Grant Dick Will N. Browne
Peter Whigham Mengjie Zhang
Lam Thu Bui Hisao Ishibuchi Yaochu Jin
Xiaodong Li Yuhui Shi Pramod Singh
Kay Chen Tan Ke Tang (Eds.)

Simulated Evolution and Learning

10th International Conference, SEAL 2014
Dunedin, New Zealand, December 15-18, 2014
Proceedings

 Springer

Volume Editors

Grant Dick, Peter Whigham, University of Otago, Dunedin, New Zealand
E-mail: {grant.dick, peter.whigham}@otago.ac.nz

Will N. Browne, Mengjie Zhang
Victoria University of Wellington, New Zealand
E-mail: {will.browne, mengjie.zhang}@ecs.vuw.ac.nz

Lam Thu Bui, Le Quy Don University, Hanoi, Vietnam
E-mail: lam.bui07@gmail.com

Hisao Ishibuchi, Osaka Prefecture University, Japan
E-mail: hisaoi@cs.osakafu-u.ac.jp

Yaochu Jin, University of Surrey, Guilford, UK
E-mail: yaochu.jin@surrey.ac.uk

Xiaodong Li, RMIT University, Melbourne, VIC, Australia
E-mail: xiaodong.li@rmit.edu.au

Yuhui Shi, Xi'an Jiaotong-Liverpool University, Suzhou, China
E-mail: yuhui.shi@xjtlu.edu.cn

Pramod Singh
Indian Inst. of Information Technology and Management, Gwalior, India
E-mail: pksingh@iiitm.ac.in

Kay Chen Tan, National University of Singapore
E-mail: eletankc@nus.edu.sg

Ke Tang
UBRI and University of Science and Technology of China, Hefei, China
E-mail: ketang@ustc.edu.cn

ISSN 0302-9743   e-ISSN 1611-3349
ISBN 978-3-319-13562-5   e-ISBN 978-3-319-13563-2
DOI 10.1007/978-3-319-13563-2
Springer Cham Heidelberg New York Dordrecht London

Library of Congress Control Number: 2014954886

LNCS Sublibrary: SL 1 –
Theoretical Computer Science and General Issues

Typesetting: Camera-ready by author, data conversion by Scientific Publishing Services, Chennai, India
Printed on acid-free paper
Springer is part of Springer Science+Business Media (www.springer.com)

Lecture Notes in Computer Science 8886

Commenced Publication in 1973
Founding and Former Series Editors:
Gerhard Goos, Juris Hartmanis, and Jan van Leeuwen

Preface

This LNCS volume contains the papers presented at SEAL 2014, the 10th International Conference on Simulated Evolution and Learning, held during December 15-18, 2014, in Dunedin, New Zealand. SEAL is a prestigious international conference series in evolutionary computation and learning. This biennial event was first held in Seoul, Korea, in 1996, and then in Canberra, Australia (1998), Nagoya, Japan (2000), Singapore (2002), Busan, Korea (2004), Hefei, China (2006), Melbourne, Australia (2008), Kanpur, India (2010), and Hanoi, Vietnam (2012).

SEAL 2014 received over 109 paper submissions from nearly 30 countries. After a rigorous peer-review process involving at least three reviewers for each paper, the best 42 papers were selected to be presented at the conference as a regular talk (acceptance rate of 39%) and an additional 29 papers as short talks (an acceptance rate of about 26%).

The papers included in this volume cover a wide range of topics in simulated evolution and learning. The accepted papers have been classified into the following main categories: (a) evolutionary optimization, (b) evolutionary multi-objective optimization, (c) evolutionary (machine) learning, (d) theoretical developments, (e) evolutionary feature reduction, (f) evolutionary scheduling and combinatorial optimization, and (g) real-world applications and evolutionary image analysis.

The conference featured three distinguished keynote speakers: Xin Yao, Kay Chen Tan, and Zbigniew Michalewicz. Prof. Xin Yao's talk was on "Learning in the Model Space." Prof. Kay Chen Tan's talk was on "Advances in Evolutionary Multiobjective Optimization and Applications." Prof. Zbigniew Michalewicz's talk was on "Some Thoughts on Complexity of Real-World Problems—Evolutionary Computation for Real-World Applications." We were very fortunate to have such internationally renowned research leaders giving talks at SEAL 2014, given their busy schedules. Their presence at the conference was yet another indicator of the importance of the SEAL conference series on the international research map.

SEAL 2014 also included six tutorials, which were free to all conference participants. Two tutorials were kindly provided by two of the keynote speakers, Prof. Xin Yao and Prof. Zbigniew Michalewicz, and in addition, we were also fortunate to have Prof. Frank Neumann, Prof. Marcus Gallagher, Prof. Hernan Aguirre, and Prof. Simon Lucas to each present a tutorial at the conference.

These six tutorials covered some of the hottest topics in evolutionary computation and learning, and their applications, including "Evolving and Designing Neural Network Ensembles Effectively" (Xin Yao), "How to develop a Killer EC-Based Application?" (Zbigniew Michalewicz), "Parameterized Complexity Analysis of Bio-Inspired Computing" (Frank Neumann), "Estimation of Distribution

Algorithms and Probabilistic Modelling in Evolutionary Computation" (Marcus Gallagher), "Advances on Evolutionary Many-Objective Optimization" (Hernan Aguirre), and "Monte Carlo Tree Search and Evolutionary Enhancements" (Simon Lucas). They provided an excellent start to the four-day conference.

The success of a conference depends on its authors, reviewers, participants and Organizing Committees. SEAL 2014 was no exception. We are very grateful to all the authors for their paper submissions and to all the reviewers for their outstanding effort in refereeing the papers within a tight schedule. We relied heavily upon a team of volunteers to keep SEAL 2014 running smoothly and efficiently. They were the true heros working behind the scene. In particular, Heather Cooper and Stephen Hall-Jones and the local organizing team from the University of Otago played an important role in supporting the running of the conference. We are most grateful to all the student volunteers for their great efforts and contributions.

We would also like to thank our sponsors for providing all the support and financial assistance to SEAL 2014, including the Department of Information Science, University of Otago, School of Engineering and Computer Science and Evolutionary Computation Research Group, Victoria University of Wellington, and IEEE Computational Intelligence Society.

December 2014   Grant Dick
                                                                   Will Browne
                                                                   Peter Whigham
                                                                   Mengjie Zhang

Organization

The 10th International Conference on Simulated Evolution and Learning (SEAL 2014) was organized and hosted by University of Otago and Victoria University of Wellington, New Zealand.

SEAL 2014 Conference Committee

General Chairs

Mengjie Zhang (New Zealand) Peter Whigham (New Zealand)

Program Chairs

Grant Dick (New Zealand) Will Browne (New Zealand)

Technical Co-chairs

Lam Thu Bui (Vietnam) Yuhui Shi (China)
Hisao Ishibuchi (Japan) Pramod Singh (India)
Yaochu Jin (UK) Kay Chen Tan (Singapore)
Xiaodong Li (Australia) Ke Tang (China)

Local Organizing Chairs

Heather Cooper (New Zealand) Stephen Hall-Jones (New Zealand)

Tutorial Chair

Mark Johnston (New Zealand)

Special Session Chair

Aaron Chen (New Zealand)

Publicity Chairs

Jing Liu (China) Kourosh Neshatian (New Zealand)
Andy Song (Australia) Nguyen Xuan Hoai (Vietnam)

International Advisory Board

Hussein Abbass   Australia
Carlos A. Coello Coello   Mexico
Kalyanmoy Deb   USA
Garry Greenwood   USA
Jong-Hwan Kim   Korea
Bob McKay   Korea
Zbignew Michalewicz   Australia
Lipo Wang   Singapore
Xin Yao   UK

SEAL 2014 Keynotes

Learning in the Model Space
  Xin Yao

**Advances in Evolutionary Multiobjective Optimization
and Applications**
  Kay Chen Tan

**Some Thoughts on Complexity of Real-World Problems —
Evolutionary Computation for Real-World Applications**
  Zbigniew Michalewicz

SEAL 2014 Tutorials

Evolving and Designing Neural Network Ensembles Effectively
  Xin Yao

How to Develop a Killer EC-Based Application
  Zbigniew Michalewicz

Parameterized Complexity Analysis of Bio-Inspired Computing
  Frank Neumann

**Estimation of Distribution Algorithms and Probabilistic Modelling
in Evolutionary Computation**
  Marcus Gallagher

Advances on Evolutionary Many-Objective Optimization
  Hernan Aguirre

Monte Carlo Tree Search and Evolutionary Enhancements
  Simon Lucas

Sponsors

Department of Information Science, University of Otago
School of Engineering and Computer Science and Evolutionary Computation
Research Group, Victoria University of Wellington
IEEE Computational Intelligence Society

Program Committee

Hussein Abbass
Hernn Aguirre
Soha Ahmed
Forrest Sheng Bao
Pasquale Barile
Luigi Barone
Urvesh Bhowan
Juergen Branke
Lam Thu Bui
Stefano Cagnoni
Zhenjiang Cai
Jinhai Cai
Allan Campbell
Ying-Ping Chen
Gang Chen
Wei-Neng Chen
Raymond Chiong
Siang Yew Chong
Vic Ciesielski
Kalyanmoy Deb
Hepu Deng
Alexander Donaldson
Haibin Duan
Daryl Essam
Wenlong Fu
Marcus Gallagher
Yang Gao
Richard Green
Steven Gustafson
Hisashi Handa
Jingsong He
Jun He
Tim Hendtlass
Wei-Chiang Hong
Zeng-Guang Hou

Daniel Howard
Yujing Hu
Rachel Hunt
Muhammad Iqbal
Hisao Ishibuchi
David Jackson
Xiuyi Jia
Licheng Jiao
Xin Jin
Mark Johnston
Nikola Kasabov
Michael Kirley
Mario Koeppen
Yun Sing Koh
Krzysztof Krawiec
Ivan Lee
Per Kristian Lehre
Xiaodong Li
Bin Li
Jinyan Li
Kunlun Li
Jiuyong Li
Wei Li
Ying Lin
Jing Liu
Cheng-Lin Liu
Wenjian Luo
Hui Ma
Syahaheim Marzukhi
Michael Mayo
Jon McCormack
Yi Mei
Kathryn Merrick
Bernd Meyer
Asad Mohammadi

Ammar Mohemmed
Ian Moser
Gul Muhammad Khan
Syed Saud Naqvi
Kourosh Neshatian
Su Nguyen
Yew-Soon Ong
Chao Qian
Kai Qin
Rong Qu
Juan Rada-Vilela
Marcus Randall
Ramesh Rayudu
Patricia Riddle
Ramn Sagarna
Muhammad Sarfraz
Mahdi Setayesh
Lin Shang
Bam Shrestha
Yuhui Shi
Josefina Sierra-Santibanez
Andy Song
Jinhua Song
Qing Song
Kay Chen Tan
Ke Tang
Chuan-Kang Ting
Krzysztof Trojanowski
Roberto Ugolotti
Yu-Xuan Wang
Yuping Wang
Dianhui Wang
Jingxuan Wei
Clinton Woodward
John Woodward

Jason Xie   Qiang Yang   Shichao Zhang
Feng Xie   Yubin Yang   Qingfu Zhang
Bing Xue   Tina Yu   Dongbin Zhao
Shengxiang Yang   Lean Yu   Ning Zhong
Ming Yang   Yang Yu   Zhihua Zhou
Peipei Yang   Sihai Zhang   Xiaofeng Zhu

Table of Contents

Evolutionary Optimization

Solving Dynamic Optimisation Problem with Variable Dimensions 1
   AbdelMonaem F.M. AbdAllah, Daryl L. Essam, and Ruhul A. Sarker

A Probabilistic Evolutionary Optimization Approach to Compute
Quasiparticle Braids.. 13
   Roberto Santana, Ross B. McDonald, and Helmut G. Katzgraber

Adaptive System Design by a Simultaneous Evolution of Morphology
and Information Processing 25
   Olga Smalikho and Markus Olhofer

Generating Software Test Data by Particle Swarm Optimization 37
   Ya-Hui Jia, Wei-Neng Chen, Jun Zhang, and Jing-Jing Li

A Steady-State Genetic Algorithm for the Dominating Tree Problem ... 48
   Shyam Sundar

Evolution of Developmental Timing for Solving Hierarchically
Dependent Deceptive Problems 58
   Kouta Hamano, Kei Ohnishi, and Mario Köppen

The Introduction of Asymmetry on Traditional 2-Parent Crossover
Operators for Crowding and Its Effects 70
   Shigeyoshi Tsutsui

The Performance Effects of Interaction Frequency in Parallel
Cooperative Coevolution 82
   Xiaofen Lu, Stefan Menzel, Ke Tang, and Xin Yao

Customized Selection in Estimation of Distribution Algorithms 94
   Roberto Santana, Alexander Mendiburu, and Jose A. Lozano

A Hybrid GP-Tabu Approach to QoS-Aware Data Intensive Web
Service Composition.. 106
   Yang Yu, Hui Ma, and Mengjie Zhang

A Modified Screening Estimation of Distribution Algorithm for
Large-Scale Continuous Optimization............................. 119
   Krishna Manjari Mishra and Marcus Gallagher

Clustering Problems for More Useful Benchmarking of Optimization
Algorithms... 131
   Marcus Gallagher

An Analysis of Differential Evolution Parameters on Rotated
Bi-objective Optimization Functions.............................. 143
    Martin Drozdik, Kiyoshi Tanaka, Hernán Aguirre, Sebastien Verel,
    Arnaud Liefooghe, and Bilel Derbel

Fuzzy Clustering with Fitness Predator Optimizer for Multivariate
Data Problems 155
    Shiqin Yang and Yuji Sato

Effects of Mutation and Crossover Operators in the Optimization of
Traffic Signal Parameters 167
    Rolando Armas, Hernán Aguirre, and Kiyoshi Tanaka

A GP Approach to QoS-Aware Web Service Composition
and Selection 180
    Alexandre Sawczuk da Silva, Hui Ma, and Mengjie Zhang

Evolutionary Multi-objective Optimization

A Novel Hybrid Multi-objective Optimization Framework: Rotating the
Objective Space 192
    Xin Qiu, Ye Huang, Jian-Xin Xu, and Kay Chen Tan

PaCcET: An Objective Space Transformation to Iteratively Convexify
the Pareto Front 204
    Logan Yliniemi and Kagan Tumer

Evolving Hard and Easy Traveling Salesman Problem Instances:
A Multi-objective Approach 216
    He Jiang, Wencheng Sun, Zhilei Ren, Xiaochen Lai, and Yong Piao

A Multi-Objective A* Search Based on Non-dominated Sorting 228
    Mohammad Haqqani, Xiaodong Li, and Xinghuo Yu

Extending AϵSϵH from Many-objective to Multi-objective Optimization. 239
    Hernán Aguirre, Yuki Yazawa, Akira Oyama, and Kiyoshi Tanaka

User Preferences for Approximation-Guided Multi-objective
Evolution 251
    Anh Quang Nguyen, Markus Wagner, and Frank Neumann

Multi-objective Optimisation, Software Effort Estimation and Linear
Models 263
    Peter A. Whigham and Caitlin Owen

Adaptive Update Range of Solutions in MOEA/D for Multi and
Many-Objective Optimization 274
    Hiroyuki Sato

Evolutionary Machine Learning

Classification of Lumbar Ultrasound Images with Machine Learning 287
  Shuang Yu and Kok Kiong Tan

Schemata Bandits for Binary Encoded Combinatorial Optimisation
Problems 299
  Madalina M. Drugan, Pedro Isasi, and Bernard Manderick

Anomaly Detection Using Replicator Neural Networks Trained on
Examples of One Class 311
  Hoang Anh Dau, Vic Ciesielski, and Andy Song

Walking Motion Learning of Quadrupedal Walking Robot by
Profit Sharing That Can Learn Deterministic Policy for POMDPs
Environments 323
  Yuya Morino and Yuko Osana

Genetic Programming for Multiclass Texture Classification Using a
Small Number of Instances 335
  Harith Al-Sahaf, Mengjie Zhang, and Mark Johnston

A Stepwise Multi-centroid Classification Learning Algorithm with GPU
Implementation 347
  Cain Cresswell-Miley and Kourosh Neshatian

Dynamic Selection of Evolutionary Algorithm Operators Based on
Online Learning and Fitness Landscape Metrics..................... 359
  Pietro A. Consoli, Leandro L. Minku, and Xin Yao

Learning Patterns of States in Time Series by Genetic Programming ... 371
  Feng Xie, Andy Song, and Vic Ciesielski

Reusing Learned Functionality to Address Complex Boolean
Functions 383
  Isidro M. Alvarez, Will N. Browne, and Mengjie Zhang

Analysis of Online Signature Based Learning Classifier Systems for
Noisy Environments: A Feedback Control Theoretic Approach 395
  Kamran Shafi and Hussein A. Abbass

Multi-objective Multiagent Credit Assignment Through Difference
Rewards in Reinforcement Learning 407
  Logan Yliniemi and Kagan Tumer

Theoretical Developments

On the Impact of Utility Functions in Interactive Evolutionary
Multi-objective Optimization 419
  Frank Neumann and Anh Quang Nguyen

Beyond the Edge of Feasibility: Analysis of Bottlenecks 431
  Mohammad Reza Bonyadi, Zbigniew Michalewicz,
  and Markus Wagner

Adaptive Risk Aversion in Social Dilemmas 443
  Michael Kirley and Friedrich Burkhard von der Osten

Fitness Landscape Analysis of Circles in a Square Packing Problems.... 455
  Rachael Morgan and Marcus Gallagher

Local Landscape Patterns for Fitness Landscape Analysis 467
  Shinichi Shirakawa and Tomoharu Nagao

Why Advanced Population Initialization Techniques Perform Poorly in
High Dimension?... 479
  Borhan Kazimipour, Xiaodong Li, and A.K. Qin

Bloat and Generalisation in Symbolic Regression 491
  Grant Dick

Evolutionary Feature Reduction

Improved PSO for Feature Selection on High-Dimensional Datasets..... 503
  Binh Tran, Bing Xue, and Mengjie Zhang

Multi-objective Feature Selection in Classification: A Differential
Evolution Approach 516
  Bing Xue, Wenlong Fu, and Mengjie Zhang

A Multi-objective Optimization Method for Product Feature Fatigue
Problem 529
  Jinze Chai, Ming Li, Yu Zheng, Liya Wang, and Fan Yu

Genetic Programming for Channel Selection from Multi-stream Sensor
Data with Application on Learning Risky Driving Behaviours 542
  Hoang Anh Dau, Andy Song, Feng Xie, Flora D. Salim,
  and Vic Ciesielski

Variable Neighbourhood Iterated Improvement Search Algorithm for
Attribute Reduction Problems 554
  Yahya Z. Arajy, Salwani Abdullah, and Saif Kifah

PSO and Statistical Clustering for Feature Selection:
A New Representation 569
  Hoai Bach Nguyen, Bing Xue, Ivy Liu, and Mengjie Zhang

Feature Selection Method with Proportionate Fitness Based Binary
Particle Swarm Optimization 582
  Zhe Zhou, Xing Liu, Ping Li, and Lin Shang

Genetic Programming for Measuring Peptide Detectability   593
Soha Ahmed, Mengjie Zhang, Lifeng Peng, and Bing Xue

Overview of Particle Swarm Optimisation for Feature Selection in
Classification   605
Binh Tran, Bing Xue, and Mengjie Zhang

Evolutionary Scheduling and Combinatorial Optimization

A Comparison between Two Evolutionary Hyper-Heuristics for
Combinatorial Optimisation   618
Richard J. Marshall, Mark Johnston, and Mengjie Zhang

Improving Efficiency of Heuristics for the Large Scale Traveling Thief
Problem   631
Yi Mei, Xiaodong Li, and Xin Yao

Load Balance Aware Genetic Algorithm for Task Scheduling in Cloud
Computing   644
*Zhi-Hui Zhan, Ge-Yi Zhang, Ying-Lin, Yue-Jiao Gong,
and Jun Zhang*

Selection Schemes in Surrogate-Assisted Genetic Programming for Job
Shop Scheduling   656
Su Nguyen, Mengjie Zhang, Mark Johnston, and Kay Chen Tan

Developing a Hyper-Heuristic Using Grammatical Evolution and the
Capacitated Vehicle Routing Problem   668
Richard J. Marshall, Mark Johnston, and Mengjie Zhang

An Organizational Cooperative Coevolutionary Algorithm for
Multimode Resource-Constrained Project Scheduling Problems   680
Lixia Wang, Jing Liu, and Mingxing Zhou

Scaling Up Solutions to Storage Location Assignment Problems by
Genetic Programming   691
Jing Xie, Yi Mei, Andreas T. Ernst, Xiaodong Li, and Andy Song

Dual Population Genetic Algorithm for the Cardinality Constrained
Portfolio Selection Problem   703
Nasser R. Sabar and Andy Song

An Evolutionary Algorithm for TD-LTE Resource Allocation Based on
Adaptive Fairness Threshold   713
*Qiang Wang, Hai-Lin Liu, Zhen-hua Li, Yiu-ming Cheung,
and Jun Zhang*

Enhancing Heuristics for Order Acceptance and Scheduling Using
Genetic Programming 723
    John Park, Su Nguyen, Mengjie Zhang, and Mark Johnston

Real World Applications and Evolutionary Image Analysis

Application of Adaptive Streaming Technology in Remotely Driven
Electric Vehicles 735
    Kyaw Ko-Ko-Htet, Arun-Shankar Narayanan, Tan Kok-Kiong,
    and Chandran Nair

Optimising Wi-Fi Installations Using a Multi-Objective Evolutionary
Algorithm.. 747
    Lyndon While and Chris McDonald

Automated Design of Architectural Layouts Using a Multi-Objective
Evolutionary Algorithm.. 760
    Darcy Chia and Lyndon While

An Approach for Real-Time Frame Size Adaptation in M-JPEG
Streams 773
    Kyaw Ko-Ko-Htet and Tan Kok-Kiong

Automatic Melody Generation Considering User's Evaluation Using
Interactive Genetic Algorithm 785
    Mio Takano and Yuko Osana

Object Recognition by Stochastic Metric Learning .... ·798
    Oliver Batchelor and Richard Green

Automatic Resolution Selection for Edge Detection Using Genetic
Programming... 810
    Wenlong Fu, Mark Johnston, and Mengjie Zhang

Evolutionary Feature Combination Based Seed Learning for
Diffusion-Based Saliency 822
    Syed S. Naqvi, Will N. Browne, and Christopher Hollitt

Analysis of Hybrid Classification Approach to Differentiate Dense and
Non-dense Grass Regions 835
    Sujan Chowdhury, Brijesh Verma, and David Stockwell

Image Segmentation: A Survey of Methods Based on Evolutionary
Computation 847
    Yuyu Liang, Mengjie Zhang, and Will N. Browne

Author Index.. 861

Solving Dynamic Optimisation Problem
with Variable Dimensions

AbdelMonaem F.M. AbdAllah, Daryl L. Essam, and Ruhul A. Sarker

School of Engineering and Information Technology, University of New South Wales Canberra
(UNSW Canberra@ADFA), Canberra 2600, Australia
a.abdallah@student.adfa.edu.au,
{d.essam,r.sarker}@adfa.edu.au

Abstract. Over the last two decades, dynamic optimisation problems (DOPs) have become a challenging research topic. In DOPs, at least one part of the problem changes as time passes. These changes may affect the objective function(s) and/or constraint(s). In this paper, we propose and define a novel type of DOP in which dimensions change as time passes. It is called DOP with variable dimensions (DOPVD). We also propose a mask detection procedure to help algorithms in solving single objective unconstrained DOPVDs. This procedure is used to try to detect ineffective and effective dimensions while solving DOPVDs. In this paper, this procedure is added to Genetic Algorithms (GAs) to be tested. The results in this paper demonstrate that GAs which use the mask detection procedure outperform GA without it especially Periodic GA 5 (PerGA5).

Keywords: Dynamic optimisation, variable dimensions, genetic algorithm, mask detection, periodic.

1 Introduction

Optimisation is an important topic that relates to several aspects in our life, such as transportation, management and industry. There are different categories of optimisation problems. Firstly, problems can be either discrete or continuous. In discrete optimisation, problems may have a finite number of objects to be ordered in their best possible way, for example, when finding an optimised path among a set of locations [1]. In contrast, continuous optimisation problems have real values to be optimised [2]. Secondly, the problems can be either single objective or multi-objective [3]. Thirdly, problems can be either unconstrained or constrained [2]. Finally, problems may either be stationary (static), where they do not change over time [4], or dynamic, where they have at least one part that changes over time [5].

In this paper, we consider dynamic optimisation problems (DOPs), in particular, a new class of problems in which effective dimensions change as time passes. We call these problems DOPs with variable dimensions (DOPVDs). Also, we develop an approach to identify active and inactive dimensions during the search process; this is called mask detection procedure. The experiments are conducted by developing test

G. Dick et al. (Eds.): SEAL 2014, LNCS 8886, pp. 1–12, 2014.

problems based on existing well-known functions. The analysis of the experimental results shows that GAs with the proposed mask detection procedure outperform GAs without it.

The rest of this paper is organised as follows. In Section 2, DOPs are addressed in general. In Section 3, DOPVD is defined and a framework is constructed to design such problems. Section 4 includes a comparison and discussion of six GAs that have been designed for solving DOPVDs. While an overall discussion of the implications of the results and suggestions for future work are concluded in this paper in Section 5.

2   Dynamic Optimisation Problems

In dynamic optimisation problems (DOPs), at least one part of each problem changes as time passes. These changes may include the objective function(s) and/or constraint(s). Various methods have been used to solve DOPs, for example, Genetic Algorithms (GAs) [6], Immune-based algorithms (IBAs) [7] and Evolutionary Algorithms (EAs) [8]. When EAs are used to solve dynamic environments, they are called evolutionary dynamic optimisation (EDO) algorithms [9]. These algorithms have attracted a lot of research effort during the last 20 years. In this section, we consider three issues; change detection, optimisation approaches for solving DOPs and DOPs test problems and generators.

2.1   Change Detection

Ignoring the dynamics of the problem is the simplest way to solve DOPs, but such an approach is not practical [5]. Hence, to deal with DOPs the methods have two main goals, to track the changes in a problem, and to locate the optimal solution [9]. In addition, some type of correlation between the problem-after-change and the problem-before-change must be considered. Otherwise, after a problem changes, the algorithm needs to solve a different problem by starting from scratch.

When a change occurs, most of the algorithms need to detect the changes. Change detection mechanisms can be categorised into: detecting change by re-evaluating solutions [9], and detecting changes based on algorithm behaviour [9].

2.2   Optimisation Approaches for Solving DOPs

This subsection briefly critically reviews two of the most typical approaches that have been proposed to solve DOPs.

I) Introducing Diversity when Changes Occur

This category contains methods that try to introduce diversity into their population when they detect a change. For example, by increasing the mutation rate as in hyper-mutation [10]. These approaches are good in solving problems with continuous changes, when the changes are small and medium [11]. However, these methods might have some disadvantages, for example, they are dependent on changes being easy to detect [12].

II) Memory Approaches

It might be useful to reuse previously found solutions, if the changes of a DOP are periodical or recurrent. In these situations, memory can be used to save computational time. The memory can also play the role of maintaining diversity by reserving place(s) for storing old solution(s). The memory can be integrated implicitly as a redundant representation in an algorithm, or be maintained explicitly as a separate memory component [9, 11]. These approaches are particularly effective in solving problems with periodically changing environments [6]. However, if the ideal solutions do not repeat, then the memory might become redundant [9].

This category is the closest to the approach proposed in this study. In this paper, we use a partial explicit memory in our approach, as ineffective dimensions are prohibited from changing.

2.3   DOPs Test Problems and Generators

There are various test problems and generators to be used to test the proposed algorithms to solve DOPs. Some test problems in continuous search space are mentioned as follows:

- Branke [13] proposed the Moving Peaks Benchmark (MPB) Problem, which has been widely used in the literature [14].
- Dynamic Composition Benchmark Generator (GDBG) the dynamic composition functions, are actually extended from the static functions that devised by Liang et al. [15].
- Dynamic test problems for the CEC 2009 Competition; The GDBG was used to construct these test problems [16]. These dynamic test problems consist of Sphere, Rastrigin, Weierstrass, Griewank and Ackley functions. The detailed of each function can be found in [16]. These problems are used in this paper.

3   Dynamic Optimisation Problems with Variable Dimensions

A dynamic optimisation problem with variable dimensions (DOPVD) is a DOP in which the effective dimensions change as time passes. In real life DOPVDs arise, because sometimes the decision variables that affect a decision changes as time passes. For example:

- Stock exchange; if a decision maker wants to optimise a group of illiquid stocks, while their availability changes as time passes.
- Production systems; consider a production process that produces multiple products based on market demand. In this case, all products do not have the same availability/requirement as time passes.

The effective dimensions are dimensions that affect a decision during the current time slot while the ineffective dimensions are those that do not. To construct a framework for designing DOPVD, benchmark function(s) containing multiple dimensions are used. In this paper, Sphere, Rastrigin, Weierstrass, Griewank and Ackley are used [16]. Without loss of generality, this paper only considered minimisation problems.

To construct a problem, the parameter, prob_change, is used; this parameter determines the probability of a problem change as time passes. The parameter, g_random, is a variable that is randomly generated every generation. The parameter, MAX_Dim, is the maximum number of dimensions that problem contains. The parameter, InEff_Dim, is the number of ineffective dimensions in the time slot. Then the problem changes as follows:

- For each generation:
  - ○ Generate a random value (g_random)
  - ○ If (g_random < prob_change)
    - Change the mask of the problem
  - ○ Else
    - Do not change

To determine which dimensions are ineffective, while all others are effective, a problem mask is randomly generated. For example, if we have a problem with ten dimensions (MAX_Dim = 10), where three are ineffective (InEff_Dim = 3), then three unique indices $\in$ [1-10] are randomly generated, for example, 1, 5 and 9, and then those dimensions are chosen to be ineffective (its mask value is equal to 0) as shown in Fig. 1. When the fitness of function is evaluated, the value used for each of the effective dimensions, is the one specified by an algorithm, however the value for the ineffective dimensions is always 0. For example, consider a simple example function Minimise (abs($x_1 + x_2$)). The minimal value for this function is obviously x_1 and $x_2 = 0$. However, consider if x_2 is ineffective. Due to mutation and crossover, x_2 will gradually diversify, because if $x_1 = 0$, and the function will have its minimal value, regardless the value of x_2 (because 0 is always used for its value).

Dimension	1	2	3	4	5	6	7	8	9	10
Problem Mask Value	0	1	1	1	0	1	1	1	0	1

Fig. 1. Example of a randomly generated problem mask

Hence, the efficiency of an algorithm for solving DOPVD depends on determining and tracking the effective dimensions to be optimised, thus saving computational power. In this paper, the optimisation approaches for solving DOPVD detect the ineffective and effective dimensions by using a mask detection procedure. This procedure is used periodically, every g generations (this parameter determines how often to detect the effective dimensions), by using a randomly selected solution as follows:

(a) A random solution is chosen from the current population.
(b) Calculate its actual fitness, let it be F1.

(c) For each dimension, a random value is generated:
  (i) The fitness is recalculated for the solution with the new random value, let it be F2.
  (ii) If F1 is equal to F2, then this dimension is assumed to be detected as ineffective (its detected mask value is equal to 0), otherwise it is assumed to be detected as effective (its detected mask value is equal to 1).

An example of a detected mask is shown in Fig. 2. In this figure, dimensions 2 and 6 are assumed to be detected as ineffective dimensions; that is they do not affect the fitness value when different values are used for them, while other dimensions are assumed to be detected as being effective dimensions.

Dimension	1	2	3	4	5	6	7	8	9	10
Detected Mask Value	1	0	1	1	1	0	1	1	1	1

Fig. 2. Example of a detected mask

4   Experimental Results and Discussion

To solve DOPVD and demonstrate the effect of the proposed procedure, six Genetic Algorithms (GAs) were implemented for experimentation with the set of unconstrained optimisation benchmark functions i.e. Sphere, Rastrigin, Weierstrass, Griewank and Ackley [16]. In this paper, the algorithms were coded in Microsoft C++, on a 3.4GHz/16GB Intel Core i7 machine. The six GAs were:

- 1) GA without the mask detection procedure.
- 2-6) Periodic GA N (where N = 1, 5, 10, 20 or 40), the mask detection is used periodically every N generations. The generated GAs are PerGA1, PerGA5, PerGA10, PerGA20 and PerGA40 respectively.

In all the implemented GAs, only the detected effective dimensions were modified by the genetic operators. Note that all genes were assumed to be effective in GA without mask detection procedure. For a fair evaluation, every algorithm ran one million fitness evaluations. To compare these algorithms, a group of points were determined for calculations over the fitness evaluations. This was done because each system ran for a differing number of generations; depending on whether the mask detection procedure was used or not and how often it was used. In this paper, twenty calculation points were determined, so the values for every $\frac{1000000}{20} = 50,000$ fitness evaluations solutions were recorded. A variation of the Best-of-Generation measure was used where the best-of-generation values were averaged over all generations at each calculation point [17], it is calculated as follows:

$$\overline{F}_{BOG} = \frac{1}{G} \sum_{i=1}^{i=G} \left(\frac{1}{N} \sum_{j=1}^{j=N} F_{BOG_{ij}} \right) \tag{1}$$

where $\overline{F}_{BOG}$ is the mean best-of-generation fitness, G is the number of generations, N is the total number of runs, and $F_{BOG_{ij}}$ is the best-of-generation fitness of generation i of run j of an algorithm to solve a problem [18].

The parameter, prob_change, was also varied from less change (0.05) to more change (0.75) to demonstrate how the implemented GAs deal with these variations. The number of dimensions, MAX_Dim, was twenty dimensions, and the number of ineffective dimensions, InEff_Dim, was randomly determined between 5 and 10 dimensions.

Table 1 shows the parameters and processes of the implemented GAs. The search space of all variables was [-5, 5] [16]. Note that all GAs had the same initial population in the beginning of each run for a fair comparison. Also, in each run the same masks are loaded for each algorithm to try to simulate the same changes in the problems.

Table 1. Parameters of experiments

Parameters	
Population size	100
Max. number of fitness evaluations / run	1000000
Probabilities of problem mask change (prob_change)	0.05, 0.25, 0.50, 0.75
Selection procedure	Tournament
Tournament size	2
Selection pressure	0.9
Elitism percentage	2
Crossover	Single-point
Crossover rate	0.9
Mutation	Uniform
Mutation rate	0.15
Number of dimensions (MAX_Dim)	20
Number of ineffective dimensions (InEff_Dim) / change	Randomly $\in$ [5, 10]

4.1   Performance Evaluation

To evaluate the performance while regarding best-of-generation values, the $\overline{F}_{BOG}$ in equation (1) was averaged over the twenty points. Tables 2 to 5 show the comparison among the algorithms; the best results that have lower values are shown in bold and shaded cells.

Table 2. Performance comparison at prob_change = 0.05

Probability of change	Function	Without mask	PerGA1	PerGA5	PerGA10	PerGA20	PerGA40
0.05	Ackley	1.144	0.513	**0.389**	0.472	0.612	0.965
	Griewank	0.114	0.020	**0.018**	0.019	0.047	0.086
	Rastrigin	13.369	6.029	**4.692**	4.942	7.237	10.119
	Sphere	1.008	0.233	**0.217**	0.409	0.521	0.792
	Weierstrass	2.454	1.165	**1.058**	1.229	1.265	2.024

Table 3. Performance comparison over prob_change = 0.25

Probability of change	Function	Without mask	PerGA1	PerGA5	PerGA10	PerGA20	PerGA40
0.25	Ackley	0.488	0.297	**0.230**	0.242	0.339	0.310
	Griewank	0.016	0.006	**0.005**	**0.005**	0.014	0.016
	Rastrigin	6.575	3.689	**2.567**	2.980	3.653	3.822
	Sphere	0.196	0.122	**0.068**	0.081	0.200	0.199
	Weierstrass	1.346	0.894	**0.494**	0.529	0.716	0.729

Table 4. Performance comparison at prob_change = 0.50

Probability of change	Function	Without mask	PerGA1	PerGA5	PerGA10	PerGA20	PerGA40
0.5	Ackley	0.514	0.239	**0.191**	0.211	0.245	0.283
	Griewank	0.016	0.004	**0.002**	0.005	0.005	0.009
	Rastrigin	6.789	3.603	**1.976**	2.535	3.205	3.235
	Sphere	0.194	0.075	**0.041**	0.059	0.086	0.144
	Weierstrass	1.627	0.785	**0.468**	0.499	0.621	0.664

Table 5. Performance comparison at prob_change = 0.75

Probability of change	Function	Without mask	PerGA1	PerGA5	PerGA10	PerGA20	PerGA40
0.75	Ackley	0.538	0.282	**0.160**	0.215	0.257	0.254
	Griewank	0.016	0.004	**0.002**	0.004	0.008	0.006
	Rastrigin	6.697	3.661	**1.924**	2.104	2.890	2.853
	Sphere	0.221	0.096	**0.041**	0.053	0.111	0.098
	Weierstrass	1.821	0.727	**0.369**	0.502	0.639	0.656

The previous tables show that GAs with periodic mask detection outperform the GA without mask. Also, the PerGA5 outperforms the other periodic GAs. The previous tables were averaged and summarised in Table 6.

Table 6. The overall comparison

Probability of change	Function	Without mask	PerGA1	PerGA5	PerGA10	PerGA20	PerGA40
All	Ackley	0.671	0.333	**0.243**	0.285	0.363	0.453
	Griewank	0.041	0.008	**0.007**	0.008	0.019	0.029
	Rastrigin	8.357	4.245	**2.790**	3.140	4.246	5.007
	Sphere	0.405	0.132	**0.092**	0.151	0.230	0.308
	Weierstrass	1.812	0.893	**0.597**	0.690	0.810	1.018

Table 6 shows that over all the functions and probabilities of change, the PerGA5 outperforms the other periodic GAs. Also, all periodic GAs outperform GA without the mask detection procedure; the mask detection procedure is therefore significant while solving DOPVD.

Here, in order to be able to compare our results more accurately, we also performed statistical significance tests. The non-parametric Friedman test that is similar to the parametric repeated measures ANOVA was used [19]. It is a multiple comparison test that aims to detect significant differences between the performances of two or more algorithms. Friedman test was performed with a confidence level of 95%

($\alpha = 0.05$) on results in Table 6 with null hypothesis that there is no significant differences among the performances of the compared algorithms. The computation of the p-value for this test was $0.002 <= 0.05$; so we reject the null hypothesis; as there is a significant difference among the performances of the compared algorithms. Table 7 shows the ranks of the algorithms based on Friedman test.

Table 7. Performances Friedman test ranks

Algorithm	Without mask	PerGA1	PerGA5	PerGA10	PerGA20	PerGA40
Rank	6	2.90	1	2.30	3.80	5

To test the stability of the compared algorithms, Table 8 shows their standard deviations. In this table, periodic GAs are more stable than GA without mask; while PerGA5 is the most stable GA.

Table 8. The overall comparison over the standard deviation

Probability of change	Function	Without mask	PerGA1	PerGA5	PerGA10	PerGA20	PerGA40
All	Ackley	0.3012	0.1852	**0.1577**	0.2070	0.3049	0.3503
	Griewank	0.0424	**0.0102**	0.0108	0.0151	0.0335	0.0463
	Rastrigin	3.7768	1.9280	**1.7171**	2.2645	3.3627	4.3457
	Sphere	0.5198	**0.1463**	0.1569	0.3086	0.4498	0.6015
	Weierstrass	0.7796	0.4741	**0.4157**	0.4856	0.6467	0.7728
	Average	1.0839	0.5487	**0.4916**	0.6562	0.9595	1.2233

In these problems, we had million fitness evaluations (10000 generations), 20 dimensions and 100 population size; the mask detection procedure must be included in this. So the periodic GAs had a number of wasted generations, in comparison to the without mask GA (normal GA). PerGA1, as there were 100 individuals and mask detection used 20 fitness evaluations, it wasted 1 of every 5 generations (about 2000 generations). Doing the same previous calculations for the other periodic GAs; PerGA5, PerGA10, PerGA20, PerGA40 wasted 400, 200, 100 and 50 generations respectively. Despite this, the periodic GAs outperformed the without mask GA which did not waste any generation. However, it can be seen that when the wastage reaches a critical level, the periodic GA could not improve its solutions; as PerGA5 outperforms PerGA1.

Lastly, Table 9 shows the effect of varying the probabilities of change. It averages the values of the previous Tables (2 to 5) over the probabilities of change. Note that the best results have lower values. From it, it can be observed that PerGA40 is worst for low probabilities of change. This presumably because a change usually is maintained for a long period of time, and so on ineffective dimension can widely diverge. For higher probability of change, such dimensions would not have as much time.

Table 9. Comparison over the different probabilities of change

Probability of change	Without mask	PerGA1	PerGA5	PerGA10	PerGA20	PerGA40
0.05	3.618	1.592	**1.275**	1.414	1.936	2.797
0.25	1.724	1.001	**0.673**	0.768	0.985	1.015
0.5	1.828	0.941	**0.536**	0.662	0.833	0.867
0.75	1.859	0.954	**0.499**	0.576	0.781	0.774

4.2 Behaviour Tracking Evaluation

In this section, we try to also evaluate the behaviour and the convergence speed of the algorithms during solving the problems over the twenty points of calculations. This tries to monitor and track how the algorithms perform when the problem changes over time. An evaluation technique is used in this paper, which is similar to the normalised scores [20]. While judging system i in terms of its average of the best solutions to test problem j at calculation point k, F_{ijk} is defined as the actual value of the average of the best solutions that the system obtained, while $BF_{jk} = \min (F_{ijk})$ and $WF_{jk} = \max (F_{ijk})$ are the overall best and worst averages of the best solutions for test problem j at a calculation point k respectively, and the score of system i (S_{ijk}) is calculated as follows:

$$S_{ijk} = \frac{|F_{ijk} - BF_{jk}|}{|BF_{jk} - WF_{jk}|} \tag{2}$$

Tables 10 to 13 show the comparison based on this evaluation approach, the best results that have lower values are shown in bold and shaded cells. The less value indicates that the algorithm converges better.

Table 10. Behaviour comparison at prob_change = 0.05

Probability of change	Function	Without mask	PerGA1	PerGA5	PerGA10	PerGA20	PerGA40
0.05	Ackley	19.131	3.650	**0.645**	2.733	6.149	14.957
	Griewank	18.944	1.344	**1.086**	1.387	6.734	14.158
	Rastrigin	19.560	3.535	**0.753**	1.240	6.816	13.376
	Sphere	17.221	1.680	**1.171**	5.940	7.575	12.856
	Weierstrass	19.834	2.721	**1.346**	3.755	4.269	14.509

Table 11. Behaviour comparison at prob_change = 0.25

Probability of change	Function	Without mask	PerGA1	PerGA5	PerGA10	PerGA20	PerGA40
0.25	Ackley	18.651	5.807	**1.285**	2.484	11.898	7.821
	Griewank	9.699	1.733	**1.340**	1.940	10.084	15.585
	Rastrigin	19.259	4.629	**0.373**	2.500	6.971	8.508
	Sphere	8.574	3.541	**1.156**	1.297	11.537	10.990
	Weierstrass	19.998	7.361	**0.448**	1.231	6.434	6.257

Table 12. Behaviour comparison at prob_change = 0.50

Probability of change	Function	Without mask	PerGA1	PerGA5	PerGA10	PerGA20	PerGA40
0.5	Ackley	19.374	2.898	**1.005**	2.327	4.923	8.703
	Griewank	18.199	1.398	**0.087**	3.394	5.042	6.301
	Rastrigin	19.982	6.120	**0.118**	2.102	5.650	5.671
	Sphere	17.675	3.771	**0.036**	2.968	7.142	10.313
	Weierstrass	20.000	4.825	**0.403**	0.680	2.911	4.682

Table 13. Behaviour comparison at prob_change = 0.75

Probability of change	Function	Without mask	PerGA1	PerGA5	PerGA10	PerGA20	PerGA40
0.75	Ackley	20	6.162	**0.025**	2.281	6.126	4.900
	Griewank	18.401	1.044	**0**	1.132	6.960	3.922
	Rastrigin	20	6.764	**0.080**	0.899	3.589	4.363
	Sphere	19.051	4.937	**0.013**	1.426	6.560	7.696
	Weierstrass	20	4.380	**0.068**	1.702	3.826	3.739

The previous tables show that GAs with periodic mask detection outperform the GA without mask in behaviour and convergence speed. Also, the PerGA5 outperforms the other periodic GAs. The previous tables were averaged and summarised in Table 14 which shows the average over all functions and probabilities of change, based on the normalised scores.

Table 14. Summary of the behaviour tracking comparison

Probability of change	Function	Without mask	PerGA1	PerGA5	PerGA10	PerGA20	PerGA40
All	Ackley	19.289	4.629	**0.740**	2.456	7.274	9.095
	Griewank	16.311	1.380	**0.628**	1.964	7.205	9.992
	Rastrigin	19.700	5.262	**0.331**	1.685	5.756	7.980
	Sphere	14.940	3.385	**1.101**	3.007	8.656	10.841
	Weierstrass	19.958	4.687	**0.948**	1.600	4.402	7.745

Again, Friedman test was performed with a confidence level of 95% ($\alpha = 0.05$) on results in Table 14 with null hypothesis that there is no significant differences among the behaviour and convergence speed of the compared algorithms. The computation of the p value for this test was $0.002 <= 0.05$; so there is a significant difference among the behaviour and convergence speed of the compared algorithms. Table 15 shows the ranks of the algorithms based on Friedman test.

Table 15. Behaviour Friedman test ranks

Algorithm	Without mask	PerGA1	PerGA5	PerGA10	PerGA20	PerGA40
Rank	6	3	**1**	2.20	3.80	5

The experimental results and statistical test show that GAs with mask detection procedure gradually improve the solutions as time goes, while GA without mask disturbs the ineffective dimensions, and so consequently this prevents GA from effectively converging. Regarding the GAs with the mask detection procedure, when it is periodically called in small periods this allows them to detect the ineffective and effective dimensions early, and therefore they can solve DOPVD efficiently; however, this consumes more fitness evaluations. However when the wastage reached a critical level, the periodic GA could not improve its solutions, thus PerGA5 outperformed PerGA1.

5   Conclusions and Future Work

In this paper, we proposed a novel DOP with variable dimensions (DOPVD), in which the ineffective and effective dimensions change as time passes. We also proposed the mask detection procedure to help algorithms to solve DOPVDs. Based on the experimental results and statistical tests; the proposed GAs with mask detection outperformed the pure GA (without mask detection procedure) in both performance and convergence speed; however, this consumes more fitness evaluations. Despite of the wastage of fitness evaluations, periodic GAs outperformed the without mask GA. However when the wastage reached a critical level, the periodic GA could not improve its solutions, thus Periodic GA 5 (PerGA5) outperformed Periodic GA 1 (PerGA1). In general, the advantages of the usage of the mask detection procedure are:

— save computational resources; algorithms deal with only the detected effective dimensions.
— does not disturb the ineffective dimensions, and this helps algorithms to effectively converge.
— can help a decision maker (user) to know which dimension(s) not affect the considered problem.

However, the disadvantages of using the mask detection procedure are that is might detect the wrong mask, which might happen when two values of a dimension gave the same fitness value while it is effective. Any such wrong detection might lead to wrong values of the fitness function and/or prevent some dimensions (those wrongly detected as ineffective) to be optimised.

Two directions for future work, the first direction is trying to enhance the mask detection procedure, for example sampling more points rather than one point only. Also, try to solve more problems with more complex function especially that have dependent variables. The second direction is the DOPVDs algorithms; as more advanced approaches will attempt to implicitly detect when changes occur, this might save computational resources and so more effectively solve DOPVDs. Also, we intend to investigate how to use local search procedure(s) in solving DOPVDs.

References

1. Gendreau, M., Potvin, J.-Y., Bräysy, O., Hasle, G., Løkketangen, A.: Metaheuristics for the Vehicle Routing Problem and extensions: A Categorized Bibliography. In: Golden, B., Raghavan, S., Wasil, E. (eds.) The Vehicle Routing Problem: Latest Advances and New Challenges. Springer (2008)

2. Nocedal, J., Wright, S.J.: Numerical Optimization, 2nd edn. Springer (2006)
3. Bandyopadhyay, S., Saha, S.: Some Single - and Multiobjective Optimization Techniques. In: Unsupervised Classification, pp. 17–58. Springer, Heidelberg (2013)
4. Dadkhah, K.: Static Optimization. In: Foundations of Mathematical and Computational Economics, pp. 323–346. Springer, Heidelberg (2011)
5. Branke, J.: Evolutionary Optimization in Dynamic Environments. In: Genetic Algorithms and Evolutionary Computation. Kluwer (2001)
6. Yang, S.: Genetic algorithms with memory- and elitism-based immigrants in dynamic environments. Evolutionary Computation 16(3), 385–416 (2008)
7. Trojanowski, K., Wierzchoń, S.T.: Immune-based algorithms for dynamic optimization. Information Sciences 179(10), 1495–1515 (2009)
8. Branke, J., Schmeck, H.: Designing evolutionary algorithms for dynamic optimization problems. In: Advances in Evolutionary Computing: Theory and Applications, pp. 239–262. Springer, Heidelberg (2003)
9. Nguyen, T.T., Yangb, S., Branke, J.: Evolutionary dynamic optimization: A survey of the state of the art. Swarm and Evolutionary Computation 6, 1–24 (2012)
10. Cobb, H.G.: An investigation into the use of hypermutation as an adaptive operator in genetic algorithms having continuous, time-dependent nonstationary environments
11. Nguyen, T.T., Yang, S., Branke, J., Yao, X.: Evolutionary Dynamic Optimization: Methodologies. In: Yang, S., Yao, X. (eds.) Evolutionary Computation for DOPs. SCI, vol. 490, pp. 39–64. Springer, Heidelberg (2013)
12. Jin, Y., Branke, J.: Evolutionary Optimization in Uncertain Environments — A Survey. IEEE Transactions on Evolutionary Computation 9(3), 303–317 (2005)
13. Branke, J.: Memory enhanced evolutionary algorithms for changing optimization problems. In: Proceedings of the 1999 Congress on Evolutionary Computation, CEC 1999 (1999)
14. Moser, I., Chiong, R.: Dynamic Function Optimization: The Moving Peaks Benchmark. In: Alba, E., Nakib, A., Siarry, P. (eds.) Metaheuristics for Dynamic Optimization. SCI, vol. 433, pp. 35–59. Springer, Heidelberg (2013)
15. Liang, J.J., Suganthan, P.N., Deb, K.: Novel composition test functions for numerical global optimization. In: Proceedings of 2005 IEEE Swarm Intelligence Symposium, SIS 2005 (2005)
16. Li, C., Yang, S., Nguyen, T.T., Yu, E.L., Yao, X., Jin, Y., Beyer, H.-G., Suganthan, P.N.: Benchmark Generator for CEC 2009, Competition on Dynamic Optimization (2009)
17. Morrison, R.W.: Performance Measurement in Dynamic Environments. In: GECCO Workshop on Evolutionary Algorithms for Dynamic Optimization Problems, pp. 5–8 (2003)
18. Yang, S., Nguyen, T.T., Li, C.: Evolutionary Dynamic Optimization: Test and Evaluation Environments. In: Yang, S., Yao, X. (eds.) Evolutionary Computation for Dynamic Optimization Problems, pp. 3–37. Springer, Heidelberg (2013)
19. García, S., Molina, D., Lozano, M., Herrera, F.: A study on the use of non-parametric tests for analyzing the evolutionary algorithms' behaviour: a case study on the CEC'2005 Special Session on Real Parameter Optimization. Journal of Heuristics 15(6), 617–644 (2009)
20. Nguyen, T.T.: Continuous Dynamic Optimisation Using Evolutionary Algorithms. In: School of Computer Science, p. 300. The University of Birmingham, Birmingham (2010)

A Probabilistic Evolutionary Optimization Approach to Compute Quasiparticle Braids

Roberto Santana[1], Ross B. McDonald[2], and Helmut G. Katzgraber[2,3]

[1]Department of Computer Science and Artificial Intelligence,
University of the Basque Country (UPV/EHU), P. Manuel de Lardizabal,
20018, Guipuzcoa, Spain
[2]Department of Physics and Astronomy, Texas A&M University, College Station,
Texas 77843-4242, USA
[3]Santa Fe Institute, 1399 Hyde Park Road, Santa Fe, NM 87501, USA
roberto.santana@ehu.es, katzgraber@physics.tamu.edu

Abstract. This paper proposes the use of estimation of distribution algorithms to deal with the problem of finding an optimal product of braid generators in topological quantum computing. We investigate how the regularities of the braid optimization problem can be translated into statistical regularities by means of the Boltzmann distribution. The introduced algorithm obtains solutions with an accuracy in the order of 10^{-6}, and lengths up to 9 times shorter than those expected from braids of the same accuracy obtained with other methods.

Keywords: topological computing, quasiparticle braids, probabilistic graphical models, EDAs, braid optimization, Fibonacci anyons.

1 Introduction

The idea of using the theory of quantum mechanics to obtain computers potentially exponentially faster for certain applications, such as the factorization of prime numbers, arouses considerable interest and research efforts from the scientific community nowadays. In quantum computation, information is represented and manipulated using quantum properties. An obstacle for the construction of large quantum computers is the problem of quantum decoherence, that can be viewed as the loss of information of the quantum system due to the interaction with the environment. One possible solution to this problem is the design of quantum systems immune to quantum decoherence on a hardware level.

Topological quantum computing (TQC) [2,14] investigates quantum computing systems that, given the properties of quasiparticles they use, are not affected by quantum decoherence. The key idea of these systems is that quantum information can be stored in global properties of the system and thus affected only by global operations but not by local perturbations such as noise. In TQC, quantum gates are carried out by adiabatically braiding quasiparticles around each other. This braiding is used to perform the unitary transformations of a quantum computation.

G. Dick et al. (Eds.): SEAL 2014, LNCS 8886, pp. 13–24, 2014.
© Springer International Publishing Switzerland 2014

One of the essential questions to design a TQC is to find a product of braid generators (matrices) that approximates a quantum gate with the smallest possible error and, if possible, as short as possible to prevent loss [8]. The relevant question of minimizing the error of a TQC design can be posed as a braid optimization problem. Some optimization approaches to this question have been proposed. Exhaustive search [2] has been applied to search for braids of manageable size (up to 46 exchanges). Other methods such as the Solovay-Kitaev algorithm [3] provide bounds on the accuracy and length of the braids. However, they do not allow the user to tune the balance between the accuracy and the length as pioneered in [8] where the use of genetic algorithms (GAs) to find optimal braids is proposed. In this paper, we build on the GA approach introduced in [8] to solve the braid optimization problem.

We use the fitness function proposed in [8] and introduce a new representation, variation operators and enhancement procedures in the framework of estimation of distribution algorithms (EDAs) [10,7]. EDAs are evolutionary algorithms (EAs) that apply learning and sampling of distributions instead of classical crossover and mutation operators. Modeling the dependencies between the variables of the problem serves to efficiently orient the search to more promising areas of the search space by explicitly capturing and exploiting potential relationships between the problem variables.

2   Braids and Anyons

Qubits play in quantum computation a role similar to that played by bits in digital computers. A braid operation can be represented by a matrix that acts on the qubit space. These matrices are referred to as generators, and the quantum gate that a braid represents is the product of the generators that encode the individual braid operations.

Let σ_1 and σ_2 represent two possible generators. σ_1^{-1} and σ_2^{-1} respectively represent their inverses. Given a braid B, $len()$ is a function that returns the braid's length l (e.g. $B = \sigma_1\sigma_1\sigma_2\sigma_1^{-1}$, $l = len(B) = 4$).

Since the product of a matrix by its inverse reduces to the identity matrix, some braids can be simplified reducing their length. Therefore, we also define function $elen()$, that has a braid as its argument and returns the braid's *effective length* which is the length of braid after all possible simplifications have been conducted. For example, the effective length values of braids $(\sigma_1\sigma_1\sigma_1\sigma_1\sigma_1^{-1} = \sigma_1\sigma_1\sigma_1)$ and $(\sigma_2^{-1}\sigma_1\sigma_1\sigma_1^{-1}\sigma_1^{-1}\sigma_2\sigma_1^{-1} = \sigma_1^{-1})$ are 3 and 1, respectively.

Let T represent the target matrix (gate to be emulated), the braid error is calculated as [8]: $\epsilon = |B - T|$ where the matrix norm used is $|M| = \sqrt{\sum_{ij} M_{ij}^2}$.

The problem of finding braiding operations that approximate gates is then reduced to finding a product chain of the reduced generators and their inverses that approximates the matrix representing the quantum gate. Two elements that describe the quality of a braid are its error ϵ and its length l.

Anyons appear as emergent quasiparticles in fractional quantum Hall states and as excitations in microscopic models of frustrated quantum magnets that harbor topological quantum liquids [11]. Fibonacci anyons are the simplest anyons with non-Abelian braiding statistics that can give rise to universal quantum computation. Fibonacci anyon braids [2] only encompasses one-qubit gates. In such systems, the braid transition operators result in a phase change for the non computational state, and therefore it can be ignored. Overall, phases in the problem can also be ignored. Therefore the transition matrices can be projected onto SU(2) by a multiplication with $e^{\frac{i\pi}{10}}$, yielding for the generators

$$
\sigma_1 = \begin{pmatrix} e^{\frac{-i7\pi}{10}} & 0 \\ 0 & -e^{\frac{-i3\pi}{10}} \end{pmatrix} \qquad \sigma_2 = \begin{pmatrix} -\tau e^{\frac{-i\pi}{10}} & -i\sqrt{\tau} \\ -i\sqrt{\tau} & -\tau e^{\frac{-i\pi}{10}} \end{pmatrix} \tag{1}
$$

where $\tau = \frac{\sqrt{5}-1}{2}$.

In this paper we address the problem of finding a product of generator matrices for Fibonacci anyon braids. Although the methodology we propose can be extended to other braids, we focus on anyon braids since they are one of the best known in TQC [8,15]. As a target gate for computing the error we use $T = \begin{pmatrix} i & 0 \\ 0 & i \end{pmatrix}$.

3   Problem Formulation

Let $\mathbf{X} = (X_1, \ldots, X_n)$ denote a vector of discrete random variables. We use $\mathbf{x} = (x_1, \ldots, x_n)$ to denote an assignment to the variables. I denotes a set of indices in $\{1, \ldots, n\}$, and X_I (respectively x_I) a subset of the variables of $\mathbf{X}$ (respectively $\mathbf{x}$) determined by the indices in I.

In our representation for the quasiparticle braids problem, $\mathbf{X} = (X_1, \ldots, X_n)$ represents a braid of length n, where X_i takes values in $\{0, 1, \ldots, 2g - 1\}$ and g is the number of generators. Given an order for the generators $\sigma_1, \sigma_2, \ldots, \sigma_g$, $X_i = j, j < g$ means that the matrix in position i is σ_{j+1}. If $X_i = j, j \geq g$, then the matrix in position i is $\sigma_{(j-g)+1}^{-1}$. For example, for generators shown in Equation (1), and $B = \sigma_1 \sigma_1 \sigma_2 \sigma_2^{-1} \sigma_1^{-1}$, the corresponding braid representation is $\mathbf{x} = (0, 0, 1, 3, 2)$. Notice that this is a fixed length representation.

We are interested in the solution of an optimization problem formulated as $\mathbf{x}^* = argmax_{\mathbf{x}} f(\mathbf{x})$, where $f : S \to R$ is called the objective or fitness function. The optimum $\mathbf{x}^*$ is not necessarily unique. To evaluate the fitness function associated to a solution $\mathbf{x}$, firstly the product of braid matrices B is computed according to $\mathbf{x}$ and then the error ϵ is calculated from B as previously defined. The fitness function [8] is defined as:

$$
f(\mathbf{x}) = \frac{1 - \lambda}{1 + \epsilon} + \frac{\lambda}{l} \tag{2}
$$

where l is the braid's length, and λ serves to balance the two conflicting goals, i.e., having short braids or low approximation error. When $\lambda = 0$, braids are optimized only for the error and the function reaches its maximum value when this error is minimized.

We define functions $\hat{f}(\mathbf{x})$ and $\bar{f}(\mathbf{x})$ as two variations of function (2). Function $\hat{f}(\mathbf{x})$ is identical to $f(\mathbf{x})$, except that the effective length $\hat{l} = elen(B)$ is used instead of the braid's length. Function $\bar{f}(\mathbf{x})$ outputs the maximum value of the function for any of the braids contained in B that start from the first position, i.e. $\bar{f}(\mathbf{x}) = max_{\mathbf{y}, \mathbf{y} \in \{(x_1),(x_1,x_2),\dots,(x_1,\dots,x_i),\dots,(x_1,\dots,x_n)\}} f(\mathbf{y})$.

4   Probabilistic Modeling of Braids

To optimize the braid problem we use EDAs, a class of evolutionary algorithms that capture and exploit statistical regularities in the best solutions. EDAs assume that such regularities exist. As a preliminary proof of concept on the existence of such regularities, we investigate the Boltzmann distribution for braids of manageable size. A similar approach has been successfully applied to investigate the dependencies that arise in the configurations of simplified protein models [13] and conductance-based neuron models [12].

4.1   Boltzmann Distribution

We use complete enumeration to define a probability distribution on the space of all possible braids for $n = 10$. Using the fitness value as an energy function, we associate to each possible braid a probability value $p(\mathbf{x})$ according to the Boltzmann probability distribution. The Boltzmann probability distribution $p_B(\mathbf{x})$ is defined as

$$p_B(\mathbf{x}) = \frac{e^{\frac{g(\mathbf{x})}{T}}}{\sum_{\mathbf{x}'} e^{\frac{g(\mathbf{x}')}{T}}}, \tag{3}$$

where $g(\mathbf{x})$ is a given objective function and T is the system temperature that can be used as a parameter to smooth the probabilities.

In our approach, $p_B(\mathbf{x})$ assigns a higher probability to braids that give a more accurate approximation to the target gate. The solutions with the highest probability correspond to the braids that maximize the objective function. We use an arbitrary choice of the temperature, $T = 1$, since our idea is to compare the distributions associated to different fitness functions with fixed T.

Using the Boltzmann distribution we can investigate how potential regularities of the fitness function are translated into statistical properties of the distribution. In particular, we are interested in the marginal probabilities associated to the variables and the mutual information between pairs of variables.

4.2   Statistical Analysis of the Braids Space

Figure 1 shows the univariate probabilities computed from the Boltzmann distribution for functions f and $\bar{f}$, and 10 variables. The search space comprises $4^{10} = 1,048,576$ braids. Univariate probabilities for function $\hat{f}$ were also computed, they are similar to probabilities obtained for f, and due to space constraints we do not include figures for this function. p_1, p_2, p_3, and p_4 respectively represent the univariate probabilities for braid generators λ_1, λ_2, λ_1^{-1}, and λ_2^{-1}. For all the functions, higher probabilities for p_3 indicate that λ_1^{-1} is more likely to be present in the best solutions. This is the type of statistical regularities that can be detected and exploited by EAs that learn probabilistic models.

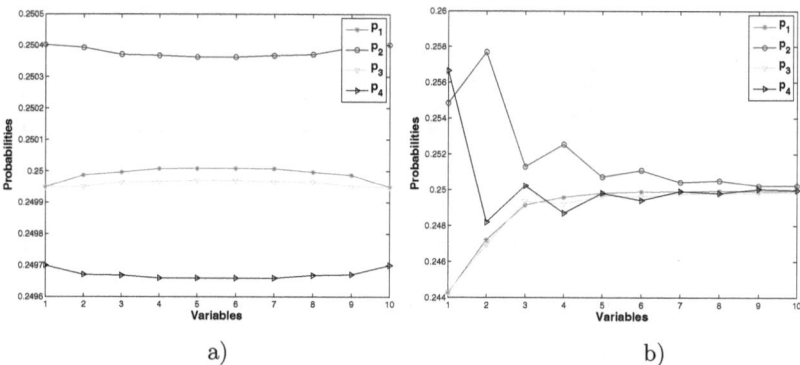

a)   b)

Fig. 1. Univariate probabilities of the Boltzmann distribution for: a) f, b) $\bar{f}$

We compute the bivariate marginal distributions between every pair of variables and derive the values of the mutual information. The mutual information is a measure of statistical dependence between the variables and can serve to identify variables that are dependent. A strong dependence between two variables may indicate that their joint effect has a strong influence on the function and the optimizer should take into account this interaction. Figure 2 shows the mutual information computed for functions f and $\bar{f}$. It can be seen in Figure 2 that for the two functions the strongest dependencies are between adjacent variables, although for function f there is also a strong dependence between the first and the last variables. It can be also seen in Figure 2b) that the dependencies between adjacent variables decreases with the index for function $\bar{f}$.

Summarizing, the statistical analysis of the Boltzmann distribution shows that there are at least two types of regularities of the braid problem that are translated into statistical features. Firstly, there are different frequencies associated to the generators in the space of the best solutions. Secondly, there are strong dependencies between the variables, particularly those that are adjacent in the braid representation.

5   Estimation of Distribution Algorithms

EDAs use samples of solutions to learn a model that captures some of the regularities that may exist in the data. The pseudocode of an EDA is shown in Algorithm 1.

We work with positive distributions denoted by p. $p(x_I)$ denotes the marginal probability for $\mathbf{X}_I = \mathbf{x}_I$. $p(x_i \mid x_j)$ denotes the conditional probability

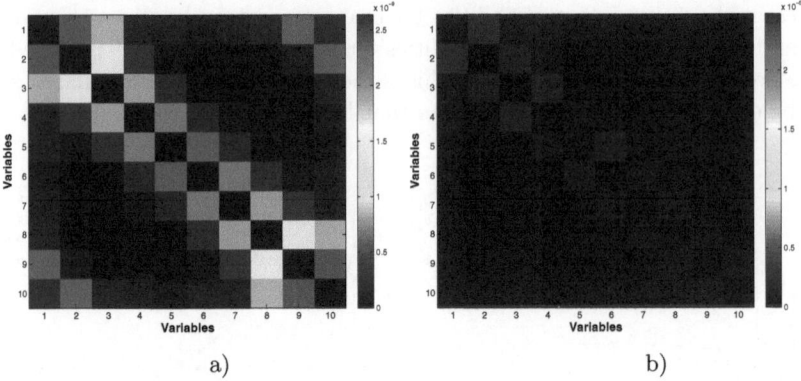

a)   b)

Fig. 2. Mutual information computed from the Boltzmann distribution for: a) f, b) $\bar{f}$

Algorithm 1. Estimation of distribution algorithm

1	Set $t \Leftarrow 0$. Generate N solutions randomly.
2	**do** {
3	Evaluate the solutions using the fitness function.
4	Select a population D_t^S of $K \leq N$ solutions according to a selection method.
5	Calculate a probabilistic model of D_t^S.
6	Generate N new solutions sampling from the distribution represented in the model.
7	$t \Leftarrow t + 1$
8	} **until** Termination criteria are met.

distribution of $X_i = x_i$ given $X_j = x_j$. Three types of probabilistic graphical models are used: 1) Univariate model. 2) 1-order Markov model. 3) Tree model.

Univariate	1-order Markov	Tree
$p_u(\mathbf{x})$	$p_{MK}(\mathbf{x})$	p_T
$\prod_{i=1}^{n} p(x_i)$	$p(x_1) \prod_{i=2}^{n} p(x_i \mid x_{i-1})$	$\prod_{i=1}^{n} p(x_i \mid pa(x_i))$

In the univariate model, variables are considered to be independent, and the probability of a solution is the product of the univariate probabilities for all variables. In the 1-order Markov model, the configuration of variable X_i depends on the configuration of its previous variable. In a probability distribution conformal with a tree, $Pa(X_i)$ is the parent of X_i in the tree, and $p(x_i \mid pa(x_i)) = p(x_i)$ when $pa(X_i) = \emptyset$, i.e. X_i is a root of the tree. We allow the existence of more than one root in the PGM (i.e. forests) although for convenience of notation we refer to the model as tree.

Univariate approximations are expected to work well for functions that can be additively decomposed into functions of order one (e.g. $g(\mathbf{x}) = \sum_i x_i$). However, other non additively decomposable functions can be easily solved with EDAs that use univariate models (e.g. $g(\mathbf{x}) = \prod_i x_i + \sum_i x_i$) [9]. Therefore, it makes sense to test the univariate approximation for the braid problem. The 1-order Markov model captures only dependencies between adjacent variables, and the tree model can represent a maximum of $n - 1$ bivariate dependencies. The computational cost of EDAs is mainly associated to the methods needed to learn and sample the models. The most complex EDA used in this paper is Tree-EDA which has a computational cost $O(n^2)$. Examples of EDAs that use univariate, 1-order Markov, and tree models are respectively presented in [10], [4] and [1,13] and details on the methods used to learn and sample the models can be obtained from these references.

5.1   Enhancements to the EDAs

We consider three enhancements to EDAs: 1) Use of a local optimizer. 2) Partial sampling. 3) Recoding.

As is the case of other EAs, EDAs can be enhanced by the incorporation of local optimizers. We use a greedy optimization algorithm that is applied during the evaluation of the population by the EDA. The algorithm starts from the solution generated by the EDA. In each iteration, the local optimizer evaluates all the $3n$ solutions that are different to the current solution in only one variable (the neighbor solutions). The next selected solution is the neighbor that improves the fitness of the current solution the most. The algorithm stops when none of the neighbors improves the fitness of the current solution.

During the sampling step of an EDA, all variables are assigned their values according to the probabilistic model and the sampling method. For the EDA that uses the univariate model, variables are independently sampled. For 1-order Markov and tree, probabilistic logic sampling (PLS) [5] is used. In both methods, all variables are assigned the new values. However, for some problems with higher-order interactions using a base-template solution can be better than generating each new solution from scratch.

In partial sampling, a solution of the population is selected and only a subset of its variables are sampled according to the model. We use two variants of partial sampling I) Partial sampling where the number of variables to be modified is randomly selected between 1 and n. II) Partial sampling, where the number of variables to be modified is randomly selected between 1 and $\frac{n}{2}$.

Recoding consists in modifying the representation of the solution after the fitness evaluation. For functions $\hat{f}$ and $\bar{f}$ it is possible to recode the solution by eliminating redundant generators (e.g., pairs $\sigma_i \sigma_i^{-1}$). The rationale of using recoding is that meaningful variables will be located closer to the beginning of the braid. Since solutions have a fixed length, the last variables will be kept unused, i.e. garbage information. Therefore, we devised two ways to fill these gaps: I) Leaving the unused variables as they were in the original solution. II) Replacing the unused variables by a reverse copy of the variables used in the evaluation. The second variant intends to replicate information that has proved to be "informative" about the problem.

6 Experiments

The main objective of our experiments is to evaluate the capacity of the EDAs to find optimal solutions to the braid problem. We run experiments for $n \in \{50, 100, 150, 200, 250\}$ in order to evaluate the scalability of the algorithms. Increasing n may lead to obtain braids with a smaller error. A second objective is to compare different variants of the problem formulation and of the algorithm.

6.1 Experimental Settings

Each EDA is characterized by 5 parameters:

- Use of local optimizer. 0: Only EDA is applied, 1: EDA is combined with greedy search as described in Section 5.1.
- Type of function and representation. 0: Function f, 1: Function $\bar{f}$ without recoding, 2: Function $\bar{f}$ with recoding I, 3: Function $\bar{f}$ with recoding II.
- λ value. 0:0.0, 1:0.01, 2:0.05, 3:0.1.
- Sampling method. 0: Normal, 1: Partial sampling I, 2: Partial sampling II.
- Type of probabilistic model. 0: Univariate, 1: 1-order Markov, 2: Tree.

The experimental settings were selected to investigate different aspects that influence the behavior of the algorithm. The total number of variants of the algorithm was $2 \times 4 \times 4 \times 3 \times 3 = 288$. All the algorithms use truncation selection, in which the best 5% of the population is selected. EDAs that do not incorporate the greedy local search use a population size $N = 10000$. For these EDAs, the number of generations was dependent on n as $N_g = 15n$. Due to the large number of evaluations spent by the greedy search method, the population size for all hybrid EDAs was $N = 100n$ and the number of generations was fixed to $N_g = 100$. For each EDA variant, 100 experiments were run.

6.2 Best Solutions Found by EDAs

Figure 3a) and Table 1 respectively show the parameters that characterize the best braids found by the EDAs for each value of n, and the braids. In Figure 3a), we also show an estimate of the length of the braids $(O[log_{10}^{3.97}(1/\epsilon)])$ that would compute the Solovay-Kitaev algorithm [3] to obtain the same error ϵ of our best solutions. The lengths of our solutions clearly outperform these estimates. Figure 3b) shows the length of all the best solutions achieved for each value of n. It can be observed in Figure 3 that EDAs are able to find several braids with different lengths for $n = 150$ and $n = 200$.

6.3 Behavior of the EDA Variants

We further investigate the behavior of the different EDA variants. Figure 4 shows the violin plots [6] with the distribution of the best values found in all the executions for: a) All EDA variants without local optimizer (14400 runs), b) All

n	l	ϵ	$log_{10}(\epsilon)$	$log_{10}^{3.97}(\frac{1}{\epsilon})$
50	44	4.8435×10^{-4}	-3.3148	116.47
100	70	8.3527×10^{-6}	-5.0782	633.37
150	64	8.3527×10^{-6}	-5.0782	633.37
200	62	8.3527×10^{-6}	-5.0782	633.37
250	124	3.5038×10^{-6}	-5.4555	841.82

a)

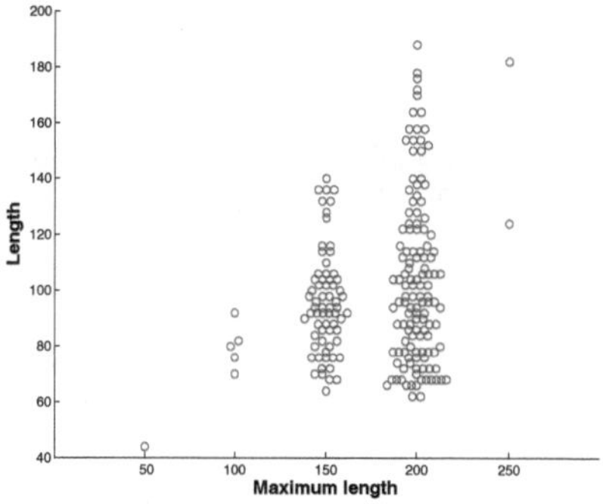

b)

Fig. 3. a) Parameters of the best braids found by the EDAs for each value of n. b) Length of the best solutions found for each value of n

EDA variants that incorporate the greedy search (14400 runs), c) EDAs with local optimizer, recoding type II, and that use partial sampling type II (300 runs). Each violin plot shows a histogram smoothened using a kernel density with Normal kernel. The mean and median are shown as red crosses and green squares, respectively.

Table 1. Best braids found by the EDAs for each value of n

n	braid
50	$\sigma_1^{-1}\sigma_2^3\sigma_1^{-2}\sigma_2\sigma_1^{-1}\sigma_2\sigma_1^{-2}\sigma_2^3\sigma_1^{-1}\sigma_2\sigma_1^{-1}\sigma_2\sigma_1^{-2}\sigma_2\sigma_1^{-2}\sigma_2\sigma_1^{-1}\sigma_2^3\sigma_1^{-1}\sigma_2^2\sigma_1^{-1}\sigma_2\sigma_1^{-1}\sigma_2\sigma_1^{-2}\sigma_2^2$ $\sigma_1^{-3}\sigma_2^2$
100	$\sigma_1\sigma_2^{-2}\sigma_1^4\sigma_2^{-1}\sigma_1\sigma_2^{-4}\sigma_1\sigma_2^{-1}\sigma_1^3\sigma_2^{-1}\sigma_1^2\sigma_2^{-2}\sigma_1^4\sigma_2^{-1}\sigma_1\sigma_2^{-4}\sigma_1\sigma_2^{-1}\sigma_1^6\sigma_2^{-2}\sigma_1\sigma_2\sigma_1\sigma_2^2\sigma_1^3\sigma_2^{-3}\sigma_1^{-1}\sigma_2$ $\sigma_1^{-3}\sigma_2\sigma_1^2\sigma_2^5$
150	$\sigma_2\sigma_1^{-1}\sigma_2\sigma_1^{-1}\sigma_2^2\sigma_1^{-1}\sigma_1\sigma_2^{-4}\sigma_1\sigma_2^{-1}\sigma_2^{-1}\sigma_1^{-1}\sigma_2^{-1}\sigma_1^4\sigma_2^{-1}\sigma_1\sigma_2^{-1}\sigma_1^2\sigma_2^{-1}\sigma_1^{-1}\sigma_2\sigma_1^{-3}$ $\sigma_2^{-1}\sigma_1\sigma_2^{-1}\sigma_1^3\sigma_2^{-1}\sigma_1\sigma_2^{-4}\sigma_2^2\sigma_1^{-4}\sigma_2^2\sigma_2^{-3}$
200	$\sigma_2\sigma_1^{-2}\sigma_2^4\sigma_1^{-1}\sigma_1\sigma_2^{-4}\sigma_1\sigma_2^{-1}\sigma_1^2\sigma_2^{-1}\sigma_1^3\sigma_2^{-1}\sigma_1\sigma_2^{-4}\sigma_1\sigma_2^{-1}\sigma_1^4\sigma_2^{-1}\sigma_1\sigma_2^{-2}\sigma_2^2\sigma_1\sigma_2^{-1}\sigma_1\sigma_2^{-5}$ $\sigma_2\sigma_1^{-1}\sigma_2^{-6}\sigma_1^{-2}\sigma_2^3$
250	$\sigma_1\sigma_2\sigma_1^{-1}\sigma_2\sigma_1^{-1}\sigma_2^{-1}\sigma_1\sigma_2^{-1}\sigma_1^{-1}\sigma_2^3\sigma_1^4\sigma_2^{-2}\sigma_2^4\sigma_1^{-1}\sigma_2\sigma_1^2\sigma_2\sigma_1^{-1}\sigma_2^4\sigma_1^{-2}\sigma_2^4\sigma_1^{-3}\sigma_2^{-1}\sigma_1\sigma_2^{-1}\sigma_1^{-1}\sigma_2^2$ $\sigma_1\sigma_2^{-1}\sigma_1^2\sigma_2^{-1}\sigma_1\sigma_2^{-4}\sigma_1^2\sigma_2^{-1}\sigma_1\sigma_2^{-1}\sigma_1^4\sigma_2^{-1}\sigma_1^{-1}\sigma_2^{-1}\sigma_1^{-1}\sigma_2^{-1}\sigma_1^4\sigma_2^{-1}\sigma_1\sigma_2^{-4}\sigma_2^{-4}$ $\sigma_1\sigma_2^{-2}\sigma_1^{-1}\sigma_2^5\sigma_1^{-1}\sigma_2\sigma_1^{-1}\sigma_2^{-4}\sigma_2^2\sigma_1^{-4}\sigma_2^2\sigma_1^{-4}\sigma_2^2\sigma_1^{-1}$

In Figure 4, the modes of the Normal distribution indicate the existence of a local optimum of the error with a very wide basin of attraction. This local optimum has value $log_{10}f(\epsilon) = -2.50785$ and the majority of the EDA runs can be trapped in this value. Differences between the EDAs due to the application of the greedy method can be appreciated for $n = 200$ and $n = 250$ (Figures 4a) and 4b)). Also, Figure 4c) reveals how a particular combination of the EDA's parameters can improve the results of the search.

There are a number of commonalities between the best EDA variants. Except in one case, all EDAs use recoding of type II. Similarly, except in one case, in all the variants $\lambda \in \{0.01, 0.05\}$. Except in two cases, the sampling method selected was partial sampling. The application of the local optimizer notably improved

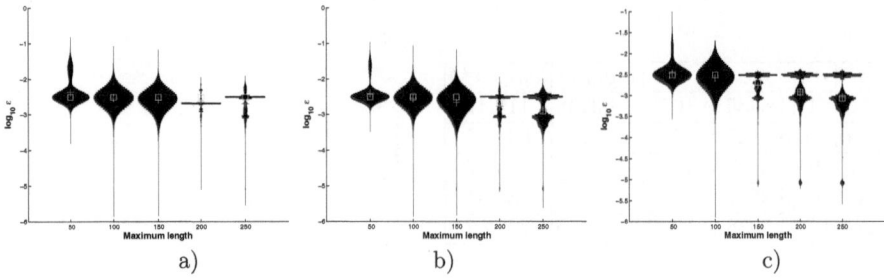

Fig. 4. Violin plots showing the distribution of the best values found in all the executions for: a) All EDAs variants without local optimizer (14400 runs), b) All EDAs variants with local optimizer (14400 runs), c) EDAs with local optimizer, recoding type II, and that use partial sampling type II (300 runs)

the results for $n \in \{150, 200\}$ but in terms of the best solution found it did not have an important influence for the other values of n.

As a summary, we recommend to use an EDA that adds the greedy search, and uses partial sampling of type II and the 1-order Markov model since it is less complex than the tree and results achieved using the two models are similar.

6.4   Improvement Over Other Search Methods

As a final validation of our method, we compare our best EDA variant with the results achieved using a random search, the greedy local optimizer, and the GA introduced in [8]. For the random search, we randomly generated 10000 solutions and selected the best solution according to function $\bar{f}, \lambda = 0.01$. The same experiment was repeated 100 times to select the 100 "best" solutions for $n \in \{50, 100, 150, 200, 250\}$.

A similar procedure was followed for the greedy local search. The local optimizer was applied to each of the 10000 solutions until no improvement was possible. For the GA, we used the results of the 100 GA runs analyzed in [8]. Since these results were obtained using solutions of different length, and with a different number of evaluations, care must be taken to interpret the differences. We only compare the GA results with the other algorithms for $n = 50$. Similarly, the results of the random search were very poor for $n > 50$ and we only include them in the comparison for $n = 50$. Results are shown in Figure 5a). In the boxplots, the central mark is the median, the edges of the box are the 25th and 75th percentiles, the whiskers extend to the most extreme data points not considered outliers, and outliers are plotted individually.

Using the 100 best solutions for each of the four algorithms, a multiple comparison test was applied to test the null hypothesis that samples corresponding to every pair of algorithms are drawn from the same population. The multiple comparison test uses the Tukey's honestly significant difference criterion. Every pair-wise comparison is based on the Kruskal-Wallis test, a nonparametric

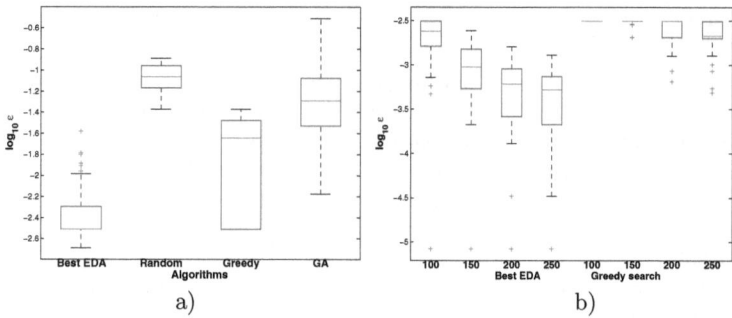

Fig. 5. a) Comparison between the Best EDA variant, the random search, the greedy local search and the GA for $n = 50$. b) Comparison between the Best EDA variant and the greedy local search $n \in \{100, 150, 200, 250\}$

version of the classical one-way ANOVA. The significance criterion was $\alpha = 0.05$. The application of the test identified statistical differences between each pair of algorithms and they were ranked as: 1) Best-EDA, 2) Greedy, 3) GA, 4) Random.

The results of the comparison between the EDA and the greedy search for $n > 50$ are shown in Figure 5b). The application of the Kruskal-Wallis test ($\alpha = 0.05$) found significant differences between the EDA and the Greedy algorithm for all n. Furthermore, it can be seen in Figure 5b) that as n increases the algorithm is able to scale and find better solutions.

7   Conclusions

In this paper we have proposed for the first time the use of probabilistic modeling of the search space to address the problem of approximating a quantum gate as a product of braid generators. We have shown that some of the problem characteristics can be translated into statistical regularities of the Boltzmann distribution. This result indicates that capturing and exploiting statistical regularities emerges as a sensible approach to the quasiparticle braid problem.

In a second step we have shown the effectiveness of EDAs to find short braids that provide accurate approximations. The best braids obtained with our EDAs have lengths up to 9 times shorter than those expected from braids of the same accuracy obtained with the Solovay-Kitaev algorithm and had not been previously reported to be found by the GA approach.

Acknowledgments. R. Santana has been partially supported by the Saiotek and Research Groups 2013-2018 (IT-609-13) programs (Basque Government), TIN2013-41272P (Ministry of Science and Technology of Spain), COMBIOMED network in computational bio-medicine (Carlos III Health Institute), and by the NICaiA Project PIRSES-GA-2009-247619 (European Commission). H. G. Katzgraber acknowledges support from the NSF (Grant No. DMR-1151387).

References

1. Baluja, S., Davies, S.: Using optimal dependency-trees for combinatorial optimization: Learning the structure of the search space. In: Fisher, D.H. (ed.) Proceedings of the 14th International Conference on Machine Learning, pp. 30–38 (1997)
2. Bonesteel, N.E., Hormozi, L., Zikos, G., Simon, S.H.: Braid topologies for quantum computation. Physical Review Letters 95(14), 140503 (2005)
3. Dawson, C.M., Nielsen, M.A.: The Solovay-Kitaev algorithm. arXiv preprint quant-ph/0505030 (2005)
4. De Bonet, J.S., Isbell, C.L., Viola, P.: MIMIC: Finding optima by estimating probability densities. In: Mozer, et al. (eds.) Advances in Neural Information Processing Systems, vol. 9, pp. 424–430. The MIT Press, Cambridge (1997)
5. Henrion, M.: Propagating uncertainty in Bayesian networks by probabilistic logic sampling. In: Lemmer, J.F., Kanal, L.N. (eds.) Proceedings of the Second Annual Conference on Uncertainty in Artificial Intelligence, pp. 149–164. Elsevier (1988)
6. Hintze, J.L., Nelson, R.D.: Violin plots: a box plot-density trace synergism. The American Statistician 52(2), 181–184 (1998)
7. Larrañaga, P., Karshenas, H., Bielza, C., Santana, R.: A review on probabilistic graphical models in evolutionary computation. Journal of Heuristics 18(5), 795–819 (2012)
8. McDonald, R.B., Katzgraber, H.G.: Genetic braid optimization: A heuristic approach to compute quasiparticle braids. Physical Review B 87(5), 054414 (2013)
9. Mühlenbein, H., Mahnig, T., Ochoa, A.: Schemata, distributions and graphical models in evolutionary optimization. Journal of Heuristics 5(2), 213–247 (1999)
10. Mühlenbein, H., Paaß, G.: From recombination of genes to the estimation of distributions I. Binary parameters. In: Ebeling, W., Rechenberg, I., Voigt, H.-M., Schwefel, H.-P. (eds.) PPSN 1996. LNCS, vol. 1141, pp. 178–187. Springer, Heidelberg (1996)
11. Read, N., Rezayi, E.: Beyond paired quantum Hall states: parafermions and incompressible states in the first excited Landau level. Physical Review B 59(12), 8084 (1999)
12. Santana, R., Bielza, C., Larrañaga, P.: Conductance interaction identification by means of Boltzmann distribution and mutual information analysis in conductance-based neuron models. BMC Neuroscience 13(suppl 1), P100 (2012)
13. Santana, R., Larrañaga, P., Lozano, J.A.: Protein folding in simplified models with estimation of distribution algorithms. IEEE Transactions on Evolutionary Computation 12(4), 418–438 (2008)
14. Sarma, S.D., Freedman, M., Nayak, C.: Topological quantum computation. Physics Today 59(7), 32–38 (2006)
15. Xu, H., Wan, X.: Constructing functional braids for low-leakage topological quantum computing. Physical Review A 78(4), 042325 (2008)

Adaptive System Design by a Simultaneous Evolution of Morphology and Information Processing

Olga Smalikho[1] and Markus Olhofer[2]

[1] Technische Universität Darmstadt, Darmstadt, Germany
Olga.Smalikho@rmr.tu-darmstadt.de
[2] Honda Research Institute Europe, Offenbach, Germany
Markus.Olhofer@honda-ri.de

Abstract. The continuous need to increase the efficiency of technical systems requires the utilization of complex adaptive systems which operate in environments which are not completely predictable. Reasons include the often random nature of the environment and the fact that not all phenomena can be explained in full detail. As a consequence the developer often gets confronted with the task to design an adaptive system with only little prior knowledge about the problem at hand. The design of adaptive systems, which react autonomously to changes in their environment, require the coordinated generation of sensors, providing information about the environment, actuators which change the current state of the system and signal processing structures thereby generating suitable reactions to changed conditions. In this work we demonstrate the applicability of a concurrent evolutionary design of the optimal morphological configuration, presented as sensory and actuation systems, and the corresponding optimal controller part of a system. We apply the process to the example of an adaptive wing design. Prior experiments for the optimization of the systems, having fixed number of sensor and actuator elements demonstrate the existence of an optimal dimensionality of the systems morphology. We show that the presented growth method is able to detect this morphological configuration and concurrently find the optimal corresponding controller autonomously during a single combined evolutionary process.

Keywords: adaptive systems, co-evolution, system growth, dynamic evolutionary optimization.

1 Introduction

The evolutionary development of higher animals can be seen as a complex process of ongoing body-brain complexification to better adjust to the environment. Since the morphology of the body is tightly coupled to the brain structure, these two functional parts of living creatures are supposed to co-evolve. Admittedly, an addition of new sensory inputs does not give an individual a performance

G. Dick et al. (Eds.): SEAL 2014, LNCS 8886, pp. 25–36, 2014.
© Springer International Publishing Switzerland 2014

advantage without the adjustment of a corresponding signal processing structure. Analog to the development of complex living systems, we assume that the design process of technical solutions with high complexity could be improved by starting the system development with an initially simple system organization while performing simultaneous complexification of its functional parts. The majority of the current engineering methods adapt isolated parts of the overall structure, which is in a strong contrast to biological body-brain co-evolution. The design of adaptive systems implies a selection of sensors and actuators, which adjust the system to a changed environment, as well as a generation of a corresponding controller, according to a predefined quality measure. The developers usually design the morphological part, defined as sensors and actuators of the system separately from the development of the corresponding controller unit. This approach has been a usual practice in a field of evolutionary robotics. First, the real mobile robots have been given fixed morphological limitations, such as fixed number and resolution of camera system, fixed joints angles range etc. Than through the following optimization of controller structure a complex behavior, like for example an obstacle avoidance tasks of a robot, can be achieved. Examples of this approach can be found in Brooks [1], Dorigo and Schnepf [2], Cliff, Husband and Harvey [3], Floreano and Mondada [4], Miglino, Nafasi, Taylor [5].

The weakness of the approach is that the controller performance strongly depends on the suitability and the amount of sensory information, as well as on the actuator resources. This causes the problem, that the optimal system performance is difficult to be achieved, if not all detailed phenomena about the system are known during the first phase, in which the hardware configuration is defined. Otherwise it can happen, that some important information about the environment or an actuator at the position in the structure, having a major impact on the system performance, is missing. As an attempt to overcome the problem, we could optimize an initially very rich system, having high number of sensory and actuation elements. This would statistically decrease a chance of missing important environmental factors during the sensors acquisition. However, the optimization progress of the system having a large scale dimensionality might be not possible due to the high number of optimization parameters. To solve the problem we implement a growth method which synchronize the design process of sensing and signal processing system parts during optimization process and additionally frees the system of early structural limitations. Therefore it gives a possibility to develop a system autonomously to optimal morphological configuration.

A variety of approaches for the co-evolutionary design of morphological and controller configurations have been developed in the field of evolutionary robotics. Early work in the field of automatic design of a systems by body-brain co-evolution has been reported by Sims [6]. He demonstrated the evolutionary development of the morphology of virtual creatures in a physical simulation fulfilling simple locomotion tasks starting from simple building blocks without any prior knowledge. Parker and Nathan [7] researched the design of sensor morphology and controller for a simulated hexapod robot. For this purpose the type of sensors, the heading angle and the range of the sensors as well as the rules for the controller are co-evolved. This method enables the system to extract information from the environment which is relevant to complete a given task by configuring a minimal controller and minimal number of sensors and actuators to increase the system's overall efficiency. Bugajska and Schutz [8] co-evolved the shape and strategies in the design of Micro Air Vehicles (MAV). The target, similar to Parker and Nathan, was to find a minimal sensor suite and reactive strategies for navigation and collision avoidance tasks. Sugiura et al. [9] also proposed a system that automatically designs the sensor morphology of an autonomous robot with two kinds of adaptation: ontogenetic and phylogenetic adaptation. Gomez, Lungarella, Eggenberger, et al. [10] extended the principle of sensor-control balance to developmental enlargement of the system by simultaneously increasing the sensor resolution, the precision of motors, as well as the size of a neural structure (controller). They showed with their experiments that a chosen system starting with a low-resolution sensory system, a low precision motor system, and a low complexity of controller, can learn a given task faster than a system with a complex configuration already in the beginning. The coordination of the growth process was realized through internal learning mechanisms where the active neural units controlling the robot were 'rewarded' or 'punished' depending on the improvement or aggravation in task fulfilling. The authors showed the advantages of concurrent optimization of the sensory and control systems as a dynamic developmental process of gradual complexification. Also Auerbach and Bongard [11] have made extensive research in the field of co-evolution of morphology and control in evolutionary robotics. In their work they implement a growth mechanism to create robots using compositional pattern-producing networks and demonstrate that the concurrent development of the morphological and controller structures of the simulated adaptive robots can give an advantage for the final system performance, compared to the approaches with separate design strategies.

The promising results of the co-evolutionary approach in evolutionary robotics motivated us to implement the biologically inspired growth process for the coordination of the fully autonomous development of sensor, actuator and control structures without dimensional limitation of sensory or actuator setup in the early stages of development. Since the final system configuration is not predefined and is the result of the concurrent optimization process, we expect an evolvable system through enlargement of the search domain and potentially increase the chances of global optima detection. Compared with the reviewed research in evolutionary robotics, we utilize the co-evolution of morphology and information processing structure for the optimal control of an adaptive wing shape. Although the generation of optimal control for adaptive wings is not in the main focus of our research we argue that this problem is a suitable test bed for the research on evolutionary design of adaptive systems. Aerodynamic problems are characterized by highly complex interactions between flow body and flow field which is in most cases difficult to understand in detail. Due to this, the manual design is generally challenging although excellent tools are available for their simulation and evaluation. In this work we demonstrate that evolutionary methods are able to generate systems which can optimally adapt to environmental conditions, while at the same time we target shedding some light on the precise synchronization of system parts during the developmental process.

In our previous work [12] we implemented the co-evolutionary growth of a sensory and controller structures given a fixed actuation system on the example of the adaptive wing. The promising results motivated us to expand the research to the combined development of all functional parts of the system. In this case the overall system configuration is fully variable during the optimization and initially minimal. Additionally we implemented cost factors for sensors and actuators, which result in a limited growth of the system dimensionality. In this case a system gets a new sensor or actuator only if it gives a significant benefit to the system performance. The target for the optimization of the adaptive wing is the reduction of the drag the airfoil generates while still creating a minimum of lift. Environmental changes are realized through the changes in the angle of attack of the airflow across a wide range of values. The co-evolutionary system development has been implemented as a structural growth process of all it functional parts, such as the number and position of the sensors and actuators and the complexity of the controller structure, defined by its input (sensors) and output (actuators) dimensionality. We show the results of the sensors-actuators-controller growth method and compare them with the optimization of the system with fixed morphological settings as well as with the results of the growth of sensors-controller systems, having a fixed actuation system.

A description of the framework of the adaptive wing with a detailed explanation of the functional set-up of the sensory, actuation and controlling systems is given in section 2. In section 3 we explain and present the results of the experiments with implemented growth method, and compare them with optimization results for a fixed morphology as well as with the results of the growth of sensory and controller systems. Finally, we conclude the paper with a summary of the main findings of our work.

2   Framework for Morphology-Controller Co-evolution

In our work we implemented a system, consisting of virtual sensors, actuators and a signal processing structure. The signal processing structure controls the adaptive system under changing environmental conditions by generating actuator signals based on sensor signals derived from the environment. The target has been to achieve a system behavior which reduces the airfoil's drag, calculated in a CFD (computational fluid dynamics) simulation of the resulting airfoil shape while maintaining specified lift values. The actuator signals correspond to changes of NURBS [13] control points and define the current airfoil shape. The virtual sensors of the system have been defined as pressure sensors, at given positions on the airfoil surface. The measurement results of the virtual sensors correspond to the surface pressure calculated in the CFD simulation and therefore depend on the blade's surface, the angle of attack and the speed of the air

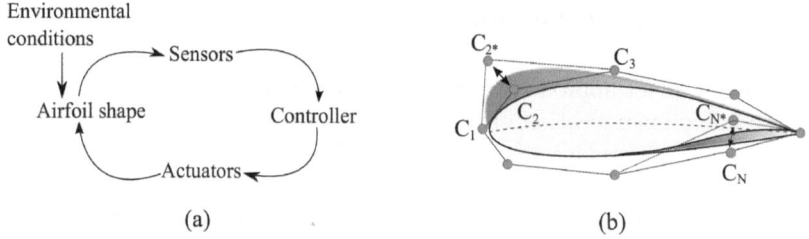

Fig. 1. (a) Adaptive airfoil framework, (b) Example of the airfoil created with NURBS. Airfoil in white, defined by the initial positions of the spline control points. The airfoil shape change (in gray) results from the movements of C_2 and C_N.

flow etc. Fig. 1 (a) shows the described relations between the single parts of the test-framework. The described setup allows us to simulate the interactions between control structure and morphology during the operation of the control structure as well as during their evolutionary development.

The two dimensional airfoil is represented by a non-uniform rational B-splines (NURBS) as shown in Fig. 1 (b). The shape of the NURBS curve is determined by the set of spline control points C_n. For the simulation of the aerodynamic airfoil characteristics and pressure distribution the computational fluid dynamic solver Xfoil[1] is used. In the simulation we change the angle of attack in order to generate variations of the airfoil environment. The Reynolds number has been fixed during the optimization ($Re = 10^7$). The pressure coefficient over the airfoil surface represent the virtual sensor data. We used Xfoil to calculate the profile of the pressure coefficients at 160 points on the airfoil surface. The more detailed description of the implemented framework can be found in [12], [14].

The presented adaptive system of a virtual wing requires a suitable controller for processing of sensor data and generating actuator signals. The actuators morph the surface of the airfoil and therefore can reduce the drag and generate the required lift. A wide variety of possible controller designs for similar purposes can be found in the literature. Parker and Nathan [7] as well as Bugajska and Schutz [8] implemented a controller as a reactive system which uses "if...then" rules to control a simulated robot. Haller, Ijspeert and Floreano [15] implemented a controller inspired by the central pattern generators underlying locomotion in animals. In this work we implemented a linear recurrent model to control the presented adaptive system. The controller input signals are the pressure coefficients s_i. The outputs of the controller are the actuator signals C_o and describe the position of the virtual actuators in the two dimensional space. The actuator adjustments ΔC_o are calculated as the sum of the signals of all sensors of the system multiplied with the corresponding linear coefficients K_{io}. The current state of the actuators C_o determines the pressure coefficients s_i by the nonlinear air flow function f, simulated with Xfoil solver.

---

[1] http://web.mit.edu/drela/Public/web/xfoil/

3   Sensor-Actuator-Controller Co-evolution

The co-evolutionary approach has been realized by the concurrent development and gradual complexification of the sensory, actuation and corresponding controller systems. Although these different functional parts of the system are coded in a single genome, the described optimization process reflects the main aspect of the co-evolutionary process, since all units are mutually co-adjusted during the entire evolutionary process. In this work we combine the co-evolutionary method with a growth process and implement it as a single dynamic optimization task. We use standard evolution strategy optimization, ES(15,100) developed by Bienert, Rechenberg and Schwefel [16] to optimize the overall system configuration. Although more sophisticated algorithms are available like CMA-ES [17], we stick to the comparable simple strategy which can also be applied in the growth process. A single genome of the individual includes four chromosomes which code the position of the sensors and actuators as well as the controller parameters and additionally the strategy parameters (mutation step sizes). Offspring are the result of the gene intermediate recombination and mutation of the parent individuals with the current mutation step size. The optimization has been implemented, using SHARK[2], open-source C++ machine learning library. One of the important advantages of the growth process is a minimal requirement on the prior knowledge about the system. The growth process shows fast convergence in small search spaces, since it starts with an initially elementary system. Thereby we free the system from the early limitations and allow it to develop autonomously into a final system, having as simple morphology as possible, optimally positioned in the structure and in the same time having an optimal signal processing structure for the given morphology.

The results of our previous work in [12] demonstrate the expected difficulties of the standard optimization strategies on large scale problems. The experiments on the optimization of the systems with different number of sensors showed the infeasibility of the standard ES for detection of the optimal solution for a systems, having more than 5 sensors. We found out, that concurrent growth method, based on the standard ES, could overcome these problems. Additionally, such a system development method resulted a structured system organization, with a strong hierarchical arrangement of the elements of sensor and controller structures. In this way the arranged system organization provides information about the importance of the present sensor elements of the system. However, the overall system configuration in [12] has been pre-defined to the system using 6 actuators at fixed positions. In this work we fully free the system of any morphological or controller system limitations on the early developmental stages and evolve all functional parts of the system simultaneously.

The system is evaluated according to its ability to reduce the drag while changes in the inflow angle of the air occur during the optimization process. The ratio of the drag coefficients before change of the inflow angle happened C_d^t and after C_d^{t+1} a change is evaluated. The fitness of an individual is defined as a sum

---

[2] http://image.diku.dk/shark/

of drag value ratios over a cascade of 16 different angles of attack. Additionally the cost functions of the sensor and actuator elements multiplied with a number of sensors and actuators respectively has been added to the fitness function.

$$Fitness(Individual) = \frac{\sum_{i=1}^{N} \frac{C_d^{t+1}(\alpha_i)}{C_d^t(\alpha_i)}}{N} + w \cdot S + v \cdot A, \tag{1}$$

where α is the angle of attack, N is the total number of angles of attack applied and on which the individual has been evaluated, C_d^t is the drag coefficient before and C_d^{t+1} - after actuator adjustments, S and A are the number of sensors and actuators, w and v are the cost factors for sensors and actuators. To get a fair comparison we use two constraint parameter derived from the parameters of the NACA 2410 [18] airfoil. NACA airfoils are the aircraft wing shapes, developed by the National Advisory Committee for Aeronautics in 1948 [18] and define since that time a set of standard airfoil shapes. The maximal thickness of the adaptive airfoil was set to the maximal thickness of the NACA 2410 airfoil which is equal to 10% of the chord length. Additionally, we put the constraint on the lift coefficient to be equal or higher than a lift of a NACA 2410 airfoil at corresponding angle of attack.

3.1   Definition of Growth Process

We defined a system growth process as an optimization through gradual enlargement of the initially minimal system by concurrent addition and adaptation of the sensors, actuators and corresponding connection weights of the controller during entire optimization process. There exists a variety of approaches for the topology optimization of the processing structures in the neuro-evolutionary domain proposed for example by Moriarty and Miikulainen [19] (SANE), Stanley and Miikulainen [20],[21] (NEAT).

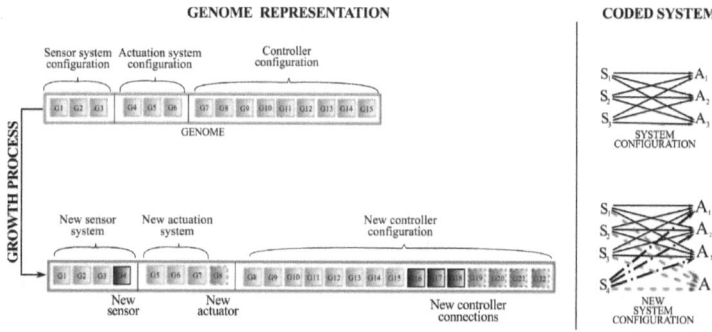

Fig. 2. Demonstration of growth process. Probability based triggering of enlargement of morphology and controller dimensionality

In this work we realize the topology optimization of the controller, consisting of a linear recurrent model, as an optimization of the input as well as output dimensionality of the control structure, while adding necessary connections to keep a fully connected network. We use a direct genotype to phenotype encoding as shown in Fig. 2. We use a neutral system mutation by initializing the weighting coefficients of a new sensor or new actuator elements with zero. The new morphological elements as well as the corresponding controller connections get individual mutation step sizes. Through this important method of system enlargement, triggered by the probability based method, we give a mutated system a possibility to evolve new elements individually while keeping the previously optimized system setup intact. Each individual in a population has a fixed probability $p = 0.2$ to get a new sensor or actuator by mutation. Compared to our previous research we implemented cost factors for the new sensory and actuation elements. The values of the cost factor has been experimentally set to $w = 0.04$ and $v = 0.04$. The cost factors allow the generation of the systems with minimal complexity of morphological and controller configurations required for fulfilling a given task.

3.2 Experimental Results of Growth Process

In this section we demonstrate the experimental results of the presented sensor-actuator-controller growth method of the adaptive wing and compare them with the results of our previous work in [14] and [12].

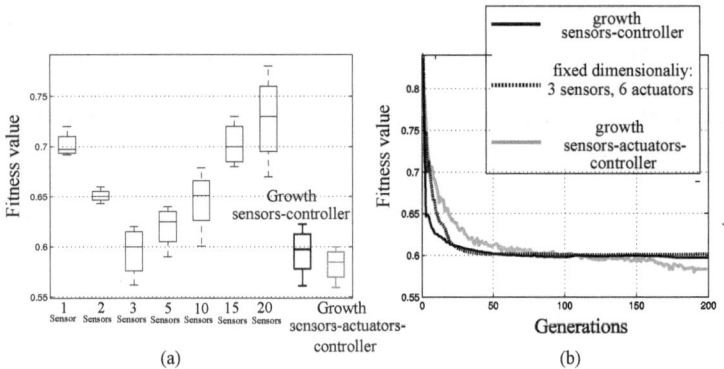

Fig. 3. (a) Final quality of sensors-controller and sensors-actuators-controller optimization runs, compared with a results of the runs with a different fixed morphological configurations. 10 optimization runs with different starting parameters, (b) Comparison of the averaged fitness curves.

Fig. 3 (a) present a comparison of the optimization results for different systems. On the one hand we have results of the systems, having different fixed number of sensory elements and equally positioned fixed 6 actuators. On the other hand we get the optimization results for the sensors-controller and sensors-actuators- controller growth method. In following we compare the results and give an explanation to the difference in the final performance. Basically as the performance of the final system is expected to get better, the more sensory information about the environment is available. Indeed Fig. 3 (a) shows a significant improvement of the final optimized system performance with an enlargement of sensory system dimensionality up to 3 sensors. However the high number of sensors and actuators leads to a high number of optimization parameters. The problems of the standard ES on the large scale problem can be seen for a higher number of sensory elements. For experiments with fixed morphological dimensionality we observe a decline of the performance for the systems, having more than 3 sensors. In this case of a morphologically rich system, an optimization has a high chance to get stuck in local optima and not reach the globally optimal solution.

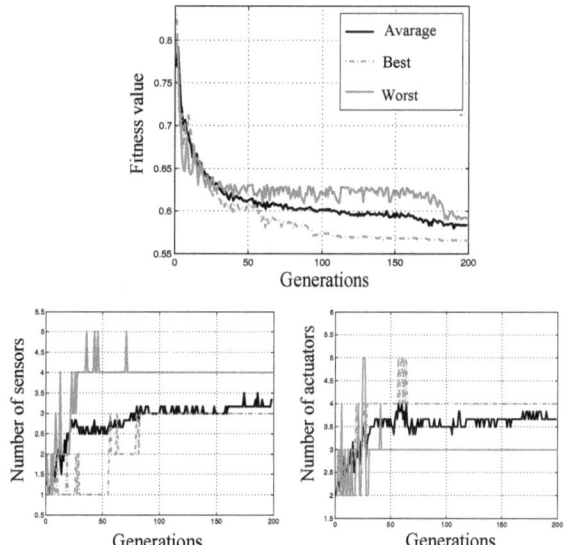

Fig. 4. (a) Average (10 runs), worst and the best fitness progress of the sensors-actuators-controller optimization, (b) Development of the number of sensors over 200 generations, (c) Development of the number of actuators

According to the optimization results, systems with 3 pressure sensors represent an optimal sufficient solution for morphological setting for given optimization strategy, since it reached the best final quality in average. The results demonstrate, that the systems developed with sensors-controller growth method show similar good performance as the systems optimized for 3 sensors, both having 6 fixed actuators. According to the average achieved quality of the fully variable system design, presented as sensor-actuators-controller growth method, we obtain a significant benefit starting the optimization with initially minimal system configuration, which evolves during the optimization through gradual step-wise complexification. The fitness value of the best and worst individual in each population as well as the average sensors-actuators-controller optimization is presented in Fig. 4. The growth method generated a systems, having in average about 3 sensors and between 3 and 4 actuators. To analyze the func-

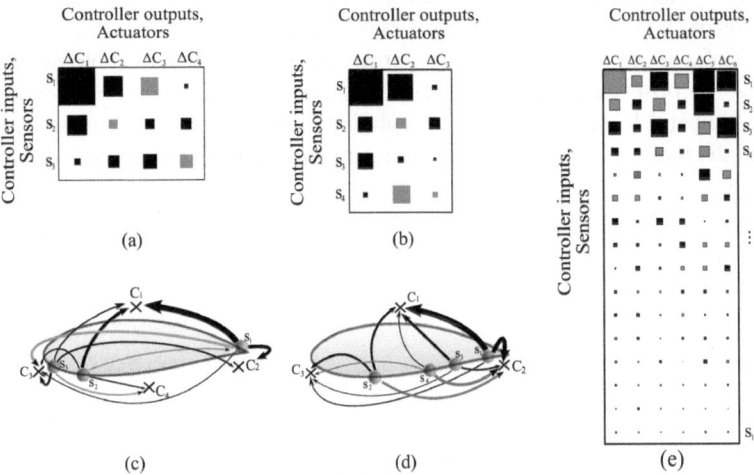

(a)

(b)

(c)

(d)

(e)

Fig. 5. Hinton diagrams of the system controller of the worst (a) and the best (b) system in Fig. 4, developed by the sensors-actuators-controller growth process at 200th generation, (c) and (d) Position of the sensors and actuators of the systems and schematic airfoil controller with the signal connections between sensors and actuators (thicker lines mean a stronger connection), (e) Example of the system controller developed with sensors-controller method without cost functions

tional configuration of the controller a Hinton diagrams has been used [22]. The size of the boxes corresponds to the value of the connection strength. The box color (gray and black) represents a positive or negative sign of the connection strength respectively. The values of the connection strengths have been scaled between minimal (no box) and maximal controller connection strength (box of maximum size). Fig. 5 demonstrates the final controller structures of the worst and the best system, developed with the growth method after 200 generations. Fig. 5 (c), (d) shows the optimized position of the sensors and actuators in both systems. A reason for the performance difference of the two systems seems to be an extra actuator of the first system. The results show in comparison to earlier work, that the actuation resources of the system have a comparable impact on the system performance than the amount of gathered sensor information about the environment. This means that the pre-definition of the configuration of each morphological structure limits a system's global evolvability gradually. The presented system growth method shows experimentally on a virtual adaptive wing design the potentials and benefits of the fully automatic globally optimal system design.

4   Conclusions

In the presented work the evolutionary generation of an optimal adaptive system which reacts autonomously to changes in its environment is described. As an example system for the evolutionary development a problem from the field of aerodynamics was utilized. A virtual wing shape had to be adapted to show minimal drag for changing unknown inflow angles based on pressure measurements on its surface. All functional parts of the systems, such as morphological configuration, defined by the sensory and actuation structures, as well as the configuration of the signal processing unit of the virtual adaptive wing are not specied in the early design stage and are the result of the evolutionary optimization process implemented as a dynamic system growth process. We showed that the proposed concurrent growth method could overcome the problems of standard evolutionary strategies on a given large scale optimization task. According to the presented results we obtain a significant benefit in starting the optimization initially as simple as possible while the system undergoes a stepwise complexification during the co-evolution process. Proposed system growth approach, combined with a cost factors for a morphological dimensionality, is able to detect a minimally possible morphological configuration required to fulfill a given task of drag reduction and maintenance of a required lift. Therefore an optimization process supports generation of preferably low dimensionality of morphological and controller units, which is still sufficient to react optimally in a simulated changing environment. Such a system development method results in a structured system organization, with a strong hierarchical arrangement of the elements of sensor and controller structures according to its importance to the system. This system organization provides an important information to the designers about the significance of sensors and actuators for the system performance.

Acknowledgments. The authors gratefully acknowledge the support of Bernhard Sendhoff and the financial support from Honda Research Institute Europe GmbH.

References

1. Brooks, R.A.: New approaches to robotics. Science 253, 1227–1232 (1991)
2. Dorigo, M., Schnepf, U.: Genetics-based machine learning and behaviour based robotics: A new synthesis. IEEE Transactions on Systems, Man, and Cybernetics 23(1), 141–154 (1993)
3. Cliff, D., Harvey, I., Husbands, P.: Explorations in Evolutionary Robotics. Adaptive Behavior 2(1), 73–110 (1993)
4. Floreano, D., Mondada, F.: Automatic creation of an autonomous agent: Genetic evolution of a neural-network driven robot. In: Proceedings of the Third International Conference on Simulation of Adaptive Behavior: From Animals to Animats 3: From Animals to Animats 3, SAB 1994, pp. 421–430. MIT Press, Cambridge (1994)

5. Miglino, O., Nafasi, K., Taylor, C.: Selection for wandering behavior in a small robot. Technical Report UCLA-CRSP-94-01, Department of Cognitive Science, University of California, Los Angeles (1994)
6. Sims, K.: Evolving virtual creatures. In: The 21st Annual Conference, pp. 15–22. ACM Press, New York (1994)
7. Parker, G.B., Nathan, P.J.: Co-evolution of sensor morphology and control on a simulated legged robot. In: International Symposium on Computational Intelligence in Robotics and Automation, CIRA 2007, pp. 516–521 (2007)
8. Bugajska, M.D., Schultz, A.C.: Coevolution of form and function in the design of micro air vehicles. In: Evolvable Hardware, pp. 154–166. IEEE Computer Society (2002)
9. Sugiura, K., Akahane, M., Shiose, T., Shimohara, K., Katai, O.: Exploiting interaction between sensory morphology and learning. In: 2005 IEEE International Conference on Systems, Man and Cybernetics, vol. 1, pp. 883–888 (2005)
10. Gomez, G., Lungarella, M., Hotz, P.E., Matsushita, K., Pfeifer, R.: Simulating development in a real robot: on the concurrent increase of sensory, motor, and neural complexity (2004)
11. Auerbach, J., Bongard, J.: 12th International Conference on the Synthesis and Simulation of Living Systems (ALife XII) (August 2010)
12. Smalikho, O., Olhofer, M.: Growth in co-evolution of sensor system and signal processing for optimal wing control. In: Proceedings of the Genetic and Evolutionary Computation Conference. ACM (2014)
13. Farin, G.E.: NURBS: From Projective Geometry to Practical Use, 2nd edn. A. K. Peters, Ltd., Natick (1999)
14. Smalikho, O., Olhofer, M.: Co-evolution of sensor system and signal processing for optimal wing control. In: Proceedings of the EvoApplication Conference. Springer (2014)
15. von Haller, B., Ijspeert, A., Floreano, D.: Co-evolution of structures and controllers for neubot underwater modular robots. In: Capcarrère, M.S., Freitas, A.A., Bentley, P.J., Johnson, C.G., Timmis, J. (eds.) ECAL 2005. LNCS (LNAI), vol. 3630, pp. 189–199. Springer, Heidelberg (2005)
16. Rechenberg, I.: Evolutionsstrategie 1994. Frommann, Stuttgart (1994), Fit via Evolutionsstrategie, Routine von Volker Tuerck vorhanden
17. Hansen, N.: The CMA Evolution Strategy: A Comparing Review (2006)
18. Jacobs, E.N., Ward, K.E., Pinkerton, R.M.: The characteristics of 78 related airfoil sections from tests in the variable density wind tunnel. Technical Report 460 (1948)
19. Moriarty, D.E., Miikkulainen, R.: Efficient reinforcement learning through symbiotic evolution. Machine Learning (AI94-224), 11–32 (1996)
20. Stanley, K.O., Miikkulainen, R.: Evolving neural networks through augmenting topologies. Evolutionary Computation 10(2), 99–127 (2002)
21. Stanley, K.O., Miikkulainen, R.: Competitive coevolution through evolutionary complexification. Journal of Artificial Intelligence Research 21, 63–100 (2004)
22. Bremner, F., Gotts, S., Denham, D.: Hinton diagrams: Viewing connection strengths in neural networks, vol. 26, pp. 215–218. Springer (1994)

Generating Software Test Data
by Particle Swarm Optimization

Ya-Hui Jia[1,3,4], Wei-Neng Chen[2,3,4,**], Jun Zhang[1,2,3,4], and Jing-Jing Li[5]

[1]School of Information Science and Technology, Sun Yat-sen University, Guangzhou, China
[2]School of Advanced Computing, Sun Yat-sen University, Guangzhou, China
[3]Key Lab. Machine Intelligence and Advanced Computing, Ministry of Education, China
[4]Engineering Research Center of Supercomputing Engineering Software, MOE, China
[5]School of Computer Science, South China Normal University, Guangzhou, China
chenwn3@mail.sysu.edu.cn

Abstract. Search-based method using meta-heuristic algorithms is a hot topic in automatic test data generation. In this paper, we develop an automatic test data generating tool named particle swarm optimization data generation tool (PSODGT). The PSODGT is characterized by the following two features. First, the PSODGT adopts the condition-decision coverage (C/DC) as the criterion of software testing, aiming to build an efficient test data set that covers all conditions. Second, the PSODGT uses a particle swarm optimization (PSO) approach to generate test data set. In addition, a new position initialization technique is developed for PSO. Instead of initializing the test data randomly, the proposed technique uses the previously-found test data that can reach the target condition as the initial positions so that the search speed of PSODGT can be further accelerated. The PSODGT is tested on four practical programs. Experimental results show that the proposed PSO approach is promising.

Keywords: Particle swarm optimization, Automatic software test case generation, Software testing,· Code coverage.

1 Introduction

With the rapid development of software industry, software is becoming bigger and more subtle. In 2002, NIST estimated the loss caused by software failure which reached 0.6 percent of GDP in America [1]. Hence, software testing, as a necessary part during the circle of software development, is more difficult than before. Software testing is also an expensive and labor-intensive work, which sometimes occupies about half of the total workload [2] and brings lots of redundant expenditure both in time and money. Hence, developing automatic test tool has important practical significance.

The basic prerequisite for automatic software testing is generating test data automatically. However, test data generation is a very challenging task, as a good data set should not only fulfill all the requirements defined by test criterion well but also be as

---

* Corresponding author.

G. Dick et al. (Eds.): SEAL 2014, LNCS 8886, pp. 37–47, 2014.

smaller as possible. As a result, more and more research effort has been attracted in software test data generation in recent years [7], [16]. In general, these studies can be classified into three classes, random, symbolic and dynamic. Random method just generates inputs at random until a useful input is found. In symbolic method, variables are assigned with symbolic values so that test data generation can be turned into a problem of solving algebraic expressions [13], [14]. In dynamic test generation, the source code is instrumented to collect information about the program when it executes. This information can help test generators to modify the program's input to satisfy the requirement heuristically. Then the problem of generating test data converted to function minimization problem. As the dynamic method is efficient and robust for different kinds of programing codes, it has been increasingly considered as a promising software test data generation technique in recent years [17]. The dynamic method is also known as search-based software testing thus several meta-heuristic optimization algorithms have been proposed for this problem, e.g. hill climbing [15], tabu search algorithm [3], genetic algorithm (GA) [5] and particle swarm optimization (PSO) [8].

Meanwhile most existing researches focused on covering paths in a program as many as possible [6]. This strategy is known as path coverage which is a coverage criterion in the field of white-box testing requiring that every path in the target program should be reached. But sometimes it is not enough just covering all paths in a program. Path coverage may also cause some conditions in the target program cannot be fully covered. In order to overcome this problem, there is also another coverage criterion called condition-decision coverage (C/DC). C/DC requires that every condition and every decision should take all possible outcomes at least once. Michael et al. [4] used C/DC as criterion in their test data generation tool GADGET but the approaches they proposed are based on GA. Though GA has strong ability in global searching, the local search capability is not good enough. Hence the convergence speed of GA often cannot satisfy the software testing requirement.

In this paper, we intend to introduce a PSO approach to search-based software testing with the C/DC criterion and further develop a PSO Data Generation Tool (PSODGT). The reason of using PSO is that PSO has a fast convergence speed [10], [18]. In addition, the self-cognitive and social-influence learning strategies of PSO make it more reliable in detecting conditions which are difficult to reach. Though PSO has been used in test data generation with the path coverage criterion in a few works like [6], [9], different from these existing approaches, our proposed PSO approach focuses on a different criterion, i.e., the C/DC criterion. In addition, we further improve the performance of PSO for test data generation by introducing a new initialization technique to PSO and adjust its parameter setting. During each optimization procedure, particles should reach the target condition before optimizing the fitness function. In the proposed initialization technique, particles are initialized according to the test data that can reach the current target condition found in previously. This modification saves time for particles to reach the target condition so that particles can early start to optimize fitness function. As for experiment, most researches just tested their approaches by simple programs like triangle classification, bubble sort. These programs are not complicated enough to simulate real situations because they are too simple and the search space is small. In this paper, four programs with different complexities of inputs and conditions are tested. Our PSO approach is compared with a

GA approach [4] which is also proposed for C/DC. Experimental results are evaluated in two aspects, conditions coverage rate and convergence rate. When observing the convergence rate, the number of executions of the target program is used as measurement instead of time consumption.

The rest of this paper is organized as follows. In section 2, we introduce the PSO algorithm and the test adequacy criteria. Section 3 shows some pivotal details about PSODGT. Then experimental results and analysis are shown in section 4. Finally in section 5 the conclusions are drawn.

2   Particle Swarm Optimization and Test Adequacy Criteria

2.1   Particle Swarm Optimization

Particle Swarm Optimization (PSO) algorithm was proposed by Russell Eberhart and James Kennedy in 1995 [10]. In PSO algorithm, each particle keeps track of a position which is the best solution it has achieved so far as *pbx* and globally optimal solution is stored as *gbest*. The basic steps of PSO are as follow:

1. Initialize N particles with random positions px_i and velocities v_i on D dimensions. Evaluate every particle's current fitness $f(px_i)$. Initialize $pbx_i = px_i$ and $gbest = i, f(px_i) = min(f(pbx_0), f(pbx_1),..., f(pbx_N))$;
2. Check whether the criterion is met. If the criterion is met, loop ends else continue;
3. Change velocities according to formula (1):

$$v_i = \omega v_i + c_1 r_1 (pbx_i - px_i) + c_2 r_2 (pbx_{gbest} - px_i); \tag{1}$$

4. Change positions according to formula (2):

$$px_i = px_i + v_i \tag{2}$$

5. Evaluate every particle's fitness $f(px_i)$; if $f(px_i) < f(pbx_i)$ then $pbx_i = px_i$;
6. Update *gbest* and loop to step 2.

Usually particle's position cannot overstep the boundary of the search space and velocity also cannot exceed one particular value which is often set as 20% of the search space's width. In formula (1), the particle velocity updating formula, ω presents inertia factor, generally obtained by formula (3) [18]

$$\omega = \omega_{max} - \frac{(\omega_{max} - \omega_{min})}{maxIt} k . \tag{3}$$

maxIt means the maximum iteration number and k means the k-th iteration; c_1, c_2 are accelerated factors which present cognition and social of the particle; r_1, r_2 are random numbers between 0 and 1.

2.2   Coverage Criteria

The goal of software test is to uncover as many faults as possible with a potent set of tests. But predicting how many faults will be uncovered by a given test set is almost impossible [11]. We need test adequacy criteria to help us judge whether a data set is good enough to accomplish the test. Regardless of whether test adequacy criteria really can represent the quality of a test suite, they do represent the thoroughness of testing.

There are several common coverage criteria in structural test like statement coverage, branch coverage, condition coverage, multiple condition coverage, condition-decision coverage (C/DC) and path coverage [2]. A condition is a leaf-level Boolean expression and cannot be broken down into a simpler Boolean expression. A decision is a Boolean expression composed of conditions and Boolean operators. A decision without any Boolean operators is a condition. In these criteria exists a hierarchy, the top one is multiple condition coverage which requires every permutation of values for the Boolean variables in every condition occurring at least once. On the contrary, function coverage only requires the execution of every function.

In many automatic testing researches, path coverage was used as their criterion. Though path coverage is applicable to a number of program's testing, it is not perfect. For example, assume there is a decision consisted by disjunction of two conditions like $(a \parallel b)$ in the program. For the true path, according to the short circuit evaluation, most programming language will check condition a first and if condition a is true then ignore the condition b; for the false path, both condition a and b need to be false. Finally we find that condition b may never be true even both paths were already covered.

In PSODGT, we use condition-decision coverage as the coverage criterion. Condition-decision coverage requires that every decision in the program has taken all possible outcomes at least once, and every condition in a decision in the program has taken all possible outcomes at least once. Although in the hierarchy, the level of C/DC is lower than multiple condition coverage, C/DC already can make sure that every piece of the program can be executed if the requirement of C/DC is fully fulfilled.

3   The PSO Data Generation Tool (PSODGT)

In this section, the PSO data generation tool is introduced. First of all, we give a brief overview of the PSODGT. Then we discuss three issues of the PSODGT in detail, including the main data structure, implementation of the fitness function and the improved PSO algorithm for this tool.

3.1   Overview of PSODGT

PSODGT is designed to work on programs written in C or C++ programing language and the architecture of PSODGT is shown in Fig. 1. There are two parts in PSODGT, automatic instrumentation and test data generation. Original source code is automatically instrumented and compiled in the automatic instrumentation part. After compiling, an instrumented executable program is generated for data generation part to work on. The data generation part also consists of two classes. Class controller maintains a condition table and a test data set, taking charge of choosing target condition and branch, storing useful inputs found by class optimizer and updating the condition table all by using a key function named runOnce. The optimizer class only focuses itself on reaching the target branch chosen by controller.

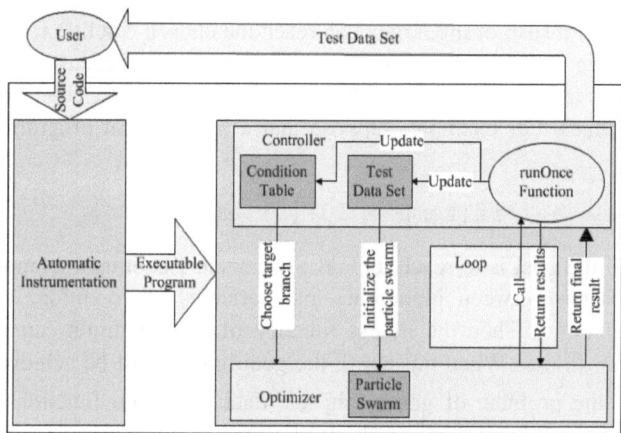

Fig. 1. Architecture of the PSODGT

3.2   Condition Table

The condition table is a vitally important data structure for the PSODGT. It is derived from the decision table proposed in [12] and modified in [4] by replacing decision with condition. Different from the condition table in [4], each branch in PSODGT's condition table has three statuses not two. A sample condition table is shown in Table 1.

Table 1. Condition table

	Branch	
Condition	TRUE	FALSE
1	1	2
2	0	0
3	1	1
4	1	0

Status 0 means this branch is not covered yet, 1 means this branch has been covered and 2 means that the algorithm has failed in optimizing this branch of the condition. Status 2 is used to avoid endless loop. When a goal branch is needed, PSODGT always chooses the condition the state of which is 1/0 or 0/1, i.e., this condition has been reached but its branches are not fully covered yet. To satisfy C/DC requirement, we take advantage of the short circuit evaluation. If every condition is fully covered (both TRUE and FALSE branches have been achieved), all decisions can be fully covered. So we can only focus on covering conditions as many as possible.

3.3   Fitness Calculation

When a condition's branch is chosen, things we need to do next is to get its fitness under test data. If a strip of input data can reach the chosen condition, variables in the chosen condition have relationship with the input data. Because different inputs should cause different value for variables in condition so that conditions may take different outcomes. For example, suppose that a hypothetical program contains the condition

$$\texttt{if(temp > 10)\{…\} else \{…\}}$$

on line 50 and the goal is to reach its FALSE branch. Denoting x as input, according to the relationship between input data and variables in condition, temp can be indicated as $line_{50}(x)$. Then the fitness function of this condition can be express as $f(x) = 10 - line_{50}(x) + 1$. When $f(x) <= 0$, the goal branch will be achieved. Using the value of $f(x)$, the problem of generating test data turns into function minimization problem. If the input data cannot reach the chosen condition, we set a very large value as fitness to represent that this condition is not related to the input data.

Table 2. Computation of the fitness function

Condition Type	Goal Branch	Fitness Calculation
$c > d$	T	$d - c + minConst$
	F	$c - d$
$c < d$	T	$c - d + minConst$
	F	$d - c$
$c >= d$	T	$d - c$
	F	$c - d + minConst$
$c <= d$	T	$c - d$
	F	$d - c + minConst$
$c == d$	T	$|c - d|$
	F	$minConst - |c - d|$
c	T	1000
	F	1000

Table 2 shows how fitness is calculated for all condition types if the condition is reached. We add a constant named minConst to some fitness so that all fitness calculations can be evaluated as a positive number no matter what the goal branch is. The goal is to reduce the fitness down to zero or negative numbers. For integer problem, minConst is set to 1; for float problem, minConst can be set to a very small float number according to the precision needed in the real situation.

3.4   Particle Swarm Optimization for PSODGT

According to the actual situation in test data generation, some improvements are applied to the PSO algorithm in initialization step and parameters setting. Initialization step in PSO can be divided into position initialization and velocity initialization.

Plenty of existing researches use a strategy in initial population generating that add the test data which can successfully reach the target condition to the population firstly; if additional inputs are needed, generate some random inputs to fill the population. However for PSO, these randomly generated inputs take a lot of time in reaching the target condition. Considering this problem, this paper proposes a new method for the initialization step of PSO using the idea of crossover operation in GA and stratified sampling for reference. Suppose there are N particles needed in the swarm and the number of the target program's input is D. The position of the i-th particle is presented as an array $P_i[1...D]$. The positions initialization steps are as follow:

1. If the number of the test data which can successfully reach the condition is bigger or equal to N, randomly add N of them into swarm and end the initialization process; else add all of them into swarm and calculate how many additional particles are needed.
2. Assume M particles are needed. If $M<=N–M$, use random M particles in the swarm as seeds and each seed generates one additional particle, else use all the particles in the swarm as seeds and each seed generates $M/(N–M)$ particles.
3. For every generation of every seed, it begins with copying the position of i-th seed to the new particle, $NewP[1...D]=P_i[1...D]$.
4. After copy, generate a number d smaller than D randomly. Then construct an d-length array $changePos[1...d]$ filling up with different numbers which are generated randomly and smaller than D.
5. Finally substitute the values in $NewP$ with random numbers according to $changePos$, $NewP[changePos[1...d]]=randomNumber$.

Research in [4] found that there are lots of serendipitous coverages during test data generation. This means some test data do cover new condition branches but these conditions are not the one the optimizer is currently working on. Serendipitous coverage requires degree of randomness in optimizer. However the directional character makes PSO perform badly in gaining serendipitous coverage. Considering this problem, particles' velocities are initialized within the same boundary as positions to obtain more randomness in early stages of iteration. This setting causes a consequence that convergence speed becomes slower than basic PSO. To make up the losses on convergence rate, ω the inertia factor is set as 0.4 down to 0.3 with the iteration growing. A lot of experiments have been done to verify this setting about inertia factor. When it is set much higher, convergence speed is too slow to meet the requirement. While, if it is smaller than 0.3, the algorithm usually fails.

4   Experimental Studies

4.1   Experimental Settings

In the experiments, we test the proposed method on four programs: triangle classification program, week calculation program, student grade judgment program and blood glucose judgment program. (Denoted as P1, P2, P3, and P4) Different from the simple programs tested in [4], these programs are practical and full of various conditions. When we test these programs, two main aspects are taken into consideration. One is the dimension (number of inputs) and size of search space; the other is the number of conditions. Specific figure about the dimension of search space and the number of conditions are shown in Table 3. And each program is tested with two different size of search space measured by bit shown in Table 4.

Table 3. Number of conditions and inputs

Program	Condition number	Inputs Number (dimension)
P1	20	3
P2	76	3
P3	16	5
P4	33	12

Our proposed PSO approach is compared with two genetic algorithms with different coding schemes which are gray code (GAG) and binary code (GAB). Also a PSO method using the same initialization way with GA is tested to verify the necessity of the proposed initialization technique, denoted as iPSO. We use 100 individuals, allow 30 generations to elapse before two GAs give up, the same as in [4], and the mutation probability of every bit is 0.01. For PSO, 20 individuals and 100 generations are allowed. Values of accelerated factors c_1 and c_2 are 2.

4.2   Experimental Results and Analysis

In this paper, experimental results are estimated in two aspects, efficiency (convergence rate) and effectiveness (coverage rate). In the course of experiment, we take five serial attempts as a group of tests. After six groups of tests for each method, the best coverage rate in each group is selected and the best, worst, average coverage rate of these six numbers are shown in Table 4, displayed in percentage.

Comparing the experimental data on different search space size for the same program, we can find that when the number of input is relative small, increasing on the search space size doesn't affect the coverage rate greatly. However when the number of input grows larger, the contrary is the case. Though both GA and PSO suffer the increasing on search space size, the PSO approach is much more stable. And the poor performance made by iPSO also demonstrates the importance of our proposed initialization tech in PSO method for software test data generation. In general, except the data of best coverage rate for P1 in 16-bit search space which is underlined in Table 4, all the rest data show the advantage of our PSO approach in effectiveness.

Table 4. Experimental results on coverage rate

Prog.		P1		P2		P3		P4	
Bit		32	16	16	12	12	8	12	8
PSO	mean	95	95	99.45	100	100	100	88.38	100
	worst	95	95	98.68	100	100	100	65.15	100
	best	95	95	100	100	100	100	100	100
GAG	mean	88.75	93.33	96.49	96.49	82.29	100	45.20	100
	worst	82.50	92.50	94.74	93.42	43.75	100	33.33	100
	best	95	<u>97.50</u>	98.03	99.34	100	100	57.58	100
GAB	mean	72.50	85.42	96.71	96.93	59.90	100	36.36	100
	worst	50	72.50	95.39	96.05	43.75	100	28.79	100
	best	85	92.50	98.03	98.68	90.63	100	40.91	100
iPSO	mean	85.42	87.50	82.46	86.95	41.67	100	26.27	94.70
	worst	80	87.50	79.61	85.53	40.63	100	24.24	71.21
	best	87.50	87.50	84.21	87.50	43.75	100	30.30	100

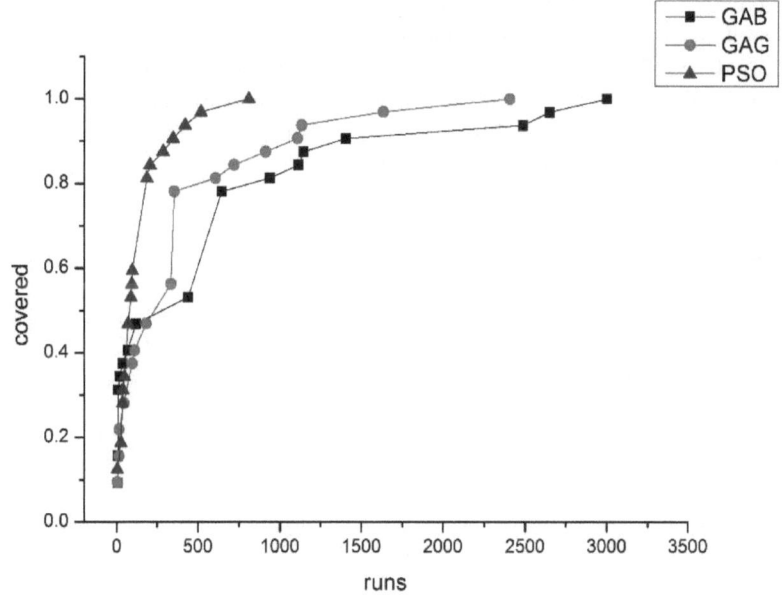

Fig. 2. Converge rate of three methods on P3

As to the convergence rate, P3 with 8-bit search space is used in comparing the efficiency because all three approaches performed well on this program and iPSO is not shown because it is too slow comparing with others. The relationship between runs and coverage rate is illustrated in Fig. 2. This plot shows that PSO begins to demonstrate its advantages after reach 50% coverage and covers all conditions first. Clearly the convergence rate of PSO is much faster than the two GA approaches. The same fact also can be verified by the other programs.

All these experimental results show that the proposed PSO approach is effective and efficient in automatic software test data generation.

5   Conclusion

In this paper, an improved PSO approach is proposed to apply to search-based test data generation. The main contributions are in two aspects. First, the PSODGT is developed by combining the PSO algorithm and C/DC. Second, a new position initialization technique is developed for PSO to adapt accommodate software testing. Experimental results show that the proposed PSO approach is very promising.

In the future research, it will be interesting to find out which method is suitable to which kind of condition so that more hybrid methods can be proposed to apply to different conditions. And how to use much higher-level coverage criterion is also a promising research topic.

Acknowledgement. This work was supported in part by the National High-Technology Research and Development Program (863 Program) of China No. 2013AA01A212, in part by the NSFC for Distinguished Young Scholars 61125205, in part by the NSFC Nos. 61379061, 61070004, in part by Natural Science Foundation of Guangdong No. S2013040014949, and in part by Programs Foundation of Ministry of Education of China No. 20130171120016.

References

1. National Institute of Standards and Technology, "Then Economic Impacts of Inadequate Infrastructure for Software Testing," Planning Report 02-3 (May 2002)
2. Myers, G.J., Sandler, C., Badgett, T.: The art of software testing. John Wiley & Sons (2011)
3. Díaz, E., Tuya, J., Blanco, R.: Automated Software Testing Using a Metaheuristic Technique Based on Tabu Search. In: Proc. 18th IEEE Int'l Conf. Automated Software Eng., pp. 310–313 (2003)
4. Michael, C.C., McGraw, G., Schatz, M.A.: Generating software test data by evolution. IEEE Trans. Software Eng. 27(12), 1085–1110 (2001)
5. Bottaci, L.: Instrumenting Programs with Flag Variables for Test Data Search by Genetic Algorithm. In: Proc. Genetic and Evolutionary Computation Conf., pp. 1337–1342 (2002)
6. Li, A., Zhang, Y.-L.: Automatic Generating All-Path Test Data of a Program Based on PSO. In: WRI World Congress on Software Eng., pp. 189-193 (2009)
7. Windisch, A., Wappler, S., Wegener, J.: Applying Particle Swarm Optimization to Software Testing. In: Proc. 9th Ann. Genetic and Evolutionary Computation Conf., pp. 1121–1128 (2007)

8. Cui, H.-H., Chen, L., Zhu, B., Kuang, H.-L.: An Efficient Automated Test Data Generation Method. In: Int'l Conf. Measuring Technology and Mechatronics Automation, vol. 1,
   pp. 453–456 (2010)
9. Zhang, S., Zhang, Y., Zhou, H., He, Q.-Q.: Automatic Path Test Data Generation Based on
   GA-PSO. In: Proc. IEEE Int'l Conf. Intelligent Computing and Intelligent Systems, pp.
   142–146 (2010)
10. Kennedy, J., Eberhart, R.: Particle swarm optimization. In: Proc. IEEE Int'l Conf. Neural
    Networks, vol. 4, pp. 1942–1948 (1995)
11. Frankl, P., Hamlet, D., Littlewood, B., Strigini, L.: Choosing a Testing Method to Deliver
    Reliability. In: Proc. Int'l Conf. Software Eng., pp. 68–78 (1997)
12. Chang, K.H., Cross II, J.H., Carlisle, W.H., Liao, S.-S.: A Performance Evaluation of Heu-
    ristics-Based Test Case Generation Methods for Software Branch Coverage. Int'l J. Soft-
    ware Eng. and Knowledge Eng. 6(4), 585–608 (1996)
13. Clarke, L.A.: A System to Generate Test Data Symbolically and Execute Programs. IEEE
    Trans. Software Eng. 2(3), 215–222 (1976)
14. Offutt, A.J.: An Integrated Automatic Test Data Generation System. J. Systems Integra-
    tion 1, 391–409 (1991)
15. Korel, B.: Automated Software Test Data Generation. IEEE Trans. Software Eng. 16(8),
    870–879 (1990)
16. Sofokleous, A.A., Andreou, A.S.: Automatic, evolutionary test data generation for dynam-
    ic software testing. J. Systems and Software 81(11), 1883–1898 (2008)
17. Harman, M., McMinn, P.: A theoretical and empirical study of search-based testing: Local,
    global, and hybrid search. IEEE Trans. Software Eng. 36(2), 226–247 (2010)
18. Shi, Y., Eberhart, R.: A modified particle swarm optimizer. In: Proc. IEEE Int'l Conf.
    Evolutionary Computation, pp. 69–73 (1998)

A Steady-State Genetic Algorithm for the Dominating Tree Problem

Shyam Sundar

Department of Computer Applications
National Institute of Technology Raipur
Raipur - 492010, India
ssundar.mca@nitrr.ac.in

Abstract. Dominating tree problem (DTP) is a recent variant of dominating set problems in graph theory and finds its root in providing virtual backbone routing in wireless sensor networks. This problem consists in finding a tree, say DT, with minimum total edge weight on an undirected, weighted and connected graph such that each vertex of the graph is either in DT or adjacent to a vertex in DT. In this paper, a steady-state genetic algorithm (SSGA) is proposed for the solution of DTP. In particular, crossover operator of SSGA is designed in such a way that it generates a DT of the child solution which not only avoids the generation of a forest of trees, but also contributes in finding a high quality of child solution. Crossover and mutation of SSGA as well as other elements such as pruning procedure for the DTP are effectively coordinated in such a way that they help in evolving high quality solutions in a less time. SSGA has been compared with the best approaches in the literature. Computational results show the superiority of SSGA over these state-of-the-art approaches in terms of both solution quality and computational time.

Keywords: Evolutionary Algorithm, Genetic Algorithm, Steady-State, Dominating Tree, Wireless Sensor Networks.

1 Introduction

In recent years, many hard combinatorial optimization problems have been encountered in the domain of wireless sensor networks (WSNs). Dominating tree problem (DTP) is one of $\mathcal{NP}$-Hard problems in WSNs. Given an undirected, weighted and connected graph $G = (V, E, w)$, where V is a set of vertices, E is a set of edges, and w is a non-negative weight function $w : E \rightarrow \Re^+$ associated with the edges of G, DTP consists in finding a tree, say DT, with minimum total edge weight on G such that for each vertex $v \in V$, v is either in DT or adjacent to a vertex in DT. Every vertex in DT is called a dominating vertex, whereas every vertex not in DT is called a non-dominating vertex. Note that vertices and nodes are used interchangeably in this paper.

A solution to the DTP offers an application in providing a virtual backbone for routing in WSNs. Since a non-dominating node is at least adjacent to one

G. Dick et al. (Eds.): SEAL 2014, LNCS 8886, pp. 48–57, 2014.

of dominating nodes of dominating tree (DT) in the WSN, the routing information can be stored only on dominating nodes of DT (solution). In this scheme,
a message can be sent from one source to destination by first forwarding this
message to its nearest dominating node of DT, then with the help of DT, this
message is further routed to one of its dominating nodes nearest to the receiver
and then, finally destined to the receiver. Each non-dominating node is required
to remember only its nearest dominating node. Advantage of this scheme is
that the total number of dominating nodes used for storing routing information
(virtual backbone) is small in comparison to the total nodes in the WSN, which
in turn, the less overhead on the size of routing table occurs.

2   Related Work

Dominating tree problem (DTP) is a recent variant of dominating set problems in
graph theory and is proven to $\mathcal{NP}$-Hard problem [6, 13]. In literature, the concept
of connected dominating set [2, 4, 10–12] has been studied for constructing a
routing backbone with minimum energy consumption in WSNs. Such papers
consider the weight on each node instead of the weight on each edge. In fact,
energy consumption in routing is directly related to the energy consumed by
edges on the route. This led to the introduction of DTP [6, 13] with the objective
of minimizing energy consumption of routing. They proved inapproximability
result and presented an approximation algorithm - quasi-polynomial ($|V|^{O(lg|V|)}$)
algorithm- to solve the DTP. Due to quasi-polynomial algorithm, both Zhang *et
al.* [13] and Shin *et al.* [6] developed a polynomial time problem-specific heuristic
for the solution of DTP. Later, Sundar and Singh [8] presented a problem-specific
heuristic and two swarm intelligence techniques – artificial bee colony algorithm
and ant colony optimization algorithm – and demonstrated the superiority of
results over the results reported in [6, 13].

This paper presents a steady-state genetic algorithm (SSGA) for the solution
of DTP. SSGA has been compared with the state-of-the-art approaches, i.e., artificial bee colony (ABC) approach and ant colony optimization (ACO) approach
[8]. Computational results show the superiority of SSGA over ABC and ACO
approaches in terms of both solution quality and computational time.

The rest of this paper is organized as follows: Section 2 describes a a brief
introduction of SSGA, whereas Section 3 describes an SSGA for the DTP. Computational results are reported in Section 4. Finally, Section 5 contains some
concluding remarks.

3   SSGA for the DTP

Genetic algorithm (GA) [3] is an evolutionary algorithm that works on the principles of natural evolution. It is one of the most powerful metaheuristic techniques
for optimization problems. This paper is focused on presenting a steady-state
genetic algorithm (SSGA) for the DTP. SSGA works on steady-state population replacement method [1]. SSGA starts iteratively with selecting two parents,

performing crossover and mutation to generate a child solution that replaces a worst individual of the population. It is quite different from generational genetic algorithm (GGA), where the whole parent population is replaced with the same number of newly generated child solutions every generation. In general, SSGA finds highly fit solutions in a faster way in comparison to GGA [1]. The notion behind this one is that highly fit solutions in the population are kept permanently and available instantly for selection and reproduction in order to generate child solutions. Also, GGA may contain multiple copies of highly fit solutions in the population which can dominate the whole population within few generations. In such a situation, crossover operator becomes completely ineffective. However, mutation can be used to improve the solution quality, but such improvement, if occurs, is quite slow. In SSGA, one can simply prevent this situation by comparing each newly generated child solution with the current individuals of the population and rejecting the newly generated child solution, in case, it is equivalent to one of current individuals of the population.

Elements of SSGA for the DTP are described as follows:

3.1   Encoding

Each chromosome (solution) is represented as a set of dominating vertices of a dominating tree.

3.2   Initial Solution Generation

Each initial solution is generated by an iterative process [8]. Initially, S and U are the two empty sets. A vertex v_1 is selected uniformly at random from V and added to S. All vertices adjacent to v_1 are added to U. At each step, an edge connecting a vertex $v_x \in S$ to a vertex $v_y \in U$ is selected, where v_x and v_y are selected uniformly at random from S and U respectively. After this, v_y is deleted from U and added to S. Each vertex, which is adjacent to v_y and is neither a member of U nor S, is added to U. This whole procedure is repeated again and again until the sum of cardinality of S and U becomes equal to the total number of vertices in G. At this juncture, a dominating tree DT is constructed.

Once a solution DT is generated, a pruning procedure is applied to DT [8]. According to this pruning procedure, a dominating vertex with degree one, say $v_p, \in DT$ is examined for pruning. It is possible only when all non-dominating vertices adjacent to v_p are also adjacent to other dominating vertices in DT. When it is possible, then only the edge incident to v_p can be deleted from DT, which in turn reduces the total edge weight of DT. This pruning procedure is applied to DT repeatedly till it is no longer possible to prune any dominating vertex with degree one. Thereafter, Prim's algorithm [5] is applied to construct a minimum spanning tree (MST) on the sub-graph of G induced by the set of dominating vertices of DT [8]. This may lead to further minimize the total weight of DT. The notion behind this one is that even after pruning, the total weight of DT may not be minimum due to the selection of incorrect edges while constructing DT. Numerous dominating trees can be constructed in G on a given a set of dominating vertices. Obviously, resultant MST or DT after applying Prim's algorithm will always be of minimum cost among all such dominating trees.

Uniqueness of each generated individual (solution) is checked against the individuals of the population generated so far and if it is unique, it is included in the initial population, otherwise it is discarded.

3.3   Fitness

Fitness of each solution is computed, where the fitness function of a solution is same as objective function of a solution.

3.4   Selection

Binary tournament selection method is used consecutively two times for selecting two different chromosomes as parents for crossover. This method starts with selecting two chromosomes uniformly at random from the current population. With probability P_b, the chromosome with better fitness is selected, otherwise, worse one is selected (with probability $1 - P_b$). This selection is also applied to select a chromosome for mutation.

3.5   Crossover

Crossover starts with selecting two chromosomes (solutions) as parents (say, p_1 and p_2) from the population with the help of binary tournament selection method, initializing an empty child solution C, and labeling all vertices $\in V$ as unmarked. With equal probabilities, first vertex $v_{p_1 1}$ (gene) is selected from the first index of ordered set of dominating vertices of p_1, otherwise first vertex $v_{p_2 1}$ (gene) is selected from that of p_2. This selected vertex is added to C. All vertices adjacent to this selected vertex and the selected vertex itself are labeled marked. Hereafter, iteratively, with equal probabilities, a *next* vertex, say $v_{p_1 i}$, is selected from the ordered set of dominating vertices of p_1, otherwise *next* vertex, say $v_{p_2 j}$, is selected from that of p_2. Here i and j are the indices of vertices in the ordered set of dominating vertices of p_1 and p_2 respectively, and *next* vertex means a vertex is next to previously selected vertex in the ordered set of dominating vertices of selected parent. In addition, it is possible that *next* vertex (suppose that this *next* vertex is $v_{p_1 i}$) already exists in C. This possibility is based on two things: the first one is that this *next* vertex may be common to both p_1 and p_2 and it may be selected from p_2 in an earlier iteration, and the second one is that this *next* vertex may be selected in an earlier iteration due to *potential path* (*potential path* is explained later). In such a situation, index i in p_1 is incremented to $i + 1$ for selecting a *next* vertex from p_1 in the next time, and the procedure starts a fresh for selecting a *next* vertex. The selected *next* vertex which will be either $v_{p_1 i}$ or $v_{p_1 j}$ will be referred to as v_1. After this, a connectivity of v_1 against all other vertices in C is checked first before adding it to C. Here checking of a connectivity means checking of existence of an edge between v_1 and one of the vertices $\in C$ in G. If an edge exists, then v_1 is added to C, and all unmarked vertices adjacent to v_1 and v_1 itself are now labeled marked. Otherwise, a path (say, $v_1 \rightsquigarrow v_2$, where $v_2 \in C$) having maximum *potential* is selected from among all candidate paths connecting v_1 to all vertices in C. A *potential path* (say, $v_1 \rightsquigarrow v_c$, where $v_c \in C$) is based on $\frac{|SP_{um(v_1 \rightsquigarrow v_c)}|}{SP_{v_1 \rightsquigarrow v_c}}$, where $|SP_{um(v_1 \rightsquigarrow v_c)}|$ is the number of unmarked vertices in the path $v_1 \rightsquigarrow v_c$, $v_c \in C$, and $SP_{v_1 \rightsquigarrow v_c}$ is the cost of shortest path $v_1 \rightsquigarrow v_c$. All vertices, say V_{sp}, constituting this selected path $v_1 \rightsquigarrow v_2$ are added to C except v_2, as v_2 is already in C. All unmarked vertices adjacent to V_{sp} and all unmarked vertices in V_{sp} are now labeled marked. This whole iterative procedure continues until all vertices are labeled marked.

It is to be noted that simply selecting a vertex each time from one of ordered set of dominating vertices of p_1 and p_2 for C is not applied in crossover operator. The reason behind this one is that such crossover would possibly lead to a forest of trees and in that situation, even after crossover it would require a repair operator to transform this forest of trees into a dominating tree that would be costly. To overcome this situation, the concepts of checking a connectivity and selecting a *potential path* are applied during crossover which not only avoid the construction of a forest of trees, but also contribute in finding a high quality of child solution.

3.6   Mutation

Mutation starts with selecting a solution from the population with the help of binary tournament selection method, and copying this solution to an empty solution, say C. A small set (V_m) of non-dominating vertices is selected uniformly at random from $V \backslash C$ set, where $V_m = P_m \times \text{minimum}\{|C|, |V \backslash C|\}$. P_m is a parameter determined empirically. Iteratively, a vertex $v_s \in V_m$ having a path of minimum cost with one (say v_d) of the vertices in C is determined. It is to be noted that the degree of v_s must be greater than one in G. All vertices constituting the selected path are added to C except v_d, as v_d is already in C. This iterative procedure continues until V_m becomes empty. Hereafter, Prim's algorithm [5] is applied to construct a minimum spanning tree on the sub-graph of G induced by the set of dominating vertices of C.

Similar to [7, 9], here crossover and mutation operators are also applied in a mutually exclusive way to generate a child solution. With probability P_c, crossover operator is selected, otherwise mutation operator is selected with the probability $(1 - P_c)$. Once a child solution C is generated either from crossover or mutation operator, similar to [8], a series of procedures on the current DT of C, i.e., pruning on DT of C, Prim's algorithm on the resultant DT, and again pruning on the resultant DT is applied to further minimize the cost of DT. Such procedures are already explained in Section 3.2.

The reason behind considering crossover and mutation operators in a mutually exclusive way is that crossover operator generates a child solution (C) based on selecting high-quality building blocks (genes) in a randomized manner either from p_1 or p_2, and greedy approach for connectivity. Whereas, mutation operator generates a child solution C based on adding some random non-dominating vertices to C. If mutation operator is applied after crossover operator, then a series of procedures – pruning, Prim's algorithm and pruning (discussed above) – would be applied on the current C to further minimize the cost of C. In that case, the resultant C may lose some potentially high-quality building blocks (genes).

Algorithm 1. Pseudo-code of SSGA for the DTP

Generate a population (*pop_size*) of solutions $s_1, s_2, \ldots, s_{pop\_size}$ randomly (see Section 3.2);
$best \leftarrow$ Best solution in the population;
while *(Termination criteria is not met)* **do**

    if *(u01 $< P_c$)* **then**

        $p_1 \leftarrow Binary\_Tournament\_Selection(s_1, s_2, \ldots, s_{pop\_size})$;
        $p_2 \leftarrow Binary\_Tournament\_Selection(s_1, s_2, \ldots, s_{pop\_size})$;
        $Child \leftarrow Crossover\_Operator(p_1, p_2)$;

    else

        $p_1 \leftarrow Binary\_Tournament\_Selection(s_1, s_2, \ldots, s_{pop\_size})$;
        $Child \leftarrow Mutation\_Operator(p_1)$;

    if *(Child is a partial DT)* **then**
        Apply repair procedure on *Child*;

    Apply pruning procedure on *DT* of *Child*;
    Apply Prim's algorithm to construct a *MST* on the sub-graph of G induced by the set of dominating vertices of *DT*;
    Apply pruning procedure on *DT* of *Child*;
    if *(Child is better than best)* **then**
        $best \leftarrow Child$;

    Apply replacement policy;

    return *best*;

3.7   Replacement Policy

In this replacement policy, uniqueness of the newly generated child solution C is examined against each individual of the population. If C is found to be different from all individuals of the population, then it replaces the worst individual of the population, irrespective of its own fitness. Otherwise, it is rejected.

Algorithm 1 presents the pseudo-code of SSGA for the DTP, where the size of the population is *pop_size*. Two procedures called *Crossover_Operator(p_1, p_2)* and *Mutation_Operator(p_1)* perform crossover and mutation operations respectively. *Binary_Tournament_Selection($s_1, s_2, \ldots, s_{s_{pop\_size}}$)* is another procedure which selects a solution from solutions $s_1, s_2, \ldots, s_{s_{pop\_size}}$ with the help of using binary tournament selection method and returns the solution selected.

4   Computational Results

The proposed SSGA for the DTP has been implemented in C and executed on a Linux with the configuration of 3.0 GHz Core 2 Duo system with 2 GB RAM. In this approach, $pop\_size = 300$, $P_b = 0.8$, $P_c = 0.8$ and $P_m = 0.4$ are considered for the DTP. SSGA has been allowed $|V| \times 500$ generations to execute. All these parameter values are chosen empirically. These parameter values give good results although they may not be optimal for all problem instances. SSGA has been compared with the best approaches such as artificial bee colony (ABC) approach and ant colony optimization (ACO) approach [8] in the literature. Configuration of a computer system used to execute both ABC and ACO approaches was similar to that of SSGA. Similar to ABC and ACO approaches, SSGA has been also executed 20 independent times for each problem instance. In subsequent subsections, a brief description of problem instances and the performance comparisons of SSGA with ABC and ACO approaches are provided.

4.1   Problem Instances

SSGA uses same problem instances for the DTP as used in [8]. These problem instances have been downloaded from `http://dcis.uohyd.ernet.in/~alokcs/dtp.zip`. Each problem instance is described as follows: each instance is a disc graph, $G = (V, E)$ where each disk represents the transmission range of each node. The weight on each edge e_{uv} in E is defined as $w(u, v) = C_v \times d_{uj}^2$, where d_{uv} is the euclidean distance between two nodes (u and v), and C_v is a random constant which is considered as 1. The assumption is that all nodes are distributed randomly in a $500m \times 500m$ area and the transmission range of each node is $100m$. Three different problem instances are created for each value of $|V| \in \{50, 100, 200, 300, 400, 500\}$, resulting a total of eighteen problem instances.

4.2   Comparison of SSGA with ABC and ACO Approaches

SSGA has been compared with ABC and ACO approaches [8] on a set of problem instances (see Section 4.1). Experimental results are reported in Table 1. The descriptions of various columns of this table are as follows: Column 1 represents the name of each instance; columns 2-6 present best known value (BKV), average solution quality (Avg.), standard deviation (SD) of solution values, average number of dominating vertices (ANDV) and average total execution time (ATET) for each test instance that are obtained through ABC; columns 7-11 and 12-16 present same information (BKV, Avg., SD, ANDV and ATET) that are obtained by ACO and SSGA respectively.

Table 1 clearly shows that SSGA is much better than both ABC and ACO approaches in terms of solution quality (BKV and Avg.) and computational time (ATET). Note that best values are presented in bold numbers. Considering its all 18 instances, comparing with ABC approach, SSGA is better on 11 problem instances and equal on 7 problem instances in terms of best solution quality

Table 1. Comparison of the results of SSGA with the results of ABC and ACO approaches [8]

Instance	ABC					ACO					SSGA				
	BKV	Avg.	SD	ANDV	ATET	BKV	Avg.	SD	ANDV	ATET	BKV	Avg.	SD	ANDV	ATET
50_1	1204.41	1204.41	0.00	19.00	25.57	1204.41	1204.41	0.00	19.00	2.41	1204.41	1204.43	0.03	18.75	0.73
50_2	1340.44	1340.44	0.00	21.00	21.46	1340.44	1340.44	0.00	21.00	4.18	1340.44	1340.44	0.00	21.00	1.31
50_3	1316.39	1316.39	0.00	19.00	22.99	1316.39	1316.39	0.00	19.00	2.50	1316.39	1317.10	0.30	18.15	0.73
100_1	1217.47	1218.15	0.69	18.45	28.64	1217.47	1217.47	0.00	19.00	12.71	1217.47	1217.98	0.27	18.00	1.74
100_2	1128.40	1128.42	0.09	17.90	27.58	1152.85	1152.85	0.00	17.00	10.86	1128.40	1128.79	0.09	16.10	1.56
100_3	1252.99	1253.14	0.23	19.70	28.39	1253.49	1253.49	0.00	19.00	8.96	1252.99	1253.26	0.25	19.45	1.98
200_1	1206.79	1209.52	2.69	18.25	84.10	1206.79	1207.61	3.58	18.05	81.13	1206.79	1207.28	0.23	18.05	4.41
200_2	1216.41	1219.74	2.15	18.90	87.78	1216.23	1217.73	2.61	17.65	78.72	1216.23	1217.63	3.34	17.15	4.30
200_3	1253.02	1258.06	3.42	22.15	90.44	1247.25	1248.94	2.99	20.90	97.93	1247.25	1251.69	2.60	20.90	4.89
300_1	1229.97	1237.47	2.89	21.75	145.17	1228.24	1243.70	9.71	22.85	352.89	1226.11	1230.06	4.25	20.15	8.91
300_2	1182.52	1200.79	7.82	19.60	162.59	1176.45	1193.95	10.51	21.10	260.30	1170.85	1173.57	7.04	18.80	7.65
300_3	1257.21	1271.20	6.74	20.50	145.75	1261.18	1276.75	9.27	24.60	251.91	1247.51	1253.73	8.40	20.35	8.35
400_1	1223.61	1241.75	7.88	21.90	263.13	1220.62	1237.45	9.50	26.05	600.74	1213.45	1224.87	8.16	21.55	13.93
400_2	1220.54	1235.29	6.97	22.45	249.39	1209.69	1246.14	21.41	24.40	591.44	1199.67	1205.69	4.79	20.55	13.04
400_3	1266.41	1276.80	4.59	22.30	216.95	1254.10	1270.34	9.42	25.85	530.58	1248.29	1265.25	9.75	20.85	12.59
500_1	1233.14	1241.60	4.56	21.40	379.72	1219.66	1240.05	9.17	26.50	1163.20	1203.34	1215.34	10.93	20.45	18.12
500_2	1245.59	1258.33	5.40	22.35	364.04	1273.86	1295.51	13.39	28.65	1031.81	1233.89	1247.42	7.13	21.90	22.84
500_3	1249.17	1278.67	11.96	21.60	338.25	1232.71	1259.08	20.03	24.35	917.73	1231.92	1251.23	16.52	20.50	17.76

(BKV), whereas SSGA is better on 13 problem instances, equal on 1 problem instances and worse on 4 instances in terms of average solution quality (Avg.). Only on small sized problem instances, solution quality obtained by ABC approach is slightly better than that of SSGA at a large computation time. Based on ATET, SSGA is much faster than ABC approach. For example, the maximum time (ATET) taken by ABC approach on the problem instance (*500_1*) is *379.72 seconds*, whereas the maximum time (ATET) taken by SSGA on the problem instance (*500_2*) is *22.84 seconds*. In a similar way, comparing with ACO approach, SSGA is better on 11 problem instances and equal on 7 problem instances in terms of best solution quality (BKV), whereas SSGA is better on 13 problem instances, equal on 1 problem instances and worse on 4 instances in terms of average solution quality (Avg.). Solution quality obtained by ACO approach is slightly better than that of SSGA on only small sized problem instances. Based on ATET, SSGA is much faster than ACO approach. For example, the maximum time (ATET) taken by ACO approach on the problem instance (*500_1*) is *1163.20 seconds*, whereas the maximum time (ATET) taken by SSGA on the problem instance (*500_2*) is *22.84 seconds*.

Since the number of dominating vertices plays an important role in the performance of any routing protocols based on virtual backbone, therefore, the performance of SSGA is also examined with ABC and ACO approaches in terms of average number of dominating vertices (ANDV). Table 1 clearly shows that SSGA obtains less ANDV than ABC and ACO approaches on most of the instances.

Overall, SSGA outperforms both ABC and ACO approaches on most of the problem instances in terms of solution quality. SSGA has found new values for 9 problem instances out of 18. In particular, the convergence of SSGA is much faster in finding high quality solutions in comparison to ABC and ACO approaches.

It is to be noted that the number of dominating vertices in the solution and the solution quality do not vary significantly with instance size for SSGA. The reason behind this one is that all problem instances consisting of vertices are randomly distributed in a $500m \times 500m$ area, and the average degree of vertices also increases with the increase of instance size, resulting no significantly changes in the number of dominating vertices of the solution and the solution quality with the increase of instance size [8].

5   Conclusions

This paper presents a steady-state genetic algorithm (SSGA) for the dominating tree problem (DTP). In particular, crossover operator is designed in such a way that it generates a *DT* of the child solution which not only avoids the generation of a forest of trees, but also contributes in finding a high quality of child solution. Crossover and mutation of SSGA as well as other elements such as pruning procedure for the DTP are effectively coordinated in such a way that they help in evolving high quality solutions in a less time. In fact, the proposed SSGA has experimentally proved that SSGA is much superior than the state-of-the-art approaches, i.e., artificial bee colony approach and ant colony optimization approach in terms of both solution quality and computational time. In particular, the convergence of SSGA is much faster in finding a high quality solution.

Since there is still room for improvement based on the results, particularly the values of standard deviation of problem instances, obtained by SSGA. Therefore, as a future work, we will intend to develop other metaheuristic techniques for this dominating tree problem.

References

1. Davis, L.: Handbook of Genetic Algorithms. Van Nostrand Reinhold, New York (1991)
2. Guha, S., Khuller, S.: Approximation algorithms for connected dominating sets. Algorithmica 20, 374–387 (1998)
3. Holland, J.H.: Adaptation in Natural and Artificial Systems: An Introductory Analysis with Applications in Biology, Control, and Artificial Intelligence. Michigan Press, MI (1975)
4. Park, M., Wang, C., Willson, J., Thai, M.T., Wu, W., Farago, A.: A dominating and absorbent set in wireless ad-hoc networks with different transmission range. In: Proceedings of the 8th ACM International Symposium on Mobile Ad Hoc Networking and Computing (MOBIHOC) (2007)
5. Prim, R.C.: Shortest connection networks and some generalizations. Bell Systems Technical Journal 36, 1389–1401 (1957)
6. Shin, I., Shen, Y., Thai, M.: On approximation of dominating tree in wireless sensor networks. Optimization Letters 4, 393–403 (2010)
7. Singh, A., Gupta, A.K.: Two heuristics for the one-dimensional bin-packing problem. OR Spectrum 29, 765–781 (2007)
8. Sundar, S., Singh, A.: New heuristic approaches for the dominating tree problem. Applied Soft Computing 13, 4695–4703 (2013)
9. Sundar, S., Singh, A.: Metaheuristic approaches for the blockmodel problem. IEEE Systems Journal (accepted, 2014)
10. Thai, M., Tiwari, R., Du, D.Z.: On construction of virtual backbone in wireless ad hoc networks with unidirectional links. IEEE Transactions on Mobile Computing 7, 1–12 (2008)
11. Thai, M., Wang, F., Liu, D., Zhu, S., Du, D.Z.: Connected dominating sets in wireless networks with different transmission ranges. IEEE Transactions on Mobile Computing 6, 721–730 (2007)
12. Wan, P., Alzoubi, K., Frieder, O.: Distributed construction on connected dominating set in wireless ad hoc networks. In: Proceedings of the Conference of the IEEE Communications Society (INFOCOM) (2002)
13. Zhang, N., Shin, I., Li, B., Boyaci, C., Tiwari, R., Thai, M.T.: New approximation for minimum-weight routing backbone in wireless sensor network. In: Li, Y., Huynh, D.T., Das, S.K., Du, D.-Z. (eds.) WASA 2008. LNCS, vol. 5258, pp. 96–108. Springer, Heidelberg (2008)

Evolution of Developmental Timing for Solving Hierarchically Dependent Deceptive Problems

Kouta Hamano, Kei Ohnishi, and Mario Köppen

Kyushu Institute of Technology,
680-4 Kawazu, Iizuka, Fukuoka 820-8502, Japan
hamano@evocomp.cse.kyutech.ac.jp, ohnishi@cse.kyutech.ac.jp,
mkoeppen@ci.kyutech.ac.jp

Abstract. Conventional evolutionary algorithms (EAs) cannot solve given optimization problems efficiently when their evolutionary operators do not accommodate to the structures of the problems. We previously proposed a mutation-based EA that does not use a recombination operator and does not have this problem of the conventional EAs. The mutation-based EA evolves timings at which probabilities for generating phenotypic values (developmental timings) change, and brings different evolution speed to each phenotypic variable, so that it can solve a given problem hierarchically. In this paper we first propose the evolutionary algorithm evolving developmental timing (EDT) by adding a crossover operator to the mutation-based EA and then devise a new test problem that conventional EAs are likely to fail in solving and for which the features of the proposed EA are well utilized. The test problem consists of multiple deceptive problems among which there is hierarchical dependency, and has the feature that the hierarchical dependency is represented by a graph structure. We apply the EDT and the conventional EAs, the PBIL and cGA, for comparison to the new test problem and show the usefulness of the evolution of developmental timing.

Keywords: developmental timing, deceptive problem, graph structure, dependency between variables, estimation distribution algorithm.

1 Introduction

Evolutionary algorithms (EAs) evolve several spatial patterns at different levels, such as genotype, phenotype and population, by using evolutionary operators. These spatial patterns, as objects of evolution, are related to the structure of optimization problems solved by EAs. Therefore, evolutionary operators have to be adapted to spatial patterns involved in optimization problems. For instance, fixed recombination operators that do not adapt linkages between variables have been shown to be inadequate and scale-up exponentially in terms of population size with increasing problem size [17].

To overcome the dependence of evolutionary operators on spatial patterns involving optimization problems, EAs must have a mechanism to reconstruct the

G. Dick et al. (Eds.): SEAL 2014, LNCS 8886, pp. 58–69, 2014.

spatial patterns for their evolutionary operators to work effectively. One such approach is to adopt a genetic code with a (position, value)-style, as used in the messy genetic algorithm [6] and the linkage learning genetic algorithm [8]. This genetic coding style allows EAs to rearrange phenotypic variables originally arranged in a fixed order. Another method that allows EAs to rearrange phenotypic variables within a genotype is the use of grammatical genetic codes. Grammatical genetic codes, as used in a genetic algorithm using grammatical evolution [16], enable EAs to provide priority to phenotypic variables by means of sequential interactions among phenotypic variables induced by grammar, even in uniformly-scaled problems, as well as to rearrange phenotypic variables within a genotype in an arbitrary order [12].

A more direct approach is to introduce a learning mechanism instead of genetic and evolutionary operators into EAs. This approach is referred to as probabilistic model building of genetic algorithms [14] or estimation of distribution algorithms [11]. This approach involves learning the distribution of selected genotypes in a genotypic space, which is a spatial pattern representing a part of the whole structure of an optimization problem, so that its performance is not influenced by linkage between variables in a genotype.

Under the background mentioned above, we previously proposed a mutation-based EA that does not use a recombination operator and does not have the problem of conventional EAs [13]. The mutation-based EA evolves timings at which probabilities for generating phenotypic values (developmental timings) change, and brings different evolution speed to each phenotypic variable, so that it can solve a given problem hierarchically. In the study, it was shown that the mutation-based EA sequentially solves sub-problems comprising a hard uniformly-scaled problem in which there is no prioritized variable. Concretely, the 4-bit trap deceptive function [5] was used as a hard uniformly-scaled problem and the scale-up of the mutation-based EA in terms of the number of function evaluations was shown to be sub-exponential.

In this paper we first propose a new EA by adding a crossover operator to the mutation-based EA. We refer to this EA as the evolutionary algorithm evolving developmental timings (EDT) hereinafter. Then, we devise a new test problem that conventional EA are likely to fail in solving and for which features of the proposed EA are well utilized. The test problem consists of multiple deceptive problems among which there is hierarchical dependency, and has a feature that the hierarchical dependency is represented by a graph structure. We apply the EDT and the conventional EAs, the population-based incremental learning (PBIL) [1] and the compact genetic algorithm (cGA) [9], for comparison to the new test problem and show the usefulness of the evolution of developmental timing. Though it is not shown in this paper due to page limitation, the EDT was shown to have better scalability for the 4-bit trap deceptive function mentioned above than the mutation-based EA proposed previously through simulations.

The present paper is organized as follows. Section 2 briefly describes related research. In Section 3, we present the evolutionary algorithm evolving developmental timings (EDT). Section 4 describes the new test problem. In Section 5, we examine the performance of the EDT on the new test problem through simulation. We summarize our results and present the conclusions in Section 6.

2   Related Work

Biological development is a complicated process in which a living organism develops into an adult from only a fertilized egg that includes a genome. In biological development, an appropriate temporal pattern of interactive developmental events, such as gene expression and cell division, produces an appropriate form. In certain species, when biological evolution occurs, the developmental system must also change. Since biological evolution is continuous, the change of a developmental system should also be continuous. Heterochrony is a biological term that refers to the change in timing of developmental events, and heterochrony has been reported to be key in explaining biological evolution [7].

One of the differences between biological development and genotype-genotype-mapping in EAs is that while living things continue to be exposed to environments (a fitness function) during the entire process of biological development and different genes are expressed at different times, individuals in EAs, in the case of problems with a fixed structure, are exposed to a fitness function (environments) only when fixed-structured phenotypes are formed by genotype-phenotype-mapping. Since only the formation of phenotypes that can be evaluated is meaningful in terms of EA functions, bringing temporal elements in biological development to genotype-phenotype-mapping in EAs is occasionally difficult or meaningless.

However, there have been several attempts to allocate temporal elements to EAs in the case of problems with fixed structures. One such attempt involves the development of grammatical genetic codes [16]. Genes in grammatical genetic codes are decoded in a certain order into phenotypic values according to grammar. In grammatical genetic codes, there are interactions among phenotypic variables induced by grammar; however, as in biological development, there is no interaction among decoding processes themselves.

In the case of optimizing variable structures such as trees and networks, the situation is somewhat different. Since all possible structures can be evaluated, temporal elements can be used by considering the growth of structures to be a process similar to biological development. Therefore, genotype-phenotype-mapping including interactions among multiple decoding processes (developmental events) [4] as well as grammatical genetic codes [10][15] can be used for structure optimization problems.

3   Evolutionary Algorithm Evolving Developmental Timing

In this section, we present the evolutionary algorithm evolving developmental timings (EDT) that adds a crossover operator described in Section 3.3 to the mutation-based evolutionary algorithm that we previously proposed in [13], which is described in other parts of this section.

3.1   Individual

In conventional evolutionary algorithms, individuals are usually equivalent to genotypes, which are mapped into phenotypes in a fixed manner. A conventional individual is evaluated only once during a generation and proceeds to a selection phase that produces a generation gap. Here, we introduce a lifespan to individuals within a generation. The individual then generates several phenotypes during its lifetime.

The individual consists of two types of vectors. In the first type, the element of the vector represents a cycle time of changing a probability to determine a phenotypic value. This vector is a primary object to which evolutionary operators are applied, and its elements can be considered as a type of information on developmental timing. We refer to this vector and its element as a genotype and a gene, respectively. In the second type, the element of the vector is a probability for determining a phenotypic value. For example, in the case of bit optimization, the probability of generating zero at a certain phenotypic position is an element of the vector. While the vector of the cycle times does not vary during a generation, the vector of probabilities does vary during a generation, as discussed later.

The lengths of the genotype and the vector of probabilities are the same as the length of the phenotype. Each position in these two vectors corresponds to the same position in the phenotype.

3.2   Individual Development

As mentioned in the previous section, the individual within a generation has a lifespan and consists of a genotype and a vector of probabilities. A time in a lifetime of the individual is denoted by $n \in [1, N]$, where N is the algorithm parameter representing the end of life of the individual. In addition, let $(t_1, t_2, \cdots, t_\ell)$ and $(p_1, p_2, \cdots, p_\ell)$ be the genotype and the vector of probabilities, respectively, where t_i is an integer within $[1, T_c]$ and p_i is a real value. $T_c (T_c \leq N)$ is the algorithm parameter representing the possible maximum cycle time. The genotype does not vary during the lifetime of the individual, but the vector of probabilities does vary during its lifetime. In this section, we explain how the vector of probabilities varies due to developmental timings composing the genotype and how N phenotypes are time-sequentially produced from the genotype and the vector of probabilities.

The initialization of the individuals is performed as follows. The genotypes are randomly generated. All elements of the vector of probabilities are set 0.5 in the case of bit optimization problems, which represents the probability with which zero is generated. A phenotype is generated using the vector of probabilities at each time during the lifetime of the individual. After every generation of a phenotype, by comparing the current time, n, and each element of the genotype, t_*, it is determined whether it is time to change the probability to generate each phenotypic value. If n is a multiple of t_*, then p_* is modified.

The modification of the elements of the vector of probabilities is performed as follows. We assume that the i-th gene is $t_i \in [1, T_c]$. The modification of the i-th element of the vector of probabilities, p_i, is carried out every t_i time step during development of the individual. When the time is $a \times t_i$, the modification is performed using the phenotypes generated within $(a-1) \times t_i$ to $a \times t_i$ and their fitness values. For example, when t_i is 2, if the phenotypic values at the i-th position in the phenotypes generated between time 1 and time 2 are 1 and 0, and if the phenotypic value at the i-th position in the phenotype with the best fitness value among these two values, pv_b, is 0, then the probability with which 0 is generated on the i-th position in the phenotype, p_i, increases in proportion to the number of 0s in the two phenotypic values generated, num_0. If pv_b is 1, then p_i decreases in proportion to the number of 1, num_1. The new probability, p_i^{new}, is determined by Equation (1).

$$
p_i^{new} = \begin{cases} p_i + C \times num_0 & if \;\; pv_b = 0, \\ p_i - C \times num_1 & if \;\; pv_b = 1, \end{cases} \tag{1}
$$

where C is the algorithm parameter. This can be considered as a type of learning process, not for the distribution of all phenotypes in the phenotypic space, but rather for the distribution of pieces of the phenotypes. Figure 1(a) also illustrates an example of modification of the vector of probabilities during the lifetime of the individual.

The best fitness value among all of the phenotypes generated during the lifespan of the individual is set as the fitness value of the individual, and the individual proceeds to the selection, crossover, and mutation phases.

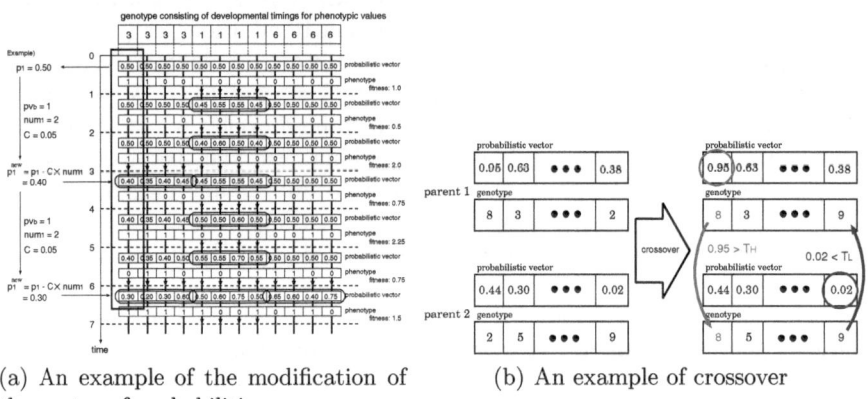

(a) An example of the modification of the vector of probabilities

(b) An example of crossover

Fig. 1. Overview of the evolutionary algorithm evolving developmental timing (EDT)

3.3   Crossover Operator

The EDT proposed in this section cannot rearrange phenotypic variables on a phenotype. Therefore, if the EDT uses a fixed crossover (recombination) operator, for example, one-point crossover, it depends on optimization problems to be solved whether it can efficiently solve the problems. So, we design a crossover operator that does not break good linkages between phenotypic variables that the EDT has identified so far but brings them to the next generation.

The crossover operator designed here estimates the dependency between phenotypic variables from element values of the vector of probabilities of the individual. When the i-th element value of the vector of probabilities of some parent individual, p_i, is close to 0 (or 1), the i-th phenotypic value becomes 1 mostly. That is to say, it can be said that the i-th phenotypic variable's value has almost converged. We consider that multiple phenotypic variables whose values have almost converged at some moment depend on each other. Then, the crossover operator copies the i-th gene of that parent individual, t_i, onto the genotype of another parent individual. The reason for copying not element values of the vector of probabilities but genes of the genotype to decide the developmental timing is that phenotypic values might have converged incorrectly.

Before applying the crossover operator, the EDT does not conduct selection for reproduction but just randomly divides all individuals of the present population into pairs throughly. When the population size is P, the number of the parent pairs is $P/2$, where P is an even number. Each pair of parent individuals produces two new individuals, so that the total number of individuals produced is P. The condition in which the i-th gene of a parent individual of focus, t_i, is copied onto the genotype of another parent individual is that the i-th element value of the vector of probabilities of the parent individual of focus, p_i, satisfies $p_i < T_L$ or $p_i > T_H$ $(T_L < T_H)$, where T_L and T_H are the algorithm parameters. An example of the crossover is shown in Figure 1(b).

3.4   Mutation and Selection Operators

A mutation operator is applied to each element of the genotype and the vector of probabilities The mutation rates for the i-th element of the genotype and the vector of probabilities, pm_i, are the same, and the rate is determined using the i-th element of the genotype, $t_i \in [1, T_c]$. The mutation rate, pm_i, is determined by Equation (2).

$$pm_i = 1 - \frac{t_i}{N+1} \tag{2}$$

This equation indicates that the smaller the cycle time, the larger the mutation rate. In addition, the mutation rate is always greater than zero.

The mutation to the genotype randomly changes its element value. The mutation to the vector of probabilities sets its element value as 0.5. Using this mutation operator, a parent-individual generates R child-individuals. When the population size is P, $R \times P$ child-individuals are generated in one generation. P and R are the algorithm parameters.

A selection operator for survival selects P individuals with better fitness values from among $R \times P$ child-individuals plus P parent-individuals as the population of the next generation.

4 Hierarchically Dependent Deceptive Problem

We devise a new test problem for which the features of the EA proposed in Section 3, the EDT, are well utilized. This is a bit optimization problem. The salient feature of the EDT is that it brings a different evolution speed to each phenotypic variable. It depends on individuals how evolution speeds are different between phenotypic variables, but in any cases, individuals that bring suitable evolution speeds to phenotypic variables for solving a given problem are selected and adapted. Therefore, we consider a new test problem that requires bringing different evolution speeds to phenotypic variables in order to to solve it efficiently.

The test problem considered here includes hierarchical dependency among phenotypic variables in terms of fitness values, and is hard to determine values of phenotypic variables at higher hierarchies correctly, and leads to incorrect determination of values of phenotypic variables at lower hierarchies if the determination at higher hierarchies is incorrect.

The general procedures to produce the test problem as mentioned above are as follows.

1. We generate a connected graph having some topology whose number of nodes is equal to the number of phenotypic variables (the length of a bit string to be optimized), ℓ.
2. We assign position numbers $(1, 2, \ldots, \ell)$ on a bit string as a phenotype to each of the ℓ nodes on the generated graph.
3. We define L m-bit deceptive problems [5] on the generated connected graph, where nodes of higher degree (the number of edges) form higher hierarchies. If multiple nodes having the same degree are included in one m-bit deceptive problem, the nodes having smaller position numbers become ones of higher degree. A solution candidate of one m-bit deceptive problem consists of m nodes (bits) sequentially connected by edges. More concretely, the solution candidate is formed by arranging the m nodes (bits) from higher to lower in terms of hierarchy. We need to prepare a way to determine a fitness value of the obtained sequence of bits. In addition, we introduce dependency among multiple m-bit deceptive problems. That is to say, multiple m-bit deceptive problems share several same nodes (bits).
4. We set the sum of fitness values of L m-bit deceptive problems to be a fitness value of a solution candidate of the entire problem. This is a maximization problem.

In simulations described in the following section, we determine the parameters of the procedures above as follows. We generate the connected graph used in the simulations by using the algorithm described in [3]. The topology of the generated graph follows a power law [2] with respect to the distribution of degree.

We use 2-bit deceptive problems. One 2-bit deceptive problem consists of two nodes connected by an edge. We introduce the dependency between two 2-bit deceptive problems by setting a node at the lower hierarchy in one 2-bit deceptive problem to be a node at the higher hierarchy in another 2-bit deceptive problem. The total number of the 2-bit deceptive problems defined on the generated graph is equal to the number of the edges of the graph.

Figure 2(a) shows an example graph representing the problem used here and Figure 2(b) shows a way to determine a fitness value of the 2-bit deceptive problem. As shown in Figure 2(b), the optimum solution for the 2-bit optimization problem is "11", and the global optimum of the entire problem is obtained when all bits (nodes) are 1. The second best solution for the 2-bit optimization problem is "00". Due to the deceptive structure of the problem, it is likely that the local optimum in which all bits are 0 is obtained.

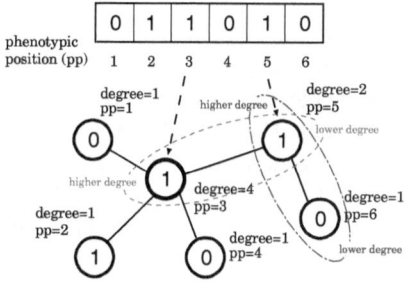

phenotypic value of higer degree node	phenotypic value of lower degree node	fitness value
0 ——→ 0		9
0 ——→ 1		7
1 ——→ 0		0
1 ——→ 1		10

(a) An example graph representing the problem

(b) A fitness value of the 2-bit deceptive problem

Fig. 2. The proposed test problem

5   Simulations

In this section, we apply the EDT presented in Section 3 to the test problem devised in Section 4. In addition, we compare the results with those obtained by other algorithms for comparison.

5.1   Algorithms for Comparison

We use the population-based incremental learning (PBIL) [1] and the compact genetic algorithm (cGA) [9] for comparison, which are the simple algorithms that belong to the probabilistic model building of genetic algorithms or the estimation of distribution algorithms. These algorithms are both for bit-optimization and represent a population as one vector of probabilities for generating phenotypes. Each element of the vector of probabilities is probability for generating "1" at the corresponding position on the phenotype and is initialized as 0.5 at the beginning.

The PBIL generates M phenotypes using the vector of probabilities within one generation and selects Q better phenotypes in terms of fitness values from the M $(Q \le M)$ phenotypes, where Q and M are the algorithm parameters. Suppose that the present generation is g. The PBIL determines each element value of the vector of probabilities for the next generation, $g + 1$, using the selected Q phenotypes. More specifically, the PBIL determines the probability for generating "1" at each position i $(i = 1, 2, \ldots, \ell)$ of the phenotype, $P_{g+1}(X_i = 1)$. This probability is given by Equation (3).

$$P_{g+1}(X_i = 1) = (1 - \alpha) \cdot P_g(X_i = 1) + \alpha \cdot \frac{1}{M} \cdot \sum_{k=1}^{M} x_i^k, \tag{3}$$

where $\alpha \in (0, 1]$ is the algorithm parameter and x_i^k is the phenotypic value at the i-th position on the k-th phenotype in the Q selected ones, which takes "0" or "1".

Next, the cGA generates two phenotypes, a and b, using the vector of probabilities. Suppose that the fitness value of the phenotype a is better than b's and the present generation is g. Then, if phenotypic values of a and b at the phenotypic position i $(i = 1, 2, \ldots, \ell)$ are different, the probability for generating "1" at the phenotypic position i for the next generation $g + 1$, $P_{g+1}(X_i = 1)$, is given as follows. If the phenotypic value of a is "1", $P_{g+1}(X_i = 1)$ is given by Equation (4). If the phenotypic value of a is "0", $P_{g+1}(X_i = 1)$ is given by Equation (5).

$$P_{g+1}(X_i = 1) = P_g(X_i = 1) + \frac{1}{K}, \tag{4}$$

$$P_{g+1}(X_i = 1) = P_g(X_i = 1) - \frac{1}{K}, \tag{5}$$

where K is the algorithm parameter.

The PBIL and cGA explained above build the vector of probabilities by learning the distribution of generated phenotypes. Meanwhile, the EDT builds the vector of probabilities by learning the distribution of generated phenotypes piece by piece at different times during the lifetime of the individual.

5.2   Settings

The test problem used here is the concrete one explained in the last part of Section 4. We use $10 \times j$ $(j=1,2\ldots,10)$ as the number of nodes (phenotypic variables). Figure 3 shows the degree distributions of the graphs representing the problems when the number of nodes is 10, 30, and 100.

All the parameter values of the EDT, PBIL, and cGA are not varied depending on the number of nodes. The termination conditions for all the algorithms are the same, which is that the number of evaluations reaches 10^6. The parameter values of the EDT are as follows. The population size, P, is 100, the lifetime length, N, is 10, the maximum possible value for the gene is 9, the parameter for updating the probability, C, is 0.05, the parameters of the crossover operator,

T_H and T_L, are 0.92 and 0.08, and the number of child individuals created from one individual by the mutation operator, R, is 5. The parameter values of the PBIL are as follows. The number of generated phenotypes, M, is 100, the number of selected phenotypes, Q, is 50, and the parameter for updating the probability, α, is 0.01. The parameter of the cGA is only the parameter for updating the probability, K, and its value is set to be 100.

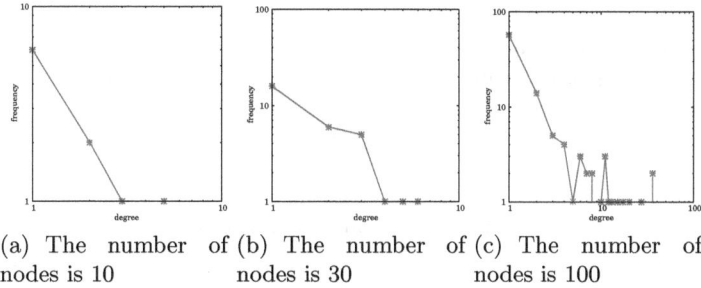

(a) The number of nodes is 10

(b) The number of nodes is 30

(c) The number of nodes is 100

Fig. 3. The degree distribution of the used test problems

5.3   Results and Discussions

Figure 4(a) shows the relationship between the number of nodes (phenotypic variables) and the best fitness value obtained by 10^6 evaluations for the EDT, PBIL, and cGA. The fitness values shown in Figure 4(a) are obtained by dividing the actual fitness values by the number of edges in the graph, which is equal to the number of the 2-bit deceptive problems defined on the graph. Therefore, when the global optimum whose bits are all "1" is obtained, the fitness value in Figure 4(a) becomes 10 and when the local optimum whose bits are all "0", the fitness value in Figure 4(a) becomes 9, no matter what number of nodes is used.

Figure 4(b) shows the result only for the EDT and the time transitions of the ratio of "1" at three phenotypic positions, 3, 16, and 20, in all generated phenotypes at the generation in case that the number of nodes is 30. The degrees of the nodes corresponding to phenotypic positions 3, 16, and 20 are six, two, and one, respectively. The degree of six is the highest in the graph.

We can observe from Figure 4(a) that the EDT obtained the global optimum 10 times out of the 10 simulation runs when the number of nodes is less than or equal to 30. However, the EDT obtained the global optimum only a few times out of the 10 simulation runs when the number of nodes is 40, 50, and 60. When the number of nodes is more than or equal to 70, the EDT obtained the local optimum whose bits are all "0" at all the simulation runs. Thus, at this moment, the EDT is not so scalable against the increasing number of nodes (size of problem) and needs the improvement of the scalability.

Meanwhile, we can observe from Figure 4(a) that the PBIL and cGA for comparison obtained the local optimum or the nearly local optimum at all the

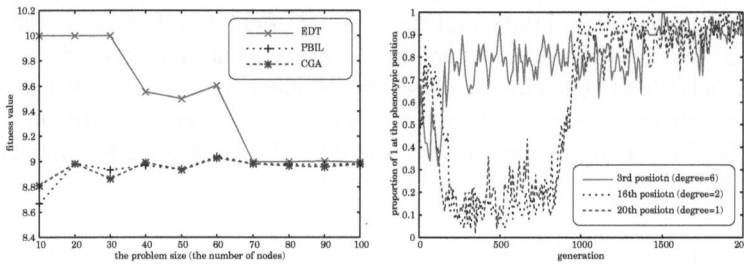

(a) The average final fitness values over 10 independent runs for different number of nodes

(b) The time transitions of the proportion of "1" at the 3rd, 16th, and 20th nodes, whose degrees are six (a hub node), two, and one, respectively

Fig. 4. The simulation results

simulation runs. Compared the EDT with the PBIL and cGA, the EDT is superior to the PBIL and cGA. Since the test problem used here is basically designed for the EDT to work efficiently, this result is expected. In addition, we can guess from Figure 4(b) that the EDT solved the problem in a way such that values at the phenotypic positions corresponding to nodes of higher degree are converged correctly earlier than those corresponding to nodes of lower degree. This is also expected and this way would contribute to the efficient problem solving.

6   Concluding Remarks

We proposed the evolutionary algorithm evolving developmental timings (EDT) based on the mutation-based evolutionary algorithm that we previously proposed. In addition, we devised the new test problem that was expected to be solved efficiently by the EDT, and applied the EDT and the conventional evolutionary algorithms for comparison to the new test problem. It was shown through the comparison that the EDT is better.

In the future work, we will improve the scalability of the EDT and look for the connection of the new test problem presented in the paper to problems in the real world. Since the presented test problem is represented by a graph structure, it might be, for example, related to realizing cooperative behaviors of nodes on a network.

Acknowledgement. This work was supported by the Japan Society for the Promotion of Science through a Grant-in-Aid for Scientific Research (C)(25330289).

References

1. Baluja, S.: Population-based incremental learning: A method for integrating genetic search based function optimization and competitive learning. Tech. rep (1994)
2. Barabasi, A.L., Albert, R.: Emergence of Scaling in Random Networks. Science 286, 509–512 (1999)
3. Bu, T., Towsley, D.: On Distinguishing between Internet Power Law Topology Generators. In: Proceedings of IEEE Infocom 2002, pp. 638–647 (2003)
4. Cangelosi, A.: Heterochrony and adaptation in developing neural networks. In: Proceedings of the Genetic and Evolutionary Computation Conference 1999, San Francisco, CA, pp. 1241–1248 (1999)
5. Deb, K., Goldberg, D.E.: Analyzing deception in trap functions. Foundations of Genetic Algorithms 2, 93–108 (1993)
6. Goldberg, D.E., Korb, B., Deb, K.: Messy genetic algorithms: Motivation, analysis, and first results. Complex Systems 3(5), 493–530 (1989)
7. Gould, S.J.: Ontogeny and Phylogeny. Harvard Univ. Press, Oxford (1977)
8. Harik, G.R., Goldberg, D.E.: Learning linkage. Foundations of Genetic Algorithms 4, 247–262 (1996)
9. Harik, G.R., Lobo, F.G., Goldberg, D.E.: The compact genetic algorithm. IEEE Transactions on Evolutionary Computation 3(4), 287–297 (1999)
10. Kitano, H.: Designing neural networks using genetic algorithms with graph generation system. Complex Systems 4(4), 461–476 (1990)
11. Larranaga, P., Lozano, J.A.: Estimation of Distribution Algorithms: A New Tool for Evolutionary Computation. Kluwer Academic Publishers (2001)
12. Ohnishi, K., Sastry, K., Chen, Y.-P., Goldberg, D.E.: Inducing sequentiality using grammatical genetic codes. In: Proceedings of the Genetic and Evolutionary Computation Conference 2004, Seattle, WA, pp. 1426–1437 (2004)
13. Ohnishi, K., Uchida, M., Oie, Y.: Evolution and Learning Mediated by Difference in Developmental Timing. Advanced Computational Intelligence and Intelligent Informatics (JACIII) 11(8), 905–913 (2007)
14. Pelikan, M., Goldberg, D.E., Lobo, F.: A survey of optimization by building and using probabilistic models. IlliGAL Report No. 99018, Illinois Genetic Algorithms Lab., Univ. of Illinois, Urbana, IL (1999)
15. Ryan, C., Collins, J.J., O'Neill, M.: Grammatical evolution: Evolving programs for an arbitrary language. In: Banzhaf, W., Poli, R., Schoenauer, M., Fogarty, T.C. (eds.) EuroGP 1998. LNCS, vol. 1391, pp. 83–96. Springer, Heidelberg (1998)
16. Ryan, C., Nicolau, M., O'Neill, M.: Genetic algorithms using grammatical evolution. In: Foster, J.A., Lutton, E., Miller, J., Ryan, C., Tettamanzi, A.G.B. (eds.) EuroGP 2002. LNCS, vol. 2278, pp. 278–287. Springer, Heidelberg (2002)
17. Thierens, D., Goldberg, D.E.: Mixing in genetic algorithms. In: Proceedings of the 5th International Conference on Genetic Algorithms (ICGA 1993), pp. 38–45 (1993)

The Introduction of Asymmetry
on Traditional 2-Parent Crossover Operators
for Crowding and Its Effects

Shigeyoshi Tsutsui

Hannan University, Matsubara Osaka 580-8502, Japan
tsutsui@hannan-u.ac.jp

Abstract. This paper proposes a novel crowding method, which we call "Crowding with Asymmetric Crossover (CAX)" that can be applied to traditional 2-parent crossover operators. The asymmetric crossover operator begins with two parents. Then two offspring individuals are created, each offspring taking more characteristics from one of the two parents. This is an easy method to perform replacement between parents and offspring individuals. Experimental results showed that CAX can increases the performance of traditional 2-parent crossover operators in finding global optimal solutions. CAX is also useful to find multiple solutions (niching).

1 Introduction

Crowding methods constitute an important research area in genetic and evolutionary computation. There are two main objectives of crowding methods: (1) one is to prevent premature convergence of a population by preserving the population diversity, and obtain one global optimal solution; (2) the other is to converge the population to multiple, highly fit, and significantly different solutions (niching).

In this paper, we propose a novel crowding method, which we call "Crowding with Asymmetric Crossover (CAX)" and show that the CAX can increases the performance of traditional 2-parent crossover operators in finding global optimal solutions. In the literature of crowding methods, the main efforts are focused on how replacement is performed between parental individuals and offspring individuals using similarity between them as a replacement criteria. Typical studies of these are the crowding factor model [1], the deterministic crowding (DC) [2], the probabilistic crowding (PC) [3], the Boltzmann crowding [4] and the generalized crowding [5].

CAX does not use the similarity metric among individuals as a criteria for replacement. Instead, we use "asymmetric crossover" for crossover operators. The asymmetric crossover operator generates offspring individuals which are each similar to one of two parent individuals. The degree of the similarity between the parents individuals and the offspring individuals is controlled by a parameter. By choosing the value of the parameter, CAX can maintain population diversity

G. Dick et al. (Eds.): SEAL 2014, LNCS 8886, pp. 70–81, 2014.

to obtain one global optimal solution, or converge the population to multiple different solutions (niching).

In the remainder of this paper, a brief review of the crowding methods are described in Section 2. Then, we describe how the CAX is configured in Section 3. The empirical analysis is given in Section 4 putting main emphasis on obtaining one global optimal solution. Finally, Section 5 concludes the paper.

2 A Brief Review of Crowding Methods

Here, we review typical crowding methods. The original idea of crowding methods was proposed by De Jong [1]. De Jong reasoned that in nature, as like individuals begin to dominate a niche, increased competition for limited resources decreases life expectancy and birthrates. Less crowded niches experience less pressure and achieve life expectancy and birthrates much closer to their potential. To enforce such a crowding pressure in artificial genetic algorithms, De Jong forced newly generated offspring to replace similar, older adults in the hope of maintaining more diversity in the population [6].

Crowding consists of two main phases: pairing and replacement. In the pairing phase, offspring individuals are paired with individuals in the current population according to a similarity metric. In the replacement phase, a decision is made for each pair of individuals as to which of them will remain in the population [5].

2.1 Crowding Factor Model

The main purpose of the crowding factor model by De Jong [1] is to maintain population diversity to find global optimal solutions. In the crowding factor model, replacement for each offspring produced is considered individually. For each such individual, a sample of CF (Crowding Factor) number of individuals are drawn from the population and searched for the most similar individual to the offspring in question. The most similar individual from the small sample is then directly replaced in the population by the offspring, without regard for fitness.

2.2 Deterministic Crowding (DC)

Since offspring are obtained by recombination of their parents, parent individuals and offspring individuals have a certain degree of similarity. Deterministic Crowding (DC) [2] uses this feature as shown in Fig. 1.

2.3 The Extension of Deterministic Crowding

Unlike DC, Probabilistic Crowding (PC) [3] uses a non-deterministic rule to establish the winner of a competition between a parent p and a child c. In PC, c and p compete in probabilistic tournaments. The probability of c winning is given by:

1. Select two parents, p_1 and p_2, randomly, without replacement
2. Cross them, yielding c_1 and c_2
3. If $|p_1, c_1| + |p_2, c_2| \leq |p_1, c_2| + |p_2, c_1|$
        If $f(c_1) > f(p_1)$, replace p_1 with c_1
        If $f(c_2) > f(p_2)$, replace p_2 with c_2
  Else
        If $f(c_2) > f(p_1)$, replace p_1 with c_2
        If $f(c_1) > f(p_2)$, replace p_2 with c_1

Fig. 1. Deterministic crowding methods

$$P_c = \frac{f(c)}{f(c) + f(p)}, \tag{1}$$

where f is the fitness function. Boltzmann Crowding (BC) is based on the well-known Simulated Annealing method, implemented with the Boltzmann acceptance rule [4] in Eq. (1).

Generalized Crowding (GC) [5] allows selective pressure to be controlled in a simple way in the replacement phase of crowding, thus overcoming limitations of the other approaches. Both DC and PC turn out to be special cases of GC. The temperature parameter used in Simulated Annealing is replaced by a parameter called scaling factor that controls the selective pressure applied.

3   Crowding with Asymmetric Crossover (CAX)

As we saw in Section 2, in usual crowding methods, offspring individuals are generated using usual crossover operators in the pairing phase and then the similarity between two parents and offspring individuals are measured in the replacement phase. In Crowding with Asymmetric Crossover (CAX), we do not use the similarity measure in the replacement phase. Instead, an "asymmetric crossover (AX)" in the pairing phase is used. AX generates two offspring individuals each which is similar statistically to one of the two parent individuals.

Although AX is not restricted to 2-point crossover, hereafter we explain the AX using 2-point crossover. Let l_x be the length between cut-point cut1 and cut2. In usual 2-point crossover, l_x distributes in $[1, n-1]$ uniformly. Thus, the average value of l_x, $E(l_x)$, is $n/2$, where n is the string length, or problem size.

In AX, we sample two cut-points so that $E(l_x)$ is bigger than $n/2$. If we choose two cut-points so that $E(l_x)$ is closer to n, then both offspring individuals c_1 and c_2 are more similar to parents p_1 and p_2, respectively as shown in Fig. 2.

To control the similarity, AX introduces a parameter $\lambda(0.5 \leq \lambda < 1)$ which controls the similarity by sampling l_x as $E(l_x) = n \times \lambda$. For probability density function (p.d.f.) of l_x, we determine in the following manner which is similar to our previous study on cAS (cunning Ant System) [7]. When we apply c cut-points $(c=2, 3, 4, \ldots)$ to a string, the string can be divided into c segments.

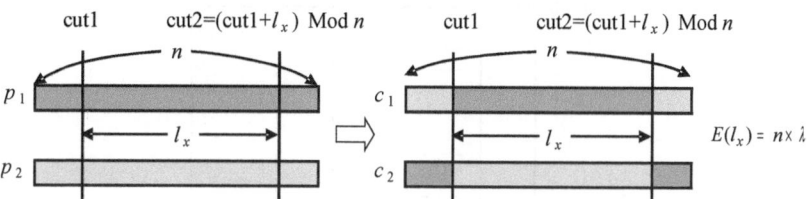

Fig. 2. Asymmetric crossover (AX). If we choose two cut-points so that $E(l_x)$ is closer to n, then both offspring individuals c_1 and c_2 are more similar to parents p_1 and p_2, respectively.

If we take any set of connected $c - 1$ segments, then the average length of those $c - 1$ segments should be $n \times (c - 1)/c$. We then use this length for l_x. The probability density function of l_x, $f_{n,\lambda}(l_x)$, can be obtained as

$$f_{n,c}(l_x) = \frac{c-1}{n} \left(\frac{l_x}{n}\right)^{c-2}, \quad 0 < l_x < n, \ c \geq 2. \tag{2}$$

Here, $E(l_x)$ is $n \times 1/2$, $n \times 2/3$, $n \times 3/4$, $\cdots$ for $c = 2, 3, 4, \ldots$, and, λ corresponds to $(c-1)/c$, i.e., λ can take only the values of 0.5, 0.667, 0.75, corresponding to $c = 2, 3, 4, \ldots$. Hereafter, we extend this function of Eq. (2) to a more flexible technique which allows for λ to take values in the rage [0.5, 1.0) by setting $(c-1)/c = \lambda$, i.e., $c = 1/(1 - \lambda)$. Then, Eq. (2)can be rewritten as

$$f_{n,\lambda}(l_x) = \frac{\lambda}{n(1-\lambda)} \left(\frac{l_x}{n}\right)^{\frac{2\lambda-1}{1-\lambda}}. \tag{3}$$

Fig. 3 shows $f_{n,\lambda}(l_x)$ for $\lambda = 0.5, 0.6, 0.7, 0.8$, and 0.9. We can see from this figure that for a bigger λ, longer lengths of l_x become dominant and thus, c_1 and c_2 become more similar to p_1 and p_2, respectively (see Fig. 2). Please note here that the case of $\lambda = 0.5$ becomes uniform distribution, i.e., usual 2-point crossover or "symmetrical crossover".Overall description of the CAX algorithm is shown in Fig. 4.

4   Empirical Analysis

In this section, we explore the effects of crowding with asymmetric crossover (CAX). Many of studies on crowding explore the performance of their algorithms to converge the population to highly fit solutions (niching). Instead, in experiments in this paper, we mainly explore the performance of CAX to find the global optima. In these experiments, we will see the performance of CAX changing values of the control parameter λ in the range [0.5, 1). Note here that the results with λ value of 0.5 are performances with "symmetric crossover", i.e., canonical two-point crossover.

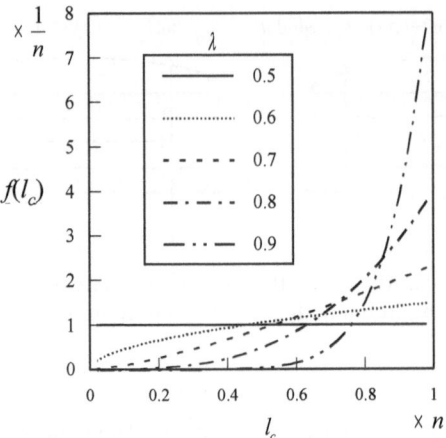

Fig. 3. $f_{n,\lambda}(l_x)$ for various λ values

We use the following three types of chromosome domains: (1) the binary domains, (2) the permutation domains, and (3) the real-value domains. In the following experiments, we run CAX 10 times on each test problem.

4.1 Results on the Binary Domains

Here we use two problems; the 0-1 Knapsack Problem (KP), and the fully deceptive problem. For the KP, we generated a problem which consists of 100 items. Value and weight of each item are obtained randomly in $[1, 100]$. Thus the length of the bit string is 100. The optimal solution was obtained by the Minknap Algorithm [8] which is based on dynamic programming. For the deceptive problem, we used the function which was proposed in [9]. In this experiment, we connected 20 3-bit fully deceptive functions tightly. Thus length of the bit string is 60. The population size N is set to 100. We run the algorithm until 100,000 function evaluations are reached or the global optima are found.

We analyze the CAX by $\#OPT$ (the number of runs in which algorithms succeeded in finding the global optimum), MNE (the mean number of function evaluations to find the global optimum in those runs where it did find the optimum), and population diversity measure. For the diversity measure of population with binary domains, we use the "$Bias$" $(0.5 \leq Bias \leq 1)$ proposed by Grefenstette in developing his GA package "GENESIS" [10]. We used this measure to see the current convergence status of the population in previous study [11] and it worked well. Here, the $Bias$ is defined as follows: Let a population be represented by $P = (p_{ij})$ where each row vector represents the string of an individual and p_{ij} is 0 or 1 where, $i = 0, 1, 2, \ldots, N-1$ and $j = 0, 1, 2, \ldots, n-1$. Then, $Bias$ is

$$Bias = \frac{1}{N \times n} \sum_{j=0}^{n-1} \left(\left| \sum_{i=0}^{N-1} p_{ij} - \frac{N}{2} \right| \right) + 0.5. \tag{4}$$

> 1. Set λ (e.g., λ=0.8)
> 2. Generate initial N size of population
> 3. Create $N/2$ number of parental pairs
> 4. For each parental pair $[p_1, p_2]$, apply AX
>   4.1 Sample cut1 in $[0, n-1]$, randomly
>   4.2 Sample l_c according to Eq. (3)
>   4.3 Set cut2 as (cut2+ l_c) Mod n
>   4.4 Cross p_1 and p_2, yielding c_1 and c_2
>   4.5 Replacement
>   • If $f(c_1) > f(p_1)$, replace p_1 with c_1
>   • If $f(c_2) > f(p_2)$, replace p_2 with c_2
> 5. Shuffle the population
> 6. If the termination criteria are met, terminate the algorithm.
>   Otherwise, go to Step 3.

Fig. 4. Description of the CAX algorithm. N is the population size.

Bias is the first-order convergence indicator that indicates the average ratio of the most prominent value in each position of the individuals of the population. Larger values mean low genotype diversity.

Fig. 5 (a) shows the variations of $\#OPT$ and MNE for various λ on KP. Here, λ values were varied starting from 0.5 to 0.95 with step 0.05. We can see that CAX with λ values within range [0.7, 0.85] finds the optimum ($\#OPT \cong 10$) showing relatively smaller values of MNE. Although with a smaller value of λ near 0.5, the results of MNE show better results, but $\#OPT$ results are rather poor. Typically with $\lambda = 0.5$, $\#OPT$ being only 1. Fig. 5 (b) shows the convergence process of the population measured by *Bias* for $\lambda = 0.5, 0.6, 0.7$, 0.8, and 0.9. Fig. 5 (b) supports the results of Fig. 5 (a), i.e., using appropriate λ values ($\lambda > 0.5$), we can maintain population diversity, and thus we can obtain optimal solution effectively, as seen in (a).

Fig. 6 shows results of the deceptive function, showing similar results to KP in Fig. 5. We can see that CAX with λ values within range [0.7, 0.85] finds the optimum ($\#OPT \cong 25$) showing relatively smaller values of MNE. Although with $\lambda = 0.5$, $\#OPT$ shows smaller value than results with λ values in [0.75, 0.9], but $\#OPT$ again results in a value of one with $\lambda = 0.5$, as with KP in Fig. 5 (a). Fig. 6 (b) shows the convergence process of the population measured with *Bias* for $\lambda = 0.5, 0.6, 0.7, 0.8$, and 0.9.

4.2 Results on the Permutation Domains

For the permutation domains, we use the quadratic assignment problem (QAP). The QAP is a problem which assigns a set of facilities to a set of locations and can be stated as a problem to find a permutation ϕ which minimizes

$$cost(\phi) = \sum_{i=0}^{n-1} \sum_{j=0}^{n-1} a_{ij} b_{\phi(i)\phi(j)}, \tag{5}$$

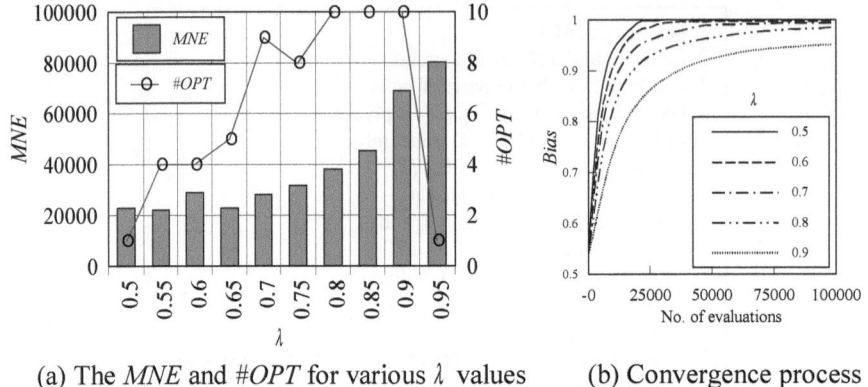

(a) The *MNE* and *#OPT* for various λ values   (b) Convergence process

Fig. 5. Results of CAX on the Knapsack Problem (KP). (a) shows *#OPT* and *MNE*, and (b) shows the convergence process measured by *Bias* of the population.

where $A = (a_{ij})$ and $B = (b_{ij})$ are two $n \times n$ matrices and ϕ is a permutation of $\{0, 1, 2, \ldots, n - 1\}$. Matrix A is a flow matrix between facilities i and j, and B is the distance between locations i and j.

We used benchmark instances in QAPLIB [12]. According to [13], benchmark instances can be classified into i) randomly generated instances, ii) grid-based instances, iii) real-life instances, and iv) real-life-like instances. In this experiment, we used the following four instances; tai30b, tai40b, tai50b, and tai60b, which are classified as real-life-like instances.

Since the performance of QAP in Eq. (5) depends mainly on absolute position of nodes in a string, we used the partially matched crossover (PMX) [6] for the base operator of AX. The QAP is considered one of the hardest optimization problems. To get high quality solutions, it is common to combine heuristic algorithms to the base algorithm [14,15]. However, we do not combine any heuristic algorithms in this experiment to see the pure effect of the AX. As for the convergence measure of the population, we used the following entropy E,

$$E = -\frac{1}{n} \sum_{j=0}^{n-1} \sum_{i=0}^{n-1} \frac{q_{ij}}{N} \log_2 \left(\frac{q_{ij}}{N}\right), \tag{6}$$

where q_{ij} represents the number of strings which have node value i at position j in the population. In this experiment, we used the following parameter setting: the population size $N = 8 \times n$, the maximum number of evaluations is $100,000 \times n$.

Fig. 7 shows results on QAP for various λ values. In the figure, performances are shown by $Error = (best\_functional\_value - optimal\_value)/optimal\_value \times 100$. From this figure, we can see CAX with λ values in $[0.7, 0.8]$ works well showing smaller *Error* values. Fig. 8 shows the convergence process of CAX on tai40b as a representative. In the figure, (a) shows the change of *Error* (%) and (b) shows the change of entropy E defined by Eq. (6). For example, with λ value of

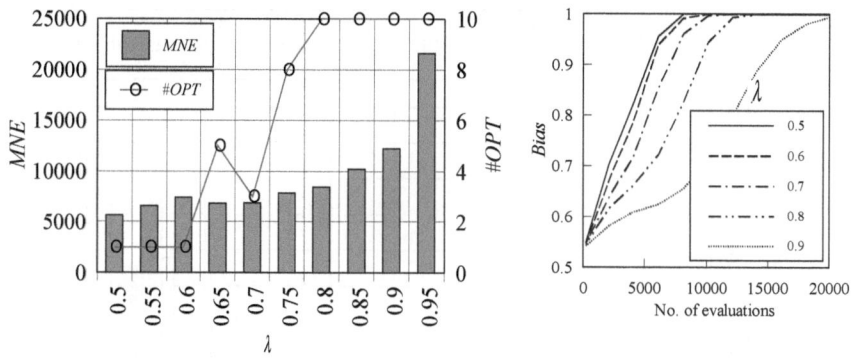

(a) The *MNE* and *#OPT* for various λ values   (b) Convergence process

Fig. 6. Results of CAX on the deceptive function. (a) shows *#OPT* and *MNE*, and (b) shows the convergence process measured with *Bias* of the population.

0.5, the value of E rapidly decrease, as seen in (b) and a premature convergence occurred, as seen in (a). In contrast to this, with λ value of 0.9, the population converges very slowly. With λ values of 0.7 and 0.8, the population converges with a degree of diversity, showing smaller *Error* values.

4.3   Results on the Real-Value Domains

A typical traditional 2-parent crossover operator for real-value domains is BLX-α proposed by Eshelman & Schaffer [16]. In this experiment, we use BLX-α to solve the following two test functions, the Sphere function (F_{Sphere}) and the Ridge function (F_{Ridge}) defined as

$$F_{Sphere}(x) = \sum_{i=0}^{n-1} x_i^2, \quad -5.12 \le x_i \le 5.11, \tag{7}$$

$$F_{Ridge}(x) = \sum_{i=0}^{n-1} \left(\sum_{j=0}^{i} x_j \right)^2, \quad -64 \le x_i \le 64. \tag{8}$$

Function F_{Sphere} is a unimodal one and has no linkage among variables. F_{Ridge} has a weak linkage among variables. We set $n = 10$.

BLX-α creates two children c_1 and c_2 of parents p_1 and p_2 which lie on the line joining the parents, but not between them, as shown in Fig. 9. Since the value of 0.5 is often used for α, we used that value.

BLX-α for AX samples using Eq. (3) as follows: Let X_1^i and X_2^i be values obtained by sampling $f_{|J^i|,\lambda}(x)$ of Eq. (3), respectively. Then, we obtain children c_1 and c_2 as follows:

$$\begin{cases} c_1^i = e_1^i + |J^i| - X_1^i \\ c_2^i = e_1^i + X_2^i \end{cases} \tag{9}$$

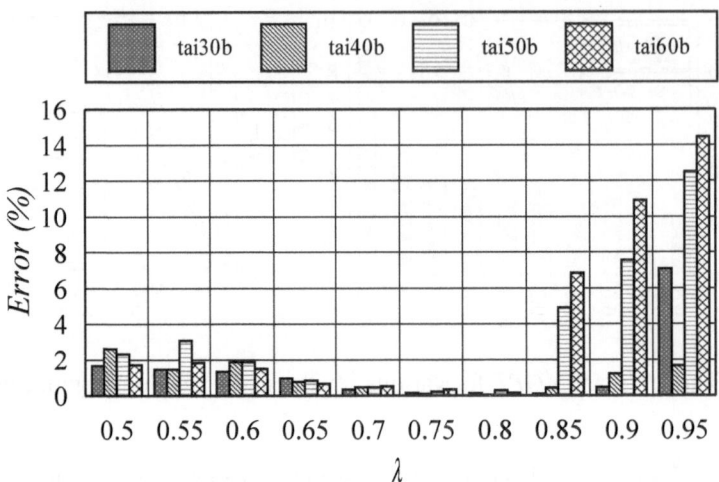

Fig. 7. Results of CAX on QAP instances. *Error* deviation from optimal values in %.

where, $i = 0, 1, 2, \ldots, n - 1$. It is clear from Eq. (9) that c_1 and c_2 are similar to p_1 and p_2, respectively, depending on the parameter value λ of CAX.

Fig. 10 shows results of CAX on functions F_{Sphere} and F_{Ridge}. On F_{Sphere}, which has no linkage among variable, the solution converges well when λ value is 0.5, i.e., the standard BLX-α is used. According to increasing λ value, the convergence speed becomes slower. In contrast to this, on F_{Ridge}, which has a linkage among valuables, the solution converges most well around λ value of 0.7.

4.4   CAX for Niching

In experiments in this section, we briefly show how CAX converges the population to multiple, highly fit, and significantly different solutions (niching). We use the following two functions [17].

$$F_1(x) = \sin^6(5\pi x), \qquad (10)$$

$$F_2(x) = e^{-2(\ln 2)\left(\frac{x-0.1}{0.8}\right)^2} \sin^6(5\pi x). \qquad (11)$$

We encoded x in the range $[0, 1]$ with 30-bit binary string, and population size was 100 as were set in [2]. As shown in Fig. 11, CAX found multiple solutions with larger value of λ than 0.5. The advantage of niching with CAX is that we do not need any similarity measure, as was needed in crowding methods such as DC.

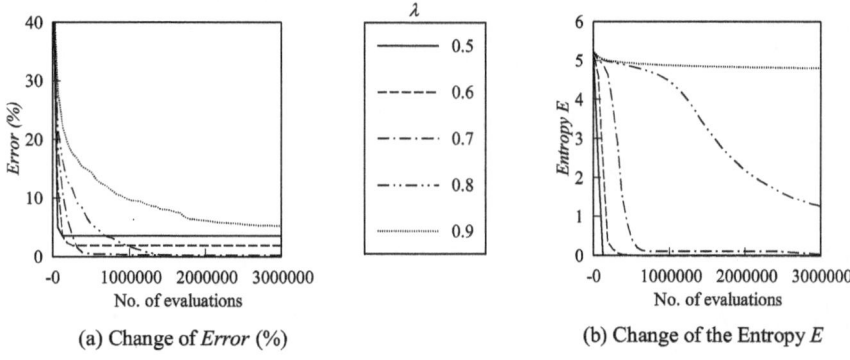

(a) Change of *Error* (%)   (b) Change of the Entropy E

Fig. 8. Convergence process of CAX on tai40b. (a) shows the change of *Error*, and (b) shows the change of the entropy E defined by Eq. (6).

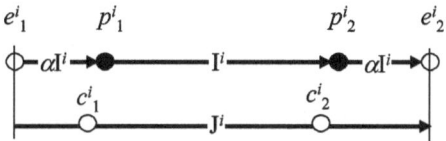

Fig. 9. BLX-α. BLX-α uniformly samples new individuals with values that lie on $\mathbf{J}^i$ $(i = 1, 2, \ldots, n)$

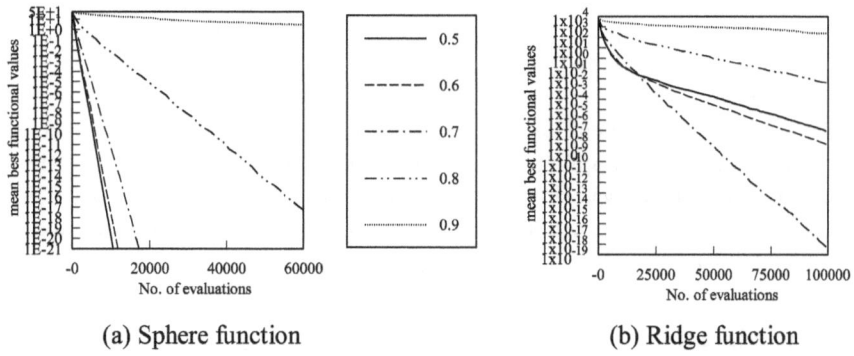

(a) Sphere function   (b) Ridge function

Fig. 10. Results of CAX on F_{Sphere} and F_{Ridge}. Each line shows functional values for evaluations.

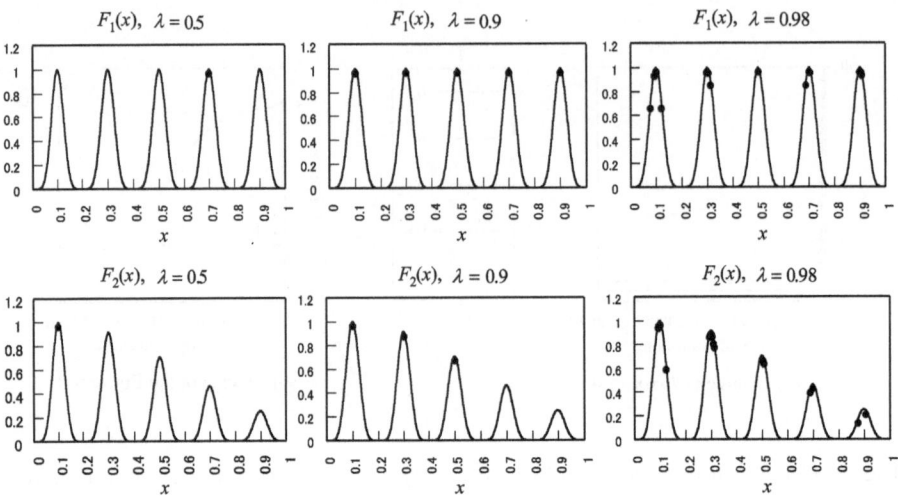

Fig. 11. Niching with CAX

5   Conclusions

In this paper, we proposed a novel crowding method, Crowding with Asymmetric Crossover (CAX). CAX can be applied to the traditional 2-parent crossover operators in binary, permutation, and real-value domains.

Through the wide range of experiments, we showed CAX can enhance the performance of traditional 2-point crossover operators, i.e., CAX can prevent premature convergence of a population by preserving the population diversity, and thus, CAX can obtain one global optimal solution efficiently. We used BLX-α for the base of AX on the real-value domains in the experiments. But BLX-α is not suitable to apply for problems which have a strong linkage among variables. To test CAX on advanced 2-parent crossover operators for real-code such as UNDX [18] or to extend CAX with multi-parent recombination operators remain for future work.

Although our main focus was finding one global optimal solution, we showed that CAX can also be applied to find multiple, highly fit, and significantly different solutions (niching). However, niching with CAX must be studied more intensively comparing to other crowding methods in the literature. This also remains for future work.

Acknowledgements. The author would like to acknowledge to the reviewers for their insightful comments on improving this paper. This research is partially supported by the Ministry of Education, Culture, Sports, Science and Technology of Japan under Grant-in-Aid for Scientific Research number 1381211400.

References

1. De Jong, K.A.: An Analysis of the behavior of a class of genetic adaptive systems. Ph.D. Dissertation, University of Michigan (1975)
2. Mahfoud, S.W.: Niching methods for genetic Algorithms. PhD Dissertation, Department of General Engineering, University of Illinois at Urbana-Champaign (1995)
3. Mengshoel, O.J., Goldberg, D.E.: Probabilistic crowding: Deterministic crowding with probabilistic replacement. In: Genetic and Evolutionary Computation Conference, pp. 409–416 (1999)
4. Mengshoel, O.J., Goldberg, D.E.: The crowding approach to niching in genetic algorithms. Evolutionary Computation 16(3), 315–354 (2008)
5. Galan, S.F., Mengshoel, O.J.: Generalized crowding for genetic algorithms. In: Genetic and Evolutionary Computation Conference, pp. 775–782. ACM (2010)
6. Goldberg, D.E.: Genetic algorithms in search, optimization and machine learning. Addison-Wesley publishing company (1989)
7. Tsutsui, S.: cAS: Ant colony optimization with cunning ants. In: Runarsson, T.P., Beyer, H.-G., Burke, E.K., Merelo-Guervós, J.J., Whitley, L.D., Yao, X. (eds.) PPSN IX. LNCS, vol. 4193, pp. 162–171. Springer, Heidelberg (2006)
8. Pisinger, D.: A minimal algorithm for the 0-1 Knapsack Problem. Operations Research 45, 758–767 (1994)
9. Goldberg, D.E., Korb, B., Deb, K.: Messy genetic algorithms: motivation, analysis, and first results. Complex Systems, 493–530 (1989)
10. Grefenstette, J.J., Davis, L., Cerys, D.: GENESIS and OOGA: Two GA Systems. TSP Publication, Melrose (1991)
11. Tsutsui, S., Fujimoto, Y., Ghosh, A.: Forking GAs: GAs with search space division schemes. Evolutionary Computation 5(1), 61–80 (1997)
12. Burkard, R.E., Çela, E., Karisch, S.E., Rendl, F.: QAPLIB - a quadratic assignment problem library (2009), http://www.seas.upenn.edu/qaplib
13. Taillard, É.D.: Comparison of iterative searches for the quadratic assignment problem. Location Science 3(2), 87–105 (1995)
14. Stützle, T., Hoos, H.: Max-Min Ant System. Future Generation Computer Systems 16(9), 889–914 (2000)
15. Tsutsui, S., Fujimoto, N.: ACO with tabu search on a GPU for solving QAPs using move-cost adjusted thread assignment. In: Genetic and Evolutionary Computation Conference, pp. 1547–1554. ACM (2011)
16. Eshelman, L.J., Schaffer, J.D.: Real-coded genetic algorithms and interval-schema. Foundations of Genetic Algorithms 2, 187–202 (1993)
17. Goldberg, D.E., Richardson, J.: Genetic algorithms with sharing for multimodal function optimization. In: Proceedings of the 2nd Inter. Conf. on Genetic Algorithms and their Application, pp. 41–49 (1987)
18. Ono, I., Kobayashi, S.: A real coded genetic algorithm for function optimization using unimodal normal distributed crossover. In: ICGA, pp. 246–253 (1997)

The Performance Effects of Interaction Frequency in Parallel Cooperative Coevolution

Xiaofen Lu[1], Stefan Menzel[2], Ke Tang[3], and Xin Yao[1,3]

[1] CERCIA, School of Computer Science, University of Birmingham,
Edgbaston, Birmingham B15 2TT, UK
[2] Honda Research Institute Europe, 63073 Offenbach/Main, Germany
[3] UBRI, School of Computer Science and Technology, University of Science and Technology
of China (USTC), Hefei, Anhui 230027, China
{xxl332,x.yao}@cs.bham.ac.uk, stefan.menzel@honda-ri.de,
ketang@ustc.edu.cn

Abstract. Cooperative coevolution (CC) employs a divide-and-conquer paradigm for tackling complex optimization problems. Its performance is influenced by many design decisions. Therefore, to beneficially use it, it is important to acquire some knowledge of the effects of different design settings on the performance of CC. In this paper, we investigate experimentally the performance effects of interaction frequency in parallel CC. The experimental results show that it is overall best for subpopulations to interact with each other as frequently as possible when communication cost is ignored; when communication cost is considered, the best interaction frequency varies from problem to problem and a dynamic change of it is desirable during the optimization process.

Keywords: Cooperative Coevolution, Interaction Frequency.

1 Introduction

Cooperative coevolution (CC) is an evolutionary method that tries to solve large-scale problems by problem decomposition. The idea of CC was firstly introduced in genetic algorithm (GA) by Potter and De Jong [11], and later a framework for using the cooperative coevolutionary algorithm (CCEA) was developed [10]. In this framework, the decision variables of an optimization problem are firstly decomposed into several subcomponents, each of which is then evolved in a population. The fitness for each subpopulation individual is assessed by assembling it with representative individuals of the other subpopulations. Experimental results in [11] have shown that this framework can achieve superior performance to traditional GA. Since then, different evolutionary algorithms (EAs) have been successfully extended under this framework due to its generality, such as particle swarm optimization (PSO) ([1], [5]), differential evolution (DE) ([13],[22]), MOEAs [16], and memetic algorithm (MA) [6].

Except combined with different EAs, the CCEA framework has different implementations in different computing environments [3]. In a sequential computing environment, subpopulations in CCEA take turns in evolving. This is called sequential CCEA. The other is parallel CCEA in which all subpopulations are evolved simultaneously. This is more natural in a parallel computing environment.

G. Dick et al. (Eds.): SEAL 2014, LNCS 8886, pp. 82–93, 2014.

As pointed out in [17] and [9], CCEA has many components that can be adjusted, such as how and when to make subpopulations interact with each other. As a result, a practitioner will face a lot of design decisions when applying CCEA to real-world problems. Therefore, it would be useful to generate the knowledge about the performance effects of different choices of these components in CCEA. In literature, various studies have been carried out along the directions of how to select representative individuals, how many representatives to select, and how to calculate the fitness of each subpopulation individual based on these representatives for both parallel CCEA ([2], [7],[19]) and sequential CCEA ([18], [8]). For when to make subpopulations interact with each other, the performance effects of interaction frequency was systematically studied for sequential CCEA [9] but little has been done for parallel CCEA in literature.

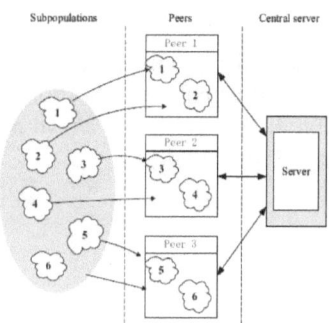

Fig. 1. The Model of Parallel CCEA, taken from [16]

In this paper, we focus on studying the performance effects of communication frequency in parallel CCEA. Different from sequential CCEA, communication cost exists in parallel CCEA. Consider one model of parallel CCEA given in Fig. 1, in which subpopulations are decomposed into several groups and these groups will be distributed over several peer computers. The interaction among subpopulations is indirectly achieved through the exchange of information between the peers and a central server. In such a case, communication overhead can not be ignored especially for distributed networks with limited communication speeds. As communication cost is involved, the performance influence of communication frequency in parallel CCEA would be more complicated than in sequential CCEA.

In CCEA, one critical step is problem decomposition. In [11], each dimension was considered as a subcomponent and evolved in a population. The experimental results have shown that this decomposition method loses its efficiency on non-separable problems. Since then, various studies ([20], [21], [12], [4], [15]) have been carried out towards grouping interdependent variables into the same subcomponent during the search process. In this paper, we consider fixed problem decomposition. In an extreme case, when the subcomponents are independent, they do not need to communicate with each other; when the subcomponents are interdependent, intuitively, it is better for them to interact with each other. This means that the best communication frequency depends on the correlation degree of subcomponents.

Considering this, our first and second hypotheses are that the performance effects of interaction frequency depends on communication cost and correlation degree of sub-components, respectively. Besides, our another hypothesis is that the performance effects relate to the evolution speed of the EA. To confirm our hypotheses, we observe the change of the best communication frequency using different EAs, different communication costs, and test functions with different subcomponent correlation degrees. Through this, we also try to answer the questions as follows.

1. Is there an optimal setting for the interaction frequency of parallel CCEA for a particular class of problems?
2. How does the best interaction frequency vary in different EAs?
3. How does the best communication frequency relate to communication cost?

The remaining part of this paper is organized as follows. Section 2 will detail the experiment methodology and experimental setup. Section 3 will give the experimental results and analysis. In Section 4, answers to the above-mentioned questions will be given as well as the directions of future work.

2   Experiment Methodology

2.1   Correlation of Subcomponents

The correlation degree of two subcomponents in this study is calculated by averaging the correlation degree of variables between them, which is calculated based on the statistical interdependence learning model proposed in [15]. That is:

- Suppose we have decision vectors $\overrightarrow{\alpha} = (..., x_i, ..., x_j, ...)$, $\overrightarrow{\beta} = (..., x'_i, ..., x_j, ...)$, and they satisfy that $f(\overrightarrow{\alpha}) \leq f(\overrightarrow{\beta})$ (f is the function to optimize). If the value of x_j is changed to x'_j, resulting in $\overrightarrow{\alpha}' = (..., x_i, ..., x'_j, ...)$, $\overrightarrow{\beta}' = (..., x'_i, ..., x'_j, ...)$, and $f(\overrightarrow{\alpha}') > f(\overrightarrow{\beta}')$, then it is called variable x_i is affected by variable x_j under context vector $\overrightarrow{c} = (..., x_{i-1}, -, x_{i+1}..., x_{j-1}, -, x_{j+1}...)$.
- The extent to which the variable x_i can be affected by x_j depends on the probability of the inequality change, i.e., $P\{f(\overrightarrow{\alpha}') > f(\overrightarrow{\beta}')\}$, which can be estimated by selecting a number of $\overrightarrow{c}$ and checking the affect of x_j on x_i under each sample $\overrightarrow{c}$.
- The correlation degree of two variables is calculated by averaging their separate probabilities of the inequality change.

Note that it has been validated in [9] that the performance effects of interaction frequency are dependent on a problem property called *best-response curves*. In this study, we instead use the correlation degree of subcomponents to depict the problem property. The reason is that the later could be much easier to estimate than the former if the information is possibly used.

2.2 The Domains

We start by using three families of test functions, each with a dimension of 2. One is taken from [9], defined as follows:

$$BR_n^\alpha(x,y) = \begin{cases} 2y + \frac{\alpha-3}{2\alpha}(x-n), & \text{if } \alpha y < x + (\alpha-1)n; \\ 2x + \frac{\alpha-3}{2\alpha}(y-n), & \text{if } y > x + (\alpha-1)n; \\ n + \frac{x+y}{2}, & \text{otherwise} \end{cases} \tag{1}$$

where $n \in \mathbf{N}, \alpha \in [0,1], x, y \in [0,n]$. The optimum for this function is $BR_n^\alpha(n,n) = 2n$. In this paper, n is set to 10.

The other two are the shifted rotated sphere function and shifted rotated rastrigin's function [14], which are typical unimodal and multimodal function, respectively, defined as:

$$F_{sphere}(x) = \sum_{i=1}^{D} z_i^2, \mathbf{z} = (\mathbf{x} - \mathbf{o}) * \mathbf{M}, \mathbf{x} = [x_1, x_2, ..., x_D], \mathbf{x} \in [-100, 100]^D \tag{2}$$

$$F_{sphere}(x) = \sum_{i=1}^{D} (z_i^2 - 10cos(2\pi z_i) + 10), \mathbf{z} = (\mathbf{x} - \mathbf{o}) * \mathbf{M},$$
$$\mathbf{x} = [x_1, x_2, ..., x_D], \mathbf{x} \in [-5, 5]^D \tag{3}$$

where D is the number of dimensions, $\mathbf{o} = [o_1, o_2, ..., o_D]$ is the shifted global optimum. In this paper, D is set to 2.

We chose these three functions because the correlation degree of x (or x_1) and y (or x_2) can be changed by changing α (or $\mathbf{M}$). For $BR_n^\alpha(x,y)$, when increasing α from 0 to 1, the correlation degree of x and y increases monotonically from 0. For $F_{sphere}(x)$ and $F_{rastrigin}(x)$, assume $\mathbf{M} = \begin{bmatrix} m_{11} & m_{12} \\ m_{21} & m_{22} \end{bmatrix}$ and the initial $\mathbf{M}$ is an identity matrix. By increasing both m_{12} and m_{21} to the same value from 0 to 1, the correlation degree of x_1 and x_2 monotonically increases from 0.

2.3 The CCEA Configuration

We use a two-population CCEA as the basic setup, in which one population is used to evolve x (or x_1) and the other is to evolve y (or x_2). For each population, we run a non-overlapping generational EA. The parameter setting for the used EA in our experiments is given in Table 1.

At first, an initial population is generated for each subcomponent. The population member of each population is evaluated by assembling it with a randomly chosen member of the other population. Then, each population is simultaneously evolved with the above-mentioned EA. The fitness of a population member is evaluated by combining it with the current best member of the other population. During the search process, they communicate with each other their current best solution every constant number of generations.

Table 1. Parameter Setting for EA

Population size (NP)	20
Mutation Probability (P_{mut})	0.9
Evolutionary Operators	Gaussian mutation, elitism of size 1, tournament selection of size 2

2.4 Experimental Setup

In our experiments, we vary the interaction frequency and observe the changes in the performance of CCEA. To confirm the hypotheses, we observe the change of the best interaction frequency by using the above-mentioned test functions with different $alpha$ and M values, the EA with different Gaussian mutation sigma, and different communication costs, respectively. We set the number of maximum fitness evaluations constant across experiments, and independently run each experiment for 100 times, and then use Wilcoxon rank-sum test at a 0.05 significance level to make comparisons.

We start from the extreme scenario in which communication cost equals 0. In this scenario, we study the relationship between the best communication frequency and each of correlation degree of subcomponents and the mutation step size. Then, we consider the scenario in which communication has a cost and study how the best communication frequency changes as each of correlation degree of subcomponents, the mutation step size, and communication cost changes.

3   Experimental Studies: Communication Without Cost

3.1   Different Correlation Degrees of Variables

In this study, the maximum number of fitness evaluations was set to 3220, and thus the maximum number of generations is 80 ($20 + 80 * (2 * 20) = 3220$). The Gaussian mutation sigma was set to 0.2, 1.0, and 0.2 for $BR_n^\alpha(x, y)$, $F_{sphere}(x)$, and $F_{rastrigin}(x)$, respectively. We varied the communication period from 1 to 80 generations and used Wilcoxon rank-sum test to make comparisons. Here, communication period means every how many generations two populations interact with each other. Table 2 summarizes the interval of best communication periods for each test function, which means communication periods in this interval performs the same best according to statistical test.

It can be seen from Table 2 that communication frequency indeed has an effect on the performance of parallel CCEA as not all values in [1,80] performed the same best, and thus worths studying . For each family of test functions, the interval of the best communication period varies as the correlation degree of variables varies, and the change tendency is similar. That is, when the variables are independent, all interaction periods performs the same; as the interdependence degree of subcomponents increases, the algorithm more and more prefers short interaction periods; when the correlation degree of variable is high enough, the upper bound of the best interaction period interval increases as correlation degree increases. This can substantiate our first hypothesis.

Table 2. The Interval of Best Interaction Periods (denoted as I_{bp}) for $BR_n^\alpha(x, y)$ with Different α Values, $F_{sphere}(x)$ and $F_{rastrigin}(x)$ with Different m_{12}/m_{21} Values, respectively (the column of 'corr' is the estimated correlation degree of x (or x_1) and y (or x_2))

$BR_n^\alpha(x, y)$			$F_{sphere}(x)$			$F_{rastrigin}(x)$		
α	corr	I_{bp}	m_{12}/m_{21}	corr	I_{bp}	m_{12}/m_{21}	corr	I_{bp}
0.0	0.0	[1,80]	0.0	0.0	[1,80]	0.0	0.0	[1,80]
0.05	0.0299	[1,10]	0.05	0.0364	[1,5]	0.01	0.0509	[1,4]∪[10,16]
0.10	0.058	[1,10]	0.10	0.0805	[1,2]	0.02	0.1024	[1,4]
0.20	0.1206	[1,4]	0.15	0.1207	[1,2]	0.03	0.1534	[1,5]
0.30	0.1669	[1,2]	0.20	0.1770	[1]	0.04	0.1997	[1,4]
0.40	0.2202	[1,2]	0.30	0.2523	[1]	0.05	0.2434	[1,2]
0.50	0.2601	[1]	0.40	0.3205	[1,2]	0.068	0.3040	[1,2]
0.60	0.3061	[1]	0.50	0.3549	[1]	0.08	0.3270	[1,4]
0.70	0.3483	[1]	0.60	0.3726	[1]	0.10	0.3456	[1,4]
0.80	0.3737	[1]	0.70	0.3907	[1]	0.20	0.3813	[1,10]
0.90	0.4025	[1]	0.80	0.3893	[1]	0.40	0.4251	[1,4]
0.95	0.4078	[1]	0.90	0.3981	[1,2]	0.60	0.4410	[1,5]
0.993	0.4090	[1,10]	0.95	0.3994	[1,80]	0.80	0.4416	[1,4]
1.0	0.4107	[1,80]	1.0	0.4017	[1,2]∪[20,80]	1.0	0.4474	[1,10]∪[70,80]

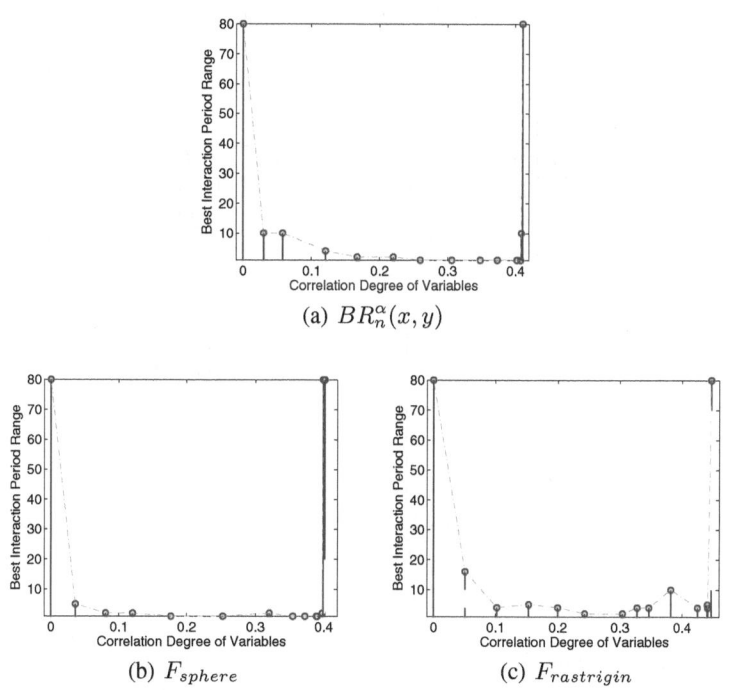

(a) $BR_n^\alpha(x, y)$

(b) F_{sphere}

(c) $F_{rastrigin}$

Fig. 2. The Curves of Best Interaction Period Range versus Correlation Degree of Variables

To get a clearer picture of how the best interaction period interval changes as the correlation of variables changes, the curves of the best communication period interval versus the correlation degree of variables for $BR_n^\alpha(x, y)$, $F_{sphere}(x)$, and $F_{rastrigin}(x)$ are plotted in Fig. 2. It can be seen from Fig. 2 that the upper bound of the interval and the correlation of subcomponents are non-linearly and complicatedly related. Take a closer look at Fig. 2, this non-linear relationship is much higher when the correlation degree is at about 0.4 than when the correlation degree is less than 0.2. When the correlation degree of subcomponents falls in the range between 0.2 to 0.35, it is highest required that communication happens frequently. Moreover, the lower bound of the communication interval is always 1, which means that overall every-generation communication is best when communication cost can be neglected.

3.2 Different Mutation Step Sizes

In this study, we varied Gaussian mutation sigma and recorded the best interaction period interval. Table 3 gives the best interaction frequency interval over $BR_n^\alpha(x, y)$ for three different sigma values (0.1, 0.2, and 0.5).

When the mutation sigma changes from 0.1 to 0.2, we cannot observe an obvious trend of the change of the best communication period interval from Table 3. But, when the sigma changes from 0.1 (or 0.2) to 0.5, it can be seen that the range of the best interaction period decreases while short periods are still preferred.

Table 3. The Interval of Best Interaction Periods (denoted as I_{bp}) for $BR_n^\alpha(x, y)$ with Different Gaussian Mutation Sigma Values

$BR_n^\alpha(x, y)$		I_{bp}		
α	corr	$\sigma = 0.1$	$\sigma = 0.2$	$\sigma = 0.5$
0.0	0.0	[1,80]	[1,80]	[1,80]
0.05	0.0299	[1,10]	[1,10]	[1,8]
0.10	0.0580	[1,2]∪[5]	[1,10]	[1,5]
0.20	0.1206	[1,5]	[1,4]	[1,4]
0.30	0.1669	[1,2]	[1,2]	[1,4]
0.40	0.2202	[1,2]	[1,2]	[1,2]
0.50	0.2601	[1]	[1]	[1]
0.60	0.3061	[1]	[1]	[1]
0.70	0.3483	[1]	[1]	[1]
0.80	0.3737	[1]	[1]	[1]
0.90	0.4025	[1]	[1]	[1]
0.95	0.4078	[1]	[1]	[1]
0.993	0.4090	[1,4]∪[16,80]	[1,10]	[1,2]
1.0	0.4107	[1,80]	[1,80]	[1,4]

In general, if communication cost can be ignored, it is best for subpopulations to communicate every generation. However, it should be noted here in previous experiments we did not reevaluate each current subpopulation after the communication happens. Considering that in some implementation of parallel CCEA (e.g. Differential

Evolution (DE) is chosen as the optimization algorithm) the current population needs to be reevaluated when information of the best individual in the other subpopulation changes, we further did experiments with reevaluation over $BR_n^\alpha(x,y)$, $F_{sphere}(x)$, and $F_{rastrigin}(x)$, and the results are summarized in Table 4. It can be seen from Table 4 that the lower bound of the best interaction interval changes to 2 or 3 over some cases of $F_{sphere}(x)$. The reason why every-generation communication did not perform best in these cases may be that reevaluation would not help improving the best solution but waste more fitness evaluations if the information does not change too much. Thus, if CCEA is implemented with reevaluation, it should be checked whether the reevaluation is needed every exchange of information.

Table 4. The Interval of Best Interaction Periods (I_{bp}) with Reevaluation for $BR_n^\alpha(x,y)$ with Different α Values, $F_{sphere}(x)$ and $F_{rastrigin}(x)$ with Different m_{12}/m_{21} Values

$BR_n^\alpha(x,y)$			$F_{sphere}(x)$			$F_{rastrigin}(x)$		
α	$corr$	I_{bp}	m_{12}/m_{21}	$corr$	I_{bp}	m_{12}/m_{21}	$corr$	I_{bp}
0.0	0.0	[1,79]	0.0	0.0	[2,79]	0.0	0.0	[1]∪[7,79]
0.05	0.0299	[1,9]	0.05	0.0364	[1,4]	0.01	0.0509	[1,19]
0.10	0.0580	[1,4]	0.10	0.0805	[1,3]	0.02	0.1024	[1,7]
0.20	0.1206	[1,3]	0.15	0.1207	[1,3]	0.03	0.1534	[1,9]
0.30	0.1669	[1,3]	0.20	0.1770	[1,3]	0.04	0.1997	[1,7]
0.40	0.2202	[1,2]	0.30	0.2523	[2,3]	0.05	0.2434	[1,7]
0.50	0.2601	[1,2]	0.40	0.3205	[3]	0.068	0.3040	[1,4]
0.60	0.3061	[1]	0.50	0.3549	[2]	0.08	0.3270	[1,4]
0.70	0.3483	[1]	0.60	0.3726	[2]	0.10	0.3456	[1,4]
0.80	0.3737	[1]	0.70	0.3907	[2]	0.20	0.3813	[1,19]
0.90	0.4025	[1]	0.80	0.3893	[1]	0.40	0.4251	[1,3]
0.95	0.4078	[1]	0.90	0.3981	[1,79]	0.60	0.4410	[1,7]
0.993	0.4090	[1,9]	0.95	0.3994	[1,79]	0.80	0.4416	[1,7]
1.0	0.4107	[1,79]	1.0	0.4017	[1]∪[39,79]	1.0	0.4474	[1,15]∪[69,79]

4 Experimental Studies: Communication with Cost

4.1 Different Communication Costs

In this study, we assume communication cost as a fixed number of fitness evaluations as fitness evaluation denotes the time elapse in the CCEA framework. We used 7 cases of $BR_n^\alpha(x,y)$. The maximum number of of fitness evaluations was set to 3000. The interaction period varies from 1 to 20 generations while the communication cost varies from $0 * NP$ to $60 * NP$ (NP is the population size) fitness evaluations. The interval of the best interaction periods for each test function and communication cost is given in Table 5.

It can be seen from Table 5 that every-generation communication no longer performed best when communication exists. For uncorrelated subcomponents, the largest

Table 5. The Interval of Best Interaction Periods (I_{bp}) over $BR_n^{\alpha}(x,y)$ with Different Communication Costs

cost	$\alpha = 0.0$ $corr = 0.0$	0.2 0.1206	0.4 0.2202	0.6 0.3061	0.8 0.3737	0.99 0.4090	1.0 0.4107
0	[1,20]	[1,4]	[1,2]	[1]	[1]	[1,10]	[1,20]
$4*NP$	[12,20]	[3,8]	[2,5]	[2]	[2]	[1,20]	[1,20]
$10*NP$	[10,20]	[3,8]	[4]	[3,4]	[2,4]	[1,20]	[1,20]
$14*NP$	[10,20]	[4,7]	[5]	[3,5]	[2,5]	[1,20]	[1,20]
$20*NP$	[14,20]	[6,8]	[6,8]	[4]	[2,8]	[1,20]	[1,20]
$30*NP$	[14,20]	[7,8]	[6,8]	[5,8]	[2,20]	[1,20]	[1,20]
$40*NP$	[16,20]	[8,14]	[7,16]	[3,16]	[2,16]	[1,20]	[1,20]
$50*NP$	[16,20]	[7,10]	[6,10]	[5,12]	[3,20]	[1,20]	[1,20]
$60*NP$	[18,20]	[6]	[5,7]	[4,7]	[2,20]	[1,20]	[1,20]

communication period performed best. For strongly-related subcomponents, all communication periods performed the same. For other functions, the best communication periods are different for different communication costs. This confirms our second hypothesis. Moreover, when the communication cost is the same, the best communication

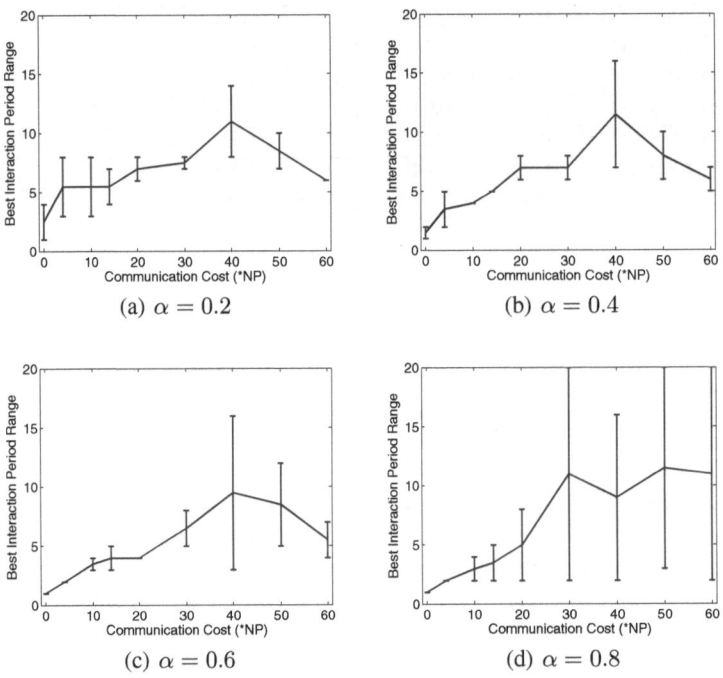

(a) $\alpha = 0.2$ (b) $\alpha = 0.4$

(c) $\alpha = 0.6$ (d) $\alpha = 0.8$

Fig. 3. The Curves of Best Interaction Period Range versus Different Communication Costs

periods are different for different correlation degrees of subcomponents. This substantiates our first hypothesis.

To get a picture of how best interaction period changes as the communication cost changes, the curves of the best communication period interval versus communication cost for $\alpha = 0.2$, $\alpha = 0.4$, $\alpha = 0.6$, and $\alpha = 0.8$ are plotted in Fig. 3. It can be seen that there is a non-linear relationship between the interval of best interaction period and communication cost, and the relationship varies for different correlation degrees of subcomponents. Overall, when communication cost becomes larger, the best interaction period is prone to larger values (note that the decreasing part of the curve as the communication cost increases is because we set the maximum number of fitness evaluations fixed across all communication costs).

4.2   Different Mutation Step Sizes

In this experiment, we change Gaussian mutation sigma and recorded the best interaction period interval over 10 cases of $BR_n^\alpha(x, y)$ with fixed communication cost $(4 * NP$ fitness evaluations). We calculate the number of fitness evaluations to achieve the function value of 19.99 for each interaction period in [1,20], and obtain the best interaction period according to the Wilcoxon rank-sum test. Table 6 shows the interval of best interaction period for three different sigma values $(0.05, 0.1,$ and $0.2)$.

Table 6. The Interval of Best Interaction Periods (denoted as I_{bp}) for Different Gaussian Mutation Sigma Values over $BR_n^\alpha(x, y)$ with Communication Cost $(4 * NP)$

$BR_n^\alpha(x,y)$		I_{bp}		
alpha	corr	$\sigma = 0.05$	$\sigma = 0.1$	$\sigma = 0.2$
0.0	0.0	[7,20]	[6,20]	[7,20]
0.05	0.0299	[5, 15]	[4,9]	[3,8]
0.10	0.058	[5,13]	[3,7]	[3,4]
0.20	0.1206	[4,9]	[3,7]	[2,4]
0.30	0.1669	[3,8]	[2,5]	[2,3]
0.40	0.2202	[3,7]	[2,4]	[2,3]
0.50	0.2601	[3,6]	[2,4]	[2]
0.60	0.3061	[2,5]	[2,3]	[2]
0.70	0.3483	[2,4]	[2,3]	[2]
0.80	0.3737	[2,3]	[1,2]	[2]

By comparing the best interaction period interval for 0.05, 0.1, and 0.2, it can be seen that when the mutation step size decreases, the best interaction period will prefer larger values. This confirms our third hypothesis. Furthermore, this indicates that a different interaction period needs to be set if the evolution speed changes during the search process.

5   Conclusion and Future Work

In this paper, we studied the communication frequency effects on the performance of parallel CCEA. We showed that the best interaction frequency depends on the correlation degree of subcomponents, the evolution speed of the optimization algorithm as well as the communication cost.

As an answer to question 1, for problems that communication cost can be comparatively ignored, it is best to communicate as frequently as possible. For problems that consider communication cost, the best interaction period is hard to set, it is not only related to the correlation of variables and communication cost but also to the evolution speed of the optimization algorithm. But, for separable variables, it is best to set the communication frequency as low as possible.

As an answer to question 2, if the evolution speed of optimization algorithm is quick, high communication frequency is needed. Actually, during the evolutionary process of a CCEA, the evolution speed would change. It would become slower in the later phase of the evolutionary process. Therefore, a dynamic change of interaction frequency is desirable during a parallel CCEA run.

As an answer to question 3, the relationship between interaction frequency and communication cost is non-linear and complicated. The trend is that the best interaction period is prone to larger values when communication cost becomes higher. For problems with a fixed communication cost, the best interaction frequency represents the best trade-off between frequent communication and increased communication overhead, which varies from problem to problem.

The answers to these questions are given based on the 2-D test functions used in our experiments. In future, we will consider high-dimensional test functions. Moreover, the studies were done using a fixed interaction frequency. In reality, a dynamic or adaptive one might be more appropriate, which is another direction of our future work.

Acknowledgments. This work was supported by Honda Research Institute Europe (HRI-EU) and European Commission FP7 grants (Nos. 257906 and 247619).

References

1. Van den Bergh, F., Engelbrecht, A.P.: A cooperative approach to particle swarm optimization. IEEE Transactions on Evolutionary Computation 8(3), 225–239 (2004)
2. Bull, L.: Evolutionary computing in multi-agent environments: Partners. In: Proceedings of the 7th International Conference on Genetic Algorithms, pp. 370–377. Morgan Kaufmann (1997)
3. Jansen, T., Wiegand, R.P.: Sequential versus parallel cooperative coevolutionary (1+ 1) eas. In: Proceedings of the 2003 IEEE Congress on Evolutionary Computation (CEC 2003), vol. 1, pp. 30–37. IEEE (2003)
4. Khare, V.R., Yao, X., Sendhoff, B.: Credit assignment among neurons in co-evolving populations. In: Yao, X., Burke, E.K., Lozano, J.A., Smith, J., Merelo-Guervós, J.J., Bullinaria, J.A., Rowe, J.E., Tiňo, P., Kabán, A., Schwefel, H.-P. (eds.) PPSN VIII. LNCS, vol. 3242, pp. 882–891. Springer, Heidelberg (2004)
5. Li, X., Yao, X.: Cooperatively coevolving particle swarms for large scale optimization. IEEE Transactions on Evolutionary Computation 16(2), 210–224 (2012)
6. Mei, Y., Li, X., Yao, X.: Cooperative co-evolution with route distance grouping for large-scale capacitated arc routing problems. IEEE Transactions on Evolutionary Computation 18(3), 435–449 (2014)

7. Panait, L., Luke, S.: Time-dependent collaboration schemes for cooperative coevolutionary algorithms. In: Proceedings of the 2005 AAAI Fall Symposium on Coevolutionary and Coadaptive Systems (2005)
8. Popovici, E., De Jong, K.: A dynamical systems analysis of collaboration methods in cooperative co-evolution. In: AAAI Fall Symposium on Coevolutionary and Coadaptive Systems (2005)
9. Popovici, E., Jong, K.D.: The effects of interaction frequency on the optimization performance of cooperative coevolution. In: Proceedings of the 2006 Genetic and Evolutionary Computation Conference (GECCO 2006), pp. 353–360. ACM (2006)
10. Potter, M.A.: The design and analysis of a computational model of cooperative coevolution. Ph.D. thesis, George Mason University (1997)
11. Potter, M.A., De Jong, K.A.: A cooperative coevolutionary approach to function optimization. In: Davidor, Y., Männer, R., Schwefel, H.-P. (eds.) PPSN III. LNCS, vol. 866, pp. 249–257. Springer, Heidelberg (1994)
12. Ray, T., Yao, X.: A cooperative coevolutionary algorithm with correlation based adaptive variable partitioning. In: Proceedings of the 2009 IEEE Congress on Evolutionary Computation (CEC 2009), pp. 983–989. IEEE (2009)
13. Shi, Y.-J., Teng, H.-F., Li, Z.-Q.: Cooperative co-evolutionary differential evolution for function optimization. In: Wang, L., Chen, K., S. Ong, Y. (eds.) ICNC 2005. LNCS, vol. 3611, pp. 1080–1088. Springer, Heidelberg (2005)
14. Suganthan, P.N., Hansen, N., Liang, J.J., Deb, K., Chen, Y.P., Auger, A., Tiwari, S.: Problem definitions and evaluation criteria for the cec 2005 special session on real-parameter optimization. KanGAL Report 2005005 (2005)
15. Sun, L., Yoshida, S., Cheng, X., Liang, Y.: A cooperative particle swarm optimizer with statistical variable interdependence learning. Information Sciences 186(1), 20–39 (2012)
16. Tan, K.C., Yang, Y., Goh, C.K.: A distributed cooperative coevolutionary algorithm for multiobjective optimization. IEEE Transactions on Evolutionary Computation 10(5), 527–549 (2006)
17. Wiegand, R.P.: An analysis of cooperative coevolutionary algorithms. Ph.D. thesis, George Mason University (2003)
18. Wiegand, R.P., Liles, W.C., De Jong, K.A.: An empirical analysis of collaboration methods in cooperative coevolutionary algorithms. In: Proceedings of the Genetic and Evolutionary Computation Conference (GECCO 2001), vol. 2611, pp. 1235–1245 (2001)
19. Wiegand, R.P., Liles, W.C., De Jong, K.A.: Analyzing cooperative coevolution with evolutionary game theory. In: Proceedings of the 2002 IEEE Congress on Evolutionary Computation (CEC 2002), vol. 2, pp. 1600–1605. IEEE (2002)
20. Yang, Z., Tang, K., Yao, X.: Differential evolution for high-dimensional function optimization. In: Proceedings of the 2007 IEEE Congress on Evolutionary Computation (CEC 2007), pp. 3523–3530. IEEE (2007)
21. Yang, Z., Tang, K., Yao, X.: Large scale evolutionary optimization using cooperative coevolution. Information Sciences 178(15), 2985–2999 (2008)
22. Yang, Z., Zhang, J., Tang, K., Yao, X., Sanderson, A.C.: An adaptive coevolutionary differential evolution algorithm for large-scale optimization. In: Proceedings of the 2009 IEEE Congress on Evolutionary Computation (CEC 2009), pp. 102–109. IEEE (2009)

Customized Selection in Estimation of Distribution Algorithms

Roberto Santana, Alexander Mendiburu, and Jose A. Lozano

Department of Computer Science and Artificial Intelligence, University of the Basque Country (UPV/EHU). P. Manuel de Lardizabal, 20018, Guipuzcoa, Spain
{roberto.santana,alexander.mendiburu,ja.lozano}@ehu.es

Abstract. Selection plays an important role in estimation of distribution algorithms. It determines the solutions that will be modeled to represent the promising areas of the search space. There is a strong relationship between the strength of selection and the type and number of dependencies that are captured by the models. In this paper we propose to use different selection probabilities to learn the structural and parametric components of the probabilistic graphical models. Customized selection is introduced as a way to enhance the effect of model learning in the exploratory and exploitative aspects of the search. We use a benchmark of over $15,000$ instances of a simplified protein model to illustrate the gains in using customized selection.

Keywords: selection, estimation of distribution algorithms, optimization, customized selection.

1 Introduction

Since their inception most of the research on estimation of distribution algorithms (EDAs) [9,10,13] has been devoted to the analysis of the learning and sampling components of these algorithms. The characteristic feature of EDAs with respect to other evolutionary algorithms (EAs) is the use of probabilistic modeling to capture the most relevant features of the selected solutions. Therefore, learning and sampling steps are critical for these algorithms and research on these methods in EAs almost began with EDAs. A different situation arises for the selection methods used by EDAs. The selection schemes traditionally applied in these algorithms are essentially those widely applied in GAs.

Different approaches investigate how selection mediates the information about the fitness function that is passed to the probabilistic models. Among the research directions explored are: 1) Explicitly using fitness information in the construction of the probabilistic models to learn more accurate models [15,17,20] and 2) Explicitly modeling fitness information as part of the probabilistic model [7,11]. These research directions are very related since it has been shown that the explicit modeling of fitness information can produce more accurate representations of the interactions between the variables. Furthermore, some research

G. Dick et al. (Eds.): SEAL 2014, LNCS 8886, pp. 94–105, 2014.

on the role of selection has gone beyond the classical aim of optimization to investigate the effect of selection in recovering the original problem structure [2].

In this paper we propose customized selection as a new way to implement selection in EDAs. We start from the assumption that the role played by selection in EDAs is two-fold. As in GAs, selection should be able to capture information about the promising areas of search space. But in addition, the selection method should contribute to a meaningful and efficient representation by the probability model. Basically, this assumption states that, in EDAs, considering the choice of the selection method in accordance with the type of probability modeling applied can contribute to a more efficient search for the optimal solutions. Customized selection takes into account this assumption by splitting the information extracted during the selection step into: 1) Information used for structural learning and 2) Information used for parametric learning. We empirically evaluate this idea for Boltzmann and truncation selections.

2   Selection and Learning in EDAs

In this section we assume the reader is familiar with EDAs. Let $\mathbf{X} = (X_1, \ldots, X_n)$ denote a vector of discrete random variables. We use $\mathbf{x} = (x_1, \ldots, x_n)$ to denote an assignment to the variables. I denotes a set of indices in $\{1, \ldots, n\}$ and X_I (respectively x_I) a subset of the variables of $\mathbf{X}$ (respectively $\mathbf{x}$) determined by the indices in I. p denotes a distribution, $p(x_I)$ the marginal probability for $\mathbf{X}_I = \mathbf{x}_I$, and $p(x_i \mid x_j)$ the conditional probability distribution of $X_i = x_i$ given $X_j = x_j$.

Algorithm 1. Tree-EDA

1	$D_0 \leftarrow$ Generate N individuals randomly
2	$t = 0$
3	**do** {
4	Evaluate the individuals using the fitness function.
5	Assign a selection probability to each individual.
6	Create a compact population D_t^S where copies of the same individual add up their probabilities p_t^S.
7	Calculate a probabilistic model using D_t^S and p_t^S.
8	Compute the univariate and bivariate marginal frequencies $p_i^s(x_i \mid D_t^s)$ and $p_{i,j}^s(x_i, x_j \mid D_t^S)$ using D_t^S and p_t^S
9	Calculate the mutual information using bivariate and univariate marginals.
10	Calculate the maximum weight spanning tree from the mutual information.
11	Compute the parameters of the model.
12	$t \leftarrow t + 1$
13	$D_t \leftarrow$ Sample N individuals from the tree and add elitist solutions.
14	} **until** A stop criterion is met

Algorithm 1 shows the pseudocode of an EDA that uses the complete population to define the selection probabilities. This EDA learns a tree model. In this section we focus on the analysis of the selection procedure and leave the analysis of the model learning step for Section 4. In terms of the selection procedure, the main difference between Algorithm 1 and the typical EDA is that in the former, instead of selecting a subset of individuals based on their fitness, a vector of selection probabilities is computed over the complete population. The probabilistic model is learned from the vector and the population. The procedure described in steps 5 to 7 of Algorithm 1 was originally introduced in [17].

Using this way to implement the selection has two advantages: 1) When possible, the fitness information of the complete population is used. 2) Although the computation of the compact population is not essential for computation of the probabilistic model, it helps to make model learning faster, particularly when there are multiple copies of the same individuals in the population. A requirement for the application of this method is that model learning could be done directly on the probabilities. This can be easily done for most of the model learning methods [5,15,17,20].

There is an extra cost in finding the compact population but notice that for detecting that two solutions are different, it is sufficient to find a variable where they differ. Therefore, although comparison between pairs of solutions can have a worst case cost of n, this cost will depend on the homogeneity of the population and the expected cost of finding the compact population will be often much smaller than $Nlog(N)n$, where N is the population size.

3   Customized Selection

Probabilistic models learned by EDAs can be classified according to the type of learning they use into two groups: 1) Models that apply non-structural (parametric) learning. 2) Models that apply structural *and* parametric learning. We extend this classification to EDAs and talk of non-structural learning and structural-learning EDAs, understanding that the second class of algorithms also applies parametric learning of the models. Among non-structural-learning EDAs are the univariate marginal distribution algorithm (UMDA) [14] and other EDAs based on marginal product models [12]. Structural-learning EDAs include algorithms based on Bayesian networks and Markov networks.

The key idea of customized selection is to learn the structure and the parameters of the model from different selection probabilities. We assume that, in terms of population diversity, non-structural learning and structural learning may have different requirements for accurately modeling. For example, in truncation selection, we may need to have a selection threshold of 0.5 (half the population) to guarantee a dataset large and diverse enough from which to learn the model structure applying statistical tests. However, once the structure is learned, we would like to make the marginal probabilities to represent the characteristics of solutions of highest fitness, for instance, those included in the best 30% of the population. In this way, we combine learning a robust structure with non-structural learning more focused on the best solutions. In all selection methods currently used by EDAs, the same selected population is used for both tasks. We aim to split this process and investigate whether customized selection has an impact in the quality of the search of EDAs.

3.1   Customized Boltzmann and Truncation Selections

We use notation introduced in Section 2. In Boltzmann selection, the probability of each solution to be part of the selected population is computed according to the Boltzmann probability distribution $\hat{p}(\mathbf{x}) = \frac{e^{\frac{f(\mathbf{x})}{T}}}{\sum_{\mathbf{x}'} e^{\frac{f(\mathbf{x}')}{T}}}$, where $\sum_{\mathbf{x}'} e^{\frac{f(\mathbf{x}')}{T}}$ is the so-called partition function, and T is the temperature of the system that can be used as a parameter to smooth the probabilities. The partition function is computed using all the solutions in the current population and a probability of selection is associated to each solution.

We use T as a way to influence the strength of selection. When $T \to \infty$, the models can not recover any information about the problem structure because all solutions are given the same probability. Similarly, when $T \to 0$ all the probability is concentrated in the point with highest function value in the population. Customized Boltzmann selection is implemented by using two different values of the temperature, Ts and Tp which will be associated to the structural and non-structural learning, respectively. Ts and Tp will bias the type and amount of information captured by the probabilistic models. In our experiments, we focus on the analysis of $Ts, Tp = 2^k$ for $k \in \{-3, -2, \cdots, 1, 2\}$.

In truncation selection, the best $M = \alpha N$ individuals of the population is selected, being $\alpha \in (0, 1]$. We define truncation selection on the complete population by associating a probability $\frac{1}{M}$ to the best M individuals and 0 to the rest. Customized truncation selection is implemented by defining two different truncation thresholds α_{T_s} and α_{T_p} for structural and non-structural learning, respectively. In usual application of truncation selection, $\alpha_{T_s} = \alpha_{T_p}$, but in customized selection these values can be different.

4   EDAs with Customized Selection

Customized selection can only be applied to EDAs that apply structural learning. We use Tree-EDA, an EDA that learns a tree probabilistic model and is similar to the ones introduced in [1] and [16]. The probability distribution of a tree is defined as $p_{\mathcal{T}}(\mathbf{x}) = \prod_{i=1}^{n} p(x_i | pa(x_i))$ where $Pa(X_i)$ is the parent of X_i in the tree, and $p(x_i | pa(x_i)) = p(x_i)$ when $pa(x_i) = \emptyset$, i.e. X_i is a root of the tree. The distribution $p_{\mathcal{T}}(\mathbf{x})$ itself will be called a tree model when no confusion is possible. We allow the existence of more than one root in the PGM (i.e. forests) although for convenience of notation we refer to the model as tree. Algorithm 1 shows the pseudocode of Tree-EDA.

We choose Tree-EDA to evaluate customized selection because it exhibits a good balance between the capacity of the probabilistic model to represent dependencies and the computational cost of learning and sampling the tree.

For comparison purposes we use UMDA. The univariate model used by this algorithm can be seen as a particular case of the tree when we have $pa(X_i) = \emptyset$ for all i.

As shown in Algorithm 1, the tree is learned using the Chow-Liu method [3] that calculates the maximum weight spanning tree from the matrix of mutual information. Notice that the mutual information is computed from the bivariate and univariate probabilities calculated upon marginalization of the selection probabilities of the compact population. When customized selection is used, the computation of the bivariate and univariate probabilities is done twice. The first time, from the selection probabilities calculated using T_s (respectively α_s for truncation selection). It is during this first step when the mutual information and the tree structures are computed. Then, during a second step, the univariate and bivariate probabilities are computed again, this time using T_p (respectively α_p for truncation selection), but only for the edges of the tree, i.e. a maximum of $n - 1$ bivariate probabilities instead of $\frac{n(n-1)}{2}$.

5   HP Functional Model Protein

As a testbed we use an optimization problem defined on a simplified protein model. The HP model considers hydrophobic (H) residues and hydrophilic or polar (P) residues. A protein is considered a sequence of these two types of residues, which are located in regular lattice models forming self-avoided paths. Given a pair of residues, they are considered neighbors if they are adjacent either in the chain (connected neighbors) or in the lattice but not connected in the chain (topological neighbors).

The functional model protein is a "shifted" HP model that can represent native states that are not maximally compact [6]. An energy function that measures the interaction between topological neighbor residues is defined as $\epsilon_{HH} = -2$ and $\epsilon_{PP} = \epsilon_{HP} = \epsilon_{PH} = 1$. The functional model protein problem consists of finding the solution that minimizes the total energy and it is a NP-hard problem.

Figure 1 shows an example of a functional model protein instance. In our solution representation, for a given sequence and lattice, X_i represents the relative move of residue i in relation to the previous two residues. Taking as a reference the location of the previous two residues in the 2D lattice, X_i takes values in

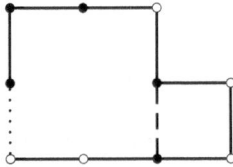

Fig. 1. One possible configuration of sequence $HHHPHPPHPP$ in the functional model protein. There is one HH interaction (represented by a dashed line), and one HP interaction (represented by a dotted line).

$\{0, 1, 2\}$. These values respectively mean that the new residue is located in one of the 3 possible directions (left, ahead, right) with respect to the previous two locations [8]. Therefore, values for X_1 and X_2 are meaningless. The locations of these two residues are fixed. A solution $\mathbf{x}$ can be seen as a walk in the lattice, representing one possible folding of the protein.

The codification corresponding to the configuration of the sequence shown in Figure 1a) is $\mathbf{x} = (0, 0, 2, 1, 2, 0, 2, 2, 1, 1)$. The objective function is computed as the opposite of the energy for feasible configurations.

In our representation there can be self-intersecting paths that correspond to unfeasible configurations. We use two ways to deal with these solutions: 1) Penalized fitness functions and 2) Repairing of solutions. We penalize self-intersecting solutions by dividing the energy by the number of the sequence's self-intersections. To repair solutions, a variation of the backtracking method introduced in [4] is applied.

As a problem benchmark, the functional model protein is a very interesting problem because, disregarding multiple representations of the same solution, the problem reaches the optimum on a unique configuration. We have selected a database of 15,575 protein sequences ($n = 23$) [8] for which, the optimal value, the closest suboptimal value, and the number of configurations where this suboptimal value is reached have been previously determined. The complexity of the optimization problem can be very different between sequences.

6 Experiments

The aim of the experiments is determining if using customized selection can help to improve the results of the EDAs that apply Boltzmann and truncation selection. A second goal is to find an appropriate choice of the selection parameters. Finally, we investigate the effect of the number of local optima in the behavior of the EDAs for the different selection methods.

6.1 Experiment Settings

The population size for all EDAs was fixed to $N = 500$ and as termination criterion we used a maximum number of generations $N_g = 50$. Experiments were run with and without repairing of the solutions. In the second case, the fitness values of infeasible solutions were penalized.

For Boltzmann selection we investigate in detail the effect of using different probability distributions to learn the parameters and structure of the model. This is done by trying all combinations of $Ts, Tp = 2^k$, for $k \in \{-3, -2, \cdots, 1, 2\}$. For truncation selection, $\alpha_s, \alpha_p \in \{0.1, 0.2, 0.3, 0.4, 0.5, 0.6\}$. When using the UMDA, the selection parameters of the structural and non-structural learning have identical values. Therefore, for each type of selection there are 36 variants of Tree-EDAs, and 6 variants of UMDA. We run the EDAs 100 times for each of the 15,575 instances. The total number of EDA runs for each type of selection method was $2 \times 100 \times 15,575 \times (36 + 6) = 130,830,000$ that were executed in a cluster of 575 computers.

6.2   Results for Customized Boltzmann Selection

We compare the algorithms in terms of the number of times that the optimum was found in 100 runs and in terms of the mean fitness value of the best solutions found in each of the 100 runs. The mean success rate of the different EDA variants from the 15, 575 instances is shown in Figure 2.

It can be seen in Figure 2a) that there are important variations in the success rate of Tree-EDA due to parameters Ts and Tp. The influence of the parameters can be critical for the behavior of the algorithm. Notably, Tree-EDA with $Tp > -1$ can not outperform the behavior of UMDA. Notice however, that UMDA is also very sensitive to the influence of parameter T. For all values of Tp, except $Tp = -3$, the number of times that Tree-EDA finds the optimum improves by selecting $Ts < Tp$. This means that, for a given selection strength applied to non-structural learning, a stronger selection applied to structural learning will likely improve the results. Figure 2b) shows how the mean success rate of Tree-EDA also increases by repairing the solutions. The same trend in the influence of the selection parameters is observed. Except for $Tp = -3$, the results of Tree-EDA improves by selecting $Ts < Tp$.

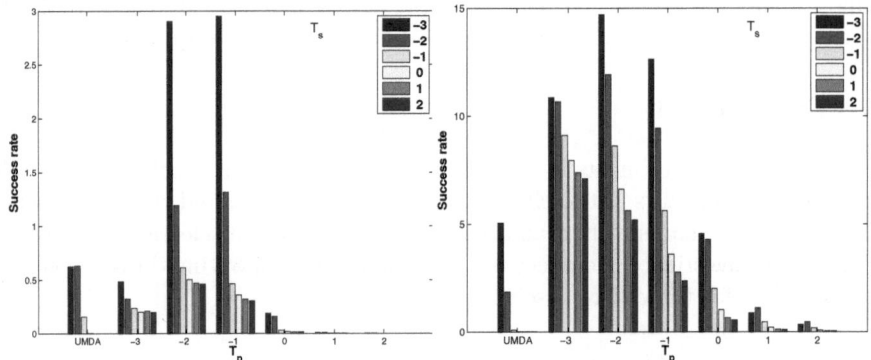

Fig. 2. Customized Boltzmann selection: Number of times the best solution was found for the different values of the temperature. a) Without repairing. b) When repairing is applied.

We also test, for each instance independently, for statistical differences between the EDAs that could be attributed to the use of customized selection. Fixing the value of Tp, we apply a multi-comparison test (p-value 0.05) for the six variants of Tree-EDA. The test uses as information the best result in each of the 100 runs for the corresponding instance. Among the six variants of Tree-EDA corresponding to the six values of Ts, we test which pairs of means are significantly different, and which are not. A test that can provide such information is called a multiple comparison procedure. Adjustment for multiple testing is applied using the Dunn-Sidak method [18], a procedure similar to, but less conservative than, the Bonferroni procedure.

Using the results of the test, we compute the number of times each Tree-EDA variant was significantly better (S_b) and significantly worst (S_w) than the Tree-EDA that uses $Ts = Tp$. For example, for $Tp = 1$, we compute the number of times that Tree-EDA $(Ts = i)$ was significantly better than Tree-EDA $(Ts = 1)$ for all $i \neq 1$. Similarly, we compute the number of times that Tree-EDA $(Ts = i)$ was significantly worse than Tree-EDA $(Ts = 1)$. The difference between these two numbers gives an idea of the appropriate choice for Ts in relation with Tp. Figure 3 shows the values of $(S_b - S_w)$ for all values of Tp. A positive value for $Ts = i$ means that Tree-EDA improves its performance when it takes this value. Conversely, a negative value indicates a poorer behavior.

Figure 3 confirms the previous results obtained from the analysis of mean success rate. For Boltzmann distribution, improvements can be achieved by using, for structural learning, selection probabilities with a higher selection strength than that used for non-structural learning. We should learn the structure from a set of very good solutions but the selection strength can be relaxed at the time of learning the model's parameters.

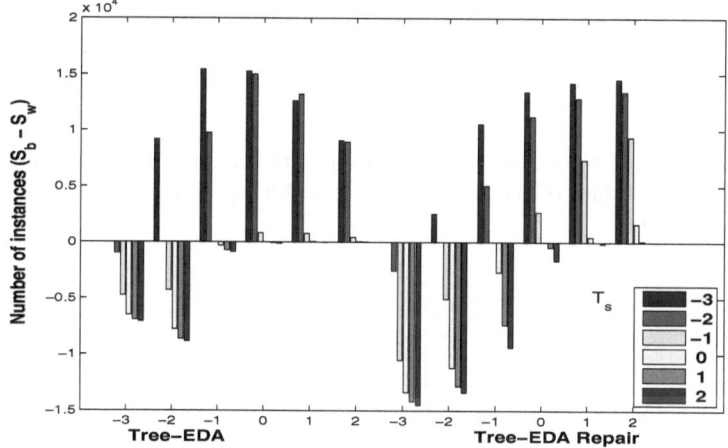

Fig. 3. Results of the statistical tests for customized Boltzmann selection

6.3   Results for Customized Truncation Selection

The analysis of customized truncation selection are conducted using the same methodology.

Figure 4 shows the mean success rate of Tree-EDA with and without repairing. The results of Tree-EDA improves by selecting $\alpha_s < \alpha_p$ for $\alpha_p > 0.1$. Also for Tree-EDA with repairing the results improve for $\alpha_s < \alpha_p$ for $\alpha_p > 0.1$, but the differences are not that clear. For truncation selection, $\alpha_s \in \{0.2, 0.3\}$ is the best choice for almost all values of α_p. Another remarkable fact that makes a difference with Boltzmann selection is that Tree-EDA is always better than UMDA when truncation selection is used.

The results of the statistical tests for customized truncation selection are shown in Figure 5. It can be seen in the figure that Tree-EDA with truncation selection exhibits a behavior similar to Tree-EDA with Boltzmann selection when the repairing procedure is not used. Nevertheless, when repairing is applied, there are fewer significant differences in the behavior of the algorithms. This fact can be also observed in Figure 5.

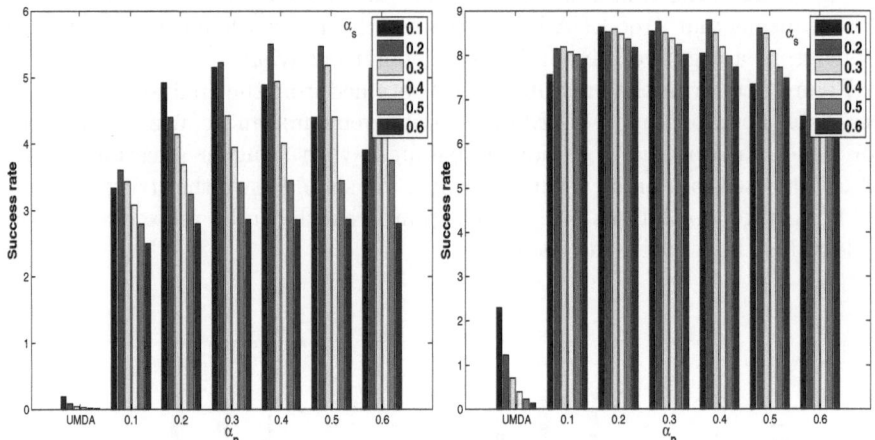

Fig. 4. Customized truncation selection: Number of times the best solution was found for the different values of the truncation parameter. a) Without repairing. b) When repairing is applied.

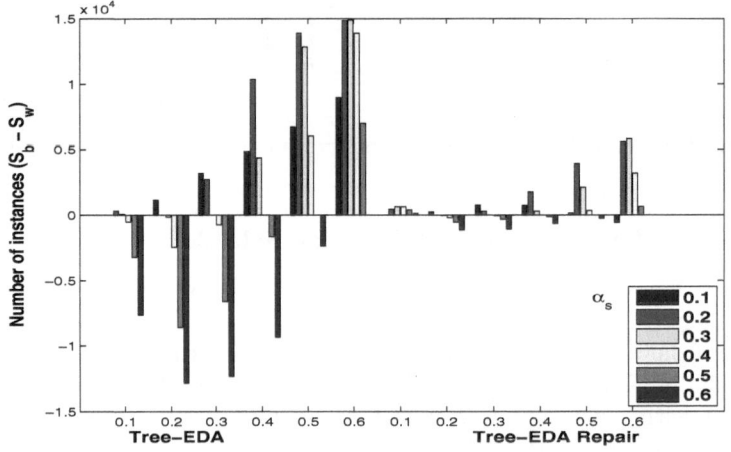

Fig. 5. Results of the statistical tests for customized truncation selection

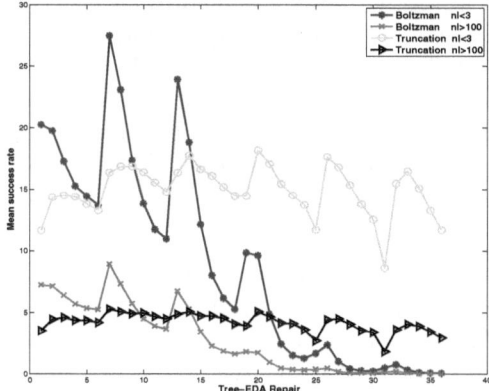

Fig. 6. Mean success rate of Tree-EDA with repairing for solutions grouped according to the number of local optima (nl)

To evaluate how sensitive are the introduced selection methods to the number of optima in the problem, we selected two sets of instances. The first group contains instances with one or two local optima (703 instances). The second group comprises instances with more than 100 local optima (685 instances). Using Tree-EDA with repairing we compute the mean success rate for the selection methods on all possible combinations of parameters Tp and Ts ($6 \times 6 = 36$). The results are shown in Figure 6 where nl is the number of local optima in addition to the global optimum. The main conclusion from the analysis of Figure 6 is that both selection methods are sensitive to the number of local optima. However, while the Boltzmann customized selection is able to outperform customized truncation selection for appropriate combination of parameters, the second selection method is more robust to the variation of the parameters.

7   Conclusions and Future Work

In this paper we have introduced customized selection in EDAs. We have shown that, by using different selection probabilities for structural and non-structural learning of the models, it is possible to increase the success rate of Tree-EDA for the functional model protein. Our results show that improvements are more important for Boltzmann selection than for truncation selection.

Beyond the improvements in optimization, customized selection opens new possibilities for research on the relationship between selection and model learning in EDAs. We can independently evaluate the effect of selection in the structural and parametric components of the graphical models. For instance, we can investigate the quality of the models as fitness surrogates by independently manipulating the different selection probabilities from which the model's components are learned.

There are optimization problems where some sets of variables make a higher contribution to the fitness. Evolutionary algorithms can fail in these situations when these *salient* building blocks converge before those with lower marginal fitness [19]. One possible extension of customized selection is the computation of marginal probabilities of different subsets of variables using different selection probabilities. In this way, marginal probabilities could be adjusted, according to different strengths of selection, to "synchronize" building blocks with different temporal-salience behaviors.

Acknowledgments. This work has been partially supported by the Saiotek and Research Groups 2013-2018 (IT-609-13) programs (Basque Government), TIN2013-41272P (Ministry of Science and Technology of Spain), COMBIOMED network in computational bio-medicine (Carlos III Health Institute), and by the NICaiA Project PIRSES-GA-2009-247619 (European Commission).

References

1. Baluja, S., Davies, S.: Using optimal dependency-trees for combinatorial optimization: Learning the structure of the search space. In: Fisher, D.H. (ed.) Proceedings of the 14th International Conference on Machine Learning, pp. 30–38. Morgan Kaufmann, San Francisco (1997)
2. Brownlee, A.E.I., McCall, J., Shakya, S.K.: The Markov network fitness model. In: Shakya, S., Santana, R. (eds.) Markov Networks in Evolutionary Computation, pp. 125–140. Springer (2012)
3. Chow, C.K., Liu, C.N.: Approximating discrete probability distributions with dependence trees. IEEE Transactions on Information Theory 14(3), 462–467 (1968)
4. Cotta, C.: Protein structure prediction using evolutionary algorithms hybridized with backtracking. In: Mira, J., Álvarez, J.R. (eds.) IWANN 2003. LNCS, vol. 2687, pp. 321–328. Springer, Heidelberg (2003)
5. Echegoyen, C., Mendiburu, A., Santana, R., Lozano, J.A.: On the taxonomy of optimization problems under estimation of distribution algorithms. Evolutionary Computation 21(3), 471–495 (2013)
6. Hirst, J.D.: The evolutionary landscape of functional model proteins. Protein Engineering 12, 721–726 (1999)
7. Karshenas, H., Santana, R., Bielza, C., Larrañaga, P.: Multi-objective optimization based on joint probabilistic modeling of objectives and variables. IEEE Transactions on Evolutionary Computation 18(4), 519–542 (2014)
8. Krasnogor, N., Blackburne, B.P., Burke, E.K., Hirst, J.D.: Multimeme algorithms for protein structure prediction. In: Guervós, J.J.M., Adamidis, P.A., Beyer, H.-G., Fernández-Villacañas, J.-L., Schwefel, H.-P. (eds.) PPSN VII. LNCS, vol. 2439, pp. 769–778. Springer, Heidelberg (2002)
9. Larrañaga, P., Karshenas, H., Bielza, C., Santana, R.: A review on probabilistic graphical models in evolutionary computation. Journal of Heuristics 18(5), 795–819 (2012)
10. Lozano, J.A., Larrañaga, P., Inza, I., Bengoetxea, E. (eds.): Towards a New Evolutionary Computation: Advances on Estimation of Distribution Algorithms. Springer (2006)

11. Miquélez, T., Bengoetxea, E., Larrañaga, P.: Evolutionary computation based on Bayesian classifiers. International Journal of Applied Mathematics and Computer Science 14(3), 101–115 (2004)
12. Mühlenbein, H., Mahnig, T., Ochoa, A.: Schemata, distributions and graphical models in evolutionary optimization. Journal of Heuristics 5(2), 213–247 (1999)
13. Mühlenbein, H., Paaß, G.: From recombination of genes to the estimation of distributions I. Binary parameters. In: Voigt, H.-M., Ebeling, W., Rechenberg, I., Schwefel, H.-P. (eds.) PPSN IV. LNCS, vol. 1141, pp. 178–187. Springer, Heidelberg (1996)
14. Mühlenbein, H., Schlierkamp-Voosen, D.: The science of breeding and its application to the breeder genetic algorithm (BGA). Evolutionary Computation 1(4), 335–360 (1994)
15. Munetomo, M., Murao, N., Akama, K.: Introducing assignment functions to Bayesian optimization algorithms. Information Sciences 178(1), 152–163 (2008)
16. Pelikan, M., Mühlenbein, H.: The bivariate marginal distribution algorithm. In: Roy, R., Furuhashi, T., Chawdhry, P. (eds.) Advances in Soft Computing - Engineering Design and Manufacturing, pp. 521–535. Springer, London (1999)
17. Santana, R.: Factorized distribution algorithms: Selection without selected population. In: Proceedings of the 17th European Simulation Multiconference ESM 2003, Nottingham, England, pp. 91–97 (2003)
18. Šidák, Z.: Rectangular confidence regions for the means of multivariate normal distributions. Journal of the American Statistical Association 62(318), 626–633 (1967)
19. Thierens, D., Goldberg, D.E., Pereira, A.G.: Domino convergence, drift, and the temporal-salience structure of problems. In: Proceedings of 1998 IEEE International Conference on Evolutionary Computation, Anchorage, AK, pp. 535–540 (1998)
20. Valdez-Peña, I.S., Hernández-Aguirre, A., Botello-Rionda, S.: Approximating the search distribution to the selection distribution in EDAs. In: Proceedings of the Genetic and Evolutionary Computation Conference GECCO 2009, pp. 461–468. ACM, New York (2009)

A Hybrid GP-Tabu Approach to QoS-Aware Data Intensive Web Service Composition

Yang Yu, Hui Ma, and Mengjie Zhang

School of Engineering and Computer Science,
Victoria University of Wellington, PO Box 600, Wellington 6012, New Zealand
{yang.yu,hui.ma,mengjie.zhang}@ecs.vuw.ac.nz

Abstract. Web service composition has become a promising technique to build powerful business applications by making use of distributed services with different functions. Due to the explosion in the volume of data, providing efficient approaches to composing data intensive services will become more and more important in the field of service-oriented computing. Meanwhile, as numerous web services have been emerging to offer identical or similar functionality, web service composition is usually performed with end-to-end Quality of Service (QoS) properties which are adopted to describe the non-functional properties (e.g., response time, execution cost, reliability, etc.) of a web service. In this paper, a hybrid approach that integrates the use of genetic programming and tabu search to QoS-aware data intensive service composition is proposed. The performance of the proposed approach is evaluated using the publicly available benchmark datasets. A full set of experimental results show that a significant improvement of our approach over that obtained by the simple genetic programming method and several traditional optimization methods, has been achieved.

1 Introduction

Service-oriented architecture (SOA) [1] is a widely accepted and engaged paradigm for the realization of complex business processes. The aim of SOA is to implement business processes covering different organisations and computing platforms in a dynamic and loosely-coupled manner. As a promising technology to implement such a service-oriented architecture, web services encapsulate software functions and make them available to anyone in the world over the network via standard interfaces and protocols (e.g., SOAP and WSDL). The advent of web services has boosted the creation of business applications by reusing existing resources on the network, rather than building new applications from scratch to fulfill business functional requirements.

Today, a large number of web services on the Internet offer identical or overlapping functionality but present various non-functional characteristics which are called *quality of service (QoS)* properties such as response time, execution cost and reliability. Therefore, how to select a suitable web service that satisfies user's requirements still remains an open question. On the other hand, when no

G. Dick et al. (Eds.): SEAL 2014, LNCS 8886, pp. 106–118, 2014.

single web service is able to respond to user's request, it is necessary to compose a range of existing services together in order to provide new value-added and complex functionality, which is referred to as *web service composition*, and the aggregated web service becomes a *composite web service.*

Aside from business processes, the service-oriented approach using web services is also of great interest for the implementation of data intensive processes such as data mining and image processing. Such web services are defined as data intensive services that generally have large amounts of data as their inputs and outputs. For example, a facial recognition solution for video and image content could be published through a web service that, given a set of original images, will provide a set of processed images with identified facial features. Over the recent years, the amount of data generated by humanities, scientific activities, as well as commercial applications from a diverse range of fields has been increasing exponentially. Data volumes used in the fields of sciences and engineering, finance, media, online information resources and so on, are expected to double every two years over the next decade and further [2]. There is no doubt in the industry and research community that the importance of data intensive computing has been raising and will continue to be the foremost research field. As a result, data intensive services based applications have become the most challenging type of applications in SOA. Also, data intensive service composition has become an appealing research area in academia and industry.

Although various approaches have been presented to solve web service composition problems [3–7], there is limited work in the literature for data intensive service composition [2, 8–10]. The authors of [8, 10] consider the data intensity of service composition, but they overlook the communication cost of mass data transfer and its effects on the performance of business processes with different structures. An ant colony based algorithm is proposed in [2, 9] to find the cost minimized data intensive service composition by considering the access cost and communication cost of mass data transfer. However, the approach focuses on minimizing the cost only without reflecting other important QoS attributes (e.g., availability and reliability). To the best of our knowledge, the use of genetic programming in data intensive service composition was not examined in the past research.

In this paper, we present a hybrid GP-Tabu approach to the problem of QoS-aware data intensive service composition. Tabu search (TS) [11] as a meta-heuristic local search is integrated into the evolutionary process of genetic programming (GP) [12] in order to overcome the downsides of GP such as its prematurity and proneness to trap in local optima. The main contributions of this paper are two-fold. First, a QoS-aware mathematical model is developed to take into account the effect of mass data transfer. Second, a hybrid approach that combines the use of GP and TS is proposed to address the QoS-aware data intensive service composition problem. The experimental results demonstrate the effectiveness and efficiency of the approach, especially it offers better performance than traditional optimization techniques.

The rest of the paper is organized as follows. Section 2 introduces the background of the QoS-aware service composition problem. Section 3 discusses the time and cost aware model used in the context of data intensive service composition. In Section 4, the details of the proposed approach are explained, and Section 5 reports on the experimental results. Finally, the conclusions and future work are outlined in Section 6.

2   Background

Standard protocols adoption and platform independent have enabled web services to be integrated together to produce a value-added complex service. Nowadays due to the explosive increase of web services that provide identical or overlapping functionality on the Web, quality of service namely QoS becomes a key factor in distinguishing these functionally equivalent services.

2.1   Atomic Web Service *vs.* Composite Web Service

A composite web service is made up of a collection of single web services each of which is referred to as *atomic web service*. The goal of a service composition is to generate the desired outputs given a set of available inputs that are described by semantic concepts. Assume the task scenario is to search for an appropriate flight as well as the weather forecast for the destination based upon the given departure date, return date, departure city and arrival city. In other terms, a web service is requested to take {*departure date, return date, departure city, arrival city*} as inputs in order to produce {*flight information, weather forecast*} as outputs. However, an atomic web service itself has limited functionality which is not sufficient to respond to the user's request. A composite web service that consists of multiple atomic web services is needed to accomplish the task. A valid service composition must guarantee that the inputs of any atomic web service are available either from the outputs of its ancestors or from the original inputs (i.e., from the user).

2.2   Quality of Service

QoS can be characterized according to various non-functional properties of web services called *QoS attributes* such as response time, execution cost, availability and reliability. Based on a selection of relevant characteristics in the field of web services, the QoS attributes considered in this research work are latency, execution cost, availability and accuracy all of which are defined as follows.

- *latency L* measures the expected delay in seconds between the moment when a request is sent and the moment when the results are received.
- *execution cost C* is the amount of money that a service requester has to pay for executing the web service.
- *availability A* is the probability that a web service is accessible.

- *accuracy* R is the measurement of the degree to which the real results produced by the web service match the desired results.

Amongest the above four QoS attributes, latency and execution cost are decreasing measures with respect to QoS. In other terms, the grades of these measures increase as their values decrease. In contrast, availability and accuracy are increasing measures of which the grades decrease as their values decrease.

QoS reflects the non-functional properties of web services which in turn have an influence on user satisfaction, thereby the QoS factors have become significant criteria that cannot be overlooked in web service composition. The problem of *QoS-aware service composition* denotes selecting a number of atomic web services while obtaining the highest possible QoS of the service composition and satisfying the global constraints posed by the user. Here, global constraints define an upper or lower bound for the aggregated QoS values of a composite service. To optimize the quality of a service composition, a method must be applied to estimate the QoS of the service composition from its constituent services. This estimation is called QoS aggregation, and the QoS aggregation formulas defined for basic composition patterns [13] and major QoS attributes are summarized in Table 1.

Table 1. Aggregation formulae for each pair QoS attribute - workflow structure

QoS attribute	Sequence	Parallel (Flow)	Choice (Switch)	Loop
Latency	$L = \sum_{i=1}^{j} l_i$	$L = MAX\{l_i \| i \in \{1, \ldots, j\}\}$	$L = \sum_{i=1}^{j} p_i * l_i$	$L = k * l$
Execution cost	$C = \sum_{i=1}^{j} c_i$	$C = \sum_{i=1}^{j} c_i$	$C = \sum_{i=1}^{j} p_i * c_i$	$C = k * c$
Availability	$A = \prod_{i=1}^{j} a_i$	$A = \prod_{i=1}^{j} a_i$	$A = \sum_{i=1}^{j} p_i * a_i$	$A = a^k$
Accuracy	$R = \prod_{i=1}^{j} r_i$	$R = \prod_{i=1}^{j} r_i$	$R = \sum_{i=1}^{j} p_i * r_i$	$R = r^k$

3 Time and Cost Aware Model for Data Intensive Service

Since a data intensive service s_i is provided by a service provider and deployed on a server, a data intensive service composition SC that consists of a set of m service servers and certain composition patterns can be denoted by $SC = \{S_1, S_2, \ldots, S_m\}$. Each atomic data intensive service s_i is associated with a QoS vector $Q_i = [l_i, c_i, r_i, a_i]$ where l_i, c_i, r_i, a_i represents the latency, execution cost, accuracy and availability of the service. Assume that a data intensive service s_i requires a set of data denoted by D_i to form part of its inputs. The latency l_i for the service s_i is made up of three parts: the queue time q_i, the processing time p_i and the transfer time t_i, as shown in Equation 1.

$$l_i = q_i + p_i(D_i) + t_i(D_i, S_j, S_i) \tag{1}$$

where q_i is the time spent in waiting in the queue for the data D_i to be processed by the server, $p_i(D_i)$ is the actual time used to process the set of data D_i, and $t_i(D_i, S_j, S_i)$ is the data transfer time for transferring the data set D_i from the server hosting the service s_j to the server hosting the service s_i.

To be detailed, the queue time q_i depends on server load, i.e., the current request queue length on the server, while the processing time p_i and the transfer time t_i can be calculated using Equation 2 and 3.

$$p_i(D_i) = size(D_i)/pr(S_i) \tag{2}$$

$$t_i(D_i, S_j, S_i) = size(D_i)/bw(S_j, S_i) + size(D_i)/ws(S_i) \tag{3}$$

where $size(D_i)$ is the size of the data set D_i, $pr(S_i)$ denotes the processing rate of the server S_i, $bw(S_j, S_i)$ is the network bandwidth between the server S_j and the server S_i, and $ws(S_i)$ is the disk write speed of the server S_i. To sum up, as described in the above equations, the processing time depends on the processing capacity of the server, and the data transfer time is determined by the network bandwidth and the amount of data to be transferred between two service servers. As each server has many requests at the same time and it serves only one request at a time, the current service request needs to wait until all requests prior to it in the queue have completed.

In addition to time, the costs generated by data intensive services as well as the movement of mass data have a significant impact on the total cost of a service composition. Consider a data intensive service s_i, similarly its execution cost c_i consists of three parts: the data access cost ac_i, the data transfer cost tc_i, and the service related cost sc_i. As can be seen from Equation 4, the data access cost ac_i is the price to be paid for writing the data D_i to the server that hosts the service s_i and reading the data in order to invoke the service. The data transfer cost tc_i is proportional to the size of the data set D_i, which depends on the available network bandwidth between two service servers. The service related cost sc_i expresses the cost to provision the service s_i as well as the cost to process the service request including data processing.

$$c_i = ac_i(D_i) + tc_i(D_i, S_j, S_i) + sc_i(D_i) \tag{4}$$

$$ac_i(D_i) = size(D_i) * wcost(S_i) + size(D_i) * rcost(S_i) \tag{5}$$

$$tc_i(D_i, S_j, S_i) = size(D_i) * tcost(S_j, S_i) \tag{6}$$

$$sc_i(D_i) = pcost(S_i) + size(D_i) * dcost(S_i) \tag{7}$$

where $size(D_i)$ denotes the size of the data set D_i, $wcost(S_i)$ is the cost of writing per unit of data to the server S_i, $rcost(S_i)$ is the cost of reading per unit of data from the disk on the server S_i, $tcost(S_j, S_i)$ is the transfer cost from the service server S_i to the service server S_j for per unit of data, $pcost(S_i)$ is the price charged to use the service s_i which is usually specified by service provider, and $dcost(S_i)$ is used to represent the expenditure for processing each unit of data on the server S_i.

For each data intensive service, the other two QoS attributes (i.e., accuracy and availability) are supposed to have fixed values which can be collected from service providers. Therefore, the QoS-aware data intensive service composition can be regarded as an optimization problem. Clearly the goal of the optimization problem is to minimize the latency and execution cost that have been defined in Equations 1 and 4, meanwhile achieving the maximum possible accuracy and availability for a composite web service.

4   The Hybrid GP-Tabu Approach

Artificial intelligence techniques have been widely used to solve many optimization problems. In recent years, GP has become increasingly popular as an alternative to more classical techniques in science and engineering disciplines. As another powerful optimization procedure, TS is capable of escaping local optimum trap by employing a flexible memory system, and it has been successfully applied to a diverse range of combinatorial optimization problems. These methods seem to be promising and are still evolving. Next, the GP and TS methods are briefly reviewed before the proposed hybrid approach is presented.

4.1   An Overview of Genetic Programming

GP [12] simulates Darwin's principles of natural selection and evolution to create a working computer program that is typically represented as a tree-like structure. Our approach employes the same tree representation as [14], that is, atomic web services invoked by the composition are denoted by leaf nodes and the workflow patterns (e.g., sequence) are expressed by intermediate nodes. In addition, GP requires a fitness function to measure the quality of each individual (i.e., service composition in our case) in the population. In general, GP involves three types of operators which are described as follows.

- *Selection.* A number of individuals in the population are selected to breed a new generation. Individuals selection is a fitness-based process, where fitter individuals are typically more likely to be selected for reproduction.
- *Crossover.* The crossover operator takes two parents and replaces a randomly chosen part of one parent with another randomly chosen part of the other in order to produce two new offsprings. For the specific service composition problem, two individuals are stochastically selected. If there is a common node that represents the compatible inputs and outputs in both individuals, the two nodes along with their subtrees are then swapped between the two individuals. This guarantees that the offspring generated is feasible.
- *Mutation.* The mutation operator takes one parent and replaces a randomly selected part of that parent with a randomly generated sequence of code. In our particular case, a random node of a service composition is selected and its subtree is replaced with a new stochastically generated one.

4.2   An Overview of Tabu Search

TS [11] is a meta-heuristic that guides a local heuristic search procedure to explore a problem's solution space with the goal of avoiding local optimum and ultimately finding the desired solution. The basic principle of TS is to avoid cycling back to previously visited solutions and allow non-improving moves whenever a local optimum is encountered. This is achieved by using a short-term memory that records the recent history of the search to prevent investigating the solution space that has been visited before. However, in some situations, TS permits backtracking to previous solutions which may ultimately lead to better solutions via a different direction. The two main components of TS are the tabu list and the aspiration criteria of the solution associated with the recorded moves.

- *Tabu list.* Certain forbidden moves (trial solutions) are maintained in the list to prevent cycling when moving away from local optimum through non-improving moves. As a result, the search is not allowed to return to a recently visited point in the search space, that is, a recent move is not allowed to be reversed. Usually the tabu list stores a fixed or fairly limited quantity of information. Empirically, the size of the list that provides good results often grows with the size of the problem, and stronger restrictions are generally coupled with smaller list size.
- *Aspiration criteria.* A key issue for tabu list is that it is sometimes so powerful to prohibit attractive moves, even cycling cannot occur, or they may lead to an overall stagnation of the search process. Hence, aspiration criteria are used to allow for exceptions from the tabu list, if such moves lead to promising solutions. The simplest and most commonly used aspiration criterion, found in almost all TS implementations, allows a tabu move when it results in a solution with an objective value better than that of the current best-known solution (since the new solution has obviously not been previously visited).

4.3   The Proposed Hybrid Approach

In order to determine a solution to the QoS-aware data intensive service composition problem, we propose a new hybrid approach where the evolutionary process of GP works along with the local TS search procedure. To be specific, the proposed hybrid approach adopts the neighbour solutions found by TS to generate part of a new population in the global search process of GP. The major steps of our approach are described in Algorithm 1. The approach starts from a randomly initialized population after setting up the necessary variables for GP and Tabu. The individuals in the current population are then evaluated using the specific fitness function. Crossover and mutation are performed on the selected individuals to produce next generation. For every t generations, the m best individuals (i.e., service compositions) in the population are selected as the initial solutions of the TS procedure. As a result, n neighbour solutions are generated by mutating a random node in the tree representation of a candidate service composition solution, and the n worst individuals in the current population are replaced. The above process is repeated until the maximum number of iterations is reached.

In our approach, the fitness function introduced to measure the performance of each individual i in the gth generation of the evolution process is defined as follows.

$$f_i = \frac{(w_1 R_i + w_2 A_i) * (w_5 I_i + w_6 O_i)}{w_3 L_i + w_4 C_i} \tag{8}$$

where R_i, A_i, L_i and C_i denote the aggregated accuracy, availability, latency and execution cost of a composite data intensive service, each of which can be

Algorithm 1. GP-Tabu for QoS-aware data intensive service composition

---

Require: available inputs, required outputs, QoS constraints and a service repository
Ensure: a service composition that meets both functional and non-functional requirements
1: Initialize the parameters of GA and TS, and set $g=1$
2: Generate an initial population P randomly
3: Evaluate each individual i in P using the fitness function
4: **while** $g < g_{max}$ **do**
5:     Select two parents from the population P. Perform crossover with rate P_c, and
        perform mutation with rate P_m to generate a new population P'
6:     **if** g mod $k = 0$ **then**
7:         Choose the m best individuals from the current population, and apply the TS
            algorithm to generate n neighbours to substitute the worst n individuals in
            the new population P'
8:     **end if**
9:     Evaluate each individual i' in P' using the fitness function
10:    Set $g=g+1$
11: **end while**
12: **return** the individual with the best fitness

---

calculated using the formulae described in Table 1 and the equations proposed in
Section 3. For example, the aggregated latency for a sequence workflow structure
is $L = \sum_{i=1}^{j} l_i$. In the function, w_1, w_2, w_3, w_4, w_5 and w_6 are real and positive
weights. A larger weight means that that particular QoS attribute is considered
more important than others from the point of view of users. Note that the fitness
function can be easily adapted to user's requirements, i.e., adding or removing
QoS attributes without affecting the performance of our approach. I_i and O_i
that indicate the degree to which a valid solution has been found are presented
in Equation 9.

$$I_i = \frac{|input_r|}{|input_r \bigcup input_a|} \qquad O_i = \frac{|output_r \bigcap output_a|}{|output_r|} \qquad (9)$$

where $input_r$ is the list of inputs available for a composite service solution, $input_a$
is the list of inputs required by the solution, $output_r$ is the list of outputs desired
by a composition task, $output_a$ is the list of outputs that are actually produced
by the solution, and $|.|$ represents the size of the list.

To guarantee incommensurable QoS attributes have fair impact on the calculation of fitness, the value of each QoS attribute is to be normalized in the
interval [0,1]. All the weights utilized in the fitness function also falls within
the range [0, 1]. As illustrated in Equation 8, QoS-aware data intensive service
composition is converted to a maximization problem, i.e., greater fitness denotes
more satisfying solution.

5   Experimental Results and Analysis

For the purpose of evaluation, we carry out a set of experiments using the test
cases provided by the public benchmark datasets, WSC2008 and WSC2009.

Each dataset consists of a great number of web services associated with randomly generated inputs and outputs. However, QoS attributes are not included in neither datasets. Therefore, the QoS values of web services are generated based on the data collected in another public dataset called QWS [15]. Each test case is made up of available inputs, required outputs, and a service repository. The complexity of the test cases is diverse in terms of the number of atomic web services and the number of workflow structures involved.

5.1   Parameter Settings

The experiments are conducted with the population size of 200 for maximum 100 generations (i.e., g_{max}=100). The crossover probability P_c=0.9, the mutation probability P_m=0.2, and the size of the tabu list is 7. In our experiments, the top 10% individuals in the current population (i.e., m=20) will be selected for applying TS for every 10 generations (i.e., t=10), so the 20 worst (i.e., n=20) individuals will be replaced with the new neighbour solutions. Assume that availability and execution cost are considered more important than accuracy and latency. The weights defined in the fitness function are $w_1 = 0.2$, $w_2 = 0.3$, $w_3 = 0.2$, $w_4 = 0.3$, $w_5 = 0.5$ and $w_6 = 0.5$ which can give better performance indicated by a large number of empirical trials. Since our approach is nondeterministic, 30 independent runs are performed for each test case.

To simulate the time and cost for data access and transfer in our experiments, the amount of input data for an atomic data intensive service is randomly generated in the interval (0, 30]MB, and the amount of output data is determined by multiplying by a random factor of 0~10. For the sake of simplicity, the queue time required by a service server is between 0~10s, all server's data write speed (i.e., $ws(S_i)$) and process rate (i.e., $pr(S_i)$) are both 10MB/s, the transfer rate between two service servers (i.e., $bw(S_j, S_i)$) is 3MB/s, and all other costs such as data access cost and transfer cost for per unit of data are all 1.

5.2   Experimental Results

The experimental results for the test cases are presented in Table 2. Each row in the table shows the fitness of the solutions found by GP and GP-Tabu for each test case. To demonstrate the superiority of our approach over the simple GP method, a significance test (z-test) is conducted to compare the solutions found by the two approaches. As illustrated in the table, for simple composition tasks (i.e., WSC2008-1, WSC2008-2 and WSC2008-4), both GP and GP-Tabu are able to make the same *optimal* service composition to achieve the high-level task. However, it is observed that for more complicated test cases, the GP-Tabu approach is capable of finding better compositions of services indicated by a significant improvement on the fitness. In summary, our hybrid approach was successful in computing a solution to each of the service composition tasks, and the results showed that it is much more effective for complicated tasks.

Table 2. The results of the test cases for GP and GP-Tabu. (↑ denotes significantly better)

Test case	GP	GP-Tabu
WSC2008-1	0.9167 ± 0.0000	0.9167 ± 0.0000
WSC2008-2	0.9206 ± 0.0000	0.9206 ± 0.0000
WSC2008-3	0.9025 ± 0.0017	0.9114 ± 0.0009↑
WSC2008-4	0.8668 ± 0.0000	0.8668 ± 0.0000
WSC2008-5	0.8126 ± 0.0014	0.8135 ± 0.0008↑
WSC2008-6	0.8461 ± 0.0003	0.8556 ± 0.0002↑
WSC2008-7	0.8988 ± 0.0008	0.9052 ± 0.0013↑
WSC2008-8	0.8825 ± 0.0011	0.8857 ± 0.0007↑
WSC2009-1	0.8276 ± 0.0002	0.8311 ± 0.0006↑
WSC2009-2	0.8844 ± 0.0001	0.8982 ± 0.0011↑
WSC2009-3	0.7846 ± 0.0023	0.7931 ± 0.0009↑
WSC2009-4	0.7024 ± 0.0006	0.7546 ± 0.0019↑
WSC2009-5	0.7290 ± 0.0008	0.8449 ± 0.0023↑

5.3   Further Analysis

To further study the effectiveness and efficiency of our approach, here we conduct a set of experiments with the same test cases on two traditional optimization methods, i.e., TS and integer linear programming (ILP) [16]. In order to evaluate the quality of the solutions found by different optimization methods, a unity function that adopts the simple additive weighting approach is used as shown in Equation 10.

$$U(SC) = w_1 \cdot R + w_2 \cdot A + w_3 \cdot L + w4 \cdot C \tag{10}$$

where w_1, w_2, w_3, and w_4 remain the same as described in Section 5.1. I and O are not included in the function as they are specified as constraints in all the optimization methods, that is, the solution found must be able to generate the desired outputs given the available inputs.

The simulation results of the experiments are shown in Table 3. The last three columns of the table present the fitness of the solutions found by GP-Tabu, ILP and Tabu, respectively. As can be observed from the table, the ILP method cannot find a valid solution for most of the service composition tasks except for tasks WSC2008-1, WSC2008-2, WSC2008-4 and WSC2008-5, and the significance test demonstrates the superiority of the GP-Tabu approach over ILP in this problem domain. In contrast, the TS method is able to find a service composition solution for each task, especially when the tasks (i.e., WSC2008-1, WSC2008-2, WSC2008-3, WSC2008-4, WSC2008-5 and WSC2009-2) are relatively simple. However, in most situations where the service composition request is very complex, our GP-Tabu approach is recommended due to its better performance implied by the significant improvement from the statistical test.

Table 3. The results of the test cases for GP-Tabu, ILP and Tabu. ($\downarrow$ denotes significantly worse, and $\uparrow$ denotes significantly better)

Test case	GP-Tabu	ILP	Tabu search
WSC2008-1	0.5946 ± 0.0000	$0.5849 \pm 0.0014\downarrow$	$0.5916 \pm 0.0019\downarrow$
WSC2008-2	0.4997 ± 0.0000	$0.4654 \pm 0.0006\downarrow$	$0.5108 \pm 0.0007\uparrow$
WSC2008-3	0.4588 ± 0.0005	n/a	$0.4836 \pm 0.0008\uparrow$
WSC2008-4	0.4738 ± 0.0000	$0.4531 \pm 0.0013\downarrow$	$0.4876 \pm 0.0006\uparrow$
WSC2008-5	0.4888 ± 0.0007	$0.3626 \pm 0.0027\downarrow$	0.4884 ± 0.0010
WSC2008-6	0.4212 ± 0.0003	n/a	$0.4192 \pm 0.0014\downarrow$
WSC2008-7	0.4177 ± 0.0006	n/a	$0.3791 \pm 0.0006\downarrow$
WSC2008-8	0.5146 ± 0.0009	n/a	$0.5033 \pm 0.0012\downarrow$
WSC2009-1	0.6666 ± 0.0008	n/a	$0.5779 \pm 0.0007\downarrow$
WSC2009-2	0.4447 ± 0.0015	n/a	0.4441 ± 0.0009
WSC2009-3	0.5612 ± 0.0008	n/a	$0.5531 \pm 0.0004\downarrow$
WSC2009-4	0.4623 ± 0.0006	n/a	$0.3145 \pm 0.0011\downarrow$
WSC2009-5	0.4844 ± 0.0019	n/a	$0.3487 \pm 0.0023\downarrow$

6   Conclusions and Future Work

In this paper, a novel combination of genetic programming and tabu search for solving the problem of QoS-aware data intensive service composition has been presented. A time and cost aware mathematical model was developed for describing the effect of the movement of mass data. In the proposed approach, the local search procedure employed by TS was integrated into the global search process of GP, in order to avoid premature convergence and getting stuck in local optima. To verify the effectiveness and efficiency of the proposed hybrid approach, it was successfully applied to two public benchmark datasets, i.e., WSC2008 and WSC2009, each of which consists of a large variety of web services as well as diverse service composition tasks. Compared to the simple GP and two traditional optimization methods, the analysis of the experimental results showed the superiority of our approach in finding more satisfying service compositions.

In this study, multiple QoS constraints are transformed into one single criterion to be optimized by our approach. This makes it difficult to simultaneously optimize multiple conflicting QoS objectives. Therefore, for future work we will investigate the use of multi-objective GP with the expectation that multiple and often conflicting QoS properties (e.g., time and cost) can be optimized simultaneously to produce a set of pareto-optimal solutions. Future work also includes applying our approach in real life situations so server load (e.g., queue time, processing time, transfer time, etc.) can be measured with real values.

References

1. Perrey, R., Lycett, M.: Service-oriented architecture. In: Proceedings. 2003 Symposium on Applications and the Internet Workshops, pp. 116–119. IEEE (2003)
2. Wang, L., Shen, J., Di, C., Li, Y., Zhou, Q.: Towards minimizing cost for composite data-intensive services. In: 2013 IEEE 17th International Conference on Computer Supported Cooperative Work in Design (CSCWD), pp. 293–298 (2013)
3. Zeng, L., Benatallah, B., Dumas, M., Kalagnanam, J., Sheng, Q.Z.: Quality driven web services composition. In: Proc. of the 12th International Conference on World Wide Web, pp. 411–421. ACM (2003)
4. Zheng, H., Zhao, W., Yang, J., Bouguettaya, A.: Qos analysis for web service compositions with complex structures. IEEE Transactions on Services Computing 6(3), 373–386 (2013)
5. Cardellini, V., Casalicchio, E., Grassi, V., Lo Presti, F.: Flow-based service selection for web service composition supporting multiple qos classes. In: IEEE International Conference on ICWS 2007, pp. 743–750 (2007)
6. El Haddad, J., Manouvrier, M., Ramirez, G., Rukoz, M.: Qos-driven selection of web services for transactional composition. In: IEEE International Conference on ICWS 2008, pp. 653–660 (2008)
7. Aversano, L., di Penta, M., Taneja, K.: A genetic programming approach to support the design of service compositions. International Journal of Computer Systems Science and Engineering 4, 247–254 (2006)
8. Bucchiarone, A., Presti, L.: Qos composition of services for data-intensive application. In: Second International Conference on ICIW 2007, p. 46 (2007)
9. Wang, L., Shen, J., Beydoun, G.: Enhanced ant colony algorithm for cost-aware data-intensive service provision. In: 2013 IEEE Ninth World Congress on Services (SERVICES), pp. 227–234 (2013)
10. Zhang, Y., Zhou, X., Gao, Y.: Optimizing the data intensive mediator-based web services composition. In: Zhou, X., Li, J., Shen, H.T., Kitsuregawa, M., Zhang, Y. (eds.) APWeb 2006. LNCS, vol. 3841, pp. 708–713. Springer, Heidelberg (2006)
11. Glover, F., Laguna, M.: Tabu search. Springer (1999)
12. Koza, J.R.: Genetic programming: on the programming of computers by means of natural selection. MIT Press, Cambridge (1992)
13. Jaeger, M., Rojec-Goldmann, G., Muhl, G.: Qos aggregation for web service composition using workflow patterns. In: Proc. Eighth IEEE International EDOC 2004, pp. 149–159 (2004)
14. Yu, Y., Ma, H., Zhang, M.: An adaptive genetic programming approach to qos-aware web services composition. In: 2013 IEEE Congress on Evolutionary Computation (CEC), pp. 1740–1747 (June 2013)
15. Al-Masri, E., Mahmoud, Q.H.: Qos-based discovery and ranking of web services. In: Proc. of 16th International Conference on ICCCN 2007, pp. 529–534. IEEE (2007)
16. Canfora, G., Di Penta, M., Esposito, R., Villani, M.L.: An approach for qos-aware service composition based on genetic algorithms. In: Proceedings of the 7th Annual Conference on Genetic and Evolutionary Computation, GECCO 2005, pp. 1069–1075. ACM, New York (2005)
17. Zheng, H., Yang, J., Zhao, W., Bouguettaya, A.: Qos analysis for web service compositions based on probabilistic qos. In: Kappel, G., Maamar, Z., Motahari-Nezhad, H.R. (eds.) Service Oriented Computing. LNCS, vol. 7084, pp. 47–61. Springer, Heidelberg (2011)

18. Ruberg, N., Ruberg, G., Manolescu, I.: Towards cost-based optimization for data-intensive web service computations. In: Proc. of SBBD, p. 283 (2004)
19. Wang, A., Ma, H., Zhang, M.: Genetic programming with greedy search for web service composition. In: Decker, H., Lhotská, L., Link, S., Basl, J., Tjoa, A.M. (eds.) DEXA 2013, Part II. LNCS, vol. 8056, pp. 9–17. Springer, Heidelberg (2013)
20. Juric, M.B., Mathew, B., Sarang, P.G.: Business Process Execution Language for Web Services: An Architect and Developer's Guide to Orchestrating Web Services Using BPEL4WS. Packt Publishing Ltd. (2006)
21. Li, Y., Lin, C.: Qos-aware service composition for workflow-based data-intensive applications. In: 2011 IEEE International Conference on Web Services (ICWS), pp. 452–459 (2011)

A Modified Screening Estimation of Distribution Algorithm for Large-Scale Continuous Optimization

Krishna Manjari Mishra and Marcus Gallagher

School of Information Technology and Electrical Engineering,
The University of Queensland, Brisbane, 4072. Australia.
{krishna,marcusg}@itee.uq.edu.au

Abstract. Continuous Estimation of Distribution Algorithms (EDAs) commonly use a Gaussian distribution to control the search process. For high-dimensional optimization problems, several practical issues arise when estimating a large covariance matrix from the selected population. Recent work in continuous EDAs has aimed to address these issues. The Screening Estimation of Distribution Algorithm (sEDA) is one such algorithm which, uniquely, utilizes the objective function values obtained during the search. A sensitivity analysis technique is then used to reduce the rank of the covariance matrix, according to the estimated sensitivity of the fitness function to individual variables in the search space.

In this paper we analyze sEDA and find that it does not scale well to very high-dimensional problems because it uses a large number of additional fitness function evaluations per generation. A modified version of the algorithm, named sEDA-lite is proposed which requires no additional fitness evaluations for sensitivity analysis. Experiments on a variety of artificial and real-world representative problems evaluate the performance of the algorithm compared with sEDA and EDA-MCC, a related, recently proposed algorithm.

Keywords: Estimation of Distribution algorithms, High-Dimensional Optimization problems, Screening Technique.

1   Introduction

Estimation of Distribution Algorithms (EDAs) are a class of evolutionary optimization algorithms, where probabilistic models play the key role in controlling the search process. In EDAs, the selected population is used to learn a probability distribution and subsequent solutions obtained by sampling from this distribution. The general procedure for an EDA is summarized in Algorithm 1. A number of different types of density estimation models have been used in EDAs for both discrete and continuous search spaces. Probabilistic models can be used to identify and represent interactions among the variables and can represent a priori information about the problem structure, which may assist the search process. EDAs have been developed in both the discrete and continuous setting, and have been successfully applied to solving a variety of problems.

G. Dick et al. (Eds.): SEAL 2014, LNCS 8886, pp. 119–130, 2014.

Algorithm 1. General pseudocode framework for an EDA

1: Initialization: set $t = 0$, Generate initial population uniformly in search space
2: Evaluate $f(\mathbf{x}')$ for each individual $\mathbf{x}'$ in the current population
3: Select promising individuals
4: Build probabilistic model $p(\mathbf{x})$ based on selected individuals
5: Generate new population by sampling from $p(\mathbf{x})$
6: $t = t + 1$
7: Goto step 2 until a stopping criterion is met

Reviews of EDAs can be found in [13,16,19]. This paper focuses on continuous EDAs ($\mathbf{x} \in \mathbb{R}^n$).

Truncation selection is typically applied in EDAs. The parameters of $p(\mathbf{x})$ are typically fitted using maximum likelihood estimation. While several different models have been considered for continuous optimization, a Gaussian distribution is most commonly used. The continuous Univariate Marginal Distribution Algorithm (UMDA$_c$)[13] uses a factorized Gaussian model (i.e. a diagonal covariance matrix) which assumes all the variables are independent. UMDA$_c$ is easy to apply and computationally robust and efficient, but the model may have difficulty in solving problems with strong dependencies between variables. Multivariate Gaussian EDAs, such as the Estimation of Multivariate Normal Algorithm (EMNA$_{global}$) address this issue by modelling dependencies between all pairs of variables using a full covariance matrix [13].

The behaviour of basic Gaussian EDAs has been shown to be sometimes undesirable. On some fitness landscapes, performance is poor due to premature shrinking of the model variance at an exponential rate (Eg., in slope-like regions of the search space, described in [1] or in an elliptical region [10]). To address such issues, a number of enhancements have been proposed. Adaptive variance scaling (AVS) provides a way to control the rate of contraction and expansion of the model and scale the variance to improve the progress of the EDA model. Anticipated Mean Shift (AMS) additionally modifies sampled solutions in the direction of mean shift of the previous generation [1,2]. Nevertheless, the task of covariance matrix estimation remains a fundamental step in state of the art Gaussian EDAs.

In practice, numerical issues can arise with estimating the full covariance matrix. The covariance matrix, Σ, is by definition positive semi-definite, but this is not guaranteed in implementation because of finite precision representation. Computational errors or numerical issues arise when the sample used to estimate the model does not adequately span all dimensions of the search space, which becomes likely when the sample size is relatively small compared to the problem dimensionality. As a result of these issues, several techniques have been proposed to avoid the covariance matrix becoming ill-posed in EDAs. The Eigenspace EDA (EEDA) [20] was one of the first modifications in this direction (see [7] for a discussion and comparison of variants). Dong et al. [6] developed the Covariance Matrix Repairing (CMR) method, where a positive value (equal to the absolute value of the smallest eigenvalue) is added to the diagonal of Σ.

The Eigen-decomposition EDA (ED-EDA) builds on this previous work with eigenvalue-based repairing strategies [7] . Experimental results show that these different covariance repairing methods can avoid numerical difficulties. The algorithms also show good performance results with respect to the best solution found and because of the improved numerical properties, a smaller population can be used.

More recently, Dong et al.[5] proposed the EDA Model Complexity Control (EDA-MCC) to scale up continuous EDAs to high-dimensional problems using a sparse covariance matrix, with reduced computational cost and a smaller population. EDA-MCC uses EMNA$_{\text{global}}$ for each subset of the variables. A set of artificial test problems were used for comparing EDA-MCC with UMDA$_c$, EMNA$_{\text{global}}$ and EEDA [20]. While EDA-MCC does not outperform traditional EDAs on low dimensional problems, EDA-MCC shows significantly better results on high dimensional problems. Other existing statistical methods have also been applied to control the amount of covariance/dependency modelling in EDAs. In [11], regularization techniques were adopted into EDAs. The resulting algorithm shows the ability to solve high dimensional problems with a comparable quality of solutions using much smaller populations.

The screening Estimation of Distribution Algorithm (sEDA), was proposed in [14] as an EDA to control the degree of covariance modelling. However unlike other approaches, this algorithm explicitly uses the objective function values obtained during the search. Using this information, a notion of variable importance is derived by adapting a screening technique from experimental design. The algorithm also improves on numerical stability in EDAs by allowing the level of dependency modelling to be controlled. It performs better than traditional EDAs on low dimensional problems. In this paper, a modified version of sEDA, called sEDA-lite is proposed which improves on the previous algorithm, specifically by allowing the algorithm to scale to higher dimensional problems using less function evaluations than sEDA.

This paper is structured as follows. In Section 2, the existing sEDA algorithm is described. We analyse the issues arising in high dimensions when using sEDA. In Section 3, we propose sEDA-lite to address these issues. In Section 4, we compare the solutions of sEDA-lite with solutions of UMDA$_c$, EEDA and EDA-MCC on a set of artificial test problems. In addition to this, the solution of sEDA-lite is also compared with UMDA$_c$, EMNA$_{\text{global}}$ and sEDA on a couple of real world problems. Conclusions of this paper are drawn in Section 6.

2   Screening Estimation of Distribution Algorithms (sEDA)

The continuous global optimization problem is to find $\mathbf{x}^*$ such that

$$f(\mathbf{x}^*) \leq f(\mathbf{x}), \forall \mathbf{x} \in S,$$

where $S \subseteq \mathbb{R}^n$ is the set of feasible solutions, $f(\mathbf{x})$ is the fitness or objective function and $\mathbf{x} = (x_1, \ldots, x_n)$ is an individual or candidate solution vector.

In the Screening EDA [14], variables are modeled based on their estimated influence on the fitness function. From this, the most important variables are then modeled using the EMNA$_{global}$ model (full covariance matrix), while the variables which are least important are modeled using the UMDA$_c$ model (no covariance), to try and capture the advantages and limit the potential problems of both approaches. Hence, the sEDA uses a multivariate Gaussian model where the covariance matrix contains some degree of sparseness. To select which variables to model using covariance, a technique from sensitivity analysis known as the Morris method is used.

The Morris method [15], is based on measuring the mean and standard deviation of changes in the fitness function value given perturbations of individual variables, calculated via so-called *elementary effects* terms. The elementary effect for the ith variable, $E_i(\mathbf{x})$, is defined as follows. Let Δ be a pre-determined amount to perturb each variable. For a given $\mathbf{x}$,

$$E_i(\mathbf{x}) = \frac{f(x_1, x_2, ..., x_{i-1}, x_i + \Delta, x_{i+1}, ..., x_n) - f(\mathbf{x})}{\Delta} \tag{1}$$

where $\mathbf{x} = (x_1, x_2, ..., x_n)$ is a given starting or "baseline" vector in the solution space. The perturbations, Δ are by default determined according to a full factorial sampling grid of some fixed resolution or increment size. In other words, for each variable x_i, over some fixed range and increment size, the value of x_i is changed and f is recalculated, producing a sample or set of values of $E_i(\mathbf{x})$.

Given a set of elementary effect values, the mean, $\overline{E}_i(\mathbf{x})$, and standard deviation, $std(E_i(\mathbf{x}))$ over the sample can be calculated. If this calculation is done over an arbitrary set of points in an arbitrary order, the absolute values, $|E_i|$ are used [4,12] and we take this approach in sEDA. A high value of $\overline{E}_i(\mathbf{x}^*)$ indicates that x_i has a large average influence on the value of f. A high value of $std(E_i(\mathbf{x}))$ indicates that variable x_i has a fluctuating influence on the value of f, which may indicate that it is involved in interactions with other variables [15].

In sEDA, the Morris method is adapted by calculating elementary effects values using the selected population on each generation of the algorithm rather than being based on a predetermined grid of points. Specifically, the mean of the selected population is calculated for each dimension x_i. A new set of candidate solutions is then generated, by creating new solution vectors where the mean value is substituted in turn for each problem variable (e.g. $\mathbf{x^i} = x_i, ..., x_{i-1}, m_i, x_{i+1}, ..., x_n$) in each individual in the selected population. This produces $n \times \tau \times M_{sel}$ new individuals, where M_{sel} is the size of the selected population[1], n is the dimension of the problem and τ $(0 < \tau < 1)$ is the selection ratio, which are evaluated using f to produce a set of elementary effect values.

Given these values and their sample mean and standard deviation, sEDA uses the concepts of dominance and Pareto optimality from multi-objective optimization (see, e.g. chapter 9 of [8]) to determine which variables are the most "important". We consider the mean and standard deviation of elementary effects as two different (aka decision-making) criteria. One solution is said to dominate

---

[1] Rounding if $M_{sel} \cdot \tau$ is not an integer.

the other if its score is at least as high for all objectives, and is strictly better for at least one. The set of all non-dominated solutions is called the Pareto set or the Pareto front.

A fixed fraction η ($0 < \eta < 1$) of the variables need to be selected for covariance modeling in the sEDA. Variables that belong to the Pareto set are selected first. If more variables are required, then those which have the minimum (Euclidean) distance to the Pareto front are selected. On the other hand, if the number of variables on the Pareto front is greater than required, then a random subset of these variables is selected. Hence at each generation, a sparse matrix is formed that has $n \times \eta$ variables[2] which are modeled using covariance parameters while the remaining variables are modelled using only variance terms. The mean vector of the selected population along with the sparse covariance matrix is then used to generate the new population for the next generation.

2.1   Scaling of sEDA

Due to the nature of the algorithm, sEDA as described above will require a relatively large number of function evaluations when applied to high-dimensional problems. This is due to the fact that for each generation, the population size is directly proportional to the dimension of the problem. Hence, the number of function evaluations per generation is $O(nM_{sel})$.

3   Scaling sEDA to High-Dimensional Problems: sEDA-lite

In this section we describe a modified version of sEDA, called sEDA-lite. The algorithm uses the same principles as sEDA but differs in the calculation of the elementary effects values. As discussed in Section 2.1, using the mean of the selected population to calculate elementary effects values in sEDA results in significant increase in the number of fitness function evaluations required per generation. In sEDA-lite, the main innovation is to instead use the *median* of each dimension in the selected population to calculate elementary effects values. Like the mean, the median is representative of the center of the selected population. However, the median is by definition located at one of the given individual points. Hence, all calculations in Equation(1) are carried out between individuals in the selected population (i.e. their fitness values have already been evaluated). Hence for each generation of sEDA-lite the number of function evaluations is reduced from $M_{sel} + (M_{sel} \times n \times \tau)$ to M_{sel}.

The median of the selected population is taken as the central/reference point for the elementary effect calculations. Other points in the selected population represent perturbations around this point and hence the elementary effect values measure the sensitivity of the objective function to changes in the solution values in the region of the search space represented by the current selected population.

---

[2] Rounding if $n \times \eta$ is not an integer.

The mean is also still calculated, since this is used as a parameter of the EDA model itself.

sEDA-lite uses the same Pareto optimal concept as in sEDA, to select the important variables in the problem as the ones that have the largest mean/standard deviation of elementary effect values. After selection, the covariance matrix for the EDA model is formed as a sparse matrix, with non-zero covariance terms for selected variables. This is used in combination with the EDA mean vector (estimated from the selected expanded population) and the model is then used to generate the new population as in a standard EDA. The process is repeated until some stopping criterion is met.

Algorithm 2. Pseudo code for sEDA-lite

1: Given: Population size M, dimensionality n, selection parameter $0 < \tau < 1$, model selection parameter $0 < \eta < 1$.

2: Begin (set $t = 0$)
   Initialize population P by generating M individuals uniformly in S.

3: **while** stopping criteria not met **do**

4:     Evaluate f for population P.

5:     Truncation selection: $P_{sel} = M_{sel}$ best individuals from P; $M_{sel} = \text{Rnd}(M \cdot \tau)$.

6:     Calculate mean, μ and median, $\tilde{m}$ of P_{sel}

7:     Calculate $\tilde{m} = \text{median}(P_{sel})$, where $\tilde{m} = \tilde{m}_1, \cdots, \tilde{m}_n$.

8:     **for** $i = 1$ to n **do**

9:         **for** $j = 1$ to M_{sel} **do**

10:            Calculate $E_{ij}(x)$ using Eqn.1, where $\tilde{m}$ is the baseline point and the perturbation value is given by j^{th} individual

11:        **end for**

12:    **end for**

13:    Calculate $mean(E)$ and $std(E)$ over M_{sel} perturbations in E_{ij}.

14:    Determine the Pareto optimal solutions/variables for objectives $abs(mean(E))$ and $std(E)$. Let this number of variables be p_o .

15:    Let $B = \text{Round}(n \cdot \eta)$.

16:    If $p_o > B$, randomly choose B variables from p_o.

17:    If $p_o < B$, select/add the next $B - p_o$ variables nearest to the Pareto front.

18:    Build Σ_t using covariance terms for the B selected variables and variance terms only for the remaining $n - B$ variables

19:    $p(\mathbf{x}) \leftarrow (\mu, \Sigma_t)$.

20:    Generate P new population by sampling from $p(\mathbf{x})$.

21: **end while**

---

4   Experimental Design

To evaluate the performance of sEDA-lite, we have carried out experiments on 3 different sets of problems. The first is a set of commonly used artificial test functions. The second set of problems are Circle in a square (CiaS) packing problems and the third set are the 50-customer Location Allocation problems with different numbers of facilities. While the artificial functions are useful for comparison with other algorithms, we also consider it important to evaluate the technique on real-world representative problems. The problems used here are all scalable in terms of dimensionality. They are also known to have features that make them difficult to solve for many algorithms, e.g. they are not everywhere differentiable and contain a large number of local optima.

4.1   Artificial Test Problems

The artificial test problems considered in this paper are taken from Dong et al. [5]. The problems are categorized into separable unimodal (F_1 and F_2), non separable problems ($F_3, \cdots, F_{10}$) and multimodal problems (F_{11}, F_{12}, F_{13}). The offset values used in the test functions are same as described in [5], with the exception of F_4 and F_6. For these 2 functions, the offset values were generated randomly. While this means the results are not precisely comparable, we generated offset values for these functions using the same formula as given in [5].

The problem sizes were 50D and 100D for each artificial test function. The maximum number of function evaluation was set at $10000 \times n$. The population sizes used in [5], were tested for sEDA-lite (i.e., 200, 500, 1000 and 2000). From this, a population size of 2000 was used for all functions except F_1 and F_2, where 200 was used since these very simple functions do not require a large population. Initially, rough experiments were conducted to determine reasonable algorithm parameter values: $\tau = \{0.1, 0.2, 0.3, 0.5\}$ and $\eta = 0.1$ were trialled, though not explored exhaustively. The parameter values that seemed to work best for each set of problems were then used. The algorithm stopped when the difference between the global optimum and the optimal values obtained from the algorithm is 1E-12 or it attained maximum number of function evaluations. The results reported are based on 25 repeated trials.

4.2   Circle in a Square Problem

Given the 2D unit square and a pre-specified number of circles, n_c, constrained to be of equal size, the circles in a square (CiaS) problem is to find an optimal packing; i.e. to position the circles and compute the radius length of the circles such that the circles occupy the maximum possible area within the square. All circles must remain fully enclosed within the square, and cannot overlap. The problem can be formulated as finding the positions of n_c points inside the unit square such that their minimum pairwise distance is maximized:

$$d_{n_c} = \max \min_{i \neq j} \| \mathbf{w}^i - \mathbf{w}^j \|_2 \tag{2}$$

$$\mathbf{w}^i \in [0, 1]^2, i = 1, \ldots, n_c \tag{3}$$

The feasible search space of a CiaS problem is defined by the unit hypercube $[0, 1]^{2n_c} \subset \mathbb{R}^{2n_c}$, and solutions outside this are infeasible. To ensure the generation of such candidate solutions by the EDAs, any value in a solution vector generated that lies outside the feasible region is reset to the (nearest) boundary.

For the (CiaS) problems, experiments were conducted on the number of circles, ranging from 2,...,50. The problem dimensionality n is equal to $2 \times n_c$. UMDA$_c$, EMNA$_{global}$, sEDA and sEDA-lite were implemented on this problem. The population size of all the algorithms was set to 2000 except sEDA, where the population was 50 times n. The value of τ for all the algorithms is set to 0.2. Since EMNA$_{global}$ is performing better than UMDA$_c$ in CiaS problem discussed in [14], the value of η as 0.5 was set for sEDA and sEDA-lite. The algorithm is stopped after 2E+06 function evaluations or if the difference between the best fitness value and the global optimum is 1E-04. The results were computed based on 25 repeated trials.

4.3   Location Allocation

In the continuous location allocation problem [18], the aim is to determine the location of n_f facilities in a 2D Euclidean space in order to serve customers at c fixed points so that the distances between each customer and the nearest facility are minimized [3]. There is no restriction on the capacity of the facilities to serve customers.

The (uncapacitated, continuous) location-allocation problem is formulated as follows:

$$\min \left(\sum_{j=1}^{c} \min_{i} d(X_i, A_j) \right)$$

where X is the vector consisting of the coordinates of the facilities. For n_f facilities problem, there are $2n_f$ variables for optimization. The $X_i th$ facility has coordinate values (x_i, x_{n_f+i}). A is the vector consisting of the given coordinates of the customers in the problem. For the $A_j th$ customer the coordinate values are represented as (a_{1j}, a_{2j}). $d(X_i, A_j)$ is the Euclidean distance from the location of facility X_i, to the location of a customer at fixed point A_j. For $n_f > 1$, this problem is known to be non convex and generally contains a large number of local minima [3].

In this paper we consider the widely used (e.g.[3,17]) 50-customer problem with a unit weight value for all the customers. The data A_i for the problem is given in [9]. For the 50-Customer problems, experiments were conducted using, $n_f = 5, 10, 15, 20, 25, 35$. The dimensionality of the problem $n = 2 \times n_f$. UMDA$_c$, EMNA$_{global}$, sEDA and sEDA-lite were compared on these problems. The population size of all the algorithms was set to 2000 except sEDA, where the population was $50 \times n$ to allow the algorithm a sufficient number of generations to converge. The value of τ for all the algorithms was 0.3, while the value of η was 0.1 for sEDA and sEDA-lite. The maximum number of function evaluations was $10000 \times n$. The results are over 25 repeated trials.

5   Results

5.1   Artificial Test Problems

The results of sEDA-lite are compared here with the values of $UMDA_c$, EEDA and EDA-MCC which are taken from [5]. The results of $EMNA_{global}$ are not repeated since it was previously found to perform worst on the test problems [5]. Comparative results for sEDA are also not reported here since it requires a prohibitive number of function evaluations for these larger-scale problems. The comparative results between $UMDA_c$, EEDA, EDA-MCC and sEDA-lite are listed in Table 5.1.

The results of the experiments for separable problems (F_1 and F_2), show that all the algorithms for 50D and 100D can solve these problems without difficulty. Functions from $F_3 \ldots F_{10}$ are non-separable problems with only a few local optima. On these functions, $UMDA_c$ and EEDA do not show the best performance. The performance of EDA-MCC is significantly better than rest of the algorithms in problems F_3 and F_5. Since the offset values are generated randomly, the previous solutions of F_4 and F_6 for EEDA and EDA-MCC are not reported here. We recomputed the results of $UMDA_c$ on these functions and compared with sEDA-lite for the same offset values. sEDA-lite clearly outperforms $UMDA_c$. The results also show EDA-MCC and sEDA-lite outperform$UMDA_c$ in 50D F_7 and F_8 functions. Solution comparison Table 5.1 shows that EEDA performs well on the 50D F_{10} function. Overall, from functions F_3 to F_{10}, the performance of sEDA-lite is similar to EDA-MCC.

Functions from F_{11} to F_{13} are multimodal functions. In these functions EDA-MCC and EEDA do not perform well. The performance of $UMDA_c$ and sEDA-lite are similar for function F_{11}, however, on the functions F_{12} and F_{13}, the performance of sEDA-lite exceeds $UMDA_c$. It is to be expected that there would be some variability in the relative performance of the algorithms. Overall, sEDA-lite is generally competitive and in some cases provides the best performance for these problems.

5.2   Results for the Circle in a Square (CiaS) Problems

The performance of the algorithms on a large set of CiaS problems (4D - 100D) is presented in Figure 1. The x-axis denotes the problem size (n_c) while the y-axis is a performance ratio given by $d_n/f(x_n)$, where d_n is the known global optimum and $f(x_n)$ is the solution found by the algorithm.

The results show that up to $n_c = 16$, $UMDA_c$, sEDA and sEDA-lite perform similarly. $EMNA_{global}$ does not perform as well, likely because it requires a larger population size and/or number of function evaluations. When $16 < n_c < 24$, sEDA actually performs slightly better than the other algorithms, but its performance then quickly degrades when $n_c > 24$. This is when the total budget of function evaluations for this experiment means that sEDA cannot perform sufficient generations, due to the requirement for calculating elementary effects values during execution. However sEDA-lite does not suffer from this, maintaining performance that is a little better than $UMDA_c$ up to $n_c = 50$.

Table 1. Solution quality comparison. Each cell contains the mean and standard deviation of the difference between the best fitness value and the global optimum. Bold font represents the best result. A "+" indicates a statistically significant difference (t-test, unequal variances, 0.05 level) when compared with sEDA-lite. A "-" sign indicates no significant difference. ζ indicates previous results for the algorithms which are incomparable due to the random values of the offset.

Prob.	Dim	UMDA$_c$	EEDA	EDA-MCC	sEDA-lite
F_1	50	0±0	0±0	0±0	0±0
	100	0±0	0±0	0±0	0±0
F_2	50	0±0	0±0	0±0	0±0
	100	0±0	5.3e-10±1.4e-09(-)	0±0	0±0
F_3	50	2.6e-04±1.5e-05(+)	1.8e-08±2.4e-09(+)	0±0(+)	4.2e-07±2.7e-08
	100	2.6e-02±8.3e-02(-)	1.5e-03±8.5e-04(+)	0±0(+)	2.5e-06±4.2e-06
F_4	50	4.1e+01±2.3e+00(+)	ζ	ζ	**3.6e+01±2.1e+00**
	100	5.3e+01±2.5e+00(+)	ζ	ζ	**4.8e+01±2.8e+00**
F_5	50	1.5e+01±4.1e+00(+)	2.4e-02±3.7e-03(+)	0±0(+)	1.0e+01±4.0e+00
	100	1.3e+02±2.7e+01(-)	3.8e-01±4.7e-02(+)	0±0(+)	1.2e+02±2.1e+01
F_6	50	6.5e+01±1.4e+01(+)	ζ	ζ	**3.8e+01±1.1e+01**
	100	1.1e+03±1.6e+02(+)	ζ	ζ	**8.6e+02±1.6e+02**
F_7	50	4.8e+01±3.4e-02(+)	5.0e+01±9.2e+00(+)	**4.7e+01±2.1e-01(-)**	**4.7e+01±2.4e-02**
	100	9.7e+01±6.4e-02(-)	9.7e+01±3.7e-01(-)	9.6e+01±7.5e-02(+)	9.7e+01±3.1e-02
F_8	50	4.1e+02±9.1e+02(+)	5.2e+02±1.0e+03(+)	**4.8e+01±1.5e-01(-)**	4.8e+01±1.4e+00
	100	9.3e+02±3.1e+03(-)	4.4e+04±4.4e+04(+)	9.6e+01±1.3e-01(+)	1.1e+02±2.8e+01
F_9	50	4.3e+07±4.1e+06(+)	4.1e+06±1.4e+06(+)	**3.6e+06±1.5e+06(+)**	4.0e+08±4.6e+07
	100	4.3e+07±3.1e+06(+)	2.2e+07±3.7e+06(+)	**9.6e+06±2.5e+06(+)**	2.5e+09±3.7e+08
F_{10}	50	4.9e+03±1.8e+02(+)	**2.0e+03±2.0e+03(+)**	3.1e+03±3.4e+02(+)	4.5e+03±1.6e+02
	100	5.9e+03±4.3e+02(+)	4.4e+03±6.0e+02(+)	**1.9e+03±3.6e+02(+)**	5.3e+03±3.4e+02
F_{11}	50	**0±0(-)**	3.1e+02±1.3e+01(+)	2.9e+02±1.4e+01(+)	0±0
	100	**0±0(-)**	7.3e+02±1.5e+01(+)	7.5e+02±1.6e+01(+)	0±0
F_{12}	50	2.1e+00±9.5e-01(+)	3.1e+02±1.7e+01(+)	3.0e+02±1.4e+01(+)	1.5e+00±9.9e-01
	100	8.6e+00±2.1e+00(+)	7.3e+02±2.5e+01(+)	7.4e+02±2.3e+01(+)	**0±0**
F_{13}	50	7.8e+00±8.3e-01(+)	2.7e+01±1.1e+00(+)	2.6e+01±9.2e-01(+)	**5.9e+00±5.4e-01**
	100	1.5e+01±2.0e+00(+)	3.8e+01±2.6e+01(+)	6.5e+01±1.6e+00(+)	**1.1e+01±7.3e-01**

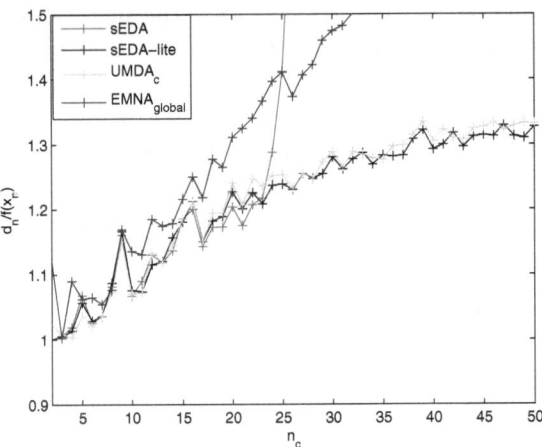

Fig. 1. Median Performance of UMDA$_c$, EMNA$_{global}$, sEDA and sEDA-lite on CiaS problems

5.3   Results for the 50-Customer Location Allocation Problem

Table 5.3 shows the results for $UMDA_c$, $EMNA_{global}$, sEDA and sEDA-lite. The performance of sEDA-lite is relatively good, particularly for high-dimensional problems. While it is possible that sEDA and $EMNA_{global}$ could give similar (or even better) performance, the amount of function evaluations required is prohibitive. Note however that there is some difference between these average performance results and the known optimal values. Also, these results are not as good as some of the previously reported results [3,17].

Table 2. Solution quality comparison (mean and standard deviation) for 50 Customer problem. Bold font represents the best result. A "+" indicates a statistically significant difference (t-test, unequal variances, 0.05 level) when compared with best result. A "-" sign indicates no significant difference.

(n_f)	Optimum	$UMDA_c$	$EMNA_{global}$	sEDA	sEDA-lite
5	72.2369	72.688±0.551(+)	77.454±3.650(+)	**72.560 ±0.591**	72.542±0.543(-)
10	41.6851	**43.401±0.823**	52.539±2.753(+)	47.930±3.809(+)	43.614±0.764(+)
15	27.6282	29.295±0.948(+)	42.991±3.171(+)	49.376±3.224(+)	**28.952±1.080**
20	19.3560	21.135±0.494(+)	36.292±2.559(+)	48.554±9.155(+)	**20.889±0.321**
25	13.3016	14.653±0.705(-)	34.402±2.772(+)	50.093 ±16.515(+)	**14.218±0.492**
30	8.7963	10.104±0.970(+)	31.237±2.173(+)	49.796±20.015(+)	**9.592±0.538**
35	5.0483	7.364±0.605(-)	29.917±2.293(+)	47.771±20.717(+)	**7.246±0.757**

6   Summary

This paper has proposed a modified version of the sEDA algorithm called sEDA-lite. Like the original algorithm, sEDA-lite is a Gaussian-EDA with a sparse co-variance matrix model, that uses a screening technique to predict the important variables which is subsequently used to control covariance modelling. However, sEDA-lite achieve this without using any additional objective function values per generation to carry out the modelling. As a result, it can be effectively applied to high-dimensional problems without a prohibitive number of function evaluations. Experimentally, sEDA-lite has been shown to be competitive with $UMDA_c$, EEDA, EDA-MCC, $EMNA_{global}$ and sEDA in various problems.

References

1. Bosman, P.A.N., Grahl, J., Thierens, D.: Enhancing the performance of maximum–likelihood gaussian eDAs using anticipated mean shift. In: Rudolph, G., Jansen, T., Lucas, S., Poloni, C., Beume, N. (eds.) PPSN X. LNCS, vol. 5199, pp. 133–143. Springer, Heidelberg (2008)

2. Bosman, P.A.N.: On empirical memory design, faster selection of Bayesian factorizations and parameter-free Gaussian EDAs. In: Raidl, G., et al. (eds.) Proceedings of the Genetic and Evolutionary Computation Conference — GECCO–2009, pp. 389–396. ACM Press, New York (2009)
3. Brimberg, J., Hansen, P., Mladenovic, N., Taillard, E.D.: Improvements and comparison of heuristics for solving the uncapacitated multisource Weber problem. Operations Research 48(3), 444–460 (2000)
4. Campolongo, F., Cariboni, J., Saltelli, A.: An effective screening design for sensitivity analysis of large models. Environmental Modelling & Software 22(10), 1509–1518 (2007)
5. Dong, W., Chen, T., Tino, P., Yao, X.: Scaling up Estimation of Distribution Algorithms for continuous optimization. IEEE Transactions 17(6), 797–822 (2013)
6. Dong, W., Yao, X.: Covariance matrix repairing in Gaussian based EDAs. In: IEEE Congress on Evolutionary Computation (CEC), pp. 415–422. IEEE (2007)
7. Dong, W., Yao, X.: Unified eigen analysis on multivariate Gaussian based Estimation of Distribution Algorithms. Information Sciences 178(15), 3000–3023 (2008)
8. Eiben, A.E., Smith, J.E.: Multimodal problems and spatial distribution. In: Introduction to Evolutionary Computing, pp. 153–172. Springer (2003)
9. Eilon, S., Watson-Gandy, C.D.T., Christofides, N.: Distribution management: mathematical modelling and practical analysis. Griffin, London (1971)
10. Hansen, N.: The CMA evolution strategy: a comparing review. In: Towards a New Evolutionary Computation, pp. 75–102. Springer (2006)
11. Karshenas, H., Santana, R., Bielza, C., Larrañaga, P.: Regularized continuous Estimation of Distribution Algorithms. Applied Soft Computing (2012)
12. King, D.M., Perera, B.J.C.: Morris method of sensitivity analysis applied to assess the importance of input variables on urban water supply yield–a case study. Journal of Hydrology 477, 17–32 (2013)
13. Larrañaga, P., Lozano, J.A. (eds.): Estimation of Distribution Algorithms: A New Tool for Evolutionary Computation. Kluwer (2001)
14. Mishra, K.M., Gallagher, M.: Variable screening for reduced dependency modelling in gaussian-based continuous estimation of distribution algorithms. In: 2012 IEEE Congress on Evolutionary Computation (CEC), pp. 1–8. IEEE (2012)
15. Morris, M.D.: Factorial sampling plans for preliminary computational experiments. Technometrics 33(2), 161–174 (1991)
16. Pelikan, M., Goldberg, D.E., Lobo, F.G.: A survey of optimization by building and using probabilistic models. Computational Optimization and Applications 21(1), 5–20 (2002)
17. Salhi, S., Gamal, M.D.H.: A genetic algorithm based approach for the uncapacitated continuous location–allocation problem. Annals of Operations Research 123(1), 203–222 (2003)
18. Scaparra, M.P., Scutellà, M.G.: Facilities, locations, customers: Building blocks of location models. a survey. Technical Report TR-01-18, Universits' degli Studi di Pisa (2001)
19. Hansen, N., Büche, D., Ocenasek, J., Kern, S., Müller, S.D., Koumoutsakos, P.: Learning probability distributions in continuous evolutionary algorithms–a comparative review. Natural Computing 3(1), 77–112 (2004)
20. Wagner, M., Auger, A., Schoenauer, M.: EEDA: A new robust estimation of distribution algorithms. Technical Report INRIA RR-5190 (2004)

Clustering Problems for More Useful Benchmarking of Optimization Algorithms

Marcus Gallagher*

School of Information Technology and Electrical Engineering,
The University of Queensland, Brisbane, 4072. Australia
marcusg@itee.uq.edu.au

Abstract. This paper analyses the data clustering problem from the continuous black-box optimization point of view and proposes methodological guidelines for a standard benchmark of clustering problem instances. Clustering problems have been used many times in the literature to evaluate evolutionary, metaheuristic and other global optimization algorithms. However much of this work has occurred independently and the various experimental methodologies used have produced results which tend to be incomparable and provide little collective wisdom as to the difficulty of the problems used, or an objective measure for comparing and evaluating the performance of algorithms. This paper surveys some of the clustering literature and results to identify issues relevant for benchmarking. A set of 27 problem instances ranging from 4-D to 40-D and based on three well-known datasets is identified. To establish some pilot results on this benchmark set, experiments are presented for the Covariance Matrix Adaptation-Evolution Strategy and several other standard algorithms. A web-repository has also been created for this problem set to facilitate better experimental evaluations of algorithms.

Keywords: Algorithm Benchmarking, Continuous Black-box Optimization, Clustering.

1 Introduction

In evolutionary computation and metaheuristic optimization, an enormous number of algorithms have been developed. Since no algorithm is superior in the theoretical, No Free Lunch sense, in practice the performance differences we observe depend on how well the mechanisms of the algorithm match the structure of the problem landscape. A key step towards understanding the matching between problems and algorithms is to develop better benchmark problems and more rigorous approaches to the experimental analysis of algorithms. Unfortunately, the dominating paradigm in the literature has been to continually develop new algorithm variants and to evaluate these techniques in isolation. For continuous

---

* The author acknowledges the contribution of the Dagstuhl Theory of Evolutionary Algorithms Seminar 13271 (http://www.dagstuhl.de/13271/) to the work in this paper.

G. Dick et al. (Eds.): SEAL 2014, LNCS 8886, pp. 131–142, 2014.

black-box optimization, artificial test functions (e.g. Sphere, Rastrigin, Rosen-brock) have been used hundereds of times, but a question such as "what is the best performance of a black-box optimization algorithm on function f, given 10^5 functions evaluations?" seems to be difficult (if not impossible) to answer using the literature. The situation is even more problematic, because subtle differences in experimental settings in different papers (e.g. using a different bound on the feasible search space) mean that results are often not strictly comparable. Recently, research has begun to focus more on such experimental issues. For example, the Black-Box Optimization Benchmarking (BBOB) problem set[1] resolves many of these issues by standardizing many aspects of the experimental setting. However, it is very important to also evaluate algorithms on real-world problems, since it is difficult to know how well artificial test problems represent real-world problems and hence to what extent algorithm performance on artificial problems is indicative of real-world performance. It can be difficult to use real-world problems for algorithm benchmarking because real problems may require expert domain knowledge to configure, or may come with additional complexities that are not part of the basic optimization algorithm (e.g. complex constraints). Ideally, problems that are real-world "representative" while being convenient for benchmarking should provide a valuable contribution to experimental research practice.

This paper examines data clustering as a useful source of continuous, black-box benchmark problems. In Section 2, the sum of squares clustering problem is defined and its key properties discussed. Clustering problems have previously been used in the literature to test optimization algorithms: Section 4 reviews some of this literature and discusses why it is difficult to compare with previously reported results. A specification is proposed to describe clustering problem instances and a set of problem instances defined (and made available via the web). To establish some baseline results for future comparison, a number of commonly-used algorithms are applied to the clustering problem sets. The experiments are described in Section 5 and Results presented in Section 6. Where possible, the results are also compared with previous results from the literature, revealing some surprising insights. The work is summarised in Section 7.

2   Clustering

The sum of squares clustering problem (see, e.g.[13]) can be stated as follows. Given a set $\mathcal{X} = \{x_1, \ldots, x_n\} \subseteq \mathbb{R}^d$ of n data points, find a set of k cluster centers $\mathcal{C} = \{c_1, \ldots, c_k\} \in \mathbb{R}^d$ to minimize:

$$f(\mathcal{C}|\mathcal{X}) = \sum_{i=1}^{n} \sum_{j=1}^{k} b_{i,j} ||x_i - c_j||^2$$

where

$$b_{i,j} = \begin{cases} 1 & \text{if } ||x_i - c_j|| = \min_j ||x_i - c_j|| \\ 0 & \text{otherwise.} \end{cases}$$

---

[1] http://coco.gforge.inria.fr/doku.php

The problem variables are the coordinates of the cluster centres in the data space. Let the d-dimensional coordinates of $\mathbf{c}_i = (y_{d(i-1)+1}, y_{d(i-1)+2}, \ldots, y_{di})$, then we have an unconstrained, continuous optimization problem of dimensionality dk:

$$\min f(\mathbf{y}), \mathbf{y} \in \mathbb{R}^{dk}$$

A clustering problem instance is therefore defined by a dataset, $\mathcal{X}$ and a value of k.

An equivalent problem from operations research is the (continuous, uncapacitated) location-allocation problem , also known as facility location or multisource Weber problem [2,9]. Given a set of *customers* to be serviced by a set of *facilities*, the problem is to position the facilities to optimize a criterion measuring overall service. Under the following conditions:

- the set of customer locations is given by $\mathcal{X}$,
- the set of facility locations be $\mathcal{C}$,
- assuming equal customer weightings, unlimited capacity of facilities to provide service and Euclidean distances between customers and facilities,

the problem then reduces to the sum of squares clustering problem.

Clustering is a fundamental task in machine learning, data analysis and operations research. Finding a global optimum is known to be NP-hard, even in the restricted cases where $d = 2$ or $k = 2$. A large number of algorithms have been proposed for clustering, though there is little doubt that the k-means algorithm is the most widely known and used [11]. From an optimization perspective, k-means is a local iterative optimization algorithm which follows a non-increasing trajectory over f. It is not a black-box algorithm, nevertheless its popularity makes it frequently used in experimental comparative studies. Note also that solving the optimization problem (i.e. locating cluster centres) is often not the final goal of clustering. Further analysis might include studying which data points are assigned to which cluster centre, or producing a classifier, where each cluster represents a class in the data set and the class label of future data points can then be predicted (e.g. using the minimum distance from the cluster centres).

3   Why Use Clustering Problems for Black-Box Optimization Benchmarking?

Clustering problems have a number of properties which suggest that they might provide an extremely useful source of benchmark problems for the evaluation and comparison of algorithms:

- They seem to be generally challenging to solve.
- They are scalable in dimensionality (via d and k).
- They are "real-world" problems in data analysis (i.e. datasets can come from real-world problems).
- They are unconstrained, meaning that black-box algorithms can be readily applied without the need for a constraint-handling mechanism.

- They can be implemented relatively simply and do not require a large amount of problem-domain-specific knowledge to understand.
- The objective function is not expensive to evaluate.

There are currently few (if any) benchmark problem sets that have all of the above properties. This suggests an exciting opportunity to improve on and increase the utility of experimental black-box algorithm evaluation and comparison by building a standardised set of clustering problem instances.

4   Black-box Optimization Approaches to Clustering

Given the fundamental nature of the clustering problem and data analysis, it is not surprising that hundreds of clustering algorithms have been proposed in the literature. At the same time, general-purpose metaheuristics and other optimization techniques have also been applied to clustering problems. This paper does not attempt an exhaustive review of all this work, but rather aims to extract the important issues to be considered in developing a specification of clustering problems for black-box optimization benchmarking.

4.1   Difficulties in Comparing with Previous Results

One of the major difficulties in trying to compare an algorithm with previous work stems from the lack of standard in the way authors present their results. Clustering results are presented in a variety of ways in the literature [13]. While the sum of squares objective function is frequently used, the actual function values (and the number of evaluations made by the algorithm) obtained are sometimes not reported. Instead, measures of cluster shape around the cluster centres produced have been used (e.g. the Rand index is used by Chang et al. to evaluate their genetic algorithm variant [3]). When the intended application is classification, measures such as classification accuracy on the data are used (e.g. Liu et al.[7] evaluate a fuzzy C-means, genetic algorithm based fuzzy C-means and an immunodominance clonal selection fuzzy C-means algorithm in this way).

While clustering problems have been widely used to compare algorithms, the datasets that have been used also vary from paper to paper. Some authors generate artificial datasets with known structure/distributions. There are a large number of benchmark datasets available in machine learning, and different authors select different datasets to use. Focussing on specific datasets would clearly improve the comparability of results for black-box optimization algorithms.

Finally, there are many experimental factors that are not specific to clustering problems that can impact on future comparisons of results. The fundamental performance results are in terms of the best objective function values found (or statistics of such values over multiple trials) and the number of function evaluations used. Presenting results in figures has several advantages, but on the other hand it is often difficult to read off numerical values from a graph for comparison. Full details of the experimental configuration (e.g. algorithm parameter settings, termination criteria, number of repeated trials) are essential to permit fair comparison and reproduction of results.

4.2   Results Selected for Comparison

The literature was further reviewed for experimental results that could be compared in a black-box optimization context. A representative number of approaches were identified:

- Maulik [8] develops a real-valued genetic algorithm (GA) for clustering and compares with k-means.
- Ye and Chen [14] apply particle swarm optimization (PSO), ant colony optimization (ACO) and a honey bees algorithm.
- Kao and Cheng [6] develop an ant colony optimization algorithm and compare it with a previous ACO clustering algorithm (due to Shelokar et al.[10]) and k-means.
- Fathian et al.[5] present a honey bee mating algorithm and compare it with ACO, a GA, Tabu search (TS) and simulated annealing (SA).
- Taherdangkoo et al.[12] propose a blind, naked mole-rats algorithm and compare it with k-means, two GA variants, PSO, ACO, simulated annealing and artificial bee colony algorithms.

These papers have each used different datasets to evaluate and compare algorithms. One problem instance is common across all the papers - these results are compared in Section 6.1.

4.3   Clustering Problem Instances

In the literature, many different datasets have been utilized to evaluate and compare clustering algorithms. Sometimes, authors generate artificial test data with known clustering structure. This can be useful, for example to visualize results. However if the exact dataset used is not available, then results can only be compared qualitatively. Benchmark datasets have also been widely used, such as those from the UCI Machine Learning Repository [1]. In particular, Du Merle et al.[4] used an interior point algorithm to compute approximate global optimum values for problem based on the Iris, Ruspini and German Towns (Spath) datasets. This is useful because we can assess the performance of algorithms relative to the optimal value on these problems. These datasets have also been used in other papers, therefore the following set of problem instances is used:

- The Iris dataset, with $d = 4$, $k = 2, \ldots, 10$ and initial search space $[0.1, 7.9]^{dk}$.
- The Ruspini dataset, with $d = 2$, $k = 2, \ldots, 10$ and initial search space $[4, 156]^{dk}$.
- The German Towns dataset, with $d = 3$, $k = 2, \ldots, 10$ and initial search space $[24.49, 1306024]^{dk}$.

Clustering Problem Specification. To be useful for black-box optimization evaluation, a clustering problem should be specified with the following elements:

1. A dataset, $\mathcal{X}$ of dimensionality d.
2. A value of k.
3. An initial bounded search space, which contains the global optimum. This can be done by using the minimum and maximum value in the dataset as the upper and lower bounds of the search space. For simplicity the overall minimum and maximum are used for every variable. A tighter search space could consider the minimum and maximum of each variable independently, however the implementation would be more complex.

To facilitate future use of these problems, a web repository has been created at http://realopt.uqcloud.net/crwr.html. The repository records the specifications of each problem instance, the global optimum (solution vector and objective function value) and a copy of the dataset. This will be extended to record results on these problems from the literature.

5   Experimental Details

To make a comparison and establish some results for the selected clustering problems, the following algorithms were evaluated:

- CMA-ES: the Matlab implementation of the Covariance Matrix Adaptation Evolution Strategy available from
  https://www.lri.fr/~hansen/cmaes_inmatlab.html was used with default parameter settings (as recommended, the initial search variance was set to 2/3 of the search space.
- CMA-ES (50,100): the same implementation of CMA-ES but with a (larger) population size of 50.
- NM: the Nelder-Mead simplex algorithm, as implemented in the Matlab `fminsearch` function. Default parameter settings were used, with termination criteria extended so that the algorith ran until a tolerance of change in variables or function values was less than 10^{-10}, or if 3×10^5 function evaluations were reached. The algorithm is initialized at a random point in the search space for each trial.
- RS: uniform random search over the search space. The algorithm was given 10^5 function evaluations.
- KM: the k-means clustering algorithm. Cluster centers were initialized to be randomly selected data points (this is probably the most common method in the literature but there are many other possibilities [11]).

The algorithms were chosen firstly because they can be applied with little setting of internal parameters. In addition: CMA-ES is a well-regarded black-box algorithm; NM is the standard Matlab solver; KM is a standard non-black-box clustering algorithm; RS provides a useful baseline. Each algorithm was run for

50 restarts. Note that the different algorithms ran for different numbers of function evaluations. The intention was not to impose a fixed budget of function evaluations across the algorithms but to allow them to use the amount of resources they request to "converge". Results on these problems can be of interest for any reasonable budget of function evaluations. Future research may choose to focus on a "limited budget" scenario or on finding high quality solutions using a possibly large number of function evaluations. Different algorithm specifications will be more suitable to different budgets of function evaluations and any result that improves upon previous results makes a worthwhile contribution.

6   Results

The experimental results are shown in Tables 1 and 2. Overall, CMA(50,100) gave the best performance, with average values that were closer to the optimal value that the other techniques. It required between 10000 and 50000 function evaluations. CMA used a smaller population size (determined automatically) and used between 2000 and 25000 function evaluations. The results were very similar on some problems (e.g. for Ruspini $n = 4, 6, 10$) but an order of magnitude worse on others (e.g. German, $n = 8 - 20$). The NM results are considerably worse across the problem sets than CMA(50,100) and worse than CMA for the Iris and Ruspini problems, but (interestingly) better on the German Towns problems. As a local search algorithm, it does however use a much smaller budget of function evaluations: between 1000 and 8000. As a completely non-local algorithm, RS outperforms the standard Matlab solver (NM) on most of the Ruspini problem instances! Finally, KM as a non-black-box solver has a considerable advantage over the other algorithms. It converges very quickly, taking less than 20 iterations/function evaluations across the problems tested. Its performance is relatively good, however CMA(50,100) still provides better performance on all problem instances! This is an impressive result for a black-box solver and experimentally demonstrates that metaheuristics are able to outperform problem-specific algorithms, and that global/population-based search would seem to lead to results that are difficult to obtain with a local/trajectory-based algorithm. With such small requirements for function evaluations, large amount of restarts could be performed for KM. Nevertheless the results here over 50 runs at least indicate that the fitness landscapes of clustering problems contain structure that causes problems for the standard solver in this problem domain.

On most problems (shown with bold), CMA(50,100) found the global optimum on at least one of the 50 trials. The exceptions were some of the larger problems on the German Towns problems. It is an open question to establish results on these problems to see how many functions evaluations are required to locate the global optimum. KM finds the global optimum for around half of the problems (lower dimensions) and CMA and NM do so for some of the smaller problem instances. Figs 1-3 compare the average fitness performance of the algorithms over the problems from each dataset. Results are given as a performance ratio with the global optimum value for each problem instance.

Table 1. Results for clustering problems. Problem instances are defined by a dataset (D), i.e. Iris (I), Ruspini (R) or German towns (G), together with the problem dimensionality (n), where $n = dk$. Shown are mean and standard deviations over 50 trials of each algorithm. The mean and average number of function evaluations $(\#f)$ is also shown for each algorithm. A result is in bold if one or more of the 50 trials located the global optimum (f^*) to at least 15 significant figures.

D	n	f^*	CMA(50,100)	CMA(50,100) #f	CMA	CMA #f
I	8	1.52348e02	**1.523480e02(0.0e00)**	1.1168e04(2.9e02)	**2.68733e02(2.2e02)**	2.41780e03(2.1e02)
	12	7.88514e01	**7.885144e01(0.0e00)**	1.7136e04(5.7e02)	1.96530e02(1.8e02)	3.77214e03(5.3e02)
	16	5.72285e01	**5.922328e01(5.0e00)**	2.1938e04(7.4e02)	1.26181e02(8.8e01)	4.98944e03(5.9e02)
	20	4.64462e01	**4.904783e01(2.3e00)**	2.5092e04(9.5e02)	1.20999e02(8.9e01)	6.58952e03(9.8e02)
	24	3.90400e01	**4.076697e01(3.5e00)**	2.7466e04(1.3e03)	1.06623e02(8.9e01)	8.39298e03(1.3e03)
	28	3.42982e01	**3.554887e01(1.4e00)**	3.0220e04(1.1e03)	8.90713e01(3.3e01)	9.90306e03(1.2e03)
	32	2.99889e01	**3.232285e01(2.1e00)**	3.3250e04(9.3e02)	8.84456e01(3.3e01)	1.20515e04(1.8e03)
	36	2.77861e01	**2.946070e01(1.9e00)**	3.6432e04(1.4e03)	8.38895e01(2.9e01)	1.46102e04(2.4e03)
	40	2.58341e01	**2.705470e01(1.3e00)**	3.9214e04(1.5e03)	8.00406e01(3.2e01)	1.67537e04(3.0e03)
R	4	8.93378e04	**8.93378e04(0.0e00)**	1.3842e04(4.2e03)	**8.93378e04(0.0e00)**	1.4188e03(3.0e02)
	6	5.10635e04	**5.11278e04(4.4e01)**	1.9382e04(2.9e03)	**5.11094e04(4.8e01)**	3.2726e03(3.2e02)
	8	1.28811e04	**1.28811e04(0.0e00)**	1.6822e04(5.1e03)	**1.66519e04(1.2e04)**	2.8380e03(7.9e02)
	10	1.01267e04	**1.10334e04(6.3e02)**	2.0862e04(4.9e03)	1.14295e04(1.2e03)	3.1720e03(2.2e02)
	12	8.57541e03	**8.84859e03(6.8e02)**	2.2022e04(1.1e03)	9.86935e03(9.6e02)	3.8949e03(2.1e02)
	14	7.12620e03	**7.55527e03(7.4e02)**	2.5912e04(2.3e03)	8.76290e03(1.4e03)	4.9135e03(7.2e02)
	16	6.14964e03	**6.43566e03(3.5e02)**	2.8792e04(2.7e03)	7.75958e03(1.5e03)	5.7632e03(6.7e02)
	18	5.18165e03	**5.64378e03(2.0e02)**	3.1892e04(4.3e03)	6.67363e03(1.1e03)	6.1772e03(7.1e02)
	20	4.446.28e03	**4.80930e03(2.5e02)**	3.1822e04(2.4e03)	6.87915e03(1.4e03)	7.2632e03(1.5e03)
G	6	6.02546e11	**6.02547e11(0.0e00)**	1.6282e04(4.7e03)	1.55257e12(8.2e11)	3.6002e03(4.5e02)
	9	2.94506e11	**3.08336e11(2.9e10)**	2.2952e04(5.5e03)	1.04493e12(8.0e11)	5.4150e03(8.9e02)
	12	1.04474e11	**1.42481e11(8.0e10)**	2.8652e04(8.1e03)	1.05834e12(7.8e11)	7.3940e03(1.7e03)
	15	5.97615e10	**7.46629e10(1.5e10)**	3.2272e04(5.0e03)	8.33722e11(7.2e11)	1.0106e04(2.4e03)
	18	3.59085e10	**4.80932e10(9.0e09)**	3.3402e04(1.5e03)	5.99661e11(6.0e11)	1.19768e04(1.4e03)
	21	2.19832e10	4.40172e10(9.4e09)	3.9092e04(2.6e03)	4.31552e11(1.5e11)	1.44593e04(1.8e03)
	24	1.33854e10	**3.11688e10(1.2e10)**	4.0962e04(2.6e03)	4.17723e11(1.6e11)	1.82163e04(1.4e03)
	27	7.80442e09	2.26611e10(1.2e10)	4.6382e04(3.5e03)	4.05634e11(1.8e11)	2.09944e04(1.9e03)
	30	6.44647e09	2.80614e10(1.2e10)	4.9872e04(5.8e03)	4.19464e11(1.7e11)	2.3522e04(3.2e03)

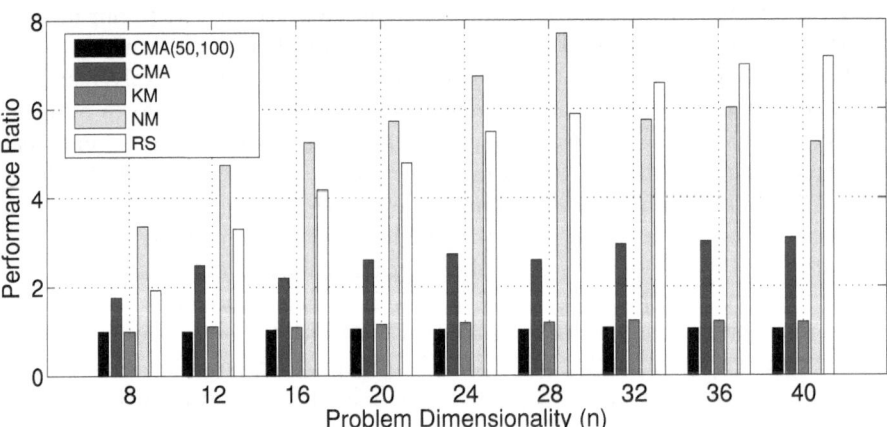

Fig. 1. Performance results (mean best fitness) for the Iris (I) dataset problems, as a ratio with the globally optimal value (e.g. a value of 2.5 means the average best solution found by an algorithm was 2.5 times the value of the global optimum

Table 2. Results for clustering problems. Problem instances are defined by a dataset (D), i.e. Iris (I), Ruspini (R) or German towns (G), together with the problem dimensionality (n), where $n = dk$. Shown are mean and standard deviations over 50 trials of each algorithm. The mean and average number of function evaluations $(\#f)$ is also shown for each algorithm. A result is in bold if one or more of the 50 trials located the global optimum (f^*) to at least 15 significant figures.

D	n	NM	NM #f	RS	KM	KM #f
I	8	**5.1208e02(2.5e02)**	1.0938e03(5.8e02)	2.93656e02(4.1e01)	**1.52348e02(0.0e00)**	4.66e00(1.4e00)
	12	3.7454e02(2.6e02)	1.6421e03(6.1e02)	2.60478e02(3.0e01)	**8.79682e01(2.3e01)**	6.94e00(2.9e00)
	16	3.0047e02(2.4e02)	2.0770e03(5.5e02)	2.39601e02(2.8e01)	**6.24359e01(6.9e00)**	8.2e00(3.7e00)
	20	2.6579e02(2.2e02)	2.6568e03(8.8e02)	2.2239e02(2.7e01)	**5.37562e01(8.8e00)**	9.18e00(3.6e00)
	24	2.6285e02(2.3e02)	3.0913e03(1.3e03)	2.1431e02(1.9e01)	4.66559e01(8.7e00)	7.86e00(2.9e00)
	28	2.6404e02(2.2e02)	3.8847e03(2.8e03)	2.0180e02(2.4e01)	4.09041e01(5.5e00)	8.82e00(3.4e00)
	32	1.7233e02(1.3e02)	4.3597e03(1.4e03)	1.9722e02(1.8e01)	3.70879e01(7.5e00)	8.34e00(2.8e00)
	36	1.6729e02(1.1e02)	5.0430e03(2.8e03)	1.9432e02(1.9e01)	3.39236e01(3.8e00)	7.78e00(2.1e00)
	40	1.3556e02(3.2e01)	5.8026e03(2.3e03)	1.8527e02(1.8e01)	3.09606e01(10.3e01)	8.52e00(3.0e00)
R	4	**8.9338e04(0.0e00)**	5.03e02(2.5e02)	9.01376e04(4.0e02)	**9.67212e04(2.3e04)**	3.82e00(1.5e00)
	6	6.6428e04(2.1e04)	8.004e02(1.3e02)	5.37281e04(1.0e03)	**5.11096e04(4.6e01)**	3.70e00(1.5e00)
	8	**3.5846e04(2.1e04)**	1.275e03(3.3e02)	1.99433e04(2.3e03)	**2.83654e04(1.8e04)**	3.99e00(1.6e00)
	10	3.5366e04(2.1e04)	1.9712e03(1.3e03)	1.65256e04(1.6e03)	**1.86163e04(1.5e04)**	4.45e00(1.6e00)
	12	2.7332e04(2.0e04)	4.203e03(2.8e03)	1.50499e04(1.0e03)	**1.46650e04(1.2e04)**	4.66e00(1.7e00)
	14	2.6147e04(2.0e04)	7.4512e03(2.8e03)	1.36176e04(1.0e03)	**1.07800e04(7.7e03)**	5.08e00(1.7e00)
	16	2.5967e04(2.1e04)	7.5234e03(4.4e03)	1.24834e04(1.0e03)	**9.58363e03(6.8e03)**	5.00e00(1.7e00)
	18	2.4995e04(2.1e04)	1.1427e04(3.9e03)	1.15749e04(7.4e02)	8.22721e03(5.1e03)	5.01e00(1.7e00)
	20	1.0728e04(1.2e03)	5.9614e03(2.1e03)	1.09593e04(7.1e02)	6.97757e03(3.9e03)	5.30e00(1.7e00)
G	6	1.5526e12(8.7e11)	9.908e02(4.1e02)	1.08704e12(1.6e11)	**6.51273e11(1.6e10)**	5.95e00(1.4e00)
	9	**4.1772e11(1.7e11)**	2.8568e03(1.4e03)	9.12653e11(1.3e11)	**3.63195e11(3.7e09)**	9.07e00(1.5e00)
	12	5.7125e11(1.7e11)	2.1122e03(6.1e02)	8.50900e11(9.5e10)	**2.73230e11(3.1e10)**	1.51e01(2.8e00)
	15	4.9316e11(1.5e11)	3.0132e03(1.8e03)	7.83531e11(8.3e10)	**2.49510e11(4.4e10)**	1.55e01(3.2e00)
	18	3.3045e11(1.8e11)	7.8998e03(7.5e03)	7.50905e11(9.0e10)	2.29057e11(6.0e10)	1.53e01(5.3e00)
	21	3.2013e11(3.5e10)	5.0868e03(4.4e02)	6.66544e11(8.9e10)	2.09148e11(8.1e10)	1.43e01(4.2e00)
	24	4.5023e11(1.8e11)	4.2644e03(1.4e03)	6.67006e11(8.7e10)	1.79267e11(9.8e10)	1.39e01(4.3e00)
	27	2.8713e11(2.1e11)	7.598e03(4.1e03)	6.50674e11(8.4e10)	1.35754e11(1.1e11)	1.33e01(3.6e00)
	30	2.7033e11(9.7e10)	6.4218e03(1.5e03)	6.34447e11(8.7e10)	1.10532e11(1.0e11)	1.32e01(4.0e00)

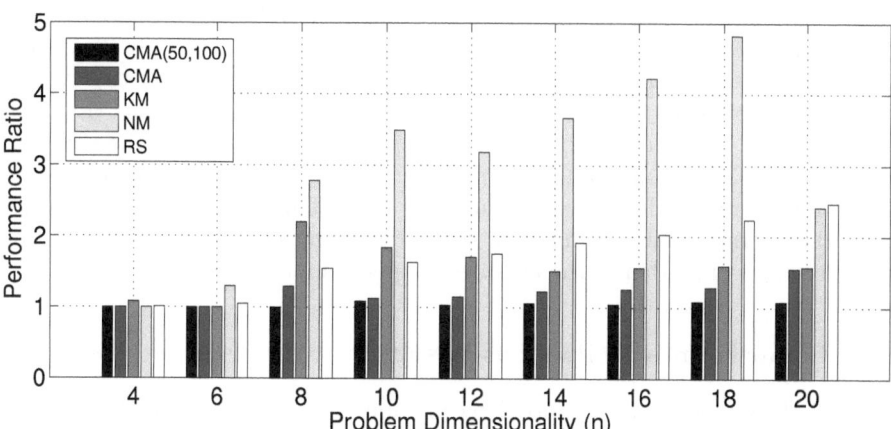

Fig. 2. Performance results (mean best fitness) for the Ruspini (R) dataset problems, as a ratio with the globally optimal value

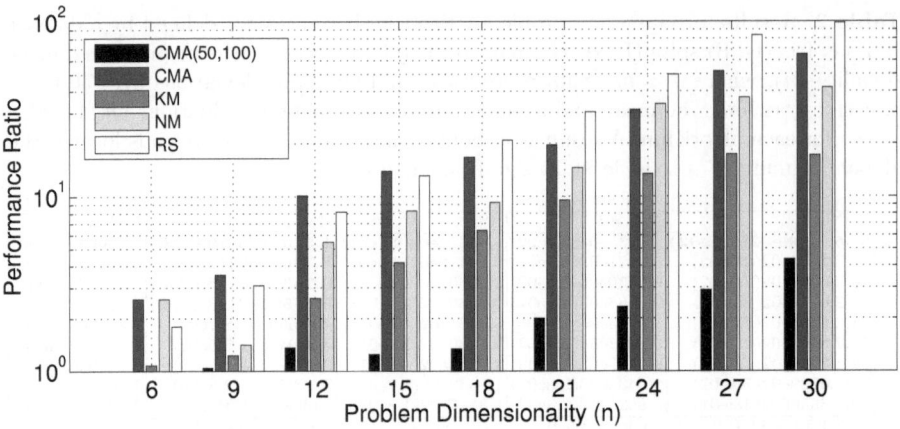

Fig. 3. Performance results (mean best fitness) for the German towns (G) dataset problems, as a ratio with the globally optimal value

The trends as problem dimensionality increases give some indication of the scaling behaviour of the performance of each algorithm. As expected, random search steadily increases (given a fixed budget of function evaluations). Some individual problem instances also appear to be particularly challenging for the algorithms tested. For example, the 8D Ruspini problem appears more difficult for KM and CMA, than both the 6D *and* the 10D Ruspini problem instances (Fig.2). Note also that the German towns problems lead to poor performance ratios for the algorithms compared to the other datasets (in Fig.3 the y-axis is on a log scale).

The comparative results in Figs 1-3 give a general indication of performance, but it is important to note that average fitness values are a relatively gross summary of the results and may hide important details. For example, the average performance of CMA on the 12D and 15D German towns problems (Fig.3) is slightly worse than random search! However, Table 1 shows that the standard deviation of CMA results on these problems is relatively large. Further investigation of these results revealed that many trials found solutions much better than the average, but a number of other trials converged to a poor solution considerably worse. Hence, the average is a poor summary of such results.

6.1   Comparison with Previous Results

Table 3 shows the results reported by previous papers for one of the problem instances tested, Iris with $k = 3$. A variety of algorithms have been tested on this problem. Comparing these results with those from above, the most striking thing is that *all* of these reported results are relatively poor. The average objective function values are far from optimal and are significantly outperformed by CMA(50,100). These results tend to be based on fewer function evaluations, but the papers do not seem to be targetting a "low budget" scenario, but rather evaluating the potential of the algorithms. Another significant anomaly is the differences between the k-means (KM, SBKM) results reported in these papers (average values between 97 and 101) and the result obtained in this paper for KM on this problem (8.79682e01(2.3e01)). There may be a difference in the initialization technique used which is not mentioned in all papers. In any case, there are clearly unresolved questions here, demonstrating the need for the development of standard specifications and experimental practice when evaluating black-box optimization algorithms.

Table 3. Previous results for the Iris (k=3, n=12) problem. A question mark (?) means that it is not clear from the paper what value was used in the experiments.

Reference	Algorithm	f_{ave}	Evals
[8]	GA	97.10077 (5 times!)	10^4
[8]	KM (best trial of 5)	97.204574	
[14]	KM	98.1872	?
[14]	Fuzzy c-means	96.9280	?
[14]	AKPSO	96.7551	?
[6]	KM	99.84	10^4
[6]	Shelokar ACO	97.78	10^4
[6]	ACOC	97.22	10^4
[5]	HBM	96.95316	11214
[5]	ACO	97.171546	10998
[5]	GA	125.197025	38128
[5]	TS	97.868008	20201
[5]	SA	97.134625	29103
[12]	SBKM	101.3672	3e04(?)
[12]	GAPS	97.3868	3e04(?)
[12]	VGAPS	96.2022	3e04(?)
[12]	PSO	96.0176	3e04(?)
[12]	ACO	99.9176	3e04(?)
[12]	SA	101.4574	3e04(?)
[12]	BNMR	95.0927	3e04(?)

7   Summary

This paper has examined sum of squares clustering problems as a source of real-world benchmark problems for the evaluation and comparison of black-box optimization algorithms. It was shown that clustering problems have many useful properties for benchmarking. To facilitate better comparisons of algorithms and experimental results, a specification was provided for clustering problems and a web repository has been created. Experimental results were presented on a set of 27 clustering problems and some comparisons made with existing results in the literature. It is intended that future work will build on and add to the problems specified here, with additional datasets. Also, future research should be able to make better use of published experimental results on clustering problems.

References

1. Blake, C., Keogh, E., Merz, C.J.: UCI repository of machine learning databases (1998), http://www.ics.uci.edu/~mlearn/MLRepository.html (retrieved)
2. Brimberg, J., Hansen, P., Mladenovic, N., Taillard, E.D.: Improvements and comparison of heuristics for solving the uncapacitated multisource Weber problem. Operations Research 48(3), 444–460 (2000)
3. Chang, D.-X., Zhang, X.-D., Zheng, C.-W.: A genetic algorithm with gene rearrangement for k-means clustering. Pattern Recognition 42(7), 1210–1222 (2009)
4. Du Merle, O., Hansen, P., Jaumard, B., Mladenovic, N.: An interior point algorithm for minimum sum-of-squares clustering. SIAM Journal on Scientific Computing 21(4), 1485–1505 (2000)
5. Fathian, M., Amiri, B., Maroosi, A.: Application of honey-bee mating optimization algorithm on clustering. Applied Mathematics and Computation 190(2), 1502–1513 (2007)
6. Kao, Y., Cheng, K.: An ACO-based clustering algorithm. In: Dorigo, M., Gambardella, L.M., Birattari, M., Martinoli, A., Poli, R., Stützle, T. (eds.) ANTS 2006. LNCS, vol. 4150, pp. 340–347. Springer, Heidelberg (2006)
7. Liu, R., Shen, Z., Jiao, L., Zhang, W.: Immunodomaince based clonal selection clustering algorithm. In: 2010 IEEE Congress on Evolutionary Computation (CEC), pp. 1–7 (2010)
8. Maulik, U., Bandyopadhyay, S.: Genetic algorithm-based clustering technique. Pattern Recognition 33(9), 1455–1465 (2000)
9. Salhi, S., Gamal, M.D.H.: A genetic algorithm based approach for the uncapacitated continuous location-allocation problem. Annals of Operations Research 123, 230–222 (2003)
10. Shelokar, P.S., Jayaraman, V.K., Kulkarni, B.D.: An ant colony approach for clustering. Analytica Chimica Acta 509(2), 187–195 (2004)
11. Steinley, D.: K-means clustering: A half-century synthesis. British Journal of Mathematical and Statistical Psychology 59, 1–34 (2006)
12. Taherdangkoo, M., Shirzadi, M.H., Yazdi, M., Bagheri, M.H.: A robust clustering method based on blind, naked mole-rats (bnmr) algorithm. Swarm and Evolutionary Computation 10, 1–11 (2013)
13. Xu, R., Wunsch II., D.: Survey of clustering algorithms. IEEE Transactions on Neural Networks 16(3), 645–678 (2005)
14. Ye, F., Chen, C.-Y.: Alternative kpso-clustering algorithm. Tamkang Journal of Science and Engineering 8(2), 165 (2005)

An Analysis of Differential Evolution Parameters on Rotated Bi-objective Optimization Functions

Martin Drozdik[1], Kiyoshi Tanaka[1], Hernan Aguirre[1], Sebastien Verel[2], Arnaud Liefooghe[3], and Bilel Derbel[3]

[1] Interdisciplinary Graduate School of Science and Technology, Shinshu University, Nagano, Japan
martin@iplab.shinshu-u.ac.jp
[2] LISIC, Université du Littoral Côte d'Opale, France
[3] Inria Lille-Nord Europe, Villeneuve d'Ascq, France

Abstract. Differential evolution (DE) is a very powerful and simple algorithm for single- and multi-objective continuous optimization problems. However, its success is highly affected by the right choice of parameters. Authors of successful multi-objective DE algorithms usually use parameters which *do not* render the algorithm *invariant* with respect to *rotation* of the coordinate axes in the decision space. In this work we try to see if such a choice can bring consistently good performance under various *rotations* of the problem. We do this by testing a DE algorithm with many combinations of parameters on a testbed of bi-objective problems with different modality and separability characteristics. Then, we explore how the performance changes when we rotate the axes in a controlled manner. We find out that our results are consistent with the single-objective theory but *only for unimodal problems*. On multi-modal problems, unexpectedly, parameter settings which do *not* render the algorithm rotationally invariant have a consistently good performance for all studied rotations.

Keywords: differential evolution, rotational invariance, multi-objective optimization, parameter analysis.

1 Introduction

Differential evolution [6] started as a simple single-objective continuous optimization heuristic. The need for a versatile multi-objective optimizer has motivated researchers to generalize the basic algorithm for multi-objective problems. Now we have a great number of multi-objective DE variants. Many of them use the same mechanism to generate new individuals. In a problem with n variables a new individual is created using a crossover variation operator which randomly selects $k; k \leq n$ variables which are perturbed. The magnitude of the mutation is generated by scaling a *difference* of randomly chosen individuals.

Many research papers on DE such as [1] or [9] provide little insight into how the authors chose the parameters for their benchmarking. We find this striking

G. Dick et al. (Eds.): SEAL 2014, LNCS 8886, pp. 143–154, 2014.

since many authors choose their parameters such that the crossover operator perturbs only a small number of variables in an existing individual. In other words, the search for the Pareto optimal set proceeds *along the coordinate axes*. Since these algorithms perform very well [1][9], we have a suspicion that this may be due to some characteristic of the problem, such as separability, that makes it easy to optimize along the axes. This would mean that if the axes are transformed, the algorithm should lose some performance.

Very strict warning against the practice of perturbing a small number of variables at a time has been raised as soon as 1996 by Salomon [8]. Salomon empirically demonstrated that the stellar performance of many popular single-objective genetic algorithms owes to the fact that most of the benchmark functions were *separable* and that the low mutation rate caused them to be optimized *one component at a time*. Once Salomon stripped the separability by *rotating* the principal axes of the benchmark functions, many algorithms were significantly slowed down, while some failed to converge completely. Salomon's theoretical results state that, in some cases, the probability of finding the global optimum can drop below that of random search. We are concerned that the same is true for the multi-objective realm since many authors perform their experiments with separable test functions.

In DE the number of variables that are perturbed is controlled by a parameter. If all variables are perturbed, the algorithm has the same performance *regardless of rotation*. Let us have a parameter setting, that perturbs only a *small proportion* of the variables, which outperforms a setting that perturbs *all* variables. In this work we attempt to answer this question: Is this *exceptionally good* performance on a problem with a particular alignment of the coordinate axes balanced by *exceptionally bad* performance on a different alignment?

We do this *empirically* by observing the performance of a simple multi-objective algorithm DEMO (Differential evolution for multi-objective optimization) [9] on a bi-objective subset of the WFG (Walking Fish Group) test suite [3]. We run all our experiments with a fixed population size and a fixed number of variables, while varying the *parameters*. Then we gradually *rotate* the problems in a controlled manner and observe the new behavior.

The answer to our question is, unexpectedly, *negative*. We find a statistically significant difference between the performance on the rotated problems and the original ones. Closer inspection reveals that a systematic performance *loss* happens when we rotate the *separable* problems, but the performance is still *significantly better* than for a rotationally invariant algorithm. We find that this happens for *multi-modal* problems, while *single-modal* problems exhibit the behavior we would expect from the work of Salomon.

In the following section we provide background information on DE and on the previous related work on DE parameters. In Section 3 we introduce the experimental design, where we explain which problems are used and why were they chosen. In addition we introduce a new performance metric called the *relative hypervolume*, and explain the controlled manner in which the rotations are generated. In Section 4 we present our data along with a discussion. Finally, in Section 5, we present the conclusion.

Algorithm 1. Modified DEMO [9] algorithm

---

1 initialize $P = \{X_1, ..., X_N\}$ uniformly randomly in the decision space
2 **for** generation := 1 *to* G_{max} **do** *Evolutionary loop*
3   **for** target := 1 *to* N **do** *Generational loop*
4   randomly generate mutually distinct $r_1, r_2, r_3 \neq$ target
5          $X_{\text{mutant}} := X_{r_1} + \text{F}(X_{r_2} - X_{r_3})$
6   randomly generate inv $\in \{1, \ldots, n\}$
7   **for** $i := 1$ *to* n **do**
8   **if** rand$(0.0; 1.0) <$ Cr *or* $i =$ inv **then**
9                  $X_{\text{trial},i} := X_{\text{mutant},i}$
10   **else**
11                 $X_{\text{trial},i} := X_{\text{target},i}$
12   **end**
13   **end**
14   project X_{trial} to decision space
15   **if** X_{target} *dominates* X_{trial} **then**
16   discard X_{trial}
17   **else if** X_{trial} *dominates* X_{target} **then**
18   replace X_{target} with X_{trial}
19   **else if** X_{target} *and* X_{trial} *are mutually non-dominated* **then**
20   add X_{trial} to the end of the population
21   **end**
22   **end**
23   Trim the P to size N using non-dominated sorting[9] and MNN diversity[4]
24 **end**

---

2   Background

2.1   Differential Evolution

In this section, we describe the variant of DE which we use in this work. It is a slightly modified version of the DEMO algorithm [9] described in Algorithm 1. The modified parts are highlighted with yellow color in lines 14 and 23. Let us look at the algorithm in detail. First, the population of the algorithm is randomly initialized (line 1). Then, the algorithm runs for a fixed number of generations (evolutionary loop). In each generation, DE iterates through the entire population generating a *trial* individual, which is compared to an existing *target* individual. The trial is generated utilizing the traditional method in lines 4 to 13. Here we introduce the parameters of DE.

To generate a new individual, three distinct individuals are selected from the population. By forming a *difference* between two of them, scaling it by a fixed parameter F, and adding to a third individual we obtain a so called mutant individual (line 5). The trial individual is created by crossover between the mutant and the target. First, a randomly chosen variable from the mutant is inherited (line 6). Next, each other variable is inherited from the mutant with a fixed crossover probability of Cr. Otherwise it comes from the target.

After the individual is generated we project it to the decision space. This is a modification of the original algorithm which did not explicitly deal with domain issues. The purpose is to keep the algorithm as simple as possible while being able to optimize problems with simple constraints. Next the trial is compared to the target. If one of them is dominated by the other, we discard the dominated one. If they are *mutually non-dominated*, we keep them both. At the end of the generation loop, the population is trimmed to size N using non-dominated sorting and the M nearest neighbor diversity estimation procedure [4]. We chose this procedure because it achieves a better distribution along the Pareto front than the original crowding distance computation.

Note that $Cr = 1$ is the *only* value of Cr for which the DE algorithm is rotationally invariant with probability 1. Rotational invariance does not by itself imply good performance. Its merit is that it allows us to *generalize* a single observation to an entire invariance class [2].

2.2 Crossover Probability and Separability

In this section we summarize what we know about the relationship between the separability of the test functions and the good choice of Cr parameter.

There are many different types of separability. One of the simplest is *additive* separability. A function $f : D \subseteq \mathbb{R}^n \to \mathbb{R}$ is called *additively* separable if:

$$\exists f_1, \ldots, f_n \text{ such that } f(x_1, \ldots, x_n) = \sum_{i=1}^{n} f_i(x_i)$$

The most important consequence of additive separability is that the n-dimensional problem can be optimized sequentially *one variable at a time*. Therefore separable problems are not subject to the curse of dimensionality [8].

Salomon [8] illustrates the problems of algorithms which vary the individuals *one variable at a time* on a quadratic function of two variables in Figure 1. The ellipses in the left part are contours of a *separable* quadratic function. We can see two individuals on one of the contours. The blue individual represents an individual in a randomized algorithm. If we mutate one variable of this individual, the probability to get an improvement in the objective function is relatively high, since the improvement intervals d_1, d_2 are long. If we rotate the coordinate axes, thus rendering the function non-separable, the improvement intervals shrink.

One more illustration of problems which arise is using a sequential deterministic algorithm which finds the optimum with respect to *one variable at a time*. The red individual illustrates the path of one such an algorithm. When the function is aligned with the axes, this algorithm achieves optimum in just two iterations, while in the rotated case the algorithm not only progresses slower, but never actually reaches the optimum.

Huband et al. from the Walking Fish Group (WFG) define separability from the optimizational standpoint[3]. A variable x_i is separable if the set of global optima of a problem:

$$\underset{x_i}{\arg\min} f(x_1, \ldots, x_n)$$

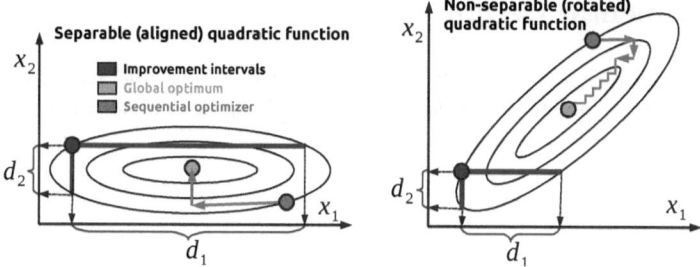

Fig. 1. Illustration of variable-wise optimization on a rotated quadratic function

is the same for any choice of the other variables $x_1, \ldots, x_{i-1}, x_{i+1}, \ldots, x_n$. For example, an additively separable function is WFG-separable, hence WFG-separability is a *generalization* of additive separability. The authors define a separable multi-objective problem as one where *each* objective is separable. The majority of the frequently used DTLZ and ZDT problems are WFG-separable [3], while their objective functions are *not* additively separable.

The multi-objective model is fundamentally different from the single-objective model because all objectives are being optimized *simultaneously*. The global optima of each optimized function constitute only a relatively small subset of the Pareto optimal set. Therefore, it is appropriate to ask if the problems of sequential algorithms which are illustrated in Figure 1 persist in multi-objective optimization. Also, while additively separable *unimodal* functions are inherently similar to the quadratic function in Figure 1, it is not clear if the intuition holds for *multi-modal* functions or for functions which are WFG-separable but not additively separable.

2.3 Variance as a Common Currency

Probably the most significant work on the theoretical properties of DE has been written by Zaharie [10]. Let us collect all the trial vectors that are generated in the course of one generational loop of Algorithm 1 into a set P_{trial}. Then the relationship between the *variance in decision space* of P and P_{trial} is given by the simple equation $\mathbb{E}[Var(P_{\text{trial}})] = c\mathbb{E}[Var(P)]$ where:

$$c = 2\mathrm{F}^2\mathrm{Cr} + \frac{\mathrm{Cr}^2 - 2\mathrm{Cr}}{N} + 1 \tag{1}$$

Zaharie omits the fact that in most DE variants the individuals which generate the trial individual are chosen *distinct* from the target individual (Algorithm 1 line 4). However her results hold unchanged also after adding this assumption.

The work of Zaharie is important since it transforms the two parameters into a single number c (common currency) which has a very intuitive interpretation. If $c < 1$ we see that the algorithm tends to contract the population while if $c > 1$ it expands the population. Based on empirical data Kukkonen concluded in [5] [7] that a good choice of parameters is one that satisfies $c \in [1.0; 1.5]$ with the upper bound not very strict.

3   Experimental Design

In this section, we describe which test problems we chose and why. We explain what we mean by *rotating the problem* and we propose a new performance metric which we use.

3.1   WFG Problems

In order to explore the relationship between the control parameters of DE and the characteristics of the problem, we chose 4 problems from the WFG test suite[3]. These problems have been chosen since they have the same Pareto front and contain all possible combinations of the *WFG-separability* and *modality* characteristics. They are summarized in Table 1. We chose the number of variables to be 10 of which one is a positional variable.

Table 1. Characteristics of the selected WFG problems

	WFG4	WFG7	WFG6	WFG9
separable	yes	yes	no	no
unimodal	no	yes	yes	no

3.2   Rotations in $\mathbb{R}^n$

As humans we have a very good intuitive understanding of rotation in 2 or 3 dimensional space. However in higher dimensions things are not as intuitive as they might seem. An elementary rotation by the angle ϕ is characterized by the matrix:

$$R^e = \begin{pmatrix} cos(\phi) & -sin(\phi) \\ sin(\phi) & cos(\phi) \end{pmatrix}$$

We can generalize this rotation to n-dimensional space by taking an n-dimensional identity matrix I and replacing $I_{i,i}, I_{i,j}, I_{j,i}, I_{j,j}$ by $R^e_{1,1}, R^e_{1,2}, R^e_{2,1}, R^e_{2,2}$ respectively. We can see that the rotation is not executed *around an axis* as we might intuitively feel, but *around an $n-2$ dimensional subspace* which is coincidentally a 1-dimensional axis in the intuitive 3-dimensional case. For our experiments, we generate the rotation matrix R by applying a rotation to *each $n-2$ dimensional subspace* in sequence, one rotation after the other.

We rotate the *entire decision space* (DS). This way the entire Pareto optimal set is always attainable since the entire decision space rotates along. In the case of WFG problems this means rotating a n-dimensional hyper-box. For example, in order to initialize the population in Algorithm 1 in the rotated DS (line 1), we first initialize the population in the original DS and then multiply by R^{-1}. Similar process is used to project the individual to the rotated DS on line 14. To evaluate the objective value of an individual we first multiply the decision vector by R and evaluate the original objective functions.

3.3 Relative Hypervolume

In our experiments, we use only one performance metric, the hypervolume (HV) [11], since it includes information on both convergence and spread of the individuals. With WFG problems, it is not easy to choose the reference point for the HV. Even if we choose the point as tight as possible, there are some individuals after the initialization of DE which dominate the reference point. Therefore the HV at the start is not zero and it is hard to say if a certain attained HV is good or bad. Moreover, it is hard to make quantitative comparisons based on HV. If some algorithm achieves HV of 100 and another one achieves a HV of 99.98, it may seem that the difference is not very big, but it all depends on the HV *at initialization*. If the algorithms started with HV = 0, the interpretation of the results would be quite different from one where HV = 99.99 at the start.

We attempt to mitigate this problem by subtracting the HV at initialization (HV_{init}) and normalizing the result using the *maximal attainable hypervolume* (HV_{max}). We define the relative hypervolume (RHV) in the following equation:

$$RHV := \frac{HV - HV_{init}}{HV_{max} - HV_{init}} \qquad (2)$$

We compute HV_{max} deterministically by integrating the space between the true Pareto front (PF) and the reference point. From (2), we have RHV $\in [1; -\infty)$. RHV = 1 implies convergence, RHV at initialization is 0 and RHV < 0 indicates an algorithm which is receding from the Pareto front.

We use RHV since its normalized nature is more intuitive and it is more robust with respect to the selection of the reference point. It may be more meaningful to compare two algorithm runs in terms of RHV. If we have two algorithm runs starting from the same randomly initialized population then the ratio of their RHVs is independent of the choice of the reference point. [1] On the other hand, two independent runs which produce the same final population may yield different relative hypervolume.

4   Results and Discussion

In our experiments we varied the parameters F $\in [0.05; 1.5]$, Cr $\in [0; 1]$ equidistantly with a resolution of 0.05. For each combination we performed 10 runs of Algorithm 1. To simplify the setup, the population size was kept constant at 100 individuals and the length of each run was 250 generations. We explored the rotations from 0 to 90 degrees with a resolution of 5 degrees. In the following we discuss our results on a subset of the experimental data. To simplify the analysis, in each section we keep either F, Cr or the rotation angle fixed.

4.1 Fixed Rotation Angle

Figure 2 shows the average RHV on *non-rotated* problems. For illustration, we plot the combinations of F and Cr which result in $c = 1.0, 1.5$ and 3.0 according

---

[1] Given that the reference point is dominated by all individuals in the population.

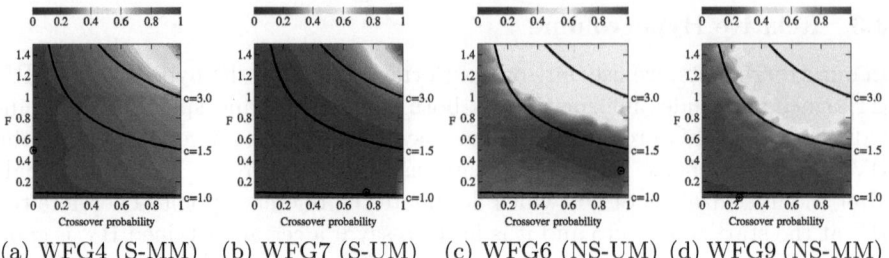

(a) WFG4 (S-MM)   (b) WFG7 (S-UM)   (c) WFG6 (NS-UM)   (d) WFG9 (NS-MM)

Fig. 2. Average RHV without rotation

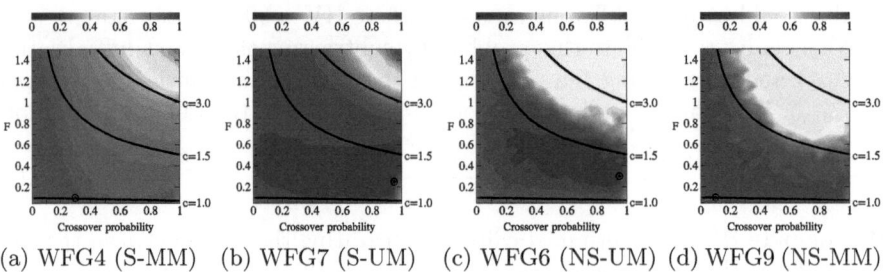

(a) WFG4 (S-MM)   (b) WFG7 (S-UM)   (c) WFG6 (NS-UM)   (d) WFG9 (NS-MM)

Fig. 3. Average RHV with rotation angle of 5 degrees

to (1). The circle marks the combination of parameters with the best RHV. The data in all our figures from now on is presented from left to right in the same order as in Table 1. The separable problems (S) are on the left, the non-separable (NS) on the right, while the unimodal (UM) are on the inside and multi-modal (MM) ones are near the page margins. For each problem, an L-shaped favorable region containing RHV of 0.8 and higher, roughly corresponds to $c \in [1; 1.5]$. Low value of Cr is more robust, since it allows for a wider interval of F values. Unexpectedly, this holds also for non-separable problems WFG6 and WFG9.

The effect of introducing a rotation by 5 degrees is shown in Figure 3. The two figures seem identical, but the *ratio* of these averages in Figure 4 reveals a difference. A value of less than 1 indicates that the rotation caused the performance to decrease. We highlighted the contour at level 1 and marked the maximal and minimal value by circles. In order to make the results most readable we chose a color scale of $[0.5; 1.2]$ for separable problems and $[0.6; 1.7]$ for non-separable problems. The separable problems on the left half exhibit a performance loss consistent with Salomon's single-objective results. Performance dropped for almost all Cr smaller than 1. Non-separable WFG6 and WFG9 do not show such a systematic decrease. In some areas we even see an *increase* of performance.

It seems that there is relatively little difference between the rotated and non-rotated data. These result may seem not as significant as Salomon's. However, there is an important methodological difference. When he mentions that the performance on the rotated benchmark is *six orders of magnitude* worse than

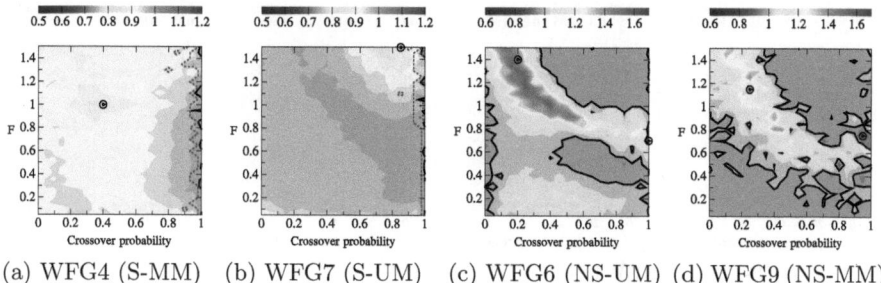

(a) WFG4 (S-MM) (b) WFG7 (S-UM) (c) WFG6 (NS-UM) (d) WFG9 (NS-MM)

Fig. 4. $\dfrac{\text{Average RHV with a rotation of 5 degrees}}{\text{Average RHV without a rotation}}$

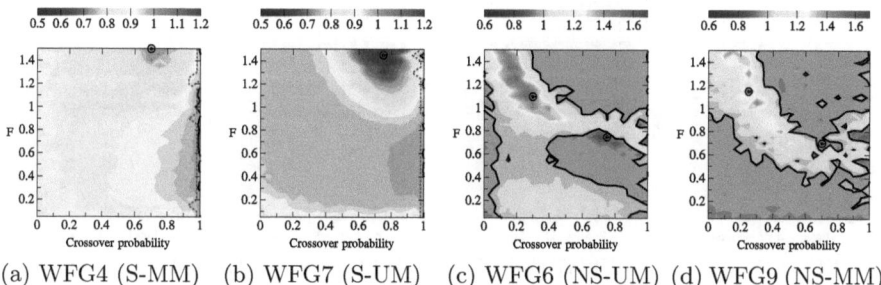

(a) WFG4 (S-MM) (b) WFG7 (S-UM) (c) WFG6 (NS-UM) (d) WFG9 (NS-MM)

Fig. 5. $\dfrac{\text{Average RHV with a rotation of 45 degrees}}{\text{Average RHV without a rotation}}$

the performance on the non-rotated benchmark ([8, p.273]), he means that *the minimal attained value* $2.65 \cdot 10^5$ is six orders of magnitude worse in absolute numbers. But the *value at initialization* was three orders of magnitude greater yet. This means that both algorithms started somewhere near $2.65 \cdot 10^8$ and the non-rotated one progressed to $2.65 \cdot 10^{-1}$ while the rotated one progressed to $2.65 \cdot 10^5$. In terms of relative hypervolume, this would be a very small difference [2]. In order to provide a scale-independent comparison, we compared all data using a two-tailed Wilcoxon signed rank test at a significance level of 0.05. For separable problems in Figures 4 and 5 we separate the parameter space with a dashed line into two areas. The area on the right is such that the rotated and non-rotated data is not significantly different, while on the left there is a *significant decrease in performance*. The data for non-separable problems contains areas of both significant decrease and significant decrease, as well as areas with no significant difference so in this case the separation cannot be plotted so compendiously.

The effects are more visible with 45 degree rotation in Figure 5. Again, there is a *systematic* decrease in performance for the separable problems for $Cr < 1$. However, this decrease does not imply that $Cr = 1$ is a good choice. Looking

---

[2] Assuming that the minimum of the given function is 0, the difference would be on the order of 10^{-3}.

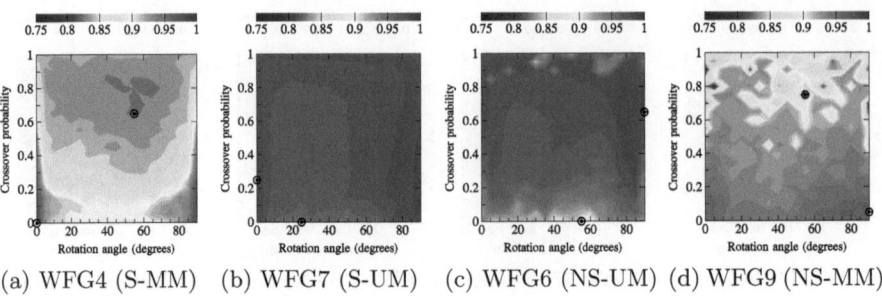

(a) WFG4 (S-MM) (b) WFG7 (S-UM) (c) WFG6 (NS-UM) (d) WFG9 (NS-MM)

Fig. 6. Average RHV for F = 0.5

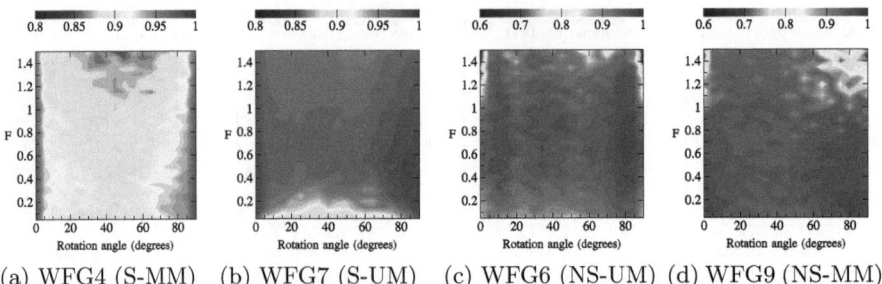

(a) WFG4 (S-MM) (b) WFG7 (S-UM) (c) WFG6 (NS-UM) (d) WFG9 (NS-MM)

Fig. 7. Average RHV for Cr = 0.1

at Figures 2 and 3, we see that $Cr = 1$ is a consistently bad choice for the *multi-modal* problems WFG4 and WFG9.

4.2 Fixed F

In Figures 2 and 3 we see that F = 0.5 is compatible with many different values of Cr and achieves consistently good performance. The average RHV for F = 0.5 is shown in Figure 6. For *multi-modal* problems WFG4 and WFG9, *very low* values of Cr are *consistently* good for all studied rotations, while for *uni-modal* problems WFG6 and WFG7 *big* values of Cr yield a *consistently* good performance. On the other hand, poor performance is achieved with *big* values of Cr for multi-modal problems and *small* values for uni-modal problems. The data for WFG4 and WFG9 suggests that the exceptionally good performance of a small Cr setting does *not* have to be balanced by an exceptionally bad performance after the problem is rotated. Based on the observation from Figure 6 we see that for each problem either Cr = 0.1 or Cr = 0.9 perform well through the observed spectrum of rotations.

4.3 Fixed Cr

In Figures 7 and 8 we see data with a fixed value of Cr = 0.1 and Cr = 0.9 respectively. For Cr = 0.1 the regions with the best performance are for rotations

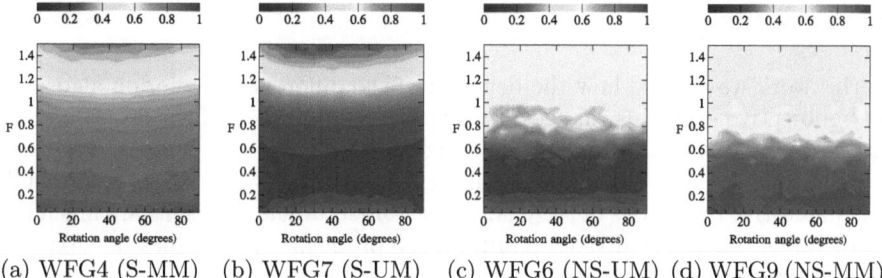

(a) WFG4 (S-MM) (b) WFG7 (S-UM) (c) WFG6 (NS-UM) (d) WFG9 (NS-MM)

Fig. 8. Average RHV for Cr = 0.9

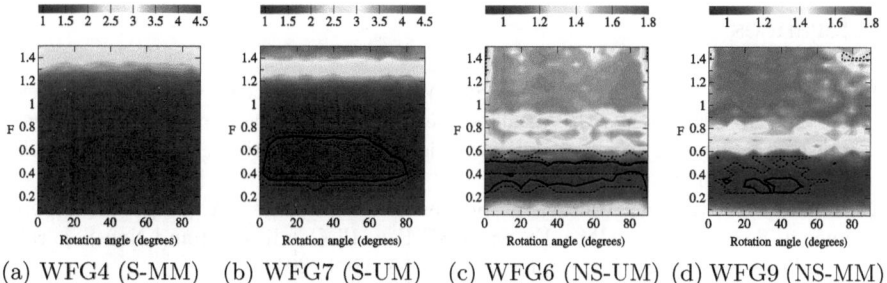

(a) WFG4 (S-MM) (b) WFG7 (S-UM) (c) WFG6 (NS-UM) (d) WFG9 (NS-MM)

Fig. 9. $\dfrac{\text{Average RHV with Cr} = 0.1}{\text{Average RHV with Cr} = 1}$

which are either close to 0 or 90 degrees. This is true also for non-separable problems, but it is more visible for separable problems. The data for Cr = 0.9 seems different. The choice of Cr close to 1 means that the algorithm is nearly rotationally invariant. The gained robustness with respect to coordinate rotation is balanced by lost robustness in the choice of F. Almost in all cases the interval with favorable values of F became shorter.

We see that for values of Cr < 1 there is a performance loss when the coordinate axes are rotated, but does the performance drop bellow that of a rotationally invariant choice of Cr = 1? The data supporting a *negative* answer is presented in Figure 9. Here we divided the average RHV with Cr = 0.1 by the average RHV attained with a rotationally invariant Cr = 1. The interpretation of the dashed and full contour lines is the same as for Figures 4 and 5. For WFG4, the setting of Cr = 0.1 statistically significantly outperformed Cr = 1 for *all rotations* and *all* values of F. This means a definitive *negative* answer to our main question. The results are similar for the second multi-modal problem WFG9. Here we see a small region in which the data for Cr = 0.1 and Cr = 1 are not significantly different and Cr = 1 is significantly better in a few isolated cases. The unimodal problems on the other hand show that Cr = 1 is significantly better for most rotations and for the best performing values of F.

5   Conclusion

In this work we showed how the behavior of the differential evolution algorithm on bi-objective problems changes when the coordinate axes of the decision space are rotated. Our findings show that the change is significant even for small rotations. There is a consistent *drop* in performance on *separable* problems while the qualitative properties of the change for *non-separable* problems are much less predictable. Unexpectedly, for *multi-modal* problems, *low* values of crossover probability perform better through the observed spectrum of rotations. As a future work we propose to see if this holds for problems other than the ones we studied and if this is the case, to find the cause of this behavior.

References

1. Denysiuk, R., Costa, L., Esprito Santo, I.: Many-objective Optimization Using Differential Evolution with Variable-wise Mutation Restriction. In: Proceedings of the Conference on Genetic and Evolutionary Computation, pp. 591–598. ACM, New York (2013)
2. Hansen, N., Ros, R., Mauny, N., Schoenauer, M., Auger, A.: Impacts of Invariance in Search: When CMA-ES and PSO Face Ill-Conditioned and Non-Separable Problems. Applied Soft Computing 11, 5755–5769 (2011)
3. Huband, S., Hingston, P., Barone, L., While, L.: A Review of Multiobjective Test Problems and a Scalable Test Problem Toolkit. IEEE Transactions on Evolutionary Computation 10(5), 477–506 (2006)
4. Kukkonen, S., Deb, K.: A Fast and Effective Method for Pruning of Non-dominated Solutions in Many-Objective Problems. In: Runarsson, T.P., Beyer, H.-G., Burke, E.K., Merelo-Guervós, J.J., Whitley, L.D., Yao, X. (eds.) PPSN IX. LNCS, vol. 4193, pp. 553–562. Springer, Heidelberg (2006)
5. Kukkonen, S., Lampinen, J.: An Empirical Study of Control Parameters for The Third Version of Generalized Differential Evolution (GDE3). In: IEEE Congress on Evolutionary Computation, pp. 2002–2009 (2006)
6. Price, K., Storn, R.M., Lampinen, J.A.: Differential Evolution: A Practical Approach to Global Optimization. Springer, New York (2005)
7. Saku, K.: Generalized Differential Evolution for Global Multi-Objective Optimization with Constraints. PhD thesis, Lappeenranta Univ. of Technology (2012)
8. Salomon, R.: Re-evaluating Genetic Algorithm Performance Under Coordinate Rotation of Benchmark Functions. Biosystems 39(3), 263–278 (1996)
9. Robič, T., Filipič, B.: DEMO: Differential Evolution for Multiobjective Optimization. In: Coello Coello, C.A., Hernández Aguirre, A., Zitzler, E. (eds.) EMO 2005. LNCS, vol. 3410, pp. 520–533. Springer, Heidelberg (2005)
10. Zaharie, D.: Critical Values for the Control Parameters of Differential Evolution Algorithm. In: Proceedings of MENDEL 2002 (2002)
11. Zitzler, E.: Evolutionary Algorithms for Multiobjective Optimization: Methods and Applications. PhD thesis, Comput. Eng. Netw. Lab. Swiss Federal Instit. Technol (ETH), Zurich, Switzerland (1999)

Fuzzy Clustering with Fitness Predator Optimizer for Multivariate Data Problems

Shiqin Yang and Yuji Sato

Graduate School of Computer and Information Sciences
Hosei University, Tokyo, Japan

Abstract. Fuzzy c-means (FCM) is the most common fuzzy clustering model and uses an objective function to measure the desirability of partitions. However, if the data sets contain several noise points, or if the data sets are very high dimensional, the iteration process of optimization the FCM model often falls into local optima solution. To avoid this problem, this paper proposes a new hybrid fuzzy clustering algorithm that incorporates the Fitness Predator Optimizer (FPO) into the FCM model. FPO is a new bionic-inspired algorithm to avoid premature convergence for the multimodal optimization problem. The excellent probability of finding the global optimum of FPO enhances the quality of fuzzy clustering. Five benchmark data sets from the UCI Machine Learning Repository are used to compare the performances of proposed FPO-FCM with FCM and a hybrid swarm algorithm based on Quantum-behaved PSO. Experimental results show that the proposed approach could demonstrate the desirable performance and avoid the minimum local value of objective function for multivariate data type clustering problems.

Keywords: Fitness Predator Optimizer, Fuzzy C-means Model, multimodal optimization problem, hybrid fuzzy clustering algorithm.

1 Introduction

The most common popular data mining techniques discussed are clustering and classification. The clustering aims at identifying and extracting significant groups in underlying data, which is an unsupervised learning method. In the field of clustering, Fuzzy c-means (FCM) is one of the most popular algorithms. Although FCM is extensively used in literature, it suffers from several drawbacks. The objective function of the FCM is the multimodal function which means that it may contain many local minima. Consequently, while minimizing the objective function, there is possibility of getting stuck at local minima or saddle points. In addition, the performance of the FCM depends on the initial selection of the cluster center.

To increase the probability of finding the global optimum, various alternative methods for the optimization of clustering models were suggested in the literature. Some researchers adopt the stochastic methods such as evolutionary or swarm-based methods to increase the global convergence ability of fuzzy clustering. In [7], authors used a Fuzzy c-means algorithm based on Picard iteration and PSO (PPSO-FCM) to improve the performance of FCM. In [8], a hybrid data clustering algorithm based on PSO and

G. Dick et al. (Eds.): SEAL 2014, LNCS 8886, pp. 155–166, 2014.

KHM is proposed, which makes full use of the merits of PSO and KHM. However, to the classical PSO, it couldn't guarantee to convergence to the global best solution. The QPSO algorithm proposed by [4] outperforms traditional PSO in search ability as well as having less parameter to control. Then, [6] proposed a new hybrid fuzzy clustering algorithm that incorporates the Quantum-behaved PSO into the FCM model.

Basically, the major problem with most of swarm intelligent algorithms in multi-modal function is premature convergence. To avoid premature convergence by maintaining diversity in the population, many kinds of optimization algorithms are proposed. However, few of the swarm intelligence techniques focus on individual competition and independent self awareness. The individual competition is more likely to reduce the rapid social collaboration process and increase the ability of being out of the local optimum. This motivated our attempt to present a new swarm intelligence algorithm, called Fitness Predator Optimizer (FPO) [15]. In this paper an application of the proposed FPO is presented in the field of fuzzy clustering. In FCM model, the probability of finding the global optimum can be increased by FPO due to its outstanding global searching ability. Consequently a new hybrid fuzzy clustering algorithm FPO-FCM is proposed in this paper.

The outline of this paper is organized as follows. In section 2, the brief introduction of the state of the art and the characteristics of the fuzzy clustering. In section 3, basic conceptions and pseudocode of FPO is introduced first. Then a new hybrid fuzzy clustering algorithm based on the FPO (FPO-FCM) is proposed. A hybrid FCM algorithm based on the QPSO is also introduced in this section. In Section 4, the FPO-FCM is verified by five widely used data sets in the pattern recognition literature. Finally, some concluding remarks and suggestions for future research are provided in Section 5.

2   Related Work

There are three main types of fuzzy clustering - fuzzy clustering based on fuzzy relation, fuzzy clustering based on objective function and fuzzy generalized k-nearest neighbour rule. The fuzzy clustering based on objective function is the most popular one, because it is quite facile, and allows the most precise formulation of the clustering criteria. The most popular version is the Bezdek's FCM model [1],[10] with the generalized objective function.

$$J_m(U,V) = \sum_{i=1}^{c} \sum_{k=1}^{n} (u_{ik})^m \mid x_k - v_i \mid^2 \tag{1}$$

Where m $(m > 1)$ is a scalar termed the 'weighting exponent' and controls the fuzziness of the resulting clusters. The FCM model partitions a data set $X = \{x_1,...,x_n, n \in N\}$ into c $(1 < c < n)$ number of fuzzy clusters with $V = \{v_1, v_2,...,v_c\}$ cluster centroids by a partition matrix U. The matrix U shows the fuzzy relation from set of data objects, which is expressed as follows:

$$U = \begin{bmatrix} u_{11} & \cdots & u_{1n} \\ \cdots & u_{ij} & \cdots \\ u_{c1} & \cdots & u_{cn} \end{bmatrix} \tag{2}$$

In which u_{ij} is the membership function of the jth data object with the ith cluster within the constraints of $u_{ij} \in [0,1]$ and $\sum_{i=1}^{c} u_{ij} = 1$. Clustering partitions a data set into subsets by finding the maximum membership grade u_{ij} of data object x_i belonging to the cluster j. This model aims to minimize the following objective function with respect to each fuzzy membership grade u_{ij} and each cluster centroid v_i. In most of the cases, the distance between x_k and v_i is assigned with the Euclidean norm and the fuzzifier $m = 2$. A popular method to optimize the FCM model is Alternating Optimization (AO) through the necessary conditions extrema of $J_m(U,V)$:

$$u_{ik} = \frac{1}{\sum_{j=1}^{c} (\frac{|v_i - x_k|}{|v_j - x_k|})^{2/(m-1)}} \tag{3}$$

$$v_i = \frac{\sum_{k=1}^{n} u_{ik}^m x_k}{\sum_{k=1}^{n} u_{ik}^m} \tag{4}$$

The subsequent computation of the partition matrix u_{ij} can be merged to $V(u(V,X),X)$ in FCM algorithm. The reformulated version of $J_m(U,V)$ [2] is obtained by inserting (3) into (1).

$$J_m(V,X) = \sum_{i=1}^{c} \sum_{k=1}^{n} \frac{|v_i - x_k|^2}{\sum_{j=1}^{c} (\frac{|v_i - x_k|}{|v_j - x_k|})^{2m/(m-1)}} \tag{5}$$

In this paper we consider a widely used FCM model with a number of cluster centers prototype. Then FCM-AO-V is described in Algorithm 1.

Algorithm 1. FCM-AO-V

Initialize data: $X = \{x_1, x_2, ..., x_n\}$
Initialize the clustering centroids $V = \{v_1, v_2, ..., v_c\}$
Initialize the maximum iterations $t_{max} \in N$
while $t \leq t_{max}$ **do**
　　Generate the partition matrix u_{ij} by (3)
　　Generate the new clustering center v_i by (4)
end while
Output U, V

Figure 1 shows $J_m(V,X)$ in 3-dimensional graph with two clustering centroids $v_1 \in [-1,1]$ and $v_2 \in [-1,1]$. The reformulated function can be visualized for the trivial data set $X = \{x_1, ..., x_{100}\}$, $x_i \in [-5,5]$ with the parameters $m = 2, c = 2$. It also shows that the objective function $J_m(V;X)$ is a non-linear multimodal function with a number of local

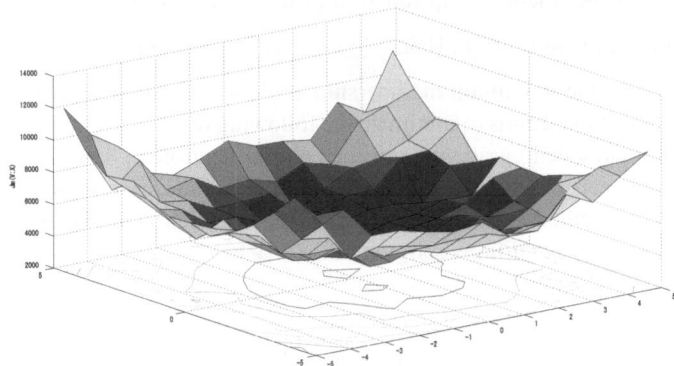

Fig. 1. Reformulated objective function $J_m(V;X)$

minima. Obviously, the alternating optimization or gradient based methods might get stuck in these local extrema. To avoid local minima, heuristic search algorithms can be applied to minimize the objective function, such as evolutionary or swarm intelligence methods.

3 Proposal of New Hybrid Fuzzy Clustering Algorithm for Multivariate Data Type Clustering Problems

In order to overcome the shortcomings of Fuzzy C-means algorithm, various computing techniques such as artificial neural networks [12] [13], hybrid fuzzy time series approach [11], genetic algorithms and PSO-based fuzzy clustering algorithms [9] have been used in FCM recently. We also proposed a new hybrid fuzzy clustering algorithm by incorporation the QPSO into the FCM model [6] in our previous work. Similarly, a new optimization technique named as IQPSO-FCM in [14] is a combination of FCM and improved QPSO to drive the clustering efficiency in standard medical and non-medical data sets. However, the diversity declines rapidly in the later iteration period, leaving the QPSO with great difficulties of escaping the local optima. In order to improve the diversity of the population, we present a new swarm intelligence algorithm called Fitness Predator Optimizer (FPO) [15].

3.1 The Fitness Predator Optimization with Competitive Predators

Fitness Predator Optimizer (FPO) is a new bionic-inspired algorithm proposed in our previous work to avoid premature convergence for the multimodal problems. All of the individuals in the FPO are defined as predators, whose purpose is to find the global optimum (seen as prey) in the search space. Each individual is depicted only by its position vector x, which determines the trajectory of the particle. Then an individual is named as a "position" which comprises the population in FPO. If position i has a higher value of fitness function, then i has more power of locomotion. If position i does not know what is the next best place, a sensible way is to dynamically adjust it according to its own experience and its companions' experience. The definition of an updated position is:

$$newx_{(i,j)} = x_{(i,j)} + (rand - 0.5) * \mu * w * (x_{(k,j)} - x_{(i,j)})$$ (6)

To the position i, $x_i = \{x_{(i,1)}, ..., x_{(i,j)}, ..., x_{(i,d)}\}$ is a vector with d $(1 \leqslant j \leqslant d)$ dimensions. $newx_{(i)}$ is the updated position. $rand$ is a positive decimal randomly produced by computer. μ is a positive constant, w is defined as inertia weight, first proposed by Y.Shi and Eberhart [3] for PSO. It is a variable parameter in the range of $[0.2, 1]$ in this paper. However, it must be said that the given method of w's selection is a trial. Generally, it is not sure whether the new position $newx_{(i)}$ is better than x_x during the optimum search process. In FPO, only a limited number of chances are released in each iteration. A position with better value of fitness function has a prior possibility to access the chance and update its next position by (6). The remaining positions keep the previous positions until they get a chance to update. The main function of FPO is described in Algorithm 2.

Algorithm 2. Main Function

Initialize population: *popsize*
Initialize x_i: $rand(x_i) \in (x_{min}, x_{max})$
while $t \leq \rho$ **do**
    if $rand_1 < rand_2 * \dfrac{fitness(x_i)}{\sum\limits_{i=1}^{n} fitness(x_i)}$ **then**
    get a chance to update its position
    $x_{(i,j)} = newx_{(i,j)}$
    Generate a new position by (6)
    For each position use the elitism strategy
    end if
end while
For all population use the elitism strategy

In Algorithm 2, the vector x_i depicts a position i, *popsize* denotes the population of particles, the amount of chances ρ equals to *popsize* on each iteration. ρ number of chances are released in each iteration, and only the competitive position could get the chance to update. The elitism strategy in FPO is to reserve the best optimal position as shown in (7).

$$f_{k+1}^* = \begin{cases} f(X) & f(X) < f_k^* \\ f_k^* & \text{others} \end{cases}$$ (7)

The f_k^* is the best optimal position fitness value after k times comparison with other positions. X is the new position which will be compared with f_k^* in the $(k+1)th$ time. The elitism strategy function is shown in Algorithm 3.

Algorithm 3. Elitism Strategy Function

$GlobalMin = find(min(f(x_i))$
$newx_{(i,j)} = x_{(i,j)} + (rand - 0.5) * \mu * w * (x_{(k,j)} - x_{(i,j)})$
$tmp = find(min(f(newx_{(i,j)})))$
if $tmp < GlobalMin$ then
$GlobalMin = tmp$
end if

3.2 The Proposal of FPO-FCM Algorithm

In FPO-FCM, the position of particle x_i represents a set of clustering centers. Which can be expressed as follows:

$$\begin{bmatrix} x_{11} & \cdots & x_{1d} \\ \vdots & \ddots & \vdots \\ x_{c1} & \cdots & x_{cd} \end{bmatrix} \tag{8}$$

To the particle x_i, each row of the center matrix $(x_{i1}, x_{i2}, ..., x_{id})$ denotes the cluster center with d dimensions. There are P number of particles that composed a swarm of FPO-FCM. In FPO-FCM algorithm, we need a function for evaluating the generalized solutions called fitness function. In this paper, equation (5) is used for the fitness function. The smaller is $J_m(V, X)$, the better is the clustering effect and the better is the individual fitness function. The FPO-FCM algorithm can be stated as follows:

Algorithm 4. FPO-FCM

Initialize population: *popsize*
Initialize the sample data
Initialize the total number of iteration t_{max}
Initialize *pbest* for each particle and *gbest* for the warm
while $t \leq t_{max}$ **do**
   if $rand < \dfrac{fitness(x_i)}{\sum\limits_{i=1}^{n} fitness(x_i)}$ **then**
     get a chance to update its position $x_{(i,j)} = newx_{(i,j)}$
     for each center **do**
       Generate a new position by (6)
     end for
     for each position of particle **do**
       Use the elitism strategy function
     end for
   end if
   For all population use the elitism strategy function
   Update the clustering center
end while
Partition the data set with the final clustering centroids
Output the final clustering centroids and classified data sets

The performance of FPO-FCM is verified by several widely used data sets in the pattern recognition literature.

4 Experiments

4.1 Experimental Method

In this section, we compare the performances of FPO-FCM with FCM and QPSO-FCM by using five different benchmark data sets that obtained from the UCI Machine Learning Repository. In Table 1, all of the data sets are multivariate data type. In particular, Lung Cancer data set is a relatively higher dimensional data than the others. The numbers of instances and attributes are shown in third and fourth column of the table 1 respectively. Class distribution reflects the number of classes and the number of instance in each class. Basically, the intra-cluster distance and inter-cluster distance are used as clustering evaluation index in this paper.

Table 1. Benchmark data set description

Data Set Name	Data Types	Instances	Attributes	Class distribution
Iris	Multivariate	150	4	(50,50,50)
Wine	Multivariate	178	13	(59,71,48)
Breast Cancer Wisconsin (BCW)	Multivariate	699	10	(458,241)
Wisconsin Diagnostic Breast Cancer	Multivariate	569	32	(357,212)
Lung Cancer	Multivariate	27	56	(8,10,9)

$$Intra-D = [\sum_{i=1}^{c} \sum_{k \in c_i} \| x_k - v_i \|^2]/c \tag{9}$$

$$Inter-D = \sum_{i,j \in c_i} \| v_i - v_j \| \tag{10}$$

$$QE = \sum_{k=1}^{n} \sum_{i=1}^{c} (u_{ik})^m d_{ik}^2(x_k, v_i) \tag{11}$$

When the value of the intra-cluster distance is decreasing, it means that the data partition is more accurate. On the contrary, when the value of inter-cluster distance is increasing, the data partition is more accurate as well. Quantum Error equals to the objective function of FCM which reflected the tightness of clustering.

In experiment A, four data sets (iris, wine, breast cancer wisonsin and Wisconsin Diagnostic Breast Cancer(WDBC)) are selected to test the performance of FPO-FCM comparing with K-means, FCM and QPSO-FCM. Each of hybrid fuzzy swarm algorithms was run with 100 iterations and a population size of 30 on each data set. The test environment and experimental execution parameters are shown in Table 2 and Table 3 respectively.

Table 2. Evaluation test environment

OS	Windows 7
Processor	Intel(R) $Core^{TM}$ i7 CPU 2.80GHz
Memory (RAM)	8.00GB
System type	64-bit operation system
Tool	MATLAB 7.10.0

Table 3. Parameters Setting of experiment A

Experiment A	Algorithm	Parameters
Population=30	FCM	m=2
Max iteration=100	QPSO-FCM	m=2, $\alpha \in [0.5, 1.0]$
Run Number=50	FPO-FCM	m=2, c=2, $\omega \in [0.2, 1.0]$

Table 4. Experimental results of A

$Dataset^{\diamond}$	Algorithms	Error Distribution	Intra-D	Inter-D	QE	Accuracy (%)
Iris	K-means	(0,3,13)	0.4483	1.0942	1.2393*	89.33
	FCM	(0,7,6)	0.3742	1.0409	1.2701	91.33
	QPSO-FCM	(0,8,4)	0.0065	0.0112	1.1957	92.00
	FPO-FCM	(0,2,9)	0.0062	0.0122	1.1953	92.67
Wine	K-means	(13,21,19)	0.0476	0.2383	0.1429*	70.22
	FCM	(13,20,23)	0.0570	0.1804	6.2100	68.54
	QPSO-FCM	(2,18,29)	0.0033	0.0117	5.6024	72.47
	FPO-FCM	(10,21,17)	0.0025	0.0150	5.5827	73.03
BCW	K-means	(11,18)	275.6728	0.9629	551.3456*	95.85
	FCM	(14,4)	271.7616	0.9212	258.0470	97.42
	QPSO-FCM	(61,17)	335.8733	0.2877	296.6230	88.84
	FPO-FCM	(5,7)	266.2934	1.1054	256.6245	98.28
WDBC	K-means	(83,0)	1.4610	0.1447	2.9219*	85.41
	FCM	(0,86)	1.7658	0.0897	1.9173	84.89
	QPSO-FCM	(0,77)	1.4462	0.1352	0.0018	86.47
	FPO-FCM	(87,0)	1.7720	0.1317	2.0719	84.71

$^{\diamond}$ Each data set is normalized within a range of $[0.1, 0.9]$

* the Quantum Error of K-means algorithm is defined as: $QE = \sum_{i=1}^{c} \sum_{k=1}^{N_i} d_{ik}^2(x_k, v_i)$

* Where c is the total number of clusters and N_i is the count of data in each cluster

4.2 Experimental Results

Table 4 resumes the clustering results of FPO-FCM, K-means, FCM and partly results of QPSO-FCM obtained from our previous work [6]. Error distribution reflects the number of instances that were wrongly assigned to each class. The intra-cluster distance, inter-cluster distance and quantum error are used as clustering evaluation indexes for all of the algorithms. The statistical results including four kinds of clustering evaluation indexes and accuracy show that FPO-FCM is able to provide very competitive results on each benchmark data set partition.

In experiment B, a relatively high dimensional data set, lung Cancer data, is selected to evaluate the performance of FPO-FCM comparing with K-Means, FCM and QPSO-FCM. The FPO-FCM and QPSO-FCM terminating condition are limited in 500 consecutive iterations and a population size of 30 for both of them. All execution parameters in experiment B are shown in Table 5.

Table 5. Parameters setting of experiment B

Experiment B	Algorithm	Parameters
Population=30	FCM	m=2
Max iteration=500	QPSO-FCM	m=2, $\alpha \in [0.5, 1.0]$
Run Number=30	FPO-FCM	m=2, c=2, $\omega \in [0.2, 1.0]$

The lung cancer data set described 3 types of pathological lung cancers. In the original data, five instances were missing some feature values, as such only 27 vectors are collected as our experimental samples. These three clusters are more likely to highly overlap according with [5], so finding the clusters is very difficult. Figure 2 shows the convergence curve of FCM with lung cancer data set. Obviously, the FCM model is trapped into local minima and cannot improve the objective function value after the fourth iteration.

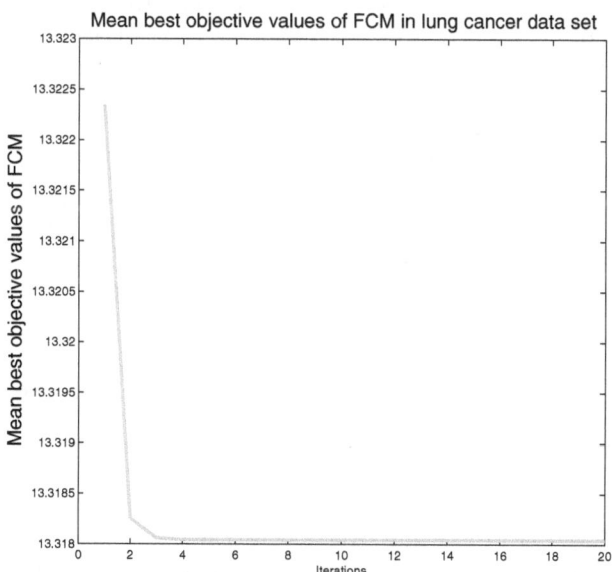

Fig. 2. Convergence curve of FCM

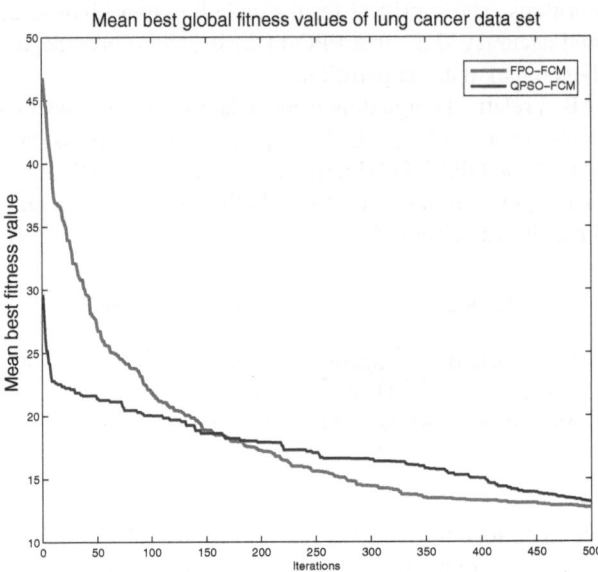

Fig. 3. Comparison of convergence curves between FPO-FCM and QPSO-FCM

Figure 3 shows the convergence curves between FPO-FCM and QPSO-FCM in lung cancer data set. All of algorithms constantly minimize objective functions of clustering within 500 times of the iteration. Compared with QPSO-FCM, the FPO-FCM has a faster speed and higher convergence rate during the iteration process.

Table 6. Experimental results of B

Dataset°	Algorithms	Error Distribution	Intra-D	Inter-D	QE	Accuracy (%)
	K-means	(2,7,4)	19.8399	3.0913	59.5198*	51.85
	FCM	(4,5,4)	23.5750	1.18e-11	13.3180	51.85
lung cancer	QPSO-FCM	(3,4,6)	23.6350	1.5895	13.1833	51.85
	FPO-FCM	(4,3,2)	22.8330	0.9592	12.7418	66.67

° The data set is normalized within $[0.1, 0.9]$

* the Quantum Error of K-means algorithm is defined as: $QE = \sum\limits_{i=1}^{c} \sum\limits_{k=1}^{N_i} d_{ik}^2(x_k, v_i)$

* Where c is the total number of clusters and N_i is the count of data in each cluster

Table 6 resumes the lung cancer data's clustering results of K-means, FCM, QPSO-FCM and FPO-FCM. Due to the highly overlapping character of clusters in lung cancer, the precision of all of the algorithms is dramatically decreased compared with experiment A. However, the FPO-FCM has a higher clustering accuracy than the others. The experimental results show that the proposed algorithm has a good robustness for the high dimensional and overlapping clusters in data set.

4.3 Discussion

According to the results of experiment A, it is easy to see that FPO-FCM shows its competitive global search ability on the three benchmark data sets except Wisconsin Diagnostic Breast Cancer (WDBC) data. The best classification results of WDBC in experiment A is the QPSO-FCM algorithm. We think that this is because QPSO has better global search ability for the clustering of space partition that resembles the shape of a completely round ball. Then QPSO-FCM demonstrates its superior performance to the WDBC that with the similar space distribution. In experiment B, the lung cancer data set is a fairly difficult clustering problem due to its high dimensions and overlapping clusters. In table 6, the intra-distance index and the inter-distance index of K-means algorithm is better than the others. However, the K-means algorithm has not worked as well as expected. One of the reasons is that using euclidean norm as similarity calculation formulation may not fit for the overlapping cluster such as lung cancer data set. Despite of it, FPO-FCM is still able to escape the trap of the suboptimal values of objective function and to find the global minimum.

5 Conclusion

In this study, a cluster optimization methodology is proposed based on the Fitness Predator Optimizer (FPO) algorithm. The proposed approach deals with the modified FPO algorithm for fuzzy clustering optimization. In the proposed new hybrid fuzzy clustering algorithm (FPO-FCM), the position of each particle represents a set of clustering centroids, a number of particles composed of a swarm of FPO-FCM. The objective function $J_m(U, V)$ of FCM is used for evaluating the generalized solutions. The experimentation is done with five benchmark data sets covered examples of data from low and high dimensions. Compared with traditional algorithms (K-means and FCM) and hybrid swarm algorithm (QPSO-FCM), FPO-FCM has a higher robustness and better global optimization ability of clustering partition.

References

[1] Bezdek, J.C.: Pattern recognition with fuzzy objective function algorithms. Kluwer Academic Publishers (1981)
[2] Hathaway, R.J., Bezdek, J.C.: Optimization of clustering criteria by reformulation. IEEE Transactions on Fuzzy Systems 3(2), 241–245 (1995)
[3] Shi, Y., Eberhart, R.: A modified particle swarm optimizer. In: IEEE World Congress on Computational Intelligence, pp. 69–73. IEEE (1998)
[4] Sun, J., Feng, B., Xu, W.: Particle swarm optimization with particles having quantum behavior. In: Congress on Evolutionary Computation, pp. 325–331 (2004)
[5] Runkler, T.A., Katz, C.: Fuzzy clustering by particle swarm optimization. In: 2006 IEEE International Conference on Fuzzy Systems, pp. 601–608. IEEE (2006)
[6] Wang, H., Yang, S., Xu, W., Sun, J.: Scalability of hybrid fuzzy c-means algorithm based on quantum-behaved pso. In: Fourth International Conference on Fuzzy Systems and Knowledge Discovery, vol. 2, pp. 854–857 (2007)
[7] Liu, H.-C., Yih, J.-M., Wu, D.-B., Liu, S.-W.: Fuzzy c-mean clustering algorithms based on picard iteration and particle swarm optimization, pp. 838–842. IEEE (2008)

[8] Yang, F., Sun, T., Zhang, C.: An efficient hybrid data clustering method based on k-harmonic means and particle swarm optimization. Expert Systems with Applications 36(6), 9847–9852 (2009)

[9] Gupta, K., Shrivastava, M.: Web usage mining clustering using hybrid fcm with ga. International Journal of Advanced Computer Research, 322–336 (2010)

[10] Bezdek, J.C.: Fuzzy c-means cluster analysis. Scholarpedia 6(7), 2057 (2011)

[11] Egrioglu, E., Aladag, C.H., Yolcu, U.: Fuzzy time series forecasting with a novel hybrid approach combining fuzzy c-means and neural networks. Expert Systems with Applications 40(3), 854–857 (2013)

[12] Ortiz, A., Palacio, A.A., Górriz, J.M., Ramírez, J., Salas-González, D.: Segmentation of brain mri using som-fcm-based method and 3d statistical descriptors. Computational and Mathematical Methods in Medicine (2013)

[13] Khan, A., Jaffar, M.A., Choi, T.-S.: Som and fuzzy based color image segmentation. Multimedia Tools and Applications 64(2), 331–344 (2013)

[14] Anusuya, S., Parthiban, L.: Efficient hybridized fuzzy clustering with fcm-iqpso for biomedical datasets. Annual Review & Research in Biology 4(17) (2014)

[15] Yang, S., Sato, Y.: Fitness predator optimizer to avoid premature convergence for multimodal problems. In: IEEE International Conference on Systems, Man and Cybernetics, pp. 264–269 (2014)

Effects of Mutation and Crossover Operators in the Optimization of Traffic Signal Parameters

Rolando Armas, Hernán Aguirre, and Kiyoshi Tanaka

Faculty of Engineering, Shinshu University
4-17-1 Wakasato, Nagano, 380-8553 Japan
rolando.armas@iplab, {ahernan,ktanaka}@shinshu-u.ac.jp

Abstract. In this work, we analyze crossover and mutation operators for traffic signals optimization aiming to understand the problem from a system level perspective. We use MATSim to simulate the transport system of the business district of Quito (Ecuador) with 20000 agents moving in a one-day congested scenario. We relax the usual assumption of common cycle length for all signals and minimize travel time focusing on the optimization of 11 consecutive signals located in a main road. We study individual and combined effects of crossover and mutation for cycle length, offset, and green times. The results of this study provide valuable insights to know better the problem, validate the mobility scenario, and understand the effects of the operators.

1   Introduction

Population growth and urbanization trends have increased the demand of road networks causing congestion. This adds substantial costs, increases gas emissions, and the risk of accidents. A way to alleviate traffic congestion is to make better use of the existing roads, which can be achieved in part by a properly setting traffic signals. Essential parameters of a traffic signal are cycle length, green times of the phases bounded by the cycle length, and the offset between the beginning of the cycle of consecutive signals. Common measures to evaluate the performance of a network are average delay, travel time through the network, number of stops, or some combination of these. The choice of a measure of performance depends on the type of traffic that should benefit (e.g. private and commercial vehicles, public transportation, pedestrians and cyclists); societal objectives (e.g. safety, priority to businesses, reduction of emissions); and cost of traffic congestion (e.g. delay, fuel consumption). Clearly there are tradeoffs among these objectives.

There are many suggestions for setting traffic signals, ranging from statistically based methods developed in the early 60's to adaptive and cooperative methods that use actual flow information supplied by traffic detectors [12]. Here we consider a system for offline optimization. Most works in the literature focus on optimization for single intersections or coordination of signals along a main road [12], where genetic algorithms are among the preferred optimizer [8,11,2,7]. A typical optimization formulation involves a common cycle length, green times

G. Dick et al. (Eds.): SEAL 2014, LNCS 8886, pp. 167–179, 2014.
© Springer International Publishing Switzerland 2014

for the phases and offsets [11,12]. Assuming a common cycle length simplifies the complexity of the problem. Nonetheless, the search space still is vast and the evaluation of a candidate solution could be very costly, especially when some form of simulation is performed to compute the objective function values. Useful knowledge about traffic management has been gained from previous studies. Unfortunately, this knowledge is not readily applicable to highly congested networks. In addition, transport networks differ substantially from one another and we need to understand the particularities of the system and validate the model used for optimization.

We consider the optimization of traffic signals as one important component in the design of a sustainable transport system, where mobility, societal and economical aspects should be considered. In this work we relax the assumption of common cycle length for all signals and focus on the optimization of 11 consecutive signals located in a main road of a real-world scenario. We use MATSim [1], a multi-agent transport system simulator, to simulate the transport system of the business district of Quito (Ecuador) with 20000 agents moving in a one-day congested scenario. We study the individual and combined effects of operators for crossover and mutation of cycle length, offset, and green times minimizing travel time. Analyzing congested scenarios is a complex task, especially when several optimization objectives related to sustainability are present. Here, our main objective is to understand the kind and range of optimal signal configurations when travel time is optimized without considering other important objectives. The results of this study provide valuable insights to know better the problem, validate the mobility scenario, and understand the effects of the operators.

2   Problem Formulation

The traffic signals optimization problem aims to coordinate traffic signals in order to provide smooth flow of traffic along streets and highways to reduce travel times, stops and delays[10]. A transport network can be represented by a directed graph $G = (N, A)$, where N represents nodes and A represents links. The travel time for a given vehicle is

$$t_{ia} = t_{ia}^x - t_{ia}^e \quad a = 1, ..., A; \quad i = 1, ..., V, \tag{1}$$

where t_{ia} represents the travel time on link a for vehicle i, t_{ia}^x denotes the time vehicle i exited link a (see Fig.2), t_{ia}^e denotes the time vehicle i entered link a, V is the number of vehicles being simulated, A is the number of links in network, e is the enter node and x is the exit node [9].

In this work, we minimize average travel time expressed by

$$\min \quad \frac{\sum_{i=1}^V \sum_{a=1}^A t_{ia}}{V} \tag{2}$$

subject to signal timing design and feasibility constraints shown in Eq.(3)- Eq.(9) [11].

The principal components of a traffic signal are **cycle length, phase, off-set, stage, green** and **inter-green** time. **Cycle length** is the time in seconds required for one complete color sequence of the signal. A **phase** is the set of movements that can take place simultaneously or the sequence of signal indicators received by such movements. **Offset** is the time lapse in seconds between the beginning of a a corresponding green phase at an intersection and the beginning of a corresponding green phase at the next intersection. One **stage** is a green and inter-green time sequence. The list of principal components are summarized in Table 1. Fig.1 illustrates a traffic signal that models two traffic flows in orthogonal directions.

Equations Eq.(3) - Eq.(5) represent the range for cycle length C_h, offset θ_h and green time $\phi_{h,r}$, respectively. C_{hmin} is determined by identifying the signal that needs the longest duration just to accommodate the inter-green times and the minimum green times as shown in Eq.(6). C_{max} is set to 120 seconds [10]. Inter-green is 3 seconds and minimum green time duration is 7 seconds for all signals as shown in Eq.(7).

$$C_{hmin} \leq C_h \leq C_{hmax} \tag{3}$$

$$0 \leq \theta_h \leq C_h - 1 \tag{4}$$

$$\phi_{h,rmin} \leq \phi_{h,r} \leq \phi_{h,rmax} \tag{5}$$

$$C_{min} = Max\left\{ \left(\sum_{r=1}^{S_h} \phi_{h,r} + \sum_{r=1}^{S_h} I_{h,r} \right) : \quad h = 1, 2..., N \right\} \tag{6}$$

$$\phi_{h,rmin} = 7 \quad sec \quad \forall h, r \tag{7}$$

$$C_h = \sum_{r=1}^{S_h} \phi_{h,r} + \sum_{r=1}^{S_h} I_{h,r} \quad \forall h \tag{8}$$

$$\phi_{h,rmax} = C_h - \sum_{r=1}^{S_h} I_{h,r} - \sum_{y=1 y \neq r}^{S_h} \phi_{h,ymin} \tag{9}$$

Table 1. Notation

Variable	Description
C_h	Cycle length at signal h
θ_h	Offset at signal h
$\phi_{h,r}$	Green time at signal h for stage r (phase)
$I_{h,r}$	Inter-green time signal h for stage r (phase)
S_h	Total number of stages (phases) at signal h

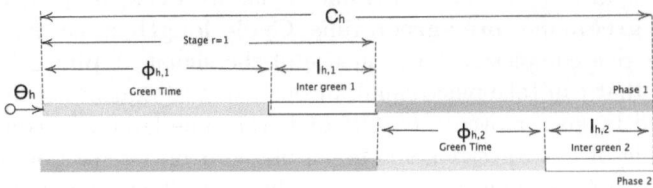

Fig. 1. Traffic Light Components

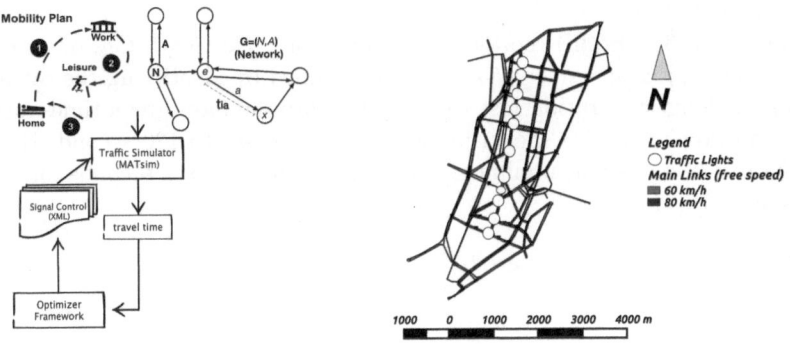

Fig. 2. Optimization System   **Fig. 3.** Area of Study

Eq.(8) ensures that the sum of the green times in a signal together with inter-green do not exceed the cycle length set for the signal. Eq.(9) establishes the maximum green time for the signal phase based on the cycle time, inter-green and minimum green time.

3   Method

3.1   Transport Simulation and Optimization Algorithm

In this work, we use evolutionary algorithms to find optimal signal settings of a transportation system in order to minimize average travel time. Fig.2 illustrates the interaction of the various components of the optimization system. The evolutionary algorithm (EA) evolves a population of candidate solutions, each solution represents a configuration of signals (signal control) for the transportation system. At each iteration, the evolutionary algorithm calls the transport simulator for each candidate solution in order to evaluate it. Once all solutions are evaluated, the evolutionary algorithm continues to the next iteration.

We use the Multi-agent Transport Simulator MATSim [6]. MATSim allows micro-simulation of the transport system producing detailed information about the behavior of the agents being modeled. MATSim receives initial mobility

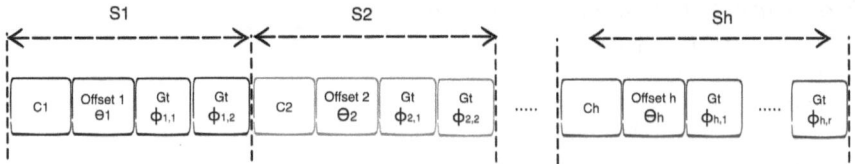

Fig. 4. Chromosome Representation

plans for a set of agents and a model of the transport infrastructure as inputs. It simulates traffic following the plans of the agents trying various routes and iterates to optimize plans and routes for all agents in order to provide a system in an equilibrium state. The *network equilibrium model* refers to a Wardrop user equilibrium (UE) condition [13]: *A stable condition is reached only when no traveler can improve his travel time by unilaterally changing routes.* To run a scenario with traffic lights, MATSim simulates traffic lights microscopically using fixed-time controls [4].

Before we run the optimizer, we first run MATSim until it reaches an equilibrium state. When the optimizer calls MATSim to evaluate a solution, MATSim starts from the equilibrium state setting its signals controls with the tentative solution provided by the optimizer and runs one additional iteration. The output collected from that iteration of the simulator is used to calculate travel time using Eq.(2), which is passed as fitness of the solution to the optimizer.

3.2 Evolutionary Algorithm (EA)

Representation. A signal S in junction h is represented by set of **integer** variables expressed by

$$S_h = (C_h, \theta_h, \phi_{h,1}, \cdots, \phi_{h,r}),\qquad(10)$$

where C_h is cycle length, θ_h is the offset, and $\phi_{h,1}, \cdots, \phi_{h,r}$ are the green times for the r phases of the signal. The ranges and constrains of these variables are given in Eq.(3)-(9). Signal S_h represents one *gene* and a set of signals form the *chromosome* of an individual; a solution with a complete specification of the signals considered for optimization. Fig.4 illustrates the representation of a solution with h signals, each one with two phases.

Algorithm. In this work we use a simple elitist evolutionary algorithm, which general flow is shown in **Procedure 1**. The main steps of the algorithm are as follows:

Initial Population. The population P can be initialized as a combination of pre-settings, mutation of the pre-settings, and randomly creating individuals. Details of how the initial population is created in this work are included in Section 4.2.

Parent Selection. Individuals are selected to reproduce using binary tournaments among randomly sampled solutions from the population P. The winner of a tournament is decided based on fitness.

Procedure 1. Evolutionary Algorithm

1: $P \leftarrow$ Initial Population()
2: Evaluation (P)
3: **repeat**
4:     $P\prime \leftarrow$ Parent selection(P)
5:     $Q \leftarrow$ Recombination and mutation $(P\prime)$
6:     Evaluation (Q)
7:     $P \leftarrow$ Survival selection $(P, nElite, Q)$
8: **until** condition is met

Recombination and Mutation. The offspring population Q is created applying crossover to the selected parents with probability P_c followed by mutation. There is one operator of mutation per variable of the signal. $P_m{}^{(Ct)}$,
$P_m{}^{(Ot)}$, and $P_m{}^{(Gt)}$ are mutation probabilities for cycle, offset and green
times, respectively. To mutate, we first decide which mutation operator will
be applied using the probabilities of the operators. Then we apply the chosen
mutation operator with probability P_m per signal.

Evaluation. To evaluate each individual we first run the traffic simulator. The
parameters of the simulation are the signals settings contained in the variables of the individual. The fitness value of the individual is calculated from
the output of the simulation using Eq.(2).

Survival Selection. The $nElite$ best individuals from the current population
P and the offspring population Q are combined. For the next generation, we
select the best $|P|$ individuals from this combined population.

Operators. To create offspring we follow the representation described above
and employ one crossover operator and three mutation operators for cycle length,
offset, and green times, respectively. In the following we explain each one of them.

Crossover. In this work we implement one point crossover taking each signal
as an atomic unit. The crossing point is selected randomly with equal probability in the range $[1, h - 1]$, where h is the number of signals. Then the
crossover operator interchanges complete signals between parents.

Cycle Length Mutator. This operator increases or decreases randomly with
equal probability the cycle length of a signal using step size $stepCt$. If the
new cycle length is out of the specified range, we adjust it accordingly to be
either C_{hmin} or C_{hmax}. After that, it is necessary to check whether offset
time violates its constraint. If offset is larger than the new cycle length, it is
reset to new cycle length - $stepOff$, where $stepOff$ is the offset step size.
Finally, for each signal phase the green times are adjusted proportionally to
the new cycle length. Due to the correlation of offset and green times to the
cycle length, this operator may act as a macro-mutation operator.

Green Time Mutator. This operator decreases the green time of one phase
and adds it to another phase using step size $stepGt$. To determine the phase
that will decrease its green time, we randomly visit the phases until we find
one in which the decrement does not violate the constraint for minimum

green time $\phi_{h,rmin}$. The phase to which the green time is added is also determined randomly among all phases, except the one in which time was reduced.

Offset Time Mutator. This operator increases or decreases randomly with equal probability the offset time of a signal using step size $stepOff$. If offset becomes negative, it is reset to 0. Likewise, if offset is greater than the maximum cycle length C_{hmax}, it is reset to C_{hmax} - $stepOff$.

4   Simulation Results and Discussion

4.1   Mobility Scenario and MatSim Simulation

The geographical area of study is the business district of Quito (Ecuador), which covers approximately 7x3 Km^2 as shown in Fig.3. For this experiment, the area takes into account only the primary and secondary pathways which free speeds are in the range from 30 to 80 Km/h. The network has 1000 links approximately and comes from Geofabrik and OpenStreetMap[3].

The number of simulated agents is 20000. The mobility plan for each agent consists of three main trips: (1) home to work, (2) work to leisure, and (3) leisure to home (see Fig.2). The plans are designed so that all agents move first from south to north, completely crossing the geographical area of study. In their second trip, the agents move from north to the central zone of the area under study and in their last trip from the central zone to the south. Eleven signal lights are located in a main two-way street with flows in south-north and north-south directions. We run the multi-agent transport simulator MATSim for 500 iterations, making sure it reaches a user equilibrium state without setting any traffic signal. The traffic simulation period is for 24 hours. It takes approximately 1 hour and 30 minutes of computation time to run MATSim for this number of iterations. Traffic signals are optimized using the equilibrium state as an initial condition.

4.2   Evolutionary Algorithm Experimental Setup

We set the number of elite individuals $nElite$=10 and use a fixed population size of 21. The initial population is created deterministically as follows. We prepare 21 cycle lengths in the range [20, 120] seconds in steps of 5. All solutions are set with a different cycle length, but all signals of a solution are set to the same cycle length. The offset times of all signals are set to zero and green times per phase are set to the same value according to the cycle length, i.e. green time = (cycle length - inter-green) /2. That is, all signals are synchronized but not coordinated.

We conduct 10 experiments using different settings for the probabilities of the operators. All experiments start with the same initial population and use the same random seed. The parameters used for each experiment are detailed in Table 2, where P_c is the probability of recombination, P_m is the probability of

Table 2. Experiment Settings (EA Parameters)

Exp.	Pm	$Pm^{(Ct)}$	$Pm^{(Ot)}$	$Pm^{(Gt)}$	Pc	Comments
E01	0	-	-	-	1	only crossover
E02	2/11	0	0	1	0	only green time mutation
E03	2/11	0	1	0	0	only offset time mutation
E04	2/11	0	0.5	0.5	0	offset and green time mutation
E05	2/11	1	0	0	0	only cycle time mutation
E06	2/11	0.5	0.5	0	0	cycle and offset time mutation
E07	2/11	0.5	0	0.5	0	cycle and green time mutation
E08	2/11	0.5	0.3	0.2	0	cycle,offset and green time mutation
E09	2/11	0.5	0.3	0.2	0	mutation 100 generations
E10	2/11	0.5	0.3	0.2	1	mutation and crossover 100 generations

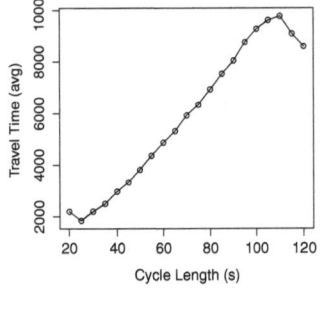

Fig. 5. Initial population evaluation

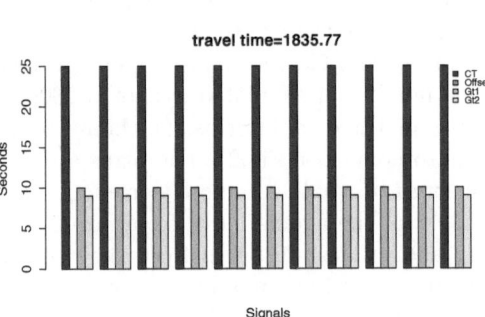

Fig. 6. Best solution in initial population

mutation per signal, and $P_m^{(Ct)}$, $P_m^{(Of)}$ and $P_m^{(Gt)}$ are mutation probability for cycle length, offset and green time, respectively. For each experiment we run the algorithm just once. The first eight experiments run for 50 generations, whereas the last 2 run for 100 generations. It takes in average 70 seconds to evaluate one individual. The mutation steps are set to $stepCt=5$, $stepOff=10$, $stepGt=3$ for cycle, offset and green time respectively.

4.3   Results

Fig.5 shows the average travel time (fitness) in seconds of all solutions in the initial population. Note that the smallest travel time is achieved when cycle length is set to 25 seconds for all traffic lights. Increasing cycle length from 25 to 110 seconds in steps of 5 linearly increases travel time. It is well known that shorter cycle lengths usually result in reduced delays [10]. The cycle length trend observed for the mobility pattern studied here is in accordance with the above statement. However, note that the minimum cycle length of 20 seconds tried

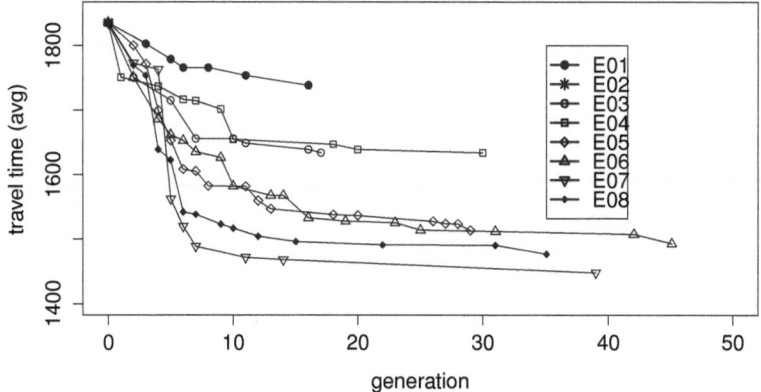

Fig. 7. Fitness transition per experiment (50 generations)

here does not lead to the smallest travel time. Also note that the travel time trend is reversed for cycle lengths above 110, i.e. travel time reduces linearly for cycle length 115 and 120. Fig.6 shows the configuration (Ct, Off, G1, G2) of all eleven signals in the solution with smallest travel time in the initial population.

Next, in order to study the impact of optimizing independently cycle length, offset, and green times, we analyze one experiment using recombination without mutation (E01) and seven experiments (E02-E08) using mutation without recombination. All experiments start with the same initial population shown in Fig.5 and use the same random seed.

Fig.7 shows the travel time transition of the best solution found so far over the generations for the eight experiments and Fig.9 shows the signal configuration of the best solution for some experiments. From these figures, it is worth mentioning the following. Experiment E01 performs only crossover without mutation ($P_c = 1$, $P_m = 0$), which has the effect of recombining the different cycle lengths of the signals in the initial population without changing their initially set offset and green times. In this case, travel time improves until generation 16 and the best solution includes signals with cycle length 20, 25, 30 and 35 as shown in Fig.9(a). This shows that a better overall system configuration can be found by having different signals set with different cycle lengths. This fact is crucial because a large number of works focus on the optimization of groups of signals that are assumed to be synchronized or coordinated, i.e. cycle length is a variable considered for optimization but all signals are assigned the same cycle length. The problem with these approaches is that it is not possible to know in advance which signals have to be coordinated and assigned the same cycle, as shown by this simple experiment.

Experiment E02 that mutates green times ($P_m^{(Gt)} = 1$) could not find a solution better than the best solution of the initial population because mutating green times

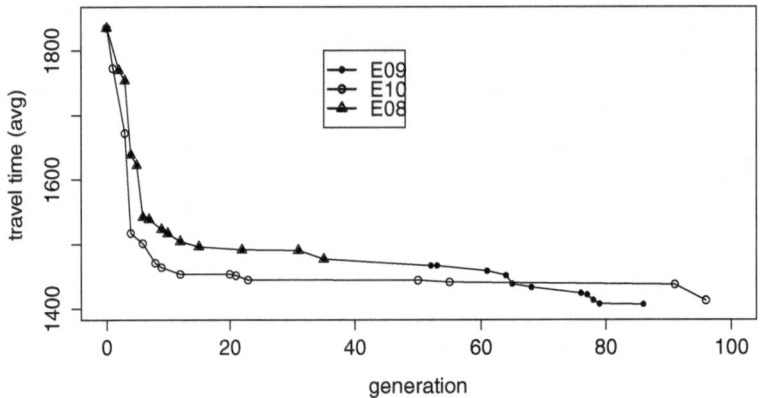

Fig. 8. Fitness transition per experiment (100 generations)

of the best and second best initial configurations (Ct, Of, G1, G2)=(25,0,10,9) and (20,0,7,7), respectively, violates the minimum and maximum allowed green times. Remember that, in this work, the green time mutation step (stepGt) is set to 3 seconds. Also, improvement in other solutions is not large enough to surpass the best configuration of the initial population.

Experiment E03 tests the effect of mutation on offset of traffic signals ($P_m{}^{(Of)} = 1$). Offset parameter is the only one that is independent of the cycle length and green times, and it is used to coordinate signals in order to induce green waves [5]. As Fig.7 shows, travel time improves until generation 17 and then it stagnates. Note however, that travel time is considerably better than by experiment E01. From Fig.9(b) note that the only difference with the best configuration of the initial population is the offset for each signal. This results is in agreement with previous efforts where improvements are achieved by signal coordination, i.e. common cycle and different offsets. Experiment E04 mutates offset and green time ($P_m{}^{(Of)} = 0.5$, $P_m{}^{(Gt)} = 0.5$), which leads to the same solution of experiment E03, where only offset was mutated. This is because green time mutation violates the minimum constraint, similar to experiment E02.

E05 mutates exclusive cycle length ($P_m{}^{(Ct)} = 1$). Note that travel time reduces substantially compared to the previous experiments as shown in Fig.9(c). Although the operator is directed to cycle length it also affects green times, because when cycle changes green times must also be adjusted accordingly, up or down, to be within the limits imposed by the newly mutated cycle length. Experiment E06 mutates cycle length and offset with same probability ($P_m{}^{(Ct)} = 0.5$, $P_m{}^{(Of)} = 0.5$), which leads to a slightly better travel time than E05 that only mutates cycle length.

E07 mutates cycle length and green times with equal probability ($P_m{}^{(Ct)} = 0.5$, $P_m{}^{(Gt)} = 0.5$). Note that this experiment leads to the best travel time. E08 mutates cycle length, offset, and green times with larger probability for cycle

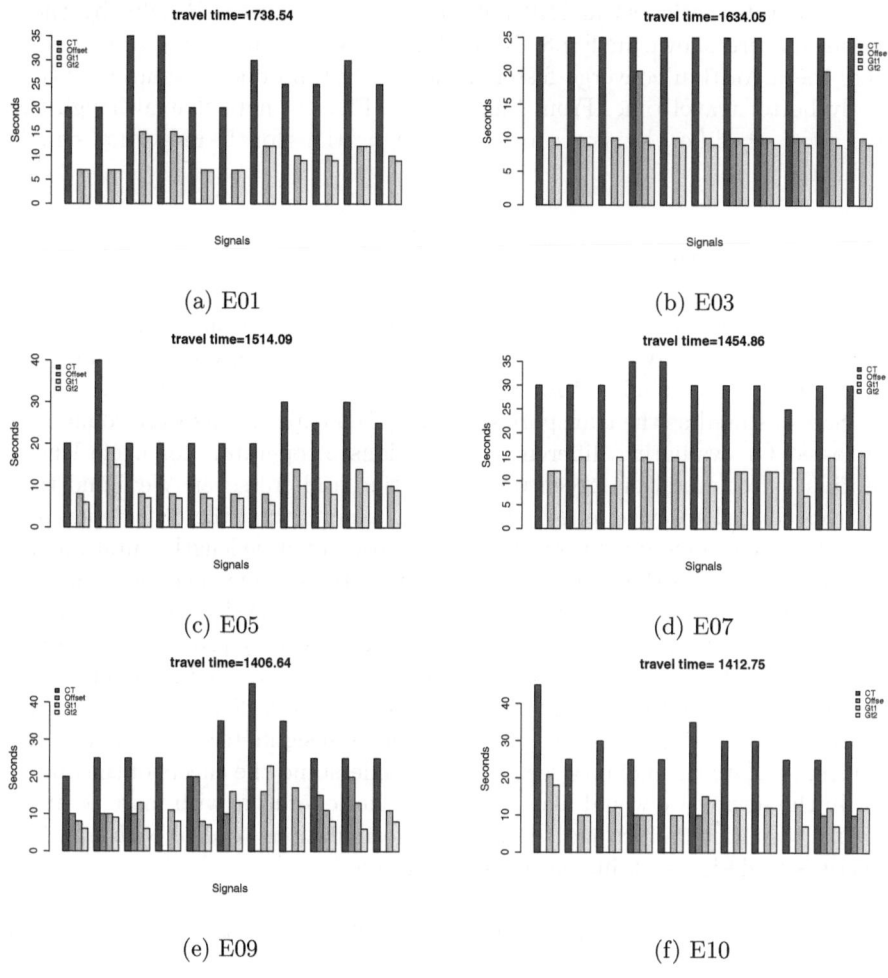

Fig. 9. Signal configuration of best solution

length ($P_m^{(Ct)} = 0.5$, $P_m^{(Of)} = 0.3$, $P_m^{(Gt)} = 0.2$), which leads to the second best travel times of the experiments included in this figure. E07 clearly shows that mutating green times in addition to cycle length contributes to reduce travel time, but the addition of offset mutation in E06 and E08 seems not to improve the effectiveness of the search in the scenario studied here. Fig.9(d) shows the signals configurations of the best solution found by experiment E07.

Next, in experiment E10 we analyze the combined effect of crossover and the three mutation operators for cycle length, offset and green times (P_c=1.0, $P_m^{(Ct)} = 0.5$, $P_m^{(Of)} = 0.3$, $P_m^{(Gt)} = 0.2$). For this experiment, we allow the algorithms to run for 100 generations instead of the 50 used in the experiments before. For comparison, we run experiment E08 again also using 100 generations and call it E09. Note that in E09 the three-mutation operators are used with the same probabilities as in E10 but no crossover is used. Results by these experiments are shown in Fig.8. Note that the inclusion of crossover in E10 makes the algorithm converge faster. However, E09 produce a solution with a slightly better travel time. From Fig.9(e) and Fig.9(f) note that although the travel time only differs by six seconds the configurations for the individual signals are different.

5   Conclusions

In this article, we have analyzed the effects of different EA operators for traffic signals optimization. We performed a system-level optimization of 11-traffic signals simulating the mobility of 20000 agents on a 21Km^2 area of Quito (Ecuador). MATSim was used as the transport simulator. Ten experiments were configured and tested for evaluating different combinations of operators for cycle length mutation, offset mutation, green times mutation and crossover. We found that operators related to cycle length show better results than the other operators related to green times and offsets. The combination of cycle length mutation and green times produces the best results in terms of travel time. The incorporation of crossover did not lead to better travel time, but it speeded up convergence, which could be important in this computationally expensive problem. An important finding of this study is that heterogeneous cycle lengths reduce significantly travel time compared to settings where a common cycle length is used for all signals. The results of this study provide valuable insights to explain better the problem, validate the mobility scenario, and understand the effects of the operators. In the future, we would like to explore other operators with more complex mobility scenarios and extend the problem formulations to deal with multiple objectives to design sustainable transport systems.

Acknowledgements. The first author gratefully acknowledges the support of National Secretariat of Higher Education, Science, Technology and Innovation of Ecuador.

References

1. Balmer, M., Rieser, M., Meister, K., Charypar, D., Lefebvre, N., Nagel, K.: Matsim-t: Architecture and simulation times. In: Multi-Agent Systems for Traffic and Transportation Engineering (2009)
2. Ceylan, H.: Traffic signal timing optimisation based on genetic algorithm approach, including drivers routing. Transportation Research 38, 329–342 (2004)
3. Frederik, R., Topf, J., Karch, C.: Geofabrik (2007), http://www.geofabrik.de (accessed: January 2014)
4. Grether, D.: Traffic light control in multi-agent transport simulations (2011)
5. Kelly, B.: A green wave reprieve. Traffic Engineering & Control 53, 55–58 (2012)
6. Matsim. Multi agent transport simulation, http://matsim.org (accessed: January 2014)

7. Park, B.: Traffic signal optimization program for oversaturated conditions: Genetic algorithm approach. Transportation Research 1683, 133–142 (2007)
8. Sanchez-Medina, J.J., Galan-Moreno, M.J., Rubio-Royo, E.: Traffic signal optimization in la almozara district in saragossa under congestion conditions, using genetic algorithms, traffic microsimulation and cluster computing. IEEE Transactions on Intelligent Transportation Systems 11(1), 10 (2010)
9. Spiegelman, C., Sug-Park, E., Rilett, L.: Transportation Statistics and Microsimulation. CRC Press (2011)
10. STM. Signal Timming Manual. Federal Highway Administration, USA (2008)
11. Teklu, F., Sumalee, A., Watling, D.: A genetic algorithm approach for optimizing traffic control signals considering routing. In: Computer-Aided Civil and Infrastructure Engineering, pp. 31–43 (2007)
12. Warberg, A., Larsen, J., Munk, R.: Green Wave Traffic Optimization - A Survey. IMM-Technical Report-2008-01. Informatics and Mathematical Modelling (2008)
13. Wardrop, J.G.: Some theoretical aspects of road traffic research. Proceedings of Institution of Civil Engineers 1(2), 325–378 (1952)

A GP Approach to QoS-Aware Web Service Composition and Selection

Alexandre Sawczuk da Silva, Hui Ma, and Mengjie Zhang

School of Engineering and Computer Science,
Victoria University of Wellington, New Zealand
{Alexandre.Sawczuk.Da.Silva,Hui.Ma,Mengjie.Zhang}@ecs.vuw.ac.nz

Abstract. Web services are independent functionality modules that can be used as building blocks for applications that accomplish more specific tasks. The large and ever-growing number of Web services means that performing this type of Web service composition manually is unfeasible, which leads to the exploration of automated techniques to achieve this objective. Evolutionary Computation (EC) approaches, in particular, are a popular choice because they are capable of efficiently handling the complex search space involved in this problem. Therefore, we propose the use of a Genetic Programming (GP) technique for Web service composition, building upon previous work that combines the identification of functionally correct solutions with the consideration of the Quality of Service (QoS) properties for each atomic service. The proposed GP technique is compared with two PSO composition techniques using the same QoS-aware objective function, and results show that the solution fitness values and execution times of the GP approach are inferior to those of both PSO approaches, failing to converge for larger datasets. This is because the fitness function employed by the GP technique does not have complete smoothness, thus leading to unreliable behaviour during the evolution process. Multi-objective GP and the use of functional correctness constraints should be considered as alternatives to overcome this in the future.

1 Introduction

With the popularisation of the Internet, Web services have become feasible building blocks for applications. Web services can be defined as independent functionality modules that are used for achieving specific tasks, and that can be accessed via a network communication protocol. The combination of services in order to create a larger application is known as *Web service composition* [11], a technique that encourages component reuse and consequently leads to the expedient development of software solutions. Even though Web service composition can be performed manually, the number of services has been growing so quickly that doing so could prove to be quite time-consuming, particularly when *selecting* between many services with the same functionality but different non-functional attributes. As a result, significant research efforts have been invested to identifying and developing techniques for the automated composition and selection of services.

G. Dick et al. (Eds.): SEAL 2014, LNCS 8886, pp. 180–191, 2014.
© Springer International Publishing Switzerland 2014

Applying Evolutionary Computation (EC) approaches to Web service composition and selection is a popular direction of investigation, since EC techniques employ non-deterministic strategies for discovering solutions and thus are capable of efficiently handling large search spaces (i.e. large numbers of possible service combinations) [13]. The goal of this work is to present an alternative QoS-aware Web service composition approach using GP, introducing the idea of a fitness function with separate ranges of values to differentiate between the composition solutions that are not fully functionally correct and those that are. This work assumes knowledge of genetic programming [6] and particle swarm optimisation techniques [8]. This paper is organised as follows: Section 2 presents a brief overview of the research on GP and PSO conducted in this area. Section 3 presents the improved GP approach. Section 4 briefly reviews the two PSO approaches used for the comparison and presents the details of the comparative evaluation. Section 5 analyses the evaluation results while section 6 briefly presents further investigations that we attempted to improved our GP approach. Section 7 concludes this paper and discusses future work possibilities.

2   Background

2.1   Problem Description

The basic idea behind Web service compositions is to meet user task requirements by combining services into a composition with appropriate functionality. However, this fundamental process only takes into account how well the inputs and outputs of the services within the composition match, meanwhile overlooking important non-functional requirements such as execution time and reliability. A more sophisticated approach is to consider the *Quality of Service* (QoS) measures [10] of each service when performing the composition, in what is known as a QoS-aware Web service composition. While many different Web service quality measures exist, four of them have appeared consistently in previous works [7,20]: the probability of a service being available for execution (A), the probability of a service conforming to previously estimated execution times (R), the estimated time limit between sending requests and receiving responses (T), and the service's financial execution cost (C). A and R are expressed as probabilities, where the highest values denote the highest quality, and T and C are absolute figures, where the lowest values denote the highest quality.

Commonly used Web service composition languages, including BPEL4WS and OWL-S, support four basic constructs for configuring the interaction between services: sequence, parallel, choice and loop [20]. A number of papers only considers sequence and parallel constructs [20,5], and the same applies to this work. These two constructs are described as follows:

- **Sequence Construct:** The services organised using the sequence construct are executed sequentially, which means that the output of the first service feeds into the input of the second in a chain. The total time (T) and cost (C)

of this construct can be calculated simply by adding the individual property values from each service in the sequence. As probabilities, the availability (A) and reliability (R) are calculated by multiplying the individual property values from each service.

– **Parallel Construct:** The services organised using the parallel construct are executed in parallel, which means that the inputs of each service must be fulfilled independently and that the outputs are consequently also produced independently. The cost, availability and reliability (C, A, R) of this construct are calculated in the same way as in the sequence construct. The total time (T), on the other hand, is calculated by selecting the highest individual execution time out of all services.

2.2   Objective Function

An objective function must be employed to perform QoS-aware service composition. This function ensures that desirable quality properties are maximised in the solutions to a composition task [2]. In accordance to the four QoS values chosen in this work, the following objective function was employed for a candidate solution i:

$$f_i = w_1 A_i + w_2 R_i + w_3(1 - T_i) + w_4(1 - C_i) \tag{1}$$

where $\sum_{i=1}^{4} w_i = 1$.

The QoS attributes used in this function are calculated according to the strategies described above for each sequential and parallel construct in the overall composition. The output of this objective function is within the range $[0, 1]$, where 1 represents the best possible composition quality and 0 represents the worst. As the function weights (w_1 to w_4) all add to 1, the T and C values must be normalised between 0 and 1 so that the overall result falls within the required range. To perform this normalisation, C is divided by the sum of costs in all the services that could possibly be in the composition, and T is divided by the sum of times of these services. The services that are possibly in the composition can be identified using a simple discovery algorithm outlined in previous works [17,16]. Finally, as the lowest possible T and C values represent the best quality, the objective function must be offset using $(1 - T)$ and $(1 - C)$.

A key aspect in the comparison performed in this work is that it utilises the same objective function for all the techniques compared. This consistency is important because it means that the results of the comparison are fair with regards to the composition quality measure.

2.3   Existing GP-Based Composition Approaches

Genetic programming is a popular EC technique for performing Web service composition. In [12,15] a GP approach is proposed that guarantees functional correctness by generating the initial population candidates according to a context-free grammar. After the initial generation of candidates, any genetic operation to the trees is also guaranteed to maintain the functional correctness by checking that inputs and outputs match. The authors claim to have supported all composition structures present in the OWL-S and BPEL4WS models, but the context-free grammar upon which this work is based does not seem to support loop constructs. Their work was tested using a collection of composition problems and large service repositories, with favourable results for all runs. The positive results demonstrate that this is a robust approach. An important limitation of this work is that it does not consider QoS measurements, instead assuming that the minimum execution path needed to achieve a solution in the measure of its quality (e.g., the depth of the tree [12]).

Similarly, [17] proposes a technique in which all initial candidate compositions are functionally correct, and any subsequent candidates must also be functionally correct. This approach is more accurate than [15], since the latter may generate candidates that are functionally correct but do not relate to the original composition task, thus requiring the imposition of additional penalties by the fitness function. In the case of [17], on the other hand, all candidates in all populations are both guaranteed to be functionally correct and also guaranteed to fulfil the original task's need. This is accomplished by utilising a greedy search algorithm that generates suitable composition candidates and subtrees during mutation. It must be noted that the fitness function in this work relies on the total number of unique web services as its measure of goodness, but does not consider QoS measurements.

[19] investigates the use of GP for Web service composition by proposing a dedicated composition framework. This framework uses a fitness function that incorporates the results from black-box testing using automatically generated use cases, as well as taking into account the overlap between inputs and outputs of each solution's subtrees. The black-box testing ensures that the behaviour of the generated candidates is correct, thus preventing compositions in which the input and output names match but the behaviour of the combined services is not logically compatible. The framework also relies on a Service Dependency Graph to ensure that all generated candidates are functionally correct when performing genetic operations. However, once again the proposed solution neglects to use QoS measurements as the criteria with which to evaluate candidate compositions, a pattern that is repeated in [4].

[20] proposes a genetic programming approach to solve the problem of Web service composition, which is unique because it achieves both the goals of functional correctness and non-functional Quality of Service through a single fitness function. In contrast to process-driven composition approaches — where only the input, output, and total number of services are considered —, this method provides the benefit of evolving the final composition based on global QoS measures. Nevertheless, while this work does consider QoS measures, it does not guarantee functional correctness for all composition candidates.

2.4   Existing PSO-Based Composition Approaches

Web service composition using QoS-aware PSO techniques has been discussed by several works, see [16]. In [3] a fitness function is proposed that considers the availability, response and execution times, successful execution rate, and reputation of each Web service to be included in the composition. This function is employed in the PSO's evolution process, where each particle dimension corresponds to a Web service with the required functionality for the composition. This approach assumes that the overall workflow in which each individual service is to be placed has already been provided, which simply leaves particles to discover the most suitable services for each workflow slot. While the preselection of a workflow considerably facilitates the composition process, it requires either a selection mechanism or a person with sufficient domain knowledge to make the appropriate decision.

[18] carries on the idea of preselecting a workflow upon which to perform PSO. Their main contribution lies in the utilisation of a multi-objective fitness function to evaluate composition candidates. Multi-objectivity is ideal when the goal is to maximise several, often conflicting, desirable attributes in a population. For example, an ideal Web service composition would incur the lowest possible cost while providing the highest possible availability. However, there is often a trade-off between these two quality measures in a composition. The advantage of a multi-objective function is that it allows the retrieval of a Pareto set of solutions that are equivalent overall, despite being different from the perspective of a single quality measure [14].

Despite once again preselecting a workflow to be optimised, [9] proposes a unique method with which to update the position of the particles in the swarm. The idea is to apply list of changes to each particle in order to update it, as opposed to performing the usual numerical calculations. Effectively, particles undergo a transformation process at every step of the PSO search, yielding new workflow configurations. As this approach can lead to stagnant particles, a technique to search solutions within the radius of a given candidate is also implemented, thus diminishing the probability of early convergence on local optima.

3   Proposed GP Approach

The GP approach proposed herein is based on [20]. Candidates are represented using trees where inner nodes consist of parallel and sequence constructs that direct the flow of the composition, and leaf nodes consist of the Web services used as basic components. Each parallel and sequence construct requires a set of inputs and produces a set of outputs according to the nodes that compose its subtree. The genetic operations employed are *crossover*, in which subtrees are swapped, and *mutation*, in which a node is randomly selected and modified. This particular tree representation is convenient because it allows the set of inputs, outputs, and QoS values for each inner node to be calculated by performing a simple depth-first tree traversal. Since this approach relies on the mutation and crossover of the trees to explore the search space, it is capable of composing workflows with varying configurations while at the same time selecting the services with the best QoS properties. However, it may generate solutions that are not fully functional, so the fitness function that evaluates each candidate must proportionally penalise those solutions with functionality issues.

3.1   Fitness Function

The novel contribution of the proposed approach is in its fitness function, which maximises desirable QoS attributes while penalising solutions that are not entirely functionally correct. This works similarly to the approach of [20], but the difference is that the function produces two separate ranges of values, an inferior range denoting partially functional solutions without considering QoS, and a superior range denoting fully functional solutions with QoS. By creating this separation, the fitness function priorities the achievement of full functional correctness before considering non-functional properties.

The range of the fitness function is $[-1, 1]$, where $[-1, 0)$ corresponds to the functional correctness of the solution (with 0 being a completely correct solution), and $[0, 1]$ corresponds to the total QoS properties of a fully functional solution, with 1 indicating the best quality. Before calculating the fitness function the candidate solution tree is traversed in a depth-first fashion, and within each node both the functional (output and input matching) and non-functional (QoS properties) aspects of the candidate are calculated.

In the case of the tree leaves, which represent atomic Web services, the non-functional properties A, R, T and C are the values of those properties in that service and the functional score is always 0 (i.e. not considered). In the case of the inner nodes, which represent workflow configurations, the values of A, R, T and C are calculated as explained in Subsection 2.1, treating each child node as an atomic service.

It is impossible to evaluate the functional correctness of isolated parallel nodes in the tree, since they simply hold services that should be simultaneously executed and are thus unaware of the outside arguments provided to them. Because of this, they do not contribute to the calculation of the functional score component of a candidate's fitness. For sequence nodes, on the other hand, the functional score can be calculated by creating an average of the output-input matches between each pair of child nodes, s_{i-1}, s_i in the sequence. This average is calculated using the equation below, and results in a value in the range $[0,1]$:

$$average = \frac{\sum_{i=2}^{n} \frac{|output_{i-1} \cap input_i|}{|input_i|}}{n} \tag{2}$$

where n is the number of children of the sequence node, $output_{i-1}, input_i$ is the output of service s_{i-1}, the input of services s_i, respectively.

This score is added to an overall running average of the candidate tree. Once the entire tree has been visited, the match for the overall task inputs and outputs is also calculated and added to the running average. Finally, this average is offset using -1 and the functional score yields a value between -1 and 0. If the value is 0, that means that the solution is fully functional, so the objective function introduced earlier (Eq. 1) is employed. Otherwise, the functional score is returned as the fitness value. Thus, the fitness function for a solution i can be expressed as follows:

$$fitness(i) = \begin{cases} f_i & \text{if } func(i) = 0 \\ func(i) & \text{otherwise} \end{cases} \qquad (3)$$

where

$$func(i) = -1 + \frac{w_5\left(\frac{|in_i \cap in_{req}|}{|in_{req}|}\right) + w_6\left(\frac{|out_i \cap out_{req}|}{|out_{req}|}\right) + treeScore(root)}{2},$$

$w_5 + w_6 = 1$ and $treeScore(root)$ is a recursive function that traverses the tree and calculates the average of the results obtained by applying Equation 2 to the sequence constructs within the structure.

4   Design of Experiments

Experiments were carried out to compare the proposed GP approach with two recent PSO approaches, graph-based PSO, and greedy-based PSO. The datasets used for the set of experiments were generated in [20] using the QWS dataset [1] as its basis, since currently no benchmark datasets are available for evaluating QoS-aware web service composition. The exception to this is dataset 6, which is based on dataset 5 but expanded with synthetically generated Web services in order to test the scalability of the approaches. The datasets contain information that has been collected online detailing the inputs, outputs, time, cost, reliability, and availability of real Web services. Four different composition tasks were used throughout this set of experiments, requiring the creation of composition solutions of various sizes and complexities. Their details are displayed in Table 1.

Table 1. Experiment tasks

Task	Inputs	Outputs	Dataset (No. of Services)
1	PhoneNumber	Address	1(20)
2	ZipCode, Date	City, WeatherInfo	2(30)
3	From, To, DepartDate, ReturnDate	ArrivalDate, Reservation	3(60)
4	From, To, DepartDate, ReturnDate	ArrivalDate, Reservation, BusTicket, Map	4(150), 5(450), 6 (4500)

4.1   Two Recent PSO Approaches

The proposed GP approach was compared to two Web composition approaches that rely on PSO. For reasons of brevity, only the key characteristics of each

PSO approach will be described here, however their full explanation can be found in the original work from which they were reproduced [16]. For both of these approaches, the fitness function employed in the evolutionary process is the unchanged objective function presented in Subsection 2.2 (Eq. 1). This function is different from that of the GP method presented in section 3 in that it does not need to constrain functionality, so it only ranges from 0 to 1.

Greedy-Based PSO Approach. The greedy-based PSO approach [16] uses a greedy algorithm, originally proposed in [17], to generate an initial Web service composition workflow where services can be executed sequentially, in parallel, or in a combination thereof. This workflow contains abstract slots for placing services, each slot presenting a different set of available inputs and required outputs. For each slot, a list of compatible services is compiled. PSO is then employed to select the best possible service for each slot in order to arrive at a solution with the best possible QoS attributes overall. Each particle is represented as having n dimensions, where n corresponds to the number of abstract slots in the workflow, and each dimension points to a Web service from its list of compatible services. In summary, in greedy-based PSO the structure of the composition is determined first, and the services to populate that structure are selected afterwards.

Graph-Based PSO Approach. The graph-based PSO approach [16] also employs the greedy composition algorithm, but this time during the evolutionary process. Initially, the discovery of all services from the repository that could possibly be used for the requested composition task is performed using a basic algorithm. Once the discovery is finished, a directed graph showing all the input-output relationships between these services is created — this is referred to as the master graph. The services in the master graph are represented as nodes, and the relationships between them as edges. Each particle has k dimensions, where k corresponds to the number of edges in the master graph. Each dimension holds a value between 0 and 1, which represents a weight associated with that edge. Since each particle only contains a series of weights, during PSO it is necessary to extract the candidate composition workflow from the master graph using the greedy algorithm. The algorithm is run aided by the weights in the particle, meaning that edges with the highest weights are selected to be in the candidate composition. After the workflow has been extracted its fitness can finally be calculated. In summary, in graph-based PSO both the structure of the composition and the services that populate it are selected simultaneously.

4.2   Parameters

Experiments were conducted on a personal computer with a 3.4 GHz CPU and 8 GB RAM. For GP, 50 independent runs were executed per dataset with a population size of 1000 — smaller populations were previously attempted with unsatisfactory convergence rates. Each run was required to continue until a fully functional result was achieved, at which point 50 more iterations would occur and the run would finish. The fitness function was configured with weights of 0.25 for all QoS properties, and of 0.5 for both w_5 and w_6. The crossover and mutation probabilities were set to 0.9 and 0.1, respectively. The single best solution in one generation was copied to the next.

For both PSO approaches, the same settings outlined in the original work were preserved [16]. 50 independent runs were executed per dataset, all of them using a swarm of 30 particles. Runs were allowed to execute a maximum of 100, but were terminated earlier if the global best fitness remained the same throughout 10 iterations. The fitness function was configured with weights of 0.25 for all QoS properties, the PSO inertia weight w was set to 1, and acceleration constants c_1 and c_2 were both set to 1. The greedy-based PSO approach was configured to choose the initial composition workflow from 50 randomly generated candidates.

5   Results and Analysis

The results of the comparison are shown in Table 2, where the first column records the dataset used and its total number of services, the second column contains the composition task employed, and the third column shows the minimum number of services from that dataset which had to be used in order to create a fully functional solution for the composition task. The fourth, fifth and sixth columns present the fitness of the greedy-based, graph-based and GP approaches, respectively; the seventh, eighth and ninth columns show the execution time of the greedy-based, graph-based and GP approaches, including setup times associated with service discovery, creation of the master graph, etc. A Wilcoxon signed-rank test at 0.95 confidence interval was carried out to verify whether there was any statistically significant time or fitness differences between the graph-based and the other two approaches. These differences are indicated in the table as ↓, ↓, ↑ and ↑ symbols denoting significantly smaller and significantly larger values, respectively.

The results show that our GP based approach has clearly worse execution time than that of graph-based PSO approach and the greedy-based PSO approach, though graph-based PSO has clearly worse execution time than that

Table 2. Average time and fitness results for each approach

Dataset (No. of Serv.)	Task	Min. Cmp. Size	Fitness			Time (ms)		
			Greedy PSO	Graph PSO	GP	Greedy PSO	Graph PSO	GP
1(20)	1	1	0.808±0	0.808±0	0.808±0	22.9±1.2	41.3±10	149.6±58.3 ↑↑
2(30)	2	2	0.713±0	0.713±0	0.639±0.04 ↓↓	9±0.1	13.8±2.8	346± 282 ↑↑
3(60)	3	2	0.634±0	0.631±0.011	0.634±0 ↑	11±0	87.2±18	180.6±68.6 ↑↑
4(150)	4	4	0.532±0	0.524±0.01	0.413±0.06 ↓↓	21.7±0.5	116.1±24.5	67689.7 ±109320.9 ↑↑
5(450)	4	4	0.532±0	0.525±0.01	–	33.6±1	60.4± 2.3	–
6(4500)	4	4	0.586±0.01	0.637±0.022	–	462.4±61.2	752.3±78.6	–

[a] ↓ / ↑ mean significant lower / higher in comparison with Greedy PSO
[b] ↓ / ↑ mean significant lower / higher in comparison with Graph PSO

of greedy-based PSO. The average fitness, on the other hand, suggests that the fitness of the GP approach becomes progressively inferior with the growth of dataset sizes, though overall performance of the greedy-based and graph-based approaches is equivalent. As it can be observed, the fitness and time values for the execution of GP using datasets 5 and 6 are missing from the table. This is because the runs using those two datasets failed to converge after a significant amount of time. In fact, the efficiency of GP is severely reduced for dataset 4, as seen by the sudden spike in the execution time and drop in the fitness value.

In hindsight, the fundamental problem with the proposed approach is in its fitness function. Specifically, the division of function values into ranges is problematic because it means that the fitness of solutions does not increase smoothly as they evolve past the threshold of functional correctness. For example, suppose that the fitness for the best solution in generation k is -0.01, i.e. not fully functionally correct. If a crossover operation occurs in generation $k + 1$ and pushes descendants of that solution to the threshold of functional correctness (0), these descendants' QoS scores will be used as their fitness values from that point onwards. However, these QoS values are likely to already be significantly higher than 0, thus causing a jump in the fitness progression of these candidates and leading to unreliable behaviour during the evolution process. In the future, this problem could be addressed by employing a multi-objective GP approach to adequately consider the independent goals of functional correctness and composition quality. Alternatively, functional correctness constraints could be enforced to determine which candidates are structurally valid before applying a fitness function that would concern itself exclusively with QoS optimisation.

6   Further Investigation

As seen from the previous section our proposed GP approach to QoS aware service composition does not perform well comparing with two PSO approaches, due to the fitness function we used. To further improve our GP approach we adjust our GP approach by considering functional correctness during the process of evolutions, i.e, when applying mutation and crossover operations. Correspondingly, we change the fitness function to only measure the aggregate QoS properties of the individuals of each generation. To show the effectiveness of our improved GP approach, ImprGP, we have conducted a further experimental evaluation using the same datasets and the same parameter settings as in Section 4. Table 3 below shows the experimental results.

The results show that the fitness for all approaches is mostly equivalent, with small variations for datasets 4 and 5, but differences are more pronounced in dataset 6. The execution time for ImprGP is higher than for both PSO-based approaches for all datasets except dataset 6, for which ImprGP takes less time than graph-based PSO. When looking at datasets 5 and 6, the increase in the number of services (from 450 to 4500) causes an increase in the execution time by a factor of 15 for graph-based PSO, while the execution time for ImprGP increases only by a factor of less than 2. Compared to graph-based PSO, ImprGP

Table 3. Average time and fitness results for the improved GP and the two PSO-based each approaches

Dataset (No. of Serv.)	Task	Min. Cmp. Size	Fitness			Time (ms)		
			Greedy PSO	Graph PSO	ImprGP	Greedy PSO	Graph PSO	ImprGP
1(20)	1	1	0.808±0	0.808±0	0.808±0	6.7±8.3	27.6±36.5	62.2±81.4 ↑↑
2(30)	2	2	0.713±0	0.713±0	0.713±0	4.4±0.5	33.4±17.4	193.5±13.5 ↑↑
3(60)	3	2	0.634±0	0.634±0	0.634±0	4.9±0.3	32.5±6.8	187.1±11.2 ↑↑
4(150)	4	4	0.532±0	0.527±0.01	0.527±0.01 ↓	9.4±0.5	60.4±3.6	340.8±36.5 ↑↑
5(450)	4	4	0.532±0	0.527±0.01	0.526±0.01↓↓	10.7±1.1	62.7±5.3	351.3±32.5↑↑
6(4500)	4	4	0.586±0.01	0.637±0.02	0.617±0.02↑	374.4±71.9	934.3±44.5	634.8±51.4 ↑↓

[a] ↓ / ↑ mean significant lower / higher in comparison with Greedy PSO
[b] ↓ / ↑ mean significant lower / higher in comparison with Graph PSO

produces a 3% lower fitness in a 30% shorter execution time for dataset 6. This indicates that there is a trade-off between fitness and execution time for larger datasets, an observation that was also made in [16]. In summary, after modifying the fitness function our improved GP approach performs better than the original GP approach. In particular, the experiment results indicate that for large data sets (such as dataset 6), ImprGP achieves better fitness than greedy PSO, and executes faster than graph-based PSO.

7   Conclusions and Future Work

This paper proposed a GP approach for QoS-aware Web service composition which builds upon previous work by employing an improved fitness function. This approach was compared through a set of experiments against two previously defined PSO techniques for QoS-aware composition, namely greedy-based and graph-based PSO. Results showed that while fitness values for GP oscillated between noteworthy and undesirable when compared with the other two approaches, its execution time was clearly higher in all instances, and convergence could not be achieved for the larger datasets. The problem was that the fitness function employed in the GP approach lacked the smoothness required for a reliable evolution process. Further investigation has attempted to improve our proposed GP approach. Experiments has shown that our further improved GP approach shows its efficiency for large datasets. For future work we will evaluate our GP approach using larger datasets to test its scalability. Finally, further work will investigate to apply evolutionary multi-objective optimization (EMO) techniques to QoS aware service composition.

References

1. Al-Masri, E., Mahmoud, Q.H.: Qos-based discovery and ranking of web services. In: 16th Int. Conf. Computer Comm. Networks, pp. 529–534. IEEE (2007)

2. Alrifai, M., Risse, T.: Combining global optimization with local selection for efficient QoS-aware service composition. In: 18th Int. Conf. World Wide Web, pp. 881–890. ACM (2009)
3. Amiri, M.A., Serajzadeh, H.: Effective web service composition using particle swarm optimization algorithm. In: 6th Int. Symposium Telecommunications, pp. 1190–1194. IEEE (2012)
4. Aversano, L., Di Penta, M., Taneja, K.: A genetic programming approach to support the design of service compositions (2006)
5. Cardoso, J., Sheth, A., Miller, J., Arnold, J., Kochut, K.: Quality of service for workflows and web service processes. Web Semantics 1(3), 281–308 (2004)
6. Cramer, N.L.: A representation for the adaptive generation of simple sequential programs. In: 1st Int. Conf. Genetic Algorithms, pp. 183–187 (1985)
7. Jaeger, M.C., Mühl, G.: Qos-based selection of services: The implementation of a genetic algorithm. In: ITG-GI Conf. Comm. Distributed Systems, pp. 1–12 (2007)
8. Kennedy, J.: Particle swarm optimization. In: Encyclopedia of Machine Learning, pp. 760–766. Springer (2010)
9. Ludwig, S.A.: Applying particle swarm optimization to quality-of-service-driven web service composition. In: IEEE 26th Int. Conf. Advanced Information Networking and Applications, pp. 613–620 (2012)
10. Menascé, D.A.: Qos issues in web services. IEEE Internet Comp. 6(6), 72–75 (2002)
11. Milanovic, N., Malek, M.: Current solutions for web service composition. IEEE Internet Comp. 8(6), 51–59 (2004)
12. Mucientes, M., Lama, M., Couto, M.I.: A genetic programming-based algorithm for composing web services. In: 9th Int. Conf. Intelligent Systems Design and Applications, pp. 379–384. IEEE (2009)
13. Rao, J., Su, X.: A survey of automated web service composition methods. In: Cardoso, J., Sheth, A.P. (eds.) SWSWPC 2004. LNCS, vol. 3387, pp. 43–54. Springer, Heidelberg (2005)
14. Rezaie, H., NematBaksh, N., Mardukhi, F.: A multi-objective particle swarm optimization for web service composition. In: Zavoral, F., Yaghob, J., Pichappan, P., El-Qawasmeh, E. (eds.) NDT 2010. CCIS, vol. 88, pp. 112–122. Springer, Heidelberg (2010)
15. Rodriguez-Mier, P., Mucientes, M., Lama, M., Couto, M.I.: Composition of web services through genetic programming. Evolut. Intell. 3(3-4), 171–186 (2010)
16. Sawczuk da Silva, A., Ma, H., Zhang, M.: A graph-based particle swarm optimisation approach to qos-aware web service composition. In: IEEE Congress on Evolutionary Computation (CEC) (2014)
17. Wang, A., Ma, H., Zhang, M.: Genetic programming with greedy search for web service composition. In: Decker, H., Lhotská, L., Link, S., Basl, J., Tjoa, A.M. (eds.) DEXA 2013, Part II. LNCS, vol. 8056, pp. 9–17. Springer, Heidelberg (2013)
18. Xia, H., Chen, Y., Li, Z., Gao, H., Chen, Y.: Web service selection algorithm based on particle swarm optimization. In: 8th IEEE Int. Conf. Dependable, Autonomic and Secure Computing, pp. 467–472 (2009)
19. Xiao, L., Chang, C.K., Yang, H.-I., Lu, K.-S., Jiang, H.-Y.: Automated web service composition using genetic programming. In: IEEE 36th Annual Conf. Computer Software and Applications, pp. 7–12 (2012)
20. Yu, Y., Ma, H., Zhang, M.: An adaptive genetic programming approach to qos-aware web services composition. In: IEEE Congress Evolutionary Computation (CEC), pp. 1740–1747 (2013)

A Novel Hybrid Multi-objective Optimization Framework: Rotating the Objective Space

Xin Qiu[1], Ye Huang[2], Jian-Xin Xu[2], and Kay Chen Tan[2]

[1] NUS Graduate School for Integrative Sciences and Engineering,
National University of Singapore, 28 Medical Drive, Singapore 117456
[2] Department of Electrical and Computer Engineering,
National University of Singapore, 4 Engineering Drive 3, Singapore 117576
{qiuxin,huangye,elexujx,eletankc}@nus.edu.sg

Abstract. Multi-objective Evolutionary Algorithms (MOEAs) are popular approaches for solving multi-objective problems (MOPs). One representative method is Non-dominated Sorting Genetic Algorithm II (NSGA-II), which has achieved great success in the field by introducing non-dominated sorting into survival selection. However, as a common issue for dominance-based algorithms, the performance of NSGA-II will decline in solving problems with 3 or more objectives. This paper aims to circumvent this issue by incorporating the concept of decomposition into NSGA-II. A grouping-based hybrid multi-objective optimization framework is proposed for tackling 3-objective problems. Original MOP is decomposed into several scalar subproblems, and each group of population is assigned with two scalar subproblems as new objectives. In order to better cover the whole objective space, new objective spaces are formulated via rotating the original objective space. Simulation results show that the performance of the proposed algorithm is competitive when dealing with 3-objective problems.

Keywords: Multi-objective evolutionary algorithm, hybrid, decomposition.

1 Introduction

Problems in reality usually have multi-objectives instead of one single objective. A Multi-Objective Optimization Problem (MOP) can be defined as follows:

$$\max/\min F(x) = \left(f_1(x), \ldots, f_m(x)\right)^T \qquad (1)$$

$$subject\ to\ x \in \Omega$$

where x refers to the decision variables which lie in the decision (variable) space Ω. The MOP consists of m objective functions and it maps the decision space Ω into an m-dimensional objective space R^m, i.e. $F: \Omega \rightarrow R^m$.

Objectives of a MOP are often conflicting with each other, meaning that the optimized solution in one objective does not produce optimal result for the other

G. Dick et al. (Eds.): SEAL 2014, LNCS 8886, pp. 192–203, 2014.
© Springer International Publishing Switzerland 2014

objectives. Thus, there are many or even infinite Pareto Optimal solutions for a MOP instead of a single solution to optimize all the objectives simultaneously. The best tradeoff among objectives is defined as the Pareto Front (PF).

In real life applications, the task of solving MOP eventually becomes a task of providing a good approximation of PF in the objective space for decision makers [1]. Therefore, it is desired to have optimization algorithms to produce a good approximation of the real PF with manageable number of Pareto optimal solutions which are evenly distributed along the real PF. NSGA-II [2] is one of the most famous evolutionary algorithms for multi-objective optimization. Non-dominated sorting and density estimation is utilized in the survival selection process to help maintain the diversity of the population. According to empirical results in literatures [3-5], NSGA-II is able to provide powerful performance in 2-objective problems. However, when the number of objectives increases, the quality of the solution set obtained by NSGA-II will impair. To seek the reason, it becomes more difficult for non-dominated sorting to decide which individual should survive to next generation as most solutions are non-dominated, which is a common issue for dominance-based approaches. To circumvent such issue, this paper proposes a new algorithm that combines the concept of decomposition used in MOEA/D [1] with the current NSGA-II framework. The whole population is divided into several groups and the MOP is decomposed into a number of scalar subproblems. Each group will then be assigned with two scalar subproblem as new objectives. Non-dominated sorting and density estimation is conducted within the group based on the new objectives. To better cover the whole objective space, new objective spaces are formulated for each group by rotating the original objective space. Simulation results demonstrate that the proposed algorithm is competitive in solving 3-objective problems.

The rest of the paper is organized as follows. Section 2 reviews some related work. Section 3 provides the details of the proposed framework. Section 4 presents the experimental results and compares the performance of the new framework with that of original NSGA-II. Conclusions are drawn in Section 5.

2   Related Work

2.1   NSGA-II

NSGA-II makes use of the important techniques of non-dominated sorting and density estimation in the survival selection process [2].

The concept of domination can be explained as follows.

In a maximization problem, let u and v be two points in an m-dimensional objective space R^m, i.e. u, v $\in R^m$, u is said to dominate v if and only if:

1. $u_i > v_i$ for at least one index i $\in \{1, 2, \dots, m\}$
2. $u_j \geq v_j$ for every index j $\in \{1, 2, \dots, m\}$

This is to say, performance of u must be better than v in at least one objective (i.e. condition 1) and cannot be worse than v in any of the m objectives (i.e. condition 2) in order for us to say u dominates v.

NSGA-II makes use of the non-dominated sorting technique to give every individual solution a rank. The improved fast non-dominated sort is a sorting method that helps to separate the combined parents and child population into different fronts of dominance level. To achieve the first non-dominated front, each individual is compared with others to see whether it is being dominated. The best set of individual which dominates over other population will made up the first non-dominated front. The process repeats itself until every individual is allocated to a non-dominated front.

Density estimation computes the distance between individual solutions. In the case that there are more than required number of individuals with the same rank in a selection process, the algorithm considers the contribution of the individuals in diversity maintenance as well. The solution set that presents better diversity in objective space will have a higher chance to be selected for the next generation.

Evaluation of the overall fitness in NSGA-II is based on the rank as well as density estimation results. From the second generation onwards, each individual is to generate its own offspring through crossover and mutation. Survival selection is conducted among parents as well as offspring. Population for the next generation is selected through non-dominated sorting as well as density estimation.

2.2   MOEA/D

Essentially, MOEA/D decomposes a MOP into a set of scalar subproblems using uniformly distributed aggregation weight vectors and optimizes all of them simultaneously. Throughout the searching process, each individual solution is assigned with a scalar subproblem as its new objective. By doing so, individual solutions are in fact assigned with specified searching directions in the objective space. Uniformly distributed aggregation weight vectors are utilized to ensure the searching directions are evenly distributed in the objective space. Thus, a good approximation of PF with individual solutions evenly distributed along the real PF can be expected. The most common decomposition approaches that have been adopted are weight sum approach and Tchebycheff approach.

Weighted Sum Approach [6]. Weighted sum approach involves a convex combination of the different objectives in a MOP. In this case, each scalar subproblem is in fact a linear combination of the original objectives in the MOP with all the coefficients to be non-negative and sum to 1. Let $\lambda = (\lambda_1, \dots, \lambda_m)^T$ be a weight vector of a MOP with m objectives, $\lambda_i \geq 0$ for $i = 0, 1, \dots, m$; and $\sum_{i=1}^{m} \lambda_i = 1$. The corresponding scalar function g(x) produced with this weight vector λ would be:

$$\min \ g(x|\lambda) = \sum_{i=1}^{m} \lambda_i \cdot f_i(x) \tag{2}$$

where $f_i(x)$ is the real objective value obtained on objective i.

Tchebycheff Approach [7]. In the Tchebycheff approach, the scalar optimization problem with weight vector λ is the difference between the current performance on objective i and the optimal result obtained on the same objective, while i is decided to be the objective producing the maximum value of such difference.

$$\min \ g(x|\lambda) = \max_{1 \leq i \leq m}\{\lambda_i|f_i(x) - z_i^*|\} \tag{3}$$

where z_i^* is the reference point storing the optimal value found so far for objective i and $|f_i(x) - z_i^*|$ gives the absolute difference between the performance of decision variable x on objective i and the optimal result on objective i stored in the reference point.

3   Proposed Framework

The new framework proposed in this paper is a grouping approach combining the non-dominated sorting technique from NSGA-II and the decomposition concept from MOEA/D. After the very first initialization of the population, the overall population is immediately grouped into N groups. The original objectives from the MOP are decomposed into a fixed number of scalar subproblems and every group will be assigned with two of the scalar subproblems. Non-dominated sorting will thus be conducted within each group with the 2 scalar functions as the new objectives.

The essence of the new approach is to assign groups of individuals to look for different sections of the real PF while expecting the overall coverage by all the sections is a good approximation of the real PF.

3.1   Basic Concepts

Optimal Points for a Scalar Function. For a specific scalar function given, there always exists a corresponding point in the objective space representing the intersection of the specified searching direction by the scalar function with the real PF. This point is referred as the optimal solution for the specific scalar function [8].

Group Solution Lines. In the 3-D objective space corresponding to a 3-objective MOP, a section of PF eventually found by a group of individual solutions with two specific scalar functions assigned is expected to be a line, which is distributed along the real Pareto surface connecting the two optimal points in the objective space corresponding to the two scalar functions given. For the rest of the paper, such lines will be referred as the group solution lines.

3.2   Distribution of Group Solution Lines

The goal of the new framework in solving 3-objective MOPs is to cover the Pareto surface with group solution lines as evenly as possible. There are quite a number of ways to do so. One of the possible assignment scheme shown in Fig. 1 is eventually chosen to implement the new framework to deal with 3-objective MOP.

As shown in Fig. 1, twelve group lines are necessary in this case implying the group number in the new framework is fixed to be twelve (i.e. N=12). Essentially, the implementation of the new framework to solve 3-objective MOPs is a reversed process in which we identify the way to cover the Pareto surface with group solution lines first. Corresponding scalar functions are in turn identified according to the points predefined in the objective space. Each group is assigned with two scalar functions as the new objective to conduct non-dominated sorting within the group.

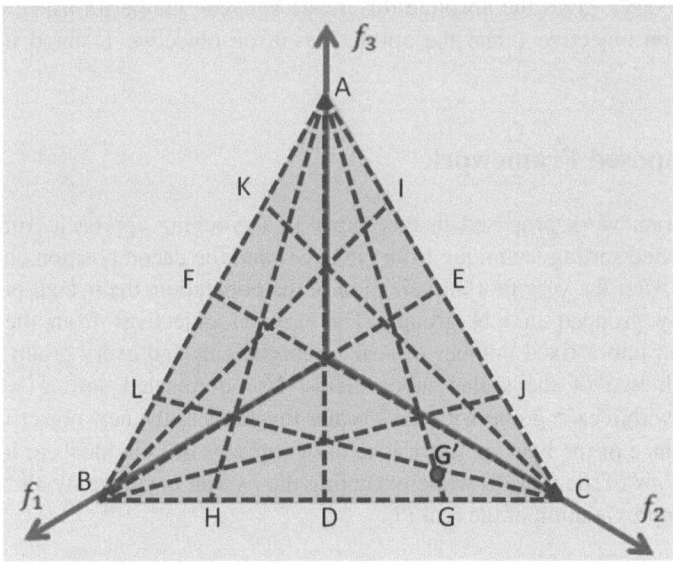

Fig. 1. Distribution of group solution lines on Pareto surface

One of the difficulties faced is about the scalar functions identification. Scalar functions corresponding to vertex points A, B and C are easy to be identified. Obviously, these three points imply the best performances on two of the objectives and the worst performance on the other. Due to the conflicting nature of the three objectives, we can interpret that the weight vectors assigned to the two objectives with best performance are 0.5 while the one assigned to the other objective with worst performance is 0 in order to generate such points in the objective space.

It is very difficult if not impossible to get the scalar functions to represent the other points in our group lines assignment scheme. The common feature for these points is that they perform extremely well on one of the three objectives. Taking points H, D and G as examples, they all have the best performance on the third objective f_3. The difficulty comes from the fact that it is hard to give an exact ratio to the importance among the three objectives for these points. Generally they could all be generated by scalar functions $0f_1 + 0f_2 + 1f_3$. However, due to the conflicting nature of the three objectives, point D would eventually be generated with this scalar function. The same idea can be applied to points E and F.

The rest of the points are approximated with the points that are extremely close to them. Taking point G as an example, approximated point G' would be a point that is extremely near point G with position slightly lifted up with respect to the third objective f_3. Weight vector assignment is found through solution mapping vector. The concept of solution mapping vector is proposed in [8]. The coordinates of the optimal point corresponding to a specific scalar function is found through the solution mapping vector λ', while λ' is expressed as follows:

$$\lambda' = \left(\frac{\frac{1}{\lambda_1}}{\sum_{i=1}^{m} \frac{1}{\lambda_i}}, \frac{\frac{1}{\lambda_2}}{\sum_{i=1}^{m} \frac{1}{\lambda_i}}, \cdots, \frac{\frac{1}{\lambda_m}}{\sum_{i=1}^{m} \frac{1}{\lambda_i}} \right) \tag{4}$$

Thus, with weight vector λ given, we are in fact able to find out the searching direction through the solution mapping vectors as calculated above. While in our case, it is a reversed process in which we predefined the points of intersection of the searching direction with PF, and calculate for the weight vector λ correspondingly. The coordinates are obtained based on the assumption that the PF of the problem is the simple case as the one show in Fig.1. Table 1 below shows the corresponding weight vectors to generate the points.

Table 1.

Point	Coordinates in Objective Space			Weight Vector Assignment		
	f1	f2	f3	f1	f2	f3
A	0	0	1	0.5	0.5	0
B	1	0	0	0	0.5	0.5
C	0	1	0	0.5	0	0.5
D	0.5	0.5	0	0	0	1
E	0	0.5	0.5	1	0	0
F	0.5	0	0.5	0	1	0
G	0.25	0.75	0	0.07	0.02	0.9
H	0.75	0.25	0	0.02	0.07	0.9
I	0	0.25	0.75	0.9	0.07	0.02
J	0	0.75	0.25	0.9	0.02	0.07
K	0.25	0	0.75	0.07	0.9	0.02
L	0.75	0	0.25	0.02	0.9	0.07

With the knowledge of the weight vectors corresponding to the points, group solution lines can be obtained by assigning designated scalar functions to the groups. The scalar function assignment scheme employed during implementation is shown in Table 2.

3.3   Rotating the Objective Space

Both weighted sum and Tchebycheff decomposition approaches are employed in the new framework to deal with 3-objective MOP. Experimental results demonstrated that weighted sum approach is good at searching for the solution lines that are at the edge of Pareto surface (i.e. solution line AD, BE and CF in Fig. 1). Tchebycheff approach is good at looking for the central lines (i.e. solution line AB, BC and CA in Fig. 1) but not able to look for those lines at the edge of the Pareto surface. Unfortunately there are no good ways to obtain the rest of the group lines with the current available decomposition schemes.

Table 2.

Group Index	Line Assignment	scalar function 1			scalar function 2		
		f1	f2	f3	f1	f2	f3
0	AB	0.5	0.5	0	0	0.5	0.5
1	AD	0.5	0.5	0	0	0	1
2	AG	0.5	0.5	0	0.07	0.02	0.9
3	AH	0.5	0.5	0	0.02	0.07	0.9
4	BC	0	0.5	0.5	0.5	0	0.5
5	BE	0	0.5	0.5	1	0	0
6	BI	0	0.5	0.5	0.9	0.07	0.02
7	BJ	0	0.5	0.5	0.9	0.02	0.07
8	CA	0.5	0	0.5	0.5	0.5	0
9	CF	0.5	0	0.5	0	1	0
10	CL	0.5	0	0.5	0.02	0.9	0.07
11	CK	0.5	0	0.5	0.07	0.9	0.02

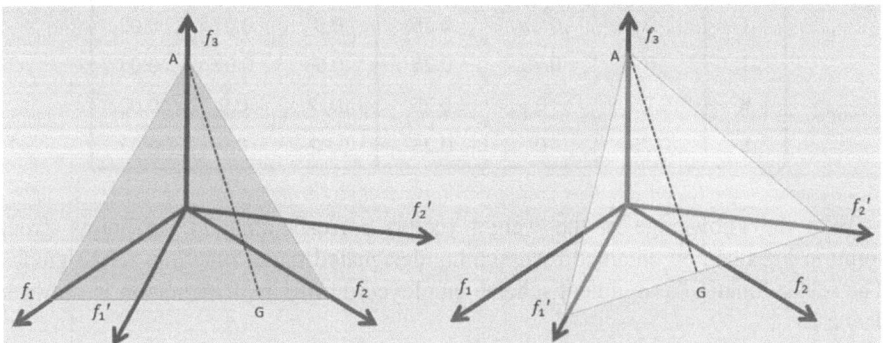

Fig. 2. Rotation of the Objective Space

However, given that Tchebycheff approach is good at searching for central lines, one possible way to tackle this problem is to rotate the coordinate system of the objective space so that desired group lines can be rotated to the position of a central line. Tchebycheff decomposition approach can thus be adopted to obtain that line easily.

The idea is illustrated with group solution line AG in Fig. 2. Basically we rotate the coordinate system and input the updated objective values into the fitness evaluation process while adopting Tchebycheff approach. Decomposition scheme in fact works in a rotated objective space in which AG is at the central position. Thus Tchebycheff is adopted to obtain AG easily. Throughout the process, the real objective space is not rotated.

In our implementation, we are actually trying to rotate the coordinate system of the objective space. Thus rotation with angle θ about f1 axis in fact means to rotate the coordinate system with an angle of θ degrees about f1 axis.

Assuming that the shape of true Pareto Front is an equilateral triangle as shown in Fig. 1, and then the angle of rotation to obtain AG at the central position is calculated to be 26.565° about f3 axis. This angle of rotation is in fact the universal angle that can be used to obtain all the missing group lines. The differences are about the axis of rotation as well as the direction of rotation. The latter could be either clockwise (i.e. θ=26.565°) or anticlockwise (i.e. θ=-26.565°).

Table 3.

Group Index	Line Assignment	Group Index /4	%4	Decomposition Scheme	Rotate Axis	Angle of rotation (degree)
0	AB	0	0	weighted sum	NA	
1	AD	0	1	Tchebycheff	NA	
2	AG	0	2	Tchebycheff with rotation	f3	26.565
3	AH	0	3	Tchebycheff with rotation	f3	-26.565
4	BC	1	0	weighted sum	NA	
5	BE	1	1	Tchebycheff	NA	
6	BI	1	2	Tchebycheff with rotation	f1	26.565
7	BJ	1	3	Tchebycheff with rotation	f1	-26.565
8	CA	2	0	weighted sum	NA	
9	CF	2	1	Tchebycheff	NA	
10	CL	2	2	Tchebycheff with rotation	f2	26.565
11	CK	2	3	Tchebycheff with rotation	f2	-26.565

The new framework targeting at 3-objective MOPs is eventually implemented as follows. The first four groups (i.e. Group 0, 1, 2, 3) look for lines starting with point A. The axis of rotation is f_3 for groups looking for AH and AG. The second four groups (i.e. Group 4, 5, 6, 7) look for lines starting with point B. The axis of rotation is f_1 for groups looking for BI and BJ. The last four groups (i.e. Group 8, 9, 10, 11) look for lines starting with point C. The axis of rotation is f_2 for groups looking for CK and BL. A summary is given in Table3.

4   Experimental Results

Performance of the proposed algorithm is evaluated with various representative benchmark problems [9][10]. The experimental setting is fixed with 150,000 evaluation times for all the 3-objective problems. Population size is taken to be 300 for all the simulation runs. The Inverted Generational Distance (IGD) [11][12] is utilized to provide a quantitative evaluation of the performance. Table 4 shows the mean of IGD values of the proposed framework and traditional NSGA-II algorithm over 30 independent runs. Best entries are marked in boldface. Same offspring reproducing mechanisms (i.e. simulated binary crossover (SBX) [13]) were adopted by all the algorithms in this evaluation of their performance. This is to avoid the effects from different offspring reproducing mechanisms on the performance, thus to ensure the performances are solely based on the frameworks.

Table 4. Mean of IGD values over 30 independent runs

Benchmark Problems (3-objective)		Average IGD values for 30 runs	
		NSGA-II(SBX)	New framework (SBX)
DTLZ cases	DTLZ1	0.01451	**0.01394**
	DTLZ2	**0.04040**	0.04143
	DTLZ3	**0.04040**	0.04309
	DTLZ4	0.78061	**0.75946**
UF cases	UF8	0.21940	**0.19853**
	UF9	0.16350	**0.15094**
	UF10	0.32360	**0.22053**

According to the experimental results, the proposed framework performs better than the original NSGA-II in 5 out 7 benchmark problems, which demonstrates that the new mechanism is able to obtain better diversity and convergence in handling 3-objective problems.

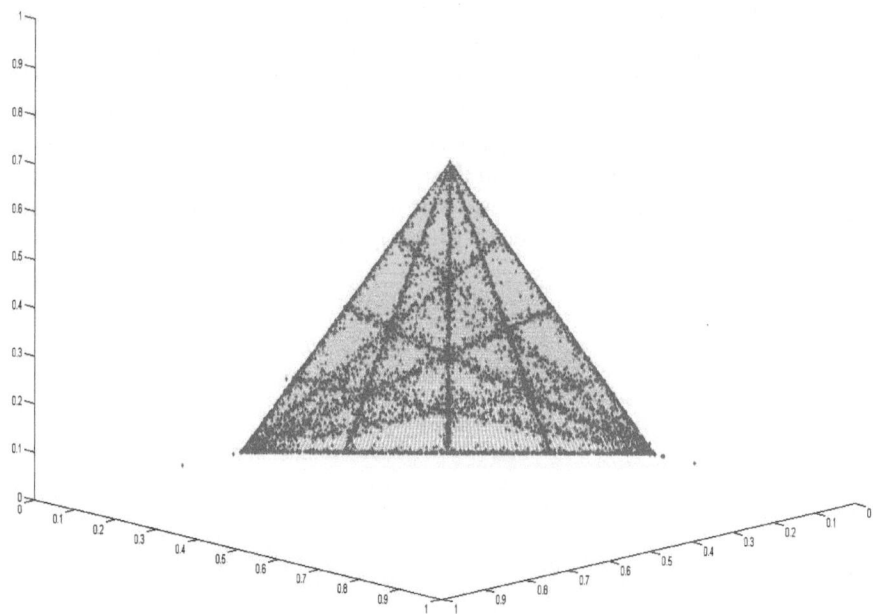

Fig. 3. Overall performance of the new framework for 30 runs on DTLZ1

DTLZ1 New Framework

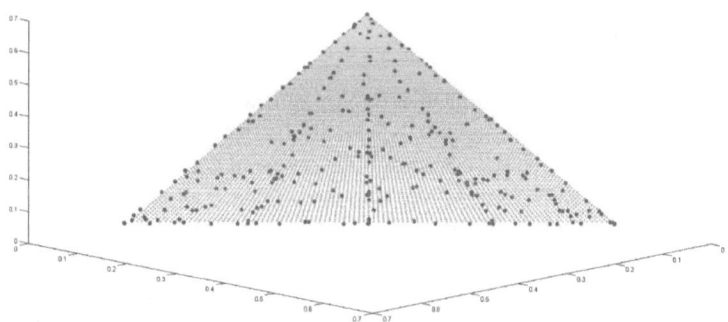

Fig. 4. Single run performance on DTLZ1 by the new framework

In order to give a more comprehensive picture of the resulted performance, Fig. 3 plots all the solutions obtained by the proposed framework for DTLZ1 over 30 runs. The green dots represent the true Pareto Front, and the red dots represent the obtained solutions. From Fig. 3, it could be observed that the new framework is able to produce the group lines to cover the Pareto surface in the expected way, given the PF of the problem is a regular surface in the 3-D objective space. To visually show the difference between the proposed framework and traditional NSGA-II method, Fig. 4 and

Fig. 5 plot the final solution sets obtained by both methods on DTLZ1 for a single run. Based on the plots, the solutions obtained by NSGA-II are relatively random with regard to the allocation pattern on the Pareto surface, while the solutions found by the new framework follows exactly the predefined group solution lines as expected. The new framework is able to find the edge lines of the Pareto surface in the 3-D objective space and thus to provide the outline of the real Pareto surface to the decision makers. From this perspective, the new framework is able to produce a better approximation of the real PF compared to the current existing algorithms in 3-objective problems.

DTLZ1 NSGA-II

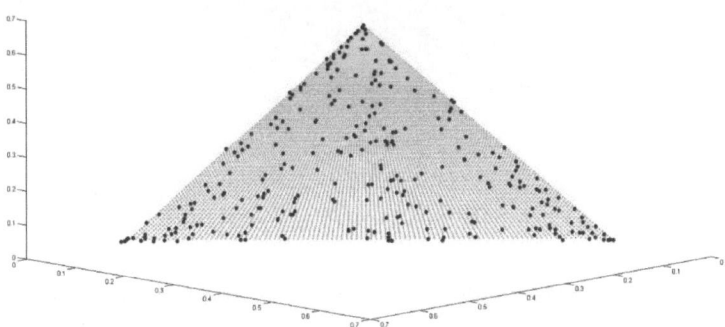

Fig. 5. Single run performance on DTLZ1 by NSGA-II

5   Conclusion

This paper proposes a new multi-objective optimization framework that incorporates the concepts of decomposition into NSGA-II. By rotating the objective space, the whole objective space is better covered by the evolved solutions in a more reasonable manner. Empirical results demonstrate that the proposed algorithm outperforms traditional NSGA-II in handling 3-objective problems. Future work could be working on MOPs with disparately scaled objectives, in which rotation angles of new objective spaces need to be calculated correspondingly. Moreover, performance comparison with MOEA/D and other state-of-the-art MOEAs are expected for further investigation.

References

1. Zhang, Q., Li, H.: MOEA/D: A multiobjective evolutionary algorithm based on decomposition. IEEE Transactions on Evolutionary Computation 11(6), 712–731 (2007)
2. Deb, K., Pratap, A., Agarwal, S., Meyarivan, T.: A fast and elitist multiobjec-tive genetic algorithm: NSGA-II. IEEE Transactions on Evolutionary Computation 6(2), 182–197 (2002)

3. Antonin, P., Antonio, L.J., Carlos, A.C.C.: A Survey on Multiobjective Evolutionary Algorithms for the Solution of the Portfolio Optimization Problem and Other Finance and Economics Applications. IEEE Transactions on Evolutionary Computation 17(3), 321–344 (2013)
4. Anirban, M., Ujjwal, M., Sanghamitra, B., Carlos, A.C.C.: A Survey of Multiobjective Evolutionary Algorithms for Data Mining: Part I. IEEE Transactions on Evolutionary Computation 18(1), 4–19 (2014)
5. Pindoriya, N.M., Singh, S.N., Kwang, Y.L.: A Comprehensive Survey on Multi-objective Evolutionary Optimization in Power System Applications. In: 2010 IEEE Power and Energy Society General Meeting, pp. 1–8 (2010)
6. Voss, T., Beume, N., Rudolph, G., Igel, C.: Scalarization versus indicator-based selection in multi-objective CMA evolution strategies. In: Proc (IEEE World Congress on Computational Intelligence). IEEE Congress on Evolutionary Computation, CEC 2008, pp. 3036–3043 (2008)
7. Miettinen, K.: Nonlinear Multiojective Optimization. Kluwer, Norwell (1999)
8. Qi, Y.T., Ma, X.L., Liu, F., Jiao, L.C., Sun, J.Y., Wu, J.S.: MOEA/D with Adaptive Weight Adjustment. Evolutionary Computation 22(2), 231–264 (2014)
9. Zhang, Q., Zhou, A., Zhao, S., Suganthan, P.N., Liu, W., Tiwari, S.: Multiobjective optimization test instances for the CEC 2009 special session and competition. Tech. Rep. CES-487, University of Essex and Nanyang Technological University (2008)
10. Deb, K., Thiele, L., Laumanns, M., Zitzler, E.: Scalable Test Problems for Evolutionary Multi-Objective Optimization, Zurich, Switzerland, Tech. Rep. 112 (2001)
11. Veldhuizen, D.A.V., Lamont, G.B.: On measuring multiobjective evolutionary algorithm performance. In: 2000 Congress on Evolutionary Compuation, vol. 1, IEEE Service Center, Piscataway (2000)
12. Veldhuizen, D.A.V., Lamont, G.B.: Multiobjective evolutionary algorithm research: A history and analysis. Technical Report TR-98-03, Department of Electrical and Computer Engineering, Graduate School of Engineering, Air Force Institute of Technology, Wright-Patterson AFB, OH (1998)
13. Deb, K., Agrawal, R.B.: Simulated binary crossover for continuous search space. Complex Syst. 9, 115–148 (1995)

PaCcET: An Objective Space Transformation to Iteratively Convexify the Pareto Front

Logan Yliniemi and Kagan Tumer

Oregon State University
Corvallis, Oregon, USA
logan.yliniemi@engr.orst.edu
kagan.tumer@oregonstate.edu

Abstract. In multi-objective problems, it is desirable to use a fast algorithm that gains coverage over large parts of the Pareto front. The simplest multi-objective method is a linear combination of objectives given to a single-objective optimizer. However, it is proven that this method cannot support solutions on the concave areas of the Pareto front: one of the points on the convex parts of the Pareto front or an extreme solution is always more desirable to an optimizer. This is a significant drawback of the linear combination.

In this work we provide the Pareto Concavity Elimination Transformation (PaCcET), a novel, iterative objective space transformation that allows a linear combination (in this transformed objective space) to find solutions on concave areas of the Pareto front (in the original objective space). The transformation ensures that an optimizer will always value a non-dominated solution over any dominated solution, and can be used by any single-objective optimizer. We demonstrate the efficacy of this method in two multi-objective benchmark problems with known concave Pareto fronts. Instead of the poor coverage created by a simple linear sum, PaCcET produces a superior spread across the Pareto front, including concave areas, similar to those discovered by more computationally-expensive multi-objective algorithms like SPEA2 and NSGA-II.

1 Introduction

Multi-objective optimization is very important in the real world [12]. Multiple competing objectives must be balanced in applications like the design of high-speed transport planes [13], the design of trusses [2], job shop scheduling [21], urban planning [1], and greywater reuse [16]. In these, the "best" solutions characterize a tradeoff between the multiple objectives. This array of solutions is known as the "Pareto optimal set", and is a commonly sought-after solution type for a multi-objective problems [3].

Successful methods function on arbitrarily-shaped Pareto fronts, as the shape is unknown before optimization. One simple method is to use a linear combination of all objectives, which has the benefits of being easy to understand and computationally cheap, but this is unable to find the concave areas of a Pareto front [3,4,11,12,15,17,18,19], because a convex part (or an extreme point) will be more desirable than the concave region [12].

G. Dick et al. (Eds.): SEAL 2014, LNCS 8886, pp. 204–215, 2014.

The primary contribution of this work is to present the Pareto Concavity Elimination Transformation (PaCcET), a novel, optimizer-independent, iterative multi-objective transformation. It transforms the objective space so that the Pareto Front is convex, and requires only a single user-defined parameter. This allows an linear combination with unit weights (in the transformed objective space) to find concave areas of the Pareto front (in the original objective space), removing the major drawbacks of a linear combination, and allowing a simple linear combination to be used instead of more computationally expensive multi-objective evolutionary algorithms, and produce similar results.

This work is organized as follows: Section 2 provides background on multi-objective problems and multi-objective methods. Section 3 describes PaCcET. Section 4 provides theoretical guarantees for PaCcET. Sections 5 and 6 describe two test domains and show results using PaCcET. Section 7 discusses the work and concludes.

2 Background

This work draws from many distinct concepts from within multi-objective research, which we introduce in this section. We assume (without loss of generality) pure minimization of k objectives $\Lambda \in \mathbb{R}^k$ through the control of the n design variables $\Omega \in \mathbb{R}^n$.

Dominance: A solution u dominates another solution v ($u \prec v$) if it scores lower on all criteria (objectives $c \in C$): $\forall c \in C[f_c(u) < f_c(v)]$. A solution u weakly dominates another solution v ($u \preceq v$) if it scores equal on some objectives, but less on others: $\forall c \in C[f_c(u) \leq f_c(v)] \wedge \exists j \in C[f_j(u) < f_j(v)]$ [20].

Pareto optimal set: A solution which is not dominated by any other feasible solution is part of the Pareto optimal set $\mathbf{P}^*$. As an incomplete optimizer solves a problem, it will approximate $\mathbf{P}^*$ with a *Pareto approximate set* P_I^* at iteration I.

Multi-objective spaces: $\Omega \in \mathbb{R}^n$ is the *design variable space* (domain). $\Lambda \in \mathbb{R}^k$ is *objective space* (range or codomain) [20]. The mapping from $\Omega \to \Lambda$ is unknown to the chosen optimizer Ξ, but is usually repeatable with some stochastic error. We also use Λ^{norm}, a normalized version of Λ, which places P_I^* elements $\in [0{:}1]$, and Λ^τ, the post-PaCcET analogue to Λ. Additionally, we break Λ into three sub-spaces, Λ_D, Λ_N, and Λ_B. Λ_D is the subspace in Λ that is strongly dominated by the current P_I^*. Λ_N is the subspace of Λ that is non-dominated. Λ_B forms the border between the two (Fig. 1), and includes P_I^* and all points weakly dominated by P_I^*.

Utopia and nadir vectors: Two important concepts in multi-objective problems are the utopia and nadir points. The utopia point takes on the best possible value for each objective, minus some small amount so that it is always infeasible. This point is difficult to find, requiring an optimization for each objective individually. Instead, we approximate:

$$\hat{u}^\circ(c) = min(P_I^*(c)) - \Delta \tag{1}$$

where $\hat{u}^\circ(c)$ is the cth element in the estimated utopia vector, $min(P_I^*(c))$ is the minimum c^{th} element of any vector in P_I^*, and Δ is a small value [5]. The nadir point takes the worst value for each objective in the Pareto optimal set, which we approximate:

$$\hat{u}^{\text{nad}}(c) = max(P_I^*(c)) \tag{2}$$

but it is very important to note that this is a distinct concept from the worst feasible vector; it is instead the upper bound of the objective values for solutions within P_I^* [5].

2.1 Multi-objective Methods

Many successful multi-objective algorithms have been developed. In this work we address linear combinations since they are a component of PaCcET, as well as NSGA-II and SPEA2, two successful multi-objective evolutionary algorithms.

Linear Combination: a simple metric that is sufficient, but not necessary, for finding Pareto optimal points [12]:

$$LC(w, v) = \sum_c w(c)v(c) \tag{3}$$

where $LC(w, v)$ is the linear combination evaluation or L_1 norm of vector v, $v(c)$ is the evaluation of vector v on the cth objective, w is the vector of weights, and $w(c)$ is the weight for the cth objective.

This method is computationally cheap when paired with typical optimizers like an evolutionary algorithm [19], but presents three primary problems. First, as the number of objectives increases, the choice of weights can become difficult. Second, this method is incapable of finding certain areas of the Pareto Front, those that are non-convex. Third, incrementing the weights evenly to converge to different parts of the Pareto front does not necessarily lead to evenly-spaced solutions along the front [4].

NSGA-II: is an evolutionary algorithm which sorts solutions into a series of successive fronts. Those solutions on the less-dominated fronts are more desirable and are kept. To break ties, a local density measure is used. Details can be found in [6].

SPEA2: is an evolutionary algorithm which assigns each vector a "strength" equal to the number of vectors in the current population it dominates. Each vector then sums the strengths of all vectors which dominate it, and this forms a raw fitness evaluation. This is altered by a local k-nearest neighbor density calculation, and the best solutions survive. Details can be found in [8].

3   Pareto Concavity Elimination Transformation (PaCcET)

Each point in the current Pareto-approximate set, P_I^* represents a tradeoff between which we are indifferent [9]. PaCcET makes each solution on Λ_B (including P_I^*) equally valuable to a linear combination in Λ^τ through a two step transformation, which first transforms from Λ to Λ^{norm}, and then transforms from Λ^{norm} to Λ^τ, where the τ superscript on any space or set denotes the transformed space or set. This means that any Pareto-approximate solution will have a linear combination evaluation of $(k-1)$ when all weights are set to 1. All solutions in Λ_N^τ will have a linear combination evaluation $< (k-1)$, and all solutions in Λ_D^τ will have a linear combination evaluation $> (k-1)$.

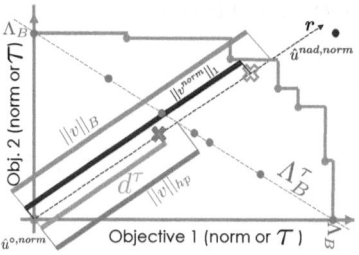
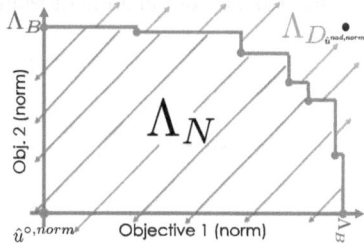

Fig. 1. Visualization of quantities used in transformation (left) and partitions in the multi-objective space (right). The vector v^{norm} is represented by the hollow green X mark, and v^τ by the solid red X mark. v^{norm} lies outside of the dominated hypervolume, so is a desirable point to discover. Green dots correspond to vectors in $P_I^{*,\text{norm}}$ (which form the border, Λ_B, between the non-dominated hyperspace Λ_N and the dominated hyperspace, Λ_D). Red correspond to their transformations in $P_I^{*\tau}$. All measurements are Manhattan Distance (L_1 norm) along r.

Algorithm: To determine the transformed evaluation for a given solution vector v, we require the current Pareto approximate set P_I^*, from which we can calculate the approximate utopia point $\hat{u}^\circ$ based on P_I^* (Eq. 1), and the matching nadir approximation $\hat{u}^{\text{nad}}$ (Eq. 2).

The first step is to normalize the target vector v such that each objective takes on a value not less than 0, transforming Λ to Λ^{norm} [5,12]:

$$v^{\text{norm}}(c) = \frac{v(c) - \hat{u}^\circ(c)}{\hat{u}^{\text{nad}}(c) - \hat{u}^\circ(c)} \tag{4}$$

By definition $\hat{u}^{\text{nad,norm}} \equiv 1$ and $\hat{u}^{\circ,\text{norm}} \equiv 0$, and each element of a member of P_I^* will be in the range [0:1].

The second step is to perform the transformation from Λ^{norm} to Λ^τ. Within this process, we use the unit vector r that points from $\hat{u}^{\circ,norm}$ toward v^{norm}:

$$r = \frac{v^{\text{norm}}}{|v^{\text{norm}}|} \tag{5}$$

All distance measurements in the transformation process are taken along the direction of r. We measure three distances for use in PaCcET:

Algorithm 1. PaCcET for iteration I

Require: Set of solutions V 6:   Find $||v^{\text{norm}}||_1$ (Eq. 6)
Require: Pareto Approximate Set P_I^* 7:   Find $||v||_\text{B}$ (Eq. 7)
1: Find $\hat{u}^\circ, \hat{u}^{\text{nad}}$ (Eq. 1–2)   8:   Find $||v||_{\text{hp}}$ (Eq. 8)
2: $\forall c \ w_c = 1$ 9:   Find d^τ (Eq. 9)
3: **for all** Solutions $i \in V$ **do**   10:   Find v_i^τ (Eq. 10)
4:    Find v_i^{norm} (Eq. 4)   11: $Fit_{\text{PaCcET}}(v_i) = LC(v_i^\tau)$ (Eq. 3)
5:    Find r (Eq. 5)   12: **end for**

- L_1 distance (linear combination or Manhattan Distance) from $\hat{u}^{\circ,norm}$ to v^{norm}:

$$||v^{norm}||_1 = \sum_i v_i^{norm} \tag{6}$$

- L_1 distance from $\hat{u}^{\circ,norm}$ to the normalized dominated border Λ_B^{norm} along r:

$$||v||_B = min(\gamma) \ni \gamma r \succeq P_I^* \tag{7}$$

- L_1 distance from $\hat{u}^{\circ,norm}$ to the normalized utopia hyperplane Λ_B^τ [14] along r:

$$||v||_{hp} = \beta \ni \sum_i \beta r_i = (k-1) \tag{8}$$

We then calculate d^τ, which determines where v^τ is located:

$$d^\tau = ||v||_{hp} \frac{||v^{norm}||_1}{||v||_B} \tag{9}$$

And finally we determine the location of v^τ, enclosing the whole process:

$$v^\tau = d^\tau r = PaCcET(v) \tag{10}$$

Choosing the Maximum Size of P_I^*, the Pareto approximate set: P_I^* is maintained in the same way as any Pareto optimality calculation. However, for computation and memory concerns, its size must be limited. The size of P_I^* is the only user-defined parameter in PaCcET, and corresponds directly to the granularity of the Pareto front estimation. In our experiments we use 250 as the size. We ran tests with a size as small as 50, in which the algorithm still functions, but provides a very coarse approximation of the true Pareto front. Once over the chosen size, we used random elimination of non-extreme elements. We also tested with nearest-neighbor elimination and k-nearest neighbor elimination, the performance of PaCcET was not sensitive.

4   Theoretical Properties of PaCcET

In this section we provide two theorems which together prove that PaCcET finds Pareto optimal solutions, even in concave areas of the Pareto front. We begin by assuming:

Assumption A1. *The system designer specifies k points that are incomparable to the Pareto front, which describe a hyper-prism that completely bounds the Pareto Front.*

Assumption A2. *Optimizer Ξ solves the PaCcET problem exactly in a single iteration.*

Assumption A3. *The feasible region has no solutions that are weakly dominated by the Pareto Front.*

Assumption A4. *The Pareto Front is continuous.*

A1 provides us vectors with which we seed P_I^*, and assures PaCcET is calculable in the whole feasible objective space. A2 allows us to use the exact solution to the PaCcET minimization problem to determine how P_I^* changes over iterations. A3 and A4 allow us to draw conclusions in k-objective space without any other restrictions on the shape of the Pareto Front.

Theorem 1. *The solution to the PaCcET optimization problem will be Pareto Optimal.*

Proof. There exists an infinite number of possible rays $r \in R$ (where R is the set of all rays originating from $\mathbf{0}$) on which the true solution may exist. This solution exists only along one of those rays, which must pass through the feasible space. We do not seek to determine which r it lies on. For any individual r, the PaCcET optimization problem takes the form (Eq. 9, reorganized):

$$\min(d^\tau) = \min\left(||v||_{\mathrm{hp}}\frac{||v^{\mathrm{norm}}||_1}{||v||_{\mathrm{B}}}\right) \tag{11}$$

And for a constant r, $||v||_{\mathrm{B}}$ and $||v||_{\mathrm{hp}}$ are constant on a given iteration:

$$\min(d^\tau) = \min(\alpha||v^{\mathrm{norm}}||_1) \tag{12}$$

where α is some positive constant. $||v^{\mathrm{norm}}||_1$ increases monotonically as distance from the origin increases, therefore d^τ does as well. The minimum of d^τ, then, will be on the border of the feasible space, a Pareto Optimal Solution or a weakly dominated solution [12]. By A3 and A4, this is a Pareto optimal solution. This can also be assured by the same logic as [4], since it is equivalent to a scaled linear combination. □

Theorem 2. *PaCcET finds solutions in concave areas of the Pareto front.*

Proof. We prove this by contradiction. Assume a globally concave search space. By theorem 1, in the worst case, the solution to the PaCcET optimization problem will lead to the k anchor points (single objective extremes) in the first k iterations. By A4, we know additional Pareto optimal points exist. We show that the d^τ calculations for those points in the current P_I^* is greater than those in Λ_N (super/sub-scripts denoting the calculation for a member of the set named in the super/sub-script):

$$d_{P_I^*}^\tau > d_{\Lambda_N}^\tau \tag{13}$$

$$||v||_{\mathrm{hp}}^{P_I^*}\frac{||v_{P_I^*}^{\mathrm{norm}}||_1}{||v||_{\mathrm{B}}^{P_I^*}} > ||v||_{\mathrm{hp}}^{\Lambda_N}\frac{||v_{\Lambda_N}^{\mathrm{norm}}||_1}{||v||_{\mathrm{B}}^{\Lambda_N}} \tag{14}$$

$$(k-1)\frac{||v_{P_I^*}^{\mathrm{norm}}||_1}{||v||_{\mathrm{B}}^{P_I^*}} > (k-1)\frac{||v_{\Lambda_N}^{\mathrm{norm}}||_1}{||v||_{\mathrm{B}}^{\Lambda_N}} \tag{15}$$

By definition $||v||_{\mathrm{hp}}^{P_I^*} = (k-1)$. Also, $\frac{||v_{P_I^*}^{\mathrm{norm}}||_1}{||v||_{\mathrm{B}}^{P_I^*}} = 1$ because $||v_{P_I^*}^{\mathrm{norm}}||_1 = ||v||_{\mathrm{B}}^{P_I^*}$, and the quantity $\frac{||v_{\Lambda_N}^{\mathrm{norm}}||_1}{||v||_{\mathrm{B}}^{\Lambda_N}} \in [0:1)$, because it is in the non-dominated subspace (so $||v_{\Lambda_N}^{\mathrm{norm}}||_1 < ||v||_{\mathrm{B}}^{\Lambda_N}$), and the inequality in Eq. 15 holds. Because of Theorem 1, we know that the solution will be Pareto Optimal, and because of the globally concave assumption, we know this point is on a concave region of the Pareto front. □

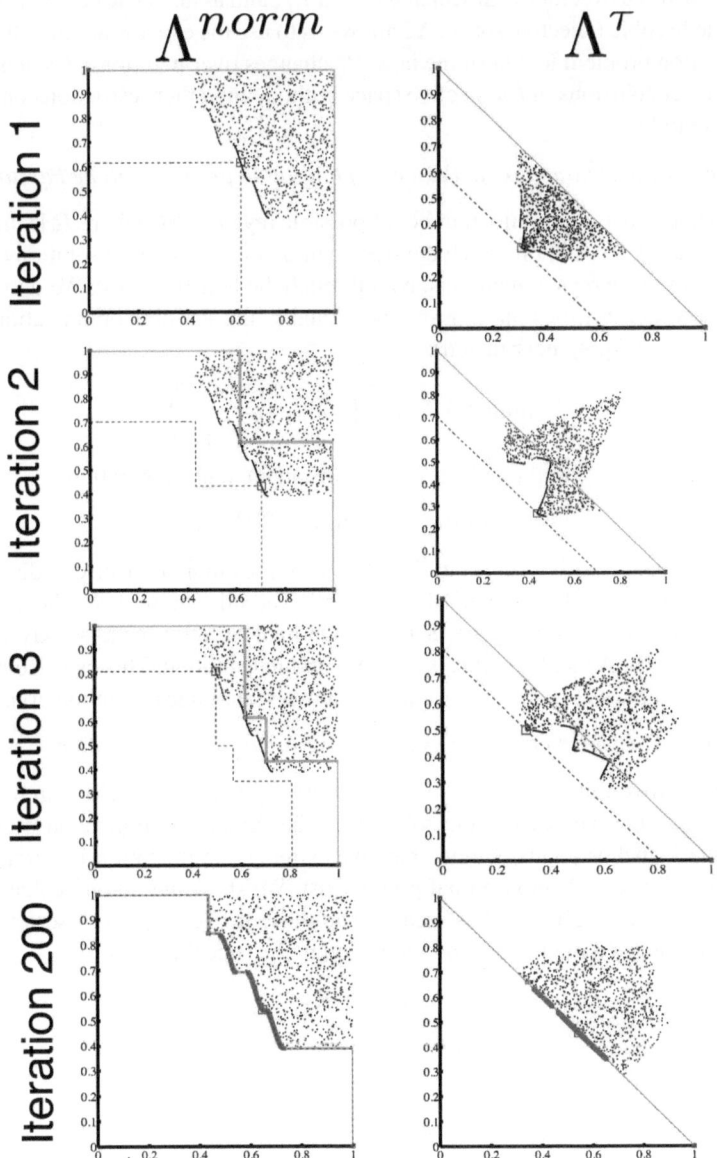

Fig. 2. Visualization of PaCcET procedure over iterations. The left column is the normalized objective space Λ^{norm}. The right column is the transformed objective space, Λ^{τ}. The rows show, in turn, the optimizer working at the 1st, 2nd, 3rd, and 200th iteration. In the left column, Black points are candidate solutions. Red points are solutions in P_I^*, the Pareto approximate set. The green solid line denotes Λ_B^{norm} The blue square denotes the true solution to the PaCcET optimization problem at that iteration. The blue dashed line is the level curve of the PaCcET evaluation on which all solutions are as valuable as the discovered solution. In the right column, the colors and symbols map to the transformed versions of the same points as described previously, in Λ^{τ}.

Implications: The significance of these two theorems is as follows: the true solution to the PaCcET problem will always be a Pareto optimal solution, and PaCcET will be able to find concave areas of the Pareto front. Because the assumptions used to generate these conclusions are restrictive, in the following empirical results sections, we take steps to violate each of the assumptions categorically, and PaCcET is still able to find good coverage over concave Pareto fronts.

5   Experiment: KUR

As a first experimental domain, we use a test problem (KUR) from multi-objective optimization with a discontinuous and locally concave Pareto front (which breaks A4) [5]:

$$f_1(\mathbf{x}) = \sum_{i=1}^{2} \left[-10 \, exp \left((-0.2) \sqrt{x_i^2 + x_{i+1}^2} \right) \right] \tag{16}$$

$$f_2(\mathbf{x}) = \sum_{i=1}^{3} \left[|x_i|^{0.8} + 5 \sin(x_i)^3 \right] \tag{17}$$

Where f_1 and f_2 are to be minimized by controlling the decision variables:

$$x_i \in [-5, 5] \quad ; \quad i \in \{1, 2, 3\} \tag{18}$$

A vector $\mathbf{x}$ is evaluated:

$$Fit_{\text{LC}}(\mathbf{x}) = w_1 f_1 + w_2 f_2 \tag{19}$$

where in this experiment, $w_1 = w_2 = 1$ (other values lead to different portions of the Pareto front being better covered, but similar overall performance). For PaCcET:

$$Fit_{\text{PaCcET}}(\mathbf{x}) = f_1^{\tau} + f_2^{\tau} \tag{20}$$

where f_1^{τ} and f_2^{τ} represent the transformed objectives, within Λ^{τ}, calculated as:

$$\{f_1^{\tau}, f_2^{\tau}\} = PaCcET(\{f_1, f_2\}) \tag{21}$$

As the optimizer Ξ, we use an evolutionary algorithm (which breaks A2), in which the population members are vectors of length 3 that meet the criteria set forth in Eq. 18. We maintain a population of 100 solutions, with the 50 worst-performing solutions removed after each generation, replaced by copies of the winner of 50 binary tournaments, with each element of the vector changed by a random number chosen by a normal distribution centered around 0 with standard deviation 0.25. We do not seed P_I^* (which breaks A1).

Figure 3 shows the Empirical Attainment Function (EAF) [10] for each method, respectively. It shows PaCcET's worst performance exceeds that of the linear combination's median performance, and PaCcET's worst performance exceeds NSGA-II's worst performance. SPEA2 and PaCcET perform comparably after 5000 generations.

Figure 4 shows the percent of dominated hypervolume by PaCcET and two successful multi-objective methods, SPEA2 and NSGA-II, as a function of number of individual fitness evaluations. PaCcET proceeds faster than the other two methods toward the Pareto front. All methods shown eventually converge to a good approximation of the Pareto front, and dominate a similar amount of hypervolume.

6 Experiment: DTLZ2

As a second experimental domain, we use one of the test problems out of the battery developed by Deb, Thiele, Laumanns and Zitzler, DTLZ2 [7]. A solution is described by a vector ($\mathbf{x} = \{x_1, x_2, \mathbf{x_M}\}$) of length 12, where 2 elements (x_1, x_2) determine at what angles in the 3 dimensional objective space evaluation v will lie and the remaining 10 elements ($\mathbf{x_M}$) determine the distance from the origin at which v will lie. The three functions to be minimized are:

$$f_1(\mathbf{x}) = (1 + g(\mathbf{x}_M)) \cos\left(x_1 \frac{\pi}{2}\right) \cos\left(x_2 \frac{\pi}{2}\right) \tag{22}$$

$$f_2(\mathbf{x}) = (1 + g(\mathbf{x}_M)) \cos\left(x_1 \frac{\pi}{2}\right) \sin\left(x_2 \frac{\pi}{2}\right) \tag{23}$$

$$f_3(\mathbf{x}) = (1 + g(\mathbf{x}_M)) \sin\left(x_1 \frac{\pi}{2}\right) \tag{24}$$

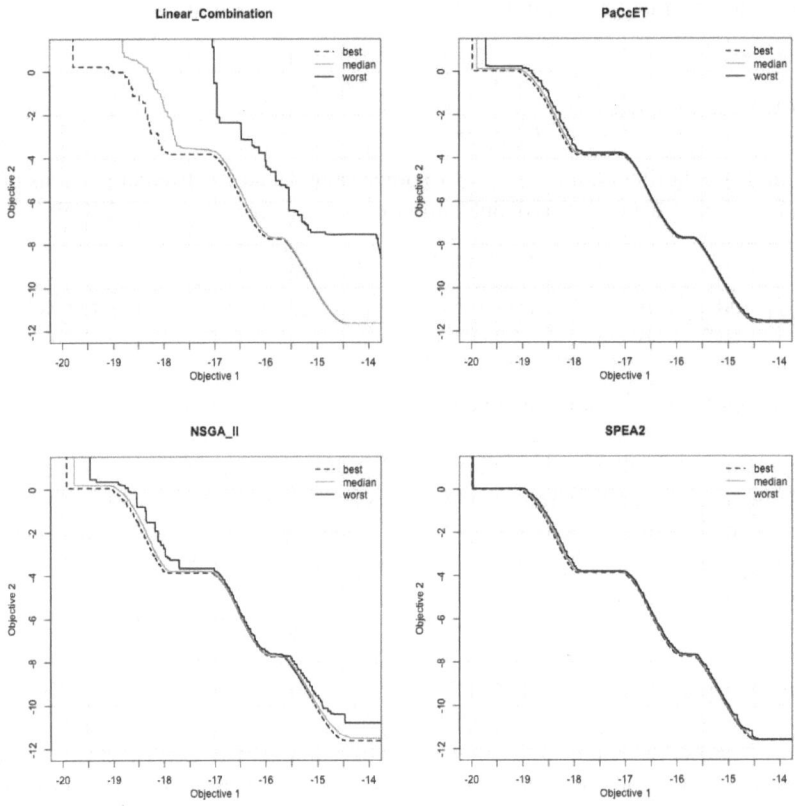

Fig. 3. KUR Empirical Attainment Functions, shown in Λ

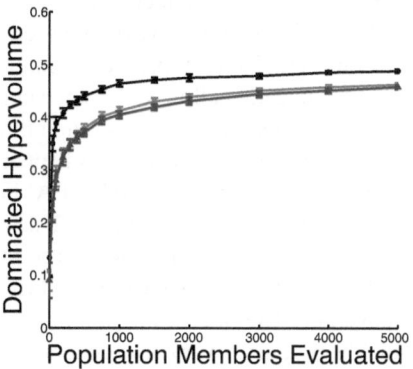

Fig. 4. Percent of hypervolume dominated in KUR, calculated using the limits of Fig. 3 over 50 statistical runs. Error in the mean ($\sigma/\sqrt{N}$ where $N = 50$), is smaller than the plotted symbols

subject to each element of **x** remaining in the range [0:1], and the evaluation $g(\mathbf{x}_M)$ calculated as:

$$g(\mathbf{x}_M) = \sum_{x_i \in \mathbf{x}_M} (x_i - 0.5)^2 \tag{25}$$

This results in a known Pareto front that can be described by the octant of a sphere of radius 1 for which f_1, f_2, and f_3 are all positive. The feasible space has a large area that is weakly dominated by the Pareto front (which breaks A3).

The fitness of a vector **x** is calculated as:

$$Fit_{LC}(\mathbf{x}) = w_1 f_1^{\text{norm}} + w_2 f_2^{\text{norm}} + w_3 f_3^{\text{norm}} \tag{26}$$

and for PaCcET,

$$Fit_{\text{PaCcET}}(\mathbf{x}) = f_1^\tau + f_2^\tau + f_3^\tau \tag{27}$$

where f_1^τ, f_2^τ, and f_3^τ represent the transformed objectives, within Λ^τ, calculated as:

$$\{f_1^\tau, f_2^\tau, f_3^\tau\} = PaCcET(\{f_1, f_2, f_3\}) \tag{28}$$

We use the same optimizer $\varXi$ for DTLZ2 as for KUR (which breaks A2), except members are vectors of length 12 with each element in the range [0:1], and the mutation

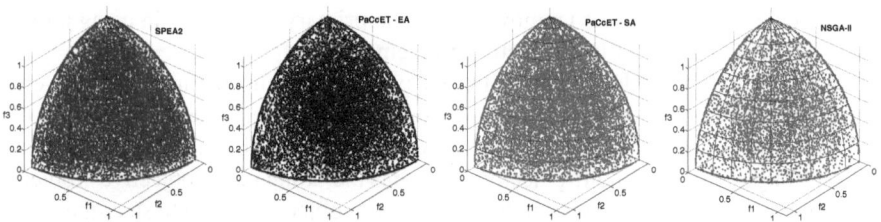

Fig. 5. All Pareto optimal points discovered in one statistical run of DTLZ2 in Λ

operator alters each element by a random number drawn from a normal distribution centered around 0 with standard deviation 0.05. We do not seed P_I^* (which breaks A1).

Figure 5 shows the results on DTLZ2 for a typical experimental run of 5000 generations for each method (simulated annealing allowed the same number of global function calls as the EAs), reporting the non-dominated points found through the entire experimental run (Note that this is distinct from P_I^*, which was kept at a size of 250). SPEA2 and PaCcET using an evolutionary algorithm (PaCcET – EA) both find a similar number of solutions spread all across the Pareto front. PaCcET using simulated annealing (PaCcET – SA) is slightly less successful but still generates good coverage, even though it is not using a population-based optimizer. NSGA-II produces fewer Pareto optimal points, but still maintains coverage. The linear combination (not shown) converges to one of the extremes very quickly, producing very poor coverage, regardless of the choice of weights.

7   Discussion and Conclusion

In this work we have presented a low computational cost way to improve the performance of a linear combination in multi-objective problems. PaCcET convexities concave regions of the Pareto front for the sake of training, and allows for solutions in these areas to be found by an optimizer using a linear combination of transformed objectives.

The primary benefits of PaCcET displayed in this work are:

1. It allows a linear combination of transformed objectives to find concave areas of the Pareto front in the original objective space.
2. It acts independently of the chosen optimizer.
3. It creates a wide spread of solutions along the Pareto front on concave or discontinuous fronts.
4. It removes the need for the system designer to choose weights.
5. It functions in higher-than-two objective problems.

The first benefit (1) allows a simple linear combination to be applied to a much broader class of multi-objective problems than it could be otherwise. Benefit (2) means that optimizers like evolutionary algorithms, A* search, simulated annealing, or particle swarm optimization can be applied to multi-objective problems through PaCcET with little alteration; it also means that future developments in single-objective optimizers are immediately useful to a large class of multi-objective problem, but comes at the cost that PaCcET is limited by the quality of the optimizer. Benefit (3) reinforces (1): Even on challenging Pareto fronts, PaCcET develops a desirable array of solutions to choose between. Benefits (4,5) remove one of the primary challenges in using a linear combination on more-than-two objective problems.

PaCcET offers a fundamentally different possible avenue for multi-objective research: the elimination of concavity as opposed to the development of methods that deal well with concave Pareto fronts. Future work in this area includes testing the PaCcET on a large testbed of multi-objective problems including many-objective problems, examining how reference points can be used in the τ-objective space to steer the search or otherwise altering the method so that it is steerable, and developing guarantees for complete Pareto front coverage.

Acknowledgements. This work was partially supported by the National Energy Technology Laboratory under grants DE-FE0011403 and DE-FE0012302.

References

1. Balling, R., Taber, J., Brown, M., Day, K.: Multiobjective urban planning using genetic algorithm. Journal of Urban Planning and Development 125(2), 86–99 (1999)
2. Coello, C.A., Christiansen, A.D.: Multiobjective optimization of trusses using genetic algorithms. Computers and Structures 75(6), 647–660 (2000)
3. Coello, C.A.C.: A comprehensive survey of evolutionary-based multiobjective optimization techniques. Knowledge and Information Systems 1(3), 269–308 (1999)
4. Das, I., Dennis, J.E.: A closer look at drawbacks of minimizing weighted sums of objectives for pareto set generation in multicriteria optimization problems. In: Structural Optimization, pp. 63–69 (1997)
5. Deb, K.: Search Methodologies, ch. 10, pp. 273–316. Springer (2005)
6. Deb, K., Pratap, A., Agarwal, S., Meyarivan, T.: A fast elitist multi-objective genetic algorithm: NSGA-II. Evolutionary Computation 6, 182–197 (2002)
7. Deb, K., Thiele, L., Laumanns, M., Zitzler, E.: Scalable test problems for evolutionary multi-objective optimization. Technical report, ETH Zurich (2001)
8. Laumanns, M., Zitzler, E., Thiele, L.: SPEA2: Improving the strength pareto evolutionary algorithm. Computer Engineering 3242(103) (2001)
9. Edgeworth, F.Y.: Mathematical Psychics: An essay on the application of mathematics to moral sciences. C. Kegan Paul and Company (1881)
10. Fonseca, C.M., Guerreiro, A.P., López-Ibáñez, M., Paquete, L.: On the computation of the empirical attainment function. In: Takahashi, R.H.C., Deb, K., Wanner, E.F., Greco, S. (eds.) EMO 2011. LNCS, vol. 6576, pp. 106–120. Springer, Heidelberg (2011)
11. Marler, R.T., Arora, J.S.: The weighted sum method for multi-objective optimization: new insights. Structural and Multidisciplinary Optimization (2009)
12. Marler, R.T., Arora, J.S.: Survey of multi-objective optimization methods for engineering. Structural and Multidisciplinary Optimization 26, 369–395 (2004)
13. Messac, A., Hattis, P.D.: Physical programming design optimization for high speed civil transport (hsct). Journal of Aircraft 33(2), 446–44 (1996)
14. Messac, A., Ismail-Yahaya, A., Mattson, C.A.: The normalized normal constraint method for generating the pareto frontier. Struct. and Multidisc. Optimization 25, 86–98 (2003)
15. Parsopoulos, K.E., Vrahatis, M.N.: Particle swarm optimization method in multiobjective problems. In: ACM Symposium on Applied Computing (2002)
16. Penn, R., Friedler, E., Ostfeld, A.: Multi-objective evolutionary optimization for greywater reuse in municipal sewer systems. Water Research 47(15), 5911–592 (2013)
17. Rosehart, W., Cañizares, C.A., Quintana, V.H.: Multi-objective optimal power flows to evaluate voltage security costs in power networks. IEEE Tr. on Power Systems (2001)
18. Vamplew, P., Dazeley, R., Berry, A., Issabekov, R., Dekker, E.: Empirical evaluation methods for multiobjective reinforcement learning algorithms. Machine Learning (2010)
19. VanVeldhuizen, D.A.: Multiobjective Evolutionary Algorithms: Classifications Analyses and New Innovations. PhD thesis, Air Force Institute of Technology (1999)
20. Van Veldhuizen, D.A., Lamont, G.B.: Multiobjective evolutionary algorithms: Analyzing the state-of-the-art. Evolutionary Computation 8(2), 125–147 (2000)
21. Zhang, G., Shao, X., Li, P., Gao, L.: An effective hybrid particle swarm optimization algorithm for multi-objective flexible job-shop scheduling problem. Computers and Industrial Engineering 56, 1309–1318 (2009)

Evolving Hard and Easy Traveling Salesman Problem Instances: A Multi-objective Approach

He Jiang, Wencheng Sun, Zhilei Ren, Xiaochen Lai, and Yong Piao

School of Software, Dalian University of Technology, Dalian, China
{jianghe,zren,laixiaochen,piaoy}@dlut.edu.cn, wencheng.sun.dlut@gmail.com

Abstract. It becomes a great challenge in the research area of meta-heuristics to predict the hardness of combinatorial optimization problem instances for a given algorithm. In this study, we focus on the hardness of the traveling salesman problem (TSP) for 2-opt. In the existing literature, two approaches are available to measure the hardness of TSP instances for 2-opt based on the single objective: the efficiency or the effectiveness of 2-opt. However, these two objectives may conflict with each other. To address this issue, we combine both objectives to evaluate the hardness of TSP instances, and evolve instances by a multi-objective optimization algorithm. Experiments demonstrate that the multi-objective approach discovers new relationships between features and hardness of the instances. Meanwhile, this new approach facilitates us to predict the distribution of instances in the objective space.

Keywords: TSP · 2-opt · multi-objective optimization algorithm · random forest.

1 Introduction

Many metaheuristics such as genetic algorithms [12], local search [1], simulated annealing [11], tabu search algorithm [7], and ant colony optimization [9] have been used to solve NP-hard combinatorial optimization problems (COPs). For a particular NP-hard problem, there exist easy instances and hard instances for distinct algorithms. Hereafter, an instance could be obtained by specifying all the problem parameters with the given problem formulations [10]. With the development of metaheuristics, it becomes a hot topic to select an appropriate algorithm to resolve a given instance of a NP-hard COP. In [18], Rice first proposed the problem of algorithm selection, which seeks to predict which algorithm is likely to perform best on one given instance.

What exactly makes an optimization problem instance hard or easy? To answer this question, Macready [14] makes it clear that the features of an instance determine its hardness for a particular algorithm. Some recent survey papers [4] [15] point out that the instance features might influence algorithm performance which is denoted as exploratory landscape analysis. Researches in [13] [17] study the problem hardness to an algorithm by analyzing the expected running time.

G. Dick et al. (Eds.): SEAL 2014, LNCS 8886, pp. 216–227, 2014.

In this study, we focus on the hardness of the Traveling Salesman Problem (TSP), which aims at finding a shortest tour visiting each of N cities once and returning to the starting city in the end. There have been a great number of metaheuristics to solve the TSP. We choose 2-opt [8], one of the most popular local search algorithms, to analyze the hardness of TSP instances based on their feature vectors. For a large-scale TSP instance, we calculate its features to predict its hardness for 2-opt, then we can know whether it is cost-effective to select 2-opt. More precisely, if the instance is hard for 2-opt, it is considerable to choose some other metaheuristics instead. However, it is still a challenge to measure the hardness of TSP instances for 2-opt. Two different approaches have been proposed to evaluate the hardness of TSP instances. One adopts the efficiency of 2-opt obtaining a local optimum to measure the hardness of TSP instances when solving these instances [19], while the other employs the effectiveness of the solutions achieved by 2-opt to evaluate the hardness of the instances [16]. Accordingly, each of them only considers one objective, either the efficiency or the effectiveness of 2-opt respectively. However, there exist some conflicts between two objectives [16]. For example, 2-opt possesses high efficiency but may achieve poor effectiveness with bad solutions on some instances, whereas it obtains desired effectiveness with low efficiency on some other instances.

To address this challenge, we evaluate the hardness of TSP instances by combining both the effectiveness and the efficiency objectives. More precisely, for 2-opt, one instance is easier than another if 2-opt achieves higher efficiency and better effectiveness on the former instance. Based on this evaluation formulation, we evolve easy and hard instances by a multi-objective optimization algorithm following NSGA-II [2]. For the purpose of straightforward illustration and significant analysis, all the instances are mapped into a 2-dimension objective space. Results show that the easy instances and the hard instances are distributed within different areas in the objective space. To study which features mainly affect the distribution of the instances in the objective space, we get the influence coefficient of each feature by training a prediction model. New relationships are discovered by the multi-objective approach that at least six features have a major influence on the hardness of TSP instances. The distribution of random TSP instances and TSPLIB instances in the objective space can be well predicted based on these relationships.

The remainder of this paper is organized as follows. Section 2 analyzes the relationships between two existing evaluation approaches to the hardness of TSP instances. Section 3 generates instances based on the multi-objective approach. Section 4 investigates the relationships between features and hardness. We evaluate the relationships on random TSP instances and sampled TSPLIB instances in Section 5. We conclude this paper in Section 6.

2   Traditional Evaluations of Hardness of TSP Instances

In this section, we demonstrate the conflicts between two existing approaches for evaluating the hardness of TSP instances. There are two approaches to evaluate the hardness of TSP instances in the literature based on single objective: efficiency or effectiveness. Smith-Miles et al. [19] measure the hardness of a given TSP instance by the efficiency of 2-opt on this instance, which is calculated by the number of 2-opt swaps to reach a local optimum. They consider that 2-opt has high efficiency on easy instances and low efficiency on hard instances. Meanwhile, Mersemann et al. [16] evaluate the hardness of a given instance for 2-opt by the effectiveness of 2-opt on this instance, which is presented by the quality of the solution achieved by 2-opt. To measure the quality of a solution obtained by 2-opt, they compare the solution against the global optimal solution achieved by the concorde solver [3]. Both researches use an evolutionary algorithm to evolve hard and easy TSP instances, and analyze the relationships between the features and the hardness. We adopt the genetic algorithm with the same crossover and mutation operators used in [16] to evolve instances based on the efficiency objective or the effectiveness objective. Moreover, we denote the corresponding collections of instances as "swaps_instances" and "quality_instances", respectively. A TSP instance is represented by a list of N (x, y) city coordinates on a 1×1 grid. To validate our finding on the instances provided on TSPLIB, we rescale the city coordinates of TSPLIB instances to a 1×1 grid as well.

2.1 Evolving TSP Instances by Traditional Evaluations of Hardness

We generate swaps_instances and quality_instances with fixed sizes of 25, respectively [16]. The size of an instance means the number of cities in the instance. We choose the 2-opt in [8] whose main idea is that making an initial solution randomly and obtaining a local optimum after a few of 2-opt swaps. Accordingly, we adopt 2-opt on each TSP instance and take the number of 2-opt swaps to reach a local optimum as the fitness of the instance for the genetic algorithm when generating swaps_instances. It is obvious that the fitness of each instance depends on the random initial solution, which makes the fitness of instances uncertain. To make the fitness of instances more reasonable, we use 2-opt to solve each instance 500 times, and take the average of the number of 2-opt swaps to reach a local optimum as the fitness of the instance.

We generate TSP instances randomly for the initial population. When evolving an easy swaps_instance, the instance which takes less 2-opt swaps for 2-opt to reach a local optimum has higher fitness. We select the instances with higher fitness from the current generation for the next generation, and the instance with the highest fitness in the last generation will be choosed as an easy swaps_instance. We repeat this process until we get the expected number of easy swaps_instances. In contrast, the instance taking more 2-opt swaps to reach a local optimum has higher fitness when evolving hard swaps_instances, and we choose the instance with the highest fitness after generations of optimization as a hard swaps_instance. Repeat this process until the desired number of hard swaps_instances are evolved.

In addition, we evolve quality_instances based on the effectiveness objective which is measured by the approximation ratio of path length that 2-opt achieves for a given TSP instance to the length of global optimal path achieved by the concorde solver. The approximation ratio equals to 1 means that 2-opt has the same effectiveness as the concorde solver when solving an instance. Therefore, the closer the approximation ratio of a given instance is to 1, the easier the instance is for 2-opt. Similar to the process of evolving swaps_instances, we generate the quality_instances by taking the approximation ratio instead of the 2-opt swaps as the fitness of instances for the genetic algorithm.

100 swaps_instances and 100 quality_instances of either easy or hard with fixed sizes of 25 are evolved. Genetic algorithm parameters are set as follows. The size of initial population is 100, and the number of generations is 1000. The uniform mutation rate is 0.001, while the normal mutation rate is 0.01. We use a 1-elitism strategy that the best individual survives and will be contained in the next population, while the other instances for the next population will be generated by uniform crossover of the instances with high fitness.

2.2   The Conflicts between Two Single Objective Approaches

To observe whether there exist conflicts between two objectives, we get the efficiency of 2-opt on quality_instances and the effectiveness of 2-opt on swaps_instances. Then each of quality_instances and swaps_instances can be mapped into a 2-dimensional objective space.

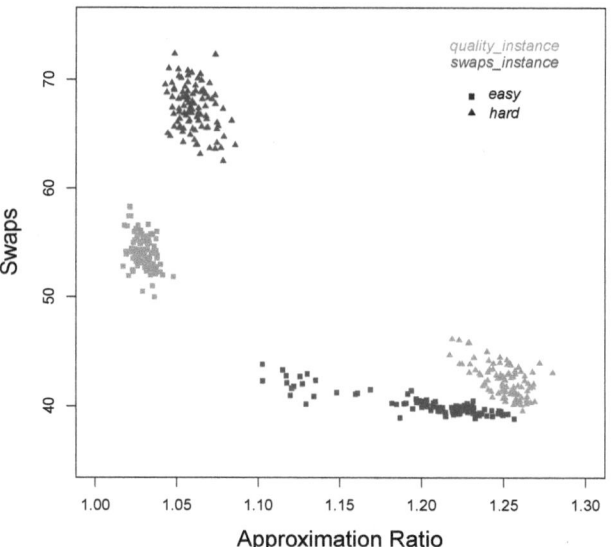

Fig. 1. The distribution of swaps_instances and quality_instances in the 2-dimensional objective space

In Fig.1, instances are denoted as points, the x-axis represents the effectiveness of 2-opt on each instance, and the y-axis indicates the efficiency of 2-opt on each instance. Swaps_instances are represented in blue color, and quality_instances are in green color. Hard instances are denoted by triangles, while easy instances are denoted as squares. It is shown in Fig.1 that 2-opt has lower efficiency on easy quality_instances than that on hard quality_instances, and has higher effectiveness on hard swaps_instances than that on easy swaps_instances. Therefore, there raise some conflicts that the instances which are considered as easy instances by one objective may be judged as hard ones based on the other objective, which implies evaluating the hardness of instances based on separate consideration of the efficiency objective or the effectiveness objective might be insufficient. To address this issue, we present a new approach to evaluate the hardness of instances which considers both the efficiency objective and the effectiveness objective. More precisely, for 2-opt, one instance is easier than another for 2-opt if 2-opt has higher efficiency and better effectiveness on the former, and vice versa. Based on the concept of Pareto optimality [6], we also evolve easy and hard instances which are denoted as "mul_instances" by a multi-objective optimization algorithm, and discover which features achieve the most influence on the hardness of instances for 2-opt.

3   Evolving TSP Instances by Multi-objective Optimization

Since we evaluate the hardness of instances based on a multi-objective approach, we evolve TSP instances by a multi-objective optimization algorithm in this section. We first impose an additional concept into the multi-objective optimization algorithm as follows.

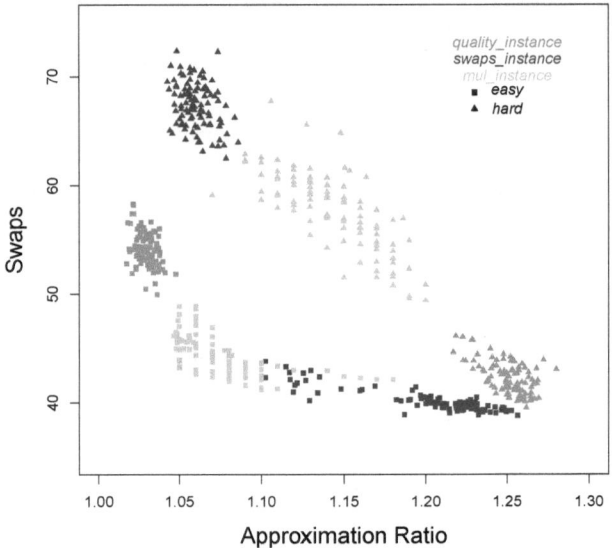

Fig. 2. The distribution of instances in the 2-dimensional objective space

Given two individuals p and q in the population Pop, p dominates q (denoted by $p \succ q$) if they satisfy the following conditions, where $f_k(*)$ is the k_{th} objective of individual $*$:

- For all the objectives, p is not worse than q, i.e., $f_k(p) \leq f_k(q)$, $(k = 1, 2)$.
- There exists at least one objective such that p is better than q. That is, $\exists l \in \{1, 2\}$, s.t. $f_l(p) < f_l(q)$.

Algorithm 1. The construction of non-dominated individual set

---
1: **for** each $p \in Pop$ **do**
2:     **for** each $q \in Pop$ **do**
3:         **if** p dominates q **then**
4:             $s_q = s_q \cup \{q\}$ //the set of individuals dominated by the individual q
5:         **end if**
6:         **if** q dominates p **then**
7:             $n_p = n_p + 1$ //the number of individuals dominating the individual p
8:         **end if**
9:     **end for**
10:   **if** $n_p = 0$ **then**
11:       $P_1 = P_1 \cup \{p\}$
12:   **end if**
13: **end for**
14: $i = 1$
15: **while** $P_i \neq \emptyset$ **do**
16:   $H = \emptyset$
17:   **for** each $p \in P_i$ **do**
18:     **for** each $q \in s_p$ **do**
19:         $n_q = n_q - 1$ //the number of individuals dominating the individual q
20:         **if** $n_q = 0$ **then**
21:             $H = H \cup \{q\}$
22:         **end if**
23:     **end for**
24:   **end for**
25:   $i = i + 1$
26:   $P_i = H$ //the set of non-dominated individuals after the ith generation
27: **end while**

2-opt is also conducted on each instance 500 times, and we take the average number of swaps as the efficiency of 2-opt on the instance, while the average of approximation ratio is taken as the effectiveness of 2-opt on the instance. The parameters of the multi-objective optimization algorithm are also the same as the genetic algorithm used in the previous section. The key difference between the genetic algorithm [16] and the multi-objective optimization algorithm proposed in this paper is that we choose two instances from a non-dominated individual set randomly to evolve new instances by uniform crossover. We obtain the easy instances or the hard instances from the non-dominated individual set in the last generation. The pseudo-code (Algorithm 1) is used to build the non-dominated individual set in each generation.

Finally, we generate 100 easy instances and 100 hard instances with fixed sizes of 25 which are mapped into the objective space as well. Fig.2 shows that the easy and the hard instances locate in different regions and the mul_instances present convex distribution as expected. Some hard mul_instances locate in those regions that hard swaps_instances or hard quality_instances locate in, which illustrates that the instances considered to be hard by the single objective approaches are also considered to be hard by the multi-objective approach.

4   The Influential Features to the Hardness of Instances

In this section, we investigate whether there are different combinations of features that affect the efficiency and the effectiveness of 2-opt most. We choose the features used in [16]. There are totally 47 features classified into 8 groups, including *Distance Features*, *Mode Features*, *Cluster Features*, *Nearest Neighbor Distance Features*, *Centroid Features*, *MST Features*, *Angle Features*, and *Convex Hull Features*. We calculate the features of all evolved instances and then conduct comparative analysis between the single objective approaches and the multi-objective approach.

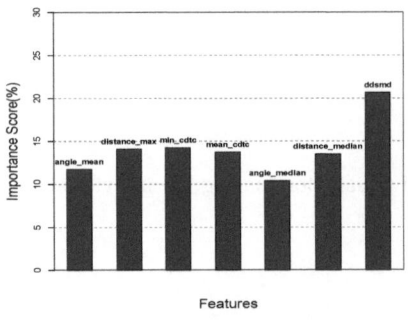

(a) Influential features discovered by mul_instances w.r.t. the effectiveness objective

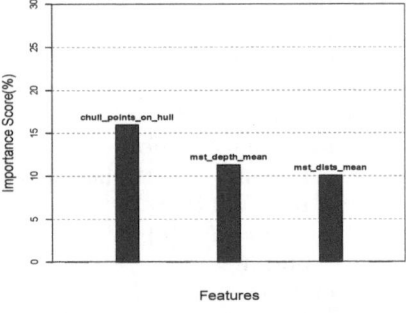

(b) Influential features discovered by quality_instances w.r.t. the effectiveness objective

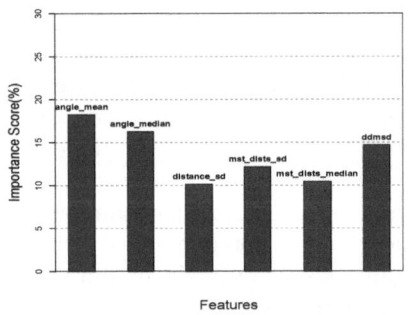

(c) Influential features discovered by mul_instances w.r.t. the efficiency objective

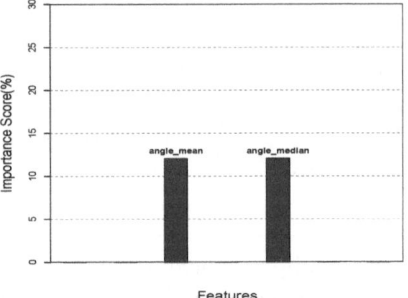

(d) Influential features discovered by swaps_instances w.r.t. the efficiency objective

Fig. 3. The importance scores of features to the hardness

We discover the features that affect the hardness of instances most by training a prediction model based on the random forest [5]. The training set consists of 75 easy and 75 hard mul_instances, and the other 25 easy and 25 hard mul_instances compose the test set. We use the features of the test set to predict the corresponding efficiency and effectiveness of 2-opt. Root Mean Squared Error (RMSE) is used to indicate prediction error, which is defined as follow:

$$RMSE = \sqrt{\frac{1}{n} \sum_{i=1}^{n} (y_i - \hat{y}_i)^2}, \tag{1}$$

where y_i is the true value and $\hat{y}_i$ is the predicted value of the ith element. Then we delete each feature in turn, and record the percentage of error increase when removing a certain feature which is denoted as the importance score of this feature to the hardness. To find the features that affect the hardness of instances most, we select the features whose importance scores to the hardness are greater than 10%. Using the same approach on swaps_instances, we can get another combination of features that influence the efficiency of 2-opt most. The combination of features that influence the effectiveness of 2-opt most can be discovered on quality_instances as well.

Considering Fig.3, we can find the features that affect the effectiveness and the efficiency of 2-opt most discovered by the multi-objective approach are quite different from those discovered by the single objective approaches, which implies that new relationships between features and hardness are discovered by the multi-objective approach. The features that affect the efficiency of 2-opt most are also different from those affect the effectiveness of 2-opt most, which also explains that separately considering the efficiency objective or the effectiveness objective to evaluate the hardness of instances might be insufficient.

5   Validating the Features on TSP Instances

To clarify whether the new relationships discovered by the multi-objective approach are practically useful, we use the prediction model trained by mul_instances to predict the effectiveness and the efficiency of 2-opt on random TSP instances and TSPLIB instances based on the feature vectors of these instances in this section. In order to compare with the relationships discovered by the single objective approaches, the prediction model trained by quality_instances is used to predict the effectiveness of 2-opt and the prediction model trained by swaps_instances is used to predict the efficiency of 2-opt on these instances, respectively. In this section, RMSE is also used to indicate prediction error.

5.1   Validating the Features on Random TSP Instances

We generate 50 TSP instances with fixed sizes of 25 randomly which compose the test set. There are three groups of training sets for building random forest:

- Mul_Training_Set: consisting of 75 easy mul_instances and 75 hard mul_instances;
- Quality_Training_Set: consisting of 75 easy quality_instances and 75 hard quality_instances;
- Swaps_Training_Set: consisting of 75 easy swaps_instances and 75 hard swaps_instances.

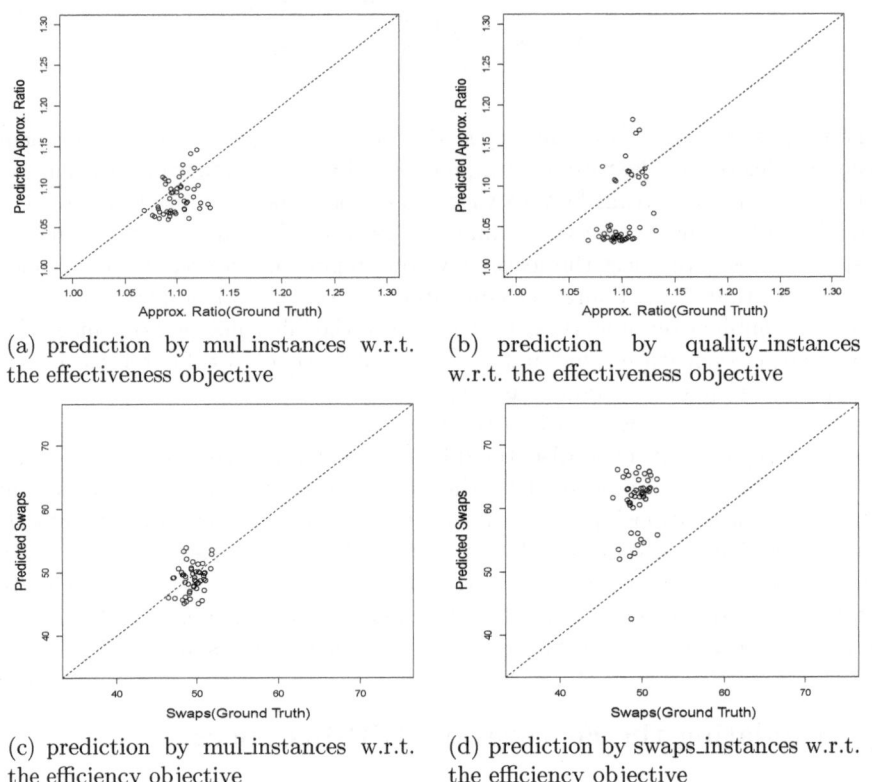

(a) prediction by mul_instances w.r.t. the effectiveness objective

(b) prediction by quality_instances w.r.t. the effectiveness objective

(c) prediction by mul_instances w.r.t. the efficiency objective

(d) prediction by swaps_instances w.r.t. the efficiency objective

Fig. 4. Prediction comparision on random TSP instances

We achieve a RMSE of 0.0253 when using the prediction model trained by Mul_Training_Set to predict the effectiveness of 2-opt on random instances, and the RMSE for the prediction model trained by Quality_Training_Set is 0.0523. The RMSE for the prediction model trained by Mul_Training_Set to predict the efficiency of 2-opt on random instances is 2.32, which is lower than that of 12.37 obtained by the prediction model trained by Swaps_Training_Set. The RMSE values in the two different objective dimensions are not in the same order of magnitude. This is because there exists a big gap between the magnitudes of the 2-opt swaps and the approximation ratio. Fig.4 shows that the prediction model trained by Mul_Training_Set is better to predict the efficiency and the effectiveness of 2-opt on random instances, which illustrates that the multi-objective approach can better discover the relationships between the features and the hardness than the single objective approaches.

5.2   Validating the Features on Sampled TSPLIB Instances

In order to further validate the new relationships between features and hardness, we use the prediction models trained by these three groups of training sets to predict the efficiency and the effectiveness of 2-opt on sampled TSPLIB instances. The instances on TSPLIB have different sizes. However, the training instances are all with the fixed sizes of 25. To be coincident with the training instances, we select 50 TSP instances from TSPLIB whose size is larger than 25 and extract 25 coordinates of cities from each of the TSPLIB instances randomly. Then we will obtain a test set with 50 sampled TSPLIB instances.

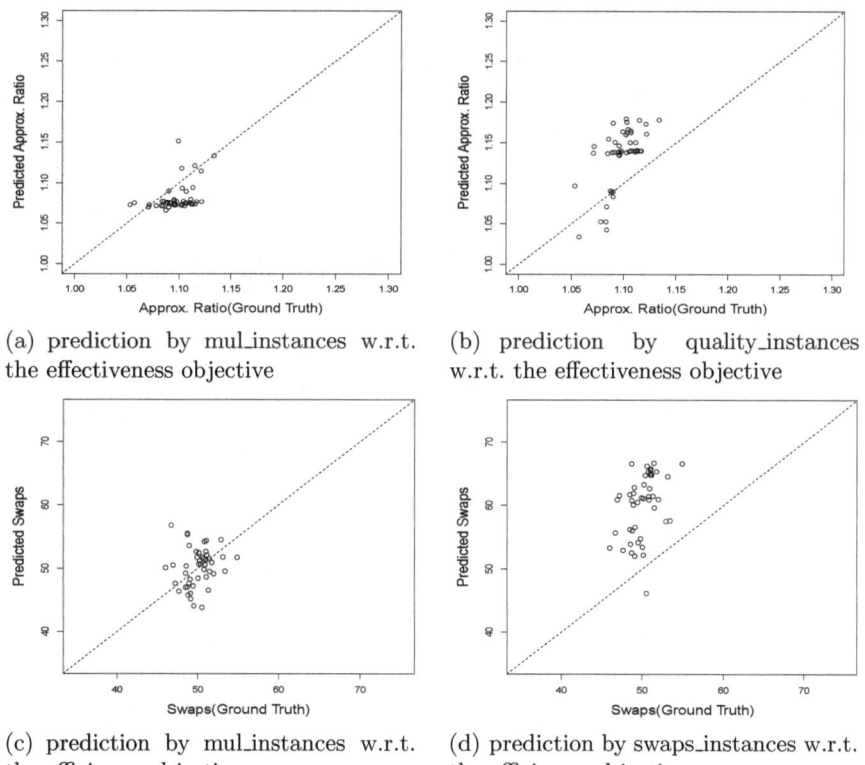

(a) prediction by mul_instances w.r.t. the effectiveness objective

(b) prediction by quality_instances w.r.t. the effectiveness objective

(c) prediction by mul_instances w.r.t. the efficiency objective

(d) prediction by swaps_instances w.r.t. the efficiency objective

Fig. 5. Prediction comparision on sampled TSPLIB instances.

The prediction results are shown in Fig.5. The RMSE is 0.0464 for the prediction model trained by Quality_Training_Set, and RMSE obtained by the prediction model trained by Mul_Training_Set is 0.0247 when predicting the effectiveness of 2-opt on sampled TSPLIB instances. The prediction model trained by Mul_Training_Set achieves a better prediction with RMSE of 3.26 than the RMSE of 10.93 obtained by the prediction model trained by Swaps_Training_Set, which also implies that the model trained by Mul_Training_Set can better predict the efficiency of 2-opt on sampled TSPLIB instances. Overall, the multi-objective approach achieves higher accuracy in predicting the distribution of TSP instances in the objective space, which illustrates that the multi-objective approach can better discover the relationships between the features and the hardness of instances for 2-opt.

In this section, we validate the features on the sampled TSPLIB instances with fixed sizes of 25. Further investigation needs to be conducted on TSPLIB instances with real sizes, which needs us to evolve training instances with different sizes.

6   Conclusion

There are two existing approaches to evaluate the hardness of TSP instances for 2-opt based on single objective. However, the objectives may conflict with each other. The instances which are considered as easy instances by one single objective may be judged as hard ones w.r.t. the other objective. To address this challenge, we propose a new evaluation approach by combining both objectives. For 2-opt, one instance is easier than another instance if 2-opt has higher efficiency and better effectiveness on the former, and vice versa.

We use a multi-objective optimization algorithm to evolve hard and easy TSP instances. Then we study the relationships between features and hardness. To find the combinations of features that affect the efficiency and the effectiveness of 2-opt most, we employ the random forest to get the importance score of each feature. Experimental results show that the relationships between features and hardness discovered by the multi-objective approach is quite different from those discovered by single objective approaches. There are at least six features with the most effect on the efficiency and the effectiveness of 2-opt discovered by the multi-objective approach. In the end, we verify our finding on random TSP instances and TSPLIB instances, and the results show that the relationships discovered by the multi-objective approach can provide more help for us to predict the distribution of TSP instances in the objective space.

Acknowledgements. This work was supported in part by the Fundamental Research Funds for the Central Universities under Grant DUT13RC(3)53, in part by the New Century Excellent Talents in University under Grant NCET-13-0073, in part by China Postdoctoral Science Foundation under Grant 2014M551083, and in part by the National Natural Science Foundation of China under Grant 61175062.

References

1. Aarts, E.H., Lenstra, J.K.: Local search in combinatorial optimization. Princeton University Press (2003)
2. Abbass, H.A., Sarker, R., Newton, C.: PDE: a pareto-frontier differential evolution approach for multi-objective optimization problems. In: Proceedings of the 2001 Congress on Evolutionary Computation, pp. 971–978. IEEE (2001)
3. Applegate, D., Bixby, R., Chvatal, V., Cook, W.: Concorde tsp solver (2011), http://www.tsp.gatech.edu/concorde.html
4. Bischl, B., Mersmann, O., Trautmann, H., Preuss, M.: Algorithm selection based on exploratory landscape analysis and cost-sensitive learning. In: Proceedings of the 14th International Conference on Genetic and Evolutionary Computation Conference, pp. 313–320. ACM (2012)
5. Breiman, L.: Random forests. Machine Learning 45(1), 5–32 (2001)
6. Censor, Y.: Pareto optimality in multiobjective problems. Applied Mathematics and Optimization 4(1), 41–59 (1977)
7. Chen, L., Bostel, N., Dejax, P., Cai, J., Xi, L.: A tabu search algorithm for the integrated scheduling problem of container handling systems in a maritime terminal. European Journal of Operational Research 181(1), 40–58 (2007)
8. Croes, G.: A method for solving traveling-salesman problems. Operations Research 6(6), 791–812 (1958)
9. Dorigo, M., Birattari, M.: Ant colony optimization. In: Encyclopedia of Machine Learning, pp. 36–39. Springer (2010)
10. Garey, M.R., Johnson, D.S.: Computers and intractability: a guide to the theory of NP-completeness. WH Freeman & Co., San Francisco (1979)
11. Goffe, W.L., Ferrier, G.D., Rogers, J.: Global optimization of statistical functions with simulated annealing. Journal of Econometrics 60(1), 65–99 (1994)
12. Goldberg, D.E.: Genetic algorithms. Pearson Education India (2006)
13. He, J., Chen, T., Yao, X.: On the easiest and hardest fitness functions (2012)
14. Macready, W.G., Wolpert, D.H.: What makes an optimization problem hard? Complexity 1(5), 40–46 (1996)
15. Mersmann, O., Bischl, B., Trautmann, H., Preuss, M., Weihs, C., Rudolph, G.: Exploratory landscape analysis. In: Proceedings of the 13th Annual Conference on Genetic and Evolutionary Computation, pp. 829–836. ACM (2011)
16. Mersmann, O., Bischl, B., Trautmann, H., Wagner, M., Bossek, J., Neumann, F.: A novel feature-based approach to characterize algorithm performance for the traveling salesperson problem. Annals of Mathematics and Artificial Intelligence 69(2), 151–182 (2013)
17. Qian, C., Yu, Y., Zhou, Z.-H.: On algorithm-dependent boundary case identification for problem classes. In: Coello, C.A.C., Cutello, V., Deb, K., Forrest, S., Nicosia, G., Pavone, M. (eds.) PPSN 2012, Part I. LNCS, vol. 7491, pp. 62–71. Springer, Heidelberg (2012)
18. Rice, J.R.: The algorithm selection problem. Advances in Computers 15, 65–118 (1976)
19. Smith-Miles, K., van Hemert, J., Lim, X.Y.: Understanding TSP difficulty by learning from evolved instances. In: Blum, C., Battiti, R. (eds.) LION 4. LNCS, vol. 6073, pp. 266–280. Springer, Heidelberg (2010)

A Multi-Objective A* Search Based on Non-dominated Sorting

Mohammad Haqqani, Xiaodong Li, and Xinghuo Yu

School of Computer Science and Information Technology, RMIT University
{mohammad.haqqani,xiaodong.li,x.yu}@rmit.edu.au

Abstract. This paper present a *Non-dominated Sorting based Multi Objective A* Search* ($NSMOA^*$) algorithm for multi-objective search problem. It is an extension of the *New Approach for Multi Objective A* Search* ($NAMOA^*$). This study aims to improve the selection phase of the $NAMOA^*$ algorithm which can affect the performance of the algorithm considerably, especially when the number of non-dominated solutions increases to a large number during the search. This research proposes a new sorting method that allows selection and expansion of the partial solutions be carried out more efficiently. The results demonstrate that our algorithm expands fewer nodes and explores a smaller region of solution space using the same heuristic.

Keywords: *A** Search, Multi-Objective Optimization, Non-dominated Sorting.

1 Introduction

Finding the shortest path in a graph is a classical optimization problem with a large number of applications. For instance, shortest path algorithms are applied to automatically find the shortest route between two different locations. In this scenario, travel time is usually considered as the main objective. However, it is sometimes necessary to have more than just one objective. For example, you may be interested in routes that are not only faster but also cheaper. People who want to use public transportation have several criteria for their journey as well (i.e. number of transfers, monetary cost, comfort of the journey etc.). The multi-objective shortest path problem considers more than one objective that need to be optimized simultaneously, and these objectives may be in conflict with each other. In multi-objective search the aim is to find the Pareto optimal solutions, i.e., paths that are not dominated by any other solutions in the search space with respect to all objective functions. Real-world multi-objective optimization problems are often NP-hard even for bi-criteria problems [10].

Selection and expansion of open nodes are the basic operations in A^* . In the original A^* , each open node is a partial solution which can be expanded. However, this cannot be directly applied to multi-objective problems; where paths may reach the same node at the different times. Moreover, solution costs in

G. Dick et al. (Eds.): SEAL 2014, LNCS 8886, pp. 228–238, 2014.
© Springer International Publishing Switzerland 2014

multi-objective search is not a scalar value and cannot be fully ordered. Therefore, selecting one of the partial solutions (which may or may not be on the same node) is one of the most important tasks in multi-objective path finding. Selecting a solution from non-dominated set is an important issue in MOA^* . If the selection is done in an efficient way the number of nodes that we have to expand decreases (due to the elimination based on solutions that we have already found) and the overall complexity of the algorithm can be reduced as well. In selection phase of the $NAMOA^*$ algorithm [9], a partial solution is selected randomly from all the non-dominated solutions to expand.

Multi-objective path finding problems have received tremendous attention in the past few decades. An extension of Dijkestras algorithm [3] to the multi-objective case is presented by Hansen[4]. In [6] Loui demonstrated that some of the stochastic search problems could be mapped to multi-objective ones. Stewart and White [2], described MOA^*, a multi-objective augmentation of A^* [5], and also provided proofs on admissibility, node expansion, as well as comparison of various heuristics' efficiencies. In [1] Dasgupta extended the MOA^* and presented versions for non-consistent heuristic A^* (MOA^{**}) with limited memory A^* ($MOMA^*$). In [10, 11] Perny and Spanjaard presented a generalization of MOA^* focusing on a specific application for a Web access problem. Mandow and Perez de la Cruz considered the extension of A^* to the multi-objective case, outlined a new algorithm ($NAMOA^*$ -New Approach to Multi-objective A^*), and briefly discussed its admissibility in [8]. More recently, Mandow presented a revision on $NAMOA^*$ and presented new proofs on its admissibility, node expansion. [9].

This paper presents an extension to the algorithm which was presented by Mandow [9]. The algorithm is fully described, and an example is presented to explain the algorithm. For proofs of admissibility and optimality of $NAMOA^*$, we highly recommend the readers to take a look at $NAMOA^*$ presented by Mandow [9].

This paper is organized as follows: Section 2 reviews previous relevant studies in scalar search and points out analogies and differences with the multi-objective search problem. Section 3 presents $NSMOA^*$ algorithm and illustrates its behaviour with an example to demonstrate its differences from $NAMOA^*$. Section 4 provides an experimental comparison of results between $NSMOA^*$ and $NAMOA^*$. Finally, conclusions are summarized in section 5.

2   Preliminaries

This section presents an overview of A^* search as well as non-dominated sorting algorithms. At first, we describe the scalar A^* search and discuss its properties. The extension to multi-Objective A^* is discussed, and then the differences are identified. We will also describe the crowding distance which is the parameter that we used for non-dominated sorting phase of our algorithm.

2.1   A* Definition

A shortest path problem can be represented by graphs which may be directed, undirected or mixed. In this article, we consider undirected graphs, $G = (V, E)$ where V and E represent the vertices and edges of the graph respectively.

Two nodes are adjacent when they both share a common edge. A path in an undirected graph is a sequence of nodes $P = (V_1, V_2, ..., V_n) \in V$ such that V_i is adjacent to V_{i+1} for $1 \leq i \leq n$. The path P is called a path of length n from V_1 to V_n.

Let $e_{i,j}$ be the edge shared between both V_i and V_j. Given a weight function $f : E \rightarrow \mathbb{R}$, and an undirected graph, the shortest path from V to V' is the path $P = (V_1, V_2, ..., V_n)$ where $V_1 = V$ and $V_n = V'$ which minimizes the sum $\sum_{i=1}^{n-1} f(e_{i,i+1})$.

A^* uses best-first search algorithm and finds a least-cost path from an initial node to the goal node. As A^* explores the graph, it expands the nodes with lowest expected total cost or distance. It uses a knowledge based heuristic cost function of node x, usually expressed as $f(x)$, to determine the order of nodes which the algorithm expands in the tree. The cost function is a summation of two functions:

- The past path-cost function, which is the associated distance between the starting node and the current node x, usually denoted as $g(x)$.
- A planned cost function which is an admissible heuristic estimate of the distance from x to the goal node and is indicated as $h(x)$.

The $h(x)$ part of the $f(x)$ function must be an admissible heuristic. Therefore, it must not overestimate the distance to the goal. If the heuristic function h is admissible, it means that it never overestimates the true cost of reaching the goal, then A^* is admissible as well. A^* is also optimally efficient, and that means no other search algorithm using the same heuristic will expand fewer nodes than A^*.

2.2   An Extension on Multi-Objective Search

In multi-objective optimization, an optimization problem is usually involved more than one objective functions that need to be optimized simultaneously [12, 13]. A multi-objective optimization problem can be expressed as:

$$\min(f_1(x), f_2(x), ..., f_k(x)) \quad s.t. \ x \in X \tag{1}$$

where $k > 2$ is the number of objectives and X is the feasible set of decision variables. The feasible set is defined by some constraint functions. In addition, the vector-valued objective function is usually represented as:

$$f : X \rightarrow R, f(x) = (f_1(x), f_2(x), ..., f_k(x))^T \tag{2}$$

An element $x \in X$ is called a feasible solution. A vector $z := f(x) \in \mathbb{R}^k$ for a feasible solution x denotes an objective vector. Usually in multi-objective

optimization, a feasible solution which can minimize all objective functions simultaneously cannot be found. Therefore, the optimizer should find a set of
solutions which are called Pareto optimal solutions, i.e., solutions that cannot
be improved in any of the objectives without degrading at least one of the other
objectives. In mathematical terms, a feasible solution $a \in X$ dominate another
solution $b \in X$, if:

$$f_i(a) \leq f_i(b) \quad for \ all \ \ i \in \{1, 2, ..., k\} \tag{3}$$

$$f_j(a) < f_j(b) \quad for \ at \ least \ one \ \ j \in \{1, 2, ..., k\} \tag{4}$$

A solution $x^* \in X$ is called Pareto optimal, if there is no other solution that
dominates it. The set of Pareto optimal solutions is called the Pareto front.

The aim of any multi-objective optimization algorithm is to find the set of
solutions that converge as closely as possible to the true Pareto front, and also
as evenly distributed along the front as possible. Therefore, the goal for MOA^*
is to find all optimal paths to the goal nodes. However, a few key differences
from the scalar search should be considered [9].

1. The search which A^* uses to record the best known path to the nodes is
   no longer applicable. As it is presented in the next section, in $NAMOA^*$ a
   directed acyclic graph is used to record the set of non-dominated paths to
   generated nodes.
2. All arcs and non-dominated cost vectors reaching each node have to be
   recorded. Therefore, the number of generated nodes may no longer be a fair
   estimation of the memory that the algorithm needs.
3. Each time a new path is generated to a known node, its cost should be tested
   for dominance with the rest of all path reaching the node.
4. The heuristic functions for all objectives should be admissible.

2.3   Crowding Distance

In A^* algorithm, there might be several non-dominated solutions at each iteration. However, only one of them should be selected to expand. Therefore, one
auxiliary parameter should be added to the A^*s Priority queue to sort the non-
dominated solutions. in this study we used crowding distance value to sort the
open solutions.

The crowding distance value of a solution gives an assessment of the density of
solutions encompassing that solution [7]. Figure 1 demonstrates the computation
of the crowding distance of the point i which is an assessment of the measure of
the biggest cuboid walling i in without including other points. Crowding distance
is computed by first sorting the set of solutions in ascending objective function
values. The crowding distance value of a specific solution is the normal distance
of its two neighbors. This procedure is done for each objective function. The
overall crowding distance for a solution is calculated by averaging the individual
crowding distance values for every objective functions.

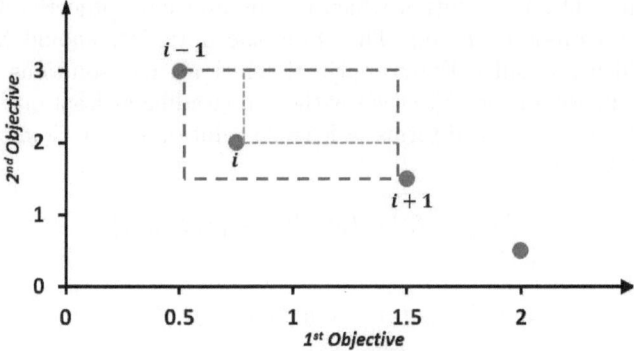

Fig. 1. Crowding distance computation

3   NAMOA* and Proposed Method

This section describes the proposed method for $NSMOA^*$. First the algorithm of $NAMOA^*$ is presented and then the method for sorting open partial solutions is explained. at the end of this section, an example is provided to illustrate the steps of the algorithm.

3.1   NAMOA* Algorithm

The $NAMOA^*$ algorithm presented in [9] can be briefly described as follows:

The algorithm uses an acyclic graph named SG to record partial solution path. For each node n in SG there is an entry in $G_{cl}(n)$ and $G_{op}(n)$ to store sets of non-dominated solutions reaching to n. G_{cl} and G_{op} illustrates that solution have or have not been explored respectively. Initially, the start node (S) is the only node in SG.

In the algorithm, the partial solutions which need to be expanded later are stored in the $OPEN$ set. The $OPEN$ is a set of triples such $(n, g_n, F(n, g_n))$ where g_s denotes the actual cost for reaching n and F is the heuristic approximation of the cost to reach V.

In the selection phase of the algorithm, it selects randomly one of the non-dominated solutions in the $OPEN$ set to expand. $GOALN$ and $COSTN$ sets record all goal nodes and corresponding non-dominated costs for reaching that goal node respectively. Once the algorithm reaches a goal node, it updates $GOALN$ and $COSTN$ set and uses the $COSTN$ set to prune the $OPEN$ set as well. The algorithm uses a backward procedure to produce the actual path for the corresponding cost based on nodes and costs in $GOALN$ and $COSTN$ sets as well as SG graph. $COSTN$ vector is also utilised for pruning the search space. When a cost vector enters to $COSTN$ all the partial solutions in the $OPEN$ set that their $F(n, g_n)$ vector is dominated by $COSTN$ will be removed from the $OPEN$ set. The algorithm only terminates when the $OPEN$ set is empty.

Notice that all the partial solutions in the $OPEN$ has to be either expanded or filtered.

3.2   Extension to NSMOA* Algorithm

As mentioned before, the main contribution in this study is on selection phase of the $NAMOA^*$ algorithm. A^* is a single individual search algorithm and it means that at each iteration of the algorithm we have to select only one solution to expand. Moreover, the number of non-dominated solutions in the $OPEN$ set can grow during the algorithm. Therefore, selecting one of the non-dominated solutions may not be wise enough in in practice especially when number of nodes and edges in the graph increase to some large number.

In $NSMOA^*$ algorithm, at each iteration, the algorithm selects the partial solutions from $OPEN$ set which has the lowest crowding distance value among all non-dominated solutions. As mentioned before the lower crowding distance value shows that the distance between a particular solution and its neighbours is close. In other words, it shows that in that region of the solution space, solution density is high. Therefore, the chance of searching for a more worthy solution in that region is higher than the other regions. In the next section, an example is provided to demonstrate steps of the algorithm.

3.3   An Example

Let us consider the graph presented in figure 2. The start node is S and the only goal node is V. The heuristic function for each node $H(n)$ is presented at table 1. Figure 3 to 7 illustrate the trace of the algorithm. Each figure shows the distribution of the $OPEN$ set in objective space. Triangles and squares represent non-dominated and dominated solutions respectively. As mentioned earlier, every solution is represented by a set of triples such $(n, g_n, F(n, g_n))$. The first two numbers (g_n show the actual cost for reaching a particular node (a number for each objective) and the last two numbers $F(n, g_n)$ represent the estimation of the cost to reach the goal node from this particular node. Finally, evolution of the SG graph through the algorithm is presented at figure 8.

At iteration 1 there is only one node in the SG (S) and one partial solution in the $OPEN$ set $(s, g_s, F(s, g_s))$. Therefore, it will be selected, and its three adjacent nodes $(n1, n2, n4)$ are added to the $OPEN$ set. At iteration 2, since the only non-dominated solution is $n2$, this solution is selected, and its neighbors are added to the $OPEN$ set. At the same time, the SG graph updates as well (figure 2 (a)). In the third iteration, since between two non-dominated solutions $([n_1, (6, 1), (8, 3)] and [n_4, (3, 5), (4, 8)])$ n_1 has the lower crowding distance value. Therefore, the solution $[n_1, (6, 1), (8, 3)]$ will be selected to expand and its adjacent nodes will be added to the $OPEN$ set as well.

At iteration 4, after the $OPEN$ set is sorted, the partial solution $[n_6, (8, 2), (9, 3)]$ is picked since it is a non-dominated solution with lowest crowding distance value and its neighbor (V) are added to the $OPEN$ set. At iteration 5, the solution $[V, (9, 3), (9, 3)]$ has the lowest crowding distance between non-dominated solutions

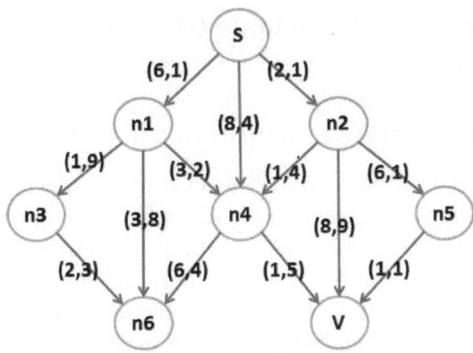

Fig. 2. A sample directed graph where edge labels present cost for each objective

Table 1. Heuristic estimates for sample graph

n	S	n_1	n_2	n_3	n_4	n_5	n_6	V
H(n)	(3,3)	(2,2)	(2,2)	(1,1)	(1,1)	(1,1)	(0,0)	(0,0)

and become selected. Since the V is the goal node, it will be added to $GOALN$ array and its corresponding cost adds to $COSTN$ array as well. At this point there is a value in $COSTN$ and can be used to prune the $OPEN$ set. As explained earlier, all the solutions where their F function is dominated by the solutions in $COSTN$ will be eliminated from the $OPEN$ set. Since the F function is admissible, these solutions cannot reach the goal nodes with a cost lower than their F function values, hence removing them from the $OPEN$ set will not eliminate any interesting solutions. The solution $[(n_4, (9, 3), (10, 4))], [(n_4, (8, 4), (9, 5)], [(n_6, (9, 9), (9, 9)]$ and $[(V, (10, 10), (10, 10)]$ will be eliminated from the $OPEN$ set and their corresponding values in G_{op} and G_{cl} will be removed as well.

At iteration 6, solution $[n_4, (3, 5)(4, 6)]$ is selected but since all the path which pass through its neighbours are dominated by the solution we have already found, just one new solution will be added to the $OPEN$ set. In iteration 7, solution $[V, (4, 10), (4, 10)]$ is selected, and since V is the goal node, we add it to $GOALN$ and $COSTN$. Now we eliminate the $OPEN$ and since all the solutions in $OPEN$ is dominated by the costs in $COSTN$, all the solutions in $OPEN$ set will be removed and the main loop of the algorithm is finished. Now the actual path related to the elements in $GOALS$ and $COSTN$ is generated based on information in SG, G_{op}, G_{cl} sets.

4   Expriments

In this section, we compare the results of $NSMOA^*$ with the original $NAMOA^*$ which was presented in [9]. Since our main contribution is in node selection phase of the algorithm, we compare the algorithms by the number of nodes which have to be expanded to find the solutions. figure 8 and 9 present the number of

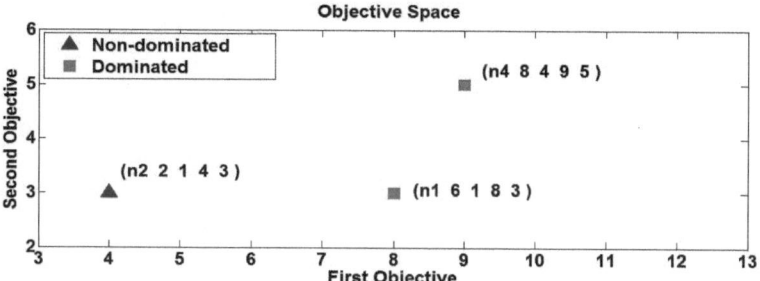

Fig. 3. Solution space and SG graph at iteration 2

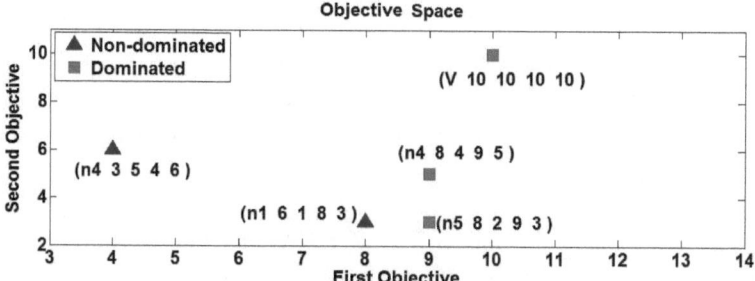

Fig. 4. Solution space and SG graph at iteration 3

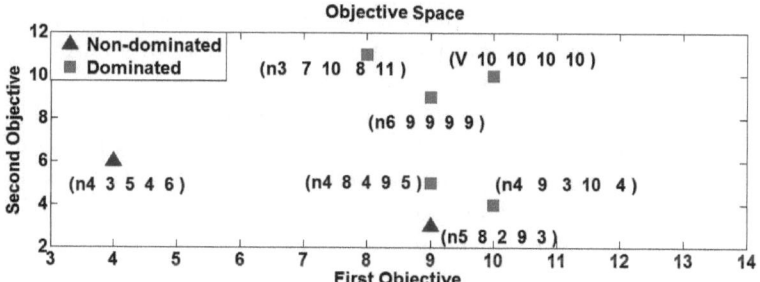

Fig. 5. Solution space and SG graph at iteration 4

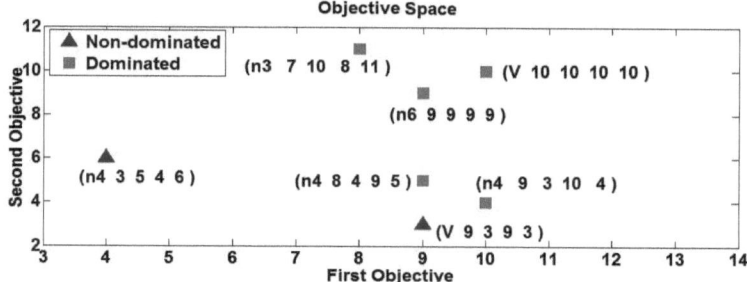

Fig. 6. Solution space and SG graph at iteration 5

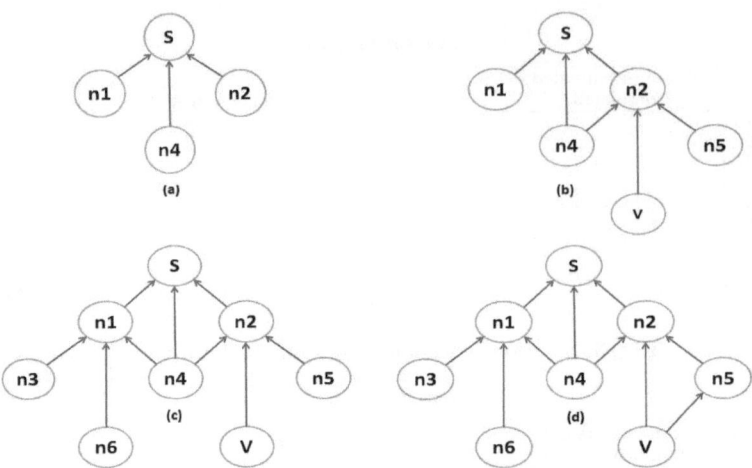

Fig. 7. (a) to (d) - SG graph at iteration 2-5 respectively

solutions in the *OPEN* set (size of the *OPEN* set) before the first solution is found as well as the number of solutions which is expanded, respectively. The information of some of the graphs and their correspondence results for *NAMOA** and *NSMOA** is presented in table 2 and 3 respectively. Finding the first solution is critical in this algorithm since we cannot prune any solution until we find at least one solution. We compare the number of solutions that the algorithms have to expand until the first solution is found.

As mentioned earlier, we select a solution from the *OPEN* solution set which has the lowest crowding distance. In other words, we picked a non-dominated solution which has a lot of solutions around; therefore, it must be in a decent region to explore. The results shows that in average, our algorithm has to expand only half of the nodes that the *NAMOA** expands.

Table 2. NAMOA* results [9]

Node#	Edge#	No. of expanded nodes before the first solution is found	Size of the OPEN set before the first solution is found
8	13	5	7
50	245	6	17
100	928	30	264
200	4008	19	373
400	4816	202	1903
500	24954	23	1134
1000	10006	99	870
2000	69875	497	4316

Table 3. NSMOA* results

Node#	Edge#	No. of expanded nodes before the first solution is found	No. of open solutions before the first solution is found
8	13	4	6
50	245	4	13
100	928	13	102
200	4008	10	201
400	4816	68	688
500	24954	9	424
1000	10006	40	380
2000	69875	240	2142

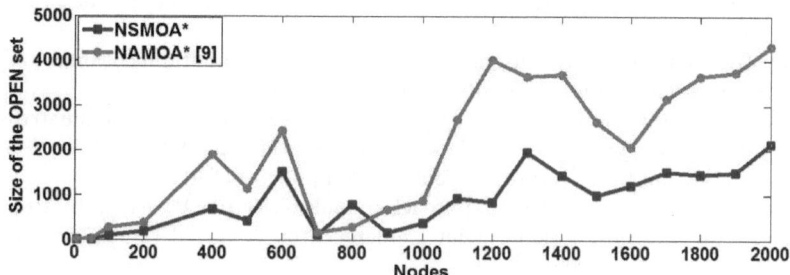

Fig. 8. Number of iterations before the first solution is found

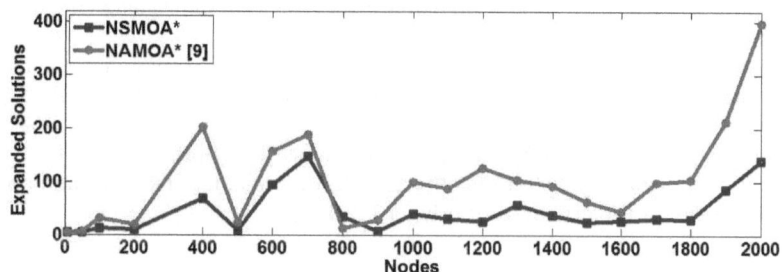

Fig. 9. Number of open solutions before the first solution is found

5   Conclusion

In this paper, a new extension on multi-objective A^* is presented. The main contribution of this paper is in the selection phase of the algorithm. The crowding distance value has been used to sort the $OPEN$ solutions set. The idea behind this method is that the solution which has a lower crowding distance is in a dense region of the solution space and therefore is superior to the other solutions for expansion. The results demonstrated a significant improvement over the amount of the solution space which the algorithm has to explore to find the Pareto front.

References

1. Dasgupta, P., Chakrabarti, P., Desarkar, S.: Utility of pathmax in partial order heuristic search. Information Processing Letters 55(6), 317–322 (1995)
2. Dasgupta, P., Chakrabarti, P., Desarkar, S.: Multiobjective heuristic search: An introduction to intelligent search methods for multicriteria optimization. Springer (1999)
3. Dijkstra, E.W.: A note on two problems in connexion with graphs. Numerische mathematik 1(1), 269–271 (1959)
4. Hansen, P.: Bicriterion path problems. In: LNEMS, vol. 177, pp. 109–127. Springer (1979)
5. Hart, P., Nilsson, N., Raphel, B.: A formal basis for the heuristic determination of minimum cost paths. IEEE Transactions on Systems Science and Cybernetics 4(2), 100–107 (1968)
6. Loui, R.P.: Optimal paths in graphs with stochastic or multidimensional weights. Communications of the ACM 26(9), 670–676 (1983)
7. Deb, K., Agrawal, S., Pratab, A., Meyarivan, T.: A fast elitist nondominated sorting genetic algorithm for multiobjective optimization: NSGA-II. In: Proc. Parallel Problem Solving from Nature VI Conference, pp. 849–858 (2000)
8. Mandow, L., Pérez, J.L.: A new approach to multiobjective A* search. In: Proceedings of the XIX International Joint Conference on Artificial Intelligence (IJCAI 2005), pp. 218–223 (2005)
9. Mandow, L., Péerez, J.L.: Multiobjective A* Search with Consistent Heuristics. Journal of the ACM 57(5), Article 27 (2010)
10. Perny, P., Spanjaard, O.: On preference-based search in state space graphs. In: Proceedings of the 18th National Conference on Artificial Intelligence, pp. 751–756. AAAI Press (2002)
11. Perny, P., Spanjaard, O.: A preference-based approach to spanning trees and shortest paths problems. European Journal of Operational Research 162(3), 584–601 (2005)
12. Milettinen, K.: Nonlinear Multiobjective Optimization. Springer (1999)
13. Hwang, C., Masud, A.: Multiple objective decision making, methods and applications: a state-of-the-art survey. Springer (1979)

Extending AεSεH from Many-objective to Multi-objective Optimization

Hernán Aguirre[1], Yuki Yazawa[1], Akira Oyama[2], and Kiyoshi Tanaka[1]

[1] Faculty of Engineering, Shinshu University
4-17-1 Wakasato, Nagano, 380-8553 Japan
[2] Institute of Space and Astronautical Science, Japan Aerospace Exploration Agency
{ahernan@,yazawa@iplab,ktanaka@}shinshu-u.ac.jp
oyama@flab.isas.jaxa.jp

Abstract. This work analyzes the dynamics of dominance based multi-objective evolutionary algorithms and extends a many-objective evolutionary algorithm so that it can also work effectively in multi-objective problems. The many-objective algorithm incorporates in its selection mechanism a density sampling approach based on ε-dominance and performs recombination within neighborhoods created by another ε-dominance based procedure. The many-objective algorithm works well during the stage of the search where there are too many non-dominated solutions and dominance is not capable of ranking solutions. Here we modify the selection mechanism of the algorithm to also work effectively during the early stage of the search where dominance can be used to bias selection. This allows the algorithm to solve multi- or many-objective problems formulations using the same framework.

1 Introduction

Evolutionary algorithms are being successfully applied to solve single-, multi- and many-objective optimization problems. In many real world application domains, like design innovation [1] and sustainability, the problems we aim to solve are usually ill-defined and complex. Thus, an important stage of the problem solving task is to gain insights about the problem itself and to collect knowledge about the trade-offs in objective and variable space in order to help the decision maker. We gain insights and knowledge by solving simple versions of the problem, use them to adjust the problem formulation and try to solve more detailed versions of it. In this process we might change, add or eliminate objective functions, variables and constraints. Even if the problem is well-defined, we still might want to solve it incrementally to better understand the outcome of the optimization, especially in the case of multi- and many-objective optimization.

There are effective algorithms for single- and multi-objective optimization [2,3]. Lately, some effective algorithms are also being proposed for many-objective optimization [4,5]. However, in most cases we need yo use different algorithms depending on whether the formulation of the problem is multi- or many-objective, small or large-scale. This imposes a heavy burden in the user. Therefore it is

G. Dick et al. (Eds.): SEAL 2014, LNCS 8886, pp. 239–250, 2014.

desirable to have scalable algorithms, within a same framework, that can be applied to any formulation of the problem.

From this standpoint, in this work, we extend the many-objective evolutionary algorithm $A\varepsilon S\varepsilon H$ [5] so that it can also work effectively in multi-objective problems. We first analyze the dynamics of dominance-based algorithms in multi- and many-objective problems showing that them usually face two clearly recognizable stages of search regardless of the number of objective functions. One stage based on Pareto dominance and the other one based on density-estimation. Then, we modify the selection mechanism of $A\varepsilon S\varepsilon H$ to work in the different stages of the search the algorithm goes through. Experimental results show that the proposed algorithm can perform significantly better than conventional multi-objective algorithms when applied to multi-objective problems while retaining its effectiveness for many-objective problems.

2   Two Search Stages in Dominance-Based Multi-objective Evolutionary Algorithms

Let us assume a conventional elitist multi-objective algorithm evolving a population P. At each generation t, individuals are selected from P_t to reproduce and create an offspring population Q_t using recombination and mutation. The surviving population for the next generation P_{t+1} is selected from the combined population $P_t \cup Q_t$. Dominance based evolutionary multi-objective algorithms use Pareto dominance to establish a primary rank among solutions. This implies that non-dominated solutions are assigned the same dominance-based rank. Thus, these algorithms also compute a secondary rank based on density estimation to discriminate among non-dominated solutions. These primary and secondary ranks are used to select parents and to select the surviving population as well.

Fig.1 (a)-(d) illustrates the dynamics of a dominance-based algorithm by showing the number of non-dominated solutions F_1 over the generations, obtained from the instantaneous combined population of parents and offspring $P_t \cup Q_t$ before survival selection. Results are shown for DTLZ2 and DTLZ3 problems [6] with 2, 3 and 4 objectives using population sizes $P = Q = 300$ and $P = Q = 1000$. In the figures, the horizontal lines at 300 and 1000 show the size of the surviving population P for the next generation.

First, looking at the dynamics in problem DTLZ2 where fitness functions are unimodal, it can be seen that the search process goes through two clearly defined stages as generations pass by. That is, a first stage in which the number of non-dominated solutions grows but is still smaller than the population size followed by a second stage in which the number of non-dominated solutions is larger than the population size. Consequently, in the first stage, only a part of the surviving population P_{t+1} after truncating $P_t \cup Q_t$ is non-dominated and parent selection can effectively use the primary rank based on dominance to favor dominant solutions over dominated ones. On the contrary, in the second stage all solutions in the surviving population P_{t+1} are non-dominated, i.e. are a subset

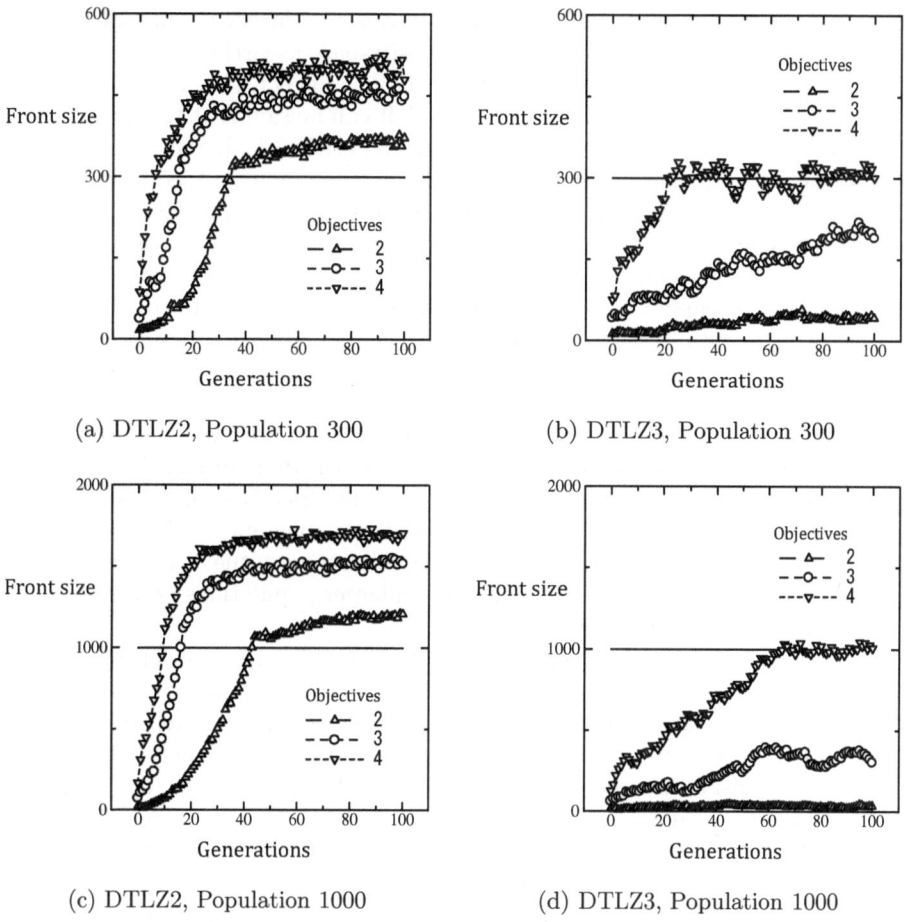

(a) DTLZ2, Population 300   (b) DTLZ3, Population 300

(c) DTLZ2, Population 1000   (d) DTLZ3, Population 1000

Fig. 1. Number of non-dominated solutions obtained from the combined population of parents and offspring for $m = 2, 3, 4$ objectives and population sizes 300 and 1000

of the non-dominated solutions present in the combined population $P_t \cup Q_t$; the primary rank becomes useless and parent selection relies exclusively in the secondary rank based on density estimation. We call these stages of the search as **dominance-based search** and **density-based search**, respectively. Note that when the number of objectives increases the transition from dominance-based to density-based search happens earlier. Also, larger populations slightly delays this transition.

In the case of DTLZ3, a problem in which each objective function is multimodal with a large number of local optima, it can be seen that for 2 and 3 objectives problems the search is governed by the dominance-based stage. In 4 objectives, the density-based stage also plays a predominant role in a large number of generations, although a clearly defined two stage search cannot be seen for the population sizes used here. Similar to DTLZ2, earlier transitions are expected when the number of objectives increase and larger populations delays the transition from dominance-based to density-based search.

Similar behavior to the one observed for DTLZ2 and DTLZ3 can be seen in other continuous and discrete problems. In general, it can be said that the dynamics of dominance-based multi-objective evolutionary algorithms is characterized by both dominance- and density-based search when applied to multi-objective problems and by density-based search when applied to many-objective problems. More precisely, it has been observed that the transition from dominance- to density-based search depends on the ratio of the population size to the number of optimal solutions of the landscape [7,8] . It is well known that the number of optimal solutions increases exponentially with the number of objectives and it also depends strongly on the correlation between objectives and the ruggedness of the landscape (multimodality) [9,10]. Usually in 2 and 3 objective problems we use population sizes that are large enough to guarantee a dominance-based stage for some generations. However, if a sufficiently small population is used, the dynamics of the algorithm will be governed by density-based search even in bi-objective problems [7]. Thus, population size is a determinant factor related to the dynamics of the algorithm in multi- and many-objective optimization. The robustness of the algorithm to small populations and its performance scalability in larger populations becomes an important factor of algorithm evaluation.

AεSεH is an effective many-objective optimizer which good performance compared to conventional multi-objective optimizers and other many-objective optimizers has been shown in problems with 4 or more objectives [8]. Its design is based on the assumption that for most generations all solutions in the instantaneous population are non-dominated, i.e. the algorithm mostly operates in a density-based search stage. In this work we extend AεSεH to search effectively also in the dominance-based stage of the search, so that the algorithm can be used for multi- and many-objective optimization.

The lack of scalability of conventional dominance-based multi-objective evolutionary algorithms is attributed to the incapability of the algorithm to rank solutions in many-objective problems, which happens during a density-based search stage. However, as we showed above, this density-based search stage also appears when we optimize multi-objective problems after evolving the population for some generations. Thus, it is expected that in multi-objective problems the improved AεSεH would perform better than widely used conventional multiobjective evolutionary algorithms such NSGA-II [11].

3   The AεSεH EMyO

Adaptive ε-Sampling and ε-Hood (AεSεH) [5] is an elitist evolutionary manyobjective algorithm that applies ε-dominance principles for survival selection and parent selection. In ϵ-dominance, the objective vector $\boldsymbol{f}(\boldsymbol{x})$ of a solution $\boldsymbol{x}$ is first mapped to another point $\boldsymbol{f}'(\boldsymbol{x})$ in objective space and dominance is calculated using the mapped point. Let us consider, without loss of generality, a maximization multi-objective problem with M objectives $\boldsymbol{f}(\boldsymbol{x}) = (f_1(\boldsymbol{x}), f_2(\boldsymbol{x}), \cdots, f_M(\boldsymbol{x}))$, a

---

Procedure 1. AɛSɛH EMyO

Require: Population size P_{size}, reference neighborhood size H_{size}^{Ref}
Ensure: $\mathcal{F}_1$, set of Pareto non-dominated solutions
1: $N_H^{Ref} \leftarrow P_{size}/H_{size}^{Ref}$ // set reference number of neighborhoods
2: $\varepsilon_s \leftarrow 0, \Delta_s \leftarrow \Delta_0$ // set ε_s-dominance factor and its step of adaptation
3: $\varepsilon_h \leftarrow 0, \Delta_h \leftarrow \Delta_0$ // set ε_h-dominance factor and its step of adaptation
4: $\mathcal{P} \leftarrow random, \mathcal{Q} \leftarrow \emptyset$ // initial populations $\mathcal{P}$ and $\mathcal{Q}$, $|\mathcal{P}| = P_{size}$
5: evaluation($\mathcal{P}$)
6: **repeat**
7:    // Parent selection
8:    $\{\mathcal{H}, N_H\} \leftarrow \epsilon$-hood creation ($\mathcal{P}, \epsilon_h$) // $\mathcal{H} = \{\mathcal{H}_j\}, j = 1, 2, \cdots, N_H$
9:    $\{\epsilon_h, \Delta_h\} \leftarrow$ adapt ($\epsilon_h, \Delta_h, N_H^{Ref}, N_H$)
10:    $\mathcal{P}' \leftarrow \epsilon$-hood mating($\mathcal{H}, Psize$)
11:    // Offspring creation
12:    $\mathcal{Q} \leftarrow$ recombination and mutation($\mathcal{P}'$) // $|\mathcal{Q}| = |\mathcal{P}| = P_{size}$
13:    // Evaluation and front sorting
14:    evaluation($\mathcal{Q}$)
15:    $\mathcal{F} \leftarrow$ non-dominated sorting($\mathcal{P} \cup \mathcal{Q}$)   // $\mathcal{F} = \{\mathcal{F}_i\}, i = 1, 2, \cdots, N_F$
16:    // Survival selection
17:    $\{\mathcal{P}, N_S\} \leftarrow \epsilon$-sampling truncation($\mathcal{F}, \epsilon_s, P_{size}$) // N_S, number of samples
18:    $\{\epsilon_s, \Delta_s\} \leftarrow$ adapt ($\epsilon_s, \Delta_s, P_{size}, N_S$)
19: **until** termination criterion is met
20: **return** $\mathcal{F}_1$

---

solution x is said to ϵ-dominate another solution y, denoted by $x \succ^\epsilon y$, if the following conditions are satisfied:

$$f(x) \mapsto^\epsilon f'(x)$$
$$\forall i \in \{1, \cdots, M\} \quad f'_i(x) \geq f_i(y) \ \wedge \tag{1}$$
$$\exists i \in \{1, \cdots, M\} \quad f'_i(x) > f_i(y),$$

where $f(x) \mapsto^\epsilon f'(x)$ is a mapping function that depends on a parameter ϵ.

The general flow of the AɛSɛH is illustrated in **Procedure 1**. The main steps of the algorithm at each generation t can be summarized as follows.

Parent Selection. For parent selection the algorithm first uses a procedure called ε-*hood creation* to cluster solutions in objective space. This procedure randomly selects an individual from the surviving population and applies ε-dominance with parameter ε_h. A neighborhood is formed by the selected solution and its ε_h-dominated solutions. Neighborhood creation is repeated until all solutions in the surviving population have been assigned to a neighborhood. Then, parents are selected by the procedure ε-*hood mating*, which sees neighborhoods as elements of a list than can be visited one at the time in a round-robin schedule. The first two parents are selected randomly from the first neighborhood in the list. The next two parents will be selected randomly from the second neighborhood in the list, and so on. When the end of the list is reached, parent selection continues with the first neighborhood in the list. Thus, all individuals have the same probability of being selected within a specified neighborhood, but due to the round-robin schedule individuals belonging to neighborhoods with fewer members have more reproduction opportunities that those belonging to neighborhoods with more members.

Offspring Creation. Once the pool of mates has been formed, recombination and mutation is applied to the selected parent individuals to create the offspring population $\mathcal{Q}_t$.

Evaluation and Front Sorting. The newly created offspring population is evaluated. Then, the current population $\mathcal{P}_t$ and its offspring $\mathcal{Q}_t$ are joined and divided into non-dominated fronts $\mathcal{F} = \{\mathcal{F}_i\}, i = 1, 2, \cdots, N_F$ using the non-dominated sorting procedure.

Survival Selection. Next, survival selection is performed using the ϵ-*sampling truncation* procedure, which applies two different procedures according to the number of non-dominated solutions. In the rare case in many-objective optimization where the number of non-dominated solutions is smaller than the population size $|\mathcal{F}_1| < \boldsymbol{Psize}$, the sets of solutions $\mathcal{F}_i$ are copied iteratively to $\mathcal{P}_{t+1}$ until it is filled; if set $\mathcal{F}_i$, $i > 1$, overfills $\mathcal{P}_{t+1}$, the required number of solutions are chosen randomly from it. On the other hand, in the common case in many-objective optimization where $|\mathcal{F}_1| > \boldsymbol{Psize}$, it calls ε-*sampling* with parameter ε_s. This procedure samples solutions randomly from the set $\mathcal{F}_1$, inserting the sample in $\mathcal{P}_{t+1}$ and eliminating from $\mathcal{F}_1$ the sample itself and the solutions ε-dominated by the sample. Sampling is repeated until there are no remaining solutions in $\mathcal{F}_1$. After sampling, if $\mathcal{P}_{t+1}$ is overfilled solutions are randomly eliminated from it. Otherwise, if there is still room in $\mathcal{P}_{t+1}$, the required number of solutions are randomly chosen from the initially ε-dominated solutions and added to $\mathcal{P}_{t+1}$. This guarantees that the size of $\mathcal{P}_{t+1}$ is exactly $\boldsymbol{Psize}$. The mapping functions $\boldsymbol{f}(\boldsymbol{x}) \mapsto^{\epsilon} \boldsymbol{f}'(\boldsymbol{x})$ used for ε-dominance in ε-*sampling* and ε-*hood creation* determine the distribution of solutions the algorithm aims to find.

Adaptation. Both epsilon parameters ε_s and ε_h used in survival selection and neighborhood creation, respectively, are dynamically adapted during the run of the algorithm. Further details about AεSεH can be found in [5].

4   Extension of AεSεH for Dominance-Based Search Stage

AεSεH was designed for many-objective optimization and assumes that in most generations all solutions in the population will be non-dominated, i.e. all solutions are equally good. Consequently, in its original formulation *ε-hood creation* randomly selects a solution to create a neighborhood with its ε-dominated solutions and *ε-hood mating* also selects parents randomly within each neighborhood. Density-estimation based on ε-dominance combined with local recombination after clustering solutions in objective space based on ε-dominance allows the algorithm an effective search in many-objective problems, where a density-based search is predominant. However, during a dominance-based search stage, there will also be dominated solutions in the population. Thus, in order to extend AεSεH to work effectively during this stage, two slight changes are required for neighborhood creation and parent selection.

First, *ε-hood creation* is modified so that solutions chosen to initiate a neighborhood are among the non-dominated solutions in the population. The rest of the procedure remains the same. In this way, we make sure that each neighborhood contains at least one non-dominated solution in it, eliminating the possibility that a neighborhood contains only dominated solutions. **Procedure 2** shows the extended *ε-hood creation*, where line 4 replaces the following in the original formulation

$$z \leftarrow x_r \in \mathcal{P} \mid r = rand(\,1,\,|\mathcal{P}|\,) \quad // \ z, \text{ a randomly chosen solution}$$

Second, to select parents, in the extended *ε-hood creation* procedure we replace random selection with binary tournaments between solutions within the neighborhood. Solutions with better dominance rank win a tournament. If two solutions have the same rank, the winner is decided randomly. Thus, selection favors dominant solutions over dominated ones within each neighborhood. **Procedure 3** shows the extended *ε-hood mating* procedure, where lines 5-8 replace the following in the original formulation

$$\{y, z\} \leftarrow \{x_{r_1}, x_{r_2} \in \mathcal{H}_i \mid r_1 \wedge r_2 = rand(\,1,\,|\mathcal{H}_i|\,), r_1 \neq r_2\}$$

5 Experimental Setup, Test Problems, and Performance Indicators

In this work we study the performance of the extended AεSεH algorithm comparing with NSGA-II, a widely used algorithm for multi-objective optimization. We use DTLZ2, DTLZ3, and DTLZ4 [6] continuous functions as benchmark problems. In our experiments, we vary the number of objectives from $m = 2$ to 4 and set the total number of variables to $n = (m - 1) + 10$. DTLZ2 has a non-convex Pareto-optimal surface that lies inside the first quadrant of the unit hyper-sphere. DTLZ3 and DTLZ4 are variations of DTLZ2. DTLZ3 introduces a large number of local Pareto-optimal fronts in order to test the convergence ability of the algorithm. DTLZ4 introduces biases on the density of solutions to some of the objective-space planes in order to test the ability of the algorithms to maintain a good distribution of solutions.

Procedure 2. Extended ε-hood creation ($\mathcal{P}$, ε_h)

Require: Population $\mathcal{P}$, ε-dominance parameter ϵ_h for neighborhood creation

Ensure: Neighborhoods $\mathcal{H} = \{\mathcal{H}_i\}$, $i = 1, 2, \cdots, N_H$

1: $\mathcal{H} \leftarrow \emptyset$
2: $i \leftarrow 0$
3: **while** $\mathcal{P} \neq \emptyset$ **do**
4:     $z \leftarrow x_r \in \mathcal{F}_1 \mid r = rand(\,1,\, |\mathcal{F}_1|\,)$ // z, a randomly chosen solution from the non-dominated set $\mathcal{F}_1 \in \mathcal{P}$
5:     $\mathcal{Y} \leftarrow \{y \in \mathcal{P} \mid z \succeq^{\varepsilon_h} y, z \neq y\}$ // solutions ε_h-dominated by z
6:     $i \leftarrow i + 1$
7:     $\mathcal{H}_i \leftarrow \{\{z\} \cup \mathcal{Y}\}$ // z and its ε_h-dominated solutions form the hood
8:     $\mathcal{H} \leftarrow \mathcal{H} \cup \mathcal{H}_i$
9:     $\mathcal{P} \leftarrow \mathcal{P} \setminus \mathcal{H}_i$ // update $\mathcal{P}$ and therefore the non-dominated set $\mathcal{F}_1 \in \mathcal{P}$
10: **end while**
11: $N_H \leftarrow i$
12: **return** $\mathcal{H}$, N_H

Procedure 3. Extended ε-hood mating ($\mathcal{H}$, P_{size})

Require: Neighborhoods $\mathcal{H} = \{\mathcal{H}_i\}$, $i = 1, 2, \cdots, N_H$, and population size P_{size}

Ensure: Pool of mated parents $\mathcal{P}'$, $|\mathcal{P}'| = 2P_{size}$

1: $\mathcal{P}' \leftarrow \emptyset$
2: $i \leftarrow 1$
3: $j \leftarrow 0$
4: **while** $j < P_{size}$ **do**
5:     $\{x_{r_1}, x_{r_2} \in \mathcal{H}_i \mid r_1 \wedge r_2 = rand(\,1,\, |\mathcal{H}_i|\,), r_1 \neq r_2\}$
6:     $y \leftarrow tournament(x_{r_1}, x_{r_2})$ // decide based on dominance rank
7:     $\{x_{r_3}, x_{r_4} \in \mathcal{H}_i \mid r_3 \wedge r_4 = rand(\,1,\, |\mathcal{H}_i|\,), r_3 \neq r_4\}$
8:     $z \leftarrow tournament(x_{r_3}, x_{r_4})$ // decide based on dominance rank
9:     $\mathcal{P}' \leftarrow \mathcal{P}' \cup \{y, z\}$
10:     $i \leftarrow 1 + (i \bmod N_H)$
11:     $j \leftarrow j + 1$
12: **end while**
13: **return** $\mathcal{P}'$

We run the algorithms 30 times and present average results. We use a different random seed in each run, but all algorithms use the same seeds. The number of generations is set to **100** generations, and population size varies from **100** to **3000**, $|\mathcal{P}| = |\mathcal{Q}|$. As variation operators, the algorithms use SBX crossover and polynomial mutation, setting their distribution exponents to $\eta_c = 15$ and $\eta_m = 20$, respectively. Crossover rate is $pc = 1.0$, crossover rate per variable $pcv = 0.5$, and mutation rate per variable is $pm = 1/n$. The reference neighborhood size H_{size}^{Ref} in AεSεH is set to 20 individuals. The mapping function $f(x) \mapsto^{\varepsilon}$

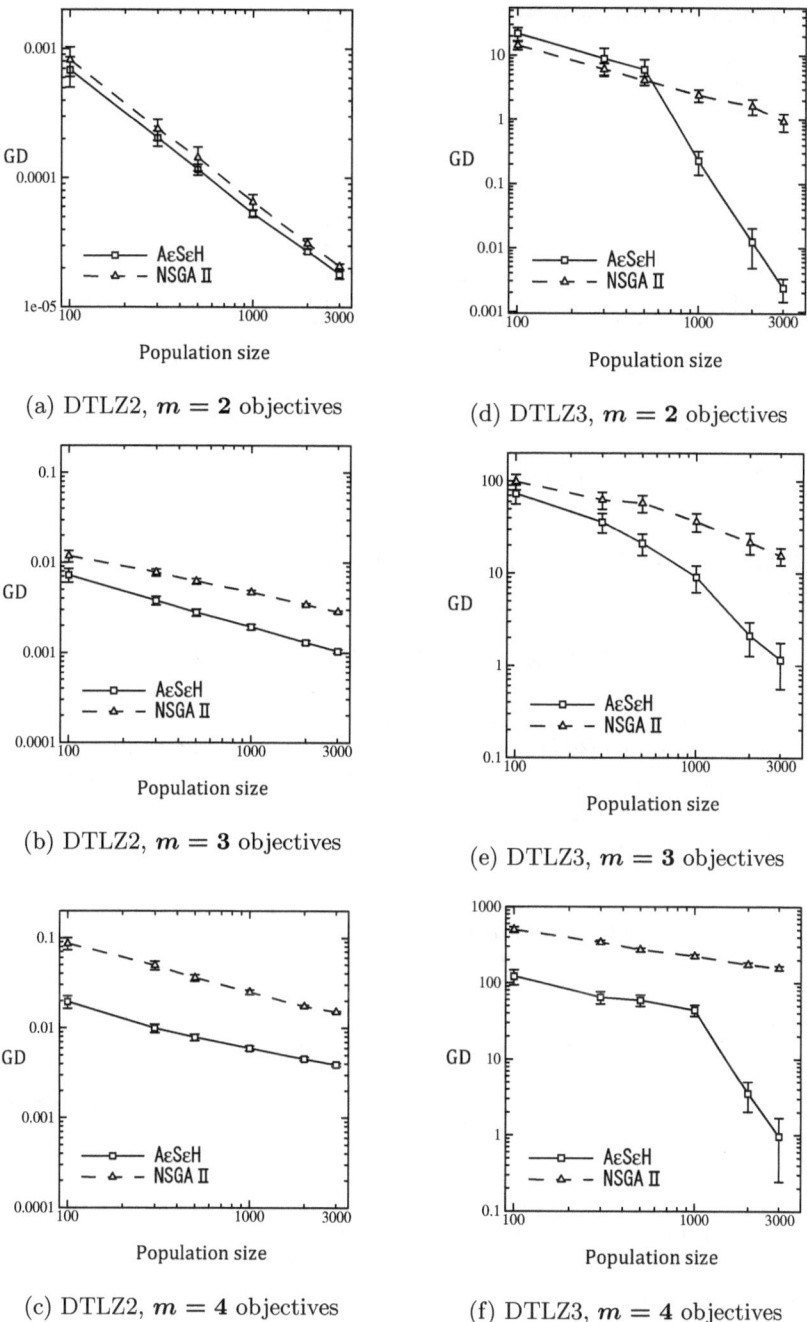

(a) DTLZ2, $m = 2$ objectives

(d) DTLZ3, $m = 2$ objectives

(b) DTLZ2, $m = 3$ objectives

(e) DTLZ3, $m = 3$ objectives

(c) DTLZ2, $m = 4$ objectives

(f) DTLZ3, $m = 4$ objectives

Fig. 2. GD at generations 100 for various population sizes, DTLZ2 and DTLZ3

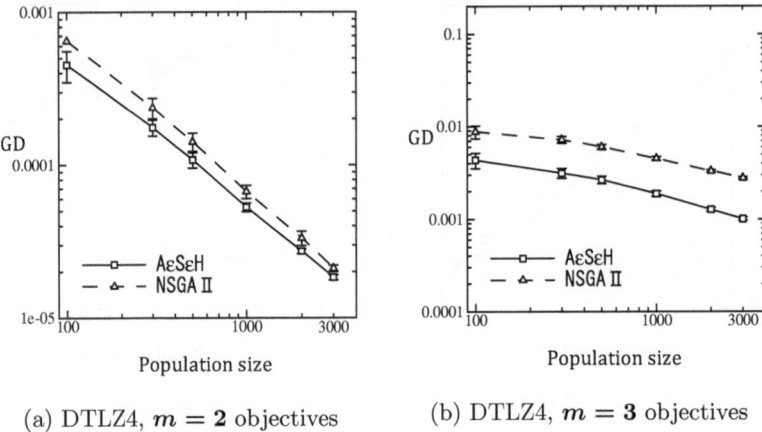

(a) DTLZ4, $m = 2$ objectives   (b) DTLZ4, $m = 3$ objectives

Fig. 3. GD at generations 100 for various population sizes, DTLZ4

$f'(x)$ used for ε-dominance in ε-*sampling truncation* and ε-*hood creation* is as follows

$$f_i'(x) = f_i(x) + (\varepsilon \times (\max_{y \in P} f_i(y) - \operatorname*{median}_{y \in P} f_i(y))), \quad i = 1, 2, \cdots, m \quad (2)$$

To evaluate the Pareto optimal solutions obtained by the algorithms we use Generational Distance (GD) [12], which measures the convergence of solutions to the true Pareto front using Eq.(3), where P denotes the set of Pareto optimal solutions found by the algorithm and x a solution in the set. Smaller values of GD indicate that the set P is closer to the Pareto optimal front.

$$GD = \operatorname*{average}_{x \in P} \left\{ \left[\sum_{i=1}^{m} (f_i(x))^2 \right]^{\frac{1}{2}} - 1 \right\} \quad (3)$$

6   Simulation Results and Discussion

Fig.2 shows the average GD by AεSεH and NSGA-II computed at the last generation for various population sizes in problems DTLZ2 and DTLZ3 with 2, 3 and 4 objectives. Similarly, Fig.3 shows results in DTLZ4 with 2 and 3 objectives. Error bars show 95% confidence intervals around the mean. From these figures it can be seen that AεSεH achieves smaller GD than NSGA-II for all population sizes and number of objectives in DTLZ2 and DTLZ4 problems. AεSεH also achieves overall smaller GD than NSGA-II in DTLZ3 problem, except for population size 500 or less in 2 objectives. These results show that convergence by AεSεH is significantly better, confirming the superiority of the proposed algorithm over NSGA-II when dominance-based search constitutes a significant part of the run of the algorithm. As expected, when density-based search becomes predominant in problems with a larger number of objectives the superiority of the proposed algorithm becomes clearer, as shown in [1,5] for an even larger number of objectives. It should be noticed that here we run the algorithms for 100 generations. Running the algorithms for a larger number of generations would make the density-based search the predominant stage and it is expected a larger performance gap in favor of the proposed algorithm.

It is worth mentioning the behavior of the algorithm in DTLZ3, a problem with multi-modal fitness functions and a large number of local optima. As illustrated in Fig.1 (b) and (d), in 2 objective DTLZ3 the number of non-dominated solutions is a small fraction of the population size when population size is 300 and 1000. Thus, in these cases the algorithm performs a dominance-based search from the start to the end of the run. NSGA-II and the proposed algorithm set the dominance rank of solutions using non-dominated sorting and apply the same operators for crossover and recombination. However, NSGA-II achieves a slightly better GD than the proposed algorithm in the 2 objective DTLZ3 for population size 500 or less, as shown in Fig.2 (b), whereas the proposed algorithm significantly outperforms NSGA-II for population size 1000 or greater. These performance differences are attributed to recombination performed locally within the neighborhood.

In general, small populations in highly multi-modal problems make the algorithm prone to get trapped in local optima. This can be seen in the large values of GD for small populations. When neighborhoods are created in the proposed algorithm, if there are very few non-dominated solutions, each one of them will most likely be placed in a different neighborhood. This reduces the chances of recombining non-dominated solutions. In addition, clustering solutions in objective space would group together solutions located in the same basin of attraction or in a near one. This will increase the chances of exploiting local optima, but it will reduce the changes of escaping them. This explains the slightly worse GD by the proposed algorithm when population sizes 500 or less is used. Using larger populations in 2 objective DTLZ3 would not increase substantially the number of non-dominated solutions. However, it increases the chances of having a better coverage of the search space with solutions located in promising basins of attraction, where recombination within the neighborhood could exploit them. It also increases the chances of having within the same neighborhood solutions located in different basins of attraction, which could be recombined to escape local optima. Once the algorithm finds promising regions, recombination within the neighborhood would quickly improve convergence. This explains the striking GD gap difference in favor of the proposed algorithm observed for population size larger than 500.

7   Conclusions

In this work we analyzed the dynamics of dominance-based multi-objective algorithms in order to clarify the search stages the algorithm goes through. We showed that dominance-based multi-objective algorithms make a transition from a dominance-based search stage to a density-based search stage. Also, we argued that the transition time depends on the ratio between population size and the number of optimal solutions of the landscape, which in turn depends strongly on the number of objectives, correlation between objectives, and ruggedness (multi-modality) of the landscape. Then the many-objective algorithm AεSεH algorithm was extended for multi-objective optimization. This algorithm works very well for density-based search, so selection was modified to make the algorithm effective also for a dominance-based search. We used multi-objective benchmark problems with 2, 3, and 4 objectives and showed that in a large range of population sizes the overall performance of the extended algorithm is superior to NSGA-II in multi-objective problems.

In the future we would like to study deeper the behavior of the algorithm in highly-multimodal landscapes to adapt neighborhood creation accordingly.

References

1. Nishio, Y., Oyama, A., Akimoto, Y., Aguirre, H., Tanaka, K.: Many-Objective Optimization of Trajectory Design for DESTINY Mission. In: Learning and Intelligent Optimization Conference. LNCS. Springer (2014)
2. Deb, K.: Multi-Objective Optimization using Evolutionary Algorithms. John Wiley & Sons, Chichester (2001)
3. Coello, C., Van Veldhuizen, D., Lamont, G.: Evolutionary Algorithms for Solving Multi-Objective Problems. Kluwer Academic Publishers, Boston (2002)
4. Hadka, D., Reed, P.: Borg: An Auto-adaptive Many-objective Evolutionary Computing Framework. Evol. Computation 2(2), 231–259 (2013)
5. Aguirre, H., Oyama, A., Tanaka, K.: Adaptive ϵ-Sampling and ϵ-Hood for Evolutionary Many-Objective Optimization. In: Purshouse, R.C., Fleming, P.J., Fonseca, C.M., Greco, S., Shaw, J. (eds.) EMO 2013. LNCS, vol. 7811, pp. 322–336. Springer, Heidelberg (2013)
6. Deb, K., Thiele, L., Laumanns, M., Zitzler, E.: Scalable Multi-Objective Optimization Test Problems. In: Proc. 2002 Congress on Evolutionary Computation, pp. 825–830. IEEE Service Center (2002)
7. Aguirre, H., Liefooghe, A., Verel, S., Tanaka, K.: A Study on Population Size and Selection Lapse in Many-objective Optimization. In: Proc. 2013 IEEE Congress on Evolutionary Computation, June 2013, pp. 1507–1514. IEEE Press (2013)
8. Aguirre, H., Liefooghe, A., Verel, S., Tanaka, K.: An Analysis on Selection for High-Resolution Approximations in Many-Objective Optimization. In: Bartz-Beielstein, T., Branke, J., Filipič, B., Smith, J. (eds.) PPSN 2014. LNCS, vol. 8672, pp. 487–497. Springer, Heidelberg (2014)
9. Aguirre, H., Tanaka, K.: Working Principles, Behavior, and Performance of MOEAs on MNK-Landscapes. European Journal of Operational Research 181(3), 1670–1690 (2007)
10. Verel, S., Liefooghe, A., Jourdan, L., Dhaenens, C.: On the structure of multi-objective combinatorial search space: MNK-landscapes with correlated objectives. European Journal of Operational Research 227(2), 331–342 (2013)
11. Deb, K., Agrawal, S., Pratap, A., Meyarivan, T.: A Fast Elitist Non-Dominated Sorting Genetic Algorithm for Multi-Objective Optimization: NSGA-II, KanGAL report 200001 (2000)
12. Zitzler, E.: Evolutionary Algorithms for Multiobjective Optimization:Methods and Applications, PhD thesis, Swiss Federal Institute of Technology, Zurich (1999)

User Preferences for Approximation-Guided Multi-objective Evolution

Anh Quang Nguyen, Markus Wagner, and Frank Neumann

Optimisation and Logistics
School of Computer Science
The University of Adelaide, Australia

Abstract. Incorporating user preferences into evolutionary multi-objective evolutionary algorithms has been an important topic in recent research in the area of evolutionary multi-objective optimization. We present a very simple and yet very effective modification to the Approximation-Guided Evolution (AGE) algorithm to incorporate user preferences. Over a wide range of test functions, we observed that the resulting algorithm called iAGE is just as good at finding evenly distributed solutions as similarly modified NSGA-II and SPEA2 variants. However, in particular for "difficult" two-objective problems and for all three-objective problems we see more evenly distributed solutions in the user preferred region when using iAGE.

Keywords: Multi-objective optimisation, approximation, user preference.

1 Introduction

Many real-world optimization problems consist of multiple objectives that conflict with each other. Solving a multi-objective optimization (MOO) problem usually means finding a set of trade-offs regarding the given objective functions. The set of all trade-offs according to the given objective functions is called the Pareto front of the underlying MOO problem. Since the size of the Pareto front can grow exponentially for discrete problems and can even be infinite for continuous problems, evolutionary algorithms on MOO problems have to restrict themselves to a smaller set of solutions which should be a good approximation of the Pareto front. There are different algorithms such as NSGA-II [4], SPEA2 [21], or IBEA [19] which try to solve two main goals of a MOO problem: find the Pareto front or a good approximation thereof by preferring a diversity of non-dominated solutions.

Motivated by the studies of multiplicative and additive approximations for multi-objective problems [3, 7, 16], the algorithm Approximation-Guided Evolution (AGE) has been introduced in [2]. AGE works with a formal notion of approximation and improves the approximation quality during its runtime without having a full knowledge about the true Pareto front. The results in [2, 17] show that, given a fixed number of evaluation, AGE outperforms state-of-the-art

G. Dick et al. (Eds.): SEAL 2014, LNCS 8886, pp. 251–262, 2014.

algorithms in terms of additive approximation and covered hypervolume. AGE has later been improved in [18] to overcome the problem of over growth archive size in high dimensional objective spaces by adapting the ϵ-dominance approach and a non-random selection of parents used for next generation of population.

Recently, great efforts have been made in order to incorporate user preferences into evolutionary multi-objective optimization (EMO) where specific regions in the objective space have higher priority than others. For NSGA-II, a reference point approach has been presented in [6]. Later on, the crowding distance assignment function has been changed in order to meet the requirement of non-even distribution of solutions along the Pareto front [10]. Zitzler et al. [22] have shown that the weighted hypervolume indicator is a good method to integrate user preferences and showed that their results are superior than the ones obtained by NSGA-II and SPEA2, where no user preference information is considered. However, all of these hypervolume-based approaches have a negative effect on the runtime of the algorithm because they require exponential runtime in the number of dimensions [1]. To overcome that problem, Friedrich et al. [9] proposed a simple approach to integrate the weight function into a wide range of EMO algorithms, including NSGA-II and SPEA2, and showed that their results now match the ones in [22] without changing the performance of the algorithms.

In relation to our series of works of integrating preferences into existing algorithms is that presented in [13–15]. There, the authors focus on reference points and on a performance metric for comparing algorithms with reference points. The preference functions that are considered in our article here, however, go beyond reference points.

We propose a new variant of AGE [2, 18] called iAGE which incorporates user preferences into the algorithm. iAGE widens the range of preference functions by using not only reference points but also preferred regions and spaces. Furthermore, we change the selection process of AGE by considering the preference functions as a factor to keep or discard solutions from the population while still keep the complexity remaining unchanged. Our experimental results show that iAGE is fast and works just as well as integrated NSGA-II and SPEA2. Furthermore, iAGE provides more evenly distributed solutions in the preferred region of the objective space.

The outline of this paper is structured as follows. In Section 2, we introduce some basic definitions of multi-objective optimization and the AGE algorithm. Section 3 shows how user preferences are incorporated into AGE and how input parameters can affect the distribution of solutions. In Section 4, we report on our experimental results, and compare them with the ones from NSGA-II and SPEA2. Finally, we finish with some conclusions.

2   Preliminaries

In this section, we give a basic introduction into the setting for multi-objective optimization, the approach of using weight function to incorporate user preferences, and the approximation-guided evolution approach.

2.1   Multi-objective Optimization

In multi-objective optimization the task is to optimize a function $f = (f_1, \ldots, f_d)$ $: S \to \mathbb{R}_+^d$ with $d \geq 2$, which assigns to each element $s \in S$ a d-dimensional objective vector. Each objective function $f_i \colon S \mapsto \mathbb{R}$, $1 \leq i \leq d$, maps from the considered search space S into the positive real values. Elements from S are called search points and the corresponding elements $f(s)$ with $s \in S$ are called objective vectors.

Throughout this paper, we consider the minimization problems of d objectives. In multi-objective optimization the given objective functions f_i are usually conflicting, which implies that there is no single optimal objective vector. Instead of this the Pareto dominance relation is defined, which is a partial order. In order to simplify the presentation we only work with the Pareto dominance relation on the objective space and mention that this relation transfers to the corresponding elements of S.

The Pareto dominance relation $\preceq$ between two objective vectors $x = (x_1, \ldots, x_d)$ and $y = (y_1, \ldots, y_d)$, with $x, y \in \mathbb{R}^d$ is defined as

$$x \preceq y :\Leftrightarrow x_i \leq y_i \text{ for all } 1 \leq i \leq d.$$

We say that x dominates y iff $x \preceq y$. If

$$x \prec y :\Leftrightarrow x \preceq y \text{ and } x \neq y$$

holds, we say that x strictly dominates y as x is not worse than y with respect to any objective, and at least better with respect to one of the d objectives.

The objective vectors x and y are called incomparable if

$$x \parallel y :\Leftrightarrow \neg(x \preceq y \lor y \preceq x)$$

holds. Two objective vectors are therefore incomparable if there are at least two (out of the d) objectives where they mutually beat each other. An objective vector x is called Pareto optimal if there is no $y = f(s)$ with $s \in S$ for which $y \prec x$ holds. The set of all Pareto optimal objective vectors is called the Pareto front of the problem given by f. Note that the Pareto front is a set of incomparable objective vectors.

Even for two objectives the Pareto front might grow exponentially with respect to the problem size. Therefore, algorithms for multi-objective optimization usually have to restrict themselves to a smaller set of solutions. This smaller set is then the output of the algorithm.

We make the notion of approximation precise by considering a weaker relation on the objective vectors called additive ϵ-dominance. It is defined as

$$x \preceq_{\epsilon+} y :\Leftrightarrow x_i + \epsilon \leq y_i \text{ for all } 1 \leq i \leq d.$$

Furthermore, we also define additive approximation of a set of objective vectors T with respect to another set of objective vectors S.

Definition 1. *For finite sets $S, T \subset \mathbb{R}^d$, the additive approximation of T with respect to S is defined as*

$$\alpha(S, T) := \max_{s \in S} \min_{t \in T} \max_{1 \leq i \leq d} (s_i - t_i).$$

We will use Definition 1 in order to judge the quality of a population P with respect to a given archive A that contains all non-dominated solutions seen so far (or an approximation thereof)—effectively, the value of $\alpha(S, T)$ is the approximation value achieved for the worst-approximated solution. In this way, we can measure how good the current population is with respect to the search points seen during the run of the algorithm.

Although, we are only using the notion of additive approximation, we would like to mention that our approaches can be easily adapted to multiplicative approximation. This can be done by adjusting the definitions accordingly.

2.2   User Preferences as Weight Functions in the Objective Space

User preferences provide information that guides the search process of the algorithm and tells the differences among incomparable solutions. In this article, we denote a weight function $w : \mathbb{R}^d \to \mathbb{R}$ which represents user preferences. In general, w can be an arbitrary function that specifies preferences to certain regions or points in the objective space.

In this article, we will use different *weight* functions for both 2- and 3-dimensional problems, which calculate the weight of a solution based on a given *scheme* value. Given a solution $x = \{x_1, x_2, \cdots, x_d\}$, the weight functions for 2-objectives problems, originally introduced in [22] and investigated in [9, 22], are defined as follows:

- *scheme* = 1: Both objectives are treated equally and the weight of a solution x is given by
$$w(x) = (e^{20x_1} + e^{20x_2})/(2 \cdot e^{20})$$

- *scheme* = 2: The user preference is based on only the second objective and the weight of a solution x is given by
$$w(x) = (e^{20x_2})(e^{20})$$

- *scheme* = 3: Given a reference point $r = \{r_1, r_2\}$, solutions closer to this point have higher user preference than the further ones. The weight of a solution x is given by
$$w(x) = \begin{cases} 10^{-5} + \frac{(3 - ((x_1 - r_1)^2 + (x_2 - r_2)^2))}{(0.001 + (2(x_1 - r_1) - 2(x_2 - r_2))^2)} \\ 10^{-5} \text{ otherwise} \end{cases}$$

For 3-objective problems, no *weight* functions had previously been defined. We extended the above-defined schemes 1 and 3:

Algorithm 1. Outline of AGE [18]

---

1 Initialize population P with μ random individuals;
2 Set ϵ_{grid} the resolution of the *approximative archive* $A_{\epsilon_{grid}}$;
3 **foreach** $p \in P$ **do**
4 $\quad$ Insert offspring *floor*(p) in the *approximative archive* $A_{\epsilon_{grid}}$ such that only non-dominated solutions remain;

5 **foreach** *generation* **do**
6 $\quad$ Initialize offspring population $O \leftarrow \emptyset$;
7 $\quad$ **for** $j \leftarrow 1$ **to** λ **do**
8 $\quad\quad$ Select two individuals from P (see Section 3.2 in [18]);
9 $\quad\quad$ Apply crossover and mutation;
10 $\quad\quad$ Add new individual to O;

11 $\quad$ **foreach** $p \in O$ **do**
12 $\quad\quad$ Insert offspring *floor*(p) in the *approximative archive* $A_{\epsilon_{grid}}$ such that only non-dominated solutions remain;
13 $\quad\quad$ Discard offspring p if it is dominated by any point *increment*(a), $a \in A$;

14 $\quad$ Add offsprings to population, i.e., $P \leftarrow P \cup O$;
15 $\quad$ **while** $|P| > \mu$ **do**
16 $\quad\quad$ Remove p from P that is of least importance to the approximation (for details on this step see [2]);

- *scheme* = 1:

$$w(x) = (e^{20x_1} + e^{20x_2} + e^{20x_3})/(3 \cdot e^{20})$$

- *scheme* = 3: Given a reference point $r = (r_1, r_2, r_3)$, the weight of a solution x is given by

$$w(x) = \begin{cases} 10^{-5} + \dfrac{(3-((x_1-r_1)^2+(x_2-r_2)^2+(x_3-r_3)^2))}{0.001+(\frac{(x_1-r_1)+(x_2-r2)+(x_3-r_3)}{3})^2} \\ 10^{-5} \text{ otherwise} \end{cases}$$

For *scheme* = 3 we selected a reference point of $(0.5, 0.6)$ for the two-dimensional problems, and $(0.5, 0.6, 07)$ for the three-dimensional ones.

How these *weight* functions will be incorporated into AGE will be shown in Section 3

2.3 Approximation-Guided Evolution

Definition 1 allows us to measure the quality of the population of an evolutionary algorithm with respect to a given set of objective vectors. AGE [2] is an evolutionary multi-objective algorithm that works with this formal notion of approximation. It stores an archive A consisting of the non-dominated objectives vectors found so far. Its aim is to minimize the additive approximation $\alpha(A, P)$ of the population P with respect to the archive A.

Algorithm 2. Function *floor* [18]

---

    input   : d-dimensional objective vector x, archive parameter ϵ_{grid}
    output: Corresponding vector v on the ϵ-grid

1 **for** $i = 1$ **to** d **do** $v[i] \leftarrow \left\lfloor \frac{x[i]}{\epsilon_{grid}} \right\rfloor$

Algorithm 3. Function *increment* [18]

---

    input   : d-dimensional vector x, archive parameter ϵ_{grid}
    output: Corresponding vector v that has each of its components increased by 1

1 **for** $i = 1$ **to** d **do** $v[i] \leftarrow o[i] + 1$

We consider the further developed version of AGE (called AGE-II in [18]). This algorithm is parametrized by the desired approximation quality $\epsilon_{grid} \geq 0$ of the archive with respect to the seen objective vectors. The algorithm is shown in Algorithm 1, and it uses the helper functions given in Algorithms 2 and 3. The latter is used to perform a relaxed dominance check on the offspring p in Line 13. A strict dominance check here would require an offspring to be not dominated by any point in the entire archive. However, as the archive approximates all the solutions seen so far (via the flooring), it might be very unlikely, or even impossible, to find solutions that pass the strict dominance test.

3   Adding User Preferences

Interactive AGE (iAGE) is a variant of AGE that considers user preferences as one of its parameters called *scheme* along with using the corresponding *weight* functions, which is mentioned in Section 3.2. The selection process of iAGE follows the same structure as the original AGE [2]. Let P be the current population where we need to remove an individual and A be the current archive. For each solution $a \in A$, we denote the best and second best approximation $\alpha_1(a)$, $\alpha_2(a)$ accordingly while $p_1(a)$ and $p_2(a)$ are solutions $p \in P$ that approximates a best and second best. In case of AGE, $p \in P$ with minimum $\beta(p)$ is removed from the population where $\beta(p)$ is known as the importance of solution p and defined as

$$\beta(p) := max_{a \in A}\{\alpha_2(a) | p_1(a) = p\}.$$

iAGE integrates the *weight* function into the selection process of the algorithm to ensure that user preference is one of the factors that decides whether a solution is removed or accepted to the next generation. We use a combination between the weight, $w(p)$, and the approximation to determine the importance of a given solution p given by expression

$$\beta(p) := max_{a \in A}\{w(p) \cdot \alpha_2(a) | p_1(a) = p\}.$$

Let $\beta_{min} := min_{p \in P} \beta(p)$ be the minimum β-value among all individuals of the population. The selection process removes a p from P for which $\beta(p) = \beta_{min}$

Algorithm 4. Outline of iAGE selection process.

---

12 See lines 12-16 in [2], Algorithm 4

17 **foreach** *solution* $p \in P$ **do**

18     $\beta(p) := max_{a \in A}\{w(p) \cdot \alpha_2(a) | p_1(a) = p\}$

19 **while** $|P| > \mu$ **do**

20   Remove an individual $p^* = \arg \min_{p \in P}(\beta(p), w(p))$ chosen uniformly at random from P.

21   See lines 21-23 in [2], Algorithm 4

---

holds. If there are multiple solutions p of value β_{min}, one with the smallest weight w value is discarded. If there are multiple solutions p with the same β_{min} and w value, then the removed solution is chosen at random. Therefore, the selection process removes an individual p from P that has the smallest vector $(\beta(p), w(p))$ according lexicographic order. The detail of the changes from Algorithm 4 in [2] is shown in Algorithm 4 where min is taken according to lexicographic order of the vector $(\beta(p), w(p))$.

The choice of iAGE's parameter ϵ_{grid} influences how well the set of solutions seen so far is approximated. Interestingly, this parameter also has a small but noticeable impact on the distribution of solutions. Some results are shown in Figure 1. As we can see, the solutions are packed more densely with decreasing grid size. The explanation is that the number of potential points in the archive increases, and consequently solutions in the population are more likely to be "responsible" for the approximation of an archive point. This, in combination with the increasing preference, results in a higher density of solutions.

We also investigate the impact of user preference on the distribution of solutions in the final population by providing different adjustments to the *weight* function. In particular, given the calculated β value for each $p \in P$, we want to study how different adjusted weight functions overwhelm the approximation and hence affect the selection process of the algorithm. In the following example, three adjustment strategies d are used:

- weight strategy 1 : $w(x) = w(x)$
- weight strategy 2 : $w(x) = sqrt(w(x))$
- weight strategy 3: $w(x) = ln(1 + w(x))$

We show some results in Figure 2. It can be seen that the choice of the adjustment strategy has hardly any impact on the distribution of points. The reason is that the weight remains its high impact even after a logarithmic scale-down. In addition, the approximation part of the adjustment only ensures that the archive points are better and better approximated; the slight change in the relative positioning of "the best population point for an archive point" (after considering the weight and the adjustment strategy) is barely noticeable in the final population and within the typical variations of results of randomized algorithms.

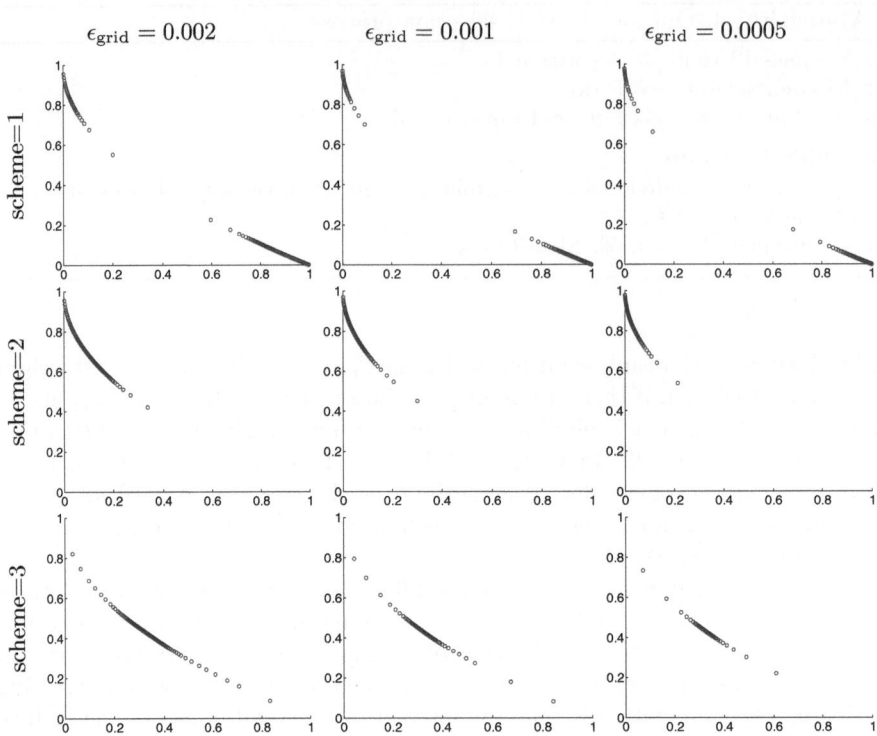

Fig. 1. Influence of ϵ_{grid} on the distributions of the solutions. The underlying problem is ZDT 1.

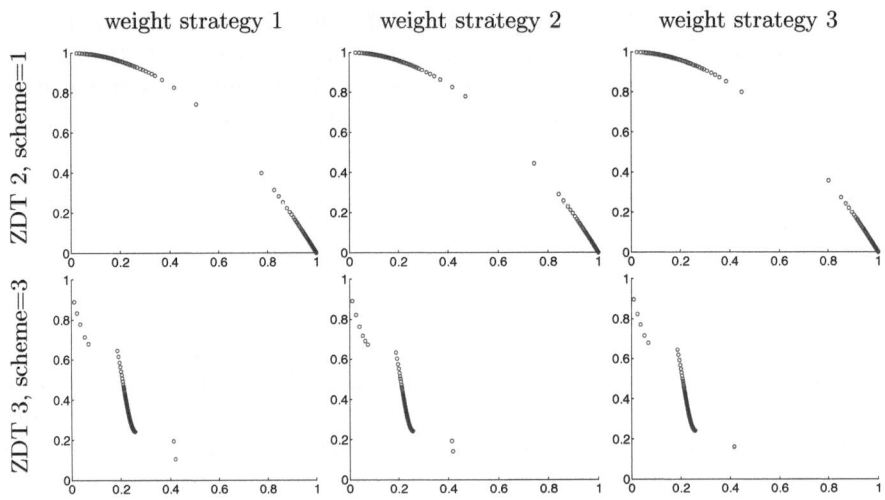

Fig. 2. Influence of different adjustment strategies to the weight function of iAGE with $\epsilon_{\mathrm{grid}} = 0.0005$

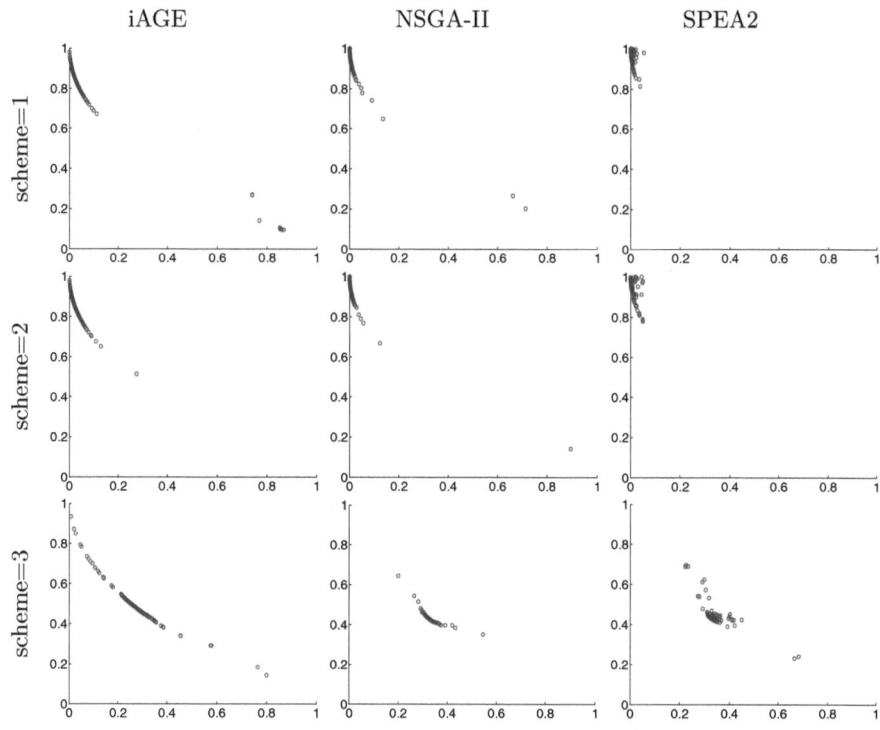

Fig. 3. LZ F5, 100.000 evaluations

4   Comparison with Other Algorithms

In our study, we investigate the performance of iAGE (using weight strategy 2) on problems with two and three objectives. We use the jMetal framework [8] to compare iAGE with the established algorithms NSGA-II [4], and SPEA2 [21]. Both algorithms are used as described in [9]: the weight functions are used multiplicatively to adjust either the crowding distance (NSGA-II) or the density (SPEA2). As benchmarks, we use the benchmark families WFG [11] and LZ [12], DTLZ [5], and ZDT [20].[1]

Note that we compare the final populations only visually. For the computation of indicator values, we would need reference sets: these are available for the true Pareto fronts in the "preference free" case, but not when non-linear preferences schemes are considered.

For many problems (mostly for the ZDT family and for DTLZ 1/2/3) we notice very few differences between the final distributions of the three algorithms. In stark contrast to this, we notice for several other problems that all algorithms would have immense problems to achieve good approximations of the true Pareto

---

[1] The code is available on our project page
http://cs.adelaide.edu.au/~optlog/research/foundations.php

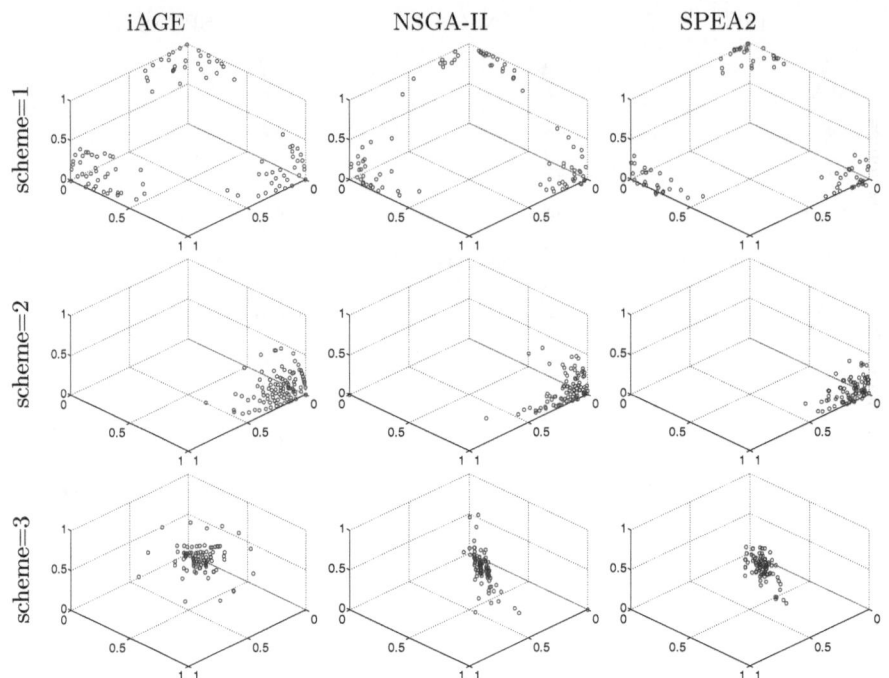

Fig. 4. DTLZ 2, d=3, 100.000 evaluations

front when a preference function was used. Some results are shown in Figure 3. The top row shows that all three algorithms have problems to cover the lower right sections of the Pareto front, even though this was a preferred region just as the top left section was. In the bottom row, we can observe that all algorithms find solutions close to the reference point. However, NSGA-II's solutions are often dominated by iAGE's, and SPEA2 itself maintains many dominated solutions.

We conjecture that the use of a preference function can restrict the diversity so much that it is not possible towards the end of the optimisation process to "rediscover" certain parts of the objective space anymore. We observe such difficulties for many functions, including DTLZ 4, many of the LZs and many of the WFGs.

As an example that preferences in objective spaces with more than two dimensions are possible, and as another extension to existing work, Figure 4 shows the results of the different algorithms on DTLZ 2, d=3. Because it is difficult to compare the outcomes using indicator values, we compare them visually. All three algorithms produce solution sets that follow the preference scheme. For iAGE, we notice "ray-like" patterns for the second scheme, and circular patterns around the reference point for the third scheme. NSGA-II and SPEA2, without their sense of an approximated archive, produce sets without any obvious visual structure. Consequently, we argue that iAGE produces the most evenly distributed solutions, even though this is in the eye of the beholder.

5 Conclusions

Evolutionary multi-objective methods are often considered in the unbiased case, where no particular area of the objective space is favored. This is in contrast to the actual decision making processes in the real world, where the decision maker typically has a preference for a particular range of non-dominated solutions.

In this article, we presented a simple and yet very effective modification to the algorithm AGE. The resulting algorithm iAGE differs from the original AGE only of the consideration of the weight function in a single step—the overall low computational complexity of the algorithm remains unchanged. Over a wide range of test functions, we observed that iAGE is just as good at finding evenly distributed solutions as similarly modified NSGA-II and SPEA2 variants. However, in particular for "difficult" two-objective problems and for all three-objective problems we have seen more evenly distributed solutions in the preferred regions of the objective space.

References

[1] Bringmann, K., Friedrich, T.: Approximating the volume of unions and intersections of high-dimensional geometric objects. In: Hong, S.-H., Nagamochi, H., Fukunaga, T. (eds.) ISAAC 2008. LNCS, vol. 5369, pp. 436–447. Springer, Heidelberg (2008)

[2] Bringmann, K., Friedrich, T., Neumann, F., Wagner, M.: Approximation-guided evolutionary multi-objective optimization. In: Proceedings of the Twenty-Second International Joint Conference on Artificial Intelligence, IJCAI 2011, vol. 2, pp. 1198–1203. AAAI Press (2011)

[3] Daskalakis, C., Diakonikolas, I., Yannakakis, M.: How good is the chord algorithm? In: Proceedings of the Twenty-First Annual ACM-SIAM Symposium on Discrete Algorithms, pp. 978–991. Society for Industrial and Applied Mathematics (2010)

[4] Deb, K., Pratap, A., Agrawal, S., Meyarivan, T.: A fast and elitist multiobjective genetic algorithm: NSGA-II. IEEE Trans. Evolutionary Computation 6, 182–197 (2002)

[5] Deb, K., Thiele, L., Laumanns, M., Zitzler, E.: Scalable test problems for evolutionary multiobjective optimization. In: Evolutionary Multiobjective Optimization, Advanced Information and Knowledge Processing, pp. 105–145 (2005)

[6] Deb, K., Sundar, J.: N. Udaya Bhaskara Rao, and S. Chaudhuri. Reference point based multi-objective optimization using evolutionary algorithms. Journal of Computational Intelligence Research 2, 273–286 (2006)

[7] Diakonikolas, I., Yannakakis, M.: Small approximate pareto sets for bi-objective shortest paths and other problems. In: Charikar, M., Jansen, K., Reingold, O., Rolim, J.D.P. (eds.) RANDOM 2007 and APPROX 2007. LNCS, vol. 4627, pp. 74–88. Springer, Heidelberg (2007)

[8] Durillo, J.J., Nebro, A.J., Alba, E.: The jMetal framework for multi-objective optimization: Design and architecture. In: IEEE Congress on Evolutionary Computation (CEC 2010), pp. 4138–4325 (2010)

[9] Friedrich, T., Kroeger, T., Neumann, F.: Weighted preferences in evolutionary multi-objective optimization. Journal of Machine Learning and Cybernetics 4, 139–148 (2013)

[10] Hu, Q., Xu, L., Goodman, E.D.: Non-even spread nsga-ii and its application to conflicting multi-objective compatible control. In: Proceedings of the First ACM/SIGEVO Summit on Genetic and Evolutionary Computation, pp. 223–230. ACM (2009)

[11] Huband, S., Barone, L., While, L., Hingston, P.: A scalable multi-objective test problem toolkit. In: Coello Coello, C.A., Hernández Aguirre, A., Zitzler, E. (eds.) EMO 2005. LNCS, vol. 3410, pp. 280–295. Springer, Heidelberg (2005)

[12] Li, H., Zhang, Q.: Multiobjective optimization problems with complicated pareto sets, MOEA/D and NSGA-II. IEEE Trans. on Evolutionary Computation 13, 284–302 (2009)

[13] Mohammadi, A., Omidvar, M.N., Li, X.: Reference point based multi-objective optimization through decomposition. In: Congress on Evolutionary Computation (CEC), pp. 1150–1157. IEEE (2012)

[14] Mohammadi, A., Omidvar, M.N., Li, X.: A new performance metric for user-preference based multi-objective evolutionary algorithms. In: Congress on Evolutionary Computation (CEC), pp. 2825–2832. IEEE (2013)

[15] Mohammadi, A., Omidvar, L.X., Nabi, M., Kalyanmoy, D.: Integrating user preferences and decomposition methods for many-objective optimization. In: Congress on Evolutionary Computation (CEC), pp. 421–428. IEEE (2014)

[16] Vassilvitskii, S., Yannakakis, M.: Efficiently computing succinct trade-off curves. In: Díaz, J., Karhumäki, J., Lepistö, A., Sannella, D. (eds.) ICALP 2004. LNCS, vol. 3142, pp. 1201–1213. Springer, Heidelberg (2004)

[17] Wagner, M., Friedrich, T.: Efficient parent selection for approximation-guided evolutionary multi-objective optimization. In: Congress on Evolutionary Computation (CEC), pp. 1846–1853 (2013)

[18] Wagner, M., Neumann, F.: A fast approximation-guided evolutionary multi-objective algorithm. In: Proceedings of the 15th Annual Conference on Genetic and Evolutionary Computation, GECCO 2013, pp. 687–694. ACM, New York (2013)

[19] Zitzler, E., Künzli, S.: Indicator-based selection in multiobjective search. In: Yao, X., et al. (eds.) PPSN 2004. LNCS, vol. 3242, pp. 832–842. Springer, Heidelberg (2004)

[20] Zitzler, E., Deb, K., Thiele, L.: Comparison of multiobjective evolutionary algorithms: Empirical results. Evol. Comput. 8, 173–195 (2000)

[21] Zitzler, E., Laumanns, M., Thiele, L.: SPEA2: Improving the strength Pareto evolutionary algorithm for multiobjective optimization. In: Evolutionary Methods for Design, Optimisation and Control with Application to Industrial Problems (EUROGEN 2001), pp. 95–100 (2002)

[22] Zitzler, E., Brockhoff, D., Thiele, L.: The hypervolume indicator revisited: On the design of pareto-compliant indicators via weighted integration. In: Obayashi, S., Deb, K., Poloni, C., Hiroyasu, T., Murata, T. (eds.) EMO 2007. LNCS, vol. 4403, pp. 862–876. Springer, Heidelberg (2007)

Multi-objective Optimisation, Software Effort Estimation and Linear Models

Peter A. Whigham and Caitlin Owen

Otago University, Dunedin, New Zealand,
peter.whigham@otago.ac.nz

Abstract. This paper examines the use of a linear model in combination with a multi-objective optimisation. A simple linear model is constructed and trained using data that has been automatically transformed based on skewness. These transformations, and their inverse, can then be used on the test data without having to make any assumptions of the underlying distribution of this data. Using *nsga2*, the coefficients of the linear model are optimised across a pareto front using 3 objective functions, representing 3 different error measurements. Although *nsga2* produces a variety of non-dominated models across the pareto front, we show that the use of these models for creating an ensemble is inappropriate. Our main conclusion is that the use of pareto modelling for creating ensemble methods does not appear to be valuable, although there is some information that can be gained from examining the change in coefficient values of a linear model across the pareto front.

Keywords: Multi-objective optimisation, nsga2, linear model, ensemble, sensitivity.

1 Introduction

The field of software effort estimation (SEE) has historically been interested in developing robust models for predicting development costs with software projects [15,2,1]. Significant early work included the constructive cost (CO-COMO) model [2] and function point analysis [1]. In recent years there has been a large number of different machine learning methods applied to the problem of effort estimation. The reader is directed to [8,4,14] for a review of machine learning methods applied to effort estimation.

This paper is motivated by a recent publication by Minku and Yao [14], where the weights of an artificial neural network were optimised using a multi-objective method to produce an effort estimation model as an ensemble of solutions from the pareto front. Our interest is in examining the use of a multi-objective method for optimising the coefficients of a linear model. Although we will show that the linear model performs at least as well as the more complex modelling approach of Minku and Yao, our real interest is in examining the properties of the linear model coefficients across the pareto front. In particular, can the variation in the coefficients be used to understand the stability and significance of the

G. Dick et al. (Eds.): SEAL 2014, LNCS 8886, pp. 263–273, 2014.
© Springer International Publishing Switzerland 2014

explanatory variables used in the model, and do the optimised linear models outperform a standard linear model? This is of general interest to the SEE community, since linear models have often been shown to not perform as well as other more complex machine-learning methods [5,18,16]. In particular, the non-linear interactions between the response and explanatory variables have been used to argue that linear models will not perform well [2]. However, there have also been arguments that linear models do perform well, but that appropriate transformations of the response and/or explanatory variables are required prior to modelling. The results from this paper will also be of general interest since we examine properties of the optimisation process, the pareto front, and the use of ensemble models in exploratory data analysis.

2   Data and Models

The dataset used in this study is the Cocomo81 data [2] obtained from the PROMISE repository [12]. The development mode (categorical variable) was removed from the dataset to allow a direct comparison with the results of Minku and Yao. This dataset describes 63 projects with 17 features (16 explanatory variables and 1 response). The response variable is the actual effort in person-months, ranging from a minimum of 5.9 to a maximum of 11400. A number of the variables in the Cocomo81 dataset are skewed, suggesting that a log or square-root transformation of these variables will be appropriate for a linear model [11,9]. Skewness is estimated using the m_1 measure [7], which has been shown to adequately characterise skewness for small samples. Since we are only interested in the magnitude of skewness, the absolute value of m_1 is taken as the measure of skewness, where larger values indicate a relatively greater divergence from a normal distribution.

2.1   Error Measures as Objective Functions

The multi-objective problem is framed by the use of three error measurements representing the objective functions. The following error measurements are commonly used in SEE and have been used in this study so that a direct comparison with the work of Minku and Yao is possible. The mean magnitude of the relative error (MMRE) measured over N examples is defined as:

$$MMRE = \frac{1}{N} \sum_{i=1}^{N} |y_i - \hat{y}_i| / y_i$$

where y_i is the ith measured value and $\hat{y}_i$ is the ith predicted value. Note that MMRE is biased towards prediction systems that under-estimate [10]. MMRE is an objective (error) measure to be *minimised*.

PRED(25) is a measure of the percentage of predictions that are within 25 percentage of the measured value, and defined as:

$$PRED(25) = \frac{1}{N} \sum_{i=1}^{N} \begin{cases} 1 & \text{if } |y_i - \hat{y}_i| / y_i \leq 0.25 \\ 0 & \text{otherwise} \end{cases}$$

PRED(25) emphasises the precision of predictions and is an objective to be *maximised*.

The logarithmic standard deviation (LSD) [6] is defined as:

$$LSD = \sqrt{\frac{\sum_{i=1}^{N}(e_i + \frac{s^2}{2})^2}{N-1}}$$

where s^2 is an estimate of the variance of the logarithmic residual e_i given $e_i = ln(y_i) - ln(\hat{y}_i)$. LSD is a measure suited to datasets that comply with a log-linear model [6] and is an objective to be *minimised*.

2.2   Linear Models and Multi-objective Optimisation

The linear model used for these experiments is the standard *lm* model from the programming environment R [17] and takes the form:

$$\hat{y}_i = \hat{\beta}_0 + \hat{\beta}_1 x_{1i} + \hat{\beta}_2 x_{2i} \ldots + \hat{\beta}_n x_{ni}$$

where the $\hat{\beta}_i$ are the estimated coefficients of the linear model for the intercept term and explanatory variables, and $\hat{y}_i$ is the estimated response. The application of any transformation to the response and explanatory variables is determined by the skewness measure m_1. Given some training data, the skewness of each variable is measured and compared with the skewness after a log (base e) and square-root transformation. If any transformation decreases the skewness this transformation is applied to the variable prior to building the linear model. When assessing the errors for a prediction (i.e. MMRE, PRED(25) and LSD), an inverse transform (if applicable) is applied to the predicted values prior to calculating the objective function errors. The multiobjective algorithm *nsga2* [3],as implemented in the R package *mco*, is used to search for the pareto front of solutions based on the 3 error objective functions. The basic outline of the algorithm is shown as Algorithm 1.

In Algorithm 1 the training data is initially analysed for transformations based on skewness, and transformed to *DT* as appropriate. A linear model is then constructed based on the (transformed) training data and a formula (F) defining the terms of the linear model. The coefficients of this initial linear model are used to determine the lower and upper bounds for the coefficients that will be searched using *nsga2*. Note that $CR = 1$ for all experiments, therefore allowing coefficient values to vary by 100% of their original values. This means that the lower bound is zero for all coefficients and the upper bound is two times the value of the initially discovered coefficient values found from the fit of the linear model (i.e. $\lceil \hat{\beta}_i \rceil = 2 * \hat{\beta}_i$). Initial experiments using larger values found that *nsga2* could not guarantee to find a pareto set, so this limitation was pragmatically set so that the algorithm converged. The result of *nsga2* is the values of the evolved coefficient values and objective function values of the linear model for each population member on the pareto front (*Solns*), along with the calculated transformation table (*Trans*). The parameters for the *nsga2* are shown in Table 1. Note that a

---

input : Training data *Data*, Formula *F*, ResponseVar *RV*,
            CoefficentFactor *CR*
output: Pareto Front Solutions *Solns*, Transformations *Trans*

1 $Trans \leftarrow$ CalcTransforms(*Data*);
2 $Trans.Data \leftarrow$ ApplyTransforms(*Trans,Data*);
3 $Linear.Model \leftarrow$ lm(*F,Trans.Data*);
4 $\hat{\beta}_i \leftarrow$ CoefficientVals(*Linear.Model*);
5 $\lfloor \hat{\beta}_i \rfloor \leftarrow \hat{\beta}_i - (\hat{\beta}_i * CR)$;
6 $\lceil \hat{\beta}_i \rceil \leftarrow \hat{\beta}_i + (\hat{\beta}_i * CR)$;
7 $Solns \leftarrow$ ApplyNSGA2(*Linear.Model,Trans.Data,Trans,$\lfloor \hat{\beta}_i \rfloor,\lceil \hat{\beta}_i \rceil$*);
8 **return** *Solns,Trans*;

Algorithm 1.   Basic algorithm for applying a linear model and nsga2

small population size was used after some initial experiments that found larger populations could not find a non-dominated set and were quite unstable between runs using the same data. The default values for *nsga2* were taken for the search parameters (crossover, mutation and the distribution indexes).

Table 1. Parameter values for *nsga2*

Parameter	Value	Parameter	Value
Pop. Size	52	Generations	500
Constraints	-	Output Dims.	3
Lower Bounds	0	Upper Bounds	$\lceil \hat{\beta}_i \rceil$
Crossover Prob.	0.7	Crossover Dist. Index	5
Mutation Prob.	0.2	Mutation Dist. Index	10

3   Experiments

The experiments are designed to show: (1) that a linear model with automatic transformations is a suitable model for SEE; (2) that an ensemble approach can be used with a least-squares linear model; (3) that the evolved coefficients for a single dataset are reasonably stable between different runs of *nsga2*; and (4) that the variation in coefficients across a pareto front allow an interpretation of explanatory variable importance.

The experimental design is based on the Minku and Yao paper so that a direct comparison is possible. Although a number of SEE papers have argued that a leave-one-out cross validation is the most appropriate format for estimating model quality (given this is the typical situation where a project manager requires an estimate of effort just for a single project), Minku and Yao used a randomised training and test split with 10 test items and the remaining examples used for training. As such we initially created 30 random sampling of training/test sets that were stored in separate files - this would allow us to reproduce the behaviour of the system and to examine the datasets if any unusual behaviour between test sets occurred.

Experiment (1) created a linear model for each of the 30 training/test splits, giving an estimate of the mean and variance of the model performance. Experiment (2) created a pareto front of models for each of the 30 training/test splits, and produced a single estimate for each test split by taking the best performing model for each objective function, producing their prediction, and taking the average effort prediction for each model and using this as the prediction of effort. This allowed a direct comparison with the results from Experiment (1). Experiment (3) used an arbitrary training set and produced 30 runs of the linear pareto model. A comparison of the coefficient values for the best objective function models for each run allowed an assessment of the variation between runs of *nsga2*. This could be used to assess whether variation observed in Experiment (4) were meaningful. Experiment (4) used the same training/test set as the previous experiment and examined the coefficient values across the pareto front to determine the stability of the coefficients as the emphasis on each objective varied.

4   Results

All variables in the Cocomo dataset were log transformed automatically due to skewness, for each training set, apart from the explanatory variables *cplx* and *turn*. These variables were left in their original form (i.e. no transform applied), apart from training set 28 where *cplx* was square-root transformed, and for training sets 7 and 11 where *turn* was also square-root transformed. In addition, *turn* was log transformed based on training set 21. Figure 1 shows the resulting test performance for the transformed linear model, and demonstrates that for this particular dataset the results are comparable or better than the more complex approach of Minku and Yao. Given the linear model with automatic transforms is a simple model it would suggest that the SEE field should be using such a model as a base measure for comparison when arguing that a new model performs well. However, this paper is not fundamentally about baseline models or showing the weakness in more complex methods (otherwise we would be comparing many datasets).

Figure 2 shows the performance using the model coefficients derived from the pareto front using the 3 error measures as objective functions. Here the best model on the front for each of the objective functions has been selected, and used to predict the test data. The quality of prediction is shown for each of the error measurements (one per graph), along with the original linear model and the combined ensemble model. The best MMRE model panel in Figure 2 shows little difference between models, however the LSD prediction model is clearer the most variable and overall the weakest. Note that the scales of each panel

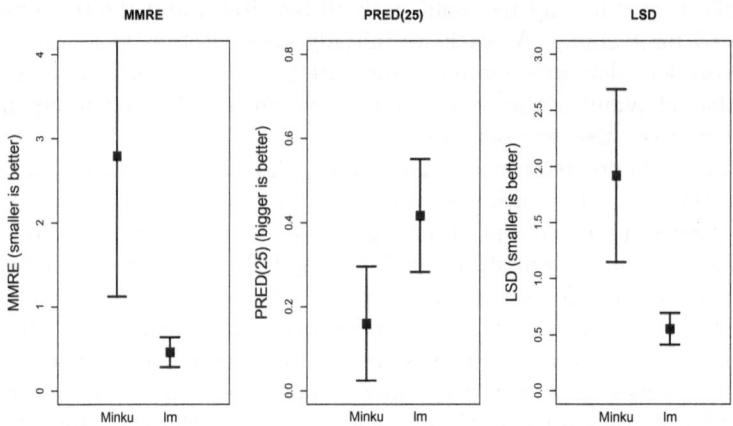

Fig. 1. Comparison of best results (mean and std. dev.) from Minku and Yao [14] versus a standard linear model with automatic transformations over 30 random training/test sets

are different since they show different error measurements. The Pred(25) model panel again shows little difference between models, although the MMRE model is the weakest. The LSD panel shows little difference between models except that the MMRE model is the most variable and has the worst median LSD. The main message from Figure 2 is that the linear model performs at least as well as the ensemble methods or any of the other best selected models from the pareto front.

Figure 3 shows the results for Experiment 3, where the variability of the $\hat{\beta}_i$ coefficients are shown for a single training set over 30 runs (lower panel). For this experiment the model with the minimum MMRE is selected for each run. The top panel shows the variability in $\hat{\beta}_i$ over the entire pareto front for the same training data. Although there is some differences in the variation between

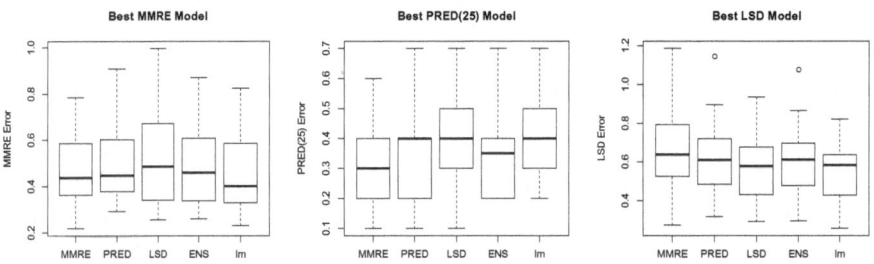

Fig. 2. Comparison of test predictions using the best training models (MMRE, PRED,LSD) from the pareto front, and the combined average ensemble (ENS). The linear model predictions are shown (labelled as *lm*) for comparison.

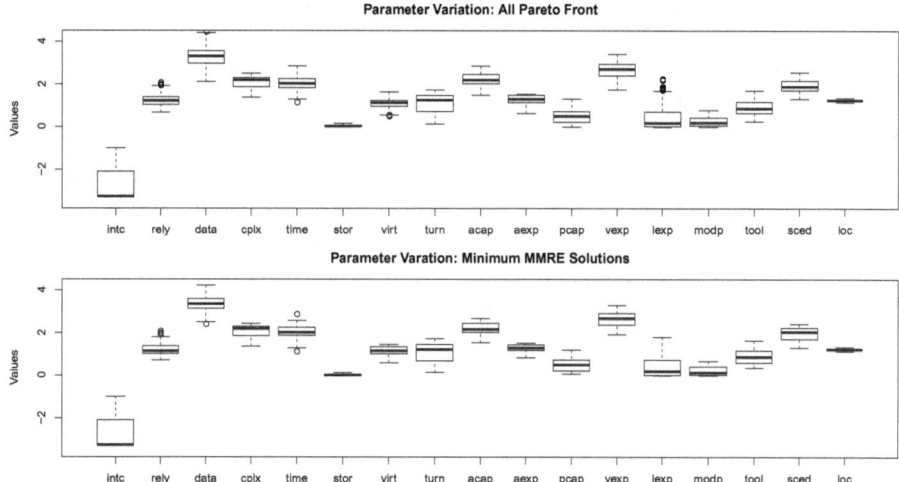

Fig. 3. Comparison of coefficients $\hat{\beta}_i$ over the entire pareto front (top panel) versus the models with the minimum MMRE model over the same training set (model built 30 times)

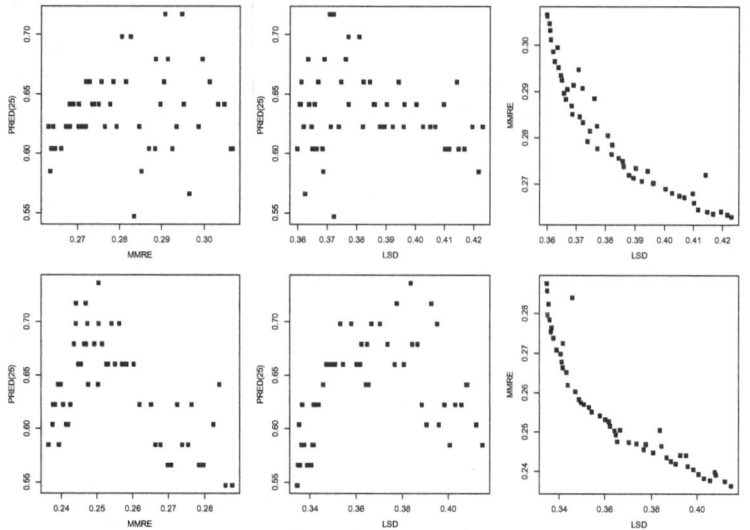

Fig. 4. Plot of pareto front objective functions for two different training sets. Note the similarity to Minku and Yao [14] (pg. 20).

panels overall the patterns are very similar, suggesting that there is significant variability between runs of *nsga2* with the same training data.

Figure 4 shows two example runs (using different training data) of the objective function values across the pareto front. Although there are differences they

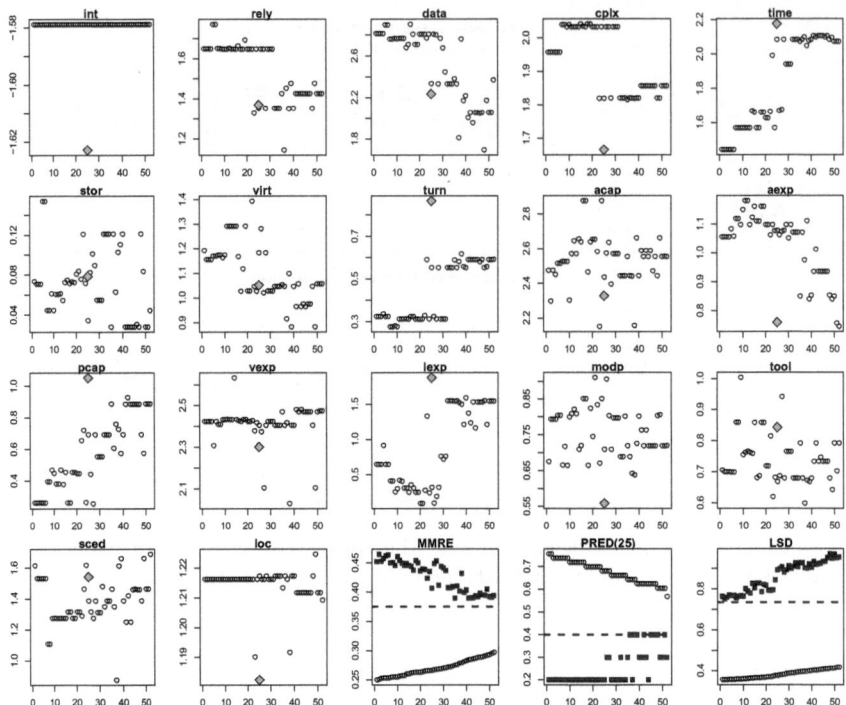

Fig. 5. Coefficient variation along the pareto front for one training run. The points are ordered from lowest to highest MMRE, with the coefficient value for the linear model is shown as a grey diamond. The final three plots show the ordered values of each objective along the pareto front, and their corresponding test predictions (black squares). The dashed line is the linear model prediction error for each objective.

are quite similar in their overall shape (especially the LSD versus MMRE). Note that since PRED(25) is an objective to be maximised the resulting plots against either LSD or MMRE (that are minimised) created a concave tradeoff shape, whereas the LSD versus MMRE plot creates a classic convex pattern as would be expected for two objectives that are minimised.

Figure 5 shows the $\hat{\beta}_i$ coefficient values of each of the 52 individuals in the population across the pareto front for one training run. Note that the models have been ordered based on increasing MMRE error. This shows that for some of the explanatory variables (and intercept term) there is little variation over the population, while other coefficients have either a clear pattern of two state values or have no discernible pattern. The lower right 3 panels show the training errors (circles) for each objective, while the black squares are the errors using the test data, indicating the generalisation error associated with the pareto front. Note that the population members have been ordered based on the objective function. The dashed line is the error for the linear model. It is clear that as we move across the front the training error increases as would be expected. For MMRE and PRED(25) the test error decreases as the training error increases, showing that for these objectives some overfitting is likely to have occurred. In contrast, the LSD test error increases across the front as the training error increases, suggesting that these models do not overfit, but also do not appear to generalise. In other words, the best LSD test error occurs with the most accurate training model for LSD. Although not considered here, this may indicate that LSD is a more robust error measurement in terms of avoiding overfitting and biasing towards outliers in the training data. Note that the linear model test predictions (dashed lines) are comparable to the best test errors across the entire front for each objective function.

5   Discussion

The results for experiments 1 and 2 (see Figures 1 and 2) indicate that the automatically transformed linear model performs at least as well as the evolved linear models on the pareto front. Note that all of the models are at least as good as the more complex models presented by Minku and Xao. In addition, examining later work using the Cocomo81 dataset and other ensemble methods [13], the linear model presented here produces a better SA measure than the best model presented in this work. This suggests that, for this particular dataset, complex modelling methods are inappropriate.

The results shown in Figure 2 suggest that constructing ensemble models based on sampling from the pareto front has little value. Since the models on the pareto front are all using the same data for training, their variation is just due to the tradeoff between objectives: although this produces different models, they are not likely to learn different aspects of the problem and therefore be valuable in combination. This implies that the use of pareto front modelling to produce ensembles is not likely to be a useful avenue of research. In addition, the variation in coefficient values found using *nsag2* (Figure 3) shows that each run of the multi-objective model (using the same training data) produces different coefficient values. This suggests that it is inappropriate to compare models between different runs, and that it is only meaningful to consider the properties of any single population along the pareto front for a particular run. This also raises some concerns regarding the stability of *nsga2*, but we cannot address this issue in this work.

A comparison between the pareto front objective functions (Figure 4) and the example figure (pg. 20) shown in [14] suggests that the overall patterns of tradeoff between objectives is similar, independent of the model. Although this is perhaps obvious, we would conjecture that *any model with enough complexity to produce a range of behaviours will produce a similar pattern for the objective tradeoffs*. We have shown this to be true when comparing a non-linear model and a simple transformed linear model, and suggest this would be true for any other models of reasonable complexity.

The coefficient variation ($\hat{\beta}_i$) along the pareto front (Figure 5) does allow some interpretation of the importance of the explanatory variables in developing a model. Those coefficients that show a random pattern of values are likely to be less important to the quality of the model prediction than those with a constant or stepped set of values. For example, the explanatory variables *stor*, *modp*, and *tool* show little pattern of consistent values across the front. A stepwise regression using the automatically transformed linear model removes all of these variables from the final model, indicating that this interpretation has some merit, although the *acap* variable was found to be important even though it has a reasonable spread of values across the front. Those variables with a consistent value (*int*, *cplx*, *vexp*, *loc*) were also found to be the most important variables using stepwise regression. Hence the pareto models do appear to have some value in assessing the quality of explanatory variables. The last three panels in Figure 5 show the training and test performance along the pareto front. The main comment is that the linear model on the test data is at least as good as the best model on the pareto front, once again supporting the conjecture that the pareto front does not contain more generalised models, and is likely to just be overfitting aspects of the training data. There is also some concern regarding the behaviour of the LSD measure: the test error increases across the front, and therefore shows no generalisation, unlike the MMRE and PRED(25) measures. This must be related to the tradeoff between the three objectives that has meant LSD is incompatible with the other error measurements, and brings into question the use of this combination of measures when constructing objective functions.

This paper has a number of weaknesses: we have not considered behaviour across multiple datasets; the ensemble method only considered the best training model for each error measurement across the pareto front, rather than considering some weighted combination of all models; and the properties and interactions between error measurements has been noted but not explored in detail. The paper has allowed some initial conjectures regarding aspects of these concepts, and they should be examined as future work to allow a more thorough assessment of ensemble models and multi-objective approaches.

6   Conclusion

This paper has shown that a linear model using automatically transformations based on the skewness of the training data can produce good quality predictions for one particular dataset in SEE. The use of a multi-objective optimisation method for tuning the coefficients of this linear model has been found to behave in a similar manner to more complex modelling methods previously published. However, we have argued that the use of a pareto front for constructing a group of ensemble models is not likely to be valuable since they all use the same training data and tend to overfit. Due to this over-fitting, unless a pareto front can be constructed that takes different samplings of the training data, the model variability will not be useful in producing an ensemble result, and therefore will be no more accurate than a single linear model.

References

1. Albrecht, A.J., John, J., Gaffney, E.: Software function, source lines of code, and development effort prediction: A software science validation. IEEE Transactions on Software Engineering 9(6), 639–648 (1983)
2. Boehm, B.W.: Software Engineering Economics. Prentice-Hall, Inc. (1981)
3. Deb, K., Pratap, A., Agarwal, S.: A fast and elitist multiobjective genetic algorithm: Nsga-ii. IEEE Trans. on Evol. Comp. 6(8), 182–197 (2002)
4. Dejaeger, K., Verbeke, W., Martens, D., Baesens, B.: Data mining techniques for software effort estimation: A comparative study. IEEE Transactions on Software Engineering 38(2), 375–398 (2012)
5. Finnie, G.R., Wittig, G.E., Desharnais, J.-M.: A comparison of software effort estimation techniques: Using function points with neural networks, case-based reasoning and regression models. The Journal of Systems and Software 39(3), 281–290 (1997)
6. Foss, T., Stensrud, E., Kitchenham, B., Myrtveit, I.: IEEE Trans. Softw. Eng. 29(11), 985–995 (2003)
7. Joanes, D., Gill, C.: Comparing measures of sample skewness and kurtosis. The Statistician 47(1), 183–189 (1998)
8. Jorgensen, M., Shepperd, M.: A systematic review of software development cost estimation studies. IEEE Transactions on Software Engineering 33(1), 33–53 (2007)
9. Kitchenham, B., Mendes, E.: Why comparative effort prediction studies be invalid. In: Proceedings of the 5th International Conference on Predictor Models in Software Engineering, pp. 1–5 (2009)
10. Kitchenham, B., Pickard, L., MacDonell, S., Shepperd, M.: What accuracy statistics really measure. IEE Proceedings Software 148(3), 81–85 (2001)
11. Kutner, M., Nachtsheim, C., Neter, J., Li, W.: Applied Linear Statistical Models, 5th edn. McGraw-Hill/Irwin (2005)
12. Menzies, T., Caglayan, B., He, Z., Kocaguneli, E., Krall, J., Peters, F., Turhan, B.: The promise repository of empirical software engineering data (2012)
13. Minku, L., Yao, X.: Ensembles and locality: Insight on improving software effort estimation. Information and Software Technology 55, 1512–1528 (2013)
14. Minku, L., Yao, X.: Software effort estimation as a multiobjective learning problem. ACM Transactions on Software Engineering and Methodology 22(4), 35–67 (2013)
15. Nelson, E.: Management Handbook for the Estimation of Computer Programming Costs. Systems Development Corporation (1966)
16. Park, H., Baek, S.: An empirical validation of a neural network model for software effort estimation. Expert Systems with Applications 35(3), 929–937 (2008)
17. R Core Team: R: A Language and Environment for Statistical Computing. R Foundation for Statistical Computing, Vienna, Austria (2014)
18. Tronto, I.B., Silva, J., Sant'Anna, N.: Comparison of artificial neural network and regression models in software effort estimation. In: IEEE International Joint Conference on Neural Networks, pp. 771–776. IEEE (2006)

Adaptive Update Range of Solutions in MOEA/D for Multi and Many-Objective Optimization

Hiroyuki Sato

The University of Electro-Communications,
1-5-1 Chofugaoka, Chofu, Tokyo 182-8585 Japan
sato@hc.uec.ac.jp

Abstract. MOEA/D, a representative multi-objective evolutionary algorithm, decomposes a multi-objective optimization problem into a number of single objective optimization problems and tries to approximate Pareto front by simultaneously optimizing each of these single objective problems. MOEA/D has several options to calculate a scalar value from multiple objective function values of a solution. In many-objective optimization problems including four or more objective functions, MOEA/D using the weighted sum scalarizing function achieves high search performance. However, the weighted sum has a serious problem that the entire concave Pareto front cannot be approximated. To overcome this problem of the weighted sum based MOEA/D, in this work we propose a method to adaptively determine update ranges of solutions in the framework of MOEA/D. The experimental results show that the weighted sum based MOEA/D using the proposed solution update method can approximate the entire concave Pareto front and improve the search performance.

Keywords: evolutionary multi-objective optimization, many-objective optimization, MOEA/D.

1   Introduction

Evolutionary algorithms are particularly suited to solve multi-objective optimization problems (MOPs) since Pareto optimal solutions (POS) approximating a trade-off among objective functions can be simultaneously searched with a population of solutions in a single run of the algorithm [1]. MOEA/D (multi-objective evolutionary algorithm based on decomposition), a representative evolutionary algorithm for solving multi-objective problems, decomposes a MOP into a number of single objective optimization problems and tries to find POS by simultaneously optimizing each of these single objective problems with a single population [2]. The decomposition of the objective space is one of promising approaches for solving many-objective optimization problems (MaOPs) involving four or more objective functions. In many-objective optimization, Pareto dominance based MOEAs such as NSGA-II [3] and SPEA2 [4] deteriorate their search performance since almost all solutions in the population become non-dominated

G. Dick et al. (Eds.): SEAL 2014, LNCS 8886, pp. 274–286, 2014.

and the proper selection pressure to improve the convergence of solutions toward Pareto front is weakened with increasing the number of objectives [5]. On the other hand, since MOEA/D uses a scalar value aggregated from multiple objective function values instead of Pareto dominance when solutions are compared, MOEA/D can easily determine the superiority of solutions even in MaOPs. Recently, NSGA-III, an improved version of NSGA-II for solving MaOPs, has also introduced the concept of decomposition of the objective space [6]. In this work we focus on MOEA/D employing the decomposition approach and aim to improve its search performance in multi and many-objective optimization problems.

MOEA/D has several options of scalarizing function to calculate a scalar value from multiple objective function values of a solution, and each of scalarizing functions has its own characteristics. The weighted Tchebycheff scalarizing function has an advantage that both convex and concave Pareto fronts can be approximated. On the other hand, the weighted sum scalarizing function has an advantage in many-objective optimization. The weighted sum based MOEA/D achieves higher search performance than the weighted Tchebycheff based MOEA/D in MaOPs. Recently, this observation was reported by Ishibuchi et al. [7]. However, the weighted sum has a serious problem that the entire concave Pareto front cannot be approximated [2]. If this problem of the weight sum based MOEA/D is solved by a modification of MOEA/D framework, the utilization of the weighted sum scalarizing function will be encouraged since the weighted sum based MOEA/D achieves high search performance in MaOPs [7].

To approximate concave Pareto fronts while using the weighted sum scalarizing function, in this work we propose a method to adaptively determine update ranges of solutions in the framework of MOEA/D. In a MOP with a concave Pareto front, solutions obtained by the weighted sum-based MOEA/D are distributed only in specific regions of the objective space. This problem is caused by a mechanism that the selection range of parents and the update range of solutions are the same in the conventional MOEA/D. Contrary, the proposed method considers the selection range of parents and the update range of solutions separately. After an offspring is generated by two parents selected from a selection range, the proposed method adaptively determines an update range of solutions based on the balance of the objective values of the generated offspring.

To verify the effectiveness of the proposed method, we use concave and convex WFG4 problems [8] with 2-6 objectives and many-objective knapsack problems [4] with 2-6 objectives and 500-1,000 bits (items) in this work.

2 MOEA/D

2.1 Algorithm

MOEA/D decomposes a MOP into a number of single-objective optimization problems. The single-objective optimization problems are defined by scalarizing functions g using uniformly distributed weight vectors $\boldsymbol{\lambda}^i$ $(i = 1, 2, \ldots, N)$. Each weight vector $\boldsymbol{\lambda}^i$ determines a search direction in the m-dimensional objective space. Each element λ_j^i $(j = 1, 2, \ldots, m)$ is one of $\{0/H, 1/H, \ldots, H/H\}$ based on the decomposition parameter H, and $N = C_{H+m-1}^{m-1}$ kinds of weight vectors satisfying $\sum_{j=1}^{m} \lambda_j^i = 1.0$ are used for the solution search. A similar idea was also proposed in Murata et al [10]. In the following, the algorithm of MOEA/D [2] is briefly described.

Step 1) Initialization:

Step 1-1) Compute the Euclidean distances between any two weight vectors and find the T nearest weight vectors to each weight vector. For each $i \in \{1, 2, \ldots, N\}$, set the parent selection and solution update range $B(i) = \{i_1, i_2, \ldots, i_T\}$ where $\boldsymbol{\lambda}^{i_1}, \boldsymbol{\lambda}^{i_2}, \ldots, \boldsymbol{\lambda}^{i_T}$ are the T nearest weight vectors to $\boldsymbol{\lambda}^{i_1}$.

Step 1-2) Randomly generate the population $\{\boldsymbol{x}^1, \boldsymbol{x}^2, \ldots, \boldsymbol{x}^N\}$.

Step 2) Solution Search:

For each $i \in \{1, 2, \ldots, N\}$, perform the following procedure.

Step 2-1) Randomly choose two indexes α and β from the selection range $B(i)$, and then generate an offspring $\boldsymbol{y}^i$ from parents $\boldsymbol{x}^\alpha$ and $\boldsymbol{x}^\beta$ by applying genetic operators.

Step 2-2) For each index k in the update range $B(i)$, if $g(\boldsymbol{y}^i|\boldsymbol{\lambda}^k)$ is better than $g(\boldsymbol{x}^k|\boldsymbol{\lambda}^k)$, then the current solution $\boldsymbol{x}^k$ is replaced by the generated offspring $\boldsymbol{y}^i$ $(\boldsymbol{x}^k = \boldsymbol{y}^i)$ [2].

Step 3) Stopping Criteria:

If the termination criterion is satisfied, then stop and pick POS from the population $\{\boldsymbol{x}^1, \boldsymbol{x}^2, \ldots, \boldsymbol{x}^N\}$ as the output of the optimization. Otherwise, go to **Step 2**.

2.2   Scalarizing Functions

In MOEA/D, there are several scalarizing approaches to aggregate m kinds of objective function values [2]. This section introduces the weighted Tchebycheff and the weighted sum scalarizing functions.

Tchebycheff

The scalar optimization problem of the weighted Tchebycheff function g^{tch} [9] is defined by

$$\text{Minimize } g^{tch}(\boldsymbol{x}|\boldsymbol{\lambda}) = \max_{1 \le j \le m} \{\lambda_j \cdot |f_j(\boldsymbol{x}) - z_j|\}, \tag{1}$$

---

[1] To approximate an entire Pareto front, MOEAs need to search a wide range of the solution space with the single population. To utilize local information of solutions having similar search directions, MOEA/D ueses the concept neighborhoods.

[2] For problems with totally different ranges of objective values, in this work, g is calculated after each objective value is normalized to the range $[0, 1]$ by the minimum and the maximum objective values in the current population.

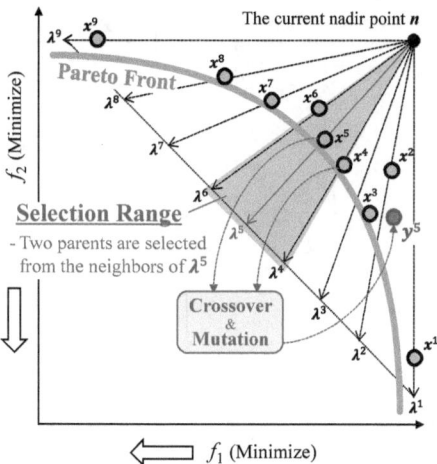

Fig. 1. The parents selection and the offspring generation in the conventional MOEA/D using the weighted sum function (**Step 2-1** in Section 2.1)

where, z is the obtained ideal point. In this work, each element z_j $(j = 1, 2, \ldots, m)$ is set to the best[3] objective function value f_i in the population. The weighted Tchebycheff approach searches a solution minimizing g^{tch} toward z. The weighted Tchebycheff has an advantage that both convex and concave Pareto fronts can be approximated.

Weighted Sum

The scalar optimization problem of the weighted sum function g^{ws} is defined by

$$\text{Maximize } g^{ws}(x|\lambda) = \sum_{j=1}^{m} \lambda_j \cdot (n_j - f_j(x)), \tag{2}$$

where, n is the current nadir point. In this work, each element n_j $(j = 1, 2, \ldots, m)$ is set to the worst[4] objective function value f_j in the population. The weighted sum approach searches a solution maximizing g^{ws} from n. A recent study [7] reported that the weighted sum approach achieves higher search performance than the weighted Tchebycheff approach in MaOPs. However, the weighted sum approach has a serious problem that the entire concave Pareto front cannot be approximated [2]. In the next section, we briefly explain the problem of the weighted sum based MOEA/D when we try to approximate concave Pareto front.

---

[3] For minimization problems, the best indicates the minimum. For maximization problems, the best indicates the maximum.

[4] For minimization problems, the worst indicates the maximum. For maximization problems, the worst indicates the minimum.

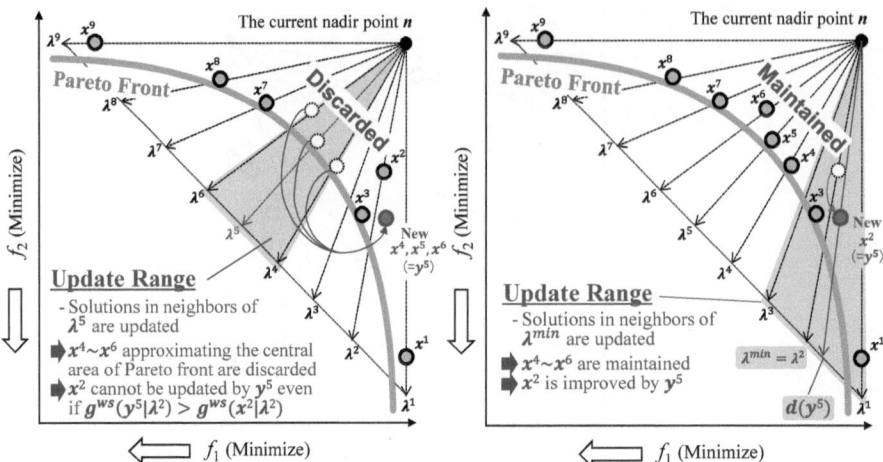

Fig. 2. The conventional solution update (**Step 2-2** in Section 2.1)

Fig. 3. The proposed solution update (**Step A~C** in Section 3.2)

2.3 Problem in the Weighted Sum Based MOEA/D

Figure 1 shows an example of the parent selection and the offspring generation in **Step 2-1** of the conventional weighted sum based MOEA/D in a $m = 2$ dimensional minimization problem with concave Pareto front. In this example, since we focus on the index $i = 5$, the selection range of parents becomes indexes $B(5) = \{4, 5, 6\}$ by considering $T = 3$ neighbors of the index $i = 5$. In this figure, $\alpha = 4$ and $\beta = 5$ are randomly chosen from the selection range $B(5) = \{4, 5, 6\}$, and the offspring $\boldsymbol{y}^5$ is generated by applying the genetic operators to the parents $\boldsymbol{x}^{\alpha=4}$ and $\boldsymbol{x}^{\beta=5}$.

Next, Fig. 2 shows the solution update in **Step 2-2** of the conventional weighted sum based MOEA/D. In the conventional MOEA/D, the update range of solutions becomes $B(5) = \{4, 5, 6\}$ which is the same as the selection range of parents shown in Fig. 1. In Fig. 2, the current solutions $\boldsymbol{x}^4$, $\boldsymbol{x}^5$ and $\boldsymbol{x}^6$ are updated (replaced) by the generated offspring $\boldsymbol{y}^5$ because the offspring $\boldsymbol{y}^5$ achieves higher g^{ws} than the current solutions for each weight vector $\boldsymbol{\lambda}^4$, $\boldsymbol{\lambda}^5$ and $\boldsymbol{\lambda}^6$, respectively. Although the solutions $\boldsymbol{x}^4$, $\boldsymbol{x}^5$ and $\boldsymbol{x}^6$ are well-converged on a central region of the concave Pareto front, they are discarded from the population since g^{ws} of the generated offspring $\boldsymbol{y}^5$ is higher than the ones of $\boldsymbol{x}^4$, $\boldsymbol{x}^5$ and $\boldsymbol{x}^6$. Thus, the weighted sum based MOEA/D has a problem that solutions approximating the central region of concave Pareto front are discarded from the population during the solution search.

In addition, for the search direction of the weight vector $\boldsymbol{\lambda}^2$, the generated offspring $\boldsymbol{y}^5$ is better than the current $\boldsymbol{x}^2$. However, the conventional MOEA/D does not update $\boldsymbol{x}^2$ by the generated offspring $\boldsymbol{y}^5$ because the index 2 is out of the update range of solutions $B(5) = \{4, 5, 6\}$. It will lead to a loss in the solution search.

If above mentioned problems are solved by a modification of MOEA/D framework, the utilization of the weighted sum function will be encouraged since the weighted sum based MOEA/D achieves high search performance in MaOPs [7]. These problems are caused by a mechanism that the selection range of parents and the update range of solutions are the same in the conventional MOEA/D. To overcome these problems in the weighted sum based MOEA/D, in this work we modify **Step 2-2** of the conventional MOEA/D.

3   Proposal: Adaptive Update Range of Solutions

3.1   Concept

To approximate concave Pareto fronts and encourage the solution search for each search direction while using the weighted sum scalarizing function, in this work we propose a method to adaptively determine update ranges of solutions in MOEA/D framework. In the conventional MOEA/D, the selection range of parents and the update range of solutions are the same. Contrary, the proposed method considers the selection range of parents and the update range of solutions separately. After an offspring is generated by two parents selected from a selection range, the proposed method adaptively determines the update range of solutions based on the balance of the objective function values of the generated offspring.

3.2   Method

The procedure of the proposed method adaptively determining the update range of solutions in MOEA/D framework is described as follows. **Step 2-2** in Section 2.1 is replaced by the following procedure **Step A~C**. For problems with totally different ranges of objective values, the following procedures are performed after each objective value is normalized to the range $[0, 1]$ by the minimum and the maximum objective values in the current population.

Step A: Calculate the balance $d(y^i)$ of the objective function values of the generated offspring y^i by

$$d_j(y^i) = \frac{n_j - f_j(y^i)}{\sum_{\ell=1}^{m} \{n_\ell - f_\ell(y^i)\}} \qquad (j = 1, 2, \ldots, m). \qquad (3)$$

Step B: Find the index min of the weight vector λ^{min} which has the minimum Euclidean distance from $d(y^i)$ among the all weight vectors $\{\lambda^1, \lambda^2, \ldots, \lambda^N\}$.

$$min = \{ q \mid \min_{q \in \{1,2,\ldots,N\}} \text{Distance}(d(y^i), \lambda^q) \} \qquad (4)$$

Step C: For each index k in the update range $B(min)$, perform the following procedure.

Step C-a: If $d(y^i)$ satisfies the following condition for the weight vector λ^k, go to **Step C-b**. Otherwise, continue **Step C**.

$$\forall j = \{1, 2, \ldots, m\} : \lambda_j^k - \frac{h}{H} < d_j(y^i) < \lambda_j^k + \frac{h}{H}. \tag{5}$$

In this work, $h = 2$ is used for this condition.

Step C-b: Calculate the balance $d(x^k)$ of the objective function values of the current solution x^k by

$$d_j(x^k) = \frac{n_j - f_j(x^k)}{\sum_{\ell=1}^{m} \{n_\ell - f_\ell(x^k)\}} \quad (j = 1, 2, \ldots, m). \tag{6}$$

Step C-c: If $d(x^k)$ does not satisfy the following condition for the weight vector λ^k, x^k is updated (replaced) by y^i ($x^k = y^i$).

$$\forall j = \{1, 2, \ldots, m\} : \lambda_j^k - \frac{h}{H} < d_j(x^k) < \lambda_j^k + \frac{h}{H}. \tag{7}$$

Also, if the both the above and the following conditions are satisfied, x^k is updated (replaced) by y^i ($x^k = y^i$).

$$g^{ws}(y^i \mid \lambda^k) \geq g^{ws}(x^k \mid \lambda^k) \tag{8}$$

3.3   Expected Effects

Figure 3 shows the expected effects of the proposed solution update method. In the proposed method, first, the balance $d(y^5)$ of the objective function values of the generated offspring y^5 is calculated. Next, the weight vector λ^{min} ($= \lambda^2$) having the minimum Euclidean distance from $d(y^5)$ is founded, then the update range of solutions becomes $B(min) = \{1, 2, 3\}$ by considering $T = 3$ neighbors of λ^{min} ($= \lambda^2$). In the update range $B(min)$ adaptively determined by the proposed method, the current solution x^2 is updated (replaced) by the generated offspring y^5. In this way, the proposed method can maintain solutions approximating the central region of the concave Pareto front. In addition, since the solution x^2 is updated by the generated offspring y^5, we can expected that the solution search for the weight vector λ^2 is encouraged.

4   Experimental Setup

4.1   Four Algorithms

In this work, we compare four algorithms. First three algorithms are MOEA/Ds. They are (i) the weighted Tchebycheff based MOEA/D using the conventional solution update method, (ii) the weighted sum based MOEA/D using the conventional solution update method and (iii) the weighted sum based MOEA/D using the proposed solution update method, respectively. The last one is (iv) NSGA-III [6] proposed by Prof. Deb. NSGA-III is an improved version of NSGA-II and tries to approximate Pareto front by decomposing the objective space similar to MOEA/D. To maintain the distribution of solutions in the objective space, NSGA-III uses distance between solutions and weight vectors instead of the crowding distance used in NSGA-II [3].

4.2 Test Problems and Parameters

As a continuous test problem, we use WFG4 [8]. Since the conventional WFG4 has a concave Pareto front, in this paper this problem is described as 'concave WFG4'. Also, as an extension of the conventional WFG4, in this work 'convex WFG4' problem is defined by

$$\text{Minimize } f_i^{\text{WFG4}}(\boldsymbol{x})^p \quad (i = 1, 2, \ldots, m), \tag{9}$$

where, f_i^{WFG4} $(i = 1, 2, \ldots, m)$ are the original objective functions of the conventional concave WFG4 [8], and p is a problem parameter. Pareto front of this problem with $p > 2$ becomes convex. In this work, $p = 5$ is used in the experiment. For both WFG4 problems, two kinds of difficulties can be controlled by the distance-related parameter L and the position-related parameter K. The difficulty of the convergence toward Pareto front is increased by increasing L. In this work, the fixed $L = 10$ is used. Also, the difficulty to cover the entire Pareto front is increased by increasing K. In this work, we verify the search performance in WFG4 problems with different K. Since assignable values of K in WFG4 is restricted, we set $K = k + (m - 1) - k \mod (m - 1)$ by varying $k = \{50, 100, 200, 400\}$. Therefore, the number of variables is $n = L + K$. Also, the number of objectives is set to $m = \{2, 4, 6\}$. To generate offspring, we adopt SBX [11] with the crossover ratio 0.8 and $\eta_c = 15$, and apply polynomial mutation [11] with the mutation ratio $4/n$ and $\eta_m = 20$. The termination criterion is set to $3,000$ generations.

As a discrete test problem, we use many-objective knapsack problems (MOKPs) [4] with $m = \{2, 4, 6\}$ objectives, $n = \{500, 750, 1000\}$ bits (items) and feasibility raio $\phi = 0.5$. Infeasible solutions are repaired by the greedy repair algorithm [4]. To generate offspring, uniform crossover with the crossover ratio 0.8 and bit-flip mutation with the mutation ratio $4/n$ are used. The termination criterion is set to $10,000$ generations.

The decomposition parameters to generate weight vectors in MOEA/D are set to $H = \{200, 9, 5\}$ for $m = \{2, 4, 6\}$ objectives, respectively. Therefore, the population sizes are $N = \{201, 220, 252\}$ for $m = \{2, 4, 6\}$, respectively. Three MOEA/Ds and NSGA-III use the same weight vectors and the population size. In MOEA/Ds, the neighborhood size is set to $T = 20$. In the following experiments, we show the average performance with 30 runs.

4.3 Performance Metric

To evaluate the search performance, Hypervolume (HV) [12] is used in this work. HV measures the m-dimensional volume of the region enclosed by the

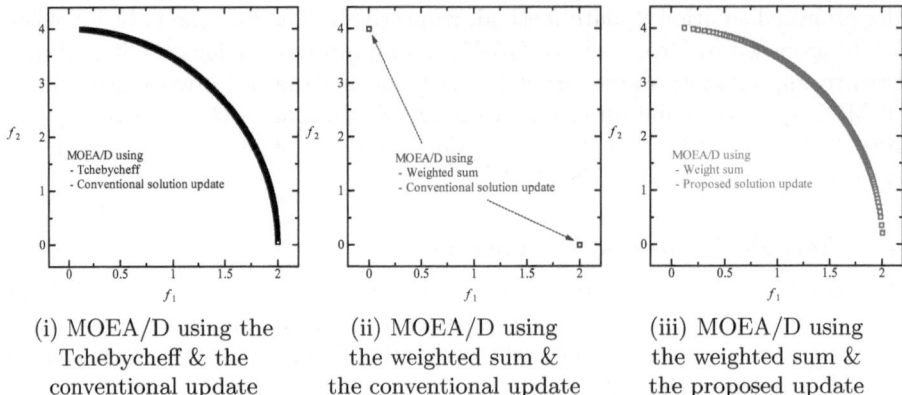

(i) MOEA/D using the
Tchebycheff & the
conventional update

(ii) MOEA/D using
the weighted sum &
the conventional update

(iii) MOEA/D using
the weighted sum &
the proposed update

Fig. 4. The obtained POS on concave WFG4 with the problem difficulty $k = 50$

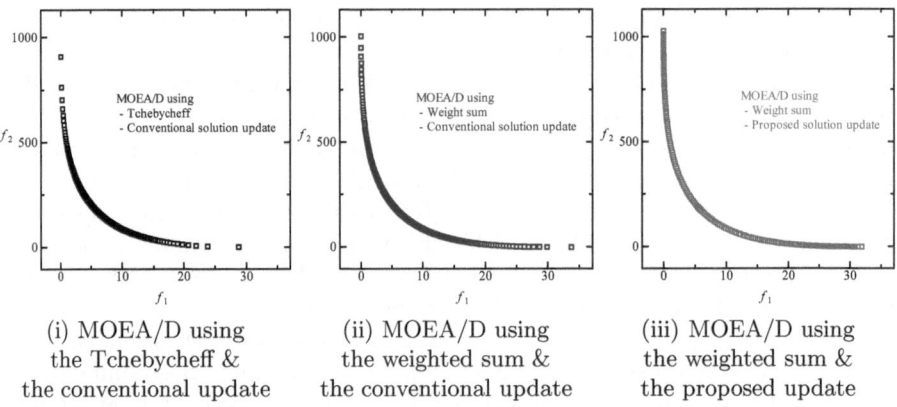

(i) MOEA/D using
the Tchebycheff &
the conventional update

(ii) MOEA/D using
the weighted sum &
the conventional update

(iii) MOEA/D using
the weighted sum &
the proposed update

Fig. 5. The obtained POS on convex WFG4 with the problem difficulty $k = 50$

obtained POS and a dominated reference point r in the objective space. A Higher HV indicates better POS in terms of both the convergence and the diversity. For MOKPs, $r = \{0, 0, \ldots, 0\}$ is used to calculate HV. For both concave and convex WFG4 problems, $r = \{1.1, 1.1, \ldots, 1.1\}$ is used to calculate HV after the objective function values are normalized as $f'_j(x) = f_j(x)/2j$ $(j = 1, 2, \ldots, m)$ for concave WFG4 and $f'_j(x) = f_j(x)/2j^p$ $(j = 1, 2, \ldots, m)$ for convex WFG4, respectively.

5 Experimental Results and Discussion

5.1 The Obtained POS on Concave and Convex WFG4s

Figure 4 shows the obtained Pareto optimal solutions by three MOEA/Ds on concave WFG4 problem with $m = 2$ objectives. Note that both objective functions should be minimized in this problem. From these results, first we can see that

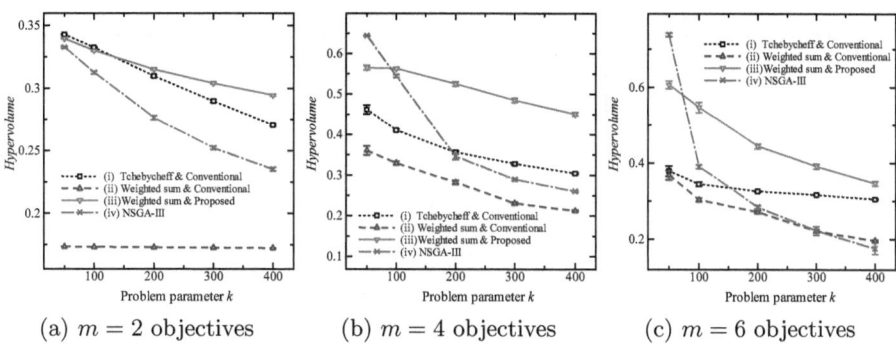

Fig. 6. Results of *HV* on concave WFG4 problems

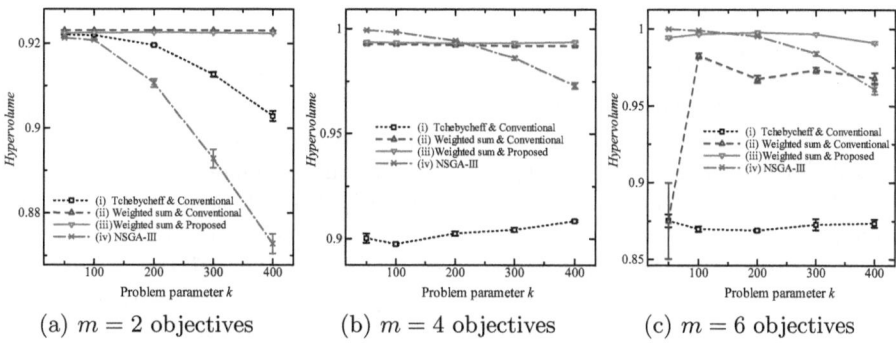

Fig. 7. Results of *HV* on convex WFG4 problems

MOEA/D using the weighted Tchebycheff function can approximate concave Pareto front. However, MOEA/D using the weighted sum and the conventional solution update method cannot approximate the central region of the concave Pareto front. On the other hand, we can see that MOEA/D using the weighted sum and the proposed solution update method can approximate the entire concave Pareto front. This result reveals that the proposed solution update method can approximate the entire concave Pareto front while using the weighted sum scalarizing function.

Similarly, Fig. 5 shows the obtained Pareto optimal solutions on convex WFG4 problem with $m = 2$ objectives. The convex WFG4 is also minimization problem. From these results, we can see that all three algorithms can obtain POS approximating the entire convex Pareto front. We can see the tendency that the distribution of POS obtained by the weighted Tchebycheff is sparse in the two extreme regions of convex Pareto front. On the other hand, the distributions of POS obtained by two MOEA/Ds using the weighted sum are dense in the two extreme regions of convex Pareto front.

5.2 The Search Performance on Concave and Convex WFG4s

Figure 6 shows results of *HV* achieved by four algorithms at the final generation on concave WFG4 problems when the problem difficulty parameter k is varied.

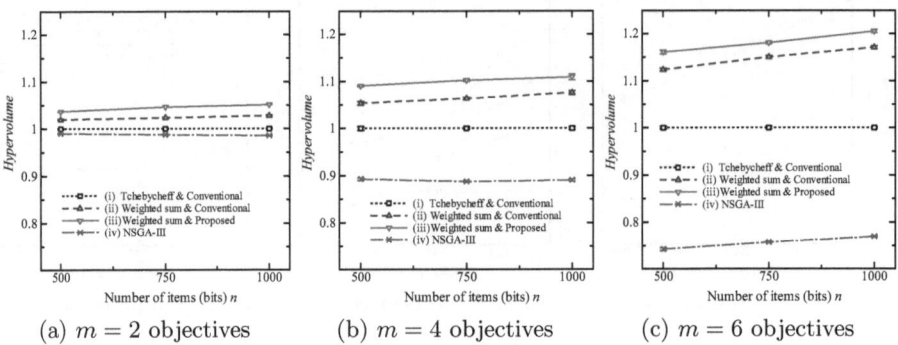

Fig. 8. Results of HV on MOKPs

The difficulty to obtain a widely spread POS is increased by increasing the problem difficulty parameter k. From these results, we can see that MOEA/D using the weighted sum and the conventional solution update method shows the lowest HV on all problems except the case of the problem with $m = 6$ and $k = 400$. This is because, as shown in Fig. 4 (ii), the solutions evolved by the weighted sum and the conventional solution update method are distributed only in specific regions in the objective space. Also, we can see that NSGA-III achieves the highest HV among four algorithms on problems with $m = \{4, 6\}$ objectives and small $k = 50$. On the other hand, in problems with large k, MOEA/D using the weighted sum and the proposed solution update method achieves the highest HV among four algorithms compared in this work. These results reveal that the proposed method can approximate concave Pareto front and improves the search performance especially in problems with the difficulty to obtain a widely spread solutions.

Similarly, Fig. 7 shows results of HV achieved by four algorithms on convex WFG4 problems. From this results, in problems with $m = \{2, 4\}$ objectives, we can see that values of HV achieved by two MOEA/Ds using the weighted sum function are comparable. However, in problems with $m = 6$ objectives, the weighted sum based MOEA/D using the proposed solution update method achieves higher HV than the one using the conventional solution update method. These results reveal that the effectiveness of the proposed method is significant especially in problems with a large number of objectives.

5.3 The Search Performance on MOKPs

Figure 8 shows results of HV achieved by four algorithms at the final generation on MOKPs. In each figure, values of HV are normalized by the results of HV achieved by MOEA/D using the weighted Tchebycheff and the conventional solution update method. From these results, we can see that MOEA/D using the weighted sum and the proposed solution update method achieves the highest HV among four algorithms on all MOKPs used in this work. Also, the effectiveness of the proposed solution update method becomes significant especially in problems with large number of objectives m and large size of solution space n.

6   Conclusions

To approximate concave Pareto fronts and encourage the solution search for each search direction in the weighted sum based MOEA/D, in this work we proposed a method adaptively determining update range of solutions in the framework of MOEA/D. Experimental results in WFG4 problems clearly showed that MOEA/D using the weighted sum and the proposed solution update method was able to approximate concave Pareto front and achieved higher search performance than other MOEA/Ds and NSGA-III especially in problems with the difficulty to obtain a widely spread Pareto optimal solutions. Also, in MOKPs, MOEA/D using the weighted sum and the proposed solution update method achieved the highest HV among four algorithms in all problems used in this work, and the effectiveness of the proposed method became significant in problems with large number of objectives and large size of solution space.

As future works, we will verify the search performance of the proposed method in problems with more large number of objectives.

Acknowledgments. This work was supported by the Okawa Foundation for Information and Telecommunications.

References

1. Coello, C.A.C., Van Veldhuizen, D.A., Lamont, G.B.: Evolutionary Algorithms for Solving Multi-Objective Problems. Kluwer Academic Publishers, Boston (2002)
2. Zhang, Q., Li, H.: MOEA/D: A Multi-objective Evolutionary Algorithm Based on Decomposition. IEEE Trans. on Evolutionary Computation 11(6), 712–731 (2007)
3. Deb, K., Pratap, A., Agarwal, S., Meyarivan, T.: A Fast Elitist Multi-Objective Genetic Algorithm: NSGA-II. IEEE Trans. on Evolutionary Computation 6, 182–197 (2002)
4. Zitzler, E., Thiele, L.: Multiobjective Evolutionary Algorithms: A Comparative Case Study and the Strength Pareto Approach. IEEE Trans. on Evolutionary Computation 3(4), 257–271 (1999)
5. Ishibuchi, H., Tsukamoto, N., Nojima, Y.: Evolutionary many-objective optimization: A short review. In: Proc. of 2008 IEEE Congress on Evolutionary Computation (CEC2008), pp. 2424–2431 (2008)
6. Deb, K., Jain, H.: An Evolutionary Many-Objective Optimization Algorithm Using Reference-point Based Non-dominated Sorting Approach, Part I: Solving Problems with Box Constraints. IEEE Trans. on EC PP 99, 1–23 (2013)
7. Ishibuchi, H., Akedo, N., Nojima, Y.: A Study on the Specification of a Scalarizing Function in MOEA/D for Many-Objective Knapsack Problems. In: Nicosia, G., Pardalos, P. (eds.) LION 7. LNCS, vol. 7997, pp. 231–246. Springer, Heidelberg (2013)

8. Huband, S., Hingston, P., Barone, L., While, L.: A Review of Multi-objective Test Problems and a Scalable Test Problem Toolkit. IEEE Transactions on Evolutionary Computation 10(5), 477–506 (2006)
9. Miettinen, K.: Nonlinear Multiobjective Optimization. Kluwer, Norwell (1999)
10. Murata, T., Ishibuchi, H., Gen, M.: Specification of Genetic Search Directions in Cellular Multi-objective Genetic Algorithms. In: Zitzler, E., Deb, K., Thiele, L., Coello Coello, C.A., Corne, D.W. (eds.) EMO 2001. LNCS, vol. 1993, pp. 82–95. Springer, Heidelberg (2001)
11. Deb, K., Goyal, M.: A Combined Genetic Adaptive Search (GeneAS) for Engineering Design. Computer Science and Informatics 26(4), 30–45 (1996)
12. Zitzler, E.: Evolutionary Algorithms for Multiobjective Optimization: Methods and Applications, PhD thesis, Swiss Federal Institute of Technology, Zurich (1999)

Classification of Lumbar Ultrasound Images with Machine Learning

Shuang Yu and Kok Kiong Tan

National University of Singapore, Mechatronics and Automation Lab, Singapore
{yushuang,kktan}@nus.edu.sg

Abstract. In this paper, we propose a feature extraction and machine learning method for the classification of ultrasound images obtained from lumbar spine of pregnant patients in the transverse plane. A set of features, including matching values and positions, appearance of black pixels within predefined windows along the midline, are extracted from the ultrasound images using template matching and midline detection. Artificial neural network is utilized to classify the bone images and interspinous images. The neural network is trained with 1000 images from 25 pregnant subjects and tested on 720 images from a separate set of 18 pregnant patients. A high success rate (96.95% on training set, 95.75% on validation set and 94.12% on test set) is achieved with the proposed method. The trained neural network further tested on 43 videos collected from 43 pregnant subjects and successfully identified the proper needle insertion site (interspinous region) in all of the cases. Therefore, the proposed method is able to identify the ultrasound images of lumbar spine in an automatic manner, so as to facilitate the anesthetists' work to identify the needle insertion point precisely and effectively.

1 Introduction

Epidural/spinal anesthesia (EA) is widely used in surgery and for post-surgical pain relief. A properly performed epidural procedure is the 'gold standard' of treatment to reduce pain during childbirth [1]. Around 50-90% of women in labour in developed countries choose EA for pain relief [2]. However, the failure rate of EA has been reported to be as high as 20% [3]. One of the key challenges for EA is the identification of the needle insertion site, which is traditionally identified by palpating the patients' lumbar spine [4]. This blind technique may require multiple needle insertion attempts, leading to complications in the process. The case is worse for patients with obesity problems, which is increasingly prevalent in the pregnant population.

Ultrasound imaging, as a non-radioactive, convenient and inexpensive medical imaging modality, has been introduced to EA to assist epidural needle insertion since the 1950s [5]. Previous researches have confirmed the effectiveness of ultrasound imaging compared with the traditional palpation method [6]. Despite the benefits of ultrasound, the effective interpretation of ultrasound images remains a challenge, especially for anesthetists who received limited training in reading

G. Dick et al. (Eds.): SEAL 2014, LNCS 8886, pp. 287–298, 2014.
© Springer International Publishing Switzerland 2014

ultrasound images [7]. The low spatial resolution and severe speckle noises of ultrasound images results in the subtle anatomical features becoming indiscernible from the surrounding background [8]. It requires professional training to fully interpret the ultrasound images and the learning curve is steep. Therefore, a large proportion of anesthetists are reluctant to adopt ultrasound imaging in the common practice.

In order to ease the ultrasound image interpretation and facilitate the applicability of ultrasound in epidural needle insertion, automatic interpretation of lumbar ultrasound images has been investigated by researchers. Train et al. utilized phase symmetry and template matching to extract the lamina and ligamentum flavum in the paramedian images [9]. Kerby et. al proposed to label the lumbar level automatically with panorama images obtained from the paramedian view [10]. Furthermore, an augmented reality system (AREA) which projected the identified lumbar vertebra levels on the patients back was developed so as to assist spinal needle insertion [11].

Although automatic interpretation of lumbar ultrasound images has been explored, it is mainly focused on the paramedian view. Ultrasound images in the transverse view, which reveal important anatomical information and frequently been used by anesthetists for precise pre-puncture localization of needle insertion site, are less researched from the automatic image interpretation perspective. In our previous research, an image processing and identification procedure was developed for the automatic interpretation of ultrasound images in the transverse view [12]. Template matching combined with position correlator (PC) was proposed to identify the interspinous images and achieved a success rate of 100% on ultrasound images obtained from lumbar spine of healthy volunteers. However, since the clarity of anatomical feature of lumbar spine might degrade during pregnancy [13], the original position correlator designed for healthy volunteer is not effectively applicable to the pregnant patient.

In order to improve the identification accuracy for pregnant patients and make the classification algorithm more generally applicable, a feature extraction and classification procedure is developed. Three contributions are achieved with this paper. First, a set of features, which are composed of important parameters, are extracted from the lumbar ultrasound images with template matching and midline detection methods. Secondly, a multi-layer neural network is utilized to classify the interspinous images and bone images with the extracted feature vector. A high success rate is achieved with the proposed feature extraction and neural network structure on images collected from the pregnant patients. Last but not least, the trained neural network model is also tested on 43 videos and it successfully identify the interspinous region and bone region on all of the cases collected, with a computational speed fast enough for real-time processing.

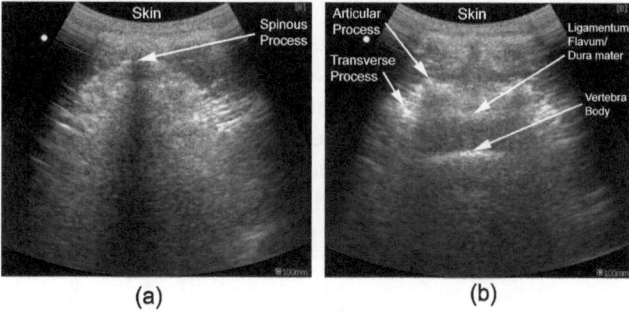

(a)   (b)

Fig. 1. Ultrasound Image of Lumbar Spine. (a) typical ultrasound image when the probe is placed above spinous process, featured by the triangular anechoic window; (b) ultrasound image when probe is placed on interspinous space, where the articular processes, epidural space and vertebra body are visible.

2 Materials and Methods

2.1 Ultrasound Image Feature of Lumbar Spine

The ultrasound images taken at different region of the lumbar spine have different features, determined by the region where the probe is placed. When the probe is placed directly on the spinous process (not proper for needle insertion), the ultrasound wave will be impeded by bones, creating a long triangular hypo-echoic acoustic shadow Fig 1(a)). The ultrasound image will be dark with a triangular dark window along the midline, which is the main feature of bony images. When the probe is moved to the interspinous region (proper for needle insertion), more details beneath the skin can be noted, as shown in Fig 1(b). The 'flying bat' alike shape on the ultrasound image indicates that the location of the probe is a suitable site for needle insertion [14].

2.2 Feature Extraction

Before feature extraction, raw ultrasound images are pre-processed with difference of Gaussian enhanced local normalization, so as to remove the speckle noises and extract the anatomical structure [12]. After pre-processing, local intensity variance induced by uneven ultrasound wave reflection rate are also eliminated. Therefore, a potential element which might deteriorate the image classification is removed.

Feature extraction procedure is extraordinarily important for image classification. Medical images generally suffer from limited training samples, thus the feature vector length shall be limited. Otherwise, the learning models will have high variance and cannot be optimally trained. In this paper, image features are extracted with two approaches, the template matching method to detect the key anatomical features and midline detection approached to obtain image features along midline.

Template Matching. The visibility of 'flying bat' shape is the criterion adopted by anesthetists to recognize interspinous images [14]. However, in computer vision, due to the variation and distribution extent of the 'flying bat' shape in the image, the recognition of the entire 'flying bat' shape is not a easy task. In our previous research, we proposed to decompose the 'flying bat' shape into three sub-features: the 'bat ear' (articular process), epidural space and vertebra body. The decomposed sub-features recognized the articular process and vertebra body with high accuracy on images obtained from volunteers [12].

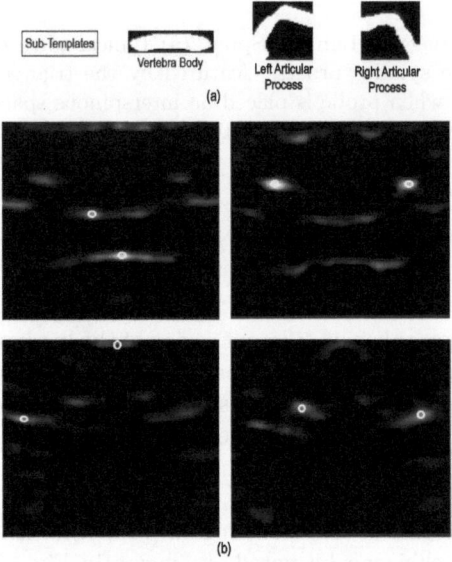

Fig. 2. Feature Extraction with Template Matching. (a) Sub-templates for anatomical features, from left to right: Vertebra body and epidural space, left articular process and right articular process; (b) Matching result of key anatomical features: The left column: matching result of vertebra body sub-template; the right column: matching result for articular process sub-templates; the upper row: interspinous image; the lower row: bone image. The optimal matching position is marked by a circle.

In this paper, similar decomposition is employed. Template matching is used to obtain the matching position and matching value between the sub-features and the images. Among the three sub-features, the appearance of the epidural space and the vertebra body both resemble a line. Thus, the same linear sub-template (as shown on Fig 2(a)) is employed for the recognition of both vertebra body and epidural space. Of the two maximum matching blobs, the one that locates lower in the image is vertebra body and the superior one is epidural space, which follows the anatomical structure of the lumbar spine. In the interspinous images, the visibility of vertebra body and epidural space is clear and both of them can be correctly recognized. While in the bone images, the maximum matching of the sub-template will occur at different regions in the image; and the matching values for both epidural space and vertebra body are low, as indicated by Fig 2(b). The situation is the same for the matching of articular processes, except that the maximum matching of articular processes will appear on the left and right side of the midline. Therefore, based on the matching position and matching value, it is possible to partially discriminate the interspinous images and bone images.

Fig 2(b) shows the template matching results of the epidural space, vertebra body and left & right articular process, with the optimal matching position been marked. The parameters obtained with template matching can be utilized to constitute part of the feature vector for the purpose of image classification, including the retrospective depth measurement of epidural space ($\mathcal{D}_1$) and vertebra body ($\mathcal{D}_2$), their matching values ($\mathcal{V}_1$ and $\mathcal{V}_2$), matching position of two articular processes ($\mathcal{P}_3$, $\mathcal{D}_3$ for left articular process and $\mathcal{P}_4$, $\mathcal{D}_4$ for right articular process) and their matching values ($\mathcal{V}_3$ and $\mathcal{V}_4$).

Midline Detection. The image features along the midline of the ultrasound image is different for interspinous images and bone images. For the bone images, ultrasound wave is impeded by the spinous process, resulting in an anechoic region along the midline (Fig 1(a)); while for interspinous images, the epidural space and vertebra body along the midline will be visible (Fig 1(b)). Therefore, the appearance of black pixels along the midline serve as an important feature for the classification of interspinous / bone images.

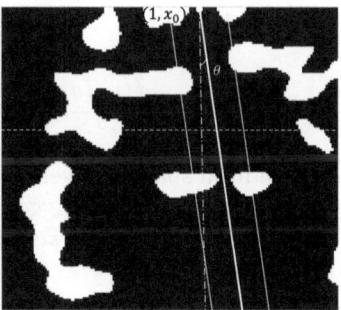

Fig. 3. Feature Extraction with Midline Detection. (Note: the background image is the pre-processed binary image; The horizontal dashed line is the depth threshold used to calculate the white pixels rate $\mathcal{R}_w$. x_0 denotes the center of scanning window along the horizontal axis and ϑ denotes the angle of scanning window against the vertical axis.).

For the detection of midline, a cost function $J(\vartheta, x_0)$ based on the summation of white pixels within a predefined scanning window is formulated. The window scanned though the entire image within [-45, +45] degrees. The position and degree that gives the minimum cost function value will locate the midline, as demonstrated by Fig 3. In order to increase the accuracy of midline detection for interspinous images, a penalty which decreases its weight as a function of depth is imposed on the cost function, so as to allow the appearance of epidural space and vertebra body to be less penalized in the cost function. The cost function is formulated in Equation 1.

$$J(\vartheta, x_0) = \sum_{i=1}^{n} \sum_{j=-C}^{C} [0.5 + exp(-0.05i)] \times f(i, itan\vartheta + x_0 + j) \times \sqrt{(1 + 0.3|\vartheta| |x_0 - n/2|)}$$

(1)

The first part of the Equation 1 is the penalty term for the appearance of white pixels at different depths. The third part is the penalty term if the detected midline is not near the middle of the ultrasound image or that it is not vertical. In equation 1, $f(i,j)$ denotes the binary image of the pre-processed ultrasound image with a dimension of $n \times m$; C represents half size of the predefined window, which can be optimally set between 5 - 10.

After optimal ϑ' and x_0' is obtained and midline is located, the rate of black pixels within the predefined scanning window can be calculated using the following equation:

$$\mathcal{R}_b = 1 - \frac{\sum_{i=1}^{n} \sum_{j=-C}^{C} f(i, itan\vartheta' + x_0' + j)}{2Cn}$$

(2)

The depth of epidural space is reported to range from 3-8 cm, indicating that the epidural space and vertebra body appear deeper than 3cm in the image. Thus, the rate of potential epidural space and vertebra body within the scanning window can be calculated with:

$$\mathcal{R}_w = \frac{\sum_{i>=3cm}^{n} \sum_{j=-C}^{C} f(i, itan\vartheta' + x_0' + j)}{2Cn}$$

(3)

Because of the presence of epidural space and vertebra body at the lower part of the interspinous image, the parameter $\mathcal{R}_w$ is bigger than 0. On the contrary, for the images obtained from bone regions, the lower part of the image is black. Thus, $\mathcal{R}_w$ approximates 0 for bone images.

After the midline is located via the cost function approach, symmetry measurement is utilized to double confirm the accuracy of midline detection. The introduction of symmetrical parameter is based on the fact that the anatomical structure of lumbar spine exhibits mirror symmetry with respect to the midline. The symmetrical parameter $\mathcal{S}$ is simply calculated with Equation 4

$$\mathcal{S} = \frac{\sum |f(x,y) - f(x',y')|}{nx_0'}$$

(4)

where $(x,y) and (x',y')$ represent the coordinates of one pair of pixels which are symmetrical to each other against the detected midline.

$\mathcal{R}_b$, $\mathcal{R}_w$ and $\mathcal{S}$ add another three parameters for the feature vector. Therefore, combining the 10 parameters obtained from template matching and 3 parameters from midline detection, a feature vector of length 13 is formulated. A detailed description of template matching and midline detection on lumbar ultrasound image processing can be further approached at [15].

2.3   Multi-Layer Neural Network

After the feature set has been extracted and normalized, the multi-layer neural network (MLP) is employed for the classification of the interspinous images and bone images. The MLP contains three or more layers, with an input layer, an output layer and one or more hidden layers of nonlinearly-activating nodes (usually sigmoid or tanh function). Theoretically, MLP can approximate any bounded continuous function [16]. In order to simplify the network structure, only one hidden layer is utilized in this paper, as shown in Fig 4.

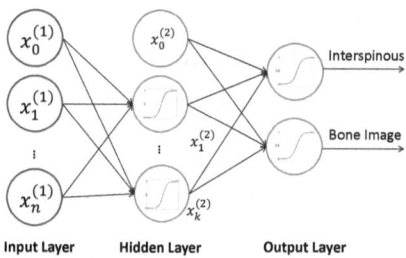

Fig. 4. Multi-layer Neural Network Structure: on input layer with 13 neurons, one hidden layer with k neurons and one output layer with 2 neurons

The learning procedure of MLP is fulfilled by changing synaptic weights of neurons after training data is processed. The most popular method for training of MLP is the back-propagation (BP) algorithm, witch provides a computationally efficient approach for MLP training [17]. The training processing using BP algorithm can be divided into two phases: the forward phase and the backward phase. In the forward phase, the synaptic weights are fixed and the training samples are propagated through the network. In the backward phase, the resulting error produced by comparing the output with desired output is propagated through the network backwards, during which successive adjustments are made to the synaptic weights [18].

2.4   Evolutionary Neural Network

Back Propagation is computationally efficient for the training of feed forward neural network. However, it is prone to be trapped in local minimum [18].

For the complex problems, evolutionary computation, e.g. genetic algorithm, can be involved in the training of neural network to obtain the global optimum by performing searches over a complex and multi-mode space [19]. For the classification problem, firstly, calculate the fitness function, usually the misclassification rate with the generated neural network; select the individuals with largest fitness and reserve them to the next generation; then perform the crossover and mutation with the current population to generate the new generation. The above procedure is repeated to evolve the inital weights until the training goal is achieved. Another commonly used evolutionary neural network training is hybrid algorithm, which combine genetic algorithm and back propagation together. The hybrid training fist use genetic algorithm to optimize the initial weight distribution and locate certain search spaces in the solution space. Then use back propagation to search the optimal solution in the small solution spaces [20].

In this paper, the problem is not very complex, since only two classes are involved and the dimension of the input vector is not high. Back propagation is good enough to train the neural network. However, in order to avoid the local minimum problem, the neural network is trained 5 times consequently.

2.5   Materials and Image Acquisition

The ultrasound video streams utilized in this research were collected from KK Women's and Children's Hospital (Singapore), with institutional review board (IRB) approval and patients' consent obtained. Pregnant women scheduled for a caesarean procedure were recruited before they were sent to the operation theater (OT). During the study, 43 ultrasound video streams are collected from 43 different subjects. After video streams are collected, the image database is obtained by extracting still images from the video streams. 40 images are randomly extracted from each of the video streams, constituting 1720 ultrasound images in the training and test database in total. The extracted images are then labelled by an experienced sonographer: '1' for interspinous images, '0' for bone images and other images not proper for needle insertion.

Table 1. Statistics of Training Set, Validation Set and Test Set

	Training Set	Validation Set	Test Set
Subject Number	21	10	12
Image Number	840	400	480
Interspinous	404	182	204
Bone Images	436	218	276

3   Results and Discussions

Of the 43 video streams collected, 21 (49%) of them are randomly selected as training set, another randomly selected 10 subjects (23%) are used for validation

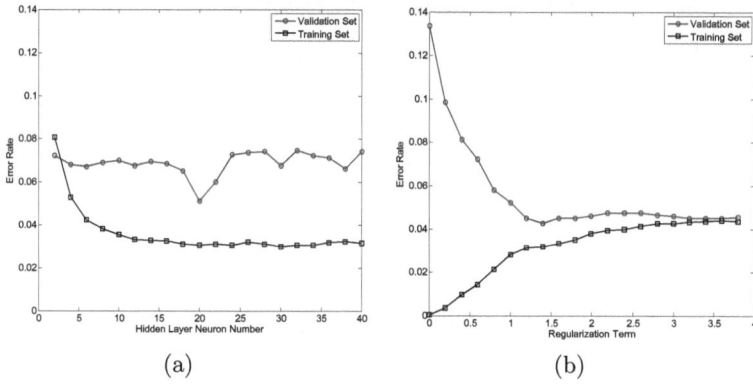

(a)   (b)

Fig. 5. Parameter Tuning of the Multi-layer Neural Network. (a). Fix the regularization term as 1 and change the hidden neuron numbers. (b). Fix the hidden neuron number as 20 and change the regularization term.

set and the reset 12 (28%) are used as test set. The training, validation and test sets are divided on the level of subjects instead of extracted images, which follows the assumption that in the clinical setting the detailed lumbar spine structure of individuals are not given neither examined with MRI or other imaging modalities before the epidural anesthesia. Since 40 images are extracted from each video, thus there are in total 840 images in the training set, 400 images in the validation set and 480 images in the test set. The detailed statistical information of the images is listed in Table 1.

3.1 Performance of Neural Network

The multi-layer neural network model is trained with the training set and then validated on the validation set to get the optimal parameters. There are two parameters involved in the neural network structure with one hidden layer: the number of neurons in the hidden layer and the regularization term to avoid over-fitting. Since the BP training might get trapped in local minimum, thus the network is trained 5 times so as to get the estimation of the mean performance.

In order to get the optimal performance, the test is conducted in two steps. First, the regularization term is fixed as 1 and then get the number of hidden layer neurons. As shown in Fig 5(a) when the regularization term is set as 1, the optimal performance is obtained when the hidden neuron number is 20. Then the hidden neuron number is set as 20 and then tune the regularization term. Minimal error rate is achieved when regularization term is 1.4.

Therefore, the optimal performance on the validation set is achieved when the hidden neuron number is 20 and the regulation term is 1.4. The trained network models with the obtained parameters is further tested on the test set. The performance of the trained model is listed on Table 2.

Table 2. Performance of Neural Network

	Training Set (%)	Validation Set (%)	Test Set (%)
Accuracy	96.95	95.75	94.12
Precision	97.11	95.00	94.35
Recall	97.02	97.71	95.51
F0.5	97.09	95.53	94.58

Fig 6 demonstrates the receiver operating characteristic curve (ROC) of the trained model on the test set. The area under the curve (AUC) of the neural network model is 0.981 for the test set, indicating that the neural network model is properly trained and has good predictability [21].

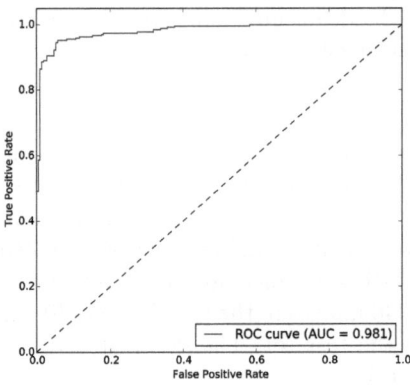

Fig. 6. Receiver Operating Characteristic Curve of the Trained Neural Network Model on Test Set

3.2 Video Processing

The trained neural network model is further tested on the ultrasound video streams collected to identify the interspinous region and bone region. In the video processing, the interspinous region is defined by the continuous appearance of more than 5 interspinous images; while for the negative detections, if it is in the interspinous region, no more than 2 bone images shall be detected by the image; vice versa for bone region. According to the definition above, the neural network model is able to identify the interspinous region and bone region correctly on all of the 43/43 video streams collected.

Table 3 lists the computation time for major operations in the pre-processing, feature extraction and classification procedure. Matlab (R2012a) was used for the implementation of the algorithm. The computation time for each frame is 67.58 ms. Given that the video is collected at the frame rate of 15 FPS, thus the computation speed is a little bit slower for real-time processing and may result in frame loss for real time processing. However, improvement in computational speed has been realized by implementing the program using Python (with OpenCV library), which shortened the computation time to 32.85 ms per frame. Therefore, the proposed image processing procedure is applicable to real time processing.

Table 3. Computation Cost of Video Processing with Matlab

Operation	Computation Cost(ms)
Preprocessing	9.10
Template Matching	17.21
Midline Detection	17.65
Symmetry Detection	15.32
Neural Network Classification	0.08
Others	8.25
Processing Time Per Frame	67.58

4   Conclusion

In this paper, we propose a feature extraction procedure for the ultrasound images collected from lumbar spine. The important anatomical features, including epidural space, vertebra body and articular processes are extracted from the ultrasound images. Moreover, the rate of black pixels along with midline are also extracted with midline detection. Based on the features extracted from training samples and test samples, neural network is used to classify the interspinous/ bone images with maximal margin. The trained neural network model is also tested on the 43 ultrasound video streams collected from pregnant patients, and successfully identified the interspinous region / bone region on all of the videos collected.

This research is part of a bigger project which aims to insert the needle for the epidural anesthesia procedure automatically under the guidance of ultrasound imaging. This paper fulfills the purpose of automatic interpretation and identification of ultrasound images, so that anesthetists are relieved from reading raw ultrasound images. It also proves that the proposed algorithms are fast enough for real-time video processing. In a future work of this research, the algorithm will be implemented in real-time manner to detect epidural needle insertion site.

References

1. Rawal, N.: Reg. Anesth. Pain Med. 37(3), 310–317 (2012)
2. Osterman, M.J.K., Martin, J.: Epidural and Spinal Anesthesia Use During Labor: 27-State Reporting Area, Centers for Disease Control and Preventnion (2008)
3. Le Coq, G., Ducot, B., Benhamou, D.: Risk factors of inadequate pain relief during epidural analgesia for labour and delivery. Can. J. Anaesth. 45(8), 719–723 (1998)

4. Paech, M.J., Godkin, R., Webster, S.: Complications of obstetric epidural analgesia and anaesthesia: a prospective analysis of 10,995 cases. Int. J. Obstet. Anesth. 7(1), 5–11 (1998)
5. La Grange, P., Foster, P.A., Pretorius, L.K.: Application of the Doppler ultrasound bloodflow detector in supraclavicular brachial plexus block. Br. J. Anaesth. 50(9), 965–967 (1978)
6. Grau, T., Leipold, R.W., Conradi, R., Martin, E., Motsch, J.: Efficacy of ultrasound imaging in obstetric epidural anesthesia. J. Clin. Anesth. 14(3), 169–175 (2002)
7. Ecimovic, P., Loughrey, J.: Ultrasound in obstetric anaesthesia: a review of current applications. Int. J. Obstet. Anesth. 19(3), 320–326 (2010)
8. Noble, J.A., Navab, N., Becher, H.: Ultrasonic image analysis and image-guided interventions. Interface Focus 1(4), 673–685 (2011)
9. Tran, D., Rohling, R.: Automatic detection of lumbar anatomy in ultrasound images of human subjects. IEEE Trans. Biomed. Eng. 57(9), 2248–2256 (2010)
10. Kerby, B., Rohling, R., Nair, V., Abolmaesumi, P.: Automatic identification of lumbar level with ultrasound. In: Conf Proc. IEEE Eng. Med. Biol. Soc., pp. 2980–2983 (2008)
11. Al-Deen Ashab, H., Lessoway, V.A., Khallaghi, S., Cheng, A., Rohling, R., Abolmaesumi, P.: An augmented reality system for epidural anesthesia (AREA): prepuncture identification of vertebrae. IEEE Trans. Biomed. Eng. 60(9), 2636–2644 (2013)
12. Yu, S., Tan, K.K., Shen, C.Y., Sia, A.: Ultrasound Guided Automatic localization of needle insertion site for epidural anesthesia. In: Proceeding of IEEE International Conference on Mechatronics and Automation, pp. 985–990 (2013)
13. Lee, Y., Tanaka, M., Carvalho, J.: Sonoanatomy of the lumbar spine in patients with previous unintentional dural punctures during labour epidurals. Reg. Anesth. Pain. Med. 33(3), 266–270 (2008)
14. Carvalho, J.C.: Ultrasound-facilitated epidurals and spinals in obstetrics. Anesthesiol Clin. 26(1), 145–158 (2008)
15. Yu, S., Tan, K.K., Sng, B.L., Li, S.J., Sia, A.: Automatic identification of needle insertion site in epidural anesthesia with a cascading classifier. Ultrasound Med. Biol. (in press)
16. Maiorov, V., Pinkus, A.: Lower bounds for approximation by MLP neural networks. Neurocomputing 25(1), 81–91 (1999)
17. Hecht-Nielsen, R.: Theory of the backpropagation neural network. In: International Joint Conference on Neural Networks, IJCNN, pp. 593–605. IEEE (1989)
18. Haykin, S.: Neural networks and learning machines (vol. 3). Pearson Education, Upper Saddle River (2009)
19. Ding, S., Li, H., Su, C., Yu, J., Jin, F.: Evolutionary artificial neural networks: a review. Artif. Intell. Rev. 39, 251–260 (2013)
20. Alba, E., Chicano, J.F.: Training Neural Networks with GA Hybrid Algorithms. In: Deb, K., Tari, Z. (eds.) GECCO 2004. LNCS, vol. 3102, pp. 852–863. Springer, Heidelberg (2004)
21. David, J., Goadrich, M.: The relationship between Precision-Recall and ROC curves. In: Proceedings of the 23rd International Conference on Machine Learning, pp. 233–240. ACM (2006)

Schemata Bandits for Binary Encoded Combinatorial Optimisation Problems

Madalina M. Drugan[1], Pedro Isasi[2], and Bernard Manderick[1]

[1] Artificial Intelligence Lab, Vrije Universitieit Brussels, Pleinlaan 2, 1050- B, Belgium
{Madalina.Drugan,Bernard.Manderick}@vub.ac.be
[2] Computer Science Department, Carlos III of Madrid University, Spain
pedro.isasi@uc3m.es

Abstract. We introduce the schemata bandits algorithm to solve binary combinatorial optimisation problems, like the trap functions and NK landscape, where potential solutions are represented as bit strings. Schemata bandits are influenced by two different areas in machine learning, evolutionary computation and multi-armed bandits. The schemata from the schema theorem for genetic algorithms are structured as hierarchical multi-armed bandits in order to focus the optimisation in promising areas of the search space. The proposed algorithm is not a standard genetic algorithm because there are no genetic operators involved. The schemata bandits are non standard schemata nets because one node can contain one or more schemata and the value of a node is computed using information from the schemata contained in that node. We show the efficiency of the designed algorithms for two binary encoded combinatorial optimisation problems.

1 Introduction

A recent trend is to transfer expertise between machine learning (ML) areas, i.e. multi-armed bandits (MAB) and evolutionary computation algorithms (ECs) [15]. There are many similarities between multi-armed bandits and evolutionary computation mainly because they are both optimisation algorithms. The main difference between the two techniques is that MAB is used to optimise stochastic environments [5], whereas the majority of EC based methods are for optimisation of very large but deterministic environments. We want to use the similarities and differences between the two techniques to design new efficient hybrid optimisation algorithms.

Schema Theorem. Genetic algorithms (GAs) [11] are powerful optimisation and search techniques that have been applied with great success to a wide range of applications [6,1]. Let's consider the standard GAs with a binary string encoding of length ℓ for each solution. There are 2^ℓ possible strings. The GA processes a population of individuals by the successive application of fitness evaluation, selection of the better individuals followed by recombination of the genotypes of the selected individuals.

The common part in the representation of several individuals is called a schema [7,11]. According to John Holland, the schema theorem [7] explains the success of genetic algorithms in general. It basically states that although the GA operates at the level of individuals, GA *implicitly* and *in parallel* processes information about schemata,

G. Dick et al. (Eds.): SEAL 2014, LNCS 8886, pp. 299–310, 2014.

subsets of the search space. Moreover, it samples the most interesting schemata called building blocks in a near-optimal way using the analogy between schemata and arms in the bandit problem. That is schemata with the fitness mean above the average are grouped as a bandit arm and the schemata with the fitness below the average are considered to be a second arm. The schema theorem shows that selection increasingly focuses on the schemata with the fitness average above the mean. This brings us to the multi-armed bandit problem (MAB).

The exploration (the search for new useful solutions) versus exploitation (the use and propagation of such solutions) trade-off is an attribute of successful adaptation. The exploration implies the evaluation of new solutions that could have low fitness and the exploitation means the usage of already known good solutions. Holland modelled the exploration vs exploitation trade-off in ECs with multi-armed bandits. The higher fitness solutions are considered (or grouped) as an arm, and the lower fitness solutions are considered as a second arm. Mixing of good building blocks in GAs, i.e. the propagation of good schema in GAs, was studied in [14].

Multi-armed bandits (MAB) problems [4] have been studied since the 1930s and they arise in diverse domains, like the online profit-seeking automated market makers [13] and yahoo recommendation system [10]. In the stochastic MAB-problem, there are K arms and each time an arm i, where the set of arms in $1, \cdots, i, \cdots, K$ is selected, a reward r_i is drawn according to the probability distribution with fixed but *unknown* mean μ_i. The goal is to maximise the total expected reward $\hat{r}_i$. If the true means of all arms where known, this task would be trivial. One selects the arm with the highest mean reward all the time. A MAB algorithm starts by uniformly exploring the K-arms, and, then, gradually focuses on the arm with the best observed performance. Since, the means are unknown one has to allocate a number of trials over the different arms so that, based on the obtained rewards, the optimal arm is identified as soon as possible and this with (very) high confidence.

To reach this goal, a tradeoff between exploration and exploitation has to be found. Exploration means that one tries a suboptimal arm to improve the estimate its mean reward while exploitation means that one tries the best observed arm which is not necessary the true best one. An arm selection policy determines what arm is selected at what time step based on the rewards obtained so far. The research question is what are (near)-optimal arm selection policies for the MAB-problem. An important heuristic that has emerged is that good policies, e.g. variants of the upper confidence bound (or UCB) policy, are *optimistic in the face of uncertainty* [12].

Thus, the trade-off between exploitation and exploration is important for both MAB and EC algorithms. MAB should pull all arms using an exploration strategy, to estimate their performance and it returns feasible, close to optimal, solutions using an exploitation strategy. Exploration in EC means to generate solutions in unexplored regions of the search space, and exploitation means to generate new solutions in promising regions of the search space using structural information about the current solutions. Selecting and using these strategies are not trivial and actually, the trade-off between them can increase the time needed to find an acceptable solution.

One MAB variant is the hierarchical bandit approach where the reward of one arm in the hierarchy is the reward of another one at one level deeper in the hierarchy [12].

Monte Carlo Tree Search (MCTS) is a recently proposed search method that builds a search tree in an incremental and asymmetric manner accordingly to a tree policy that selects the node with the highest priority to expand [3]. The tree policy needs to balance exploration versus exploitation, which for MCTS methods resembles the same trade-off in ECs. Exploration means to search in areas not sampled yet, whereas exploitation means to search in promising areas. Each round in MCTS consists of four steps: selection, expansion, simulation, and back-propagation.

Selection starts from the root, it selects successive expandable child nodes down to a leaf node. It selects child nodes that expand the tree towards most promising moves, which is the essence of MCTS. A node is expandable if it represents a non-terminal state and is unvisited. *Expansion:* unless a stopping criteria is met, MCTS creates one or more child nodes and choses from them a node, designated as the current node, using a tree policy. If no child was created, the simulation starts from a leaf node. *Simulation* plays at random from the current node using a default policy. *Backpropagation* uses the results from the previous steps to update the information in the nodes on the path from the current node to the root. MCTS is a statistical anytime algorithm for which more computing power means better results.

MCTS using *upper confidence bound* (UCB1) [2] as arm selection policy is called Upper Confidence Trees (or UCT). UCB1 is a very simple and efficient stochastic MAB with appealing theoretical properties, i.e. UCB1 is upper bounds the lost in choosing non-optimal arms. Each promising node of MCTS is evaluated accordingly to the UCB1 policy. UCT builds incrementally a search tree using random samples in the search space by expanding the nodes selected by the arm selection policy [3]. This approach is largely responsible for the success of Monte Carlo Tree Search (MCTS) where other methods fail, e.g. the game of GO [3].

The Designed Algorithms. We want to steer the optimisation in ECs using MCTS by making an analogy with schema representation of solutions. In Section 2, the search space of ECs is structured as a schemata net with 3^ℓ possible schemata. We reveal some of the properties of the schemata net. Section 3 presents a baseline schemata bandits algorithm where each node in the net, thus each schema, is an arm. This bandit searches in a 3^ℓ dimensional search space for the optimal solution.

Section 4 proposes a condensed representation of the net for the schemata bandits that searches over a reduced search space of 2^ℓ. Each node is itself a bandit of schemata and we denote this algorithm as the bilevel schemata bandits.

Experimental Results. Section 5 tests the performance of the proposed schemata bandits on two binary encoded problems: i) a deceptive trap function, and ii) a version of NK landscape that uses the deceptive trap functions and has best known solution. We show that the baseline schemata bandits performs better in terms of best found so far solution but the bilevel schemata bandits perform better in terms of minimising the expected regret and the computational efficiency. Alternative parameters are considered for the UCB1 algorithms in the schemata nets. Section 6 concludes the paper.

2   A Schemata Net Structure

In this section, we present a schemata net structure and its properties.

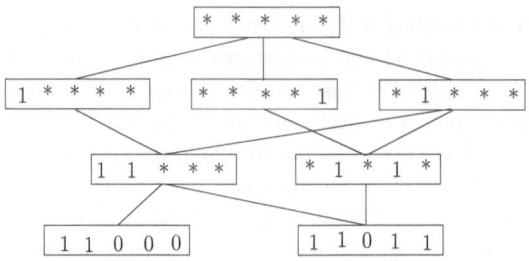

Fig. 1. An example of schemata net for a 5 dimensional strings

We focus on *search spaces* that are ℓ-dimensional hypercubes, i.e. $\mathbb{B}^\ell$ where $\mathbb{B} = \{0, 1\}$ is the set of booleans and ℓ is the length of the bitstrings in the search space. A schema H represented as $H \in \{0, 1, *\}$ is a subspace of $\mathbb{B}^\ell$ that is also a hypercube. The don't care symbol $*$ can take on any value in $\mathbb{B}$. The *order* $o(H)$ of a schema H is the number of instantiated values, i.e. either 0 or 1. Here, we will also use the *dimension* $d(H)$ of a schema H: it is the number of don't care symbols and $d(H) = \ell - o(H)$. There are in total 3^ℓ schemata of which the most general schema $** \cdots **$ has dimension ℓ and of which all the 2^ℓ fully instantiated schemata, i.e. bit strings of the search space, have dimension 0. Note that the intersection of 2 schemata is again a schema.

Example 1. Let $\mathbb{B}^5$ be the 5-dimensional hypercube, see Figure 1. Then the schema $H_1 = 11 * **$ has order 2 and represents the 3-dimensional hypercube of all bit strings of length 5 starting with 11, i.e. H_1 contains 8 elements including 11001. And, $H_2 = 1 * * * *$ has dimension 4 and the schemata H_1 and H_2 share the element 11001.

Each node in the schemata net has the following attributes: i) a value, ii) children, and iii) parents. Let H be a schema of dimension $d = d(H)$, where d are the number of symbols $*$, and order $o(H) = \ell - d(H)$, where $o(H)$ is the number of positions where the value of bits is fixed to either 1 or 0.

Children: If we replace any don't care symbol $*$ by either 0 or 1 then we obtain one of the $2 \cdot d$ children of schema H. Each child has dimension $d - 1$. The leave nodes have no children.

In Example 1, schema $H_2 = 1 * * * *$ has $2 \cdot 4 = 8$ children, that are $10 * **, 11 * **, 1 * 0 **, 1 * 1 **, 1 * *0*, 1 * *1*, 1 * * *0$ and $1 * * *1$. The schema $H_1 = 11 * **$ that has dimension $d - 3$ has $2 \cdot 3 = 6$ children, that are $110 * *, 111 * *, 11 * 0*, 11 * 1*, 11 * *0$ and $11 * *1$.

Parents: A parent for the schema H has one fixed position replaced with the symbol $*$. If we replace any of the instantiated values 0 or 1 by a don't care $*$ then we obtain one of the o parents of H. Each parent has the order $o - 1$. The fully uninstantiated schemata have no parents.

For example, the schema H_2 has 1 parent, that is the fully uninstantiated schema, because it has the order $o = 1$, and the schema H_1 has two parents, i.e. $1 * * * *$ and $*1 * **$ with the order 1.

The value of a node: Let $b_1, \cdots, b_i, \cdots, b_n \in H$ be the bit strings, or individuals, evaluated for their common schema H. The values of function f (to be optimised) for the set of bitstrings b_i are $f(b_i)$, where $i = 1, \cdots, n$. Then

$$\overline{f}(H) = \frac{1}{n} \sum_{i=1}^{n} f(b_i) \tag{1}$$

is the *estimated mean* value of f on the hypercube H, or simpler the value of H, and depends on the samples b_i used. The variance of $\overline{f}(H)$ on H will depend on its dimension $d(H)$: the higher the dimension the higher the variance and if the dimension is 0 then the variance is also 0.

There are two special types of nodes: i) the root, and ii) leaves.

The root is the fully uninstantiated schema, i.e. with the symbol $*$ everywhere. This node has no parents and it has $2 \cdot \ell$ children corresponding with the replacement of each position with a fixed value, 1 or 0.

A *leaf* is the node H that instantiate all its children, where each leaf has a fixed dimensionality $d \leftarrow |leaf|$. There are 2^d bitstrings for leaf schema corresponding with the d symbols $*$. The value of a leaf is fixed to the mean fitness values of the bitstrings generated for that schema. Let $b_1, \cdots, b_i, \cdots, b_{2^d} \in H$ be the bitstring evaluated for the leaf schema H. Then

$$\overline{f}(H) = \frac{1}{2^d} \sum_{i=1}^{2^d} f(b_i) \tag{2}$$

is the value of the leaf node on the hypercube H.

Because the size of a complete net is usually two large to be of a practical use, we expand the net given new solutions up to a given dimension $d > threshold$. The schemata with a higher dimension $d \leq threshold$ are structured in a tree where only the children but not the entire set of parents are further investigated and stored. For $threshold \ll \ell$, there are considerable more schemata in a schemata bandits, 3^ℓ, than total number of individual solutions, 2^ℓ. If $threshold$ is close to 1 the schemata net is small and the learning properties of the algorithm, e.g. generate bits strings from the best schema, are limited.

3   A Baseline Schemata Bandits Algorithm

The *baseline schemata bandits algorithm* builds a tree where each node is a schema. The starting point for each iteration of this algorithm is the root that is the most general schema. The pseudo-code for this algorithm is presented in Algorithm 1.

Considering the steps specific for the Monte Carlo tree search algorithm, cf MCTS, the schemata bandits algorithm consists of three steps:

Selection: Starting from the root, select successively child nodes down to a leaf node. As in UCT, we select each time the child node that expands the tree towards the most promising parts of the search space. A node is expandable if it is unvisited. A popular policy to select the next node to expand is $UCB1$ [2] that upper bounds the loss resulting from choosing non-optimal arms. The UCB1 goal is to play often the optimal arm, in our case the optimal schema.

Let H be the selected node. The reward corresponding with each schema H, the arms or bandits in UCT, is the estimated mean $\overline{f}(H)$ over H based on all bitstrings

---

Algorithm 1. A baseline schemata bandits algorithm

Initialise the schemata bandits algorithm with n random individuals
for a fixed number of schemata net iterations **do**
    Select the root schemata, $H \leftarrow root$
    while a leaf node is *NOT* reached **do**
        Select the most promising child H_i of the current schema H using the UCB1
        algorithm, cf. $argmax_i \overline{f}(H_i) + C \cdot \sqrt{\frac{2 \log(t)}{t_i}}$
        Update the counters of the current and selected child schemata, cf $t \leftarrow t + 1$,
        $t_i \leftarrow t_i + 1$
        Update the current schema $H \leftarrow H_i$
    end while
    if the leaf node L was not expanded before **then**
        Expand all the individual solutions $2^{d(L)}$ in the leaf node L
        Update value of the parents of the schemata on the path between the root and the
        leaf node with the $2^{d(L)}$ bitstrings
    end if
end for
return the best found solution

$b \in H$ generated so far as in Equation 1. If a schema H has dimension d, then H has $2 \cdot d$ child schemata denoted as $H_i, i = 1, \cdots, 2 \cdot d$, and t_i is the number of times that H_i was evaluated so far. In order to play UCB1, we initialise the child schemata H_i as follows: for each child schema H_i the number of trials is set to one, $t_i \leftarrow 1$, and the estimated mean value is set to its minimum value, $\overline{f}(H_i) \leftarrow 0.01$. Thus, the number of trials t of the parent schema H is set to $t \leftarrow \sum_{i=1}^{2d} t_i$. $UCB1$ selects the child node H_i with the maximum index

$$\overline{f}(H_i) + C \cdot \sqrt{\frac{2 \log(t)}{t_i}} \tag{3}$$

where the second term $C \cdot \sqrt{\frac{2 \log(t)}{t_i}}$ represents the confidence term and encourage exploration of suboptimal schemata, and C is a constant scalar value, $C > 0$, usually set to 1. A larger value for $C > 1$ would encourage the exploration of new (possible) suboptimal schemata, and a smaller value $C < 1$ would promote almost exclusively the schema with the optimal expected mean even though this expected mean depends on the already generated bitstrings.

In the beginning, all children will be played equally often. As the number of samples t increase towards ∞, the confidence term increases and the suboptimal schemata that were not visited for a long time will attain the maximum value. The schema with the maximum expected mean is played the most. Note that the mean value of a more general schema vary less than the mean value of a less general one.

Expansion: If in the selection step, a child node H_i that is not in the schemata graph is selected, i.e. $t_i \leftarrow 1$, a node in the net is created. If the selected child node H_i is already in the schemata graph, the counters are incremented, $t_i \leftarrow t_i + 1$ and $t \leftarrow t + 1$, and a child of this schema is selected. The expansion finishes with the generation of a leaf node L. When a leaf node is reached, all the corresponding $2^{d(L)}$ bitstrings are generated and evaluated. Thus, the set of bitstrings $\{b_1, \ldots, b_{2^{d(L)}}\}$ is evaluated to $\{f(b_1), \ldots, f(b_{2^{d(L)}})\}$. These bitstrings update the expected mean of each matching child and parent of the schemata on the path between the root and the leaf L schemata.

Propagation: Using each of the generated solutions, b_i where $1 \leq i \leq 2^{d(L)}$, we update the mean values of all the schemata in the schemata graph that contain that solution. Thus, if b_i belongs to the schema H, $b_i \in H$, than its expected mean is updated $\overline{f}(H) \leftarrow \frac{n\overline{f}(H)+f(b_i)}{n+1}$ and the sample counter n is also updated, $n \leftarrow n+1$. This means that, in the propagation step, the schemata (and thus the inner nodes) that contain an individual solution are created if they do not already exists in the schemata bandits and if their dimension is smaller than *threshold*. The root schema is updated for all bitstrings. A higer dimensional schema is updated more often than a lower dimensional schema because there are more bitstrings that match to a higher dimensional schema than to a lower dimensional schema.

Related Work. There are some important differences between the schemata bandits algorithm and the standard UCT-algorithm [8,3]. We actually define a graph where each node is the child of several parents. Because of the strong overlap between schemata, the rewards of the corresponding nodes are strongly correlated while UCT assumes that the rewards independent. The creation of a schema can occur both during expansion and propagation. Therefore one way to improve the performance of the proposed algorithm is to prune the unpromising branches of the graph.

The schemata bandits algorithm also relates to Estimation Distribution Algorithms [9] since no genetic operator is needed to generate new individuals. In addition, the schemata bandits approach could offer theoretical guaranties on the convergence to the optimal solution.

Discussion. The version of schemata bandits introduced here is designed for bistrings. The extension of this algorithm to a representation with k-valued strings is straightforward, where $k \leq 2$. However, the dimensionality of such schemata net would grow exponentially to k^ℓ, where ℓ is the size of the string as before. Although the number of schemata in the net increases, the sparsity of the net decreases because a schema has the same number of children regardless of the value of k.

Note that, in principle this algorithm is a *parameter free* optimisation algorithm. However, to increase the practical implications of this algorithm, we consider several parameters that decrease the computational complexity of the algorithm or tune the exploration / exploitation trade-off of the UCB1 algorithm. To decrease the search space of the schemata net, we have introduced the parameter *threshold* that separates schemata net, which is very densely connected, to the schemata tree that is sparse. The parameter $d(L)$ decreases the computational time to run a schemata net by generating more that one bitstrings from a good schema. This problem is also recognisable in the standard UCT algorithms with alternative solutions proposed in [12]. The constant value C gives the exploration / exploitation component of the UCB1 algorithm, but in the standard setting of UCB1 is set to 1.

A alternative version of this algorithm could select a random schema from the net to start the net iteration. UCB1 could be again considered to select a schema, but this

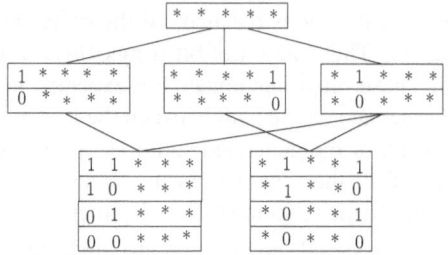

Fig. 2. An example of a bilevel schemata net for a 5 dimensional strings

time the number of arms is much larger means that the algorithm would require a longer time to learn good schemata. New schemata added recently to the net are more often selected than the schemata that were created close the initialisation of the net even though the former schemata have a better expected mean compared with the newly generated schemata. Note that in this case we have two UCB1 algorithms, one UCB1 algorithm over all schemata in the net and another UCB1 algorithm only for the children of a schema.

4   A Bilevel Schemata Bandits Algorithm

The dimensionality of the schemata net, i.e. the number of total schemata in the net, is a problem to a search algorithm as it is much larger than the dimensionality of the problem itself, i.e. a schemata net has a 3^ℓ search space whereas the search space of the initial problem is only 2^ℓ. The reduction of the dimensionality of the schemata net could mean the increase in this algorithm efficiency. In this section, we propose a version of the schemata bandits that groups the schemata with the same indifferent $*$ positions such that the search space is 2^ℓ.

A Schemata Group Net. Consider G a set of schemata with dimension $g \leftarrow d(G)$ corresponding the number of symbols $*$. There are $o(G) \leftarrow 2^{\ell-g}$ schemata contained in this node corresponding to the all schemata where the fixed positions are assigned one of the values 0 or 1. The root node contains only one schema, the most general schema with only symbols $*$. The root's children contains 2 schemata corresponding with one fixed position, and the children for the root children contains 4 schemata corresponding with the 2 fixed positions.

Example 2. Figure 2 shows with a simple example of such schemata bandits. A schemata group of order $o(G) \leftarrow 2$, i.e. there are two bits that can be fixed, has $2^2 = 4$ component schemata. Thus $00**, 01**, 10**$, and $11**$ are schemata in the same schemata group. Similarly, the schemata $*0**0, *0**1, *1**0$ and $*1**1$ also belong to the same schemata group.

There are fundamental differences between the two schemata bandits approaches. With this approach, each bit string will match exactly one schema from a schemata group. Thus all the bitstrings match all schemata group. Each schema has an expected mean value given in Equation 1. The value of a schemata group is given by the best expected mean value for that group of schemata. Each group schemata of dimension g has g children and g parents.

Note that the leaf nodes contains $2^{\ell-L}$ schemata, which for $L = 1$ means $2^{\ell-1}$ schemata where only one position has the symbol $*$. Therefore, for computational efficiency reasons, we set $L > 2$ on a high value. The variance in the expected mean values of schemata is larger for the leaf nodes than for the root node. In general, a schemata group with a higher dimension has a larger variance then a schemata group from a lower dimension.

The Algorithm. In the following, we introduce the bilevel schemata bandits algorithms that has only two steps unlike the baseline schemata bandits introduced in Section 3 which has three steps. The expansion step is not necessarily in this implementation because each bitstring matches each node in the net. We then generate upfront all the nodes of the net up to leaves. There will be $\sum_{i=0}^{L} g^i$ group nodes, each node containing 2^i schemata. Thus, in total, there will be a fixed number of $\sum_{i=0}^{L} g^i \cdot 2^i$ schemata in the net.

Selection: Each iteration, the search in the net starts with the root node. Using an UCB1 algorithm, a child is selected from the current node and the selection process continues until a leaf node. When a leaf node is reached, there are sampled 2th bitstrings from that leaf node proportional with the expected mean value of each schema, where $th > 0$ is a constant.

Propagation: We update the matching schema in each node with each generated bitstring, and the expected mean value for the matching schema is updated.

Performance of Schemata Bandits. The goal of the schemata bandits algorithms is to generate the optimal solution. The schemata bandits is a combination of evolutionary algorithms and multi-armed bandits thus measuring the performance of these algorithms is a complex task. To assess the quality of a schemata bandits algorithm, we evaluate the number of times each algorithm found the optimal solution and the mean of the generated solutions. Since storing and generating a schemata net is computational expensive, we need adequate performance metrics to compare the standard GAs and the schemata bandits algorithms. To measure its computational complexity, we take into account the number of schemata generated and the number of function evaluations.

To measure the performance of the $UCB1$-algorithm, we evaluate the expected regret of each schema that is the loss resulting from selecting suboptimal children of that schema. Each schema s has a regret that is calculated as

$$R_s = \sum_{i=1}^{g} \mathbb{E}[T_i(N)] \cdot (\overline{f}^*(N) - \overline{f}_i(N))$$

where $T_i(N)$ is the number of times the child node i was selected in N tree iterations and $\overline{f}_i(N)$ is the performance of a child node as before. We denote with $\overline{f}^*(N)$ the performance of the optimal child node, i.e. the node with the maximum expected mean value. We argue that the performance of the schemata bandits algorithms is an issue that needs further investigation.

Table 1. Performance of the schemata bandits for 10 deceptive trap functions where the block size is 5

Nr block	ℓ	best sol	mean	nr schemata	fun eval	regret
1	5	1.00 ± 0	0.34 ± 0	210 ± 0	32 ± 0	58 ± 0
2	10	1.00 ± 0	0.34 ± 0	35870 ± 70	1024 ± 0.45	3598 ± 0
3	15	1.00 ± 0	0.34 ± 0	171104 ± 0	31818 ± 0	169063 ± 0
4	20	0.95 ± 0.05	0.34 ± 0.01	127019 ± 5	276017 ± 46	1926978 ± 8507
5	25	0.92 ± 0.04	0.34 ± 0.01	179329 ± 32	318433 ± 54	2786122 ± 10451
6	30	0.88 ± 0.02	0.34 ± 0.01	230715 ± 6	319954 ± 4	3312440 ± 31333
7	35	0.86 ± 0.03	0.33 ± 0.01	281795 ± 0	319998 ± 0	3865973 ± 35264
8	40	0.85 ± 0.05	0.33 ± 0.01	332736 ± 20	320000 ± 0	4434108 ± 35609
9	45	0.76 ± 0.01	0.34 ± 0.01	383740 ± 14	320000 ± 0	5025488 ± 26731
10	50	0.73 ± 0.01	0.34 ± 0.01	434779 ± 9	320000 ± 0	5483765 ± 48880

5   Experimental Results

In this section, we experimentally compare the performance of the two versions of schemata bandits on two binary encoded functions: i) deceptive trap functions and ii) an NK problem.

Deceptive Trap Functions. As test functions, we concatenate deceptive trap functions of 5 bits. The maximum value is for all bits 1s is 5, and the deceptive local maximum for all bits 0s is 4. If there is only a single bit 1, the value is 3, for two bits 1, the value is 2, for three bits 1 the value is 1, and for four bits 1 the value is 0. Let b be a bitstring, where $b = b_1, b_2, \ldots, b_\ell$ and ℓ is a multiple of 5. We have that $f(1) = 3$, $f(2) = 2$, $f(3) = 1$, $f(4) = 0$, the deceptive optimum is $f(0) = 4$ and the global optimum for 5 bits is $f(5) = 5$. Let b is the number of deceptive blocks and $b \in \{1, \ldots, 10\}$ and the size of a deceptive block $k = 5$. Then, the normalised value of the fitness function is

$$f(b) = \frac{1}{b * max_{1 \leq j \leq k} f(j)} \sum_{i=0}^{b-1} f(\sum_{j=1}^{k} b_{i \cdot b + j})$$

The trap functions are considered a difficult test problem for GAs because the large basin of attraction of the deceptive local optimum.

Table 1 gives the values of the above enumerated performance measures for the deceptive trap function. We run each experiment for 30 times and the schemata bandits is iterated, i.e. selection, expansion and propagation, for 10^4 times. A leaf node evaluates $2^5 = 32$ solutions, i.e. $leaf \leftarrow 32$. Because of computational reasons related to memory usage, we set $threshold \leftarrow 3$. For example, for $\ell = 30$, there are $(2 \cdot 30)^3 = 216000$ schemata in the net to be stored and evaluated in the propagation step. A value 0.8 of mean fitness means that the algorithm reaches the deceptive optima, and value 0.9 of mean fitness means that 50% of the component tap functions found the global optimum. Note that the deceptive optimum is (almost always) reached in less than 320.000 function evaluations, and for $\ell \leq 40$ at least a quarter of the trap functions reach their optimal value.

For the bilevel schemata bandits, we set $th \leftarrow 5$ and $leaf \leftarrow 5$. We repeat each experiment 30 times and we iterate each schemata bandits for 10^3 times. In Table 2, the results in terms of finding the global optimum are not drastically deteriorated by the

Table 2. Performance of the bilevel schemata bandits for the deceptive trap function where the block size is 5

Nr block	ℓ	best sol	mean	nr nodes	fun eval	regret
3	15	0.98 ± 0.03	0.61 ± 0.05	2523 ± 10	32000 ± 0	162305 ± 9551
4	20	0.94 ± 0.02	0.59 ± 0.05	6957 ± 11	32000 ± 0	341496 ± 19586
5	25	0.92 ± 0.04	0.49 ± 0.01	16128 ± 10	32000 ± 0	635886 ± 40376
6	30	0.86 ± 0.02	0.44 ± 0.01	32824 ± 10	32000 ± 0	3312440 ± 31333
7	35	0.79 ± 0.02	0.44 ± 0.01	60465 ± 9	32000 ± 0	3865973 ± 35264
8	40	0.73 ± 0.05	0.42 ± 0.01	103039 ± 3	32000 ± 0	2752405 ± 401730

Table 3. Performance of the baseline (top half) and bilevel (bottom half) schemata bandits for the deceptive NK problem where the block size is 5 and for 10^3 schemata net iterations

Alg	ℓ	best sol	mean	nr nodes	fun eval	regret
Baseline	15	1.0 ± 0.0	0.35 ± 0.0	114426 ± 15	32000 ± 0	184406 ± 4140
schemata	20	0.84 ± 0.03	0.35 ± 0.00	560380 ± 12	32000 ± 0	245091 ± 9037
Bilevel	15	0.99 ± 0.03	0.63 ± 0.01	2434 ± 5	32000 ± 0	141655 ± 12068
schemata	20	0.83 ± 0.03	0.65 ± 0.01	8295 ± 10	32000 ± 0	374702 ± 41451

10 times less iterations of the bilevel schemata bandits. Furthermore, the mean fitness value, which is computed over the best mean fitness value for all the schemata in a node, is much higher than the mean fitness value of the baseline schemata bandits. Also the regret is smaller and a reason might be the smaller number of children arms for a single node. The number of fitness evaluation is also 10 smaller than in Table 1 and the number of nodes is about three times smaller.

The Deceptive NK Problem. We propose to use a NK problem that overlaps several deceptive trap functions. The fitness function is a sum of deceptive trap functions that overlap in $k - 1$ positions, here $k = 5$. Let b be a bitstring, where $b = b_1, b_2, \ldots, b_\ell$. Then, the fitness function of the deceptive NK problem is

$$f(b) = \frac{1}{(\ell - k) \cdot max_{1 \leq j \leq k} f(j)} \sum_{i=1}^{\ell-k} f(\sum_{j=1}^{k} b_{i+j})$$

Note that the resulting NK problem is more complex that the initial deceptive trap functions since all the bits should be set at once in order to obtain the global maximum. This fact is reflected in the experimental results by seldom identifying the optimal solution, see Table 3. Comparing the two algorithms, the performance is similar, but the number of nodes generated is 2 orders smaller for the bilevel schemata bandits than for the standard algorithm.

In conclusion, the baseline and the bilevel schemata bandits have a similar performance on the two tested combinatorial problems. The advantage of baseline schemata bandits is the good performance in terms of the best found solution, but the bilevel schemata bandits have the advantage of grouping similar schemata and thus of a more compact net structure.

6 Conclusions

We combine techniques from evolutionary computation and Monte Carlo Tree Search paradigms in order to create new efficient optimisation algorithms. The schemata bandits algorithm combines the schema theory with multi-armed bandits and its goal is to generate the optimum solution. The baseline schemata bandits algorithm considers that each schema is a node in the schemata net which is connected both with schemata that are more general that the current schema, denoted as parents, and with schemata that are more specialised than the current schema, denoted as children. The main drawback of this schemata bandits is the dimensionality of the net, i.e. 3^ℓ where ℓ is the size of the bitstrings, for a smaller search space 2^ℓ. The bilevel schemata bandits is the second proposed algorithm with a reduced dimension net of 2^ℓ nodes, where each node is the group of schemata with the same positions for the symbol $*$. We test and compare the two proposed algorithms on two binary combinatorial optimisation problems. The experimental showed that the baseline schemata bandits is better in finding optimal solutions whereas the bilevel schemata bandits are performing better in terms of optimising the regret. We conclude that schemata bandits is a viable alternative for the genetic algorithms that deserve further investigation towards a very efficient optimisation algorithm.

References

1. Affenzeller, M., Winkler, S., Wagner, S., Beham, A.: Genetic Algorithms and Genetic Programming: Modern Concepts and Practical Applications. CRC Press (2009)
2. Auer, P.: Using confidence bounds for exploitation-exploration trade-offs. J. of Machine Learning Res. 3, 397–422 (2002)
3. Browne, C., Powley, E., Whitehouse, D., Lucas, S., Cowling, P.I., Rohlfshanger, P., Tavener, S., Perez, D., Samothrakis, S., Colton, S.: A survey of monte carlo tree search methods. IEEE Trans. on Comp. Intel. and AI in Games 4(1), 1–46 (2012)
4. Bubeck, S., Cesa-Bianchi, N.: Regret Analysis of Stochastic and Nonstochastic Multi-armed Bandit Problems. Foundations and Trends in Machine Learning, vol. 5 (2012)
5. Gutjahr, W.J., Pichler, A.: Stochastic multi-objective optimization: a survey on non-scalarizing methods. Ann. Oper. Res (2013)
6. Haupt, R.L., Haupt, S.E.: Practical genetic algorithms. Wiley (2004)
7. Holland, J.H.: Adaptation in Natural and Artificial Systems. University of Michigan Press (1975)
8. Kocsis, L., Szepesvári, C.: Bandit based monte-carlo planning. In: Fürnkranz, J., Scheffer, T., Spiliopoulou, M. (eds.) ECML 2006. LNCS (LNAI), vol. 4212, pp. 282–293. Springer, Heidelberg (2006)
9. Larrañaga, P., Lozano, J.A.: Editorial introduction special issue on estimation of distribution algorithms. Evolutionary Computation 13(1) (2005)
10. Li, L., Chu, W., Langford, J., Schapire, R.E.: A contextual-bandit approach to personalized news article recommendation. In: Proc. of International Conference on World Wide Web (2010)
11. Mitchell, T.: Machine Learning. McGraw Hill (1997)
12. Munos, R.: From bandits to monte-carlo tree search: The optimistic principle applied to optimization and planning. Foundations and Trends in Machine Learning 7(1), 1–129 (2014)
13. Penna, N.D., Reid, M.D.: Bandit market makers. In: NIPS (2013)
14. Thierens, D.: Analysis and design of genetic algorithms. PhD thesis, Katholieke Universiteit Leuven, Belgium (1995)
15. Zhang, J., Zhan, Z.-H., Lin, Y., Chen, N.: Evolutionary computation meets machine learning: A survey. IEEE Computational Intelligence Magazine 6(4), 68–75 (2011)

Anomaly Detection Using Replicator Neural Networks Trained on Examples of One Class

Hoang Anh Dau, Vic Ciesielski, and Andy Song

RMIT University, Melbourne 3000, Victoria, Australia
http://www.rmit.com.au

Abstract. Anomaly detection aims to find patterns in data that are significantly different from what is defined as normal. One of the challenges of anomaly detection is the lack of labelled examples, especially for the anomalous classes. We describe a neural network based approach to detect anomalous instances using only examples of the normal class in training. In this work we train the net to build a model of the normal examples, which is then used to predict the class of previously unseen instances based on reconstruction error rate. The input to this network is also the desired output. We have tested the method on six benchmark data sets commonly used in the anomaly detection community. The results demonstrate that the proposed method is promising for anomaly detection. We achieve F-score of more than 90% on 3 data sets and outperform the original work of Hawkins et al. on the Wisconsin breast cancer set.

Keywords: artificial neural networks, replicator neural network, autoencoder, anomaly detection, one-class learning.

1 Introduction

The term anomaly or outlier has come from the field of statistics, wherein anomaly detection has been studied since at least the 19th century. Anomalies are also referred to as an outlier in the literature. The most quoted definition of anomaly comes from Hawkins' 1980 book: "An outlier is an observation which deviates so much from other observations as to arouse suspicions that it was generated by a different mechanism" [1]. Many real world problems can be formulated as an anomaly detection task. For example, one way to detect possible network intrusion is to look for abnormal patterns in network traffic. Detecting anomalies is important because anomalous data often implies negative or even destructive consequences. Alternatively it may represent a positive thing such as rich mineral deposit in magnetic field data. Detecting and then removing anomalies can improve the performance of classification, clustering and regression algorithms, because even a single anomalous value can significant bias these algorithms. For example, Chen recently showed that a single anomalously smooth exemplar will condemn semi-supervised time series classification algorithms to fail [2].

G. Dick et al. (Eds.): SEAL 2014, LNCS 8886, pp. 311–322, 2014.

Different names for the anomaly detection problem as used in literature are: outlier detection, surprise detection, discord detection, deviation detection, exception detection and one-class classification. The machine learning version of anomaly detection aims to build a model that can differentiate anomalous instances from the rest of the data. Techniques can operate in one class or multi-class setting. In one class setting, all anomalous instances are treated as one class. In contrast, in multi-class setting, the goal is to learn different classes of an anomalous nature. Various methods have been proposed for anomaly detection, using concepts drawn from multiple disciplines.

Anomaly detection remains a challenging research topic. It is often difficult to specify a concrete definition of normal regions as opposed to abnormal regions. For that reason, techniques in anomaly detection are usually domain-specific. The lack of labelled data is a common problem. This gives rise to the popularity of techniques operating in unsupervised or semi-supervised mode. Unsupervised anomaly detection does not require training data. However, it makes implicit assumption about the unbalanced distribution of the data set. Thus, the techniques report high false alarm rate if this condition does not hold. On the other hand, techniques that operate in semi-supervised mode require only normal examples for training. In practice, it is often more feasible to collect instances of the normal class than to obtain a comprehensive set of anomalies. This is the approach that our method uses.

Our method is inspired by studies of Hawkins et al., who first proposed Replicator Neural Network (RNN) for anomaly detection [3], [4]. The idea is to train a network capable of replicating the input as its output through a data compression and decompression process. The technique can detect most of the outliers in large multi-variate data sets and can work directly on raw data. However, the network that Hawkins et al. employed consists of 3 hidden layers and this increases the number of parameters to tune considerably. Furthermore, the method only produces an anomaly score, which gives an estimation of the anomalous degree of the pattern. It, however, does not enable direct decision-making without specific domain knowledge. In our study, we show that a network with only one single hidden layer can perform just as well or even better. The outputs of our method offers both an anomaly score and a corresponding threshold value for label assignment.

Our method considers the detection of a point anomaly in a single class setting. The goal is to decide whether an individual instance is anomalous with respect to the rest of the data. To achieve that objective, we train the neural networks to reconstruct instances of the normal class. In testing, we expose the trained network to examples of both classes. We expect the network to replicate the normal instances with marginal error while handling anomalous instances poorly, indicating by the high error rate. Therefore the error value can guide the judgement on the anomalous degree of each observation. We then apply cut-off on sorted error values to assign an anomaly label. We tested the technique on a collection of 6 data sets published in anomaly detection literature. The result shows that our method is competitive or better than other published results.

1.1 Research Questions

1. How can replicator neural network with only one hidden layer be used to detect anomalies?
2. How well does this technique work on different data sets?

Our original intention was to apply the replicator neural network technique to data sets used in previously published works on anomaly detection and compare the performance with that of the previously published methods. This proved to be surprisingly difficult. Many data sets are not available or poorly described. Measures of performance were ad hoc and inconsistent with authors choosing metrics that favour their technique.

2   Related Work

2.1 Anomaly Detection

The broader spectrum of anomaly detection techniques employs concepts from various disciplines Each technique has relative strengths and weaknesses. Anomaly detection techniques can be categorised in different ways. In terms of outputs, some techniques produce an anomaly score while the others produce an anomaly label. In terms of data type, broadly speaking, there are two groups, one works on categorical data whilst the other works on continuous data. There exists techniques to convert data between symbolic and continuous, which enable anomaly detection methods which are not originally designed for one kind of data to work on the other. Almost all anomaly detection techniques require some kinds of distance measures. An instance is assessed based on its relative distance from other instances. A comprehensive survey of research on anomaly detection is presented in [5].

2.2 One-Class Learning

Moya et al provided one of the first papers on one-class learning [6]. One-class classification is similar to binary classification, in which the goal is to categorize the instances into two distinct groups. The difference is that in one-class learning, the training set contains the objects of one class only. Training set of traditional classification problem, on the other hand, must include objects from all the classes. The machine learning version of anomaly detection problem is fundamentally one-class learning. The detector is trained to learn the normal examples and everything else can be classified as outliers.

2.3 Neural Network for Anomaly Detection

Artificial Neural networks (ANN) have been used for classification-based anomaly detection [5]. One of the main advantages is that the technique makes no assumption about the data distribution. Moreover, it is capable of inducing a predictive non-linear model and handles well data sets of high dimension. ANN has been used for for both one-class and multi-class anomaly detection. A basic anomaly detection technique using ANN in one-class setting operates in two steps. A neural network is first trained on normal examples to learn the different normal classes. Each test instance is then provided as input to this trained network. The network is supposed to reject an input if it is an outlier.

2.4   Replicator Neural Network for Anomaly Detection

The term replicator neural networks (RNN) is intended for all auto-associative neural networks with compressed internal representations. RNN is also known in literature as autoencoder. The basic idea of RNN is that the input vector is also the target vector; we simply duplicate the input to be the output. The output layer therefore has equally many nodes as the input layer. Instead of training to predict a target value y given an input x, the network is trained to reconstruct its own input x. The weights of the RNN is driven by the goal to minimize the reconstruction error rate. Fig.1 displays a visualisation of a RNN network. One of the early studies of RNN is done by Hecht-Nielsen almost two decades ago [7], in which the author discusses two theorems regarding the data compression capability of the 3-hidden-layer RNN. Hecht-Nielsen himself credited the first serious study of RNN to Kohonen with the work on Self-Organizing Map.

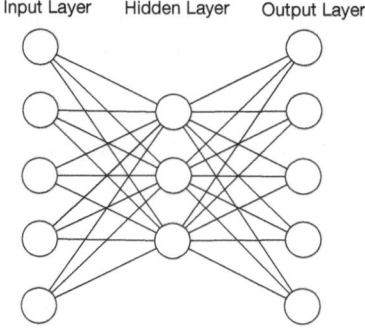

Fig. 1. A replicator neural network with one hidden layer consisting of 3 units. The input and output layer have 5 units each.

RNN was first proposed for outlier detection by Hawskin et al [3]. Their network is a feed-forward multi-layer perceptron with three hidden layers sandwiched between an input layer and an output layer. They use the *tanh* activation function for the two outer hidden layers and a staircase-like activation function for the middle hidden layer, which fit the data points into a number of clusters. They train the net to reconstruct the normal instances and reason that the trained net should replicate the unseen normal instances relatively well. Therefore if an input pattern is poorly reconstructed, indicating by the high reconstruction error, it is likely to be an anomaly. Their method produces an anomaly score for each record, which is the reconstruction error value. They test the technique on two public data sets, namely the Wisconsin breast cancer set and the 1999 KDD Cup network intrusion detection set.

Toth et al. used RNN for outlier modelling in speech recognition [8]. Without RNN, they face the lack of labelled anomaly for training and therefore have to generate synthetic outlier examples. RNN eliminates this task and still offers comparable performance. They point out that using three hidden layers is unnecessary and that the 3 and 4-layer version RNN produce similar result. They find the traditional sigmoid activation function converge the best and use it for all layers.

In a study on texture retrieval but can be easily formulated as an anomaly detection problem, Ciesielski et al. [9] use RNN to find regions of a texture of interest in arbitrary images. This study also finds that using only 1 hidden layer gives smallest train error and test error. The number of units in the hidden layer gave varied accuracies; generally choosing approximately the same number of hidden nodes as the number of inputs gave a good trade off between accuracy and training time.

3 Learning an Anomaly Detector

3.1 Problem Definition

We formulate anomaly detection with RNN as a one-class learning problem. The network is trained to learn what is normal in order to detect the abnormal. The trained network should have little problem reconstructing normal examples, showing by the low reconstruction error. On the other hand, it should encounter difficulty in reconstructing anomalous instances, indicating by the high reconstruction error. The reconstruction error therefore can be used as the anomaly score for each instance. The maximum error at the end of the training process can be used as a threshold value for outputting binary predicate. All the instances with corresponding outlier score below this cut-off value belong to the normal class and the rest is anomalous.

3.2 Data Preparation

We use raw data directly without any feature extraction. For some public data sets, the data may have gone through some kinds of pre-processing. All the attributes are normalized to the range $[0, 1]$ using the Weka software tool-kit [10]. A pattern to be given to the network is created by representing each instance as a feature vector and duplicate the input as output. Instances with missing values and data label entries are removed.

For each data set, the normal instances and anomalous instances are separated to form a normal set and a anomalous set. A training set is created by randomly sampling a number of instances from the normal set. A validation set is created by randomly sampling a number of instances from the normal set (excluding the ones used in training). A test set is created by randomly sampling a number of instances from the normal set (excluding the one used for training and validation) and a number of instances from the anomalous set. Ground-truth is only used after the testing phase to evaluate the performance.

3.3   The Training Algorithm

Generating an anomaly detector involves training a neural network and then finding a suitable threshold. The methodology is:

1. Generate a training set of N normal examples.
2. Generate a validation set of M normal examples.
3. Create a three-layer feed-forward network with random initial weights. The number of units in the input-output layer is equal to the number of variables in the data set. The number of hidden neurons is determined empirically.
4. Use back-propagation to train the network. Training ceases when the error on the validation set begins to rise.
5. Choose the maximum error in training to be the threshold.

4   Experiments and Results

We use the JavaNNS system (Java Neural Network Simulator) [11] for all the experiments. We leave most of the default parameter setting untouched.

4.1   Evaluation Metrics

For each data set, we report a range of evaluation metrics including the F-score, accuracy, true positive rate, true negative rate and top-p accuracy. The choice of F-score is to take into account the possible unbalanced distribution of some data sets. Top-p accuracy is a metric used in [5] to evaluate 10 state-of-the-art anomaly detection techniques. Take p to be the number of true anomalies in the entire test set and t to be the number of true anomalies in the top p records based on anomaly score, top-p accuracy is equal to $\frac{t}{p}$.

4.2   Data Sets

One of our research goals is to see how the technique performs on a wide range of real data sets, and provide a relative comparison with other techniques. For this reason, we selected data sets that have been used in published anomaly detection literature. Table 2 lists the statistic summary for each data set and Table 3 displays the test result.

Table 1. Summary of each data set including name, number of attributes, attribute type, total number of instances, number of instances in training set, validation set and test set respectively. All data sets are of numeric value. These data sets were used in published anomaly detection works.

Name	Patterns	Attributes	Anomalies	Training	Validation	Test
Breast Cancer	683	9	239	150	50	483
Ionosphere	351	34	126	100	30	221
Musk	7074	166	1224	3000	1000	3074
Biomed	194	4	67	50	50	94
Shape1	30	162	10	7	3	20
Shape2	70	162	10	20	10	40

Winscosin Breast Cancer Data Set [12]. The data set consists of 699 instances, of which 458 are benign (65.5%) and 241 malignant (34.5%). The data has 9 real-valued features. In the context of this paper, we consider benign to be of normal class and malignant to be of the anomalous class.

Ionosphere Data Set [13]. This is radar data collected by a system in Goose Bay, Labrador. "Good" radar returns shows some types of structure in the ionosphere while "bad" radar returns do not. We label 251 good radar as normal instances and 126 bad radar as anomalies. All 34 attributes are numeric.

Musk Data Set [14]. The data set consists of 7074 instances. Each is a 166-dimensional feature vector representing a conformation of a molecule. We label the musk instances as anomalies (1255) and non-musk ones as normal (5857).

Biomed Data Set [15]. This data set consists of 194 examples with 4 attributes corresponding to measurements made on blood samples. Normal examples are observations from healthy patients. Abnormal examples are from patients with a rare genetic disease.

Shape Anomaly Data Set [16]. The full data set is originally from the UCR time series database [17]. We only consider two subsets used in [5] and available at [16]. Each instance represents a shape converted into one dimensional time series. The normal time series correspond to one or more shapes, while the anomalous time series correspond to other shapes. **Shape2** is considered more complex than **Shape1**. Each shape time series is 1614 data-point long. We decide to down sample the features as their number is too large (more than 1000). Instead of using all 1614 features, we choose every 10th data points, resulting in 161 attributes left.

Table 2. Result for all data sets: F-score, Accuracy, True Positive Rate, True Negative Rate, the threshold value based on maximum error, Top-p Accuracy and the network architecture used. The numbers in bold indicate high performance.

Name	F-score	Accuracy	TPR	TNR	Thresh.	Top-p A.	Network
Cancer	**94.29%**	94.36%	93.30%	95.42%	0.0634	94.53%	9-9-9
Ionosphere	**92.00%**	90.74%	91.27%	90.00%	0.2138	91.27%	34-34-34
Musk	41.93%	67.76%	28.76%	94.28%	0.4453	54.49%	166-166-166
Biomed	55.67%	54.25%	40.30%	88.89%	0.0238	74.63%	4-7-4
Shape1	**95.24%**	95.00%	100.00%	90.00%	1.1901	100.00%	162-162-162
Shape2	51.28%	52.50%	100.00%	36.67%	2.3580	70.00%	162-162-162

4.3   Choosing the Number of Units for the Single Hidden Layer

The selection of the number of hidden neurons has a couple of implications. If it is too large, the system will be over specified. Conversely, if it is too small, the system can become over-generalized and therefore poorly infers specific cases.

We find that the choice of hidden unit quantity significantly affect the technique accuracy. Table 4 shows varied measurement metrics for different network architectures on the breast cancer data set. For each scenario, we run the experiments 5 times and notice the result stays consistent. Generally, having approximately the same number of hidden units as the number of input-output units give high performance (F-score of more than 90%). Interestingly, further experiments on different data sets show that having the number of hidden units equal to the input-output units consistently yields good detection rate even though depending on the data set, it may not be the optimum architecture.

Table 3. Test result for the breast cancer data set with different network architectures, training uses standard back-propagation and stops at 2000 epochs. The numbers in bold indicate high performance.

Network	F-score	Accuracy	TPR	TNR	Top-p A.
9-0-9 net	78.16%	81.21%	67.37%	95.00%	76.00%
9-1-9 net	61.85%	72.44%	44.77%	100.00%	96.00%
9-3-9 net	89.00%	90.00%	81.17%	98.75%	95.80%
9-9-9 net	**94.29%**	94.36%	93.30%	95.42%	94.34%
9-12-9 net	**93.53%**	93.53%	93.73%	93.33%	93.70%

4.4   Comparison with Other Works

We find it hard to systematically compare different techniques because the community has not converged on a single quality metric. For comparison purpose, we compute the same metrics as the original authors when possible.

When the comparison is impossible, we refer to the baseline accuracy used in [18]. Baseline accuracy for a given data set is the ratio $\frac{n}{(m+n)}$, in which n is the number of anomalies and m is the number of normal instances in the test set respectively. For the method to be considered effective, its top-p accuracy must be significantly better than its corresponding baseline accuracy.

Breast Cancer Data Set Used by Hawkins et al. [3]. We recreate their version of the breast cancer data by choosing one in every six malignant records to form an unbalanced distribution. The resultant data set has 39 malignant (8%) and 444 benign (92%). The authors, however, do not mention the specific number of instances used for training, validation and test respectively. For our experiment, we use 150 normal patterns for training, 50 normal patterns for validation. The test set includes 244 normal patterns and all 39 anomalous patterns. Overall, our method is better at detecting anomalies for this data set. We report the best result in Table 6. Our top 16 records are all anomalies while Hawkins' only contains 11 true anomalies. Our technique captures all the malignant in the top 48 records while Hawkins' method only detects 89.74%. They instead need to examine the top 112 to cover all the malignant cases present.

The Musk and Ionosphere Data Set Used by Aggarwal et al. [19]. The authors' original purpose is to compare performance of brute-forte search and evolutionary search in detecting anomaly. Only the search time for each algorithm is reported and no other detection rate was given. We therefore turn to baseline accuracy described in the Evaluation Metrics section to assess the technique's effectiveness. We achieved 54.49% top-p accuracy for the Musk set given the baseline accuracy of 39.81%. Our technique captures most of the anomalies in the top ranked records of the Ionosphere set, gaining 91.27% top-p accuracy given the baseline accuracy of 57.01%.

The Biomed Data Set Used by Bennett et al. [20]. The authors report performance as a variable dependent on the σ value. The best case is when $\sigma = 1.1$, for which the technique correctly labels 2 out of 27 normal instances and 57 out of 67 anomalous instances. According to our calculation, this gives a F-score of 76.50%. Our technique scores 55.67%. However, it does not necessarily means their technique is better than ours. Reporting performance for different σ value is no different from reporting performance for different threshold values. We can simply adjust the threshold to have a better detection rate. Note that our threshold is a product of the training process, not a value tuned in testing.

Shape1 and Shape2 Data Set Used by Varun [18]. These are among 19 data sets used to evaluate 10 existing state of the art anomaly detection techniques for time series data. Shape1 is relatively easy and most techniques give 100% top-p accuracy. The worst result is around 90%. Our method also obtains 100% top-p accuracy for this set. Shape2 is harder and most techniques

Table 4. A comparison of Hawkins et al. result and our method's result on the breast cancer data set. The first and the second column show the number of malignant present in the top ranked records. The third column displays the corresponding percentage of malignant out of all malignant there are in the whole data set. Our network is trained by standard backpropagation through 500 iterations, 9-9-9 network architecture.

Hawkins' method result			Our method's result		
Malignant	Record	% Malignant	Malignant	Record	% Malignant
0	0	0.00%	0	0	0.00%
3	4	7.69%	4	4	10.25%
6	8	15.38%	8	8	20.51%
11	16	28.21%	16	16	41.02%
18	24	46.15%	23	24	58.97%
25	32	64.10%	29	32	74.36%
30	40	76.92%	36	40	92.30%
35	**48**	**89.74%**	**39**	**48**	**100.00%**
36	56	92.31%	39	56	100.00%
36	64	92.31%	39	64	100.00%
38	72	97.44%	39	72	100.00%
38	80	97.44%	39	80	100.00%
38	100	97.44%	39	100	100.00%
39	112	100.00%	39	112	100.00%

struggle. The best performance is around 90% and the worst is only around 20%. Our method stays competitive with 70% top-p accuracy. Note that our method is not specifically designed for detecting anomaly in a time series database. Time series data has the temporal correlation between data points which does not exist in normal record data.

4.5   Discussion

We notice the lack of benchmark data for anomaly detection. Each paper uses distinct data sets and evaluation metrics, making a fair assessment challenging. One recent paper describes an anomaly detection benchmark from real data [21]. With 19 data sets from the UCI repository, the authors generate 4,369 problem set replicates of various difficulty levels. Unfortunately, these problem sets are not publicly available.

The experiment result demonstrates that using RNN with only one hidden layer is a promising approach for anomaly detection. Even though the optimum number of hidden neurons is dependent on the data dimensionality, we have managed to narrow down to an optimum range. We suggest having this number slightly fewer than, equal to or slightly higher than the number of input-output units are all reasonably good options. When an exhaustive search is impossible, we recommend using the same number of units for all three RNN layers. This finding is somewhat surprising because the intuition is that the number of units in the hidden layer should be smaller than that of the two outer layers to enable data compression and helps the network generalises unseen examples.

In terms of training iteration, we note that most of the time the network converges at 2000 epochs or less. The maximum reconstruction error in training varies for each data set. Therefore, there is no universal threshold value for all. At the same time, we see no correlation between the performance and the corresponding threshold value. We speculate that the threshold intensity may just reflect the diversity of the training examples. We find that threshold alone cannot effectively distinguish all the anomalies because some atypical instances consistently has low reconstruction error.

5   Conclusions

We have shown how replicator neural networks can be used for anomaly detection. This can be done with a three layer feed-forward network with one single hidden layer. The input to this network is also the desired output. The number of units in the input and output layer corresponds to the number of the data attributes. Only normal instances are used for training. The output of the training process is a predictive model and a corresponding threshold value. Our method not only gives a ranked estimation of the anomalous degree of each instance but also provides anomaly label for direct decision-making.

While our original intention was to examine the performance of the method on a wide range of previously used data sets, this proved to be very difficult. We, however, were able to perform comparisons on 6 data sets published in the anomaly detection literature. The technique achieves high F-score of above 90% on 3 out of 6 data sets. Comparison with other works shows that our method is competitive and sometimes better. Our method outperforms the Hawkins et al. original work on using RNN for anomaly detection with the Wisconsin breast cancer set.

Future work would be to investigate other alternatives of choosing a threshold value. For the data sets that the technique does not work quite well, we plan to investigate whether increasing the number of hidden layers can help. We also would like to have the method tested on more data sets of different dimensions and application domains.

References

1. Hawkins, D.M.: Identification of outliers, vol. 11. Springer (1980)
2. Chen, Y., Hu, B., Keogh, E., Batista, G.E.: Dtw-d: time series semi-supervised learning from a single example. In: Proceedings of the 19th ACM SIGKDD International Conference on Knowledge Discovery and Data Mining, pp. 383–391. ACM (2013)
3. Hawkins, S., He, H., Williams, G.J., Baxter, R.A.: Outlier detection using replicator neural networks. In: Kambayashi, Y., Winiwarter, W., Arikawa, M. (eds.) DaWaK 2002. LNCS, vol. 2454, pp. 170–180. Springer, Heidelberg (2002)

4. Williams, G., Hawkins, S., Gu, L., Baxter, R., He, H.: A comparative study of rnn for outlier detection in data mining. In: IEEE International Conference on Data Mining, pp. 709–709. IEEE Computer Society (2002)

5. Chandola, V., Banerjee, A., Kumar, V.: Anomaly detection: A survey. ACM Computing Surveys (CSUR) 41(3), 15 (2009)

6. Moya, M.M., Hush, D.R.: Network constraints and multi-objective optimization for one-class classification. Neural Networks 9(3), 463–474 (1996)

7. Hecht-Nielsen, R.: Replicator neural networks for universal optimal source coding. Science-New York Then Washington, 1860–1860 (1995)

8. Tóth, L., Gosztolya, G.: Replicator neural networks for outlier modeling in segmental speech recognition. In: Yin, F.-L., Wang, J., Guo, C. (eds.) ISNN 2004. LNCS, vol. 3173, pp. 996–1001. Springer, Heidelberg (2004)

9. Ciesielski, V., Ha, V.P.: Texture detection using neural networks trained on examples of one class. In: Nicholson, A., Li, X. (eds.) AI 2009. LNCS, vol. 5866, pp. 140–149. Springer, Heidelberg (2009)

10. Weka, http://www.cs.waikato.ac.nz/ml/weka/downloading.html/ (online; accessed July 23, 2014)

11. Javanns: Java neural network simulator,
http://www.ra.cs.uni-tuebingen.de/software/JavaNNS/ (online; accessed July 12, 2014)

12. Breast cancer wisconsin (prognostic) data set,
http://archive.ics.uci.edu/ml/machine-learning-databases/
breast-cancer-wisconsin/ (online; accessed July 9, 2014)

13. Ionosphere data set, https://archive.ics.uci.edu/ml/datasets/Ionosphere/ (online; accessed July 17, 2014)

14. Musk (version 2) data set,
https://archive.ics.uci.edu/ml/datasets/Musk+Version+2/ (online; accessed July 17, 2014)

15. Musk (version 2)data set, http://lib.stat.cmu.edu/datasets/ (online; accessed July 18, 2014)

16. Chandola time series data sets,
http://www-users.cs.umn.edu/~chandola/timeseries/ (online; accessed July 9, 2014)

17. The ucr time series classification/clustering page,
http://www.cs.ucr.edu/~eamonn/time_series_data/ (online; accessed July 9, 2014)

18. Chandola, V.: Anomaly detection for symbolic sequences and time series data. Ph.D. dissertation, University of Minnesota (2009)

19. Aggarwal, C.C., Yu, P.S.: Outlier detection for high dimensional data. ACM Sigmod Record 30(2), 37–46 (2001)

20. Campbell, C., Bennett, K.P.: A linear programming approach to novelty detection. In: Proceedings of the 2000 Conference on Advances in Neural Information Processing Systems 13, vol. 13, p. 395. MIT Press (2001)

21. Emmott, A.F., Das, S., Dietterich, T., Fern, A., Wong, W.-K.: Systematic construction of anomaly detection benchmarks from real data. In: Proceedings of the ACM SIGKDD Workshop on Outlier Detection and Description, pp. 16–21. ACM (2013)

Walking Motion Learning of Quadrupedal Walking Robot by Profit Sharing That Can Learn Deterministic Policy for POMDPs Environments

Yuya Morino and Yuko Osana

Tokyo University of Technology,
1401-1 Katakura, Hachioji, Tokyo, Japan
osana@stf.teu.ac.jp

Abstract. In this paper, walking motion learning of quadrupedal walking robot is realized by the Profit Sharing that can learn deterministic policy for POMDPs environments. In this research, we used the Profit Sharing that can learn deterministic policy for POMDPs environments which can obtain the deterministic policy by using the history of observations. We carried out a series of experiments using quadrupedal walking robot, and confirmed that walking motion learning can be realized by the Profit Sharing that can learn deterministic policy for POMDPs environments.

1 Introduction

Reinforcement learning is a sub-area of machine learning concerned with how an agent ought to take actions in an environment so as to maximize some notion of long-term reward[1]. Reinforcement learning algorithms attempt to find a policy that maps states of the world to the actions the agent ought to take in those states. Recently, we have proposed the Profit Sharing that can learn deterministic policy for POMDPs environments[2]. This method can obtain the deterministic policy by using the history of observations. However, in this method, robustness for noise does not guaranteed.

In this paper, walking motion learning of quadrupedal walking robot is realized by the Profit Sharing that can learn deterministic policy for POMDPs environments. In this research, we used the Profit Sharing that can learn deterministic policy for POMDPs environments which can obtain the deterministic policy by using the history of observations.

2 Profit Sharing That Can Learn Deterministic Policy for POMDPs Environments

Here, the Profit Sharing that can learn deterministic policy for POMDPs environments [2] which is used in this research. In the method, if the observation is judged as perceptual aliasing, the action is selected based on the history of observations. In the observation judged as perceptual aliasing, if enough observation

G. Dick et al. (Eds.): SEAL 2014, LNCS 8886, pp. 323–334, 2014.

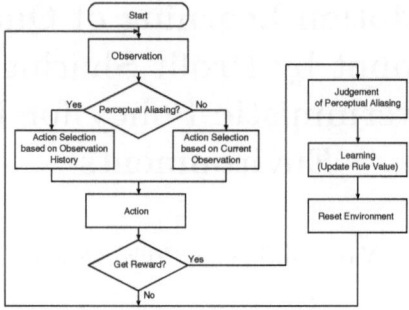

Fig. 1. Flow of Profit Sharing that can Learn Deterministic Policy for POMDPs Environments

sequences are not considered when the action is selected, observation sequences including past observation are also considered. Moreover, in this method, the deterministic rate of actions of each observation and the progress of learning in order to detect perceptual aliasing. Figure 1 shows the flow of the method.

2.1   Action Selection

In this method, for the state which is not judged as perceptual aliasing, the action is selected based on the ratio of rule values of the current observation by the Boltzmann selection. For the state which is judged as perceptual aliasing, the action is selected based on the ratio of rule values of observation sequences by the Boltzmann selection.

The action a in the observation o_x at the time x is selected based on the probability $P(o_x, a, x)$ and it is given by

$$P(o_x, a, x) = \begin{cases} \dfrac{\exp(q_n(o_x, a)/T(o_x))}{\displaystyle\sum_{b \in C^A} \exp(q_n(o_x, b)/T(o_x)),} & (o_x \notin C^{PA}) \\ \dfrac{\exp(q_n(o_x, a)/T(o_x)) + Q(o_x, a, x)}{\displaystyle\sum_{b \in C^A} \left(\exp(q_n(o_x, b)/T(o_x)) + Q(o_x, b, x)\right),} \\ \hfill (o_x \in C^{PA}) \end{cases} \quad (1)$$

where $q_n(o_x, a)$ shows the normalized value for the rule (o_x, a), $T(o_x)$ shows the temperature in the observation o_x, C^{PA} is the set of observations and observation sequences which are judged as perceptual aliasing, and C^A is the set of actions. And, $Q(o_x, a, x)$ is the summation of values for the rules on the observation o_x at the time x considering observation sequences, and is given by

$$Q(o_x, a, x) = \sum_{o_i \to O \in C^{ref}(x)} \exp(q_n(o_i \to O, a)/T(O)) \quad (2)$$

where $o_i \to O$ is the observation sequence which includes the observation o_i before the observation sequence O, and $C^{ref}(x)$ is the set of observations and observation sequences which are used for the action selection at the time x.

2.2   Judgment of Perceptual Aliasing

(1) Judgment for Each Observation in Episode

In this method, it is judged whether each observation in perceptual aliasing using the deterministic rate of actions in each observation and the progress of learning.

(a) Deterministic Rate of Actions in Each Observation

In the learning process, if the plural actions to be selected in order to obtain the reward in the same observation, the action is selected stochastically. In this method, the deterministic rate of actions in each observation which is used in the Extended On-line Profit Sharing with Judgment (EOPSwJ)[3] is used in order to detect perceptual aliasing.

The deterministic rate of actions in the observation o_x at the time x, $d(o_x, x)$ ($0 \leq d(o_x, x) \leq 1$) is given by the following equation:

$$d(o_x, x) = \frac{\sum\limits_{a \in C^A} (P(o_x, a, x) - P_{ini})^2}{N} \tag{3}$$

where C^A shows the set of actions, $P(o_x, a, x)$ is the action selection probability for the action a in the observation o_x at the time x, P_{ini} is the initial action selection probability, and N is the normalization constant.

The deterministic rate of actions in each observation $d(o_x, x)$ is close to 1 when the action selection in the observation o_x at the time x is deterministic. In contrast, the deterministic rate of actions in each observation $d(o_x, x)$ is close to 0 when the action selection in the observation o_x at the time x is stochastic.

(b) Progress of Learning

The progress of learning in the observation o_x at the time x, $l(o_x, x)$ is given by the following equation:

$$l(o_x, x) = \begin{cases} I(o_x), & (o_x \notin C^{PA}) \\ \min\{I(O)|O \in C^{ref}(x)\}, & (o_x \in C^{PA}) \end{cases} \tag{4}$$

where $I(o_x)$ is the update times of the value of the rules for the observation o_x, $I(O)$ is the update times of the value of the rules for the observation sequence O. And $C^{ref}(x)$ shows the set of observations and observation sequences which are used for the action selection at the time x in the last episode.

(c) Total Judgment

The observation whose progress of learning is high and deterministic rate of actions is low (that is, the action selection is stochastic) is judged as perceptual aliasing. That is, for the observation o_x at the time x in the last episode which satisfies the following condition:

$$\phi(l(o_x, x))(1 - d(o_x, x)) > \theta^{PA}, \tag{5}$$

the observation o_x is regarded as perceptual aliasing. Here, θ^{PA} means the threshold for judgment of perceptual aliasing. And $\phi(\cdot)$ is given by

$$\phi(u) = \frac{1}{1 + \exp(-(u - \theta^l))} \tag{6}$$

where θ^l is the threshold.

(2) Update of Set of Observations and Observation Sequences Which Are Judged as Perceptual Aliasing C^{PA}

The set of observations and observation sequences which are judged as perceptual aliasing C^{PA} is updated based on the information of the observations which are judged as perceptual aliasing.

(a) Decision of Observation Sequence Which Is Added to C^{PA}

If the condition given by Eq.(7) is satisfied, the observation sequence which is added to the set C^{PA} is determined.

$$o_x \in C^{PA} \cap C^{PA\_E} \tag{7}$$

Here, C^{PA_E} is the set of observations which are judged as perceptual aliasing in the last episode.

The observation sequence which is added to the set C^{PA} for the observation o_x, $O^{PA}(x)$ is given by

$$O^{PA}(x) = \operatorname*{argmax}_{(o \to O) \in \{C^{ref}(x) \cap \overline{C^{PA}}\}} \{d(o \to O)\} \tag{8}$$

where $C^{ref}(x) \cap \overline{C^{PA}}$ shows the set union of the set of observation sequences which are considered in the action selection at the time x and the set of the observations which are not judged as perceptual aliasing. $d(o \to O)$ is the deterministic rate of actions for the observation sequence $o \to O$ and it is given by

$$d(o \to O) = \frac{\sum\limits_{a \in C^A} (P(o \to O, a) - P_{ini})^2}{N} \tag{9}$$

where P_{ini} is the action selection probability and N is the normalized constant. $P(o \to O, a)$ shows the probability that the action a is selected for the observation sequence $o \to O$ and it is given by

$$P(o \to O, a) = \frac{\exp(q_n(o \to O, a)/T(O))}{\sum\limits_{b \in C^A} \exp(q_n(o \to O, b)/T(O))} \tag{10}$$

where $q_n(o \to O, a)$ is the normalized value of the rule $(o \to O, a)$ and $T(O)$ is the temperature in the observation sequence O.

(b) Update Set of Observations and Observation Sequences Which Are Judged as Perceptual Aliasing C^{PA}

Next, the observation sequences determined in *(a)* and the observations judged as perceptual aliasing are added to the set C^{PA}.

$$C^{PA} \leftarrow C^{PA} \cup C^{PA\_E}$$
$$\cup \{O^{PA}(x) | x : o_x \in C^{PA} \cap C^{PA\_E}\} \tag{11}$$

2.3   Learning

When the agent obtains the reward, the rule values are updated after the judgment of perceptual aliasing.

(1) Update of Value of Rule $q(o, a)$

The update times of value of the rules (o, a) $(I(o, a))$ and the value of rule $q(o, a)$ are updated as follows.

$$q(o, a) \leftarrow \left(1 - \frac{1}{I(o, a)}\right) q(o, a) + \frac{r \cdot F(o)}{I(o, a)} \tag{12}$$
$$((o, a) \in \{(o_x, a_x) | x = 1, \cdots, W\})$$

Here, r is the reward, and $F(o)$ is the reinforcement value for the rules in the observation o and it is given by

$$F(o) = \frac{1}{W - x_o} \tag{13}$$

where W is the episode length, and x_o is first time when the observation o is observed.

(2) Update of Value of Rule $\acute{q}(O, a)$

If there are some observations judged as perceptual aliasing in the last episode, values of the rules for these observations and the observation sequences including these observations are updated.

If it is considered that the observation sequence $O \to o_x (\in C^{ref}(x))$ is used for only action selection at the time x, the update time of value of the rule $(O \to o_x, a_x)$ $(I(O \to o_x, a_x))$ and the value of the rule $q(O \to o_x, a_x)$ are updated as follows.

$$q(O \to o_x, a_x) \leftarrow \left(1 - \frac{1}{I(O \to o_x, a_x)}\right) q(O \to o_x, a_x)$$
$$+ \frac{r \cdot F(o_x)}{I(O \to o_x, a_x)} \tag{14}$$

where r is the reward. $F(o_x)$ is the reinforcement value for the observation sequence $(O \to o_x)$ whose last is the observation o_x.

If it is considered that the observation sequence $O \to o_x (\in C^{ref}(x))$ is used for action selection at plural times, the update time of the value of the rule $(O \to o_x, a_y)$ $(I(O \to o_x, a_y))$ and the value of the rule $q(O \to o_x, a_y)$ are updated as follows.

$$q(O \to o_x, a_y) \leftarrow \left(1 - \frac{1}{I(O \to o_x, a_y)}\right) q(O \to o_x, a_y)$$
$$+ \frac{r \cdot F(o_x)}{I(O \to o_x, a_y)} \tag{15}$$

Here, a_y is the action at the time y which satisfy

$$o_x = o_y \tag{16}$$

and

$$\operatorname*{argmin}_{x=1,\cdots,W} \{(O \to o_x) \in (C^{ref}(x) \cap C^{ref}(x')) \tag{17}$$

$$|x \neq x'\} \leq y.$$

If the action a_y appears plural times in that episode, the values of all rules for the action are updated one time.

If the observation sequence $O \to o_x$ appears in plural times in an episode, the values of all rules for the pair of the action and the observation o_x after the observation sequence $O \to o_x$ appears first in the episode are updated equally.

2.4 Observation Sequences Which Are Used for Action Selection

In this method, the set of observations and observation sequences $C^{ref}(x)$ are used when the action is selected, the judgment of perceptual aliasing and the update of the rule values.

If $o_x \in C^A$, $C^{ref}(x)$ is determined by the following procedure.

(1) The set $C^{ref}(x)$ is set to

$$C^{ref}(x) = \{o_x\}. \tag{18}$$

(2) The set C^{add} whose elements are the observation sequences to be added to the set $C^{ref}(x)$ is given by

$$C^{add} = \{o_i \to o_x | h(x) \leq i < x\} \tag{19}$$

where $h(x)$ is the last time when the observation o_x appears before the time x and is given by

$$h(x) = \begin{cases} 1, & (\min\{i | o_i = o_x\} = x) \\ \max\{i | o_i = o_x \text{ and } i < x\}, & (\text{otherwise}). \end{cases} \tag{20}$$

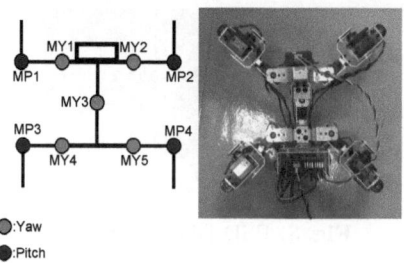

Fig. 2. Quadrupedal Walking Robot

(3) The set C^{past} of the observation sequences which have to consider more past observation is obtained by

$$C^{past} = C^{add} \cap C^{PA}. \tag{21}$$

(4) The elements of the set C^{add} are added to the set $C^{ref}(x)$.

$$C^{ref}(x) \leftarrow C^{ref}(x) \cup C^{add} \tag{22}$$

(5) If there is no observation or observation sequence which has to consider more past observation $((C^{past} = \phi))$, the procedure is finished. If $C^{past} \neq \phi$, go to (6).

(6) The set C^{add} is obtained by

$$C^{add} = \{o_i \to o_j \to O$$
$$|h(j) \leq i < j \text{ and } o_j \to O \in C^{past}\}. \tag{23}$$

(7) Back to (3).

3   Walking Motion Learning of Quadrupedal Walking Robot

Here, we explain the proposed walking motion learning of quadrupedal walking robot.Y

3.1   Qaudrupedal Walking Robot

Figure 2 shows the quadrupedal walking robot which has nine axes. In Fig.2, red circles show yaw axes and blue circles show pitch axes. The robot has PSD (Position Sensitive Detector) distance sensor as shown in Fig.3.

3.2   Observations

In this research, PSD distance sensor value is used as the observation. The used PSD sensor can be detected 10~ 80 cm, and sensor value is between 60 and 170. So, the quatized value shown in Table 1 is used as the observation.

Fig. 3. PSD Distance Sensor

Table 1. Value of PSD Distance Sensor and Observation Number

Observation No.	Sensor Value
0	~60
1	61~90
2	91~120
3	121~150
4	151~

Table 2. Angle of Servomotors.

	MP1	MY1	MP2	MY2	MY3	MP3	MY4	MP5	MY5
(a)	−20	±10, ±20	0	0	−20	0	0	0	±10, ±20
	−40	±30, ±40							±30, ±40
	MP1	MY1	MP2	MY2	MY3	MP3	MY4	MP5	MY5
(b)	0	0	20	±10, ±20	20	0	±10, ±20	0	0
			40	±30, ±40			±30, ±40		

3.3   Actions

In this research. only angles of servomotors are obtained by the learning. Walking motion is composed of following four steps.

(a) Move left forward leg and right rear leg.
(b) Back to original porision.
(c) Move right forward leg and left rear leg.
(d) Back to original porision.

Table 2 shows the angles of servomotors.

3.4   Episode

In this research, (a)~(d) described in **3.3** are treated as one episode. The quadrupedal walking robot moves from the initial position to the goal until it rearches the goal or it is off its own course.

(a) Environment (1)   (b) Environment (2)
(Wood)   (Rubber))

Fig. 4. Environments (1) & (2)

Table 3. Experimantal Conditions

Initial Temperture	T_{ini}	1.0
Minumum Temperture	T_{min}	0.005
Decay Rate of Temperture	γ	0.92
Threshold	θ^{PA}	0.6
Thresshold	θ^l	30
Parameter in Perceptual Aliasing Judgement p		13
Initial Value of Rules		5

3.5 Reward

In each episode, the robot obtained the reward. The reward r is calculated based on the difference between the PSD sensor valued before and after each episode.

$$r = \begin{cases} d_a - d_b & (d_a - d_b > 5) \\ 0 & (\text{otherwise}) \end{cases} \qquad (24)$$

where d_a is the PSD sensor value after episode and d_b is the PSD sensor value before episode.

4 Experiment Results

Here, we examined walking motion learning of quadrupedal walking robot is realized by the Profit Sharing that can learn deterministic policy for POMDPs environments.

In this experiment, euqadrupedal walking robot starts the initial position located from the goal in 80 cm position. Experimants were carried out in two environments (1) on woody plate and (2) on robber plate. Table 3 shows the parameters in the Profit Sharing that can learn deterministic policy for POMDPs environments.

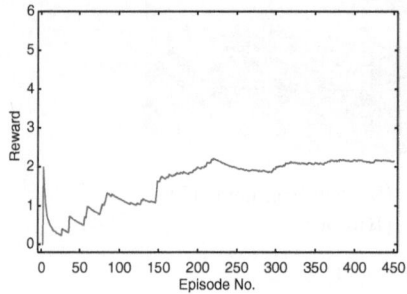

Fig. 5. Reward Transition in Environment 1

Table 4. Obtained Rules (Environment (1))

Obs. No.	Action No.	Rule Value
1	44	8.25833
1	62	9.83333
1	137	9.00000
1	149	8.50000
1	228	8.58333

Table 5. Angle of Servomotors in Each Action (Table 4)

Act. No.	44	62	137	149	228
MP1	−40	−40	0	0	0
MY1	20	40	0	0	0
MP2	0	0	40	40	20
MY2	0	0	−20	−30	10
MY3	−20	−20	20	20	20
MP3	0	0	0	0	0
MY4	0	0	−10	30	30
MP5	0	0	0	0	0
MY5	−30	−30	0	0	0

4.1   Experment in Environment (1)

Figure 5 shows the transition of the obtained reward in the environment (1). This figure shows the average reward per 30 episodes.

As shown in Fig.5, the quadrupedal walking robot can be obtained the walking motion.

Table 4 shows the observation number and the action number in the rules whose value is high after 450 episodes. And angles of servomoters are shown in Table 5.

4.2   Experment in Environment (2)

Figure 6 shows the transition of the obtained reward in the environment (2). This figure shows the average reward per 30 episodes.

As shown in Fig.6, the quadrupedal walking robot can be obtained the walking motion.

Table 6 shows the observation number and the action number in the rules whose value is high after 450 episodes. And angles of servomoters are shown in Table 7.

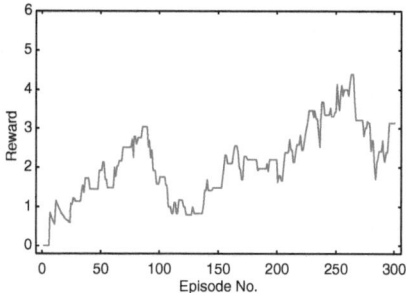

Fig. 6. Reward Transition in Environment (2)

Table 6. Obtained Rules (Environment (2))

Obs. No.	Action No.	Rule Value	Obs. No.	Action No.	Rule Value
1	7	9.00000	1	48	13.00000
1	13	8.08333	1	53	10.50000
1	17	11.72917	1	63	9.58333
1	27	10.50000	1	179	8.08333
1	31	10.00000	1	181	7.91667
1	33	9.71667	1	183	9.58333
1	35	8.50000	1	191	8.50000
1	43	8.10417			

Table 7. Angle of Servomotors in Each Action (Table 6)

Act. No.	7	13	17	27	31	33	35	43	48	53	63	179	181	183	191
MP1	−40	−40	−40	40	40	40	−40	−40	−20	−20	−20	0	0	0	0
MY1	10	20	30	40	40	10	10	30	20	20	20	0	0	0	0
MP2	0	0	0	0	0	0	0	0	0	0	0	40	40	40	40
MY2	0	0	0	0	0	0	0	0	0	0	0	−30	−30	−30	−40
MY3	−20	−20	−20	20	20	20	−20	−20	−20	−20	−20	20	20	20	20
MP3	0	0	0	0	0	0	0	0	0	0	0	0	0	0	0
MY4	0	0	0	0	0	0	0	0	0	0	0	20	30	40	40
MP5	0	0	0	0	0	0	0	0	0	0	0	0	0	0	0
MY5	−40	30	10	20	40	10	−20	−20	−30	−40	−40	0	0	0	0

4.3   Learning in Envirnonment (2) After Environment (1)

Figure 7 shows the transition of the obtained reward in the environment (2). In this experiment, the robot which learns in the environment (1) is used. This figure shows the average reward per 30 episodes.

As shown in Fig.7, the quadrupedal walking robot can be obtained the walking motion in this experiment.

Table 8 shows angles of servomoters in rules whose value are high.

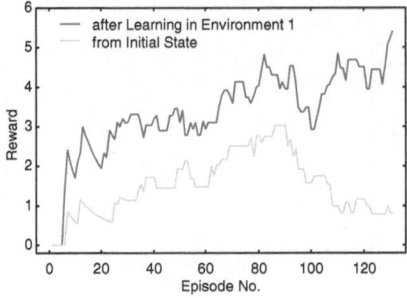

Fig. 7. Reward Transition in Environment (2) After Learning in Environment (1)

Table 8. Angle of Servomotors in Each Action)

Act. No.	54	56	98	111	137	142	149	156	172	189
MP1	−20	−20	−20	40	0	0	0	0	0	0
MY1	30	40	−10	20	0	0	0	0	0	0
MP2	0	0	0	0	0	20	40	20	20	40
MY2	0	0	0	0	20	20	−40	−40	−20	−40
MY3	−20	−20	−20	20	20	20	20	20	20	20
MP3	0	0	0	0	0	0	0	0	0	0
MY4	0	0	0	0	10	40	−30	−30	20	30
MP5	0	0	0	0	0	0	0	0	0	0
MY5	−40	−10	−20	40	0	0	0	0	0	0

5   Conclusions

In this paper, walking motion learning of quadrupedal walking robot has been realized by the Profit Sharing that can learn deterministic policy for POMDPs environments. In this research, we used the Profit Sharing that can learn deterministic policy for POMDPs environments which can obtain the deterministic policy by using the history of observations. We carried out a series of experiments using quadrupedal walking robot, and confirmed that walking motion learning can be realized by the Profit Sharing that can learn deterministic policy for POMDPs environments.

References

1. Sutton, R.S., Barto, A.G.: Reinforcement Learning: An Introduction. The MIT Press (1998)
2. Takamori, Y., Osana, Y.: Profit sharing that can learn deterministic policy for POMDPs environments. In: Proceedings of IEEE International Conference on System, Man and Cybernetics, Anchorage (2011)
3. Saito, K., Masuda, S.: Profit Sharing Introducing the Judgement of Incomplete Perception. Transactions of the Japanese Society for Artificial Intelligence 19(5), 379–388 (2004)

Genetic Programming for Multiclass Texture Classification Using a Small Number of Instances

Harith Al-Sahaf, Mengjie Zhang, and Mark Johnston

Evolutionary Computation Research Group,
Victoria University of Wellington, PO Box 600, Wellington, New Zealand
{harith.al-sahaf,mengjie.zhang}@ecs.vuw.ac.nz
mark.johnston@msor.vuw.ac.nz

Abstract. The task of image classification has been extensively studied due to its importance in a variety of domains such as computer vision and pattern recognition. Generally, the methods developed to perform this task require a large number of instances in order to build effective models. Moreover, the majority of those methods require human intervention to design and extract some good features. In this paper, we propose a Genetic Programming (GP) based method that evolves a program to perform the task of multiclass classification in texture images using only two instances of each class. The proposed method operates directly on raw pixel values, and does not require human intervention to perform feature extraction. The method is tested on two widely used texture data sets, and compared with two GP-based methods that also operate on raw pixel values, and six non-GP methods using three different types of domain-specific features. The results show that the proposed method significantly outperforms the other methods on both data sets.

Keywords: Genetic Programming, Texture Classification, Multiclass.

1 Introduction

In the fields of computer vision and pattern recognition, *image classification* represents one of the most important tasks. However, developing a program that is capable of performing image classification with good performance is a very challenging task, particularly for difficult problems. Even discriminating between instances of two classes (binary classification) can be difficult. The difficulty of this task increases with a large number of classes due to the increased complexity of detecting a good set of features. The majority of the proposed methods for multiclass classification in the literature suffer one or both of the following issues: (1) some human intervention to design and extract a good set of features is required prior to the training phase [6]; and (2) a large number of instances are required in order to reach a suitable level of performance. In terms of the first issue, an expert with background knowledge of the domain is required to perform the task of detecting a set of distinctive features. It is not always feasible to find such a person or it can be very expensive. Similarly, in terms of the second

G. Dick et al. (Eds.): SEAL 2014, LNCS 8886, pp. 335–346, 2014.

issue, acquiring a sufficiently large number of instances can be expensive, hard, or infeasible in many cases.

Genetic Programming (GP) is an evolutionary search method inspired by the principles of natural selection [12]. GP aims at evolving a computer program for a user-defined problem. Starting from a randomly generated initial population of solutions, GP uses genetic operators and a fitness function to evolve a solution. The fitness function is used to measure the goodness of each program, which reflects the performance level or the ability of that program to solve the problem. The genetic operators allow the system to explore or introduce different combinations of the genetic materials.

In tree-based GP [12], an evolved program is represented as a tree where all non-leaf nodes are drawn from the function set, while leaves are taken from the terminal set. The program evolved by GP produces a single value from the root node for each instance. For binary classification, the resulting value can be naturally translated to a class label such that all negative values represent one class and all non-negative values represent the other class.

There are at least four approaches that can be adopted to extend GP to perform multiclass classification tasks. In the first, a wrapper approach can be adopted where a multiclass classifier (e.g. nearest-neighbour) is used to perform the classification task, while GP is used to perform feature selection, extraction, or construction [15]. The second approach is to change the mapping scheme such that the real number line is divided into more than two intervals and the single value resulting from the root node is mapped to one of those intervals [13]. The third approach is to use a different program representation that produces multiple values instead of only a single value obtained from the root node [18]. Building a composite solution via breaking the multiclass classification task into a number of binary classification problems is the fourth approach [7].

Song et al. [13] utilised GP to perform multiclass texture classification by using *static range selection* (SRS) [16] and *dynamic range selection* (DRS) [7]. In SRS, the real number line is divided into a predefined number of equally sized intervals, each of which represents one class. This method requires $N - 1$ thresholding values, where N is the number of classes. Selecting a good set of threshold values introduces an extra complexity that is an interesting research topic itself. In DRS, the idea is to dynamically divide the real number line during the program evolving phase (training). Their experiments reveal that both methods are capable of handling the multiclass classification task on texture images using the raw pixel values. SRS and DRS methods are used as baseline methods in this study.

A method that decomposes the multiclass classification task into a number of binary classification tasks was used by Loveared et al. [7]. The idea is to evolve a single program for each of the pairwise class combinations. The cost of evolving a complete set of sub-classifiers represents the main drawback of this approach.

Changing the representation of the GP program to produce multiple values from the internal nodes rather than the single value of the root node represents another approach that has been adopted to tackle the problem of multiclass classification in [18]. The idea is to use a special type of node that make decisions to discriminate between instances of the different classes, and then a voting approach is used to predict the class label of the instance being evaluated.

In this paper, a wrapper-based approach is adopted via combining a nearest-neighbour classifier and GP to evolve a model. The main idea of the proposed method is to evolve a program that applies different operations on the instance being evaluated such that the response to a bank of filters (i.e. set of convolution masks) will be different depending on the textures of the different class instances. Therefore, a single program is evolved to handle the multiclass classification task rather than breaking the task into a number of sub-tasks.

The overall goal of this paper is to use GP to evolve a program for the task of multiclass image classification using a small number of instances. Precisely, we are interested in addressing the following questions:

- what GP representation and fitness function can be used to tackle the limited number of instances for multiclass image classification;
- whether the evolved program can compete with other GP-based methods that were also utilised to automatically handle the multiclass image classification task; and
- whether the evolved program can achieve better performance than the use of domain-specific features with commonly used classification methods.

The remainder of this paper is organised as follows. Section 2 provides a detailed discussion of the proposed method. The data sets, baseline methods, and evaluation procedure are presented in Section 3. The results of the experiments are presented and discussed in Section 4. Finally, the conclusions and recommendations for future research directions are given in Section 5.

2   The New Method

For presentation convenience, we call the proposed GP method *Tree-of-Filters* (ToFs). This section describes the terminal and function sets, fitness function, and the procedure to measure the fitness of an evolved program.

2.1   Terminal Set

The proposed method operates directly on the raw pixel values of the image, and does not require a prior step to perform feature detection and extraction. Therefore, the first component of the terminal set is the instance (image) represented as a 2D matrix. The second type in the terminal set is a list of filters that can be used to transform the image into another image via convolution. The major issue here is which filters to select as the literature shows that there are many filters of different types and sizes. We have decided to limit our scope to filters that can help in revealing the texture primitives such as lines, as presented in Figure 1. In addition to those filters, the Gaussian and Laplacian-of-Gaussian

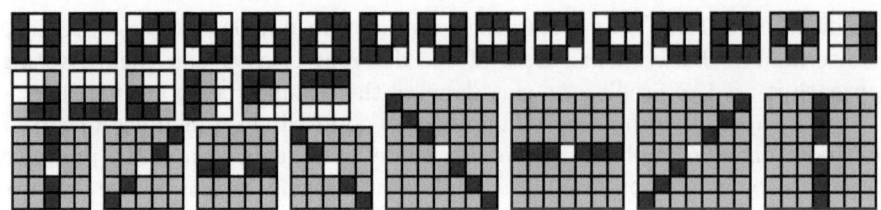

Fig. 1. The 29 filters used in this study. The blue cells are those cells having the value −1, grey cells are 0, and white cells are set to a positive value equal to the number of blue cells in order to make the sum of the filter coefficients equal to 0.

(LoG) filters were also used. The size of the Gaussian and LoG filters depends on a σ value. The size is calculated using:

$$size = (2\lceil 3\sigma \rceil) + 1 \qquad (1)$$

where $\lceil \cdot \rceil$ returns the smallest integer value greater than or equal to the argument. The value of σ is randomly chosen from the set $\{0.5, 1, 1.5, 2, 2.5, 3, 3.5, 4\}$.

The third type of terminal is a randomly generated constant value in the closed interval $[-10, +10]$.

In summary, the terminal set is made up of $\{I, F_i, G_\sigma, L_\sigma, C\}$, where I is the image being evaluated, F_i is the ith predefined filter, G_σ and L_σ are respectively the Gaussian and LoG filter of the specified σ value, and C are constant values.

2.2   Function Set

The function set consists of twelve operators that can be categorised into five groups based on the number and type of the input arguments. The first group consists of the $\{Add_I(\cdot, \cdot),\ Sub_I(\cdot, \cdot),\ Mul_I(\cdot, \cdot),\ Div_I(\cdot, \cdot),\ Min_I(\cdot, \cdot),\ Max_I(\cdot, \cdot)\}$ functions that operates on two images and return an image. The first four functions of this group are the regular mathematical operators $\{+, -, \times, \div\}$, where the $\div$ is protected; it returns zero when the second value (divisor) is zero. The $Min_I(\cdot, \cdot)$ and $Max_I(\cdot, \cdot)$ functions respectively return the minimum and maximum of the arguments. The functions of this group work in a pixel-by-pixel fashion; therefore, the input and returned images are of the same size. Similarly, the second group of functions, $\{Add_C(\cdot, \cdot),\ Sub_C(\cdot, \cdot),\ Mul_C(\cdot, \cdot),\ Div_C(\cdot, \cdot)\}$, operate on an image and a constant value. The third group is the $Conv(\cdot, \cdot)$ function that convolves the first argument (image) using the second argument (filter). The fourth group is the single function $Coder(\cdot)$, which takes only one argument of type image, and returns a feature vector. This is only used at the root of the evolved program. Each function of the first three groups normalises the resulting image to have pixel values between 0 and 255 (inclusive) before passing it to the parent node. Moreover, those functions, apart from $Coder(\cdot)$, can form long chains in different orders due to type matching between their outputs and at least one of the input arguments. Figure 3 shows two examples of programs evolved by the ToFs method.

Algorithm 1. Distance ratios (DR_{diff} and DR_{same})

    Input: Ω : list of lists of Imprints
    Output: Set of double-precision values

1   $\Sigma_{\text{differ}} \longleftarrow 0$
2   $\Sigma_{\text{same}} \longleftarrow 0$
3   $\xi \longleftarrow 0;$ // ξ is a counter
4   **foreach** $\omega_1 \in \Omega$ **do**   // ω_1 is a list of Imprint objects
5   **foreach** $\nu_1 \in \omega_1$ **do**   // ν_1 is an Imprint object
6       $D_{\text{differ}} \longleftarrow 2$
7       $D_{\text{same}} \longleftarrow -1$
8   **foreach** $\omega_2 \in \Omega$ **do**   // ω_2 is a list of Imprint objects
9   **foreach** $\nu_2 \in \omega_2$ **do**   // ν_2 is an Imprint object
10           $\lambda \longleftarrow \text{dist}(\nu_1, \nu_2)$
11   **if** $(\omega_1 \neq \omega_2) \wedge (\lambda < D_{\text{differ}})$ **then** $D_{\text{differ}} \longleftarrow \lambda$
12   **else if** $(\nu_1 \neq \nu_2) \wedge (\lambda > D_{\text{same}})$ **then** $D_{\text{same}} \longleftarrow \lambda$
13   **end**
14   **end**
15       $\Sigma_{\text{differ}} \longleftarrow \Sigma_{\text{differ}} + D_{\text{differ}}$
16       $\Sigma_{\text{same}} \longleftarrow \Sigma_{\text{same}} + D_{\text{same}}$
17   **end**
18     $\xi \longleftarrow \xi + |\omega_1|$
19   **end**
20   **return** $\left\{ \Sigma_{\text{differ}}/\xi, \Sigma_{\text{same}}/\xi \right\}$

2.3 Fitness Measure

The fitness function of the ToFs performs multiple tasks simultaneously as shown in Equation (2).

$$Fitness = DR_{\text{same}} + \left(2 - (DR_{\text{diff}} + Accuracy) \right) \qquad (2)$$

$$Accuracy = \frac{Hits}{Total} \qquad (3)$$

Here *Accuracy* measures the performance (accuracy) of the individual on the training set, and *Hits* and *Total* are respectively the number of correctly classified instances and total number of instances. The DR_{same} and DR_{diff} are the distance ratio to instances of the same and different class respectively. Algorithm 1 presents the procedure for calculating the DR_{same} and DR_{diff} values.

The dist$(\cdot, \cdot)$ function in Algorithm 1 calculates the distance between two feature vectors of the same length:

$$\text{dist}(a, b) = 1 - \left(\frac{2 \left(\sum_{i=1}^{|a|} min\,(a_i, b_i) \right)}{\sum_{i=1}^{|a|} a_i + \sum_{i=1}^{|b|} b_i} \right) \qquad (4)$$

where a and b are the two feature vectors, the $|\cdot|$ function returns the length (i.e. number of elements or items) of a vector, $min(\cdot, \cdot)$ returns the minimum value of the two arguments, a_i and b_i are the value of the ith feature of a and b respectively. This distance measure returns a value between 0 and 1, where a smaller value means a higher similarity between the two feature vectors.

2.4 Fitness Measuring Procedure

The training phase aims at evolving a program that has high accuracy on the training set, high distance ratio between the instances of different classes, and low distance ratio between instances of the same class. The evaluation of the program starts from the terminal nodes as they represent the inputs of the program's tree, and ends at the root node (i.e. *Coder*). For each instance in the training set, the system applies the operations in a bottom-up order. At each non-terminal node, an image is generated depending on the inputs and the specified operation. Then this image is normalised (to have values between 0 and 255) and passed to the parent node. The *Coder* node then receives the final image and transforms it to a feature vector. The *Coder* node has a filter-bank (list of filters) identical to those of the terminal set apart from the Gaussian and LoG filters. This node constructs a new feature vector (all elements have zero value) which consists of a number of elements equivalent to the number of filters in the filter-bank (one element for each filter). A dot product is then performed at each pixel of the image argument, along with the neighbouring pixels, with each of the filters in the filter-bank, and the responses (resulting values) are reported. The corresponding element of the filter that has the highest response (largest value) is incremented by one. Finally, the generated feature vector is normalised to have values between 0 and 1.

The system uses the feature vector generated by the *Coder* node along with the actual class label of the instance being evaluated to construct an *Imprint* object. The constructed imprint objects of the training set instances are stored in a list named *knowledge base* that will be used for two tasks: (1) to measure the fitness of the evolved program; and (2) to serve as a knowledge base during the testing phase. The DR_{same} and DR_{diff} values are first calculated using the procedure presented in Algorithm 1. Meanwhile, the accuracy is measured using *the-nearest-neighbour* (1NN) [2] method. Using those three values (DR_{same}, DR_{diff}, and accuracy), the fitness of the evolved program can be measured using the formula presented in Equation (2).

2.5   Performance Measuring Procedure

The aim of the testing phase is to measure the generalisation ability of the best evolved program on the unseen data. Therefore, the best evolved program at the end of the training phase is tested on the instances of the test set (unseen data). For each instance, a feature vector is generated from the *Coder* node in a similar way to that in the training phase. The distance between the generated feature vector and each imprint object of the *knowledge base* list is calculated. The class label of the closest imprint object is returned to serve as the predicted class label for the instance being evaluated. Then the accuracy formula (Equation (3)) is used to measure the generalisation ability of the best evolved program.

3   Experimental Design

In this section, discussions of the data sets, data set preparation, baseline methods, parameter settings, and the evaluation process are provided.

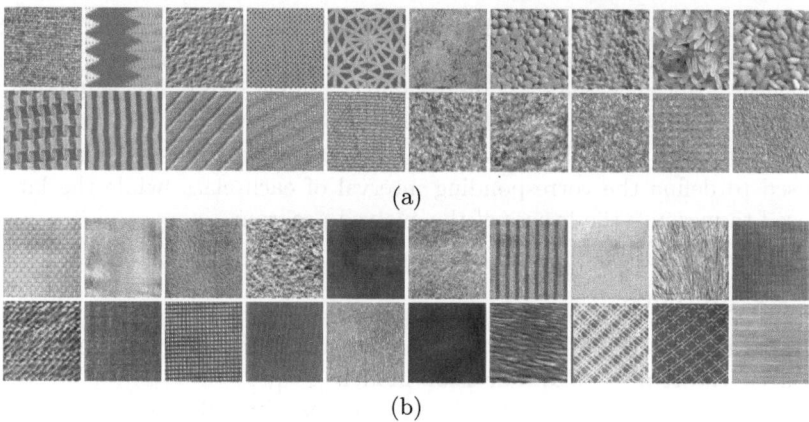

(a)

(b)

Fig. 2. Samples of the (a) Textures-1 data set; and (b) Textures-2 data set

3.1 Data Sets

In this study, two data sets are used to evaluate the performance of the ToFs method. The first data set is taken from the *Kylberg Texture* data set [5], which is made up of 28 grey-scale texture classes. Each class consists of 160 instances of size 576×576 pixels. The instances of this data set are fixed in terms of rotation and scale, but not illumination. This data set comes in another flavour where the instances are captured under different rotation angles (not used in this study). Only 20 classes of the 28 have been selected to form the first data set of this study *Textures-1* as presented in Figure 2(a). The instances of this data set have been resampled (i.e. resized) to 115×115 pixels to reduce the computation costs.

In computer vision and signal processing, the *Brodatz textures* data set [1] is one of the mostly used data sets. This data set is made up of 112 classes of different textures that each consists of only one grey-scale instance of size 640×640 pixels. Similar to *Textures-1*, 20 classes have been selected to form the second data set in this study *Textures-2* as presented in Figure 2(b). The original image of each class has been divided into 16 distinct sub-samples each of size 160×160 pixels.

3.2 Baseline Methods

In this study, two GP-based and four non-GP methods have been used as the baseline methods for comparison purposes.

GP Methods

– Static Range Selection (SRS) [16]: in the SRS method, the real number line is divided into a number of equally and fixed size intervals, where each interval is allocated for one classes. The SRS method uses the accuracy function to measure the fitness of an evolved program.

- Dynamic Range Selection (DRS) [7]: the DRS method is similar to SRS in terms of program representation, terminal and function sets, and fitness measure. However, the real number line is divided into intervals dynamically rather than using predefined intervals. Moreover, in the DRS method, the training set is divided into *segmentation* and *evaluation* sets. The former is used to define the corresponding interval of each class, while the latter is used to measure the fitness of the evolved program.

Non-GP Methods

- Naive Bayes (NB) [14]: NB is a simple, yet powerful, classifier that uses Bayes' theorem to build a decision model.
- Support Vector Machines (SVM) [14]: SVM is a broadly used classifier in the literature. A SVM is trained using algorithm of Platt [11] named *sequential minimal optimisation* (SOM).
- Naive Bayes / Decision Trees (NBTree) [14]: Combines Decision Trees (DT) with NB method to form a hybridised method that inherits the characteristics of the two methods. DT is used to build the tree, whilst NB is used at the leaves of the tree.
- K* (KStar) [14]: similar to 1NN, KStar predicts the class label of an instance based on the similarity to the closest instance in the training set. The KStar method uses an entropy-based distance measure to calculate the distance between two instances.
- Non-nested generalized (NNge) [14]: similar to 1NN, NNge is an instance-based classifier that operates based on similarity measure. Moreover, NNge uses a non-nested exemplar.
- Multilayer Perceptron (MLP) [14]: A artificial multilayer neural network trained using the back-propagation algorithm.

3.3   Data Sets Preparation

In both data sets, the total number of instances of each class has been divided equally between the training and test sets. Moreover, the instances of the two data sets are *standardised* to have zero mean and unit standard deviation. The standardised images then are *normalised* to have values between 0 and 255. The three GP-based methods do not require feature detection and extraction as they were designed to operate directly on raw pixel values. However, all non-GP methods require performing feature detection and extraction in a prior stage. Three different feature extraction methods have been used in this study: (1) domain-independent features (DIF) [17]; (2) Haralick texture features [4]; and (3) local binary patterns (LBP) [10]. In the first method, each instance has been divided into five regions that are the four quadrants and the center area of the image. The mean and standard deviation values of each of these five regions and the entire image are calculated to form the feature vector. Therefore, the feature vector of each image consists of twelve values. The second method is based on the use of the *grey-level co-occurrence matrix* (GLCM), which represents a

Table 1. The GP Parameters of all experiments

Parameter	Value	Parameter	Value	Parameter	Value
Crossover Rate	0.80	Generations	30	Selection Type	Tournament
Mutation Rate	0.19	Population Size	100	Tournament size	7
Elitism Rate	0.01	Tree depth	2-10	Initial Population	Ramped half-and-half

very popular method to extract texture features. In this study, the matrices are generated using the four orientations $\{0°, 45°, 90°, 135°\}$, of one pixel distance, and a full range (8-bits) of grey-levels. Hence, each matrix is of size 256×256. The third method, LBP, is a dense-based feature descriptor that has been used extensively in the literature to extract texture features. In our experiments, each instance has been transformed into a histogram of uniform $LBP_{8,1}$ codes [10].

3.4 Parameter Settings and Implementation

To draw fair conclusions, all experiments have been conducted under the same conditions. The parameter settings of GP-based methods are shown in Table 1.

The three GP-based methods (one proposed and two baseline) have nodes that vary in the types of inputs and outputs, and the number of input arguments. Therefore, *Strongly-typed Genetic Programming* (STGP) [9] is required to implement these methods. The *Evolutionary Computation Java-based* (ECJ) package [8] is used to implement STGP based methods. The *Waikato Environment for Knowledge Analysis* (WEKA) package [3] has been used to evaluate the non-GP methods on the two data sets.

3.5 Evaluation

Only two instances of each class are randomly selected to form the training set. Similar to other stochastic search methods, GP produces different results based on the seed to the random number generator. Hence, the process of evolving a program has been repeated 30 times independently and using different random seeds. The average performance of the best evolved programs on the test set at the end of the 30 runs is then reported. The non-GP methods, apart from MLP, that were used in this study are all deterministic. Therefore, each of them has been tested only one time; while the average performance for 30 runs of MLP is reported. The selected instances forming the training set have a great impact on the final result. Hence, and as only two instances are used, the same procedure of 30 independent runs has been further repeated 10 times using different instances in the training set each time. The average performance of those 10 repetitions along with the standard deviation is reported.

4 Results and Discussions

The results of the experiment are presented and discussed in this section. The two-tailed unpaired t-test is used to check whether the difference between the performance of the proposed method compared to that of the baseline methods is significant or not. The significance level of the t-test is set to 5%. The superscript "*" appears if ToFs has significantly outperformed the other method.

Table 2. Accuracies of the Textures-1 data set

	SRS		DRS		ToFs	
	$5.02 \pm 0.21^*$		$9.84 \pm 8.01^*$		94.31 ± 1.39	
Features	NB	SVM	NBTree	KStar	NNge	MLP
DIF	$26.07 \pm 3.51^*$	$36.27 \pm 4.20^*$	$31.06 \pm 5.60^*$	$27.73 \pm 2.39^*$	$40.63 \pm 5.12^*$	$34.71 \pm 3.72^*$
Haralick	$70.77 \pm 9.12^*$	$84.43 \pm 3.85^*$	$71.81 \pm 4.10^*$	$84.11 \pm 4.21^*$	$86.48 \pm 2.86^*$	$81.86 \pm 3.26^*$
LBP	$75.80 \pm 6.57^*$	$82.17 \pm 3.58^*$	$78.52 \pm 7.17^*$	$86.82 \pm 2.92^*$	$88.68 \pm 2.14^*$	$85.54 \pm 3.09^*$

Table 3. Accuracies of the Textures-2 data set

	SRS		DRS		ToFs	
	$5.80 \pm 1.34^*$		$2.28 \pm 0.21^*$		95.74 ± 1.90	
Features	NB	SVM	NBTree	KStar	NNge	MLP
DIF	$46.64 \pm 5.80^*$	$58.21 \pm 6.36^*$	$53.67 \pm 10.22^*$	$61.56 \pm 5.49^*$	$59.77 \pm 4.67^*$	$56.49 \pm 7.42^*$
Haralick	$84.22 \pm 4.31^*$	$90.78 \pm 1.73^*$	$80.00 \pm 5.19^*$	$89.61 \pm 4.01^*$	$92.97 \pm 2.79^*$	$92.58 \pm 3.97^*$
LBP	$83.68 \pm 3.76^*$	$89.53 \pm 3.16^*$	$84.46 \pm 6.89^*$	$90.86 \pm 3.24^*$	$90.71 \pm 3.49^*$	$92.19 \pm 3.36^*$

Each of the tables presented in this section is divided vertically into two blocks. The upper block presents the results of the GP-based methods, while the results of the non-GP methods are presented in the lower block. Moreover, three values are listed under each of the non-GP methods that each corresponds to one of the three features extraction methods were discussed in Section 3.3.

4.1   Overall Results

The results on the Textures-1 data set are presented in Table 2. The statistical test shows that ToFs has significantly outperformed all other methods on this data set. Both of the GP-based baseline methods show very poor performance on this data set. The use of hand crafted features with the six non-GP methods shows a good level of performance. A lower level of performance has been achieved when the DIF features are used by all those methods compared to LBP and Haralick. In most of the cases, the use of LBP features results in a slightly better performance than that of the Haralick features.

Table 3 presents the results on the Textures-2 data set. Similar to Textures-1, ToFs has significantly outperformed all other methods on this data set. SRS and DRS have achieved the lowest accuracies amongst all other methods. Similar to Textures-1, the non-GP methods show poor performances when the DIF features are used. Moreover, those methods achieved a good level of performance when LBP features or Haralick features are used.

4.2   Analysis

The results show that the simple domain-independent features are not sufficiently powerful for these data sets. Moreover, GP with SRS and DRS are not suitable for classification when the number of classes is large. These two methods simply translate the single floating number output into a set of class labels, while the proposed method evolves a program that implicitly performs feature extraction and generates a powerful feature vector.

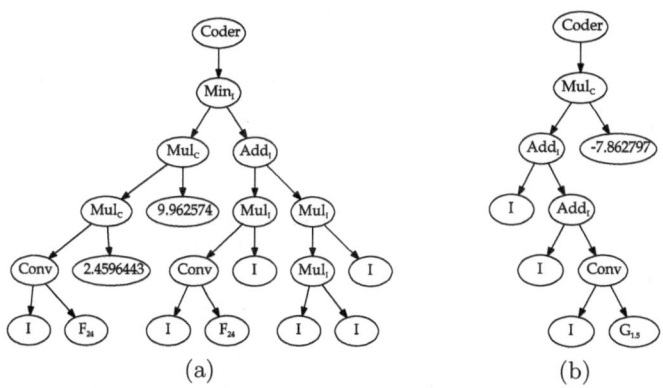

Fig. 3. Sample programs evolved on (a) Textures-1, and (b) Textures-2 data sets

Figure 3(a) shows a program that was trained on the Textures-1 data set. This program has scored 95.50% accuracy on the unseen data. Meanwhile, a program evolved by the proposed method on the Textures-2 data set is presented in Figure 3(b). This program has scored 100% accuracy on the unseen data. It convolves the image with a Gaussian filter with $\sigma = 1.5$, then adds it to the original image twice. The resulting image is then multiplied by a constant value (-7.862797) and passed over to the root node to generate the feature vector.

5   Conclusions

In this paper, a GP-based method has been proposed for the task of multiclass classification in texture images. The proposed method uses only two instances of each class to evolve a program that operates on raw pixel values. Two well-known data sets have been used to evaluate the performance of the proposed method. Moreover, the performance achieved has been compared to that of two GP-based and six non-GP methods. Similar to the proposed method, the two GP baseline methods operate on raw pixel values to perform multiclass texture classification. The non-GP methods, on the other hand, require a set of pre-extracted features to build a model. Therefore, three different feature extraction methods have been used and the performances obtained have been compared to that of the proposed method. The results of the experiments show that the proposed method significantly outperformed all other methods on both of the data sets used.

In the future, we would like to test the ability of the method to handle rotation and scale variants, and on different domains other than textures. Analysing some of the evolved programs to highlight some important patterns and to investigate the costs (i.e. speed and memory) is another objective to investigate in the near future.

References

1. Brodatz, P.: Textures: A Photographic Album for Artists and Designers. Dover Publications (1999)
2. Fix, E., Hodges, J.: Discriminatory analysis-nonparametric discrimination: Consistency properties. Technical Report 4, USAF School of Aviation Medicine, Randolf Field, Texas (1951)
3. Hall, M., Frank, E., Holmes, G., Pfahringer, B., Reutemann, P., Witten, I.H.: The WEKA data mining software: An update. SIGKDD Explorations Newsletter 11(1), 10–18 (2009)
4. Haralick, R., Shanmugam, K., Dinstein, I.: Textural features for image classification. IEEE Transactions on Systems, Man and Cybernetics 3(6), 610–621 (1973)
5. Kylberg, G.: The Kylberg texture dataset v. 1.0. External report (Blue series) 35, Centre for Image Analysis, Swedish University of Agricultural Sciences and Uppsala University, Uppsala, Sweden (2011)
6. Levner, I., Bulitko, V., Li, L., Lee, G., Greiner, R.: Automated feature extraction for object recognition. In: Proceedings of the Image and Vision Computing, New Zealand, pp. 309–314. Massey University (2003)
7. Loveard, T., Ciesielski, V.: Representing classification problems in genetic programming. In: Proceedings of the IEEE Congress on Evolutionary Computation, vol. 2, pp. 1070–1077. IEEE Press (2001)
8. Luke, S.: Essentials of Metaheuristics. Lulu, 2nd edn. (2013)
9. Montana, D.J.: Strongly typed genetic programming. Evolutionary Computation 3(2), 199–230 (1995)
10. Ojala, T., Pietikäinen, M., Mäenpää, T.: Multiresolution gray-scale and rotation invariant texture classification with local binary patterns. IEEE Transactions on Pattern Analysis and Machine Intelligence 24(7), 971–987 (2002)
11. Platt, J.C.: Fast training of support vector machines using sequential minimal optimization. In: Schölkopf, B., Burges, C.J.C., Smola, A.J. (eds.) Advances in Kernel Methods, pp. 185–208. MIT Press (1999)
12. Poli, R., Langdon, W.B., McPhee, N.F.: A Field Guide to Genetic Programming. Lulu (2008) (With contributions by J. R. Koza)
13. Song, A., Loveard, T., Ciesielski, V.: Towards genetic programming for texture classification. In: Stumptner, M., Corbett, D.R., Brooks, M. (eds.) Canadian AI 2001. LNCS (LNAI), vol. 2256, pp. 461–472. Springer, Heidelberg (2001)
14. Witten, I.H., Frank, E.: Data Mining: Practical Machine Learning Tools and Techniques, 2nd edn. Morgan Kaufmann, San Francisco (2005)
15. Zhang, L., Jack, L., Nandi, A.: Extending genetic programming for multi-class classification by combining k-nearest neighbor. In: Proceedings of the IEEE International Conference on Acoustics, Speech, and Signal Processing, vol. 5, pp. v/349–v/352 (2005)
16. Zhang, M., Ciesielski, V.: Genetic programming for multiple class object detection. In: Foo, N.Y. (ed.) AI 1999. LNCS, vol. 1747, pp. 180–192. Springer, Heidelberg (1999)
17. Zhang, M., Ciesielski, V.B., Andreae, P.: A domain-independent window approach to multiclass object detection using genetic programming. EURASIP Journal on Advances in Signal Processing 2003(8), 841–859 (2003)
18. Zhang, M., Johnston, M.: A variant program structure in tree-based genetic programming for multiclass object classification. In: Cagnoni, S. (ed.) Evolutionary Image Analysis and Signal Processing. SCI, vol. 213, pp. 55–72. Springer, Heidelberg (2009)

A Stepwise Multi-centroid Classification Learning Algorithm with GPU Implementation

Cain Cresswell-Miley and Kourosh Neshatian

Department of Computer Science and Software Engineering
University of Canterbury, Christchurch, New Zealand
cjc167@uclive.ac.nz
kourosh.neshatian@canterbury.ac.nz

Abstract. By viewing the training of classifiers as an optimisation problem, we have developed a method in this paper to train a new type of nearest centroid classifier with multiple centroids per class, using Particle Swarm Optimisation (PSO). The developed method has been compared to an earlier PSO classification algorithm, and was found to have partial success. Additionally, both the developed algorithm, and the earlier PSO algorithm have been implemented on the GPU, with results showing at least one order of magnitude difference between speeds of the GPU and sequential CPU implementations on most data sets.

1 Introduction

The training of classifiers is an optimisation problem. That is, learning a classifier from a set of labelled training examples can be seen as optimising a given performance metric (objective) by modifying the parameters or structure of the classifier.

Particle Swarm Optimisation (PSO) is a biologically-inspired metaheuristic optimisation technique that is based on simulating the behaviour of swarms, such as schools of fish or bird flocks[8]. Conceptually, the algorithm iteratively moves a population of 'particles' throughout a search space, where the direction of movement for each particle is influenced by its own memory and particles in the local neighbourhood.

PSO is simple to implement, and does not require any derivative information, or any additional problem structure other than an objective function. Because of these properties, it has been applied to many different tasks[12] including the training of classifiers. Applying PSO to classification has been achieved in multiple ways, such as using PSO to train neural networks[5,3,6], learning sets of hierarchical classification rules[14], or finding centroids in the feature space[2].

Sousa *et al.*[14] introduce a rule based approach for classifying categorical data using PSO. A hierarchical set of *if-then* rules are created, where the condition of each rule is a conjunction of feature values. An example meets the condition if each relevant feature has the same feature value. The set of rules is created by iteratively discovering new rules that classify the predominant class in the current data set, with the current data set being updated in each iteration

G. Dick et al. (Eds.): SEAL 2014, LNCS 8886, pp. 347–358, 2014.

by removing examples that match the newly created rule. PSO is used as the underlying algorithm to find the conjunction of feature values that most optimally matches examples of the given class. While the above paper considered problems consisting of categorical features, Falco *et al.*[2] describe a classification algorithm using PSO for continuous data sets. For a classification problem with C output classes, a classifier is represented by C centroids in the feature space, where each centroid is mapped to a single output class. An example is assigned a class by the closest centroid (in terms of euclidean distance). Given this representation, PSO is used to optimise the position of the centroids, such that the accuracy over the training set is maximised. However, by using only one centroid per class, this classifier representation is limited to representing linear boundaries between each class, which limits the types of data this classifier can accurately classify.

While PSO has shown some success in its use for classification, convergence of the algorithm can be slow. By taking advantage of the GPU, a processor designed for massively-parallel computations, PSO may be more useful. The GPU is a processor specialised for highly-parallel and compute intensive tasks, and since the introduction of general GPU computing platforms such as CUDA and OpenCL utilising the GPU to parallelise algorithms has become a popular research area. In order to achieve high parallelism, the GPU devotes more transistors to data processing rather than caching and flow control. This tradeoff limits the types of algorithms that can benefit from the GPU, and requires that algorithms be specifically designed with these tradeoffs in mind.

Two methods of implementing PSO for the GPU have been implemented in the literature. Zhou and Tan[16] implement PSO by parallelising each individual stage of the PSO algorithm. This approach scaled well to more particles and larger dimensions, although each stage of the algorithm required reading particle data from the GPUs global memory, which has high latency. Mussi *et al.*[10] implemented PSO as a single GPU function , where each thread on the GPU performed a single step of the algorithm for a single particle. However, due to synchronisation limitations of the GPU, a swarm was limited to a single thread block, and hence a single GPU multiprocessor, which limits the utilisation of the GPU if only a single swarm is being used. Additionally, because each particle is mapped to a single thread, the number of particles in a swarm is limited by the maximum number of threads in a thread block.

A single centroid approach to classification using PSO is limited to representing simple linear boundaries between each class in the feature space. By allowing more than one centroid to map to each class, more complex boundaries may be represented. Therefore, we develop a learning algorithm capable of training multi centroid classifiers, so more complex classification problems may be handled. Additionally, because the execution time of PSO can be prohibitive, we develop a GPU implementation of the algorithm, and evaluate the implementation by comparing the performance to an equivalent CPU implementation.

The organisation of the paper is as follows. The second section describes the background. The third section describes the algorithm design. The fourth section describes the experimental setup, and the parameters and data sets used in the experiments. The fifth section provides the results, and some analysis and discussion. Finally, the sixth section gives some concluding remarks about the developed algorithm.

2   Background

2.1   Particle Swarm Optimisation

Particle Swarm Optimisation is a population based optimisation technique introduced by Kennedy and Eberhart[7]. A population of particles (a swarm) is iteratively moved through a multi-dimensional search space until a sufficiently 'fit' solution is attained (or a maximum number of iterations is reached). Each particle is defined by two multi-dimensional vectors p and v, representing the position and velocity respectively. At each iteration the position and velocity of each particle is updated according to the following mathematical equations, where i is the current iteration index:[13]

$$v_{i+1} = w\mathbf{v_i} + c_1\mathbf{r_1} \otimes (\mathbf{p_i}^{\text{best}} - \mathbf{p_i}) + c_2\mathbf{r_2} \otimes (\mathbf{p_i}^{\text{nbest}} - \mathbf{p_i})$$
$$p_{i+1} = p_i + v_{i+1}$$

The velocity is updated such that the previous velocity, the best known position of the current particle so far, $\mathbf{p}^{\text{best}}$, and the best known position across the particles neighbourhood, $\mathbf{p}^{\text{nbest}}$, all influence the particle's movement. w is known as the inertia weight, which controls the influence of the previous velocity, and c_1 and c_2 control the influence of the current particle's last best position and the neighbourhoods best known position respectively. Finally, $\mathbf{r_1}$ and $\mathbf{r_2}$ are vectors of random numbers uniformly distributed in the interval $[0, 1]$, and are intended to introduce randomness into the search. The operator $\otimes$ performs component-wise multiplication of two vectors.

The neighbourhood of a particle is defined topologically—by viewing each particle as a node in a graph, a particle's neighbourhood is the set of particles that it is connected with. In the *gbest* topology, the particles form a fully connected graph, and thus every particle influences each other. Conversely, for *lbest* topologies, each particle is influenced by a strict subset of particles in the whole population. A popular *lbest* topology is the ring topology, where each particle has two distinct neighbours[9]. The way the neighbourhood is defined affects how the best position is communicated throughout the swarm. This appears to affect the convergence properties of the algorithm, with experiments showing that a ring-like topology appears to be better at exploration, whereas the star topology converges faster (although not necessarily to a globally optimal result)[15].

Algorithm 1 gives the basic outline of the PSO algorithm as described above. The fitness calculation is defined by the application utilising PSO, however the other steps are independent of the problem.

1 *Initialise* the position and velocity of all particles;
2 **while** *max iterations not reached* **do**
3   calculate fitness of each particle;
4   update each particle's personal and neighbourhood best;
5   update each particle's position and velocity;
6 **end**

Algorithm 1. Basic PSO Pseudocode

2.2 CUDA - GPU Programming

CUDA is a parallel computing platform and programming model developed by NVIDIA for general purpose computing on the GPU. The GPU is specifically designed for data parallel algorithms that can take advantage of a high degree of parallelism without a high degree of reliance on flow control or memory caching. More specifically, the GPU is well suited to programs which can be specified in terms of multiple threads executing the same code across multiple different pieces of data in parallel[11].

In the CUDA programming model, code is executed across a number of parallel threads, where each thread is distributed among a number of thread blocks. Each thread block is mapped to a single multiprocessor on the GPU, which allows threads that share a thread block to have the ability to synchronise with each other, and access shared memory, but separate thread blocks are executed independently from each other.

Within this model, coding for the GPU consists of writing kernels, which are special C functions written to be executed in parallel on the GPU. The kernel is executed across a number of thread blocks with each thread block consisting of a chosen number of threads. The kernel has access to the runtime constants *threadIdx*, *blockIdx*, *blockDim*, *gridDim*, specifying information about the current thread executing the kernel. This gives the ability to map different pieces of data to different threads, and have a more fine grained control over the execution path of each thread.

A thread block is executed by a GPU's multiprocessor by first partitioning the threads into groups of 32 parallel threads, called *warps*, and then scheduling each warp for execution. A warp executes one common instruction at a time for every thread, so if threads in a warp diverge via data-dependent conditional branching, the warp executes each branch serially, disabling threads which are not on that path, and converges back to a single path once each branch has been executed. Full efficiency in a warp is therefore achieved when all 32 threads agree on the same execution path.

Threads have access to multiple memory spaces during execution. Figure 1 illustrates the layout of the memory spaces. Thread local memory and global memory (along with the constant and texture memory spaces) are both stored on-device, whereas registers and shared memory are both stored on-chip (the multiprocessor). Shared memory and the registers therefore have much higher bandwidth and lower latency than local and global memory, but are not as large as the global or local memory spaces.

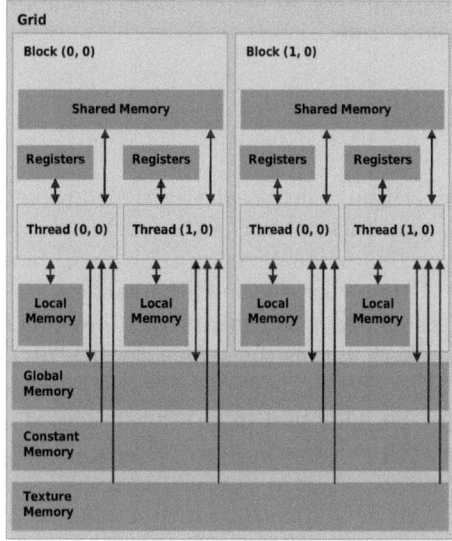

Fig. 1. CUDA Memory Hierarchy[11]

3 Design

In this section we propose a nearest centroid classification algorithm with multiple centroids per class, and introduce a stepwise learning algorithm based on PSO. This has been developed in order to evaluate the use of PSO for finding classifiers capable of classifying more complex types of data. Additionally, a GPU implementation of the algorithm is described.

3.1 Classifier Representation

A continuous, multiple centroid classifier representation is used, in order to allow the centroids to define more complex boundaries. That is, a classifier is represented by a set of centroids, K, in the feature space. Each centroid κ_i is assigned a class $c_j \in C$ by the surjective map $class$, where $|K| \geq |C|$.

$$C = \{c_1, c_2, \ldots, c_m\}$$
$$K = \{\kappa_1, \kappa_2, \ldots, \kappa_n\}$$
$$class \colon K \mapsto C$$
$$\kappa^* = \arg\min_{\kappa \in K} d(u, \kappa)$$
$$h(u) = class(\kappa^*)$$

The function $h(u)$ defines how an example is classified given the classifier represented by K and the map *class*. In particular, an example u is assigned the class of the centroid κ^*, which is the closest centroid to the given example.

By allowing multiple centroids to map to a single class, the classifier is capable of representing complex decision boundaries between individual classes of a problem.

3.2 GPU PSO

PSO is a sequential algorithm, where each step depends on the previous steps to have been completed, as in Algorithm 1. To parallelise this, the GPU is used as a coprocessor. That is, each individual step of the algorithm is implemented to be ran in parallel on the GPU, whereas the CPU manages the overall control flow of the algorithm by scheduling when to run each individual step on the GPU. The approach outlined here follows the same structure as Zhou and Tan[16].

The initialisation, position and velocity updating, and updating of each particles personal and neighbourhood best steps are parallelised in similar ways. Each particle is mapped to B thread blocks, where $B \geq 1$, such that $B \times X$ is greater than the number of dimensions of the particle, where X is the number of threads per block. In this way, an individual thread is mapped to a single dimension of a single particle, where the *block id* of the thread block identifies the particle, and the *thread id* identifies the particle dimension mapped to the current thread. This ensures the algorithm can scale to larger systems of particles without being limited to a fixed number of threads or thread blocks.

Each thread block maps to a single particle, which means the position of a single particle may be stored in *shared* memory. This means individual threads in a thread block may read the particle position data from the lower latency shared memory, rather than global memory.

Currently, the performance of single precision arithmetic on the GPU far outweighs the GPUs performance with double-precision arithmetic. Because of this, PSO has been implemented to use single precision floating point arithmetic. This trades (arithmetic) accuracy for speed. This should have limited effect on the effectiveness of the algorithm, assuming the magnitudes of values used in PSO are approximately equal.

3.3 Stepwise Centroid Algorithm

To learn a nearest centroid classifier with multiple centroids per class, we have developed an iterative algorithm which is capable of utilising multiple centroids per class in order to learn complex decision boundaries. In order to learn such a classifier, we have developed an algorithm which iteratively adds new centroids to an initial classifier, using PSO to reoptimise the centroid positions, until some stopping condition is met.

```
 1  Scale features of the training data to the range [0, 1];
 2  Partition training data into a training set (66%), and validation set;
 3  Optimise position of |C| centroids, one for each class, by using PSO;
 4  Calculate accuracy of classifier over validation set;
 5  set i to 0;
 6  while i < n do
 7      Add new centroid for a randomly chosen class;
 8      Reoptimise positions of all centroids with current centroid added;
 9      Calculate accuracy of classifier over validation set;
10      if validation accuracy increased then
11          set i to 0;
12          store currently learned classifier as the current best;
13      end
14      else
15          set i = i + 1;
16      end
17  end
18  Return the current best classifier
```

Algorithm 2. Stepwise Centroid

Step (1) scales the training data such that every example is contained within a hypercube with lower and upper bounds of 0 and 1. This is performed in order to ensure the distance measure used for finding the closest centroid is not biased by the scaling of different features. Additionally, this also bounds the search space over which PSO finds the positions of each centroid.

To begin with, step (3) initialises a nearest centroid classifier with a single centroid for each class by optimising the centroid positions with PSO. The position of a particle represents the positions of each centroid in a classifier, and the fitness of a particle is defined as the accuracy of the classifier it represents.

New centroids are iteratively added to the classifier where the new centroid is assigned a randomly chosen class. Once a new centroid is added in step (7), PSO reoptimises the position of every centroid in the classifier, where previously found centroids are initialised to their previous positions. This is done to allow the boundaries within the feature space, as defined by the centroids, to potentially adapt given the additional centroid. The while loop continues until the accuracy (fitness) over the validation set has not increased for a number of iterations n, and once complete, the classifier found throughout the process with the highest accuracy is returned.

The fitness of a particle is calculated by iterating over every training example, finding the closest centroid in the particle for each example, and then summing the total number of examples correctly classified by the closest centroid. This total is then divided by the total number of examples.

3.4   GPU Stepwise Centroid Implementation

The overall structure GPU implementation of the above algorithm is very similar. The GPU is used as a coprocessor in order to parallelise individual steps, while the CPU is still used for the scheduling and overall control flow. In particular, the GPU PSO implementation is used for optimisation, with the fitness function optimized by PSO also being implemented for the GPU, and the computation of accuracy over the validation set also utilises the GPU fitness function.

To calculate the fitness of every particle on the GPU, each particle is mapped to one or more thread blocks, where each thread in the thread block is mapped to a single example in the data set. Once the number of examples exceeds the number of threads in a block, additional thread blocks are used. This allows for the GPU implementation to scale as the number of training examples used increases.

Each thread classifies its example by iterating over each centroid of the current particle, and finds the centroid closest to the example, using Euclidean distance as the distance measure. If the class of the closest centroid matches the example, the value 1 is written to a 1d array indexed by the current thread ID, where the array is defined in shared memory, and each element is initialised to 0. Once each thread in a thread block has completed, the thread block performs a *parallel reduction* over the array to calculate the total number of correctly classified examples in the thread block. The parallel reduction algorithm finds the sum of all values in the array, by having each thread in the thread block cooperatively sum the values together. This is achieved by the first half of the threads adding the values of the second half in the array, therefore halving the number of values left to sum. This same process repeats for the remaining half of values, until the final result is stored at the first position of the array. Once the number of correctly classified examples is known, the value is divided by the total number of examples, and is atomically added to the output fitness in global memory, where the atomic add ensures serialised access to global memory across multiple thread blocks.

4   Experimental Setup

4.1   Experiment Design

We have designed an experiment to evaluate two different aspects of the proposed algorithm: accuracy and speed. The accuracy of the designed algorithm is compared to a simple nearest centroid PSO classification algorithm, where PSO is used to find a single centroid for each class. To meaningfully compare these two algorithms in terms of accuracy, a paired t-test with a p-value of 0.05 or less is ran on the results of 30 different runs of both algorithms. The GPU implementation of both classification algorithms is also compared to a sequential CPU implementation in terms of execution time over each data set. The following equation is used to calculate the speedup achieved from the GPU implementation:

$$\gamma = \frac{T_{\mathrm{CPU}}}{T_{\mathrm{GPU}}}$$

where T_{CPU} and T_{GPU} are the CPU and GPU times respectively.

4.2   Datasets

Multiple different data sets consisting of continuous features were used to evaluate the performance of the developed algorithm, where each data set has a varying number of training examples, features, and classes in order to understand the performance of both the CPU and GPU implementations with respect to different aspects of the problem. The # train and # test columns describe the distribution of examples used when evaluating the algorithms.

Table 1. Properties of each examined dataset[1]

Dataset	# features	# classes	# examples	# train	# test
Iris	4	3	150	90	60
Breast Cancer Wisconsin	9	2	683	409	274
wdbc	30	2	569	341	228
Banknote Authentication	4	3	1372	823	549
Letter Recognition	16	26	20000	12000	8000
Statlog (Shuttle)	9	7	58000	26100	17400
Magic gamma telescope	11	2	19020	11412	7608
Skin Segmentation	4	2	245057	147034	98023
SPECTF	44	2	267	80	187

4.3   Configuration Settings

The experiment has been run using a NVIDIA GTX 780 GPU, and an Intel i5 2500k CPU. The parameters of the PSO algorithm are described in Table 2, where the values for w, c_1, and c_2 have been chosen based on a paper by Eberhart et al.[4]. Additionally, the v_{max}, v_{min} and p_{max}, p_{min} values are used to confine each particles velocity and position respectively, to ensure particles stay within the search space.

Table 2. Parameters of the PSO algorithm

Setting	Value
Number of Particles	500
Number of Iterations	100
Particle Topology	Ring
w	0.7968
c_1	1.4962
c_2	1.4962
v_{max}	1.0
v_{min}	0.0
p_{max}	1.0
p_{min}	0.0

5   Results

From Table 3, we can see on most data sets, the GPU implementation is at least an order of magnitude faster than the sequential CPU implementation for both algorithms. This trend fails for two data sets, due to the number of examples in each training set being very small. The speedups achieved differ between data sets, mostly due to the size of each training set, where larger training sets correspond to larger speedups being achieved by the GPU. This is caused by how the calculation of the fitness of particles is parallelised, and computed on the GPU. As the training set grows, the number of thread blocks used for the fitness calculation increases. This means increasing the number of examples in the data set increases the parallelism exhibited by the GPU, therefore giving a larger speedup when compared to a sequential implementation.

A noticeable outlier to this pattern is the speedup achieved over the Letter Recognition data set (12000 training examples) for the stepwise centroid implementation. This could be caused by the number of classes in the problem increasing the number of centroids being added, which means the CPU is doing more work overall compared to other datasets, and additionally, having a lot of centroids increases the amount of memory reads for a particle on the GPU.

It should be noted that the speedups of the stepwise and single centroid algorithms differ on the same datasets. The reason for this is that the stepwise centroid algorithm can run a PSO optimize multiple times to learn the positions for new centroids, and the amount of times this optimize stage is called depends on the difficulty of the classification problem. If the classes in the problem are easily discriminated, the algorithm will terminate earlier, and add less centroids than for a difficult problem. This is in contrast to the single centroid algorithm, which just performs a single PSO optimize and returns the result.

Table 3. GPU and CPU times in seconds for both the stepwise centroid, and single centroid algorithms

Data set	Stepwise GPU	Stepwise CPU	Speedup	Single GPU	Single CPU	Speedup
Iris	0.8332	6.8957	8.28	0.0720	0.2137	30.86
Breast Cancer	1.5550	51.1669	32.90	0.0954	1.1767	12.33
wdbc	11.1900	137.4110	12.28	3.1090	0.2526	12.33
Banknote	1.4490	55.6737	38.42	0.0913	1.7202	18.84
Letter Recog.	804.3550	10125.2000	12.59	21.7417	696.4370	32.03
Shuttle	195.9270	5236.1800	26.73	6.9730	244.9380	35.13
Magic	75.6590	1620.7900	21.42	1.1908	36.7500	30.86
Skin Segment.	214.8290	7152.6900	33.29	4.8352	182.1390	37.67
SPECTF	15.9339	49.7974	3.13	0.1974	1.1007	5.58

Table 4 displays the achieved accuracy over the test set of both the developed stepwise classifier and the single centroid classifier approach. Average accuracy on the training set is shown in brackets, and a boldface is used to highlight the best result for each data set. The presence or absence of * indicates whether or not the result is statistically significant, according to a paired t-test over 30 runs, with a p-value of 0.05 or less.

Table 4. Average accuracies of the stepwise centroid classifier and the single centroid classifier

Data set	Stepwise Centroid Classifier	Single Centroid Classifier
Iris	0.9163 (0.9808)	**0.9294*** (1.0000)
Breast Cancer	0.9624 (0.9796)	0.9654 (0.9811)
wdbc	0.9636 (0.9846)	0.9685 (0.9841)
Banknote	**0.9950*** (0.9992)	0.9866 (0.9969)
Letter Recog.	**0.3838*** (0.3885)	0.1923 (0.1920)
Shuttle	**0.9955*** (0.9958)	0.9714 (0.9713)
Magic	**0.8313*** (0.8399)	0.7908 (0.8045)
Skin Segment.	**0.9940*** (0.9940)	0.9476 (0.9483)
SPECTF	0.7589 (0.8988)	0.7686 (0.9192)

The results indicate that the developed stepwise centroid algorithm has had some success on the chosen data sets, achieving the same, or better results on all except one dataset. The performance on the iris dataset could be explained by the use of a validation set during training for the stepwise algorithm. The number of examples used for training with iris is relatively small, only 90, and 34% of this is used for validation. This can mean that the algorithm cannot learn how to correctly discriminate between the classes, due to the training data not being representative of the problem.

6 Conclusion

This paper proposed a new learning algorithm for a nearest centroid classifier with multiple centroids, using PSO to optimise the model, and also implemented the algorithm on both the CPU and GPU in order to evaluate the effectiveness of using the GPU to speed up the designed algorithm.

The developed algorithm achieved some success in terms of classification, performing better or equal than the single centroid algorithm on all but one of the chosen data sets.

In terms of the GPU implementation, results found for most of the data sets tested, the GPU implementation was at least an order of magnitude faster than the CPU implementation for both the stepwise, and single centroid classifier learning algorithms. This indicates that a GPU implementation of PSO for classification can be useful for lowering the amount of time necessary to train a classifier.

References

1. Bache, K., Lichman, M.: UCI machine learning repository (2013)
2. De Falco, I., Della Cioppa, A., Tarantino, E.: Facing classification problems with Particle Swarm Optimization. Applied Soft Computing 7(3), 652–658 (2007)
3. Eberhart, R.C., Hu, X.: Human tremor analysis using particle swarm optimization. In: Proceedings of the 1999 Congress on Evolutionary Computation, CEC 1999, vol. 3. IEEE (1999)

4. Eberhart, R.C., Shi, Y.: Comparing inertia weights and constriction factors in particle swarm optimization. In: Proceedings of the 2000 Congress on Evolutionary Computation, vol. 1, pp. 84–88. IEEE (2000)
5. Gudise, V.G., Venayagamoorthy, G.K.: Comparison of particle swarm optimization and backpropagation as training algorithms for neural networks. In: Proceedings of the 2003 IEEE Swarm Intelligence Symposium, SIS 2003, pp. 110–117. IEEE (2003)
6. Juang, C.-F.: A hybrid of genetic algorithm and particle swarm optimization for recurrent network design. IEEE Transactions on Systems, Man, and Cybernetics, Part B: Cybernetics 34(2), 997–1006 (2004)
7. Kennedy, J., Eberhart, R.: Particle swarm optimization. In: Proceedings of the IEEE International Conference on Neural Networks, vol. 4, pp. 1942–1948 (1995)
8. Kennedy, J., Kennedy, J.F., Eberhart, R.C.: Swarm intelligence. Morgan Kaufmann (2001)
9. Kennedy, J., Mendes, R.: Population structure and particle swarm performance. In: Proceedings of the 2002 Congress on Evolutionary Computation, CEC 2002, vol. 2, pp. 1671–1676 (2002)
10. Mussi, L., Nashed, Y.S., Cagnoni, S.: GPU-based asynchronous particle swarm optimization. In: Proceedings of the 13th Annual Conference on Genetic and Evolutionary Computation - GECCO 2011, p. 1555 (2011)
11. Nvidia, C.: NVIDIA CUDA C Programming Guide. Changes, pp. 173 (2011)
12. Poli, R.: Analysis of the Publications on the Applications of Particle Swarm Optimisation. Journal of Artificial Evolution and Applications 2008(2), 1–10 (2008)
13. Poli, R., Kennedy, J., Blackwell, T.: Particle swarm optimization. Swarm Intelligence 1(1), 33–57 (2007)
14. Sousa, T., Silva, A., Neves, A.: Particle Swarm based Data Mining Algorithms for classification tasks. Parallel Computing 30(5-6), 767–783 (2004)
15. Suganthan, P.N.: Particle swarm optimiser with neighbourhood operator. In: Proceedings of the 1999 Congress on Evolutionary Computation, CEC 1999, vol. 3. IEEE (1999)
16. Zhou, Y.Z.Y., Tan, Y.T.Y.: GPU-based parallel particle swarm optimization. In: 2009 IEEE Congress on Evolutionary Computation (2009)

Dynamic Selection of Evolutionary Algorithm Operators Based on Online Learning and Fitness Landscape Metrics

Pietro A. Consoli, Leandro L. Minku, and Xin Yao

CERCIA
School of Computer Science
University of Birmingham
Birmingham, West Midlands
B15 2TT, UK
{p.a.consoli,l.l.minku,x.yao}@cs.bham.ac.uk

Abstract. Self-adaptive mechanisms for the identification of the most suitable variation operator in Evolutionary meta-heuristics rely almost exclusively on the measurement of the fitness of the offspring, which may not be sufficient to assess the optimality of an operator (e.g., in a landscape with an high degree of neutrality). This paper proposes a novel Adaptive Operator Selection mechanism which uses a set of four Fitness Landscape Analysis techniques and an online learning algorithm, Dynamic Weighted Majority, to provide more detailed informations about the search space in order to better determine the most suitable crossover operator on a set of Capacitated Arc Routing Problem (CARP) instances. Extensive comparison with a state of the art approach has proved that this technique is able to produce comparable results on the set of benchmark problems.

1 Introduction

Parameter Setting has recently become one important area of research in the Evolutionary Computation field. Since an a-priori identification of the optimal configuration of the parameters is always time-consuming and often not practicable, one must employ a dynamic selection strategy of the optimal configuration which is performed while the search is being executed. In addition, a static set of parameters is not always the optimal choice for a large number of problems where self-adapting techniques have proven to be more effective[8].

The problem of identifying the most suitable variation operator among several, also known as Adaptive Operator Selection (AOS), can be divided into two sub-tasks: the Credit Assignment (CA) mechanism, used to evaluate the performance of the operators, and the Operator Selection (OS) Rule, necessary to determine the most suitable operator using the information provided by the CA mechanism. The majority of the Credit Assignment approaches in literature are based on the evaluation of the fitness of the offspring generated by the operator, which is compared either to the current best solution [6], to the median fitness [14] or to

G. Dick et al. (Eds.): SEAL 2014, LNCS 8886, pp. 359–370, 2014.
© Springer International Publishing Switzerland 2014

the parents' fitness[2]. A different strategy evaluating both fitness and diversity of the offspring was proposed for multi-modal optimization in [18]. The reward has been mostly considered as the value assessed during the last evaluation (*Instantaneous* reward), as the average reward over a window of the last N evaluations (*Average* reward), and as the biggest improvement achieved over a window of the last N evaluations (*Extreme* reward)[9]. A different approach for population based meta-heuristics, proposed in [4], assesses the reward as the proportion of solutions generated by each operator which have been selected by the ranking phase of the evolutionary algorithm. Credit Assignment mechanism are coupled with Operator Selection rules such as Probability Matching[10], Adaptive Pursuit[24] or Multi Armed Bandit solvers (MAB)[5].

From the analysis of the existing literature, it is clear how almost all the existing CA strategies rely exclusively on the mere evaluation of the fitness of the offspring. However, the information provided by the fitness may not be sufficient to assess the optimality of an operator (e.g. in a landscape with a high degree of neutrality). The purpose of this work is therefore to develop a new dynamic CA mechanism which considers a suite of measures, and that can be adopted also as an Operator Selection Rule. We consider the Memetic Algorithm with Extended Neighborhood Search (MAENS*)[4] algorithm as a case study and for comparison purposes. More specifically, we aim to answer to the following research questions:

- **RQ1:** What kind of additional information we can provide to the Credit Assignment technique for a more "aware" calculation of the reward and does this information effectively help to improve the prediction ability of the algorithm?
- **RQ2:** What technique would be useful to handle this data and to select the most suitable operator in such a dynamic environment? Would the prediction ability of the technique be better than that of MAENS*? Would the use of this technique improve the optimization ability of MAENS*?

The contributions of this work are:

- An ensamble of four different online Fitness Landscape Analysis techniques, performed during the execution of the MAENS* algorithm in order to give a more accurate description the current population (RQ1);
- A Credit Assignment technique based on the use of a online learning algorithm to predict the reward of the most suitable operator (RQ2).

The results of the experiments carried out show that the proposed approach is able to produce results comparable to a state-of-the-art strategy and reveal how in some cases the presence of a set of measures have a beneficial effect on the optimization ability of the AOS.

The rest of the paper is divided as follows. Section 2 introduces the case scenario and the base MAENS* algorithm. Section 3 describes the ensamble of Fitness Landscape Techniques used in conjunction with the CA mechanism of the MAENS* algorithm. Section 4 describes the online Learning algorithm that has been used and adapted for the CA system. Section 5 includes a description of the proposed MAENS*-II algorithm. Section 6 describes the experiments that have been carried out to verify the assumptions of this research and their results. Finally, section 8 includes the conclusions and some future work ideas.

2   Background

2.1   Capacitated Arc Routing Problem

The Capacitated Arc Routing Problem (CARP) [11] is the problem of minimizing the total service cost of a routing plan, given a set **T** of tasks (which correspond to a subset of the arcs of a graph) and a fleet of **m** vehicles with capacity **C**. Each task **t** has a *service cost* **sc**, a *demand* **d** (the load of the vehicle necessary to service the task), a unique **id**, a reference to its head and tail vertices, and must be served once and entirely within the same route $\mathbf{R_j}$. Solutions are represented by a permutation of the tasks, divided into several routes, which must start and end in a specific vertex called *depot*. The service cost of a single route is calculated adding the service cost of all the tasks in the route plus the cost of the shortest path **sp** between each task. The problem can be formally defined as follows:

$$\min TC(S) = \sum_{i=1}^{\mathrm{length}(S)-1} (sc(t_i) + sp(t_i, t_{i+1}))$$

subject to the constraints

$$load(R_i) \leq C \ , app(t_i) = 1 \text{ and } \forall t_i \in T, m <= nveh$$

$$load(R_k) = \sum_{i=1}^{\mathrm{length}(R_k)} d(t_{ik})$$

where $\mathbf{app(t_i)}$ gives the number of appearances of tasks $\mathbf{t_i}$ in the sequence **S** and **nveh** is the number of available vehicles.

2.2   A Case Study: MAENS*

Among the several approaches for CARP involving Evolutionary Algorithms existing in literature, one of the most competitive is MAENS [23], a memetic algorithm which makes use of a crossover operator, a local search combining three local move operators and a novel long move operator called MergeSplit, and a ranking selection procedure called Stochastic Ranking (SR)[21]. The algorithm was recently refined into the MAENS*[4] algorithm. The major differences with the original algorithm are: (a) the crossover operator is replaced by a set of four operators, namely GSBX, GRX, PBX, SPBX, (b) a dynamic MAB mechanism (dMAB) [9] is adopted as an AOS rule, (c) a novel CA mechanism assigns a reward to the operators which is proportional to the number of solutions generated by each operator that "survived" the ranking phase, (d) the Stochastic

Ranking is improved considering also the diversity of the solutions (dSR) using a (e) novel diversity measure for the CARP search space.

The dMAB [9] approach, adopted in this work, combines the UCB1 algorithm [1] with the Page-Hinckley (PH) statistical test [13] to detect changes in the environment. When the PH test is triggered the MAB system is restarted and the information gathered in the previous generations is discarded. The MAENS* algorithm represents the case study of this research, as the presence of a suite of crossover operators allows the study of several AOS approaches.

3   Online Fitness Landscape Analysis

The existing Fitness Landscape Analysis (FLA) techniques have been analysed with the purpose to identify those that can be used in the CARP context. Such selection has been driven by both the necessity to reduce at most the computational effort, by exploiting some calculations that are already performed by the algorithm and the necessity to identify measures able to "capture" different features of the landscape. We identified a set of four FLA techniques, namely an evolvability measure, two neutrality measure and a fitness distribution measure, as they describe different features of the landscape and do not considerably increase the computational effort. The FLA techniques are then employed during each generation, and their results, in combination with the CA technique of MAENS, are used to create a more accurate and informative "snapshot" of the current population which eventually might lead to a more aware selection of the crossover operator. A final remark is necessary about the constraints handling and how it affects the fitness of the individuals. The landscape in which MAENS* operates is that of solutions which can potentially violate the capacity constraints of the vehicles. Therefore, we consider the following fitness function, adopted from [23]:

$$f(S) = TC(S) + \lambda * TV(S)$$

where λ is an adaptive parameter depending on the cost, on the violation and on the best feasible solution found so far, $TC(S)$ is the total cost of the solution and $TV(S)$ its total violation.

The rest of the section will introduce the four FLA techniques that have been considered in this work and how they have been integrated in the MAENS* algorithm.

Accumulated Escape Probability. The Accumulated Escape Probability[16] (aep) is a technique which aims to measure the evolvability, which can be defined as the capacity of the solutions to evolve into better solutions. We obtain the Accumulated Escape Probability (aep) by averaging the mean escape rate[19] (the proportion of solutions with equal or better fitness in the neighbourhood) of each fitness level with the formula:

$$aep = \frac{\sum_{f_i \in F} P_j}{|F|}, \text{ where } F = f_0, f_1, ..., f_L$$

where f_i is a fitness level (subset of all the solutions with fitness equal to f_i), P_j is the average Escape Rate of all samples belonging to the f_j fitness level and L is the number of possible fitness levels. Being the mean value of a set of probabilities, the aep will be equal to 0 when the instance is hard and higher (up to 1) in the opposite case. The calculation of the aep requires the analysis of the neighbourhood of each solution in order to identify how many individuals have a equal or better fitness than the original individual. We analyse therefore the evolvability of the solutions which have been selected (with probability equal to 0.2) for the local search. Since the calculation of the neighbourhood of each solution corresponds to the first step of the local search, no significant additional cost is required to compute the aep.

Dispersion Metric. The analysis of the distribution of the solutions within the landscape can be sometimes used to understand more about the difficulty that a "jump" between fitness levels requires and to gain some information on the global structure of the landscape. In this context, the Dispersion Metric (dm) [17] is a technique to obtain information about the global structure of the landscape, by measuring the dispersion of the best solutions. Ideally, if the best solutions are very close we might be in presence a single funnel structure. If, on the contrary, solutions get more distant when their fitness improves, the landscape might be more like a multi funnel structure. The technique can be described as follows:

1. A sample S of solutions is taken from the search space;
2. the best S_{best} solutions are selected from the S (using a threshold value);
3. the average pairwise distances in S ($\overline{d}(S)$) and in S_{best}($\overline{d}(S_{best})$) are calculated;
4. the dm is obtained as the difference between $\overline{d}(S_{best})$ and $\overline{d}(S)$.

The calculation of the pairwise distance between all the individuals of the sample is already performed during the dSR and therefore requires no additional cost. Thus, the dm can computed on the set of all the *popsize*offset* individuals created during each generation of MAENS*. More information about the distance measure can be found in [4]. Finally, it is possible to rely on the ranking performed by the dSR operator and choose these solutions as the subset of the best ones.

Neutrality Measures. Neutrality is the study of the width, distribution and frequency of neutral structures within the landscape (e.g. plateaus, ridges). A set of several neutrality measures was defined in [25]. Among these, we selected the following:

1. average neutrality ratio ($\overline{r}$): can be obtained averaging the neutrality ratio (e.g. the number of solutions with equal fitness) of each individual with respect to its neighbourhood;
2. average Δ−fitness of the neutral networks ($\Delta(\overline{f})$): can be defined as the average fitness gain after one mutation step of each individual belonging to a neutral network.

In the same fashion as in the case of the aep, the computation effort of this technique can be absorbed by the generation of the neighbourhood of the initial solution during the local search.

4   Online Learning

The AOS model followed in MAENS* is that of the Multi Armed Bandit scenario, where the UCB1 [1] algorithm is used to balance the exploration and the exploitation of the crossover operators and the Page-Hinckley [13] test is used to detect when a different operator has become the most suitable.

In this work, we propose the adoption of a different model. The abrupt and scarcely predictable changes of the most suitable operator which might happen during the search show many similarities to the notion of concept drift [22][20] in machine learning. Thus, in such a context, we might adopt a online learning algorithm capable of (a) predicting a reward for each operator using the online Fitness Landscape Analysis measures and (b) detecting the changes of the environment, relying only on a limited number of training instances. We employ the Dynamic Weighted Majority (DWM) [15] algorithm as our online learning algorithm, which has proved to be one of the most effective techniques in the task of tracking the concept drift. The DWM algorithm can be described as follows. A set of learners (called experts) are used to classify the incoming instances $\{\vec{x}, y\}$, where $\vec{x}$ is the vector of the n input features and y is the output feature. Each expert e_j has a own weight w_j, and operates a classification λ of the instance. The global prediction is identified as the prediction with the largest sum of weights. All the experts which have failed to classify correctly the instance have their weights reduced of a β factor. Moreover, every p instances, all the experts with a weight below a certain threshold θ, are deleted and a new expert is created if the global prediction is wrong.

DWM for the Regression Task. As the DWM algorithm was originally conceived for a classification purpose it is necessary to adapt and modify some of its mechanism for the regression task of predicting the reward of a given operator based on the FLA techniques. A pseudocode of the revised DWM algorithm for the regression task (rDWM) is available in table 1, where the grey lines indicate the novelties introduced with respect the original algorithm previously described. The modifications introduced are:

1. The global prediction σ_i is obtained calculating the weighted average of all predictions (line 10);
2. we consider correct the predictions if the difference with the output feature is less than a threshold τ (lines 5-6);
3. a new expert is created if the difference between the global prediction and the output feature is less than a t factor (lines 17-18);
4. we introduce a window containing the last n instances wTS, which is used to train the new experts upon creation (line 2).

Table 1. Dynamic Weighted Majority algorithm for the regression task

```
1   for (each instance {⃗x_i, y_i}) do
2   |     update wTS(⃗x_i);
3   |     for (each expert e^j) do
4   |     |     λ^j =classify(e^j, ⃗x_i);
5   |     |     if ( |λ_i^j − y_i| > tau ) then
6   |     |     |   w_j = β * w_j;
7   |     |     end
8   |     end
9   |     normalize weights;
10  |     σ_i = weighted average of the prediction of all the experts;
11  |     if (p mod i = 0) then
12  |     |     for (each expert e^j) do
13  |     |     |     if (w_j < θ) then
14  |     |     |     |   delete expert;
15  |     |     |     end
16  |     |     end
17  |     |     if ( |σ_i − y_i| > t ) then
18  |     |     |   create new expert and train using wTS;
19  |     |     end
20  |     end
21  |     for (each expert e^j) do
22  |     |     train(e^j, ⃗x_i);
23  |     end
24  end
```

5   MAENS*-II

The revised version of the algorithm adopting the rDWM as an AOS mechanism, named MAENS*-II, is shown in the pseudocode included in table 2, where the grey lines highlight the modifications over the MAENS* algorithm previously introduced in section 2.2. Further information about MAENS* can be found in [4]. A set of four (one for each crossover operator) rDWM instances are created upon initialization of the algorithm (line 2). During each generation, one new training example is created for each rDWM instance by using the current set of FLA metrics as input features, and the reward associated to the operator

Table 2. MAENS*-II pseudocode

```
1   initialize a population pop of popsize individuals;
2   initialize a set of four rDWM_i instances and a set of rewards rw_i (one for each crossover operator)
3   while (termination condition not met) do
4   |     choose the crossover operator op_i with largest rw_i
5   |     generate a population pop_x of popsize*offset individuals, choosing the parents from pop ∪ pop_x;
6   |     generate pop_{ls}(i) for each individual pop_x(i) with probability = 0.2;
7   |     if (pop_{ls}(i) is better than pop_x(i)) then
8   |     |     overwrite pop_x i;
9   |     end
10  |     calculate aep, r̄, Δ(f̄) and the dm measures
11  |     use d-Stochastic Ranking and overwrite pop;
12  |     use the MAENS* CA approach to calculate the output feature out_i for each op_i;
13  |     for each crossover operator op_i do
14  |     |     rw_i = rDWM_i([aep, r̄, Δ(f̄), dm_i], out_i)
15  |     end
16  end
```

Table 3. Parameters of the FLA-MAENS* algorithm

Name	Description	Value	Name	Description	Value
psize	population size	30	p	expert removal period	5
ubtrial	maximum attempts to generate a solution	50	β	decrease factor for expert weights	0.75
opsize	size of the offspring during each generation	6*psize	τ	expert weight reduction threshold	0.05
P_{l_s}	probability of performing the local search	0.2	θ	threshold for expert removal	0.05
pMS	routes selected during MergeSplit	2	t	threshold for expert creation	0.10
G_{max}	maximum generations	500			
SR_{r1}	probability of sorting solutions using diversity	0.25			
SR_{r2}	probability of sorting solutions using fitness	0.70			

Table 4. Experimental results. The first two columns show the instance name (*inst*) and the best known result (*BK*). Further columns show the average fitness of the best solution (*avg*), the standard deviation (*std*), the best solution (*best*) achieved by the four different versions of the MAENS* algorithm. Instances in boldface show results statistically significant between MAENS* and MAENS*-II with $p < 0.05$ according to the Wilcoxon rank sum test. The avg row shows the average value of each column. Bottom row shows the number of comparisons won (W), drawn (D), and lost (L) to MAENS*-II in terms of average fitness of the best solution.

inst	BK	MAENS*-rw avg	std	best	MAENS*-II avg	std	best	MAENS* avg	std	best	oracle avg	std	best
C01	1590	1668.67	13.16	1660	1670.00	17.75	1660	1671.67	19.38	1660	1665.33	13.03	1660
C05	2410	2483.33	18.36	2470	2474.00	5.83	2470	2471.00	2.00	2470	2470.00	0.00	2470
C06	855	905.17	3.98	895	901.00	4.90	895	902.00	4.58	895	896.67	3.73	895
C09	1775	1840.33	20.41	1820	1824.00	10.12	1820	1830.00	16.73	1820	1829.00	15.08	1820
C10	2190	2277.33	11.53	2270	2275.17	9.53	2270	2272.17	6.54	2270	2270.67	3.59	2270
C11	1725	1832.00	27.37	1815	1817.33	2.49	1815	1816.33	3.14	1805	1815.17	2.41	1805
C18	2315	2407.17	6.91	2390	2402.76	9.27	2385	2403.67	7.74	2385	2401.17	7.82	2390
D01	725	734.83	8.99	725	742.17	5.11	725	742.83	4.02	725	739.50	6.87	725
D07	735	836.33	3.40	835	835.00	0.00	835	835.00	0.00	835	835.00	0.00	835
D08	615	692.00	4.58	685	687.67	4.42	685	685.67	2.49	685	685.67	2.49	685
D11	920	937.67	6.42	920	936.72	3.48	930	936.50	3.91	935	934.67	4.99	920
D21	695	818.67	11.47	810	814.00	4.16	805	814.83	5.24	805	810.17	3.98	805
D23	715	772.83	12.23	745	**767.67**	7.39	755	**769.83**	12.28	740	758.17	8.51	740
E01	1855	1941.00	6.11	1935	1936.50	2.93	1935	1936.17	2.11	1935	1935.50	1.50	1935
E09	2160	2266.33	25.26	2230	2249.17	21.64	2225	2252.00	21.16	2230	2250.83	21.26	2230
E11	1810	1878.00	25.68	1850	1858.00	15.03	1840	1857.00	13.52	1835	1853.83	7.71	1845
E12	1580	1741.00	17.63	1710	1722.50	14.59	1695	1717.33	13.15	1695	1719.50	11.50	1705
E15	1555	1608.67	5.91	1595	**1604.33**	5.59	1595	**1602.50**	6.68	1590	1599.50	6.24	1590
E19	1400	1444.67	1.80	1435	**1442.00**	4.58	1435	**1442.67**	4.23	1435	1438.33	4.71	1435
E21	1700	1707.67	2.49	1705	1708.10	2.39	1705	1708.00	2.45	1705	1705.67	1.70	1705
E23	1395	1440.50	7.34	1435	1435.50	1.98	1430	1435.50	1.50	1435	1434.00	2.00	1430
F01	1065	1071.43	2.54	1065	1072.59	3.32	1065	1071.83	2.73	1065	1069.50	3.73	1065
F04	930	954.67	5.31	940	954.00	4.16	940	953.67	3.64	940	951.17	3.80	940
F09	1145	1165.54	12.34	1145	1163.45	8.48	1145	1161.00	11.79	1145	1157.00	8.12	1145
F11	1015	1026.96	13.56	1015	1027.07	12.35	1015	1030.00	11.11	1015	1021.00	6.88	1015
F12	900	940.71	32.32	910	931.83	26.94	910	925.00	23.42	910	917.33	13.09	910
F14	1025	1035.83	13.17	1025	1034.50	11.86	1025	1037.33	12.23	1025	1033.00	13.52	1025
F19	685	737.67	8.73	725	**732.50**	9.64	725	**735.17**	9.35	725	726.67	3.73	725
F24	975	997.00	9.36	980	998.83	10.38	975	999.33	8.63	980	990.50	11.28	975
e1-B	4498	4509.17	11.68	4498	4504.79	10.42	4498	4501.20	8.33	4498	4502.60	8.50	4498
e2-B	6305	6329.83	13.35	6317	6323.86	9.41	6317	6323.67	9.58	6317	6320.37	6.36	6317
e4-A	6408	6464.07	5.39	6446	6463.83	5.07	6446	6462.50	3.04	6450	6462.77	2.58	6456
e4-B	8884	9023.47	16.23	8992	9021.10	17.84	8990	9022.50	16.39	8988	9011.20	11.79	8993
s1-B	6384	6407.30	19.35	6388	**6397.59**	12.70	6388	**6399.90**	16.38	6388	6399.70	14.50	6388
s2-A	9824	9943.43	32.78	9889	9934.80	29.49	9881	9931.63	26.62	9889	9928.37	27.01	9885
s2-B	12968	13217.13	44.41	13159	**13171.41**	29.10	13123	**13179.07**	26.11	13124	13179.20	29.61	13124
s2-C	16353	16516.03	46.02	16430	16505.83	51.89	16434	16510.10	43.05	16430	16498.00	41.64	16433
s3-A	10143	10293.87	29.07	10242	10290.67	25.78	10251	10282.63	29.41	10221	10276.50	26.39	10221
s3-B	13616	13874.37	59.29	13736	13821.50	47.04	13747	13820.13	57.75	13736	13823.37	60.51	13750
s3-C	17100	17325.90	46.56	17237	17309.87	37.46	17221	17289.73	42.75	17220	17296.10	33.42	17240
s4-A	12143	12403.37	47.36	12316	12388.59	41.42	12316	12400.87	47.91	12283	12382.93	41.71	12304
s4-B	16093	16454.30	42.73	16351	16437.60	54.52	16281	16421.17	50.46	16325	16414.67	47.18	16344
avg	4266.16	4355.38	17.91	4327.16	4347.37	14.58	4323.88	4346.69	14.60	4322.95	-	-	-

		W	D	L				W	D	L			
		6	0	36				18	2	20			

as output feature (lines 10, 13-14). The set of four rDWM instances are then used predict the reward of each operator (line 4). The algorithm adopts an Instantaneous Reward mechanism to choose among the options, to limit the bias constituted by the performances of the operator in the previous generations and facilitate, in this way, the tracking of the concept drift. All the experiments were performed using the *weka* [12] implementation of REPTrees as base learners.

6   Experimental Studies

A set of experiments was designed to verify the behaviour of MAENS*-II. As a first step, an oracle was implemented with the purpose of analysing a set of CARP instances in order to obtain optimal crossover operator selection rates and to exclude those instances where all the crossover operators achieve the same results. The oracle can be briefly described as follows. Four different populations are obtained during each generation by using each crossover operator. All the individuals of the four generations are merged into a single population which is sorted using the MAENS* ranking operator. The Credit Assignment mechanism is therefore used to assess the best operator. The results achieved by the oracle show that the predictions operated by the dMAB are not optimal, as better results can be achieved. Besides, these results should be considered "optimal" only when the MAENS* reward measure is considered, while they might not be anymore when in presence of a set of multiple measures, as in the case of MAENS*-II. The experiments were performed on instances taken from the known benchmark test sets proposed in [7] and [3], named *egl* and *C,D,E,F*. The analysis of the results achieved by the oracle allowed to identify a subset of 42 instances. The set of parameters adopted in the MAENS*-II algorithm, included in table 3, was identified with a series of test-and-trial attempts and might not correspond to the most optimal one. All the values were obtained by averaging the results of 30 independent runs and all the experiments are performed on the instances selected from the two different datasets.

Effectiveness of the FLA Measures (RQ1). A first experiment was designed to understand whether the use of the online FLA techniques has a beneficial effect on both the optimization ability and the prediction capacity of the algorithm. Therefore, the performances of the MAENS*-II were compared to that of a version of the algorithm which only makes use of the original reward of MAENS* as an input feature of the learning algorithm, named MAENS*-rw. In this context, we are not interested in the results achieved by the algorithm but rather we want to verify that the results are significantly different and prove, as a consequence, a certain sensibility of the rDWM algorithm to the presence of the FLA measures. The results are included in table 4 in columns MAENS*-rw and MAENS*-II. The results have been tested for significance using the Wilcoxon signed-rank test across the problem instances, which confirmed that the two algorithms produce significantly different results (respectively $W = 26$ with $p < 0.05$ and $W = 54.5$ sample size: 42). The comparison of the average fitness shows that MAENS*-rw produced slightly better results in only 6 instances out of 42 and considerably worse ones in all the rest. This can be interpreted as a signal that the rDWM is concretely affected by the FLA measures, which influence (in a beneficial way) the decisions made by the algorithm.

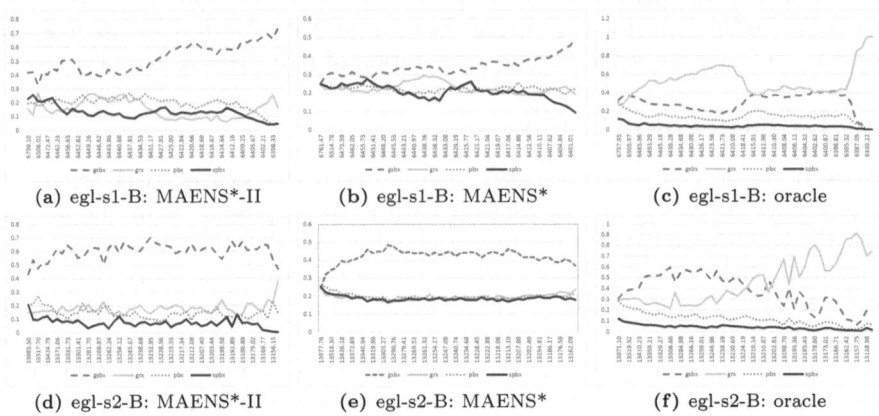

(a) egl-s1-B: MAENS*-II   (b) egl-s1-B: MAENS*   (c) egl-s1-B: oracle

(d) egl-s2-B: MAENS*-II   (e) egl-s2-B: MAENS*   (f) egl-s2-B: oracle

Fig. 1. Crossover operator selection rates on two CARP instances of MAENS* (first column), MAENS*-II (central column) and the oracle (right column)

MAENS*-II vs MAENS* (RQ2). The second research question focuses on the performance of the proposed approach with respect to the existing one. Therefore, the MAENS*-II was tested against the MAENS* algorithm and the oracle. A Wilcoxon signed-rank test performed on the dataset proved that the differences between the results achieved by the two algorithms are not statistically significant ($W = 375$ with $p > 0.05$ and sample size: 40). The results are similar also in terms of mean average fitness over all the instances, standard deviation and best solution. The online learning system is therefore able to achieve results comparable to those achieved by the bandit solver. Despite this result, it is possible to notice some significant differences between the results in some of the instances. A Wilcoxon rank-sum test was performed on each couple of results and 6 instances (highlighted in boldface in table 4) showed statistically significant results. A comparison of the selection rates of such instances is included in figure 1. The ordinates axis in the figure refers to the selection rate of each crossover operators, while the abscissas corresponds to the average fitness of the population discretised into 50 intervals. We study, therefore, how the selection rate of the four operator changes while the search is carried on and the average fitness of the population decreases. In the first instance, *egl-s1-B*, it is possible to notice three phases in the oracle prediction. A first phase where the GRX operator is preferred over the others, an intermediate phase where the GRX and GSBX operators have nearly equal selection rates and a last phase characterized by a rise of the selection rate of the GRX operator which reaches 1 in the last moments of the search. Both MAENS* and MAENS*-II award the GSBX operator with the highest selection rate for the whole search, missing the prediction of the oracle. It is possible to see, however, how MAENS*-II increases the selection rate of GSBX more rapidly than MAENS*. In the second instance, the oracle clearly identifies a change in the environment halfway through the search.

The concept drift is not detected by both MAENS* and MAENS*-II, which, however shows an higher exploitation of the GSBX operator. The performance of MAENS*-II instances suggests the hypothesis that despite the not enhanced prediction ability, the availability of more than one measures has led to better results in some instances, outperforming even the oracle, based only on the use of the Credit Assignment system of MAENS*.

7 Conclusions and Future Work

In this work we proposed the adoption of a novel Adaptive Operator Selection scheme to identify the optimal crossover operator. The AOS is tested against the Multi Armed Bandit approach employed in the MAENS* algorithm for the Capacitated Arc Routing Problem. The AOS proposed combines a set of four Fitness Landscape Analysis measures in conjunction with the existing Credit Assignment measure of MAENS* and an online learning algorithm, to predict the most suitable crossover operator. The results achieved by MAENS*-II show that this technique is able to compete with the state-of-the-art techniques and can, in some cases, exploit the multiple measures to outperform the alternative strategy. This work leaves space for interesting directions that can be explored, such as the adoption of an Average or Extreme Reward strategy, the use of different base learners or the combined use of this Credit Assignment strategy with existing Operator Selection Rules and vice versa.

Acknowledgments. This work was supported by EPSRC (Grant Nos. EP/I010297/1 and EP/J017515/1). Xin Yao was supported by a Royal Society Wolfson Research Merit Award.

References

1. Auer, P., Cesa-Bianchi, N., Fischer, P.: Finite-time analysis of the multiarmed bandit problem. Machine learning 47(2-3), 235–256 (2002)
2. Barbosa, H.J., Sá, A.: On adaptive operator probabilities in real coded genetic algorithms. In: Workshop on Advances and Trends in Artificial Intelligence for Problem Solving, SCCC 2000 (2000)
3. Beullens, P., Muyldermans, L., Cattrysse, D., Van Oudheusden, D.: A guided local search heuristic for the capacitated arc routing problem. European Journal of Operational Research 147(3), 629–643 (2003)
4. Consoli, P., Yao, X.: Diversity-driven selection of multiple crossover operators for the capacitated arc routing problem. In: Blum, C., Ochoa, G. (eds.) EvoCOP 2014. LNCS, vol. 8600, pp. 97–108. Springer, Heidelberg (2014)
5. DaCosta, L., Fialho, A., Schoenauer, M., Sebag, M.: Adaptive operator selection with dynamic multi-armed bandits. In: Proceedings of the 10th Annual Conference on Genetic and Evolutionary Computation, pp. 913–920 (2008)
6. Davis, L.: Adapting operator probabilities in genetic algorithms. In: International Conference on Genetic Algorithms 1989, pp. 61–69 (1989)

7. Eglese, R.W.: Routeing winter gritting vehicles. Discrete applied mathematics 48(3), 231–244 (1994)
8. Eiben, A.E., Hinterding, R., Michalewicz, Z.: Parameter control in evolutionary algorithms. IEEE Transactions on Evolutionary Computation 3(2), 124–141 (1999)
9. Fialho, Á., Da Costa, L., Schoenauer, M., Sebag, M.: Dynamic multi-armed bandits and extreme value-based rewards for adaptive operator selection in evolutionary algorithms. In: Stützle, T. (ed.) LION 3. LNCS, vol. 5851, pp. 176–190. Springer, Heidelberg (2009)
10. Goldberg, D.E.: Probability matching, the magnitude of reinforcement, and classifier system bidding. Machine Learning 5(4), 407–425 (1990)
11. Golden, B.L., Wong, R.T.: Capacitated arc routing problems. Networks 11(3), 305–315 (1981)
12. Hall, M., Frank, E., Holmes, G., Pfahringer, B., Reutemann, P., Witten, I.H.: The weka data mining software: an update. ACM SIGKDD Explorations Newsletter 11(1), 10–18 (2009)
13. Hinkley, D.V.: Inference about the change-point from cumulative sum tests. Biometrika 58(3), 509–523 (1971)
14. Julstrom, B.A.: What have you done for me lately? adapting operator probabilities in a steady-state genetic algorithm. In: Proceedings of the 6th International Conference on Genetic Algorithms, Pittsburgh, PA, USA, July 15-19 (1995)
15. Kolter, J.Z., Maloof, M.: Dynamic weighted majority: A new ensemble method for tracking concept drift. In: Third IEEE International Conference on Data Mining, ICDM 2003, pp. 123–130. IEEE (2003)
16. Lu, G., Li, J., Yao, X.: Fitness-probability cloud and a measure of problem hardness for evolutionary algorithms. In: Merz, P., Hao, J.-K. (eds.) EvoCOP 2011. LNCS, vol. 6622, pp. 108–117. Springer, Heidelberg (2011)
17. Lunacek, M., Whitley, D.: The dispersion metric and the cma evolution strategy. In: Proceedings of the 8th Annual Conference on Genetic and Evolutionary Computation, pp. 477–484. ACM (2006)
18. Maturana, J., Saubion, F.: A compass to guide genetic algorithms. In: Rudolph, G., Jansen, T., Lucas, S., Poloni, C., Beume, N. (eds.) PPSN 2008. LNCS, vol. 5199, pp. 256–265. Springer, Heidelberg (2008)
19. Merz, P.: Advanced fitness landscape analysis and the performance of memetic algorithms. Evolutionary Computation 12(3), 303–325 (2004)
20. Minku, L.L., White, A.P., Yao, X.: The impact of diversity on online ensemble learning in the presence of concept drift. IEEE Transactions on Knowledge and Data Engineering 22(5), 730–742 (2010)
21. Runarsson, T.P., Yao, X.: Stochastic ranking for constrained evolutionary optimization. IEEE Transactions on Evolutionary Computation 4(3), 284–294 (2000)
22. Schlimmer, J.C., Granger, R.H.: Beyond incremental processing: Tracking concept drift. In: AAAI, pp. 502–507 (1986)
23. Tang, K., Mei, Y., Yao, X.: Memetic algorithm with extended neighborhood search for capacitated arc routing problems. IEEE Transactions on Evolutionary Computation 13(5), 1151–1166 (2009)
24. Thierens, D.: An adaptive pursuit strategy for allocating operator probabilities. In: Proceedings of the 2005 Conference on Genetic and Evolutionary Computation, pp. 1539–1546. ACM (2005)
25. Vanneschi, L., Pirola, Y., Collard, P.: A quantitative study of neutrality in gp boolean landscapes. In: Proceedings of the 8th Annual Conference on Genetic and Evolutionary Computation, pp. 895–902. ACM (2006)

Learning Patterns of States in Time Series by Genetic Programming

Feng Xie, Andy Song, and Vic Ciesielski

RMIT University, Melbourne, VIC 3001, Australia
{feng.xie,andy.song,vic.ciesielski}@rmit.edu.au
http://www.rmit.edu.au/compsci

Abstract. A state in time series can be referred as a certain signal pattern occurring consistently for a long period of time. Learning such a pattern can be useful in automatic identification of the time series state for tasks like activity recognition. In this study we showcase the capability of our GP-based time series analysis method on learning different types of states from multi-channel stream input. This evolutionary learning method can handle relatively complex scenarios using only raw inputs requiring no features. The method performed very well on both artificial time series and real world human activity data. It can be competitive comparing with classical learning methods on features.

Keywords: Genetic Programming, Pattern Recognition, Time Series.

1 Introduction

Time series pattern refers to certain regularities in time series that may be of user interests. There are in general two types of time series patterns. One is event patterns which indicate the occurrence of an event, for example a heart beat on EEG readings. Another type is patterns of states which indicate the time series reading stabilizing during a relatively short period of time.

The main three differences between *state* or an *event* are that:

1. Occurrence: A *state* is usually a reflection of a stable condition, for example, a person being standing or sitting. An *event* usually happens when transiting from one condition to another, for example, a person sitting down (changing from standing to sitting).
2. Data Characteristic: A *state* may show a certain form of repetition of segments over a time period, for example, a person can be in a walking state with the repetition of leg movement. The data is often homogenous. On contrary, an *event* is heterogeneous along the time axis. Figure 1 demonstrates the transition from standing to sitting. The 3 regions divided by 2 dotted grey line show the subject be standing, sitting down and be sitting sequentially. We can see that in both two states, the readings are similar but it is not the case in transitions.

G. Dick et al. (Eds.): SEAL 2014, LNCS 8886, pp. 371–382, 2014.

3. Detection: The detection mechanism would be different as the occurrence of an event can not be decided before the completion of the event while that is not the case for detecting a state. An event usually has a certain duration while a state may last indefinitely.

The boundary between *state* or an *event* are however subtle. As shown in Fig 2, the short walking period is composed of 4 steps. Each step can be viewed as an event. The repetition of such "step" events forms the a walking state.

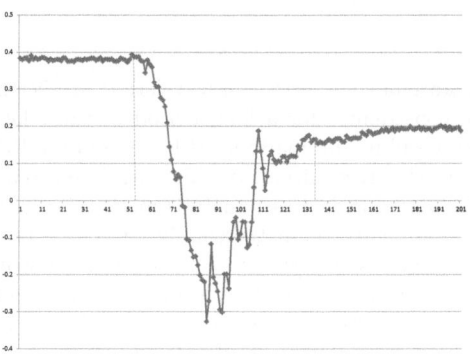

Fig. 1. A Person transiting from Standing to Sitting

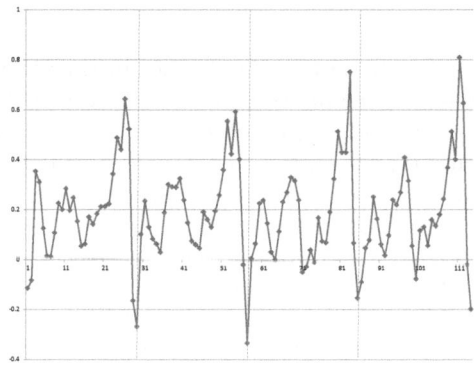

Fig. 2. Illustration of a walking state (4 steps)

A drawback of existing works on classifying time series patterns including events and states is that they often require to know the pattern size in advance [10, 8, 5, 6]. In case of state detection, the pattern size refers to the *state length*, the minimum period that a state can be existing. [1] Such information is not

---
[1] State length will be used with *pattern size* interchangeably through this paper.

always available. Moreover, a suitable set of features has to be defined for each particular task, which makes the solution usually domain-dependent. In addition, a great number of techniques cannot work well on multi-channel time series.

Given the above complexity, a method that can automatically search the state length and extract useful features is highly beneficial. In [12], Genetic Programming (GP) has been shown its effectiveness in solving event detection problem for raw, multi-channel time streams. We therefore propose to use this GP based method for state detection problems. In particular, the three research questions we are addressing in this paper are:

1. Is the GP-based event detection method applicable to state detection problems?
2. How does GP perform on a range of multivariate synthetic state detection tasks?
3. How does GP perform when applied to real scenario e.g. Activity Recognition tasks?

2   Related Work

There are mainly two categories of methods for classifying time series patterns: 1) Similarity-based techniques and 2) Features-based techniques. In the first category, the class of a time series segment can be determined by its similarity between segments from all classes. Nearest Neighbour, a typical distance-based classifier is the most popular similarity-based time series classifier [10]. Another popular choice is decision tree which uses similarity measure for the partition of trees [8]. The key factor affecting the performance of such classifiers is the effectiveness of that similarity measure [13]. The commonly used measures include Euclidean Distance [7, 3] and Dynamic Time Warping (DTW) [1]. Similarity-based methods assume that a time series pattern always appears similarly which may not be true. Feature-based methods carry out classification based on time series features. However feature extraction may be very time consuming and are often highly problem-specific [5, 6, 9, 4, 2].

Methods of both categories mentioned above have to know the pattern size beforehand and use it to define the window size for sampling segments. Our approach is different as it does not require such information. Moreover, the majority of aforementioned methods are designed for single channel time series. Patterns over several parallel time series are very difficult to be captured by those methods, because redundant or irrelevant channels have to be ruled out from decision process and the dependencies between relevant channels are sometimes complex. Our proposed method can handle multiple channel of time series.

3   A GP-Based State Detection Methodology

In this section, we present our methodology which is based on Genetic Programming. The description mainly focuses on the function set which includes the window function, temporal difference function and multi-channel function. These functions are internal nodes on a GP constructed program tree which in this case a classifier. The higher classification accuracy the better fitness the tree will receive.

3.1   Window Function

The Window Function defines the incoming sequential inputs, selects data points inside the window, and applies the operations on the select data points. It takes three parameters: i, `temporal index` and `operation`.

The first parameter (i.e. i) is the input of this function which samples a data point at every time step. It keeps the subsequence of historical values of that input in memory. The length of this subsequence is called "Window" function size (denoted as S), which is manually adjustable. The reserved data points are marked from the earliest point to the most recent one as $t_0, t_1, ..., t_{S-1}$. The value of S is set as 8 in this study. Greater values are not used so that the evolved programs can be less complex for analysis. Moreover, this value does not deteriorate the performance.

The second parameter is from terminal "`temporal index`" which returns an random integer within the range of $[1, 2^S - 1]$. First the integer is converted into its binary form. In case that the binary is shorter than S and not sufficient to mark all elements in the subsequence, it will be left padded with 0. For example, assuming S is 8 and the parameter value is 5 then the binary string should be 00000101, in which the first five 0s come from padding. This binary is then mapped to the subsequence of time series data under the window. A bit with "1" indicated the data point with the same index will be selected while a bit with "0" will be discarded.

The third parameter (i.e. operation) is a randomly generated integer valued from a range [1, 4]. Each value corresponds to one of the four operations: AVG, STD, DIF and SKEWNESS. They are used for calculating the average, the standard deviation, the sum of absolute differences and the skewness of the selected points under the window. The return value is the final output of the Window Function.

3.2   Temporal-Difference Function

Temporal-Difference Function (noted as *Temporal_Diff*) is introduced to capture temporal change between adjacent points as it is obviously important for identifying the occurrence of events.

It only takes one double value parameter i which defines the input. It stores the value t_{i-1} which is one time stamp earlier and returns the difference between t_{i-1} and the current value t_i. It consequently can be considered to have an effective window size of 2. Eventually, it calculates the first derivative of the time series, as temporal changes can be more revealing. Higher order derivatives can be achieved as well if this function is used iteratively.

3.3   Multi-channel Function

The two functions mentioned earlier only handle the temporal dependence, that is, they only work on a sequence along time axis by themselves. They can hardly be aware of any relationship cross channels at a particular time point. Consequently, a state occurring in multiple channels would not be captured by those functions. To address this problem, Multi-Channel Function is introduced. The function selects arbitrary collection of channels and computes characteristics of these channels. It takes two integers as its parameters: `channel index` and `channel operation`. No input parameter needs to be specified as the whole set of channels are treated as input. The parameter `channel index` works in a similar way as to `temporal index` in the Window Function, except its range is from 1 to $2^M - 1$, where M is the total number of channels. So assuming the channel number is 6 in total, a binary form of 13, 001101, would tell the function to operate on the 3rd, the 4th and the 6th channels. The parameter for channel operation also returns an integer from 1 to 4, which corresponds to the following functions: median value which is the middle value of the selected variables (MED), their average (AVG), their standard derivation (STD) and the distance between the maximum and minimum values (RANGE).

The Window Function can be integrated with Multi-Channel Function by taken the output of the latter as input data. Such combination enables GP to find both temporal relationships and variable dependence simultaneously.

4   Synthetic State Detection

In this session, we introduce six synthetic data sets with increasing complexity. They are used to verify the capability of the proposed method for states detection. These data sets vary in the state size and the number of channels.

4.1   Single-Channel Time Series

In single-channel time series tasks, there is only one channel involved in the time series. The time series data and the tasks are explained in following.

Box Functions. The task is to identify a state of signal at certain level. An example is shown in Figure 4.1. The starting point and the end point of a state are marked with red dots. It is the same in all other graphs in this section. This stimulates voltage or temperature maintaining at a certain level with minor fluctuations.

Oscillation. In a range of applications, constant oscillation may be viewed as a certain state, such as vibration of a spring, which may indicate the normal working condition of the spring. In this time series, a state is defined as consecutive peaks of which the top value is above 180 and the bottom value is bellow 10 (shown in Figure 4.1). Note that the state should last at least for a period of p samples ($p = 4$) given each sample taking 12 time points.

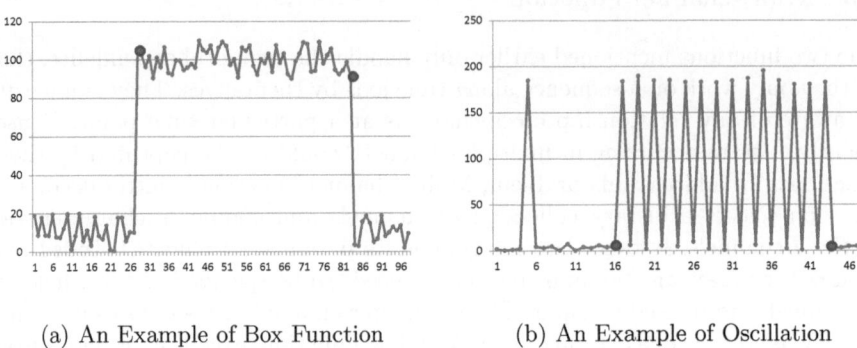

(a) An Example of Box Function   (b) An Example of Oscillation

Fig. 3. Two Non-periodical Synthetic States

Sine Wave vs. Random Numbers. A state can be not just a constant value. It can also be a regular signal. This task is to distinguish a signal being generated through a periodical function versus at random. The regulated signal is produced $y = |100 * sin(x)|$ which is sampled at every $\frac{\pi}{30}$. We define the state size as 8. An example is given in Figure 4.1.

Sine Waves. The positive state is defined the same as last problem. The negatives are however consisted of other sine waves, instead of random numbers. These variants are generated by several similar periodical function $y = |100 * sin(x * f)|$ (where $f = 2, 3, 4, 5, 6$), sampled at the same rate as target function (shown if Figure 4.1). The aim is to investigate whether our method can discriminate similar regularities. The state size is also set as 8 for this task.

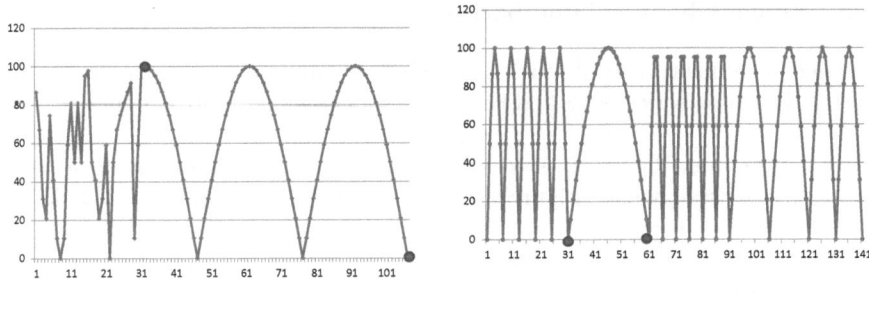

(a) Sine Wave VS. Random Numbers   (b) Sine Wave VS. Other Sine Waves

Fig. 4. Two Periodical Synthetic States

4.2 Multi-channel Time Series

In the following two tasks, there are more than one channels in the time series. The time series data and the tasks are explained in following.

Sine Wave in Two Channels. The sine wave is again $y = 100 * sin(x)$, sampled at every $\frac{\pi}{7}$. However, in this task, positives are only when time series in both channels are sine waves. If one channel is random numbers the state will be considered as negative. As shown in Figure 5, only the middle section is considered positive.

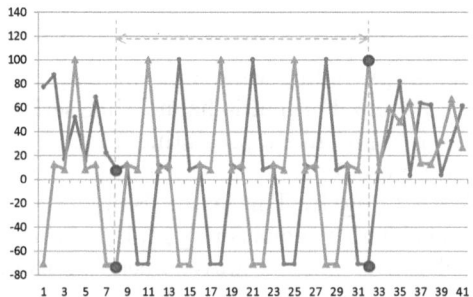

Fig. 5. An Example of Two-Channel Sine Waves

Box Functions in Four Out of Five Channels. The time series is in the positive state if there are more than four channels receiving signal value above 90 for at least 8 points. There is no constrain that at which channels the high reading may occur.

5 Experiments and Results

GP was applied on the six synthetic tasks described in Section 4. For comparison purposes five non-GP classifiers were also applied on those tasks, including *OneR*, *J48*, *Naive Bayes* and *IB1*. In addition AdaBoost was used to combine multiple classifiers as an ensemble to boost accuracy. For each task, the best conventional classifier from the four was selected as the base classifier in AdaBoost. The experimental settings for GP and Non-GP classifiers are shown in Section 5.1 and Section 5.2. The experimental results are presented in Section 5.3

5.1 The Experimental Settings for GP

The GP runtime parameter setting for synthetic data sets is: Population (300), Generation (50), Maximum Depth (8), Minimum Depth (2), Mutation Rate (5), Crossover Rate (85), Elitism Rate (10) and Window Function Size (8). In activity recognition task, a larger population size of 1000 and a greater Window Function Size of 12 are used due to the complexity of the problem. Each run is repeated 10 times and the best run is taken as GP's result.

5.2   The Experimental Settings for Non-GP Classifiers

The time series streams are converted into a list of segments as inputs for non-GP classifiers. For each tasks, the segments are extracted by a sliding window of which the size equals to the state length. This ensures that each segment contains sufficient amount of information while redundant information is eliminated. The segments containing raw data can be used as inputs to classifiers directly. We call such a segment raw input vector. Alternatively, a set of features can be extracted based on each raw input vector and called feature set. We use both types of inputs for non-GP classifiers. The processes of obtaining these inputs are demonstrated in following by an example where the time series has 2 channels and the pattern size is 3.

Raw Input Vector: A. A sliding window is moving through time series to extract raw input vectors. For multi-channel time series, all the readings are flattened into one row just like representing a matrix in a one-dimensional array as shown in Figure 6.

Feature Set B: Wave Length. This feature is uniquely designed for sine functions, which is calculated by equation $\sum_{i=1}^{3} |t_i - t_{i-1}|$. This features is not effected by the phases of the sine wave. Therefore the feature at any time point should have the identical feature values, hence a good feature for finding a state of waves.

Feature Set C: Temporal Average and Variance. The feature set provides the average and standard deviation over the length of a pattern. So the size of this feature set is the double of the number of channels.

Feature Set D: Channel Average. This feature takes averages calculations on different points at each single channel. This feature set enumerate the average value of all channels at each time point. The number of features should be equal to the pattern size.

Table 1 summarizes, for each task, the state length, the numbers of attributes in raw input vector, the type of feature set used in that task, and the numbers of attributes in the feature set.

Stream Data	Converted Vectors
$C_0(t_0)$ $C_1(t_0)$	$C_0(t_0), C_0(t_1), C_0(t_2), C_1(t_0), C_1(t_1), C_1(t_2)$
$C_0(t_1)$ $C_1(t_1)$	$C_0(t_1), C_0(t_2), C_0(t_3), C_1(t_1), C_1(t_2), C_1(t_3)$
$C_0(t_2)$ $C_1(t_2)$	$C_0(t_2), C_0(t_3), C_0(t_4), C_1(t_2), C_1(t_3), C_1(t_4)$
...	...
$C_0(t_n)$ $C_1(t_n)$	$C_0(t_{n-2}), C_0(t_{n-1}), C_0(t_n), C_1(t_{n-2}), C_1(t_{n-1}), C_1(t_n)$

Fig. 6. An Example showing converting a Two-Channel Time Series Stream To Raw Data Vectors for Conventional Classifiers (Pattern Size: 3)

Stream Data	Feature Vector B
$C_0(t_0)$ $C_1(t_0)$	$\mid(C_0(t_1) - C_0(t_0)) + (C_0(t_2) - C_0(t_1))\mid, \mid(C_1(t_1) - C_1(t_0)) + (C_1(t_2) - C_1(t_1))\mid$
$C_0(t_1)$ $C_1(t_1)$	$\mid(C_0(t_2) - C_0(t_1)) + (C_0(t_3) - C_0(t_2))\mid, \mid(C_1(t_2) - C_1(t_1)) + (C_1(t_3) - C_1(t_2))\mid$
$C_0(t_2)$ $C_1(t_2)$	$\mid(C_0(t_3) - C_0(t_2)) + (C_0(t_4) - C_0(t_3))\mid, \mid(C_1(t_3) - C_1(t_2)) + (C_1(t_4) - C_1(t_3))\mid$
...	...
$C_0(t_n)$ $C_1(t_n)$	$\mid(C_0(t_{n-1}) - C_0(t_{n-2})) + (C_0(t_n) - C_0(t_{n-1}))\mid, \mid(C_1(t_{n-1}) - C_1(t_{n-2})) + (C_1(t_n) - C_1(t_{n-1}))\mid$

Fig. 7. Illustration of Extraction Feature Type B (Pattern Size: 3)

Stream Data	Feature Vector C
$C_0(t_0)$ $C_1(t_0)$	$Average\{C_0(t_0), C_0(t_1), C_0(t_2)\}, Average\{C_1(t_0), C_1(t_1), C_1(t_2)\},$ $STD\{C_0(t_0), C_0(t_1), C_0(t_2)\}, STD\{C_1(t_0), C_1(t_1), C_1(t_2)\}$
$C_0(t_1)$ $C_1(t_1)$	$Average\{C_0(t_1), C_0(t_2), C_0(t_3)\}, Average\{C_1(t_1), C_1(t_2), C_1(t_3)\},$ $STD\{C_0(t_1), C_0(t_2), C_0(t_3)\}, STD\{C_1(t_1), C_1(t_2), C_1(t_3)\}$
$C_0(t_2)$ $C_1(t_2)$	$Average\{C_0(t_2), C_0(t_3), C_0(t_4)\}, Average\{C_1(t_2), C_1(t_3), C_1(t_4)\},$ $STD\{C_0(t_2), C_0(t_3), C_0(t_4)\}, STD\{C_1(t_2), C_1(t_3), C_1(t_4)\}$
...	...
$C_0(t_n)$ $C_1(t_n)$	$Average\{C_0(t_{n-2}), C_0(t_{n-1}), C_0(t_n)\}, Average\{C_1(t_{n-2}), C_1(t_{n-1}), C_1(t_n)\},$ $STD\{C_0(t_{n-2}), C_0(t_{n-1}), C_0(t_n)\}, STD\{C_1(t_{n-2}), C_1(t_{n-1}), C_1(t_n)\}$

Fig. 8. Illustration of Extraction Feature Type C (Pattern Size: 3)

Stream Data	Feature Vector D
$C_0(t_0)$ $C_1(t_0)$	$Average\{C_0(t_0), C_1(t_0)\}, Average\{C_0(t_1), C_1(t_1)\}, Average\{C_0(t_2), C_1(t_2)\}$
$C_0(t_1)$ $C_1(t_1)$	$Average\{C_0(t_1), C_1(t_1)\}, Average\{C_0(t_2), C_1(t_2)\}, Average\{C_0(t_3), C_1(t_3)\}$
$C_0(t_2)$ $C_1(t_2)$	$Average\{C_0(t_2), C_1(t_2)\}, Average\{C_0(t_3), C_1(t_3)\}, Average\{C_0(t_4), C_1(t_4)\}$
...	...
$C_0(t_n)$ $C_1(t_n)$	$Average\{C_0(t_{n-2}), C_1(t_{n-2})\}, Average\{C_0(t_{n-1}), C_1(t_{n-1})\}, Average\{C_0(t_n), C_1(t_n)\}$

Fig. 9. Illustration of Extraction Feature Type D (Pattern Size: 3)

5.3   Experimental Results on Synthetic Tasks

Table 2 shown the results of 6 classifiers on six state detection tasks. All the results are from test only. Considering the overall performance, GP outperformed other classifiers. In particular, in Task 3 and 5, GP significantly outperformed other counterparts. The performance gaps between GP and other classifiers are not as wide as what we found in event detection.

Table 3 presents the results of conventional classifiers on pre-defined feature sets B, C, D. The results from GP runs on **raw data** are also listed. Obviously these well designed features can help the classifiers to achieve better results. However the superior performance of GP can still be observed.

Table 1. Training and Test Data of the Six Synthetic State Detection Tasks

Tasks	Training	Test	State Size	Numbers of Attributes (No Features)	Feature Set	Numbers of Attributes (Features)
1. Box Functions	263:249	122:133	3	3	C	2
2. Oscillation	217:280	88:178	7	7	B	1
3. Sine Wave vs. Random Numbers	279:150	112:133	8	8	B	1
4.Sine Waves	219:201	69:141	8	8	B	2
5. Sine Waves in Two Channels	140:276	46:159	8	16	B	2
6. Box Functions in Four out of Five Channels	203:309	92:163	8	40	D	8

Table 2. Synthetic State Detection: Comparing GP with Non-GP Methods on Raw Input Vector%

Tasks	OneR	J48	NB	IB1	AdaBoost	GP
1	92.09 TP: 98.4 TN: 86.3	100 TP: 100 TN: 100	100 TP: 100 TN: 100	100 TP: 100 TN: 100	100 TP: 100 TN: 100	**100** TP : **100** TN : **100**
2	59.84 TP: 63.7 TN: 56.8	98.46 TP: 99.1 TN: 97.9	90.73 TP: 100 TN: 83.6	99.23 TP: 100 TN: 98.6	99.23 TP: 100 TN: 98.6	**100** TP : **100** TN : **100**
3	56.3 TP: 92.9 TN: 23.8	91.18 TP: 100 TN: 83.3	61.76 TP: 74.1 TN: 50.8	92.86 TP: 100 TN: 86.5	91.18 TP: 100 TN: 83.3	**99.58** TP : **99.11** TN : **100**
4	66.5 TP: 65.2 TN: 67.2	96.55 TP: 100 TN: 94.8	66 TP: 82.6 TN: 57.5	98.52 TP: 100 TN: 97.8	97.04 TP: 100 TN: 95.5	**98.52** TP: **100** TN: **97.76**
5	65.15 TP : 100 TN : 54.6	87.88 TP : 56.5 TN : 97.4	82.83 TP : 100 TN : 77.6	74.75 TP : 100 TN : 97.4	85.86 TP : 43.5 TN : 98.7	**100** TP: **100** TN: **100**
6	74.6 TP: 51.1 TN: 88.5	96.77 TP: 97.8 TN: 96.2	90.73 TP: 79.3 TN: 97.4	99.19 TP: 100 TN: 98.7	97.18 TP: 98.9 TN: 96.2	**100** TP : **100** TN : **100**

The results shown in Table 2 and Table 3 demonstrate that GP has the capability to extract features that can distinguish a state from the rest of time series, even when a state pattern is relying in several channels. This is because with the given functions and terminals, GP is able to combine and operate on raw numeric values. This is actually an implicit feature construction process.

5.4 Experimental Results on a Real-world Task

To further evaluate the performance of our method, we tested it on a benchmark data set [11] for mobile-based activity recognition [2], which includes 21-channel sensory data collected from 5 subjects. There are four state detection tasks: sitting, walking, running and lying flat. Note that the walking state includes different gaits, including going upstairs and going downstairs. A leave-one-person-out validation is conducted in this study. That is, for each detection task, the

---

[2] Data can be download at
http://yallara.cs.rmit.edu.au/~s3268719/AR/data.html

Table 3. Synthetic State Detection: Comparing GP with Non-GP Methods on Feature Sets %

Tasks	OneR	J48	NB	IB1	AdaBoost	GP
1	100 TP: 100 TN:100	100 TP: 100 TN:100	100 TP: 100 TN:100	100 TP: 100 TN:100	100 TP: 100 TN:100	100 TP : 100 TN : 100
2	100 TP: 100 TN:100	100 TP: 100 TN:100	100 TP: 100 TN:100	100 TP: 100 TN:100	100 TP: 100 TN:100	100 TP : 100 TN : 100
3	100 TP: 100 TN: 100	100 TP: 100 TN: 100	99.58 TP: 74.1 TN: 99.2	100 TP: 100 TN: 100	100 TP: 100 TN: 100	A:99.58 TP: 99.11 TN: 100
4	93.6 TP: 100 TN: 90.3	98.52 TP: 100 TN: 97.8	91.13 TP: 100 TN: 86.6	98.52 TP: 100 TN: 97.8	98.52 TP: 100 TN: 97.8	**98.52** TP: **100** TN: **97.76**
5	77.78 TP: 100 TN: 71.1	98.99 TP: 100 TN: 98.7	100 TP: 100 TN: 100	99.49 TP: 100 TN: 99.3	100 TP: 100 TN: 100	100 TP: 100 TN: 100
6	87.1 TP: 93.5 TN: 83.3	98.79 TP: 97.8 TN: 99.4	100 TP: 100 TN: 100	100 TP: 100 TN: 100	100 TP: 100 TN: 100	100 TP : 100 TN : 100

Table 4. Leave-one-person-out: Accuracies, true Positive and true Negative rates (trained and tested on data from the right front pant pocket)(%)

	Sitting	Walking	Running	Lying
Subject 1	91.5 TP: 100 TN: 90.1	93.7 TP: 97.5 TN: 88.5	99.7 TP: 96.9 TN: 99.9	99.6 TP: 99.2 TN: 99.6
Subject 2	97.1 TP: 79.8 TN: 99.3	94.1 TP: 96.9 TN: 87.5	99.5 TP: 94.2 TN: 99.7	97.7 TP: 98.2 TN: 97.7
Subject 3	88.7 TP: 93.4 TN: 88.1	91.2 TP: 97.8 TN: 83.6	86.0 TP: 90.3 TN: 83.3	99.6 TP: 96.4 TN: 99.9
Subject 4	95.9 TP: 94.3 TN: 96.3	93.4 TP: 95.4 TN: 91.6	96.4 TP: 94.0 TN: 96.6	98.1 TP: 94.7 TN: 98.6
Subject 5	98.5 TP: 96.3 TN: 98.9	91.0 TP: 91.2 TN: 91.0	96.0 TP: 97.5 TN: 95.7	99.7 TP: 99.1 TN: 99.8

classification is conducted for five times. For each time, the records from one subjects are used for testing and the rest for training.

Table 4 shows the results from all four tasks on 5 subjects. Our method did achieve consistently good accuracy over different scenarios of state detection. These results show that GP can detect states not only from synthetic time series but also in a complex, real-world scenario.

6   Conclusions

State and event are two main types of time series patterns. In this study, we proposed a GP-based method for state detection from multi-channel time series. This method requires no manual feature extraction. Our experiments show GP-based method can achieve significantly better results on raw inputs and competitive results when non-GP methods are provided with pre-defined features. The good performance of the proposed method is consistent on a set of synthetic problems as well as on real-world activity recognition problems. We conclude that GP based time series classification method is suitable for state detection.

References

1. Bernad, D.J.: Finding patterns in time series: a dynamic programming approach. In: Advances in Knowledge Discovery and Data Mining (1996)
2. Brooks, R.R., Ramanathan, P., Sayeed, A.M.: Distributed target classification and tracking in sensor networks. Proceedings of the IEEE 91(8), 1163–1171 (2003)
3. Chan, K.-P., Fu, A.W.-C.: Efficient time series matching by wavelets. In: Proceedings of the 15th International Conference on Data Engineering, pp. 126–133. IEEE (1999)
4. Englehart, K., Hudgins, B., Parker, P.A., Stevenson, M.: Classification of the myoelectric signal using time-frequency based representations. Medical Engineering & Physics 21(6), 431–438 (1999)
5. Garrett, D., Peterson, D.A., Anderson, C.W., Thaut, M.H.: Comparison of linear, nonlinear, and feature selection methods for eeg signal classification. IEEE Transactions on Neural Systems and Rehabilitation Engineering 11(2), 141–144 (2003)
6. Nanopoulos, A., Alcock, R., Manolopoulos, Y.: Feature-based classification of time-series data. International Journal of Computer Research 10(3) (2001)
7. Ralanamahatana, C., Lin, J., Gunopulos, D., Keogh, E., Vlachos, M., Das, G.: Mining time series data. Data Mining and Knowledge Discovery Handbook, 1069–1103 (2005)
8. Rasoul Safavian, S.: David Landgrebe. A survey of decision tree classifier methodology. IEEE Transactions on Systems, Man, and Cybernetics 21(3), 660–674 (1991)
9. Subasi, A.: Eeg signal classification using wavelet feature extraction and a mixture of expert model. Expert Systems with Applications 32(4), 1084–1093 (2007)
10. Way, M.J., Scargle, J.D., Ali, K.M., Srivastava, A.N.: Advances in Machine Learning and Data Mining for Astronomy. CRC Press, Boca Raton (2012)
11. Xie, F., Song, A., Ciesielski, V.: Genetic programming based activity recognition on a smartphone sensory data benchmark. In: 2014 IEEE Congress on Evolutionary Computation (CEC), pp. 2917–2924. IEEE (2014)
12. Xie, F., Song, A., Ciesielski, V.: Event detection in time series by genetic programming. In: 2012 IEEE Congress on Evolutionary Computation (CEC), pp. 1–8. IEEE (2012)
13. Xing, Z., Pei, J., Keogh, E.: A brief survey on sequence classification. ACM SIGKDD Explorations Newsletter 12(1), 40–48 (2010)

Reusing Learned Functionality
to Address Complex Boolean Functions

Isidro M. Alvarez, Will N. Browne, and Mengjie Zhang

Springer-Verlag, Computer Science Editorial,
Tiergartenstr. 17, 69121 Heidelberg, Germany
{alfred.hofmann,ursula.barth,ingrid.haas,frank.holzwarth,
anna.kramer,leonie.kunz,christine.reiss,nicole.sator,
erika.siebert-cole,peter.strasser,lncs}@springer.com
http://www.springer.com/lncs

Abstract. Although it is possible to identify building blocks of knowledge created by a learning classifier system in order to reuse them to solve larger scale problems, a scaling limit was still reached in certain domains. Furthermore, it was not possible to transfer functionality from one domain to another. Initial investigations have shown that it is possible and practical to reuse learned rule sets as functions in very simple problems in the same domain. The novel work here seeks to reuse learned knowledge and functionality to scale to complex problems in the same domain and to a related domain for the first time. The past work showed that the reuse of knowledge through the adoption of code fragments, GP-like sub-trees with a depth of at most two, into the XCS learning classifier system framework could provide dividends in scaling; the technique made it possible to solve until then intractable problems like the 135 bit multiplexer. The main contribution of this investigation is that a growing set of learned functions reused in the inner nodes of a code fragment tree can be beneficial. This is anticipated to lead to a reduced search space and increased performance both in terms of instances needed to solve a problem and classification accuracy. We show that through the reuse of learned functionality at the root and leaf nodes of code fragment trees, it is possible to solve complex problems such as the 18 bit hidden multiplexer problem.

Keywords: Learning Classifier Systems, Learning, Hidden Multiplexer, XCS, Code Fragments.

1 Introduction

Learning can be defined as, "the improvement of performance in some environment through the acquisition of knowledge resulting from experience in that environment" [4]. The proposed work aims to show that by continually learning useful information in a domain, problems could be solved by reusing the previously learned knowledge and relevant functionality.

G. Dick et al. (Eds.): SEAL 2014, LNCS 8886, pp. 383–394, 2014.
© Springer International Publishing Switzerland 2014

The hypothesis is that there exists a method for training a learning system where useful patterns in a simple domain can be transferred to more complex domains in order to benefit learning [9]. By presenting a learning agent with problems in a domain the system should be able to learn smaller building blocks which will help it to solve more complex problems in the same domain or in a related domain. This is similar to the threshold concept in which human learning progresses much more quickly after having learned certain crucial concepts [5].

Since their introduction by Holland 35 years ago, learning classifier systems (LCS) have provided a beneficial platform on which to perform research about learning [16]. Holland's original design was the Michigan style LCS. This type of agent evolves a population of cooperating classifiers that together constitute the solution to the problem. The reason that the Michigan style LCS is appropriate for this study is because the technique gathers sub-rules and combines them to produce groups of rules that together represent the solution.

With the development of Wilson's XCS, new improvements to the Michigan technique made it the most popular learning classifier system to date. Some of these improvements included a simpler architecture as well as new developments in reinforcement learning [3]. In XCS the resulting rule sets in the solution to the problem tend to be accurate and maximally general. Hence they encode much of the environment information without having to list each specific rule.

The XCSCFC learning classifier system, which is an extension of XCS, is an ideal tool for conducting this study because it is capable of learning building blocks of information in the form of Code Fragments (CF). Code fragments are GP-like sub-trees with a depth of at most two. The leaf nodes of the sub-trees contain either features from the message string or previously discovered code fragments. The condition can contain as many code fragments as there are condition bits, however, it is also possible to have less and still solve the problem. The first investigation involving code fragments was the introduction of CFs to represent condition bits in a classifier rule; this was named *code-fragment conditions* CFC [10]. In this approach the condition bit in a classifier was directly replaced with a code fragment. Initially, there was a separate population of code fragments used. Currently the code fragments are housed simply within the rules [11], [It is now the aim to once again make use of a separate code fragment population]. This means that the number of code fragments to be reused from a particular level was governed by the unique code fragments in good classifiers.

According to [10], this investigation showed that the multiple genotypes to a single phenotype issue in feature-rich encoding disabled the subsumption deletion function. Further, the additional methods and increased search space also led to much longer training times. However, this was compensated by the code fragments containing useful knowledge; such as the importance of the address bits in the multiplexer problems. Code fragments have enabled XCSCFC to solve previously intractable problems, e.g. the 135 bit multiplexer problem, however, it does not reuse information at the root nodes and this can lead to a limit in scalability in certain domains.

The proposed system $XCSCF^2$ is capable of reusing the previously learned code fragments as well as any function rule-sets learned, see figure 1 for more details. A rule-set can act as a function as it maps inputs (encoded as conditions) to outputs (actions dependent on the input). For example, the system could be trained sequentially with the boolean operators NAND, OR, AND, XOR, and NOR. By the time this training is complete the system will have a cache of learned function rule-sets as well as the code fragments that will have been learned. During any subsequent runs the system can make use of these learned function rule-sets along with their associated code fragments. To the system, these two components constitute one object in the form of a function. During subsequent runs the code fragments get reused at the leaf (terminal) nodes, while the functions get reused at the root (inner) nodes. It must be clarified that any code fragments branching from any of the root nodes would have been created from the group of code fragments associated with the function. This provides a tight linkage between the available rule-sets and their corresponding code fragments. This is anticipated to produce an increase in scalability and should enable the solution of problems in related domains [3].

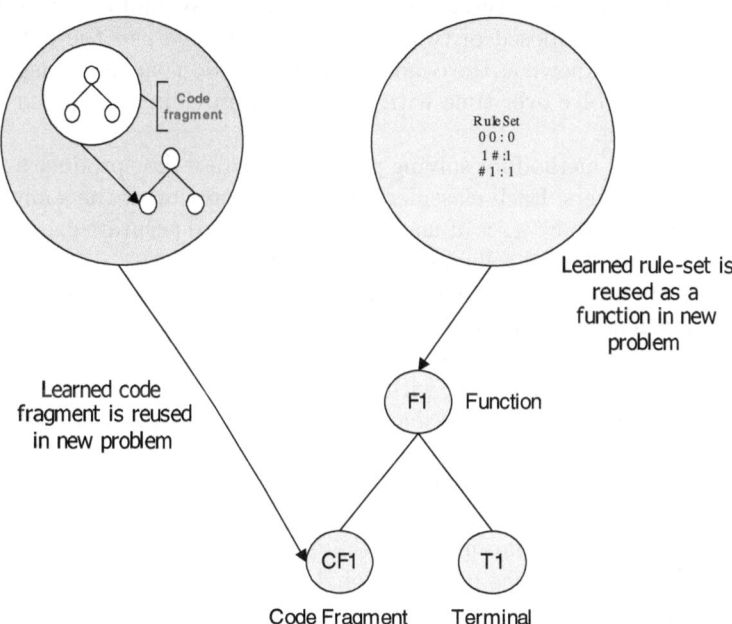

Fig. 1. Code fragment and rule set reuse - adapted from [3]

The technique will be tested against problems from the boolean domains such as the boolean operators: NAND, AND, OR, XOR, NOR, even parity, odd parity, multiplexer, and hierarchical multiplexer (hidden multiplexer). These domains were chosen because they are complex domains having a large number of relevant features, and epistatic properties [13],[15].

The long term goals are as follows:

* **Determine** the feasibility of reusing rule-sets as functions.
* **Determine** if learned knowledge can be transferred to a related domain.

The specific research objectives here are as follows:

* **Determine** if learned rule sets can provide scalability.
* **Compare** performance overhead or degradation on benchmark problems with equivalent techniques.

The benefit of the research presented here is that it explores alternative training paths in order to discover potential threshold functions from one domain, which may be used to solve problems in a different but related domain.

2   Background

A learning classifier system (LCS) is an evolutionary system first proposed by Holland in 1976 [12]. It was originally composed of four main modules: a set of resource reservoirs, a detector array, a classifier array, and an effector array. The classifiers are composed of two parts: the input side and the output side. These are currently known as the condition and the action part of a classifier. The classifiers would evolve over time with the goal of providing a suitable solution to the problem [12].

LCSs are a good method for solving problems because they produce a solution composed of classifiers. Each classifier represents a portion of the solution. The final classifiers tend to be a set of maximally general and accurate classifiers that are easily interpretable by a human.

XCS presents numerous benefits for the research community such as: providing maximally general, accurate rules, and a human readable output. The fitness of the individuals is gauged on how accurately they can predict their reward and not the reward itself. This, along with niche mutation ensures that the solution will be as general as possible while maintaining its accuracy. XCS also introduced the notion of numerosity and subsumption deletion. Numerosity measures the number of copies of a particular classifier in the population. when a new classifier is created, if it already exists in the population, it is not added to the population but the numerosity parameter of the existing classifier or macroclassifier is increased by one. This has computational benefits because less classifiers are processed [14]. It also dampens performance in terms of deletion effects. Subsumption deletion takes place during the explore phase and consists of deleting overly specific classifiers by more general ones. This has the benefit of reducing the number of classifiers in the final population [10].

In spite of the aforementioned benefits, XCS contains certain limitations in terms of scalability. One interesting characteristic of the XCS technique is the cover and delete cycle. During this phase it is possible to register a recurring cycle of covering and deletion during the match set creation. This is due to the fact that the system lacks the resources or individuals to successfully cover all of the actions without losing valuable individuals during the delete phase. This tends to occur when the maximum population size has been set too low [15].

The hierarchical multiplexer (hidden multiplexer) is a problem in which XCS exhibits the cover and delete cycle. The hidden multiplexer requires effective building block processing because they are designed in a two level hierarchy. The lower-level is evaluated by one set of boolean functions and the output of the lower-level is then fed as input to the higher level. In this novel work the 3 bit parity problems form the lower-level while the 6 bit multiplexer forms the higher level of the problem giving 18 bits in total. According to Butz, the XCS system was unable to solve the 18 bit hidden multiplexer problem with uniform crossover. The reason is that XCS is unable to process the lower level building blocks but merely disrupts them [15].

Another type of limitation in XCS is a direct result of the ternary alphabet used for representing the condition in the classifiers. The tight location correspondence between condition bits and the message string render XCS unable to capture repeated abstracted patterns in problems. As a consequence it was unable to solve certain problems like the 135 bit multiplexer without graduated rewards.

XCSCFC used CFs instead of ternary bits in the condition of classifiers, which created a store of knowledge reusable in the future problems of the same type. However, the inner nodes of the CF tree used predefined functions, which required human knowledge, this did not address repeated patterns in the data and still encountered a scaling limit.

XCSCFC has made it possible to solve previously intractable problems such as the 135 bit multiplexer. Its other siblings have also facilitated interesting solutions. For example, XCSCFA replaced the action part of the classifier with a code fragment while keeping the ternary alphabet in the condition part [10]. This tends to produce sets of rules where the optimal and sub-optimal classifiers are grouped separately, making it easier to extract the optimal classifiers. Another technique is XCSSMA, in which the static binary action is replaced by a Moore state machine while retaining the ternary alphabet in the condition of a classifier rule. This helped to discover repeated abstracted patterns in the data. The XCSSMA technique produced compact, easily understandable solutions for any n-bits even-parity problem. It also produced compact, easily understandable, and general solutions for any n+n carry problem [13].

In spite of having benefits, XCSCFC contains certain limitations as well. The fact that learned code fragments can only be reused at the leaf nodes means that at some point the system will be incapable of solving a large problem in a reasonable amount of time. More importantly, the functions are hard coded and once the run ends the rule-sets learned are forgotten prior to the next problem.

The XCS with code fragment functions, $XCSCF^2$, is based on the XCSCFC and aims to overcome the limitation mentioned above. Unlike XCSCFC where the inner nodes make use of predefined functions, which are chosen at random when the code fragment is created, $XCSCF^2$ opts to learn new functions online and catalogs them along with their rule-sets and their associated code fragments.

This newly learned functionality is then available when the next problem is processed. It is expected that this methodology will provide dividends in increased scalability as well as facilitating the transfer of building blocks of knowledge to related domains.

3   The Method

The domain used in the final experiments will be the hidden multiplexer problem. As solutions can be comprised of boolean functions it is at least feasible to learn, albeit a complex task [8]. They are also difficult to solve due to their two-stage nature, non-redundant features and high epistasis; and have commonly been used in this type of experimentation.

The first step is to determine the initial functions that the system should be given, e.g. the atomic knowledge from which to build. The more of these that are included, the more domain bias is also included without the ability to learn the linkage between functions and discovered building blocks. For example, for boolean domains, NAND gates are building blocks with which it is possible to build other gates such as the OR, AND or XOR [1], [2], [3].

The system was initially trained with the NAND function, in order to produce the relevant rule set and the associated code fragments. Once this phase was successful, the system had accumulated a population of rules that would replicate the NAND function. The next step was to have the system learn the OR function by using the rules belonging to the NAND function. At the end of this new run, there existed the additional OR function as a rules set and also a set of code fragments. The process then proceeded to learn the AND, XOR and NOR functions in a similar manner. This work builds on the system after it has learned the base boolean functions with associated code fragments (relevant building blocks of knowledge for these functions). Thus the system addresses the hidden multiplexer problem with a readily available cache of functions along with their rules sets and Code Fragments [3]. The next stage consists of three branches, see figure 2 for more details. The parity branch continues to learn the 2-bit even parity, 2-bit odd parity, 3-bit even parity, and the 3-bit odd parity problems, in that order. Once this phase was completed the system attempts to learn the 18 bit hidden multiplexer. The second branch also consists of learning the boolean operators, the parity problems but then included the 6 bit multiplexer problem. At that point the system was given the 18 bit hidden multiplexer problem. The third, and last, branch consists of learning the boolean operators as mentioned above and then first learning the 6 bit multiplexer problem prior to the 18 bit hidden multiplexer problem.

Moreover all of the systems utilized two-point crossover for the rule discovery since according to Butz uniform crossover tends to have deleterious effects on the population and XCS is unable to solve the 18 bit hidden multiplexer problem. This is because XCS is not able to process the lower level building blocks, but rather disrupts them. However if two point cossover is used, XCS is able to learn the 18 bit hidden multiplexer [15].

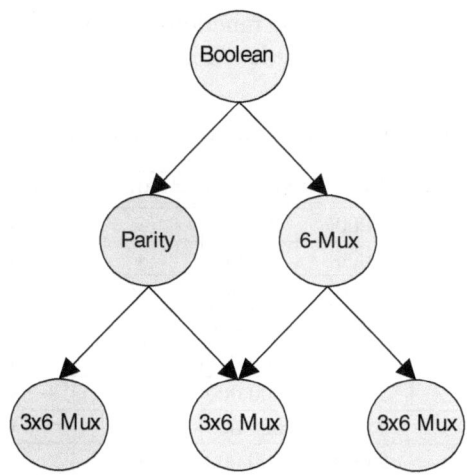

Fig. 2. Training flow of learned functions for the system.

4   Results

4.1   Experimental Setup

The experiments were run 30 times and compared with an XCSCFC and a standard XCS System. The XCSCFC system was chosen because it is the state of the art with respect to multiplexer problems. XCS is the standard benchmark algorithm in the field and therefore was an obvious choice for the role of the control in the experiments. The settings for the single step experiments were as follows: Payoff 1,000; the learning rate $\beta = 0.2$; the Probability of applying crossover to an offspring $\chi = 0.8$; the probability of using a don't care symbol when covering $P_{don'tCare} = 0.33$; the experience required for a classifier to be a subsumer $\Theta_{sub} = 20$; the initial fitness value when generating a new classifier $F_I = 0.01$; the fraction of classifiers participating in a tournament from an action set 0.4.

4.2   Boolean Problems

Boolean-Parity. The first problem presented to the system was the NAND operator. This process continued next to learning the OR, AND, XOR and NOR operators in a serial manner. By this time the system had recorded this small set of learned functions. The population sizes for the boolean and parity operators were much larger than the ones for the 6 multiplexer as they had to be increased to provide enough resources for the system to solve the problem. During this phase the system was experiencing the cover and delete cycle.

The following problems addressed by the system were the 2-bit even parity, 2-bit odd parity, 3-bit even parity, and then the 3-bit odd parity problems. Once these experiments had completed, the systems addressed the 18 bit hidden multiplexer.

The population sizes used for these functions as well as the number of training instances used for the $XCSCF^2$ are listed in table 1:

Table 1. Number of classifiers and training instances for $XCSCF^2$

Boolean	Classifiers	Instances	Branch 1	Branch 2	Branch 3
NAND	1,000	700,000	✓	✓	✓
OR	2,000	700,000	✓	✓	✓
AND	2,000	700,000	✓	✓	✓
XOR	2,000	800,000	✓	✓	✓
NOR	2,000	900,000	✓	✓	✓
2-Bit Even Parity	2,000	900,000	✓	✓	✗
2-Bit Odd Parity	2,000	1,000,000	✓	✓	✗
3-Bit Even Parity	2,000	1,100,000	✓	✓	✗
3-Bit Odd Parity	2,000	1,200,000	✓	✓	✗
6 Mux	500	500,000	✗	✓	✓

Hidden Multiplexer Problems. The population size for XCS was twice as much as for the other two systems (table 2 depicts the number of classifiers used in the populations as well as the different training sizes for each of the systems). The reason for this is because it was suggested in the equation by Butz [15], that the optimal population size should be large for XCS. Our experiments utilized a very conservative number for the population of the XCS. According to Butz, in the k-parity-$\acute{k}$-multiplexer combination, effectively the optimal population is of size:

$$||[O]|| = 2(2^{k(k'+1)}) . \tag{1}$$

$XCSCF^2$ was able to learn the 18 bit hidden multiplexer in all three branches of the experimentation, however, according to table 3, it took up to approximately 31 hrs mostly due to matching rule-sets in the learned functions to the inputs being provided by the leaf nodes in the code fragment trees. It is theorized that this was due to the fact that while learning the new function rules, the system performs sub-sampling and more importantly, the system must code all of the possible inputs for each function. Although $XCSCF^2$ was able to solve the hidden multiplexer problems utilizing less training instances than both of the other systems, this limitation must be overcome for the system to exhibit much greater scalability than is currently achievable by XCSCFC.

Figure 3 shows the results for the 18-bit hidden multiplexer experiments using just the boolean and parity learned functions. According to the graphs, the $XCSCF^2$ system was able to solve the problem with approximately 300,000 instances while the XCSCFC system achieved this with approximately 2,400,000 instances. The XCS system converged with approximately 800,000 instances. An explanation for this performance difference is the fact that the $XCSCF^2$ system

Table 2. Number of classifiers and training instances for $XCSCF^2$, XCSCFC, and XCS

System	Problem	Classifiers	Instances
$XCSCF^2$	18 Bit Hidden Multiplexer	50,000	6,000,000
XCSCFC	18 Bit Hidden Multiplexer	50,000	6,000,000
XCS	18 Bit Hidden Multiplexer	100,000	6,000,000

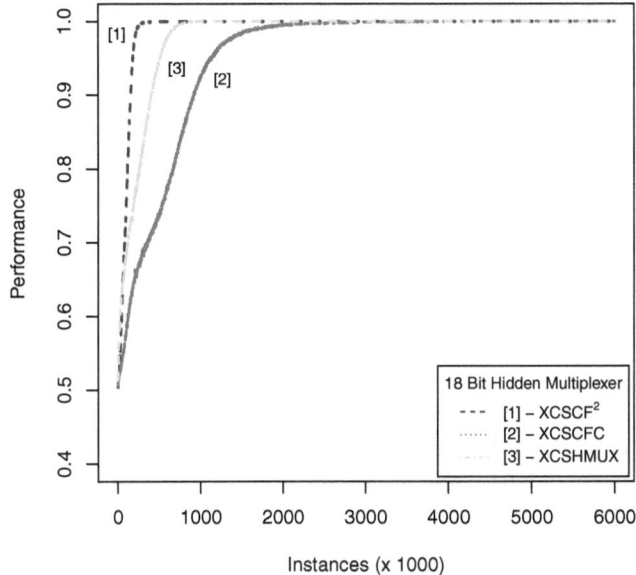

Fig. 3. 18-bit Hidden Multiplexer problem. (boolean, parity)

Table 3. Time taken for XCS, XCSCFC, and $XCSCF^2$

System	XCS	XCSCFC	$XCSCF^2$
Time Taken	8 hrs +/- 45 min	13 hrs +/- 18 min	30 hrs +/- 57 min

had a tool-set of learned functions by this time; the NAND, OR, AND, XOR, NOR, 2-bit Even Parity, 2-bit Odd Parity, 3-bit Even Parity, and the 3-bit Odd Parity functions, while the XCSCFC only had five hard coded boolean functions. The XCS system required less instances to solve the problem than the XCSCFC system and this can be attributed to the fact that the XCSCFC system did not have any previously learned code fragments at this time; it had not encountered a hidden multiplexer problem previously.

Figure 4 shows the results for the 18-bit hidden multiplexer experiments for the $XCSCF^2$ system using all three training branches. According to the graphs,

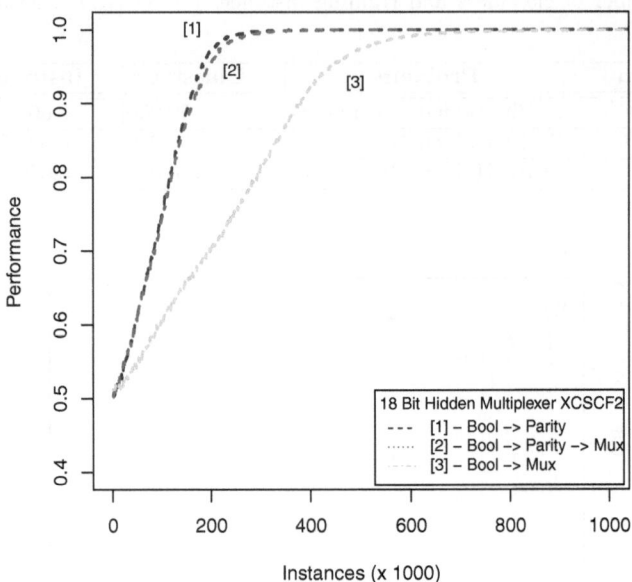

Fig. 4. 18-bit Hidden Multiplexer problem. (boolean, parity, 6mux)

there exists a marginal advantage to using the parity functions as opposed to using a combination of parity and 6 multiplexer. The former branch converged at approximately 300,00 instances and the later at approximately 320,000 training instances. This difference in performance is even more stark when compared to the branch that uses only the boolean operators and the 6 multiplexer. In this case the system converged at approximately 750,000 instances.

5   Discussion

Some of the results were as hypothesized. For example, it was likely that the $XCSCF^2$ system would do well in terms of instances needed. The differences in performance between the other two systems and $XCSCF^2$, however, were more exaggerated than anticipated; it required eight fold less iterations than XC-SCFC. The results show that XCS also outperformed XCSCFC, which was not an expected outcome. It is hypothesized that the fact that the system counted on only five hard-coded functions means that it lacked certain building blocks exhibited by $XCSCF^2$. Also, XCSCFC utilized the NOT operator as opposed to $XCSCF^2$ which was trained with the XOR operator instead. It has been shown in informal experiments that the XOR operator has certain benefits in helping XCS converge with less problem instances. The better than expected performance by $XCSCF^2$ could be attributed to the fact that by the time the system attempted the problem, it had accumulated a number of useful function rule-sets along with code fragments. It is hypothesized that upon arriving at an optimal solution for a problem using $XCSCF^2$, the newly discovered functionality could be used to seed the XCS or XCSCFC to help them arrive at a solution as well.

The results emphasize the fact that certain learned functions contain more useful building blocks than others. For instance, figure 4 clearly shows a degradation in performance when the parity functions are not utilized. Moreover, it is possible that by training the system with the 6 bit multiplexer, one is introducing possibly irrelevant building blocks for the combined problem. It is also important to take notice that the 6 bit multiplexer learned function would contain more inputs than any of the other functions being used. This could have had an impact in the number of instances needed to converge as the system would have been dealing with larger building blocks than if it had been using the other functions.

Due to the inherent nature of each of the systems, a different amount of domain knowledge is included *a priori*. The XCS system has no functions that it can use, the XCSCFC has user defined functions while $XCSCF^2$ has the ability to learn new functions based on the problems given to it by the user. The latter is a more flexible approach at the cost of the larger search space and training times.

The goal was to compare the new work with established benchmarks such as XCS and XCSCFC. Keeping this in mind, the system parameters such as payoff, learning rate and mutation rate were kept the same for all the systems. However, the setting for maxPopSize in XCS was doubled from the setting for both: XCSCFC and $XCSCF^2$. The reason behind this is because XCS would have difficulty in solving this type of problem otherwise [15].

6   Conclusions

The results show that learning the building blocks from the boolean and parity problems did translate into improved scalability for solving the 18 bit hidden multiplexer problem. There is also an indication that omitting the parity problems can lead to a greater number of training instances needed in order to solve the problem. It has been shown that learning rule-sets can provide scalability in the same and related domain for a complete boolean task.

7   Future Work

Future work will involve testing on larger problem domains and scaling within the hidden multiplexer problem as well as the sensitivity analysis with parameter settings like payoff and learning rate.

References

1. Tocci, R.J., Neal, S.: Widmer and Gregory L. Moss: Digital Systems: Principles and Applications. Prentice Hall, Upper Saddle River (2011)
2. Nisan, N., Schocken, S.: The Elements of Computing Systems: Building a Modern Computer from First Principles. MIT Press, Cambridge (2008)
3. Alvarez, I.M., Browne, W.N., Zhang, M.: Reusing Learned Functionality in XCS: Code Fragments with Constructed Functionality and Constructed Features. In: Proceedings of the 2014 Conference Companion on Genetic and Evolutionary Computation Companion, pp. 969–976. ACM, Vancouver (2014)
4. Urbanowicz, R.J., Moore, J.H.: Learning Classifier Systems: A Complete Introduction, Review, and Roadmap. Journal of Artificial Evolution and Applications (2009)
5. Falkner, N.J.G., Vivian, R.J., Falkner, K.E.: Computer Science Education: The First Threshold Concept. In: LaTiCE, pp. 39–46. IEEE Computer Society (2013)
6. Iqbal, M., Browne, W.N., Zhang, M.: Evolving Optimum Populations with XCS Classifier Systems. Soft Computing 17, 503–518 (2013)
7. Koza, J. R.: Digital Systems : Principles and Applications. Prentice Hall, Upper Saddle River (2011)
8. Koza, J.R.: Hierarchical automatic function definition in genetic programming. In: Foundations of Genetic Algorithms 2, pp. 297–318. Morgan Kaufmann (1992)
9. Thrun, S.: Is Learning The n-th Thing Any Easier Than Learning The First? In: NIPS, pp. 640–646. MIT Press (1995)
10. Iqbal, M., Browne, W.N., Zhang, M.: Reusing Building Blocks of Extracted Knowledge to Solve Complex, Large-Scale Boolean Problems. IEEE Transactions on Evolutionary Computation 99, 1–16 (2013)
11. Iqbal, M., Browne, W.N., Zhang, M.: Learning Overlapping Natured and Niche Imbalance Boolean Problems Using XCS Classifier Systems. In: Proceedings of the IEEE Congress on Evolutionary Computation, pp. 1818–1825 (2013)
12. Holland, J.: Adaptation. In: Rosen, R., Snell, F. (eds.) Progress in Theoretical Biology. Academic Press (1976)
13. Iqbal, M., Browne, W.N., Zhang, I.M.: Extending Learning Classifier System with Cyclic Graphs for Scalability on Complex, Large-Scale Boolean Problems. In: GECCO Proceedings of the 15th Annual Conference on Genetic and Evolutionary Computation, pp. 1045–1052 (2013)
14. Wilson, S.W.: Classifier Fitness Based on Accuracy. Evolutionary Computation, 149–175 (1995)
15. Butz, M.V.: Rule-Based Evolutionary Online Learning Systems. Springer, Berlin (2006)
16. Lanzi, P.L.: Learning Classifier Systems: Then and Now. Evol. Intel. 1, 63–82 (2008)

Analysis of Online Signature Based Learning Classifier Systems for Noisy Environments: A Feedback Control Theoretic Approach

Kamran Shafi and Hussein A. Abbass

School of Engineering & Information Technology
University of New South Wales,
Canberra, Australia
http://seit.unsw.adfa.edu.au/index.php

Abstract. Post training rule set pruning techniques are amongst one of the approaches to improve model comprehensibility in learning classifier systems which commonly suffer from population bloating in real-valued classification tasks. In an earlier work we introduced the term signatures for accurate and maximally general rules evolved by the learning classifier systems. A framework for online extraction of signatures using a supervised classifier system was presented that allowed identification and retrieval of signatures adaptively as soon as they are discovered. This paper focuses on the analysis of theoretical bounds for learning signatures using existing theory and the performance of the proposed algorithm in noisy environments using benchmark synthetic data sets. The empirical results with the noisy data show that the mechanisms introduced to adapt system parameters enable signature extraction algorithm to cope with significant levels of noise.

Keywords: Learning Classifier Systems, LCS, UCS, Online rule reduction, Signature based LCS, Noise, Adaptive control.

1 Introduction

Learning classifier systems (LCS) [11] are leading rule-based evolutionary machine learning techniques. Accuracy based Michigan style LCS in general evolve a single population of rules using a niche GA. The fitness of individual rules is based on their accuracy which in supervised classification context is a simple measure of the number of correctly classified instances covered by a particular rule. Population bloating leading to inferior system performance, in terms of model size and accuracy, in real-valued classification tasks is a common problem in these systems. One of the main approaches explored to mitigate this issue in LCS are the rule set reduction techniques which aim to prune post training rule sets in order to find the smallest subset of rules without compromising system performance. A number of rule set reduction techniques for LCS, specifically Michigan style LCS, have been introduced by researchers over the years. The focus of most of these algorithms is to find a subset of the post training rule

G. Dick et al. (Eds.): SEAL 2014, LNCS 8886, pp. 395–406, 2014.

population that performs equivalent to the actual population on the training set [12][6][5][13][10].

Many recent data mining applications [7], however, require dealing with data as it arrives such as network intrusion and fraud detection, web and network management, etc. These data streams are often time varying in nature and a data instance once processed is considered virtually lost because the unprecedented amount of data makes it impractical to either fully store or visit it multiple times. Due to the reliance on post training processing, such techniques are thus not well suited for online rule reduction which may be necessary for such applications. Earlier, in [9][8], the authors introduced a framework to extract accurate and maximally general rules, referred to as *signatures*, from Michigan style LCS. The framework allows extracting the *best* rules learnt by the underlying LCS dynamically, that is during the online learning as soon as they are discovered by LCS, and adaptively, that is the reduced model is updated as better signatures are learnt and existing signatures deteriorate. Experiments with the binary multiplexer problem showed that the algorithm successfully retrieved all maximally general classifiers in real time and provided a mechanism for early stopping of the learning process. Further, it was found that for continuous-valued feature spaces, realizing the precise decision boundaries is difficult. A modified subsumption operator was presented to achieve minimality and reduce partial overlapping in the signatures set. Experiments with a 2-dimensional real-valued *checkerboard* problem showed that the algorithm was able to retrieve near optimal decision boundaries with the help of a modified subsumption operator. The signature extraction system was implemented on top of UCS – a Michigan style LCS designed specifically for supervised classification tasks – and it reduced the processing time of UCS by more than 50%.

This paper focuses on the analysis of theoretical bounds for learning signatures using existing theory and the performance of the proposed algorithm in noisy environments using benchmark binary class checkerboard problem. The empirical results with the noisy data show that the mechanisms introduced to adapt system parameters enable signature extraction algorithm to cope with significant levels of noise.

The rest of the paper is organized as follows: A brief overview of the online rule reduction or signature extraction system is provided in Section 2. The analysis of theoretical bounds for learning signatures is provided in Section 3. In Section 4 we analyze the performance of UCS and the signature extraction algorithm in noisy environments and the paper is concluded in Section 5.

2   Online Signature Extraction System

The online signature extraction algorithm is implemented on top of an LCS. Its aim is to automatically detect the presence of optimal classifiers as they are discovered by an LCS and terminate the search process as soon as a complete maximally general solution is found. Figure 1 shows a block diagram of the proposed online signature extraction system using UCS [2]. For ease this system is referred to as UCSSE denoting a UCS with online Signature Extraction. A summary of UCSSE operation is provided below.

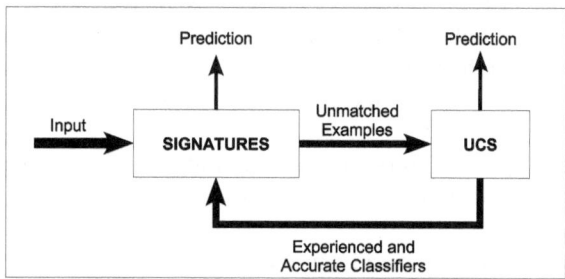

Fig. 1. UCS with Real time *Signature* Extraction System (UCSSE)

In Figure 1, *Signatures* (or a signature set referred as [S]) is essentially a subset of [P] (rule population in underlying UCS) and consists of optimal classifiers extracted during the operation of UCS. An input from the environment is first presented to [S] whereby a matchset [M] is generated using the current input label and the accuracy of those signatures participating in [M] are updated similar to UCS. The discovery component of UCS, i.e. the GA, however is bypassed when the system is run through [S]. Thus, signatures do not preserve their numerosity, fitness and nichesize parameters that they used to have in UCS. The input is escalated to UCS only if no match is available in [S], in which case a standard UCS takes over and runs its performance and discovery components using [P] for a certain number of trials.

Meanwhile, the extraction process of accurate and experienced classifiers from [P] to [S] is triggered in parallel periodically. Initially, [S] is empty and the system runs mainly through [P] getting enough exploration opportunities. The operation is shifted gradually to [S] as it starts getting populated. The transition completes when the system discovers the best map of the input space. This can be determined simply by measuring the percentage of input instances handled through [S]. For non-noisy test problems, a complete transition occurs when 100% of input instances are correctly classified by rules in [S], at which point the evolutionary search is completely halted and the system is made to run from [S]. A pruning step in [S] is carried out when the average experience of the signatures in [S] reaches a threshold. In the pruning step, the signatures that are inaccurate and have a below average experience are deleted from [S]. If the deletion causes a covering gap, control is handed back to [P] and the process is repeated until the system stabilizes to run from [S] at which point the learning process can be stopped.

A description of different parameters used in the online signature extraction algorithm is provided below. For further details interested readers are referred to [9][8].

- $N_{[P]}$: Number of standard UCS explore trials since the last extraction step.
- T_{ext} : Number of time steps since the last extraction step.
- T_{opt}: Minimum number of time steps for which control is switched to UCS when a $[M]$ through $[S]$ is found empty.
- θ_{xacc}, θ_{xexp}: The accuracy and experience thresholds for extracting a signature to $[S]$.
- C: A constant positive integer used in controlling the deletion experience threshold in $[S]$. Deletion in $[S]$ occurs when the average set experience exceeds $C.\theta_{xexp}$.
- θ_{dacc}, θ_{dexp}: The accuracy and experience thresholds for deleting a signature from $[S]$.

3   Analysis of Signature Learning Bounds

The essence of the signature extraction algorithm is based on the assumption that an LCS (XCS or UCS in particular) is able to evolve optimal rules during its search process and that these optimal rules or signatures can be extracted from the population successfully. This obviously requires a careful choice for when an extraction needs to occur. The second important decision in the signature extraction system is that of switching from $[P]$ to $[S]$ and back; that is, to decide how much search opportunities should be given to LCS that will be enough to evolve an optimal representation of the problem. Both of these decisions can be controlled by the T_{opt} parameter in UCSSE. To reiterate, T_{opt} corresponds to the number of time steps the control is switched to the normal UCS when the match set $[M]$ formed out of $[S]$ is found empty.

T_{opt} attempts to find a balance between providing enough exploration opportunities to the classifier system using $[P]$ and switching back to $[S]$ as soon as the signatures are discovered. Ideally, we wish to run the search process using $[P]$ for the entire duration of finding at least one optimal classifier before carrying out an extraction process and switching back to $[S]$. Hence T_{opt} can be formulated as a sum of the expected time to discover an optimal classifier and the expected time it will be evaluated θ_{xexp} times (the experience thresholds for extracting a signature to $[S]$).

Butz et al. provided a time bound for finding an optimal classifier by XCS using a domino convergence model [3]. They showed that learning time in XCS scales polynomially in problem length and exponentially in problem complexity. Since the evolutionary dynamics – the basis for time bound computation – in both XCS and UCS are similar we argue that the same bounds can be applied to both systems. Further UCS has generally shown to converge faster than XCS thus the bounds only provide an upper bound for UCS. Using insights from Butz et al., we derive the bounds for T_{opt} as follows:

Given that

$$T_{opt} = E(\text{Time to generate an optimal classifier}) +$$
$$E(\text{Time to evaluate an optimal classifier } \theta_{xexp} \text{ times}) \qquad (1)$$

From [3], for an equally probable input distribution, the time bound to generate an optimal classifier is given by

$$E(\text{Time to generate an optimal classifier})$$
$$= \frac{1}{P(\text{generation of an optimal classifier})} < \frac{n2^{o+s([P])l}}{\mu(1-\mu)^{l-1}} \qquad (2)$$

where n is the number of classes, o is the schema order, $s([P])$ is the average specificity of the population, l is the length of the string and μ is the mutation rate.

Considering an equally probable distribution, the expected time that this classifier will match an input θ_{xexp} times is given by:

$$E(\text{Time to match } \theta_{xexp} \text{ times})$$
$$= \frac{1}{P(\text{matching an input by the optimal classifier } \theta_{xexp} \text{ times})} \qquad (3)$$
$$= \frac{N}{N/2^o} . \theta_{xexp} = 2^o \theta_{xexp}$$

where N is the total number of instances in the feature space.

Substituting Equation 3 and the adjusted time bound ($O(l2^{o+n})$) for generating an optimal classifier from [3] in Equation 4, the expected time to generate an optimal classifier becomes:

$$E(\text{Time to generate an optimal classifier})$$
$$< \frac{n2^{o+s([P])l}}{\mu(1-\mu)^{l-1}} + 2^o \theta_{xexp} < l2^{o+n} + 2^o \theta_{xexp} \qquad (4)$$
$$= \gamma(l2^{o+n} + 2^o \theta_{xexp})$$

where γ is a constant between 0 and 1.

To test the validity of the above expression we experimented with the binary multiplexer problem of length 6, 11, 20 and 37. The theoretical values of T_{opt} can be calculated by substituting the values of l and o in Equation 4 for each of the above mentioned lengths of multiplexer and keeping $\gamma=1$. This gives us values of 352, 1024, 3200 and 10752 for 6, 11, 20 and 37 bit multiplexer problems respectively. The experimental values are obtained by recording the actual time when a member of the best action map (BAM) [2] is found with an experience equal to θ_{xexp} (set to 20 for these experiments) for different lengths of multiplexer problems.

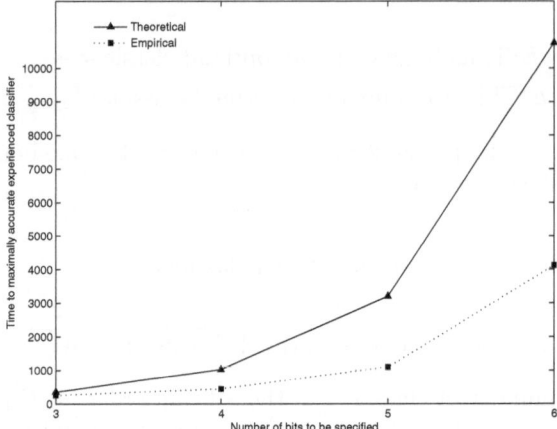

Fig. 2. Theoretical and experimental bounds for T_{opt} in the binary multiplexer problem

Figure 2 shows a comparison between the theoretical and empirical bounds. Both curves show a similar increasing trend, although the difference between the two bounds increases with the increase in the number of bits to be specified or the schema order. This is somewhat expected. In contrast to the theoretical bound which is derived assuming same parameter values for all lengths, we are using larger population sizes and $P_\#$ values for higher length multiplexer. The value of T_{opt} can be tuned using γ. A value of γ closer to 1 could delay the extraction process and thus increase the processing time. On the other hand, a value closer to 0 could lead to early switching to $[S]$ thereby losing important exploration opportunities to discover optimal rules.

For binary multiplexer problem, it was shown in [9] that using a value of 0.65 for γ, UCSSE was able to retrieve all optimal classifiers in real time as they are discovered. The system also switches completely to $[S]$ based operation and is able to reduce the processing time by more than 50%.

4   Signature Extraction in Noisy Environments

From a classification viewpoint noise refers to the distortion or error in attribute values or in the classification signs, i.e., the class of a data instance. In the presence of noise or imbalance class distribution in the training data, the accuracy of a classification algorithm is expected to drop. In addition, it is often not possible to determine a priori the best possible accuracy that a classification algorithm can achieve on a real world data set. For these reasons, setting the extraction and deletion thresholds (see Section 2) manually in UCSSE is not an appropriate strategy for effective signature extraction in such environments. In this section we first discuss the techniques used for online adaptation of these parameters and then show the results of experiments run under noisy problem domain.

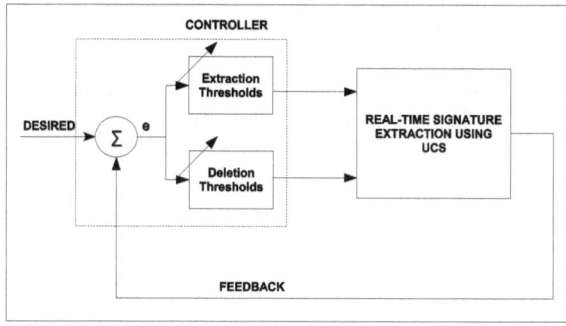

Fig. 3. Adaptive control for extraction and deletion thresholds in UCSSE

The block diagram of UCSSE control scheme for online adaptation of extraction and deletion accuracies is shown in Figure 3. Here, in control-theoretic terminology, e refers to the setpoint error and is computed as the difference between the *desired* control value and the actual output value (*feedback*). In our case, the control value refers to the percentage of inputs handled through signatures, which is ideally 100%. In other words, the extraction and deletion thresholds are adjusted based on the load on signatures. The basic idea is to keep the thresholds low when not many signatures can be supplied by UCS and increase them to appropriate levels when enough signatures are available. First let us look at the adaptation of accuracy thresholds (i.e. θ_{xacc}, θ_{dacc}). A simple procedure to adapt the extraction accuracy is given in Algorithm 1:

Algorithm 1. Update Extraction Accuracy

1: $\theta_{xacc} \leftarrow ACC0$
2: **for each** Class c **do**
3:    $\Delta = GetSupplyAccuracy(c)$
4:    **if** $\Delta > \overline{[S]_c.accuracy}$ **then**
5:       $\theta_{xacc}[c] = \Delta$
6:    **end if**
7: **end for**

At each extraction step the update procedure searches for the most accurate rule in $[P]$ which has enough experience ($> \theta_{xexp}$) to be extracted as a signature. If this accuracy value (referred to as *SupplyAccuracy*) is higher than the average accuracy in $[S]$ then θ_{xacc} is adjusted to the new value. Notice that the accuracy for each class is computed independently. This allows the system to handle varying levels of noise in different classes. Similarly, the deletion accuracy threshold θ_{dacc} is adjusted at each deletion step as follows:

$$\theta_{dacc}(c) = \overline{[S]_c.accuracy} - \Delta_c \qquad (5)$$

where Δ varies between $[0,\overline{[S]}.accuracy]$ based on the *error* signal between the desired $[S]$ based trials and the current feedback, i.e., the number of $[S]$ based trials since the last deletion step. Note that *error* here corresponds to the set point error (as commonly used in controller notations) and it has nothing to do with the classification error. At the beginning, when $[S]$ is empty, all inputs are sent to standard UCS and the error is maximum, hence Δ is set to minimum. The error starts dropping as the signatures are extracted to $[S]$ and some of the inputs are blocked by the signatures. Accordingly, Δ is increased based on controller's response. Since we are dealing with a single independent variable, the response can be given by a simple relationship of the form $y = f(x)$, where y corresponds to the controlled parameter and x is the current error signal. To gain control over the rate of change of the controlled parameter, $f(x)$ can be modelled as a simple linear exponential function. For an upper and lower bound of the controlled parameter the function can be written as:

$$y = y_{min} + (1/\exp^{ax} - 1/\exp^{x_{max}})\frac{y_{max} - y_{min}}{\|1/\exp^{x_{max}} - 1/\exp^{x_{min}}\|}$$

where y_{min} and y_{max} corresponds to the lower and upper bounds of the controlled parameter respectively, the error range is given by x_{min} and x_{max} and a (set to 2 and 5 respectively for accuracy and deletion thresholds in our experiments) is a constant which controls the slope of the exponential function.

Other schemes for adapting these parameters are possible and various other techniques were tried, however, the update combination mentioned above yielded the best outcome in terms of the test set accuracy and work load convergence.

4.1 Experiments with Noisy Checkerboard

The effect of noise is generally studied by simulating noise in the data e.g. by introducing random classification errors according to some distribution [1]. Following this practice, we simulate noise by introducing false positives (FP) and false negatives (FN) in the standard checkerboard problem. To create a noisy-checkerboard, data instances are sampled online from the feature space according to a uniform distribution and assigned a class 0 or 1 depending on their respective coordinates on the checkerboard. A FP or FN is introduced by inverting the correct class of an instance based on the noise level η. Four different noisy environments are created with varying degrees of noise in one or both classes listed in the results tables below.

To ensure a fair comparison between UCSSE and UCS we also implemented Dixon et al.'s [5] rule reduction algorithm to prune the post training rule populations evolved by UCS. Dixon's algorithm has shown to be an order of magnitude faster than Wilson's rule reduction algorithm [12] while achieving equivalent performance in terms of test accuracy. The actual algorithm is proposed and tested for XCS but we adopted it easily for UCS.

In our analysis, we compare the performance of three systems i.e. UCS, UCS with Dixon's rule reduction algorithm, referred to as UCSD, and UCSSE.

The following common practice parameter setting is used for UCS (please refer to [4] for naming convention of the parameters):

$$\alpha = 0.1,\ \beta = 0.2,\ \upsilon = 10,\ \theta_{GA} = 25,\ \chi = 0.8,\ \mu = 0.04,\ \theta_{del} = 20,\ \delta = 0.1,$$
$$ACC0 = 0.99,\ \theta_{sub} = 20,\ m_0 = 0.2,\ r_0 = 0.4,\ \text{GASubsumption} = \text{YES},$$
$$\text{ASSubsumption} = \text{NO},\ \text{Specify} = \text{NO}$$

The signature extraction algorithm parameters were set to:

$$T_{opt} = 1200,\ C = 10,\ \theta_{xacc} = 0.99,\ \theta_{xexp} = 100,\ \theta_{dacc} = 0.97,\ \theta_{dexp} = 0.5\ ,$$
$$\theta_{ol} = 0.9.$$

For Dixon's rule reduction algorithm any classifiers with experience less than 15 and accuracy less than $ACC0$ is considered as non-qualified. This setting was chosen carefully to give the best result for Dixon across the datasets and to be consistent with UCSSE.

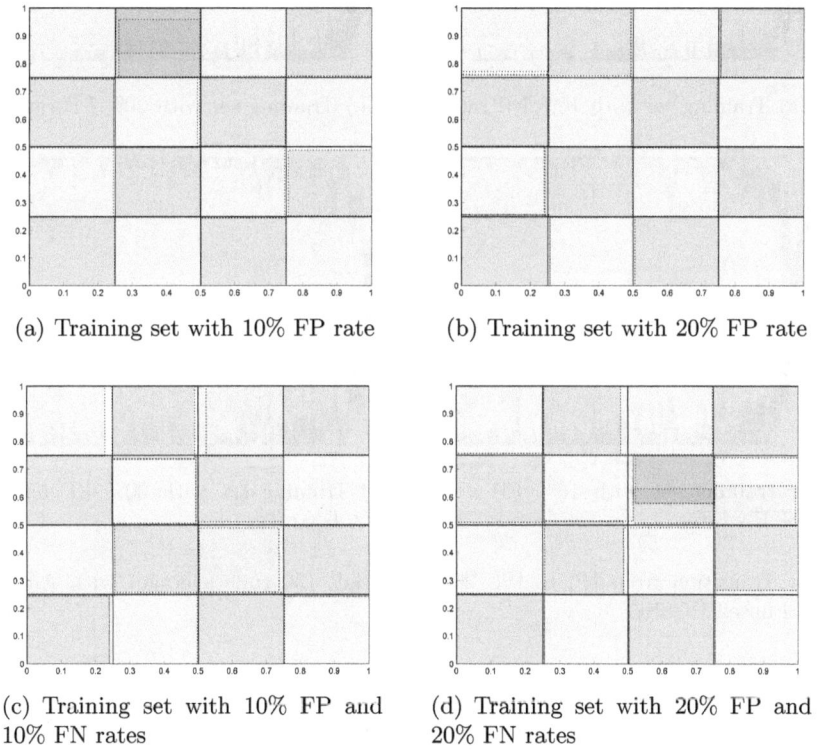

(a) Training set with 10% FP rate

(b) Training set with 20% FP rate

(c) Training set with 10% FP and 10% FN rates

(d) Training set with 20% FP and 20% FN rates

Fig. 4. Decision boundaries obtained by signatures in four noisy-checkerboard problems

Table 1 compares the test accuracy, number of rules and rules coverage of UCSSE and UCSD on 4 noisy problems. The results are averaged over 30 runs.

A ▲ is used if UCSD or UCSSE is significantly better (higher in accuracy and coverage and lower in number of rules) than UCS. A ♦ shows that UCSD or UCSSE is significantly better than both UCS and the other system. Similarly a △ denotes that UCSD or UCSSE is significantly worse (lower in accuracy and coverage and higher in number of rules) than UCS. A ◊ denotes that UCSD or UCSSE is worse than the other two systems. The significance is tested using a pairwise t-test at a significance level of 99%.

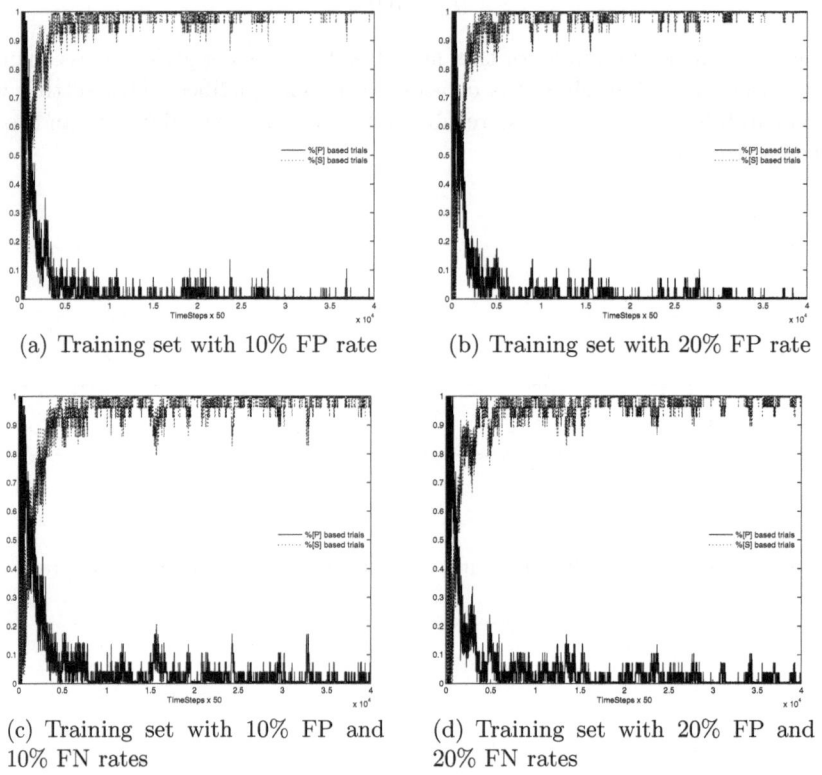

(a) Training set with 10% FP rate

(b) Training set with 20% FP rate

(c) Training set with 10% FP and 10% FN rates

(d) Training set with 20% FP and 20% FN rates

Fig. 5. Transition from $[P]$ to $[S]$, 2000000 trials (30 runs average) with Adaptive Control based UCSSE

First it can be noticed that the performance of UCS on noisy training data is quite impressive. The test set accuracy does not degrade proportional to the increasing noise levels. Given a uniform distribution of noise, the expected accuracy on a class with 10% noise is 90% at most. However UCS performs far better thanks to its fitness weighted voting policy during prediction. Contrary to test accuracy which is not affected severely, the number of rules evolved by UCS grows quickly with increasing levels of noise. Next note that UCSSE is able to retrieve approximately correct number of signatures in all four noisy problems which are also significantly less than both UCS and UCSD (see Section 4 for symbol notation). It also achieves more than 99% accuracy in all four problems with a high test set coverage. The correctness of signatures is further verified by Figures 4 and 5, which show the decision boundaries realized by signatures and the load transition curves in the four noisy problems respectively.

Table 1. Comparison of UCS, UCSD and UCSSE performance on noisy checkerboard problems. N corresponds to the noise level; 1=10% FP 0% FN , 2=10% FP 10% FN, 3=20% FP 0% FN, 4=20% FP 20% FN. Results are averages of 30 runs. See text for the explanation of notations used in significance tests.

	Test Accuracy (%)								
	UCS			UCSD			UCSSE		
N	Class 0	Class 1	Overall	Class 0	Class 1	Overall	Class 0	Class 1	Overall
1	99.83	100.00	99.93	97.10$^\triangle$	99.45$^\triangle$	98.20$^\diamond$	99.36	99.53$^\triangle$	99.53$^\triangle$
2	99.47	99.49	99.43	98.57$^\diamond$	98.83$^\diamond$	98.67$^\diamond$	99.70	99.93	99.93
3	98.32	99.96	99.17	95.92$^\diamond$	99.38$^\triangle$	97.70$^\diamond$	99.40$^\blacklozenge$	99.13$^\triangle$	99.23
4	97.64	97.73	97.70	95.43$^\diamond$	96.80$^\diamond$	96.07$^\diamond$	99.14	99.26	99.17
	Number of Rules								
1	155.20	44.67	199.87	33.20$^\blacktriangle$	11.20$^\blacktriangle$	44.40$^\blacktriangle$	8.67$^\blacklozenge$	8.47$^\blacklozenge$	17.13$^\blacklozenge$
2	154.33	154.87	309.20	32.27$^\blacktriangle$	32.87$^\blacktriangle$	65.13$^\blacktriangle$	8.77$^\blacklozenge$	9.10$^\blacklozenge$	17.87$^\blacklozenge$
3	147.67	58.00	205.67	33.10$^\blacktriangle$	13.57$^\blacktriangle$	46.67$^\blacktriangle$	8.63$^\blacklozenge$	8.57$^\blacklozenge$	17.20$^\blacklozenge$
4	163.63	159.37	323.00	30.13$^\blacktriangle$	32.07$^\blacktriangle$	62.20$^\blacktriangle$	9.07$^\blacklozenge$	9.33$^\blacklozenge$	18.40$^\blacklozenge$
	Test Set Coverage (%)								
1	99.99	99.99	99.99	99.92	99.91$^\triangle$	99.91$^\triangle$	99.20	97.50$^\triangle$	98.35
2	100.00	100.00	100.00	99.94	99.91$^\triangle$	99.93$^\triangle$	98.43	99.55$^\diamond$	98.99$^\diamond$
3	99.99	99.97	99.98	99.88	99.82$^\triangle$	99.85$^\triangle$	99.16	99.05	99.10
4	100.00	100.00	100.00	99.93$^\triangle$	99.88	99.90$^\triangle$	97.79$^\diamond$	99.20	98.49$^\diamond$

5   Conclusions

In a previous work we presented a framework for online signature extraction (or rule reduction) from LCS. An algorithm was presented to automatically identify and extract maximally general rules during the learning of LCS, which we referred to as *signatures*. In this work we have shown the validity of the signature extraction algorithm using learning time bounds in UCS and investigated the performance of UCS and UCSSE in noisy problems. With adaptive threshold control of accuracy parameters, UCSSE is able to retrieve near optimal decision boundaries for noisy checkerboard problems. We also extended a leading rule reduction algorithm for XCS to UCS and compared its performance with UCSSE. We noted that the rule reduction algorithm suffers from noise in data because it also uses preselected fixed thresholds to prune post training rule sets. We replaced fixed thresholds with per class average rule set accuracy. In noisy checkerboard problems UCSSE outperforms both UCS and UCSD in the number of rules while achieving similar test accuracy. In future, we would like to test the generalisation of signature extraction concept to other LCS including the Pittsburgh style LCS.

References

1. Angluin, D., Laird, P.: Learning from noisy examples. Machine Learning 2(4), 343–370 (1988)
2. Bernadó-Mansilla, E., Garrell, J.M.: Accuracy-Based Learning Classifier Systems: Models, Analysis and Applications to Classification Tasks. Evolutionary Computation 11(3), 209–238 (2003)
3. Butz, M.V., Goldberg, D.E., Lanzi, P.L.: Bounding Learning Time in XCS. In: Deb, K., Tari, Z. (eds.) GECCO 2004. LNCS, vol. 3103, pp. 739–750. Springer, Heidelberg (2004)
4. Butz, M.V., Wilson, S.W.: An algorithmic description of XCS. Soft Computing-A Fusion of Foundations, Methodologies and Applications 6(3), 144–153 (2002)
5. Dixon, P.W., Corne, D.W., Oates, M.J.: A Ruleset Reduction Algorithm for the XCS Learning Classifier System. In: Lanzi, P.L., Stolzmann, W., Wilson, S.W. (eds.) IWLCS 2003. LNCS (LNAI), vol. 2661, pp. 20–29. Springer, Heidelberg (2003)
6. Fu, C., Davis, L.: A modified classifier system compaction algorithm. In: Proceedings of the Genetic and Evolutionary Computation Conference (GECCO 2002), pp. 920–925. Morgan Kaufmann (2002)
7. Gaber, M.M., Zaslavsky, A., Krishnaswamy, S.: Mining data streams: a review. ACM Sigmod Record 34(2), 18–26 (2005)
8. Shafi, K., Abbass, H.: An adaptive genetic-based signature learning system for intrusion detection. Expert Systems With Applications 36(10), 12036–12043 (2009)
9. Shafi, K., Abbass, H.A., Zhu, W.: Real Time Signature Extraction From A Supervised Classifier System. In: Proceeding of the IEEE Congress on Evolutionary Computation, CEC 2007, September 25-28, pp. 2509–2516 (2007)
10. Tan, J., Moore, J., Urbanowicz, R.: Rapid rule compaction strategies for global knowledge discovery in a supervised learning classifier system. In: Advances in Artificial Life, ECAL, vol. 12, pp. 110–117 (2013)
11. Urbanowicz, R.J., Moore, J.H.: Learning classifier systems: a complete introduction, review, and roadmap. Journal of Artificial Evolution and Applications 2009, 1 (2009)
12. Wilson, S.W.: Compact Rulesets from XCSI. In: Lanzi, P.L., Stolzmann, W., Wilson, S.W. (eds.) IWLCS 2001. LNCS (LNAI), vol. 2321, pp. 197–210. Springer, Heidelberg (2002)
13. Wyatt, D., Bull, L., Parmee, I.: Building compact rulesets for describing continuous-valued problem spaces using a learning classifier system. In: Parmee, I. (ed.) Adaptive Computing in Design and Manufacture VI, pp. 235–248. Springer (2004)

Multi-objective Multiagent Credit Assignment Through Difference Rewards in Reinforcement Learning

Logan Yliniemi and Kagan Tumer

Oregon State University
Corvallis, Oregon, USA
logan.yliniemi@engr.orst.edu
kagan.tumer@oregonstate.edu

Abstract. Multiagent systems have had a powerful impact on the real world. Many of the systems it studies (air traffic, satellite coordination, rover exploration) are inherently multi-objective, but they are often treated as single-objective problems within the research. A very important concept within multiagent systems is that of credit assignment: clearly quantifying an individual agent's impact on the overall system performance. In this work we extend the concept of credit assignment into multi-objective problems, broadening the traditional multiagent learning framework to account for multiple objectives. We show in two domains that by leveraging established credit assignment principles in a multi-objective setting, we can improve performance by (i) increasing learning speed by up to 10x (ii) reducing sensitivity to unmodeled disturbances by up to 98.4% and (iii) producing solutions that dominate all solutions discovered by a traditional team-based credit assignment schema. Our results suggest that in a multiagent multi-objective problem, proper credit assignment is as important to performance as the choice of multi-objective algorithm.

1 Introduction

Cooperative multiagent systems focuses on producing a set of autonomous agents to achieve a system-level goal [12]. Multiagent frameworks have been used to study complex, real-world systems like air traffic [10], teams of satellites [3], and extra-planetary rover exploration [1]. In each case, the goal is to optimize a single, well-defined objective function.

But, in many of these cases, the problems lend themselves more naturally to multiple objectives: for example, air travel should be as safe and as expedient as possible. Satellites may need to make observations for multiple separate institutions. Extra-planetary rovers should acquire multiple different types of scientific data. However, most research in multiagent systems does not take a multi-objective viewpoint: they typically seek to find a single usable solution, without considering the tradeoffs between potential alternatives that would increase one objective's value at the cost of another. These tradeoff solutions, which form the *Pareto front*, are a key solution concept in multi-objective problems.

Developing successful agent policies in multiagent systems can be challenging. One successful approach is to use adaptive agents with tools like reinforcement learning.

G. Dick et al. (Eds.): SEAL 2014, LNCS 8886, pp. 407–418, 2014.

Each agent seeks to maximize its own reward; with a properly designed reward signal, the whole system will attain desirable behaviors. This is the science of credit assignment: determining the contribution each agent had to the system as a whole. Clearly quantifying this contribution on a per-agent level is essential to multiagent learning. This is an issue that has not been studied within the context of multiple objectives. In this work we address the challenges that arise when multiagent systems are combined with multi-objective problems.

The primary contribution of this work is to develop the concept of credit assignment for multi-objective problems. This broadens the traditional multiagent learning framework to account for the multiple objectives present in many real world problems. This improves system-level performance by (i) increasing learning speed by up to 10x (ii) reducing sensitivity to unmodeled disturbances by up to 98.4% and (iii) producing solutions that dominate all solutions discovered by a traditional team-based credit assignment schema.

The remainder of this work is organized as follows: Section 2 describes the necessary background. Section 3 describes a stateless coordination domain, the multi-objective bar problem (MOBP), and presents the results in this domain. Section 4 describes a stateful, time-extended coordination domain, the collective transport domain (CTD), and presents the results in this domain. Finally, Sec. 5 draws the conclusion to this work and identifies future directions for this line of research.

2   Background

We limit the scope of this work to consider reinforcement learners using difference rewards as a feedback signal with "a priori" scalarization of objectives. This allows us to examine the performance of multi-objective difference rewards in two scenarios in which they have been shown to out-perform a global "team" reward in a single-objective case. "A posteriori" methods, such as multi-objective evolutionary algorithms, though more generally successful, are explicitly out of the scope of this work.

2.1   Multi-objective Problems

In a multi-objective problem, there is typically not one "best" solution, but instead an array of optimal tradeoffs that are *incomparable*. For example, a man with 1 kilogram of bread and 1 kilogram of wine might be just as happy as a man with 0.9 kilograms of bread and 1.2 kilograms of wine. The two are incomparable [6]. However, both of these solutions are strictly better than a man with no bread and no wine, which is *dominated* by both of the others.

Non-dominated set (NDS). The NDS is the set of discovered feasible solutions that are not dominated by any other solution. The process for calculating the NDS is illustrated in Figure 1. Any optimizer or search will develop a set of non-dominated solutions; the globally best-possible NDS is known as the Pareto front, and the goal of any multi-objective approach is to develop an NDS that is a close approximation to the Pareto front.

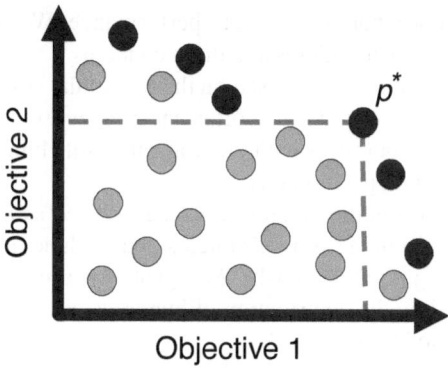

Fig. 1. Domination. The point p^* is a point in the NDS, and all points which score worse on all objectives than (below and to the left of) p^* are dominated by p^*. The three grey points not dominated by p^* are dominated by other (black) points in the NDS.

Scalarization of objectives. Within the class of *a priori* methods for multi-objective problems, there are many different ways to scalarize the objectives into a single reward signal. In this work we examine two: a linear combination and a hypervolume calculation. In each case we normalize the objectives to the range [0:1] before combining them in one of two ways:

$$R_+ = \sum_{c \in C} w_c f_c^{\text{norm}} \quad \Big| \quad R_\lambda = \prod_{c \in C} f_c^{\text{norm}} \tag{1}$$

where R_+ is the linear combination reward delivered to the reinforcement learner, R_λ is the hypervolume reward delivered to the reinforcement learner, C is the set of all criteria or objectives, and f_c^{norm} is the normalized score on objective c. In each case we give all agents either R_+ or R_λ, but never any combination of the two. The form which f_c^{norm} takes varies depending on the credit assignment schema used, which is discussed in the following section. Other types of scalarizations do exist, like an exponentially weighted set of objectives or distance from a target point, but we limit the scope of this work to consider only these two.

2.2 Reinforcement Learning

In this work we use a team of independent reinforcement learners (Action-value learners for the MOBP and Q-learners for the CTD [5,9]), with the standard notation of a learning rate α, a discount factor γ, and a reward R.

2.3 Multiagent Credit Assignment Structures

In a multiagent system, it is important to reward an agent based on its contribution to the system. This is difficult due to the other agents acting in the environment, obscuring the agent's individual contribution to system performance. We consider three popular *credit assignment structures* for addressing these concerns.

A **local reward** (L_i) is the reward based on the part of the system that an agent i can directly observe. Using this reward signal often encourages "selfish" behavior, in which the agent may act at cross-purposes with other agents while blindly increasing its own reward, causing poor system performance.

The **global reward** (G) is the system performance used as a learning signal. This encourages the agent to act in the system's interest, but includes a large amount of noise from other agents acting simultaneously. An agent's own contribution to the global reward may be dwarfed by the contribution of hundreds of other agents, resulting in a low "signal to noise ratio" [11].

The **Difference reward** (D_i) is a shaped reward signal that helps an agent learn the consequences of its actions on the system objective by removing a large amount of the noise created by the actions of other agents active in the system [11]. It is defined as

$$D_i(z) = G(z) - G(z_{-i}) \tag{2}$$

where $G(z)$ is the global system performance for the system considering the joint state-action z, and $G(z_{-i})$ is $G(z)$ for a theoretical system without the contribution of agent i. Any action taken to increase D_i simultaneously increases G, while agent i's impact on its own reward is much higher than its relative impact on G [11].

3   Multiobjective Bar Problem (MOBP)

The first domain we consider in this work is an extension of the El Farol Bar Problem originally introduced by Arthur [2]. In this extension, a group of agents **A** are each assigned a static type m or f and must independently choose to attend one of several bars. There are multiple objectives: first, the agents wish to attend a bar that is not too crowded, and not too empty. Second, the agents wish to attend a bar with an even mixing of agents of type m and f.

The first "capacity" objective for each bar is modeled as a smooth curve that takes on a value of 0 with no agents attending, near 0 with many agents attending, and a maximum at the ideal capacity ψ. This models the enjoyment of the agents (of quantity x_b) attending bar b. The second "mixture" objective for each bar is maximized when $M_b = F_b$, where these are the number of agents of type m and f attending bar b, *regardless of the number of agents at the bar*. Formally:

$$L_{cap_b} = x_b \cdot e^{\frac{-x_b}{\psi}} \quad \bigg| \quad L_{mix_b} = \frac{\min(M_b, F_b)}{(M_b + F_b)W} \tag{3}$$

where W is the number of bars available for the agents to choose from. L_{mix_b} evaluates to 0 if the agents are all of the same type, and $0.5/W$ if there is an equal mixture of types. The number of bars, W, is a constant (and therefore does not change the reinforcement learning process), and serves to limit G_{mix} to values in the range [0:0.5] for easier interpretation of results.

The global rewards for each of these objectives are simply the sum of the local rewards across all bars:

$$G_{\text{cap}} = \sum_{b \in B} L_{\text{cap}_b} \qquad \Bigg| \qquad G_{\text{mix}} = \sum_{b \in B} L_{\text{mix}_b} \qquad (4)$$

And the Difference rewards for each are calculated by Equation 2 as the global reward minus the global reward in a fictional world had agent i never attended any of the bars:

$$D_{\text{cap}_i} = x_a \cdot e^{\frac{-x_a}{\psi}} - (x_a - 1) \cdot e^{\frac{-(x_a - 1)}{\psi}} \qquad (5)$$

$$D_{\text{mix}_i} = \begin{cases} \frac{\min(M_a, F_a)}{(M_a + F_a)W} - \frac{\min((M_a - 1), F_a)}{(M_a + F_a - 1)W} & : i \in m \\[2ex] \frac{\min(M_a, F_a)}{(M_a + F_a)W} - \frac{\min(M_a, (F_a - 1))}{(M_a + F_a - 1)W} & : i \in f \end{cases} \qquad (6)$$

where x_a is the attendance in the bar attended by agent i, and M_a and F_a are the number of agents of types m or f respectively that attended the same bar as agent i. D_{mix_i} depends on the type of the agent; the second term represents the system with agent i removed from bar b.

Procedure. The procedure for running the MOBP is simple. Each agent simultaneously selects a bar to attend based on no sensory information. The local rewards L_{cap_b} and L_{mix_b} are calculated for each bar b. Then the global rewards G_{cap} and G_{mix} are calculated. Finally, D_{cap_i} and D_{mix_i} are calculated for each agent i. Once these are calculated, the selected reward type (local, global, or difference) is normalized and put through Equation 1 depending on the desired scalarization. The result is then provided to the agent as the reward R, calculated with a value of $\gamma = 0$, because the problem is only a single step.

Tradeoffs and independence of objectives. We take measures to prevent a trivial solution for either objective, or a single dominating solution:

- G_{cap}: There are many more agents (100) than capacity across all bars (a capacity of 5 for 7 bars).
- G_{mix}: Agent types are 70% type m, 30% type f.
- *Tradeoff*: L_{cap_i} is maximized at 5 agents; L_{mix_i} is maximized only when an even number of agents attend a bar.
- *Tradeoff*: A maximum G_{mix} case involves many bars with one agent of each type and the rest attending a single bar, which conflicts with G_{cap}.

We calculate the coefficient of determination (R^2) value for the correlation between the two objectives across 10^6 random Monte Carlo trials using a linear, exponential, and polynomial fit. The maximum value was the linear fit at 0.0034, which reinforces that the objectives are distinct, though they are coupled through the actions of the agents.

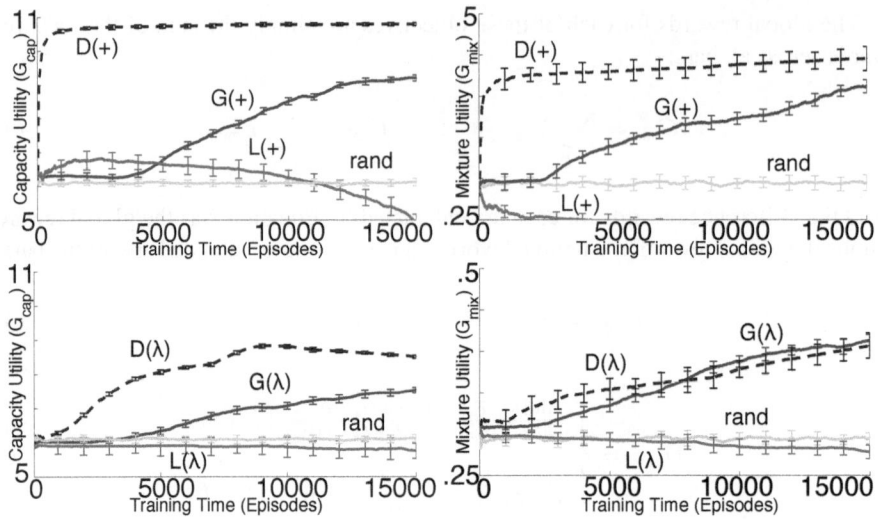

Fig. 2. Performance on G_{cap} (left) and G_{mix} (right), for agents trained on the linear scalarization (+,top) and hypervolume calculation (λ,bottom) of the three reward structures (D,G,L) and the random baseline (rand). Each of these objectives is to be maximized.

3.1   MOBP Results

To exhibit the benefits of Difference rewards in multi-objective problems, we examine 4 types of results:

- Average system performance on both system objectives (Figure 2)
- Dominance and NDS (Figure 3)
- Impact of training time (Figure 3)
- Robustness to disturbances (Figure 4)

Simulation information. We execute 30 statistical runs of the MOBP for seven independent experiments: training all agents on each structure-scalarization combination in turn (D+, Dλ, G+, Gλ, L+, Lλ), and on a random policy (rand).

Each agent selects an action using an ϵ-greedy mechanism, with an initial $\epsilon = 0.05$ for local and difference rewards, and $\epsilon = 0.1$ for global rewards[1] (both multiplied by a factor of 0.999 every episode to reduce exploration), with a learning rate $\alpha = 0.10$.

We performed a full sweep through w_c values, but due to the large effect each agent has on the overall system performance near the Pareto front, we found that an even weight, combined with the natural exploration, resulted in a spread of solutions discovered along the Pareto front.

In Fig. 2, a 100-episode moving average (across 30 statistical runs) of system performance was used. Error bars report the error in the mean, calculated as $\frac{\sigma}{\sqrt{N}}$, where N

---

[1] These values were chosen through a parameter sweep to create the best performance for each reward, though the results are not very sensitive to ϵ or α values.

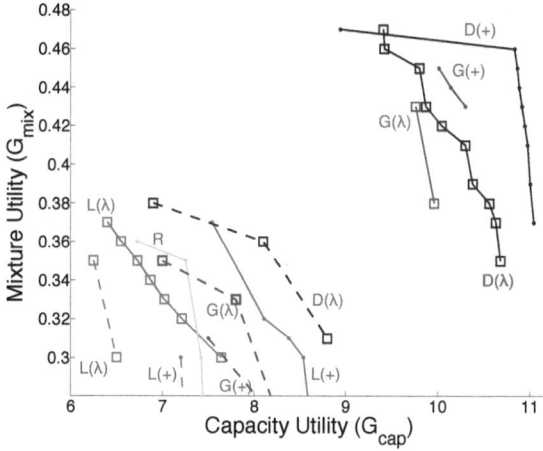

Fig. 3. The set of non-dominated episodes created over the entire training process through using hypervolume (λ) or a linear combination ($+$); dotted lines show the NDS after 1500 learning episodes; solid lines, the NDS after 15,000 episodes. Agents trained on D($+$) peak in performance before 1500 episodes, so both D($+$) NDSs are identical.

is the number of statistical trials. We identify the NDS for each structure-scalarization combination (e.g. "Hypervolume of Global Reward", Gλ), across all 30 statistical runs, and aggregate these into a single non-dominated set, for clarity [4].

3.2 Average Performance on System Objectives

It is informative to look at the performance of the system on each objective individually (Figure 2), as this performance drives the behavior of the non-dominated set. For both the linear combination and hypervolume scalarization, the local reward (L+,Lλ) performs poorly; the agents work at cross-purposes, undermining each other's efforts by all trying to attend low-attendance days. This leads to low performance, and will never lead to good system behavior, even with an extreme amount of training time. The global reward (G+,Gλ) does learn, slowly. For the hypervolume scalarization, Dλ increases system performance at a slightly higher rate than Gλ. The linear combination of difference rewards, D+, performs at a very high level very quickly, and reaches near its final performance after only 1500 episodes.

3.3 Non-Dominated Sets (NDS) and Training Time

In addition to performing well on the individual objectives, solutions produced by D+ or Dλ produce superior NDS compared to the global and local rewards with the same scalarization. The NDS are shown in Figure 3. In fact, every solution produced by the local or global rewards is dominated by a solution produced by the difference reward.

The dotted lines in Figure 3 represent the NDS produced in the first 1500 episodes (10% of the training). In all cases the NDS improve between 1500 and 15000 episodes,

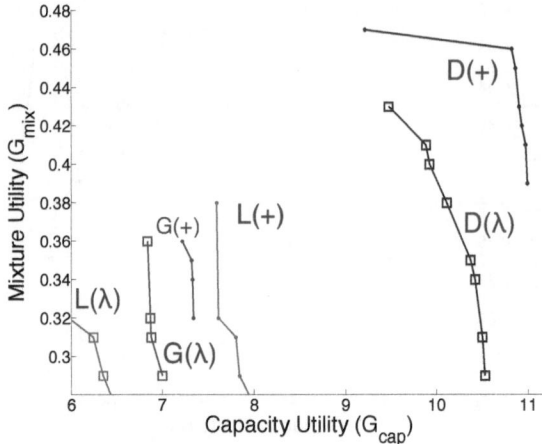

Fig. 4. The NDS produced in the 5000 training episodes after the conversion of 20% of agents to "selfish" behavior. Compare with solid lines in Figure 3: D+ and Dλ recover well; G+ and Gλ suffer a disastrous drop in performance.

except D+, which has already produced its best episodes (dominating all other credit assignment/scalarization combinations). Solutions produced by Dλ dominate the solutions produced by other methods in the same time, except D+.

3.4   Robustness to Disturbances

To model outside disturbances, after 15000 episodes 14 agents of type m and 6 of type f "fail". They have their Q-table reset to zero values and continue learning using the local reward policy regardless of the learning signal they were using previously (acting selfishly). The remaining 80 agents continue learning using the same signal they were using previously. Additional exploration was found to be necessary in this case, so we reset ϵ to initial values. All agents continue the learning process as before.

Figure 4 shows that D+ maintains its dominant NDS. G+ and Gλ are affected catastrophically by the selfish agents, while D+ loses 98.4% less dominated hypervolume. Dλ only loses performance on G_{mix}.

4   Collective Transport Domain

We additionally performed experiments in a collective transport domain, modeled after [7], in which a team of small robots must cooperate to transport an item (which we also refer to as a body or load) across a surface in much the same way that ants transport objects.

We formulate this as a time-extended, stateful reinforcement learning problem in which the robot agents try to (i) collectively transport the object as quickly as possible to the goal, while (ii) expending minimum effort.

Each robot is given discretized state information about the load's position and velocity, and is allowed to take one of nine actions, applying a force to the load in a cardinal direction (N,S,E,W), an intermediate direction (NE,SE,SW,NW), or no force. The robots are assumed to be attached to the object, and receive discretized state information based on the object's current location and speed.

The body's acceleration (acc), velocity (vel), position (pos) at time t are found with particle kinetics:

$$acc(t) = \sum_{i \in \mathbf{A}} [F_x(i)\hat{i}] + \sum_{i \in \mathbf{A}} [F_y(i)\hat{j}] - F_f \hat{f} \tag{7}$$

$$vel(t) = vel(t-1) + acc(t) \cdot t_{\text{step}} \tag{8}$$

$$pos(t) = pos(t-1) + vel(t) \cdot t_{\text{step}} \tag{9}$$

where $\mathbf{A}$ is the set of all agents, $F_x(i)$ is the force applied by agent i in the $\hat{i}$ direction, $F_y(i)$ is the force applied by agent i in the $\hat{j}$ direction, and F_f is the force of friction, which acts in the $\hat{f}$ direction, which points opposite the direction of motion of the body. We omit mass from this calculation of Newton's second law because we assume the mass of the body and transporting robots to be 1 unit total. In this context a local reward loses some meaning as all agents are collectively acting to move the same object, so we only look at global and difference reward in this case.

The first objective (proximity) is to move the load close to the goal as quickly as possible. This takes the form:

$$G_{\text{prox}}(t) = -Tdist(t) \tag{10}$$

$$D_{\text{prox}_i}(t) = -Tdist(t) + Tdist_{-i}(t) \tag{11}$$

where $Tdist()$ is a function that returns the body's Euclidian distance from the target at time t, and $Tdist_{-i}()$ returns the distance from the target if agent i took no action during timestep t.

The second objective is to minimize the effort exerted by the team to move the load to the desired target location:

$$G_{\text{effort}}(t) = \sum_{i \in \mathbf{A}} [1 - E_{i,t}] \tag{12}$$

$$D_{\text{effort}_i}(t) = 1 - E_{i,t} \tag{13}$$

where $E_{i,t}$ is 1 if the agent applied a force to the object at time t, and 0 if the agent did not apply a force.

We perform a Q-update at every time step. To visualize the performance, we aggregate these into one point for each time the load reaches the goal state. For the purpose of learning, however, we use the distance to the goal state after each time step, as this provides a smoother gradient for learning [5]. The process for conducting this experiment is described in Algorithm 1. For each credit assignment schema and scalarization combination, step 19 would use the proper evaluation (one of L_i, G, or D_i), and use the desired scalarization from Equation 1.

In this domain the two objectives are in conflict with one another: minimizing the time to deliver the load will maximize the effort required, and minimizing effort will lead to a longer time.

4.1   CTD Results

In the collective transport domain, we examine two types of results:

- Dominance and NDS (Figure 5, Left)
- Impact of training time (Figure 5, Right)

Simulation Information. We perform 4 different trials following Algorithm 1; one each for G+, Gλ, D+, and Dλ. For each, we conduct 30 statistical runs of 5000 time steps for teams of 50 agents attempting to transport a load across a surface with maximum static force of friction $F_f = 8$ units and kinetic force of friction of $F_f = 2$ units. The body's starting state is initialized as $(x, y) = (1, 1)$, with the goal as a square at $\{x_{min}, x_{max}, y_{min}, y_{max}\} = \{900, 1000, 900, 1000\}$. The boundaries are a larger square at $\{x_{min}, x_{max}, y_{min}, y_{max}\} = \{0, 1000, 0, 1000\}$. Though the calculations of the body's velocity and position are continuous, we us an approximation via tile coding [9] and discretize into 10 states each for $(x_{vel}, y_{vel}, x_{pos}, y_{pos})$ creating 10,000 states. In this

---

Algorithm 1. Collective Transport Domain using Difference Reward of Dominated Hypervolume (Dλ)

1: initialize Q-values to zero: $Q(s, a) = 0 \ \forall \ s, a$
2: initialize body position to starting location
3: initialize velocity and acceleration to **0**.
4: **for** $timestep = 1 \rightarrow max\_timesteps$ **do**
5:    **for** $i = 1 \rightarrow total\_agents$ **do**
6:        choose an action to take with ϵ-greedy action selection:
7: {none,N,NE,E,SE,S,SW,W,NW}
8:        add force contribution to body $(F_x(i), F_y(i))$
9:    **end for**
10:    evaluate body acceleration (Equation 7)
11:    evaluate body velocity (Equation 8)
12:    evaluate body position (Equation 9)
13:    **if** body position is out of bounds **then**
14:        set body position to nearest in-bounds position
15:        set body velocity to **0**
16:    **end if**
17:    evaluate global reward (Equations 10, 12)
18:    **for** $i = 1 \rightarrow total\_agents$ **do**
19:        evaluate difference rewards (Equations 11, 13)
20:        evaluate $R \leftarrow R_\lambda$ (Equation 1)
21:        update $Q(s, a)$ values
22:    **end for**
23:    **if** body is in goal state **then**
24:        set body to starting location
25:        set velocity and acceleration to **0**.
26:    **end if**
27:    reduce ϵ
28: **end for**

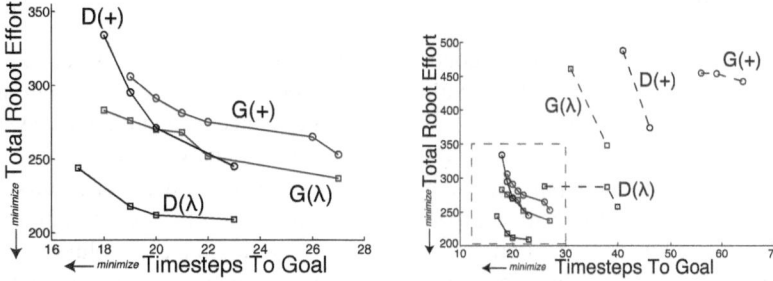

Fig. 5. (Left) Collective Transport Domain results. $D(\lambda)$ creates solutions that dominate all other methods (solutions below and to the left are superior in this domain). D+ outperforms G+, and creates an overlapping Pareto front with $G(\lambda)$. (Right) Dotted lines denote early system performance after 500 time steps. The denoted highlighted area is the range of the figure on the left.

domain, we find the best performance when we vary the weights for the objectives as a function of learning step, starting by with a value of $\{w_{\mathrm{prox}}, w_{\mathrm{effort}}\} = \{1, 0\}$ changing linearly to $\{0, 1\}$ at the final learning step. This produces policies which do find the goal state, and learn to reduce effort over time. This produces better initial performance and a spread of solutions along the NDS. Initial weights favoring the effort objective led to policies of inaction, never reaching the goal.

4.2 Dominance and NDS

Figure 5 shows the final NDS for each method. The teams of agents trained on the scalarizations of the difference reward (D+, Dλ) outperform their global counterparts in the final produced NDS. In this domain, however, the hypervolume calculations (λ) perform better than the linear combinations (+). We find nearly equivalent performance between Gλ and D+, suggesting that using the proper multi-objective scalarization is as important as proper multiagent credit assignment. The Dλ result shows that these benefits can be symbiotic.

4.3 Impact of Training Time

We also identify the NDS produced by each solution after 10% of the training time in Figure 5. Again, the difference reward using the preferable scalarization attains performance close to its final performance very quickly, while the global methods are not as near their final performance values. Dλ dominates all solutions formed by other scalarizations.

Additionally, in this domain we noticed that the performance of the global reward signals was sensitive to the learning parameters, while the difference reward signals were robust to these changes. In additional trials we found the agents trained with the difference reward to be robust to noisy actuators, noisy sensors, failing agents, and unmodeled disturbances (externally applied forces) as well.

5   Conclusion

Multiagent systems are a powerful concept for dealing with complex systems. Many multiagent systems are intrinsically multi-objective, but this has received scant attention. In this work we explicitly addressed one of the key concerns in multiagent systems — credit assignment — under the conditions of a multi-objective problem. We found that credit assignment is important under multi-objective conditions: our results show (i) a 10x increase in learning speed, (ii) a 98.4% increase in robustness to unmodeled disturbances, and (iii) the production of solutions which dominate all solutions found by a traditional global reward. These results show that proper credit assignment is of paramount importance in a multiagent multi-objective system. However, the choice of multi-objective algorithm is still extremely important. Difference rewards boosted performance in both domains, for both scalarizations. The gains from credit assignment through difference rewards were independent of the scalarization used and the domain.

Difference rewards are not limited to reinforcement learning or a priori methods, however. Future work on this topic includes an examination of the effects that credit assignment can have on multiagent implementations of well-established *a posteriori* multi-objective evolutionary algorithms.

Acknowledgements. This work was partially supported by the National Energy Technology Laboratory under grants DE-FE0011403 and DE-FE0012302.

References

1. Agogino, A.K., Tumer, K.: Analyzing and visualizing multi-agent rewards in dynamic and stochastic domains. JAAMAS 17(2), 320–338 (2008)
2. Arthur, W.B.: Inductive reasoning and bounded rationality (the El Farol Problem). American Economic Review 84(406) (1994)
3. Damiani, S., Verfaillie, G., Charmeau, M.-C.: An earth watching satellite constellation: How to manage a team of watching agents with limited communications. In: AAMAS (2005)
4. Fonseca, C.M., Fleming, P.J.: On the performance assessment and comparison of stochastic multiobjective optimizers. In: Ebeling, W., Rechenberg, I., Voigt, H.-M., Schwefel, H.-P. (eds.) PPSN 1996. LNCS, vol. 1141, pp. 584–593. Springer, Heidelberg (1996)
5. Kaelbling, L.P., Littman, M.L., Moore, A.W.: Reinforcement learning: A survey. Journal of Artificial Intelligence Research (1996)
6. Pareto, V.: Manual of Political Economy. MacMillan Press Ltd. (1927)
7. Rubenstein, M., Cabrera, A., Werfel, J., Habibi, G., McLurkin, J., Nagpal, R.: Collective transport of complex objects by simple robots: Theory and experiments. In: AAMAS (2013)
8. Sherstov, A.A., Stone, P.: Function approximation via tile coding: Automating parameter choice. In: Zucker, J.-D., Saitta, L. (eds.) SARA 2005. LNCS (LNAI), vol. 3607, pp. 194–205. Springer, Heidelberg (2005)
9. Sutton, R., Barto, A.G.: Reinforcement Learning: An Introduction. MIT Press (1998)
10. Tomlin, C., Pappas, G.J., Sastry, S.: Conflict resolution for air traffic management: A study in multiagent hybrid systems. IEEE Transactions on Automatic Control 43(4), 509–521 (1998)
11. Wolpert, D.H., Wheeler, K., Tumer, K.: Collective intelligence for control of distributed dynamical systems. Europhysics Letters 49(6) (2000)
12. Wooldridge, M.: An Introduction to MultiAgent Systems. John Wiley and Sons (2002)

On the Impact of Utility Functions in Interactive Evolutionary Multi-objective Optimization

Frank Neumann and Anh Quang Nguyen

Optimisation and Logistics
School of Computer Science
The University of Adelaide
Adelaide, SA 5005, Australia

Abstract. Interactive evolutionary algorithms for multi-objective optimization have gained an increasing interest in recent years. As multi-objective optimization usually deals with the optimization of conflicting objectives, a decision maker is involved in the optimization process when encountering incomparable solutions. We study the impact of a decision maker from a theoretical perspective and analyze the runtime of evolutionary algorithms until they have produced for the first time a Pareto optimal solution with the highest preference of the decision maker. Considering the linear decision maker, we show that many multi-objective optimization problems are not harder than their single-objective counterpart. Interestingly, this does not hold for a decision maker using the Chebeyshev utility function. Furthermore, we point out situations where evolutionary algorithms involving a linear decision maker have difficulties in producing an optimal solution even if the underlying single-objective problems are easy to be solved by simple evolutionary algorithms.

1 Introduction

Evolutionary algorithms (EAs) are frequently used for tackling multi-objective optimization problems [5,4]. Multi-Objective problems usually allow for an exponential number of trade-offs with respect to the given objective functions. In the usual setting, solutions representing the different trade-offs according to given objective functions are presented to the decision maker and he then has to decide on one of these solutions for implementation.

In order to let an EA focus on regions in the objective space that are preferable to a decision maker, one can add the possibility of interacting with the algorithm. In particular, the decision maker can make the decision which solution to prefer in the case that two solutions are incomparable with respect to the classical Pareto dominance relation which drives most evolutionary multi-objective algorithms.

Interactive evolutionary multi-objective optimization has gained increasing attention during the last years [10,6]. The goal is to involve the decision maker into the optimization process and gain knowledge about his preferences in order to focus on the regions that he prefers during the optimization run. It should

G. Dick et al. (Eds.): SEAL 2014, LNCS 8886, pp. 419–430, 2014.

be mentioned that the preferences of the decision maker are usually not known in advance as he does not know the different possibilities of solutions and their corresponding objective vectors before starting the run of the algorithm.

The runtime analysis of interactive evolutionary multi-objective optimization has been started recently by Brockhoff et al. [3]. The authors considered the algorithms iRLS and (1+1) iEA which are interactive versions of randomized local search and the (1+1) EA [8]. The algorithms iRLS and (1+1) iEA work on the Pareto dominance relation and use the knowledge of a decision maker to decide between incomparable search points. The influence of a linear decision maker using the weighted sum and a decision maker working with the Chebyshev utility function has been analyzed for two well known example problems called LOTZ and COCZ [3].

In this paper, we investigate the setting of Brockhoff et al. [3]. Our aim is to give a general characterization of problems where the use of a linear decision maker makes a multi-objective optimization problem as easy as the optimization of its single-objective functions. Here, we assume that the multi-objective problem consists of single-objective problems of the same type, e.g. a minimum spanning tree problem or a shortest path problem. We show that the linear decision maker turns such problems from a structural point of view into single-objective problems. This implies that we can translate known runtime results of RLS and (1+1) EA to their interactive versions in the multi-objective setting. For a decision maker using the Chebyshev utility function, we show that there are instances of the multi-objective setting of the knapsack problem where the interactive algorithms have an exponential expected optimization time.

After having examined multi-objective problems with linear objective functions, we turn our attention to the LeadingOnes problem. We examine multi-objective versions motivated by recent studies in the area of black box complexity [7]. Our results point out situations where iRLS and (1+1) iEA have difficulties in obtaining optimal solution according to the linear decision maker.

The outline of the paper is as follows. In Section 2 we introduce the setting for interactive multi-objective optimization and the algorithms that are subject to our analysis. In Section 3, we show how the decision maker may prevent Deteriorative Cycles where a new produced solution is worst than the previously obtained ones .In Section 4, we present a general study on a linear decision maker for multi-objective problems having linear objective functions. For a decision maker using the Chebyshev utility function we show in Section 5 that there are instances of knapsack problem leading to an exponential optimization time. Finally, we consider the general LeadingOnes problems in Section 6 in order to point out the situations where the linear decision maker runs into difficulties and finish with some concluding remarks.

2   Interactive Multi-objective Optimization

Throughout this paper, we investigate the impact of a decision maker who is involved in the optimization process for a given multi-objective problem.

A multi-objective optimization problem is given by a function $f: X \to \mathbb{R}^d$ that assigns to each element $x \in X$ of the considered search space X a vector $f(x) = (f_1(x), \ldots, f_d(x))$ consisting of d objective values. If not otherwise stated we assume that each of the d objectives should be minimized. A search point x weakly dominates a search point y $(x \preceq y)$ iff $f_i(x) \le f_i(y)$, $1 \le i \le d$. We say that x strongly dominates y $(x \prec y)$ iff $f_i(x) \le f_i(y)$, $1 \le i \le d$ and there exists an $j \in \{1, \ldots, d\}$ with $f_j(x) < f_j(y)$. Often the different objectives are in conflict with each other which means that there is no single solution which gives the minimal value for all objectives at the same time. We say that x and y are incomparable $(x \parallel y)$ if neither $x \preceq y$ nor $y \preceq x$ holds. The set $X^* = \{x \in X \mid \nexists y \in X \text{ with } y \prec x\}$ is called the Pareto optimal set and the set of corresponding objective vectors $PF = \{f(x) \mid x \in X^*\}$ is called the Pareto front.

The classical goal in multi-objective optimization is to compute a set of solutions that contains for each element of PF a corresponding solution. An alternative to computing such a set of trade-offs first and presenting it later on to a decision maker who picks one of the solutions for implementation, is to involve the decision maker in the optimization process. Asking the decision maker can be in particular very helpful when making decisions between solutions that are incomparable according to the Pareto dominance relation. Our goal is to study such approaches from a theoretical perspective and examine the influence of different types of decision makers on the optimization time.

Algorithm 1. ((1+1) iEA)

1. *Choose $x \in \{0,1\}^n$ uniformly at random*
2. *Repeat*
   - *Obtain y by flipping each bit of x with probability $1/n$.*
   - *If $y \preceq x$ then $x := y$*
   - *else if $x \parallel y$ then $x := D(x,y)$.*

In this practice, we consider the interactive version of the classical (1+1) EA. This algorithm called (1+1) iEA has been introduced in [3] and is shown in Algorithm 1. (1+1) iEA starts with a solution chosen uniformly at random from the search space $X = \{0,1\}^n$. In each iteration, a new solution y is produced by flipping each bit of the current solution x with probability $1/n$. The search point x is replaced by y if y weakly dominates x $(y \preceq x)$. If y is dominated by x $(x \prec y)$ then x remains unchanged. If x and y are incomparable $(x \parallel y)$ then the decision maker decides. The decision maker is a function $D : X \times X \to X$ which takes two search points x and y and returns one of them.

We study our algorithm with respect to the number of fitness evaluations until for the first time a Pareto optimal solution with the highest preference of the decision maker has been obtained. We call this the *optimization time* of the algorithm on a given problem. The expected number of fitness evaluations until this goal has been achieved is called the *expected optimization time*. When considering single-objective optimization problems, the *expected optimization time* is defined as the expected number of fitness evaluations until the algorithm has produced for the first time an optimal solution with respect to the given objective function.

2.1 Decision Makers

To model the decision maker, we have to specify the function $D : X \times X \to X$. We examine the two decision makers modelled in [3]. In the following, we assume that all objective functions should be minimized, but the setting can be easily adjusted in the case that some of the given objectives should be maximized.

The first is the weighted sum approach. For a given problem, the decision maker chooses a parameter $\lambda_i \in [0,1]$, $1 \le i \le d$ with $\sum_{i=1}^{d} \lambda_i = 1$, and sets $D(x,y) = y$ if

$$\sum_{i=1}^{d} \lambda_i f_i(y) \le \sum_{i=1}^{d} \lambda_i f_i(x),$$

and $D(x,y) = x$ otherwise. We call this the linear (or weighted sum) decision maker.

We also consider a decision maker using the Chebyshev utility function

$$u_c(f(x)) = \max_{i \in \{1,2,\dots,d\}} \{\lambda_i \cdot |z_i^* - f_i(x)|\}$$

where $z^* = (z_1^*, z_2^*, \dots, z_d^*)$ is a pre-defined utopian point and $\lambda_i \in [0,1]$, $1 \le i \le d$ with $\sum_{i=1}^{d} \lambda_i = 1$ are the weights determined by the decision maker.

We have $D(x,y) = y$ for the decision maker using the Chebyshev utility function iff $u_c(f(y)) \le u_c(f(x))$, and $D(x,y) = x$ otherwise.

3   Deteriorative Cycles

During the optimization run evolutionary algorithms for multi-objective optimization may produce solutions that are worse than solutions obtained previously with respect to the Pareto dominance relation [9]. Evolutionary algorithms for multi-objective optimization problems often encounter the problem of such deteriorative cycles. This is, in particular, the case if the algorithm has already obtained solutions that are close to the Pareto front. In this section, we study how the decision maker may prevent such behaviour. For a detailed discussion on the underlying principles of deteriorative cycles in the context of evolutionary multi-objective optimization we refer the reader to [2].

The decision maker does not necessarily impose a total order on the search space as it is the case for single-objective problems. The reason is that the order among the search points may not be transitive.

As an example (see Figure 1) considers three search points a, b, c with objective vectors $f(a) = (5,5)$, $f(b) = (4,4)$, $f(c) = (3,6)$ and let the preference of the decision maker be $D(b,c) = c$ and $D(c,a) = a$. In this way, an algorithm could move from a to b to c and back to a. Hence, an arbitrary decision maker does not prevent the presence of deteriorative cycles.

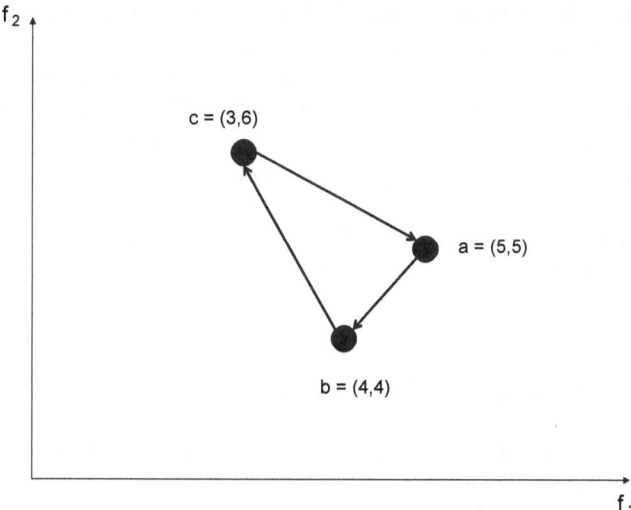

Fig. 1. Deteriorative cycle $a \to b \to c \to a$

3.1 Linear Decision Maker and Deteriorative Cycles

In the following, we show that the linear decision maker imposes a total ordering on the underlying search space which means that such an algorithm does not encounter deteriorative cycles.

Theorem 2. *The weighted sum decision maker induces a total order on the search space X.*

Proof. We show that an algorithm working with the Pareto dominance relation when considering comparable search points and working with the linear decision maker when encountering incomparable search points leads to a total order on the search space X.

Let $f: X \to \mathbb{R}^d$ and $x \preceq y$ iff $f_i(x) \leq f_i(y)$, $1 \leq i \leq d$. We define the order ($\preceq_L$) given by the Pareto dominance relation ($\preceq$) and the one by the linear decision maker as

$$x \preceq_L y \Leftrightarrow \sum_{i=1} \lambda_i f_i(x) \leq \sum_{i=1} \lambda_i f_i(y)$$

$\preceq_L$ is a total order as each search point is assigned a real value that is given by the weighted sum of its objectives.

Obviously, if $x \parallel y$ (according to the Pareto dominance relation) then the decision maker decides according to $\preceq_L$.

If $x \preceq y$ holds, then $f_i(x) \leq f_i(y)$, $1 \leq i \leq d$ and as a consequence we have

$$\sum_{i=1} \lambda_i f_i(x) \leq \sum_{i=1} \lambda_i f_i(y)$$

and hence $x \preceq_L y$.   □

The previous theorem shows that the linear decision maker prevents the presence of deteriorative cycles when working with algorithms such as (1+1) iEA. For (1+1) iEA, it also ensures convergence to the set of optimal solutions as the mutation operator has a positive probability of sampling any point in the search space $\{0,1\}^n$. Having produced a solution that is minimal with respect to $\preceq_L$ implies that (1+1) iEA will never accept a solution that is not minimal with respect to $\preceq_L$. We refer the reader to [8] for an n^n upper bound on any function defined on the search space $\{0,1\}^n$. Note, that there may be more than one optimal solution with respect to the utility function of the linear decision maker.

4 Linear Decision Maker and Linear Objective Functions

Many combinatorial optimization problems have a linear objective function that has to be optimized under a given set of constraints. This includes well known problems such as the knapsack problem or the minimum spanning tree problem. In this section, we study binary optimization problems that have linear objective functions.

4.1 Linear Objective Functions

Brockhoff et al. [3] have already made the observation that if the objective functions are linear and the underlying utility function of the decision maker is the weighted sum, then the expected optimization time of iRLS and (1+1) iEA is $\Theta(n \log n)$ (see Observation 2 in [3]).

Within this section, we want to examine this observation in greater detail by studying problems with d linear objective functions and some additional constraints. Our goal is to fit classical combinatorial optimization problems into this framework. Many combinatorial optimization problems have linear objective functions, but some additional constraints. Because of the presence of constraints the optimization time is usually not $\Theta(n \log n)$. However, we are able to relate the expected optimization time to the corresponding single-objective variants with using a decision maker working with the weighted sum.

Let P be a binary optimization problem, i.e. a problem consisting of r components where the ith component is chosen iff $x_i = 1$. We say that a binary problem P with r components has a linear objective function iff the fitness of a feasible search point x is given by $f(x) := \sum_{i=1}^{r} w_i x_i$.

Note, that we currently don't assume any restrictions on the constraints that have to be met in order to obtain a feasible solution.

W.l.o.g. we consider the case where we minimize d functions $f_1, f_2, \ldots, f_d$. Cases where at least one of the objectives has to be maximized can be treated in a similar way.

In the following, we assume that a solution x is either feasible for all objective functions or feasible for none of them. We consider evolutionary algorithms where feasible solutions are always better than infeasible solutions. For iRLS and (1+1) iEA this implies that after the algorithms have obtained a feasible solution for the first time, they will never accept an infeasible one. For the following theorem, we assume that the algorithms have already obtained a feasible solution.

Theorem 3. *Let P be a binary problem with a linear objective function and T be an upper bound on the expected optimization time of (1+1) EA on any input instance I of P when starting with an arbitrary feasible solution. Then the expected optimization time of (1+1) iEA using a linear decision maker is upper bounded by T when starting with a feasible solution.*

Proof. Let $I_1, I_2, \ldots, I_d$ be the single objective problems with objective functions

$$f_j(x) := \sum_{i=1}^{r} w_i^j x_i \qquad 1 \le j \le d$$

For a given fixed λ_j, $1 \le j \le d$, with $\sum_{j=1}^{d} \lambda_j = 1$, let

$$g(x) = \sum_{j=1}^{d} \lambda_j f_j(x) = \sum_{j=1}^{d} \lambda_j \left(\sum_{i=1}^{r} w_i^j x_i \right)$$

$$= \sum_{i=1}^{r} \left(\sum_{j=1}^{d} \lambda_j w_i^j \right) x_i = \sum_{i=1}^{r} g_i x_i$$

where $g_i = \sum_{j=1}^{d} \lambda_j w_i^j$. Note that g_i is completely determined by the input and the linear preference of the decision maker expressed by the choice of λ_j, $1 \le j \le d$.

We claim that (1+1) EA working on g accepts an offspring y of x iff (1+1) iEA working on $(f_1, f_2, \ldots f_d)$ accepts the offspring y of x.

We first assume that x and y are incomparable ($x \parallel y$). In this case, the decision maker involved in (1+1) iEA accepts y iff $g(y) \le g(x)$. Hence, y is accepted iff it is accepted by (1+1) EA working on g.

Secondly, we assume that x and y are comparable. If $y \preceq x$ then $g(y) \le g(x)$ and y is accepted by (1+1) EA and (1+1) iEA. If $x \prec y$, then $g(x) < g(y)$ and y is rejected by the (1+1) EA and (1+1) iEA. $\qquad\square$

4.2   The Knapsack Problem

In the knapsack problem the input is given by n items $1, \ldots, n$ where each item has a positive profit p_i and a positive weight w_i.

We consider the multi-objective setting for the problem where the goal is to maximize the overall profit and minimize the overall weight of the set of chosen items. We consider the search space $\{0,1\}^n$. For a bit-string x, item i is chosen iff $x_i = 1$. The fitness function $f : \{0,1\}^n \to \mathbb{R}^2$ is given by

$$f(x) = (p(x), w(x))$$

with

$$p(x) = \sum_{i=1}^{n} p_i x_i \text{ and } w(x) = \sum_{i=1}^{n} w_i x_i.$$

In the multi-objective setting, our goal is to maximize p and minimize w which introduces a partial order on the search points. $x \preceq y$ holds iff $p(x) \geq p(y)$ and $w(x) \leq w(y)$.

In order, to put it into our framework of minimizing all objectives, we can consider the case where we minimize w and minimize $-p$. If $x \parallel y$, the decision maker decides whether the new solution is accepted. For a fixed $\lambda \in [0, 1]$, $D(x,y)=y$ holds iff

$$(1 - \lambda)w(y) - \lambda p(y) \leq (1 - \lambda)w(x) - \lambda p(x)$$

and $D(x, y) = x$ otherwise.

The multi-objective formulation of the knapsack problem consists of two linear functions without any additional constraints. It is well-known that RLS and (1+1) EA optimize each linear function in time $O(n \log n)$ [8]. Together with Theorem 3, we get the following result.

Theorem 4. *Using the weighted sum utility function, the expected optimization of (1+1) iEA for the Knapsack problem is $O(n \log n)$.*

Using Theorem 3, similar results can be obtained for other multi-objective versions of classical combinatorial optimization problems having linear objective functions. For example, the runtime results on minimum spanning trees [11] and single-source shortest paths [1] can be transferred to the corresponding multi-objective problems when considering a linear decision maker.

5   Chebyshev Utility Function and the Knapsack Problem

In the following, we examine the use of a decision maker using the Chebyshev utility function. Our goal is to show that this decision maker makes it much more difficult to find the solution with the optimal preference even if there are two linear objective functions without any additional constraints.

We consider the following trap instance called KNAP2 which has been introduced in [13] in the context of the runtime analysis of evolutionary algorithms for constraint optimization. Let $p_1 = n, p_2 = \cdots = p_n = 1$ and $w_1 = n - 1, w_2 = \cdots = w_n = 1$. For the weight bound $W = n - 1$ has been chosen in [13] which implies that in the optimal solution only the first item is chosen.

For the multi-objective setting and the Chebyshev utility function we set the utopian point to $z^* = (2n, -2)$. This meets the requirement of an utopian point as $\sum_{i=1}^{n} p_i = 2n - 1 < 2n$ and each weight is positive and therefore greater than -2. Furthermore, we set $\lambda_1 = \lambda_2 = 1/2$.

The optimal solution is the string $x^* = (1, 0 \ldots 0)$ where $f(x^*) = (n, n-1)$ and $u_c(f(x^*)) = \max\{\frac{1}{2} \cdot (2n - n), \frac{1}{2} \cdot | -2 - (n-1)|\} = \frac{n+1}{2}$. x^* dominates the search point $x_l = (0, 1 \ldots 1)$ with $f(x_l) = (n-1, n-1)$ and $u_c(f(x_l)) = \max\{\frac{1}{2} \cdot (2n - (n-1)), \frac{1}{2} \cdot | -2 - (n-1)|\} = \frac{n+1}{2}$. Furthermore, x^* and x_l are incomparable to any other search point $y \in \{0, 1\}^n \setminus \{x^*, x_l\}$.

Consider a search point y where $y = (0, y_1)$ which starts with a 0-bit and has i, $0 \le i \le n-2$, ones in the remaining part y_1. Clearly $f(x) = (i, i)$ and $u_c(f(y)) = \max\{\frac{1}{2} \cdot (2n - i), \frac{1}{2} \cdot (i + 2)\} = \frac{2n-i}{2} \ge \frac{n+2}{2} > \frac{n+1}{2}$.

Consider a search point y where $y = (1, y_1)$ which starts with a 1-bit and has i, $1 \le i \le n-1$, ones in the remaining part y_1. Clearly $f(x) = (n+i, n+i-1)$ and $u_c(f(y)) = \max\{\frac{1}{2} \cdot (n-i), \frac{1}{2} \cdot (n+i-1+2)\} = \frac{n+i+1}{2} \ge \frac{n+2}{2} > \frac{n+1}{2}$.

Theorem 5. *Using the weighted Chebyshev utility function u_c with $z^* = (2n, -2)$ and $\lambda_1 = \lambda_2 = 1/2$, the optimization time of the (1+1) iEA on KNAP2 is $e^{\Omega(n)}$ with probability $\alpha = \Omega(1)$.*

Proof. The first bit is set with probability $1/2$ to 1 and with probability $1/2$ to 0 in the initial solution. We claim that this decides on whether the algorithm ends up in the local optimum x_l or the global one x^*.

Let $x^i = (x_1^i y_1)$ be the initial solution. Suppose that $x_1^i = 0$ holds (which happens with probability $1/2$). The part y_1 has at least $n/2 - \epsilon n$, $\epsilon > 0$ a constant, 1-bits with probability $1 - e^{-\Omega(n)}$ using Chernoff bounds.

Consider a phase of $T = cn$ steps where c is an appropriate constant. We claim that the number of 1-bits in y_1 is at least $n/2 + \epsilon n$ and that the bit x_1 has not been flipped during this phase. A solution with a 0 at the first bit and i 1-bits in the y_1 part has fitness (i, i) and utility value $\frac{(2n-i)}{2}$. Hence, a solution increasing the number of 1-bits in y_1 is accepted.

As long as y_1 does not contain at least $n/2 + \epsilon n$ 1-bits, the probability of increasing the number of 1-bits in y_1 is at least

$$(n/2 - \epsilon n) \frac{1}{n} \cdot (1 - 1/n)^{n-1} \ge (n/2 - \epsilon n)/(en).$$

The expected time to have obtained a solution with at least $n/2 + \epsilon n$, $\epsilon > 0$ a small constant, 1-bits is at most

$$2\epsilon n(en/(n/2 - \epsilon n)) \le 2\epsilon 3en = 6e\epsilon n.$$

The probability that the bit x_1 has not been flipped in T steps is

$$(1 - 1/n)^T = (1 - 1/n)^{cn} > \left(\frac{1}{2e}\right)^c.$$

We set $T = \alpha_1 \cdot 6e\epsilon n$. This implies that the probability of not having obtain at least $n/2 + \epsilon n$ 1-bits in y_1 is upper bounded by $1/(\alpha_1)$ using Markov's inequality and the probability that x_1 has not been flipped is at least

$$\left(\frac{1}{2e}\right)^{\alpha_1 \cdot 6e\epsilon} = \Omega(1).$$

Having obtained this solution starting with 0 and having at least $n/2 + \epsilon n$ 1-bits in the y_1 part, the utility value is at most $\frac{1.5n-\epsilon n}{2}$. A solution starting with a 1-bit and having at least $n/2$ 1-bits in the y_1-part has utility value at least $\frac{1.5n+1}{2}$ and is therefore not accepted as an offspring. Hence, only a solution having a 1-bit at the first position is accepted if at least ϵn bits flip at the same time in a single mutation step. The probability that ϵn bits flip in a single mutation steps is asymptotically Poisson distributed with parameter 1 and therefore $e^{-\Omega(\epsilon n)}$. This implies that the optimization time of (1+1) iEA is $e^{\Omega(n)}$ with probability $\Omega(1)$.   □

6   The Multi-Objective Leading Ones Problem

In this section, we investigate when using a linear utility function to model the decision maker leads to problems in the optimization process. To do this, we consider a generalization of the classical Leading Ones problem and present exponential lower bounds for the considered multi-objective problem.

Leading Ones (LO) problem was first introduced in [12] and counts the number of leading ones in a given bitstring. It is defined as

$$LO(x) = \sum_{i=1}^{n} \prod_{j=1}^{i} x_j$$

Motivated by the work in [7], where the complexity of black-box optimization on LO was analyzed, we introduce a new problem similar to the traditional Leading Ones. Given a predefined vector $a \in \{0,1\}^n$,

$$LO_a(x) = \sum_{i=1}^{n} \prod_{j=1}^{i} (1 - |x_j - a_j|)$$

counts the number of leading bits of the given solution x that agrees with a. Given two vectors $a, b \in \{0,1\}^n$, we consider a bi-objective maximization problem with objective function $MLO_{a,b}(x) = (LO_a(x), LO_b(x))$. The goal is to maximize both objective functions. Obviously it is possible to generalize the problem to d objectives by having d bitstrings and measuring the agreement of a given solution with respect to them such that d objective values are computed. In this section, we are interested in showing lower bounds for iRLS and (1+1) iEA when working with the linear decision maker. We will investigate the bi-objective problem $MLO_{a,b}$ with $a = 1^t 0^{n-t}$ and $b = 1^n$ for a given fixed value t and show when the algorithms are not able to obtain a solution with the maximal preference of the decision maker. Note that the problem has two Pareto optimal solutions, namely the strings $a = 1^t 0^{n-t}$ and $b = 1^n$ and that the weightening of the objectives decides on which one is the string with the maximum preference according to the linear decision maker.

As we are deadling with bi-objective problems, the weightening is decided by one parameter λ, $0 \le \lambda \le 1$, and utility value according to the decision maker is given by

$$\lambda \cdot LO_a(x) + (1 - \lambda) \cdot LO_b(x).$$

As we are dealing with maximizing problems the utility value should be maximized as well. Note, that if $\lambda = 1/2$ then both Pareto optimal solutions have maximum utility, whereas $\lambda > 1/2$ implies that a is the optimal solution and $\lambda < 1/2$ implies that b is the optimal solution.

In the following, we assume that $0 < \lambda < 1$ as $\lambda = 0$ or $\lambda = 1$ implies that one of the objectives can be neglected and the expected optimization time would be $\Theta(n^2)$. Furthermore, we assume $\lambda > 1/2$ such that the algorithm favours a as the optimal solution. The case $\lambda < 1/2$ can be handled in a symmetric way.

Theorem 6. *Let $\lambda > 1/2$, $a = 1^t0^{n-t}$, and $b = 1^n$. Then the optimization time of (1+1) iEA on $MLO_{a,b}$ is at least $n^{\frac{k(1-\lambda)}{2\lambda}}$ with probability $2^{-k} \cdot (1 - n^{-1/2})$ where $t = n/2$ and $t + k < n$ holds.*

Proof. If $LO_a(x) = LO_b(x) < t$ holds, the probability of increasing $LO_a(x)$ and $LO_b(x)$ is at least $1/n$. Hence, after an expected number of $O(nt)$ steps, $LO_a(x) \geq t$ and $LO_b(x) \geq t$.

Let x be the first solution in the run of the algorithm for which $LO_a(x) \geq t$ and $LO_b(x) \geq t$ holds. Since all bits at positions greater than t are still uniformly at random in x, we have $x = 1^t1^k0[0, 1]^{n-k-t-1}$ for $n-t \geq k \geq 2$ with probability 2^{-k}. In order to reach the optimal solution a, the algorithm has to accept an offspring y of x that is incomparable to x. Consider a potential offspring $y = 1^t0^c1[0, 1]^{n-t-c-1}$ of x such that x and y are incomparable. The solution y is accepted iff

$$g(x) - g(y) = k - \lambda k - \lambda c = k(1 - \lambda) - \lambda c \leq 0 \Leftrightarrow k \leq \frac{\lambda c}{1 - \lambda}.$$

This implies that in one single mutation step, $c \geq \frac{k(1-\lambda)}{\lambda}$ specific bits of the current solution x must be flipped. The probability for such a mutation is at most $n^{-\frac{k(1-\lambda)}{\lambda}}$. Let $T = n^{\frac{k(1-\lambda)}{2\lambda}}$, then the probability to obtain such a solution in T steps is at most $n^{-1/2}$. Hence, with probability $2^{-k} \cdot (1 - n^{-1/2})$, the runtime of (1+1) iEA on $MLO_{a,b}$ is at least $n^{\frac{k(1-\lambda)}{2\lambda}}$. $\square$

The previous result shows that even multi-objective versions of simple problems such as Leading Ones can become difficult to solve when using a decision maker with a linear utility function.

7   Conclusions

Incorporating the decision maker into the optimization process of evolutionary multi-objective optimization has become a very popular approach. In this paper, we have studied simple evolutionary algorithms from a theoretical perspective. Our studies show that important multi-objective combinatorial optimization problems such as the multi-objective formulation of the knapsack problems, multi-objective minimum spanning trees or multi-objective shortest paths become as easy for iRLS and (1+1) iEA as their single-objective counterparts when working with a linear decision maker. Furthermore, we have pointed out for the knapsack problem that this is in general not the case when working with the Chebyshev utility function. Our studies for the multi-objective LeadingOnes problem show situations where the algorithms using the linear decision maker fail to obtain a solution of maximal preference in expected polynomial time.

References

1. Baswana, S., Biswas, S., Doerr, B., Friedrich, T., Kurur, P.P., Neumann, F.: Computing single source shortest paths using single-objective fitness functions. In: Jansen, T., Garibay, I., Wiegand, R.P., Wu, A.S. (eds.) Proceedings of the 10th International Workshop on Foundations of Genetic Algorithms (FOGA 2009), Orlando, USA, pp. 59–66. ACM Press (2009)
2. Berghammer, R., Friedrich, T., Neumann, F.: Convergence of set-based multi-objective optimization, indicators and deteriorative cycles. Theor. Comput. Sci. 456, 2–17 (2012)
3. Brockhoff, D., López-Ibáñez, M., Naujoks, B., Rudolph, G.: Runtime analysis of simple interactive evolutionary biobjective optimization algorithms. In: Coello, C.A.C., Cutello, V., Deb, K., Forrest, S., Nicosia, G., Pavone, M. (eds.) PPSN 2012, Part I. LNCS, vol. 7491, pp. 123–132. Springer, Heidelberg (2012)
4. Coello Coello, C.A., Van Veldhuizen, D.A., Lamont, G.B.: Evolutionary Algorithms for Solving Multi-Objective Problems. Kluwer Academic Publishers, New York (2002)
5. Deb, K.: Multi-objective optimization using evolutionary algorithms. Wiley, Chichester (2001)
6. Deb, K., Sinha, A., Korhonen, P.J., Wallenius, J.: An interactive evolutionary multiobjective optimization method based on progressively approximated value functions. IEEE Trans. Evolutionary Computation 14(5), 723–739 (2010)
7. Doerr, B., Winzen, C.: Black-box complexity: Breaking the $o(n \log n)$ barrier of leadingones. In: Hao, J.-K., Legrand, P., Collet, P., Monmarché, N., Lutton, E., Schoenauer, M. (eds.) EA 2011. LNCS, vol. 7401, pp. 205–216. Springer, Heidelberg (2012)
8. Droste, S., Jansen, T., Wegener, I.: On the analysis of the (1+1) evolutionary algorithm. Theor. Comput. Sci. 276, 51–81 (2002)
9. Hanne, T.: On the convergence of multiobjective evolutionary algorithms. European Journal of Operational Research 117(3), 553–564 (1999)
10. Jaszkiewicz, A., Branke, J.: Interactive multiobjective evolutionary algorithms. In: Branke, J., Deb, K., Miettinen, K., Słowiński, R. (eds.) Multiobjective Optimization. LNCS, vol. 5252, pp. 179–193. Springer, Heidelberg (2008)
11. Neumann, F., Wegener, I.: Randomized local search, evolutionary algorithms, and the minimum spanning tree problem. Theor. Comput. Sci. 378(1), 32–40 (2007)
12. Rudolph, G.: Convergence properties of evolutionary algorithms. Kovač, Hamburg (1997)
13. Zhou, Y., He, J.: A runtime analysis of evolutionary algorithms for constrained optimization problems. IEEE Trans. Evolutionary Computation 11(5), 608–619 (2007)

Beyond the Edge of Feasibility:
Analysis of Bottlenecks

Mohammad Reza Bonyadi, Zbigniew Michalewicz, and Markus Wagner

Optimisation and Logistics
The University of Adelaide, Australia
http://cs.adelaide.edu.au/~optlog/

Abstract. The productivity of real-world systems is often limited by so-called bottlenecks. Hence, usually companies are not only interested in finding the best ways to schedule their current resources so that their benefits are maximized (optimization), but, in order to increase the productivity, they also conduct some analysis to find bottlenecks in their system and eliminate them in the most efficient way (e.g., with the lowest investment). We show that the current frequently used analysis (based on average shadow price) for identifying bottlenecks has some limitations: (1) it is limited to linear constraints, (2) it does not consider all potential sources for bottlenecks in a system, and (3) it does not provide adequate tools for decision makers to find the best way of investment to eliminate bottlenecks and maximize the profit they can gain. We propose a more comprehensive definition of bottlenecks that covers these limitations. Based on this new definition, we propose a multi-objective model for the benefit and investment. The solution for this model provides the best way of investment in resources to achieve maximum profit. As the proposed model is multi-objective and non-linear, it opens an important opportunity for the application of evolutionary algorithms, which can subsequently have a significant impact on the decision making process of companies.

Keywords: Constraints, bottlenecks, what-if analysis, feasibility.

1 Introduction

Usually real-world optimization problems contain constraints in their formulation. The definition of constraints in management sciences is "anything that limits a system from achieving higher performance versus its goal" [5]. In general, a constrained optimization problem (COP) is defined as:

$$\text{find } x \in S \text{ s.t. } z = max\{f(x)\} \text{ subject to} \tag{1}$$

$$g_i(x) \leq 0, \text{ for } i = 1 \text{ to } q$$

$$h_i(x) = 0, \text{ for } i = q+1 \text{ to } m$$

where f, g_i, and h_i are real-valued functions on the search space S, q is the number of inequalities, $m - q$ is the number of equalities, and s.t. is the short form of "such that" [2, 12]. Hereafter, the term COP refers to this formulation.

G. Dick et al. (Eds.): SEAL 2014, LNCS 8886, pp. 431–442, 2014.

It is believed that the optimal solution of most real-world optimization problems is found on the edge of a feasible area of the search space of the problem [15]. This belief is not limited to computer science, but it is also found in operational research (linear programming, LP) [3] and management sciences (theory of constraints) [11, 14] articles. The reason behind this belief is that, in real-world optimization problems, constraints usually represent limitations of availability of resources. As it is usually beneficial to utilize the resources as much as possible to achieve a high-quality solution (in terms of the objective value, f), it is expected that the optimal solution is a point where a subset of these resources is used as much as possible, i.e., $g_i(x^*) = 0$ for some i and a particular high-quality x^* in the general formulation of COPs [1]. Thus, the best feasible point is usually located where the value of these constraints achieve their maximum values (0 in the general formulation). The constraints whose values are maximized at the optimum point are called *active* constraints. The constraints that are active at the optimum solution can be thought of as *bottlenecks* that constrain the achievement of a better objective value [11, 13].

Decision makers in industries usually use some tools, known as decision support systems (DSS) [8], as a guidance for their decisions in different areas of their systems. Probably the most important areas that decision makers need guidance from DSS are: (1) optimizing schedules of resources to gain more benefit (accomplished by an optimizer in DSS), (2) identifying bottlenecks (accomplished by analyzing constraints in DSS), and (3) determining the best ways for future investments to improve their profits (accomplished by an analysis for removing bottlenecks[1], known as what-if analysis in DSS). Such support tools are more readily available than one might initially think: for example, the widespread desktop application Microsoft Excel provides these via an add-in.[2]

Identification of bottlenecks and the best way of investment is at least as valuable as the optimization in many real-world problems from an industrial point of view because: "An hour lost at a bottleneck is an hour lost for the entire system. An hour saved at a non-bottleneck is a mirage" [6]. Industries are not only after finding the best schedules of their resources (optimizing the objective function), but they are also after understanding the tradeoffs between various possible investments and potential benefits. During the past 30 years, evolutionary computation methodologies (e.g., evolutionary algorithms) have provided appropriate tools for optimization. However, the last two areas (identifying bottlenecks and removing them) that are needed in DSSs seem to have remained untouched by evolutionary computation methodologies while it has been an active research area in management and operations research.

In this article, we review some existing studies on identifying and removing bottlenecks. We investigate the most frequently used bottlenecks removing analysis (the so-called average shadow price [3]) and identify its limitations.

---

[1] The term *removing a bottleneck* refers to the investment in the resources related to that bottleneck to prevent those resources from constraining the problem solver to achieve better objective values.

[2] http://tinyurl.com/msexceldss, last accessed 29th March 2014.

We argue that the root of these limitations can be found in the interpretation of constraints and the definition of bottlenecks. In particular, the previous studies have assumed only linear constraints and they have related bottlenecks only to one specific property of resources (usually the availability of resources). Further, they have not provided appropriate tools to guide decision makers in finding the best ways of investments in their system so that their profits are maximized by removing the bottlenecks. We propose a more comprehensive definition for bottlenecks that not only leads us to design a more comprehensive model for determining the best investment in the system, but also addresses all mentioned limitations. Because the new model is multi-objective and may lead to the formulation of non-linear objective functions/constraints, evolutionary algorithms have a good potential to be successful on this proposed model. In fact, by applying multi-objective evolutionary algorithms to the proposed model, the found solutions represent points that optimize the objective function and the way of investment with different budgets at the same time.

This article is structured as follows. We explain the relevant concepts in Section 2. In Section 3, we highlight limitations of a well-known bottleneck definition. In Section 4, we present our model of bottlenecks that addresses these limitations. In Section 5 we present two first evolutionary approaches that consider investments in order to remove bottlenecks. We conclude the paper in Section 6, where we also provide directions for future research.

2 Background

In this section, we provide background information on linear programming, the concept of shadow price, and bottlenecks in general. Let us begin with linear programming. A Linear Programming (LP) problem is a special case of COP (as defined in eq. 1), where $f(x)$ and $g_i(x)$ are linear functions:

$$\text{find } x \text{ such that } z = max\{c^T x\} \text{ subject to } Ax \leq b \qquad (2)$$

where A is a $m \times d$ dimensional matrix known as *coefficients matrix*, m is the number of constraints, d is the number of dimensions, c is a d-dimensional vector, b is a m-dimensional vector known as Right Hand Side (RHS), $x \in \mathbb{R}^d$, and $x \geq 0$. The shadow price (SP) for the ith constraint of this problem is the value of z when b_i is increased by one unit. This in fact refers to the best achievable solution if the RHS of the ith constraint was larger, i.e., that more resources were available [10].

The concept of SP in Integer Linear Programming (ILP) is different from the one in LP [13]. The definition for ILP is similar to the definition of LP, except that $x \in \mathbb{Z}^d$. In ILP, the concept of Average Shadow Price (ASP) was introduced in [9]. Let us define the *perturbation function* $z_i(w)$ as follows:

$$\text{find } x \text{ s.t. } z_i(w) = max\{c^T x\} \text{ subject to } a_i x \leq b_i + w, \ a_k x \leq b_k, \forall k \neq i \quad (3)$$

where a_i is the ith row of the matrix A and $x \geq 0$. Then, the ASP for the ith constraint is defined by $ASP_i = sup_{w>0}\{(z_i(w) - z_i(0))/w\}$. ASP_i represents that if adding one unit of the resource i costs p and $p < ASP_i$, then it is beneficial (the total profit is increased) to buy w units of this resource. This information is very valuable for the decision maker as it is helpful for removing bottlenecks. Although the value of ASP_i refers to "buying" new resources, it is possible to similarly define a selling shadow price [9]. The concept of ASP was extended in a way that a set of resources were considered [4] rather than only one resource at a time. Note, however, that this set is predefined by the user and then the analysis is conducted [4]. There, it was also shown that ASP can be used in mixed integer LP (MILP) problems.

Now, let us take a step back from the definition of ASP in the context of ILP, and let us see how it fits into a bigger picture of resources and bottlenecks. As we mentioned earlier, constraints usually model availability of resources and limit the optimizers to achieve the best possible solution which maximizes (minimizes) the objective function [10, 11, 14]. Although finding the best solution with the current resources is valuable for decision makers, it is also valuable to explore opportunities to improve solutions by adding more resources (e.g., purchasing new equipment) [9]. In fact, industries are after the most efficient way of investment (removing the bottlenecks) so that their profit is improved the most.

Let us assume that the decision maker has the option of providing some additional resource of type i at a price p. It is clearly valuable if the problem solver can determine if adding a unit of this resource can be beneficial in terms of improving the best achievable objective value. It is, however, not necessarily the case that adding a new resource of the type i improves the best achievable objective value. As an example, consider there are some trucks that load products into some trains for transportation. It might be the case that adding a new train does not provide any opportunity for gaining extra benefit because the current number of trucks is too low and they can not fill the trains in time. In this case, we can say that the number of trucks is a bottleneck. Although it is easy to define bottleneck intuitively, it is not trivial to define this term in general.

There are a few different definitions for bottlenecks [13]. A definition for bottlenecks was proposed in [13] which was claimed to be the most comprehensive definition: "a set of constraints with positive average shadow price". In fact, the average shadow price in a linear and integer linear program can be considered as a measure for bottlenecks in a system [11].

3   Limitations of the Existing Bottleneck Definition

Although ASP can be useful in determining the bottlenecks in a system, it has some limitations when it comes to removing bottlenecks. In this section, we discuss some limitations of removing bottlenecks based on ASP.

Obviously, the concept of ASP has been only defined for LP and MILP, but not for problems with non-linear objective functions and constraints. Thus, using the concept of ASP prevents us from analyzing bottlenecks in a non-linear system.

Let us consider the following simple problem[3] (the problem is extremely simple and it has been only given as an example to clarify limitations of the previous definitions): in a mine operation, there are 19 trucks and two trains. Trucks are used to fill trains with some products and trains are used to transport products to a destination. The rate of the operation for each truck is 100 tonnes per hour (tph) and the capacity of each train is 2,000 tonnes. What is the maximum tonnage that can be loaded to the trains in one hour? The ILP model for this problem is given in eq. 4:

$$\text{find } x \text{ and } y \text{ s.t. } z = max\{2000y\} \text{ subject to} \qquad (4)$$
$$g_1 : 2000y - 100x \leq 0, \qquad g_2 : x \leq 19, \qquad g_3 : y \leq 2$$

where $x \geq 0$ is the number of trucks and $y \geq 0$ is the number of loaded trains (y can be a floating point value which refers to partially loaded trains). The constraint g_1 limits the amount of products loaded by the trucks into the trains (trucks can not overload the trains). The solution is obviously $y = 0.95$ and $x = 19$ with objective value 1,900. ASPs for the constraints are as follows:

- ASP for g_1 is 1: by adding one unit to the first constraint ($2000y - 100x \leq 0$ becomes $2000y - 100x \leq 1$) the objective value increases by 1,
- ASP for g_2 is 100: by adding 1 unit to the second constraint ($x \leq 19$ becomes $x \leq 20$) the objective value increases by 100,
- ASP for g_3 is 0: by adding 1 unit to the third constraint ($y \leq 2$ becomes $y \leq 3$) the objective value does not increase.

Accordingly, the first and second constraints are bottlenecks as their corresponding ASPs are positive. Thus, it would be beneficial if investments are concentrated on adding one unit to the first or second constraint to improve the objective value. Adding one unit to the first constraint is meaningless from the practical point of view. In fact, adding one unit to RHS of the constraint g_1 means that the amount of products that is loaded into the trains can exceed the trains' capacities by one ton, which is not justifiable. In the above example, there is another option for the decision maker to achieve a better solution: if it is possible to improve the operation rate of the trucks to 101 tph, the best achievable solution is improved to 1,919 tonnes. Thus, it is clear that the bottleneck might be a specification of a resource (the operation rate of trucks in our example) that is expressed by a value in the coefficients matrix and not necessarily RHS. The commonly used ASP, which only gives information about the impact of changing RHS in a constraint, cannot formulate such bottlenecks.

Figure 1 illustrates this limitation. The value of ASP represents only the effects of changing the value of RHS of the constraints (Figure 1, left) on the objective value while it does not give any information about the effects the values in the coefficients matrix might have on the objective value (constraint g_1 in Figure 1,

---

[3] We have made several such industry-inspired stories and benchmarks available:
   http://cs.adelaide.edu.au/~optlog/research/bottleneck-stories.htm

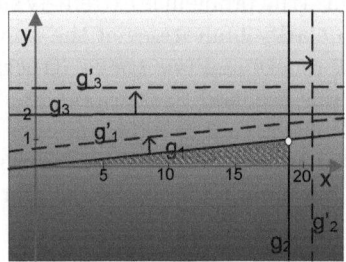 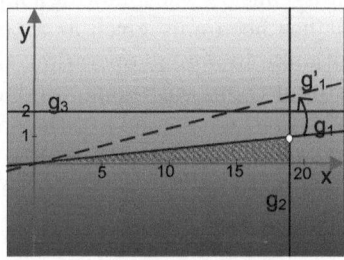

Fig. 1. x and y are number of trucks and number of trains respectively, gray gradient: indication of objective value (the lighter the better), shaded area: feasible area, g_1, g_2, g_3 are constraints, the white point is the best feasible point

right). However, as we showed in our example, it is possible to change the values in the coefficient matrix in order to achieve better solutions.

The value of ASP does not provide any information about the best strategy of selecting bottlenecks to remove. In fact, it only provides information about the benefit of elevating the RHS in each constraint or a given set of constraints and does not say anything about the order of significance of the bottlenecks. It remains the task of the decision maker to compare different scenarios by selecting different subset of constraints (also known as *what-if* analysis[4]). Of course one can analyze all possible subsets of constraints to find which subset is the most beneficial one to invest on. However, this strategy potentially leads to solving another hard problem, that is a subset selection. From a managerial point of view, it is important to answer the following question: is adding one unit to the first constraint (if possible) better than adding one unit to the second constraint (purchase a new truck)? Note that in real-world problems, there might be many resources and constraints, and a manual analysis of different scenarios might be prohibitively time consuming. Thus, a smart strategy is needed to find the best set of to-be-removed bottlenecks in order to gain maximum profit with lowest investment. In summary, the limitations of identifying bottlenecks using ASP are:

Limitation 1. ASP is only applicable if objective and constraints are linear.

Limitation 2. ASP does not evaluate changes in the coefficients matrix (the matrix A) and it is only limited to RHS.

Limitation 3. ASP does not provide information about the strategy for investment in resources, and the decision maker has to manually conduct analyses to find the best investment strategy.

---

[4] In the operational research community, there are related terms such as sensitive analysis and post-optimality [7].

4   A New Model for Bottleneck

In this section a new definition for bottleneck and a new model for removing bottlenecks (investment) is proposed that addresses limitations listed in Section 3.

Each constraint g_i in a real-world optimization problem usually models not only the availability of resources, but also other *specifications* of resources such as rates and capacities. Each of these specifications is encoded in a coefficient in the constraints. Accordingly, we propose a new definition for bottleneck:

Definition 1. *A bottleneck is a modifiable specification of resources that by changing its value, the best achievable performance of the system is improved.*

Note that this definition is a generalization of the definition of bottleneck in [13]: "a set of constraints with positive average shadow price is defined as a bottleneck". In fact, the definition in [13] concentrated on RHS only (it is just about the average shadow price) and it considers a bottleneck as a set of constraints, while our definition is based on any modifiable coefficient in the constraints (from capacity, to rates, or availability) and it introduces each specification of resources as a potential bottleneck. As an example, based on our definition, the operational rate of trucks can be a bottleneck, while according to the definition in [13], this is not possible [5] (see Limitation 2 in Section 3). Of course the set of all modifiable specifications need to be provided by the user.

According to the proposed bottlenecks definition, in order to invest on a part of a system to achieve maximum improvement of the objective of that system, not only RHS of all constraints should be assessed, but also all modifiable specifications in constraints need to be processed (e.g., tuning up trucks rather than buying new trucks) for potential changes. Hence, it is clear that the earlier methodologies based on ASP were not able to process all opportunities for removing bottlenecks and investments to maximize improvement in the objective function (see Limitation 3 in Section 3).

We propose a new model to address the limitations of ASP for removing bottlenecks and finding the best way of investment. We define the vector l_i which contains all modifiable specifications in the constraint g_i. For any COP in the form of eq. 1, we define a Bottleneck COP (BCOP) as follows:

$$\text{find } x \text{ and } l \text{ s.t. } z = \begin{cases} max(f(x, l)) \\ min(B(l)) \end{cases} \quad \text{subject to } g_i(x, l_i) \leq 0 \text{ for all } i \quad (5)$$

where l is a vector (l might contain continuous or discrete values) which contains l_i for all i and $B(l)$ is a function that calculates the cost of modified specifications of resources coded in the vector l. Note that in the linear case, $l_i = \{a_i, b_i\}$ where a_i is the ith row of the matrix A in eq. 2. If we consider $l_i = \{b_i\}$ and

---

[5] One can argue that the operational rate is another constraint that can be modeled by a new variable ($g_4 : v = 100$ in eq. 4). However, if this constraint is added to the definition of the problem then constraint g_1 becomes non-linear ($g_1 : 2000y - vx \leq 0$) which then is not suitable for ASP.

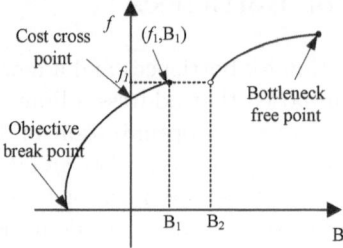

Fig. 2. The impact of an investment (B) on the achievable objective value f (maximization is the goal): positive/negative investments (reduction of resources) usually results in an increase/decrease in the best achievable objective value

$g_i(x) = a_i x - b_i$ (linear constraints) then eq.5 can express ASP_i. Figure 2 shows that, if all pieces of equipment are sold, the best achievable objective value is zero ($f = 0$) because nothing can be produced anymore (this is called "objective break" point in the figure), that is the same as the selling shadow price [9]. The "cost cross" point shows the point where the best objective value is achieved with the current specification of resources ($B = 0$). The point "bottleneck free" is the point where the optimum solution of the search space is inside the feasible region. From a practical point of view, in this situation, no matter how the decision maker invests, the profit is not improved any more. Note also that sometimes the amount of investment up to some value might not change the best achievable objective value. As an example, any investment from B_1 to B_2 does not result in any improvement in the objective value.

Let us assume that the associated solution to the point (f_1, B_1) is x' and l'. This solution can be interpreted as if the decision maker invests B_1, the best way to spend this budget is to change the specifications values to l' (which costs B_1) and the best achievable objective value in this case is f_1. Note that:

- a BCOP can be formulated for linear and non-linear systems,
- any modifiable specification of resources can be formulated in a BCOP in the vector l and the values for this vector are examined by the solver,
- solutions for a BCOP contain best investment strategies for various budgets.

Any multi-objective optimization algorithm can be applied to solve a BCOP. Also, as the specifications can be coefficients in the constraints, the constraints become non-linear, which makes the problem non-linear so that linear programming methodologies are not useful in solving this problem.

Let use assume that, in the example from Section 3, the decision maker can budget $500,000 to improve the maximum loaded products per hours into the trains. Also, the specifications that can be altered in the system are:

- the operation rate of the trucks can be increased up to 120 (i.e., $l_1 = \{100, 101, ..., 120\}$) for $100 per tph per truck,
- the capacity of trains can be increased to 2100, with the step size 20 (i.e., $l_2 = \{2000, 2020, 2040, 2060, ..., 2100\}$ tonnes) for $200 per ton per train,

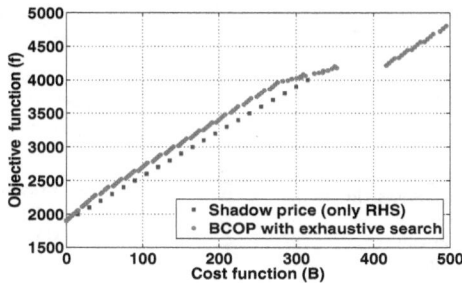

Fig. 3. Impact of an investment on the achievable objective value

– the number of trucks can be increased up to 40 (i.e. $l_3 = \{19, 20, ..., 40\}$),
  each truck costs \$15,000,
– the number of trains can be increased up to 5 (i.e $l_4 = \{2, 3, 4, 5\}$), each train
  costs \$100,000.

The BCOP for this example is written as:

$$\text{find } x \text{ and } l \text{ such that } z = \begin{cases} max\{l_2y\} \\ min\{0.1l_3(l_1 - 100) + 0.2l_4(l_2 - 2000)+ \\ 15(l_3 - 19) + 100(l_4 - 2)\} \end{cases} \quad (6)$$

$$\text{subject to} \qquad l_2y - l_1x \leq 0, \ x \leq l_3, \ y \leq l_4$$

Note that $B(l) = 0$ for $l = \{100, 2000, 19, 2\}$ (the current specification of
the resources). We can solve this problem by using two methods: changing the
value of RHS according to the ASP in eq. 4, or performing an exhaustive search
algorithm that solves BCOP in eq. 6. Note that only practically possible were
added to the list of solutions. Figure 3 shows the results.

It is clear that when BCOP is used more opportunities for investment are
examined, which can potentially result in higher benefits with smaller investments. As an example, with \$135,000 investment, a solution with $f = 2950$ is
found by solving BCOP with $l = \{118, 2000, 25, 2\}$. However, the best solution
found based on the shadow price for \$135,000 investment was only $f = 2800$ with
$l = \{100, 2000, 28, 2\}$. According to the solution of BCOP, if the decision maker
is going to invest \$135,000, the best way of investment (leading to maximum
improvement in the objective function) is to buy 6 new trucks (\$6 × 15000) and
upgrade all trucks to carry 118 tph (18 tph tune up, which costs \$18 × 100 × 25),
which all together costs \$135,000. However, by using the shadow price calcula-
tions, the decision maker needs to buy 9 new trucks (\$9 × 15000), which improves
the objective value to 2800. It is clear that better objective values can be achieved
by investing the same amount if we use BCOP.

5   An Evolutionary Algorithm for BCOP

In this section we propose two methods to solve BCOPs based on a multi-
objective genetic algorithm. As the first algorithm, we use a basic algorithm

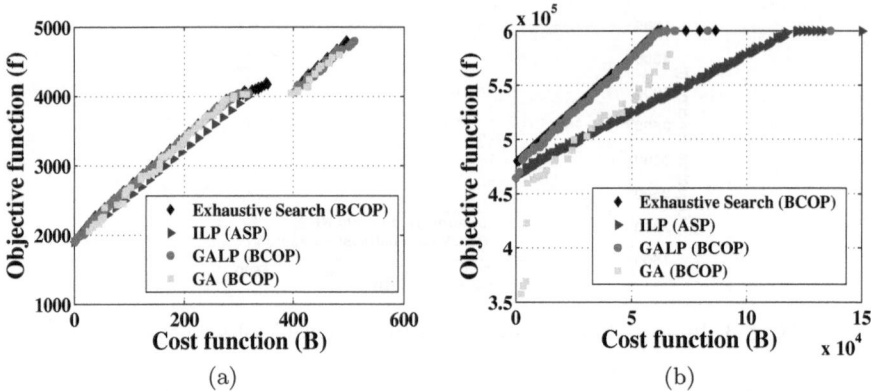

Fig. 4. Impact of an investment on the achievable objective value (a) train loading and (b) tomato farm

with tournament selection, one point crossover ($p_c = 0.9$, set via some trials), and uniform random mutation ($p_m = 0.3$, set via some trials). Each individual contains the vector l and all decision variables x. We call this first approach GA.

To handle multiple objectives, we use the following simple approach. Two solutions x_0 and x_1 are compared based on $G(x) = \sum_{i=1}^{m} max(g(x), 0)$, which is known as constraint violation value. If $G(x_0) = G(x_1)$ or $G(x_0) \leq 0$ and $G(x_1) \leq 0$, then we use the dominance relation in multi-dimensional spaces (if both are equal or non-dominating select one randomly, otherwise select the dominating one). Otherwise, we select the solution that is better in terms of constraint violation value (preferring smaller constraint violation values).

The second multi-objective algorithm GALP is based on GA. Here, the individuals contain only the vector l, and linear programming is used to find the best vector x for each generated l.

We applied both methods to the problem defined in eq. 6 with 100 individuals for 100 iterations (all non-dominating solutions found are reported in Fig. 4(a)). It is clear that both evolutionary methods have found good approximations of the Pareto front (computed by exhaustive search). It is also clear that (1) the performance of the GALP is slightly better than that of GA, and (2) our basic approaches outperform the established ASP based approach.

This means that both our approaches can be used to better plan the best investment for industries. In the following, let us consider a second example, this time from agriculture, to illustrate that our formulation is straight-forward and that it can support the decision making processes in the real-world.

Second Scenario: Agricultural Allocation.[3] A farmer owns 1,000 acres of more or less homogeneous farmland. His options are to breed cattle, or to grow wheat, corn, or tomatoes. Annually, 12,000 hours of labor are available. For simplicity, we will assume here that these can be used at any time during the year, i.e., through hiring casual labor during seasons of high need, e.g., for harvesting.

In the following, we list information regarding the profit, yield, and labor needs for the four economic activities:

- cattle: $1,600/head profit, 0.25 heads/acre yield, 40 h/head annual labor
- wheat: $5/bushel profit, 50 bushels/acre yield, 10 h/acre annual labor
- corn: $6/bushel profit, 80 bushels/acre yield, 12 h/acre annual labor
- tomatoes: 50 cent/lb profit, 1,000 lbs/acre yield, 25 h/acre annual labor

It is required that at least 20% of the farmland that is cultivated in the process is used for the purpose of cattle breeding, at most 30% of the available farmland can be used for growing tomatoes, and the ratio between the amount of farmland assigned to growing wheat and that left uncultivated should not exceed 2 to 1.

Now (and this is the challenging bit), the farmer can make certain investments that can possibly increase the overall profit per year: (1) additional acres of farmland can be rented at $200 per year, (2) additional labor can be hired at $20 per hour, (3) a tomato packing machine can be rented for $5,000 per year, which reduces 25 h/acre to 20 h/acres, (4) a "tomato grower's licence" can be bought for $10,000 per year, which increases the max ratio from 30% to 35%, and (5) the value 0.25 heads/acre can be improved up to 0.7 heads/acres for $10,000 (0.25 needs $0, 0.7 needs $10,000, and anything in between is linear).

The question now is: should the farmer invest, and if so, how? In Figure 4(b) we show the results of the different approaches.[6] Just as in the previous train loading example, our evolutionary approach GALP that makes use of the BCOP formulation clearly outperform the approach based on average shadow price. The results of GALP are close to those of found by an exhaustive search. Note that the approach based on average shadow price is not able to assess all cases for investment, which makes GALP more appropriate to find best investment plan. Note that in this example the methods were run for 1000 iterations.

6   Conclusions and Directions

In this paper we proposed a new definition for bottlenecks and a new model to guide decision makers to make the most profitable investment. We did this in order to narrow the gap between what is being considered in academia and industry. Our definition for bottlenecks and model for investment overcomes several of the drawbacks of the model that is based on average shadow prices:

1. It can work with non-linear constraints and objectives.
2. It offers changes to the coefficient matrix.
3. It can provide a guide towards optimal investments.

This more general model can form the basis for more comprehensive analytical tools as well as improved optimization algorithms. In particular for the latter application, we conjecture that nature-inspired approaches are adequate, due to the multi-objective formulation of the problem and its non-linearity.

---

[6] The construction of the BCOP formulation is straight-forward, and we omit it due to space constraints. It is available on the above-mentioned website of our *stories*.

Bottlenecks are ubiquitous and companies make significant efforts to eliminate them to the best extent possible. To the best of our knowledge, however, there seems to be very little published research on approaches to identify bottlenecks—research on optimal investment strategies in the presence of bottlenecks seems to be even non-existent. In the future, we will push this research further, in order to improve decision support systems. We will design nature-inspired single-objective and multi-objective algorithms with the goal to support decision makers to make the best possible investments in their constrained systems.

References

[1] M. R. Bonyadi and Z. Michalewicz. On the edge of feasibility: a case study of the particle swarm optimizer. In *CEC*, pp. 3059–3066. IEEE, 2014.

[2] M. R. Bonyadi, X. Li, and Z. Michalewicz. A hybrid particle swarm with velocity mutation for constraint optimization problems. In *Genetic and Evolutionary Computation Conference (GECCO)*, pp. 1–8. ACM, 2013.

[3] A. Charnes and W. W. Cooper. *Management models and industrial applications of linear programming*, Vol. 1. John Wiley and Sons, 1961.

[4] A. Crema. Average shadow price in a mixed integer linear programming problem. *European Journal of Operational Research*, 85:625–635, 1995.

[5] E. M. Goldratt. *Theory of constraints*. Nrth. River Cro.-on-Hud., NY, 1990.

[6] E. M. Goldratt and J. Cox. *The goal: a process of ongoing improvement*. Gower, 1993.

[7] P. A. Jensen and J. F. Bard. *Operations research models and methods*. John Wiley & Sons Incorporated, 2003.

[8] P. G. W. Keen. Value analysis: Justifying decision support systems. *Management Information Systems Research Center Quarterly*, 5:1–15, 1981.

[9] S. Kim and S.-c. Cho. A shadow price in integer programming for management decision. *Europ. Journal of Operational Research*, 37:328–335, 1988.

[10] T. C. Koopmans. Concepts of optimality and their uses. *The American Economic Review*, pp. 261–274, 1977.

[11] R. Luebbe and B. Finch. Theory of constraints and linear programming: a comparison. *Int. Journal of Production Research*, 30:1471–1478, 1992.

[12] Z. Michalewicz and M. Schoenauer. Evolutionary algorithms for constrained parameter optimization problems. *Evolutionary Computation*, 4:1–32, 1996.

[13] S. Mukherjee and A. Chatterjee. Unified concept of bottleneck. IIMA Working Papers WP2006-05-01, Indian Institute of Management Ahmedabad, Research and Publication Department.

[14] S. Rahman. Theory of constraints: A review of the philosophy and applications. *Int. Jour. of Oper. and Prod. Manag.*, 18:336–355, 1998.

[15] M. Schoenauer and Z. Michalewicz. Evolutionary computation at the edge of feasibility. In *Parallel Problem Solving from Nature (PPSN)*, Vol. 1141 of *LNCS*, pp. 245–254. Springer, 1996.

Adaptive Risk Aversion in Social Dilemmas

Michael Kirley and Friedrich Burkhard von der Osten

Department of Computing and Information Systems
The University of Melbourne
mkirley@unimelb.edu.au, fvon@student.unimelb.edu.au

Abstract. Risks are ubiquitous in many social dilemmas. A decision to
cooperate (or invest in a common pool) is inherently "risky" when the
reward received depends on the frequency of actions displayed by other
social group members. In this paper, we study the evolutionary dynamics
of the multiple-group public goods game where the population of mobile
individuals is divided into fixed-sized groups. Agents use continuous investment strategies. This approach allows for the evolution of an intermediate level of investments correlated with specific risk aversion levels.
Agents are also presented with an opportunity to switch groups at the
end of each round based on an environmental trigger. Detailed simulation
experiments using an evolutionary game theory framework show that investment levels can be maintained within groups. Over time, the mean
level of the risk aversion trait increases especially for larger groups. In
the conditional migration scenarios, levels of investment consistent with
risk aversion emerge.

1 Introduction

Social dilemmas, that is situations in which collective and private interests conflict, have been studied extensively in many domains, including evolutionary biology, social psychology, economics, statistical physics, and multi-agent systems
[9,25,21,4]. Evolutionary games, such as the multi-player public goods game, are
often used as an abstract mathematical framework to investigate mechanisms
used to promote and maintain cooperative behaviour in social dilemmas. Here,
groups of cooperators outperform groups of non-cooperators, but selfish individuals always do better than cooperators in their group [19,21].

A number of hypotheses explaining why individuals might contribute to a
public good, including reciprocal altruism; reputation; punishment; spatial selection and multi-level selection, have been proposed [18,19]. However, one important direction that requires further research is the role of *risk* on population
dynamics and emergent behaviour. Uncertainty and risk, including risk aversion, risk seeking and risk preference, are key features of many social dilemmas
[16,20,26]. Genetic factors have been used to help explain this observation in
human experiments, together with environmental variables and individual be-
havioural histories [1,2]. From a social psychological science perspective, people
tend to be risk averse when dealing with outcomes that are gains relative to

G. Dick et al. (Eds.): SEAL 2014, LNCS 8886, pp. 443–454, 2014.

their reference point – they chose sure smaller gains over larger, riskier gains, but became risk seeking when dealing with losses [15]. This is in contrast to the widely accepted economic perspective, where a rational decision-maker is an individual who attempts to maximize their expected utility in any decision-making scenario.

Recently, agent-based models investigating the evolution of risk taking [24] and risk aversion [12] have appeared. Starting from an assumption that risk seeking behaviour is not favoured by natural selection, inherent biases guiding decision-making strategies have been examined. Hintze and co-workers [12] report that beneficial risk aversion adaptations emerge in small populations, particularly when a larger population is divided into smaller groups, with limited migration between groups. We take this particular model as the starting point for our investigation.

In this paper, we use an evolutionary game theory framework to investigate dynamics in the public goods game, where the population of mobile agents is divided into fixed-sized interaction groups. Agents use continuous investment strategies based on their risk aversion trait. Individuals with a high trait value are considered to be risk averse, corresponding to a low chance of investing a small amount. In contrast, individuals with a low trait value are considered to have a risk seeking strategy, corresponding to a high chance of investing a large amount. At each round of the game, an individual can update their risk trait by imitating successful (fitter) individuals within their current group. We also introduce mobility modes or group switching mechanisms into the model: (a) random migration and (b) conditional migration, where each individual monitors its current environmental condition and attempts to migrate to another group if the condition is found to be undesirable.

We use Monte Carlo simulation experiments to investigate model behaviour, as it is not practical to study this group-structured model analytically. Key model parameters include the population size, the number of groups, and mobility mode. The use of a real-value for the risk investment trait generates rich dynamics. Instead of reporting results based on the "levels of cooperation" as is the norm in many social dilemma papers, we report results illustrating the evolutionary trajectory of the risk aversion trait across the population, and indirectly the average investment level. Across the population as a whole, the average value of the risk aversion trait is significantly lower in smaller groups and in the random migration model. We find that lower levels of investment are typically maintained in the conditional migration model suggesting that individuals who make decisions based on whether they are satisfied with their local neighbourhood tend to be more risk averse.

The remainder of this paper is organized as follows. In Section 2, the public goods game with continuous strategies is formally described. Related work describing mobility mechanisms in social dilemmas is also introduced. Our model is introduced in Section 3. In Section 4, the simulation experiments are described and results presented. We discuss the results, conclusions and avenues for future work in Section 5.

2   Background and Related Work

2.1   Public Goods Game Using Continuous Investment Strategies

In the continuous version of the public goods game, the traditional notion of discrete strategies, is extended by considering specific circumstances and risk orientation of individuals [13,10]. At an abstract level, this has some similarities with the probabilistic participation framework introduced in [22].

Assume that n individuals each make an investment x_i in a public good, where each $x_i \in (i = 1, \ldots, n)$ is a real number between 0 and some positive maximum value V. The payoff to individual i is given by:

$$\pi_i = (k/n) \sum_{j=1}^{n} x_j - x_i \tag{1}$$

where k is a positive constant (which can be viewed as a synergy factor or interest rate). When $1 < k < n$ every individual will maximize their payoff by making a zero investment, irrespective of the investments made by the other individuals (ie act as a free-rider). If all the players make a zero investment, they each receive a payoff 0. If all the players invest V, each would receive $(k - 1) \times V$

The evolutionary dynamics of the game can be described by the adaptive replicator dynamics for a continuous strategy space [7,6]. Under such conditions, the replicator equation assumes that the population state is described by a Borel probability measure P over $[0, V]$. The expected payoff of an individual playing x_i in a group of n players, where the other $n - 1$ players are chosen at random is:

$$\pi(x_i, P) = V + (\frac{k}{n} - 1)x_i + \frac{k}{n}\bar{x}(n - 1) \tag{2}$$

where $\bar{x} = \int_{[0,V]} x_j P(dx_j)$ is the average contribution of an individual in the population.

2.2   Mobility in Social Dilemma Models

It is now widely accepted that mobility plays an important role in affecting evolutionary dynamics in group structured social dilemmas (see [8,19]). In addition to random mobility – or random exchange of individuals between groups – the effectiveness of alternative mobility mechanisms have been investigated in spatial models. Notable examples include success-driven migration [11], adaptive migration [14] and aspiration-induced migration [28]. Success-driven migration is based on the idea that individuals can elect to move to sites with higher expected payoffs. The challenge of success-driven migration, however, is to determine in advance the potential payoff of the "non-local" site. In contrast, in the adaptive migration model individuals only make use of local information when attempting a move. In the study presented by Jiang et al. [14], adaptive migration took place probabilistically in proportion to the number of defectors in the neighbourhood.

Aspiration-induced migration is a third alternative, where individuals move to a new site if their payoff is below a certain aspiration level. Each of these migration schemes has been shown to enhance the extent of cooperative behaviour considerably, even in a noisy environment [11] or in an environment dominated by defectors [14].

In the context of multi-player social dilemma games, the efficacy of conditional mobility mechanisms have been reported in recent work by Chiong and Kirley [4,5]. However, that work was restricted to spatial models (regular lattice). Notable work investigating evolutionary dynamics in populations divided into subgroups focussed on the public goods game include the work of Killingback et al. [17] and Janssen and Goldstone [13]. These studies report on the effects of variable groups sizes and the impacts of random migration mechanisms on the emergent cooperation levels.

3   Model

Our model consists of a population of agents playing a multiple-group version of the continuous public goods game. The population is composed of g disjoint fixed-sized interaction groups, each of size n. Each agent must decide (a) how much to invest into the public pool, and (b) whether to switch groups or stay in their current group. Each of these decisions is inherently risky.

Figure 1 provides a schematic overview of the multiple group public goods game. Details of the model are discussed below.

3.1   Investment Strategy and Payoff

Each individual is defined by real value genetic trait (or strategy) χ, used to determine the level of investment in a given round of the public goods game. The level of investment x_i depends on χ as follows:

$$x_i = \begin{cases} (1 - \chi) * V & p = 1 - \chi \\ 0 & p = \chi \end{cases} \tag{3}$$

where the probability p is any value between 0 and 1, including the two endpoints, thereby capturing purely deterministic behaviour; and V is a parameter of the model controlling the maximum possible investment. The strategy x_i is thus defined by a Bernoulli random variable representing the expressed behaviour or "phenotype" of an individual given the underlying "genotype" χ. Here, χ acts as an indicator of risk aversion. When $\chi > 0.5$, there is a low chance of investing a small amount – risk averse behaviour, whereas when $\chi < 0.5$, there is a high chance of investing a large amount – risk seeking behaviour.

After each individual i has made an investment decision x_i using equation 3, the accumulated investments of the n group members are multiplied by a factor k and evenly redistributed amongst the members of the group based on equation 1. This payoff π_i is added to an individual's assets (initialized as V),

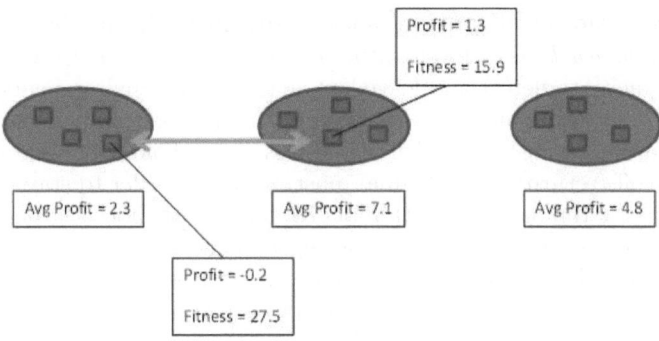

Fig. 1. The multiple group public goods game. Each agent (red square) is allocated to a fixed size group. Agent investment is based on equation 3. Payoff is determined by equation 1. Fitness corresponds to cumulative payoffs. At each round, every agent is presented with an opportunity to change groups based on their "satisfaction" or fitness level. The migration direction is illustrated via the arrow.

which serve as a function of the individual's fitness. The cumulative asset of an individual is simply the sum over all π_i values in multiple rounds of the public goods game.

The fitness of an individual is determined relative to the performance of their current group. The more successful (fitter) individuals will be imitated by others, so that the number of individuals adopting a given risk trait χ will evolve over time. At each time step, an individual i can adopt the strategy χ_j from individual j within its current groups with probability

$$W(\chi_i \leftarrow \chi_j) = \frac{1}{1 + exp[(\pi_i - \pi_j)/T]} \tag{4}$$

where T quantifies the uncertainty by strategy adoptions (without loss of generality we use $T = 0.1$). All individuals are also subject to mutation. That is, with probability μ the offspring mutates to a random strategy; otherwise its risk trait χ is identical to its parent.

3.2   Migration: Inter-group Mobility

Agent mobility is a key step in our model. Here, the overarching goal is to mediate the gradual "migration" of risk aversion traits from one group to another [27,12]. We consider three migration models:

In the **random migration** model, a fixed proportion of individuals (migration rate λ) are randomly selected to migrate to a different group. For each migrant, a randomly selected individual from a randomly selected group is simply nominated to "swap" groups. This group-based model approximates the scenario of individuals evolving in small groups with some level of inter-group mobility. Such an approach is the typical mechanism used in island-based evolutionary algorithms [3].

In the **conditional migration** model, individuals playing the game use *environmental feedback* as a mechanism to trigger migration. Here, individuals have the capacity to detect and leave low-quality social environments and share specific information about past group performance. We implement a form of the so-called *walk-away-rule*, which introduces a threshold value that defines the minimal payoff (return) an individual must receive in order to stay in the same group, otherwise the individual attempts to migrate to another group. Whether the attempted move is successful or not depends on the relative fitness of a randomly selected individual in a nominated destination group, a technique adopted in [4,5]. Two different techniques are used to select the destination group: (a) select the "best" group, and (b) select the "worst" group, where we use the group payoff at the previous game iteration to determine the nominated group. The fitness value of the migrant is then compared with the randomly selected individual from the nominated group, and if it has a higher fitness value, the swap is complete. The aim of this assortment mechanism is to reduce exploitation by selfish group members.

4   Simulation Experiments

A series of Monte Carlo simulation experiments were performed to examine population dynamics in our model. Two questions guided the experimental design:

1. Is it possible to evolve risk aversion behaviour in the multiple group public goods game?
2. Does conditional migration offer greater opportunities for positive assortment when compared to random migration?

In all simulation experiments, the value of χ was initialized randomly from a uniform distribution at the beginning of a trial. The value of the maximum investment V and payoff multiplier k/n were 5.0 and 0.8 respectively.[1] In the reproduction stage, the mutation rate μ was 0.01. The random migration rate λ was set to 0.10, based on empirical search of appropriate values. In conditional migration, the threshold value used to determine whether an agent attempts to switch groups was simply the payoff value in the current round – a loss corresponds to a low-quality social environments. The payoff performance from each group was made available to all individuals.

All results listed below are mean values over 30 simulation trials with differing seeds for the random number generators. Error bars have been omitted from time-series plots as they are smoothed values and the standard errors are very small. Key model parameters investigated are: group size (n), the number of groups (g), and the mobility mechanism (random, conditional best, conditional worse).

---

[1] We have also run a large number of simulations using a range of V and k values. The results were qualitatively the same across a range of values, thus we do not include the results in this paper.

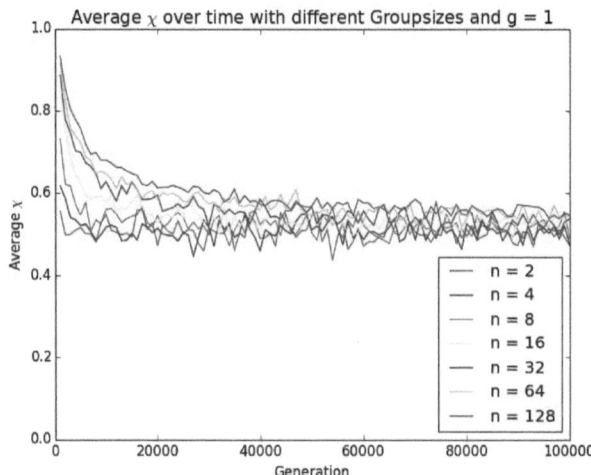

Fig. 2. The average value of χ across the population for varying population size n when $g = 1$

4.1 Evolutionary Trajectory of χ in a Single Group

We start by considering the population dynamics of a single group ($g = 1$) with varying population sizes $n \in \{2, 4, 8, 16, 32, 64, 128\}$. Figure 2 plots time series values for the average strategy χ *vs* generation number for different n values. For all values of n, the trajectory of χ settles around the mid-range value of 0.5 corresponding to the no risk/bias preference. The number of steps required to reach this value is correlated with the population size. Thus, the size of the group must be considered when analyzing perceptions of risk in the continuous public goods game.

4.2 Evolutionary Trajectory of χ in Multiple Groups

In the second set of simulation experiments, we investigate the underlying population dynamics in each of the migration models. Figure 3 plots the average value of χ, calculated over the final 2500 generations of a simulation trial, for each of the different group sizes considered $n \in \{2, 4, 8, 16, 32, 64, 128\}$ with $g = 32$ groups. Errors bars have been included on the plot. Based on the Wilcoxon rank sum test, there are significant differences between each of the migration modes considered ($p < 0.01$) for larger group sizes. When the group size is smaller ($n = 2$ and $n = 4$) the differences are not as clear cut. This is to be expected as the game is relatively "easy" (even though there is still a social dilemma) under this condition. It is important to note that as the size of the group increases there is transition to higher average χ values.

Figure 4 plots trends in the trajectory of the risk aversion trait χ. When $n = 4$, the differences between the random migration and conditional worst migration

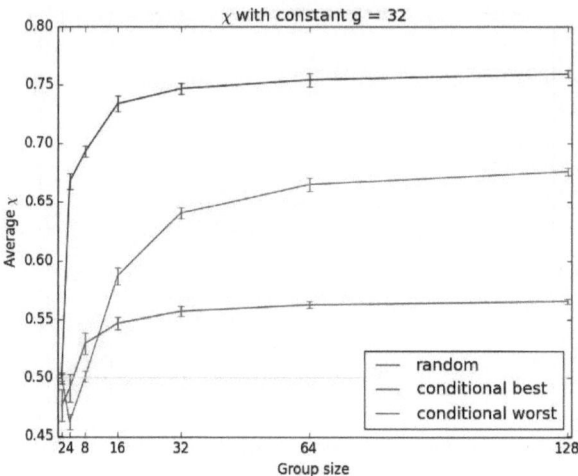

Fig. 3. The average value of χ across the population for varying groups sizes calculated over the last 2500 generations for each of the migration model. The yellow line indicates the expected value of 0.5 for unbiased evolution; i.e. no risk bias/preference.

strategy are not significant, with χ fluctuating around the 0.5 value. However, in the case of conditional migration, χ fluctuates around a value of 0.7, indicating relative high levels of average risk aversion across the population. However, when the group size is larger ($n = 32$), the evolutionary trajectory of the χ changes considerably. In each of the migration modes, the average value of χ is greater than the expected unbiased value of 0.5. The selection pressure embedded in the conditional migrations results in higher χ values, with average values around 0.8 emerging in the conditional best model.

The plots for average profit (Figures 5) reinforce the effectiveness of adopting a particular risk aversion/seeking investment level. Once again the effects are magnified in the larger groups. However, the differences between the conditional best and conditional worse modes are apparent in both group sizes. The results suggest that this transition to risk averse behaviour, especially in larger groups, is caused by the presence of "free-riders." Since the investment decision is based on both risk aversion trait and the expectation of other group members, we observe a qualitatively different relationship between risk averse investment levels and group sizes.

4.3   A Closer Look at Mobility

In the final set of simulation experiments, we investigate the effectiveness of the conditional migration mode in more detail. Conditional migration provides opportunities to taking advantage of potentially more beneficial social environments, while allowing for the possibility of leaving a group when that particular

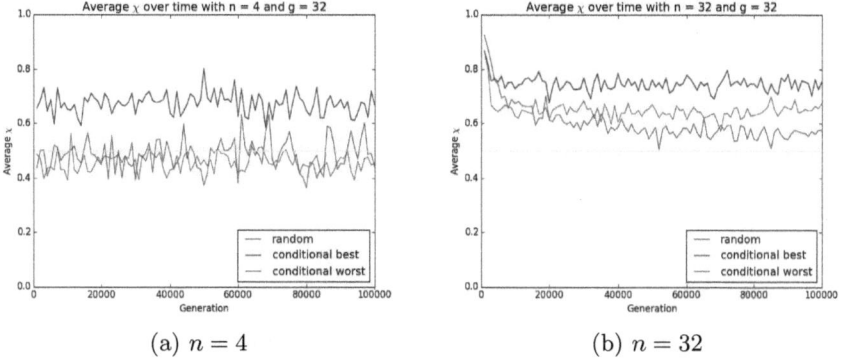

Fig. 4. The average value of χ *vs* generation number for each of the migration models. The expected value of 0.5 for unbiased evolution is also plotted.

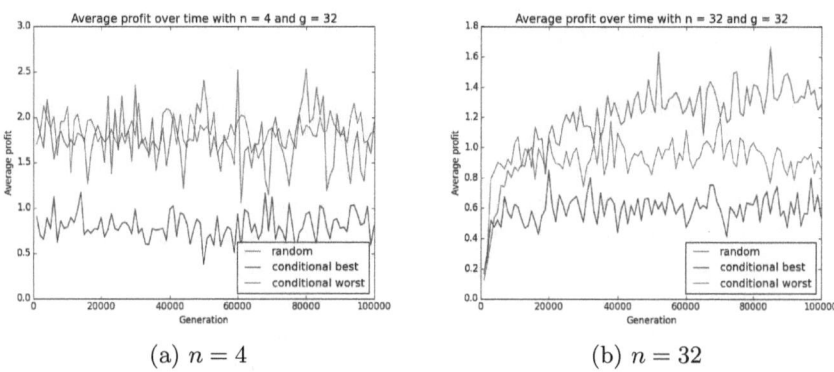

Fig. 5. The average profit *vs* generation number for each of the migration models

environment degrades (corresponding to low reward values, thus an individual is being exploited). However, a constraint on conditional movement was that the migrant must have a higher fitness value than the randomly selected individual from the nominated group (either the best or worse group depending on the model). This constraint was imposed to mediate the assortment effects.

Figures 6 plots time series values for the proportion of migrant "desired" and "successful" moves respectively for a typical population size ($g = 32$ and $n = 32$ in this case). Here, desired is simply when an individual flags that they wish to swap groups as the payoff received in the current game was not satisfactory. A successful move is when the individual actual switches groups. As expected, the proportion of successful moves is significantly lower than desired moves in the conditional modes, with lowest values apparent in the conditional best mode. This result is consistent with the expectation that potential risky moves may have a positive effect on assortment.

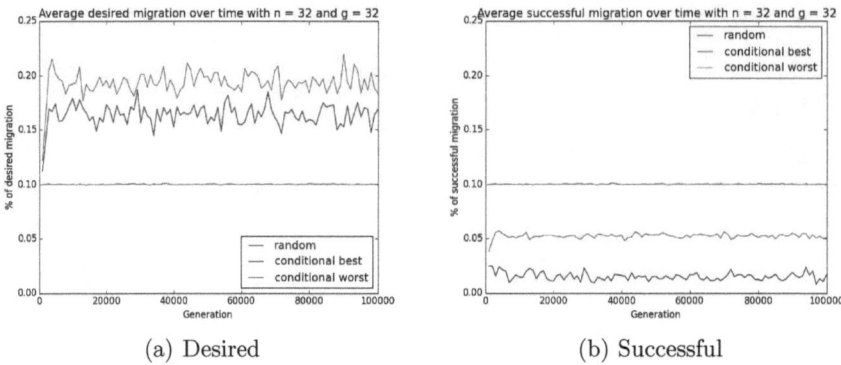

(a) Desired   (b) Successful

Fig. 6. Proportions of (a) desired and (b) successful moves *vs* generation number for each of the migration models when group size n=32 and $g = 32$

5   Discussion and Conclusions

The public goods game is a canonical group-based social dilemma game, requiring individuals to decide simultaneously whether they wish to contribute to the common pool or not. In this study, we have introduced two extensions to the game: an individual's investment strategy was based on a real value genetic trait corresponding to their level of risk aversion; and it was possible for individuals to switch between groups. Although risky, this change in group membership may be the preferred action to take if a satisfactory level of performance (payoff or return on investment) was not achieved in the current location.

Our model allows us to characterize differences in risk-taking behaviour between groups and individuals and among groups with different distributions of risk preferences. The simulation results listed in Section 4 clearly show that over time, the mean level of the risk aversion trait χ increase over time, especially for larger groups. Lower investment levels corresponding to risk adverse behaviour induces higher survivability, a result consistent with the work of [23] in social dilemma games and the work of [12] focussed on the adaptive benefits of risk aversion in an evolutionary context. The corresponding investment levels and payoff values are consistent with trends in cooperation levels typically reported in the literature. The random migration mode introduces a limited level of assortment. In contrast, conditional migration significantly affects population assortment. It is reasonable to expect that in the conditional worse mode, the proportion of successful switches is greater as, on average, the fitness of individuals in the worse pool will be relatively small, thus the chance of a successful move increases. However, the group may contain many free-riders, which adversely affect future performance. In the conditional best mode, the chance of a successful switch is less, but if successful the positive assortment leads to higher future returns, given the risk adverse behaviour. Regardless of the migration mode, the group size and not the total population size determines whether agents evolve risk averse strategies. In our model, a dynamic perception of risk is used to guide both investment levels and mobility.

There are a number of directions to explore in future work: In our model, we did not impose additional costs when changing groups. Exploring the effect of such costs and limiting the number of possible moves is an interesting direction for future empirical research. Despite the robustness of our findings, further applications in real social psychology settings may be problematic. If individuals are regularly faced with fluctuations of their general environment – variable risky situations – the proposed model should be tested experimentally to validate its conclusions.

References

1. Bateson, M.: Recent advances in our understanding of risk-sensitive foraging preferences. Proceedings of the Nutrition Society 61(04), 509–516 (2002)
2. Bell, A.M.: Approaching the genomics of risk-taking behavior. Advances in Genetics 68, 83–104 (2009)
3. Cantú-Paz, E.: A survey of parallel genetic algorithms. Calculateurs Parallèles, Réseaux et Systòmes Répartis 10(2), 141–171 (1998)
4. Chiong, R., Kirley, M.: Effects of iterated interactions in multiplayer spatial evolutionary games. IEEE Transactions on Evolutionary Computation 16(4), 537–555 (2012)
5. Chiong, R., Kirley, M.: A multi-agent based migration model for evolving cooperation in the spatial N-player snowdrift game. In: Boella, G., Elkind, E., Savarimuthu, B.T.R., Dignum, F., Purvis, M.K. (eds.) PRIMA 2013. LNCS, vol. 8291, pp. 70–84. Springer, Heidelberg (2013)
6. Cressman, R., Song, J.W., Zhang, B.Y., Tao, Y.: Cooperation and evolutionary dynamics in the public goods game with institutional incentives. Journal of Theoretical Biology 299, 144–151 (2012)
7. Doebeli, M., Dieckmann, U., Hordijk, L.: Adaptive dynamics of speciation: spatial structure. In: Adaptive Speciation, pp. 140–167 (2004)
8. Fu., F., Nowak, M.: Global migration can lead to stronger spatial selection than local migration. J. Stat. Phys. (3-4), 637–653 (2013)
9. Fu, F., Chen, X., Liu, L., Wang, L.: Social dilemmas in an online social network: the structure and evolution of cooperation. Physics Letters A 371(1-2), 58–64 (2007)
10. Hauert, C.: Cooperation, Collectives Formation and Specialization. Advances in Complex Systems 09(04), 315–335 (2006)
11. Helbing, D., Yu, W.: The outbreak of cooperation among success-driven individuals under noisy conditions. Proceedings of the National Academy of Sciences of the United States of America 106, 3680–3685 (2009)
12. Hintze, A., Olson, R.S., Adami, C., Hertwig, R.: Risk aversion as an evolutionary adaptation. arXiv preprint arXiv:1310.6338 (2013)
13. Janssen, M., Goldstone, R.: Dynamic-persistence of cooperation in public good games when group size is dynamic. Journal of Theoretical Biology 243(1), 134–142 (2006)
14. Jiang, L.L., Wang, W.X., Lai, Y.C., Wang, B.H.: Role of adaptive migration in promoting cooperation in spatial games. Physical Review E 81(3), 036108 (2010)
15. Kahneman, D., Tversky, A.: Prospect theory: An analysis of decision under risk. Econometrica: Journal of the Econometric Society, 263–291 (1979)

16. Kanagaretnam, K., Mestelman, S., Nainar, K., Shehata, M.: The impact of social value orientation and risk attitudes on trust and reciprocity. Journal of Economic Psychology 30(3), 368–380 (2009)

17. Killingback, T., Bieri, J., Flatt, T.: Evolution in group-structured populations can resolve the tragedy of the commons. Proceedings of the Biological Sciences / The Royal Society (1593), 1477–1481 (June)

18. Nowak, M.: Five rules for the evolution of cooperation. Science (5805), 1560–1563 (December)

19. Nowak, M.: Evolving cooperation. Journal of Theoretical Biology 299, 1–8 (2012)

20. Parks, C.D.: Risk preference as a predictor of cooperation in a social dilemma. In: Contemporary Psychological Research on Social Dilemmas, pp. 58–70 (2004)

21. Perc, M., Gómez-Gardeñes, J., Szolnoki, A., Floría, L.M., Moreno, Y.: Evolutionary dynamics of group interactions on structured populations: a review. Journal of The Royal Society Interface 10(80) (2013)

22. Sasaki, T., Okada, I., Unemi, T.: Probabilistic participation in public goods games. Proceedings of the Royal Society B: Biological Sciences 274(1625), 2639–2642 (2007)

23. Sella, G., Lachmann, M.: On the dynamic persistence of cooperation: How lower individual fitness induces higher survivability. Journal of Theoretical Biology 206(4), 465–485 (2000)

24. Stern, M.D.: Patrimony and the evolution of risk-taking. PloS One 5(7), e11656 (2010)

25. Szolnoki, A.: Promoting cooperation in social dilemmas via simple coevolutionary rules. The European Physical Journal B-Condensed Matter and Complex Systems 67(3), 337–344 (2009)

26. Van Assen, M., Snijders, C.: Effects of risk preferences in social dilemmas: a game-theoretical analysis and evidence from two experiments. In: Contemporary Psychological Research on Social Dilemmas, pp. 38–65 (2004)

27. West, S.A., Griffin, A.S., Gardner, A.: Social semantics: altruism, cooperation, mutualism, strong reciprocity and group selection. Journal of Evolutionary Biology 20(2), 415–432 (2007)

28. Yang, H.X., Wu, Z.X., Wang, B.H.: Role of aspiration-induced migration in cooperation. Physical Review E 81(6), 065101 (2010)

Fitness Landscape Analysis of Circles in a Square Packing Problems

Rachael Morgan and Marcus Gallagher

School of Information Technology and Electrical Engineering,
University of Queensland, Brisbane 4072, Australia
r.morgan4@uq.edu.au, marcusg@itee.uq.edu.au

Abstract. Fitness landscape analysis provides insight into the structural features of optimization problems. Landscape analysis techniques have been individually shown to capture specific continuous landscape features. However, results are typically for benchmark and artificial problems, and so the ability of techniques to capture real-world problem structures remains largely unknown. In this paper we experimentally examine and compare the ability of length scale analysis, dispersion, fitness distance correlation, information content, partial information content and information stability to characterise and distinguish instances of circle packing problems. Circle packing problems are an important abstraction of many real-world problems such as container loading, facility dispersion and sensor network layout problems. Experiments on incrementally scaled packings show that while all of the techniques provide some problem insight, only length scale analysis and information stability were able to clearly differentiate problem instances.

Keywords: Continuous optimization, fitness landscape analysis, circle packing in a square.

1 Introduction

Metaheuristics research has historically been dominated by the development of algorithms, but a recent focus has been to better understand both the relationship between algorithms and problems, and the nature of the problems themselves. To this end, fitness landscape analysis has produced theoretical frameworks and techniques for studying optimization problems. A considerable number of problem-specific and generalised analysis techniques have been proposed for combinatorial and discrete problems (see [15] for a review). In contrast, few techniques have been developed specifically for continuous problems, and practitioners often adapt discrete techniques to the continuous setting in an effort to analyse problems. Landscape analysis techniques have been shown to capture specific continuous landscape features on some example problems [8,12], however systematic comparisons of the techniques are largely missing from the literature. Furthermore, benchmark and artificial problems are typically utilised, and so the ability of the techniques to capture real-world problem structures remains unexplored.

G. Dick et al. (Eds.): SEAL 2014, LNCS 8886, pp. 455–466, 2014.

In this paper we examine and compare the ability of length scale analysis, dispersion, fitness distance correlation, information content, partial information content and information stability to characterise and distinguish instances of packing equally sized circles into a unit square. The circle-in-a-square (CiaS) packing problem is a convenient abstraction of real-world problems including container loading, facility dispersion and wireless network layout problems. Sec. 2 reviews the landscape analysis techniques used in this paper, and the CiaS problem is defined in Sec. 3. Analysis of the problem instances is conducted in Sec. 4, and Sec. 5 concludes the paper.

2   Fitness Landscape Analysis Techniques

The notion of an objective function as a 'fitness' landscape defined over a search space, $\mathcal{S}$, has been widely used as a model in evolutionary biology and computation. Formally, a fitness landscape is defined using a set of candidate solutions, $\mathbf{x} \in \mathcal{S}$, an objective function, f, and a distance function, d, between two arbitrary solutions. Fitness landscape analysis typically uses random, statistical or other sampling methods to obtain points of interest (and/or their fitness values) from a landscape. Properties of the landscape are then defined in this framework, e.g. a (strict) local optimum is a point $\mathbf{x}'$ where all neighbouring solutions have a fitness worse than $f(\mathbf{x}')$.

Length scale, denoted r, is defined as the change in objective function with respect to a step between two solutions. The length scales from all solution pairs capture a landscape's structural information and are amenable to statistical and information theoretic analysis [8,9]. Length scale was developed specifically for continuous problems, but is equally applicable in the discrete case. The distribution of r uniquely characterises problems and can be used to infer problem structure. Landscape features can be defined using measures such as the distribution's entropy, mean or median. Divergence measures such as KL-divergence can also be used to directly compare problems. Previously, 1D toy problems and 2D, 5D, 10D and 20D problems from the Black-Box Optimization Benchmarking (BBOB) problem set [5] have been analysed [8,9] using length scale.

Dispersion measures the degree to which high quality solutions are concentrated/clustered, where quality is determined by sampling m solutions and using truncation selection to retain the fittest tm solutions, where $t \in (0, 1]$ [6]. Dispersion was originally proposed for continuous problems, and has been shown to be a useful metric in studying the performance of population-based and local search algorithms relative to particular problems and their structures [6,20,14]. The dispersion values of the BBOB problem set have also been used in the feature-set of an algorithm prediction model [12]. Recently, we showed that the distance function and sampling methodology underlying dispersion can lead to convergent values in high dimensions, and adjustments to the methodology were made [10].

Landscape features conceptually similar to the discrete case can be defined mathematically (as suggested in [15]) for continuous problems, but defining neighbourhood relations and appropriate distance functions can be problematic. Despite these issues, some discrete landscape analysis techniques have been adapted to the continuous setting. *Fitness distance correlation* (FDC) measures the correlation between the fitness of solutions (from a finite sample) and their distance to the global optimum. The distance metric is a crucial component of FDC, and while the original specification used Hamming distance, Euclidean distance is typically used in the continuous setting. FDC has been used to analyse multi-layer perceptron neural networks [2], continuous NK-landscape models, the CEC 2005 benchmark functions [13] and the BBOB problem set [12].

Information content, partial information content and *information stability* aim to characterise ruggedness by analysing local neighbourhood structures and their interactions [19]. More specifically, information content is an entropic measure of the variety of six simplistic landscape structures, partial information content estimates the modality by analysing fitness fluctuations and information stability is defined as the largest fitness fluctuation between neighbours. The measures were originally defined in the discrete setting, where the notion of a neighbourhood is well-defined. Information content and related methods have been adapted for the continuous setting and applied to common benchmark problems in 1 and 30 dimensions [7] and 2D BBOB problems [11].

3   Circle Packing Problems

Packing problems have been widely studied in the mathematical and operations research literature, and in this paper we are interested in finding the optimal packing of n equally sized circles into a unit square. The problem can be formulated as finding the positions of n points inside the unit square such that their minimum pairwise distance, d_n, is maximized:

$$\max d_n = \max \min_{i \neq j} \| \mathbf{w}^i - \mathbf{w}^j \|_2 \tag{1}$$

$$\mathbf{w}^i \in [0, 1]^2 \qquad\qquad i = 1, \ldots, n \tag{2}$$

From the point of view of evaluating metaheuristic optimization algorithms, the problem given by (1) simply requires placing a set of n (2D) points within the unit square (hence the optimization problem is over $2n$ continuous variables). A candidate solution is then a vector of the circle coordinates, i.e. $\mathbf{x} = [w_1^1, w_2^1, \ldots, w_1^n, w_2^n]$.

CiaS packing problems can be considered as a simplified version of a number of different real-world problems and have received a large amount of attention in the mathematical, optimization and operations research literature (see [1] for an overview). For most values of n below 60 and for certain other values, provably optimal packings have been found using either theoretical or computational approaches [18]. The Packomania website [17] maintains a large list of the optimal (or best known) packings for many values of n from 2 up to 10000, along with references and other related resources.

CiaS packing problems represent a challenging class of optimization problems. In general, they cannot be solved using analytical approaches or via gradient-based mathematical optimization. These problems are also believed to generally contain an extremely large number of local optima. For the related problem of packing equal circles into a larger circular region, [4] estimate that the number of local optima grows unevenly but steadily, with at least 4000 local optima for $n = 25$ and more than 16000 local optima for $n = 40$. Special-purpose metaheuristics designed for the problem [1,18] have been applied, and recent results using basic continuous Estimation of Distribution algorithms ($UMDA_C$ and $EMNA$) on problems up to $n = 30$ achieved fairly modest average performance values that were between 5 and 25% worse than the global optimum values [3].

4   Experimental Results

In the following experiments we analyse the landscapes of CiaS packing problems and evaluate the ability of landscape features to (robustly) characterise these problems. In particular, we are interested in how well length scale analysis, dispersion, FDC, information content, partial information content and information stability can capture distinguishing landscape structures in CiaS instances for an increasing number of circles. Matlab source code is available at https://github.com/RachM/cias.

Given a solution of n circles, the ordering of the circles in the solution vector may be permuted without affecting d_n. Hence, for any given solution, there are $n!$ symmetric equivalent solutions. Generating $n!$ permutations for each solution sampled is computationally infeasible for large n, and so in these experiments we do not consider permutations of candidate solutions. We believe this is a reasonable design choice - algorithms are unlikely to generate permuted solutions, and so the landscapes we analyse here are the landscapes metaheuristics would typically search.

In these experiments, each feature is calculated using $2000n$ solutions sampled from $\mathcal{U}[0,1]^{2n}$ where $n = 2, 3, \ldots, 100$. To evaluate the robustness of each feature, 30 different samples are taken for each n. In the following we analyse length scale distributions, $p(r)$, and their respective entropies, $h(r)$. $p(r)$ is estimated using a kernel density estimator from length scales calculated using all combinations of solution pairs the sample. A Gaussian kernel is used where the bandwidth is chosen by the 'solve-the-equation-plug-in' method [16], and entropy is estimated numerically from $p(r)$. Dispersion is typically calculated using thresholds of 5% or less [6,12,10], and so in these experiments we use a 5% threshold and normalise values using bound-normalisation [10]. FDC is calculated using the globally optimal solution [17], as well as the best solution in the sample (we denote each estimator as $FDC_{\mathbf{x}^*}$ and $FDC_{\hat{\mathbf{x}}^*}$ respectively). The latter gives insight into how well FDC performs when the problem is treated as a black-box. FDC is typically interpretted in the context of minimisation, so in our FDC analysis, d_n was multiplied by -1. Many of the problems have multiple global optima resulting from the ability to rotate and/or reflect the circles in the global solution without affecting d_n [17], and for such packings, $FDC_{\mathbf{x}^*}$ is estimated using the distance between solutions and their *closest* global optima. Information content and its variants are based on random walks, and since we aim to compare features based on how effectively they characterise problems given the *same* sample of solutions, we treat the uniform random sample as a random walk. That is, the first solution sampled corresponds to the first step in the walk, the second solution sampled corresponds to the second step and so forth. This is similar to a random walk where steps are of size $\mathcal{U}[0,1]^{2n}$ in an isotropic, random direction, with proposed solutions 'accepted' if they are within the problem bounds. Information content and partial information content are estimated with $\epsilon = 0$, meaning transitions in objective function are 'neutral' if and only if the change in objective function value is 0. All results are reported using the mean and standard deviation (error bars) for each feature over 30 trials.

4.1   Length Scale Analysis

The entropy of the length scale distribution, shown in Fig. 1, clearly characterises and differentiates problems of different n. In addition, $h(r)$ has very low standard deviation across the different samples of the packings, which suggests that for these problems, it is a highly robust landscape feature. Since $h(r)$ is decreasing as n increases, the information required to describe r is decreasing, meaning that there is an increasing frequency of 'similar' length scales. This indicates that as n increases, the diversity of the changes in objective function between two random solutions is decreasing. Fig. 2(a) and 2(b) confirm this; Fig. 2(a) has a much heavier tail than 2(b). In particular, we see that for $n = 100$, $p(r)$ favours 'small' length scales compared to $p(r)$ for $n = 2$. To examine this more closely, the ratio of mode and 99th percentile is shown in Fig. 3. As n increases, both the mode and 99th percentile of $p(r)$ decrease, with the 99th percentile decreasing at a faster rate. The range of r in Fig. 2(b) is considerably smaller than the range in Fig. 2(a), indicating that the magnitude of objective function values for random solutions decreases as n increases. For CiaS problems, the objective function

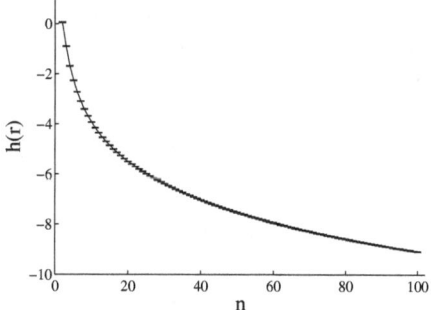

Fig. 1. $h(r)$ for CiaS problems as n increases

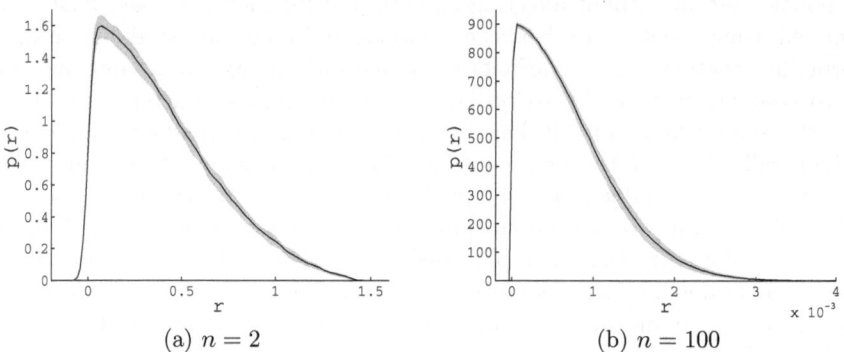

(a) $n = 2$ (b) $n = 100$

Fig. 2. The change in shape of $p(r)$ as n increases (note the change of scale in $p(r)$)

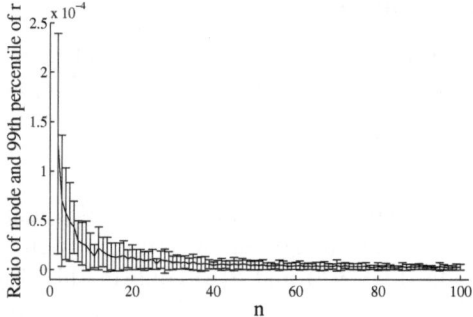

Fig. 3. Ratio of the mode and 99th percentile of r, confirming the significant change in the shape of $p(r)$ as n increases

value assigned to a solution is the maximum of the minimum distance between any two circle centres. As n increases, the radius of the circles decreases, and so for a random solution, we expect the minimum distance between two (uniform) randomly distributed circle centres to also decrease. However, our analysis of $p(r)$ and $h(r)$ provides compelling evidence that the decrease is complex and non-uniform across solutions. If the decrease across solutions was uniform, the length scale values would merely be scaled by a factor and we would see no change in the shape of $p(r)$. Hence, length scale analysis has uncovered two valuable insights into the nature of CiaS packing problems; as we increase the number of circles we pack, we can expect that 1) the packing of a (uniform) random configuration gets better and that 2) moving from one (uniform) random configuration to another will produce increasingly less significant changes in quality. The latter insight is extremely useful; potential applications include the design of more effective restart strategies for this problem and detecting algorithm convergence.

4.2   Dispersion

The bound-normalised dispersion values for the CiaS problems is shown in Fig. 4 and reveals a relatively large decrease in dispersion from 2 circles to 10 circles. A decrease in dispersion indicates that high quality solutions are increasingly closer together. Hence, Fig. 4 indicates that high quality solutions for $n = 10$ are closer together in S than high quality solutions for $n = 2$. Following the initial decrease in dispersion values, the dispersion increases slightly from $n = 10$ to $n = 40$. Unfortunately, subsequent analysis (not shown here) of the distance distributions for the fittest 5% of solutions for $n = 20, 40$ and 80 did not give insight into the increase. Fig. 4 also shows that for $n > 40$, dispersion is decreasing, meaning that high quality solutions are closer together as n increases. However, the decrease is very slight (approximately 0.0124), indicating an insignificant change in the landscape structure captured by dispersion. Overall, the bound-normalised dispersion has small variability between samples, however because the values are very similar and non-unique across problems $n > 10$, dispersion has extremely limited discriminative ability.

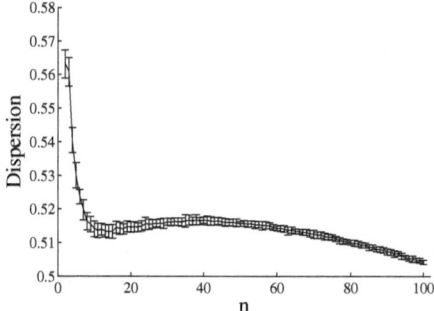

Fig. 4. Bound-Normalised dispersion for CiaS problems as n increases

4.3   Fitness Distance Correlation

In general, both estimators of FDC (shown in Fig. 5(a) and 5(b)) have small standard deviation (errorbars) over trials, which decreases as the number of circles increases. FDC values are typically small and negative for small n, and as n increases, values increases towards 0. However, an exception of this trend occurs at the transition from $n = 2$ to $n = 3$, where the FDC decrease dramatically; $FDC_{\mathbf{x}^*}$ transitions from 0.1048 to -0.0649, while $FDC_{\hat{\mathbf{x}}^*}$ transitions from -0.0265 to -0.0633. In the case of $FDC_{\hat{\mathbf{x}}^*}$, values then steadily increase towards 0 as n increases. On the other hand, $FDC_{\mathbf{x}^*}$ shows erratic fluctuations in FDC as n increases, although from approximately $n = 40$, the values steadily increase towards 0 as n increases. The fluctuations generally correlate with problems where symmetrical global solutions exist for at least one of the problems (e.g. the transition from $n = 9$ to $n = 10$). Furthermore, for $n \geq 40$ packings (where the FDC

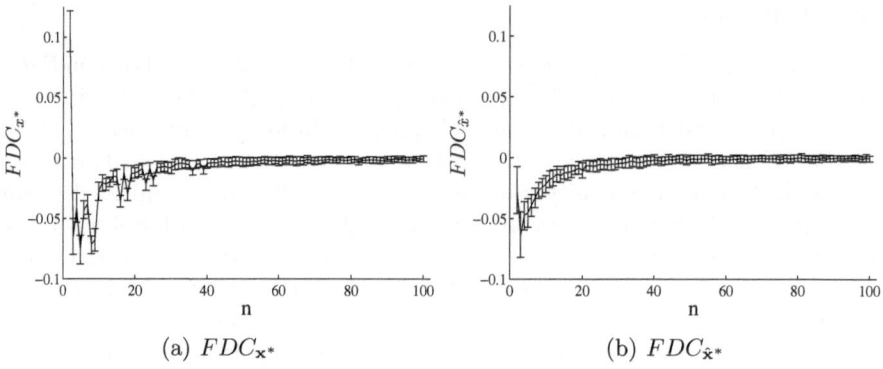

Fig. 5. FDC for CiaS problems. Lines show the mean of the 30 trials, while error bars indicate one standard deviation

values are rather stable) approximately 85% of the problems have assymetrical global solutions. Fig. 5(a) and 5(b) generally indicate that for low numbers of circles (i.e. $n < 20$), the fitness of random solutions is negatively correlated with their distance to the global optimum, however, as the number of circles increases, the fitness of random solutions has no correlation with their distance to the global optimum. Here, a negative value of FDC indicates that in general, the fitness of the sampled solutions gets better as their distance to the nearest global optimum increases. Such deceptive structure can be caused by many factors (and their interactions), including the presence of many local optima and multiple global optima, which CiaS problems are known to have [1]. The FDC values alone give no further insight into such factors, nor do they adequately differentiate between problems of varying n (particularly for $n > 40$).

Summarising the complex interaction between fitness and distance with a correlation coefficient may obviously lose important structural information, and so fitness-distance scatter plots can be used to visualise and better understand FDC and the landscape structure. The landscape at $n = 2$ has a positive $FDC_{\mathbf{x}^*}$ value and yet a negative $FDC_{\hat{\mathbf{x}}^*}$ value, and their fitness-distance scatter plots at $n = 2$ are shown in Fig. 6(a) and 6(b) respectively. Fig. 6(a) shows a general lack of correlation between fitness and distance to $\mathbf{x}^*$, however there are a few subsets of solutions that appear to be correlated. In particular, there is one area of strong positive correlation (i.e. solutions in Fig. 6(a) where $-d_n > -0.8$), indicating that the fitness gets worse as the distance from $\mathbf{x}^*$ increases. There are also two areas of weak negative correlations (i.e. solutions where $-d_n < -1$), where the fitness gets better as the distance from $\mathbf{x}^*$ increases. The positive correlation is much stronger than the two weaker correlations, and thus overall there is a positive $FDC_{\mathbf{x}^*}$ value (albeit a small one). Fig. 6(b) shows quite different structure compared to Fig. 6(a), in particular, there are much larger distances between solutions and $\hat{\mathbf{x}}^*$. This is likely due to the presence of multiple global optima; Fig. 6(a) shows the distance between solutions and their *closest* (out of

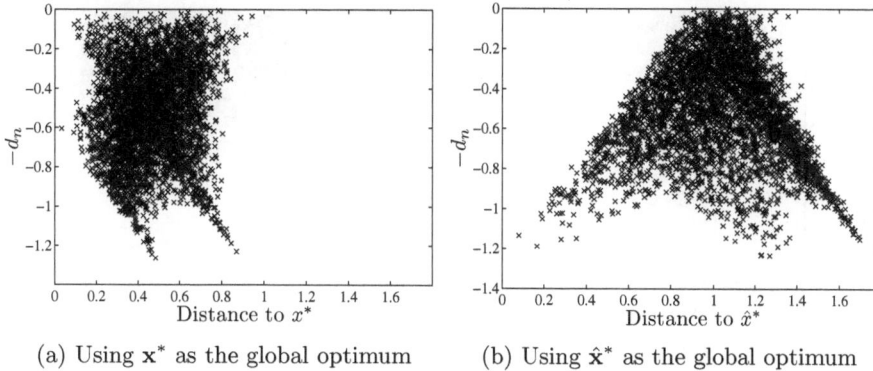

(a) Using $\mathbf{x}^*$ as the global optimum   (b) Using $\hat{\mathbf{x}}^*$ as the global optimum

Fig. 6. Typical fitness-distance scatter plots for $n = 2$ circles

8 possible) global optimum, while Fig. 6(b) shows the distance between solutions and the single best solution sampled. We expect solutions with 'more choice' in their closest global optimum to have both smaller distances and a smaller range of distances. The overall shape and trends in the data are also substantially different. Fig. 6(b) shows little evidence of the weak negative correlations that are present in Fig. 6(a). While there is perhaps a small subset of solutions with positive correlation (i.e. solutions in Fig. 6(b) where the distance to $\hat{\mathbf{x}}^*$ is less than 1), there is a prominent subset of solutions with a negative correlation (i.e. solutions where the distance to $\hat{\mathbf{x}}^*$ is greater than 1), thus explaining why $FDC_{\hat{\mathbf{x}}^*}$ is negative overall (albeit small).

In general, the interesting structure shown in Fig. 6(a) and 6(b) is absent for $n \geq 5$. For example, the fitness-distance scatter plots for $n = 9$ and $n = 10$ are shown in Fig. 7(a) and 7(b) respectively. While there are no obvious trends in either Fig. 7(a) or 7(b), the overall shape of the data is different. Solutions in Fig. 7(a) have a smaller distance to $\mathbf{x}^*$ as well as a smaller range in the distances to $\mathbf{x}^*$ compared to solutions in Fig. 7(b). This is not surprising; global solutions for $n = 9$ have 8 symmetries, compared with 1 for $n = 10$. Thus, while there are no obvious differences in the trends in Fig. 7(a) and 7(b), the difference between the number of global optima affects the value and range of distances obtained, which is evidence for the fluctuation between $FDC_{\mathbf{x}^*}$ values.

4.4   Information Content, Partial Information Content and Information Stability

The information content and partial information content were found to be highly correlated (with a sample correlation coefficient of 0.999). Consequently, we only report information content, shown in Fig. 8(a). The value of information content is roughly constant over all of the CiaS problems, with small fluctuations as indicated by the scale on the information content axis in Fig. 8(a). Comparisons with the information content and partial information content of highly rugged landscapes in [19] suggest that the values (and fluctuations) we obtained

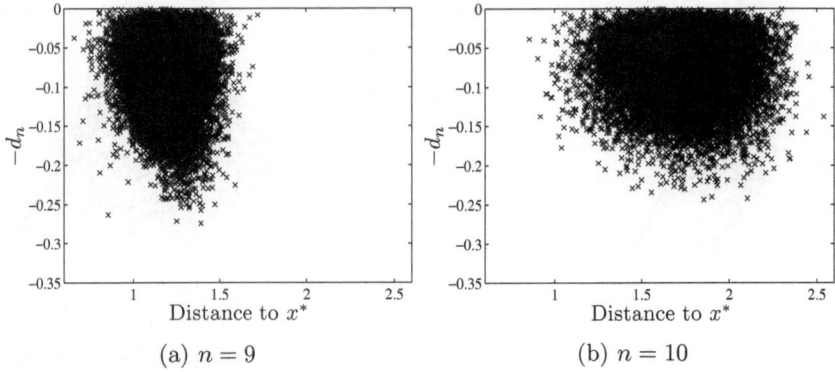

(a) $n = 9$ (b) $n = 10$

Fig. 7. Typical fitness-distance scatter plots using $\mathbf{x}^*$ as the global optimum

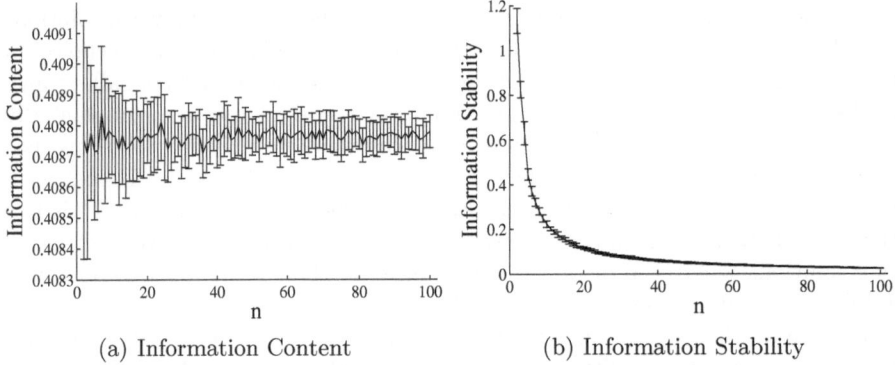

(a) Information Content   (b) Information Stability

Fig. 8. Information-Theoretic Analysis for CiaS problems as n increases

are reasonable. Similar to other features, the standard deviation decreases as n increases. The information content indicates that the problems do not significantly change in ruggedness, however the technique has no discriminative ability across the CiaS problems. Contrary to information content and partial information content, the information stability, shown in Fig. 8(b), shows a clear trend as n increases and has very small standard deviation between trials. In particular, as n increases, the information stability exponentially decays towards 0. Information stability is simply the largest change in objective function value between two steps in the walk, and because the expected objective function values (for uniform random solutions) are generally decreasing as n increases, it is no surprise that the information stability is also decreasing as n increases. Thus, while information stability can robustly differentiate the problem for changing n, it is an artifact of the objective function and sampling methodology, rather than the landscape structure. Furthermore, analysis of individual information stability values gives limited insight into landscape structure. For example, at $n = 2$, the average information stability over the 30 trials is approximately 1.13. This merely indicates that the largest change in objective function values (between a step in the walk) is 1.13; no information regarding other changes in objective function values, the distribution of objective function values or the interaction of solutions and objective function values is captured.

5 Conclusions

All the landscape features analysed in this paper provided some insight into the nature of CiaS packing problems, however dispersion, both estimators of FDC, information content and partial information content were quite limited in their ability to characterise and differentiate the CiaS packing problems. While information stability was able to differentiate between problem instances, no further insights into the structural nature of the problems were gained. In contrast, length scale analysis provided valuable insights into the nature of the problem, was statistically robust and is adept at characterising and differentiating between packings at a given value of n. Because length scale data is amenable to a variety of statistical and information theoretic techniques, potentially even more problem insights can be made from further analysis. Areas for future work include exploring other length scale analysis techniques, incorporating length scale into algorithm prediction models and linking the structural insights gained to particular algorithm behaviours.

References

1. Castillo, I., Kampas, F.J., Pintér, J.D.: Solving circle packing problems by global optimization: numerical results and industrial applications. European Journal of Operational Research 191(3), 786–802 (2008)
2. Gallagher, M.: Multi-layer perceptron error surfaces: visualization, structure and modelling. Ph.D. thesis, Dept. Computer Science and Electrical Engineering, University of Queensland (2000)
3. Gallagher, M.: Beware the parameters: Estimation of distribution algorithms applied to circles in a square packing. In: Coello, C.A.C., Cutello, V., Deb, K., Forrest, S., Nicosia, G., Pavone, M. (eds.) PPSN 2012, Part II. LNCS, vol. 7492, pp. 478–487. Springer, Heidelberg (2012)
4. Grosso, A., Jamali, A.R.M.J.U., Locatelli, M., Schoen, F.: Solving the problem of packing equal and unequal circles in a circular container. Journal of Global Optimization 47(1), 63–81 (2010)
5. Hansen, N., Auger, A., Finck, S., Ros, R.: Real-Parameter Black-Box Optimization Benchmarking: Experimental Setup. Technical report, INRIA (2013), http://coco.lri.fr/downloads/download13.09/bbobdocexperiment.pdf
6. Lunacek, M., Whitley, D.: The dispersion metric and the CMA evolution strategy. In: Genetic and Evolutionary Computation (GECCO), pp. 477–484. ACM (2006)
7. Malan, K.M., Engelbrecht, A.P.: Quantifying ruggedness of continuous landscapes using entropy. In: IEEE Congress on Evolutionary Computation (CEC), pp. 1440–1447 (2009)

8. Morgan, R., Gallagher, M.: Length scale for characterising continuous optimization problems. In: Coello, C.A.C., Cutello, V., Deb, K., Forrest, S., Nicosia, G., Pavone, M. (eds.) PPSN 2012, Part I. LNCS, vol. 7491, pp. 407–416. Springer, Heidelberg (2012)
9. Morgan, R., Gallagher, M.: Analysing and characterising optimization problems using length scale. IEEE Transactions on Evolutionary Computation (2014) (under review)
10. Morgan, R., Gallagher, M.: Sampling techniques and distance metrics in high dimensional continuous landscape analysis: Limitations and improvements. IEEE Transactions on Evolutionary Computation 18(3), 456–461 (2014)
11. Muñoz, M.A., Kirley, M., Halgamuge, S.K.: Landscape characterization of numerical optimization problems using biased scattered data. In: IEEE Congress on Evolutionary Computation (CEC), pp. 1180–1187 (2012)
12. Muñoz, M.A., Kirley, M., Halgamuge, S.K.: A meta-learning prediction model of algorithm performance for continuous optimization problems. In: Coello, C.A.C., Cutello, V., Deb, K., Forrest, S., Nicosia, G., Pavone, M. (eds.) PPSN 2012, Part I. LNCS, vol. 7491, pp. 226–235. Springer, Heidelberg (2012)
13. Müller, C.L., Sbalzarini, I.F.: Global characterization of the CEC 2005 fitness landscapes using fitness-distance analysis. In: Di Chio, C., et al. (eds.) EvoApplications 2011, Part I. LNCS, vol. 6624, pp. 294–303. Springer, Heidelberg (2011)
14. Müller, C., Baumgartner, B., Sbalzarini, I.: Particle swarm CMA evolution strategy for the optimization of multi-funnel landscapes. In: IEEE Congress on Evolutionary Computation (CEC), pp. 2685–2692 (2009)
15. Pitzer, E., Affenzeller, M.: A comprehensive survey on fitness landscape analysis. In: Fodor, J., Klempous, R., Suárez Araujo, C.P. (eds.) Recent Advances in Intelligent Engineering Systems. SCI, vol. 378, pp. 161–191. Springer, Heidelberg (2012)
16. Sheather, S.J., Jones, M.C.: A reliable data-based bandwidth selection method for kernel density estimation. Journal of the Royal Statistical Society. Series B (Methodological) 53(3), 683–690 (1991)
17. Specht, E.: Packomania (2012), http://www.packomania.com
18. Szabó, P.G., Markót, M.C., Csendes, T.: Global optimization in geometry – circle packing into the square. In: Essays and Surveys in Global Optimization, pp. 233–265. Springer (2005)
19. Vassilev, V.K., Fogarty, T.C., Miller, J.F.: Information characteristics and the structure of landscapes. Evolutionary Computation 8, 31–60 (2000)
20. Whitley, D., Lunacek, M., Sokolov, A.: Comparing the niches of CMA-ES, CHC and pattern search using diverse benchmarks. In: Runarsson, T.P., Beyer, H.-G., Burke, E.K., Merelo-Guervós, J.J., Whitley, L.D., Yao, X. (eds.) PPSN 2006. LNCS, vol. 4193, pp. 988–997. Springer, Heidelberg (2006)

Local Landscape Patterns
for Fitness Landscape Analysis

Shinichi Shirakawa[1] and Tomoharu Nagao[2]

[1] College of Science and Engineering, Aoyama Gakuin University
Sagamihara, Kanagawa, Japan
shirakawa@it.aoyama.ac.jp
[2] Graduate School of Environment and Information Sciences,
Yokohama National University, Yokohama, Kanagawa, Japan
nagao@ynu.ac.jp

Abstract. Almost all problems targeted by evolutionary computation are black-box or heavily complex, and their fitness landscapes usually are unknown. Selection of the appropriate search algorithm and parameters is a crucial topic when the landscape of a given target problem could be unknown in advance. Although several landscape features have been proposed in this context, examining a variety of landscape features is useful for problem understanding. In this paper, we propose a novel feature vector for characterizing the fitness landscape using the local landscape patterns (LLP). The proposed feature vector is composed by the histogram of the fitness patterns of the local candidate solutions. We extract the proposed LLP feature vector from well-known continuous optimization benchmark functions and BBOB 2013 benchmark set to investigate the properties of the proposed landscape feature and discuss about its effectiveness.

Keywords: Fitness Landscape Analysis, Local Feature, Problem Understanding, Continuous Optimization Problem

1 Introduction

The main target problems of evolutionary computation are black-box or heavily complex, and the landscape and characteristics of a target search space are usually unknown. Therefore, the users of the evolutionary computation method have to decide which algorithm is most suitable for their target problem by trial and error or heuristic knowledge. If we are able to know the characteristics of the landscape for the target problem in advance or during the search, it is helpful for the selection of the appropriate algorithm or parameter setting. In this context, the field of fitness landscape analysis [7,19] have been developed, and several features for characterizing the landscape are proposed such as ruggedness [24], fitness distance correlation (FDC) [3], evolvability [1,21], fitness cloud [17], neutrality [20], and dispersion metric [6].

FDC is a global feature which examines the correlation between fitness values and the distance to the global optimum. Let x_g be a global optimum, $d(\cdot, \cdot)$

G. Dick et al. (Eds.): SEAL 2014, LNCS 8886, pp. 467–478, 2014.

be an appropriate distance function, $d_i = d(x_i, x_g)$, and λ candidate solutions $X = \{x_1, \cdots, x_\lambda\}$ are given, then FDC is defined as

$$\text{FDC} = \frac{\frac{1}{\lambda} \sum_{i=1}^{\lambda} (f(x_i) - \bar{f})(d_i - \bar{d})}{\sigma_F \sigma_D}, \tag{1}$$

where $\bar{f}$ and $\bar{d}$ indicate the average values of fitness and distance, respectively, and σ_F and σ_D denote the standard deviation of fitness and distance, respectively. Note that the global optimum is usually unknown, although it can be given in the original definition of FDC. Therefore the global optimum is usually approximated by using finite candidate solutions such as $x_g = \text{argmin}_{x \in X} f(x)$[1]. FDC is one of the useful landscape features and applied various problems such as combinatorial [10] and continuous [14] ones.

Dispersion metric [6] is defined by average pairwise distance between the q best points.

$$\text{DISP} = \frac{1}{q(q-1)} \sum_{i=1}^{q} \sum_{j=1, j \neq i}^{q} d(x_{\text{rk}_i}, x_{\text{rk}_j}), \tag{2}$$

where x_{rk_i} denotes the i-th best point among λ samples. Dispersion metric is mainly applied to analyze the continuous optimization problems [6,13,15]. In order to compare the values between different search space scales, the candidate solutions are normalized to the $[0, 1]^n$ when the dispersion is computed in practice [15], where n is the problem dimension.

Fitness cloud [17] is represented as a scatter plot of the fitness values of parents against those of their offsprings (or neighbors), and negative slope coefficient (nsc) [23] is a measure of the problem hardness which is extracted from a plot of fitness cloud. Motif difficulty (MD) is introduced in [4] as a predictive difficulty measure for evolutionary algorithms by extracting motif properties from directed fitness landscape networks (FLNs). Concretely, the subgraphs of FLNs are classified into three types of classes, neutral, guide and deceptive, and then the predictive difficulty measure is calculated based on the number of these motifs in FLN. Recently, Morgan and Gallagher [11] have proposed a landscape feature called length scale and applied it to the analysis of BBOB 2010 benchmark functions. Length scale is defined by dividing the difference of the fitness of two candidate solutions by their distance. They conclude length scale is one of the promising features for the continuous optimization problems. Mersmann et al. propose an approach called exploratory landscape analysis which cheaply and automatically extracts problem properties from a concrete problem instance [9], and they investigate the relationship between low-level features and expert knowledge for the benchmark problems [8]. Smith-Miles and Tan [22] extract various features from traveling salesman problems (TSP) and investigate the relationship between the performance of several search algorithms and the problems. Muñoz et al. [13] present a neural network model for predicting the performance measure of the search algorithm. The model is input the landscape features and the algorithm parameters and outputs the predicted number of function evaluations for solving a given problem.

---

[1] This is the case of the minimization problem.

We can combine several landscape features to investigate the characteristic of a target problem. We, however, believe that it is an attractive approach to directly extract the landscape feature as a vector form which represents various characteristics of the problem. In this paper, we propose a novel feature vector for characterizing the fitness landscape, which uses the fitness patterns of the local candidate solutions. Then we extract the proposed feature from the well-known benchmark functions and BBOB 2013 benchmark set for continuous optimization problem to confirm its properties and effectiveness.

The next section of this paper presents our proposed feature, local landscape patterns (LLP), for characterizing the fitness landscape. Then, in Section 3, we extract the LLP feature vector from several continuous optimization problems and discuss about the experimental results. Finally, in Section 4, we describe our conclusions and future work.

2   Landscape Feature Using Local Landscape Patterns

Definition of Fitness Landscape. In this paper, we refer to the definition of the fitness landscape in [19]. Let $\mathscr{S}$ be a set of all candidate solutions (or a search space), $f : \mathscr{S} \to \mathbb{R}$ be a fitness function, and $d : \mathscr{S} \times \mathscr{S} \to \mathbb{R}$ be a distance function between the candidate solutions. Then the fitness landscape $\mathscr{F}$ is defined as a pair of functions of f and d in the candidate solutions $\mathscr{S}$, namely $\mathscr{F} = (\mathscr{S}, f, d)$. Let consider the continuous optimization without constraints, then the candidate solutions are $\mathscr{S} = \mathbb{R}^n$ where n is a dimension of the search space, and the Euclidean distance is usually used as the distance function d. In the context of evolutionary computation, it is impossible to get the analytical form of f or check for the fitness values and distance of all candidate solutions. We therefore consider to estimate the fitness landscape or its characteristics by only using a given finite sample. Let X be a set of sampled candidate solutions from $\mathscr{S}$ by the specific sampling method, then the approximated fitness landscape can be represented as $\bar{\mathscr{F}} = (X, f, d)$.

Note that $\bar{\mathscr{F}}$ heavily depends on the sampling method of X and may be quite different from $\mathscr{F}$ if the biased sampling is used. For example, the fitness landscape is viewed as a unimodal function when all candidate solutions are sampled from one basin area even if f is a multi-modal function. In other words, $\bar{\mathscr{F}}$ represents the fitness landscape from the viewpoint of the finite samples. In practice, we have to estimate the landscape using sampled candidate solutions and exploit it for selection of algorithm or parameters. Morgan and Gallagher discuss about the sampling issue in dispersion metric in [12].

Local Landscape Patterns. Most conventional landscape features such as FDC and dispersion metric are summarized as a scalar value. It may be difficult to represent the exhaustive features of the fitness landscape, and the multiple different landscape features are used to characterize it [8,13,22]. It is, therefore, attractive to directly extract the landscape feature as a vector which represents various characteristics of the problem. We consider to characterize the fitness

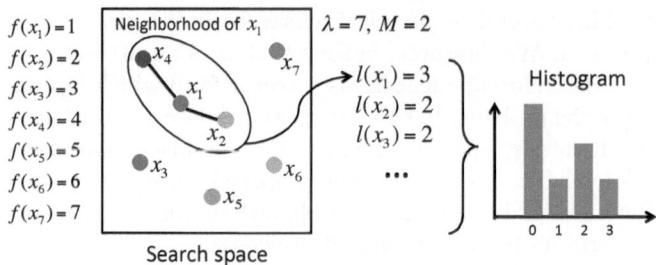

Fig. 1. Conceptual image of the procedure for calculating the LLP feature. λ and M indicate the size of candidate solution sample X and the neighborhood size, respectively.

landscape in a form of a feature vector by using the finite samples. The relationship between the fitness values and the distance of the candidate solutions is the most important characteristic, and the neighborhood structure is also an important property for almost all evolutionary algorithms. From this observation, our proposed feature focusses on the fitness patterns of the local candidate solutions.

Let $X = \{x_1, \cdots, x_\lambda\}$, $x_i \in \mathscr{S}$ be a set of the λ sampled candidate solutions, and $x_{i:j}(1 \leq j \leq \lambda - 1)$ denotes the j-th nearest neighbor candidate solution of the i-th candidate solution x_i measured by the distance function d. Then the local landscape pattern (LLP) $l(x_i)$ around x_i is defined by using the fitness values of the $M(> 0)$ nearest candidate solutions from x_i as

$$l(x_i) = \sum_{j=1}^{M} \delta_{ij} 2^{j-1}, \tag{3}$$

$$\delta_{ij} = \begin{cases} 1 & (\text{if } f(x_i) \text{ better than } f(x_{i:j})) \\ 0 & (\text{otherwise}). \end{cases} \tag{4}$$

where $f(x_{i:j})$ means the fitness value of the j-th nearest neighbor candidate solution of x_i. The value of $l(x_i)$ is an integer within the range of $[0, 2^M]$ and indicates the pattern number of x_i. The pattern number corresponds to the binary number of the fitness pattern. Let $\mathcal{L} = \{l(x_i)|1 \leq i \leq \lambda\}$ be a set of the local landscape patterns. To summarize these local landscape patterns, we construct the histogram of $\mathcal{L}$. As we employ the histogram of $\mathcal{L}$ as a feature vector, the value of $l(x_i)$ is irrelevant for the eventual feature vector. Consequently, the LLP feature vector $\text{LLP}(X)$ is given by the normalized histogram of $\mathcal{L}$, and the k-th element of $\text{LLP}(X)$ is computed by $H_k(\mathcal{L})/\lambda$, where $H_k(\mathcal{L})$ denotes the k-th element of the histogram of $\mathcal{L}$. The elements of $\text{LLP}(X)$ are divided by λ to normalize the scale of the bin values caused by the different sample size. Note that the number of the bins in the histogram is 2^M, namely the LLP feature is not a scalar but a vector form for representing the landscape features.

The conceptual image of the calculation of the LLP feature is shown in Fig. 1. In this figure, the size λ of candidate solution sample X and the neighborhood size M are set to 7 and 2, respectively. The two nearest neighbors of x_1 are x_2 and x_4, namely $x_{1:1} = x_2$, $x_{1:2} = x_4$, and $f(x_{1:1}) = f(x_2) > f(x_1)$, $f(x_{1:2}) = f(x_4) > f(x_1)$. Then the local landscape pattern of x_1 is computed by $l(x_1) = 1 \cdot 2^0 + 1 \cdot 2^1 = 3$, and it is added to the histogram bin of 3. Analogously, other local landscape patterns $l(x_2) \cdots l(x_7)$ are calculated, then the histogram feature can be obtained.

In order to extract the LLP feature vector from the given finite candidate solutions, it is only necessary to define the fitness function f and the distance function d among the candidate solutions. It means the LLP feature can be extracted from various problem domains such as continuous or combinatorial optimization problems. One of the advantages of the LLP feature is its invariant property under monotonicity-preserving transformations of f because the LLP feature only uses the magnitude relation of the fitness values. For example, the LLP feature vectors on such as $f(x) = x^T x$ and $f(x) = \exp(x^T x)$ are same without any normalization of the fitness function values.

The concept of the proposed method which uses the histogram of the local patterns has succeeded in the computer vision community, e.g. bag-of-features using SIFT descriptors [2]. The LLP feature shares the common concept with the image feature of the local binary patterns (LBP) [18] with respect to focus on the local relationship between the fitness values. LBP can be, however, applicable only for the two dimensional pixel spaces but the fitness landscape is always high dimensional and it may not even be Euclidean space.

3 Experiments and Results

To investigate the properties and the effectiveness of the LLP feature, we apply it to the fitness landscape of the continuous minimization problems defined as $\text{argmin}_{x \in \mathbb{R}^n} f(x)$ and use the Euclidean distance as the distance function between the candidate solutions. Namely, the search space $\mathscr{S}$ described in Section 2 is $\mathscr{S} = \mathbb{R}^n$ and the distance function d is the Euclidean distance in the experiments.

3.1 Basic Properties of LLP

First of all, we consider two dimensional problems whose landscape can be easily visualized. We generate $\lambda = 100$ candidate solutions by drawing random numbers uniformly from $[-3, 3]^2$ and then extract the LLP feature by the setting of the neighborhood size $M = 4$. Figure 2 (a) and (b) show the visualized two dimensional landscapes, the points of the candidate solutions, and the extracted histogram of the LLP on the sphere and Rastrigin functions, respectively. From these figures, we can see the shapes of the histogram are different between the unimodal (sphere) and the multi-modal (Rastrigin) functions. The pattern 0 or 15 indicates all fitness values of the neighborhood solutions are worse or better than that of the focused candidate solution. In Rastrigin function, the patterns of 0 and 15 (which corresponds to the binary number 0000 and 1111, respectively) have a large value because the points on the basin and ridge tend to

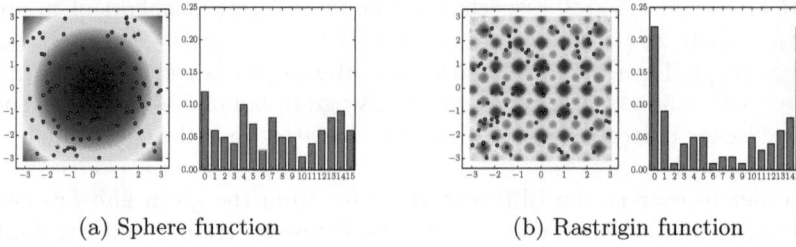

(a) Sphere function   (b) Rastrigin function

Fig. 2. Visualized fitness landscapes and the sampled candidate solutions for calculating the LLP feature (left), and the histogram of the LLP (right) on (a) Sphere and (b) Rastrigin functions, respectively

become the best and the worst point in the local region, respectively. On the other hand, the patterns which have intermediate value such as 7 (which corresponds to the binary number 0101) mean that there are both better and worse neighborhood candidate solutions than the focused one. As a lot of the local regions in the sphere function are the monotonous landscape (i.e. the fitness value monotonically decreases or increases toward a certain direction), the patters which have intermediate value are increased. The shape of the histogram of the sphere function becomes like in Fig. 2 (a) in the end.

Next, we extract the LLP feature vectors from well-known six continuous benchmark functions, Sphere, Ellipsoid, Rosenbrock, Ackley, Schaffer, and Rastrigin functions. In this experiment, the problem dimension of each function n is 20, and the 100 different sets of the candidate solutions are generated for calculating the proposed feature. The candidate solution set is sampled by drawing random numbers uniformly from $[-3, 3]^n$ for all benchmark functions. And the neighborhood size of $M = 4$ is employed in this experiment. Since our proposed feature is a form of vector, principal component analysis (PCA) is applied to reduce the dimension from $2^M = 16$ to 2 which is easy to be visualized.

Figure 3 illustrates the two dimensional plots using first and second principal components with the varying sample size $\lambda = 100, 500, 1000, 2000$. When the sample size is 100, it is difficult to discriminate each function. We can, however, observe each function is clustered when the sample size becomes larger. Both the sphere and ellipsoid functions are unimodal but the ellipsoid is ill-conditioning function. Therefore, we can regard that the difference between these functions is the range of the landscape because the LLP feature uses only the rank of fitness values, and the size of locally monotonous (or basin) regions becomes different if the same sampling region is used. This is the reason why the sphere and ellipsoid functions become different plots. From this result, we can conclude the LLP feature vector has a potential to discriminate the fitness landscape.

A large number of the candidate solutions cause the increase of computational cost. The LLP feature uses the distance between the candidate solutions, and their computational order is $O(\lambda^2)$. Of course, it requires the fitness evaluations for λ candidate solutions. Obviously, there is a tradeoff between the computational costs

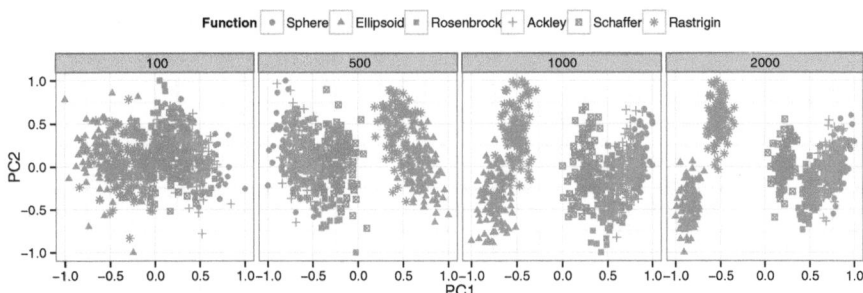

Fig. 3. Two dimensional plots using first and second principal components of the LLP feature vectors extracted from the 20 dimensional benchmark functions with the varying sample size $\lambda = 100, 500, 1000, 2000$. The 100 different sets of candidate solutions with the same size are generated for each function.

and the knowledge gain. We can observe that it seems to be sufficient to discriminate the landscapes by the 2000 samples in this case. Note that it may have additional information in the hidden dimension because Fig. 3 shows the reduced two dimensional feature by PCA.

3.2   Comparison with Conventional Landscape Features

To confirm the effectiveness of the LLP feature, we compare it with the conventional landscape features, fitness distance correlation (FDC) [3] and dispersion metric [6]. The parameter of dispersion metric is set to $q = \lfloor 0.1\lambda \rfloor$. The sampling method of the candidate solution sample X is same as the previous experiment. Figure 4 illustrates the plots of FDC and dispersion metric extracted from the 20 dimensional benchmark functions with the varying sample size $\lambda = 100, 500, 1000, 2000$. The tendency of the plots is similar to that of the LLP feature. When the sample size is small, the boundary of each function is not clear. The ellipsoid and Rastrigin functions are clustered far away from the other functions. Further the relative position of each plotted function in the two dimensional space is very similar between Fig. 3 and 4. From this result, at least the LLP feature vector has the same ability of characterizing and discriminating the fitness landscape as that of FDC and dispersion metric.

We then conduct the experiment of clustering to quantitatively evaluate the quality of each feature. We use k-means as the clustering method and 2^M dimensional feature vector is used in the proposed LLP feature. The number of the clusters is set to 6 which is the same number of the benchmark functions. To evaluate the clustering quality, we compute purity and adjusted rand index (ARI) [16] as the performance measure by referring to an ideal clustering result. Purity focuses on the frequency of the most common category in each cluster, and ARI is based on the similarity between two data clustering results. The range of both measures is $[0, 1]$ and the higher value of them indicates the better

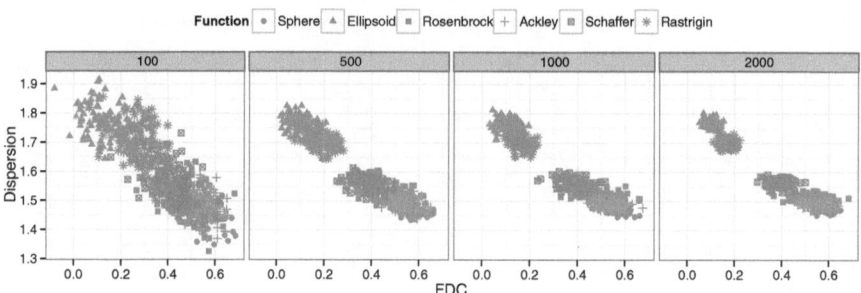

Fig. 4. Two dimensional plots using FDC and dispersion metric extracted from the 20 dimensional functions with the varying sample size $\lambda = 100, 500, 1000, 2000$. The 100 different sets of candidate solutions with the same size are generated for each function.

clustering performance. Table 1 shows the purity and ARI using the LLP and conventional features with the varying problem dimension of $n = 10, 20, 30$, the sample size of $\lambda = 500, 1000, 2000$, and the neighborhood size of $M = 3, 4, 5, 6$. In all settings, the LLP feature outperforms the conventional features, FDC and dispersion metric, in terms of purity and ARI. From this table, we can see the most stable and best setting is $M = 4$ and $\lambda = 2000$. The performance basically improves as the sample size becomes larger. This result is intuitive understandable because a large number of samples are useful to grasp the detailed landscape structure. Consequently, we can consider the LLP feature vector is one of the promising landscape features for discriminating the continuous optimization problems.

3.3   Fitness Landscape Analysis of BBOB 2013 Functions

Finally, we apply our LLP feature to analyze one of the major benchmark function sets, BBOB 2013[2], which contains the 24 noiseless benchmark functions. In this experiment, the problem dimension of each function is 20 and the candidate solutions are sampled by drawing random numbers uniformly from $[-5, 5]^n$. The range of the search space is based on the definition of BBOB 2013 benchmark functions. The neighborhood size $M = 4$ and the sample size $\lambda = 2000$ are used for extracting the LLP features.

Table 2 shows the clustering result by using k-means clustering, which the number of the clusters is set to 6. The cluster 1 and 2 contain the unimodal functions except Schwefel function (f_{20}). The reason why the multi-modal Schwefel function is assigned to the cluster 1 is that its landscape is macroscopically similar to the sphere function with the exception of near the optimum. The cluster 3 consists of the multi-modal functions such as Rastrigin and Schaffers variants and highly conditioned functions (f_6, f_7, and f_{12}), and the cluster 5 is composed

---

[2] http://coco.gforge.inria.fr/doku.php?id=bbob-2013

Table 1. Comparison of purity and ARI for the clustering results. Each feature is extracted from the 10, 20, and 30 dimensional benchmark functions with the varying sample size $\lambda = 500, 1000, 2000$. The neighborhood size varies as $M = 3, 4, 5, 6$ for the LLP feature, and the 100 different sets of the candidate solutions with the same setting are generated for each function. The bold value indicates best performance among the varying parameters.

	LLP (M=3)			LLP (M=4)			LLP (M=5)			LLP (M=6)			FDC & DISP		
λ	500	1000	2000	500	1000	2000	500	1000	2000	500	1000	2000	500	1000	2000
Problem dimension = 10															
Purity	0.74	0.75	0.84	0.75	0.76	0.87	0.77	0.76	0.84	0.75	0.78	**0.88**	0.60	0.65	0.68
ARI	0.59	0.64	0.74	0.61	0.64	**0.77**	0.62	0.64	0.72	0.58	0.66	**0.77**	0.39	0.43	0.45
Problem dimension = 20															
Purity	0.70	0.81	0.86	0.67	0.82	**0.88**	0.70	0.82	**0.88**	0.71	0.80	**0.88**	0.65	0.72	0.75
ARI	0.50	0.66	0.75	0.46	0.70	**0.79**	0.51	0.68	**0.79**	0.52	0.66	**0.79**	0.46	0.53	0.61
Problem dimension = 30															
Purity	0.74	0.80	0.91	0.76	0.83	**0.93**	0.72	0.83	0.92	0.73	0.83	0.91	0.67	0.72	0.74
ARI	0.54	0.66	0.82	0.58	0.70	**0.85**	0.53	0.69	0.84	0.53	0.68	0.83	0.45	0.55	0.59

by the multi-modal functions and Rosenbrock family. These two clusters contain both the unimodal and multi-modal functions. All the functions in the cluster 4 and 6 are highly rugged multi-modal function. Original Rosenbrock (f_8) and its rotated version (f_9) are assigned different clusters because the sampling region is not rotated along with the rotation of the fitness function and then the local landscales of the samples become different. This clustering result is similar with the qualitative grouping by the human observation.

Table 3 shows the expected running time (ERT) of BIPOP-aCMA [5] for reaching precision $\Delta f = 10^{-7}$ in the BBOB 2013 result[3] for 20 dimension in BBOB 2013. BIPOP-aCMA has only succeeded to reach the target precision at least once all benchmark functions. From Table 2 and 3, the functions in the cluster 2 are relatively easy to solve for BIPOP-aCMA because the ERT of these functions is less than 2.0×10^4. However, other clusters are not so easily understandable with respect to the ERT.

At the end of this section, we investigate whether our proposed LLP feature has the ability to predict the ERT in the manner of the supervised learning. We construct a linear regression model and predict the values which are taken the common logarithm of the ERT ($\log_{10}$ ERT). Each feature is calculated using the 2000 candidate solutions, and the neighborhood size of the LLP is 4. The LLP feature is reduced to two dimensions by PCA. To evaluate the generalized error, we adopt leave-one-out cross-validation. The generalization error of the root mean squared error (RMSE) of the LLP feature is 0.961, while that of the conventional landscape features (FDC and dispersion metric) is 1.13. Although the LLP feature outperforms the conventional feature, it is not so large

---

[3] http://coco.gforge.inria.fr/doku.php?id=bbob-2013-results

Table 2. Clustering result for the BBOB 2013 benchmark functions by k-means clustering using the LLP feature. Each feature is extracted from the 20 dimensional benchmark functions with the sample size $\lambda = 2000$. The neighborhood size M is set to 4.

Cluster 1	Cluster 2	Cluster 3
Sphere (f_1)	Ellipsoid (f_2)	Rastrigin (f_3)
Original Rosenbrock (f_8)	Linear Slope (f_5)	Attractive Sector (f_6)
Sharp Ridge (f_{13})	Rotated ellipsoid (f_{10})	Step Ellipsoid (f_7)
Different Powers (f_{14})	Discus (f_{11})	Bent Cigar (f_{12})
Schwefel (f_{20})		Non-separable Rastrigin (f_{15})
		Schaffers F7 (f_{17})
		Ill-conditioned Schaffers (f_{18})

Cluster 4	Cluster 5	Cluster 6
Büche-Rastrigin (f_4)	Rotated Rosenbrock (f_9)	Weierstrass (f_{16})
Gallagher's Gaussian 101-me Peaks (f_{21})	Composite Griewank-Rosenbrock (f_{19})	Katsuura (f_{23})
Gallagher's Gaussian 21-hi Peaks (f_{22})	Lunacek bi-Rastrigin (f_{24})	

Table 3. Expected running time (ERT) of BIPOP-aCMA [5] for reaching the precision $\Delta f = 10^{-7}$ in the BBOB 2013 result. The problem dimension is 20.

	f_1	f_2	f_3	f_4	f_5	f_6	f_7	f_8	f_9	f_{10}	f_{11}	f_{12}
ERT	6.9e2	1.1e3	4.1e3	1.5e4	4.1	9.3e3	2.5e4	4.9e3	1.7e4	1.3e4	7.7e3	2.1e4

	f_{13}	f_{14}	f_{15}	f_{16}	f_{17}	f_{18}	f_{19}	f_{20}	f_{21}	f_{22}	f_{23}	f_{24}
ERT	3.3e4	1.1e4	3.5e5	2.6e5	8.9e4	2.0e5	6.3e6	5.3e6	7.6e5	6.5e6	1.3e6	1.4e8

improvement. We, therefore, conclude at least the LLP feature has the comparable ability for predicting the problem difficulty to the conventional features.

4 Conclusion and Future Work

In this paper, we propose a novel landscape feature vector, LLP, which focusses on the local fitness patterns. The proposed method constructs the histogram of the local landscape patterns and extracts the feature as a vector form. The advantage of the LLP feature is that it is possible to extract landscape features by unified procedure, i.e. it has sufficient characterizing performance of the fitness landscape without combining the multiple landscape features. We extract the proposed LLP feature vector from the well-known benchmark functions, and show the effectiveness of it through the comparison with existing landscape features. Then we extract the LLP feature vector from the BBOB 2013 benchmark functions and present the results of the clustering and the prediction of ERT. The clustering results are understandable from the viewpoint of the qualitative properties of each function, and predicting the performance of the ERT outperforms the conventional landscape features.

In this paper, we compared the proposed LLP feature with FDC and dispersion metric. The comparison with other landscape features such as length scale [11], evolvability [1,21], and fitness cloud [17] should be conducted to verify the effectiveness of the LLP feature. Further, overall the results depend on the data analysis and the sampling methods used. It may be interesting to investigate the compatibility between the LLP feature and the data analysis techniques. We should attempt to use more sample size because the sample sizes λ used up to 2000 in the experiments might be insufficient for the 20 dimensional problems. As we noted in Section 2, the sampling method and the sample size have a big impact to the landscape features. One possible work is to generate the concatenated LLP feature using different sample sizes or sampling methods to extract the landscape characteristics from various viewpoints.

In the future, we plan to apply the proposed LLP feature to another type of problems such as combinatorial problems and real-world problems, and investigate the properties and effectiveness of it. We are able to use and investigate the several standard statistics as a local feature such as the number of the improving solutions and the probability to improve instead of the fitness patterns of the local candidate solutions. Furthermore, we will develop an efficient algorithm which switches the strategy parameters based on the LLP feature.

References

1. Altenberg, L.: The evolution of evolvability in genetic programming. In: Kinnear Jr., K.E. (ed.) Advances in Genetic Programming, pp. 47–74. MIT Press, Cambridge (1994)
2. Csurka, G., Dance, C.R., Fan, L., Willamowski, J., Bray, C.: Visual categorization with bags of keypoints. In: Workshop on Statistical Learning in Computer Vision, ECCV, pp. 1–22 (2004)
3. Jones, T., Forrest, S.: Fitness distance correlation as a measure of problem difficulty for genetic algorithms. In: Proceedings of the Sixth International Conference on Genetic Algorithms, pp. 184–192. Morgan Kaufmann (1995)
4. Liu, J., Abbass, H.A., Green, D.G., Zhong, W.: Motif difficulty (MD): a predictive measure of problem difficulty for evolutionary algorithms using network motifs. Evolutionary Computation 20(3), 321–347 (2012)
5. Loshchilov, I., Schoenauer, M., Sèbag, M.: Bi-population CMA-ES agorithms with surrogate models and line searches. In: Proceedings of the 15th Annual Conference Companion on Genetic and Evolutionary Computation (GECCO 2013) Companion, pp. 1177–1184. ACM, New York (2013)
6. Lunacek, M., Whitley, D.: The dispersion metric and the CMA evolution strategy. In: Proceedings of the 8th Annual Conference on Genetic and Evolutionary Computation (GECCO 2006), pp. 477–484. ACM, New York (2006)
7. McClymont, K.: Recent advances in problem understanding: Changes in the landscape a year on. In: Proceeding of the 15th Annual Conference Companion on Genetic and Evolutionary Computation Conference Companion (GECCO 2013) Companion, pp. 1071–1078. ACM, New York (2013)
8. Mersmann, O., Bischl, B., Trautmann, H., Preuss, M., Weihs, C., Rudolph, G.: Exploratory landscape analysis. In: Proceedings of the 13th Annual Conference on Genetic and Evolutionary Computation (GECCO 2011), pp. 829–836. ACM, New York (2011)

9. Mersmann, O., Preuss, M., Trautmann, H.: Benchmarking evolutionary algorithms: Towards exploratory landscape analysis. In: Schaefer, R., Cotta, C., Kołodziej, J., Rudolph, G. (eds.) PPSN XI. LNCS, vol. 6238, pp. 73–82. Springer, Heidelberg (2010)
10. Merz, P.: Advanced fitness landscape analysis and the performance of memetic algorithms. Evolutionary Computation 12(3), 303–325 (2004)
11. Morgan, R., Gallagher, M.: Length scale for characterising continuous optimization problems. In: Coello, C.A.C., Cutello, V., Deb, K., Forrest, S., Nicosia, G., Pavone, M. (eds.) PPSN 2012, Part I. LNCS, vol. 7491, pp. 407–416. Springer, Heidelberg (2012)
12. Morgan, R., Gallagher, M.: Sampling techniques and distance metrics in high dimensional continuous landscape analysis: limitations and improvements. IEEE Trans. Evol. Comput. 18(3), 456–461 (2014)
13. Muñoz, M., Kirley, M., Halgamuge, S.: A meta-learning prediction model of algorithm performance for continuous optimization problems. In: Coello, C.A.C., Cutello, V., Deb, K., Forrest, S., Nicosia, G., Pavone, M. (eds.) PPSN 2012, Part I. LNCS, vol. 7491, pp. 226–235. Springer, Heidelberg (2012)
14. Müller, C.L., Sbalzarini, I.F.: Global Characterization of the CEC 2005 Fitness Landscapes Using Fitness-Distance Analysis. In: Di Chio, C., et al. (eds.) EvoApplications 2011, Part I. LNCS, vol. 6624, pp. 294–303. Springer, Heidelberg (2011)
15. Müller, C., Baumgartner, B., Sbalzarini, I.: Particle swarm cma evolution strategy for the optimization of multi-funnel landscapes. In: 2009 IEEE Congress on Evolutionary Computation (CEC 2009), pp. 2685–2692 (2009)
16. Murphy, K.P.: Machine Learning: A Probabilistic Perspective. The MIT Press (2012)
17. Philippe, C., Vérel, S., Manuel, C.: Local search heuristics: Fitness Cloud versus Fitness Landscape. In: Proceedings of the 16th Eureopean Conference on Artificial Intelligence (ECAI 2004), pp. 973–974. IOS Press (2004)
18. Pietikäinen, M., Zhao, G., Hadid, A., Ahonen, T.: Computer Vision Using Local Binary Patterns. Computational Imaging and Vision, vol. 40. Springer (2011)
19. Pitzer, E., Affenzeller, M.: A comprehensive survey on fitness landscape analysis. In: Fodor, J., Klempous, R., Suárez Araujo, C.P. (eds.) Recent Advances in Intelligent Engineering Systems. SCI, vol. 378, pp. 161–191. Springer, Heidelberg (2012)
20. Reidys, C.M., Stadler, P.F.: Neutrality in fitness landscapes. Applied Mathematics and Computation 117(2-3), 321–350 (2001)
21. Smith, T., Husbands, P., Layzell, P., O'Shea, M.: Fitness landscapes and evolvability. Evolutionary Computation 10(1), 1–34 (2002)
22. Smith-Miles, K., Tan, T.: Measuring algorithm footprints in instance space. In: 2012 IEEE Congress on Evolutionary Computation (CEC 2012), pp. 1–8 (2012)
23. Vanneschi, L., Tomassini, M., Collard, P., Vérel, S.: Negative Slope Coefficient: A Measure to Characterize Genetic Programming Fitness Landscapes. In: Collet, P., Tomassini, M., Ebner, M., Gustafson, S., Ekárt, A. (eds.) EuroGP 2006. LNCS, vol. 3905, pp. 178–189. Springer, Heidelberg (2006)
24. Weinberger, E.: Correlated and uncorrelated fitness landscapes and how to tell the difference. Biological Cybernetics 63(5), 325–336 (1990)

Why Advanced Population Initialization Techniques Perform Poorly in High Dimension?

Borhan Kazimipour, Xiaodong Li, and A.K. Qin

School of Computer Science and Information Technology, RMIT University,
Melbourne, 3000, Victoria, Australia
{borhan.kazimipour,xiaodong.li,kai.qin}@rmit.edu.au

Abstract. Many advanced population initialization techniques for Evolutionary Algorithms (EAs) have hitherto been proposed. Several studies claimed that the techniques significantly improve EAs' performance. However, recent researches show that they cannot scale well to high dimensional spaces. This study investigates the reasons behind the failure of advanced population initialization techniques in large-scale problems by adopting a wide range of population sizes. To avoid being biased to any particular EA model or problem set, this study employs general purpose tools in the experiments. Our investigations show that, in spite of population size, uniformity of populations drops dramatically when dimensionality grows. The observation confirms that the uniformity loss exist in high dimensional spaces regardless of the type of EA, initializer or problem. Therefore, we conclude that the weak uniformity of the resulting population is the main cause of the poor performance of advanced initializers in high dimensions.

Keywords: Population Initialization, Large-Scale Optimization, Evolutionary Algorithm, Uniformity.

1 Introduction

Evolutionary Algorithms (EAs), like all other population-based algorithms, start with an initial population. There is a common belief that having better starting points may help algorithms to achieve better final results [2]. Based on this intuition, several advanced population initialization techniques have been proposed. Some of these techniques employ domain knowledge [7]. In many others, however, no expert knowledge is involved in the algorithms. This makes the later category more promising in dealing with general black-box problems [9]. In such a case, researchers aim to improve existing techniques to produce more uniform populations enhancing coverage over entire search space.

A large body of literature is devoted to compare different population initialization techniques on a variety of benchmarks [8]. Although there is still little agreement on which techniques are superior in dealing with unforeseen problems, almost all previous studies admit that advanced techniques perform at least better than simple random number generators. It is strongly suggested

G. Dick et al. (Eds.): SEAL 2014, LNCS 8886, pp. 479–490, 2014.

that the performance of EAs can be significantly improved by simply switching to more advanced population initialization techniques [2].

Recently, some researches have been published on the scalability of population initializers to large-scale domains [8]. The studies investigate whether the claimed advantages of using recent techniques are still significant when dimensionality of problems is beyond hundred variables. In [7], for example, authors show that even in high dimensional spaces some advanced population initialization techniques improve EAs performance. However, the effectiveness of these techniques degrades when dimensionality increases from 100 to 1,000. Furthermore, some of the studied techniques perform significantly worse than simple random number generators [7]. Another recent study indicates that when EA parameters are set properly, the improvement coming from advanced initializers becomes very marginal [8]. The study also states that size of initial population plays a more important role than the way it is generated.

Experiments in previous studies are well conducted and some advanced statistical tests confirmed validity of the results. Nevertheless, two questions still remain unanswered: 1) Whether the previous findings can be generalized to other EAs and problem sets? 2) What causes the poor performance of advanced initializers in large-scale spaces?

In this study we aim to answer the above mentioned questions. In particular, we adopt general purpose tools to measure uniformity of populations generated by different techniques to study the effect of dimensionality on the performance of initializers. The employed measures are carefully selected to guarantee the generality of the findings regardless of the type of EA or optimization problem.

The rest of the paper is as follows. Section 2 briefly introduces some of the widely used initialization techniques in the EA domain. It also shortly presents a few performance measures. Section 3 discusses the experimental setup and provides a detailed discussion about obtained results. Finally, Section 4 concludes the paper.

2   Background

This section briefly discusses advanced population initialization techniques in EAs, as well as general measures assessing their performance. Considering the wide variety of initialization techniques, we select some of the most widely cited methods from each category to maximize the generality of the study. Furthermore, to avoid being biased to any particular EA model or problem set, our main focus is on general measures. As a consequence, the findings can be easily generalised to all EAs and real-value optimization problems.

A recent survey on population initialization techniques categorized them based on three aspects: *generality, compositionality* and *randomness* [9]. From the generality point of view, population initialization techniques fall into *generic* and *application-specific* categories. In this paper, we only study generic techniques for the sake of generality. Based on the compositionality aspect, the techniques are divided into *non-compositional*, only having one core component, and *compositional*, e.g., hybrid and multi-step techniques [10]. For the sake of simplicity and clarity, we focus more on non-compositional techniques. However, we expect almost similar results from compositional techniques as they inherit the advantages and disadvantages of their parent techniques. Finally, initialization techniques are categorized as *stochastic* or *deterministic* based on the level of randomness exist in the resulting populations. The next two subsections discuss these categories in details. The third subsection, however, describes quality measures we chose for the evaluation of all initialization categories.

2.1   Stochastic Initial Populations

Following the criterion proposed in [9], we consider an initializer as a stochastic technique if its output depends on an initial seed. The most well-known subcategory of stochastic techniques is *Pseudo-Random Number Generator* (PRNG) [15]. These techniques are widely used in computer programs and simulations to generate point sets reflecting some attributes of truly random sequences, e.g., high independence and low correlation.

Many variations of PRNGs have been implemented. Indeed, at least one instance of them is available in every programming language. In addition to their versatility, the scalability of PRNGs make them the most widely used population initialization techniques in the EA domain. WELL [6], KISS [12] and Mersenne twister [13] are just a few examples of PRNGs to name.

The family of *Chaotic Number Generators* (CNGs) is another category of stochastic point generators which has recently attracted a lot of attention in the EA field [9]. Technically, *chaos* is a property of some dynamic systems which makes them very sensitive to initial condition and hardly predictable. Generally speaking, CNGs are iterative/recursive algorithms that generate sequences reflecting the properties of chaotic systems.

The algorithms of CNGs are very simple. They start with a randomly chosen initial seed and apply a function (so called *chaotic map*) on it. Then, the map is applied on the resulting numbers several times to generate a chaotic sequence. More formally, a one dimensional chaotic map works as follows:

$$\begin{cases} x_{i,j}^0 = \text{an arbitrary initial seed,} \\ x_{i,j}^{k+1} = f(x_{i,j}^k; \boldsymbol{\mu}) \end{cases} \tag{1}$$

where $x_{i,j}^k$ is the jth variable of the point number i in the kth iteration of the algorithm. In Equation (1), $\boldsymbol{\mu}$ denotes set of parameters of chaotic map f.

2.2   Deterministic Initial Populations

Based on the criterion proposed in [9], we consider an initializer as deterministic if its output does not depend on any initial seed. As a result, deterministic algorithms are those which always produce exactly the same population. Unlike stochastic techniques, these algorithms do not adopt any random element. Instead, they aim to generate evenly distributed points all over space.

There are many ways of producing uniform population. A large family of these techniques called *Quasi-Random Sequence* is widely used in Monte Carlo integration [19]. They aim to generate numbers in a unit hypercube such that when the size of population increases, its uniformity converges to the optimum value. Since discrepancy is the main measure used in evaluation of quasi-random sequences, they generally known as *low-discrepancy* point generators. Halton [4] and Sobol [1] sets are some of the most well-known low-discrepancy techniques. Subsection 2.3 provides more details about discrepancy measures.

Another family of space filling algorithms, generally called *Uniform Experimental Design*, is also widely used in initialization [17]. These techniques usually sample a small number of points from a large number of possibilities such that the uniformity of the sampled population is maximized. Beside the uniformity of population, some of these algorithms can produce point sets having extra properties such as orthogonality, i.e., meaning points are dimensionally independent [11]. These algorithms are very popular in computer simulations both in discrete and continuous spaces [3]. Latin hypercube [14], good lattice points [18], uniform design [16] and orthogonal design [11] are examples of this family of initializers. Similar to quasi-random sequences, the performance of uniform experimental designs can be assessed using discrepancy measures.

2.3   Quality Measures

Without a set of general and practical measures, population initialization techniques cannot be evaluated or compared. Therefore, many different tools have been proposed to measure quality of a given population from different aspects, e.g., uniformity or randomness. However, many of them are not applicable in general studies due to the following reasons:

1. Some of the quality measures are highly subjective. The values of these methods are sensitive to various factors such as the properties of benchmarked problems, the employed EA models and other parameters (e.g., the maximum number of function evaluations) [8]. For example, final fitness value and success rate are both of this kind. The sensitivity of the measures to those factors causes the generality of the findings to be limited by the number of studied problems, employed EAs and levels of variation in parameter settings. As a result, the findings from a limited number of experiments can hardly be generalised to other situations.
2. Some quality measures are only applicable to specific algorithms. For example, there are well-known measures of randomness, unpredictability and incompressibility which can be used to evaluate stochastic techniques [9]. However, these measures cannot be used to assess the performance of deterministic techniques due to the lack of random elements in their algorithms.
3. Some measures are computationally expensive [3]. These methods may not be applicable in large-scale domains. Since the ultimate goal is to study the performance of initializers in producing many high dimensional points, we need to find measures which are efficient in terms of memory usage and time complexity.

Considering the above limitations, a sub-class of *discrepancy measures* which has analytic formulas is studied in this research. They are selected because of three reasons: 1) Their values are not affected by the features of benchmarked problems, employed EAs or their parameters. 2) They can be easily applied to all kinds of real-value populations. 3) they are faster than similar iterative/recursive algorithms which makes them ideal for large high dimensional populations.

Discrepancy value, in general, indicates the level of non-uniformity of a point set. This means more uniform populations have smaller discrepancy values. Note that uniformity is a desirable property of initial population which plays an important role in the performance of EAs when dealing with black-box problems. Therefore, many researchers try to develop new techniques to have higher uniformity or less discrepancy when no prior knowledge is available about the landscape.

One of the early versions of discrepancy measures is widely known as L_p-discrepancy [3]. The first variants of L_p-discrepancy need many axillary random samples to be able to measure the discrepancy of a given population. This limitation makes them computationally expensive to calculate the discrepancy of large populations in high dimensional spaces. Warnock [3], for the first time, gave a novel analytic formula to compute L_2-discrepancy which resulted in a much faster algorithm. The formula is as follows:

$$(D_2(\mathcal{P}))^2 = 3^{-D} - \frac{2^{1-D}}{N} \sum_{k=1}^{N} \prod_{l=1}^{D} (1 - x_{k,l}^2) + \frac{1}{D^2} \sum_{k=1}^{N} \sum_{j=1}^{N} \prod_{i=1}^{D} [1 - \max(x_{k,i}, x_{j,i})] \quad (2)$$

where D and N are dimensionality and size of population $\mathcal{P}$, respectively.

To improve the sensitivity and accuracy of L_2-discrepancy, several expansions, such as star, modified, symmetric, wrap-around and centred L_2-discrepancies, have been proposed [3]. In this study, we focus on centred L_2-discrepancy (CD) because it is more accurate in identification of differences between populations [5]. The analytic formula of CD is as follows:

$$(CD(\mathcal{P}))^2 = \left(\frac{13}{12}\right)^D - \frac{2}{N} \sum_{k=1}^{N} \prod_{j=1}^{D} \left(1 + \frac{1}{2}|x_{k,j} - 0.5| - \frac{1}{2}|x_{k,j} - 0.5|^2\right)$$
$$+ \frac{1}{N^2} \sum_{k=1}^{N} \sum_{j=1}^{N} \prod_{i=1}^{D} \left[1 + \frac{1}{2}|x_{k,i} - 0.5| + \frac{1}{2}|x_{j,i} - 0.5| - \frac{1}{2}|x_{k,i} - x_{j,i}|\right], \quad (3)$$

where D and N are dimension and population size of population $\mathcal{P}$, respectively.

3   Experiments

This section discusses the experiment studies, then reports the obtained results and finally analyzes the findings.

3.1   Experiments Setup

This study comprises of two parts. In the first part, we study the trend of population uniformity when generated by common PRNG, i.e., the simple *rand* function. Since population size plays an important role in the uniformity of population, we also investigate its effects. In fact, we aim to answer two questions in this part: 1) How much the uniformity of a population can be affected by dimensionality? 2) Is it possible to enhance the uniformity of initial population in high dimensional spaces by increasing the population size?

In the second part of the experiments, we compare the performance of advanced initialization techniques with a commonly used PRNG technique. We repeat this experiment in a variety of low, medium and high dimensional spaces. The questions to be answered in this part are: 1) Can adopting advanced initialization techniques significantly improve population uniformity? 2) How sensitive are the advanced initializers to the variation in population size?

In the second part, we study three stochastic and three deterministic population initialization techniques. Table 1 lists the chosen techniques. More information on these techniques are provided in Section 2.

Table 1. Selected Population Initialization Techniques

Abbreviation	Name	Category	Subcategory	Ref.
RNG	Mersenne twister	stochastic	pseudo-random	[13]
SIN	sine chaotic map	stochastic	chaotic number	[20]
TNT	tent chaotic map	stochastic	chaotic number	[7]
HLT	Halton set	deterministic	low-discrepancy	[4]
SBL	Sobol set	deterministic	low-discrepancy	[19]
GLP	good lattice point	deterministic	uniform design	[18]

As mentioned earlier, each time a stochastic technique is executed, it generates a different population due to its sensitivity to the initial random seed. Therefore, the quality of the resulting population may slightly differ from time to time. To achieve a more solid conclusion, we run each stochastic technique 25 times and average their quality. Note that each run of an algorithm is independent from the other runs due to the use of explicit initial seeds.

To have similar procedures for both categories, but avoiding the production of exactly the same populations for several times, we follow *skip* scheme for deterministic techniques. This scheme generates $25 \times N$ points, but only uses ith N points in the ith run. Indices of points in the ith run is easily calculated using the following formula:

$$l_i = (i-1)N + 1, \quad \text{and} \quad u_i = iN. \tag{4}$$

where l_i and u_i are the lower and upper bounds on the indexes of points in run i, respectively. Note that, in all parts of the experiments, 20 different dimension sizes ($2 \leq D \leq 1,000$) and 20 population sizes ($10 \leq N \leq 10,000$) were examined.

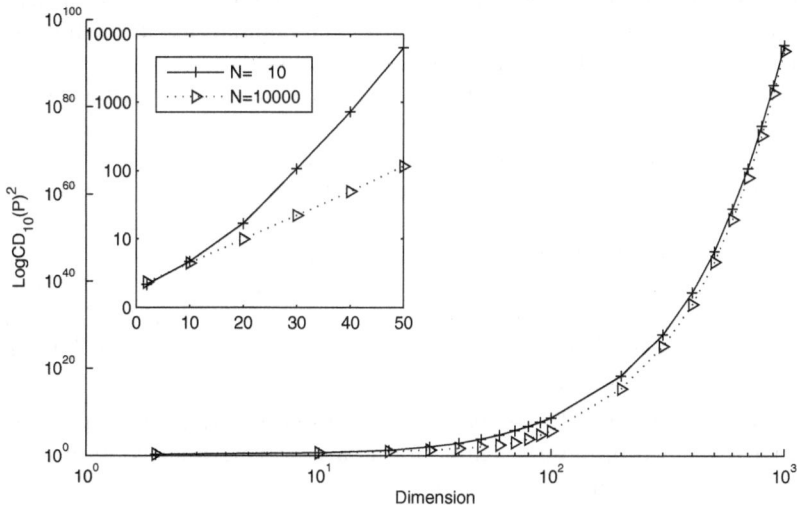

Fig. 1. Trend of $(CD(\mathcal{P}))^2$ of RNG for $2 \leq D \leq 1000$. A part of plot (i.e., $2 \leq D \leq 50$) is zoomed for better demonstration.

3.2   Results and Discussions

Experiment Part (A): In this part, we compute and compare CD values of populations generated by simple RNG to study the effects of dimensionality and population size on the uniformity.

As Figure 1 shows, discrepancy grows (i.e., uniformity drops) exponentially when the dimensionality increases. A closer look at the low dimensional part of the plot (zoomed in the figure) reveals that a large population size may lessen the undesirable effect of high dimensionality. However, the improvement may not be very significant. For example, the discrepancy of 10,000 points in 50 dimensions is comparable with the discrepancy of 10 points in 30 dimensions. In other words, 66% growth in dimensionality demands 100,000% increase in population size to recover the uniformity. This issue, widely known as curse of dimensionality, is even more severe in large-scale problems.

Figures 2-4 illustrates the effect of population size on the uniformity in small, medium and large-scale problems, respectively. As Figure 2 indicates, population size has no considerable effect on the uniformity of very small-sized problems, i.e., $D \leq 10$. For higher dimensions, specially for $30 \leq D \leq 50$, population size has a significant effect on uniformity such that it can be improved 10 to 20 times in the CD scale. However, the magnitude of improvements falls rapidly such that increasing population size beyond 1,000 points shows only a minimal improvement. In other words, it is reasonable to increase population size for the problems of size 20 to 50, while keeping it around 1,000 points.

Figure 3 demonstrates similar pattern for medium-sized spaces. The only difference is the slopes of the curves which are slower for the medium-sized spaces.

Therefore, a considerable improvement is expected even for population size beyond 3,000 points. In other words, having larger populations in medium-sized problems is reasonable when computationally feasible.

Uniformity in a high dimensional space is much worse than small or medium-sized spaces. As Figure 4 reveals, uniformity of populations in spaces of above 100 dimensions is so weak that increasing population size from 1,000 to 10,000 cannot recover it. Having a closer look at the plot shows that the reasonable population size for such large-scale problems is surprisingly less than 1,000 points. Note that this does not imply the population size has no effect in large-scale problems. Instead, it means the population size must be astronomically large to achieve a significant enhancement in uniformity. Since evaluating high dimensional populations in that magnitude is currently computationally infeasible, keeping it around 1,000 points is more practical and reasonable.

Experiment Part (B): To compare advanced initialization techniques with a common PRNG, we propose a simple formula reflecting relative improvement achieved from each advanced technique:

$$\% \text{ improvement} = \frac{log_{10}(CD(\mathcal{P}_c))^2 - log_{10}(CD(\mathcal{P}_i))^2}{log_{10}(CD(\mathcal{P}_c))^2} \times 100 \qquad (5)$$

where $\mathcal{P}_c$ is the population generated by the control technique, PRNG, and $\mathcal{P}_i$ is the population produced by the ith advanced initialization technique.

As Figure 5 reveals, some techniques are successful in improving the common initializer, although the biggest improvement in $2 \leq D \leq 50$ is less than 20%. Another observation from this plot is that whilst some techniques e.g.,

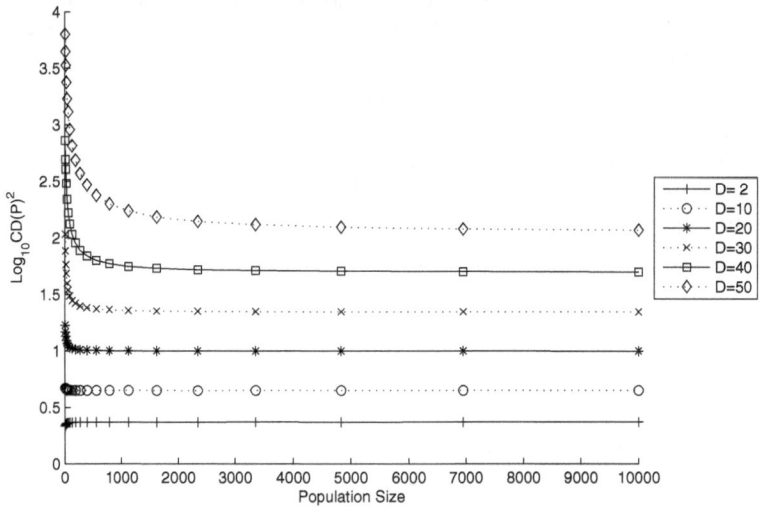

Fig. 2. Effect of population size on $(CD(\mathcal{P}))^2$ of RNG in low dimensions ($D \leq 50$)

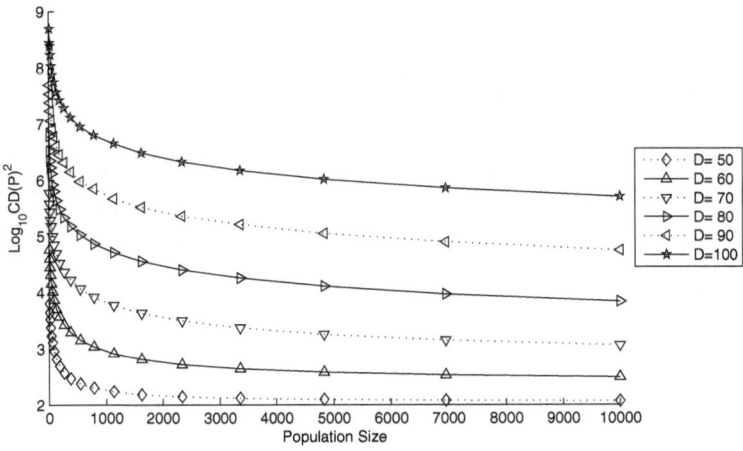

Fig. 3. Effect of population size on $(CD(\mathcal{P}))^2$ of RNG in medium dimensions

GLP are very sensitive to population size, others including SBL are more stable. However, with no exception, all techniques work relatively better when population size increases. Figure 5 also shows that mixed good and bad results can be expected from both categories of initialization techniques which confirms the findings from [7]. Note that the negative numbers indicate detrimental effects.

Figure 6 depicts the improvements gained from advanced techniques in medium and high dimensional spaces. As can be seen in the plot, all trends converge to one of the three values: 0%, -25% and -80%. This clearly shows that employing advanced

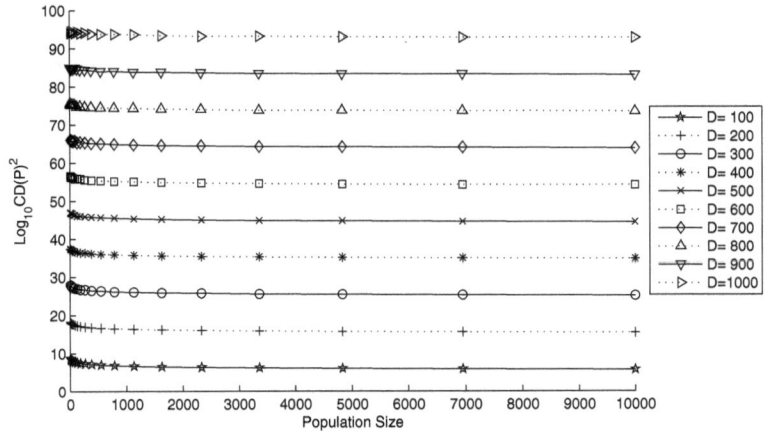

Fig. 4. Effect of population size on $(CD(\mathcal{P}))^2$ of RNG in high dimensions

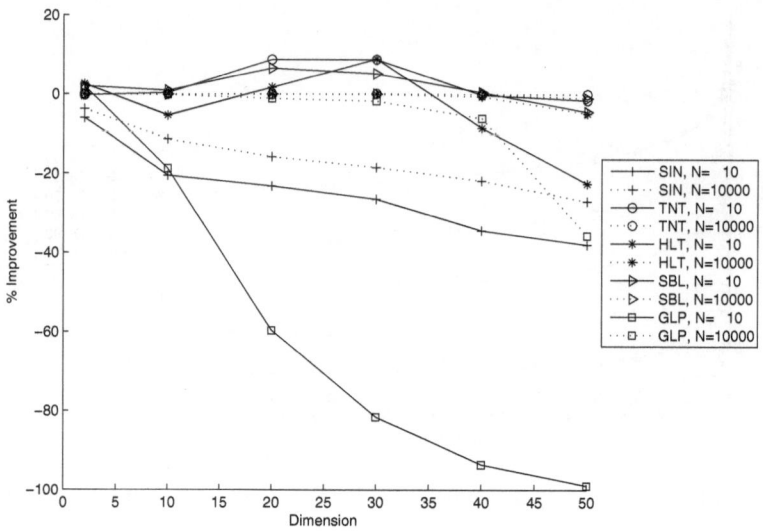

Fig. 5. Improvements gained from advanced techniques in low dimensions

initialization techniques provides no significant improvement in high dimensions, at least in terms of uniformity. Even increasing population size from 10 to 10,000 does not result in a significant improvement. Figure 6 also shows SBL with 10 and

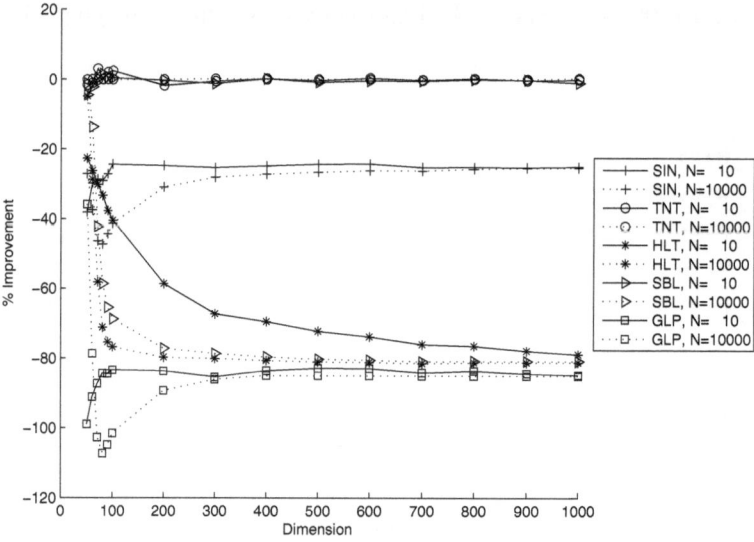

Fig. 6. Improvements gained from advanced techniques in medium and high dimensions

TNT with both 10 and 10,000 population sizes perform almost the same as PRNG. The others, however, perform poorly in comparison with a PRNG with the same population size.

In some cases, including SBL, increase in population size decreases relative improvement. This should not be mistakenly interpreted as an adverse effect of population size on those algorithms. Instead, it should be noted that the plot illustrates relative improvements which are calculated based on the performance of PRNG (see Equation 5). It means, enlarging population has a greater impact on PRNG than those algorithms such that the relative improvement from them drops when population size rises.

4   Conclusions

In this paper we investigate the reasons that causes several advanced population initialization techniques to perform poorly in high dimensional spaces. We use a general purpose measure to study the effect of dimensionality on common and advanced initializers where population size varies from 10 to 10,000 points. Our investigations show that the uniformity of initial population drops exponentially when dimensionality rises linearly. Low uniformity, which can also be interpreted as weak coverage or low diversity, degrades the quality of initial populations dramatically. Except for some small and medium-sized spaces, even increasing population size up to a computationally feasible bound cannot maintain uniformity.

Our experimental results reveal that the advanced initializers are as vulnerable to the curse of dimensionality as simple random number generators. Therefore, adopting advanced initializers in medium and large-scale spaces does not result in any significant improvement. In this regard, some advanced techniques are even more susceptible to the adverse effect of dimensionality than the simple pseudo-random number generators. Accordingly, we only recommend the use of advanced techniques when the population and dimension sizes are very small. In higher dimensional spaces or when the population size is relatively large, no significant improvement is excepted from advanced techniques.

References

1. Bratley, P., Fox, B.L.: Algorithm 659: Implementing Sobol's quasirandom sequence generator. ACM Transactions on Mathematical Software (TOMS) 14(1), 88–100 (1988)
2. Clerc, M.: Initialisations for particle swarm optimisation (2008), http://clerc.maurice.free.fr/pso
3. Fang, K.T., Lin, D.K.: Uniform experimental designs and their applications in industry. Handbook of Statistics 22, 131–170 (2003)
4. Halton, J.H.: On the efficiency of certain quasi-random sequences of points in evaluating multi-dimensional integrals. Numerische Mathematik 2(1), 84–90 (1960)
5. Hickernell, F.: A generalized discrepancy and quadrature error bound. Mathematics of Computation of the American Mathematical Society 67(221), 299–322 (1998)

6. Jun, B., Kocher, P.: The Intel random number generator. Cryptography Research Inc. white paper (1999)
7. Kazimipour, B., Li, X., Qin, A.: Initialization methods for large scale global optimization. In: 2013 IEEE Congress on Evolutionary Computation (CEC), pp. 2750–2757. IEEE (2013)
8. Kazimipour, B., Li, X., Qin, A.: Effects of population initialization on differential evolution for large scale optimization. In: 2014 IEEE Congress on Evolutionary Computation (CEC), pp. 2404–2411. IEEE (2014)
9. Kazimipour, B., Li, X., Qin, A.: A review of population initialization techniques for evolutionary algorithms. In: 2014 IEEE Congress on Evolutionary Computation (CEC), pp. 2585–2592. IEEE (2014)
10. Kazimipour, B., Omidvar, M.N., Li, X., Qin, A.: A novel hybridization of opposition-based learning and cooperative co-evolutionary for large-scale optimization. In: 2014 IEEE Congress on Evolutionary Computation (CEC), pp. 2833–2840. IEEE (2014)
11. Leung, Y.W., Wang, Y.: An orthogonal genetic algorithm with quantization for global numerical optimization. IEEE Transactions on Evolutionary Computation 5(1), 41–53 (2001)
12. Marsaglia, G., Zaman, A.: The KISS generator. Tech. rep., Department of Statistics, University of Florida (1993)
13. Matsumoto, M., Nishimura, T.: Mersenne twister: a 623-dimensionally equidistributed uniform pseudo-random number generator. ACM Transactions on Modeling and Computer Simulation (TOMACS) 8(1), 3–30 (1998)
14. McKay, M., Beckman, R., Conover, W.: A comparison of three methods for selecting values of input variables in the analysis of output from a computer code. Technometrics 42(1), 55–61 (2000)
15. Park, S.K., Miller, K.W.: Random number generators: good ones are hard to find. Communications of the ACM 31(10), 1192–1201 (1988)
16. Peng, L., Wang, Y., Dai, G.: UDE: differential evolution with uniform design. In: 2010 Third International Symposium on Parallel Architectures, Algorithms and Programming (PAAP), pp. 239–246. IEEE (2010)
17. Peng, L., Wang, Y., Dai, G., Cao, Z.: A novel differential evolution with uniform design for continuous global optimization. Journal of Computers 7(1), 3–10 (2012)
18. Sloan, I.H., Joe, S.: Lattice methods for multiple integration. Oxford University Press (1994)
19. Sobol, I.M.: On quasi-monte carlo integrations. Mathematics and Computers in Simulation 47(2), 103–112 (1998)
20. Zheng, H., Zheng, Y., Li, P.: Sine-map chaotic pso-based neural network predictive control for deployable space truss structures. In: 2013 IEEE International Symposium on Industrial Electronics (ISIE), pp. 1–5 (May 2013)

Bloat and Generalisation in Symbolic Regression

Grant Dick

Department of Information Science,
University of Otago,
Dunedin, New Zealand
grant.dick@otago.ac.nz

Abstract. Symbolic regression is a common application of genetic programming (GP). Increasingly, the GP community is identifying the need to measure the generalisation performance of the models evolved in symbolic regression, and consequently the need to design operators and methods that promote generalisation. In this paper, we explore the use of bloat control methods as a means of promoting generalisation. The results suggest that bloat control methods effectively reduce the computational requirements of symbolic regression, but do not significantly improve generalisation performance. Additionally, we compare the symbolic regression models traditional machine learning techniques, and find that the traditional methods produce models that generalise more effectively than their GP counterparts, while also using fewer computational resources. The results highlight the importance of contextualising GP performance with methods outside of evolutionary computation.

1 Introduction

The field of genetic programming (GP) specialises in the evolutionary search of solutions of arbitrary size and shape [13]. The flexibility of the representation in GP makes it a desirable approach to solving problems where the shape of the solution is not known *a priori* to search. One such domain is symbolic regression, in which a model is developed to map a set of known observation data to a prediction of an unknown response value. Symbolic regression has received a lot of attention in GP, with many papers identifying shortcomings in the standard GP approach, and proposing alternative methods to make GP more suitable to search in this domain. A key issue in any regression approach is the idea of *generalisation*, that is the ability of the produced model to adequately respond to previously unseen data. Generalisation is typically correlated to model complexity, and in GP, the size of the evolved program is one measure of its complexity. Therefore, bloat, and the required bloat management, is considered to be an important factor in promoting generalisation performance in GP regression.

This paper examines the use of bloat control methods to promote generalisation in symbolic regression. Results suggest that they do promote the development of smaller programs, but that they are neither more or less prone to over-fitting than compared to standard GP. The results are also contextualised

G. Dick et al. (Eds.): SEAL 2014, LNCS 8886, pp. 491–502, 2014.
© Springer International Publishing Switzerland 2014

with two traditional machine learning approaches to regression: a simple linear regression and random forests [4]. On all but one of the tested problems, the traditional methods offer equal or better generalisation performance, suggesting that more work is needed to improve the search characteristics of GP in the context of symbolic regression.

The remainder of this paper is structured as follows: §2 provides a brief overview of related work; §3 outlines the methods and problems used for experimentation; §4 examines the behaviour of bloat control methods in relation to GP and existing regression methods; finally §5 concludes the paper and suggests avenues for future work to explore.

2   Generalisation in Symbolic Regression

One of the key applications of genetic programming is symbolic regression, where the programs being evolved are expressions that map a set of observations to a prediction of an unknown response. By its very nature, GP appears naturally aligned with the goals of symbolic regression, so it is understandable that it receives a significant proportion of attention in GP literature. In the process, this research has raised several issues with the way that 'standard' GP handles symbolic regression, and in particular the manner in which the evolved models generalise to unseen data. While the majority of early work on symbolic regression did not examine the generalisation performance of the evolved models, it is becoming an increasingly important aspect of the relevant research [15,18,6].

It is relatively simple to incorporate modifications into GP that promote generalisation. One solution is to take a multi-objective approach and split available data into training, validation and testing. The error on the training set is used to select parents for recombination, while the validation set is used to select the representative model from the final set of candidate solutions. The final testing set is then used to externally evaluate the evolved models [9].

In its basic form, GP requires exception handling for certain operators, such as division. The solution is typically to create 'protected' versions of these operators, that return some predetermined value when error-producing inputs are provided. As a by-product of this, GP can exploit these arbitrary values to fit the resulting function to the observed data points [11]. This can cause problems when the function is applied to unseen data. A typical solution is to remove these protected operators from the available function set, but this limits the possible expressiveness of the evolved programs. Alternatively, one can use interval arithmetic to reduce the need to protect mathematical operators [12].

More recently, methods that exploit statistics about tree subtree behaviour have been introduced, such as semantically similar crossover. The idea in this approach is that crossover points are first tested to ensure that both subtrees exhibit a sufficiently similar behaviour over the data set before they can be swapped between parents. This approach was shown to outperform GP using standard subtree crossover in a number of problems [19]. Additionally, it was later shown that trees evolved using this new crossover approach are able to generalise better to unseen data [23]. Recent work has also used a semantics-driven approach to produce crossover methods that offer improvements in search and generalisation over standard GP [14].

2.1 Bloat Control and Generalisation

Very early work on GP identified the tendency for programs to grow in size at a rate greater than the observed improvement in fitness [13]. This phenomenon, later known as *bloat* has subsequently received considerable focus in the GP literature. Many theories as to the underlying cause of bloat have led to methods designed to control unreasonable growth of programs within the population. Simplest among these it to place an upper bound on the size or depth of the trees produced through crossover and mutation [13]. Alternatively, *parsimony* pressure can be incorporated into the fitness of programs to punish excessively large individuals and encourage breeding between smaller programs [25]. Later research dealt with the concept of bloat more explicitly by designing new selection operators that acknowledge size [16,17], adding size as an objective in multi-objective search, random elimination of individuals of above-average size (the so-called Tarpeian method) [20], applying a 'waiting time' to offspring in proportion to their size that delays their entry into the population [17], and rejecting offspring that do not align with a target size distribution for the population [8,22,21]. Additionally, recent work has focused upon the use of spatially-structured populations that implicitly control bloat[24,10,7].

There is a strong connection between the idea of bloat and generalisation. Typically, one expects a correlation between generalisation and model size — a smaller model typically incorporates fewer assumptions about the domain, and subsequently will be less prone to over-fitting [1]. Proactively reducing model size is a common practice in machine learning (e.g., pruning decision trees, reducing the topology and size of neural networks, eliminating high specificity rules in rule bases). Therefore, bloat control methods have received some attention as candidates for increasing the generalisation abilities of genetic programming. For example, the operator equalisation (OpEq) approach has been used to explore regression of a bio-availability problem containing hundreds of variables [22]. Interestingly, this approach managed to produce smaller programs, but also exhibited *over-fitting* to training data. The authors argue that this was a consequence of the increased number of fitness evaluations required by the approach. Subsequent work compared OpEq to a spatially-structured co-evolutionary approach (called SCALP), where results suggested that it was able to discover smaller solutions to the bio-availability problem without producing over-fitting [10].

3   Experimental Framework

The primary aim of this work is to explore the abilities of bloat control methods to promote generalisation in models evolved through symbolic regression. In particular, research into the spatially-structured plus elitism (SS+E) method is to be examined, as its generalisation performance has not been explored previously. In order to assess the abilities of this method, it needs to be placed in the context of existing bloat control methods. Therefore, this work becomes more of a general examination of bloat control methods on regression problems not typically explored in previous work. For point of reference, we adopted the Double Tournament approach [17], and dynamic operator equalisation [21] as comparative bloat control methods.

We consider five test problems from previous work: the first is the well-known quartic problem [13]. This is a very simple problem, and is included primarily for reference and to demonstrate 'best case' performance for the methods. The second problem is a bio-availability data set from previous work, with 241 variables, and 359 instances [22]. Finally, two problems from the UCI Machine Learning Repository were selected [2]. These were the Auto MPG data set, with 7 features and 398 instances, and the Wine Quality data set, with 11 features and 6498 instances [5]. The Wine Quality data set is actually supplied as two data sets, one for white wine examples (4898 instances), and another for red wine examples (1599 instances) — for this work, we found that there was very little difference in the results from investigating either the red or white subsets, so we report only the results of the red data set. The choice of these two data sets from the repository was rather arbitrary and does not reflect any bias in their selection — the former was the first suitable regression problem in the list when sorted alphabetically, while the latter was the first regression problem appearing in the 'most used' list of the web site.

For each problem, we needed to develop a training and test data set. For the quartic problem, we generated 20 uniform random samples from [-1,1] for both the training and test sets. For the remaining problems, a 70-30 split was adopted: for each run, a random sample of 70% of the available data was used as training data, while the remaining 30% was used for testing.

The common parameters are presented in Table 1. The SS+E approach requires a square toroidal population; we used a population size of 484, as it is the nearest square number to 500 as used in previous work [21]. We also used depth limiting in all the experiments: interestingly, we could not get any of the bloat control methods to improve upon depth limiting alone, but they were able to further improve the bloat properties when the two methods were combined, in line with recommendations from previous work [17]. As the SS+E approach uses elitist offspring replacement, there is no need to use a reproduction operator, therefore all SS+E experiments used full crossover, while the remaining approaches used a typical 90-10 split between crossover and reproduction. In line with previous bloat studies, no mutation was used in any experiments. Finally, the Double Tournament selection was calibrated using the recommended parameters from previous work: a parsimony tournament size of 1.4 was used, and a then fitness tournament of 7 was applied [17].

For the operator equalisation experiments, we used the dynamic variant (DynOpEq) with a bin size of 1. This method rejects individuals unless they fit a target histogram derived from the current population. If they do not fit

Table 1. The common GP parameters used for all experiments

Parameter	Value
Population Size	484
Generations	100
Selection Method	Tournament
Tournament Size	7
Crossover Rate	0.9
Reproduction Rate	0.1
Maximum Tree Depth	17
Crossover Node Selection	Koza-Style
Initialisation	Ramped-Half & Half
Minimum Initialisation Depth	2
Maximum Initialisation Depth	6
Function Set	$+$, $\times$, $-$, $\div$, $\log_e$, sin, cos
Terminal Set	*Independent variables from data set*

the distribution a further test is applied: if they are the best individual seen in the current run, or failing that are the best in generation for the target histogram bin, then they are accepted into the population. This requires a large number of evaluations per run, sometimes an order of magnitude more than the population size [10]. This was argued in previous work as a potential source of over-fitting [21], so in this work we limit DynOpEq to use the same number of total evaluations as the other approaches. If the number of evaluations is reached before the specified number of generations, then the run terminates early. All other experiments run for 100 generations after the initial random population.

Most symbolic regression experiments are presented and compared internally, that is without a comparison to traditional methods from statistics and machine learning. To provide a suitable context, we present the results from GP experimentation with two simple regression techniques — linear regression, and the random forest approach. The latter is an ensemble approach that uses resampling techniques to create a family of regression trees that combine to determine the required response prediction [4].

With two exceptions, all data was presented in an unaltered fashion to the given regression models. The first exception applied to the linear regression approach when modelling the bio-availability data set: due to the large number of 'empty' features (variables with a value of zero for every instance of the data set), linear regression produced singular fit models frequently. Therefore, before linear regression was performed, a simple feature selection was performed to eliminate such features from the data set. The second exception was in the Auto MPG data set, in which three instances contained missing data. To simplify the experiments, these instances were removed. No other feature selection or transformation of data was applied.

3.1   Spatial-Structure with Elitism

Our main point of examination in this work is the SS+E approach, as it has not been tested with respect to producing good generalisation performance. In particular, we use the *spatially-structured with lexicographic parsimonious elitism*

(SS+LPE) variant [7]. This method works in a similar manner to other spatially-structured evolutionary algorithms: each individual in the population is located in a cell in a two-dimensional torus. To produce a new generation, each location in the torus is visited, two parents from the Moore neighbourhood are selected using binary tournament selection, and crossed to produced an offspring. Once the entire offspring population has been bred, a competition takes place within the individuals of the current population: at each location, the offspring must be strictly fitter, or equal in fitness but strictly smaller in size, than the current occupant in order to survive. If the offspring fails this test, then the current individual is copied into the new generation in its place. The reader is directed to previous work for a more complete examination of the method [7].

4   Results

The results of applying each bloat control method to each problem are shown in Figures 1-4. Where appropriate, each trend in the given plot is presented with a 95% confidence interval of the measured mean. The results are summaries produced through 100 independent runs, each using a different random number generator seed.

The evolution of fitness, in terms of root-mean-square training error (RMSE) over time is shown in Figure 1. In general, a common theme is shown across all the tested problems: with the exception of Double Tournament, all methods eventually converge at a similar fitness, and have evolved long enough to have essentially plateaued in terms of improvement. The evolution of mean population size (Figure 2) shows a similar, and somewhat expected, trend: all the implemented bloat control methods offer an improvement over standard GP in terms of mean population program size.

The behaviour of double tournament was interesting, as on all problems (including the quartic problem) it struggled to find good general solutions. This appears to be a consequence of bad parameter choice; despite using the 'recommended' parameter settings for this method, it appears that, across all tested problems, there was a tendency for double tournament to overemphasise selection for smaller programs over selection for fitness.

The operator equalisation results suggest that it was able to effectively control the population bloat, relative to standard GP. On all the tested problems, the average program size in the population was around half of the equivalent standard GP results, without sacrificing fitness. The early fitness performance of OpEq requires further comment: at first glance, it would appear that OpEq is slower than the other methods at finding fitter solutions. However, the graphs shown are in *evaluations*, not generations, and the observed trend is a consequence of the increased number of evaluations that dynamic OpEq uses in each generation. Unlike previous work, we did not encounter an over-fitting phase when using OpEq. Given that we used an early halting process when sufficient evaluations had been used, this would support the claims of previous work that this over-fitting process is a result of the additional evaluations that OpEq typically consumes.

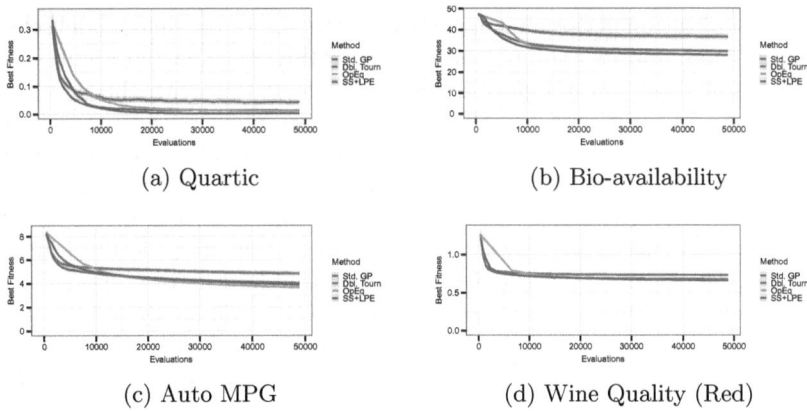

Fig. 1. Evolution of fitness of the best individual over time

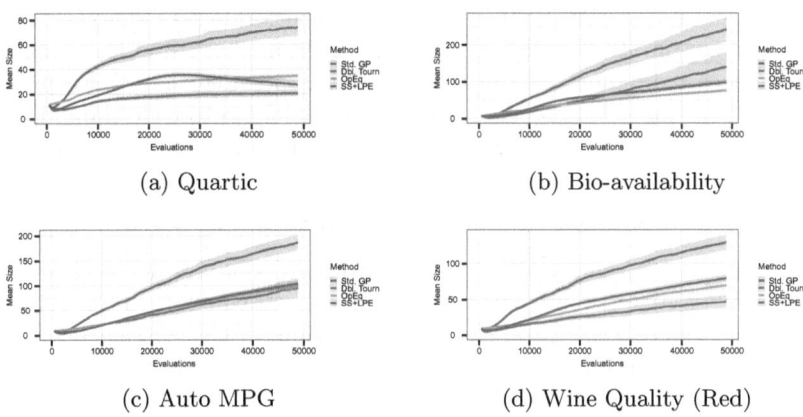

Fig. 2. Evolution of mean program growth of the population over time

Finally, the results of SS+LPE follow a similar trend to that shown by the approach in previous work. While the average size of programs in the population was slightly higher than that produced by OpEq, it was still around half of that observed in standard GP. The result on the quartic problem, while being a simple problem, is of interest here, and highlights a typical feature of SS+LPE behaviour. In the early stages of the run, SS+LPE places a strong emphasis on search for fitter solutions. However, once the populations begins to converge upon individuals of similar fitness, the pressure towards size increases, and we can observe a gradual drop in program size. Previous work suggests a possible explanation for this [7]: the programs present in the population could maintain proportions of code that is essentially non-functional (for example, containing trees resembling $(+ \, x \, (\times (- \, x \, x) \, (\Delta)))$, where the result of the subtree (Δ) is nullified). Once fitness converges, SS+LPE can place emphasis on implicitly editing out such examples by replacing them with smaller subtrees during crossover.

The generalisation performance, in terms of the size of the final model selected by each method, and its training and testing RMSE are shown in Figures 3 and 4. For each method, the program in the final population with the best fitness (or in the case of a tie, the smallest size) was selected as the representative model. As can be seen in Figure 3, all the examined bloat control methods were able to produce final models that were typically smaller than by using standard GP alone. In comparison to the mean population sizes, where the bloat control methods effectively halved the mean population size, the resulting final models using bloat are smaller, but not quite by the same margin. As mentioned earlier, Double Tournament was possibly incorrectly configured, and placed too much emphasis on size over fitness. This led to it producing runs with very small representative models. Finally, with the exception of the bio-availability problem, the models produced using SS+LPE are also slightly smaller than those produced by OpEq, although the difference is not likely to be significant.

In terms of generalisation performance, none of the bloat control methods was able to produce models that consistently outperformed standard GP by any meaningful margin. As mentioned earlier, SS+LPE appears to work by reducing the amount of redundant code in trees, and it is possible that OpEq presents a similar behaviour. This process is somewhat analogous to code simplification, and may explain why the programs are smaller, but functionally similar.

4.1 Performance of GP Relative to Traditional Methods

Comparisons with 'traditional' approaches to regression are not typically performed in GP symbolic regression research. This is in contrast to GP classification research, where it is typical to see the behaviour of GP classifiers compared with traditional methods, such as decision trees, support vector machines, or Bayesian methods [3]. The results presented so far are relevant in the context of standard GP, but it is difficult to assess the effectiveness of the approaches relative to other machine learning approaches.

To provide context, the training and testing results presented in Figure 4 include reference to two machine learning techniques: random forests and simple linear modelling. For each of these approaches, a similar experimental configuration was used to that for the GP experiments: 100 separate runs where performed, each using a random sampling process to split the data into training and testing sets, as described earlier. The regression model was trained and subsequently tested using the selected data sets, and training and testing errors were recorded. The results are shown in Figure 4.

The results presented raise some interesting questions about the performance of GP — in all but the quartic problem, GP is outperformed by traditional methods. It is not surprising that linear regression is outperformed by GP on the quartic problem, given the very nature of the relationship in the data and that the regression did not have access to any polynomial terms. However, the relative performance of GP on the other problems is surprising, and worthy of further investigation. While there is considerable effort present in the literature exploring improvements to GP with respect to regression, it is typically evaluated

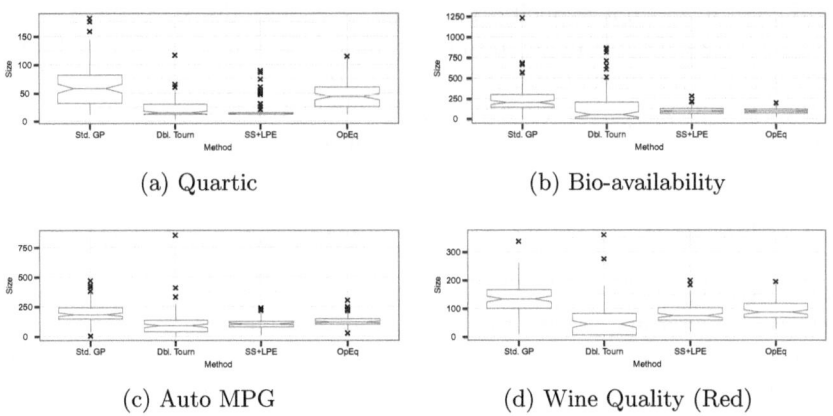

Fig. 3. Box-plots of the size of the final, best trained model evolved by each method

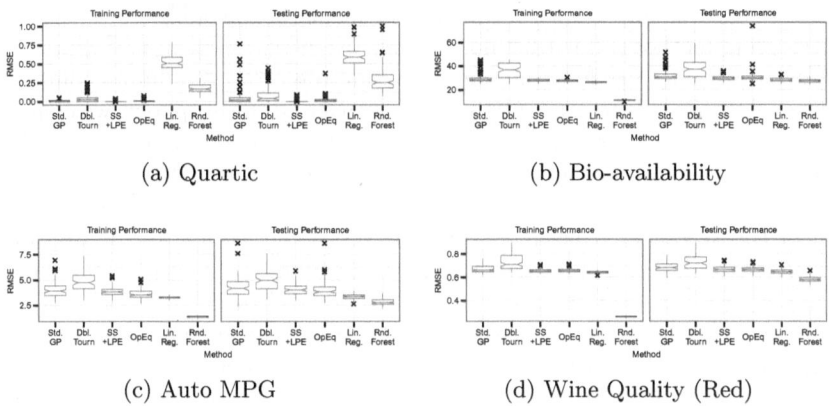

Fig. 4. Box-plots of the testing error of the final, best trained model evolved by each method

solely within in the context of 'standard GP'. The results presented here suggest that work in GP should not limit comparisons to other GP-based methods, and that traditional regression methods should be included to provide context.

An argument may be made in GP's favour that its representation provides explanatory power through direct analysis of the parse tree. [1] However, the model size evolved through GP on the tested problems typically contains hundreds of nodes, and this may require significant post-processing and analysis in order to extract the required knowledge. Additionally, most machine learning techniques provide means of knowledge extraction, so this advantage is not limited to GP.

---

[1] As suggested in the "Open Issues in Genetic Programming" tutorial presented at GECCO 2013 (Slide 21):
http://ncra.ucd.ie/papers/gecco2013_openissuesinGP_tutorial.pdf

Although it was not explicitly measured, it should also be pointed out that the computational effort required to build both the linear regression and random forest models was considerably less than for any of the GP approaches. In fact, the time required to complete all 100 runs of the random forest approach was typically less than the time required to perform a single GP run.

5 Conclusion

Generalisation performance is important in any regression domain. Although recent work increasingly acknowledges the need to measure generalisation, GP has a history of not testing generalisation of the evolved programs. One proposed approach to promoting generalisation is through the use of bloat control methods to promote the evolution of simpler parse trees. This paper has explored the used of several bloat control methods with an emphasis on promoting generalisation within the symbolic regression domain. Several problems of varying complexity were investigated, and it was shown that bloat methods could reduce the size of the resulting methods, but this did not improve generalisation performance.

A typical framework for symbolic regression research often limits comparison of the evolved models to other GP-based approaches. In this paper, we compared the results of models obtained through GP with models produced using linear regression and random forests. The results demonstrated that, on the tested problems, the performance of GP-evolved models tends to lag behind that of both of the traditional methods. This suggests that standard GP appears to struggle with the symbolic regression problem, and that other operators are needed in order to produce a better search mechanism. Most importantly, the results suggest that bloat control methods cannot be relied upon to improve the generalisation performance of GP regression methods. Rather, if they are to be used, it is likely in conjunction with other specialist operators designed to manipulate expressions in the symbolic regression domain. The results also highlight the need for GP research to contextualise its methods in relation to the wider machine learning community.

5.1 Future Work

The work presented here raises several areas for future consideration. In terms of the contribution of bloat control methods to generalisation, a recent method called approximate geometric crossover has been introduced [14]. While results using this crossover method suggest it to be more effective than standard crossover, it also tends to produce very large trees. It would be interesting to integrate this crossover method into SS+LPE, and see if the resulting models are smaller and provide better generalisation characteristics.

In order to keep the analysis simple, no attempt was made at in-depth data analysis and transformation. In other words, no feature selection was performed on the data sets, and no attempt to normalise the feature scale to reduce issues of scale were performed. This is likely to have impacted on the behaviour of all the methods examined, both traditional and GP-based. It would be an interesting piece of work to revisit the experiments performed here, only this time with data sets that have been fully-prepared for analysis.

Previous work has argued the need for harder test problems for GP [18]. The data sets presented here from the UCI Machine Learning Repository certainly appear to fulfil this requirement, and have presented challenging environments for GP to model through symbolic regression. They would certainly form a solid basis from which future work should perform its analysis.

References

1. Alpaydin, E.: Introduction to machine learning, 2nd edn. MIT Press (2010)
2. Bache, K., Lichman, M.: UCI machine learning repository (2013), http://archive.ics.uci.edu/ml
3. Bhowan, U., Johnston, M., Zhang, M., Yao, X.: Evolving diverse ensembles using genetic programming for classification with unbalanced data. IEEE Transactions on Evolutionary Computation 17(3), 368–386 (2013)
4. Breiman, L.: Random forests. Machine Learning 45(1), 5–32 (2001)
5. Cortez, P., Cerdeira, A., Almeida, F., Matos, T., Reis, J.: Modeling wine preferences by data mining from physicochemical properties. Decis. Support Syst. 47(4), 547–553 (2009)
6. Costelloe, D., Ryan, C.: On improving generalisation in genetic programming. In: Vanneschi, L., Gustafson, S., Moraglio, A., De Falco, I., Ebner, M. (eds.) EuroGP 2009. LNCS, vol. 5481, pp. 61–72. Springer, Heidelberg (2009)
7. Dick, G., Whigham, P.A.: Controlling bloat through parsimonious elitist replacement and spatial structure. In: Krawiec, K., Moraglio, A., Hu, T., Etaner-Uyar, A.Ş., Hu, B. (eds.) EuroGP 2013. LNCS, vol. 7831, pp. 13–24. Springer, Heidelberg (2013)
8. Dignum, S., Poli, R.: Operator equalisation and bloat free gp. In: O'Neill, M., Vanneschi, L., Gustafson, S., Esparcia Alcázar, A.I., De Falco, I., Della Cioppa, A., Tarantino, E. (eds.) EuroGP 2008. LNCS, vol. 4971, pp. 110–121. Springer, Heidelberg (2008)
9. Gagné, C., Schoenauer, M., Parizeau, M., Tomassini, M.: Genetic programming, validation sets, and parsimony pressure. In: Collet, P., Tomassini, M., Ebner, M., Gustafson, S., Ekárt, A. (eds.) EuroGP 2006. LNCS, vol. 3905, pp. 109–120. Springer, Heidelberg (2006)
10. Harper, R.: Spatial co-evolution: Quicker, fitter and less bloated. In: Proceedings of the Fourteenth International Conference on Genetic and Evolutionary Computation Conference, GECCO 2012, pp. 759–766. ACM, New York (2012)
11. Howard, D., Roberts, S.C.: Genetic programming solution of the convection-diffusion equation. In: Spector, L., Goodman, E.D., Wu, A., Langdon, W.B., Voigt, H.M., Gen, M., Sen, S., Dorigo, M., Pezeshk, S., Garzon, M.H., Burke, E. (eds.) Proceedings of the Genetic and Evolutionary Computation Conference (GECCO 2001), July 7-11, pp. 34–41. Morgan Kaufmann, San Francisco (2001)
12. Keijzer, M.: Improving symbolic regression with interval arithmetic and linear scaling. In: Ryan, C., Soule, T., Keijzer, M., Tsang, E., Poli, R., Costa, E. (eds.) EuroGP 2003. LNCS, vol. 2610, pp. 70–82. Springer, Heidelberg (2003)
13. Koza, J.R.: Genetic Programming: On the Programming of Computers by Means of Natural Selection. MIT Press, Cambridge (1992)

14. Krawiec, K., Pawlak, T.: Approximating geometric crossover by semantic back-propagation. In: Proceeding of the Fifteenth Annual Conference on Genetic and Evolutionary Computation Conference, GECCO 2013, pp. 941–948. ACM, New York (2013)
15. Kushchu, I.: An evaluation of evolutionary generalisation in genetic programming. Artif. Intell. Rev. 18(1), 3–14 (2002)
16. Luke, S., Panait, L.: Lexicographic parsimony pressure. In: Proceedings of the Genetic and Evolutionary Computation Conference, GECCO 2002, pp. 829–836. Morgan Kaufmann Publishers Inc., San Francisco (2002)
17. Luke, S., Panait, L.: A comparison of bloat control methods for genetic programming. Evol. Comput. 14(3), 309–344 (2006)
18. McDermott, J., White, D.R., Luke, S., Manzoni, L., Castelli, M., Vanneschi, L., Jaskowski, W., Krawiec, K., Harper, R., De Jong, K., O'Reilly, U.M.: Genetic programming needs better benchmarks. In: Proceedings of the Fourteenth International Conference on Genetic and Evolutionary Computation, GECCO 2012, pp. 791–798. ACM, New York (2012)
19. Nguyen, Q.U., Nguyen, X.H., O'Neill, M.: Semantic aware crossover for genetic programming: The case for real-valued function regression. In: Vanneschi, L., Gustafson, S., Moraglio, A., De Falco, I., Ebner, M. (eds.) EuroGP 2009. LNCS, vol. 5481, pp. 292–302. Springer, Heidelberg (2009)
20. Poli, R.: A simple but theoretically-motivated method to control bloat in genetic programming. In: Ryan, C., Soule, T., Keijzer, M., Tsang, E., Poli, R., Costa, E. (eds.) EuroGP 2003. LNCS, vol. 2610, pp. 204–217. Springer, Heidelberg (2003)
21. Silva, S., Dignum, S., Vanneschi, L.: Operator equalisation for bloat free genetic programming and a survey of bloat control methods. Genetic Programming and Evolvable Machines 13(2), 197–238 (2012)
22. Silva, S., Vanneschi, L.: Operator equalisation, bloat and overfitting: A study on human oral bioavailability prediction. In: Proceedings of the 11th Annual Conference on Genetic and Evolutionary Computation, GECCO 2009, pp. 1115–1122. ACM, New York (2009)
23. Uy, N.Q., Hien, N.T., Hoai, N.X., O'Neill, M.: Improving the generalisation ability of genetic programming with semantic similarity based crossover. In: Esparcia-Alcázar, A.I., Ekárt, A., Silva, S., Dignum, S., Uyar, A.Ş. (eds.) EuroGP 2010. LNCS, vol. 6021, pp. 184–195. Springer, Heidelberg (2010)
24. Whigham, P.A., Dick, G.: Implicitly controlling bloat in genetic programming. IEEE Transactions on Evolutionary Computation 14(2), 173–190 (2010)
25. Zhang, B.T., Mühlenbein, H.: Balancing accuracy and parsimony in genetic programming. Evolutionary Computation 3(1), 17–38 (1995)

Improved PSO for Feature Selection
on High-Dimensional Datasets

Binh Tran, Bing Xue, and Mengjie Zhang

Victoria University of Wellington, PO Box 600, Wellington 6140, New Zealand
{binh.tran,bing.xue,mengjie.zhang}@ecs.vuw.ac.nz

Abstract. Classification on high-dimensional (i.e. thousands of dimensions) data typically requires feature selection (FS) as a pre-processing step to reduce the dimensionality. However, FS is a challenging task even on datasets with hundreds of features. This paper proposes a new particle swarm optimisation (PSO) based FS approach to classification problems with thousands or tens of thousands of features. The proposed algorithm is examined and compared with three other PSO based methods on five high-dimensional problems of varying difficulty. The results show that the proposed algorithm can successfully select a much smaller number of features and significantly increase the classification accuracy over using all features. The proposed algorithm outperforms the other three PSO methods in terms of both the classification performance and the number of features. Meanwhile, the proposed algorithm is computationally more efficient than the other three PSO methods because it selects a smaller number of features and employs a new fitness evaluation strategy.

Keywords: Particle swarm optimisation, Feature selection, Classification, High-dimensional data.

1 Introduction

Learning from examples is a successful approach in machine learning and data mining. Classification is a typical task of learning from training examples to predict the class labels of unseen examples/instances based on given attributes or features. Many classification algorithms have been successfully applied to automatically learn classifiers in a variety of problems, such as image classification, text categorisation and disease classification. Recently, there are more and more classification datasets with hundreds or even thousands features, which causes the "curse of dimensionality". This makes the classifier learning process become difficult because not all features are relevant to the class labels and often contain redundant information. Such data typically needs feature selection (FS) to remove irrelevant and redundant features [9].

Existing FS methods can be classified into wrapper approaches and filter approaches depending on whether a classification algorithm is used to evaluate the goodness of the feature subsets [5]. The criteria used in wrapper methods include the classification performance of a predefined classification algorithm using only the selected feature subsets. On the other hand, filter methods rely on various measures of the general characteristics of the training data, such as distance, information, dependence and consistency

G. Dick et al. (Eds.): SEAL 2014, LNCS 8886, pp. 503–515, 2014.

measures [5] to evaluate the classification capability of the feature subsets. Filter methods are argued to be more general to different classification algorithms than wrapper methods. Wrapper approaches usually obtain feature subsets with higher classification accuracy than filter methods [14] because they directly use the classification performance to guide the search. However, wrapper approaches are computationally expensive because each evaluation involves a training process of the classification algorithm. This work focus mainly on developing a new wrapper FS algorithm.

An optimal feature subset is the smallest subset, which maximises the classification accuracy. However, finding the optimal feature subset is an NP-hard problem [14]. The size of the search space grows exponentially with the number of features in the dataset. Therefore, it is necessary to have an efficient global search technique to tackle FS problems. Evolutionary Computation (EC) techniques are well-known for their global search potential. Particle swarm optimisation (PSO) [13,22] is a relatively recent EC technique that has been successfully applied in many areas such as function optimisation [22] and feature selection [15,29,27,30,28]. Comparing to other EC techniques, PSO has some advantages such as simplicity, fewer parameters, lower computational cost, and fast convergence [8].

In PSO, a swarm of candidate solutions are encoded as particles. During the searching process, each particle remembers the best solution it obtained so far, i.e. the personal best called *pbest*. By sharing *pbest* with neighbours, each particle knows the best solution that the whole population has found so far, i.e. the global best called *gbest*. PSO searches for the optimal solutions based on the information from *pbest* and *gbest*. However, if *gbest* is near a local optimum, there is a high probability that the swarm will be stuck in this area, especially in high-dimensional/large-scale problems, e.g. FS on gene expression data, which has thousands or even more than ten thousands features.

To overcome this limitation, Chuang et al [4] proposed an improved PSO algorithm (PSO-RG) in which *gbest* will be restarted whenever it is not improved in a number of iterations. The proposed PSO achieved better performance than standard PSO. Resetting *gbest* could avoid being stuck in local optima by encouraging the exploration of the search, but it may limit the algorithm further exploit the surroundings of the already found good solutions, i.e. *gbest*. Therefore, a new PSO algorithm is still needed to better solve high-dimensional feature selection problems on gene expression data.

1.1 Goals

The overall goal of this paper is to develop a new PSO approach to feature selection on high-dimensional gene expression data to significantly reduce the number of features and increase the classification performance over using all features. To achieve this goal, the *gbest* reset mechanism is used to encourage the global search (exploration) and a new local search strategy is proposed to facilitate the exploitation of the algorithm to further improve the performance. The local search is also designed to utilise the characteristics of a simple classification algorithm, k-Nearest-Neighbour (KNN), to avoid heavy computational cost. The proposed algorithm is examined and compared with standard PSO, PSO only using the proposed local search (PSO-LS), and PSO-RG only using the *gbest* reset mechanism on five gene datasets with more than ten thousands of features. Specifically, we will investigate:

- whether the proposed algorithm can reduce the number of features and achieve similar or better classification performance than using all features,
- whether the proposed algorithm can outperform than standard PSO, PSO-LS and PSO-RG in terms of the number of features and the classification performance, and
- whether the proposed algorithm can be more efficient than standard PSO, PSO-LS and PSO-RG.

2 Background

2.1 Particle Swarm Optimisation (PSO)

PSO is an EC technique developed by Kennedy and Eberhart [13], which is inspired by social behaviours found in birds flocking or fish schooling. In PSO, a swarm consists of many individuals called particles communicating through iterations to search for optimal solutions when moving in the search space.

In PSO, each particle has a position and a velocity. The position is a candidate solution of the problem and is usually an n-dimension vector of numerical values. Velocity also has the same structure as position, which represents the speed and direction that the particle should move in the next iteration. In each iteration, the velocity of a particle is updated based on the personal best (*pbest*) which is the best position it has been explored so far and the global best (*gbest*) which is the best position it has been communicated from other particles. Formulae (1) and (2) are used to update the velocity and position of each particle.

$$v_{id}^{t+1} = w * v_{id}^t + c_1 * r_{1i} * (p_{id} - x_{id}^t) + c_2 * r_{2i} * (p_{gd} - x_{id}^t) \qquad (1)$$
$$x_{id}^{t+1} = x_{id}^t + v_{id}^{t+1} \qquad (2)$$

where v_{id}^t and x_{id}^t are velocity and position of particle i in dimension d at time t, respectively. p_{gd} and g_{gd} are *pbest* and *gbest* positions in dimension d. c_1 and c_2 are acceleration constants, and r_1 and r_2 are random values. w is the inertia weight used to control the impact of the last velocity to the current velocity. The velocity values are usually limited by a predefined maximum velocity, v_{max} to the range $[-v_{max}, v_{max}]$.

2.2 Related Works on Feature Selection

Traditional Methods for Feature Selection. Two typical wrapper FS methods are sequential forward selection (SFS)[25] and sequential backward selection (SBS) [17], which employs a greedy search method. SFS (SBS) starts with an empty (full) feature subset, then gradually adds (removes) features until the classification accuracy is not improved. However, both methods suffer from the so-called "nesting effect" because a feature which is selected or removed cannot be removed or selected in later stage. As a compromise between these two approaches, "plus-l-take-away-r" [24] applies SFS l times and then SBS r times. This strategy can avoid nesting effect, but it is hard to determine appropriate values for l and r. To avoid this, Pudil et al [21] introduced two corresponding methods: sequential backward floating selection(SBFS) and sequential forward floating selection (SFFS). These floating search methods are claimed to be better than the static sequential methods, but they are still facing the problem of stagnation in local optima.

EC Techniques for Feature Selection. Many EC techniques have been applied to FS problems such as Genetic algorithms (GAs) [3], Genetic programming (GP) [19], and Ant colony optimisation (ACO)[7]. Among these, GA is probably the first popular EC technique that has been applied in FS. Guided by Darwinian evolution principles, GAs start with a population of candidate solutions, represented as chromosomes, and evolved better solutions by using genetic operators like crossover, mutation. Many GA based FS algorithms have been proposed either in filter or wrapper approaches. In the former, feature subsets are evaluated by using inconsistency rates [16], or fuzzy sets [3]; while in the latter, different approaches proposed with different classification algorithms for fitness evaluation, such as ID3 [2], and artificial neural network [20]. GA based hybrid FS algorithms that combine both filter and wrapper approaches have also been proposed to improve the performance [10].

Using the same principles, however, instead of evolving bit strings, GP evolves computer programs to generate solutions. Each program is a tree consisting of internal nodes which are usually arithmetic operators, and leaf nodes which are constants or variables. When using GP for FS, the variables are chosen from the original features. Selected features are the ones used as leaf nodes of a GP tree. GP has been used in both filter FS methods [19] and wrapper FS methods [6]. ACO is another EC technique that stands in the same umbrella to PSO, swarm intelligence. ACO is inspired by the special communication system using pheromone between real ants about favorable paths to food. The shortest path will be the one that has most pheromone. When using ACO for FS, each feature is considered as a node, and paths between nodes represent the choices for next features. Many ACO algorithms are used for both filter [11,18] and wrapper [12] feature selection. PSO has recently gained more attention in addressing FS tasks, but most of the existing PSO based FS algorithms focus mainly on problems with a few hundreds of features [4,26].

3   Proposed Approach

In this section, a new PSO approach (named PSO-LSRG) is proposed for wrapper feature selection, where a new local search method is applied to *pbest* to exploit better solutions and a reset mechanism is applied to *gbest* to avoid stagnation in local optima. These two techniques are combined to see whether they can help PSO balance between global search and local search to improve the performance.

3.1   Overall Algorithm

PSO-LSRG mainly follows the basic steps of standard PSO. A particle represents a feature subset, where the dimensionality is the total number of features in the dataset. Each particle is encoded by a string of floating numbers in [0,1]. A threshold θ is used to determine whether a feature is selected or not. If the position value is larger than the threshold, the corresponding feature is selected. Otherwise, the corresponding feature is not selected. The fitness function of PSO-LSRG is to maximise the classification accuracy of the selected features, which is shown by Formula (3).

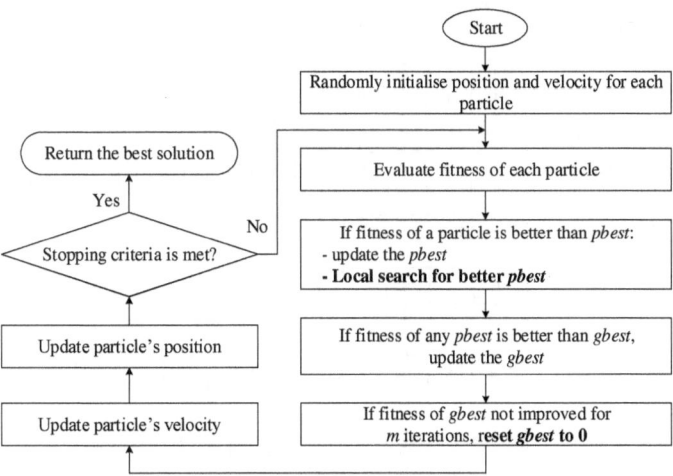

Fig. 1. Flowchart of the proposed algorithm

$$fitness = \frac{number\_of\_instances\_correctly\_classified}{Total\_number\_of\_instances} \qquad (3)$$

Fig. 1 illustrates the overall steps of PSO-LSRG. The two techniques are highlighted in the figure and will be described in the remaining of this section.

3.2 Reset $gbest$

$gbest$ plays an important role in leading the search direction of all particles in the swarm. Resetting $gbest$ if it does not change for a number of iterations can help avoid premature convergence [4], which is a limitation of PSO. Following [4], the current $gbest$ is reset to a vector of all "0"s if it does not improve in a few iterations.

3.3 Local Search on $pbest$

The local search proposed in this algorithm is basically a loop in which a predefined number of dimensions in the $pbest$ position will be flipped, i.e. a feature from being "selected" to "not selected" or from being "not selected" to "selected". To focus the search on the area surrounding current $pbest$, flipping is applied to a low percentage of the total dimensions and 2% is chosen in this work. After flipping, the new solution is evaluated. If it has better fitness, $pbest$ is updated. The search will stop after a predefined number of steps.

Fig. (2a) shows an example of the flipping procedure, where $pbest$ is converted to a binary string using the threshold θ, where "1" in the $pbest$ array means the corresponding feature is selected while "0" means the feature is not selected. Based on the randomly chosen flip dimensions positions, for example $(1, 2, 5, 6, 9)$, the current $pbest$ can be flipped to obtain a new $pbest$ position (flipped $pbest$).

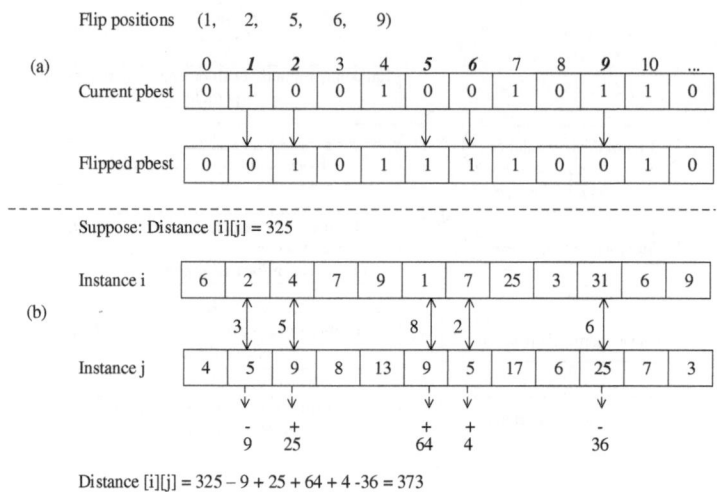

Fig. 2. An example of one local search step and re-calculating instances' distance

Local search usually brings more computation to the algorithm because of more fitness evaluations. Moreover, each evaluation in a wrapper FS method is usually expensive because it involves a training process of a classification algorithm. To avoid this problem, we proposed a new strategy to calculate the distances in KNN to find the nearest neighbours. This strategy utilises the characteristics of the local search and KNN. In each local search step, just a small percentage of the total dimensions (2%) will be changed and 98% of them in the flipped *pbest* remain the same as in the original *pbest*. In standard KNN, to calculate the distance between two instances, their differences (e.g. the squared difference) in all dimensions need to be calculated, where 98% of the calculation is repeated because 98% of the dimensions in the flipped *pbest* is the same as in the original *pbest*. Therefore, in the new evaluation strategy, only 2% of the dimensions are calculated. To achieve this, at the beginning of each local search run, all the cross distances between instances will be calculated regarding to the features selected in this given *pbest* and stored in a square matrix (distance$[i][j]$). Since this matrix is symmetric, with m instances in the dataset, the algorithm actually calculates $\frac{m(m+1)}{2}$ times. By using the distances stored in the matrix, it can speed up the computation of finding the nearest neighbours of a certain instance by calculating only 2% of the dimensions in each evaluation.

Fig. (2b) shows an example of distance re-calculation between two illustrated instances: $instance_i$ and $instance_j$, where the distance is determined by summing the squared difference of $instance_i$ and $instance_j$ in all dimensions. Suppose the distance between $instance_i$ and $instance_j$ regarding to the original *pbest* ($distance[i][j]$) is 325, the new distance regarding to the flipped *pbest* can be re-calculated as follows.

Table 1. Datasets

Data set	Number of Features	Number of Instances	Number of Classes
DLBCL	5469	77	2
9 Tumors	5726	60	9
Prostate Tumor	10509	102	2
11 Tumors	12533	174	11
Lung Cancer	12600	203	5

With 5 chosen flipped dimensions, it only need to subtract from 325 the square differences of the features' values that were selected (2 features: 1 and 9) and add those that were not selected (3 features: 2, 5 and 6). The new distance therefore is 373. Having all distances calculated, the algorithm can quickly find the new nearest neighbours for a given instance. As a result, the proposed strategy can save a significant amount of time.

4   Experimental Design

To examine the proposed approach, four different PSO algorithms are used for feature selection, which are the standard PSO (PSO), PSO with reset $gbest$ only (PSO-RG) [4], PSO with local search on $pbest$ only (PSO-LS) and PSO with both the reset $gbest$ strategy and the local search on $pbest$ (PSO-LSRG). Five datasets (Table 1) with a large number of features are chosen to test the performance of the algorithms, which are gene expressions profiles download from http://www.gems-system.org.

Since the datasets include a small number of instances, KNN (K=1) with leave one out cross validation (LOOCV) is used to calculate the classification accuracy, which is the same as in [4]. The acceleration coefficients are set as are $c_1 = c_2 = 2.0$ and inertia weight linearly decreases from 0.9 to 0.4 [23]. The swarm consists of 30 particles. The maximum number of iterations is 70 and fully connected communication topology is used here. Maximum velocity is 6.0. The threshold $\theta = 0.6$ is used to determine the selection of features. Whenever the local search is applied on a given $pbest$, it will try 100 times to find better $pbest$. In each time, 2% of the dimensions will be flipped to create a new candidate solution. Meanwhile, if $gbest$ is not improved for three iterations, it is reset to all 0, which is the same as in [4] for comparison purposes. The experiment is conducted for 30 independent runs with different random seeds. A statistical significance test, pairwise Student's T-test, is performed between the classification performance of different algorithms, where the significance level is set as 0.05.

5   Results and Discussions

Table 2 show the experimental results of the four PSO algorithms: PSO, PSO-RG, PSO-LS, PSO-LSRG. In this table, "Ave-Size" means the average number of features selected by each method over the 30 runs. "Best", "Mean" and "StdDev" respectively are the best, the average and standard deviation of the classification accuracies returned by 1NN with LOOCV in the 30 independent runs. The "All" row shows the original number of features and its classification accuracy when using all features. The highest average accuracies and the smallest size of all methods in each dataset are the bold ones.

Table 2. Experimental Results

The more "-", the better PSO-RG, PSO-LS or PSOLSRG.

Dataset	Method	Ave-Size	Best	Mean±StdDev	T_{RG}	T_{LS}	T_{LSRG}
DLBCL	All	5469	87.01		–	–	–
	PSO	2625.70	98.70	97.70±1.01	–	–	–
	PSO-RG	1766.53	98.70	98.44±0.53		=	=
	PSO-LS	2094.70	98.70	98.40±0.56			–
	PSO-LSRG	**1690.13**	98.70	**98.66**±0.24			
9 Tumors	All	5726	53.33		–	–	–
	PSO	2808.20	78.33	72.72±2.78	–	–	–
	PSO-RG	2720.63	78.33	74.67±2.68		–	–
	PSO-LS	2139.10	86.67	80.72±2.13			=
	PSO-LSRG	**2114.57**	86.67	**81.39**±1.76			
Prostate Tumor	All	10509	76.47		–	–	–
	PSO	5143.20	91.18	88.53±1.77	–	–	–
	PSO-RG	2353.17	98.04	92.42±2.74		=	–
	PSO-LS	3825.17	95.10	92.09±1.06			–
	PSO-LSRG	**2148.47**	98.04	**94.94**±1.18			
11 Tumors	All	12533	84.48		–	–	–
	PSO	6138.33	92.53	90.92±0.90	–	–	–
	PSO-RG	5623.87	95.40	91.51±1.08		–	–
	PSO-LS	4671.07	95.40	**93.87**±1.03			=
	PSO-LSRG	**4293.63**	95.40	93.79±0.70			
Lung Cancer	All	12600	90.15		–	–	–
	PSO	6144.03	96.55	95.78±0.50	–	–	–
	PSO-RG	4792.83	97.54	96.40±0.61		–	–
	PSO-LS	4641.17	97.54	97.03±0.50			=
	PSO-LSRG	**3426.43**	98.03	**97.19**±0.43			

The "T_{RG}" column shows the results of T-tests between PSO-RG and other methods, where "+" ("−") means the corresponding method achieves significantly better (worse) classification performance than PSO-RG. "=" means they are similar. Similarly, "T_{LS}" or "T_{LSRG}" are the results of T-tests comparing the classification performance achieved by other methods and PSO-LS or PSO-LSRG, respectively.

5.1   The Standard PSO

As shown in Table 2, the standard PSO obtained feature subsets with higher classification performance and smaller size than all features on all the five datasets. Using feature subsets evolved by PSO, the 1NN classifier increases its average classification accuracy about 20% on the 9 Tumor dataset, 10% on the DLBCL and the Prostate Tumors, and 5% on the other two datasets. The number of features selected by PSO is about 50% of the original number of features on all the five datasets. The results show that PSO is a suitable tool for FS problems. For the rest of this work, we will consider PSO as a baseline to compare with other methods.

5.2   Effect of Reset $gbest$ Technique (PSO-RG)

From all the "−"symbols in T_{RG} column of Table 2, we can state that the classification accuracies of feature subsets selected by PSO-RG is significantly better than using all features and those of PSO on all the five datasets. Furthermore, the subset size of the PSO-RG solutions is also smaller than that of PSO. It is about half on two datasets, which are the 9 Tumor and the 11 Tumor datasets, and one-third on the DLBCL and

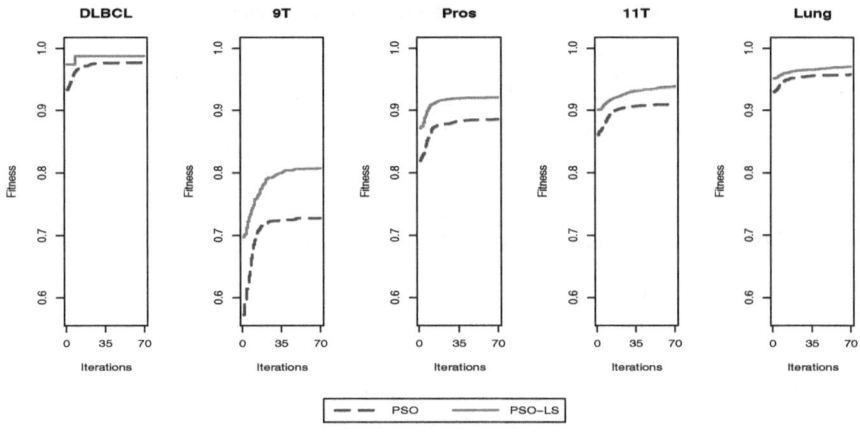

Fig. 3. Average *gbest* fitness of PSO and PSO-LS on five datasets

Lung Cancer datasets. On the Prostate Tumor dataset, this number even further reduces, which is less than a quarter of the original feature set size. This is possibly because setting *gbest* to 0 attracts all other particles moving toward this direction, resulting to smaller subsets. The results indicate that the reset *gbest* technique is useful in directing particles to other promising areas when they seems to get stuck in a local optimum.

5.3   Effect of Local Search on *pbest* (PSO-LS)

According to the results of "PSO-LS" in Table 2, the feature subsets evolved by PSO with local search on *pbest* can achieve significantly higher classification performance than using all features and standard PSO. PSO-LS obtained significantly better classification performance than PSO-RG on three datasets and they are similar on the other two datasets, which are the DLBCL and the Prostate Tumor. The overall results show that the proposed local search technique on *pbest* gives particles more chances to reach better positions in their local areas.

To have a better view of the local search effect on *pbest*, it is worth to observe how *gbest* changes during the searching process. Fig. 3 contains five graphs for the five datasets, where each graph shows the average fitness value of *gbest* over the 30 runs in each iteration. Each graph has a dashed line and a solid line representing PSO and PSO-LS, respectively. The figure shows that just after the first iteration, the average value of *gbest* of PSO-LS shows a significant improvement comparing to the standard PSO algorithm on all the five datasets. This indicates that the local search helps the population reach better solution regions and obtain feature subsets with higher classification performance.

Table 3. Average computation time (in minutes)

	PSO	PSO-RG	PSO-LS	PSO-LSRG
DLBCL	34.81	23.37	28.23	**18.99**
9 Tumors	23.64	19.79	18.85	**17.40**
Prostate Tumor	160.13	99.85	134.50	**83.77**
11 Tumors	452.14	361.68	352.26	**310.06**
Lung Cancer	558.17	435.01	446.47	**363.59**

5.4 Combination of Local Search and Reset $gbest$ in PSO (PSO-LSRG)

As can be seen from Table 2, PSO-LSRG selected the smallest number of features on all the five datasets, which is only around 20% of the original features on the Prostate Tumor dataset. Meanwhile, PSO-LSRG achieved the highest classification accuracies on four of the five datasets. The results of the significance T-tests in T_{LSRG} show that PSO-LSRG outperformed PSO-RG and PSO-LS on four and three of the five datasets, respectively. For the remaining datasets, they achieved similar results.

From Table 2, we can see an interesting pattern shown in the bold pair of T-test values in T_{LS} and T_{LSRG} on each dataset. The pattern of "– =" shows that for those datasets where PSO-LS outperformed PSO-RG, which are the 9 Tumors, 11 Tumors and Lung Cancer datasets, PSO-LSRG achieved similar classification performance to PSO-LS. On the other hand, the "= –" shows that for those datasets where PSO-LS only evolved similar results to PSO-RG, which are the DLBCL and Prostate Tumors, PSO-LSRG made an improvement. From this observation, we can conclude that the combination of reset $gbest$ and local search on $pbest$ can overcome the limitations of the two techniques and help PSO balance its exploration and exploitation abilities. Therefore, the feature subsets selected by PSO-LSRG that combines two techniques generally have a smaller size and equal or better classification performance than the case where only one technique is applied.

5.5 Computational Time

Table 3 shows the average CPU time used by each method in the 30 independent runs on the five datasets, where the numbers are expressed in minutes.

From Table 3, it can be seen that PSO-RG consumes less time than the standard PSO. Since all these PSO methods have the same settings in term of the number of particles and iterations, they have the same number of evaluations in one run. Therefore, the factor makes their computation time different is that the evolved feature subsets in the former are smaller than the latter. This again confirms the big influence of the feature subset size on the computation time in wrapper FS approaches. As a consequence, in PSO-LS and PSO-LSRG, although the local search part adds more running time to the standard PSO in every update of $pbest$, it does not make the total computation time of one run longer. By contrast, by reaching solutions with smaller subsets, it can significantly reduce the total computation time. Another important factor is the computation time saved by using the cross distance matrix of the instances to evaluate a new $pbest$, which was explained in Section 3.3. This new strategy successfully reduces the running time of KNN classifier. The quick evaluation time and selecting smaller subsets make the computation time of PSO-LSRG be the shortest in all the five different datasets.

5.6 Further Discussions

Note that in the experimental design, this paper uses the re-substitution estimator to evaluate the performance of the feature subsets, which is the same as in [4] and many other existing papers. The re-substitution estimator, in other words, means the whole dataset is used during the evolutionary feature selection process. There is no separated unseen data to test the generality of the selected features. There is a feature selection bias here, so we cannot claim that the selected features can be used for future unseen data for classification.

According to Ambroise et al [1], the feature selection bias effect can be reduced by using cross validation or bootstrap estimators in which a fraction of the original dataset are held out for testing the performance of the selected features. We will further investigate this in our future work.

6 Conclusions and Future Work

The goal of this paper was to develop a new PSO approach to feature selection on high-dimensional gene datasets with thousands of features. The goal has been successfully achieved by developing a new efficient local search on *pbest* and applying a reset *gbest* mechanism in PSO for feature selection, where KNN with LOOCV was used to evaluate the classification performance. The performance of the new PSO approach with both local search and reset *gbest* (PSO-LSRG) is examined and compared with standard PSO, PSO with the proposed local search only (PSO-LS), and PSO with the reset *gbest* mechanism only (PSO-RG). The experiments on five gene datasets of varying difficulty show that the feature subsets returned by PSO-LS are smaller and achieved better classification performance than those of PSO, and better or at least similar to those of PSO-RG. The change of *gbest* during searching processes also indicates that the proposed local search on *pbest* is an effective strategy for PSO to improve its search ability. The results of PSO-RG show that the reset *gbest* technique helped particles divert their search to other promising regions when they seem to get stuck in a near local optima. However, the *gbest* reset mechanism might also prevent particles to better exploit their findings. Meanwhile, applying local search on *pbest* gives particles more chances to obtain better solutions. Therefore, the combination of these two techniques in PSO-LSRG further increased the performance. The results confirm that PSO-LSRG could overcome their limitations to achieve even better results than both PSO-LS and PSO-RG in terms of the classification performance and the number of features. Meanwhile, the proposed strategy for the fitness evaluation in local search has successfully saved the computation time for KNN, enabling PSO-LSRG being more efficient than the other three methods.

The proposed algorithm significantly reduced the size of the feature set, but it can be seen that the numbers are still large. Further reducing the number of features is still an important and challenging task. In future work, we intend to develop a new EC approach to further reduce the number of features and improve the classification performance without increasing the computational cost.

References

1. Ambroise, C., McLachlan, G.J.: Selection bias in gene extraction on the basis of microarray gene-expression data. Proceedings of the National Academy of Sciences 99(10), 6562–6566 (2002)
2. Bala, J., Huang, J., Vafaie, H., Dejong, K., Wechsler, H.: Hybrid learning using genetic algorithms and decision trees for pattern classification. In: The 14th International Joint Conference on Artificial Intelligence, vol. 1
3. Chakraborty, B.: Genetic algorithm with fuzzy fitness function for feature selection. In: IEEE International Symposium on Industrial Electronics (ISIE 2002), vol. 1, pp. 315–319 (2002)
4. Chuang, L.Y., Chang, H.W., Tu, C.J., Yang, C.H.: Improved binary pso for feature selection using gene expression data. Computational Biology and Chemistry 32(1), 29–38 (2008)
5. Dash, M., Liu, H.: Feature selection for classification. Intelligent Data Analysis 1, 131–156 (1997)
6. Davis, R.A., Charlton, A.J., Oehlschlager, S., Wilson, J.C.: Novel feature selection method for genetic programming using metabolomic 1h NMR data. Chemometrics and Intelligent Laboratory Systems 81(1), 50–59 (2006)
7. Dorigo, M., Di Caro, G.: Ant colony optimization: a new meta-heuristic. In: IEEE Congress on Evolutionary Computation, vol. 2, pp. 1470–1477 (1999)
8. Engelbrecht, A.P.: Computational intelligence: an introduction, 2nd edn. Wiley (2007)
9. Guyon, I., Elisseeff, A.: An introduction to variable and feature selection. J. Mach. Learn. Res. 3, 1157–1182 (2003)
10. Huang, J., Cai, Y., Xu, X.: A hybrid genetic algorithm for feature selection wrapper based on mutual information. Pattern Recognition Letters 28(13), 1825–1844 (2007)
11. Jensen, R., Shen, Q.: Finding rough set reducts with ant colony optimization. In: Proceedings of the 2003 UK Workshop on Computational Intelligence, pp. 15–22 (2003)
12. Kanan, H.R., Faez, K.: An improved feature selection method based on ant colony optimization (ACO) evaluated on face recognition system. Applied Mathematics and Computation 205(2), 716–725 (2008), Special Issue on Advanced Intelligent Computing Theory and Methodology in Applied Mathematics and Computation
13. Kennedy, J., Eberhart, R.: Particle swarm optimization. In: IEEE International Conference on Neural Networks, vol. 4, pp. 1942–1948 (1995)
14. Kohavi, R., John, G.H.: Wrappers for feature subset selection. Artificial Intelligence 97(1-2), 273–324 (1997), relevance
15. Lane, M.C., Xue, B., Liu, I., Zhang, M.: Gaussian based particle swarm optimisation and statistical clustering for feature selection. In: Blum, C., Ochoa, G. (eds.) EvoCOP 2014. LNCS, vol. 8600, pp. 133–144. Springer, Heidelberg (2014)
16. Lanzi, P.L.: Fast feature selection with genetic algorithms: a filter approach. In: IEEE International Conference on Evolutionary Computation, pp. 537–540 (1997)
17. Marill, T., Green, D.M.: On the effectiveness of receptors in recognition systems. IEEE Transactions on Information Theory 9(1), 11–17 (1963)
18. Ming, H.: A rough set based hybrid method to feature selection. In: International Symposium on Knowledge Acquisition and Modeling, KAM 2008, pp. 585–588 (December 2008)
19. Neshatian, K., Zhang, M.: Pareto front feature selection: Using genetic programming to explore feature space. In: The 11th Annual Conference on Genetic and Evolutionary Computation, GECCO 2009, pp. 1027–1034 (2009)
20. Oliveira, L., Sabourin, R., Bortolozzi, F., Suen, C.: Feature selection using multi-objective genetic algorithms for handwritten digit recognition. In: 16th International Conference on Pattern Recognition (ICPR 2002), vol. 1, pp. 568–571 (2002)

21. Pudil, P., Novovičová, J., Kittler, J.: Floating search methods in feature selection. Pattern Recogn. Lett. 15(11), 1119–1125 (1994)
22. Shi, Y., Eberhart, R.: A modified particle swarm optimizer. In: IEEE International Conference on Evolutionary Computation (CEC 1998), pp. 69–73 (1998)
23. Shi, Y., Eberhart, R.: Empirical study of particle swarm optimization. In: IEEE Congress on Evolutionary Computation (CEC 1999), vol. 3, pp. 1945–1950 (1999)
24. Stearns, S.D.: On selecting features for pattern classifiers. In: Proceedings of the 3rd International Conference on Pattern Recognition (ICPR 1976), Coronado, CA, pp. 71–75 (1976)
25. Whitney, A.: A direct method of nonparametric measurement selection. IEEE Transactions on Computers C-20(9), 1100–1103 (1971)
26. Xue, B.: Particle Swarm Optimisation for Feature Selection in Classification. Ph.D. thesis, Victoria University of Wellington, Wellington, New Zealand (2014)
27. Xue, B., Cervante, L., Shang, L., Browne, W.N., Zhang, M.: A multi-objective particle swarm optimisation for filter based feature selection in classification problems. Connection Science 24(2-3), 91–116 (2012)
28. Xue, B., Cervante, L., Shang, L., Browne, W.N., Zhang, M.: Binary PSO and rough set theory for feature selection: A multi-objective filter based approach. International Journal of Computational Intelligence and Applications 13(02), 1450009 (2014)
29. Xue, B., Zhang, M., Browne, W.N.: Particle swarm optimization for feature selection in classification: A multi-objective approach. IEEE Transactions on Cybernetics 43(6), 1656–1671 (2013)
30. Xue, B., Zhang, M., Browne, W.N.: Particle swarm optimisation for feature selection in classification: Novel initialisation and updating mechanisms. Applied Soft Computing 18, 261–276 (2014)

Multi-objective Feature Selection in Classification: A Differential Evolution Approach

Bing Xue[1], Wenlong Fu[2], and Mengjie Zhang[1]

[1] School of Engineering and Computer Science
[2] School of Mathematics, Statistics and Operations Research
Victoria University of Wellington, PO Box 600, Wellington 6140, New Zealand
{Bing.Xue,Mengjie.Zhang}@ecs.vuw.ac.nz
Wenlong.Fu@msor.vuw.ac.nz

Abstract. Feature selection is an important pre-processing step in classification tasks. Feature selection aims to minimise both the classification error rate and the number of features, which are usually two conflicting objectives. This paper develops a differential evolution (DE) based multi-objective feature selection approach. The multi-objective approach is compared with two conventional methods and two DE based single objective methods, where the first algorithm is to minimise the classification error rate only while the second algorithm combines the number of features and the classification error rate into a single fitness function. Their performances are examined on nine different datasets and the results show that the proposed multi-objective algorithm successfully evolved a number of trade-off solutions, which reduce the number of features and keep or reduce the classification error rate. In almost all cases, the proposed multi-objective algorithm achieved better performance than all the other four methods in terms of both the classification accuracy and the number of features.

Keywords: Differential evolution, Feature selection, Multi-objective optimisation, Classification.

1 Introduction

In machine learning and data mining tasks, such as classification, feature selection (FS), also called dimensionality reduction, is a process of selecting a small subset of features from a large set of original features. FS can effectively increase the classification performance, speed up the training process, reduce the dimensionality of the data, and simplify the built classifiers/models [11].

FS has been of interest for many decades [6]. One of the main challenges in FS is the large search space. For a dataset including n features, the size of the search space is 2^n. Therefore, exhaustive search is impractical in most situations because of the long computational time. Although many different search techniques have been applied to FS tasks [6], most of them still have the limitations of high computational cost and being stuck in local optima [6]. Evolutionary computation (EC) includes a group of global search techniques in which differential evolution (DE) [20] is a simple yet powerful algorithm. DE has been successfully used to solve problems in a variety of fields [5], including FS [1].

G. Dick et al. (Eds.): SEAL 2014, LNCS 8886, pp. 516–528, 2014.

FS aims to minimise the number of features and maximise the classification accuracy (minimise the classification error rate). These two objectives are conflicting to each other in most cases, which makes FS a multi-objective problem. However, there are only a limited number of multi-objective FS algorithms and most of them are based on EC techniques [3, 12, 27]. Those EC techniques are population based algorithms, which are particularly suitable for multi-objective optimisation because they can produce multiple solutions in a single run [4]. Multi-objective DE gains more and more attention to solve complex multi-objective problems. Recently, Wang et al. [21] showed that DE can achieve better performance than many other EC algorithms on single objective FS, but there is only one initial work [24] on DE for multi-objective FS. This paper will further investigate this topic by significantly extend the work in [24].

1.1 Goals

The overall goal is to develop a DE based multi-objective FS approach to searching for a set of non-dominated feature subsets, which include a small subset of features and achieve similar or better classification performance than using all features. To achieve this goal, we propose a multi-objective FS approach (DEMOFS) by using a multi-objective DE algorithm to simultaneously minimise the number of features and the classification error rate. DEMOFS is compared with two conventional methods and two DE based single objective algorithms, where the first DE algorithm aims to minimise the classification error rate only and the second DE algorithm combines the classification error rate and the number of features into a single fitness function. Specifically, we will investigate:

– whether the two single objective DE algorithms can successfully reduce the number of features and maintain or even improve the classification performance over using all features,
– whether DEMOFS can achieve a set of non-dominated feature subsets, which can further reduce the number of features and improve the classification performance, and
– whether DEMOFS can outperform the two conventional FS algorithms in terms of the number of features and the classification performance.

2 Background

2.1 Differential Evolution (DE)

Differential evolution (DE) was first developed by Storn and Price [20] in 1997. Due to its simplicity, robustness and effectiveness, DE has attracted more and more attention of researchers from different fields. In DE, a candidate solution is encoded as an individual in the population. Considering there are P individuals in the population, the individual i $(1 \leq i \leq P)$ can be shown by a vector $(x_{i1}, x_{i2}, ...x_{iD})$, where D is the dimensionality of the problem or search space. There is a vector $(x_{max1}, x_{max2}, ...x_{maxD})$ to define the upper bound of the search space and a vector $(x_{min1}, x_{min2}, ...x_{minD})$ to limit the lower bound of the search space. A DE algorithm starts with randomly generated initial individuals. Then DE employs the mutation operation to produce a mutant candidate solution C_i for each individual x_i, which is also called the parent x_i.

There are different types of mutation strategies [19]. Equation (1) takes *DE/rand/1/bin* as an example to how C_i is generated.

$$C_{id} = \begin{cases} x_{id}^{i,r1} + F * (x_d^{i,r2} - x_{id}^{i,r3}), & \text{if } rand() < CR \\ x_{id}, & otherwise \end{cases} \tag{1}$$

where $x^{i,r1}$, $x^{i,r2}$, and $x^{i,r3}$ are randomly selected from the population. x_i, $x^{i,r1}$, $x^{i,r2}$, and $x^{i,r3}$ are different from each other. $F \in (0,1)$ is a scale factor, which controls the rate at which the population evolves. $rand()$ is a random number uniformly distributed in (0,1). CR is the crossover probability. If C_{id} falls out of the lower and upper bounds, a constraint method is usually applied to handle it. A simply way is to replace the value in C_{id} that exceeds the boundary value with the closest boundary value.

2.2   Related Work on Feature Selection

EC algorithms have been applied to FS problems, such as DE [14], genetic algorithms (GAs) [28], genetic programming (GP) [17], and particle swarm optimisation (PSO) [2, 22, 23, 25–27]. Typical EC based FS algorithms are reviewed in this section. Zhu et al. [28] proposed a FS method incorporating GA with local search (i.e. forms a memetic algorithm). Meanwhile, this algorithm combines filter ranking measure into a wrapper framework to take advantage of both filter and wrapper approaches. Fdhila et al. [8] applied multi-swarm PSO to solve FS problems. However, the computational cost of the proposed algorithm is high because it involves parallel evolutionary processes and multiple sub-swarms with a relative large number of particles. He et al. [14] applied a binary differential evolution (BDE) algorithm to filter FS, where mutual information is used to evaluate the goodness of the selected feature subsets. However, the proposed algorithm is not compared with any other algorithm and the datasets used in the experiments include a relatively small number (maximum 56) of features. Al-Ani et al. [1] also proposed a DE based FS method, where features are distributed to a set of wheels and DE is employed to select features from each wheel. This algorithm can significantly reduce the number of features and improve the classification performance.

EC algorithms have been applied to multi-objective FS. Hamdani et al. [12] developed a multi-objective FS algorithm using non-dominated sorting based multi-objective GA II (NSGAII). Neshatian and Zhang [17] proposed a GP based filter model as a multi-objective algorithm for FS in binary classification problems. Ke et al. [3] developed a Pareto-based multi-objective ant colony optimisation (ACO) for FS based on rough set theory. Xue et al. [27] proposed a PSO based multi-objective approach for wrapper FS, which shows that the PSO based algorithm outperforms three other commonly used EC based multi-objective algorithms. Wang et al. [21] showed that DE can achieve better performance than GA, PSO, ACO, and harmony search on single objective FS, but the use of DE for multi-objective FS has not been investigated to date.

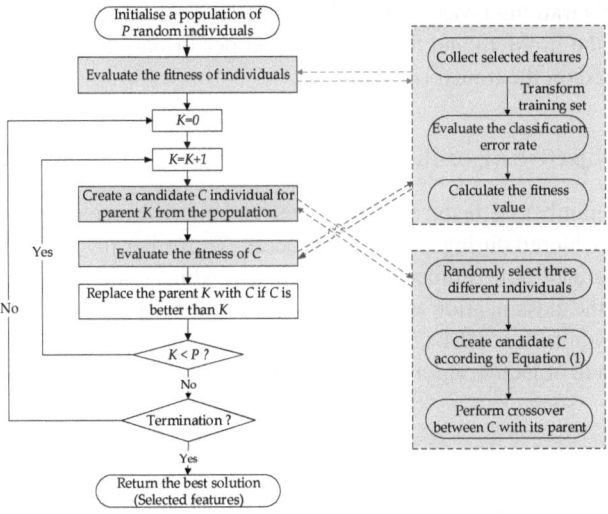

Fig. 1. Training Process of DEFS and DEFS2

3   Proposed Approach

3.1   Single Objective Algorithm 1: DEFS

To investigate the performance of DE for FS, DE is firstly used to optimise the classification performance of the selected features to form the algorithm DEFS. DEFS uses Equation (2) as the fitness function, which is to minimise the classification error rate.

$$Fit_1 = ErrorRate = \frac{FP + FN}{TP + TN + FP + FN} \tag{2}$$

where FP, FN, TP and TN mean false positives, false negatives, true positives, and true negatives, respectively.

Since FS needs to consider both the classification performance and the number of features, the number of features is not considered in the fitness function but during the evolutionary process of DEFS. If there are multiple solutions with the smallest error rate, the one with the smallest size will be reserved and others are discarded. The representation of DE follows the continuous encoding scheme since the original DE algorithm was developed for continuous problems. In DEFS, each individual in DEFS is a vector of real numbers, $x_i = (x_{i1}, x_{i2}, .., x_{id}, .., x_{iD})$, where D is the dimensionality and also the total number of features in the dataset. $0 \leq x_{id} \leq 1$ shows the probability of the dth feature being selected. A threshold θ is used to determine whether this feature is selected. If $\theta \leq x_{id}$, the dth feature is selected. Otherwise, the dth feature is not selected.

Fig. 1 shows the training or search process of DEFS, where the DE scheme *DE/rand/ 1/bin* [19] is used following [18]. The blue and green dash rectangles show the detailed steps of evaluating the fitness value and creating a new candidate individual, respectively. After this training process, the selected feature subset will be used to transform the test set of the problem and the classification performance of the selected features will be evaluated on the transformed test set.

3.2   Single Objective Algorithm 2: DEFS2

In DEFS, the number of features is not directly included in the fitness function, although it is considered during the search process. To further investigate the use of DE for FS, an integrated fitness function, Equation (3), which combines the two objectives of minimising the classification error rate and the number of features, is used to develop another algorithm named DEFS2. Although DEFS2 considers the two objectives, it is treated as a single objective algorithm since it follows a single objective search process.

$$Fit_2 = \alpha * ErrorRate + (1 - \alpha) * \frac{\#Size}{D} \tag{3}$$

where $\alpha \in (0, 1]$ is a weight parameter showing the relative importance of the classification error rate. $\#Size$ represents the number of selected features and D is the total number of features in the dataset. $\frac{\#Size}{D}$ is used to scale the value to $(0,1]$ to be in the same range of $ErrorRate$. α should be larger than 0.5 to make sure $\alpha > (1 - \alpha)$, i.e. the classification performance is more important than the number of features.

Equation (3) is used to investigate whether directly considering the number of selected features can further reduce the number of features without significantly reducing the classification accuracy. The search or training process of DEFS2 can also be shown by Fig. 1. The main difference between DEFS and DEFS2 is the fitness function.

3.3   Multi-objective Algorithm: DEMOFS

Similar to most of other EC approaches, DE was proposed to solve single objective problems. Based on a popular evolutionary multi-objective algorithm, i.e. non-dominated sorting based genetic algorithm II (NSGAII), Robič and Bogdand [18] developed a multi-objective DE algorithm named DEMO [18] to use DE for multi-objective optimisation. DEMO has shown promising performance on some problems, but it has never been applied to FS problems. In this paper, we develop a multi-objective FS algorithm named DEMOFS based on DEMO to investigate the use of DE for multi-objective FS. DEMOFS aims to minimise the classification error rate and the number of features. The representation of DEMOFS is the same as DEFS and DEFS2.

Algorithm 1 shows the pseudo-code of DEMOFS, where the two key steps are the decision of the newly constructed individual and the update of the population. In Algorithm 1, Line 8 to Line 16 show the decision on the newly constructed individual. After this procedure, the population exceeds the pre-defined maximum number of individuals. To update the population, a truncation step is needed, which is shown in Line 19. This is similar to that in NSGAII [7], which involves the use of the non-dominated sorting and crowding distance metric. Specifically, the non-dominated solutions in the population are called the first non-dominated front, which are excluded from the population. Then the non-dominated solutions in the new population are called the second non-dominated

Algorithm 1. Pseudo-Code of DEMOFS

```
 1  begin
 2  |    randomly initialise individuals;
 3  |    while Stopping Criterion is not met do
 4  |    |    evaluate the number of features selected by each individual and its Training
    |    |    error rate;
 5  |    |    for i=1 to Number of individuals do
 6  |    |    |    create candidate C from parent i ;      /* details as shown in
    |    |    |    Fig. 1 */
 7  |    |    |    evaluate the two objective values of C;
 8  |    |    |    if C dominates i then
 9  |    |    |    |    use C to replace i;
10  |    |    |    end
11  |    |    |    else if i dominates C then
12  |    |    |    |    C is discarded;
13  |    |    |    end
14  |    |    |    else if i and C are non-dominated to each other then
15  |    |    |    |    C is added to the population;
16  |    |    |    end
17  |    |    end
18  |    |    if the population size exceeds the maximum value then
19  |    |    |    truncate the population according to non-dominated sorting
20  |    |    end
21  |    |    randomly enumerate the individuals in the population;
22  |    end
23  end
24  calculate the testing classification error rate of the non-dominated solutions;
25  return the non-dominated solutions and their training and testing error rates.
```

front. The following levels of non-dominated fronts are identified by repeating this procedure. For the next generation, solutions (individuals) are selected from the top levels of the non-dominated fronts to form a new/updated population, starting from the first front. If the number of solutions needed is larger than the number of solutions in the current non-dominated front, all the solutions are added into the population for the next generation. Otherwise, the solutions in the current non-dominated front are ranked according to the crowding distance and the highest ranked (least crowded) solutions are added into the next generation.

4   Design of Experiments

4.1   Benchmark Techniques

The three DE based FS algorithms (i.e. DEFS, DEFS2 and DEMOFS) are examined and compared with each other on nine datasets shown in Table 1, which were selected from the UCI machine learning repository [9]. For each dataset, the instances are randomly

Table 1. Datasets

Dataset	NO. of Features	NO. of Classes	NO. of Instances
Wine	13	3	178
Australian	14	2	690
Zoo	17	7	101
Vehicle	18	4	846
German	24	2	1000
Lung Cancer	56	3	32
Sonar	60	2	208
Hillvalley	100	2	606
Musk Version 1 (Musk1)	166	2	476

divided into two sets: 70% as the training set and 30% as the test set. The nine datasets are chosen to have different numbers of features, classes and instances to be used as representatives of problems that the proposed algorithms can address.

All the algorithms are wrapper approaches, i.e. requiring a classification algorithm to evaluate the classification error rate of the selected features. A commonly used classification algorithm, K-nearest neighbour (KNN), is used here and K=5. During the training process, KNN with 10-fold cross-validation is employed to evaluate the classification error rate of the selected feature subset on the training set, and then the selected features are evaluated on the test set to obtain the testing classification error rate [15].

Two traditional wrapper FS methods are used to compare with that of the DE based algorithms, which are linear forward selection (LFS) [10] and greedy stepwise backward selection (GSBS), which were derived from two typical greedy search based FS, i.e. SFS and SBS, respectively. Details about LFS and GSBS can be seen from [10] and [16]. Weka [13] is used to run the experiments of LFS and GSBS. All the settings in LFS and GSBS are kept to the defaults. The parameters of the three DE based algorithms are set as follows. The population size is 80 and the maximum number of generations is 100. The crossover rate is set as 0.3. All the three DE based algorithms share the same representation. The threshold θ is set as 0.6 [27]. The parameter α in DEFS2 is set as 0.95, which means that the classification performance is much more important than the number of features. LFS and GSBS are deterministic methods, which produce a unique solution. The DE based algorithms are stochastic methods and each of them has been performed for 30 independent runs on each dataset.

5   Results and Discussions

In this section, we first compare the performance of DE based stochastic algorithms, DEFS, DEFS2 and DEMOFS, then compare their performance with that of the two traditional (deterministic) methods, LFS and GSBS.

DEFS and DEFS2 are single objective methods producing 30 solutions for each dataset from the 30 independent runs. DEMOFS is a multi-objective algorithm producing 30 sets of solutions for each dataset from the 30 independent runs. To compare their results, the 30 sets of solutions are combined together to extract two sets of solutions, which are the "best" set and the "average" set. The "best" set means the non-dominated solutions achieved by DEMOFS across the 30 independent runs. The average set contains the solutions with different numbers of features. For a certain number (e.g. m), its

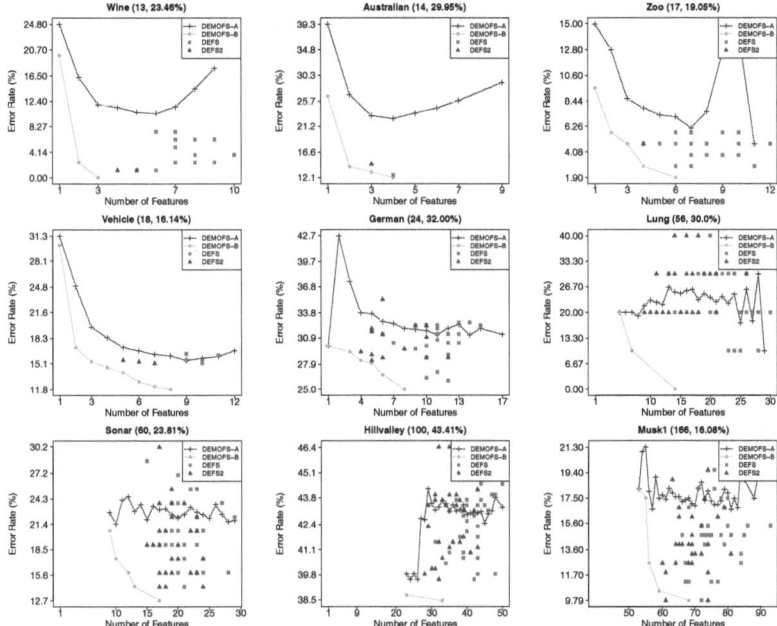

Fig. 2. Experimental Results of DEFS, DEFS2 and DEMOFS

classification error rate is the average error rate of all the available feature subsets that include m features. Fig. 2 shows the results of DEFS, DEFS2 and DEMOFS, where "DEMOFS-A" shows the "average" set and "DEMOFS-B" shows the "best" set. The nine charts correspond to the nine datasets used in the experiments. In each chart, the numbers in the bracket are the total number of features and the classification error rate achieved by using all features. For DEFS and DEFS2, there might be fewer than 30 distinct dots shown in a chart. The reason is that many different solutions may have the same number of features and the same classification error rate and they are plotted in the same dot.

5.1   Results of DEFS and DEFS2

From Fig. 2, it can be seen that on all the nine datasets, DEFS reduced around half of the features and reduced or achieved similar classification error rate to using all the available features. The results show that DEFS uses DE as the search technique can effectively search the solution space to reduce the number of features and maintain or even improve the classification performance.

Fig. 2 shows that DEFS2 selected around one third of the available features and achieved a similar or even lower classification error rate than using all features. For example, on the Australian dataset, DEFS2 selected only three features from the 14 available features and reduced the classification error rate from 29.95% to 14.5%. Comparing DEFS2 with DEFS, DEFS2 which directly considers the number of features in

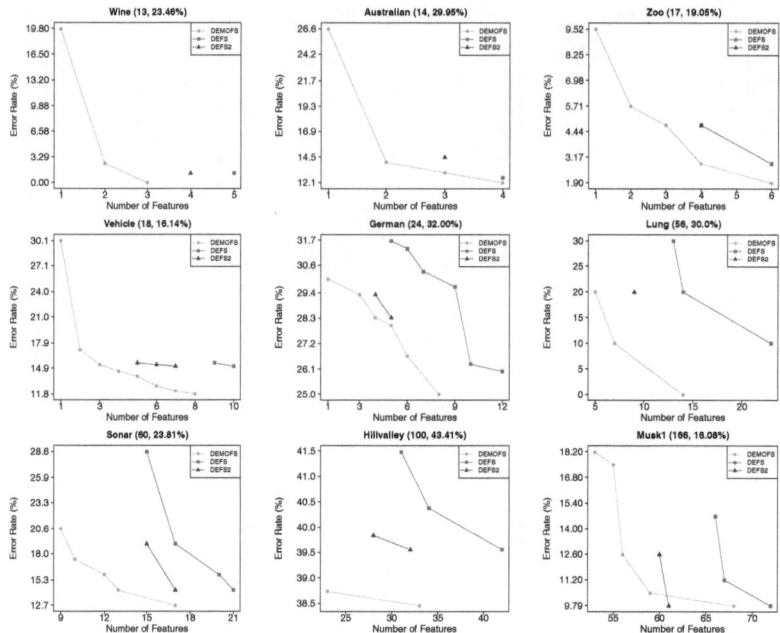

Fig. 3. Further Comparisons Between DEFS, DEFS2 and DEMOFS: Non-Dominated Solutions

the fitness function can further reduce the number of features without increasing the classification error or evening reducing it over DEFS. The main reason is that the number of features is considered but with a very small weight in the fitness function, which on one hand results in the reduction of the number of features. On the other hand, slightly compromising the classification performance with the number of features can also help avoid the over-fitting problem, which may achieve better performance on the test set.

5.2 Results of DEMOFS

From Fig. 2, it can be observed that for all cases, at least one feature subset in the "average" set included a smaller number of features and achieved similar or lower classification error rate than using all features. Note that DEMOFS reports a set of non-dominated solutions in each run, but when combining the solutions from multiple runs, some solutions may be dominated by others. Therefore, some of the solutions in the "average" set dominate others. In all datasets, the "best" set included a significantly smaller number of features and increased the classification accuracy over all features.

Fig. 2 suggests that by employing a multi-objective search mechanism, DEMOFS can effectively explore the search space to obtain a number of non-dominated feature subsets, which significantly reduced the number of features and improve the classification performance over using all features.

Table 2. Results of LFS ang GSBS

	Wine		Australian		Zoo		Vehicle		German	
	# Features	Error (%)	# Features	Error (%)	# Features	Error (%)	# Features	Error (%)	# Features	Error (%)
LFS	7	25.93	4	29.95	8	20.95	9	16.93	3	31.33
GSBS	8	14.81	12	30.43	7	20.0	16	24.21	18	35.67
	Lung		Sonar		Hillvalley		Musk1			
	# Features	Error (%)	# Features	Error (%)	# Features	Error (%)	# Features	Error (%)		
LFS	6	10.0	3	22.22	8	42.31	10	14.69		
GSBS	33	10.0	48	31.75	90	50.55	122	23.78		

5.3   Comparisons between DEFS, DEFS2 and DEMOFS

Fig. 2 also shows that the classification performance of the "average" set is often slightly worse or similar to that of DEFS and DEFS2, but the solutions in the "best" set is always better than DEFS and DEFS2. This is not surprised because DEFS, DEFS2 and DEMOFS share the same (total) number of evaluations, but DEFS and DEFS2 focus on the optimisation of the classification accuracy and return only one single solution from each run. DEMOFS returns a set of feature subsets with trade-off between the accuracy and the number of features. Therefore, when the error rates of feature subsets from different runs are averaged, it may be slightly worse the DEFS and DEFS2.

The "average" set gives an overall idea of the solutions achieved by DEMOFS, especially for the number of features, but it has a potential limitation because the solutions in the "average" set are not meaningful solutions. The reason is that in FS problems, the solutions themselves cannot be averaged because the each solution involves a number of features. Such individual features cannot be averaged to get an "average" solution, although the numbers of features and their classification error rates can be averaged to show the general performance. This is not a problem for the "best" set involving the original non-dominated solutions only. Therefore, in order to further compare the performance of DEFS, DEFS2 and DEMOFS, the non-dominated solutions obtained by DEFS and DEFS2 over the 30 independent runs are also collected and compared with that of DEMOFS, where the results are shown in Fig. 3.

From Fig. 3, it can be observed that DEFS2 generally achieved similar or better performance than DEFS. This is generally consistent with the results in Fig. 2, but shows a clearer pattern. On all the nine datasets, the solutions of DEMOFS dominate that of DEFS. On eight of the nine datasets, the solutions of DEMOFS dominate that of DEFS2. The only exception is the Musk1 dataset, where one of the solutions from DEFS2 achieved a similar classification performance to that of DEMOFS, but selected a smaller number of features. The results from Fig. 3 further show that DEMOFS has the potential to obtain better feature subsets than DEFS and DEFS2, which included a smaller number of features and a lower classification error rate.

5.4   Comparisons with Traditional Methods

Table 2 shows the results of the two traditional FS algorithms. Both LFS and GSBS are deterministic algorithms that produce a unique solution.

Comparing the results in Table 2 to that in Fig. 2 and 3, it can be seen that DEFS, DEFS2 and DEMOFS were able to outperform both LFS and GSBS in terms of both the classification performance and the number of features on eight of the nine datasets. Only on the Lung dataset, LFS outperformed DEFS and DEFS2, but DEMOFS achieved better performance than LFS. The results show that DEFS, DEFS2 and DEMOFS employ DE as the search technique can better explore the solution space to obtain better results than LFS and GSBS.

6   Conclusions and Future Work

This paper investigated the use of DE for multi-objective FS in classification. The algorithm DEMOFS was proposed to simultaneously minimise the number of features and the classification error rate. The experiments on the nine different datasets show that DEMOFS successfully evolved a set of trade-off solutions to reduce both the number of features and the classification error rate. The results show that DEMOFS outperformed two commonly used conventional FS methods (LFS and GSBS) in terms of both the classification performance and the number of features. DEMOFS was also compared with two DE based single objective FS algorithms (DEFS and DEFS2), where DEFS aimed to minimise the classification error rates and DEFS2 combined both the classification error rate and the number of features into a single fitness function. DEMOFS outperformed both DEFS and DEFS2 by employing a multi-objective search mechanism. All the three DE based algorithms achieved better classification accuracy than the two traditional algorithms.

This work discovers that DE can be successfully used for multi-objective FS. It also provides motivations for further investigating EC particularly DE methods for multi-objective FS. There are many future research directions, which can be seen as follows:

1. The performance of DEMOFS needs to compare with other EC based multi-objective FS algorithms, such as PSO and GAs, which was not conducted in this paper due to the page limit;
2. A new multi-objective DE algorithm needs to be developed to further improve the performance of DE for multi-objective FS;
3. DE was originally for continuous problems, but FS is a binary task. Therefore, a binary DE is demanded to better solve the problem;
4. This paper focuses on wrapper based algorithms and the investigation of DE for filter based multi-objective FS is still an open issue;
5. Classification on datasets with over a thousand or a few thousands of features is still a challenge. Investigating effective and efficient multi-objective FS approaches on such large-scale problems can help address this challenge; and
6. To investigate the trade-off between the number of features and the classification performance and decide how to select a single solution from a set of non-dominated solutions in multi-objective FS.

Acknowledgment. This work is supported in part by the National Science Foundation of China (NSFC No. 61170180), the Marsden Funds of New Zealand (VUW1209 and VUW0806) and the University Research Funds of Victoria University of Wellington (203936/3337).

References

1. Al-Ani, A., Alsukker, A., Khushaba, R.N.: Feature subset selection using differential evolution and a wheel based search strategy. Swarm and Evolutionary Computation 9, 15–26 (2013)
2. Cervante, L., Xue, B., Zhang, M., Shang, L.: Binary particle swarm optimisation for feature selection: A filter based approach. In: IEEE Congress on Evolutionary Computation (CEC 2012), pp. 881–888 (2012)
3. Chen, Y., Miao, D., Wang, R.: A rough set approach to feature selection based on ant colony optimization. Pattern Recognition Letters 31(3), 226–233 (2010)
4. Coello Coello, C., Veldhuizen, L.A.G., Evolutionary Algorithms, D.: Evolutionary Algorithms for Solving Multi-Objective Problems. Genetic and Evolutionary Computation Series. Springer (2007)
5. Das, S., Suganthan, P.: Differential evolution: A survey of the state-of-the-art. IEEE Transactions on Evolutionary Computation 15(1), 4–31 (2011)
6. Dash, M., Liu, H.: Feature selection for classification. Intelligent Data Analysis 1(4), 131–156 (1997)
7. Deb, K., Pratap, A., Agarwal, S., Meyarivan, T.: A fast and elitist multiobjective genetic algorithm: NSGA-II. IEEE Transactions on Evolutionary Computation 6(2), 182–197 (2002)
8. Fdhila, R., Hamdani, T., Alimi, A.: Distributed mopso with a new population subdivision technique for the feature selection. In: International Symposium on Computational Intelligence and Intelligent Informatics (ISCIII 2011), pp. 81–86 (2011)
9. Frank, A., Asuncion, A.: UCI machine learning repository (2010)
10. Gutlein, M., Frank, E., Hall, M., Karwath, A.: Large-scale attribute selection using wrappers. In: IEEE Symposium on Computational Intelligence and Data Mining, pp. 332–339 (2009)
11. Guyon, I., Elisseeff, A.: An introduction to variable and feature selection. The Journal of Machine Learning Research 3, 1157–1182 (2003)
12. Hamdani, T.M., Won, J.-M., Alimi, A.M., Karray, F.: Multi-objective feature selection with NSGA II. In: Beliczynski, B., Dzielinski, A., Iwanowski, M., Ribeiro, B. (eds.) ICANNGA 2007. LNCS, vol. 4431, pp. 240–247. Springer, Heidelberg (2007)
13. Hastie, T., Tibshirani, R., Friedman, J., Franklin, J.: The elements of statistical learning: data mining, inference and prediction. The Mathematical Intelligencer 27, 83–85 (2005)
14. He, X., Zhang, Q., Sun, N., Dong, Y.: Feature selection with discrete binary differential evolution. In: International Conference on Artificial Intelligence and Computational Intelligence (AICI 2009), vol. 4, pp. 327–330 (2009)
15. Kohavi, R., John, G.H.: Wrappers for feature subset selection. Artificial Intelligence 97, 273–324 (1997)
16. Marill, T., Green, D.: On the effectiveness of receptors in recognition systems. IEEE Transactions on Information Theory 9(1), 11–17 (1963)
17. Neshatian, K., Zhang, M.: Pareto front feature selection: using genetic programming to explore feature space. In: The 11th Annual Conference on Genetic and Evolutionary Computation (GECCO 2009), pp. 1027–1034 (2009)
18. Robič, T., Filipič, B.: DEMO: Differential evolution for multiobjective optimization. In: Coello Coello, C.A., Hernández Aguirre, A., Zitzler, E. (eds.) EMO 2005. LNCS, vol. 3410, pp. 520–533. Springer, Heidelberg (2005)
19. Storn, R.: On the usage of differential evolution for function optimization. In: 1996 Biennial Conference of the North American Fuzzy Information Processing Society (NAFIPS), pp. 519–523 (1996)
20. Storn, R., Price, K.: Differential evolution - a simple and efficient heuristic for global optimization over continuous spaces. Journal of Global Optimization 11(4), 341–359 (1997)

21. Wang, L., Ni, H., Yang, R., Pappu, V., Fenn, M.B., Pardalos, P.M.: Feature selection based on meta-heuristics for biomedicine. Optimization Methods and Software, 1–17 (2013)
22. Xue, B., Cervante, L., Shang, L., Browne, W.N., Zhang, M.: A multi-objective particle swarm optimisation for filter based feature selection in classification problems. Connection Science 24(2-3), 91–116 (2012)
23. Xue, B., Cervante, L., Shang, L., Browne, W.N., Zhang, M.: Binary PSO and rough set theory for feature selection: A multi-objective filter based approach. International Journal of Computational Intelligence and Applications 13(02), 1450009 (2014)
24. Xue, B., Fu, W., Zhang, M.: Differential evolution (DE) for multi-objective feature selection in classification. In: Proceedings of the 2014 Conference Companion on Genetic and Evolutionary Computation Companion, GECCO Comp 2014, pp. 83–84 (2014)
25. Xue, B., Zhang, M., Browne, W.N.: Multi-objective particle swarm optimisation (PSO) for feature selection. In: Genetic and Evolutionary Computation Conference, pp. 81–88 (2012)
26. Xue, B., Zhang, M., Browne, W.N.: Particle swarm optimisation for feature selection in classification: Novel initialisation and updating mechanisms. Applied Soft Computing (2013)
27. Xue, B., Zhang, M., Browne, W.N.: Particle swarm optimization for feature selection in classification: A multi-objective approach. IEEE Transactions on Cybernetics 43(6), 1656–1671 (2013)
28. Zhu, Z.X., Ong, Y.S., Dash, M.: Wrapper-filter feature selection algorithm using a memetic framework. IEEE Transactions on Systems, Man, and Cybernetics, Part B: Cybernetics 37(1), 70–76 (2007)

A Multi-objective Optimization Method for Product Feature Fatigue Problem

Jinze Chai[*], Ming Li, Yu Zheng, Liya Wang, and Fan Yu

School of Mechanical Engineering,
Shanghai Jiao Tong University, Shanghai 200240, China
{sjtuchai,imelm,yuzheng,wangliya,fanyu}@sjtu.edu.cn

Abstract. Product feature fatigue is a common problem in practice. At the moment of purchasing, customers prefer to choose products with more features. After having used these high-feature products, customers become frustrated or dissatisfied with the usability problems caused by too many features. To deal with product feature fatigue problem, this paper introduces a novel model in which capability and complexity are regarded as two conflicting objects, and NSGA-II is adopted to search for a set of Pareto solutions for this multi-objective optimization problem. Then, this paper establishes piecewise linear membership functions based on decision maker's preferences, and a priority list of non-dominated solutions can be provided according to the membership function values. The list can make it easier for decision makers to make final selection. A smart phone case study shows that the proposed method is a powerful decision-aid tool for product designers when dealing with feature fatigue problem.

Keywords: product feature fatigue, multi-objective optimization, decision support.

1 Introduction

Feature fatigue problem is a difficult problem that a company must face in the process of product development. In the fierce competitive market environment, it is a common strategy of product development to add more features for the product. Companies hope that the feature-rich products could be more attractive to customers, so as to expand the customer groups, increase the sales, and eventually maximize the earnings obtained. Although research and actual situation shows that the products with more features have higher overall capability, it can not be ignored that the increase of features will lead to the increase of the product complexity. Research indicates that, at the moment of purchasing, the main consideration of a customer is whether the product is attractive or not, therefore customers tend to choose products with more features [1]. In the purchase process, consumers are less concerned about product usability. But after actual use, consumers will gradually realize the problem brought by the operation complexity, which can lead to dissatisfaction with the products, complaint or even returns, namely feature fatigue problem [2]. In general, although adding more features for product can increase the initial sales, feature fatigue problem caused by too many features will be harmful to the long-term returns of companies.

---

[*] Corresponding author.

G. Dick et al. (Eds.): SEAL 2014, LNCS 8886, pp. 529–541, 2014.
© Springer International Publishing Switzerland 2014

In recent years, feature fatigue problem has attracted wide attention in both academia and industry. To solve this problem, the causes of feature fatigue problem have been analyzed from the perspective of customer perception, consumer psychology and so on, and qualitative proposals have been put forward specifically to deal with feature fatigue problem, such as providing training guidance, reducing quantities, launching the product with a single feature, offering trials, gradual promotion, etc[3,4]. But these qualitative researches can not provide quantitative solutions for product designers, such as how many features a product should include, and selecting which specific feature can eliminate or reduce the negative effects brought by feature fatigue. For feature fatigue problem, too many features of a product will lead to feature complexity problem, however too few will reduce the capability and competitiveness of a product, so the solution to this problem is essentially in need of balancing the increase of capability, and the decrease of usability brought by adding features, which can come down to a multi-objective problem. Some quantitative models existing in the field of feature fatigue are mainly for the study of capability or usability respectively, without proposing quantitative solutions considering two objectives synthetically[3], [5], [6]. Although researchers have explored the problem from the perspective of multi-objective optimization, there are still shortcomings of the existing models and methods. At present, there are two main methods solving feature fatigue problem from the perspective of multi-objective, one of which is to merge multiple objectives into a single objective, the other is Pareto-optimal method. Taking the classical Thompson model for example[2], this model establishes a feature fatigue model from a revenue perspective by the way of transforming capability and usability into a synthetic objective, and eventually the optimal solution can be obtained. The model focuses on the number of product features, which means that only the effects of the feature number on feature fatigue are analyzed, while ignoring the differences among the various product features. The other problem is that it's difficult for decision makers to determine the relationship between the two objectives when merging objectives[7]. Different from the method of merging multiple objectives and obtaining a unique optimal solution, the literature[8] proposed a feature fatigue multi-objective model based on Pareto optimization, this model does not need to set weight for each objective, and it can get a set of optimal solutions for decision makers to select from after the optimization computation. However, there're still some deficiencies of this method. Firstly, the existing models focus on product usability, while ignoring the effects of the feature number on product capability. Secondly, the relation functions of the feature number and product usability in existing models are mainly in the way of taking square directly, without considering the diversity of the relation functions in actual situation. Thirdly, the method based on Pareto optimization often gets too many candidate solutions, which is a burden for decision-makers to make decisions, and this method is completely dependent on artificial selection with too much subjectivity.

In this paper, on the basis of previous studies, a multi-objective optimization model for feature fatigue problem is established based on Pareto optimization. The model introduces a relation function of the feature number, product capability and complexity, in which decision-makers can choose different forms of relation functions by adjusting the parameters. Using NSGA-II algorithm to solve the model, we can get a set of candidate solutions. According to the fuzzy set theory, this paper proposes a selection method based on piecewise linear membership function. By the way of sorting candidate solutions, the proposed method can provide a powerful decision support for decision-makers to make final choice.

Before getting the final optimization result, there are several parameters to be determined: the complexity value and capability score of each feature which can be obtained from questionnaire, auxiliary parameters which can be set by decision-makers, and the acceleration and decreasing effect of the increase of feature number. Then using the proposed method in this paper, the optimal product feature combination for decision-makers can be obtained.

2 Multi-objective Optimization Model for Feature Fatigue Problem

2.1 General Form for Feature Fatigue Multi-objective Optimization Model

In product feature fatigue problem, more features often means the increase of product capability, at the same time the complexity results in the decrease of usability. In this paper, product capability and complexity are two optimization objectives of the problem. According to the definition of multi-objective optimization problem, feature fatigue problem can be described as follows:

$$
\begin{aligned}
&\text{minimize: } FXW \\
&\text{maximize: } FAW \\
&\text{s.t. } g_j(x) \leq 0, \qquad j = 1, 2, ..., p; \\
&\qquad h_k(x) = 0, \qquad k = 1, 2, ..., q.
\end{aligned}
\tag{1}
$$

Where the two objective functions represent the minimization of product complexity FXW and the maximization of product capability FAW. x represents the decision variable, and for each feature F_i, the corresponding values x_i can be 0 or 1. $x_i = 0$ indicates that the product feature combination does not include feature F_i, while $x_i = 1$ indicates that the feature F_i is included. $g_j(x) \leq 0$ represents the p inequality constraints that the product feature combination needs to satisfy and $h_k(x) = 0$ represents the q equality constraints of the product feature combination. The purpose of solving problem under the premise of considering the two contradictory objective functions FXW and FAW, is to search for a set of solutions that satisfy the Pareto optimization for decision makers to choose from.

2.2   The Objective Functions Considering the Impact of Feature Difference and Feature Number

For a product, the major negative impact of adding more features is the increase of complexity (that is the decrease of usability). In fact, every additional feature increases an item to learn, an item that is likely to cause misunderstanding, and increases an item to be searched when looking for the desired feature at the using time [9]. Research shows that the relationship between the feature number and product complexity is not linear. For a product of which the feature number is increasing, the complexity will have acceleration effect as the feature number increases. The earlier models in the feature fatigue field mainly take square of the feature number to represent the acceleration effect [2, 3]. But actually there is diversity of the relation function between feature number and complexity. At the same time, considering the feature difference in terms of complexity, this paper uses the following equation to model product complexity:

$$FXW = \sum_{i=1}^{n} a_i x_i + \alpha (\sum_{i=1}^{n} x_i - \beta)^{\omega} + \gamma \tag{2}$$

Where x_i takes the value 1 or 0 which represents whether the feature F_i is included in the feature combination or not. n represents the total number of features for choosing. The first item in the equation a_i is the complexity value of feature F_i, and the value of a_i represents the difference of each feature. The larger the value of a_i is, the greater contribution the corresponding feature makes to product complexity. The second item in the equation $\sum_{i=1}^{n} x_i$ represents the number of features which are actually included in product feature combination. $\omega > 1$ represents that the increase of feature number has acceleration effect on the increase of product complexity[2]. α 、 β and γ in the equation are auxiliary parameters which can be set by decision-makers.

The existing models in the feature fatigue field, mainly define the relationship as a linear correlation when considering the impact of feature number on product capability [2]. However, research indicates that, in practice there is decreasing effect of the contribution that the increase of feature number makes to the increase of product capability[1]. At the same time, considering the different effects of the various features on product capability, this paper proposes the model for the product capability as follows:

$$FAW = \sum_{i=1}^{n} b_i x_i + \alpha '(\sum_{i=1}^{n} x_i - \beta ')^{\omega'} + \gamma ' \tag{3}$$

The first item in the above equation b_i is the capability score of feature F_i, and the higher the value of b_i is, the lager contribution this feature makes to product capability. The second item in the above equation $0 < \omega' < 1$, indicates that there is decreasing effect of the contribution that the increase of feature number makes to the increase of product capability. α' 、 β' and γ' are auxiliary parameters that decision-makers can set.

3   Multi-objective Optimization Based on NSGA-II

To solve multi-objective optimization problem, a number of methods are proposed in academia, among which multi-objective genetic algorithms are widely used to solve multi-objective optimization problem with the advantage of simple generality, overall search, strong flexibility, fast operation speed, etc. [10]. According to the existing literature, non-dominated sorting genetic algorithm II(NSGA-II ,proposed by Deb, etc. [11]) is one of the best solving methods. NSGA-II adopts succinct and clear non-dominated sorting mechanism to make algorithm possess the superior ability to approach the Pareto front, and the adoption of crowding distance sorting mechanism ensures that the Pareto solutions have a good distribution. According to the characteristics of feature fatigue problem, in this paper, the steps using NSGA-II to solve multi-objective optimization problems are as follows:

Step 1 Feature identification and chromosome coding. For the feature fatigue problem to be studied, we must first determine which features need to be studied specifically. Assuming that there are n optional features in the problem, then for multi-objective optimization problem, a solution can be represented as a chromosome of length n. Each gene on the chromosome represents an optional product feature, and the value of each gene can be 0 or 1. 0 indicates that the individual (product feature combination) does not include the corresponding feature , and 1 indicates that the corresponding feature is included in the individual. For feature fatigue problem, the advantage of using 0-1 encoding is that it can make the problem solutions one-to-one correspondence with the individuals in genetic space. According to the value of each gene on a chromosome in genetic space, bringing which into fitness equation directly will calculate the fitness function value for optimization problem.

Step 2 Determine the objective functions. The general form for feature fatigue multi-objective optimization model is shown in Equation (1), and the objective is minimizing the product complexity FXW , meanwhile, maximizing the product capability FAW . The 0-1 value of the i_{th} gene in each chromosome is corresponding to the value of x_i in Equation (2) and Equation (3). Therefore, taking each chromosome as the input, we will obtain the objective value of product complexity considering the impact of feature difference and feature number by Equation (2), and will obtain the objective value of product capability by Equation (3). Each feature score and parameter in objective function can be obtained through customer surveys, experiments or given by experts, see Case Study for details.

Step 3 Generate the initial population. To start the genetic algorithm, we should create a random parent population P_0 with N individuals.

Step 4 Selection. The selection method adopted in this paper is a binary tournament selection. The concrete method is that by selecting 2 individuals in the parent population randomly, determine whether it is good or bad in terms of the selection criteria, and then put the better individual into the crossness-pool, until the individuals in the crossness-pool reach the number needed. For any 2 individuals in the initial population P_0 , the one with lower non-domination rank will be the priority selection. And for the non-initial population, we will select in terms of crowded-comparison operator $\prec_n$. According to the definition of $\prec_n$, for individuals in two different non-domination ranks, the one with lower rank is better. And for 2 individuals in the same non-domination rank, the one with larger crowded distance is better.

Step 5 Crossover and mutation. In this paper, the crossover and mutation is performed based on the crossover and mutation rate. For each individual S_i in the crossness-pool, a random probability rate r_i is generated. If $r_i < r_c$, a single-point crossover is performed for individual S_i based on a crossover random point. If $r_i \geq r_c$, generate a random value r_k for each gene of individual S_i randomly, if $r_k < r_m$, the mutation is achieved by reversing the value of the corresponding gene.

Step 6 Generate the next generation of parent. For the t_{th} generation of the algorithm, a parent population P_t of size N and an offspring population of size N is formed, and the combined population is $R_t = P_t \cup Q_t$. Then the individuals of R_t are sorted according to non-domination. The sorted individuals of R_t are added into the next parent population P_{t+1} in the increasing order of their ranking until the solution number of some front exceeds N . To choose N population members, all the individuals of the last front will be sorted based on crowding distance, and add individuals into P_{t+1} based on crowding distance in descending order until the number reaches N .

Step 7 If it reaches the termination condition, then take the output P_{t+1} as an optimal solution. If not, go to Step 4 to continue the iteration.

4   Pareto Optimization Based on Piecewise Linear Membership

NSGA-II is adopted to solve the multi-objective optimization problem and can get a set of Pareto-optimal product feature combinations for decision makers to select from. But if the number of the obtained Pareto-optimal solutions is large, it will be a burden for decision makers, and it carries too much subjectivity depending on artificial selection completely. To deal with this problem, this paper establishes piecewise linear membership functions based on fuzzy set theory to assist Pareto optimization [12].

It is assumed that there are m optimal solutions in Pareto solution set, and S_i is the i_{th} solution, then the corresponding complexity value of S_i is FXW_i , and the capability value is FAW_i . For complexity, the lower the value of FXW_i is, the more satisfied the decision maker will be with S_i . According to the definition of membership, the value of membership is 1 at the most satisfied time, and the value is 0 at the most dissatisfied time. When we calculate FXW using Equation(2), the acceleration effects caused by feature difference and feature number are considered comprehensively, and in this case, the relationship between FXW value and membership is difficult to be represented by a single linear function. When decision makers propose the membership function, they can set membership values for some subpoints to describe their preferences[13]. For complexity, the membership function μ_i^x can be expressed as follows:

$$\mu_i^X = \begin{cases} 1, & FXW_i \le FXW_{min} \\ \dfrac{1}{g_r - g_{r-1}}[(f_r - f_{r-1})FXW_i + (f_{r-1}g_r - f_r g_{r-1})], \\ & g_{r-1} < FXW_i \le g_r \\ 0, & FXW_i \ge FXW_{max} \end{cases} \tag{4}$$

Where FXW_{max} is the maximum value of complexity, FXW_{min} is the minimum value. g_r and f_r are the complexity score of the r_{th} endpoint and the corresponding value of membership function respectively, and g_{r-1} and f_{r-1} are respectively the complexity score and the value of membership function of the one point before, where $r \ge 1$.

For product capability, the higher the value of FAW_i is, the more satisfied the decision maker will be with S_i, so the membership function of capability can be obtained according to the following equation:

$$\mu_i^A = \begin{cases} 1, & FAW_i \ge FAW_{max} \\ \dfrac{1}{g_r' - g_{r-1}'}[(f_r' - f_{r-1}')FAW_i + (f_{r-1}'g_r' - f_r'g_{r-1}')], \\ & g_{r-1}' < FAW_i \le g_r' \\ 0, & FAW_i \le FAW_{min} \end{cases} \tag{5}$$

After getting the membership μ_i^X and μ_i^A of each solution S_i by computing on both objectives, the total membership μ_i can be obtained through the following equation:

$$\mu_i = \frac{\mu_i^X + \mu_i^A}{\sum_{j=1}^{m}(\mu_j^X + \mu_j^A)} \tag{6}$$

Where m is the solution number in Pareto solution set, the solutions with lager value of μ_i will possess the higher comprehensive satisfactory degree. The candidate solutions are sorted based on the value of μ_i in descending order, which provides Pareto-optimal solutions with priority order to decision makers, and this will reduce the burdens of decision when searching for optimal solutions.

5   Case Study

5.1   Case Description

In this paper, a case study of smart phone is presented to illustrate how the proposed multi-objective optimization method can be applied in real-world. In this case, we select ten typical features for the feature combination study: Music, Video, Radio, TV, Video Conferencing, WiFi, Payment, Camera, Email, Bluetooth. Each feature and the corresponding score of capability and complexity are shown in Table 2. The capability score of each feature is based on a questionnaire survey, and the questionnaire is shown in Fig.1. Respondents were asked to choose the degree of preference points for each feature on a scale of 1 to 9, and the preference is normalized to value from 0 to 1, then the obtained evaluation data of each capability can be shown in Table 1. The average of the preference value of all 173 copies of reasonable questionnaires is the capability value and shown in Table 2[13]. In this case, the using complexity score is represented using the number of steps to finish related tasks. For a feature, more steps means more complicate to use[14].

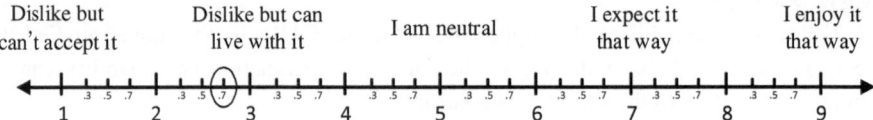

Fig. 1. The questionnaire about Functional/Dysfunctional question

Table 1. Evaluation data of capability

Index	F_1	F_2	F_3	F_4	F_5	F_6	F_7	F_8	F_9	F_{10}
C_1	0.778	0.333	0.222	0.333	0.222	0.333	0.222	0.222	0.111	0.333
C_2	0.111	1.000	0.333	0.667	0.556	0.556	0.111	0.333	0.333	0.556
C_3	1.000	0.444	1.000	0.444	0.333	0.556	0.111	0.556	0.778	0.778
...	...	...	...	...	...	...	...	...	...	...
C_{173}	0.333	0.556	0.556	0.778	0.556	0.556	0.556	0.222	0.333	0.333

Table 2. Capability and complexity scores of the product features

Index	Features	Capability score	Complexity score
F_1	Bluetooth	0.643	8
F_2	Music	0.587	16
F_3	WiFi	0.560	8
F_4	Email	0.500	12
F_5	Video	0.490	11
F_6	Camera	0.419	13
F_7	Radio	0.356	14
F_8	Payment	0.314	6
F_9	TV	0.276	10
F_{10}	Video Conferencing	0.250	12

5.2   Problem Solving

In this case, each solution for the multi-objective optimization model is represented by a chromosome with ten genes. Each gene is a product feature combination, and the value of 1 means the corresponding feature is included in the solution while 0 means not. In the process of establishing the objective function, decision makers need to set the parameters to reflect the impact of the feature number on the product. In this case, the objective function can be obtained after computation based on Equation (2) and Equation (3), and the scores in Table 1 are used as the complexity score a_i and capability score b_i of each feature, and set the constant term $\gamma = \gamma' = 0$. The second items in Equation (2) and Equation (3) represent the effect of feature number. In this case, two sets of parameter combinations are set aiming at the effect of number to illustrate the difference of effect curve, which are parameter combination A (PCA) and parameter combination B (PCB), and the parameter values are shown in Table 3. Fig.2 (a) and (b) represent the effect of feature number on capability and complexity in the case of PCA, and Fig.2(c) and (d) are two effect curves in the case of PCB.

Table 3. Parameters combination

Parameters•	α	β	ω	α'	β'	ω'
PCA	0.5	0.9	0.1	0.01	0.9	3.5
PCB	0.3	0.5	0.05	0.1	0.5	2

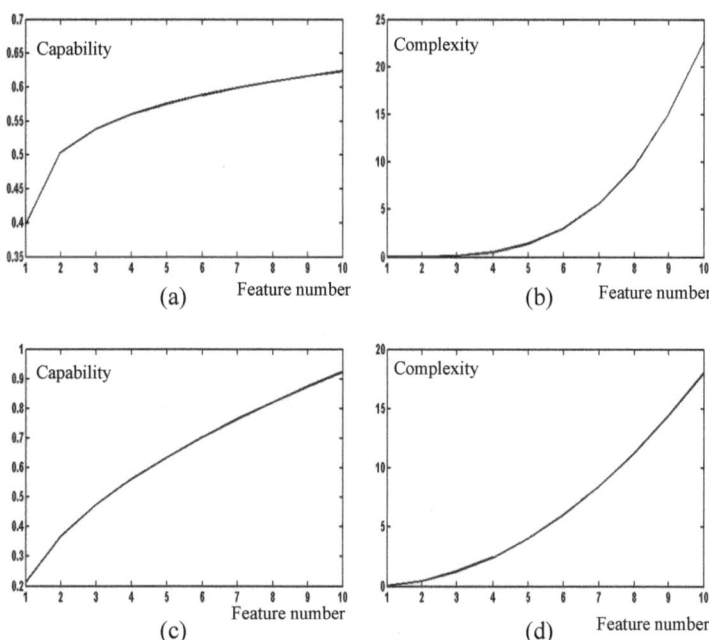

Fig. 2. The effect of feature number on objective function

After the completion of chromosome coding and the confirmation of objective function, we can run NSGA-II algorithm to search for optimal solutions. The detailed parameters of NSGA-II algorithm are as follows: Each population has 150 chromosomes (the initial population is generated randomly), and the crossover rate is 0.8, the mutation rate is 0.1 and the iteration number is 500.In the case of PCA, 37 Pareto-optimal solutions can be obtained. In Fig.3, optimal solutions are represented by hollow dots, and all dots constitute the Pareto fronts.

Based on Pareto-optimal solutions, decision makers can establish piecewise linear membership function to search for optimal solutions. In this case, for each objective function, decision makers need to give the objective function values where membership is 0, 0.5 and 1.If the two extreme points of the objective value correspond to the extreme points of membership, the decision maker just needs to give one intermediate point to establish membership function. Taking the computation of μ_i^x as an example, we take three quarters of the largest value of each objective function as the corresponding objective value where the membership is 0.5, then:

$$g_0 = 0 \bullet f_0 = 1$$
$$g_1 = 132.732 \times 0.75 = 99.549 \bullet f_1 = 0.5$$
$$g_2 = 132.732 \bullet f_2 = 0$$

According to Equation (4), if $g_0 < FXW_i \leq g_1$, then:

$$\mu_i^x = \frac{0.5 - 1}{99.549 - 0} FXW_i + \frac{1 \times 99.549 - 0.5 \times 0}{99.549 - 0}$$
$$= -0.005 FXW_i + 1$$
If $g_1 < FXW_i \leq g_2$, then:
$$\mu_i^x = \frac{0 - 0.5}{132.732 - 99.549} FXW_i + \frac{0.5 \times 132.732 - 0 \times 99.549}{132.732 - 99.549}$$
$$= -0.015 FXW_i + 2$$

In the same way, we can get the value of the membership function μ_i^A which corresponds to the capability of each solution, and then the total membership function value μ_i can be obtained based on Equation (6).The optimal solutions obtained are sorted according to μ_i in descending order, and the top ten solutions which are identified by arrows in Fig.3 are shown in Table 4.

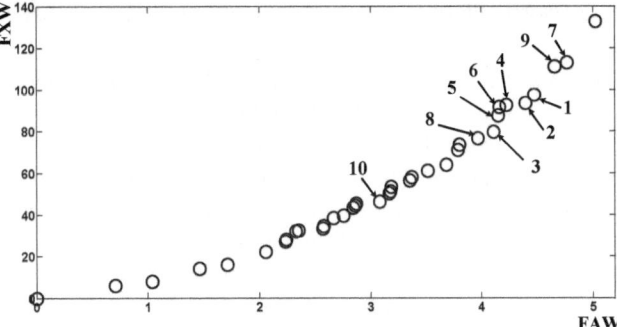

Fig. 3. Pareto front and sorting identification

Table 4. The sorting result

Index	Feature combination	FAW_i	FXW_i	μ_i
S_1	1111111100	4.477	97.537	0.03009
S_2	1111110110	4.397	93.537	0.02981
S_3	1111110100	4.112	79.606	0.02879
S_4	1111100111	4.228	92.537	0.02836
S_5	1111111000	4.154	87.606	0.02825
S_6	1011111110	4.166	91.537	0.02791
S_7	1111111110	4.761	113.125	0.02785
S_8	1111100110	3.969	76.606	0.02782
S_9	1111110111	4.655	111.125	0.02757
S_{10}	1011100100	3.083	46.396	0.02739

5.3   Discussion

The top ten solutions are listed in Table 4 according to the value of μ_i in the Pareto-optimal solution set, of which each solution represents a product feature combination. For example, the coding form of solution S_7 is 1111111110, which means that nine out of the ten features are included in this combination. The capability score of the product with these features is 4.761 and the complexity score is 113.125. Compared with the other Pareto solutions in Table 4, S_7 has the highest capability score and the highest complexity score as well. S_7 is easier to stimulate the customer's purchasing interest compared with other feature combinations, and it is also easier to cause customer dissatisfaction after use because of complexity. In contrast, the coding of S_6 is 1011111110, which means one feature less than S_7, and correspondingly the capability score is reduced to 4.166 and the complexity score to 91.537. Both of S_6 and S_7 are non-domination solutions and it is difficult to decide which one is better.

Although S_6 is worse than S_7 according to capability, it is better according to complexity.

The model mainly focuses on feature number in literature[2], which means that only the effect of feature number on feature fatigue is analysed while the difference among each feature is ignored, and that's to say, the effect of a_i and b_i hasn't been considered in the model above. The method proposed in this paper takes the effect of feature number and the difference of product features into consideration comprehensively. The curve in Fig. 2 mainly reflects the different effects of feature number on capability and complexity, while Table 4 reflects that the feature difference can result in the difference of final assessment result. For example, for S_3, S_5 and S_8 with seven features, S_8 has the lowest capability and complexity score, and its membership value is also the lowest of the three. S_3 is moderate in both aspects, but its membership value is the highest. There are only three solutions whose feature number is seven in Table 4, which means a solution with seven features is either dominated by the member of Pareto-optimal solution set or lower than the listed solutions in Table 4 according to the value of membership. So the method proposed in this paper can get the optimal feature combination when the feature number is the same.

The relation function of the feature number and product usability established in literature [8] is mainly in the way of taking square directly, but in this case the variety of relation function in practice has been considered. In addition, methods similar to the method in literature [8] will result in too many results, which is a burden for decision. One problem of using multi-objective optimization method to get Pareto-optimal solutions is that the obtained optimal solutions may be too many, which will be a burden for decision makers. As shown in Fig.3, there are 37 optimal solutions in the case of PCA, which is difficult for decision makers to select from directly. Using the method proposed in this paper, decision makers only need to give some information of the key membership value to establish membership functions, and obtain the sorted Pareto-optimal solutions. Decision makers can select the solution with the highest membership value in Table 4 (that is the solution with the best comprehensive effect) directly or take solutions of high ranking as a new set of candidate solutions for further analysis and selection to achieve the purpose of reducing the number of candidate solutions.

6   Conclusions

This paper mainly studies multi-objective optimization methods for feature fatigue problem, for which capability and complexity are two conflicting objectives. In the objective functions established in this paper, the differences between features are represented by the different feature scores. The diminishing effect of feature number on product capability and the acceleration effect on complexity are described by establishing models. For the multi-objective optimization model established in this paper, NSGA-II is adopted to search for solutions, and based on the Pareto-optimal solutions ,the candidate solutions are sorted comprehensively using piecewise linear membership functions which represent the customer preferences. The feasibility of the proposed method is evaluated by using a smart phone case study. The data result shows that, the proposed method can reduce the decision making burden and provide powerful decision support for product developers when dealing with feature fatigue problem.

Acknowledgements. This research is supported by the National Natural Science Foundation of China (Grant no.71072061/G020801) and Innovation Action Program of Science and Technology Commission of Shanghai Municipality (Grant no. 12dz1125200 /12dz1125201/12dz1125202).

References

1. Nowlis, S.M., Simonson, I.: The Effect of New Product Features on Brand Choice. Journal of Marketing Research 33(1), 36–46 (1996)
2. Thompson, D.V., Hamilton, R.W., Rust, R.T.: Feature Fatigue: When Product Capabilities Become Too Much of A Good Thing. Journal of Marketing Research 42(4), 431–442 (2005)
3. Fang, F., Xu, X.: An Analysis of Consumer Training for Feature Rich Products. Decision Support Systems 52, 169–177 (2011)
4. Lakshmanan, A., Krishnan, H.S.: The Aha! Experience: Insight and Discontinuous Learning in Product Usage. Journal of Marketing 75(6), 105–123 (2011)
5. Kumar, V., Gordon, B.R., Srinivasan, K.: Competitive Strategy for Open Source Software. Marketing Science 30(6), 1066–1078 (2011)
6. Li, M., Wang, L.: Feature Fatigue Analysis in Product Development Using Bayesian Networks. Expert Systems with Applications 38(8), 10631–10637 (2011)
7. Guan, X.S., Wang, Y.Q., Tao, L.Y.: Maching Scheme Selection of Digital Manufacturing Based on Genetic Algorithm and AHP. Journal of Intelligent Manufacturing 20, 661–669 (2009)
8. Li, M., Wang, L., Wu, M.: A multi-objective genetic algorithm approach for solving feature addition problem in feature fatigue analysis. Journal of Intelligent Manufacturing (published online ahead of print May 8, 2012)
9. Nielsen, J.: Usability Engineering. Academic Press, San Diego (1993)
10. Gen, M., Cheng, R.: Genetic Algorithm and Engineering Optimization. Tsinghua University Press, Beijing (2003)
11. Deb, K., Pratap, A., Agarwal, S., et al.: A fast and elitist multi-objective genetic algorithm: NSGA-II. IEEE Transactions on Evolutionary Computation 6(2), 182–197 (2002)
12. Fuller, R., Carlsson, C.: Fuzzy multiple criteria decision making: recent developments. Fuzzy Set Systems 78, 139–153 (1996)
13. Wu, M., Wang, L.: A continuous fuzzy kano's model for customer requirements analysis in product development. Proceedings of the Institution of Mechanical Engineers, Part B: Journal of Engineering Manufacture 226(3), 535–546 (2012)
14. Klockar, T., Carr, D.A., Hedman, A., Johansson, T., Bengtsson, F.: Usability of mobile phones. In: Proceedings of the 19th International Symposium on Human Factors in Telecommunication, Berlin, Germany, pp. 197–204 (2003)

Genetic Programming for Channel Selection from Multi-stream Sensor Data with Application on Learning Risky Driving Behaviours

Hoang Anh Dau, Andy Song, Feng Xie, Flora Dilys Salim, and Vic Ciesielski

RMIT University, Melbourne 3000, Victoria, Australia
http://www.rmit.com.au

Abstract. Unsafe driving behaviours can put the driver himself and other people participating in the traffic at risk. Smart-phones with built-in inertial sensors offer a convenient way to passively monitor the driving patterns, from which potentially risky events can be detected. However, it is not trivial to decide which sensor data channel is relevant for the task without domain knowledge, given the growing number of sensors readily available in the phone. Using too many channels can be computationally expensive. Conversely, using too few channels may not provide sufficient information to infer meaningful patterns. We demonstrate Genetic Programming (GP) technique's capability in choosing relevant data channels directly from raw sensor data. We examine three risky driving events, namely harsh acceleration, sudden braking and swerving in the experiment. GP performance on detecting these unsafe driving behaviours is consistently high on different channel combinations that it decides to use.

Keywords: feature selection, channel selection, Genetic Programming, risky driving behaviours.

1   Introduction

Most road accidents are caused by human error with the highest accident risk group being young male drivers under 25 [1] coupled with those impaired by alcohol or other drugs [2]. This group is known for risk-taking either due to inexperience, peer pressure or sensory and mental degradation. Most drivers, even those in the groups given, do not want to crash, and will take evasive manoeuvres to avoid a collision. These groups consequently are much more likely to put themselves into dangerous situations by driving inappropriately on road conditions, most commonly speeding. To be specific, risky driving behaviour involves the practice of a set of risky manoeuvres. Within the scope of this paper, we however refer to risky driving event or behaviour interchangeably as being deviated from normal driving behaviours. Typical risky driving can be characterized by a pattern of sudden braking, swerving and excessive acceleration. We address these risky driving events in this study. The potential of this research is however open to classifying other driving events as well.

G. Dick et al. (Eds.): SEAL 2014, LNCS 8886, pp. 542–553, 2014.

An onboard inertial measurement unit, or IMU, can provide a data rich enough to classify this behaviour. They are however expensive and require extra installation. We find smart phones a much better option for these reasons. Smartphones with built-in sensors are ubiquitous and affordable these days. They come with many standard built-in sensors, originally driven by user experience. For this study we have chosen to use the iPhone5 as it is one the most popular smartphones in Australia. We solve the problem using the sensors in this phone alone.

The iPhone5 comes with three built-in sensors: accelerometer, gyroscope and magnetometer, each with three channels representing three orthogonal axis. These sensors therefore collectively offer nine channels of data, from which another twelve channels can be derived. The continuous reading of data provided by each sensor over time provides a collection for observation, which is a time series. As data is coming from more than one channel, so-called multi-channel time series, the task of detecting risky driving behaviours can be considered a multi-stream time series classification. Mining of this data stream can provide interesting information. This sensor data is very close and relevant to motion caused by abnormal manoeuvres. For example, the accelerometer reads user acceleration and gravitational acceleration. The gyroscope measures the rotation rates and angles of the device, which are clear indications of the vehicle turn movement. An advantage of this approach is that the data is readily available and the device can conveniently travel with the driver all the time, hence, being handy and intrusive.

Traditional approach requires a good feature sets or sometimes manually designed complicated mathematical model. Finding a good set of features for this particular problem is not straightforward. A suitable feature set for one type of risky driving event may not be relevant to another type. Moreover, deciding which channels to use out of 21 data channels available on the iPhone is not easy. Both of these tasks require domain knowledge. In this study, we leave this task for GP to handle automatically. On the one hand, traditional machine learning algorithm may not be suitable for time series data because they ignore the temporal dependency between data points. On the other hand, most techniques developed for time series classification only concerns single stream time series, hence, do not handle multi-stream time series effectively.

This study is the development of our preliminary work on detecting risky driving behaviours [3], in which we established that this problem can be well represented as a time series classification problem and that GP is a good approach. Our GP-based methodology works directly on raw sensor data in a supervised learning manner without any explicit feature extraction. In this study, we focus on exploring the capability of GP in data channel selection. Our research gives insight into the question: Can GP choose relevant sensor reading channels for learning risky driving behaviours?

2   Related Work

Intelligent transport system has been an active area of research. Many studies have been done on modelling, classifying and predicting driver behaviour for better road safety. Based on data for analysis, most of these studies fall into vision-based approach, sensor-based approach or a mixture of two. Hidden Markov models (HMM), a popular stochastic tool for studying time series data, is one of the favourite techniques in this problem domain. HMM is a borrowed technique from speech recognition research. It is naturally suitable for driver behaviour analysis due to the capability to model stochastic events over time.

Lee et al. identifies driving patterns from the machine vision point of view [4]. Their system uses two cameras, one to capture the driver's image and the other to capture the front road image. From these pictures, the orientation information such as the car heading direction, the driver's sight line and the lane path can be obtained and mapped, which facilitates the calculation of the driver and lane correlation coefficients. HMM is used to train the sequences of lane correlation coefficients. Their system can recognize four common driving patterns: driving in a straight lane, driving in a curve lane, driving of changing lanes and driving of making a turn. It works reasonably well in ideal conditions but does not generalize well to some real world scenarios. For example, the lane path detection can suffer from traffic letters and signs painted on the road. Also, the system fails the case when the car stands for traffic light or briefly stops before making a turn.

Mitrovic [5] uses discrete HMM to develop a driving event recognition system. The results shows that HMMs could accurately and reliably recognize various driving events. The data he used was collected from real vehicles in normal driving conditions through a number of sensors. These are accelerometers, airbag sensor, and the GPS receiver as a velocity sensor. The limitation of this approach is that data has to go through pre-processing (filtering and normalization, waveform segmentation, vector quantization, and event segmentation). Also, the author had to probably put a lot of effort into manually marking hundreds of events for training, which is time consuming.

Imkamon et al. [6] propose a system to detect hazardous driving behaviour using fuzzy logic. Their system measures the unsafe behaviour from three perspectives. The first is the passenger's point of view, for this they use an accelerometer mounted to the passenger's seat to detect heavy jolts caused by sudden turn or brake. The second is the driver's point of view, for which they use a camera mounted on the car's console to emulate the driver's vision. The third perspective is the vehicle's status, for which they use an On-Board Diagnosis II reader to obtain the velocity and the engine speed of the vehicle. All the data are sent to a fuzzy logic systems for classification, which outputs a driving risk level ranging from 1 to 3. The test results show that the system can be competitive to human opinion. However, the limitation of this system is day-time operation requirement due to the constraints of the image processing algorithm they use.

Smartphones are ubiquitous today, with sales exceeded those of feature phones in early 2013.[1] Not just a communication tool, they have found their way into behaviour and activity recognition research. Johnson et al. [7] propose a system called MIROAD, that uses Dynamic Time Warping (DTW) algorithm and smartphone based sensor-fusion to classify driving styles as aggressive or non-aggressive. The iPhone5 is used to collect sensor data. The fusion of data from the accelerometer and gyroscope are analyzed. The system can identify correctly nearly 97% of the aggressive events. They conclude that the combination of three channels, x-axis rotation rate, y-axis acceleration and pitch are the signals best suited for DTW algorithm in this problem domain. The limitation of this study is the requirement of manual feature reduction.

3 Learning a Risky Driving Behaviour Classifier with GP

3.1 GP Methodology

GP needs a way to initialize the population and evaluate the individuals in this population. Each individual is an executable program tree represented in Lisp S-expression. Terminal sets and function sets are the ingredients to create the program trees. Functions fit in the internal nodes (branch) of the tree whilst terminals can only serve as the outer node (leaf). A GP run is a competitive search for all the possible combinations of terminal and function sets favouring the fittest. The programs evolve towards complexity over generations.

3.2 GP Representation

Regarding function set, we use a set of basic arithmetic functions. Apart from that, three functions are specially designed for multi-channel time series problem: $Window$, $Temporal\_Diff$ and $Multi\_Channel$. Function $Window$ samples data points from a time series for analysis. Function $Temporal\_Diff$ is similar to the standard temporal difference function. It can be used by other functions such as function $Window$. Function $Multi\_Channel$ is designed to capture the dependency between channels. It is very similar to function $Window$. The difference is that it operates on data channels rather than data points. The complete function set is displayed in table IV.

In terms of terminal set, terminal Channel_$[m]$ is available to all functions (m is the index of the channel, which starts at 0). Terminal Temporal-Index and Temporal-Operation are designed specifically for the function $Window$. Terminal Channel-Index and Channel-Operation are only for the function $Multi -$ $Channel$. More details of our GP representation can be found in [3].

To address the bias towards the majority class, we used AUC (Area under the ROC (Receiver Operating Characteristics) Curve) as the fitness measure. It is considered as a better alternative over accuracy for unbalanced data [12].

---

[1] http://www.3news.co.nz/Smartphones-now-outsell-dumb-phones/
tabid/412/articleID/295878/Default.aspx

Table 1. GP Function Set

Function	Parameter		Parameter	
			Type	Value
+	1		Double	[DOUBLE_MIN,DOUBLE_MAX]
	2		Double	[DOUBLE_MIN,DOUBLE_MAX]
-	1		Double	[DOUBLE_MIN,DOUBLE_MAX]
	2		Double	[DOUBLE_MIN,DOUBLE_MAX]
	1		Double	[DOUBLE_MIN,DOUBLE_MAX]
	2		Double	[DOUBLE_MIN,DOUBLE_MAX]
/	1		Double	[DOUBLE_MIN,DOUBLE_MAX]
	2		Double	[DOUBLE_MIN,DOUBLE_MAX]
Window	1		Double	[DOUBLE_MIN,DOUBLE_MAX]
	2	Temporal-Index		$[1, 2^{window-size} - 1]$
	3	Temporal-Operation		AVG, STD, DIF, SKEWNESS
Temporal_Diff	1		Double	[DOUBLE_MIN,DOUBLE_MAX]
Multi_Channel	1		Double	[DOUBLE_MIN,DOUBLE_MAX]
	2	Channel-Index		$[1, 2^{num-of-channels} - 1]$
	3	Channel-Operation		AVG, STD, MED, RANGE

4   Experiment

4.1   Data Sets

Data Collection. All data is from one driver controlling the 2006 Mazda 6 in a real environment. The driver drives normally through most of the sessions with the occasional performance of an hypothesised unsafe event. Care was taken so that no actual unsafe conditions were presented to other road users during the process.

An iPhone application was developed for data recording and labelling is done in real time. The driver uses voice command to confirm when one of the pre-registered risky driving behaviours has just occurred. The phone is attached to the wind-shield throughout the journey. This is to ensure that the phone is relatively fixed with respect to the car, hence any abnormal in sensor readings actually reflects the driving patterns, not the effects from the phone free movement.

The Apple CoreMotion Framework provides access to both raw and processed sensor data. In total there are 21 data channels available (Table 2). 9 channels are raw readings of the accelerometer, gyroscope and magnetometer (each provides 3 channels in x, y and z axis), from which the other 12 channels are derived. We sample data 10 times every second.

Data Preparation. We consider three risky driving behaviours, namely harsh acceleration, sudden stop and swerving. For each of these behaviours, we perform

Table 2. 21 data channels provided by the iPhone5

Channel No.	Channel
1-3	Raw Acceleration X, Y, Z (raw accelerometer reading)
4-6	Gravity X, Y, Z
7-9	User Acceleration X, Y, Z
10-12	Yaw, Pitch, Roll
13-15	Raw Rotation Rate X, Y, Z (raw gyro reading)
16-18	Unbiased Rotation Rate X, Y, Z
19-21	Magnetic Heading X, Y, Z

two separate driving sessions, one to collect the training data and one for the test set. Summary of the data sets used are presented in Table 3. We use all the 21 data channels available from the Apple framework and do not perform any extra feature extraction.

Table 3. Training set and test set for three types of risky driving behaviours

Driving Event	Training set			Testing set		
	Total	Positive	Negative	Total	Positive	Negative
Harsh Acceleration	1182	12	1170	690	9	681
Sudden Stop	1688	9	1679	947	6	941
Swerving	1206	12	1194	828	6	822

The data is highly unbalanced. The number of positive instances is minor compared to the massive number of negative examples. This reflects real-life where in most circumstances a driver does not hit things or have to swerve or brake suddenly.

4.2 Runtime Parameters

The process is terminated when a solution with 100% accuracy is found or the 50th generation is reached. The best programs evolved are then selected for testing. We favour the program that has both high accuracy and high true positive rate. This is because we do not want the classifier to miss any risky driving event whilst a false positive can still be informative. The sliding window of size 12 and step-size of 1 is used to sample data along the y axis. The GP runtime parameters are listed in Table 4.

Table 4. GP runtime parameters

Population	1000
Generation	50
Maximum Depth	8
Minimum Depth	2
Mutation Rate	5%
Crossover Rate	85%
Elitism Rate	10%
Number of Runs	10

4.3 Evaluation Metrics

We report the F-score, which is a harmonic balance of precision and recall. This is to take into account the imbalance aspect of the data, for which, accuracy is not a suitable metric. A classifier that declares all the instances to be of the majority class can still achieve high accuracy even though it fails to detect any risky driving event. On the other hand, with F-score, the technique is heavily punished for false alarm and mis-detection. We also report true positive rate and true negative rate.

4.4 Experiment Design

All the experiments are done in a binary classification setting. We use the strategy one-vs-all, where a single classifier is trained per class to distinguish that class from the rest.

GP-21-Channels. We train GP on raw sensor data of the full 21 channels. We call this experiment GP-21-channels.

GP-Selected-Channels. We analyse the best classifiers evolved from GP-21-channels to see which channels GP picked up for each individual task. We then re-train GP on these selected channels only. We call this experiment GP-selected-channels.

GP-9-Channels. We manually select 9 data channels that we think are relevant to the problem at hand and train GP on these channels only. The nine channels selected here are: Channel 1-3, 10-12 and 19-21. The motivation behind this selection is as follow. The choice of channel 1-3 (raw accelerometer reading) represents domain knowledge involvement. In the past, most driving pattern recognition systems have used data from the accelerometers only. The inclusion of channel 10-12 (yaw, pitch, roll) and 19-21 (magnetic heading) is considered noisy elements. When we visualize the data, we notice the pattern of swerving is hardly observable. However, if we remove the six channels mentioned above, the pattern stands out. We would like to investigate if GP can choose relevant features whilst ignoring the irrelevant input. We call this experiment GP-9-channels.

4.5 Result

Table 5 displays the test result of the GP classifiers. In experiments with three different data versions, GP manages to detect all targeted risky driving behaviours (100% TPR). Even though both the TPR and TNR are reasonably high (above 90%), the F-score gives varying performance. This is because the data is highly unbalanced as illustrated in Table 3. GP achieves 0 false negative but is punished heavily by a false alarms. Note that the number of false alarms is minor given the rareness of the positive class.

Table 5. Test result for GP trained on 21 channels, GP trained on channels previously selected by GP-21-channels and GP trained on 9 manually selected channels

		GP-21-channels	GP-selected-channels	GP-9-channels
Harsh acceleration	F-score	54.54%	21.95%	60.00%
	TPR	100%	100%	100%
	TNR	97.76%	90.05%	98.20%
Sudden stop	F-score	80.00%	75.00%	25.53%
	TPR	100%	100%	100%
	TNR	99.68%	99.56%	96.23%
Swerving	F-score	60.00%	60.00%	85.71%
	TPR	100%	100%	100%
	TNR	99.01%	99.01%	99.75%

For purpose of comparison, we test some popular machine learning algorithms on the same data sets tested with GP. We use the Weka implementation of these classifiers and keep default configuration settings. They are Random Forest, Naïve Bayes, k-Nearest Neighbour (IB1), Support Vector Machine (SMO) and AdaBoost. We use the sliding window of size 12 to sample the data, the same with the window size used by GP. The result is displayed in Table 6. Overall, all Weka classifiers perform poorly on the three recognition task. They fail to detect actual unsafe driving events and tend to classify every instances as negatives.

For all the experiments, the performance of GP is reliably consistent. In all cases, GP successfully detects all risky driving events. This is not the case with other traditional classifiers. Their performance is low in general and quite sensitive to dataset (channels) used. The number of positive instances identified is quite low. The classifiers obtain high accuracy mainly due to the bias toward the majority class.

Table 6. Test result of 5 traditional classifiers: Random Forest, Naïve Bayes, k-Nearest Neighbour, Support Vector Machine and AdaBoost

21 original channels						
		R. Forest	Naïve Bayes	kNN	SVM	AdaBoost
Acceleration	F-score	30.07%	7.36%	33.33%	0.00%	0.00%
	TPR	22.22%	66.70%	33.30%	0.00%	0.00%
	TNR	99.70%	77.90%	99.10%	99.90%	100%
Sudden stop	F-score	0.00%	0.00%	0.00%	0.00%	0.00%
	TPR	0.00%	0.00%	0.00%	0.00%	0.00%
	TNR	100%	100%	100%	100%	100%
Swerving	F-score	28.57%	19.23%	61.53%	0.00%	18.00%
	TPR	16.70%	83.30%	66.70%	0.00%	16.70%
	TNR	100%	94.90%	99.60%	98.50%	99.80%
Channels selected by GP trained on 21 original channels						
		R. Forest	Naïve Bayes	kNN	SVM	AdaBoost
Acceleration	F-score	0.00%	6.45%	35.29%	0.00%	33.33%
	TPR	0.00%	55.60%	33.30%	0.00%	22.20%
	TNR	100%	78.90%	99.30%	99.90%	99.90%
Sudden stop	F-score	0.00%	0.00%	0.00%	0.00%	0.00%
	TPR	0.00%	0.00%	0.00%	0.00%	0.00%
	TNR	100%	100%	100%	100%	100%
Swerving	F-score	0.00%	12.30%	20.00%	18.18%	0.00%
	TPR	0.00%	66.70%	16.70%	16.70%	0.00%
	TNR	100%	93.20%	99.60%	99.50%	99.80%
9 channels manually selected based on the domain knowledge						
		R. Forest	Naïve Bayes	kNN	SVM	AdaBoost
Acceleration	F-score	0.00%	7.27%	31.57%	0.00%	0.00%
	TPR	0.00%	66.70%	33.30%	0.00%	0.00%
	TNR	100%	77.60%	99.00%	99.30%	100%
Sudden stop	F-score	0.00%	80.00%	60.00%	0.00%	80.00%
	TPR	0.00%	66.70%	50.00%	0.00%	66.70%
	TNR	100%	100%	99.90%	100%	100%
Swerving	F-score	25.00%	18.18%	52.63%	0.00%	40.00%
	TPR	16.70%	100%	83.30%	0.00%	33.30%
	TNR	99.90%	93.30%	99.00%	100%	99.80%

5   Discussion

5.1   GP's Capability in Channel Selection

The less channels used to fulfil the task, the better it is for the following reasons:

– It is directly translated into less computational cost.

- It is easier to analyse the program evolved.
- It gives insight into the time series pattern.

Selecting the relevant channels from multi-stream sensor data requires domain knowledge of the problem at hand. Instead of turning to a human expert, we delegate this task to GP. Our experiment result has shown that GP is selective in choosing which data channel to use for detecting each risky driving behaviours.

Table 7 shows the number of channels GP actually selected out of all the channels given. In all cases, we notice the GP-evolved program do not use all channels that it was given (trained on). The classifier for sudden stop is a good example. It uses only 4 channels when given 21 channels. It uses only 2 channels when given 4 channels. It uses only 1 channel when provided 9 channels.

Re-training GP on channels selected by the previous GP training proves to be beneficial. When trained on channels selected by GP trained on the original full set of 21 channels, GP only needs to use 2 channels to handle each detecting tasks. Still, it does not miss any unsafe driving event and report little false alarms.

Table 7. Number of channels GP selected out of the number of channels GP was given

GP Training	Harsh Acceleration	Sudden Stop	Swerving
GP-21-channels	12 out of 21	4 out of 21	5 out of 21
GP-selected-channels	2 out of 12	2 out of 4	2 out of 5
GP-9-channels	6 out of 9	1 out of 9	7 out of 9

5.2   Analysis of the GP-evolved Programs

The output of each GP classifiers is a numerical value. The program takes reading from sensor channel, do come calculation. Each evolved program comes with a threshold. The program output is then compared with this threshold. The discrimination threshold is chosen to be the closest point from the AUC curve to the optimum curve. If the output is greater than the threshold, the risky event is confirmed. If it is less than the threshold, the risky event is negative. Examples of GP-evolved classifiers are shown in Figure 1.

The lest number of channels selected belongs to GP-selected-channels. It only needs 2 channels for each classification task. For harsh acceleration, it is x-axis gravity and y-axis gravity. For sudden stop, it is y-axis user acceleration and z-axis user acceleration. For swerving, it is x-axis raw acceleration and y-axis raw rotation rate. The number of channels used is less than all other cases, except for GP-9-channels on sudden braking. It needs only 1 channel, which is z-axis raw acceleration.

The choice of this single channel for observing braking patterns indicates there is a pattern of this data channel associated with sudden stop behaviour. The use of z-axis raw acceleration to detect sudden stop makes sense in reality. When a harsh brake happens, the car is decelerating. Everything in the car will tend to stay in motion in accordance with Newtonian physics. There is also a second effect, which is the air imbalance between the front and the back of the car. This will push people and everything in the car backward. However, the first effect is bigger. Consequently, when a car is braking, anything goes forward in the car if the density of the object is greater than air. As a result, the phone will move forward because of its inertia. This causes the changes in the z-axis of raw accelerometer reading.

```
[1] Sudden Acceleration
(d/ (Window toper-STD tid-438 Channel_3) (Window toper-
DIF tid-1631 (dDiff Channel_2)))

[2] Sudden Braking
(Window toper-STD tid-39 (d- (d- (Multi_Channel soper-STD vid-6) 1) Channel_2))

[3] Swerving
(d+ (d* (d+ Channel_0 Channel_0) Channel_4) (Window toper-
STD tid-3116 Channel_0))
```

Fig. 1. Best Programs Evolved for Three Tasks with GP-selected Channels Data

6   Conclusion and Future Work

We have shown that Genetic Programming is capable of selecting relevant time series channel for detecting risky driving behaviour task. The problem is challenging due to the stochastic nature of different driving events, the need to find suitable feature set and select relevant data channels. We trained GP on the full set of 21 raw channels, channels that it previously selected and on 9 manually selected channels. We found that GP always choose less data channels than it is given. Still, its performance is consistently high. On all three tasks, GP achieves 100% true positive rate and low false alarms.

We tested 5 traditional classifiers on three versions of the same data set: the original version, the GP-selected channel version and the 9 manually selected channel version. We found the performance is not consistent and low in general. Most of the time, the classifier obtain high true negative rate but fail to detect instances of the rare class, which is the risky driving event.

Future work would be as follows. We would like to obtain a richer and more comprehensive data set for training and testing. We would also like to investigate alternatives to extract the knowledge learnt by GP for better feature selection and channel selection.

References

1. Clarke, D.D., Ward, P., Truman, W.: Voluntary risk taking and skill deficits in young driver accidents in the uk. Accident Analysis & Prevention 37(3), 523–529 (2005)
2. Johnston, L.D., O'Malley, P.M., Bachman, J.G., Schulenberg, J.E.: Monitoring the future national survey results on drug use, 1975-2010, p. 744. Institute for Social Research, The University of Michigan, Ann Arbor (2011)
3. Xie, F., Song, A., Salim, F., Bouguettaya, A., Sellis, T., Bradbrook, D.: Learning risky driver behaviours from multi-channel data streams using genetic programming. In: Cranefield, S., Nayak, A. (eds.) AI 2013. LNCS, vol. 8272, pp. 202–213. Springer, Heidelberg (2013)
4. Lee, J.-D., Li, J.-D., Liu, L.-C., Chen, C.-M.: A novel driving pattern recognition and status monitoring system. In: Chang, L.-W., Lie, W.-N. (eds.) PSIVT 2006. LNCS, vol. 4319, pp. 504–512. Springer, Heidelberg (2006)
5. Mitrovic, D.: Reliable method for driving events recognition. IEEE Transactions on Intelligent Transportation Systems 6(2), 198–205 (2005)
6. Imkamon, T., Saensom, P., Tangamchit, P., Pongpaibool, P.: Detection of hazardous driving behavior using fuzzy logic. In: 5th International Conference on Electrical Engineering/Electronics, Computer, Telecommunications and Information Technology, ECTI-CON 2008, vol. 2, pp. 657–660. IEEE (2008)
7. Johnson, D.A., Trivedi, M.M.: Driving style recognition using a smartphone as a sensor platform. In: 2011 14th International IEEE Conference on Intelligent Transportation Systems (ITSC), pp. 1609–1615. IEEE (2011)
8. Kaboudan, M.: Spatiotemporal forecasting of housing price by use of genetic programming. In: The 16th Annual Meeting of the Association of Global Business (2004)
9. Wagner, N., Michalewicz, Z.: An analysis of adaptive windowing for time series forecasting in dynamic environments: further tests of the dyfor gp model. In: Proceedings of the 10th Annual Conference on Genetic and Evolutionary Computation, pp. 1657–1664. ACM (2008)
10. Hetland, M.L., Sætrom, P.: Temporal rule discovery using genetic programming and specialized hardware. Applications and Science in Soft Computing 24, 87 (2004)
11. Xie, F., Song, A., Ciesielski, V.: Event detection in time series by genetic programming. In: 2012 IEEE Congress on Evolutionary Computation (CEC), pp. 1–8. IEEE (2012)
12. Ling, C.X., Huang, J., Zhang, H.: Auc: a statistically consistent and more discriminating measure than accuracy. In: IJCAI, vol. 3, pp. 519–524 (2003)

Variable Neighbourhood Iterated Improvement Search Algorithm for Attribute Reduction Problems

Yahya Z. Arajy, Salwani Abdullah, and Saif Kifah

Data Mining and Optimisation Research Group (DMO)
Centre for Artificial Intelligence Technology
Universiti Kebangsaan Malaysia, 43600 Bangi Selangor, Malaysia
{yahya.arajy,saifkc}@gmail.com, salwani@ukm.edu.my

Abstract. Attribute reduction is one of the main contributions in Rough Set Theory (RST) that tries to find all possible reducts by eliminating redundant attributes while maintaining the information of the problem in hand. In this paper, we propose a meta-heuristic approach called a Variable Neighbourhood Iterated Improvement Search (VNS-IIS) algorithm for attribute reduction. It is a combination of the variable neighbourhood search with the iterated search algorithm where two local search algorithms i.e. a random iterated local search and a sequential iterated local search algorithm are employed in a parallel strategy. In VNS-IIS, an improved solution will always be accepted. The proposed method has been tested on the 13 well-known datasets that are available in the UCI machine learning repository. Experimental results show that the VNS-IIS is able to obtain competitive results when compared with other approaches mentioned in the literature in terms of minimal reducts.

Keywords: Attribute Reduction, Variable Neighbourhood Search, Iterated Search.

1 Introduction

Rough Set Theory (RST) is a part of the most useful methods for data mining and knowledge discovery, machine learning and imprecise knowledge[1]. The main goals of the rough set are: induction of approximations of concepts and offers mathematical tools to discover hidden patterns in data. It can be used for feature selection, feature extraction, attribute reduction, decision rule generation and approximate reasoning [2, 3].

Attribute reduction is one of the most important subjects of RST. It is a process of finding an optimal subset from a system to effectively represent the giving dataset. It plays an important role in reducing the problem size for clustering and classification problems. According to the complexity of real life data, finding all minimal attribute reductions is considered as an NP-hard problem[4]. Over the past years, researches gave a great interest in an attribute reduction domain by applying meta-heuristic algorithms to find the optimal solution and show some successful signs, for instance: ant colony [5, 6], genetic algorithm and simulated annealing [7], scatter search[8], tabu search[9], composite neighbourhood structure [10], hyper-heuristic [11], and great

G. Dick et al. (Eds.): SEAL 2014, LNCS 8886, pp. 554–568, 2014.

deluge algorithm [12, 13]. For further reading, readers can find other approaches and surveys about rough set attribute reduction in [7, 14, 15].

Variable neighbourhood search algorithm (VNS) is a meta-heuristic algorithm for solving combinatorial optimization problems that has been developed by Mladenović and Hansen [16]. It is based on a strategy of changing neighbourhood structures systematically from a pre-set list during a local search process. This helps the VNS to explore neighbourhoods which are distant from the current solution and jump to a new solution if the new version has better quality. In this paper, a VNS-based algorithm, called a variable neighbourhood iterated improvement search algorithm (VNS-IIS) for attribute reduction is developed. VNS-IIS extends the basic idea of Hybrid Variable Neighbourhood Search algorithm for attribute reduction (HVNS) [17]. The experimental results have shown that HVNS is a promising approach. Nevertheless, VNS has a good share of the researcher's interest which it went through a rapid development, with several published works where it shows the ability to solve several complicated NP-hard problems [18]. An overview of the recent attributes reduction methods shows that most of the meta-heuristic approaches applied to this problem used a small number of neighbourhood structures. This motivates the investigation of using a large number of neighbourhood structures in this work to solve this problem, which can explore the search space differently during the search process to obtain a minimal reduct.

The paper is organised as follows: Section 2 presents the preliminaries of the research. In Section 3, the description of VNS-IIS algorithm is presented. Section 4 shows the experimental results, followed by the concluding remarks and future work in Section 5.

2   Preliminaries

This section emphasizes the main concept of rough set theory for attribute reduction.

2.1   Rough Set Theory (RST)

Rough set theory it considered as a mathematical approach to analyse the vagueness of information for an object [1]. Note that in this work, RST is used to measure a minimal reduct.

Table 1. Example of dataset

U/A	a	b	c	d
u0	0	1	0	0
u1	1	0	1	2
u2	0	1	0	0
u3	0	2	2	1
u4	0	2	1	2
u5	1	0	2	1
u6	1	0	2	1

Definition 1. Let an information system be $Is = (U, A)$, where U is a non-empty set of a finite object (the universe) and A is a non-empty finite set of attributes such that a: $U \rightarrow Va$ for every $a \in A$. V_a represents the domain of an attribute a [19]. Let C be the set of condition attributes and D is the set of decision attributes. Hence, $C \subset A$, $C \cup D = A$, and $C \cap D = \emptyset$.

Definition 2. The intersection of all equivalence relations in P for any P,A is denoted by $IND (P)$ and is called as an indiscernibility relation over P. If $(x,y) \in IND(P)$, then x and y are indiscernible by attributes from P. The $IND(P)$ relation can be defined as:

$$IND\ (P) = \{(\ x, y) \in U^2 \mid \forall\ a \in P\ a(x) = a(y)\} \tag{1}$$

Definition 3. Attribute reduction in rough set theory rely on two basic concepts the P-*lower* and P-*upper* approximations of X. Let $W \subseteq U$. W can be approximated using only the information contained within P:

$$\underline{P}W = \{x \mid [x]_p \subseteq W\} \tag{2}$$

$$\overline{P}W = \{x \mid [x]_p \cap W \neq \varphi\} \tag{3}$$

Using the datasets in Table 1, we have $U = \{u0, u1, u2, u3, u4, u5, u6\}$, $A = \{a, b, c, d\}$, $C = \{a, b, c\}$, $D = \{d\}$. The equivalence classes of the P-indiscernibility relation are denoted as $[x]_p$. For example, if $P = (a,b)$ and $Q = (d)$, then objects u0 and u2 are indiscernible, objects u1, u5 and u6 are indiscernible, as are u3 and u4, with respect to P. Then:

$$U/IND\ (P) = \{\{u0, u2\}, \{u3, u4\}, \{u1, u5, u6\}\}.$$
$$U/IND\ (Q) = \{\{u0, u2\}, \{u1, u4\}, \{u3, u5, u6\}\}.$$

By calculating the lower approximation as follows:
- If $W = \{u0, u2\}$ then $\underline{P}W = \{u0, u2\}$.
- If $W = \{u1, u4\}$ then $\underline{P}W = \emptyset$.
- If $W = \{u3, u5, u6\}$ then $\underline{P}W = \emptyset$.

Definition 4. Let P and Q be an equivalence relation over U, then the positive regions can be defined as:

$$POS_P(Q) = \bigcup_{W \in U/Q} \underline{P}W \tag{4}$$

The positive region contains all objects of U that can be classified as classes of U/Q using the information in attributes P. For example, let $P = (a,b)$ and $Q = (d)$, then:

$$POS_P(Q) = \bigcup \{\emptyset, \{u0, u2\}, \emptyset\ \} = \{u0, u2\}.$$

Definition 5. It is easily to notice that objects u0 and u2 are the only one can certainly be classified as belong to a class in attribute d, when considering attributes a and b. Discovering dependencies between attributes considered as one of the major issues in RST. The dependency degree is calculated as follows:

$$k = \gamma_P(Q) = \frac{|POS_P(Q)|}{|U|}$$

(5)

If $k = 1$, Q totally depends on P; if $0<k<1$, Q partially depends (in a degree of k) on P; and if $k = 0$ then Q does not depend on P. In the above example, the dependency degree of attribute $\{d\}$ from the attributes $\{a,b\}$ is calculated as:

$$k = \gamma_{\{a,b\}}(\{d\}) = \frac{|POS_{\{a,b\}}(\{d\})|}{|U|} = \frac{|\{u0, u2\}|}{|\{u0, u1, u2, u3, u4, u5, u6, u7\}|} = \frac{2}{7}$$

Definition 6. A reduct is defined as a subset of minimal Cardinality of the conditional attribute set C *such:* $\gamma R\ (D) - \gamma C\ (D)$

$$R = \{X : X \subseteq C, \gamma_{X(D)} = \gamma_{C(D)}\}$$

(6)

$$R_{min} = \{X : X \in R, \forall Y \in R, |X| \le |Y|\}$$

(7)

The Core is defined as an intersection of all the sets in R_{min}

$$Core(R) = \bigcap_{X \in R} X$$

(8)

The elements of the core are those attributes that not possible to omit without introducing more contradiction to the data set.

For example all possible subset of C can be calculated as follow:

$$\gamma_{\{b\}}(\{d\}) = \frac{|2|}{|7|} \quad \gamma_{\{a,b\}}(\{d\}) = \frac{|2|}{|7|} \quad \gamma_{\{c\}}(\{d\}) = 1$$

$$\gamma_{\{a\}}(\{d\}) = 0 \quad \gamma_{\{b,c\}}(\{d\}) = 1 \quad \gamma_{\{a,c\}}(\{d\}) = 1$$

So, the minimal reduct set for this example is:
$R_{min} = \{c\}$.

3   Variable Neighbourhood Iterated Improvement Search for Attribute Reduction (VNS-IIS)

VNS is an efficient meta-heuristic [16]. It has been employed in a wide variety of combinatorial optimisation problems. However, VNS, like other meta-heuristics, suffers from the slow convergence that brings about a high computational cost. Recently, many approaches have been tried to speed up the convergence of local search methods by developing a modified version of VNS algorithms. Most of these approaches hybridise local search methods with VNS to obtain more efficient methods with relatively has a faster convergence[20, 21]. Parallelisation strategy is also considered as one of the most considerable policies to overcome the problem of reaching a good solution within short or reasonable times [22]. Parallelisation helps in expanding the search space region for the local search. In this work, we attempt to utilise the idea of hybridising the variable neighbourhood search with two iterated local search algorithms that are employed in parallel.

This section illustrates the construction of initial solution and shows the evaluation of the solution quality. The basic components of VNS are discussed by presenting the combination of VNS with iterated search to deal with the attribute reduction problem.

3.1   Construction and Representation of the Solution

In this work, the construction of the initial solution is done by equalising the number of attributes $|N|$ in the original datasets to a one dimensional vector, A, with a binary representation by randomly assigning "1" or '0". If a cell A_i of A, (where $i= 1...N$), has the value "1" then the represented attribute is contained in the attribute subset. Otherwise, the cells with the value "0" are not contained in the attribute subset.

3.2   Quality Measurement and Acceptance Conditions

The solution quality is measured based on the dependency degree (i.e. Section 2.1, definition 5), denoted as γ. Given two solutions i.e. current solution, x, and trial solution, x'. The trial solution x' is accepted if there is an enhancement in the dependency degree (i.e. if $\gamma(x') > \gamma(x)$). If the dependency degree for both solutions are same (i.e. $\gamma(x') = \gamma(x)$), then the solution with the less number of the attribute will be accepted.

3.3   Neighbourhood Structures

The following neighbourhood structures are employed at the local search level:

- NS_1: Randomly add one attribute to the current solution.
- NS_2: Randomly add two attributes to the current solution.
- NS_3: Randomly add three attributes to the current solution.
- NS_4: Intelligently add one attribute to the current solution (i.e. the attribute that has the highest priority value).
- NS_5: Randomly remove one attribute from the current solution.
- NS_6: Randomly remove two attributes from the current solution.
- NS_7: Randomly remove 20% of the attributes from the current solution.
- NS_8: Randomly remove 10% of the attributes from the current solution.
- NS_9: Intelligently remove one attribute from the current solution (i.e. the attribute that has the lowest priority value).
- NS_{10}: Intelligently remove two attributes from the current solution (i.e. the attributes that have lowest priority values).

- NS_{11}: Randomly swap one attribute from the current solution with another from the original dataset (with respect to duplication avoidance).
- NS_{12}: Randomly swap two attributes from the current solution with others from the original dataset (with respect to duplication avoidance).
- NS_{13}: Intelligently swap one attribute with another (i.e. swap the attribute that has the lowest priority value from the current solution with one that has the highest priority value from the original dataset).
- NS_{14}: Mix swap one attribute with another (i.e. swap the one that has the lowest priority value from the current solution with a random attribute from the original dataset).

For a better understanding, the following illustration will explain the process of applying the intelligent selection mechanism. Let us assume that S1 is a set of all attributes from the original dataset. Let S2 be a set of attributes for the current solution, where:

S1 = {A1,A2,A3,A4,A5,A6,A7}.

S2 = {A1,A3,A4,A7}.

Let NS_4 be the selected neighbourhood structure to be applied on S2. Let P_{list} be the priority list of attributes, where each number in the list represents the total number of sets taken from calculating the intersections between objects (i.e. indiscernibility relationship in Section 2) for each individual attribute.

P_{List} = {2,3,1,4,2,1,1}, every number represents the priority value of each attribute from S1.

The highest priority attribute will be selected and added to the current solution. In this case, attribute **A2** from the original dataset is selected and added to S2, then update the current solution to be a new solution, where

New S2 = {A1,**A2**,A3,A4,A7}.

In the case of all attributes have the same number of sets, in other words, if they have the same priority number, then a random mechanism will be used to select the attributes.

3.4   Basic Variable Neighbourhood Search Algorithm (Basic-VNS)

The basic VNS method combines deterministic and stochastic changes of neighbourhood [16]. The basic procedure of VNS is presented in Algorithm 1. In the initial step, a set of neighbourhood structures (i.e. NS_1, NS_2, NS_3,..., NS_{13}) is defined as N_k (where $k=1,...,K$), and K is the total number of neighbourhood structures to be used in the search. Note that in the Basic-VNS, 13 neighbourhood structures are used as in Arajy and Abdullah (2010). Let $f(x)$ be the quality measurement (as presented in Section 3.2) of the solution x. The number of attributes and the dependency degree for the initial solution are calculated.

```
Algorithm 1: Basic-VNS

- INITIALIZATION:
Select the set of neighbourhood structures NS , k=1,…,K,
                                            k
Set initial solution as x;
Choose a termination condition
Repeat until termination condition is met
Set k← 1;
 Repeat until (k = K or Dependency degree =1);
  - SHAKING: Randomly generate a solution x' from the k
                                                        th
neighbourhood of
      x(x'∈NS (x));
            k
  - LOCAL SEARCH: Apply local search method on x' until local
optima x" is
      obtained;
  - MOVE OR NOT: If the f(x") is better than f( x) then
x←x";k←1;                   else k←k+1;

End.
```

The following steps explain the procedure of the Basic-VNS algorithm.

Step 1) Randomly generate an initial solution (x).

Step 2) Implementing shaking procedure with randomly initialising neighbouring solution (x') based on the current solution (x) from the kth neighbourhood.

Step 3) Local search visits all neighbourhood structures (consider x' as an input) to obtain the local optima (x'').

Step 4) Move or not the procedure is carried out for comparing the quality of new solution (x'') with the current solution (x). If there is an improvement in the quality of the solution, then replace (x) with (x'') ($x \leftarrow x''$). Then start the next iteration with the first neighbourhood structure. Otherwise, the algorithm employed the next neighbourhood structure from the list.

The process is repeated until the termination criterion is met. In this work, we set the termination criterion as a number of iterations or the dependency degree = 1.

3.5   Hybrid Variable Neighbourhood Search Algorithm for Attribute Reduction (HVNS)

The basic procedure of HVNS as presented by Arajy and Abdullah [17] as follows: The employed algorithm is divided into two phases. In the first phase, a basic variable neighbourhood search (Basic-VNS) algorithm is applied, where the number of neighbourhood structures is reduced from thirteen to six (i.e. NS_1, NS_2, NS_3, NS_4, NS_{11} and NS_{13}) that involve the 'add' and 'swap' neighbourhood structures (as presented in Section 3.3) to obtain a solution with a dependency degree = 1. While in the second phase, a random iterated search technique with nine neighbourhood structures is applied with an aim to further improve the quality of the solution in terms of the number of minimal reducts. The idea of the second phase is based on a random-restart hill climbing algorithm. The pseudo code for the iterated search is presented in Algorithm 2.

```
Algorithm 2: HVNS Second phase


Select the set of neighbourhood structures NSₖ, k=1,…,K, where
K= 9 (i.e. NS₅, NS₆, NS₇, NS₈, NS₉, NS₁₀, NS₁₁, NS₁₂ and NS₁₃);
Set initial solution as x;
Choose a termination condition
Repeat until termination condition is met
 - Randomly generate a solution x' from the kᵗʰ neighbourhood of
   x(x'∈NSₖ(x));
 - Move or not: If the f(x') is better than f(x)  then x←x';

End.
```

Algorithm 2 shows the process for the iterated search algorithm that is employed in the second phase of HVNS. Given a list of neighbourhood structures (i.e. NS_5, NS_6, NS_7, NS_8, NS_9, NS_{10}, NS_{11}, NS_{12} and NS_{13}) as NS_k (where $k=1,..,K$), where K is a maximum number of structures, the following steps will take place.

Step 1) Get the solution from the first phase and treated as an initial solution coded as x.

Step 2) Randomly generate a new solution x' based on the current solution x from the kth neighbourhood.

Step 3) Compare the quality of the new solution x' with the quality of the current solution x. If there is an improvement in the quality of the solution, then replace x with x' ($x \leftarrow x'$). Then the search is continued with another randomly selected neighbourhood structure from the list.

The process is repeated until the termination criterion is met (i.e. based on the number of iterations). The neighbourhood structures used in this phase are based on 'remove' and 'swap' operations only. The maximum number of iterations was set to 250, where 20 iterations are set for the first phase and the rest of iterations are set for the second phase. Note that, throughout our experiments, we are able to obtain solutions with a dependency degree = 1 in all the cases from the first phase.

3.6   Proposed Algorithm (VNS-IIS)

Parallelisation strategy is one of the most considerable policies to overcome the problem of reaching a good solution within short or reasonable times [22]. The employed strategy in this section consists in applying independent two search algorithms using a shared memory multiprocessor (SMP). SMP consists of using two or more central processing units (CPUs) within a single computer system. This method is considered as a parallelisation strategy that adapts several searches simultaneously to explore the search space, start from the same initial solution and select the best solution obtained by all searches at the end. It is being described by Crainic and Toulouse [23] as:

"Independent multi-search methods turn out to be effective, simply because of the sheer quantity of computing power they allow one to apply to a given problem"

This method was presented effectively by several papers, i.e. solving QAP using tabu search [24] and simulated annealing for graph partitioning problems [25].

The proposed approach presented here is an extension to HVNS as presented by Arajy and Abdullah [17] where the algorithm is divided into two phases. In the first phase, Basic-VNS is employed (as in Section 3.5). In the second phase, two iterated search algorithms that work in parallel are implemented i.e. a random iterated search (coded as RIS) and a sequential iterated search (coded as SIS). The idea of implementing the proposed method is to use two different iterative mechanisms that work in parallel then exchange the information (solutions in this case) between them during their work. Two identical lists of neighbourhood structures are created (i.e. NS_5, NS_6, NS_7, NS_9, NS_{10}, NS_{11}, NS_{12}, NS_{13} and NS_{14}) that only contain the 'remove' and 'swap' neighbourhood structures.

The RIS and SIS algorithms start with the same initial solution obtained from the first phase. For the RIS algorithm, a random neighbourhood structure to be employed at every iteration, whilst for the SIS algorithm the neighbourhood structures are selected based on a pre-determined sequence (i.e. NS_7, NS_{12}, NS_6, NS_{10}, NS_{11}, NS_{14}, NS_5, NS_{13}, NS_9) for each iteration. Note that from our preliminary experiments, the best sequence of neighbourhood structures is to order them by decreasing size. At each iteration, the better solution obtained from each iterated search later will be swapped with another iterated search algorithm with an aim to allow the algorithms to work on different regions of the solution space. At the end of the iteration, the final solution for RIS and SIS will be compared and the best solution among them will be chosen.

Figure 1 shows the flowchart of the presented description above. Again, the total number of iterations is set to 250, where 20 iterations are set for the first phase (based on our preliminary experiments) and the rest of the iterations are set for the second phase.

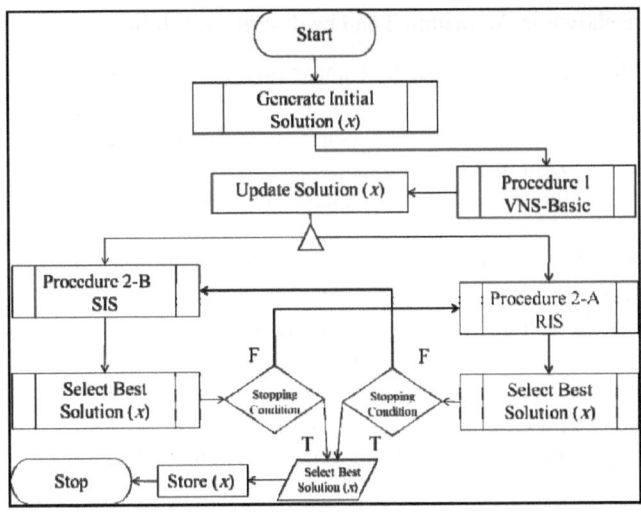

Fig. 1. VNS-IIS Flowchart

Algorithm 3: VNS-IIS second phase

Pre-define 1st List of neighbourhood structures $NS1_{k,}$ $k=1,...,K,$
where $K= 9$;
Pre-define 2nd List of neighbourhood structures $NS2_{k},$ $k=1,...,K,$
where $K= 9$;
Set initial solution as x; Set $x2=x1=x$;
Choose a termination condition
 Repeat
   For each Processer $Pro_{(c)}$ where C = number of processes;
     - **GENERATING:**
         $Pro_{(1)}$: Randomly generate a solution $x1'$ from the kth
neighbourhood of
             $x1(x1' \in NS1_{k}(x1))$;   //from the 1st List.
         $Pro_{(2)}$: Generate a solution $x2'$ from the next k^{th}
neighbourhood of
             $x2(x2' \in NS2_{k}(x2))$;   //from the 2nd List.
     - **MOVE OR NOT:**
         $Pro_{(1)}$: If the $f(x1')$ is better than $f(x1)$ then $x1 \leftarrow x1'$;
         $Pro_{(2)}$: If the $f(x2')$ is better than $f(x2)$ then $x2 \leftarrow x2'$;
     - **SWAPPING:**
         Swap the solutions between $(x1)$ and $(x2)$;
  Until termination condition is met
SELECT BEST SOLUTION:
   If the $f(x1)$ is better than $f(x2)$ then, $x \leftarrow x1$;
   else $x \leftarrow x2$;
End.

The procedure shown in Algorithm 3 can be described as follows.

Step 1) Get the solution from the first phase and treat it as an initial solution coded as x.

Step 2) Apply RIS on solution x by employing a random selected neighbourhood structure to obtain $x1'$.

Step 3) Get the solution from the first phase and treated as an initial solution coded as x.

Step 4) Apply SIS on solution x by selecting the neighbourhood structure from the pre-determined sequence to obtain $x2'$.

Step 5) Update the solution if there is an improvement i.e. $(x1 \leftarrow x1'; x2 \leftarrow x2')$.

Step 6) Exchange the solutions between the two algorithms $(x1 \leftrightarrow x2)$ to let the algorithms work on different search spaces.

Step 7) Repeat Step (1) to (4) until the termination criterion is met. Choose the best solution between $x1$ and $x2$ and return as a better solution.

4   Experimental Results

The algorithms were programmed in Java, and simulations were performed on the 2.1 GHz CPU with 3 GB of RAM. We use 13 standard benchmark UCI datasets that can be downloaded from http://archive.ics.uci.edu/ml/. For each dataset, the algorithm was run 20 times also the maximum number of iterations was set to 250. These values were set based on most of other researchers' work. So we can have an almost equivalent environment like others. Be noted that some researchers didn't obtain the 20 runs for some of the datasets (i.e. GenRSAR applied to M-of-N) for unexplained reasons.

Table 2. Comparison on minimal reducts between VNS-IIS and other approaches

Datasets	Att.	VNS-IIS	TSAR	HVNS	IS-CNS	NLGD-RSAR	GenRSAR	ACOAR	CHH-RSAR
M-of-N	13	6	6	6	6	6	$6^{(6)}7^{(12)}$	6	$6^{(11)}7^{(9)}$
Exactly	13	6	6	6	6	6	$6^{(10)}7^{(10)}$	6	$6^{(13)}7^{(7)}$
Exactly2	13	10	10	10	10	10	$10^{(9)}11^{(11)}$	10	10
Heart	13	6	6	6	6	9	$6^{(18)}7^{(2)}$	6	6
Vote	16	8	8	8	8	$10^{(14)}11^{(6)}$	$8^{(2)}9^{(18)}$	8	8
Credit	20	$8^{(10)}9^{(4)}10^{(6)}$	$8^{(13)}9^{(5)}10^{(2)}$	$8^{(7)}9^{(6)}10^{(7)}$	$8^{(10)}9^{(9)}10^{(1)}$	11	$10^{(6)}11^{(14)}$	$8^{(16)}9^{(4)}$	$8^{(10)}9^{(7)}10^{(3)}$
Mushroor	22	4	$4^{(17)}5^{(3)}$	4	4	4	$5^{(1)}6^{(5)}7^{(14)}$	4	4
LED	24	5	5	5	5	$7^{(15)}8^{(5)}$	$6^{(1)}7^{(3)}8^{(16)}$	5	5
Letters	25	8	$8^{(17)}9^{(3)}$	8	8	9	$8^{(8)}9^{(12)}$	8	8
Derm	34	6	$6^{(14)}7^{(6)}$	$6^{(16)}7^{(4)}$	$6^{(18)}7^{(2)}$	$11^{(17)}12^{(3)}$	$10^{(6)}11^{(14)}$	6	6
Derm2	34	$8^{(8)}9^{(11)}10^{(1)}$	$8^{(2)}9^{(14)}10^{(4)}$	$8^{(5)}9^{(12)}10^{(3)}$	$8^{(4)}9^{(16)}$	$11^{(15)}12^{(5)}$	$10^{(4)}11^{(16)}$	$8^{(4)}9^{(16)}$	$8^{(5)}9^{(5)}10^{(10)}$
WQ	38	$12^{(1)}13^{(6)}$ $14^{(12)}15^{(1)}$	$12^{(1)}13^{(13)}$ $14^{(6)}$	$12^{(3)}13^{(6)}14^{(}$ $^{8)}15^{(3)}$	$12^{(2)}13^{(8)}14^{(}$ $^{0)}$	$15^{(11)}16^{(9)}$	16	$12^{(1)}13^{(}$ $^{12)}14^{(4)}$	$12^{(13)}14^{(7)}$
Lung	56	4	$4^{(6)}5^{(13)}6^{(1)}$	$4^{(16)}5^{(4)}$	$4^{(17)}5^{(3)}$	4	$6^{(8)}7^{(12)}$	4	$4^{(10)}5^{(7)}6^{(3)}$

We carried out a comparison with state-of-the art approaches as well the originated version of this work in terms of minimal reducts. The first two columns in Table 2 show the dataset name and its original number of attributes (Att.). Rest of the entries in Table 2 represents the number of attributes in the minimal reducts obtained by each method. The superscripts in parentheses represent the number of runs that achieved the minimal reducts. The number of the attribute without superscripts means that the method able to obtain this number of attribute for all runs.

The comparison included two categories, a comparison with single solution-based and population-based approaches. Single solution-based approaches in comparison are: Hybrid Variable Neighbourhood Search algorithm (HVNS) [7], Tabu search (TSAR) [9], Intelligent selection composite neighbourhood structure (IS-CNS) [10], Hyper-heuristic (CHH-RSAR) [11], and Great deluge algorithm (NLGD-RSAR) [12] . Population-based approaches in comparison as provided are Genetic algorithm (GenRSAR) [7] and Ant colony optimisation (ACOAR) [26]. From Table 2, we can discern that VNS-IIS can obtain one best result on Derm2 dataset and ties on ten datasets with other best known results in the literature. Note that we did not obtain any worst results in this case.

In order to investigate the robustness of the proposed algorithm, it is required to do a further comparison with other methods from the literature. However, due to the lack of some information, especially on the quality of the solutions, complexity and computational cost make it hard to do an equitable comparison between methods. Therefore, we applied some statistical studies To represent the statistical significance of independent variables, we examined the approaches using T-Test method [27]. The results of *t-value* for VNS-IIS approach is reported in Table 3 with comparison to state of the art methods studied here. The purpose of this test is to assess whether the means of two groups are *statistically* different from each other. The quality of the produced solutions gets higher when the *t*-value is greater. Lower *t*-values indicate

Table 3. T-test comparison with state-of-the art approaches

	t-value / p-value VNS-IIS $_{(\alpha\,Level\,=\,0.05)}$						
Datasets	**TSAR**	**HVNS**	**ACOAR**	**GenRSAR**	**IS-CNS**	**NLGD-RSAR**	**CHH-RSAR**
M-of-N	-	-	-	6.66/1.1E-6	-	-	3.9/4.3E-4
Exactly	-	-	-	4.36/0.0001	-	-	2.2/0.002
Exactly2	-	-	-	4.82/5.9E-5	-	-	-
Heart	-	-	-	1.45/0.081	-	N/A	-
Vote	-	-	-	13.08/2.9E-11	-	21.8/3.09	-
Credit	-1.39/0.086	0.72/0.237	0.005	8.41/1.4E-9	-1.04/0.154	11/5.5E-10	-0.58/0.283
Mushroom	1.83/0.041	-	-	20.18/1.3E-14	-	-	-
LED	-	-	-	22.36/2.0E-15	-	22.64/1.64E-15	-
Letters	1.83/0.041	-	-	5.34/1.8E-5	-	N/A	-
Derm	2.85/0.005	2.18/0.021	-	44.71/5.1E-21	1.45/0.081	62.86/8.22E-24	-
Derm2	2.50/0.008	1.29/0.103	0.177	13.42/1.9E-15	0.93/0.177	0.936/0.177	2.60/0.006
WQ	-2.06/0.023	-0.38/0.35	0.001	15.67/1.2E-12	-1.17/0.125	11.14/2.32E-13	-3.58/5.2E5
Lung	6.1/3.6E-6	2.18/0.021	-	23.13/1.1E-15	1.83/0.041	-	3.9/4.8E-4

the low reliability, and it means that the two samples are not significantly different from each other. Table 3 also shows the exploratory test to check the probability of error (*p-value*) involved in accepting our research hypothesis about the existence of a difference. The *p-value* is compared with the actual significance level (α) of our test, where $\alpha = 0.05$. The general rule is that a small *p-value* (where $p < 0.05$) makes the quality of the solutions produced with higher evidence against the null hypothesis. A large *p-value* (where $p > 0.05$) means that the null hypothesis has no evidence against it. The sign "-" in Table 3 represents no difference between two groups compared.

The sum up from the previous comparison shows that VNS-IIS outperforms GenRSAR on all datasets and 3 datasets in comparison with ACOAR as population-based methods. Moreover it can produces better outcomes than IS-CNS, HVNS, CHH-RSAR , TSAR and NLGD-RSAR on 3, 5, 5, 5 and 8 datasets, respectively (when compared to single solution-based approaches). The performance of ACOAR and VNS-ILS is comparable where our approach is able to obtain better results on Derm2 dataset, while ACOAR outperforms our approach on Credit and WQ datasets. We believe that the hybridization of variable neighbourhood search and parallelising iterated search has the ability to find better results due to the aptitude of the algorithm to explore the search space differently when applying an altered type of neighbour-hood search, and later further improve the quality of the solution through an iterated search that acts as an exploitation mechanism.

5   Conclusion and Future Work

An attribute reduction method which is based on the variable neighbourhood search algorithm is proposed in this work. The overall goal is to examine the behaviour of a hybrid approach that combines the variable neighbourhood search with diverse iterated search algorithms with applying a parallel strategy. In the presented work, the iterated search algorithm works in parallel where it depends on two different mechanisms in choosing the neighbourhood structures making the search work into different regions of solution space, without accepting any worst solutions.

Preliminary comparisons indicate that the hybrid approach is better than a basic variable neighbourhood search alone. Numerical and statistical experiments on 13 well-known datasets demonstrate the strength of our work. Further comparison shows that our hybrid approach is competitive with other approaches in the literature and produced one new best solution. Our future work aims to examine different datasets from other machine learning repositories and consider other quality measurement tools to enhance the accuracy of the results. We believe that improving the solutions can be done by further enhancing the neighbourhood structures with applying advanced method of selection.

References

1. Pawlak, Z.: Rough Sets. In: International Joint Conference on Information Sciences, pp. 341–356 (1982)
2. Jensen, R., Shen, Q.: Fuzzy-rough sets for descriptive dimensionality reduction. In: Proceedings of the 2002 IEEE International Conference on FUZZ-IEEE 2002, vol. 1, pp. 29–34 (2002)
3. Pawlak, Z.: Rough Sets: Theoretical Aspects of Reasoning about Data. Springer (1991)
4. Engle, M.L., Burks, C.: Artificially Generated Data Sets for Testing DNA Sequence Assembly Algorithms. Genomics 16, 286–288 (1993)
5. Jensen, R., Shen, Q.: Finding Rough Set Reducts with Ant Colony Optimization. In: Proceedings of the 2003 UK Workshop on Computational Intelligence, vol. 1, pp. 15–22 (2003)
6. Xu, Z., Zhao, H., Min, F., Zhu, W.: Ant Colony Optimization with Three Stages for Independent Test Cost Attribute Reduction. Mathematical Problems in Engineering (2013)
7. Jensen, R., Shen, Q.: Semantics-preserving dimensionality reduction: rough and fuzzy-rough-based approaches. IEEE Transactions on Knowledge and Data Engineering 16, 1457–1471 (2004)
8. Wang, J., Hedar, A.R., Guihuan, Z., Shouyang, W.: Scatter Search for Rough Set Attribute Reduction. In: Computational Sciences and Optimization, pp. 531–535 (2009)
9. Hedar, A.-R., Wang, J., Fukushima, M.: Tabu search for attribute reduction in rough set theory. Soft Computing - A Fusion of Foundations, Methodologies and Applications 12, 909–918 (2006)
10. Jihad, S.K., Abdullah, S.: Investigating composite neighbourhood structure for attribute reduction in rough set theory, pp. 1183–1188 (2010)
11. Abdullah, S., Sabar, N.R., Nazri, M.Z.A., Turabieh, H., McCollum, B.: A constructive hyper-heuristics for rough set attribute reduction, pp. 1032–1035 (2010)
12. Abdullah, S., Jaddi, N.S.: Great Deluge Algorithm for Rough Set Attribute Reduction. In: Zhang, Y., Cuzzocrea, A., Ma, J., Chung, K.-i., Arslan, T., Song, X. (eds.) DTA and BSBT 2010. CCIS, vol. 118, pp. 189–197. Springer, Heidelberg (2010)
13. Jaddi, N.S., Abdullah, S.: Nonlinear great deluge algorithm for rough set attribute reduction. Journal of Information Science and Engineering 29, 49–62 (2013)
14. Zhang, W., Qiu, G., Wu, W.: A general approach to attribute reduction in rough set theory. Science in China Series F: Information Sciences 50, 188–197 (2007)
15. Düntsch, I., Gediga, G.: Uncertainty measures of rough set prediction. Artificial Intelligenc 106, 109–137 (1998)
16. Mladenović, N., Hansen, P.: Variable neighborhood search. Computers and Operations Research 24, 1097–1100 (1997)
17. Arajy, Y.Z., Abdullah, S.: Hybrid variable neighbourhood search algorithm for attribute reduction in Rough Set Theory, pp. 1015–1020 (2010)
18. Hansen, P., Mladenović, N., Moreno Pérez, J.: Variable neighbourhood search: methods and applications. Annals of Operations Research 175, 367–407 (2010)
19. Wang, Y.: Research on Optimization Algorithm for Attribute Reduction of Decision System. In: Second International Conference on Genetic and Evolutionary Computing, WGEC 2008, pp. 315–318. IEEE (2008)
20. Hansen, P., Mladenović, N., Moreno Pérez, J.: Variable neighbourhood search: methods and applications. 4OR: A Quarterly Journal of Operations Research 6, 319–360 (2008)
21. Han, H., Ye, J., Lv, Q.: A VNS-ANT Algorithm to QAP. In: Third International Conference on Natural Computation, vol. 3, pp. 426–430 (2007)

22. Yazdani, M., Amiri, M., Zandieh, M.: Flexible job-shop scheduling with parallel variable neighborhood search algorithm. Expert Systems with Applications 37, 678–687 (2010)
23. Crainic, T., Toulouse, M.: Parallel Meta-heuristics. In: Handbook of Metaheuristics, pp. 497–541 (2010)
24. Taillard, E.: Parallel taboo search techniques for the job shop scheduling problem. ORSA Journal on Computing 6, 108–108 (1994)
25. Banos, R., Gil, C., Ortega, J., Montoya, F.: A parallel multilevel metaheuristic for graph partitioning. Journal of Heuristics 10, 315–336 (2004)
26. Liangjun, K., Zuren, F., Zhigang, R.: An efficient ant colony optimization approach to attribute reduction in rough set theory. Pattern Recognition Letters 29, 1351–1357 (2008)
27. Montgomery, D.C., Runger, G.C.: Applied Statistics and Probability for Engineers, 5th Edition Binder Ready Version. Wiley (2010)

PSO and Statistical Clustering for Feature Selection: A New Representation

Hoai Bach Nguyen[1], Bing Xue[1], Ivy Liu[2], and Mengjie Zhang[1]

[1] School of Engineering and Computer Science
[2] School of Mathematics, Statistics and Operations Research
Victoria University of Wellington, PO Box 600, Wellington 6140, New Zealand
{nguyenhoai2,Bing.Xue,Mengjie.Zhang}@ecs.vuw.ac.nz,
Ivy.Liu@msor.vuw.ac.nz

Abstract. Classification tasks often involve a large number of features, where irrelevant or redundant features may reduce the classification performance. Such tasks typically requires a feature selection process to choose a small subset of relevant features for classification. This paper proposes a new representation in particle swarm optimisation (PSO) to utilise statistical clustering information to solve feature selection problems. The proposed algorithm is examined and compared with two conventional feature selection algorithms and two existing PSO based algorithms on eight benchmark datasets of varying difficulty. The experimental results show that the proposed algorithm can be successfully used for feature selection to considerably reduce the number of features and achieve similar or significantly higher classification accuracy than using all features. It achieves significantly better classification accuracy than one conventional method although the number of features is larger. Compared with the other conventional method and the two PSO methods, the proposed algorithm achieves better performance in terms of both the classification performance and the number of features.

Keywords: Particle swarm optimisation, Feature selection, Classification, Representation.

1 Introduction

In recent years, with the advances of data collection techniques, machine learning and data mining tasks such as classification often include a large number of features/variables. This causes the problem of "the curse of dimensionality" and leads to many issues, e.g. learning/classification algorithms fail to achieve satisfactory accuracy, the classification process is time-consuming, and the trained classifier is too complicated to understand/interpret. Feature selection can address these issues by removing irrelevant/redundant features and selecting only a small subset of relevant features for classification [8].

Feature selection is a challenging task due to the *large search space* and *feature interaction* problems. The size of the search space is 2^n for a dataset with n features [8]. Existing feature selection algorithms, such as greedy search based algorithms [11], suffer from stagnation in local optimal and/or high computational cost. Therefore, an efficient

G. Dick et al. (Eds.): SEAL 2014, LNCS 8886, pp. 569–581, 2014.

global search technique is needed to address feature selection problems. Evolutionary computation (EC) techniques are a group of powerful "global" search algorithms and have been successfully applied to a variety of fields [9]. Particle swarm optimisation (PSO) [13, 19] is an EC technique based on social intelligence, which has fewer parameters and is computationally less expensive than other EC techniques, such as genetic programming (GP) and genetic algorithms (GAs). PSO has been recently used to address feature selection problems and shown a certain level of success [27].

Feature interaction is a common and complex problem in classification tasks [8]. Because of feature interaction, an individually relevant feature may become less useful or redundant when combined with other features. On the other hand, a weakly relevant feature may become highly useful when used together with other features. In an "optimal" subset, features are expected to be complementary to each other and can work together to increase the classification performance. Therefore, during the feature selection process, the removal or addition of features needs to consider the appearance or absence of other features, which increases the difficulty of feature selection tasks. Finding a way to cope with feature interaction problems is expected to increase the performance of a feature selection algorithm. Meanwhile, feature interaction is also an important issue being considered in statistical data analysis. We generalise the statistical clustering method [15, 17] by taking feature interaction into account to group relatively homogeneous features into clusters. Intuitively, these ideas could be useful to address feature interaction problems in feature selection, but this has not been seriously investigated. The main challenge is how to incorporate the statistical clustering information in the feature selection process.

1.1 Goals

The overall goal of this paper is to develop a new representation scheme to incorporate the statistical clustering information in PSO for feature selection. To achieve this goal, a statistical clustering method as a preprocessing step is performed on the training set to group features into different clusters. A new representation scheme is developed to utilise such statistical clustering information to improve the performance of PSO for feature selection. A new algorithm using the new representation is then developed and compared with two existing PSO based feature selection algorithms and two conventional algorithms on eight datasets with different numbers of features, classes and instances. Specifically, we will investigate:

- whether the new algorithm can be used to reduce the number of features and increase the classification performance,
- whether the new algorithm can utilise the statistical clustering information to achieve better performance than the two existing PSO based feature selection algorithms, and
- whether the new algorithm can achieve better performance than the two conventional feature selection algorithms.

2 Background

2.1 Particle Swarm Optimisation (PSO)

Particle swarm optimisation (PSO) [13, 19] is an evolutionary computation method, which is inspired by social behaviours such as birds flocking and fish schooling. In PSO, candidate solutions are represented by a population or a swarm of particles. In order to find the optimal solutions, each particle moves around the search space by updating its position as well as its velocity. Particularly, the current position of particle i is represented by a vector $x_i = (x_{i1}, x_{i2}, \ldots, x_{iD})$, where D is the dimensionality of the search space. These positions are updated by using another vector, called velocity $v_i = (v_{i1}, v_{i2}, \ldots, v_{iD})$. During the search process, each particle maintains a record of the position of its previous best performance, called *pbest*. The best position of its neighbours is also recoreded, which is called *gbest*. The position and velocity of each particle are updated according to the following equations:

$$v_{id}^{t+1} = w * v_{id}^t + c_1 * r_{i1} * (p_{id} - x_{id}^t) + c_2 * r_{i2} * (p_{gd} - x_{id}^t) \tag{1}$$

$$x_{id}^{t+1} = x_{id}^t + v_{id}^{t+1} \tag{2}$$

where t denotes the tth iteration in the search process, d is then d^{th} dimension in the search space, i is the index of particle, w is inertia weight balancing the global and local search abilities, c_1 and c_2 are acceleration constants, r_{i1} and r_{i2} are random values uniformly distributed in [0,1], p_{id} and p_{gd} represent the position value of *pbest* and *gbest* in the dth dimension, respectively.

2.2 Related Work on Feature Selection

Existing feature selection algorithms can be generally classified into two categories, filter approaches and wrapper approaches [8, 28]. Their main difference is whether a classification/learning algorithm is used during the feature selection process. A wrapper algorithm typically includes a classification algorithm to measure the classification performance of the selected features to evaluate the goodness of the selected features. Filter approaches are independent of any classification algorithm. Filter approaches are argued to be computationally cheaper and more general than wrappers, but wrapper approaches can usually achieve better classification performance than filters due to the interaction between the selected features and the classification algorithm. This work focuses mainly on wrapper feature selection. In this section, typical wrapper feature selection algorithms and the use of statistics in feature selection are briefly reviewed.

Traditional Feature Selection Methods. Sequential forward selection (SFS) [22] and sequential backward selection (SBS) [14] are two commonly used wrapper feature selection algorithms. Both of them use a greedy hill-climbing search strategy to search for the optimal feature subset. However, both SFS and SBS suffer from the so-called nesting effect, which means that once a feature is selected (discarded) it cannot be discarded (selected) later. Therefore, both SFS and SBS are easily trapped in local optima. In addition, both SFS and SBS require long computational time when the number of features is large. In order to avoid nesting effect, Stearns [20] proposed a "plus-l-take away-r" method in which SFS was applied l times forward and then SBS was applied for r back tracking steps. However, determining the best values of (l, r) is a challenging task.

Later, Pudil et al. [18] proposed two floating selection methods, sequential backward floating selection (SBFS) and sequential forward floating selection (SFFS) to automatically determine the values of (l, r). In addition, the values of (l, r) in SBFS and SFFS that denotes the number of forward and backtracking steps are dynamically controlled instead of being fixed in the "plus-l-take away-r" method. Although the floating methods are claimed to be at least as good as the best sequential method, they are still likely to become trapped in a local optima even the criterion function is monotonic and the scale of the problem is small. Meanwhile, based on the best-first algorithm and SFFS, Gutlein et al. [11] proposed a linear forward selection (LFS) in which the number of features considered in each step is restricted. Experiments show that LFS improves the computational efficiency of sequential forward methods while maintaining comparable accuracy of the selected feature subset. However, LFS starts with ranking all the individual features without considering the presence or absence of some other features, which in turn limits the performance of the LFS algorithm in problems where there are interactions between features.

EC Approaches to Feature Selection. EC algorithms have been applied to feature selection problems, such as PSO, GAs [31], GP [16], ant colony optimisation (ACO) [12] and differential evolution (DE) [1]. Zhu et al. [31] proposed a feature selection method using a memetic algorithm that is a combination of local search and GA. Experiments show that this algorithm outperforms GA alone and other algorithms. Neshatian and Zhang [16] proposed a GP relevance measure (GPRM) to evaluate and rank feature subsets in binary classification tasks. Experiments show that the proposed method detected subsets of relevant features in different situations, where other methods had difficulties. Based on ACO, Kanan and Faez [12] developed a wrapper feature selection algorithm, which outperforms GA and other ACO based algorithms on a face detection dataset, but its performance has not been tested on other problems. Al-Ani et al. [1] also proposed a DE based feature selection method, where features are distributed to a set of wheels and DE is employed to select features from each wheel. This algorithm can significantly reduce the number of features and improve the classification performance.

Recently, BPSO has been applied to feature selection problems. Yang et al. [30] proposed two BPSO based wrapper feature selection approaches based on two inertia weight setting methods. The results show that the two algorithms can outperform SFS, SFFS, sequential GA and different hybrid GAs. Fdhila et al. [10] applied a multi-swarm PSO algorithm to solve feature selection problems. However, the computational cost of the proposed algorithm is high because it involves parallel evolutionary processes and multiple sub-swarms with a relative large number of particles. Xue et al. [27] proposed a PSO based two-stage feature selection algorithm to optimise the classification performance in the first stage and consider the number of features in the second stage. Chuang et al. [7] applied the so-called catfish effect to PSO for feature selection, which is to introduce new particles into the swarm by re-initialising the worst particles when *gbest* has not improved for a number of iterations. The authors claimed that the introduced

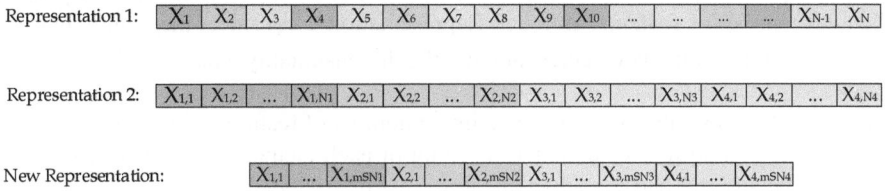

Fig. 1. Example of N features that are grouped into 4 clusters with N_1, N_2, N_3 and N_4 features, respectively, then $N = N_1 + N_2 + N_3 + N_4$. mSN_j is the predefined maximum number of features selected from cluster j and $mSN_1 < N_1, ..., mSN_4 < N_4$.

catfish particles could help PSO avoid premature convergence and lead to better results than sequential GA, SFS, SFFS and other methods. Xue et al. [29] developed new initialisation and *pbest* and *gbest* updating mechanisms in PSO for feature selection, which can increase the classification accuracy and reduce both the number of features and the computational time. Other PSO based feature selection methods can be found from [4–6, 24–26, 29].

Many statistical methods can be used to reduce the dimensionality of a dataset, such as principal component analysis, linear discriminant analysis, or canonical correlation analysis [3], but most of them are not feature selection approaches because they create new features. Clustering analysis is an important topic in statistics which aims to group features/variables to a number of clusters. We use the statistical clustering method [15, 17] to find relatively homogeneous feature groups by taking feature interactions into account. Therefore, the statistical grouping information could be used to develop a good feature selection algorithm.

3   Proposed Algorithm

In this section, a new representation scheme is proposed in PSO for feature selection to utilise the statistical clustering information to reduce the number of features selected and increase the classification performance. A newly developed clustering method based on statistical models proposed by Pledger and Arnold [17] and Matechou et. al. [15] is used to group features into different clusters. Features in the same cluster are considered similar and features in different clusters are dissimilar to each other. The technical detail of statistical clustering methods is not described here due to the page limit and the scope of this paper.

Fig. 1 shows three different types of representations, where a dataset with N features which can be grouped into 4 clusters is used as an example. N_1, N_2, N_3 and N_4 show the numbers of features in the 4 clusters, respectively. Representation 1 shows the traditional way of using PSO for features selection without considering the feature clustering information. Representation 2 and the proposed new representation consider the feature clustering information. Representation 2 is different from Representation 1 by putting features in the same cluster together. The new representation is different from Representations 1 and 2 in two main aspects. The first is the dimensionality of the particles (search space). In Representations 1 and 2, the dimensionality equals to the total number of features, although Representation 2 considers the feature clustering information. In the new representation, the dimensionality equals to $\sum\limits_{1 \leq j \leq 4} mSN_j$, where mSN_j shows the predefined maximum number of features selected from the jth cluster. The second difference is the meaning of each element in the position vector. In Representations 1 and 2, each element (e.g. x_i or $x_{j,k}$) determines whether the corresponding feature is selected or not. In the new representation, each element shows which feature is selected from the corresponding cluster. Therefore, in this new representation, two important tasks are how to determine the value of mSN for each cluster and how to determine which features are selected from a cluster. They will be described as follows.

3.1   How to Determine mSN_j

Since the features from the same cluster are similar features, a small proportion of these features can be used as the representatives of this cluster. However, it is difficult to determine how many features should be selected from each cluster. Selecting a large number of features may contain redundant information while selecting a small number of features may deteriorate the classification performance. Therefore, in the new representation, we propose the use of mSN_j, which means the maximum number of features selected from the jth cluster, to limit the number of features selected. The algorithm is expected to search for a feature subset which contains fewer than mSN_j features from cluster j, but can achieve better performance than using all features in cluster j. Since the sizes of clusters are usually different, the value of mSN_j should vary in different clusters.

$$mSN_j = \sqrt{N_j} \qquad (3)$$

Fig. 2 compares three different ways to determine mSN_j, which are a square root function of N_j shown as Eq. 3, a constant value, and a linear function of N_j. As can be seen from the figure, Eq. 3 allows selecting more features from a cluster that contains a larger number of features, which cannot be done by the constant function. On the other hand, Eq. 3 is preferred over the linear scaling function, since it leads to a smaller number of selected features from large feature clusters, which is more likely to include redundant features. The smaller mSN_j in Eq. 3 may reduce the chance of selecting those redundant features. Therefore, in this work, Eq. 3 is used to determine the value of mSN_j.

3.2   How to Select Features

In traditional representation, the position value determine whether a feature is selected or not, which is usually determined by a threshold. If the position value is larger than the threshold, the corresponding feature is selected. Otherwise, it is not selected. In the new representation, the position value in a dimension determines which feature is

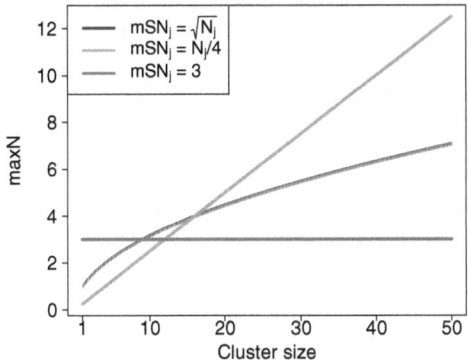

Fig. 2. Three different ways of determining mSN_j

selected from a certain cluster. To achieve this, the position value is limited to [0,1]. For the dimensions corresponding to the jth cluster, [0, 1] is equally divided into ($N_j +$ 1) intervals, where N_j is the total number of features in the jth cluster. Each interval corresponds to one feature in the cluster, which ensures that features in the same cluster has the equal chance to be selected. A feature is selected if the position value falls into its corresponding interval. There are ($N_j + 1$) intervals rather than N_j intervals because a virtual feature, called "Null" feature, is introduced to each cluster. The "Null" feature allows the selection of zero feature from a cluster if all features in that cluster are irrelevant or redundant.

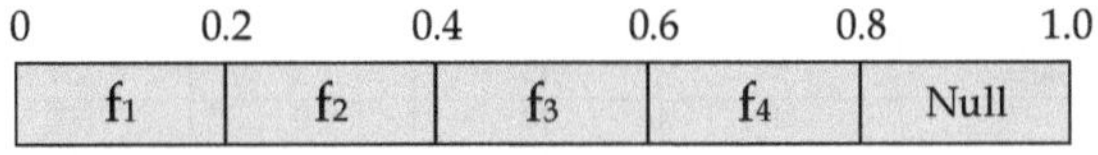

0	0.2	0.4	0.6	0.8	1.0
f_1	f_2	f_3	f_4	Null	

Fig. 3. Intervals for selecting features (Not PSO positions)

Fig. 3 takes a cluster with four features (f_1, f_2, f_3, f_4) showing the intervals for selecting features. As can been seen in Fig. 3, the interval [0,1] is further divided into five intervals, where four of them corresponds to the four features while the last interval corresponds to the "Null" feature, i.e. no feature is selected. Suppose that its $mSN_1 = 2$ and the position values are $\{x_{1,1} = 0.5, x_{1,2} = 0.96\}$. As $x_{1,1} \in [0.4, 0.6]$, which is the interval of Feature f_3, f_3 will be selected. Similarly, $x_{1,2} \in [0.8, 1.0]$ that belongs to Null feature, which means that no feature is selected. So the values are interpreted as selecting only feature f_3 from the cluster. Eq. 4 shows a general case of how to determine which feature or no feature is selected from cluster j, where x is the position value in a dimension.

$$Feature = \begin{cases} f_k, \text{if } x \in [\frac{k-1}{N_j+1}, \frac{k}{N_j+1}], \text{where } k \in [1, N_j] \\ \text{Null feature, if } [\frac{N_j}{N_j+1}, 1] \end{cases} \quad (4)$$

Algorithm 1. Pseudo-code of PSOR

---

begin
  indexing features in each cluster;
  define mSN for each cluster according to Eq. 3;
  randomly initialise the position and velocity of each particle;
  while *Maximum iterations is not reached* **do**
    Collect the features selected by each particle;
    evaluate the fitness of each particle according to its classification accuracy;
    for $i = 1$ *to Population size* **do**
      update *pbest* and *gbest* of particle i;

    for $i = 1$ *to Population size* **do**
      update v_i of particle i according to Eq. 1;
      update x_i of particle i according to Eq. 2;

  calculate the training and testing classification accuracy of the selected feature
  subset on the test set;
  return the position of *gbest*, the training and testing classification accuracies;

3.3 Pseudo-code of the Algorithm

By using the proposed representation, a new feature selection algorithm is proposed, which is named PSOR. The pseudo-code of PSOR is shown in Algorithm 1. The fitness function of PSOR is to maximise the classification accuracy of the selected features.

Table 1. Datasets

Dataset	NO. of features	NO. of clusters	NO. of classes	No of instances
Wine	13	6	3	178
Vehicle	18	6	4	846
Ionosphere	34	11	2	351
Sonar	60	12	2	208
Musk1	166	14	2	476
Arrhythmia	279	15	16	452
Madelon	500	11	2	4400
Multiple Features	649	15	10	2000

4   Experimental Design

To examine the performance of the proposed algorithm PSOR, two traditional feature selection methods, which are linear forward selection (LFS) [11] and greedy stepwise backward selection (GSBS), and two existing PSO based feature selection algorithms (PSOFS [27] and PSO42 [29]) are used for comparison purposes in the experiments. LFS and GSBS were derived from two typical feature selection algorithms, i.e. sequential forward selection (SFS) and sequential backward selection (SBS), respectively. LFS [11] restricts the number of features that are considered in each step of the forward selection. The greedy stepwise feature selection algorithm implemented in Weka [23] can move either forward or backward. Given that LFS performs a forward selection, a backward search is chosen in greedy stepwise search to form a greedy stepwise backward

Table 2. Experimental Results

Dataset	Method	Ave-Size	Best	Ave-Test-Acc	Std-Test-Acc	T
Wine	All	13	76.54			-
	PSOFS	7.93	98.77	95.6	1.7953	-
	PSO42	6.73	98.77	94.86	1.8628	-
	PSOR	4.75	100	96.70	3.10	
Vehicle	All	18	83.86			-
	PSOFS	9.5	87.01	85.03	0.8899	=
	PSO42	10.33	87.01	85.44	0.8372	+
	PSOR	5.87	86.22	84.72	0.8720	
Ionosphere	All	34	83.81			-
	PSOFS	12.47	93.33	88.41	2.3079	=
	PSO42	3.13	91.43	86.69	1.6444	-
	PSOR	9.7	91.43	88.63	1.6765	
Sonar	All	60	76.19			-
	PSOFS	26.1	84.13	77.3	3.5765	-
	PSO42	11.23	84.13	77.94	3.2104	=
	PSOR	14.33	84.13	78.94	4.0185	
Musk1	All	166	83.92			=
	PSOFS	85.93	88.81	84.61	2.0568	=
	PSO42	77.3	89.51	84.87	2.7042	=
	PSOR	35.03	90.21	83.12	3.4196	
Arrhythmia	All	279	94.46			-
	PSOFS	118.73	95.14	94.56	0.3517	=
	PSO42	69.77	95.59	94.77	0.4495	-
	PSOR	44.17	95.59	94.96	0.38	
Madelon	All	500	70.9			-
	PSOFS	259.07	78.97	76.35	1.0909	-
	PSO42	206.57	84.23	78.81	3.1171	-
	PSOR	54.39	85.13	83.40	2.0368	
Multiple features	All	649	98.63			-
	PSOFS	297.07	99.2	99.0	0.0934	+
	PSO42	314.5	99.2	99.0	0.0935	+
	PSOR	51.07	99.23	98.84	0.1751	

selection (GSBS). The algorithm PSOFS [27] selects features by using continuous PSO. The other PSO based algorithm, PSO42 [29], introduced a new initialisation strategy and $pbest$ and $gbest$ updating mechanism.

Eight datasets (Table 1) chosen from the UCI machine learning repository [2] are used in the experiments. These datasets have a different number of fetures, classes and instances. For each dataset, all instaces are randomly divided into a training set and a test set, which contains 70% and 30% of the instances, respectively. Up to 500 training instances are used in the statistical clustering method to group features into different clusters, where the number of clusters are listed in the second column in Table 1. In the experiments, the classification/learning algorithm is K-nearest neighbour (KNN) where K = 5. The parameters of PSO are set as follows [21]: w =0.7298, $c_1 = c_2 =$ 1.49618, $v_{max} = 6.0$, population size is 30, the maximum number of iterations is 100. The fully connected topology is used. All the PSO based algorithms have been run for 30 independent times on each dataset. A statistical significance test, Wilcoxon signed-rank test, is performed to compare the classification accuracies of different algorithms. The significance level was set as 0.05.

5 Experimental Results

Table 2 shows the experimental results of the PSO based algorithms, where "All" means that all the available features are used for classification. "Ave-size" shows the average number of selected features over the 30 runs. "Best", "Ave-Test-Acc", "Std-Test-Acc" illustrate the best, average and standard deviation of the testing accuracies over the 30 independent runs. T shows the results of the statistical significance tests between the accuracy of PSOR and other algorithms. "+" or "-" means that the algorithm achieved significantly better or worse classification performance than PSOR (the more "-", the better PSOR is). "=" means there is no significant difference between them.

Table 3. Results of GSBS and LFS

Method	Wine		Vehicle		Ionosphere		Sonar	
	# Features	Accuracy(%)	# Features	Accuracy(%)	# Features	Accuracy(%)	# Features	Accuracy(%)
GSBS	8	85.19	16	75.79	30	78.1	48	68.25
LFS	7	74.07	9	83.07	4	86.67	3	77.78

Method	Musk1		Arrhythmia		Madelon		Multiple Features	
	# Features	Accuracy(%)	# Features	Accuracy(%)	# Features	Accuracy(%)	# Features	Accuracy(%)
GSBS	122	76.22	130	93.55	489	51.28	-	-
LFS	10	85.31	11	94.46	7	64.62	18	99.0

From Table 2, it can be observed that the number of features selected by PSOR is significantly smaller than the total number of features, but using the selected features only, the 5NN classification algorithm achieved significantly better or similar classification accuracy. For example, on the Multiple Features dataset, PSOR selected on average 51 features from the original 649 features, but significantly increased the classification accuracy. The results suggest that PSOR can be successfully used for feature selection to reduce the dimensionality of the data and significantly increase the classification performance over using all features.

Comparing PSOR with PSOFS, the feature subsets selected by PSOR are smaller than that of PSOFS on all the eight datasets. In terms of the classification performance, PSOR achieved similar or significantly better classification accuracy than PSOFS on seven of the eight datasets. Comparing PSOR with PSO42, it can be observed that PSOR selected smaller feature subsets and achieved similar and significantly better classification performance than PSO42 on six of the eight datasets. The results suggest that PSOR using the new representation can effectively utilising the statistical clustering information to improve the classification performance over PSOFS and PSO42 and further reduce the number of features.

5.1 Further Comparisons with Traditional Methods

The results of LFS and GSBS are shown in Table 3. Since LFS and GSBS are deterministic algorithms, each of them produces only a single solution on each dataset. Since the experiment of using GSBS on the Multiple Features dataset cannot finish within two days, the results are not listed in the table.

Comparing the results of PSOR in Table 2 with the results in Table 3, it can be seen that LFS selected a smaller number of features than PSOR, but achieved significantly worse classification accuracy than PSOR. PSOR outperformed GSBS in terms of both the number of features and the classification performance on all datasets. The results show that PSOR, which is based on PSO and the feature clustering information, can better explore the solution space to obtain better feature subsets than LFS and GSBS.

6   Conclusions and Future Work

The goal of this paper was to develop a new approach to using the statistical clustering information in PSO for feature selection. The goal was successfully achieved by developing a new representation scheme in PSO. By using the new representation, the dimensionality of the search space is reduced over the traditional representation scheme and the statistical clustering information can be incorporated in the feature selection process. We have conducted the experiments to compare the new algorithm with two conventional methods and two existing PSO algorithms without using statistical clustering information on eight datasets of varying difficulty. The results show that the proposed algorithm can effectively utilise the statistical clustering information in PSO for feature selection, which results in smaller feature subsets and better classification accuracy than the existing methods.

In future work, new search mechanisms will be investigated in PSO and statistical clustering for feature selection to further increase the classification accuracy and reduce the number of features. Meanwhile, it will be interesting to split the data multiple times to test the stability of the feature selection algorithms.

References

1. Al-Ani, A., Alsukker, A., Khushaba, R.N.: Feature subset selection using differential evolution and a wheel based search strategy. Swarm and Evolutionary Computation 9, 15–26 (2013)
2. Asuncion, A., Newman, D.: Uci machine learning repository (2007)
3. Bach, F.R., Jordan, M.I.: A probabilistic interpretation of canonical correlation analysis. Tech. rep. (2005)
4. Cervante, L., Xue, B., Shang, L., Zhang, M.: Binary particle swarm optimisation and rough set theory for dimension reduction in classification. In: 2013 IEEE Congress on Evolutionary Computation (CEC), pp. 2428–2435 (2013)
5. Cervante, L., Xue, B., Shang, L., Zhang, M.: A multi-objective feature selection approach based on binary pso and rough set theory. In: Middendorf, M., Blum, C. (eds.) EvoCOP 2013. LNCS, vol. 7832, pp. 25–36. Springer, Heidelberg (2013)
6. Cervante, L., Xue, B., Zhang, M., Shang, L.: Binary particle swarm optimisation for feature selection: A filter based approach. In: IEEE Congress on Evolutionary Computation (CEC 2012), pp. 881–888 (2012)
7. Chuang, L.Y., Tsai, S.W., Yang, C.H.: Improved binary particle swarm optimization using catfish effect for feature selection. Expert Systems with Applications 38, 12699–12707 (2011)

8. Dash, M., Liu, H.: Feature selection for classification. Intelligent Data Analysis 1(4), 131–156 (1997)
9. Engelbrecht, A.P.: Computational intelligence: an introduction, 2nd edn. Wiley (2007)
10. Fdhila, R., Hamdani, T.M., Alimi, A.M.: Distributed mopso with a new population subdivision technique for the feature selection. In: International Symposium on Computational Intelligence and Intelligent Informatics (ISCIII 2011), pp. 81–86 (2011)
11. Gutlein, M., Frank, E., Hall, M., Karwath, A.: Large-scale attribute selection using wrappers. In: IEEE Symposium on Computational Intelligence and Data Mining (CIDM 2009), pp. 332–339. IEEE (2009)
12. Kanan, H.R., Faez, K.: An improved feature selection method based on ant colony optimization (aco) evaluated on face recognition system. Applied Mathematics and Computation 205(2), 716–725 (2008)
13. Kennedy, J., Eberhart, R.: Particle swarm optimization. In: IEEE International Conference on Neural Networks, vol. 4, pp. 1942–1948 (1995)
14. Marill, T., Green, D.: On the effectiveness of receptors in recognition systems. IEEE Transactions on Information Theory 9(1), 11–17 (1963)
15. Matechou, E., Liu, I., Pledger, S., Arnold, R.: Biclustering models for ordinal data. Presentation at the NZ Statistical Assn. Annual Conference, University of Auckland (2011)
16. Neshatian, K., Zhang, M.: Genetic programming for feature subset ranking in binary classification problems. In: Vanneschi, L., Gustafson, S., Moraglio, A., De Falco, I., Ebner, M. (eds.) EuroGP 2009. LNCS, vol. 5481, pp. 121–132. Springer, Heidelberg (2009)
17. Pledger, S., Arnold, R.: Multivariate methods using mixtures: correspondence analysis, scaling and pattern detection. Computational Statistics and Data Analysis (2013), http://dx.doi.org/10.1016/j.csda.2013.05.013
18. Pudil, P., Novovicova, J., Kittler, J.V.: Floating search methods in feature selection. Pattern Recognition Letters 15(11), 1119–1125 (1994)
19. Shi, Y., Eberhart, R.: A modified particle swarm optimizer. In: IEEE International Conference on Evolutionary Computation (CEC 1998), pp. 69–73 (1998)
20. Stearns, S.: On selecting features for pattern classifier. In: Proceedings of the 3rd International Conference on Pattern Recognition, pp. 71–75. IEEE Press, Coronado (1976)
21. Van Den Bergh, F.: An analysis of particle swarm optimizers. Ph.D. thesis, University of Pretoria (2006)
22. Whitney, A.: A direct method of nonparametric measurement selection. IEEE Transactions on Computers C-20(9), 1100–1103 (1971)
23. Witten, I.H., Frank, E.: Data Mining: Practical Machine Learning Tools and Techniques, 2nd edn. Morgan Kaufmann (2005)
24. Xue, B., Cervante, L., Shang, L., Browne, W.N., Zhang, M.: A multi-objective particle swarm optimisation for filter based feature selection in classification problems. Connection Science 24(2-3), 91–116 (2012)
25. Xue, B., Cervante, L., Shang, L., Browne, W.N., Zhang, M.: Binary PSO and rough set theory for feature selection: A multi-objective filter based approach. International Journal of Computational Intelligence and Applications 13(02), 1450009 (2014)
26. Xue, B., Zhang, M., Browne, W.N.: Multi-objective particle swarm optimisation (PSO) for feature selection. In: Genetic and Evolutionary Computation Conference (GECCO 2012), Philadelphia, PA, USA, pp. 81–88. ACM (2012)
27. Xue, B., Zhang, M., Browne, W.N.: New fitness functions in binary particle swarm optimisation for feature selection. In: IEEE CEC 2012, pp. 2145–2152 (2012)
28. Xue, B., Zhang, M., Browne, W.N.: Particle swarm optimization for feature selection in classification: A multi-objective approach. IEEE Transactions on Cybernetics 43(6), 1656–1671 (2013)

29. Xue, B., Zhang, M., Browne, W.N.: Particle swarm optimisation for feature selection in classification: Novel initialisation and updating mechanisms. Applied Soft Computing 18, 261–276 (2014)
30. Yang, C.S., Chuang, L.Y., Li, J.C.: Chaotic maps in binary particle swarm optimization for feature selection. In: IEEE Conference on Soft Computing in Industrial Applications (SMCIA 2008), pp. 107–112 (2008)
31. Zhu, Z.X., Ong, Y.S., Dash, M.: Wrapper-filter feature selection algorithm using a memetic framework. IEEE Transactions on Systems, Man, and Cybernetics, Part B: Cybernetics 37(1), 70–76 (2007)

Feature Selection Method with Proportionate Fitness Based Binary Particle Swarm Optimization

Zhe Zhou, Xing Liu, Ping Li, and Lin Shang

State Key Laboratory of Novel Software Technology,
Department of Computer Science and Technology,
Nanjing University, Nanjing 210023, China
{zhouzhenjucs,lipingnju}@gmail.com
liu.xing@outlook.com
shanglin@nju.edu.cn

Abstract. Particle swarm optimization(PSO) has been applied on feature selection with improved results. Traditional PSO methods have some drawbacks when dealing with binary space, which may bring negative effects on the results. In this paper, an algorithm based on fitness proportionate selection binary particle swarm optimization(FPSBPSO) will be discussed in detail aiming to overcome the problems of traditional PSO methods. FPSBPSO will be utilized in the feature subset selection domain. The performance of feature selection will be compared in a benchmark dataset, and experimental results prove that the FPSBPSO-based feature selection methods can avoid premature convergence and improve the classification accuracy at the same time.

1 Introduction

Feature Selection refers to the process of selecting a subset of relevant features in model construction, which is often seen in classification problems. A well selected feature subset can boost the classifier's performance and reduce the time cost at the same time. Feature selection has been widely studied during past years and much work has been done trying to find the most informative features while maintaining the performance of the classifier. Current feature selection methods include frequency-based methods, information-gain-based methods, lexicon-based methods and so on [14] [20] [2]. Besides, some evolutionary algorithms can be utilized for feature selection as well, such as generic algorithms [1] and particle swarm optimization [17].

Particle swarm optimization(PSO) is an optimization technique firstly introduced by Kennedy and Eberhart in 1995 [7], which was inspired by the social behavior of birds flocking or fish schooling. Since then, PSO has attracted significant attention and it has been reported that compared with other evolutionary algorithms like genetic algorithms(GA), PSO is computationally more efficient [5]. The basic idea of PSO can be described as follow: a swarm is made up by

G. Dick et al. (Eds.): SEAL 2014, LNCS 8886, pp. 582–592, 2014.

many particles, each of which represents a candidate solution. In each iteration, particles update their position in the search space according to their own historical information and their neighbours' information. After the iterations reach the end, an optimal solution is supposed to be found. PSO was originally designed to solve continuous optimization problems. In order to deal with discrete problems, Kennedy and Eberhart proposed binary particle swarm optimization(BPSO) [8] in 1997 where the candidate particles in the search space are presented as a binary string.

The basic idea of PSO is that a particle can exploit its own historical position and performance information, as well as the historical position and performance information of the whole swarm. A particle tends to move closer to its own historical best position and the best position found by the whole swarm during iterations. This idea enables PSO to find an optimal solution, and it is also the idea shared by most variations of PSO. However, as we will explain in following sections, we found that BPSO does not necessarily follow the same idea as PSO, or at least not in the way other PSO variations do. In fact, a particle in BPSO will probably move away from the best position found by itself or the best position found by the whole swarm. Therefore it is necessary to pay attention to the theoretical basis of BPSO rather than only focusing on its applications.

In this work, we will discuss two major problems of PSO in binary space and try to prove that the traditional way of calculating velocities is the main cause of the two problems. A new way to calculate velocities based on fitness values is then proposed and based on that we will propose a new binary version of PSO called FPSBPSO. After that, we will perform feature selection in a binary classification problem with FPSBPSO-based methods to test its efficiency. If the selected feature subset with FPSBPSO can return better classification results than that with traditional PSO, our proposed scheme can be regarded as an efficient one.

The remainder of this paper is organized as follows. In section 2, we will introduce some popular feature selection methods as well as the development of PSO and BPSO. In section 3, we will show two problems PSO will face in discrete space, and introduce a modified BPSO method called FPSBPSO to deal with them. Then we will utilize FPSBPSO to deal with the task of feature subset selection. In section 4, different experimental results will be presented to prove FPSBPSO's efficiency both in optimization problems and in feature subset selection. Conclusions and future work are included in section 5.

2   Background

2.1   Feature Selection

With the rapid development of information age, there usually exists a large amount of potentially useful features in current classification tasks. It will be time-consuming to use all those features in the model building process, so the feature selection process is needed to select some informative features from the original feature set without sacrificing the quality of the classifier.

Many feature selection methods give every candidate feature a "goodness" value to label its usefulness in classification. Features with higher value are considered more informative and will be selected. Yang and Pederson [20] compared five different goodness-based feature selection methods including document frequency threshold(DF), information gain(IG), mutual information(MI), χ^2 statistic(CHI) and term strength(TS). Results show that DF, IG and CHI can all bring satisfying classification accuracy and three scores of a term calculated by them are strongly correlated. In recent years, feature selection has also caused much interest. Basu and Murthy [2] used the similarity between a feature and a particular class as the goodness value. Song et al. [14] introduced a novel clustering-based feature selection method for high dimensional data.

Apart from goodness-based methods, evolutionary algorithms are frequently used in feature selection as well. Abbasi et al. [1] proposed a feature selection method based on generic algorithm. Wang et al. [17] used rough set and binary PSO to perform feature selection. Due to the potential limitations of BPSO, Xue et al. [19] used PSO in continuous space to perform feature selection.

2.2   Continuous Particle Swarm Optimization

PSO was proposed by Kennedy and Eberhart in 1995 [7]. It was inspired by the social behavior of birds flocking or fish schooling. A swarm has some particles, each particle has a position component representing a specific solution, and a velocity component representing the direction of a particle's movement in the solution space. PSO is an iterative optimization algorithm with three main steps. The first step is to initialize the population by generating each particle's velocity component and position component randomly. The second step is to evaluate solutions represented by particles' positions. The final step is to update particles' velocities and then update particles' positions using the following formulas.

$$v_i^{t+1} = w * v_i^t + c_1 * rand * \left(pbest - x_i^t\right)$$
$$+ c_2 * rand * \left(gbest - x_i^t\right) \tag{1}$$

$$x_i^{t+1} = x_i^t + v_i^{t+1} \tag{2}$$

where v_i^{t+1} and x_i^{t+1} represents the velocity component and the position component of particle p_i at the $(t+1)_{th}$ iteration respectively, c_1 and c_2 are two fixed confidence coefficient, rnd is a uniformly distributed random variable ranging from 0 to 1, w is the inertia weight. $pbest$ means the position of particle p_i's personal best while $gbest$ means the position of particle p_i's global best.

The right side of equation (1) can be divided into three parts. $w * v_i^t$ represents the previous direction, $c_1 * rand * (pbest - x_i^t)$ represents the tendency of moving towards a particle's personal best, $c_2 * rand * (gbest - x_i^t)$ represents the tendency of moving towards the swarm's global best. The three parts together guide a particle's movement. The following figure shows the effect of their cooperation.

Since the basic PSO was proposed, researchers have proposed many variations aiming to improve its performance. In fact, there was no inertia weight in the

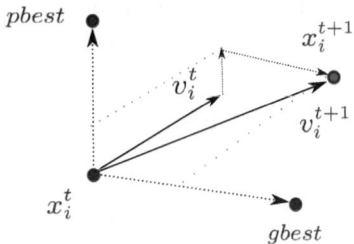

Fig. 1. The effect of pbest and gbest

basic PSO, Shi and Eberhart introduced inertia weight into PSO in 1998 [13], and this parameter has been adopted in almost every PSO variation. Other important modifications to the basic PSO model include Bare Bones PSO proposed by Kennedy [6], and the idea to clamp velocities [4] proposed by Eberhart et al.. Ioan gave a good analysis of PSO's convergence and parameter selection problem in [16].

2.3 Binary Particle Swarm Optimization

In order to use PSO to solve discrete problems, Kennedy proposed BPSO [8], which became the most widely used PSO model dealing with combinatorial optimization problems. Kennedy mainly made two changes to the continuous PSO, and the first is the representation of a particle's position. Unlike in the continuous PSO where a particle's position is a set of real values, in BPSO, a particle's position becomes a binary string like '1001101'. The second change is that the velocity of a particle no longer represents the direction of a particle's movement, rather, it means the possibility of choosing 1 at a specific bit of a particle's position component. And the update formula of the velocity component becomes the following one.

$$x_{id} = \begin{cases} 1, rand < S(v_{id}) \\ 0, otherwise. \end{cases} \tag{3}$$

where

$$S(v_{id}) = \frac{1}{1 + e^{-v_{id}}} \tag{4}$$

There are many applications of BPSO, including using BPSO to solve feature subset selection problems [11] [18], the travelling salesman problem [12], the lot sizing problem [15], flow shop scheduling problems [10], and many other combinatorial optimization problems.

3   Feature Selection with FPSBPSO

In this section, we will propose a modified BPSO algorithm called FPSBPSO, which suits the nature of PSO in discrete space better. Then we will perform a feature subset selection with the concept of FPSBPSO and test whether our proposed method can be beneficial to the classification process.

3.1 Problems of Traditional BPSO

One main problem in traditional BPSO is that the new position of a particle is solely decided by its velocity while the particle's current position hardly makes any influence in updating its next position. This phenomenon is to some extent contradictory to the basic idea of PSO which is inspired by the behaviour of bird flocking. As a result, we need to define a new way to calculate a particle's new position in which the current position of the particle can play an important role as well.

There is another problem PSO will have to face in discrete space. In traditional BPSO the velocity means the possibility that a particle's value changes between 0 and 1 respectively in every bit. But consider the following situation: a particle's value on the i_{th} bit is 0 as well as the i_{th} bit of the personal best and the global best, and the current velocity on this bit is 0 too. In this case the i_{th} bit will have equal possibility to be 0 or 1 in the next iteration as the velocity on the i_{th} bit is 0 according to equation (1). However, if the personal best, global best and the particle itself all have the value of 1 in the i_{th} bit, then the particle will also have equal possibility of being 0 or 1 in the next iteration, which is not consistent with our intuitions because the bit should have far bigger chance to remain the value of 1 when every related particle does. Thus in BPSO, we not only need a new way to update the position of a particle (like in equation (3)), but also have to change the velocity updating formula to make a particle's position constantly moving towards the best solution while maintaining the general diversity to avoid premature convergence at the same time.

3.2 FPSBPSO: Fitness Proportionate Selection BPSO

Based on the two issues mentioned in the previous subsection that may reduce the performance of BPSO, we now describe the newly designed update formulas for position component and velocity component which are the essential part of the new model. We adopt fitness proportionate selection to update a particle's position component, and that is why we call the new model FPSBPSO. The newly designed update formulas are as follows. In the following content, when we refer to the concept "involved particles", we actually mean the three particles p_i, p_i's personal best particle p_i^{pb} and p_i's global best particle p_i^{gb}.

$$v_{id}^{t+1} = \begin{cases} mr, & if \ n_0 = 0 \\ 1 - mr, & if \ n_1 = 0 \\ \frac{f_1}{f_1 + f_0}, & otherwise. \end{cases} \tag{5}$$

$$x_{id}^{t+1} = \begin{cases} 1, rand < v_{id}^{t+1} \\ 0, otherwise. \end{cases} \tag{6}$$

where n_0 is the number of involved particles with $x_{id} = 0$ and n_1 is the number of involved particles with $x_{id} = 1$. f_1 and f_0 are computed as follows: First, involved particles are divided into two sets S_1 and S_0 based on whether they select 1 or 0 at the d_{th} bit. Then, f_1 and f_0 are calculated by averaging the fitness values of particles in S_1 and S_0 respectively. v_{id}^{t+1} is the probability of setting x_{id}^{t+1} to 1. The higher f_1 is, the higher probability x_{id}^{t+1} will be 1.

In this new method, the update formula for velocities and positions is both very different from that of traditional BPSO. First, in the proposed method, the current position of a particle plays an important role in deciding its new state by taking part in the vote-like mechanism. This solves the first problem mentioned in the previous subsection. Besides, the vote-like mechanism ensures that if all three involved particles have the same value in a specific bit, then the new position will have a large possibility keeping that value unchanged (but still have a smaller possibility to mutate), which is more consistent with people's intuitions and the basic ideas of PSO. By this way the second problem mentioned in the previous subsection is also solved.

Our proposed FPSBPSO method has another advantage, which is that there is only one free parameter in FPSBPSO mr. Fewer parameters means that the model will be easier to be tuned because the difficulty of finding a good combination of parameter settings will increase exponentially as the number of parameter increases. The parameter mr is introduced to prevent the problem of premature convergence. In the proposed model, if $x_{id}^{pb} = x_{id}^{gb} = x_{id}^t$, then x_{id}^{t+1} will be fixed. For example, if $x_{id}^{pb} = x_{id}^{gb} = x_{id}^t = 1$, then v_{id}^{t+1} will be 0, which means x_{id}^{t+1} will be set to 1 with probability 1. This will lead to premature convergence. In this case, the parameter of mr will force the bit to mutate at times, which is often seen in generic algorithms.

How to select a proper value of mr is an important problem in the new model. By comparing the results of FPSBPSO using different settings of mr, we found two basic regularities. First, if mr is set to 0 or a too small value, premature convergence will happen. Second, if mr is set to a large value, then it will be very difficult for the swarm to convergence. Experimental results show that 0.01 is a good choice for many different problems. However, we suggest that it is better to use early-exploration-late-exploitation strategy, which means setting mr to a much larger value in the early stage so that particles can explore more space and gradually decreasing mr to ensure that the particles can pay more attention to exploiting the neighbourhood information in the later stage.

3.3   FPSBPSO Based Feature Subset Selection

In most classification tasks, the process of feature selection must be executed, during which a smaller set of all features is gained in order to reach better classification performance. There are many feature selection methods, such as information gain based method, lexicon based method, generic algorithm based method and so on.

PSO can be used to perform feature selection as well. Similar to the generic algorithm based method, every feature selection result is transformed into a bit string in PSO based method. The length of string is equal to the size of all different features and the value of each bit represents whether the corresponding feature is selected or not. The detailed algorithm is shown in Table 1. As the main aim of feature selection is to raise the classification accuracy, the fitness value of a particle is represented by the classifier's accuracy on a validation set using the corresponding feature subset. Our aim is to use FPSBPSO to perform the feature selection process to obtain a feature subset which is supposed to get better classification results than that brought by a traditional-BPSO-based one.

Table 1. FPSBPSO Based Feature Selection

FPSBPSO Based Feature Selection
1: Initialize parameters of BPSO
2: Randomly initialize swarm
3: **WHILE** stopping criterion not met **DO**
4:      calculate each particle's fitness function
6:      **For** $i = 1$ to $swarmSize$ **DO**
7:          update the lbest of P_i
8:          update the pbest of P_i
9:      **END**
10:     **FOR** $i = 1$ to $swarmSize$ **DO**
11:         **FOR** $j = 1$ to $dimension$ **DO**
12:             update the velocity of P_i according
13:             to equation (5)
14:             update the position of P_i according
15:             to equation (6)
16:         **END**
17:     **END**
18: **END**
19: Return the best feature subset found by the swarm

4   Experimental Results

In this section, we aim to prove FPSBPSO's effectiveness in feature selection. This is done through two steps. We first compare the performance of traditional BPSO and FPSBPSO on the MKP problem to show that our modification on traditional BPSO is reasonable. Then we utilize FPSBPSO into the feature selection in a binary classification problem to test whether the feature subset can improve the accuracy of classification.

4.1   Experiments for the Problem of MKP

First we want to prove that FPSBPSO works better than traditional BPSO in binary space by comparing the two algorithms on the famous problem of Multidimensional Knapsack Problem(MKP). MKP is an intensively studied discrete optimization problem because many real world problems can be formulated as a MKP, for example, cargo loading, cutting stock problems and so on. MKP is a constrained optimization problem, which means that there exists illegal solution space. In our tests, we will not try to ignore or avoid illegal solution space, instead, we add a penalty term into the fitness function of FPSBPSO. The final fitness function we used was proposed by Khuri et al. in [9].

$$f(x) = \sum_1^n p_i x_i - s * maxp_i \tag{7}$$

where

$$s = |j| \sum_1^m w_{ij} x_i > c_j| \tag{8}$$

where p_i is the value of item i, w_{ij} is the cost of item i for knapsack j, and c_j is the size of knapsack j. n is the number of items and m is the number of constraints. The goal is to maximize $\sum_1^n p_i x_i$ with the constrains $\forall j, \sum_1^n w_{ij} x_i <= c_j$.

We used 7 benchmarks named mknap1 in OR-Library [3] to test FPSBPSO's performance on MKP. Optimums of all the 7 benchmarks are all already known. Detailed information of the 7 benchmark problems are given in Table 2, where n is the number of items and m is the number of constraints and the column titled 'Best' shows the known best of the corresponding problem.

For standard BPSO, we use the following parameter settings: $w = 0.689343$, $c = 1.42694$, and $V_{max} = 4$. For FPSBPSO, mr decreases from 0.1 to 0.001 as iteration process goes on using formula given follows.

$$mr = (1 - iter/Max\_iter) * 0.1 + 0.01 \tag{9}$$

Where $iter$ is the current number of iterations and Max_iter is the maximum number of iterations.

Max iteration times and population size are 1000 and 40 respectively. In order to reduce the time cost of training, the iteration process will immediately stop if the value of global best has kept unchanged for over 120 times.

For each problem, we run BPSO and FPSBPSO for 20 times. We record the average and best value found during the 20 times. Experimental results are given in Table 3. As can be seen, BPSO succeeded in finding the best solution in 4 out of 7 problems while FPSBPSO succeeded in all 7 problems. Besides, FPSBPSO has the best average performance in most of the problems, and its advantage over BPSO will expand as the dimension of problems increases. In a word, FPSBPSO has better performance than BPSO, especially in high-dimensional problems.

4.2   Feature Selection with FPSBPSO

Our next goal is to evaluate FPSBPSO's performance in feature selection domain. To achieve this, we use the famous benchmark dataset of Madelon to perform the binary classification. The Madelon dataset consists of 4400 instances.

Table 2. MKP Benchmark

n m Best Known	n m Best Known
1 6 10 3800	5 28 10 12400
2 10 10 8706.1	6 39 5 10618
3 15 10 4015	7 50 5 16537
4 20 10 6120	

Table 3. MKP Results

	BPSO		FPSBPSO	
	Best	Avg.	Best	Avg.
1	3800	3800	3800	3800
2	8706.1	8706.1	8706.1	8706.1
3	4015	4014.5	4015	4014.5
4	6120	6107.5	6120	**6117.5**
5	12380	12298.5	**12400**	**12386.5**
6	10490	10431.3	**10618**	**10564**
7	16217	16150.55	**16537**	**16481**

Among them 2000 are used as training set, 600 are used as validation set and the remaining 1800 are used as test set. In all of the three sets positive instances and negative instances both make up 50% of the whole set. The original dataset has 500 features in all.

Table 4. Experimental results by BPSO

time	accuracy	iteration times
1	0.8133	230
2	0.8200	219
3	0.8200	206
4	0.8233	310
5	0.8233	233
average	0.8200	239.6

Table 5. Experimental results by FPSBPSO

time	accuracy	iteration times
1	0.8417	428
2	0.8817	1000
3	0.8467	373
4	0.8600	652
5	0.8433	231
average	0.8546	536.8

We use basic BPSO method and FPSBPSO method to get two different feature subsets, then the two subsets are separately used on the dataset to test the classification accuracy. In both cases, the classification algorithm is CART decision tree which is used both in particle fitness evaluation and final test. In this experiment, we run each algorithm for 5 times with different initialization settings.

The experimental results are shown in Table 4 and Table 5. From Table 4, we can know that the average accuracy of results by BPSO is 0.82. From Table 5, we can see the average accuracy of results by FPSBPSO is 0.8546, which has increased the accuracy by 3%. However, It may take more iteration time for FPSBPSO to return the feature selection result, which can be explained as that FPSBPSO is more precious than traditional BPSO and can avoid the "premature convergence".

5   Conclusions and Future Work

In this paper, we discussed the feasibility of utilizing PSO based methods to select feature subset for classification problems. First, we investigated the detail of traditional BPSO model and argue that BPSO is not a good binary version of PSO as it does not follow the essence of PSO in some cases. By analysing two major problems PSO will face in solving discrete problems, we proposed a new binary version of PSO called FPSBPSO. We redefined the velocity component in FPSBPSO to make it suit the character of PSO better and more computationally efficient. Experimental results showed that the feature selection result from our proposed FPSBPSO based method is superior to the one from traditional BPSO, as the former one can reach better classification results than the latter one.

Although preliminary results show that the new model can provide solid performance, there are still many problems to be dealt with. Firstly, in the proposed scheme we used only three particles' information to calculate the average performance of 1 and 0 at a specific bit, it remains to be seen if more particles' information could improve the model's performance. Secondly, we have only evaluated the model's performance on only one benchmark dataset, which is not sufficient enough. More experiments need to be taken to prove the validity and stability of the FPSBPSO based feature selection method. In the future, we will mainly focus on the two problems.

References

1. Abbasi, A., Chen, H., Salem, A.: Sentiment analysis in multiple languages: Feature selection for opinion classification in web forums. ACM Transactions on Information Systems (TOIS) 26(3), 12 (2008)
2. Basu, T., Murthy, C.: Effective text classification by a supervised feature selection approach. In: 2012 IEEE 12th International Conference on Data Mining Workshops (ICDMW), pp. 918–925. IEEE (2012)
3. Beasley, J.E.: Or-library: distributing test problems by electronic mail. Journal of the Operational Research Society 41(11), 1069–1072 (1990)

4. Eberhart, R., Simpson, P., Dobbins, R.: Computational intelligence PC tools. Academic Press Professional, Inc., San Diego (1996)
5. Hassan, R., Cohanim, B., de Weck, O.: A comparison of particle swarm optimization and the genetic algorithm. In: Proceedings of the 1st AIAA Multidisciplinary Design Optimization Specialist Conference (April 2005)
6. Kennedy, J.: Bare bones particle swarms. In: Proceedings of IEEE Swarm Intelligence Symposium, pp. 80–87 (April 2003)
7. Kennedy, J., Eberhart, R.: Particle swarm optimization. In: Proceedings of IEEE International Conference on Neural Networks, vol. 4, pp. 1942–1948 (1995)
8. Kennedy, J., Eberhart, R.: A discrete binary version of the particle swarm optimization. In: Proceedings of IEEE International Conference on Systems, Man, and Cybernetics, Computational Cybernetics and Simulation, vol. 5, pp. 4104–4108 (October 1997)
9. Khuri, S., Back, T., Heitkotter, J.: The zero/one multiple knapsack problem and genetic algorithms. In: Proceedings of the 1994 ACM Symposium on Applied Computing, pp. 188–193 (April 1994)
10. Liao, C.-J., Tseng, C.-T., Luarn, P.: A discrete version of particle swarm optimization for flowshop scheduling problems. Computers & Operations Research 34(10), 3099–3111 (2007)
11. Liu, X., Shang, L.: A fast wrapper feature subset selection method based on binary particle swarm optimization. In: Proceedings of IEEE Congress on Evolutionary Computation, pp. 3347–3353 (June 2013)
12. Shi, X., Liang, Y., Lee, H., Lu, C., Wang, Q.: Particle swarm optimization-based algorithms for tsp and generalized tsp. Information Processing Letters 103(5), 169–176 (2007)
13. Shi, Y., Eberhart, R.: A modified particle swarm optimizer. In: Proceedings of IEEE World Congress on Computational Intelligence, pp. 69–73 (May 1998)
14. Song, Q., Ni, J., Wang, G.: A fast clustering-based feature subset selection algorithm for high-dimensional data. IEEE Transactions on Knowledge and Data Engineering 25(1), 1–14 (2013)
15. Tasgetiren, M.F., Liang, Y.-C.: A binary particle swarm optimization algorithm for lot sizing problem. Journal of Economic and Social Research 5(2), 1–20 (2004)
16. Trelea, I.C.: The particle swarm optimization algorithm: convergence analysis and parameter selection. Inf. Process. Lett. 85(6), 317–325 (2003)
17. Wang, X., Yang, J., Teng, X., Xia, W., Jensen, R.: Feature selection based on rough sets and particle swarm optimization. Pattern Recognition Letters 28(4), 459–471 (2007)
18. Xue, B., Zhang, M., Browne, W.N.: Novel initialisation and updating mechanisms in PSO for feature selection in classification. In: Esparcia-Alcázar, A.I. (ed.) EvoApplications 2013. LNCS, vol. 7835, pp. 428–438. Springer, Heidelberg (2013)
19. Xue, B., Zhang, M., Browne, W.N.: Particle swarm optimization for feature selection in classification: a multi-objective approach. IEEE Transactions on Cybernetics 43(6), 1656–1671 (2013)
20. Yang, Y., Pedersen, J.O.: A comparative study on feature selection in text categorization. In: ICML, vol. 97, pp. 412–420 (1997)

Genetic Programming for Measuring Peptide Detectability

Soha Ahmed[1], Mengjie Zhang[1] , Lifeng Peng[2], and Bing Xue[1]

[1] School of Engineering and Computer Science
[2] School of Biological Sciences
Victoria University of Wellington, PO Box 600, Wellington 6140, New Zealand
{soha.ahmed,mengjie.zhang,bing.xue}@ecs.vuw.ac.nz, lifeng.peng@vuw.ac.nz

Abstract. The biomarker discovery process usually produces a long list of candidates, which need to be verified. The verification of protein biomarkers from mass spectrometry data can be done through measuring the detection probability from the mass spectrometer (Peptide detection). However, the limited size of the experimental data and lack of a universal quantitative method make the identification of these peptides challenging. In this paper, genetic programming (GP) is proposed to measure the detection of the peptides in the mass spectrometer. This is done through measuring the physicochemical chemicals of the peptides and selecting the high responding peptides. The proposed method performs both feature selection and classification, where feature selection is adopted to determine the important physicochemical properties required for the prediction. The proposed GP method is tested on two different yeast data sets with increasing complexity. It outperforms five other state-of-the-art classification algorithms. The results also show that GP outperforms two conventional feature selection methods, namely, Chi Square and Information Gain Ratio.

1 Introduction

Biomarkers are indicators of a specific biological or disease state [18]. They are important for many clinical applications and classification of different stages of diseases. Biomarker detection methods usually produce many biomarkers [11], and it is necessary to verify those biomarkers before passing them to clinical validation [18]. The peptide detection (helps in verifying candidate biomarkers) is a classification problem where the task is to classify peptides as flyers or non-flyers [12]. The detectable peptides (referred to as quantifiable surrogates) are the peptides that are characterised to be high responding in the body fluids (e.g. blood) [9]. This process of discovering the quantifiable surrogates is called verification, which is a necessary process to bridge the gap between the biomarker discovery and the clinical-validation experiments [24]. The verification process is typically done through the absolute quantification of peptides [1]. The verification of biomarkers is a hard problem due to the high dynamic range of proteins [18], the complexity of the data and the lack of a universal quantitative method.

G. Dick et al. (Eds.): SEAL 2014, LNCS 8886, pp. 593–604, 2014.
© Springer International Publishing Switzerland 2014

Mass spectrometry (MS) is capable of sensitive detection, identification and quantification of proteins. Mass spectrometer measures the molecular weight of the peptides (with respect to a charge ratio) and its corresponding intensity. The product spectrum is composed of the mass to charge ratio (m/z) and the intensity of peptides. Mostly, MS is accompanied with liquid chromatography for separation of the sample which helps decrease the complexity of the sample. The produced LC-MS spectrum contains the m/z, the intensity and the retention time of the peptides. MS-based quantification faces the problem of selecting the best quantifiable peptides that can give detectable MS peak. Therefore, machine learning methods can be useful to automatically predict the high responding peptides.

The physicochemical properties of the peptides can represent the feature vector for predicting the detectability of peptides. Mostly, the peptide detection data sets are composed of a large number of features (properties) some of which can be irrelevant to the classification task. Hence, an effective and powerful method is needed to perform two tasks. Firstly, feature selection is needed to select important physiochemical properties. Secondly, the classification of the data aims to determine the detection probability.

Genetic programming (GP) is an evolutionary technique which has been used successfully for feature selection and classification [5,8]. GP solves a problem by evolving computer programs (functions) [22]. It usually starts with a population of random programs then modifies these programs using its genetic operators [30]. The GP algorithm consists of the following steps [22]:

1. Initialize a random population of programs;
2. Calculate the goodness of each program through the fitness function;
3. If the stopping criteria are not met, do the following:
    – Select some good programs through the selection method;
    – Use the genetic operators to perform the changes on the selected programs;
    – Pass the new programs to the following generation;
    – Calculate the fitness of the programs in the new generation;
4. Return the program with the best fitness as the designed solution.

GP has the potential to perform feature selection and classification at the same time [26], and due to the high dimensionality of peptide data, GP is a good choice for solving peptide detection problem. This paper represents one of the few attempts to use GP for selecting important features required for peptide detection.

Goals. The main goal of this paper is to develop a new GP method for measuring peptide detectability. The proposed method performs two important tasks. Firstly, feature selection that helps in determining the important physiochemical properties for detection of peptides. Secondly, prediction of flyers (detectable) and non-flyers (non-detectable) peptides which will be useful for both verification of biomarkers candidates and at the same time *absolute quantification* of peptides. Precisely, we will investigate the following:

1. What is the appropriate fitness measure which can make GP reduces the number of selected features with preserving the maximum classification performance?
2. Can GP outperform conventional feature selection and classification methods?
3. What are the important physiochemical properties selected?

Organisation. The rest of the paper is organised as follows. Section 2 discusses the related work on using GP for feature selection and classification and also the previous work done on peptide detection. Section 3 describes the proposed GP approach for peptide detection. Section 4 presents the experiment setup, the data sets description and the feature vector production process. Section 5 reports the full experiment results and discussions. Section 6 concludes the paper and gives some directions for future work.

2   Related Work

2.1   GP for Feature Selection and Classification

GP has been successfully used to select features in either filter or embedded approaches [27, 28]. The advantage of GP for building classification models (without the need to be wrapped to another classifier) makes it a perfect choice for performing both classification and feature selection tasks, especially in high dimensional data such as in [2–4]. GP has been also used to solve the problem of classification of unbalanced data such as in [6–8]. The success of GP in feature selection and classification has encouraged us to use it in prediction of peptide detectability.

2.2   Peptide Detection

Previous studies have been adopted for the use of machine learning techniques for peptide detection [32]. Decision trees and artificial neural networks (ANN) have been used in [15] and [31] to relate the physicochemical properties of proteins to their MS detectability. Evolutionary algorithms were also used in a small number of studies to solve the peptide detection prediction problem in MS data. For example, in [33], genetic algorithms (GAs) have been used to solve this problem where the aim was to reach the optimum experimental conditions for protein detection in MS. GP has been used only in two studies [12, 34] with promising results. Most of these studies were focused on the maximisation of the flyers peptides without taking into account the overall accuracy of prediction both flyers and non-flyers peptides. Moreover, previous studies were mostly focused in determining detectability of peptides based on the whole set of peptides' properties. The advantage of GP to perform both feature selection and classification has not been fully investigated in those studies. In this paper, the determination of the important physicochemical properties for detection prediction is investigated. Moreover, the use of GP system as a prediction system for peptide detectability is also investigated here.

3   The New GP Method for Peptide Detection

3.1   Overall Structure

The proposed GP method is performing two tasks, firstly feature selection, in order to select important physicochemical properties required for accurate prediction, and secondly classification. The method first starts with data set preparation and generation of feature vectors. This is done through search of MS/MS through SEQUEST, which produces a data set containing peptides where the length of each peptide is chosen to be between 5-24 residues. Afterwards the feature vectors are generated, which are composed of the physicochemical properties of each peptide in the data set. For each peptide, 544 properties are extracted from *AAindex* database [19]. The data sets are divided into half for training and half for testing. Only the training set is passed to GP to build a classifier model. The produced model automatically selects features in the terminal nodes of the tree. The selected features are used to form new training and test sets. Finally, the algorithm is applied to the unseen test set to measure the detectability of the peptides.

3.2   Feature Selection

The search space using all of the 544 features (physicochemical property) is extremely large and hence, feature selection should be performed. GP can automatically select features during the evolution process [26]. The terminal nodes of evolved trees contain the selected features for building the classification model. Therefore, GP has the advantage of selecting the features, which have the potential to produce a classifier with better classification performance.

3.3   Peptides Detection (Classification)

Prediction of the detectability of a peptide is a non-trivial classification task which involves complicated relationships between the classification rules and also between the input features [12, 17]. The proposed GP method performs classification by setting a threshold value (as a decision stump) by which the classification decision is taken. If the GP tree output is less than or equal to this threshold, the peptide is classified as detectable (flyer) otherwise, it is classified as non-detectable (non-flyer).

3.4   Improved Fitness Function

The typical standard classification accuracy of the training set may be inappropriate for the peptide data sets due to the large number of features. We aim to select only the most important features. Hence, the fitness function used is designed to take into account feature selection and classification tasks.

The classification of the data as, true or false has four outcomes: true positive, false positive, true negative, false negative. These outcomes are represented using

Table 1. Binary Confusion Matrix

	Positive class	Negative class
Positive prediction	True Positive (TP)	False Positive(FP)
Negative prediction	False Negative (FN)	True Negative (TN)

a confusion matrix which is shown in Table 1. The first task is to maximise the classification accuracy, the classification accuracy is given by:

$$Accuracy = \frac{TP + TN}{TP + FN + FP + FN}$$

The second task is to reduce the number of features selected by each genetic program. Therefore, we used the following fitness function which is inspired by [25].

$$Fitness = (1 + a * exp^{\frac{n}{N}}) \times Accuracy \tag{1}$$

where n is the number of features selected by GP and N is the original number of features. The exponential factor decreases with increasing n to give more fitness to the program with less features. a is a factor used to measure the importance between reducing the number of features and increasing the classification accuracy. a is equal to the following:

$$a = (1 - \frac{\text{Current}}{\text{Total}}) \tag{2}$$

Current is the index of the current generation while *Total* is the maximum number of generations. Therefore, the fitness function used in Equation (1) will achieve the two tasks which are reducing the number of features, increasing classification performance.

4   Experiments Setup

This section outlines the data sets acquisition, feature vector production, program representation and evolutionary parameters.

4.1   Peptide Data Sets and Feature Vectors Production

Data Sets. Two tryptic peptide data sets are used to test the new method. Both data sets were analysed using LC-ESI-MS and obtained from [12]. The peptides of first data set were generated from 13 proteins. The proteins were searched against NCBInr database [29] using Mascot server [20](Matrix science) to confirm the identity and elution time. Extracted ion chromatograms were generated for the peptides that did not yield tandem MS data. Each peptide contains at least five amino acids and generated with either 0 or 1 missed cleavage and the m/z values range from 300 and 1800. The class label as a flyer or non-flyer was set by cross referencing the peptides with the generated peptides in the lab. This data set (DS_1) contains 931 peptides (501 in flyer and 430 in non-flyer class)

The second data set (DS_2.) was downloaded from *PeptideAtlas* [10] and originally produced from 24 yeast experiments. The total number of proteins is 2733.The peptides' length (number of amino acids) ranges from 6 to 42 residues with 0-2 missed cleavage. Each peptide was assigned a flyer's class label if it was observed in the 24 experiments otherwise, it was assigned a non-flyer class label. The total number of peptides examples is 21515 in which 2121 peptides are in the flyers' class and 19394 are in the non-flyers' class. More details about the data sets can be found in [12].

Feature Vectors. The data sets were obtained in the form of peptide sequences (amino acids) and the class label. Hence, in order to use those peptides with the machine learning techniques they should transformed to numerical feature vectors. The physicochemical properties of the peptides have shown to be related to their detectability [1]. Therefore, for each peptide, 544 properties were calculated to transform the peptide data into numerical feature vectors. The 544 properties were extracted from *AAindex* database [19] and for each peptide sequence the average of the property value of each individual amino acid is calculated over the whole peptide. The physicochemical properties include, for example, *mass, alpha-helical* which is the predicted percentage of the secondary structure, *hydrophobicity, gasphase basicity* and *isoelectric point*. Therefore, each peptide is an instance used for training and testing the GP algorithm which modeled by 544 feature values and either flyer or non-flyer class label.

4.2   GP Program Representation

The tree structure is used in the experiments as a representation of the GP program [21]. The features and also a randomly generated constant terminal are used in the terminal set. The function set contains the four standard mathematical operators $+, -, \%, \times$ and a conditional operator *if*, a *max* operator and a *Abs* operator. The $+, -, \times$ take two arguments and return the addition, subtraction or multiplication of the two arguments. The $\%$ is the usual division , which takes two arguments, but it is protected which returns zero if the division is by zero. *max* returns the maximum of two arguments while *if* takes three arguments and returns the second argument if the first is negative otherwise, returns the third one. The *Abs* operator takes only one argument and returns the absolute value of this argument. The classification is performed by taking a threshold value of zero in which if the output of the genetic program is less than or equal to zero the peptide is classified as belonging to the flyer class. Otherwise, it is classified as belonging to the non-flyer class.

4.3   GP Evolutionary Parameters

The initial population is generated using the ramped half and half method [21]. The number of individuals in the population is 1024. Crossover, mutation and elitism rates are 70%, 29% and 1% , respectively. The maximum tree depth is set

Table 2. GP evolutionary parameters

Initialization method	Ramped Half-and Half
Tree Depth	8
Number of Generations	100
Mutation rate	29%
Crossover rate	70%
Elitism rate	1%
Population Size	1024
Selection type	Tournament
Tournament Size	4

to 8 in order to avoid bloating. The method of selection used is the tournament method and its size is set to 4. The evolution runs for 100 generations. 50% of the data is randomly selected for training GP and the other 50% is kept as a test set. These parameters are selected based on the literature [3]. Table 2 shows the evolutionary parameters used.

4.4   Methods for Comparison

The proposed GP is compared with several state-of-the-art feature selection and classification algorithms. The Waikato Environment for Knowledge Analysis (WEKA) package [16] is used to run the feature selection and classification algorithms.

Benchmark Classification Methods. Five different classifiers are used (with both the original features and the GP's selected features) and compared to GP classifier. The five classifiers are commonly used in classification tasks.

1. Naive Bayes (NB): NB belongs to the category of Bayesian classifiers which captures the behavior of the data on probability distributions. NB makes an assumption that all the features are conditionally independent [35].
2. Support Vector Machines (SVM): SVM forms a number of hyperplanes and classifies the instances according to the side of the hyperplane to which the instance belongs to [35].
3. Decision Tree (J48): J48 classifies instances through sorting them in a tree which is composed of a hierarchy of nodes. The root node first test the value of the feature and then moves to the child nodes until the label node is reached [35].
4. Conjunctive Rule (CR): CR builds a single conjunctive rule to predict the class labels. It uses the "AND" logical operator to determine correlation of features and classes [35].
5. Voted Perceptron (VP): VP is based on the perceptron algorithm and uses kernel functions to build hyperplanes as decision boundaries [14].

4.4.1   Benchmark Feature Selection Methods

We selected two common feature selection methods to compare the impact of the GP's selected features on the classifiers to the impact of those benchmark methods' features.

1. Chi Square (χ^2) feature evaluation: In statistical analysis methods, χ^2 test is used to measure the in dependency of two events. χ^2 as a feature selection measure the association between the features and classes. A score is given for each feature, according to its χ^2 statistics with respect to the class [13].
2. Information Gain Ratio (IGR) feature evaluation: The features are evaluated by measuring the gain ratio with respect to the class [13]. The gain ratio is the ratio between the total entropy of the features and the intrinsic value.

5   Results and Discussions

Several sets of experiments were performed to test the effectiveness of the proposed GP method. Firstly, GP was run with all the 544 physicochemical properties of the peptides. Secondly, the same GP algorithm was run with the features selected in the terminal nodes of the GP program. The feature selection phase resulted in an average of 5 physicochemical properties for the data set DS_1 and 14 property for data set DS_2. The selected features are fed to the other benchmark classifiers. Moreover, the benchmark feature selection methods (χ^2 and IGR) were used to select features and the top 5 and 14 features from both methods are fed to the same classifiers. In Table 3, the second column gives the performance of the new GP method (annotated as GP). The mean ($\bar{x}$), best and the standard deviation (s) of the 30 runs are reported in the table. The rest of the columns give the results of using the other benchmark classifiers. As these classifiers are deterministic methods only one result is given for each data set. The best performance for each data set is made bold. Table 4 gives the results of using the GP's, χ^2's and IGR 's selected features with the five benchmark classifiers. As the average number of features selected by GP for DS_1 and DS_2 is 5 and 14, respectively, we used the top 5 and 14 features from both χ^2 and IGR to make the comparison. When using the GP's selected features, each of the 30 runs' features are used with each of the classifiers and the average ($\bar{x}$), best and standard deviation are given in Table 4. A statistical T-test (Z-Test) with 0.05 degrees of freedom (95% significance level) is performed to check the significance of the results between the proposed GP method and the methods of comparison. In Tables 3 and 4 the mark $^-$ ($^+$) means the method of comparison is significantly worse (better) than GP, while the mark $^=$ means that there is no significant difference between them. For running GP, the Java-based Evolutionary Computation research system ECJ [23] package was used. All the experiments were executed on a machine with an Intel(R) Core(TM) i7-3770 CPU @ 3.40GHz, running Ubuntu 4.6 and Java 1.7.0_25 with a total memory of 8GByte .

5.1   GP as a Classifier

As shown in Table 4, the best of GP managed to outperform NB, SVM, J48, CR and VP for data set DS_1. The best of GP is better than these classifiers by 2.37-9.48%, while the mean of the 30 runs is better than SVM, J48, CR and VP

Table 3. Comparison of the performances of GP to benchmark classifiers.

Dataset	GP		NB	SVM	J48	CRt	VP
	Best	$\bar{x} \pm s$					
DS_1	**59.48**	56.62±1.00	57.11$^+$	56.03$^=$	56.03$^=$	50.00$^-$	53.23$^-$
DS_2	**90.15**	**90.14**±0.01	67.07$^-$	90.13$^-$	87.91$^-$	89.14$^-$	90.13$^-$

by 0.61-6.62%. For data set DS_2, the average and the best of GP outperformed all other classifiers. The results of T-test also show that GP is significantly better than the five classifiers in the data set DS_2. Furthermore, GP is significantly better than CR and VP for data set DS_2. However, there is no significant difference between GP and SVM and J48 in DS_1. The only exception is NB in DS_1, where the performance of NB is slightly better than that of GP, although the best of GP outperforms NB.

5.2   GP as a Feature Selection Method

For each GP run, we used the selected features in the terminal nodes of the GP evolved tree with the other classifiers. The purpose is to test the capability of GP to select the important features in addition to its capability for classification. GP selected an average number of features of 5 for DS_1 and 14 for DS_2 and hence, for both χ^2 and IGR the top 5 and 14 features were used. It can be seen from Table 4 that GP managed to select the features, which achieve better performance with most the classifiers than both χ^2 and IGR on both DS_1 and DS_2. The significance test shows that for DS_1, GP selected features that have a significant better performance than those of IGR when used with all the classifiers. Moreover, it significantly outperformed χ^2 when used with NB, SVM and VP and they were similar when used with J48 and CR. For DS_2, GP features made a significantly better performance when used with NB and equal performance when used with most of the rest of the classifiers. The only exception

Table 4. Comparison of the Performances of GP, χ^2 and IGR Selected Features

Data set	Classifier	GP		χ^2	T-test	IGR	T-test
		Best	$\bar{x} \pm s$	Best		Best	
DS_1	NB	**57.11**	**53.87**±2.34	50.96	−	52.50	−
	SVM	**60.56**	**54.71**±2.19	52.90	−	52.04	−
	J48	**57.11**	**54.60**±1.61	54.19	=	52.04	−
	CR	**57.32**	**53.00**±2.14	52.68	=	50.04	−
	VP	**60.56**	**54.35**±2.32	52.04	−	52.04	−
DS_2	NB	**85.77**	**84.59**±0.55	71.24	−	71.05	−
	SVM	90.15	90.15±0.0	90.14	=	**90.22**	=
	J48	**90.34**	89.95±0.29	90.06	+	90.13	+
	CR	90.15	90.15±0.0	**90.22**	=	**90.22**	=
	VP	**90.22**	90.15±0.20	**90.22**	=	**90.22**	=

here is with J48 which has a slightly better accuracy with χ^2 and IGR more than the average of GP. This is perhaps because J48 also uses IGR to further select features, and therefore might be biased to IGR.

6   Conclusions and Future Work

The objective of this paper was to investigate the performance of GP capability to reduce the number of redundant properties with preserving the maximum accuracy for measuring peptide detectability. This goal was successfully achieved by developing a GP system which selects features and at the same time performs detection. The proposed method works by maximising the classification accuracy and minimising the number selected features in the terminal nodes of the GP tree, and therefore, the system is a multi-objective system. The new method is tested against five other classifiers namely, NB, SVM, J48, CR, VP. Moreover, in order to compare the feature selection capability of the proposed method, it is tested against two well known feature selection methods, namely, χ^2 and IGR. The results show that GP outperformed most of these state of art feature selection and classification algorithms.

There are many other investigations that need to be done in the future. Firstly, the peptide data sets are mostly characterized by being unbalanced data which means that peptides in one class (mostly flyer's class) is much less than the peptides in the other class. This makes the classifiers biased towards the majority class, and hence, the specificity rate will be much higher than the sensitivity rate. This means that the overall classification accuracy is not the only evaluation criteria that should be used for measuring the peptide detectability and the imbalance problem should be taken into account. The use of GP to solve the imbalance problem will be the first future direction. Another future direction is the verification of the candidate biomarkers detection in MS data through the linkage of the detectability of the biomarkers in the mass spectrometer. Finally, the absolute quantification of proteins using GP through peptide detection will be performed in the future.

References

1. Abbatiello, S., Mani, D., Keshishian, H., Carr, S.: Automated Detection of Inaccurate and Imprecise Transitions in Peptide Quantification by Multiple Reaction Monitoring Mass Spectrometry. Clinical Chemistry 56, 291–305 (2010)
2. Ahmed, S., Zhang, M., Peng, L.: Feature Selection and Classification of High Dimensional Mass Spectrometry Data: A Genetic Programming Approach. In: Vanneschi, L., Bush, W.S., Giacobini, M. (eds.) EvoBIO 2013. LNCS, vol. 7833, pp. 43–55. Springer, Heidelberg (2013)
3. Ahmed, S., Zhang, M., Peng, L.: Genetic programming for biomarker detection in mass spectrometry data. In: Thielscher, M., Zhang, D. (eds.) AI 2012. LNCS, vol. 7691, pp. 266–278. Springer, Heidelberg (2012)
4. Ahmed, S., Zhang, M., Peng, L.: Enhanced feature selection for biomarker discovery in LC-MS data using GP. In: Proceedings of 2013 IEEE Congress on Evolutionary Computation, pp. 584–591 (2013)

5. Augusto, D.A., Barbosa, H.J.C., Ebecken, N.F.F.: Coevolutionary multi-population genetic programming for data classification. In: Proceedings of the 12th Annual Conference on Genetic and Evolutionary Computation, GECCO 2010, pp. 933–940. ACM, New York (2010)
6. Bhowan, U., Johnston, M., Zhang, M.: Developing new fitness functions in genetic programming for classification with unbalanced data. IEEE Transactions on Systems, Man, and Cybernetics, Part B 42(2), 406–421 (2012)
7. Bhowan, U., Johnston, M., Zhang, M., Yao, X.: Evolving diverse ensembles using genetic programming for classification with unbalanced data. IEEE Trans. Evolutionary Computation 17(3), 368–386 (2013)
8. Bhowan, U., Zhang, M., Johnston, M.: Genetic programming for classification with unbalanced data. In: Esparcia-Alcázar, A.I., Ekárt, A., Silva, S., Dignum, S., Uyar, A.Ş. (eds.) EuroGP 2010. LNCS, vol. 6021, pp. 1–13. Springer, Heidelberg (2010)
9. Cho, C.-K.J., Drabovich, A.P., Batruch, I., Diamandis, E.P.: Verification of a biomarker discovery approach for detection of Down syndrome in amniotic fluid via multiplex selected reaction monitoring (SRM) assay. J Proteomics, 2052–2059 (2011)
10. Desiere, F., Deutsch, E.W., King, N.L., Nesvizhskii, A.I., Mallick, P., Eng, J., Chen, S., Eddes, J., Loevenich, S.N., Aebersold, R.: The PeptideAtlas project. Nucleic Acids Research 34(suppl 1), D655–D658 (2006)
11. Domon, B., Aebersold, R.: Options and considerations when selecting a quantitative proteomics strategy. Nat. Biotechnology 28, 710–721 (2010)
12. Eyers, C.E., Lawless, C., Wedge, D.C., Lau, K.W., Gaskell, S.J., Hubbard, S.J.: CONSeQuence: prediction of reference peptides for absolute quantitative proteomics using consensus machine learning approaches. Molecular & Cellular Proteomics 10(11) (2011)
13. Forman, G.: An extensive empirical study of feature selection metrics for text classification. J. Mach. Learn. Res. 3, 1289–1305 (2003)
14. Freund, Y., Schapire, R.E.: Large margin classification using the perceptron algorithm. Mach. Learn. 37(3), 277–296 (1999)
15. Gay, S., Binz, P.-A., Hochstrasser, D.F., Appel, R.D.: Peptide mass fingerprinting peak intensity prediction: Extracting knowledge from spectra. PROTEOMICS 2(10), 1374–1391 (2002)
16. Hall, M., Frank, E., Holmes, G., Pfahringer, B., Reutemann, P., Witten, I.H.: The WEKA data mining software: an update. In: SIGKDD Explorer Newsletter, pp. 10–18 (2009)
17. He, H., Garcia, E.A.: Learning from Imbalanced Data. IEEE Transactions on Knowledge and Data Engineering 21(9), 1263–1284 (2009)
18. Huttenhain, R., Malmstrom, J., Picotti, P., Aebersold, R.: Perspectives of targeted mass spectrometry for protein biomarker verification. Curr. Opin. Chem. Biol. 13, 518–525 (2009)
19. Kawashima, S., Kanehisa, M.: AAindex: Amino Acid index database. Nucleic Acids Research 28(1), 374 (2000)
20. Koenig, T., Menze, B.H., Kirchner, M., Monigatti, F., Parker, K.C., Patterson, T., Steen, J.J., Hamprecht, F.A., Steen, H.: Robust Prediction of the MASCOT Score for an Improved Quality Assessment in Mass Spectrometric Proteomics. Journal of Proteome Research 7(9), 3708–3717 (2008)
21. Koza, J.R.: Genetic Programming: On the Programming of Computers by Means of Natural Selection. MIT Press, Cambridge (1992)
22. Koza, J.R.: Introduction to genetic programming: tutorial. In: GECCO (Companion), pp. 2299–2338 (2008)

23. Luke, S.: Essentials of Metaheuristics. In: Lulu, 2nd edn. (2013),
    http://cs.gmu.edu/$\sim$sean/book/metaheuristics/
24. Mallick, P., Schirle, M., Chen, S., Flory, M., Lee, H., Martin, D., Ranish, J., Raught,
    B., Schmitt, R., Werner, T., Kuster, B., Aebersold, R.: Computational Prediction
    of Proteotypic Peptides for Quantitative Proteomics. Nat. Biotechnol. 25(1), 125–
    131 (2007)
25. Muni, D., Pal, N., Das, J.: Genetic programming for simultaneous feature selection
    and classifier design. IEEE Transactions on Systems, Man, and Cybernetics, Part
    B: Cybernetics 36(1), 106–117 (2006)
26. Neshatian, K., Zhang, M.: Unsupervised Elimination of Redundant Features Us-
    ing Genetic Programming. In: Proceedings of 22nd Australasian Conference on
    Artificial Intelligence, pp. 432–442 (2009)
27. Neshatian, K., Zhang, M.: Using genetic programming for context-sensitive feature
    scoring in classification problems. Connect. Sci. 23(3), 183–207 (2011)
28. Neshatian, K., Zhang, M.: Improving relevance measures using genetic program-
    ming. In: Moraglio, A., Silva, S., Krawiec, K., Machado, P., Cotta, C. (eds.) Eu-
    roGP 2012. LNCS, vol. 7244, pp. 97–108. Springer, Heidelberg (2012)
29. Pruitt, K.D., Tatusova, T., Maglott, D.R.: NCBI Reference Sequence (RefSeq): a
    curated non-redundant sequence database of genomes, transcripts and proteins.
    Nucleic Acids Research 33(suppl 1), D501–D504 (2005)
30. Smart, W., Zhang, M.: Using Genetic Programming for Multiclass Classification
    by Simultaneously Solving Component Binary Classification Problems. In: Keijzer,
    M., Tettamanzi, A.G.B., Collet, P., van Hemert, J., Tomassini, M. (eds.) EuroGP
    2005. LNCS, vol. 3447, pp. 227–239. Springer, Heidelberg (2005)
31. Tang, H., Arnold, R.J., Alves, P., Xun, Z., Clemmer, D.E., Novotny, M.V., Reilly,
    J.P., Radivojac, P.: A computational approach toward label-free protein quantifica-
    tion using predicted peptide detectability. Bioinformatics 22(14), e481–e488 (2006)
32. Timm, W., Scherbart, A., Bocker, S., Kohlbacher, O., Nattkemper, T.: Peak in-
    tensity prediction in MALDI-TOF mass spectrometry: A machine learning study
    to support quantitative proteomics. BMC Bioinformatics 9(1), 443 (2008)
33. Vaidyanathan, S., Broadhurst, D.I., Kell, D.B., Goodacre, R.: Explanatory Op-
    timization of Protein Mass Spectrometry via Genetic Search. Analytical Chem-
    istry 75(23), 6679–6686 (2003)
34. Wedge, D.C., Gaskell, S.J., Hubbard, S.J., Kell, D.B., Lau, K.W., Eyers, C.: Pep-
    tide detectability following ESI mass spectrometry: prediction using genetic pro-
    gramming. In: GECCO 2007: Proceedings of the 9th Annual Conference on Genetic
    and Evolutionary Computation, vol. 2, pp. 2219–2225 (2007)
35. : In: Witten, I.H., Frank, E. (eds.) Data Mining: Practical Machine Learning Tools
    and Techniques, 2nd edn. Morgan Kaufmann Series in Data Management Systems,
    Morgan Kaufmann Publishers Inc., San Francisco (2005)

Overview of Particle Swarm Optimisation
for Feature Selection in Classification

Binh Tran, Bing Xue, and Mengjie Zhang

Victoria University of Wellington, PO Box 600, Wellington 6140, New Zealand
{tran.binh,bing.xue,mengjie.zhang}@ecs.vuw.ac.nz

Abstract. Feature selection is a process of selecting a subset of relevant features from a large number of original features to achieve similar or better classification performance and improve the computation efficiency. As an important data pre-processing technique, research into feature selection has been carried out over the past four decades. Determining an optimal feature subset is a complicated problem. Due to the limitations of conventional methods, evolutionary computation (EC) has been proposed to solve feature selection problems. Particle swarm optimisation (PSO) is an EC technique which recently has caught much interest from researchers in the field. This paper presents a review of PSO for feature selection in classification. After describing the background of feature selection and PSO, recent work involving PSO for feature selection is reviewed. Current issues and challenges are also presented for future research.

Keywords: Particle swarm optimisation, feature selection, evolutionary computation, classification.

1 Introduction

In many fields such as data mining and machine learning, data sets may contain a large number of features. However, the redundant or irrelevant features may reduce the classification performance. In order to solve this problem, feature selection is proposed to pick a subset of features that are relevant to the target concept [11]. By removing the irrelevant and redundant features, feature selection could significantly shorten the running time, improve the classification accuracy, and/or simplify the structure of the learned classifiers or models [11]. However, feature selection is a difficult problem, especially when the number of features is large [49,28]. Therefore, the optimal solution cannot be guaranteed to be acquired except when an exhaustive search is performed. However, an exhaustive search often takes a long time [50]. In real-world applications, obtaining good solutions in a reasonable amount of time is more interested than being obsessed with optimal solutions.

EC techniques are population-based techniques with a set of genetically motivated operations. These operations are used by a population of candidate solutions to obtain the optimal or near-optimal solution of the problem. Recently, different EC algorithms have been applied to feature selection problems such as particle swarm optimisation (PSO) [3,23,27], genetic algorithms (GAs) [30,41], genetic programming (GP) [29,32], and ant colony optimisation (ACO) [16,35]. As a relatively new EC technique, PSO is

G. Dick et al. (Eds.): SEAL 2014, LNCS 8886, pp. 605–617, 2014.

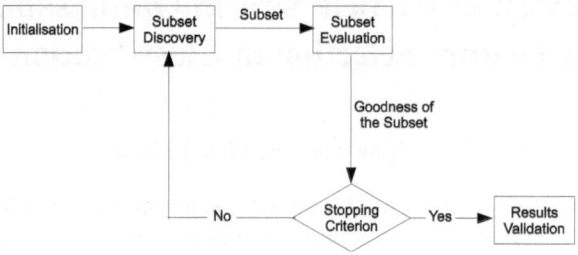

Fig. 1. General Feature Selection Process [11]

inspired by social behaviour such as birds flocking and fish schooling. Compared with other EC algorithms such as GAs, PSO is easier to implement and can converge more quickly [19]. It has been shown to be an effective method for feature selection problems [3,15,23,27]. This goal of this paper is to review recent work about PSO and its binary version called binary PSO (BPSO) [18] for feature selection in classification and to find the need for future research. The remainder of this paper is organised as follows. Feature selection is introduced in section 2 and Section 3 describes the standard PSO and BPSO algorithms. Section 4 reviews recent studies about PSO for feature selection in classification. The final section discusses current issues and challenges.

2   Feature Selection

Feature selection attempts to select the minimally sized subset of features that are necessary and sufficient to describe the target concept [11]. Its purposes include reducing the amount of data needed for learning, shortening the running time, improving the system accuracy, and increasing the comprehensibility of the learned model [24]. Fig. 1 shows the process of a typical feature selection method [11], which consists of five basic steps:

1. Initialisation: A feature selection algorithm starts with an initialisation procedure based on all the original features.
2. Subset discovery: A discovery procedure to generate candidate subsets. It is a search procedure [22], which can start with no features, all features, or a random subset of features. Many search techniques including conventional search methods and EC techniques are applied in this generation step to search for the best subset of features.
3. Subset evaluation: An evaluation function to measure the goodness of the generated feature subsets.
4. Stopping criterion: The algorithm will stop according to a given criterion, which can be based on the generation procedure or the evaluation function. The former can be a predefined number of features selected or a predetermined maximum number of iterations reached. The latter includes whether an optimal feature subset according to a certain evaluation function is obtained or whether addition or deletion of any feature does not produce a better subset.
5. Results validation: The validity of the selected subset is tested by carrying out tests on unseen data.

Table 1. Taxonomy of the reviewed papers

PSO for feature selection			
		Wrapper approach	**Filter approach**
Continuous PSO	Sing.Obj.	[3,15,23,27,45]	[14]
	Multi.Obj.	[48]	
Binary PSO	Sing.Obj.	[1,9,10,20,21,25,26,39,47,49,50,51]	[4,6,5,7,8,33,36,38,40]
	Multi.Obj.	[44]	[42,43]

Based on whether the subset evaluation process includes a learning algorithm or not, Langley [22] grouped different feature selection methods into two broad categories: filter approaches and wrapper approaches. Filter approaches are utilized to select features based on the evaluation criterion without using a learning algorithm. They are argued to be computationally less expensive and more general [7,50]. On the other hand, wrapper approaches implement a learning algorithm to construct a classifier in the evaluation procedure. They add or delete features to produce various feature subsets, and then measure the subsets depending on the performance of the developed classifier. Compared with filter approaches, wrapper approaches usually produce better results, especially when the classifier is designed to solve a particular problem. However, they are computationally expensive when the number of features is large [32]. In order to take advantage of both wrapper and filter approaches, recent studies proposed a hybrid approach in which filter methods were first used to select informative features before transferring to wrapper methods.

Naturally, an optimal feature subset is the smallest one that can obtain the highest classification quality, which makes feature selection a multi-objective problem [37]. Single-objective approaches can only produce one subset of features. Feature selection as a multi-objective problem producing several trade-off subsets can meet different user requirements in real-world applications. This paper briefly reviewed different PSO based feature selection algorithms, which can be seen in Table 1.

3   Particle Swarm Optimisation

3.1   Continuous Particle Swarm Optimisation

Particle swarm optimisation (PSO) is an evolutionary computation technique proposed by Kennedy and Eberhart in 1995 [12,17]. In PSO, each potential solution is called a bird or particle with no weight and no volume. The ith particle flies in a D-dimensional search space to find the optimal solution. There is a vector $x_i = (x_{i1}, x_{i2}, ..., x_{iD})$ presenting the position of particle i, where $x_{id \in [l_d, u_d]}$, $d \in [1, D]$, l_d and u_d are the lower and upper bounds of the dth dimension. The velocity of the ith particle is represented as $v_i = (v_{i1}, v_{i2}, ..., v_{iD})$. The best previous position of any particle is recorded as the personal best called *pbest*. The best solution visited by the whole swarm so far is the global best called *gbest*. The swarm is initialised with a population of random solutions. According to the *pbest* and the *gbest*, the algorithm searches for the best solution by updating particles' positions and velocities using the following formulae:

$$v_{id}^{t+1} = w * v_{id}^t + c_1 * r_1 * (p_{id} - x_{id}^t) + c_2 * r_2 * (p_{gd} - x_{id}^t) \tag{1}$$

$$x_{id}^{t+1} = x_{id}^t + v_{id}^{t+1} \tag{2}$$

where t means that the algorithm is going on the tth iteration. c_1 and c_2 are acceleration constants. r_1 and r_2 are random values uniformly distributed in $[0, 1]$. p_{id} presents the *pbest* while p_{gd} presents the *gbest*. w is the inertia weight first introduced by Shi and Eberhart [34]. w can make a balance between the global search and the local search to improve the performance of PSO. The velocity v_{id}^t is limited by a maximum velocity, $v_{id}^{t+1} \in [-v_{max}, v_{max}]$ and v_{max} is predefined based on the problem to be solved. Eberhart and Shi [13] suggested v_{max} to be set at about $10 - 20\%$ of the dynamic range of the variable in each dimension.

3.2 Binary Particle Swarm Optimisation

PSO was originally proposed for continuous optimisation. Kennedy and Eberhart [18] developed a binary PSO (BPSO), which can be used for discrete problems. In BPSO, the position of each particle is encoded by a binary string. x_{id}, p_{id} and p_{gd} are restricted to 0 or 1. The velocity in BPSO represents the probability of an element in the position taking value 1. Equation (1) is still applied to update the velocity. A sigmoid function $s(v_{id})$ is introduced to transform v_{id} into the range of 0 and 1. BPSO updates the position of the particle according to the following formulae:

$$x_{id} = \begin{cases} 1, & \text{if } rand() < \dfrac{1}{1 + e^{-v_{id}}} \\ 0, & otherwise \end{cases} \tag{3}$$

where $rand()$ is a random number selected from a uniform distribution in $[0,1]$. v_{id} is transformed to $[0,1]$ by a sigmoid limiting function.

4 PSO for Feature Selection

PSO shares many similarities with EC algorithms like GAs, but compared with GAs, PSO has its own advantages such as converging quickly and computationally inexpensive [52]. Both continuous PSO and binary PSO have been used for feature selection. Generally, when a continuous PSO algorithm is applied to feature selection problems, a particle in the swarm is formed by a vector of n real numbers, where n is the total number available features. In order to determine whether a feature will be selected or not, a threshold is needed to compare with the value in the vector. In BPSO, the representation of a particle is a n-bit binary string. The feature mask is Boolean that "1" represents that the feature is selected and "0" otherwise. A short review of recent work on PSO for feature selection will be presented in this section.

4.1 PSO Based Wrapper Feature Selection

Azevedo et al. [3] proposed a wrapper feature selection algorithm using PSO and support vector machines (SVM) for personal identification in a keystroke dynamic system. Experimental results showed that the proposed approach produced better performance than a GA with SVM regarding the classification error, processing time and the feature reduction rate. However, the false acceptance rate of the program was still high.

As a relatively new EC technique, PSO cannot avoid to have some disadvantages. Its high possibility to get stuck in local optima can be a typical example. Different strategies have been proposed to solve this problem. Yang et al. [50] proposed two BPSO based algorithms using two chaotic maps, a logistic map and a tent map to determine the inertia weights dynamically. The K-nearest neighbor (KNN) method with leave-one-out cross-validation (LOOCV) was applied in the wrapper models to evaluate the classification accuracies. Experiments showed that the proposed methods, especially BPSO with tent map, produced slightly higher classification accuracy than other methods, including sequential forward search (SFS), plus and take away (PTA), sequential forward floating search (SFFS), sequential genetic algorithm (SGA) and different hybrid genetic algorithms (HGAs).

Yang et al. [49] constructed a strategy for *gbest* using Boolean operator to improve BPSO. When *gbest* fitness was identical after three iterations, a Boolean operator ' and(.)' would 'and' each bit of the *pbest* of all particles to create a new binary string. This new binary string would replace the old *gbest*. KNN with LOOCV was also applied to evaluate the classification accuracies in the experiments. Results illustrated that the proposed method usually achieved higher classification accuracy with fewer features than GA and BPSO. However, proposed BPSOs were not compared with other variations of PSO, which might produce better results. Chuang et al. [9] developed another resetting strategy for *gbest* to improve the performance of BPSO for feature selection. *gbest* would be reset to zero if the *gbest* fitness maintained the same value after several iterations. The fitness of each particle was evaluated by KNN with LOOCV. Experiments were conducted on gene expression data sets. Results showed that this method effectively reduced the number of needed features and got the higher classification accuracy than the method created by Yang et al [49] in most cases.

Chuang et al [10] also applied another resetting strategy called catfish effect to improve BPSO. Similar to [9], if the fitness of *gbest* stayed the same for a predefined number of iterations, ten percent of the population with the worst fitness would be forced to extreme positions which were either all 0s or all 1s randomly. The reported results of this method were better than BPSO and those such deterministic algorithms as SFS, PTA, SFFS and current stochastic algorithms for feature selection including simple GA and hybrid GAs on all data sets. However, only 5 runs were conducted for stochastic algorithms. Another improved BPSO algorithm was proposed in [26] for gene expression data. Speed concept was introduced to update particles' positions instead of velocity to increase the probability of not choosing a feature. In this way, PSO was able to find much smaller feature subsets than [9] and other compared methods. However, this method also reduced the accuracy in the cases, which might require a higher number of features in order to create a good prediction. Therefore, there were only six out of the ten data sets which had a higher average accuracy than [9]. Furthermore, except for some data sets which have 100% accuracy in all the runs, with a 10-run experiment, the standard deviations of the results on the remaining data sets were quite high (ranging from 0.3 to 0.9). There was also no statistical significance test done for the results.

In order to simultaneously maximise the classification accuracy and minimize the number of features selected, an aggregate fitness function was proposed for BPSO in [47]. This BPSO evolved in two stages. Classification error rate was the only measure used in the fitness function of the first stage. In the second stage, the subset size was added into the fitness function with an adaptive weight. This method evolved smaller feature subsets and higher classification performance than the standard BPSO and the PSO using only stage one. Alba et al. [1] combined geometric BPSO with SVM for feature selection where current position, *pbest* and *gbest* were used as three parents in a three-parent mask-based crossover operator to determine the new position for each particle. SVM with 10-fold cross-validation was applied in the fitness evaluation process. An aggregate fitness function was used to simultaneously maximise the classification accuracy and minimize the number of features with different weights. Compared to the second algorithm proposed in this paper which combined GA with SVM, this method performed slightly better with smaller feature subsets. Experiments also showed that the initialisation of the PSO produced a great influence in the performance, since it introduced an early subset of acceptable solutions in the evolution process.

Based on this motivation, Xue et al [45] proposed three mechanisms to initialize particles in PSO for feature selection in classification. While small initialisation method generated particles with a small number of roughly 10% of features selected, the large initialisation method generated particles with a random large number of more than 50% of features selected. The mixed method used the small initialisation for most of the particles (about two-thirds) and the large initialisation for the remainders. The mixed initialisation gave the best results of the three methods. Although achieving as a good classification performance as that of standard PSO, the feature subsets evolved by the mixed method are smaller than those of PSO in eight out of 14 data sets, thereby, reducing the computation time. Additionally, the paper also introduced three new updating mechanisms for *pbest*, *gbest*. Combining the mixed initialisation method with the new updating method achieved much smaller subsets and better or at least similar accuracy as the standard PSO and the two-stage algorithm [46,47].

Unler et al. [39] proposed a wrapper feature selection method for binary classification problems based on a modified BPSO and a logistic regression model. In this study, BPSO was modified by extending social learning to update the velocity of the particles. An adaptive feature subset selection strategy was developed, where the features were selected not only according to their independent likelihood calculated by BPSO, but also according to their contribution to the subset of features already selected. Meanwhile, this strategy maintained a list of features, which had already been considered for feature addition. Only a limited number of features were considered to be selected in the feature subset, thus the computational effort for the classification learning was reduced. Experimental results indicated that the proposed method outperformed tabu search and scatter search algorithms.

Based on a statistical clustering method and BPSO, Lane et al. [20] developed a new wrapper feature selection algorithm, where features are grouped into different clusters by the statistical clustering method based on their similarity, i.e. relatively homogeneous features in the same group. The proposed algorithm aimed to select one representative features from each cluster. The results show that by selecting only a very small number of features, the algorithm can achieve similar or even better classification performance than using all features. Later, multiple or zero features are allowed to be selected from each cluster in [21] to further improve the classification performance.

A wrapper multi-objective PSO was proposed by Xue et al [48] using the non-dominated sorting concept (NSPSOFS) and the crowding, mutation and dominance concept (CMDPSOFS) to evolve non-dominated solutions for feature selection in classification. The results showed that both algorithms achieved more and better solutions than existing deterministic feature selection algorithms, i.e. linear forward selection (LFS), greedy stepwise backward selection (GSBS), and stochastic algorithms as the standard PSO, and the two-stage PSO [47]. By using the strategies of maintaining the diversity of the swarm, CMDPSOFS outperformed NSPSOFS and other three well-known evolutionary multi-objective algorithms, namely non-dominated sorting-based multi-objective GA II, strength Pareto evolutionary algorithm 2 and Pareto archived evolutionary strategy on 12 benchmark data sets. The performance of continuous PSO and binary PSO for multi-objective feature selection is compared in [44]. The results show that continuous PSO generally achieved better performance than binary PSO. More PSO based filter feature selection algorithms can be seen from [2,25,51].

4.2   PSO Based Filter Feature Selection

Wang et al [40] proposed a filter feature selection approach based on an improved BPSO and rough sets theory. In this BPSO method, velocity was used to determine the number of bits should be changed in the binary position of the particle. The fitness function combined the dependency degree of classes on features calculated according to the rough sets theory and the proportion of the selected features. Experimental results showed that the improved BPSO was computationally less expensive than a GA using rough sets in terms of both memory and running time. However, the classification performance of the feature subsets was only tested on LEM2 algorithm, which had some bias for rough set based algorithms. A fuzzy sets based fitness function was introduced by Chakraborty [7] to build a BPSO based filter feature selection algorithm. Feature evaluation index [31] was used in the fitness function. It aimed to find a feature subset which had minimum intraclass ambiguity and maximum interclass ambiguity. Experiments with the Iris and Sonar data sets illustrated that the proposed BPSO performed better than GA did. However, both Iris and Sonar include a relatively small number of features.

Cervante et al. [6] developed two filter based approaches using BPSO. In the first algorithm, mutual information was used to measure the relevance between features and the class labels and the redundancy between a pair of features. Meanwhile, entropy was employed in the second method to measure the relevance of a group of features to the class labels and the redundancy within a group of features. Both algorithms used an aggregate fitness function combining the relevance level and the redundancy level with different weights. The results showed that the first algorithm evolved smaller subsets while the second one produced better classification accuracy. However, the classification performance using the feature subsets evolved by both algorithms was just as good as using all the features in three out of four data sets. This confirms one of the drawbacks of filter based approaches. Xue et al [43] further explored the effectiveness of mutual information and entropy in multi-objective feature selection. These two measures were combined with non-dominated sorting concept and the crowding, mutation and dominance concept as in [48] to form four different multi-objective BPSO algorithms: NSfsMI, NSfsE, CMDfsMI, CMDfsE. The results showed that the proposed multi-objective approaches achieved better solutions than the single-objective BPSO using the same measures [6]. The algorithms using entropy achieved better classification performance than those using mutual information. Although algorithms using mutual information selected a smaller number of features than those using entropy in single-objective algorithms, this observation did not appear in multi-objective algorithms. CMDfsMI and CMDfsE outperformed all the other methods in terms of both the number of features and the classification performance.

Some studies not only used PSO for feature selection but also employed PSO to optimize parameters for the classification algorithm used to evaluate the feature subsets. Lin et al. [23] proposed a wrapper feature selection approach (PSO+SVM), which simultaneously determined the parameters and picked a subset of features using continuous PSO. Radial basis kernel function (RBF) with two parameters was used in SVM. In PSO+SVM, each particle with $(n + 2)$ variables represents n features and 2 parameters. Experiments with 10-fold crossover validation demonstrated that the classification accuracy of PSO+SVM outperformed that of a grid search, a newton SVM and a Lagrangian SVM. The PSO+SVM approach could simultaneously determine the parameter values and find a subset of features without the lowering the classification accuracy.

Similarly, Huang et al. [15] also developed a wrapper method (PSO-SVM) for feature selection and parameters determination in one process. The difference between PSO-SVM and PSO+SVM [23] was that PSO-SVM used binary PSO and continuous PSO to simultaneously optimize the feature subset and SVM kernel parameters, respectively. Experiments with 10-fold crossover validation showed that PSO-SVM could determine the parameters, search the discriminating feature subset simultaneously and also achieve high classification accuracy.

Mohemmed et al. [27] proposed a hybrid method (PSOAdaBoost), which incorporated PSO with an AdaBoost framework for face detection. The PSOAdaBoost algorithm picked the best feature subset and determined the decision thresholds of AdaBoost simultaneously, so it could speed up the process of the training and increase the accuracy of weak classifiers. The method encoded the particle with the feature parameters and two centroids which were used to label an instance into the positive or negative class. Experimental results showed that PSOAdaBoost could be trained in a much shorter time and improve the performance of feature selection. This method used different learning algorithms and test problems with PSO+SVM [23] and PSO-SVM [15], but all of them could optimize the feature subset and parameters in one process.

A filter-based PSO was introduced by Guan et al [14] for feature selection in microarray data sets. Two informativeness metrics constructed based on ANOVA statistics were used to evaluate feature subsets. The experiment results on two binary-class data sets were compared with six methods including the Two-Phase EA/KNN, the SVM, the GA-SVM, the EA, the Redundancy based and the PSO-SVM. Although the proposed method always evolved the smallest subsets, it can only achieved the best accuracy on one dataset. More than ten independent runs should be conducted and statistical significant test should be done in order to have a stronger conclusion. A two-stage based filter feature selection algorithm using BPSO was proposed in [33], where k-means technique was used in the first stage to cluster features into k clusters. Then, signal-to-noise ratio score was used as a filter approach to select the best feature from each cluster. The k selected features were transfered into the second stage for PSO to search for the optimal subset. SVM, KNN and Probabilistic Neural Network were used to evaluate the goodness of the selected features. The results showed that the proposed method achieved higher accuracy with much smaller subsets than using all features. However, the experiment was applied on only four binary-class gene data sets. The efficiency of the proposed method were not discussed and compared with any other EC techniques. The different setting values of k used in k-means technique for different data sets might require some expert knowledge about microarray data. The work in [33] is similar to that in [20,21], but use different clustering methods to group features. It will be interesting to investigate their advantages and disadvantages by comparing with each other. More PSO based filter feature selection algorithms can be seen from [4,5,8,36,38,42].

5   Conclusions: Current Issues and Challenges

Many different EC algorithms have been applied to feature selection problems. In recent work, both filter and wrapper approaches were developed and feature selection was also regarded as multi-objective problems. This paper mainly reviewed recent PSO based feature selection approaches. In conclusion, some discussions about current issues and challenges as well as some possible research directions of PSO for feature selection are given as follows:

- Feature selection for *large-scale* classification with thousands or tens of thousands of features is still a challenging task, since most of the existing methods have difficulties to scale up to such high dimensions. These problems typical require a novel search mechanism and evaluation methods;
- High computational cost is one of the main problems in feature selection. Efficient evaluation or fitness *measures* can significantly speed up the feature selection process. This can be achieved by developing new filter measures, such as information theory measures, consistency measures, statistical measures, and fuzzy or rough sets based measures;
- In PSO or other EC based feature selection methods, the traditional *representation* is one of the issues limiting their performance on large-scale complex problems. Developing a novel representation is also an open issue but with very few existing works. The novel representation can be continuous or discrete (or binary) and fixed-length or variable length;
- Feature selection is an NP hard problem which requires a powerful global search technique. To improve the performance of an EC based approach, new *search mechanisms* are needed to develop, which may involve local search, hybrid different EC search mechanisms and so on;

- The number of features and the classification performance are often conflicting objectives, but only a small number of multi-objective feature works have been conducted. Meanwhile, developing new evaluation metrics and further selection method to choose a single solutions from a set of trade-off solutions are also an interesting topics, and
- Feature selection is not only important in classification, but also for other problems, such as clustering and regression problems. Feature selection in such domains will also be interesting.

Acknowledgment. This work is supported in part by the National Science Foundation of China (NSFC No. 61170180), the Marsden Funds of New Zealand (VUW1209 and VUW0806) and the University Research Funds of Victoria University of Wellington (203936/3337).

References

1. Alba, E., Garcia-Nieto, J., Jourdan, L., Talbi, E.G.: Gene selection in cancer classification using PSO/SVM and GA/SVM hybrid algorithms. In: IEEE Congress on Evolutionary Computation (CEC 2007), pp. 284–290 (2007)
2. Aneesh, M., Masand, A.A., Manikantan, K.: Optimal feature selection based on image pre-processing using accelerated binary particle swarm optimization for enhanced face recognition. Procedia Engineering 30(5), 750–758 (2012)
3. Azevedo, G., Cavalcanti, G., Filho, E.: An approach to feature selection for keystroke dynamics systems based on PSO and feature weighting. In: IEEE Congress on Evolutionary Computation (CEC 2007), pp. 3577–3584 (2007)
4. Cervante, L., Xue, B., Shang, L., Zhang, M.: Binary particle swarm optimisation and rough set theory for dimension reduction in classification. In: IEEE Congress on Evolutionary Computation, pp. 2428–2435 (2013)
5. Cervante, L., Xue, B., Shang, L., Zhang, M.: A multi-objective feature selection approach based on binary pso and rough set theory. In: Middendorf, M., Blum, C. (eds.) EvoCOP 2013. LNCS, vol. 7832, pp. 25–36. Springer, Heidelberg (2013)
6. Cervante, L., Xue, B., Zhang, M., Shang, L.: Binary particle swarm optimisation for feature selection: A filter based approach. In: IEEE Congress on Evolutionary Computation (CEC 2012), pp. 881–888 (2012)
7. Chakraborty, B.: Feature subset selection by particle swarm optimization with fuzzy fitness function. In: 3rd International Conference on Intelligent System and Knowledge Engineering (ISKE 2008), pp. 1038–1042 (2008)
8. Chakraborty, B., Chakraborty, G.: Fuzzy consistency measure with particle swarm optimization for feature selection. In: 2013 IEEE International Conference on Systems, Man, and Cybernetics (SMC 2013), pp. 4311–4315 (2013)
9. Chuang, L.Y., Chang, H.W., Tu, C.J., Yang, C.H.: Improved binary PSO for feature selection using gene expression data. Computational Biology and Chemistry 32(29), 29–38 (2008)
10. Chuang, L.Y., Tsai, S.W., Yang, C.H.: Improved binary particle swarm optimization using catfish effect for feature selection. Expert Syst. Appl. 38(10), 12699–12707 (2011)

11. Dash, M., Liu, H.: Feature selection for classification. Intelligent Data Analysis 1(4), 131–156 (1997)
12. Eberhart, R., Kennedy, J.: A new optimizer using particle swarm theory. In: Proceedings of the Sixth International Symposium on Micro Machine and Human Science, MHS 1995, pp. 39–43 (October 1995)
13. Eberhart, R., Shi, Y.: Particle swarm optimization: developments, applications and resources. In: IEEE Congress on Evolutionary Computation, vol. 1, pp. 81–86 (2001)
14. Guan, J., Han, F., Yang, S.: A new gene selection method for microarray data based on PSO and informativeness metric. In: Huang, D.-S., Jo, K.-H., Zhou, Y.-Q., Han, K. (eds.) ICIC 2013. LNCS, vol. 7996, pp. 145–154. Springer, Heidelberg (2013)
15. Huang, C.L., Dun, J.F.: A distributed PSO-SVM hybrid system with feature selection and parameter optimization. Appl. Soft Comput. 8, 1381–1391 (2008)
16. Ke, L., Feng, Z., Xu, Z., Shang, K., Wang, Y.: A multiobjective ACO algorithm for rough feature selection. In: Second Pacific-Asia Conference on Circuits,Communications and System (PACCS 2010), vol. 1, pp. 207–210 (2010)
17. Kennedy, J., Eberhart, R.: Particle swarm optimization. In: IEEE International Conference on Neural Networks, vol. 4, pp. 1942–1948 (1995)
18. Kennedy, J., Eberhart, R.: A discrete binary version of the particle swarm algorithm. In: IEEE International Conference on Systems, Man, and Cybernetics, vol. 5, pp. 4104–4108 (1997)
19. Kennedy, J., Spears, W.: Matching algorithms to problems: an experimental test of the particle swarm and some genetic algorithms on the multimodal problem generator. In: IEEE World Congress on Computational Intelligence, pp. 78–83 (1998)
20. Lane, M., Xue, B., Liu, I., Zhang, M.: Particle swarm optimisation and statistical clustering for feature selection. In: Cranefield, S., Nayak, A. (eds.) AI 2013. LNCS, vol. 8272, pp. 214–220. Springer, Heidelberg (2013)
21. Lane, M., Xue, B., Liu, I., Zhang, M.: Gaussian based particle swarm optimisation and statistical clustering for feature selection. In: Blum, C., Ochoa, G. (eds.) EvoCOP 2014. LNCS, vol. 8600, pp. 133–144. Springer, Heidelberg (2014)
22. Langley, P.: Selection of relevant features in machine learning. In: AAAI Technique Report FS-94-02, pp. 127–131 (October 1994)
23. Lin, S.W., Ying, K.C., Chen, S.C., Lee, Z.J.: Particle swarm optimization for parameter determination and feature selection of support vector machines. Expert Systems with Applications 35(4), 1817–1824 (2008)
24. Liu, H., Dougherty, E., Dy, J., Torkkola, K., Tuv, E., Peng, H., Ding, C., Long, F., Berens, M., Parsons, L., Zhao, Z., Yu, L., Forman, G.: Evolving feature selection. IEEE Intelligent Systems 20(6), 64–76 (2005)
25. Mohamad, M., Omatu, S., Deris, S., Yoshioka, M., Abdullah, A., Ibrahim, Z.: An enhancement of binary particle swarm optimization for gene selection in classifying cancer classes. Algorithms for Molecular Biology 8(1), 15 (2013)
26. Mohamad, M., Omatu, S., Deris, S., Yoshioka, M.: A modified binary particle swarm optimization for selecting the small subset of informative genes from gene expression data. Information Technology in Biomedicine 15(6), 813–822 (2011)
27. Mohemmed, A., Zhang, M., Johnston, M.: Particle Swarm Optimization based Adaboost for face detection. In: IEEE Congress on Evolutionary Computation, pp. 2494–2501 (2009)
28. Narendra, P., Fukunaga, K.: A Branch and Bound Algorithm for Feature Subset Selection. IEEE Transactions on Computers C-26(9), 917–922 (1977)
29. Neshatian, K., Zhang, M., Andreae, P.: A filter approach to multiple feature construction for symbolic learning classifiers using genetic programming. IEEE Transactions on Evolutionary Computation 16(5), 645–661 (2012)

30. Oliveira, L., Sabourin, R., Bortolozzi, F., Suen, C.: Feature selection using multi-objective genetic algorithms for handwritten digit recognition. In: 16th International Conference on Pattern Recognition, vol. 1, pp. 568–571 (2002)
31. Pal, S., Chakraborty, B.: Fuzzy Set Theoretic Measure for Automatic Feature Evaluation. IEEE Transactions on Systems, Man and Cybernetics 16(5), 754–760 (1986)
32. Purohit, A., Chaudhari, N., Tiwari, A.: Construction of classifier with feature selection based on genetic programming. In: IEEE Congress on Evolutionary Computation, pp. 1–5 (2010)
33. Sahu, B., Mishra, D.: A Novel Feature Selection Algorithm using Particle Swarm Optimization for Cancer Microarray Data. Procedia Engineering 38(0), 27–31 (2012)
34. Shi, Y., Eberhart, R.: A modified particle swarm optimizer. In: IEEE Congress on Evolutionary Computation, pp. 69–73 (1998)
35. Sivagaminathan, R.K., Ramakrishnan, S.: A hybrid approach for feature subset selection using neural networks and ant colony optimization. Expert Systems with Applications 33(1), 49–60 (2007)
36. Stevanovic, A., Xue, B., Zhang, M.: Feature selection based on pso and decision-theoretic rough set model. In: IEEE Congress on Evolutionary Computation, pp. 2840–2847 (2013)
37. Subbotin, S., Oleynik, A.: The multi objective evolutionary feature selection. In: International Conference on Modern Problems of Radio Engineering, Telecommunications and Computer Science, pp. 115–116 (2008)
38. Unler, A., Alper Murat, R.B.C.: mr2pso: A maximum relevance minimum redundancy feature selection method based on swarm intelligence for support vector machine classification. Information Sciences 20, 4625–4641 (2011)
39. Unler, A., Murat, A.: A discrete particle swarm optimization method for feature selection in binary classification problems. European Journal of Operational Research 206(3), 528–539 (2010)
40. Wang, X., Yang, J., Teng, X., Xia, W., Jensen, R.: Feature selection based on rough sets and particle swarm optimization. Pattern Recognition Letters 28(4), 459–471 (2007)
41. Waqas, K., Baig, R., Ali, S.: Feature subset selection using multi-objective genetic algorithms. In: 13th International Conference on INMIC, pp. 1–6 (2009)
42. Xue, B., Cervante, L., Shang, L., Browne, W.N., Zhang, M.: Binary PSO and rough set theory for feature selection: A multi-objective filter based approach. International Journal of Computational Intelligence and Applications 13(02), 1450009 (2014)
43. Xue, B., Cervante, L., Shang, L., Browne, W., Zhang, M.: A multi-objective particle swarm optimisation for filter-based feature selection in classification problems. Connect. Sci. 24(2-3), 91–116 (2012)
44. Xue, B., Zhang, M., Browne, W.N.: Multi-objective particle swarm optimisation for feature selection. In: Genetic and Evolutionary Computation Conference (GECCO 2012), Philadelphia, PA, USA, pp. 81–88. ACM (2012)
45. Xue, B., Zhang, M., Browne, W.N.: Particle swarm optimisation for feature selection in classification: Novel initialisation and updating mechanisms. Applied Soft Computing 18(0), 261–276 (2014)
46. Xue, B., Zhang, M., Browne, W.N.: Novel initialisation and updating mechanisms in PSO for feature selection in classification. In: Esparcia-Alcázar, A.I. (ed.) EvoApplications 2013. LNCS, vol. 7835, pp. 428–438. Springer, Heidelberg (2013)
47. Xue, B., Zhang, M., Browne, W.: New fitness functions in binary particle swarm optimisation for feature selection. In: IEEE Congress on Evolutionary Computation, pp. 1–8 (2012)
48. Xue, B., Zhang, M., Browne, W.: Particle swarm optimization for feature selection in classification: A multi-objective approach. IEEE Transactions on Cybernetics 43(6), 1656–1671 (2013)

49. Yang, C.S., Chuang, L.Y., Ke, C.H., Yang, C.H.: Boolean binary particle swarm optimization for feature selection. In: IEEE Congress on Evolutionary Computation (CEC 2008), pp. 2093–2098 (2008)
50. Yang, C.S., Chuang, L.Y., Li, J.C., Yang, C.H.: Chaotic maps in binary particle swarm optimization for feature selection. In: IEEE Conference on Soft Computing in Industrial Applications, pp. 107–112 (2008)
51. Yong, Z., Dunwei, G., Ying, H., Wanqiu, Z.: Feature selection algorithm based on bare bones particle swarm optimization. Neurocomputing 148(0), 150–157 (2015)
52. Yu, X.M., Xiong, X.Y., Wu, Y.W.: A PSO-based approach to optimal capacitor placement with harmonic distortion consideration. Electric Power Systems Research 71(1), 27–33 (2004)

A Comparison between Two Evolutionary Hyper-Heuristics for Combinatorial Optimisation

Richard J. Marshall[1], Mark Johnston[1], and Mengjie Zhang[2]

[1] School of Mathematics, Statistics and Operations Research
[2] School of Engineering and Computer Science
Victoria University of Wellington, New Zealand
{richard.marshall,mark.johnston}@msor.vuw.ac.nz
mengjie.zhang@ecs.vuw.ac.nz

Abstract. Developing and managing a general method of solving combinatorial optimisation problems reduces the need for expensive human experts when solving previously unseen variations to common optimisation problems. A hyper-heuristic provides such a method. Each hyper-heuristic has its own strengths and weaknesses and we research how these properties can be managed. We construct and compare simplified versions of two existing hyper-heuristics (adaptive and grammar-based), and analyse how each handles the trade-off between computation speed and quality of the solution. We test the two hyper-heuristics on seven different problem domains using the HyFlex framework. We conclude that both hyper-heuristics successfully identify and manipulate low-level heuristics to generate "good" solutions of comparable quality, but the adaptive hyper-heuristic consistently achieves this in a shorter computational time than the grammar based hyper-heuristic.

1 Introduction

Suppose you are given the task of solving a combinatorial optimisation problem within a time limit with only outline knowledge of the problem domain, and with only high-level details of the heuristics and operators you have at your disposal. Traditional methods of solving combinatorial optimisation problems use algorithms and heuristics, such as a branch-and-bound algorithm [6] or meta-heuristic search, e.g., tabu search [7]. In general, these methods achieve good results but often require detailed domain information and can be complex and time consuming to design and execute. A hyper-heuristic is useful where a more general (domain independent) method is required. It requires only outline knowledge of the problem domain and is particularly useful when dealing with specific variations to common optimisation problems.

The term hyper-heuristic was defined by Cowling et al. [5] as "heuristics to choose heuristics". Ochoa et al. [15] note that the focus in hyper-heuristic research is to adaptively find a solution *method* rather than producing a solution for the particular problem instance at hand. They repeat the observation by Ross [16] that the difference between hyper-heuristics and (meta-)heuristics is

G. Dick et al. (Eds.): SEAL 2014, LNCS 8886, pp. 618–630, 2014.

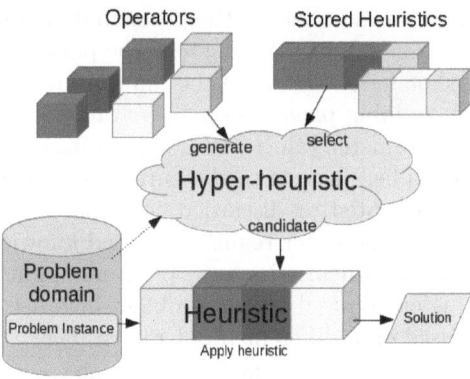

Fig. 1. The relationship between a hyper-heuristic and heuristics

that it is the search space of heuristics, rather than the search space of problem solutions, that is traversed.

The motivation for our research is to understand the major design properties required when developing or selecting a hyper-heuristic. The goal of this paper is to identify and report the strengths and weaknesses of two different evolutionary hyper-heuristics when solving previously unseen problem instances in one of several possible problem domains. Each hyper-heuristic is required to select and execute operators from an unseen set of low-level (domain specific) operators (heuristics), which in turn incrementally build and/or modify a solution to each problem instance. Understanding what makes a particular hyper-heuristic efficient and effective enables the trade-off between computational speed and quality of the result to be managed when faced with larger problem instances and more complex problem domains.

The remainder of this paper is organised as follows. A brief overview of hyper-heuristic research is given in Section 2 followed by our method in Section 3. The design, results and discussion of our experiments are detailed in Sections 4 and 5 followed by our conclusions in Section 6.

2   Background

A heuristic can be described as a method that iteratively applies one or more operators which build or modify a (possibly empty) solution, which is either accepted for the next iteration or discarded.

Burke et al. [2,4,3] analyse the various hyper-heuristic approaches published in the literature. They note that in some respects, genetic programming (GP) [8] can be regarded as a hyper-heuristic to select or generate heuristics. They identified two general hyper-heuristic categories (see Figure 1):

1. **Selection:** Hyper-heuristics which dynamically select one or more existing heuristics or operators (heuristic components) from a set of candidate

heuristics and operators suitable for the problem domain. Should a heuristic or operator require one or more parameters, the parameters can be adjusted dynamically based on feedback during the run (on-line learning) or fixed during a preliminary training phase using a separate set of problem instances to determine the parameter values (off-line learning). Each selected heuristic or operator is applied in turn to the current solution to generate a new solution, which is accepted or discarded. A hyper-heuristic differs from a meta-heuristic in that the latter requires detailed knowledge of the problem domain. The sequence of heuristics and operators applied during an on-line learning process is customised to the problem instance and is not usually effective when reused on a different problem instance.

2. **Generation:** Hyper-heuristics which generate new heuristics by recombining the operators (or components) of existing heuristics. An on-line learning process adjusts the operator combinations as well as setting any parameters. An on-line learning process can also be applied to a set of training instances to determine a reusable sequence of one or more heuristics and any relevant parameters. The sequence of heuristics is stored for future application on unseen problem instances in the same problem domain. If application of the heuristic does not involve further training (e.g. parameter modification), this two-stage combination of training and application is usually referred to as off-line learning.

A comparison of hyper-heuristics was undertaken in 2011 as part of the First Cross-domain Heuristic Search Challenge (CHeSC) [13]. The winning entry of the first CHeSC was the hyper-heuristic developed by Misir et al. [12]. The first hyper-heuristic used in this paper (see Section 3.2) is a simplified version of the multi-phase adaptive approach used by Misir et al. [12]. The proposed adaptive hyper-heuristic (AdaptiveHH) is a selection approach using on-line learning. Firstly, the hyper-heuristic generates a heuristic/operator selection vector, which is updated at the start of the second and subsequent phases based on the performance of each low-level heuristic or operator in the previous phase. The second step iteratively chooses and applies a heuristic/operator to a solution selected from a small population of solutions. The new solution replaces an existing solution in the population. The third step adjusts operator parameters depending on the frequency of improving solutions.

The second hyper-heuristic we use in this paper is based on the grammar guided GP [22] approach taken by Sabar et al. [18] which uses GP [8] in the manner described by Burke et al.[2]. The grammar guided genetic programming (GGGP) [22,11] hyper-heuristic (GrammarHH) is a generation approach using on-line learning during training and no learning when applied to a new problem instance (see Section 3.3), i.e., it is a generation approach with off-line learning. GrammarHH evolves a single heuristic, with parameter values, suitable for solving comparable problem instances. We use a separate set of six different sized problem instances for training.

The two hyper-heuristics are almost complete opposites in terms of hyper-heuristic classification [2,4]. Our choice of two different hyper-heuristics enables us to test the hypothesis that, when compared to AdaptiveHH, the knowledge gained (i.e. which operators are productive) during the GrammarHH training phase leads to a reduced computational time to achieve (or better) a target solution for a given problem instance.

3   Method

3.1   HyFlex Framework

The HyFlex (**Hy**per-heuristic **Flex**ible) framework [14] was originally developed in 2011 for the First Cross-domain Heuristic Search Challenge (CHeSC) [13]. The framework includes six in-built optimisation problem domains, to which we add a standard Capacitated Vehicle Routing Problem (CVRP) [20] domain:

1. Maximum satisfiability (MAX-SAT).
2. One-dimensional bin packing.
3. Permutation flow shop.
4. Personnel scheduling.
5. Travelling salesman (TSP).
6. Capacitated vehicle routing (CVRP) with time windows [19].
7. Standard CVRP [20] (added domain).

Associated with each problem domain is a set of between 8 and 15 unseen low-level operators (heuristics). Each set contains at least one operator belonging to each of the four defined operator types: mutation, ruin-recreate, local search and crossover. A crossover operator takes part of one solution and uses non-duplicating parts of another solution to complete a new solution. The HyFlex mutation and crossover operator types should not be confused with the identically named functions used with Genetic Programming [8].

Each operator can use (if appropriate) the two HyFlex parameters α and β, where ($0 \leq \alpha, \beta \leq 1$). The Intensity of Mutation parameter, α, affects the scale of any mutation or ruin operation, e.g., 0.5 would mean half the current solution would be altered by an operator using this parameter. The Depth of Search parameter, β, defines a range or number of repetitions an operator will undertake to find an improved solution in a single execution of the operator.

To enable a more detailed investigation, we implement a new CVRP (without time windows) problem domain compatible with the HyFlex framework. This additional problem domain helps us to undertake more detailed analysis on whether performance outcomes are due to the quality of the hyper-heuristic or the effectiveness of the chosen low-level operators. For the CVRP domain we develop twelve appropriate low-level operators (see Table 3) similar (but not identical) to those proposed by Walker et al. [21].

3.2   Adaptive On-line Learning Hyper-Heuristic (AdaptiveHH)

This is a simplified variation of the adaptive on-line learning approach developed by Misir et al. [12]. This hyper-heuristic is similar to a meta-heuristic process, and manipulates the unseen low-level operators to find a solution to a single problem instance. AdaptiveHH is only suitable for on-line learning since the process is dynamically customised for a given problem instance.

The original multi-phase approach, and our modification thereof, is as follows:

1. **Operator Selection Probability Vector.** Selection of operators is based on a probability vector. In the first phase, all operators have an equal chance of being randomly selected at each iteration. After a defined number of iterations, AdaptiveHH starts a new phase and the probability vector is recalculated and normalised based on each operator's success rate (number of improving solutions ÷ operator calls) in the previous phase. The modified vector improves the chances of operators with a higher success rate being selected during the new phase. Misir et al. [12] use more complex formulae to recalculate the vector, incorporating operator execution time and time remaining. The selection probability of an operator which consistently fails to improve the solution is gradually reduced to a specified minimum.

2. **Operator Selection and Execution.** To facilitate the crossover operator type (see Section 3.1), which uses two parent solutions, a small population of solutions is maintained at any one time. Initially the population consists of, say, 6 solutions generated using the unseen construction method defined for the relevant problem domain. An operator is selected and executed on a solution in the current population of solutions. Misir et al. [12] use the best solution found so far as the primary parent solution whereas we randomly select a parent solution from the population at each iteration.

3. **Reinitialisation.** If no improved solutions are achieved for a specified number of iterations (i_1) then the parameters α,β (see Section 3.1) are increased in steps of 0.025 to a maximum of 0.8. If no improved solutions are achieved for a larger specified number of iterations (i_2, where $i_2 \gg i_1$), then the population of solutions (other than the best found so far) is reinitialised using the construction method defined for the relevant problem domain. In the latter case, the parameters α and β are reset to the defined minimum values.

4. **"Relay" Hybridisation.** Misir et al. [12] include a step which selects and executes operators in pairs. We omit this step in the interests of simplicity and to enable later evaluation of the usefulness of this step.

5. **Adaptive Move Acceptance.** In the event no improving solutions are found for a defined number of iterations, Misir et al. [12] incorporate a step which enables selection of a solution other than the best found so far as the primary parent solution for the next iteration. We omit this step as it is unnecessary when randomly selecting the primary parent solution.

The hyper-heuristic is repeated until a pre-determined time limit or early termination condition (i.e. attainment of a target value) is reached.

3.3   Grammar Guided Genetic Programming Hyper-Heuristic (GrammarHH)

GrammarHH develops a reusable heuristic which can be applied to any problem instance in the same domain. The GP population consists of low-level heuristics

Table 1. Grammar used with GrammarHH on the CVRP domain. The HyFlex parameters, α and β, are set by the $< global >$ production rule.

LHS	Options
$< heuristic >$	$< global >< alter >< localsearch >$
$< global >$	$< range >< range >$
$< alter >$	$< mutate >\ \|\ < mutate >< alter >$
$< localsearch >$	$< search >\ \|\ < search >< crossover >\ \|\ < crossover >< search >$
$< range >$	$0.2\|\ 0.25\|\ 0.3\|\ 0.35\|\ 0.4\|\ 0.45\|\ 0.5\|\ 0.55\|\ 0.6\|\ 0.65\|\ 0.7\|\ 0.75\|\ 0.8$
$< mutate >$	$m0\|\ m1\|\ m2\|\ m3\|\ r0\|\ r1$
$< crossover >$	$x0\|\ x1$
$< search >$	$s0\|\ s1\|\ s2\|\ s3$

evolved from component operators in one of the sequences in the "language" defined by the domain independent grammar (see Table 1). Each heuristic consists of one or more operators from the *mutation* and/or *ruin-recreate* operator types (see Section 3.1) which are applied in sequence. This is followed by a *search* operator which may be preceded or succeeded by a *crossover* operator.

Each type of operator in the grammar is identified by a letter prefix followed by a identification number. For convenience, and to avoid use of modular arithmetic, the $< mutate >$, $< crossover >$ and $< search >$ rules are customised for each problem domain to match the number of operators of each operator type. A domain independent grammar can be generated by omitting this customisation and using modular arithmetic to map the selected rule to the operator.

Tournament selection is used at each new GP generation to choose the heuristics on which GP crossover and mutation operators are to be applied to create the next generation. Sabar et al. [18] used Grammatical Evolution [17] to manage the GP process, whereas we use a generic Grammar Guided Genetic Programming (GGGP) [22,11] approach.

The evolved heuristic is repeatedly applied to each problem instance until a time limit or early termination condition is reached. During the training phase we apply the evolved heuristic to six problem instances selected in advance.

With GrammarHH, only improving or equal solutions are accepted, whereas Sabar et al. [18] include a choice of eight different acceptance options. We repeat training 30 times and store every generated heuristic that delivers a fitness on the six training instances within a specified target threshold (within 1.5% of the aggregate of the best found solutions to each instance).

Both hyper-heuristics were implemented using the HyFlex HyperHeuristic [14] template. GGGP is implemented as strongly-typed GP within the ECJ [9,10] software package.

4   Experiment Design

We evaluate the operator selection process and the quality of the solution achieved when the chosen operators are applied. The quality of the solution is measured against two target values set for each of 58 CVRP instances [1].

Table 2. Parameters used for experiments

Parameter	AdaptiveHH	GrammarHH
Time limit	2 mins.	2 mins.
Parameter value range	$0.2 \leq \alpha, \beta \leq 0.8$	$0.2 \leq \alpha, \beta \leq 0.8$
Parameter adjustment threshold	500 calls	n/a
Parameter increment on threshold	0.025	n/a
No progress reinitialisation threshold	10,000 calls	10,000 calls
Hyflex Solution Population	6	6
GP Heuristic Population	n/a	25
GP Generations	n/a	8
GP Crossover probability	n/a	0.8
GP Mutation probability	n/a	0.2

Both hyper-heuristics will endeavour to select the best performing operators from an unseen set of operators applicable to the problem domain. The problem domains we use contain between 8 and 15 operators. We analyse the frequency with which each operator is selected by AdaptiveHH in the seven problem domains, and, in the case of the CVRP domain, compare the frequency with the number of times the operator is selected as part of a "good" heuristic generated by GrammarHH.

To test our hypothesis we measure the speed of each hyper-heuristic in finding (if possible) a solution equal to, or better than, one of two targets set at 0.5% and 1.5% above the best solution found during preliminary runs on each of 58 CVRP instances. The instances range in size between 32 and 262 customers. Due to the small number of instances in the six in-built HyFlex domains, we do not extend this test to those domains.

After some preliminary experimentation we use the parameter settings shown in Table 2.

5   Results and Discussion

We evaluate performance of the operator selection process and the attainment of the solution target values. Preliminary runs established a best found solution (bfs) for each CVRP instance (minimisation objective) and target values established, being bfs + 0.5% and bfs + 1.5%. The preliminary runs used a fixed computational time of 2 minutes per run. We omit the results from the fixed computational time runs from this paper as the two hyper-heuristics successfully manipulated the low-level operators to deliver solutions of comparable quality to all instances.

5.1   Operator Selection and Performance

We compare the number of times a particular operator is called, its relative success rate, and failed run rate (i.e. zero improving solutions in a run) when using

Table 3. CVRP Domain: Operator call (selection) and execution record. Runs: 1,920 (30 runs × 64 instances). Total operator calls (AdaptiveHH): 724.7 million. A *failed run* is recorded if an individual operator fails to deliver a single improved solution during an entire run.

Operator Name	Operator Type	Calls per Improvement	Failed Run	AdaptiveHH Call Rate	GrammarHH Selection
Swap within route	mutation	10,669	54.9%	3.0%	5.8%
Move within route	mutation	18,117	83.9%	0.9%	1.9%
Swap routes	mutation	14,332	42.4%	3.6%	6.7%
Move to route	mutation	18,028	85.6%	0.7%	4.8%
Band ruin	ruin rec.	6,727	0.5%	16.1%	15.4%
Ring ruin	ruin rec.	5,926	0.7%	17.6%	10.6%
Move if better	search	4,280	2.5%	16.4%	13.5%
Swap if better	search	9,688	38.1%	4.4%	8.6%
2-opt swap	search	6,360	20.2%	9.5%	8.6%
Move to best	search	6,610	13.6%	8.3%	4.8%
Random combine	crossover	6,572	7.9%	14.0%	13.5%
Longest combine	crossover	20,586	0.6%	5.4%	5.8%

AdaptiveHH and GrammarHH. We conduct this experiment using a maximum computation time of 2 minutes per run to achieve two target solution values. If, during the run, improving solutions cease to be achieved for 10,000 consecutive operator executions, the population of solutions (other than the best solution found so far) is reinitialised. The run continues with a fresh set of solutions, generated using the unseen construction method defined for the problem domain.

We observe each operator's average execution time but, unlike Misir et al. [12], we do not use the execution time to adjust the operation selection vector. In the CVRP domain the *local search* operators have an execution time approximately 36 times longer, and the *ruin-recreate* and *crossover* operators approximately 12 times longer, than the *mutation* operators.

Table 3 records the results from AdaptiveHH across 1,920 runs (30 repetitions × 64 CVRP instances (58 + 6 GrammarHH training instances)) with the CVRP problem domain using a maximum computation time of 2 minutes per run. We record a *success* against an operator each time the operator delivers an improved solution when executed (which may not be on the best solution to date). A *failed run* is recorded against an operator if the operator fails to deliver a single improved solution during the entire run.

The GrammarHH selection results in Table 3 are based on the number of times an operator is contained in the 47 different heuristics that achieved a fitness within a defined threshold during training.

A total of 724.7 million operator executions were made during the 1,920 runs with AdaptiveHH when allowing a maximum computation time of 2 minutes per run. A run was terminated early if the current solution for the CVRP instance was within a target value of 0.5% above the best solution found during preliminary runs. Overall computation time for the 1,920 runs was 30.5 hours.

Table 4. Low-level operator type performance across 300 or 360 runs using Adaptive HH on in-built HyFlex [14] domains. A *failed run* is recorded if an individual operator fails to deliver a single improved solution during an entire run. Note that the in-built crossover operator in the Bin Packing domain fails to capture usage data.

Domain	Op.Type	Num. Ops.	Call Rate	Calls/improvement	Failed Run
CVRP+TW	mutation	3	22.5%	299	29.3%
	ruin-rec.	2	20.2%	503	1.7%
	search	3	39.3%	69	6.2%
	crossover	2	18.0%	229	0.0%
TSP	mutation	5	7.7%	34,482	99.6%
	ruin-rec.	1	1.0%	∞	100%
	search	3	87.1%	5,881	46.0%
	crossover	4	4.2%	75,578	99.9%
MaxSAT	mutation	5	62.4%	88	6.3%
	ruin-rec.	1	3.2%	1,508	81.9%
	search	2	28.3%	84	0.0%
	crossover	2	6.1%	2,565	89.2%
Bin Packing	mutation	3	n/a	146	39.7%
	ruin-rec.	2	n/a	113	11.0%
	search	2	n/a	56	1.5%
	crossover	1	n/a	n/a	35.3%
Flow Shop	mutation	5	8.4%	62,472	99.8%
	ruin-rec.	2	16.9%	770	81.0%
	search	4	68.0%	1,265	5.9%
	crossover	4	6.7%	20,305	99.7%
Psnl Sched	mutation	1	8.9%	∞	100%
	ruin-rec.	3	25.9%	9	76.6%
	search	5	36.8%	2	34.5%
	crossover	3	28.4%	∞	100%

Comparable operator call results for the six in-built Hyflex problem domains, (see Section 3.1), grouped by operator type, are shown in Table 4. The in-built domains each contain 10 or 12 problem instances. Each problem instance is solved 30 times using different random number generator seeds. The data is therefore based on 300 or 360 runs for each domain.

5.2   Speed to Target Solution

We assess the speed at which each hyper-heuristic reaches (if possible) two target solution values for each of 58 CVRP instances ranging in size between 32 and 262 customers. A maximum time of 2 minutes is allowed to achieve the target value, after which the run is classified as out of time. Each instance is run 30 times, and the time to reach (if possible) a solution equal to, or better than, the two target values recorded. The two target values are based on the best solution to each CVRP instance found during previous experiments. GrammarHH uses the best performing heuristic obtained from the training phase. The results are shown in Table 5.

Table 5. Time to achieve (if possible) target solution on 58 CVRP instances run 30 times. Maximum 2 minute time limit. Target based on best found solution (bfs) to each instance during preliminary test runs.

Time (seconds)	AdaptiveHH bfs + 0.5%	GrammarHH bfs + 0.5%	AdaptiveHH bfs + 1.5%	GrammarHH bfs + 1.5%
$0 \rightarrow 1$	491	266	726	453
$1 \rightarrow 5$	150	192	228	311
$5 \rightarrow 10$	93	98	104	138
$10 \rightarrow 20$	91	94	101	122
$20 \rightarrow 30$	43	47	60	50
$30 \rightarrow 60$	71	77	92	91
$60 \rightarrow 120$	99	87	88	102
out of time	702	879	341	473
≤ 5 secs.(all)	37%	26%	55%	44%
.. small instances	79%	62%	94%	79%
.. mid size inst.	48%	35%	68%	62%
.. large instances	6%	4%	26%	20%
out of time (all)	40%	51%	20%	27%
.. small instances	11%	23%	1%	8%
.. mid size inst.	34%	47%	15%	20%
.. large instances	71%	77%	40%	50%

5.3 Discussion

Because AdaptiveHH adjusts the probability of an operator being selected based on its performance, selections become biased towards operators which have a higher success rate. Table 3 illustrates that both hyper-heuristics select operators in roughly the same proportion. However, AdaptiveHH uses the full range of operators during each run, whereas GrammarHH uses only one generated heuristic (of 47) containing a small subset of operators. The low success rate of mutation operators is offset by the relative speed of execution. However, some mutation operators have a very high failed run rate meaning additional calls of this type of operator may not achieve improving solutions regardless of the number of calls made. By comparison, the two ruin-recreate type operators in the CVRP domain have a relatively high success rate and a very low failed run rate, suggesting additional calls of either of these two operators might achieve improved solutions faster. The call rate shown in Table 3 illustrates that AdaptiveHH has progressively biased the selection towards the ruin-recreate operators and the first of the local search operators.

Table 4 illustrates that operators of a particular type do not perform consistently across domains and in some cases individual operators may have a very high failed run rate (100% in some cases). This leaves some domains with only a few productive operators, e.g., the Flow Shop domain. The challenge for the hyper-heuristic is to quickly identify and focus on applying those operators which perform well. The call rate results in Table 4 indicates that AdaptiveHH has, in general, successfully identified the best performing operators and biased selection accordingly.

Generally, AdaptiveHH outperforms GrammarHH in both speed and number of times the target value is attained. This relative performance between the two hyper-heuristics is consistent across all sizes of problem instance. Repeat experiments using the next three best heuristics developed during GrammarHH training achieve similar outcomes. Allowing more computational time enables some out-of-time runs to achieve the target value.

The hypothesis that knowledge gained (i.e. which domain-specific operators are productive) during the training phase of GrammarHH can improve the computation speed compared to AdaptiveHH is shown to be incorrect. There is evidence that the flexibility of AdaptiveHH outperforms the more rigid structure of GrammarHH when applied to new problem instances. This is possibly due to the variable performance of the mutation operators, which periodically fail for the entire run. In such cases, AdaptiveHH applies alternative operators, whereas GrammarHH is unable to adjust its approach. However the difference in performance may also be due to numerous factors ranging from the structure and content of the GrammarHH grammar to the arbitrary setting of the parameters both hyper-heuristics require.

Deciding whether to terminate a run early, or allow execution to continue until the time limit is reached, is a trade-off between speed and quality. In our research we use attainment of a target value to trigger early termination. However it may not always be possible to specify a realistic target when dealing with previously unseen problem instances. As shown in Table 5, some runs find a "good" solution in under 1 second. With AdaptiveHH at least one (in some cases, all) of the thirty runs with each instance attained the target solution within one second with 50 of the 58 CVRP instances.

6   Conclusions

These experiments indicate that both hyper-heuristics can successfully manipulate unseen low-level operators in different problem domains to deliver solutions of a reasonable quality. With hindsight, there appears to be a good case for including execution time in the vector adjustment process when using AdaptiveHH. The flexibility of AdaptiveHH outperforms the more rigid structure of GrammarHH in both computational speed and solution quality.

Our future research will investigate ways to improve the efficiency of the operator selection process. We shall also investigate ways in which the different hyper-heuristics can assist each other to automatically and intelligently set the hyper-heuristic environmental parameters and manage the trade-off between computational speed, solution quality, knowledge transferability, and scalability.

References

1. Capacitated Vehicle Routing Problem Instances (October 2013),
   http://neo.lcc.uma.es/vrp/vrp-instances/capacitated-vrp-instances/
2. Burke, E.K., Gendreau, M., Hyde, M., Kendall, G., Ochoa, G., Özcan, E., Qu, R.:
   Hyper-heuristics: a survey of the state of the art. Journal of the Operational Research
   Society 64(12), 1695–1724 (2013)
3. Burke, E.K., Hyde, M.R., Kendall, G., Ochoa, G., Ozcan, E., Woodward, J.R.:
   Exploring hyper-heuristic methodologies with genetic programming. In: Mumford,
   C.L., Jain, L.C. (eds.) Computational Intelligence. ISRL, vol. 1, pp. 177–201.
   Springer, Heidelberg (2009)
4. Burke, E.K., Hyde, M., Kendall, G., Ochoa, G., Özcan, E., Woodward, J.: A classification of hyper-heuristic approaches. In: Gendreau, M., Potvin, J.-Y. (eds.)
   Handbook of Metaheuristics, vol. 146, pp. 449–468. Springer (2010)
5. Cowling, P.I., Kendall, G., Soubeiga, E.: A hyperheuristic approach to scheduling
   a sales summit. In: Burke, E., Erben, W. (eds.) PATAT 2000. LNCS, vol. 2079,
   pp. 176–190. Springer, Heidelberg (2001)
6. Fisher, M.L.: Optimal solution of vehicle routing problems using minimum k-trees.
   Operations Research 42(4), 626–642 (1994)
7. Glover, F.: Tabu search: Part I. ORSA Journal on Computing 1(3), 190–206 (1989)
8. Koza, J.R.: Genetic Programming: On the Programming of Computers by Means
   of Natural Selection. MIT Press (1992)
9. Luke, S.: Essentials of Metaheuristics, 2nd edn. Lulu (2013),
   http://cs.gmu.edu/~sean/book/metaheuristics/
10. Luke, S., Panait, L., Balan, G., Paus, S., Skolicki, Z., Kicinger, R., Popovici, E.,
    Sullivan, K., Harrison, J., Bassett, J., Hubley, R., Desai, A., Chircop, A., Compton,
    J., Haddon, W., Donnelly, S., Jamil, B., Zelibor, J., Kangas, E., Abidi, F., Mooers,
    H., O'Beirne, J., Talukder, K.A., McDermott, J.: Evolutionary Computation in
    Java (May 2014), http://cs.gmu.edu/~eclab/projects/ecj/
11. McKay, R.I., Hoai, N.X., Whigham, P.A., Shan, Y., O'Neill, M.: Grammar-
    based genetic programming: A survey. Genetic Programming and Evolvable Ma-
    chines 11(3-4), 365–396 (2010)
12. Misir, M., Verbeeck, K., De Causmaecker, P., Vanden Berghe, G.: A new hyper-
    heuristic as a general problem solver: an implementation in HyFlex. Journal of
    Scheduling 16, 291–311 (2013)
13. Ochoa, G., Hyde, M.: Cross-domain Heuristic Search Challenge (2011),
    http://www.asap.cs.nott.ac.uk/external/chesc2011/
14. Ochoa, G., Hyde, M., Curtois, T., Vazquez-Rodriguez, J.A., Walker, J., Gendreau,
    M., Kendall, G., McCollum, B., Parkes, A.J., Petrovic, S., Burke, E.K.: HyFlex: A
    benchmark framework for cross-domain heuristic search. In: Hao, J.-K., Middendorf,
    M. (eds.) EvoCOP 2012. LNCS, vol. 7245, pp. 136–147. Springer, Heidelberg (2012)
15. Ochoa, G., Qu, R., Burke, E.K.: Analysing the landscape of a graph based hyper-
    heuristic for timetabling problems. In: Proceedings of the Genetic and Evolutionary
    Computation Conference, pp. 341–348 (2009)
16. Ross, P.: Hyper-heuristics. In: Burke, E.K., Kendall, G. (eds.) Search Method-
    olgies: Introductory Tutorials in Optimization and Decision Support Techniques,
    pp. 529–556. Kluwer (2005)
17. Ryan, C., Collins, J.J., Neill, M.O.: Grammatical evolution: Evolving programs for
    an arbitrary language. In: Banzhaf, W., Poli, R., Schoenauer, M., Fogarty, T.C.
    (eds.) EuroGP 1998. LNCS, vol. 1391, pp. 83–96. Springer, Heidelberg (1998)

18. Sabar, N.R., Ayob, M., Kendall, G., Qu, R.: Grammatical evolution hyper-heuristic for combinatorial optimization problems. IEEE Transactions on Evolutionary Computation 17(6), 840–861 (2013)
19. Solomon, M.M.: Algorithms for the vehicle routing problem with time windows. Transportation Science 29(2), 156–166 (1995)
20. Toth, P., Vigo, D.: The Vehicle Routing Problem. SIAM (2002)
21. Walker, J.D., Ochoa, G., Gendreau, M., Burke, E.K.: Vehicle routing and adaptive iterated local search within the *HyFlex* hyper-heuristic framework. In: Hamadi, Y., Schoenauer, M. (eds.) LION 2012. LNCS, vol. 7219, pp. 265–276. Springer, Heidelberg (2012)
22. Whigham, P.A.: Grammatically-based genetic programming. In: Proceedings of the Workshop on Genetic Programming: From Theory to Real-World Applications, pp. 33–41 (1995)

Improving Efficiency of Heuristics for the Large Scale Traveling Thief Problem

Yi Mei[1], Xiaodong Li[1], and Xin Yao[2]

[1] School of Computer Science and Information Technology,
RMIT University, Australia
{yi.mei,xiaodong.li}@rmit.edu.au
[2] School of Computer Science, University of Birmingham, UK
x.yao@cs.bham.ac.uk

Abstract. The Traveling Thief Problem (TTP) is a novel problem that combines the well-known Traveling Salesman Problem (TSP) and Knapsack Problem (KP). In this paper, the complexity of the local-search-based heuristics for solving TTP is analyzed, and complexity reduction strategies for TTP are proposed to speed up the heuristics. Then, a two-stage local search process with fitness approximation schemes is designed to further improve the efficiency of heuristics. Finally, an efficient Memetic Algorithm (MA) with the two-stage local search is proposed to solve the large scale TTP. The experimental results on the tested large scale TTP benchmark instances showed that the proposed MA can obtain competitive results within a very short time frame for the large scale TTP. This suggests the potential benefits of designing intelligent divide-and-conquer strategies that solves the sub-problems separately while taking the interdependence between them into account.

1 Introduction

The Traveling Thief Problem (TTP) is a novel composite problem proposed by Bonyadi *et al.* [1] in order to investigate the effect of interdependence between sub-problems in a complicated problem. TTP is a combination of two well-known combinatorial optimization problems, i.e., the Travelling Salesman Problem (TSP) and Knapsack Problem (KP). Specifically, a thief is to visit a set of cities and pick some items from the cities to put in a rented knapsack. Each item has a value and a weight. The knapsack has a limited capacity that cannot be exceeded by the total weight of the picked items. In the end, the thief has to pay the rent for the knapsack, which depends on the travel time. TTP aims to find a tour for the thief to visit all the cities exactly once, pick some items along the way and finally return to the starting city, so that the profit of the visit, which is the total value of the picked items minus the rent of the knapsack, is maximized. Since the TSP and KP have been intensively investigated, the TTP facilitates studies on analysing the interdependence between sub-problems.

A potential real-world applications of TTP is the capacitated arc routing problem with service profit, where each customer has a demand and a service

G. Dick et al. (Eds.): SEAL 2014, LNCS 8886, pp. 631–643, 2014.

profit, and the travel cost (e.g., the petrol consumption) depends on the load of the vehicle. This novel form of problem is more practical and closer to reality than the models that have been studied intensively [2] [3] [4] [5].

In this paper, the challenges caused by the interdependence between the TSP and KP components are discussed. In particular, the computational complexities of the search-based heuristics are analyzed in the context of TTP and compared with that for TSP and KP, respectively. It is found that the combination of TSP and KP in TTP can lead to a much higher computational complexity of the search. Such issue becomes much more severe for large scale instances with thousands or tens of thousands of cities and items.

To address the computational complexity issue and to speed up the algorithm, a two-stage local search with novel fitness approximation schemes for TTP is proposed, and embedded into a Memetic Algorithm (MA) framework to solve the large scale TTP. The resultant MA is tested on the CEC'2014 TTP competition benchmarks and the results showed that it managed to achieve competitive solution quality with a limited computational budget.

The rest of the paper is as follows: In Section 2, TTP is described in details. In Section 3, the computational complexities of the commonly used heuristics such as 2-opt operator for TSP and single flip for KP are analyzed in the context of TTP. Then, the complexity reduction strategies for TTP are derived from that for TSP and KP in Section 4. The proposed two-stage local search that further improves the efficiency is described in Section 5.

2   Travelling Thief Problem

The TTP is a combination of TSP and KP. In TSP, n cities with the distance matrix of $D_{n \times n}$ are given, where $d(i, j)$ is the distance from city i to j. In KP, there are m items. Each item i has a weight w_i, a value b_i and an available city a_i. A thief aims to visit all the cities exactly once, pick items on the way and finally come back to the starting city. The thief rents a knapsack to carry the items, which has a capacity of Q. The rent of the knapsack is R per time unit. The speed of the thief decreases linearly with the increase of the total weight of carried items, and is computed by the following formula:

$$v = v_{\max} - \nu \bar{w}, \tag{1}$$

$$\nu = \frac{v_{\max} - v_{\min}}{Q}, \tag{2}$$

where $0 \leq \bar{w} \leq Q$ is the current total weight of the picked items. When the knapsack is empty ($\bar{w} = 0$), the speed is maximized ($v = v_{\max}$). When the knapsack is full ($\bar{w} = Q$), the speed is minimized ($v = v_{\min}$). Then, the profit gained by the thief is defined as the total value of the picked items minus the rent of the knapsack. Let a TTP solution be represented by a tour $\mathbf{x}$ and a picking plan $\mathbf{z}$, where $\mathbf{x} = (x_1, \ldots, x_n)$ is a permutation of the cities and $\mathbf{z} = (z_1, \ldots, z_m)$

is a 0-1 vector of the items. z_i takes 1 if item i is picked, and 0 otherwise. Then, the TTP problem can be stated as follows:

$$\max \sum_{j=1}^{m} b_j z_j - R \cdot \left(\sum_{i=1}^{n-1} \frac{d(x_i, x_{i+1})}{v(i)} + \frac{d(x_n, x_1)}{v(n)} \right), \tag{3}$$

$$s.t.: v(i) = v_{\max} - \nu \bar{w}(i), \ \ 1 \le i \le n, \tag{4}$$

$$\bar{w}(i) = \sum_{k=1}^{i-1} \bar{w}(k) + cw(x_i), \ \ 1 \le i \le n, \tag{5}$$

$$cw(i) = \sum_{j=1}^{m} w_j z_j [a_j = i], \ \ 1 \le i \le n, \tag{6}$$

$$x_i \ne x_j, \ \ 1 \le i \ne j \le n, \tag{7}$$

$$\sum_{j=1}^{m} w_j z_j \le Q, \tag{8}$$

$$x_i \in \{1, \dots, n\}, z_j \in \{0, 1\}. \tag{9}$$

Eq. (3) is the objective function, which is to maximize the profit. $v(i)$ is the velocity at x_i, and is calculated by Eq. (4). $\bar{w}(i)$ is the accumulated weight from the beginning of the tour to x_i. It is computed by Eq. (5), where $cw(x_i)$ indicates the total weight of the items picked at x_i, and is obtained by Eq. (6). $[a_j = i]$ takes 1 if $a_j = i$, and 0 otherwise. Eqs. (7) and (9) ensures that $\mathbf{x}$ is a valid tour, i.e., each city appears exactly once in the permutation. Eq. (8) indicates that the total weight of the picked items cannot exceed the capacity.

Fig. 1 illustrates an example of a TTP solution that travels through the path A-B-C-D-A, picking items 1 and 2 at cities A and C, respectively. The weights and values of the items are $w_1 = b_1 = 2$ and $w_2 = b_2 = 1$. The numbers associated with the arcs indicate the distances between the cities. The total value of the picked items is $b_1 + b_2 = 3$. The travel speeds between each pair of cities are $v_{AB} = v_{BC} = 4 - 3 \cdot 2/3 = 2$ and $v_{CD} = v_{DA} = 1$. Then, the total travel time is $(2 + 1.5)/2 + (1 + 1.5)/1 = 4.25$. Finally, the profit of the travel is $3 - 1 \cdot 4.25 = -1.25$ (a loss of 1.25).

3 Complexity of Heuristics in TTP

It is obvious that TTP is NP-hard, as it can be reduced to TSP when there are no items. Therefore, the exact methods are not applicable in practice, where the problem has a large size. The heuristics are commonly used for they can provide near-optimal solutions in a short time.

The heuristics can be categorized into two types. The former includes the so-called constructive heuristics that construct a solution based on domain knowledge. For example, the Christofides's heuristic [6] is the best-so-far TSP heuristic that can construct tours whose lengths are no larger than 1.5 times of the optimum. The latter includes the search-based algorithms that start from one or

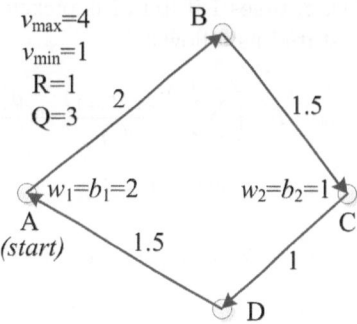

Fig. 1. An example of a TTP solution

more initial solutions, and then modify it/them by some operators to search for improvements. The well-known search-based heuristics for TSP include the chained Lin-Kernighan (LK) heuristic [7], the efficient LK Heuristic [8], and Ant Colony Optimization [9]. Obviously, the search-based heuristics can perform no worse than the constructive heuristics, since they can take the solutions generated by the constructive heuristics as the initial solutions. Thus, we focus on the search-based heuristics in this paper. Specifically, we focus on the local-search-based heuristics, which iteratively move the current solution to one of its neighbors (e.g., the LK heuristic for TSP). To facilitate the analysis, the simplest 2-opt operator for the tour and flip operator for the picking plan is chosen as an example. The 2-opt operator selects two cities and reverse the sub-tour between them. The flip operator selects an item and flip its status (from picked to unpicked, or from unpicked to picked).

Assuming that the local search has T steps, and enumerates all the neighbors at each step. The neighborhood consists of the TSP neighborhood generated by the 2-opt operator and the KP neighborhood generated by the flip operator. Then, the TSP neighborhood size is $n(n-1)/2$ and the KP neighborhood size is m, where n and m are the number of cities and items, respectively. The fitness evaluation of a TTP solution is given in Algorithm 1. Without any efficiency optimization, the complexity of the evaluation of a neighbor is $O(n+m)$. The complexity of moving the current solution to the next one (modification) can be ignored compared with the enumeration of the neighborhood. Therefore, the complexity of the entire local search is:

$$O(LS) = T\left(\left(\frac{n(n-1)}{2} + m\right)O(m+n)\right) \tag{10}$$

$$= O(T(n^2 + m)(n + m)) \tag{11}$$

$$= O(T(n^3 + n^2 m + m^2)). \tag{12}$$

Although the complexity is polynomial, the preliminary studies showed that it still makes the algorithm slow in practice, especially when n and m are large.

Algorithm 1. The evaluation of a TTP solution

1: **for** $i = 1 \rightarrow n$ **do**
2:     $cw(i) = 0$, $cb(i) = 0$;
3: **end for**
4: **for** $i = 1 \rightarrow m$ **do**
5:     **if** item i is picked in city a_i **then**
6:         $cw(a_i) \leftarrow cw(a_i) + w_i$;
7:         $cb(a_i) \leftarrow cb(a_i) + b_i$;
8:     **end if**
9: **end for**
10: Set $\bar{w} = 0$, $\bar{t} = 0$, $\bar{b} = 0$;
11: **for** $i = 1 \rightarrow n - 1$ **do**
12:     $\bar{w} \leftarrow \bar{w} + cw(tour(i))$;
13:     $\bar{b} \leftarrow \bar{b} + cb(tour(i))$;
14:     $\bar{t} \leftarrow \bar{t} + \frac{d(tour(i), tour(i+1))}{v_{\max} - \nu \bar{w}}$;
15: **end for**
16: $\bar{t} \leftarrow \bar{t} + \frac{d(tour(n), tour(1))}{v_{\max} - \nu \bar{w}}$;
17: **return** $\bar{b} - R \cdot \bar{t}$;

---

4   Complexity Reduction Strategies for TTP

The following strategies are commonly used to reduce the complexity of the local search:

Neighborhood Size. It is obvious that there are a large number of poor neighbors which are not worth evaluating during the local search. For example, it is highly unlikely to link the cities that are far away from each other in the tour. Substantial efforts have been done to reduce the neighborhood size (e.g., the kNN strategy, K-d tree [10] [11], Delaunay triangulation [12] and Delaunay candidate set [13]). There are a number of works on dynamic neighborhood structures to adaptively choose the neighborhood that is the most suitable for the current region [14] [15] [16] [17]. In this paper, the Delaunay triangulation is chosen to reduce the TSP neighborhood, since it has been empirically shown that the edges in the optimal TSP tour are highly likely to be in the Delaunay triangulation [18]. In the Delaunay triangulation, the number of edges is reduced from $n(n-1)/2$ to at most $3n - 6$, and each city has on average 6 surrounding triangles. The Delaunay triangulation is obtained by an efficient 2-D sweepline algorithm [19], which has a complexity of $O(n \log n)$.

Incremental Evaluation. In Algorithm 1, lines 1–9 compute the total weight and value of the items picked in each city, and lines 10–16 computes the total travel time of the tour. It is clear that the complexity of the evaluation is $O(m + n)$. However, when evaluating neighbors during the local search, most information required for the evaluation is maintained, and thus does not have to be re-calculated. The incremental evaluation is thus designed to prevent such redundant computations. It is obvious that the incremental evaluation for TSP and KP both reduces the complexity to $O(1)$. In TTP, all the $cw(a_i)$'s and

$cb(a_i)$'s are the same for the neighbors generated by the 2-opt operators (and all the operators that only modify the tour), and only the $cw(a_i)$ and $cb(a_i)$ of the city of the flipped items (and all the items whose status are changed) need to be updated. Therefore, it is easy to reduce the complexity of the lines 1–9 to $O(1)$, and thus reduce the complexity of the evaluation to $O(n)$. However, the complexity of computing the travel time can hardly be reduced, since one needs to recalculate $\bar{w}$, $\bar{b}$ and $\bar{t}$ for each city behind the cities moved in the tour. To the best of our knowledge, one can only keep the information of the unchanged subtour, which does not reduce the complexity. In summary, the incremental evaluation of a TTP solution has a complexity of $O(n)$.

Efficient Solution Modification. The modification of the tour has a worst-case complexity of $O(n)$. However, when changing the data structure of the tour, the complexity can be reduced. For example, the splay tree [20] reaches the amortized time of $O(\log n)$.

When applying all the above strategies to TTP, the complexity of the local search is reduced to:

$$O(LS) = T(O(S_1 + S_2)O(n) + O(\log n)) \tag{13}$$
$$= O(T(S_1 + S_2)n), \tag{14}$$

where S_1 and S_2 are the size of the reduced neighborhood of the 2-opt and flip operators, respectively.

Without any neighborhood reduction for the flip operator, it is known that $S_2 = O(m)$, where m is the number of items. After reducing the TSP neighborhood with the Delaunay triangulation, S_1 is reduced to $O(n)$. Therefore, we have

$$O(LS) = O(T(m + n)n) = O(T(n^2 + mn)) \tag{15}$$

It can be seen that the reduced complexity is much lower than the original one ($O(T(n^3 + n^2m + m^2)$ shown in Eq. (12)).

5   A More Efficient Two-Stage Local Search

The reduced complexity of the local-search-based heuristics for TTP shown in Eq. (15) is still high for the large scale problems encountered in practice. It is no less than n times of the complexity of the TSP and KP heuristics due to the incremental evaluation complexity of $O(n)$ ($O(1)$ for the TSP and KP heuristics). This will lead to a much longer computational time. For example, assuming that the LK heuristic takes 30 seconds to obtain a near-optimal for a $30,000$-city large scale TSP instance. With the TTP incremental evaluation, the computational time will become roughly $30,000 \times 30 = 900,000$ seconds, i.e., 250 hours. Therefore, the complexity needs to be further improved. To this end, a two-stage local search is proposed. It simply divides the entire search into two stages. The first stage is the TSP search for improving the TSP tour, and the second stage is the KP search for finding the improved picking plan under the given TSP tour.

To further reduce the complexity, the approximated fitness evaluations are designed for the two stages, respectively. During the first stage, the objective of maximizing the profit is approximated by minimizing the tour length. The approximation is motivated by maximizing the profit under the condition that no item is picked, i.e., $z_j = 0, \forall j = 1, \ldots, m$. In this situation, Eqs. (3)–(9) can be simplified as follows:

$$\max \quad - R \cdot \frac{\sum_{i=1}^{n-1} d(x_i, x_{i+1}) + d(x_n, x_1)}{v_{\max}}, \tag{16}$$

$$s.t. : x_i \neq x_j, \ 1 \leq i \neq j \leq n, \tag{17}$$

$$x_i \in \{1, \ldots, n\}. \tag{18}$$

It is obvious that the objective Eq. (16) is equivalent to minimizing the tour length $\sum_{i=1}^{n-1} d(x_i, x_{i+1}) + d(x_n, x_1)$.

The TSP search provides a TTP solution with a promising empty tour (without picking any item), which is a lower bound of the optimal TTP solution. Then, during the second stage, the items are inserted into the empty tour to improve the profit. The insertion of the items in TTP is different from that is in KP, because the profit of the insertion depends not only on the value of the inserted item, but also on the location of its inserted city in the tour. Specifically, given a tour $\mathbf{x}$ and the current picking plan $\mathbf{z}$, the change of profit caused by inserting an item j is given as follow:

$$\Delta p(j, \mathbf{x}, \mathbf{z}) = b_j - R \cdot \Delta t(j, \mathbf{x}, \mathbf{z}), \tag{19}$$

$$\Delta t(j, \mathbf{x}, \mathbf{z}) = \sum_{i=loc(a_j)}^{n-1} \left(\frac{d(x_i, x_{i+1})}{v'(i)} - \frac{d(x_i, x_{i+1})}{v(i)} \right) + \frac{d(x_n, x_1)}{v'(n)} - \frac{d(x_n, x_1)}{v(n)}, \tag{20}$$

$$v'(i) = v_{\max} - \nu(\bar{w}(i) + w_j) = v(i) - \nu w_j, \ 1 \leq i \leq n, \tag{21}$$

where $loc(a_j)$ is the location index of the city a_j in the tour. From Eq. (19), one can see that the insertion of an item j leads to a profit gain of its value b_j and a profit loss of $R \cdot \Delta t(j, \mathbf{x}, \mathbf{z})$ due to the increased travel time. $\Delta t(j, \mathbf{x}, \mathbf{z})$ depends on the $\bar{w}(i)$'s along the subtour after $loc(a_j)$, i.e., all the items picked in the subtour. Therefore, it is impossible to precisely evaluate each item separately, as it is in KP.

To address the above issue, three approximations to $\Delta t(j, \mathbf{x}, \mathbf{z})$ are proposed to separately estimate the priority of the items to be inserted into the tour. The first one is the called the empty-tour increased time, which is the increased time when inserting the item into an empty tour. It can be calculated as follows:

$$\Delta t_1(j, \mathbf{x}) = L(\mathbf{x}, loc(a_j)) \left(\frac{1}{v_{\max} - \nu w_j} - \frac{1}{v_{\max}} \right), \tag{22}$$

where

$$L(\mathbf{x}, loc(a_j)) = \sum_{i=loc(a_j)}^{n-1} d(x_i, x_{i+1}) + d(x_n, x_1) \qquad (23)$$

is the length of the subtour of $\mathbf{x}$ from $loc(a_j)$ to the end.

The second approximation is called the worst-case increased time under a given total weight. It is known that given a total weight W, the worst-case increased time by inserting the item j is when all the weights are picked before its picked city a_j. Therefore, the worst-case increased time under a given total weight W is calculated as follows:

$$\Delta t_2(j, \mathbf{x}, W) = L(\mathbf{x}, loc(a_j)) \left(\frac{1}{v_{\max} - \nu(W + w_j)} - \frac{1}{v_{\max} - \nu W} \right). \qquad (24)$$

When inserting an item into the tour $\mathbf{x}$ with the total weight of W, it is clear that $\Delta t_1(j, \mathbf{x}) \le \Delta t(j, \mathbf{x}, \mathbf{z}) \le \Delta t_2(j, \mathbf{x}, W)$. In other words, $\Delta t_1(j, \mathbf{x})$ and $\Delta t_2(j, \mathbf{x}, W)$ provide the lower and upper bounds of $\Delta t(j, \mathbf{x}, \mathbf{z})$ for each item.

The last approximation is called the expected increased time under a given total weight. This approximation is based on the assumption that the total weight W is uniformly distributed along the tour. Under this assumption, given a location l, where l indicates the distance from the location to the end of the tour, then the accumulated weight $\bar{w}(l)$ at location l can be calculated as follows:

$$\bar{w}(l) = \left(1 - \frac{l}{L(\mathbf{x})} \right) W, \qquad (25)$$

where $L(\mathbf{x})$ is the length of the tour $\mathbf{x}$. Therefore, the expected increased time under the given total weight of W is calculated as follows:

$$\Delta t_3(j, \mathbf{x}, W) = \int_{L(\mathbf{x})-L(\mathbf{x}, loc(a_j))}^{L(\mathbf{x})} \left(\frac{1}{v_{\max} - \nu(\bar{w}(l) + w_j)} - \frac{1}{v_{\max} - \nu\bar{w}(l)} \right) dl \qquad (26)$$

$$= \frac{1}{a} \ln \left(\frac{(aL(\mathbf{x}) + b_1) \cdot (a(L(\mathbf{x}) - L(\mathbf{x}, loc(a_j))) + b_2)}{(a(L(\mathbf{x}) - L(\mathbf{x}, loc(a_j))) + b_1) \cdot (aL(\mathbf{x}) + b_2)} \right), \qquad (27)$$

where

$$a = \frac{W}{L(\mathbf{x})}, b_1 = v_{\max} - \gamma(W + w_j), b_2 = v_{\max} - \gamma W. \qquad (28)$$

Based on the above three approximations, a heuristic of inserting the items is designed and described in Algorithm 2.

The idea of Algorithm 2 can be explained as follows: At first, the priorities of the items are determined by $\frac{b_j - R \cdot \Delta t_1(j, \mathbf{x})}{w_j}$, which is the profit change when the tour is empty (Line 1. This approximation is more precise when the total weight of the tour is not large relative to the capacity, and works better in the early stage of the insertion. Then, as more and more items are inserted and the total weight of the tour increases, it becomes more and more important to evaluate

---

Algorithm 2. The insertion heuristic of the items given a tour

1: Sort the items by the decreasing order of $\frac{b_j - R \cdot \Delta t_1(j,\mathbf{x})}{w_j}$;
2: Let the sorted item list be $(s(1), \ldots, s(m))$, and set $W = 0$;
3: **for** $j = 1 \to m$ **do**
4:     **if** $W + w_{s(j)} \leq Q$ **then**
5:         **if** $b_{s(j)} > R \cdot \Delta t_2(s(j), \mathbf{x}, W)$ **then**
6:             $z_{s(j)} = 1, W \leftarrow W + w_{s(j)}$;   ▷ Insert this item
7:         **else if** $b_{s(j)} > R \cdot \Delta t_3(s(j), \mathbf{x}, W)$ **then**
8:             $z_{s(j)} = 1, W \leftarrow W + w_{s(j)}$;   ▷ Insert this item
9:         **end if**
10:     **end if**
11: **end for**

---

the items given the total weight of the tour. First, the worst-case approximation is evaluated (Line 5). If it is positive, then the item is inserted since its insertion must lead to a profit gain. Otherwise, the expected approximation is evaluated (Line 7). If it is positive, then the item should be more likely to have a profit gain rather than a profit loss. Thus, it is inserted into the tour as well. The capacity constraint is imposed as a global constraint. That is, any insertion that leads to the violation of the capacity constraint will be abandoned.

After Algorithm 2, a subsequent local search is conducted on the generated picking plan by applying the flip operator only to the inserted items. The preliminary studies showed that the local search was very fast in practice because the generated picking plan is usually good enough and there is nearly no space for the improvement by local search.

The complexity of the two-stage local search consists of that of the TSP search and the KP search. The TSP search has a complexity of $T_1 O(n)$, where T_1 is the number of steps of the search, since the evaluation approximation reduces the complexity of evaluation from $O(n)$ to $O(1)$. Then, the complexity of the KP search is $O(m) + T_2 O(nm)$, where the former term is the complexity of Algorithm 2, and the latter one is the complexity of the subsequent local search. Therefore, the complexity of the proposed two-stage local search is as follows:

$$O(TSLS) = T_1 O(n) + T_2 O(nm) \tag{29}$$

Since T_2 is usually small ($T_2 = 1$ for most of the cases in the experimental studies), the practical complexity of the two-stage local search is usually $T_1 O(n) + O(nm)$.

6   A Memetic Algorithm with the Two-Stage Local Search

The proposed Memetic Algorithm with the Two-stage Local Search (MATLS) starts from an initial population consisting of individuals with tours generated

Table 1. The performance of the compared algorithms on the large scale TTP benchmark instances with bounded strongly correlated item weights. The best results are marked in bold.

Name	n	m	RLS	EA	MATLS
brd14051	14051	140500	-5.50e+7(1.22e+6)	-6.30e+7(2.52e+6)	**2.66e+7**(2.07e+5)
d15112	15112	151110	-6.42e+7(2.52e+6)	-7.01e+7(1.63e+6)	**2.85e+7**(5.35e+5)
d18512	18512	185110	-8.18e+7(1.27e+6)	-8.89e+7(3.98e+6)	**3.07e+7**(2.73e+5)
pla33810	33810	338090	-1.71e+8(1.49e+6)	-1.80e+8(1.64e+6)	**6.34e+7**(4.59e+5)
rl11849	11849	118480	-4.38e+7(7.58e+5)	-5.05e+7(2.76e+5)	**1.97e+7**(8.29e+4)
usa13509	13509	135080	-5.45e+7(7.57e+5)	-6.17e+7(2.95e+5)	**2.92e+7**(2.60e+5)

by the chained LK heuristic, and picking plans by the insertion heuristic (Algorithm 2). In each generation, two parents are randomly selected and undergo the ordered crossover [21] to obtain the tour of the offspring, which is then improved by the TSP search. After that, the picking plan of the offspring is initialized by Algorithm 2 and improved by the KP search. Finally, the offspring is added into the population to replace the worst individual.

7   Experimental Studies

To evaluate the proposed MATLS, it is tested on a subset of large scale TTP benchmark instances developed by Polyakovskiy *et al.* [22] and compared with the RLS and EA proposed in the same paper. RLS generates a decent tour by the chain LK heuristic, and then does a simple local search on the picking plan with the single flip operator. EA takes the same search process except that the status of each item has a probability of $1/m$ (m is the number of items) to be flipped in each generation. Since the investigation of TTP is still in its infancy, the two compared algorithms are the state-of-the-art published algorithms although they are relatively simple. The time budget of MATLS is set to 10 minutes, the same as that of RLS and EA. The population size is set to 30. The tours of 10 individuals are initialized by the more time-consuming chained LK heuristic, and tours of the remaining 20 individuals are initialized by the minimal spanning tree heuristic.

Tables 1–3 show the average performance (mean and standard deviation) of the compared algorithms on the selected benchmark problems, where n and m stand for the number of cities and items, respectively. It can be seen that the selected benchmark instances all have more than 10,000 cities, and 100,000 items. From the tables, it can be seen that for all the large scale instances, MATLS performed significantly better than RLS and EA. In fact, both RLS had highly negative profits, while MATLS managed to obtain highly positive profits for all the instances. This verifies the advantage of MATLS and thus the efficacy of the proposed two-stage local search. Since all the compared algorithms generated the tours of the solutions by the chained LK heuristic, the efficacy of the proposed item insertion heuristic is verified by the outperformance of MATLS over RLS and EA.

Table 2. The performance of the compared algorithms on the large scale TTP benchmark instances with bounded uncorrelated item weights. The best results are marked in bold.

Name	n	m	RLS	EA	MATLS
brd14051	14051	140500	-4.12e+7(7.68e+5)	-4.73e+7(1.64e+5)	**1.69e+7**(2.09e+5)
d15112	15112	151110	-4.56e+7(7.44e+5)	-5.18e+7(1.08e+5)	**1.83e+7**(2.43e+5)
d18512	18512	185110	-5.98e+7(6.12e+5)	-6.62e+7(8.82e+5)	**1.94e+7**(2.47e+5)
pla33810	33810	338090	-1.22e+8(8.40e+5)	-1.28e+8(1.01e+6)	**4.10e+7**(2.38e+5)
rl11849	11849	118480	-3.23e+7(1.79e+6)	-3.81e+7(2.12e+6)	**1.29e+7**(4.47e+4)
usa13509	13509	135080	-4.00e+7(5.79e+5)	-4.55e+7(1.29e+5)	**1.88e+7**(1.69e+5)

Table 3. The performance of the compared algorithms on the large scale TTP benchmark instances with bounded uncorrelated but similar item weights. The best results are marked in bold.

Name	n	m	RLS	EA	MATLS
brd14051	14051	140500	-3.88e+7(1.86e+6)	-4.32e+7(1.82e+5)	**1.31e+7**(2.28e+5)
d15112	15112	151110	-4.30e+7(1.43e+6)	-4.74e+7(7.51e+5)	**1.38e+7**(2.64e+5)
d18512	18512	185110	-5.50e+7(5.86e+5)	-6.00e+7(1.09e+6)	**1.42e+7**(3.39e+5)
pla33810	33810	338090	-1.10e+8(7.47e+5)	-1.15e+8(6.16e+5)	**3.10e+7**(3.38e+5)
rl11849	11849	118480	-3.10e+7(4.54e+5)	-3.51e+7(1.70e+5)	**9.15e+6**(4.31e+4)
usa13509	13509	135080	-3.77e+7(1.18e+6)	-4.20e+7(6.38e+5)	**1.52e+7**(1.70e+5)

8   Conclusion

In this paper, the complexity of large scale Traveling Thief Problem (TTP) is analyzed, and the complexity reduction strategies are proposed to solve the complex problem within a given limited computational budget. The resultant Memetic Algorithm with Two-stage Local Search (MATLS) is evaluated on the large scale TTP benchmark instances with more than 10,000 cities and 100,000 items, and our results demonstrate the efficacy of MATLS on large scale TTP.

The success of MATLS implies the importance of complexity reduction for solving large scale problems to speed up the search process significantly without losing much of its accuracy, e.g., by means of neighborhood filtering and surrogate models. However, this needs to exploit domain knowledge, and the strategy may vary from problem to problem. In the case of TTP, the efficacy of the two-stage local search gives an insight that an intelligent divide-and-conquer approach taking into account the interdependence between the sub-problems (TSP and KP) can perform better than the extreme approaches, i.e., solving the problem as a whole and solving the sub-problems separately.

References

1. Bonyadi, M., Michalewicz, Z., Barone, L.: The travelling thief problem: the first step in the transition from theoretical problems to realistic problems. In: Proceedings of the 2013 IEEE Congress on Evolutionary Computation, Cancun, Mexico, pp. 1037–1044 (2013)
2. Mei, Y., Tang, K., Yao, X.: Improved memetic algorithm for capacitated arc routing problem. In: Proceedings of the 2009 IEEE Congress on Evolutionary Computation, pp. 1699–1706 (2009)
3. Mei, Y., Tang, K., Yao, X.: Decomposition-based memetic algorithm for multi-objective capacitated arc routing problem. IEEE Transactions on Evolutionary Computation 15(2), 151–165 (2011)
4. Mei, Y., Tang, K., Yao, X.: A Memetic Algorithm for Periodic Capacitated Arc Routing Problem. IEEE Transactions on Systems, Man, and Cybernetics, Part B: Cybernetics 41(6), 1654–1667 (2011)
5. Mei, Y., Li, X., Yao, X.: Cooperative co-evolution with route distance grouping for large-scale capacitated arc routing problems. IEEE Transactions on Evolutionary Computation 18(3), 435–449 (2014)
6. Christofides, N.: The optimum traversal of a graph. Omega 1(6), 719–732 (1973)
7. Applegate, D., Cook, W., Rohe, A.: Chained lin-kernighan for large traveling salesman problems. INFORMS Journal on Computing 15(1), 82–92 (2003)
8. Helsgaun, K.: An effective implementation of the lin–kernighan traveling salesman heuristic. European Journal of Operational Research 126(1), 106–130 (2000)
9. Dorigo, M., Gambardella, L.: Ant colony system: A cooperative learning approach to the traveling salesman problem. IEEE Transactions on Evolutionary Computation 1(1), 53–66 (1997)
10. Bentley, J.: K-d trees for semidynamic point sets. In: Proceedings of the Sixth Annual Symposium on Computational Geometry, pp. 187–197. ACM (1990)
11. Bentley, J.L.: Fast algorithms for geometric traveling salesman problems. ORSA Journal on computing 4(4), 387–411 (1992)
12. Delaunay, B.: Sur la sphere vide. Izv. Akad. Nauk SSSR, Otdelenie Matematicheskii i Estestvennyka Nauk 7, 793–800 (1934)
13. Reinelt, G.: Fast heuristics for large geometric traveling salesman problems. ORSA Journal on Computing 4(2), 206–217 (1992)
14. Yao, X.: Simulated annealing with extended neighbourhood. International Journal of Computer Mathematics 40(3), 169–189 (1991)
15. Yao, X.: Dynamic neighbourhood size in simulated annealing. In: Proceedings of International Joint Conference on Neural Networks (IJCNN 1992), vol. 1, pp. 411–416 (1992)
16. Mei, Y., Tang, K., Yao, X.: A Global Repair Operator for Capacitated Arc Routing Problem. IEEE Transactions on Systems, Man, and Cybernetics, Part B: Cybernetics 39(3), 723–734 (2009)
17. Tang, K., Mei, Y., Yao, X.: Memetic Algorithm with Extended Neighborhood Search for Capacitated Arc Routing Problems. IEEE Transactions on Evolutionary Computation 13(5), 1151–1166 (2009)
18. Krasnogor, N., Moscato, P., Norman, M.G.: A new hybrid heuristic for large geometric traveling salesman problems based on the delaunay triangulation. In: Anales del XXVII Simposio Brasiliero de Pesquisa Operacional, Citeseer, pp. 6–8 (1995)

19. Žalik, B.: An efficient sweep-line delaunay triangulation algorithm. Computer-Aided Design 37(10), 1027–1038 (2005)
20. Fredman, M.L., Johnson, D.S., McGeoch, L.A., Ostheimer, G.: Data structures for traveling salesmen. Journal of Algorithms 18(3), 432–479 (1995)
21. Goldberg, D., Lingle, R.: Alleles, loci, and the traveling salesman problem. In: Proceedings of the First International Conference on Genetic Algorithms and their Applications, pp. 154–159. Lawrence Erlbaum Associates, Publishers (1985)
22. Polyakovskiy, S., Bonyadi, M.R., Wagner, M., Michalewicz, Z., Neumann, F.: A comprehensive benchmark set and heuristics for the traveling thief problem. In: Proceedings of Genetic and Evolutionary Computation Conference (GECCO), pp. 477–484 (2014)

Load Balance Aware Genetic Algorithm
for Task Scheduling in Cloud Computing[*]

Zhi-Hui Zhan[1,**], Ge-Yi Zhang[5], Ying-Lin[2,6], Yue-Jiao Gong[3], and Jun Zhang[4]

[1] Department of Computer Science, Sun Yat-sen University, Guangzhou, 510006, China
[2] Key Lab. Machine Intelligence and Advanced Computing, Ministry of Education, China
[3] Engineering Research Center of Supercomputing Engineering Software, MOE, China
[4] Key Lab. Software Technology, Education Department of Guangdong Province, China
[5] School of Sofware Engineering, Sun Yat-sen University, Guangzhou, 510006, China
[6] Department of Psychology, Sun Yat-sen university, 510275, China
zhanzhh@mail.sysu.edu.cn

Abstract. This paper proposes to solve the task scheduling problem in cloud computing by using a load balance aware genetic algorithm (LAGA) with Min-min and Max-min methods. Task scheduling problems are of great importance in cloud computing, and become especially challenging when taking load balance into account. Our proposed LAGA algorithm has several advantages when solving this kind of problems. Firstly, by introducing the time load balance (TLB) model to help establish the fitness function with makespan, the algorithm benefits from the ability to find the solution that performs best on load balance among a set of solutions with the same makespan. More importantly, the interaction between makespan and TLB helps the algorithm to minimize makespan in the same time. Secondly, Min-min and Max-min methods are used to produce promising individuals at the beginning of evolution, leading to noticeable improvement of evolution efficiency. We evaluated LAGA on several task scheduling problems and compared with a Min-min, Max-min improved version of genetic algorithm (MMGA), which does not use the TLB strategy. The results show that LAGA can obtain very competitive results with good load balancing properties, and outperform MMGA in both makespan and TLB objectives.

Keywords: Genetic Algorithm, Cloud Computing, Load Balance, Task Scheduling.

---

[*] This work was supported in part by the National High-Technology Research and Development Program (863 Program) of China No.2013AA01A212, in part by the National Natural Science Fundation of China (NSFC) with No. 61402545, 61332002, and 61300044, and in part by the NSFC for Distinguished Young Scholars with No. 61125205.
[**] Corresponding author.

G. Dick et al. (Eds.): SEAL 2014, LNCS 8886, pp. 644–655, 2014.
© Springer International Publishing Switzerland 2014

1   Introduction

In today's society, one of the hottest emerging fields in the information technology is cloud computing [1]. Cloud computing is a concept that reorganizes the physical resources, platforms, and software applications through Internet as a kind of service to satisfy users requests [2].

Task scheduling is the critical problem in cloud computing. The number of users of a cloud system could be billions and the users may come from all over the world. Therefore, large-scale task scheduling happens frequently among the cloud providers and the requesting users, which is becoming an urgent problem. Whether the schedule is efficient or not will significantly impact the performance of a cloud system. Therefore, scheduling is one of the most important concerns when establishing cloud computing systems. Task scheduling is to find a way to assign a certain number of tasks to the appropriate resources, and make the total completion time as small as possible, which is an NP-complete problem [3]. In this paper, we consider the problem of scheduling a large amount of independent tasks in the heterogeneous collection of cloud resources, i.e., the virtual machines (VMs), so as to reduce the total completion time of the tasks.

Some heuristic methods like evolutionary computation (EC) algorithms are very useful to solve NP-complete problems, such as task scheduling problems. EC algorithms are referred to a kind of heuristic stochastic search methods used for optimization problems, and typical EC algorithms include genetic algorithm (GA) [4], ant colony optimization (ACO) [5], and particle swarm optimization (PSO) [6][7]. Among these methods, we choose GA for extensive study due to its simple concept, potential parallelism and strong searching capability.

Many researchers have already applied various EC algorithms for task scheduling problems. For example, Kumar et al. [8] proposed an improved GA based on Min-min, Max-min techniques for cloud computing scheduling. Their approach greatly improves the evolutionary speed and shortens the generations of GA process. However, their approach has short comes when considering load balancing. Ying et al. [9] proposed an energy-aware GA to find a compromise solution between makespan and energy consumption. And the comparison between List heuristics and GA made by Loukopoulos et al. states [10] that an improved version of GA can better deal with task computation when involved large data transfers. Their experiments analyzed from three view of completion time, including makespan, average completion time of the tasks and the nodes. There is also an ACO with service flow model [11] to handle QoS requirements from users. Liu et al. [12] used ACO for VM scheduling on physical machine for reducing energy consumption. Moreover, a simulated annealing (SA) algorithm considered scheduling for different types of tasks [13][12], and another mutation based SA [14][12] was developed to minimize makespan and meanwhile improves utility of resources.

There are many different scheduling goals set by users and cloud providers, besides the most common makespan criterion, objectives like efficient utilization of resources, load balance, quality of service (QoS) [15], and minimizing total cost with a budget constrain [16][16] are also considered. Among these, load balance is especially an objective of great importance. Therefore, this paper takes load balance into account with the makespan criterion. As for the implement of load balance evaluation, most of the existing researches focus on CPU and memory utilities as indicators of load balance. For example, Multi-agent GA [17] is a GA-based heuristic merged with load balance model, which mimicked the competition and learning ability of multi-agent environment to improve the performance, and also select more balanced schedules through the observation of CPU and memory utilization rate. However, because of the fundamental role of the completion time in the practical application of scheduling, to calculate the degree of load balance from the view of completion time is a more convenient and meaningful way to conduct. Hence, the time load balance (TLB) model is developed and adopted in the proposed Load Balance Aware Genetic Algorithm (LAGA) of this paper.

LAGA uses the TLB model to optimize the makespan while considering the load balance simultaneously, for that the TLB model is introduced to help establishing a new fitness function so as to discover suitable schedules that are more load balanced while maintains the makespan. Moreover, the Min-min, Max-min methods, which can increase the evolution efficiency of GA are used in the population initialization [18][19]. We conducted experiments to test the performance of LAGA, and compared it with a Min-min, Max-min improved GA (MMGA) [8], the results show that LAGA can successfully pick up the more balancing schedules among those fair schedules with similar makespan, and also outperforms MMGA in the optimization of makespan.

The rest of the paper is organized as follows. In Section 2, the background including problem description and TLB model are described. Section 3 presents LAGA algorithm in details. Experiments and comparisons are conducted in Section 4, followed by conclusions in Section 5.

2   Background

2.1   Task Scheduling Problem

Task scheduling problem is a kind of problem that map n tasks to m resources for completion, each task is independent and there is an expected completion time for every possible resource that can accomplish it. The same task has different completion time for different resources when considering heterogeneous system, which makes the scheduling problem more complex.

The scheduling aims to minimize the total completion time of the task set, which is defined as makespan:

$$makespan = \max_{j \in m} completion(j) \tag{1}$$

where $completion(j)$ denotes the total time that resource R_j need to perform the assigned tasks.

The makespan is the fundamental objective for most scheduling problems. Moreover, we also consider load balance as the other objective and evaluate it from the view of time load balance.

The $n \times m$ task scheduling problem is represented by the corresponding Resource-Task Model, and the characteristics of this problem can be described as an *Expected Time of Completion (ETC)* matrix which contains completion time of each task with each resource. There is another matrix called *Expected Scheduling to Compute (ESC)* matrix, which each describes a solution to the task scheduling problem by recording the matching of tasks and resources. The three concepts mentioned above are defined as follows:

2.1.1   Resource-Task Model

Define $T=\{T_1, T_2, T_3,...,T_n\}$ as a set of independent tasks, where T_i is the i-th task, $0 \leq i \leq n$, n is the total number of tasks. And $R=\{R_1, R_2, R_3, ...,R_m\}$ as a set of resources, where R_j is the j-th resource, $0 \leq j \leq m$, m is the total number of resources.

Then, we define $C=\{T, R\}$ as a $n \times m$ Resource-Task model with n tasks and m resources, where T is a task set and R is a resource set.

2.1.2   Expected Time of Completion (ETC)

Define $ETC=\{ETC_{ij}\}_{n \times m}$ as a matrix of expected time of completion for each tasks in every resources corresponding to a $n \times m$ Resource-Task model, where ETC_{ij} is the expected time of completion for task T_i to be executed by resource R_j.

For each Resource-Task Model, there is an *ETC* serving as fundamental attribute, and helps to describe the problem in mathematics view.

2.1.3   Expected Scheduling to Compute (ESC)

Define $ESC=\{ESC_{ij}\}_{n \times m}$ as a matrix of expected scheduling to compute for a $n \times m$ Resource-Task model, also known as a solution to the $n \times m$ task scheduling problem, where $ESC_{ij}=1$ if task T_i is scheduled on resource R_j, otherwise $ESC_{ij}=0$.

Each *ESC* represents a possible solution to a $n \times m$ task scheduling problem, and every matrix can be different from each other. To find the optimal *ESC* for a given *ETC* is the process of solving task scheduling problem.

2.2   Task Load Balance Model

Load balance refers to a situation that most of the resources are expected to be occupied according to a schedule, and finally contributes to a balanced system, instead of most resources being idle or overload. In other words, only when every resource is occupied and load balancing, the ideal load balance situation comes true, results in adequate and efficient use of computing resources, and the minimum total completion time.

Load balance model aims to build a measurable model to determine the load balancing degree of a cloud system. In this paper we consider load balancing from the view of task completion time, so the new model is called time balance load (TLB) model. To be noted, the balance value computed by this model is mentioned below as TLB value (TLBV). The comparison between TLBVs can easily tell which schedule is better load balanced.

Assume that for a $n \times m$ Resource-Task model, every task can only be assigned to one resource and can never be interrupted. The total time load of resource R_j is defined as:

$$L_j = \sum_{i=0}^{n} ETC_{ij} \times ESC_{ij} \tag{2}$$

where ETC_{ij} is the matrix denotes the completion time of each task assigned to each resource, ESC_{ij} is the matrix decides which task should be assigned to which resource, $0 \leq i \leq n$ and $0 \leq j \leq m$.

The average time load of all the resources is defined as:

$$EL_j = \frac{1}{m} \sum_{j=0}^{m} L_j \tag{3}$$

where L_j is defined by Eqs. (2), $0 \leq j \leq m$.

The difference between time load of resource R_j and the average time load is: $|L_j - EL_j|$. After that we add up the differences, the time load balance value (TLBV) can be defined as follows:

$$time\ load\ balance\ value = \sum_{j=0}^{m} |L_j - EL_j| \tag{4}$$

where L_j and EL_j are defined by Eqs. (2) and (3), $0 \leq j \leq m$.

3   Proposed LAGA Algorithm

LAGA is an improved version of GA that takes not only makespan but also load balance into consideration. First we adopt Min-min, Max-min methods for population initialization, makes the algorithm more efficient. Then, the TLB model is introduced and combined with makespan to establish new fitness function. The chromosome coding, fitness function, LAGA evolutionary operators, and the whole LAGA flowchart are described as follows.

3.1   Chromosome Coding

We use resource-task representation for coding, the length of chromosome is the number of tasks, and each gene in the chromosome represents the index of the resource that is used to accomplish the corresponding task, shown as:

$$X_i = [X_{i1}, X_{i2}, \ldots\ldots, X_{in}] \tag{5}$$

where i is the chromosome index, n is the number of tasks, and X_{ij} ($0 \leq j \leq n$) is an integer to indicate which VM the task j is scheduled on.

3.2   Fitness Function

As defined, the individual with larger fitness is better to survive, and because we aim to find individuals with smaller makespan and smaller time load balance value (TLBV), the fitness function is formulated as follows:

$$f = \frac{1}{makespan} + \frac{1}{time\ load\ balance\ value} \tag{6}$$

3.3   Population Initialization

A population is a set of individuals, every individual has its own chromosome, and each represents a schedule in the scheduling problem. Individuals are randomly generated within the scope of gene value, which make sure they are valid schedules.

Moreover, in LAGA, both the Min-min method and Max-min method are used in the initialization to enhance the initialization quality.

Min-min first finds the minimum completion time for each task in each resource, and among these minimum times for all tasks, the minimum value is selected and the corresponding task should be scheduled to the resource which can complete it in minimum time. Then, the expected completion time matrix is updated for every other task in this resource, and the task which is already assigned will not be considered again. The same process repeats until all tasks are scheduled.

Max-min is almost the same as Min-min, except it selects the maximum value among the minimum completion times for all tasks, then schedule the task to the resource, update the completion time matrix in the same manner, and repeat until all tasks are scheduled.

Therefore, in the population initialization process, we use the Min-min method to generate one chromosome, and use the Max-min methods to generate another chromosome, while all the other chromosomes are generated randomly. This way, the average fitness of the initial population is improved significantly, helping the efficiency in evolution.

3.4   Selection

Selection operator determines whether the individual will enter the next generation or not, we use proportion selection to calculate the probability which is proportional to individual's fitness in the population.

3.5   Crossover

We use single-point crossover operator, which means the intersection only happens in one single point and the part of chromosome is exchanged, the crossover point is chose in random.

3.6   Mutation

Mutation is a process that value of some genes was replaced by random value accidentally, which will generate new individuals for the next generation.

3.7   Flowchart

The flowchart of LAGA is shown in Fig. 1. In the step that first apply fitness function, the best individuals of current population is saved, and then the best individual is updated in the whole evolution process.

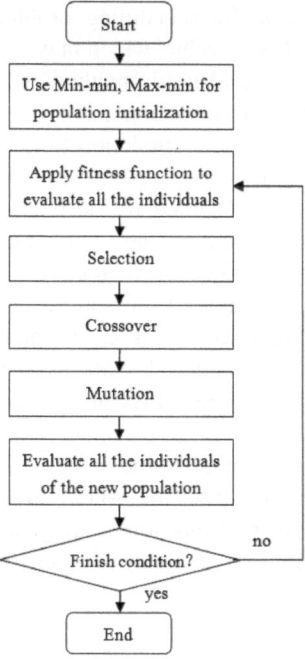

Fig. 1. The flowchart of LAGA

4   Experiments and Analysis

4.1   Parameter Analysis

Before the main experiment, we conduct parameter analysis for two main parameters named crossover rate and mutation rate in GA [20][20]. Therefore, a scheduling problem which has 20 tasks and 10 resources is used as instance to test the performance of LAGA and MMGA when the parameter values are different. For test of crossover rate, there are 11 sets of experiment and the value varying from 0.4 to 0.9 with a difference of 0.05, while the mutation rate is fixed as 0.01. For test of mutation rate, there are 9 sets of experiment, the value varying from 0.01 to 0.1 and being separated as octiles, the crossover rate is fixed as 0.5. In both tests, other parameters remain the same and set as Table I shows. And to be noted, each algorithm repeated to run 30 times for the given scheduling problem and the average of 30 independent runs is showed as results.

Table 1. Parameter Settings of Parameter Ananlysis

Name	Value
Population	50
Evolution Generation	200
Crossover Rate	[0.4,0.9]
Mutation Rate	[0.001,0.1]
Chromosome Length	20
Gene Value	[0,9]

For different crossover rates, the results of makespan and time load balance value are shown in Fig. 2.

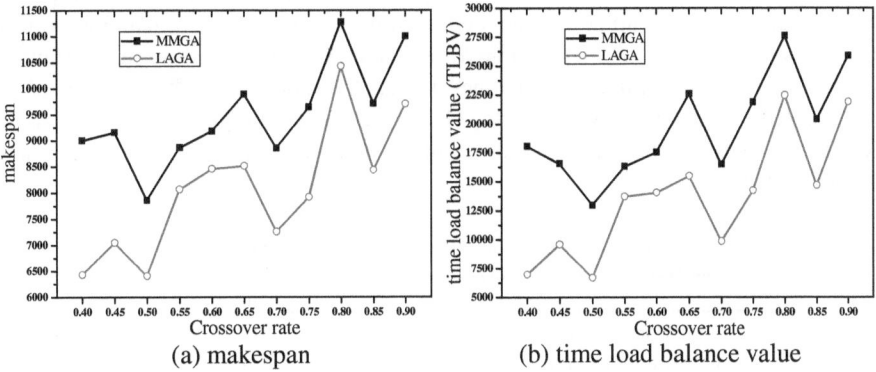

(a) makespan   (b) time load balance value

Fig. 2. Comparison of different crossover rates

For different mutation rate, the results of makespan and time load balance value are shown in Fig. 3.

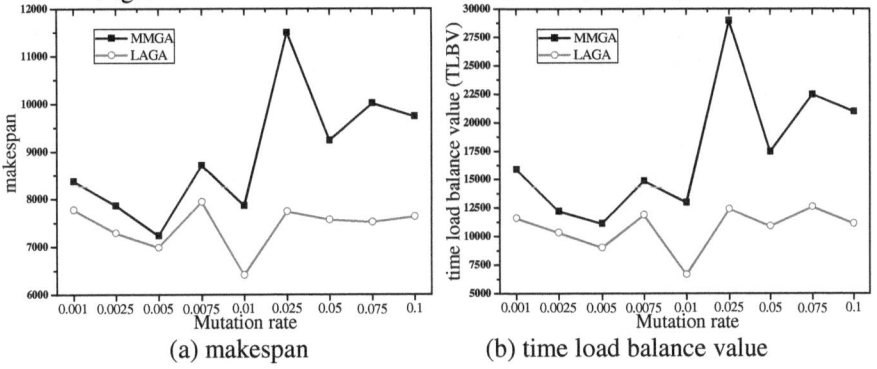

(a) makespan   (b) time load balance value

Fig. 3. Comparison of different mutation rates

From the two figures, it can be observed that the makespan and time load balance value of LAGA are both less than that of MMGA in two tests. Moreover, the best value of crossover rate for both algorithms is 0.5, the best value of mutation rate for LAGA is 0.01, and for MMGA is 0.005.

4.2   Experimental Settings

To test the performance of the proposed algorithm, we use 2 groups of task scheduling problems. In the first group, we fix the number of resources as 10, while the number of tasks varying from 10 to 40 with a different of 5. In the second group, we fix the number of tasks as 40 and vary the number of resources from 10 to 40 with a difference of 5. Thus, each group contains 7 different scheduling problems, and in total there are 14 problems in the two groups. All these problems are generated randomly, the setting of the scope of completion time for each tasks, and the heterogeneous degree of the cloud resources are showed in Table 2.

In order to show the advantages of the proposed LAGA, we use the MMGA proposed in [8] as comparison. The other parameters such as parameters of genetic algorithms, range of completion time, degree of heterogeneity, are remain the same as to make fair comparisons. The detailed parameter settings are showed in Table 2. Moreover, each algorithm run 30 times for each problem and the average of 30 independent runs is showed as results. The terminate condition is reached when the evolution process is carried to the maximum generation, which is 200 generation as defined in Table 2.

Table 2. Parameter Settings of Algorithm Experiment

Name	Value
Population	50
Evolution Generation	200
Crossover Rate	0.5
Mutation Rate	0.01
Completion Time	[200, 4426]
Heterogeneous Degree	[0, 198]

4.3   Experimental Results

For the first test group, by using LAGA and MMGA, the results of makespan and time load balance value for different number of tasks are shown in Fig. 4.

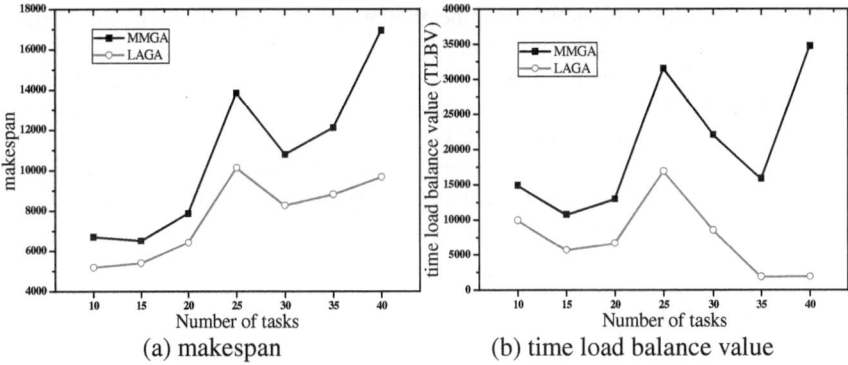

Fig. 4. Comparison of LAGA and MMGA on the first test group with different number of tasks. (a) makespan. (b) time load balance value.

For the second test group, by using LAGA and MMGA, the results of makespan and time load balance value for different number of resources are shown in Fig. 5.

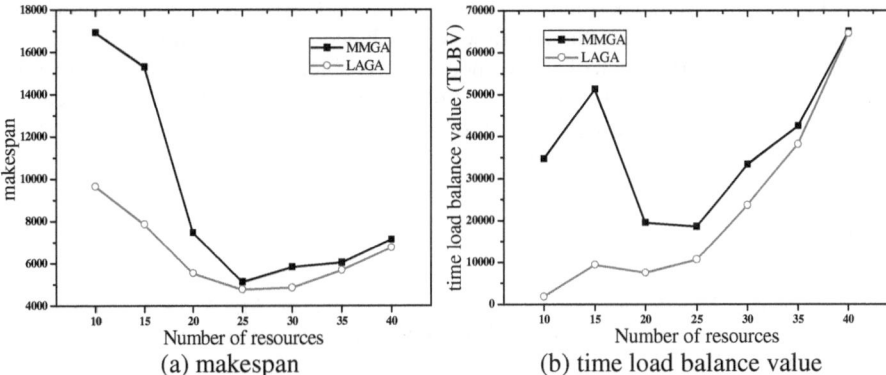

Fig. 5. Comparison of LAGA and MMGA on the second test group with different number of resources. (a) makespan. (b) time load balance value.

The results in the four figures show that both makespan and time load balance value that LAGA obtained are much better than that of MMGA, especially the load balancing of systems improves significantly, which indicates that LAGA is an effective algorithm for solving task scheduling problem.

Besides, LAGA has better performance than MMGA as load balance aware method, so the greatly more balancing results are as expected, and also indicates the advantage of load balance model. Moreover, LAGA better minimizes the makespan than MMGA does, claiming the advantage of combining makespan with load balance in the time view. In conclusion, LAGA outperforms MMGA and it obtains best results of makespan and load balance in most of the cases.

There is another observation that the performance of LAGA gets to be much better as the task-resource ratio get closer to 1. In other words, while the number of tasks that need to be scheduled is much more than the number of resources that can perform them, the proposed LAGA tends to perform better. This advantage makes LAGA suitable for complex situations. Imagine that when the ratio is high, the system is more likely to be seriously unbalanced if the algorithm cannot deal with load balancing scheduling, which might be the reason why LAGA is prominent there. It should be noted that no matter the task-resource ratio is high or low, the performance of LAGA remain better than that of MMGA.

To sum up, LAGA performs better than MMGA in cases with different characteristics. The makespan and load balance objectives are both achieved in the same time, the performance is much better especially when dealing with the load balance.

5   Conclusions

This paper mainly proposes a LAGA, which is developed to solve task scheduling problems in cloud computing environment. As an improved version of existing algorithms, it is suitable for complex problems and is capable of finding more load balancing solutions while maintain good performance of the makespan. Through the introduced time load balance model to modify the fitness function, and the Min-min, Max-min methods used for population initialization, the algorithm is proved to be efficient and useful. Experimental results show that this proposed algorithm can successfully optimize makespan and time load balance at the same time.

In the future work, we will try to model the problem by other code schemes for real-world application [21]. Moreover, multi-objective model and approach can be considered [22]. Another promising future work direction is to apply other EC related approach to solve the problem [23][24].

References

[1] Vaquero, L.M., Rodero-Merino, L., Caceres, J., Lindner, M.: A break in the clouds: towards a cloud definition. ACM SIGCOMM Computer Communication Review 39(1), 50–55 (2009)

[2] Buyya, R., Yeo, C.S., Venugopal, S., Broberg, J., Brandic, I.: Cloud computing and emerging IT platforms: Vision, hype, and reality for delivering computing as the 5th utility. Future Generation Computer Systems 25(6), 599–616 (2009)

[3] Ibarra, O.H., Kim, C.E.: Heuristic Algorithms for Scheduling Independent Tasks on Nonidentical Processors. Journal of the ACM 24(2), 280–289 (1977)

[4] Zhan, Z.H., Zhang, J., Fan, Z.: Solving the optimal coverage problem in wireless sensor networks using evolutionary computation algorithms. In: Deb, K., Bhattacharya, A., Chakraborti, N., Chakroborty, P., Das, S., Dutta, J., Gupta, S.K., Jain, A., Aggarwal, V., Branke, J., Louis, S.J., Tan, K.C. (eds.) SEAL 2010. LNCS, vol. 6457, pp. 166–176. Springer, Heidelberg (2010)

[5] Zhan, Z.H., Zhang, J., Li, Y., Liu, O., Kwok, S.K., Ip, W.H., Kaynak, O.: An efficient ant colony system based on receding horizon control for the aircraft arrival sequencing and scheduling problem. IEEE TransIntell. Transp. Syst. 11(2), 399–412 (2010)

[6] Zhan, Z.H., Zhang, J., Li, Y., Chung, H.: Adaptive particle swarm optimization. IEEE Trans. Syst., Man, and Cybern. B 39(6), 1362–1381 (2009)

[7] Zhan, Z.H., Zhang, J., Li, Y., Shi, Y.H.: Orthogonal learning particle swarm optimization. IEEE Trans. Evol. Comput. 15(6), 832–847 (2011)

[8]   Kumar, P., Verma, A.: Independent Task Scheduling in Cloud Computing by Improved Genetic Algrithm. International Journal of Advanced Research in Computer Science and Software Engineering 2(5), 111–114 (2012)

[9]   Ying, C., Yu, J.: Energy-Aware Genetic Algorithms for Task Scheduling in Cloud Computing. In: ChinaGrid Annual Conference (ChinaGrid), pp. 43–48 (2012)

[10] Loukopoulos, T., Lampsas, P., Sigalas, P.: Improved Genetic Algorithms and List Scheduling Techniques for Independent Task Scheduling in Distributed Systems. In: Proc.IEEE International Conference on Parallel and Distributed Computing, Applications and Technologies, pp. 67–74 (2007)

[11] Liu, H., Xu, D., Miao, H.: Ant colony optimization based service flow scheduling with various QoSrequirements in cloud computing. In: Proc. ACIS International Symposium on Software and Network Engineering (SSNE), pp. 53–58 (2011)

[12] Liu, X.F., Zhan, Z.H., Du, K.J., Chen, W.N.: Energy aware virtual machine placement scheduling in cloud computing based on ant colony optimization approach. In: Proc. Genetic Evol. Comput. Conf., pp. 41–47 (2014)

[13] Gan, G.N., Huang, T.L., Gao, S.: Genetic simulated annealing algorithm for task scheduling based on cloud computing environment. In: Proc. IEEE International Conference on Intelligent Computing and Integrated Systems (ICISS), pp. 60–63 (2010)

[14] Abdulal, W., Jabas, A., Ramachandram, S., Jadaan, O.A.: Mutation based simulated annealing algorithm for minimizing Makespan in Grid Computing Systems. In: Proc. International Conference on Electronics Computer Technology (ICECT), pp. 90–94 (2011)

[15] Gkoutioudi, K., Karatza, H.D.: A Simulation Study of Multi-criteria Scheduling in Grid Based on Genetic Algorithms. In: Proc. IEEE International Symposium on Parallel and Distributed Processing with Applications (ISPA), pp. 317-324 (2012)

[16] Shiand, W.M., Hong, B.: Resource Allocation with a Budget Constraint for Computing Independent Tasks in the Cloud. In: Proc. IEEE International Conference on Cloud Computing Technology and Science (CloudCom), pp. 327-334 (2010)

[17] Zhu, K., Song, H., Liu, L., Gao, J., Cheng, G.: Hybrid Genetic Algorithm for Cloud Computing Applications. In: Proc. IEEE Asia-Pacific Services Computing Conference (APSCC), pp. 182–187 (2011)

[18] Brauna, T.D., Siegelb, H.J., Beckc, N., Bölönid, L.L., Maheswarane, M., Reutherf, A.I., et al.: A Comparison of Eleven Static Heuristics for Mapping a Class of Independent Tasks onto Heterogeneous Distributed Computing Systems. Journal of Parallel and Distributed Computing 61(6), 810–837 (2001)

[19] Doulamis, N., Varvarigos, E., Varvarigou, T.: Fair Scheduling Algorithms in Grids. IEEE Trans. Parallel and Distributed Systems 18, 1630–1648 (2007)

[20] Holland, J.H.: Outline for a logical theory of adaptive systems. Journal of the ACM 9(3), 279–314 (1962)

[21] Shen, M., Zhan, Z.H., Chen, W.N., Gong, Y.J., Zhang, J., Li, Y.: Bi-velocity discrete particle swarm optimization and its application to multicast routing problem in communication networks. IEEE Trans. Ind. Electron. 61(12), 7141–7151 (2014)

[22] Zhan, Z.H., Li, J., Cao, J., Zhang, J., Chung, H., Shi, Y.H.: Multiple populations for multiple objectives: A coevolutionarytechnique for solving multiobjectiveoptimization problems. IEEE Trans. Cybern. 43(2), 445–463 (2013)

[23] Li, Y.H., Zhan, Z.H., Lin, S., Wang, R.M., Luo, X.N.: Competitive and cooperative particle swarm optimization with information sharing mechanism for global optimization problems. Information Sciences (in press, 2014)

[24] Zhan, Z.H., Zhang, J., Shi, Y.H., Liu, H.L.: A modified brain storm optimization. In: Proc. IEEE Congr. Evol. Comput., pp. 1–8 (2012)

Selection Schemes in Surrogate-Assisted Genetic Programming for Job Shop Scheduling

Su Nguyen[1,2], Mengjie Zhang[1], Mark Johnston[1], and Kay Chen Tan[3]

[1]Victoria University of Wellington, Wellington, New Zealand
[2]International University - VNU HCMC, Vietnam
[3]National University of Singapore, Singapore
{su.nguyen,mengjie.zhang}@ecs.vuw.ac.nz
mark.johnston@msor.vuw.ac.nz
eletankc@nus.edu.sg

Abstract. Designing effective dispatching rules is particularly important for dynamic job shop scheduling (JSS) problems. Recently, genetic programming (GP) and computer simulation have been combined to automatically design effective dispatching rules for different JSS problems. Although the literature has shown some success, expensive performance assessments or fitness evaluations still cause difficulty for design tasks, especially for very complicated and large-scale manufacturing systems. Therefore, it is important to effectively utilise the computational budget. The goal of this paper is to investigate the influence of surrogate models and the use of simulation replications on the performance of GP. The results show that the combination of the two techniques can enhance the quality of evolved dispatching rules. Analyses also show the advantages and disadvantages of different selection schemes in surrogate-assisted GP.

Keywords: genetic programming, job shop scheduling, heuristic.

1 Introduction

Dispatching rules are a simple and efficient approach to dealing with scheduling problems in dynamic job shops. The goal of dispatching rules is to prioritise jobs waiting in the manufacturing system based on the attributes of jobs (e.g. processing time, due date) and machines (e.g. workload, position). Jobs will be processed based on their assigned priorities. As dispatching rules are only applied at the moment machines decide which jobs to process next, they can take into account the current system status and cope with dynamic changes more easily. Many studies on dispatching rules, particularly for job shops, have been conducted in the scheduling literature to investigate the behaviour of different rules and to propose new ways to improve their performance. In the current highly competitive market, manufacturing systems need to adapt quickly to cope with changes and dispatching rules also need to be customised to deal with special requirements and production processes. Unfortunately, designing effective dispatching rules is not an easy task and involves a lot of trial and error. To avoid wasting manufacturing resources, computer simulation has often been used to assess

G. Dick et al. (Eds.): SEAL 2014, LNCS 8886, pp. 656–667, 2014.

the quality of dispatching rules before applying them to the real manufacturing systems. However, the design process is still difficult and requires extensive problem domain knowledge. Recent advances in machine learning and optimisation have offered different ways to facilitate the design process. Genetic programming (GP) is currently one of the most promising approaches to automatically design of competitive and robust dispatching rules for different scheduling problems, especially for job shop-like environments.

GP is an evolutionary computation (EC) method which is usually used to evolve programs to solve particular computational problems. In recent years, GP has been successfully used to evolve dispatching rules or scheduling heuristics for job shop scheduling (JSS) [1–3]. Different variants of GP such as tree-based GP [4], grammar-based GP [4], and gene expression programming (GEP) [5] are capable of evolving rules that outperform most rules manually designed by human experts in the literature [6]. As compared to other machine learning methods, GP has some advantages for these design tasks: (1) dispatching rules can be represented easily with GP's flexible representations; (2) GP can explore very complex combinations of attributes and discover unknown and effective dispatching rules; (3) many advanced techniques in EC can be applied to GP to enhance the quality of obtained dispatching rules. One of the key drawbacks of GP for evolving dispatching rules is long computational times. Because GP has to rely on discrete event simulation, which is computationally expensive, to assess the quality of evolved rules, the automatic design process is time-consuming.

Some strategies have been proposed to improve the efficiency of GP, particularly by focusing on effectively utilising the computation budget (i.e. the number of simulation replications). Hildebrandt et al. [2] investigated the balance between the number of simulation replications and the number of generations in GP. In their experiments, the fitness of evolved rules is measured using a predefined number of replications with fixed or *changing* random seeds (per generation) and the best individual of a generation will be fully evaluated. The analysis shows that using one simulation replication with different random seeds at different generations produces the best results. Nevertheless, there are two main drawbacks with this approach. First, because of the changes of random seeds at different generations, GP can explore diverse solutions (exploration ability) but it may have difficulty to fine-tune potential solutions (exploitation ability). This might prevent GP from finding high-performance rules. Second, GP tends to evolve large programs in this case because more genetic materials are needed to help evolve rules that can cope with new simulation replications. The large evolved dispatching rules will be more difficult to analyse and interprete.

Another way to cope with expensive fitness evaluations is to use surrogate models. Hildebrandt and Branke [7] investigated two surrogate models for evolving dispatching rules to minimise mean flowtime. In their approach, a large number of individuals are generated through genetic operations and the fitness of these rules is *approximated* by using the fitness of the most similar rules generated in the previous generations. Then, only rules with the top approximated fitness are *selected* for the next generation and receive *real* fitness evaluations.

The experimental results showed that surrogate-assisted GP (SGP) is more effective than the simple GP method for evolving dispatching rules. Specifically, SGP can converge to good dispatching rules faster than GP given that the same computational budget is used.

However, they have used fixed replications (across all generations) for fitness evaluations and have not taken the advantage of changing replications [2] to further improve the effectiveness of SGP. Also, how to build the population of rules selected based on their approximated fitness has not been investigated. In this paper, we explore two research questions regarding SGP. Firstly, can SGP be combined with changing replications to enhance the performance of GP? Secondly, what is the influence of selection scheme in SGP for choosing rules for the next generation? Moreover, this paper also investigates the effectiveness of SGP when dealing with different scheduling performance measures.

The remainder of this paper is organised as follows. Section 2 gives a brief description of job shop scheduling (JSS) and the existing studies on GP for JSS. Section 3 presents SGP methods and the two selection schemes investigated in our experiments. The results and analyses are shown in Section 4. Section 5 provides the conclusions and discusses future work directions.

2   Background

A brief introduction to JSS is provided here to show its key characteristics and the traditional methods to deal with this problem. Then, an overview of GP for JSS is presented.

2.1   Job Shop Scheduling

The general JSS problem could be simply defined as the scheduling of different jobs to be processed on different machines to satisfy certain objectives. In this case, a job is a sequence of operations, each of which is to be performed on a particular machine. In JSS, the routes of jobs are fixed, but not necessarily the same for each job [8]. For the static JSS problem, the shop (or the working/manufacturing environment) includes a set of m machines and n jobs that need to be scheduled. Each job has its own pre-determined route through a sequence of machines to follow and its own processing time at each machine it visits. In static JSS, processing information of all jobs is available. In the dynamic JSS problem, jobs arrive randomly over time and the processing information of jobs is unknown before their arrival.

A majority of past research focuses on developing effective and efficient optimisation methods for static JSS problems. As JSS is NP-hard, heuristics and meta-heuristics such as tabu search [9], evolutionary algorithms [10], and particle swarm optimisation [11] are usually employed to find acceptable solutions in reasonable computational times. For dynamic JSS, dispatching rules are more popular due to their ability to cope with dynamic changes and ease of implementation. A large number of rules have been developed in the literature to deal with both general job shops as well as specialised job shops (e.g. assembly, batch processing). Dispatching rules are usually classified based on the information used to make scheduling decisions (e.g. static, dynamic, local, global) [8] and how these pieces of information are combined. The effectiveness of a dispatching rule depends on how it reacts to dynamic changes of the shops and the ability to take into account different factors that can affect the considered objective to be optimised. Due to specific characteristics of each manufacturing system, there is no universal dispatching rule that can dominate in all situations. Thus, practitioners have to manually adapt their dispatching rules to deal with their specific operating conditions [12]. This suggests the need for automatic approaches to facilitating the selection or design of dispatching rules.

2.2   GP for JSS

Recently, GP has also been applied to evolve dispatching rules for JSS problems [1, 13]. Miyashita [14] proposed a hyper-heuristic that is based on a predetermined classification of machines into bottlenecks and non-bottlenecks and evolves one rule for each class of machine. Similarly, Jakobović and Budin [15] designed a hyper-heuristic that optimises the classification of machines while searching for good dispatching rules for each class. More specifically, each individual consists of three functions, where one of them is a discriminating function of attributes relating to the workload of a machine that determines which of the two dispatching rules, encoded by the other two (priority) functions, to apply. The best rule sets evolved by these hyper-heuristics are generally shown to outperform single benchmark rules. Nguyen et al. [16] investigated different representations of dispatching rules with GP. The experiments showed that a mixed representation, based on decision-tree like representation and arithmetic representations, provided the best results. However, they mainly examined their rules on static scheduling instances.

Tay and Ho [17] performed a study on using GP for multi-objective JSS problems. In their method, three objectives are linearly combined (with the same weights) into an aggregate objective, which is used as the fitness function in the GP method. The experiments showed that the evolved rules are quite competitive as compared to simple rules but still have trouble dominating the best rule for each single objective. Hildebrandt et al. [2] evolved dispatching rules by training them on different simulation scenarios and only minimised the mean flow time. Some aspects of the simulation models were also discussed in their study. The experimental results showed that the evolved rules were quite effective (although complicated) as compared to other existing rules. Moreover, Nguyen et al. [3] proposed a cooperative coevolution GPHH for multi-objective dynamic JSS problems. In that work, two scheduling rules (dispatching rule and due date assignment rule) are simultaneously considered in order to develop effective scheduling policies. More details about GP for JSS can be found in [7, 3].

3   Surrogate-Assisted Genetic Programming (SGP)

This section first presents the simulation model used in this paper. Then, we describe the overall algorithm for SGP and details for each components in SGP such as representation, genetic operations, and selection schemes.

3.1   Simulation Model

All experiments in this paper are based on the simulation model of a symmetrical job shop which has been used in previous studies on dispatching rules [18, 6, 19]. Here are the simulation configurations:

- 10 machines
- Each job has 2 to 14 operations (re-entry is allowed)
- Processing times follow discrete uniform distribution $U[1, 99]$
- Job arrivals follow Poisson process
- Due date = current time + allowance factor × total processing time (allowance factor of 4 is used in our experiments)
- Utilisation of the shop is 95%
- No machines break-down; preemption is not allowed
- Weights of jobs are assigned based on the 4 : 2 : 1 rule [20, 21].

In each simulation replication, we start with an empty shop and the interval from the beginning of the simulation until the arrival of the 500th job is considered as the warm-up time and the statistics from the next completed 2000 jobs [22] will be used to calculate performance measures. Three scheduling performance measures examined in our experiments are (1) mean flowtime, (2) mean tardiness, (3) total weighted tardiness. Although this simulation model is relatively simple, it still reflects key issues of real manufacturing systems such as dynamic changes and complex job flows. This section only considers a shop with high utilisation (95%) and tight due date (allowance factor of 4) because scheduling in this scenario is more challenging, and therefore easier to demonstrate the usefulness of GP. In order to reliably measure the effectiveness of evolved rules, a large number of simulation replications are usually needed (e.g. 30 to 50 simulation replications are usually needed to accurately estimate the performance of rules in the scenario described here). However, using simulation to evaluate the fitness of the evolved rules is also the most time-consuming part in GP for JSS. Therefore, only a small number of replications are usually used for fitness evaluations during the *training* process. As suggested by Hildebrandt et al. [2], we will use only one replication (corresponding to one random seed) for each fitness evaluation; however, we will change the replication (use a different random seed for simulation) when moving to a new generation. This strategy has been shown to be beneficial to prevent GP from overfitting to certain situations (replications). Two sets, each with 50 simulation replications, are used for determining *full training performance* and *testing performance*. The *full training set* is used to verify effectiveness of evolved rules and the *test set* is used to examine

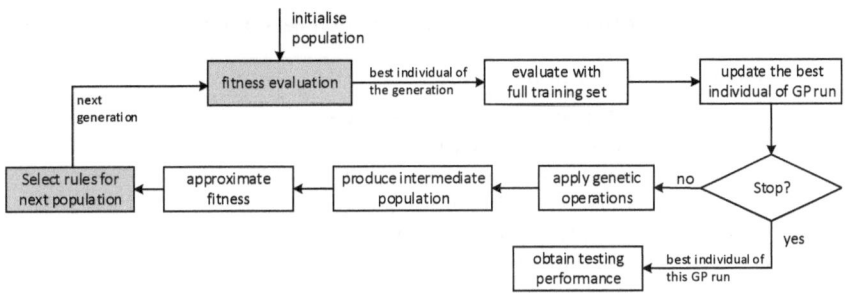

Fig. 1. Overall algorithm of SGP

the performance of rules on unseen situations. In this dynamic case, we cannot identify the best rule with the training performance or fitness functions, because they are changing across generations and they cannot accurately estimate the effectiveness of evolved rules.

3.2 Overall Algorithm

Fig. 1 shows how SGP can evolve dispatching rules for a particular simulation scenario. Similar to most EC methods, SGP starts by randomly initialising a population (ramped-half-and-half). Each rule will be evaluated using simulation. The fitness of an evolved rule depends on the performance measure achieved by the rule when applied to the training replication (e.g. mean flowtime). The rule with the best fitness will be evaluated with the full training set (refer to [2] for detailed discussions of this strategy). If the best rule of the generation has better full training performance than the current best rule, it will be assigned as the current best rule of the run. If the stopping condition is met (i.e. maximum generation in GP), SGP will stop; otherwise, we perform the next steps to build the population for the next generation.

First, a intermediate population is created using genetic operations. This intermediate population has a larger population size as compared to the original population to increase the diversity in the population as well as improve the chance to find better rules. The fitness of all rules in the intermediate population is *approximated* by using the surrogate model proposed in [7], which estimates the fitness of an evolved rule by using the fitness of the most similar rules generated in the previous generations. SGP in [7] used fixed simulation replications and utilised individuals in the last two generations to approximate fitness of newly generated rules. For the surrogate model, the behaviour of an evolved rule is characterised by a *decision vector* based on a reference rule (2PT+WINQ+NPT) and the similarity of two rules is measured by the distances of their corresponding decision vectors (see [7] for a detailed description). Different from [7], because fitness of rules in different generations is not compatible in our paper (as different replications are used), we only use rules in the most recent generation for fitness approximation. A selection scheme is

Table 1. Terminal and function sets of GP

Symbol	Description
rJ	job release time (arrival time)
RJ	operation ready time
RO	number of remaining operation within the job.
RT	work remaining of the job
PT	operation processing time
DD	due date of the job
RM	machine ready time
SL	slack of the job $= DD - (t + RT)$
WT	is the current waiting time of the job $= \max(0, t - RJ)$
#	Random number from 0 to 1
NPT	processing time of the next operation
WINQ	work in the next queue
APR	average operation processing time of jobs in the queue
Function set	$+, -, \times, \%$, min, max

*t is the time when the sequencing decision is made.

then used to select rules in the intermediate population for the next generation (see details in Section 3.4).

3.3 Representation and Genetic Operations

This paper uses the traditional GP tree to represent dispatching rules, the same as previous studies [1, 2, 16]. Table 1 shows the terminal set and function set used by GP to construct priority functions. The attributes in the tables have been extensively used in the existing dispatching rules as well as GP for JSS. For the function set, four basic arithmetic operators and *min/max* are used to construct composite dispatching rules (the protected division is similar to normal division but returns a value of 1 when division by 0 is attempted). For genetic operations, we employ the standard subtree crossover and subtree mutation. The subtree crossover creates new individuals for the next generation by randomly recombining subtrees from two selected parents. Meanwhile, the subtree mutation is performed by selecting a node of a chosen individual and replacing the subtree rooted by that node with a newly randomly-generated subtree.

3.4 Selection Scheme

One key aspect of SGP is how can we decide which rules in the intermediate population should be put into the next generation. Of course we want rules with top approximated fitness to be in the next generation to increase the chance to find more competitive rules. However, we also need to think of how diversity can be maintained. In our experiments, we will examine two selection schemes:

a. *No duplication* [7]: we will sweep through all rules in the intermediate population and eliminate duplicate rules. Two rules are considered the same if

Table 2. Parameter settings

Parameter	Description
Initialisation	ramped-half-and-half
Crossover/mutation/reproduction rates	80%/15%/5%
Maximum depth	8
Number of generations	100
Population size	500
Size of intermediate population	500×5=2500
Selection	tournament selection (size = 5)

they have exactly the same decision vector. The rationale of this scheme is to promote the diversity in the population.

b. *No previous match*: this scheme is little bit less extreme than the previous scheme. Basically, we only ignore rules which are exactly the same as the one in the previous population (used for fitness approximation). This scheme tries to create a balance between the diversity strategy and elitism. Therefore, the same rules can still exist in the population and have higher chances to pass their genetic materials to the next generation.

3.5   Experimental Settings

The parameters used in GP are presented in Table 2. In the experiments, we compare three GP methods: (1) simple GP with changing replications, (2) SGP with *no-duplicate* scheme (SGP-NODUP), and (3) SGP with *no-previous-match* scheme (SGP-NOMAT). The results for each GP method are based on 30 independent runs. All methods have the computational budgets of $(500 \times 1 + 50) \times 100 = 55000$ simulation replications. It is noted that the best rule of a generation and the current best rule based on full training performance are always copied to the next generation.

4   Results

This section compares the effectiveness of three GP methods based on their testing performance measures, lengths of obtained rules, and computational times. The comparisons are shown in Fig. 2 to Fig. 4. For each figure, the first and second boxplots respectively show the testing performance measures and lengths (number of nodes) of rules obtained by the three GP methods. For presentation purposes, the total weighted tardiness is normalised in Fig. 4 by dividing it by the number of (recorded) jobs. The third boxplot shows the computational times for GP runs. The behaviours of the three methods are shown in Fig. 5 to Fig. 7 where the plots show the progress of the best evolved rules and the average length of programs in the population over generations (averaged across all independent runs).

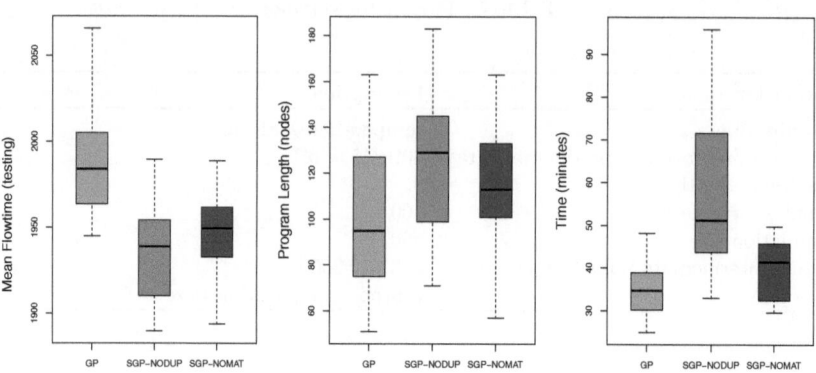

Fig. 2. Performance of GP methods - Minimise mean flowtime

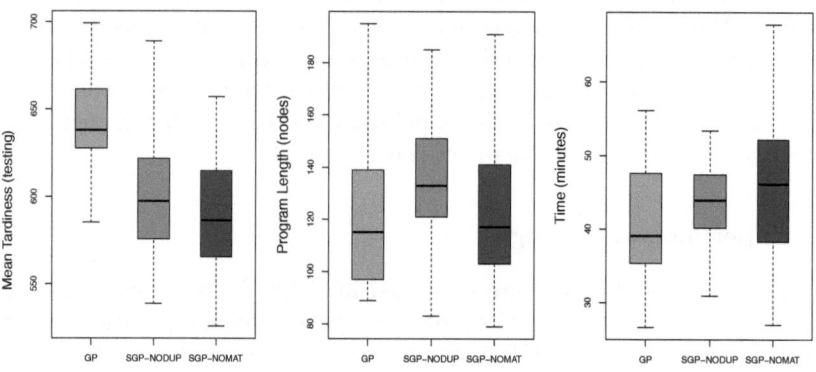

Fig. 3. Performance of GP methods - Minimise mean tardiness

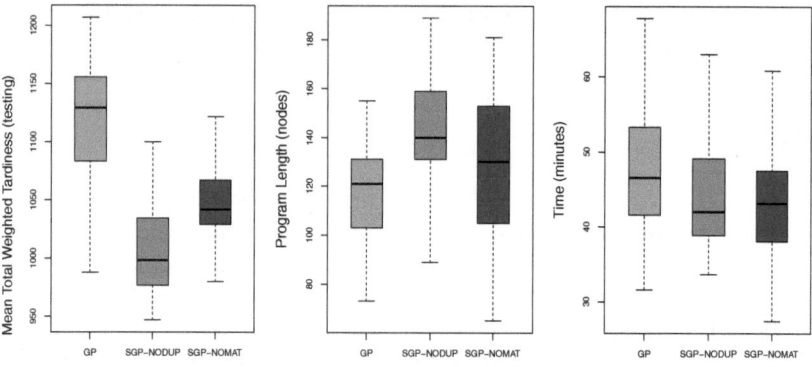

Fig. 4. Performance of GP methods - Minimise total weighted tardiness

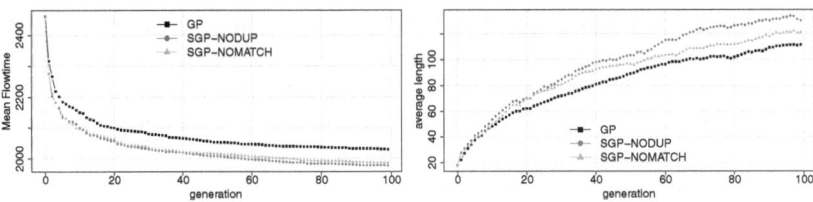

Fig. 5. Behaviours of GP methods - Minimise mean flowtime

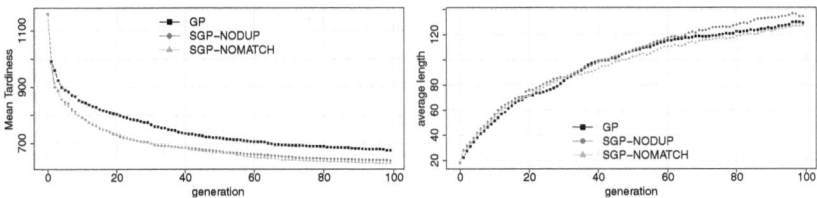

Fig. 6. Behaviours of GP methods - Minimise mean tardiness

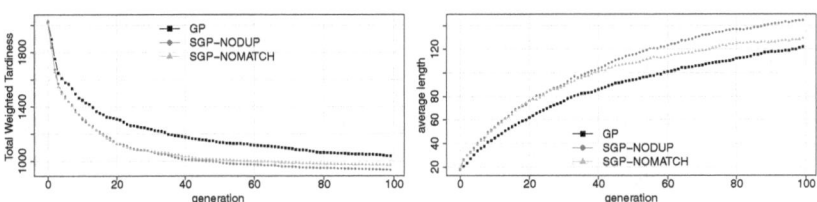

Fig. 7. Behaviours of GP methods - Minimise total weighted tardiness

From the results, it is obvious that SGP methods can find more competitive dispatching rules as compared to GP for all performance measures. As we observe the performance of the best rules over generations, we can see that SGP methods can find good rules significantly faster than GP given that they all start with the same populations. In the early generations, the gap between the GP and SGP methods widen very quickly. Then, the improvement rates of SGP decrease but they still continue to find better rules rather than prematurely converging. For the three performance measures, SGP methods tend to evolve effective rules twice as fast than GP (e.g. SGP methods only need 20 generations to find the best rules obtained by GP in 40 generations). These results suggest that SGP can successfully incorporate changing replications to further enhance the effectiveness of GP.

For the two selection schemes, there is no significant difference between them regarding the testing performance for minimising mean flowtime and mean tardiness. SGP-NODUP is only significantly better than SGP-NOMAT when total weighted tardiness is used as the performance measure. However, SGP-NODUP tends to evolve larger rules and it is also slightly slower than SGP-NOMAT in most cases. In the case with mean flowtime as the performance measure, the running time of SGP-NODUP is quite unpredictable as compared to GP and SGP-NOMAT.

The detailed results from Fig. 5 to Fig. 7 indicate that both SGP-NODUP and SGP-NOMAT make quite similar progress during the evolution. SGP-NODUP tends to perform slightly better than SGP-NOMAT in the later generations, especially for the case with total weighted tardiness as the performance measure. This observation shows that the no-duplicate scheme can be more effective when we deal with complex scheduling problems where sophisticated rules are needed and diversity is required to help SGP finding those rules. Nevertheless, the downside of the no-duplicate scheme is that it can also significantly increase the lengths of evolved rules as shown in the right plot of Fig. 5 to Fig. 7. In the three cases, SGP-NODUP tends to evolve larger rules than GP and SGP-NOMAT. For minimising mean flowtime, it is very clear that SGP-NODUP evolves rules much larger than they should be, which make the fitness evaluations slower as discussed. SGP-NOMAT performs quite well in most cases and it is able to maintain the length of evolved rules reasonably well. For minimising mean tardiness, the evolved rules of SGP-NOMAT are even slightly smaller than those of GP. Therefore, no-previous-match is a promising selection scheme in SGP; however, it still needs to be further improved to cope better with complex environments. Probably a hybrid between the two selection schemes will provide us the most favourable results.

5   Conclusions

In this paper, we have investigated different techniques to deal with expensive fitness evaluations which is one of the most critical aspects of evolving dispatching rules with GP. The experimental results indicate that surrogate models can be applied successfully along with changing replications to significantly enhance the performance of GP. The combination of the two techniques allows GP to evolve more effective rules using the same computation budget. We also investigate the influence of selection schemes in this study to see how they govern the behaviours as well as the performance GP. The results suggest that diversity is one of the key factors to help GP explore better rules, especially for complicated JSS problems. However, care must be taken to ensure that rules will not grow too large, which can slow down fitness evaluations in GP.

For future work, we want to further build better surrogate models for GP to cope with different scheduling problems. Furthermore, it would be interesting to see how surrogate models can be used to cope with multiple conflicting objectives. We also want to develop smarter selection schemes in order to help GP cope better with complicated situations and hopefully generate more compact rules.

References

1. Geiger, C.D., Uzsoy, R., Aytuğ, H.: Rapid modeling and discovery of priority dispatching rules: an autonomous learning approach. J. of Sched. 9(1), 7–34 (2006)
2. Hildebrandt, T., Heger, J., Scholz-Reiter, B.: Towards improved dispatching rules for complex shop floor scenarios — a genetic programming approach. In: GECCO 2010: Proceedings of the 12th Annual Conference on Genetic and Evolutionary Computation, pp. 257–264 (2010)

3. Nguyen, S., Zhang, M., Johnston, M., Tan, K.C.: Automatic design of scheduling policies for dynamic multi-objective job shop scheduling via cooperative coevolution GP. IEEE Transactions on Evolutionary Computation 18(2), 193–208 (2014)
4. Banzhaf, W., Nordin, P., Keller, R., Francone, F.: Genetic Programming: An Introduction. Morgan Kaufmann (1998)
5. Ferreira, C.: Gene Expression Programming: Mathematical Modeling by an Artificial Intelligence, 2nd edn. Springer, Germany (2006)
6. Nguyen, S., Zhang, M., Johnston, M., Tan, K.C.: Dynamic multi-objective job shop scheduling: A genetic programming approach. In: Uyar, A.S., Ozcan, E., Urquhart, N. (eds.) Automated Scheduling and Planning. SCI, vol. 505, pp. 251–282. Springer, Heidelberg (2013)
7. Hildebrandt, T., Branke, J.: On using surrogates with genetic programming. Technical report, Warwick Business School (2014)
8. Pinedo, M.L.: Scheduling: Theory, Algorithms, and Systems, 3rd edn. Springer, New York (2008)
9. Nowicki, E., Smutnicki, C.: A fast taboo search algorithm for the job shop problem. Management Science 42, 797–813 (1996)
10. Cheng, V.H.L., Crawford, L.S., Menon, P.K.: Air traffic control using genetic search techniques. In: Proceedings of the 1999 IEEE International Conference on Control Applications, vol. 1, pp. 249–254 (1999)
11. Sha, D., Hsu, C.Y.: A hybrid particle swarm optimization for job shop scheduling problem. Computers & Industrial Engineering 51(4), 791–808 (2006)
12. Philipoom, P., Russell, R.S., Fry, T.D.: A preliminary investigation of multi-attribute based sequencing rules for assembly shops. International Journal of Production Research 29(4) (1991)
13. Jakobović, D., Jelenković, L., Budin, L.: Genetic programming heuristics for multiple machine scheduling. In: Ebner, M., O'Neill, M., Ekárt, A., Vanneschi, L., Esparcia-Alcázar, A.I. (eds.) EuroGP 2007. LNCS, vol. 4445, pp. 321–330. Springer, Heidelberg (2007)
14. Miyashita, K.: Job-shop scheduling with GP. In: GECCO 2000: Proceedings of the Genetic and Evolutionary Computation Conference, pp. 505–512 (2000)
15. Jakobović, D., Budin, L.: Dynamic scheduling with genetic programming. In: Collet, P., Tomassini, M., Ebner, M., Gustafson, S., Ekárt, A. (eds.) EuroGP 2006. LNCS, vol. 3905, pp. 73–84. Springer, Heidelberg (2006)
16. Nguyen, S., Zhang, M., Johnston, M., Tan, K.C.: A computational study of representations in GP to evolve dispatching rules for the job shop scheduling problem. IEEE Transactions on Evolutionary Computation 17(5), 621–639 (2013)
17. Tay, J.C., Ho, N.B.: Evolving dispatching rules using genetic programming for solving multi-objective flexible job-shop problems. Computers & Industrial Engineering 54(3), 453–473 (2008)
18. Jayamohan, M.S., Rajendran, C.: New dispatching rules for shop scheduling: a step forward. International Journal of Production Research 38, 563–586 (2000)
19. Branke, J., Hildebrandt, T., Scholz-Reiter, B.: Hyper-heuristic evolution of dispatching rules: A comparison of rule representations. Evolutionary Computation (in press, 2014), doi:10.1162/EVCO_a_00131
20. Kreipl, S.: A large step random walk for minimizing total weighted tardiness in a job shop. Journal of Scheduling 3, 125–138 (2000)
21. Pinedo, M., Singer, M.: A shifting bottleneck heuristic for minimizing the total weighted tardiness in a job shop. Naval Research Logistics 46(1), 1–17 (1999)
22. Holthaus, O., Rajendran, C.: Efficient jobshop dispatching rules: Further developments. Production Planning & Control 11(2), 171–178 (2000)

Developing a Hyper-Heuristic Using Grammatical Evolution and the Capacitated Vehicle Routing Problem

Richard J. Marshall[1], Mark Johnston[1], and Mengjie Zhang[2]

[1] School of Mathematics, Statistics and Operations Research
[2] School of Engineering and Computer Science
Victoria University of Wellington, New Zealand
{richard.marshall,mark.johnston}@msor.vuw.ac.nz
mengjie.zhang@ecs.vuw.ac.nz

Abstract. A common problem when applying heuristics is that they often perform well on some problem instances, but poorly on others. We work towards developing a hyper-heuristic that manages delivery of good quality solutions to Vehicle Routing Problem instances with only limited prior knowledge of the problem domain to be solved. This paper develops a hyper-heuristic, using Grammatical Evolution, to generate and apply heuristics that develop good solutions. Through a series of experiments we expand and refine the technique, achieving good quality results on 40 well known Capacitated Vehicle Routing Problem instances.

1 Introduction

A common problem when applying heuristics to an optimisation problem is that they often perform well on some problem instances, but poorly on others. Helping to identify which heuristic to apply to a particular problem instance is one of the objectives of hyper-heuristic research. Here we develop a hyper-heuristic, using Grammatical Evolution (GE) [25], to generate heuristics for the Vehicle Routing Problem (VRP). The VRP has wide ranging application in the transport and logistics industry. Generating a good solution, and possibly several alternatives, to a VRP instance can have significant operational and cost benefits to the industry.

The Capacitated Vehicle Routing Problem (CVRP) [28] contains a single depot holding a fleet of identical vehicles. A set of customers, each at a known location and with a known demand, are to be serviced. The objective is to service all customers while travelling the shortest possible total distance. Each customer must be serviced only once (split deliveries across multiple routes are not permitted), and the capacity of each vehicle must not be exceeded at any time.

The goal in this paper is to use a hyper-heuristic to evolve heuristics that progressively construct and improve a partial solution until a complete solution to a CVRP instance is achieved. We demonstrate that a compact heuristic can

G. Dick et al. (Eds.): SEAL 2014, LNCS 8886, pp. 668–679, 2014.

be generated from a small set of operators (heuristic components) which consistently deliver "good" quality solutions to CVRP instances in a reasonable computation time.

In the remainder of this paper we discuss the background and prior research in Section 2, followed by details of the experimentation method and design in Sections 3 and 4. The results of this research and discussion are presented in Section 5 followed by our conclusions in Section 6.

2   Background

In this section we review the relevant background literature.

2.1   Vehicle Routing Problem

The VRP was introduced in 1959 by Dantzig and Ramser [8], which Lenstra and Rinnooy Kan [17] show is a NP-hard combinatorial optimisation problem. The problem has been well studied and comes in many variations [11,13,16,28]. A VRP normally includes constraints on the capacity of each vehicle, and/or the duration or distance a vehicle may travel. Further features may be added, such as time windows for servicing a particular customer, or allowing multiple depots, split deliveries or interchanges to be considered.

Solving the VRP often requires a trade-off between computation speed and achieving the best possible solution. Only relatively small CVRP instances are able to be solved optimally in reasonable computation time (e.g. using branch-and-bound [10]). Consequently heuristics are used to quickly find an acceptable solution. Because of its speed and simplicity, the Clarke and Wright Savings (CWS) heuristic [28] is a frequently used heuristic. Where extra computational effort is acceptable, the solutions generated by the CWS heuristic can often be improved by applying a search technique such as Tabu Search [12] or Iterated Local Search [18].

2.2   Meta-Heuristics and Hyper-Heuristics

Meta-heuristics have been developed over the last 40 years using different techniques which seek improvements to an initial solution by searching the adjacent and/or wider solution space. Many of these achieve good results but are often complex and time consuming to design and execute. More recent research has looked at hyper-heuristics which Cowling et al. [6] define as "heuristics to choose heuristics". Ross [23] modifies this definition to say hyper-heuristics are heuristics which search a space of heuristics, as opposed to searching the space of solutions directly.

Developing a hyper-heuristic, using GE, for the VRP is a relatively new field of research. Existing approaches have focussed on either building a VRP solution step by step from scratch [3] or, more commonly, commence with an initial solution developed using an existing fast heuristic (such as the CWS heuristic) and evolve a search heuristic to improve the solution [9,26]. Here we use a hyper-heuristic to progressively build and search for improvements in parallel. This enables application of a search operation to a partially built solution.

We follow the example of Burke et al. [4,3] and generate new heuristics from the operators (or components) of existing heuristics. The choice of operators are guided by work in recent years by Bader el Den and Poli [2], Drake et al. [9] and Sabar et al. [26]. Unlike Drake et al. [9], who study several variations of VRP to evolve the components for a variable neighbourhood search (VNS) framework, we use operators that both construct and modify a solution.

2.3   Grammatical Evolution

Grammatical Evolution (GE) is an evolutionary computation technique pioneered by Ryan et al. [25]. A key feature of GE is the separation between the search engine and the problem. This enables different classes of problem to be solved using the same search engine, and conversely, an alternative engine can be employed to create and evolve a 'genotype'. The linking element is a grammar relevant to the class of problem (i.e. CVRP in this case) which is applied through a mapper to the output of the search engine. Defining a good grammar requires a degree of inspiration and experimentation as the structure and content of the grammar can influence the quality of the result, much in the way the grammar of a natural language defines the richness of that language.

McKay et al. [20] note that GE belongs to a wider family of Grammar Guided Genetic Programming (GGGP) approaches which emerged in the mid 1990s [29]. They note that in GE the genotype is linear, as opposed to the tree structure used in both standard Genetic Programming (GP) [15] and other grammar based GP. The linear genotype enables a range of theory and practice applicable to Genetic Algorithms and Evolution Strategies to be employed. The benefits of GGGP over standard GP include the ability of the grammar to restrict the search space and reduce the likelihood of generating semantically meaningless output.

Over the last decade there has been research into the benefits and challenges of using GE. Rothlauf and Oetzel [24] have investigated the mapping of the genotype to the output (phenotype) in GE and find genotype neighbours in the population do not correspond well with phenotype neighbours produced by the grammar, a phenomenon they refer to as a low degree of locality. The consensus among researchers, notably in the works of Rothlauf et al. [24,27], is that there needs to be a close correlation (high locality) between genotype neighbours and phenotype neighbours for an efficient search process. Sub-Tree crossover and Sub-Tree mutation [14] are regarded as the best means of achieving this in GE. This requires a reverse mapping of the phenotype to the genotype to ensure a crossover is only performed between compatible points. Thorhauer and Rothlauf [27] note that the single point crossover operation has a very different effect when applied in GE than when applied in standard GP, resulting in GE swaps involving, on average, half of the tree structure as illustrated in Figure 1.

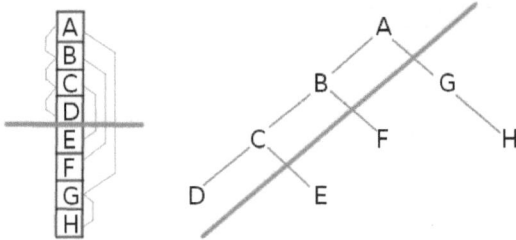

Fig. 1. The effect of GE Single Point Crossover on a Derivation Tree

Another known feature of GE is redundancy in both the encoding and length of the genotypes. GE uses variable length integer (or bit) strings (codon strings) which are mapped with the grammar using modular arithmetic. This means numerous different encodings will map to the same sentence in the grammar. Also, since the length of genotype can be longer (possibly by a considerable amount) than that needed by the grammar mapper, the population may comprise of many different individuals who only differ from each other in the unused portion of the genotype.

The use of tree-adjunct or tree-adjoining grammars as suggested by McKay et al. [20] and Murphy et al. [21] may remedy some of these issues, although the technique does not scale well and a grammar such as the one we use would generate an unmanageable number of tree elements.

3   Method

In this section we describe how a heuristic is developed by the proposed hyper-heuristic and illustrate in Algorithm 1 how the operators are selected and processed.

A CVRP solution consists of a set of one or more vehicle routes, beginning and ending at the depot, with each route listing the customers in the order in which they are visited. The solution is complete if all customers are visited exactly once, and feasible if each vehicle's capacity is never exceeded. Each operator we use performs an action on the current partial solution that selects a customer and/or modifies one or more routes.

A successful heuristic need not be intuitive, so a range of build, modify and destroy operators are enabled with few constraints on the selection of an operator or the number of times it is executed.

Each heuristic consists of four distinct elements:

1. A *strategy* which defines how the heuristic is to be developed. A build strategy starts with an empty solution and routes are developed step by step. An improvement strategy starts with a complete and feasible, but possibly sub-optimal, solution developed using a fast heuristic.

Algorithm 1. Heuristic structure

$T \leftarrow$ set strategy (*build* or *improve*)
$k \leftarrow$ randomly select number of operators
if T is *build* **then**
    $i \leftarrow$ select initial seeding method
    $s_0^i \leftarrow$ initialise (partial) solution
    for $n \leftarrow 1$ to k **do**
        $op_n^{type} \leftarrow$ select build or improve operator
        $op_n^{repeat} \leftarrow$ select number of repetitions
        $op_n^{param} \leftarrow$ select parameter
    end
    end if
else if T is *improve* **then**
    $s_0 \leftarrow$ initialise (complete) solution (out-and-back routes)
    for $n \leftarrow 1$ to k **do**
        $op_n^{type} \leftarrow$ select improve operator
        $op_n^{repeat} \leftarrow$ select number of repetitions
        $op_n^{param} \leftarrow$ select parameter
    end
    end if
$search^{type} \leftarrow$ select local search operator
$search^{param} \leftarrow$ select parameter(s)
$r \leftarrow 0$
while $(r < 1)$ or $(s_r < s_{r-1})$ or (s_r is incomplete) **then**
    increment r
    $s_r \leftarrow s_{r-1}$
    for $n \leftarrow 1$ to k **do**
        $s_r \leftarrow s_r +$ **execute** op_n^{type}
    end
    $s_r \leftarrow s_r +$ **execute** $search^{type}$
    evaluate s_r
    repeat
return s_{r-1}
end

2. A sequence of one or more *operators* (excluding a search operator) to construct or modify the current solution.
3. A *search* operator to improve the current solution.
4. The number of times the whole sequence of operators (including the search operator) is *repeated* to deliver a complete and feasible solution. We refer to each repetition of the sequence of operators as a *cycle*.

3.1 Operators

An operator manipulates the current partial solution. Application of some operators may result in a customer being returned to the pool of unallocated customers.

Build operators define the method of selecting the next customer from the pool of unallocated customers. The chosen customer is inserted into the least-cost feasible route and location in the current partial solution. A new route is created if the customer cannot be inserted into an existing route and retain feasibility. A build operator has two parameters:

1. How many times execution of the operator is repeated.
2. Whether replacement of one existing customer in a route is permitted when considering insertion of a new customer into that route. A replaced customer is blocked from being reinserted in the same route for the remaining repetitions of the current operator. However the customer may be reinserted by another operator or by the same operator in a later cycle.

Perturbation operators manipulate all or part of the solution. We use a selection of simple move, swap, and ruin-recreate operators. With two exceptions, individual perturbation operators are executed no more than once in each cycle.

The *local search operator* we use is a deterministic variation of Iterated Local Search (ILS) [18]. A brief description of the search is shown in Algorithm 2. We limit the search operator to one execution per cycle to keep the computational time within reasonable bounds.

Algorithm 2. Deterministic Iterated Local Search

---

let N be the set of customers allocated to a route in the current solution.
range ← parameter sets the range of search
pair each n_i and $n_j \in N$ **where** $distance_{i,j} \leq range$
sort $pair(n_i, nj)$ by ascending $distance_{i,j}$
while queue of $pair(n_i, n_j)$ is not empty.
    let R_a and R_b be the routes containing n_i and n_j respectively
    if $R_a \neq R_b$ **then** $R_a \leftarrow$ **concatenate** R_a, R_b (including *depot*)
    swap positions of n_i and n_j in R_a
    improve R_a using either a 2opt [7] or 3opt algorithm
    if *depot* appears in intermediate position within R_a, divide R_a into two routes
    if better feasible routes result, update current solution.
    repeat
return solution

3.2   GE Grammar

We use GE to select the operators and their respective parameters. To achieve this we define a grammar that maps the output of the search engine (the genotype) to a syntactically correct and semantically meaningful sequence of operators.

The only element of a heuristic that is not specified in the grammar is the number of cycles. Instead, the sequence of operators (up to a pre-defined maximum number of cycles) are repeated until a cycle ends with a complete solution to the CVRP instance.

The grammar is structured so a strategy is specified as the first element. Thereafter selected operators, and any required parameters, are added in sequence, terminating with a search operator. The Backus Naur Form [22] grammar we use is detailed below.

```
<strategy>   ::= build, <seed>; <action1> <search>; | improve; <action2> <search>;
<action1>    ::= <build>; | <build>; <action1> | <build>; <action2>
<action2>    ::= <improve>; | <improve>; <action2>
<build>      ::= <select>,<num>,<replace>
<improve>    ::= <improve1>,<num> | <improve2>
<improve1>   ::= mergeBestSaving | mergeNextNearest
<improve2>   ::= splitRoutes, 1, <num> | redoRoute, 1 | removeLowestDemand, 1
<search>     ::= 2opt | 3opt | ILSearch,<num>,<optType>
<select>     ::= cheapest | largestDemand | farthest | nearest | remotest
<seed>       ::= blank | cheapest | largestDemand | farthest | nearest
<num>        ::= 1 | 2 | 3 | 4 | 5 | 6 | 7 | 8 | 9 | 10
<replace>    ::= 0 | 1
<optType>    ::= 2 | 3
```

A typical phenotype from the mapper takes the following form.

build, farthest; largestDemand, 3, 0; cheapest, 6, 1; ...; ILSearch, 8, 3;

The output is passed to a parser application which interprets and processes it. If a complete solution is not achieved within the pre-defined maximum number of cycles, the distance is set to ∞. The total distance of the solution is passed back to the fitness evaluation function.

We adopt the requirement that applying a given heuristic to a particular CVRP instance will always generate the same solution. To this end, all random number generation occurs within the search engine and is passed to the parser as a parameter(s) with each operator. This includes the local search operator which follows a deterministic sequence when seeking improvements.

4   Experiment Design

We use 30 replications for each of 40 CVRP instances developed by Augerat et al. and Eilon et al. [1]. The experiments are repeated twice, firstly using an on-line learning process to dynamically customise the heuristic for the problem instance being solved, and secondly using off-line learning with a separate set of six problem instances for training, and applying the resulting heuristic to the 40 CVRP instances. We test different numbers of GP generations (between 10 and 1,000) and population size (between 40 and 100). Additionally we test different combinations of crossover and mutation functions. The best heuristic from each experiment is accepted and the performance compared to the best known published solution for each problem instance.

Table 1. Results from on-line learning and off-line learning experiments compared to published best solution (recalculated to ensure rounding consistency)

CVRP	Off-line	On-line	Best [1]	CVRP	Off-line	On-line	Best [1]
A-n32-k5	**787**	**787**	788	A-n33-k5	668	**662**	663
A-n33-k6	**743**	**743**	**743**	A-n34-k5	790	**781**	**781**
A-n36-k5	**802**	**802**	**802**	A-n37-k5	**672**	**672**	673
A-n37-k6	958	**951**	952	A-n38-k5	742	**734**	**734**
A-n39-k5	**829**	**829**	**829**	A-n39-k6	835	**833**	**833**
A-n44-k7	**938**	**938**	n/a	A-n45-k6	954	950	**945**
A-n45-k7	1160	**1147**	**1147**	A-n46-k7	**918**	**918**	**918**
A-n48-k7	1093	**1074**	**1074**	A-n53-k7	1020	**1012**	1013
A-n54-k7	1186	**1172**	**1172**	A-n55-k9	1075	1075	**1074**
A-n60-k9	1356	**1356**	**1356**	A-n61-k9	1051	1048	**1039**
A-n62-k8	1303	1302	**1294**	A-n63-k9	1638	1634	**1622**
A-n63-k10	1323	1319	**1314**	A-n64-k9	1428	1416	**1401**
A-n65-k9	1192	1185	**1182**	A-n69-k9	1180	**1166**	**1166**
A-n80-k10	1792	1770	**1767**	E-n13-k4	**290**	**290**	n/a
E-n22-k4	**375**	**375**	**375**	E-n23-k3	**569**	**569**	**569**
E-n30-k4	**505**	**505**	n/a	E-n31-k7	1239	**1205**	n/a
E-n33-k4	**838**	**838**	839	E-n51-k5	543	**525**	**525**
E-n76-k7	697	**688**	**688**	E-n76-k8	762	**741**	n/a
E-n76-k10	842	**837**	**837**	E-n76-k15	1047	**1038**	n/a
E-n101-k8	852	**829**	n/a	E-n101-k14	1105	**1095**	n/a

5   Results and Discussion

We now detail and discuss the results from these experiments. The labelling of the test instances [1] can be identified as being from the set produced by Augerat et al. (prefixed A) or Eilon et al. (E), with n nodes ($n-1$ customers plus a depot) and a minimum of k routes required.

5.1   Overall Results

Since generation of a CWS heuristic from individual operators is enabled by the grammar, the solution generated by the CWS heuristic should, if discovered by the search engine, represent an upper bound on generated solutions.

Reducing the number of GP generations means the resulting heuristic contains significantly fewer operators. When using 1,000 GP generations, the resulting heuristic can contain a sequence of up to 200 operators, whereas limiting the run to 10 generations means a heuristic rarely contains more than a dozen operators. The heuristics with only a few operators prove to be just as effective as those with numerous operators indicating that increasing the number of operators in a sequence does not necessarily improve performance.

The solution distances from our research are given in Table 1 and compared to the best known results [1] (recalculated to ensure rounding consistency).

5.2 Discussion

Our research shows that it is possible to apply a hyper-heuristic to select operators to develop a heuristic for a CVRP instance that delivers a high quality solution. As regards a general CVRP heuristic capable of consistently delivering a "good" solution to any CVRP instance, our research shows that the CWS heuristic followed by the application the 2-opt algorithm will produce a reasonable result that cannot be consistently matched or beaten by any single heuristic discovered in our research. The failure to locate a previously undiscovered solve-all heuristic from these experiments is consistent with the results observed by Bader el Den and Poli [2], who applied a similar technique on exam timetabling problems. Applying the best outcome from a small number of general heuristics achieves much better results, but still falls short of what can be achieved by applying the proposed hyper-heuristic to generate a heuristic for an individual CVRP instance.

Analysis of the results from our research indicate that some of the operators, such as the *splitRoutes* and *redoRoute*, can be simplified or omitted. Also, interpreting the number of operator repetitions relative to the size of the problem instance rather than as an absolute number may improve the computation time of the heuristic on larger CVRP instances.

The decision to use GE has provided challenges, including many of those discussed in Section 2. While we have achieved good results using GE it is difficult to avoid the impression that the search process is not as efficient as it could be. Analysis suggests the problem lies in the nature of the feedback provided from the fitness evaluation function. A CVRP contains two interdependent sub-problems: vehicle loading and travel distance. The single score based on distance alone does not adequately assist the search engine. Further, as identified by Rothlauf and Oetzel [24], the low degree of locality between genotype and phenotype hinders an efficient search process. Applying the proposed hyper-heuristic using GE on the CVRP reveals there are a number of barriers to achieving high locality, which we discuss further in Section 5.4.

5.3 Crossover and Mutation Operators

When using the sub-tree crossover and mutation operators recommended by Manrique et al. [5,19], we observe an early and rapid decline in the diversity of the population. This favours those heuristics that develop a good solution using relatively few operators. Equally it means the best solution that the search engine is likely to deliver in this run is found in relatively few generations. Although this is not necessarily a bad situation, it means that more complex combinations of operators, which may produce a better solution, are prematurely pushed out of the population by pseudo-clones of the same good solution found so far. The *mergeBestSaving* operator (which is the core operator of the CWS heuristic) is a case in point as this develops a reasonably good solution using a minimal number of operators.

We also trialled three different combinations of crossover and mutation operators. As illustrated in Figure 2 the performance of these operators is quite

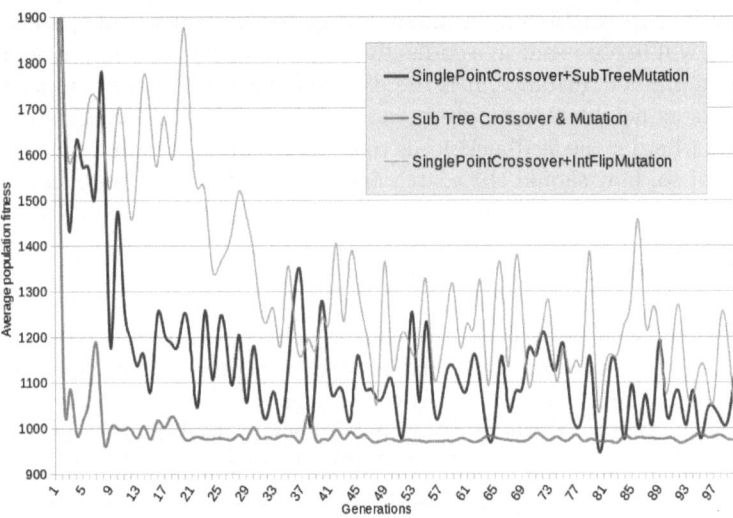

Fig. 2. Comparison between Sub-Tree and Single Point crossover and Sub-Tree and IntFlip mutation operators. Showing average population fitness per generation on the A-n37-k6.vrp [1] problem instance. Population size = 80.

different although all eventually arrived at similar results in every case. The difference between the operators lies in the speed of execution and the average fitness of the current population of the search engine.

Figure 2 illustrates the difference in the performance of the crossover and mutation operators by measuring the average fitness of the individuals in the population in the first 100 generations of a typical example (CVRP instance A-n37-k6.vrp [1]). The upper line records the average fitness from using single point crossover and intFlip mutation operators. This results in a wider variance in the fitness of the individuals in the population compared to sub-tree crossover and sub-tree mutation operators (lowest line). The population created by the latter combination rapidly converges towards the best fitness found so far and thereafter shows little diversity. The hybrid single point crossover and sub-tree mutation combination retains diversity in the population for longer, but then converges on the best fitness found so far.

5.4   Hyper-Heuristics and Capacitated Vehicle Routing

If, as discussed in Section 2, high locality is a desirable feature, then we need to be able to define what makes two solutions neighbours. With hyper-heuristics the elements used are operators to be processed rather than the raw data of the problem to be solved. If we were solving a CVRP directly we could identify neighbouring solutions by the similarity of routes or commonality of arcs between customers.

With a hyper-heuristic, the sequence of operators and parameters cannot be easily identified as neighbours. A minor change to the sequence of operators, or a parameter, will likely result in a radically different solution when the heuristic is applied to a problem instance. It is therefore inappropriate to refer to a sequence of operators as neighbours simply because they appear similar. This raises the question of whether the feedback needs to be more complex than a single numeric score and, if so, how should the search engine interpret such feedback.

6   Conclusions

We have shown that good heuristics can be delivered from a GE-based hyper-heuristic that use operators that enable both construction and improvement of a solution to a CVRP instance. However, elements of the hyper-heuristic search process show scope for considerable improvement. Indeed, a different evolutionary computation approach may be desirable. Further research will examine to what extent those improvements can be found in modifying the feedback provided to the search engine, streamlining the search operator, and/or using a different evolutionary search process.

References

1. Capacitated Vehicle Routing Problem Instances (October 2013), http://neo.lcc.uma.es/vrp/vrp-instances/capacitated-vrp-instances/
2. Bader-el-Den, M.B., Poli, R.: Grammar-based genetic programming for timetabling. In: Proceedings of the IEEE Congress on Evolutionary Computation (CEC 2009), pp. 2532–2539 (2009)
3. Burke, E.K., Hyde, M., Kendall, G., Ochoa, G., Özcan, E., Woodward, J.: Exploring hyper-heuristic methodologies with genetic programming. In: Mumford, C.L., Jain, L.C. (eds.) Computational Intelligence. ISRL, vol. 1, pp. 177–201. Springer, Heidelberg (2009)
4. Burke, E.K., Hyde, M.R., Kendall, G.: Grammatical evolution of local search heuristics. IEEE Transactions on Evolutionary Computation 16(3), 406–417 (2012)
5. Couchet, J., Manrique, D., Ríos, J., Rodríguez-Patón, A.: Crossover and mutation operators for grammar-guided genetic programming. Soft Computing 11(10), 943–955 (2007)
6. Cowling, P., Kendall, G., Soubeiga, E.: A hyperheuristic approach to scheduling a sales summit. In: Burke, E., Erben, W. (eds.) PATAT 2000. LNCS, vol. 2079, pp. 176–190. Springer, Heidelberg (2001)
7. Croes, G.A.: A method for solving traveling salesman problems. Operations Research 6, 791–812 (1958)
8. Dantzig, G.B., Ramser, J.H.: The truck dispatching problem. Management Science 6, 80–91 (1959)
9. Drake, J.H., Kililis, N., Özcan, E.: Generation of VNS components with grammatical evolution for vehicle routing. In: Krawiec, K., Moraglio, A., Hu, T., Etaner-Uyar, A.Ş., Hu, B. (eds.) EuroGP 2013. LNCS, vol. 7831, pp. 25–36. Springer, Heidelberg (2013)

10. Fisher, M.L.: Optimal solution of vehicle routing problems using minimum k-trees. Operations Research 42(4), 626–642 (1994)
11. Gendreau, M., Laporte, G., Potvin, J.-Y.: Metaheuristics for the vehicle routing problem. In: Les Cahiers du GERAD G-98-52, Montréal, Canada (1999)
12. Glover, F.: Tabu search: Part I. ORSA Journal on Computing 1(3), 190–206 (1989)
13. Goel, A., Gruhn, V.: A General Vehicle Routing Problem. Elsevier Science, Germany (2006)
14. Harper, R., Blair, A.: A structure preserving crossover in grammatical evolution. In: Proceedings of the IEEE Congress on Evolutionary Computation, vol. 3, pp. 2537–2544 (2005)
15. Koza, J.R.: Genetic Programming: On the Programming of Computers by Means of Natural Selection. MIT Press (1992)
16. Laporte, G.: The vehicle routing problem: An overview of exact and approximate algorithms. European Journal of Operational Research 59, 345–358 (1992)
17. Lenstra, J.K., Rinnooy Kan, A.H.G.: Complexity of vehicle routing and scheduling problems. Networks 11, 221–227 (1981)
18. Lourenço, H.R., Martin, O.C., Stützle, T.: Iterated local search. In: Handbook of Metaheuristics, pp. 321–354 (2003)
19. Manrique, D., Ríos, J., Rodríguez-Patón, A.: Grammar-guided genetic programming. In: Rabuñal, J.R., Dorado, J., Pazos, A. (eds.) Encyclopedia of Artificial Intelligence, pp. 767–773. Information Science Reference (2008)
20. McKay, R.I., Hoai, N.X., Whigham, P.A., Shan, Y., O'Neill, M.: Grammar-based genetic programming: A survey. Genetic Programming and Evolvable Machines 11(3-4), 365–396 (2010)
21. Murphy, E., O'Neill, M., Galaván-Lopéz, E., Brabazon, A.: Tree-adjunct grammatical evolution. In: Proceedings of the IEEE Congress on Evolutionary Computation, pp. 1–8 (2010)
22. Naur, P.: Revised report on the algorithmic language ALGOL 60. Communications of the ACM 3, 299–314 (1960)
23. Ross, P.: Hyper-heuristics. In: Burke, E.K., Kendall, G. (eds.) Search Methodolgies: Introductory Tutorials in Optimization and Decision Support Techniques, pp. 529–556. Kluwer (2005)
24. Rothlauf, F., Oetzel, M.: On the locality of grammatical evolution. In: Collet, P., Tomassini, M., Ebner, M., Gustafson, S., Ekárt, A. (eds.) EuroGP 2006. LNCS, vol. 3905, pp. 320–330. Springer, Heidelberg (2006)
25. Ryan, C., Collins, J.J., Neill, M.O.: Grammatical evolution: Evolving programs for an arbitrary language. In: Banzhaf, W., Poli, R., Schoenauer, M., Fogarty, T.C. (eds.) EuroGP 1998. LNCS, vol. 1391, pp. 83–96. Springer, Heidelberg (1998)
26. Sabar, N.R., Ayob, M., Kendall, G., Qu, R.: Grammatical evolution hyper-heuristic for combinatorial optimization problems. IEEE Transactions on Evolutionary Computation 17(6), 840–861 (2013)
27. Thorhauer, A., Rothlauf, F.: Structural difficulty in grammatical evolution versus genetic programming. In: Proceeding of the Fifteenth Annual Conference on Genetic and Evolutionary Computation Conference, pp. 997–1004 (2013)
28. Toth, P., Vigo, D.: The Vehicle Routing Problem. SIAM (2002)
29. Whigham, P.A.: Grammatically-based genetic programming. In: Proceedings of the Workshop on Genetic Programming: From Theory to Real-World Applications, pp. 33–41 (1995)

An Organizational Cooperative Coevolutionary Algorithm for Multimode Resource-Constrained Project Scheduling Problems

Lixia Wang, Jing Liu, and Mingxing Zhou

Key Laboratory of Intelligent Perception and Image Understanding of Ministry of Education,
Xidian University, Xian 710071, China

Abstract. In recent years, the resource-constrained project scheduling problem (RCPSP) with multiple execution modes is becoming more and more popular. In this paper, a new cooperative coevolutionary algorithm based on the concept of organizations, namely Organizational Cooperative Coevolutionary Algorithm for MRCPSPs (OCCA-MRCPSPs), is proposed for solving this problem. The objective is to find a schedule of activities together with their execution modes so that the makespan is minimized. In the OCCA-MRCPSPs, the population is divided into two subpopulations, for activities execution modes, respectively. The two subpopulations evolve independently, and each subpopulation is composed of organizations. During the evolutionary process, the global searching and the local searching are combined efficiently by conducting different operators. At first, each subpopulation searches the whole space of its domain through the splitting operator, the annexing operator, and the cooperation operator. Afterwards, the two subpopulations are combined to form complete solutions, and a local search operator is performed. In the experiments, the performance of OCCA-MRCPSPs is validated on benchmark problem sets J10, J12, J14, and J16 from the PLPSIB, and the experimental results show that the OCCA-MRCPSPs obtains a good performance not only in terms of the optimal solutions found but also in terms of the average deviations from optimal solutions.

Keywords: Resource-constrained project scheduling problems, Multimode, Cooperative coevolutionary algorithms, Organizations.

1 Introduction

The resource-constrained project scheduling problem (RCPSP) is a popular problem that interests lots of researchers. In RCPSPs, a project is consisted of several activities, and each activity needs a related duration and several different resources to complete. The resources can be divided into two kinds, namely renewable resources (RR) and nonrenewable resources (NR). In addition to the resource capacity constraints, there are precedence constraints among these activities. Usually, once an activity starts, it can not be interrupted.

The multimode RCPSP (MRCPSP) is an extension of the RCPSP [1]. The most important difference between them is that there are several execution modes for each activity in the MRCPSP and each execution mode may need different types of and

G. Dick et al. (Eds.): SEAL 2014, LNCS 8886, pp. 680–690, 2014.

amounts of resources, while in the RCPSP, each activity has only one execution mode. The objective is to find a schedule of activities and modes with minimum makespan which also satisfy both resource constraints and precedence constraints.

The MRCPSP is strongly NP-hard. Researchers tried to solve it with different kinds of methods. Jozefowska *et al.* [2] and Bouleimen *et al.* [3] proposed the algorithms based on simulated annealing. Slowiński *et al.* [4] proposed a single-pass approach, a multi-pass approach as well as a simulated annealing algorithm. A biased random sampling approach was proposed by Drexl and Grünewald [5]. A local search strategy was proposed by Kolisch and Drexl [6]. Hartmann *et al.* used genetic algorithms to solve the MRCPSP [7,8]. Sprecher *et al.* [9] proposed an exact algorithm based on the branch-and-branch strategy. A new mathematical formulation for the MRCPSP is proposed by Maniezzo and Mingozzi and two new lower bounds were derived [10]. Boctor presented heuristics to solve the MRCPSP without the nonrenewable resources [11–13]. Tseng *et al.* proposed a two-phase genetic local search algorithm in [1], which first search the whole solution space to find a set of elite solutions, and then most solutions in the elite set are selected to construct the initial population of the second phase. This method improves the search efficiency and performs well on the benchmark problems.

Coase explained the sizing and formation of organization from the framework of transaction cost in economic [14]. The basic idea is that the organization exists because it reduces the overhead transaction costs associated with exchanging goods and services. Wilcox introduced this concept into learning classifiers based on genetic algorithms [15]. Inspired by the idea of organizations, in our previous work, we proposed an organizational coevolutionary algorithm for classification [16], which achieved a higher predictive accuracy and lower computational cost, and an organizational evolutionary algorithm (OEA) for numerical optimization problems [17], which showed a good performance in terms of both the solution quality and the computational cost. In [18], we also applied OEA to general floorplanning problems, which can obtain high quality solutions for various and large-scale problems.

All our previous work shows that OEA has a huge potential in solving complex problems. Therefore, in this paper, with the intrinsic properties of MRCPSPs in mind, we combine the OEA with cooperative coevolutionary algorithm to solve MRCPSPs, and the algorithm is named as OCCA-MRCPSPs. During the evolutionary process of OCCA-MRCPSPs, the global searching and the local searching are combined efficiently by conducting different operators. The performance of OCCA-MRCPSPs is tested upon the benchmark problem sets J10, J12, J14, and J16, and the results show that OCCA-MRCPSPs obtains a good performance not only in the optimal solutions found but also in the average deviations from optimal solutions or critical path lower bounds.

The rest of the paper is organized as follows. Section II gives a brief introduction on MRCPSPs. Section III describe the proposed algorithm in details. The experiments are given in Section IV. Finally, Section V summarizes the work in this paper.

2   Problem Definition

In MRCPSPs, the activities of a project are marked by $1, 2, \ldots, n$. For an arbitrary activity i, its execution modes are represented by a set $EMi = \{1, 2, \ldots, M_i\}$, and a mode means a group of different kinds and quantities of resources, and it costs a related duration. If activity i executes in mode m_i, then it needs $r^{RR}_{im_ik}$ units of renewable resource k, and $r^{RR}_{im_ik}$ units of nonrenewable resource k, d_{im_i} indicates its related duration.

As indicated above, there are two kinds of constraints in the MRCPSP, one is the precedence constraint, and the other is the resource constraint. The precedence constraint can be described by a graph in Fig. 1, where each node represents an activity, and if there is an arrow from node i to j, it means that activity i is the immediate predecessor of activity j, similarly, activity j is called the immediate successor of activity i. An activity can not be arranged before all of its immediate predecessors have been completed. Image resource k belongs to renewable resources, and its gross is a constant Q^{RR}_k, which means no matter at which moment, the total amount of resource k is equal to Q^{RR}_k, while if k is a nonrenewable resource and its gross is a constant Q^{RR}_k, which means in the overall process, the total amount of resource k is stationary.

The objective of MRCPSPs is to optimize the start time and the execution mode for each activity so that the makespan of the whole project is minimized. At the same time, the precedence and resource constraints must be satisfied. In Fig. 1., there are two dummy activities called fictitious initial activity and fictitious finish activity that do not need any resources and their durations are zeros. Suppose each activity has two execution modes except the dummy activities, and there are one type of renewable resources and one type of nonrenewable resources. The data above node i mean the duration, the renewable resource cost, and the nonrenewable resource cost for activity i when it is executed under mode 1. Similarly, the data below node i mean the duration, the renewable resource cost, and the nonrenewable resource cost for activity i when it is executed under mode 2.

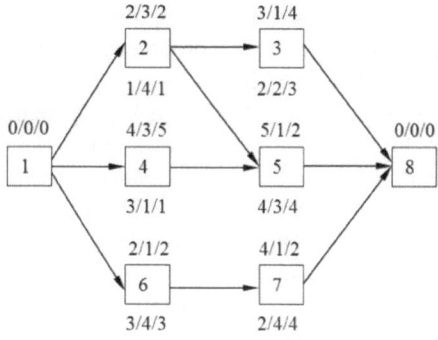

Fig. 1. An instance of the MRCPSP

3   Organizational Cooperative Coevolutionary Algorithm for MRCPSPs

3.1   Preprocessing

In order to reduce the search space, a reduction procedure for MRCPSP introduced by Sprecher *et al.* in [19] is first used. If an execution mode exceeds the renewable resource constraints, it is called a non-executable mode. When executing a mode, if its duration is not shorter than another mode, and its costs of resources are not less than this mode, then it is an inefficient mode. Besides, if a nonrenewable resources total amount is not less than the sum of the largest requirements of each activity, then it is said to be redundant. Obviously, the non-executable and inefficient modes as well as the redundant resources can be deleted from the data, and this not only has no effect on the optima, but also can reduce the search space.

As described in [19] and [7], the remove of modes and nonrenewable resources can affect each other. For example, deleting a nonrenewable resource may cause some modes inefficient, while removing a mode can make some resources be redundant. Thus, the data can be recomposed by the following steps. First, remove all of the non-executable modes from data. Second, delete all of the redundant nonrenewable resources and, subsequently, all inefficient modes. Conduct the second step repeatedly until no redundant nonrenewable resource is left. After these steps, both the number of feasible solutions and the search space is reduced.

3.2   Populations and Organizations for MRCPSPs

In the OCCA-MRCPSPs, the population is divided into two subpopulations, labeled as pop_a and pop_o. pop_a means the population of activities and pop_o means the population of modes. Besides, each subpopulation is composed by several organizations, and an organization is composed by members. A complete solution is constructed by combining a member in pop_a and a member in pop_o. Thus, we first introduce the definition of members and then define organizations.

Definition 1. A member in pop_a is labeled as $M\langle A\rangle$, and a member pop_o is labeled as $M\langle O\rangle$. $M\langle A\rangle$ is a permutation of all activities, and all activities in this permutation are arranged according to the precedence constraints,

$$M\langle A\rangle = \{\pi_1, \pi_2, \ldots, \pi_n\} \tag{1}$$

where $(\pi_1, \pi_2, \ldots, \pi_n)$ is a permutation of $(1, 2, \ldots, n)$, and for $\pi_i, i = 1, 2, \ldots, n - 1$, no immediate predecessors of π_i is in $\{\pi_{i+1}, \pi_{i+2}, \ldots, \pi_n\}$. $M\langle O\rangle$ is the set of execution modes for each activity,

$$M\langle O\rangle = \{o_1, o_2, \ldots, o_n\} \tag{2}$$

where $o_j \in EM_{\pi_i}, j = 1, 2, \ldots, n$, and stand for the execution mode of activity π_i.

After the preprocessing, the execution modes that violate renewable resource constrains are already removed. In addition, the permutation of activities in a member of pop_a strictly obeys the precedence constraints. Therefore, a solution $M = \{M\langle A\rangle, M\langle O\rangle\}$ is an infeasible solution only if it violates the nonrenewable resource constraints. In this paper, we use the fitness function proposed in [8] to evaluate the quality of a solution, which is defined as follows,

$$f = \begin{cases} mak(M) & \text{if } M \text{ is feasible} \\ max\_fea\_pop\_mak + mak(M) \\ -min\_project\_CC + SFT(M) & \text{Others} \end{cases} \tag{3}$$

where mak(M) indicates the makespan of M. $max\_fea\_pop\_mak$ expresses the maximum makespan among all feasible solutions in current population, which should be updated in every generation. $min\_project\_CC$ is the critical path using the minimum durations of activities, which is smaller than the makespan of any infeasible scheduling. $SFT(M)$ represents the amount of the nonrenewable resources that exceeds the capacities. For an infeasible solution, $SFT(M)$ is undoubtedly larger than zero, and the length of critical path is smaller than any of the makespans of feasible solutions. Therefore, the makespans of infeasible solutions are greater than that of feasible ones.

Here, a simple example is given to show the meaning of (3). For the instance in Fig. 1, let the maximum makespan of all feasible solutions in current population be 12. $min\_project\_CC$ is 7. Suppose there is a solution M, where $M\langle A \rangle$={1, 2, 3, 4, 5, 6, 7, 8} and $M\langle O \rangle$={1, 1, 2, 1, 2, 1, 2, 1}, and the total capacity of nonrenewable resources is 16. Then, we can see that the solution is infeasible, because the cost of nonrenewable resources is 20, which is larger than 16. $SFT(M)$ is 4 and $mak(M)$ is 10. Therefore, $f(M)$ is 19.

Based on the members, an organization is defined as follows,

Definition 2. An organization org_a in pop_a is a set of $M\langle A \rangle$, while an organization org_o in pop_o is a set of $M\langle O \rangle$. The best member of an organization is the *leader*. The fitness value of an organization is equal to that of its leader.

When two organizations are compared, only their leaders are considered; that is, one organization is better than another when its leader is better than another ones leader.

3.3   Evolutionary Operators for Members

There are three types of operators which can be conducted on members directly, namely crossover, mutation, and local search operators.

1)$CrossoverOperator$: The crossover operator is used by the activity subpopulation pop_a. This operator is based on two-point crossover operators that operator on two parent members $M_{p1}\langle A \rangle$ and $M_{p2}\langle A \rangle$, and generate two child members $M_{c1}\langle A \rangle$ and $M_{c2}\langle A \rangle$.

First, two cut-points r_1 and r_2 are randomly drawn. Then, the first r_1 elements of $M_{c1}\langle A \rangle$ are inherited from the first r_1 elements of $M_{p1}\langle A \rangle$. The following $r_2 - r_1$ elements of $M_{c1}\langle A \rangle$ are inherited from the first $r_2 - r_1$ elements of $M_{p2}\langle A \rangle$ which do not appear in the first section of $M_{c1}\langle A \rangle$. The remaining elements of $M_{c1}\langle A \rangle$ are inherited from the remaining elements of $M_{p1}\langle A \rangle$ which do not appear in the first two sections of $M_{c1}\langle A \rangle$. All elements of $M_{c1}\langle A \rangle$ inherited from the parents must keep the order as they appear in $M_{p1}\langle A \rangle$ or $M_{p2}\langle A \rangle$. $M_{c2}\langle A \rangle$ is constructed in the same way with the roles of $M_{p1}\langle A \rangle$ or $M_{p2}\langle A \rangle$ being exchanged.

2)$MutationOperators$: Since the ways of mutation for the two subpopulations are different, there are two kinds of mutation operators used. The mutation operator that employed by the activity subpopulation pop_a is labeled as $Mutation_1$. First, n_1 activities are randomly selected, where n_1 is smaller than one thirds of the total number of activities. Then, each activity of the n_1 ones is randomly moved to another position without violating the precedence constrains. That is, an activity can move to any position as long as it is behind all of its immediate predecessors at the same time before all of its immediate successors.

A simple example is given here to further explain $Mutation_1$. Suppose $Mutation_1$ is performed on $M_p\langle A\rangle$, where $M_p\langle A\rangle$={1, 2, 3, 4, 5, 6, 7, 8}, and n_1 is equal to 1. Suppose the activity selected is 4, then its positions is changed; that is, $M_p\langle A\rangle$ can be changed to {1, 4, 2, 3, 5, 6, 7, 8} or {1, 2, 4, 3, 5, 6, 7, 8}.

The mode subpopulation pop_o uses a different kind of mutation method, which is labeled as $Mutation_2$. Firstly, n_2 activities are randomly selected, where n_2 is smaller than one thirds of the total number of activities. Then for each of the n_2 activities, change its execution mode to another one.

3)$LocalSearchOperator$: In order to improve the search efficiency, a local search operator is designed based on $Mutation_1$ and $Mutation_2$. This operator works as follows. First, the best members among pop_a and pop_o are selected and combined to form a complete solution. Then, $Mutation_1$ and $Mutation_2$ are alternatively executed on this complete solution. If the new solution is better than the previous one, it is used to replace the previous one. The above process is repeated until the upper bound times T is reached.

3.4 Evolutionary Operators for Organizations

In the real-world situation, there exists a severe competition among organizations, and the weak ones are always annexed by the strong ones. Besides, the organizations also have a cooperative relationship. As the strength of an organization is represented by the fitness value of its leader, the purpose of each organization is to make the fitness of its leader be better. Based on these, three evolutionary operators, namely the splitting operator, the annexing operator, and the cooperating operator which are performed on organizations directly are designed.

1)$SplittingOperator$: In human societies, when the size of an organization is too large, it will be split into small ones in order to be managed easily. In the OCCA-MRCPSPs, if one organization is too large, it will affect the search efficiency. Therefore, an upper bound size, labeled as max_os (>1), is set to prevent an organization being too large. When the size of an organization exceeds the upper bound or satisfies the other condition in (4), the organization will be split.

$$(|org| > max_{os} \ or \ ((|org| \le max_{os}) \ and \ (U(0,1) < |org|/N_0))) \qquad (4)$$

where $U(0,1)$ is a uniformly distributed random number between 0 and 1, and N_0 is the initial number of organizations. Then, randomly generate an integer M satisfying $|org|/3 < M < 2|org|/3$, then divide the individuals randomly into the two sub-organizations until one of them has M members, and the other one has $|org| - M$ members. After the two child organizations are generated, their best members are selected to be their leaders, and the original organization is removed from the population.

2)$AnnexingOperator$: The annexing operator reflects the competition between two organizations. The organization with stronger strength will defeat the weak one and annex it to construct a larger organization. Suppose the annexing operator is performed on two parent organizations, org_{p1} and org_{p2}, which come from pop_a or pop_o, and $f(Leader_{org_{p1}}) < f(Leader_{org_{p2}})$. That is, the fitness of org_{p1} is worse than that of org_{p2}. The method for computing the fitness of members in each subpopulation will be introduced in the next subsection. Then, org_{p2} will be annexed by org_{p1} in the following way to generate a new organization org_{new}. First, all members of org_{p1} are moved to org_{new} without any change. Then, if the two organizations come from pop_a, $Mutation_1$ is performed on $Leader_{org_{p1}}$ to generate $|org_{p2}|$ new members for org_{new}; otherwise, $Mutation_2$ is used. Finally, org_{new} is added to the subpopulation and org_{p1} and org_{p2} are removed. After performing the annexing operator, the two organizations become a larger organization and the leader of the new organization needs to be selected again.

3)$CooperatingOperator$: The cooperating operator reflects the cooperative relationship between two organizations. Suppose the cooperating operator is performed on two parent organizations, org_{p1} and org_{p2} , which come from pop_a or pop_o . If the two organizations come from org_a , then crossover operator is performed on $Leader_{org_{p1}}$ and $Leader_{org_{p2}}$; otherwise, $Mutation_2$ is conducted on $Leader_{org_{p1}}$ and $Leader_{org_{p2}}$ separately. After performing the cooperating operator, two new members are generated, and the two old leaders are replaced by the new members.

3.5   Implementation of OCCA-MRCPSPs

In the OCCA-MRCPSPs, the population is divided into two subpopulations, pop_a and pop_o, and each subpopulation is composed of organizations. In the initialization, each subpopulation has N0 organizations and each organization has only one member. Members in pop_a are generated used the method in [19], and members in pop_o are generated randomly based on the activities after the preprocessing. After the initialization, the fitness of each member in the two subpopulations are calculated as follows. For each member in pop_a, randomly choose a member from pop_o, and combine the two members to form a complete solution, then calculate its fitness according to (3). The result will be the fitness of the member from pop_a. The fitness of each member in pop_o is calculated in the same way.

Next, in each generation, the two subpopulations evolve independently. For each subpopulation, the size of each organization is first checked. If it satisfies (4), the organization will be split. Then, two organizations are randomly selected from the corresponding subpopulation, and the annexing operator or the cooperating operator will conduct on them with the same probability. Afterwards, update the fitness of each member. To be different from the process of calculate fitness in the initialization, now, for each member in pop_a, first combine it with the best member in pop_o to form a complete solution, then combine it with a random member in pop_o to form another complete solution, choose the better one to be the fitness of this member. The fitness of members in pop_o are updated in the same way.

Then, choose two members from pop_a and pop_o separately and form a complete solution, and conduct the local search operator on it. The population evolves generation by generation. During the evolutionary process, the number of organizations increases or decreases through the splitting operator or the annexing operator. Although the number of organizations in the population varies from generation to generation, the total number of members remains constants. The details of OCCA-MRCPSPs are summarized in Algorithm 1.

Algorithm 1. Organizational cooperative coevolutionary algorithm for MRCPSPs

---

Conduct the preprocessing;

Initialize the subpopulations with N_0 organizations separately, and each organization has one member;

Calculate the fitness value of each member;

while *the termination criteria are not satisfied* **do**

    for *each subpopulation* **do**

        For each organization, if it satisfies (4), then conduct the splitting operator on it;

        Randomly select two organizations from the current subpopulation, and conduct the annexing operator or the cooperating operator on them with the same probability;

    end

    Update the fitness of each member;

    Conduct the local search operator on the current population;

end

---

4   Experiments

In this section, benchmark problem sets J10, J12, J14 and J16 from the PSPLIB [20] are used to test the performance of OCCA-MRCPSPs. The total number of instances in J10, J12, J14 and J16 are 536, 547, 551, and 550, respectively. In each instance of J10 data set, there are 10 non-dummy activities; for J12, each instance has 12 non-dummy activities, and so on. Each instance in these sets has two kinds of renewable resources and two kinds of nonrenewable resources.

The OCCA-MRCPSPs is tested upon the four data sets, for each instance, 30 independent runs are conducted. The optimal solutions of these instances are known, so we use the two indexes namely the percentage of the optimal solution found and the average deviation from the optimal solutions to evaluate the performance of the algorithm. The average deviation from the optimal solutions is calculated according to (5).

$$Average\ deviation = \frac{\sum_{instance_i} \left(\frac{best\ makespan_i}{optimal\ makespan_i} - 1\right)}{number\ of\ feasible\ instances} \tag{5}$$

Tables 1 and 2 show the results for J10 and J12 while the maximum number of schedules evaluated varies from 5000 to 50000. As can be seen, with the increasing of the maximum number of schedules evaluated, the performance of the OCCA-MRCPSPs is getting better. While the maximum schedules become 50000, the percentages of optimal solutions found for both J10 and J12 is large than 98% and the average deviation from optimal solutions is smaller than 0.1%.

In existing literature on MRCPSPs, we find that most of them show the performance by calculating the average deviation from optimal solutions evaluated being a fixed number, while majority of these set the maximum schedules to 5000. Therefore, Table 3 compares the performance of OCCA-MRCPSPs with that of another method for J10 to J16 under 5000 schedules.

Table 1. The percentage of optimal solutions found (%) for J10 and J12

Set\Size	5000	10000	20000	50000
J10	95.44	97.76	99.03	99.67
J12	86.15	92.00	96.01	98.59

Table 2. The average deviation (%) from optimal solutions for J10 and J20

Set\Size	5000	10000	20000	50000
J10	0.275	0.141	0.070	0.035
J12	0.786	0.408	0.199	0.076

Table 3. Comparison in terms of average deviation (%) from optimal solutions between OCCA-MRCPSP and another Method under 5000 schedules

Method\Set	J10	J12	J14	J16
OCCA-MRCPSPs	**0.28**	**0.79**	**1.18**	**2.75**
Jozefowska et al.[2]	1.16	1.73	2.6	4.07

Table 4. The comparison in terms of optimal solutions found (%) for J10 and J12 between OCCA-MRCPSP and another Method under 5000 schedules

Set	Method\Size	5000	10000	20000	50000
J10	OCCA-MRCPSPs	**95.44**	**97.76**	**99.03**	**99.67**
	Jozefowska et al.[2]	85.60	93.70	96.60	97.20
J12	OCCA-MRCPSPs	**86.15**	**92.00**	96.01	**98.59**
	Jozefowska et al.[2]	80.30	91.60	**96.70**	97.60

Tables 4 and 5 compare the performance of OCCA-MRCPSPs and another method with the maximum schedules evaluated fixed to 5000, 10000, 20000, and 50000. As can be seen, the OCCA-MRCPSPs outperforms the algorithm proposed by Jozefowska *et al.* [2].

Table 5. The comparison in terms of average deviation (%) from optimal solutions for J10 and J12 between OCCA-MRCPSP and another Method

Set	Method\Size	5000	10000	20000	50000
J10	OCCA-MRCPSPs	**0.275**	**0.141**	**0.070**	**0.035**
	Jozefowska et al.[2]	1.160	0.470	0.270	0.230
J12	OCCA-MRCPSPs	**0.786**	**0.408**	**0.199**	**0.076**
	Jozefowska et al.[2]	1.730	0.740	0.420	0.370

5   Conclusion

The OCCA-MRCPSPs, a new organizational cooperative coevolutionary algorithm for solving project scheduling problems with multiple modes, is proposed in this paper. In the OCCA-MRCPSPs, the population is divided into two subpopulations, namely the activity subpopulation and mode sub-population. Each subpopulation is composed by several organizations and organizations are composed by members. In the earlier stage of each generation, the two subpopulations evolve independently through the splitting operator, the annexing operator, and the cooperating operator, which realize the global search. Then in the later stage of each generation, combine the individuals of the two subpopulations to form complete solutions, and a local search strategy is executed in order to further improve the quality of members.

In the experiments, we have tested the OCCA-MRCPSPs upon the benchmark problem sets J10, J12, J14, and J16. The results show that when the number of maximum schedules evaluated increases, the percentages of optimal solution found increase obviously and the average deviations from optimal solutions decrease obviously, while compare with other algorithms, the OCCA-MRCPSPs obtains a better performance.

Acknowledgements. This work is partially supported by the National Natural Science Foundation of China under Grants 61271301 and 61103119, the Research Fund for the Doctoral Program of Higher Education of China under Grant 20130203110010, and the Fundamental Research Funds for the Central Universities under Grant K5051202052.

References

1. Tseng, L.-Y., Chen, S.-C.: Two-phase genetic local search algorithm for the multimode resource-constrained project scheduling problem. IEEE Transactions on Evolutionary Computation 13(4), 848–857 (2009)
2. Józefowska, J., Mika, M., Różycki, R., Waligóra, G., Weglarz, J.: Simulated annealing for multi-mode resource-constrained project scheduling. Annals of Operations Research 102(1-4), 137–155 (2001)
3. Bouleimen, K., Lecocq, H.: A new efficient simulated annealing algorithm for the resource-constrained project scheduling problem and its multiple mode version. European Journal of Operational Research 149(2), 268–281 (2003)
4. Słowiński, R., Soniewicki, B., Weglarz, J.: Dss for multiobjective project scheduling. European Journal of Operational Research 79(2), 220–229 (1994)
5. Drexl, A., Grünewald, J.: Nonpreemptive multi-mode resource-constrained project scheduling. IIE Transactions 25(5), 733–750 (1993)
6. Kolisch, R., Drexl, A.: Local search for nonpreemptive multi-mode resource-constrained project scheduling. IIE Transactions 29(11), 987–999 (1997)
7. Hartmann, S.: Project scheduling with multiple modes: a genetic algorithm. Annals of Operations Research 102(1-4), 111–135 (2001)
8. Alcaraz, J., Maroto, C., Ruiz, R.: Solving the multi-mode resource-constrained project scheduling problem with genetic algorithms. Journal of the Operational Research Society 54(6), 614–626 (2003)
9. Sprecher, A., Drexl, A.: Multi-mode resource-constrained project scheduling by a simple, general and powerful sequencing algorithm. European Journal of Operational Research 107(2), 431–450 (1998)

10. Maniezzo, V., Mingozzi, A.: A heuristic procedure for the multi-mode project scheduling problem based on benders decomposition. In: Project Scheduling, pp. 179–196. Springer (1999)
11. Boctor, F.F.: Heuristics for scheduling projects with resource restrictions and several resource-duration modes. The International Journal of Production Research 31(11), 2547–2558 (1993)
12. Boctor, F.F.: A new and efficient heuristic for scheduling projects with resource restrictions and multiple execution modes. European Journal of Operational Research 90(2), 349–361 (1996)
13. Boctor, F.F.: Resource-constrained project scheduling by simulated annealing. International Journal of Production Research 34(8), 2335–2351 (1996)
14. Coase, R.H.: The firm, the market, and the law. University of Chicago press (2012)
15. Wilcox, J.R.: Organizational learning within a learning classi er system. Urbana 51, 61801 (1995)
16. Jiao, L., Liu, J., Zhong, W.: An organizational coevolutionary algorithm for classification. IEEE Transactions on Evolutionary Computation 10(1), 67–80 (2006)
17. Liu, J., Zhong, W., Jiao, L.: An organizational evolutionary algorithm for numerical optimization. IEEE Transactions on Systems, Man, and Cybernetics, Part B: Cybernetics 37(4), 1052–1064 (2007)
18. Liu, J., Zhong, W., Jiao, L., Li, X.: Moving block sequence and organizational evolutionary algorithm for general floorplanning with arbitrarily shaped rectilinear blocks. IEEE Transactions on Evolutionary Computation 12(5), 630–646 (2008)
19. Sprecher, A., Hartmann, S., Drexl, A.: An exact algorithm for project scheduling with multiple modes. Operations-Research-Spektrum 19(3), 195–203 (1997)
20. Kolisch, R., Sprecher, A.: Psplib-a project scheduling problem library: Or software-orsep operations research software exchange program. European Journal of Operational Research 96(1), 205–216 (1997)

Scaling Up Solutions to Storage Location Assignment Problems by Genetic Programming

Jing Xie[1], Yi Mei[1], Andreas T. Ernst[2], Xiaodong Li[1], and Andy Song[1]

[1] School of Computer Science and IT, RMIT University, Melbourne, Australia
{jing.xie,yi.mei,xiaodong.li,andy.song}@rmit.edu.au
[2] Commonwealth Scientific and Industrial Research Organisation (CSIRO),
Melbourne, Australia
Andreas.Ernst@csiro.au

Abstract. The Storage Location Assignment Problem (SLAP) is to find an optimal stock arrangement in a warehouse. This study presents a scalable method for solving large-scale SLAPs utilizing Genetic Programming (GP) and two sampling strategies. Given a large scale problem, a sub-problem is sampled from the original problem for our GP method to learn an allocation rule (in the form of a matching function). Then this rule can be applied to the original problem to generate solutions. By this approach, the allocation rule can be obtained in a much shorter time. When sampling the problem, the representativeness is a key factor that can largely affect the generalizability of the trained allocation rule. To investigate the effect of representativeness, two sampling strategies, namely the random sampling and filtered sampling, are proposed and compared in this paper. The filtered sampling strategy adopts more information about the problem structure to increase the similarity of the sampled problem and the entire problem. The results show that the filtered sampling performs significantly better than the random sampling in terms of both solution quality and success rate (i.e., the probability of generating feasible solutions for the large problem). The good performance of filtered strategy indicates the importance of sample representativeness on the scalability of the GP generated rules.

Keywords: Scalability, Storage Location Assignment Problem, Sampling Strategy, Genetic Programming.

1 Introduction

The Storage Location Assignment Problem (SLAP) [1] is an important optimization problem in warehouse management. It improves the overall operational efficiency (e.g. space utilization [2], total picking effort [3], relocation effort [4] or peak picking load [5]) by rearranging the inventory layout in warehouses. There are many factors to be considered in this problem. The popularity of products is crucial and has been used to determine whether a product can be placed to locations close to a loading zone or not. Also, products frequently ordered simultaneously are considered to have stronger demand dependencies and should be assigned to closer locations [6]. Other factors such as picking strategies [7], resource availability [5] or warehouse maintenance cost [8] may also be included and the problem can be extremely hard to solve.

G. Dick et al. (Eds.): SEAL 2014, LNCS 8886, pp. 691–702, 2014.
© Springer International Publishing Switzerland 2014

There are many approaches for solving SLAPs in literature. Comprehensive surveys are presented in [1] and [9]. The majority of the work in this area is focused on finding the exact location for each product in a warehouse. Deterministic methods such as branch-and-bound have been successfully applied to small instances [10]. Stochastic approaches such as Simulated Annealing [11], Tabu Search [4] or Genetic Algorithms [12] are more widely used for larger problems to get near-optimal results. These methods usually find a good trade-off between solution quality and computational budget. However, the solution found is only applicable to particular scenarios. In reality, the scenario can change dramatically and frequently. To cope with such changes, one often has to repeat the whole optimization procedure and get a completely new solution. Undoubtedly, this is expensive in both time and resource. As a result, a Genetic Programming (GP) approach has recently been proposed, which searches in the space of allocation rules in the form of matching functions instead of solutions for this problem [13]. In this way, one can optimize the matching functions for the past or current scenarios, and then apply it to any future scenario to generate decent solutions efficiently. In contrast to the traditional Integer Linear Programming (ILP), where the size of the search space becomes prohibitive when the problem size reaches 400, the GP method can still achieve good optimization performance. However, the average elapsed time increases from around 300 seconds to about 1700 seconds when the number of items increases from 400 to 900. In other words, the efficiency of this method deteriorates rapidly with increasing problem size. Based on the above preliminary studies, large scale problems are defined as the problems with more than 1000 items.

Unlike the solutions obtained by other methods in literature, the matching function obtained by this GP approach is reusable and efficient and in most of the cases it can get feasible and good solutions for similar sized unseen problems. In this paper, we attempt to extend the GP approach to problems with distinct problem sizes and investigate the re-usability of the obtained matching function in this case. Specifically, a subset of items is sampled from the entire item set to represent the original problem. Then, the GP method is applied to the representative problem so as to obtain the allocation rule in a much shorter time. Finally, the obtained allocation rule is applied to the original problem to generate the corresponding solution for it. A similar approach has been explored in [14] for bin packing problems on data drawn uniformly from one distribution. This experimental setup is not applicable when the representative problems are sampled from the original problem. To investigate the representativeness of these smaller problems, two sampling methods are developed and compared, namely the random sampling and filtered sampling. As the name indicates, the random sampling is a pure random sampling technique for the items. The filtered sampling, on the other hand, filters the items based on some criteria before the sampling to obtain a more representative subset of items. The comparative results show that the filtered sampling performs much better than the random sampling. This demonstrates the importance to consider the characteristics of the representative problem when sampling the subset of items.

The rest of the paper is organized as follows: first, the problem description of SLAP and the recently proposed GP approach are briefly introduced in Section 2. Section 3 discusses the two proposed sampling techniques that are used to generate the representative problem. The experimental studies are carried out in Section 4. Finally, the conclusions and future work are described in Section 5.

2   Background

Our research is conducted in a warehouse storing garments. This section presents a brief description of the problem and the recently proposed GP approach, which will be used to solve the large-scale SLAPs in this paper.

2.1   Problem Description

Suppose we have a set of products, each consisting of a number of items in different colors and sizes with their own picking frequencies, to be assigned to a set of locations in the given warehouse. Intuitively, the same products are preferred to be stored together, while putting items with huge popularity difference together can lead to inefficient solutions. The grouping constraint was introduced to get rid of this dilemma by allowing the split of one product into several subgroups so that they can be stored to different areas. The total picking-frequency-weighted distance is used as the approximation of the overall picking effort. Frequently picked items are expected to be assigned to locations that are closer to the loading zone of the warehouse. The Integer Linear Programming (ILP) model [15] of the problem is stated in Eqs. (1) – (9).

$$\min\ \varsigma(x) = \sum_{i=1}^{N}\sum_{l=1}^{N} 2\,P_i\,(V_l + H_l)\,x_{il} \tag{1}$$

$$s.t.\ \sum_{i=1}^{N} x_{il} = 1, \quad l = 1,...,N \tag{2}$$

$$\sum_{l=1}^{N} x_{il} = 1, \quad i = 1,...,N \tag{3}$$

$$\sum_{l=1}^{N} y_{sl} \le 2, \quad s = 1,...,S \tag{4}$$

$$x_{il} \le \sum_{s=1}^{S} A_{is}\left(y_{sl} + \sum_{j=1}^{N} A_{js}x_{j,l-1} \right), \quad i,l = 1,...,N \tag{5}$$

$$\sum_{s=1}^{S} y_{sl} = 1, \quad \forall\, l\ mod\ C = 1 \tag{6}$$

$$\sum_{s=1}^{S} B_{is}x_{il} \le \sum_{s=1}^{S} B_{is}\left(\sum_{s=1}^{S} A_{is}y_{sl} \right), \quad i = 1,...,N \tag{7}$$

$$\sum_{i=1}^{N} A_{is}x_{il} \ge y_{sl}, \quad s = 1,...,S, \quad l = 1,...,N \tag{8}$$

$$x_{il}, y_{sl} \in \{0,1\}. \tag{9}$$

Eq. (1) is the objective function, which is to minimize the total picking frequency-weighted distance. Eqs. (2) and (3) indicate that each location is occupied by a unique item. Eq. (4) means that each product can be split into at most two subgroups, where y_{sl} takes 1 if location l is the starting point of a subgroup of the product s, and 0 otherwise. Eq. (5) ensures that items from the same subgroup are stored to adjacent locations, where A_{is} equals 1 if item i belongs to product s, and 0 otherwise. Eq. (6) indicates that the first bin of a shelf must be the staring point of a product. Eq. (7) is a tightening constraint that states that the most popular item of a product is always at the start of a subgroup. Finally Eq. (8) states that the start of a product group should actually have an item from that product (this constraint is not strictly speaking necessary to get a valid bound, but ensures that the y variables have the correct meaning). Table 1 lists all the notations involved in this ILP model.

Table 1. Notations for the ILP Model

Notation	Meaning	Notation	Meaning
i	Index of items, $i = 1, ..., N$	l	Index of locations, $l = 1, ..., N$
s	Index of products, $s = 1, ..., S$	P_i	Picking frequency of item i
M	Number of shelves	C	Number of bins on a shelf
V_l	Vertical distance of location l to the loading zone	H_l	Horizontal distance of location l to the loading zone
A_{is}	equals to 1 if item i is in product s; 0, otherwise.	B_{is}	equals to 1 if item i is the most popular item in product s; 0, otherwise
x_{il}	equals 1 if item i is assigned to location l, and 0 otherwise. ($i = 1, ..., N; l = 1, ..., N$)	y_{sl}	equals 1 if location l is a starting point of product s, and 0 otherwise. ($s = 1, ..., S; l = 1, ..., N$)

2.2   A Genetic Programming Approach

The general idea of the GP approach is relatively simple. GP is used to evolve matching functions which help to identify the most suitable subsets of product to an equally sized set of locations. A matching function takes the set of items and consecutive locations as input, and returns a value to reflect the degree of suitability to assign the given items to the corresponding locations. Given a matching function, the solution is constructed from scratch. At each step, the best-fit set of items to the nearest available location is identified and allocated to the set of locations starting from the nearest available location. The solution generation is completed after all the items have been assigned into the locations. When evolving the matching functions, the fitness of a matching function is set to the objective function value of the solution generated by it.

A standard tree-based representation is adopted by the GP. To handle item sets with arbitrary size, statistical data of the picking frequencies of these items and distances of locations to the loading zone are used as the terminal set of the GP method. Simple arithmetic and logic operators are applied to connect these terminals to form the GP tree. Details of this method can be found in [13].

3 Methodology

The complexity of the optimization process of the allocation rule by GP depends on the problem size, and is time consuming for large scale problems. Thus, it is impractical to apply the proposed GP directly on the large scale problems. To address this scalability issue, smaller sized problems are generated by sampling subsets of items out of the item set of the original large scale problem as its representatives. The allocation rules are evolved on the smaller sized representative problems, and then applied to the original problems to obtain the solution. For example, given an 1000-item problem, one can sample 200 items out of the total 1000 items to form a 200-item problem, and evolve the allocation rules for this smaller problem by the GP. In this way, the allocation rule can be obtained in a much shorter time.

As demonstrated in [13], the evolved allocation rules managed to obtain promising performance on unseen problems with the same sizes. However, it is unknown whether such property is maintained for those have different sizes, since the difference in problem size may lead to distinct problem structures, and thus different desirable properties of the allocation rules. Therefore, one cannot guarantee that an allocation rule optimized on smaller problems can be generalized well to larger problems. To evolve a more generic allocation rule with better re-usability regardless of the problem size, we propose and compare two sampling techniques, namely the *random sampling* and *filtered sampling*, to generate representative problems for the original problem. Here, the re-usability of an allocation rule is defined as its ability of getting feasible and high quality solutions when applied to problems with different sizes.

3.1 Random Sampling

Random sampling is the most intuitive way of sampling. Given a list I of the items, the algorithm of random sampling is as described in Algorithm 1. For example, given the 8-item list shown in Table 2, a subset $\{1, 5\}$ consisting of the first and fifth item is randomly sampled. Ease of implementation is the main advantage of this method. It simply places all the items in a giant sequence and picks the items randomly until the size limitation $Limit$ has been reached. The algorithm starts with an empty list X, the number of sampled items $Limit$ and the original list of items. In each iteration, a random number r is generated, and the r_{th} item in the current item list is selected. Then, the selected item is removed from the item list and is inserted into list X. The procedure terminates after $Limit$ iterations.

3.2 Filtered Sampling

In our previous study [13], we have observed the following behaviours of the proposed GP:

1. It consistently obtained feasible solutions during the optimization.
2. It can obtain feasible and good solutions on most of the unseen problems.

Table 2. Example Data Set

Item No.	SKU	Color	Size	Picking Frequency
1	$AK001$	$BLACK$	S	221
2	$AK001$	$BLACK$	M	1070
3	$AK001$	$BLACK$	L	293
4	$AK001$	$WHITE$	S	15
5	$AK001$	$WHITE$	M	2200
6	$AK001$	$WHITE$	L	378
7	$BL78$	$BLUE$	XS	735
8	$BL78$	$BLUE$	S	467
-	-	-	-	-

Algorithm 1. Random Sampling Procedure

1: Initialize an empty item list X;

2: Initialize $Limit$ to a positive integer;

3: **repeat**

4:     Assign r with a random number between $[0, size(I)]$;

5:     Add I_r to item list X;

6:     Remove I_r from I;

7:     $Limit := Limit - 1$;

8: **until** $Limit == 0$;

9: **return** X;

---

3. When there exists a product containing a large number of items, and if a smaller number of items in this product is firstly picked and the rest is not allowed to be split. Then none of the shelves is capable of holding the rest items without violating any constraint. As a result, an infeasible solution occurs.

Based on the above observations, we deduce that the number of items in the products plays an important role in the generalizability of the obtained allocation rule. In order to increase the probability of obtaining an allocation rule that can deal with products with a large number of items, a filtered sampling method is proposed to remove the products with single item. It is described in Algorithm 2. In Step 3, products with one item are firstly deleted as those products are not required to be split and thus can hardly provide information for the learning procedure. In the algorithm, at each step, a set of λ consecutive items are picked instead of only one item so as to increase the probability of selecting the items in the same product. The parameter λ can be specified to a sufficiently large integer, e.g., 20 or 30, in practice.

4   Experiments and Results

The test problems are generated from two raw data sets, each with more than $10,000$ items. Each raw data set is randomly split into five exclusive subsets of items, each

---

Algorithm 2. Filtered Sampling Procedure

 1: Initialize X to an empty item list and I to the list of all items;
 2: Initialize $Limit$ to a positive integer;
 3: Delete from I those products with only one item;
 4: Initialize parameter λ;
 5: $inter = \lambda$;
 6: **repeat**
 7:     Assign r with a random number between $[0, size(I) - inter]$;
 8:     Add $I_r, ..., I_{r+inter}$ to item list X;
 9:     Remove $I_r, ..., I_{r+inter}$ from I;
10:     $Limit := Limit - inter$;
11:     $inter := \min\{inter, \ Limit\}$;
12: **until** $Limit == 0$;
13: **return** X;

---

consisting of over $2,000$ items. Overall, 10 test problems have been generated for experimental studies, labelled from p_1 to p_{10} ($p_1 \sim p_5$ for the first raw data set, and $p_6 \sim p_{10}$ for the second). When sampling the subsets to form the representative problem for each test problem, the number of sampled items is set to 400. The sampling is repeated five times and thus five representative problems are generated. During the optimization process, the fitness of an allocation rule is defined as the average fitness of the solutions generated by applying the rule to the five representative problems. Finally, the obtained best-fit allocation rule is evaluated by applying it to the original problem and calculating the objective value of the corresponding solution. Thirty independent runs were conducted for each test problem, and the best, worst and average results are recorded.

4.1   Random Sampling

Table 3 shows the experimental results for the random sampling method. It lists the number of infeasible solutions obtained in the 30 independent runs and the best, worst and average fitnesses achieved when applying the allocation rule obtained from the representative problems to the original problem. For each test problem, the best known fitness α is obtained by optimizing the allocation rules on the full problem without using representative subproblems. It can be seen that on $p_6 \sim p_8$, the obtained allocation rules failed to get feasible solutions in all the 30 runs. p_{10} shows the best performance in terms of feasibility. In terms of the percentage deviation of the best fitness from the best-known fitness $Diff_{Min}^{\alpha}$, it can be observed that the random sampling obtained promising results on p_1, p_3 and p_5, which are less than 1%. Due to the high success rate, the random sampling achieved the best average results on p_{10}.

Table 3. The Results of Random Sampling

Set	α	β	Min	$Diff^{\alpha}_{Min}$	Max	$Diff^{\alpha}_{Max}$	$Avg \pm (Stdv)$	$Diff^{\alpha}_{Avg}$
p_1	3059169	23	3078208	0.62%	6673130	118.14%	$3716687.90 \pm (3.48E05)$	21.49%
p_2	2977428	24	3029585	1.75%	3877362	30.23%	$3133622.43 \pm (2.73E06)$	5.25%
p_3	3092760	25	3120909	0.91%	4463370	44.32%	$3254232.40 \pm (5.10E06)$	5.22%
p_4	2912623	27	2943177	1.05%	4259747	46.25%	$3036241.17 \pm (1.91E06)$	4.24%
p_5	3114503	12	3130649	0.52%	10086239	223.85%	$3530287.10 \pm (1.42E05)$	13.35%
p_6	1900213	30	2337416	23.01%	3898660	105.17%	$2788313.30 \pm (9.90E05)$	46.74%
p_7	3265076	30	4135615	26.66%	17004364	420.80%	$7389082.73 \pm (1.92E05)$	126.31%
p_8	3686151	30	4339916	17.74%	23698704	542.91%	$7331767.53 \pm (2.32E05)$	98.90%
p_9	3400729	27	3436107	1.04%	13076871	284.53%	$4313234.23 \pm (2.38E05)$	26.83%
p_{10}	3006363	8	3059604	1.77%	3844230	27.87%	$3118322.03 \pm (1.32E06)$	3.72%

[1.] α is the best fitness known so far, obtained by optimizing the set directly.

[2.] β is the number of infeasible solutions achieved in total 30 runs.

[3.] $Diff^B_A = \frac{A-B}{B} \times 100\%$.

4.2 Filtered Sampling

Tables 4 and 5 give the results using the filtered sampling method. Four experiments with $\lambda = 10, 20, 50$ and 100 are conducted for each set to compare with random sampling. The tables record the number of infeasible solutions in 30 runs for each set and for each λ configuration. The λ value of the row for random sampling is denoted by "–". The minimum, maximum and average fitness values and the differences of these fitnesses to the best known result for the corresponding test problem are also calculated. In addition, some statistical data related to the representative problems are presented for further discussion. Each representative problem is to allocate 400 items into a warehouse with 8 shelves, each consisting of 50 bins.

To have a clearer understanding of the results in Table 4 and 5, we firstly compare the overall performance of the random sampling method and the filtered sampling method. Table 6 shows the average number of infeasible solutions, the average difference of the best, worst and average fitness obtained by the filtered sampling with different λ settings. It shows that the filtered sampling can get more feasible solutions in general. For $p_1, p_2, p_3, p_4, p_5, p_6$ and p_9, the number of infeasible solutions obtained by the filtered sampling is nearly half of that of random sampling. In addition, the solution quality is much better when using filtered sampling. This can be observed by columns $Diff^{\alpha}_{Min}$, $Diff^{\alpha}_{Max}$ and $Diff^{\alpha}_{Avg}$ as the filtered method consistently achieved smaller percentage deviation from the best known results.

Then we investigate the impact of parameter λ on the performance of the method. This parameter is used in Algorithm 2 to determine the size of chunks taken after randomly selecting a start point in the data set. The larger the λ is set, the more likely it can generate representative problems with products containing larger number of items. This can be observed in Table 4 and 5 where γ_3 increases with λ for most of the test problems.

Table 4. Performance Comparison Using Different Representative Problems Generated using Different λ Configurations for Filtered Sampling Method

			Results for the Original Sets						Representative Problems				
set	λ	β	Min	$Diff^\alpha_{Min}$	Max	$Diff^\alpha_{Max}$	Avg	$Diff^\alpha_{Avg}$	γ_1	γ_2	γ_3	γ_4	γ_5
p_1	10	19	3049702	−0.31%	3256170	6.44%	3144771.80	2.80%	123.00	3.26	16.60	21	13
	20	19	3044269	−0.49%	4719413	54.27%	3261829.43	6.62%	115.40	3.49	21.20	25	18
	50	7	3031050	−0.92%	3106600	1.55%	3066181.73	0.23%	103.60	3.97	25.40	31	22
	100	5	3046582	−0.41%	3100820	1.36%	3062909.17	0.12%	64.80	6.20	34.00	37	32
	−	23	3078208	0.62%	6673130	118.14%	3716687.90	21.49%	210.20	1.92	8.20	9	8
p_2	10	1	3004222	0.90%	3064512	2.92%	3037839.73	2.03%	130.80	3.07	16.20	21	11
	20	4	2996131	0.63%	3051311	2.48%	3025666.50	1.62%	129.00	3.14	18.80	25	15
	50	26	3020389	1.44%	3745966	25.81%	3145886.40	5.66%	125.40	3.29	18.40	22	15
	100	7	2989703	0.41%	3072195	3.18%	3034130.20	1.90%	60.00	6.69	34.60	38	28
	−	24	3029585	1.75%	3877362	30.23%	3133622.43	5.25%	202.40	1.99	8.20	10	7
p_3	10	13	3109296	0.53%	4278922	38.35%	3223621.93	4.23%	134.40	2.99	13.40	17	11
	20	10	3106868	0.46%	7285031	135.55%	3342338.00	8.07%	130.60	3.09	19.40	23	17
	50	20	3099309	0.21%	3261410	5.45%	3177372.20	2.74%	124.40	3.24	18.40	28	15
	100	13	3102046	0.30%	3756533	21.46%	3171127.93	2.53%	62.00	6.51	32.80	35	30
	−	25	3120909	0.91%	4463370	44.32%	3254232.40	5.22%	204.80	1.97	8.80	11	7
p_4	10	22	2926350	0.47%	3307064	13.54%	2990174.07	2.66%	134.60	2.97	15.60	18	12
	20	22	2919410	0.23%	3177614	9.10%	2964727.20	1.79%	123.80	3.29	17.40	22	15
	50	11	2926303	0.47%	3257099	11.83%	2962987.50	1.73%	122.80	3.30	22.20	28	17
	100	3	2925870	0.45%	2967114	1.87%	2942840.63	1.04%	63.00	6.38	31.00	34	27
	−	27	2943177	1.05%	4259747	46.25%	3036241.17	4.24%	213.60	1.89	8.60	10	7
p_5	10	7	3129402	0.48%	3446347	10.65%	3161727.47	1.52%	132.80	3.05	15.60	18	12
	20	7	3128977	0.46%	3403266	9.27%	3168708.33	1.74%	124.40	3.23	20.80	26	16
	50	5	3131274	0.54%	3732355	19.84%	3166259.27	1.66%	127.60	3.20	21.00	33	16
	100	1	3131965	0.56%	3165643	1.64%	3148795.97	1.10%	65.00	6.18	34.60	37	28
	−	12	3130649	0.52%	10086239	223.85%	3530287.10	13.35%	211.60	1.91	8.00	9	7

[1.] For each representative problem, the number of products, the average number of items a product has and the maximum number of items a product contains are already known and can be denoted as ϕ, φ and Ω correspondingly. γ_1 is the average of ϕ for 5 representative problems. γ_2 is the average φ for 5 representative problems. γ_3, γ_4 and γ_5 are respectively the average, maximum and minimum of Ω.
[2.] The definition for α, β and $Diff$ are the same as in Table 3.

The assumption of proposing filtered sampling is that the GP method can learn to deal with grouping constraints if enough information is provided in representative problems. In other words, when the representative problems include products containing a larger number of items, the GP method can handle the grouping constraint better. The results shown in Tables 4 and 5 are consistent with this assumption. For each of the test problems, we have four rows for different λ configuration in filtered sampling. In terms of β, $Diff^{Min}_\alpha$, $Diff^\alpha_{Max}$ and $Diff^\alpha_{Avg}$, the row with the smallest value is considered as the winner. The comparative results are shown in Table 7. It is obvious that $\lambda = 100$ obtained the best performance, as it was the winner for 7 out of the total 10 test problems in terms of β, $Diff^\alpha_{Max}$ and $Diff^\alpha_{Avg}$.

Table 5. Performance Comparison Using Different Representative Problems Generated using Different λ Configurations for Filtered Sampling Method (Cont.)

Set	λ	β	Min	$Diff^{\alpha}_{Min}$	Max	$Diff^{\alpha}_{Max}$	Avg	$Diff^{\alpha}_{Avg}$	γ_1	γ_2	γ_3	γ_4	γ_5
						Results for the Original Sets					Representative Problems		
	10	30	2015539	6.07%	3338767	75.70%	2853695.37	50.18%	74.40	5.39	25.40	31	21
	20	20	1911381	0.59%	2709338	42.58%	2008070.67	5.68%	66.40	6.07	28.40	40	23
p_6	50	15	1923527	1.23%	2034104	7.05%	1953522.73	2.81%	57.20	7.28	42.20	52	29
	100	4	1924827	1.30%	4206497	121.37%	2024823.67	6.56%	25.60	15.77	69.40	77	57
	–	30	2337416	23.01%	3898660	105.17%	2788313.30	46.74%	107.60	3.75	16.60	19	15
	10	30	3386434	3.72%	13010004	298.46%	5235240.37	60.34%	60.20	6.66	33.00	38	29
	20	30	3348193	2.55%	8896970	172.49%	4538060.60	38.99%	59.20	6.86	39.60	53	28
p_7	50	30	3285883	0.64%	6158601	88.62%	3900587.93	19.46%	50.60	8.25	52.80	66	41
	100	24	3325634	1.85%	5276951	61.62%	3624195.83	11.00%	21.80	18.51	85.20	100	70
	–	30	4135615	26.66%	17004364	420.80%	7389082.73	126.31%	83.80	4.81	23.80	31	19
	10	30	4004036	8.62%	26155280	609.56%	6869958.07	86.37%	47.00	8.52	35.40	47	27
	20	30	3908860	6.04%	23733476	543.86%	6821483.63	85.06%	47.20	8.55	35.40	43	30
p_8	50	30	3797211	3.01%	14658422	297.66%	5387505.03	46.16%	41.60	9.91	50.60	55	44
	100	18	3873500	5.08%	6886889	86.83%	4343174.63	17.82%	17.40	23.07	82.60	92	76
	–	30	4339916	17.74%	23698704	542.91%	7331767.53	98.90%	54.40	7.41	20.80	22	20
	10	7	3428812	0.83%	4178877	22.88%	3485899.07	2.50%	85.40	4.69	22.80	30	19
	20	10	3415128	0.42%	3690611	8.52%	3454856.63	1.59%	76.00	5.33	29.00	39	23
p_9	50	10	3414876	0.42%	3830894	12.65%	3472938.93	2.12%	91.80	4.38	29.40	33	23
	100	6	3425275	0.72%	3606875	6.06%	3460712.57	1.76%	33.40	12.22	65.40	74	54
	–	27	3436107	1.04%	13076871	284.53%	4313234.23	26.83%	128.00	3.15	16.20	21	14
	10	6	3041814	1.18%	3476412	15.64%	3083405.33	2.56%	127.60	3.15	15.00	21	12
	20	12	3028730	0.74%	3555192	18.26%	3097883.93	3.04%	128.60	3.17	16.00	20	14
p_{10}	50	14	3027124	0.69%	3709570	23.39%	3089759.90	2.77%	106.20	3.90	22.00	27	18
	100	11	3022651	0.54%	3097094	3.02%	3049117.07	1.42%	63.60	6.34	31.60	38	30
	–	8	3059604	1.77%	3844230	27.87%	3118322.03	3.72%	213.00	1.89	8.60	10	8

[1.] For each representative problem, the number of products, the average number of items a product has and the maximum number of items a product contains are already known and can be denoted as ϕ, φ and Ω correspondingly. γ_1 is the average of ϕ for 5 representative problems. γ_2 is the average φ for 5 representative problems. γ_3, γ_4 and γ_5 are respectively the average, maximum and minimum of Ω.

[2.] The definition for α, β and $Diff$ are the same as in Table 3.

In summary, the experimental studies demonstrate that the filtered sampling strategy performs much better than random sampling on the large scale test problems in terms of both the capability of generating feasible solutions and the quality of the generated solutions. The improvement of the filtered sampling with increasing λ values indicates the importance of keeping products with larger numbers of items in the representative problems. In other words, the consistency in the product size is playing an important role in retaining the re-usability of the allocation rules obtained by the GP approach.

Table 6. Comparison of Overall Performance of Random and Filtered Sampling

No.	Average of Filtered Sampling ($\lambda = 10, 20, 50, 100$) / Random Sampling			
	β	$Diff_{Min}^{\alpha}$	$Diff_{Max}^{\alpha}$	$Diff_{Avg}^{\alpha}$
p_1	12.5 / 23	−0.53% / 0.62%	15.91% / 118.14%	2.44% / 21.49%
p_2	9.5 / 24	0.85% / 1.75%	8.60% / 30.23%	2.80% / 5.25%
p_3	14 / 25	0.38% / 0.91%	50.20% / 44.32%	4.39% / 5.22%
p_4	14.5 / 27	0.41% / 1.05%	9.08% / 46.25%	1.80% / 4.24%
p_5	5 / 12	0.51% / 0.52%	10.35% / 223.85%	1.50% / 13.35%
p_6	17.25 / 30	2.29% / 23.01%	61.68% / 105.17%	16.30% / 46.74%
p_7	28.5 / 30	2.19% / 26.66%	155.30% / 420.80%	32.45% / 126.31%
p_8	27 / 30	5.69% / 17.74%	384.48% / 542.91%	58.85% / 98.90%
p_9	8.25 / 27	0.60% / 1.04%	12.53% / 284.53%	2.00% / 26.83%
p_{10}	10.75 / 8	0.79% / 1.77%	15.07% / 27.87%	2.45% / 3.72%

[1.] The definition for α, β and $Diff$ are the same as in Table 3.

Table 7. Numbers of Wins for Different λ Configuration

λ	β	$Diff_{Min}^{\alpha}$	$Diff_{Max}^{\alpha}$	$Diff_{Avg}^{\alpha}$
10	2	0	0	0
20	1	3	1	2
50	0	5	2	1
100	7	2	7	7

[1.] The definition for α, β and $Diff$ are the same as in Table 3.

5 Conclusions and Future Work

This paper extends the GP approach with two sampling methods to solve a SLAP with grouping constraints. It adopts a relatively simple concept of generalization from sub-problem to the original problem for the purpose of using the GP approach for handling large-scale SLAPs efficiently. The major idea is to evolve the allocation rules on smaller sized representative problems sampled from the original problem. Two sampling methods, random sampling and filtered sampling, are developed to generate the representative problems. This paper conducts a comprehensive experimental study of these two sampling methods and the results demonstrate that the filtered sampling performs better in terms of both feasibility and solution quality. It also shows that the GP method can learn to deal with hard constraints regardless of the problem size if adequate critical information is provided.

In future, several possible extensions of this study could be developed. The size of products in representative problems has been identified as an important factor for the re-usability of the allocation rules obtained by the GP method in this paper. More factors are to be determined to develop an automatic procedure for generating better representative problems. For example, the size of representative problems affects the efficiency of the training procedure and the performance of the evolved allocation rules. Besides, the proposed method has only applied to a SLAP with grouping constraints. More realistic models are to be developed by including the constraints that have not been considered in the current model. For example, more accurate approximations of operational effort rather than the simple picking-frequency-weighted distance will be designed in the future. This can be achieved by using order information instead of picking frequency. We may also consider routing problems in warehouse instead of simply Manhattan distance between two points. In addition, developing new techniques for solving this problem in dynamic environments is another promising direction.

References

1. Gu, J., Goetschalckx, M., McGinnis, L.F.: Research on warehouse operation: A comprehensive review. European Journal of Operational Research 177(1), 1–21 (2007)
2. Chen, L., Langevin, A., Riopel, D.: The storage location assignment and interleaving problem in an automated storage/retrieval system with shared storage. International Journal of Production Research 48(4), 991–1011 (2010)
3. Önüt, S., Tuzkaya, U.R., Doğaç, B.: A particle swarm optimization algorithm for the multiple-level warehouse layout design problem. Computers & Industrial Engineering 54(4), 783–799 (2008)
4. Chen, L., Langevin, A., Riopel, D.: A tabu search algorithm for the relocation problem in a warehousing system. International Journal of Production Economics 129(1), 147–156 (2011)
5. Montulet, P., Langevin, A., Riopel, D.: Minimizing the peak load: an alternate objective for dedicated storage policies. International Journal of Production Research 36(5), 1369–1385 (1998)
6. Frazele, E.A., Sharp, G.P.: Correlated assignment strategy can improve any order-picking operation. Industrial Engineering 21(4), 33–37 (1989)
7. Lin, C.-H., Lu, I.-Y.: The procedure of determining the order picking strategies in distribution center. International Journal of Production Economics 60, 301–307 (1999)
8. Kofler, M., Beham, A., Wagner, S., Affenzeller, M., Achleitner, W.: Re-warehousing vs. healing: Strategies for warehouse storage location assignment. In: 2011 3rd IEEE International Symposium on Logistics and Industrial Informatics (LINDI), pp. 77–82. IEEE (2011)
9. Rouwenhorstp, B., Reuter, B., Stockrahm, V., Van Houtum, G.J., Mantel, R.J., Zijm, W.H.M.: Warehouse design and control: Framework and literature review. European Journal of Operational Research 122(3), 515–533 (2000)
10. Adil, G.K., et al.: A branch and bound algorithm for class based storage location assignment. European Journal of Operational Research 189(2), 492–507 (2008)
11. Adil, G.K., et al.: Efficient formation of storage classes for warehouse storage location assignment: a simulated annealing approach. Omega 36(4), 609–618 (2008)
12. Zhang, G.Q., Xue, J., Lai, K.K.: A genetic algorithm based heuristic for adjacent paper-reel layout problem. International Journal of Production Research 38(14), 3343–3356 (2000)
13. Xie, J., Mei, Y., Andries, E., Li, X., Song, A.: A genetic programming-based hyper-heuristic approach for storage location assignment problem. In: Proceedings of the 2014 IEEE Congress on Evolutionary Computation, pp. 3000–3007. IEEE Computational Intelligence Society (2014)
14. Burke, E.K., Hyde, M.R., Kendall, G., Woodward, J.R.: The scalability of evolved on line bin packing heuristics. In: 2007 IEEE Congress on Evolutionary Computation, pp. 2530–2537. IEEE Computational Intelligence Society (2007)
15. Wolsey, L.A.: Integer programming, vol. 42. Wiley, New York (1998)

Dual Population Genetic Algorithm for the Cardinality Constrained Portfolio Selection Problem

Nasser R. Sabar[1] and Andy Song[2]

[1] The University of Nottingham Malaysia Campus, Jalan Broga, 43500 Semenyih,
Selangor, Malaysia
Nasser.Sabar@nottingham.edu.my
[2] The School of Computer Science and Information Technology, RMIT University, Australia
andy.song@rmit.edu.au

Abstract. Portfolio Selection (PS) is to allocate a given amount of investment fund across a set of assets in such a way that the return is maximized and the risk is minimized. PS is a challenging financial engineering problem and optimization problem. GA is well known for its effectiveness in solving optimization problems. However it may experience slow convergence especially when dealing with constrained optimization problems. To address this issue, we propose a variation of genetic algorithm (GA), which utilizes dual populations to solve PS problems. The first population is responsible for exploration in the search space, whilst the second one is for exploitation to speed up the convergence process. These two populations share individuals periodically. The proposed algorithm has been tested on the standard PS benchmark instances. The results reveal that our method can obtain very good results compared to the state of the art methods. More importantly, this dual population method is much faster than other methods.

Keywords: genetic algorithm, financial marketing, portfolio selection.

1 Introduction

The amount of investment in the world financial market every year is enormous. Investors are usually seeking for maximizing their returns which naturally associate with high risks [1], [2], [3]. Thus, a tradeoff between maximizing the return and minimizing the risk should be considered. That is the challenge faced by portfolio managers in providing high quality services for their customers, the investors. An efficient optimization method which can find the best point would be highly desirable [2].

Portfolio Selection (PS) is a key task of portfolio managers. It can be defined as the problem of how to select a subset of assets to form a single portfolio that could maximize the return and minimize the risk [2]. PS is a challenging and important problem that plays a pivotal role in financial engineering. PS seeks for allocate a given amount of money across a set of assets in such a way that the return is maximized and the risk is minimized. Manually selecting a good portfolio of assets is usually intractable especially when considering the complexity of the asset and the fast changing market

G. Dick et al. (Eds.): SEAL 2014, LNCS 8886, pp. 703–712, 2014.
© Springer International Publishing Switzerland 2014

dynamics. Formulate an appropriate model first and then try to solve the model based on a good optimization algorithm is critical in PS.

The Markowitz mean–variance PS model [1], [4] has been widely used to form a single portfolio as it captures the expected return and the risk of the formed portfolio. Although this model has been the core of PS theory, it has been criticized for being impractical in real world application as it is based on assumptions that might not realistic. Among many extensions and improvements that aims to make Markowitz mean–variance PS model more practical, cardinality and boundary constraints are notable and have been widely used in the literature [2]. Cardinality constraint limits the number of assets to be included in the formed portfolio, while the boundary constraint determines lower and upper bounds for each asset of the formed portfolio. In this paper, we consider the extended Markowitz mean–variance PS model that includes cardinality and boundary constraints which is known as cardinality constrained portfolio selection problem.

Traditionally, Markowitz mean–variance PS model has been formulated as a quadratic programming problem and thus, despite being an NP-hard problem [2], a model with a small number of variables can be optimally solved using exact methods. However, when the number of variables is increased or some constraints are introduced, it becomes much more difficult, if not impossible, to use exact methods to find an optimal solution within a realistic amount of time frame. Therefore, researchers adopted meta-heuristic algorithms instead of exact methods because a good quality solution can be generated within a reasonable period of time [5]. Examples of meta-heuristic algorithms that have been adopted in the literature for PS are: Genetic Algorithm [2], Tabu Search [2], Simulated Annealing [2], Particle Swarm Optimization [6], Harmony Search [7] and hybrid algorithms [8], [9].

In this study, we propose a variation of genetic algorithm (GA) for PS with cardinality and boundary constraints. GA is a nature inspired population based meta-heuristic that simulates the process of natural selection [5], [10]. It has been proven as an efficient and effective solution method for various optimization problems. However, despite its success on solving many hard optimization problems, GA suffers from the slow convergence problem, which may prevent it from getting good results for constrained optimization problems such as PS. This tendency is usually due to the use of a single population and the lack of using an appropriate exploitation operator [11]. To enhance the convergence process of GA, we propose a GA that utilizes dual populations (denoted as DPGA). The first population is responsible for exploring the search space, whilst the second one acting as a local search operator to speed up the convergence process. The second population exploits around the explored area. These two populations periodically exchange individuals to share promising solutions. The performance of the proposed algorithm has been evaluated over the classic PS benchmark widely used in literature. The results demonstrate the effectiveness of the proposed DPGA compared to the existing algorithms that are popular among researchers and PS practitioners.

2   Problem Description

This paper focuses on the cardinality constrained portfolio selection problem which is an extension of the Markowitz mean–variance model. Two constraints are added to the model, cardinality and boundary constraints. These constraints are mainly to reduce the transaction cost and avoid holdings that are too small or too large. Cardinality constraint limits the number of assets to be included in the formed portfolio. Boundary constraint set a lower bound and upper bound for each asset of the formed portfolio. The formulation of the PS model is proposed by [2], [12]:

$$\text{minimize } \lambda \left[\sum_{i=1}^{n} \sum_{j=1}^{n} w_i w_j \alpha_{ij} \right] + (1-\lambda) \left[-\sum_{i=1}^{n} w_i \mu_i \right] \tag{1}$$

$$\text{Subject to } \sum_{i=1}^{n} w_i = 1 \tag{2}$$

$$\sum_{i=1}^{n} s_i = K \tag{3}$$

$$\varepsilon_i s_i \le w_i \le \delta_i s_i, \quad i = 1,...,n \tag{4}$$

$$s_i \in \{0,1\}, \quad i = 1,...,n \tag{5}$$

where n is the total number of assets, w_i is the proportion of the budget invested in the i-th asset, α_{ij} is the covariance between i-th and j-th assets, λ is the risk aversion, λ in [0, 1], μ_i is the expected return of the i-th asset, K is is the number of assets to be invested in assets in a portfolio, s_i is a decision variable represents whether the i-th asset has been selected or not, and ε_i and δ_i respectively are the upper and lower bounds. The cardinality constrained PS model involves two sub-problems: (1) the selection problem that seeks to select a subset of assets and (2) the allocating problem which aims at determining the proportion for each of the selected asset. In the literature, PS formulations are treated as a mixed integer programming [2]. Thus, when the number of variables is increased or some constraints are introduced, whether the optimal solution can be found in a realistic amount of time would be unknown. Therefore, in this paper, we propose a meta-heuristic based method (genetic algorithm) for the PS problem.

3   The Proposed Method

In this section, we first briefly describe the basics of genetic algorithm followed the proposed dual population variation DPGA.

3.1   Genetic Algorithm

Genetic algorithm (GA) is a meta-heuristic introduced by [10]. GA is a well-known nature inspired population based meta-heuristic that simulates the process of natural selection. It operates on a population of solutions and iteratively improves them by invoking the selection procedure, crossover procedure, mutation procedure and update procedure for a certain number of generations [5]. The pseudocode of the canonical GA is presented in Algorithm 1.

Algorithm 1. The canonical GA
1	*Set the number of generations, MaxG, population size, Ps, Crossover rate,*
2	*CR, Mutation rate, MR, G=0;*
3	*Population← initializepopulation(Ps)*
4	*FitnessCalculation (Population)*
5	**while** *(G< MaxG)* **do**
6	*parent₁← SelectParents (Ps)*
7	*parent₂← SelectParents (Ps)*
8	*child₁←Crossover (parent₁, parent₂, CR)*
9	*child₂←Crossover (parent₂, parent₁, CR)*
10	*child₁ₘ← Mutation (child₁, MR)*
11	*child₂ₘ←Mutation (child₂, MR)*
12	*population ← populationUpdate (child₁ₘ, child₂ₘ)*
13	*G=G+1;*
14	end **while**
15	Return the best solution

It starts by setting GA parameters (Lines 1 and 2), generates a population of solutions (Line 3) and then assigns a fitness value for each solution in the population (Line 4). Next, the while-loop is executed for a certain number of generations (Lines 5 to 14). At each generation, a selection mechanism is invoked to select two parents (Lines 6 and 7). The selected parents are passed together with the crossover rate to the employed crossover operator in order to generate two children (Lines 8 and 9). Next, the generated children are mutated by the utilized mutation operator according to the assigned mutation rate (Lines 10 and 11). The updating procedure (Line 12) replaces the new children with the worse solutions in the population if they are better in term of fitness value. If the stopping condition is satisfied, stop (Line 14) and return the best solution (Line 15). Otherwise, starts a new generation.

3.2   The Proposed Dual-Population Genetic Algorithm

The basic GA has proven as an effective algorithm for dealing with various optimization problems. However, when dealing with constrained and high dimensional problems, basic GA suffers from the slow convergence tendency, which may due to the use of population of solutions and thus it may produce uncompetitive results. Thus, to improve the effectiveness of the basic GA as well as the convergence process, the basic GA has been hybridized with one or a few local search algorithms, as local search algorithms are very good in exploiting the search process [11]. This kind of hybridization is called memetic algorithms, in which the local search take places at every generation to further improve the generated solutions [5], [11]. However, calling the local search algorithm at every generation would be computationally expansive and may lead to a premature convergence. To address this issue, in this paper, we propose dual-population GA (DPGA) that utilizes two types of populations (denoted as POP1 and POP2). POP1 is responsible for exploring the search space. POP2 is act as a local search method that focuses the search around the explored area. These two populations are periodically updated to share the search experience and exchange promising solutions. The flowchart of the proposed dual populations GA is shown in Figure 2.

In this paper, the solution is represented by a two-dimensional array where the array size is equal to the total number of assets. Figure 1 shows an example of a solution representation.

Asset index	1	2	3	.	.	.	n
The cardinality	1	1	0	0	0	1	1
The boundary value	0.3	0.5	0	0	0	0.4	0.1

Fig. 1. An example of solution representation

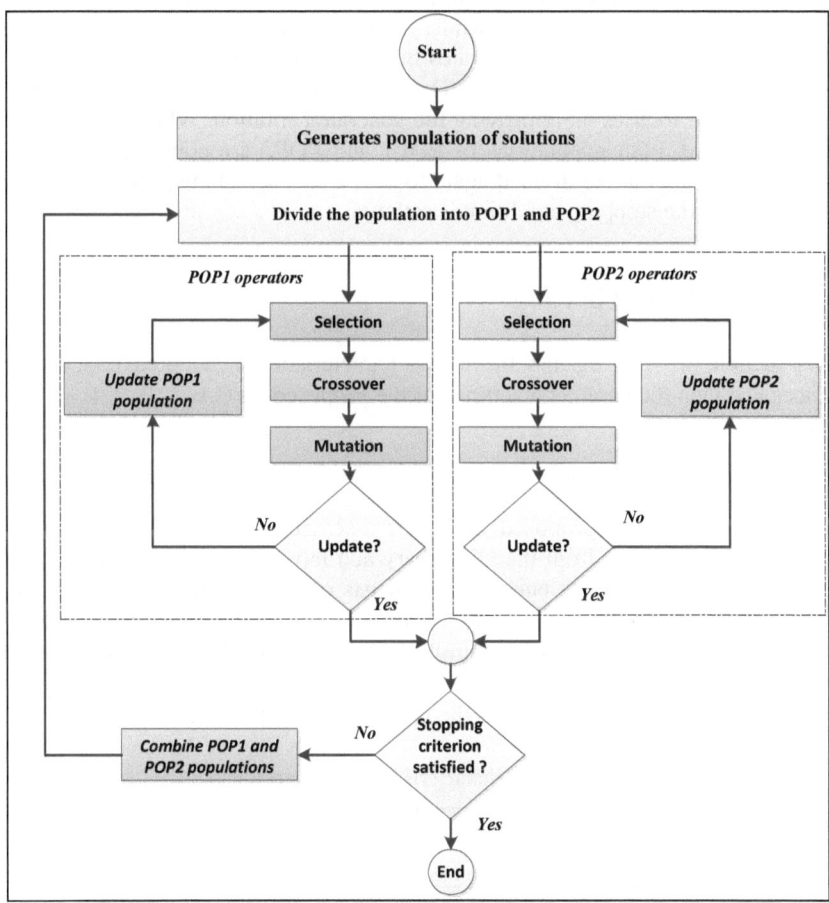

Fig. 2. The Flowchart of DPGA

In Figure 1, first row represents the cardinality that takes either "0" or "1", where "1" means the corresponding asset is selected and "0" is not. The second row represents the boundary value of the chosen asset, which takes a real value within the prescribed lower and upper boundary constraint. In this paper, the population of solutions is randomly created by assigning for each array cell of the first row either "0" or "1" while makes sure that the cardinality constraint is not violated. Next, the selected assets are randomly assigned a real number within the predefined boundary constraints that is represented by the second row of the array. The fitness value of each solution is calculated using Equation (1).

The proposed dual populations GA first divide the population into POP1 and POP2 where 30% of the worst solutions from the main population are assigned to POP2, while the rest are assigned to POP1. Then, each one is independently executed for a certain number of generations. Both POP1 and POP2 use roulette wheel as a selection mechanism, one point crossover operator, one point mutation operators and steady state updating rule [5]. POP1 follow the general procedure of the basic GA. While, POP2 is acting as a local search method as follows: first the crossover operator is associated with low crossover rate, whilst the mutation operator is assigned a higher probability. Crossover operator generates only one child and the best among the selected parents is the dominated one. The mutation operator accepts only positive changes and thus focuses on improving the generated solution. When the update condition is satisfied, both populations of POP1 and POP2 are combined to share the search outcome and then re-divided again to start a new searching period. The process is repeated until the stopping condition is satisfied.

4   Experimental Settings

In this section, we first discuss the main characteristics of adopted PS benchmark instances and then the parameter settings of the proposed DPGA.

4.1   Benchmark Instances

The performance of the proposed DPGA is assessed using the PS benchmark instances that are available from the OR-library and have been used by other researchers in the literature [13]. The benchmark consists of five instances that represent the weekly prices for five different countries. Table 1 shows the main characteristics of these five instances [13]. In this table, n is the number of the assets, K is the maximum number of assets in a formed portfolio (cardinality), ε_i $(i=1,...,n)$ is the lower bound of the asset and δ_i $(i=1,...,n)$ is the upper bound of the asset.

Table 1. The characteristics of PS benchmark

	Name	*Country*	*n*	*K*	*ε*	*δ*
1-	Hang Seng	Hong Kong	31	10	0.01	1
2-	DAX 100	Germany	85	10	0.01	1
3-	FTSM 100	UK	89	10	0.01	1
4-	S&P 100	USA	98	10	0.01	1
5-	Nikkei	Japan	255	10	0.01	1

4.2   Parameter Settings

The parameter setting of the proposed DPGA is presented in Table 2. These parameters were set based on the empirical results from a set of preliminary experiments. The parameter λ of Equation (1) was tested using 51 different values and each value is tested for $1000*n$ fitness evaluations same as [2] and [6].

Table 2. DPGA parameter settings

#	Parameter name	POP1	POP2
		Value	Value
1-	Population size, Ps	70	30
2-	Crossover rate, CR	0.8	0.2
3-	Mutation rate, MR	0.3	0.85
4-	POP1 and POP2 populations update	Every 100 fitness evaluations	
5-	Maximum number of generations, MaxG	1000*n fitness evaluations	

5   Results and Comparisons

Two sets of experiments were conducted. The goal of the first experiment is to evaluate the effectiveness of using dual populations within GA instead of single population. The second one compares the results of the proposed DPGA against the best known methods in the literature.

5.1   Evaluation of Effectiveness

In this section, we evaluate the effectiveness of using dual populations within GA by comparing the results of DPGA against POP1 as the canonical single population GA and using POP2 in single population GA. Each algorithm in this comparison (namely DPGA, POP1 and POP2) has been executed 51 independent runs with different random seeds using the same population of solutions, stopping condition and computer resources.

A Wilcoxon test with 0.05 critical level is conducted to statistically compared the results of the DPGA against POP1 and POP2. The p-value results are shown in Table 3, where a p-value < 0.05 indicates that the DPGA is statistically better than the compared methods (indicated in bold font).

Table 3. Comparison between DPGA and using POP1 and POP2 as single-population GA

	DPGA vs.	POP1	POP2
#	Name	p-value	p-value
1-	Hang Seng	**0.047**	0.063
2-	DAX 100	**0.001**	**0.024**
3-	FTSM 100	**0.004**	**0.045**
4-	S&P 100	**0.001**	**0.037**
5-	Nikkei	**0.006**	**0.041**

The *p*-values listed in Table 3 reveal that the proposed DPGA is statistically better than POP1 on all tested instances (as shown in bold font) and statistically better than POP2 on four out of five tested instances. The results verify that the use of dual populations within GA significantly improves the performance of the basic GA and improve the convergence process.

5.2   The Computational Results of DPGA against the State-of-Art Methods

In this section, we compare the computational results of DPGA with the following algorithms:

- Tabu search based algorithm (TS) proposed in [2]
- Simulated annealing based (SA) proposed in [2]
- Genetic algorithm based method (GA) proposed in [2]
- Particle swarm optimization based approach (PSO) proposed in [6]

The minimum mean percentage error (*MP%*) of 51 independent runs of the DPGA is compared against TS, SA, GA and PSO. The results are presented in Table 4, where bold font indicates the best results. As can be seen from Table 4, DPGA is better than the compared algorithms on all tested instances. Furthermore, the overall average result (last row in Table 4) of DPGA is lower than the compared algorithms. The comparison did not involve statistical tests, as the aim is not to claim the superior performance of DPGA, but to show the capability of this method. The key comparison is on computational time.

Table 4. The results of DPGA compared to GA, SA, TS and PSO

#	Name	DPGA	GA	SA	TS	PSO
1	Hang Seng	**1.0810**	1.0974	1.0957	1.1217	1.0953
2	DAX 100	**2.2124**	2.5424	2.9297	3.3049	2.5417
3	FTSM 100	**1.0457**	1.1076	1.4623	1.6080	1.0628
4	S&P 100	**1.6269**	1.9328	3.0696	3.3092	1.6890
5	Nikkei	**0.6771**	0.7961	0.6732	0.8975	0.6870
Average overall		**1.32862**	1.49526	1.8461	2.04826	1.41516

The computational time (seconds) of DPGA as well as GA, SA, TS and PSO are presented in Table 5. Result in bold indicates the best computational time. The figures in Table 5 show that, on all test instances, the computational time of DPGA is significantly lower than the running time required by GA, SA, TS and PSO.

Table 5. Computation time (in seconds) of DPGA compared to GA, SA, TS and PSO

#	Name	DPGA	GA	SA	TS	PSO
1-	Hang Seng	**0.86**	172	79	74	4.8
2-	DAX 100	**15.41**	544	210	199	26.8
3-	FTSM 100	**27.28**	573	215	246	31.4
4-	S&P 100	**28.63**	638	242	225	36.6
5-	Nikkei	**70.95**	1964	553	545	75.8

The set of comparisons of DPGA with GA, SA, TS and PSO reveal that the proposed DPGA is an effective solution method for portfolio selection, as it can produce good quality solutions on all test instances. More importantly good results can be obtained within shorter period of time.

6   Conclusion and Future Work

This work proposed a variation of genetic algorithm for the cardinality constrained portfolio selection problem. In order to improve the convergence of the search process, the proposed genetic algorithm utilizes dual populations, first population for exploring the search space, whilst second acting as a local search method. To evaluate the effectiveness of the propped algorithm, the cardinality constrained portfolio selection problem benchmark instances is used. The results show that the proposed algorithm produced very good results on all test instances when compared to existing algorithms. More importantly the computational time of the proposed algorithm is much lower than existing algorithms. We conclude that the proposed DPGA is effective method for the cardinality constrained portfolio selection problem. The use of dual population can significantly improve search performance. This work is the first step of establishing dual population GA for portfolio selection. There is a long list of issues that we would like to investigate and to improve. For example we will study further on memetic algorithms of PS and its similarity with DPGA approach. Multi-objective approaches will be also included in our next study.

References

1. Markowitz, H.: Portfolio selection. The Journal of Finance 7(1), 77–91 (1952)
2. Chang, T.-J., Meade, N., Beasley, J.E., Sharaiha, Y.M.: Heuristics for cardinality constrained portfolio optimisation. Computers & Operations Research 27(13), 1271–1302 (2000)
3. Varian, H.: A portfolio of Nobel laureates: Markowitz, Miller and Sharpe. The Journal of Economic Perspectives 7(1), 159–169 (1993)
4. Markowitz, H.: Portfolio selection: efficient diversification of investments. John Wiley and Sons, New York (1959)
5. Gendreau, M., Potvin, J.-Y.: Handbook of metaheuristics. Springer (2010)
6. Deng, G.-F., Lin, W.-T., Lo, C.-C.: Markowitz-based portfolio selection with cardinality constraints using improved particle swarm optimization. Expert Systems with Applications 39(4), 4558–4566 (2012)
7. Sabar, N.R., Kendall, G.: Using Harmony Search with Multiple Pitch Adjustment Operators for the Portfolio Selection Problem. Paper presented at the In Proceedings of the 2014 IEEE Congress on Evolutionary Computation (CEC 2014) (2014)
8. Kendall, G., Su, Y.: Imperfect evolutionary systems. IEEE Transactions on Evolutionary Computation 11(3), 294–307 (2007)
9. Fernández, A., Gómez, S.: Portfolio selection using neural networks. Computers & Operations Research 34(4), 1177–1191 (2007)
10. Holland, J.: Adaptation in Natural and Artificial Systems. MIT Press, Cambridge (1992)

11. Neri, F., Cotta, C.: Memetic algorithms and memetic computing optimization: A literature review. Swarm and Evolutionary Computation 2, 1–14 (2012)
12. Moral-Escudero, R., Ruiz-Torrubiano, R., Suarez, A.: Selection of optimal investment portfolios with cardinality constraints. Paper presented at the IEEE Congress on Evolutionary Computation, CEC 2006 (2006)
13. Beasley, J.E.: OR-Library: distributing test problems by electronic mail. Journal of the Operational Research Society 41(11), 1069–1072 (1990)

An Evolutionary Algorithm for TD-LTE Resource Allocation Based on Adaptive Fairness Threshold

Qiang Wang[1], Hai-Lin Liu[1, *], Zhen-hua Li[1],
Yiu-ming Cheung[2,3], and Jun Zhang[4]

[1]Guangdong University of Technology, Guangzhou, China
[2]Department of Computer Science
Hong Kong Baptist University, Hong Kong SAR, China
[3]United International College, Beijing Normal University –
Hong Kong Baptist University, Zhuhai, China
[4]Department of Computer Science
Sun Yat-Sen University, Guangzhou, China
{hlliu@gdut.edu.cn}

Abstract. In Time Division Long Term Evolution(TD-LTE) system, decreasing the cross time slot interference is very important to improve the system throughput. Accordingly, we will propose a multi-cells Orthogonal Frequency Division Multiplexing(OFDM) resource allocation model for TD-LTE system in this paper, which considers the cross time slot interference and the fairness. Subsequently, a TD-LTE resource allocation evolutionary algorithm based on adaptive fairness threshold is proposed to solve this model through introducing adaptive fairness threshold. Empirical studies show that the proposed model and algorithm can effectively decrease the cross time slot interference and improve the transmit rate of OFDM system under the same fairness.

Keywords: TD-LTE, Evolutionary Algorithm, Cross Slot, Adaptive Fairness Threshold.

1 Introduction

Recent years, TD-LTE system has received much attention in the wireless communication community. Along with the increase of the number of users, the demand for wireless system capacity inevitably grows. Therefore, it is becoming a major issue to allocate wireless resource for TD-LTE system, which could greatly improve the utilization of radio resources and thus improve the user's experiences.

During the scheduling process, the major resources of OFDM in TD-LTE system involve subcarriers, power, and time slot resources. Most of the current OFDM resource scheduling models focus on the relationship between power

---

* Corresponding author.

G. Dick et al. (Eds.): SEAL 2014, LNCS 8886, pp. 713–722, 2014.

and the system transmission rate, dealing with the scheduling of subcarrier and power. For example, they optimize the transmission power under the limit of the transmission rate of the system [8-19], or optimize the system capacity under the constrain of the case [11]. In the literature, paper [12] puts the transmission power and rate into an objective function using the weighted sum and then solves the model using an improved particle swarm algorithm. Paper [13] sets the transmission rate and the total system transmission power as a multi-objective model and NSGA-II is used to solve it. Furthermore, [14] formulates the OFDM resource allocation problem as a graph model. However, these models only allocate subcarrier and power without considering the resource allocation under certain conditions.

To this end, papers [14-15] have built the models in the case of limited bit error rate for the OFDM resource allocation problem. Also, [14] takes the limited bit error rate into account by adding a constraint about bit error rate. Furthermore, [15] builds a model for OFDM resource allocation problem under the condition of forward delay. Although these models take some specific transmission conditions into account, they are not very effective for the scheduling time slots.

Cross time slot interference of TD-LTE is due to different time slots ratios between different channels and along with the gradual increase of the network size, the system cross-slot interference becomes more and more serious. The above OFDM models focus on subcarrier and power scheduling problem. Thus, these models hardly solve the problem of resource scheduling in TD-LTE systems. Currently, there are some researches on slot resource allocation for TD-LTE system. For instance, paper [3] proposes a new slot allocation strategy for each cell using different transmission modes. [4] adopts a dispersion slot allocation strategy. Also, a fixed time slot allocation ratio is used in [1-7], which fixes the slots ratio between uplink and downlink within each cell. This strategy cannot respond flexibly to changing environments. That is, when a user's business requirements change, the slot resource is wasted if the slot allocating ratio cannot change in time. Most of the current researches estimate the transmission conditions for some time in a cell, and then fix the uplink and downlink slot ratio according to the estimation rather than adopting a dynamic strategy slot resources allocation. If the transmission conditions change within a short time and the slot allocation cannot respond in time, this will inevitably result in an increase in cross time slot interference.

In recent years, the fairness issue of OFDM resource allocation has attracted much more attention. In communication system, to maximize system throughput, the higher the quality of a radio channel is, the more resources it will get, and vice verse. This will affect the channel quality of other users.

This paper proposes a dynamic model dealing with time slot allocation issues of OFDM resources in TD-LTE system. The cross time slot interference among different cells is taken into account in this model. In order to facilitate the calculation of cross time slot interference, a time slot interference ratio corresponding mapping table is created. Furthermore, the model can allocate subcarrier, power and time slot simultaneously, which will significantly improve the allocation efficiency for the entire system. In order to solve the fairness issue described above, this paper presents an adaptive fairness strategy which can be combined with evolutionary algorithm. In the proposed algorithm, an adaptive fairness constraint value is set initially. In each generation, the best individual's fairness degree is calculated. If it meets the constraint, the algorithm will continue. Otherwise, it adopts the best individual on the direction of increasing fairness degree, and the constraint value will increase with the evolution generations. The motivation is that we want to focus on optimizing systems transmission capacity in the early phase and optimizing system's fairness which avoids the algorithm's performance shortcomings caused by fixed fairness threshold. In order to improve computational efficiency of the algorithm, we also design a local search strategy to improve the performance of the algorithm. Comparison experiments are implemented on cross-slot interference and the fairness among users. The results show that the proposed model and algorithm can effectively reduce system cross-slot interference and improve system fairness.

The remainder of the paper is organized as follows: OFDM resource allocation mathematic model in TD-LTE system and Evolutionary algorithm for TD-LTE resource allocation based on adaptive fairness threshold are described in Section II and Section III, respectively. Simulation results are given in Section IV. Finally, we draw a conclusion in Section V.

2 OFDM Resource Allocation Model in TD-LTE System

For TD-LTE system, we assume that there are I cells, K users, N subcarriers, and each eNodeB's total power is P_{max}. The channel gain on the kth channel in j^{th} cell to the nth user in i^{th} cell is $g_{ji,k,n}$. T_i is the time slot type of ith cell, $i = 1, 2, 3, ..., I$. In order to get the system throughput, every sub-channel SNR should be firstly calculated. For TD-LTE system, since cross-time slot interference needs to be taken into account, the SNR of nth channel for k^{th} user which belongs to i^{th} cell can be calculated as follows:

$$SNR_{i,k,n} = \frac{g_{ji,k,n} \times p_{i,k,n}}{\Delta} \tag{1}$$

where $p_{i,k,n}$ is the power of the nth channel on k^{th} user of ith cell;

$$\Delta = \sum_{j=1, j \neq i}^{I} \sum_{k=1}^{K} g_{ji,k,n} p_{j,k,n} + \sum_{j=1, j \neq i}^{I} \sum_{k=1}^{K} g_{ji,k,n} p_{ji,k,n}^{cts} + \sigma^2 \tag{2}$$

$p_{ji,k,n}^{cts}$ is the cross time slot interference, which is a function between T_i and T_j. The detail of $p_{ji,k,n}^{cts}$ can refer to literature [9]; σ^2 is additional white Gaussian noise.

The kth user's capacity in i^{th} cell can be expressed as follows:

$$\sum_{n=1}^{N} w_{i,k,n} \frac{B}{N} log_2(1 + SNR_{i,k,n}) \tag{3}$$

where $w_{i,k,n} \in \{0,1\}$. The model of TD-LTE system resource allocation is:

$$max \sum_{i=1}^{I} \sum_{k=1}^{K} \sum_{n=1}^{N} \frac{w_{i,k,n} B}{N} log_2(1 + \frac{SNR_{i,k,n}}{\Gamma}) \qquad (4)$$

$s.t$

$$\sum_{k=1}^{K} \sum_{n=1}^{N} p_{i,k,n} \leq P_T, i = 1, 2, 3, \cdots, I \qquad (5)$$

$$\sum_{k=1}^{K} w_{i,k,n} \leq 1, i = 1, 2, \cdots, I; n = 1, 2, \cdots, N \qquad (6)$$

$$p_{k,n} \geq 0, w_{k,n} \geq 0 \qquad (7)$$

where Γ is SNR gap, which is a constant; The more details of this model can be referred to [9].

3   Evolutionary Algorithm for TD-LTE Resource Allocation Based on Adaptive Fairness Threshold

In our proposed TD-LTE resource allocation model, subcarrier and time slot allocation are discrete and power allocation is continuous. Firstly, subcarriers are allocated using the criteria of [10]. Then, there are only power and time slot allocation variables in the model. Considering the system fairness and the time slot allocation are discrete variables, a TD-LTE resource allocation evolutionary algorithm based on adaptive fairness threshold (EAAF) is proposed to optimize the model. In this section, we firstly introduce the definition of system fairness, and then EAAF's framework will be shown.

In proposed algorithm, firstly, the adaptive fairness threshold φ in each generation is obtained (In the next subsection, we will show the details of the adaptive fairness threshold). In each generation, before mutation operation is implemented, the fairness of some selected individuals for mutation is obtained. If fairness is larger than current adaptive fairness threshold, perform ordinal mutation operation on these individuals. Otherwise, fixed mutation operation is performed. The fixed mutation operation is to find the user with maximal throughput, then decrease its power, meanwhile increase equal power to the user whose throughout is minimal. During the optimization process, the adaptive fairness threshold will become larger. The motivation of adaptive fairness threshold is that we hope to improve the system throughput quickly in the beginning of the algorithm, and at the last, we focus more attention on the issue of system fairness.

3.1   Definition of Fairness

Assume $\{\gamma_1, \gamma_2, \cdots, \gamma_k\}$ is a group of proportional constants about users throughput requirement, and $\{R_1, R_2, \cdots, R_k\}$ represents each users practical rate.

The mathematical definition of fairness is given by literature [13]:

$$fairness = \frac{(\sum\limits_{k=1}^{K} \frac{R_k}{\gamma_k})^2}{K \sum\limits_{k=1}^{K} (\frac{R_k}{\gamma_k})^2}$$

The range of $fairness$ is (0,1]. When $fairness$ equal to 1, it is clear that the system is the most fair. A lot of works consider the fairness as a fixed constrain in the model. However, this method is infeasible. For example, if the constraint value is large, this method will make the improvement system throughput ability change weak, and if the constraint value is small, the fairness of the system will be poor. To overcome this shortcoming, we propose an adaptive fairness threshold method in this paper. The details of adaptive fairness threshold will be shown in the following subsection.

3.2 Adaptive Fairness Threshold

In the EAAF, the adaptive fairness threshold is set as follows:

$$\varphi = exp(-\frac{1}{\alpha \times gen})$$

where α and gen are regulate factor and operation generation number respectively. Obviously, adaptive fairness threshold is positive corresponding with operation generation number, and it is close to 1 with increasing of operation generation number.

3.3 The Framework of the Evolutionary Algorithm Based on Adaptive Fairness Threshold for TD-LTE Resource Allocation

The framework of evolutionary algorithm based on adaptive fairness threshold for TD-LTE resource allocation is given as follows:

Step 1. Initialize population and adaptive fairness threshold φ;

Step 2. Perform crossover operation to population;

Step 3. Select individuals to perform mutation operation, and calculate these individuals' fairness; if a certain individual's fairness is larger than the adaptive fairness threshold, execute ordinary mutation operation to this individual; otherwise, this individual executes fixed mutation operation;

Step 4. Perform select operation to update population;

Step 5. If the stopping criteria is unsatisfied, go to Step 2; otherwise, stop.

4   Experimental Results and Performance Evaluation

In this section, the experimental results of fixed time slot types method and our proposed algorithm about cross time slot interference are shown firstly. Then we choose round robin algorithm, EA with fixed fairness constrain (EAFC) to compare with our proposed algorithm and analyze the experiment results. In the simulation, Slow-varying Rayleigh fading channel has been used and it has been assumed to be known to the resource allocator at base-station. The number of users ranges from 2 to 20, and the number of cells is 7. There are 100 time slots. The total transmit power and bandwidth have been taken as 1W and 5MHZ.

In the first simulation experiment, we simulate the ability of decreasing cross time slot interference of our proposed model. Fig.1 shows the different cross time slot interference of our proposed model and fixed time slot type method with the number of users ranging from 2 to 20. It is shown by fig.1 that the cross time slot interference of these two method are close when the number of users is small. However, with the increasement of user number, the cross time slot interference created by fixed time slot types method is much larger than our proposed dynamic update time slot types model.

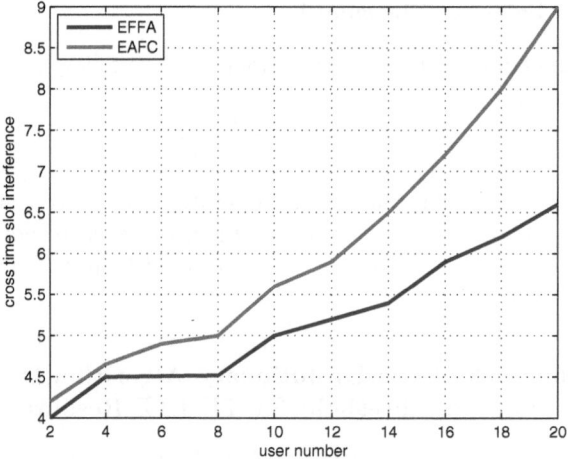

Fig. 1. Cross time slot interference of our proposed dynamic update time slot types model and fix time slot type

At the beginning of proposed algorithm, the adaptive fairness threshold is small so that it is beneficial to solve the objective function. At the end of the algorithm, we hope the system fairness can be enhanced. Therefore, the adaptive fairness threshold is adjusted larger. As a result, we can achieve a good tradeoff between system throughput and fairness.

Fig.2 shows the different system throughput of round robin, EA with fixed fairness constrain and our proposed algorithm. From fig.2, our proposed algorithm has better search ability than other two algorithms. At the beginning of our algorithm, the algorithm's search ability is improved since the fairness threshold is small.

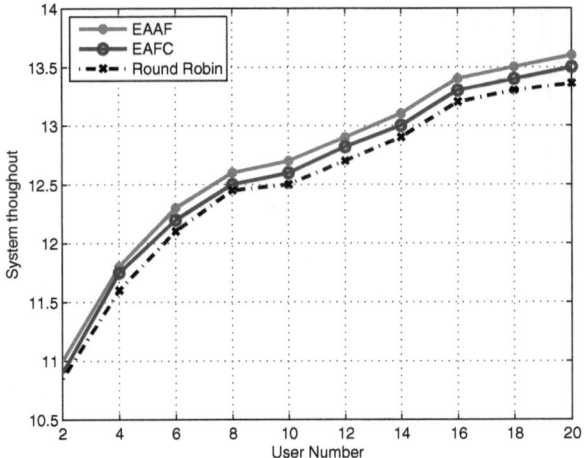

Fig. 2. System throughput vs. number of users

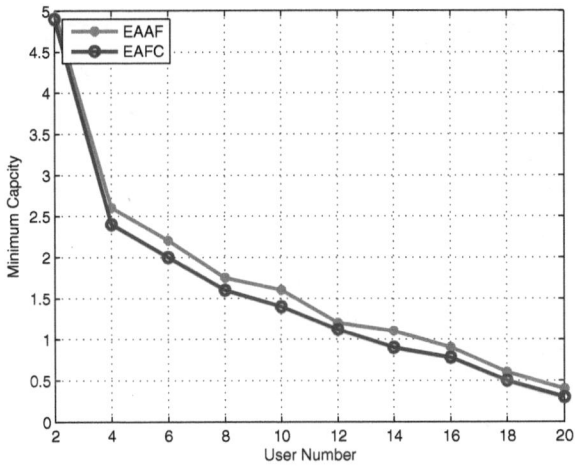

Fig. 3. Minimum user capacity vs. number of users

Minimum user capacity is also very important to wireless communication system. If the minimum user capacity is large, the total capacity is large, and the system fairness is good. Fig.3 shows the different minimum user capacity using EAAF, EAFC respectively. From Fig.3, EAAF has better minimum user capacity than EAFC.

We also compare the minimum transmit power generated by EAAF and EAFC. Fig.4 shows the minimum transmit power calculated by EAAF is less than EAFC with the increasement of user number.

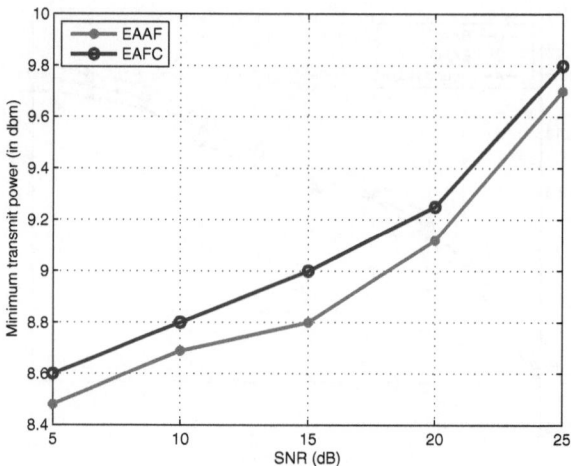

Fig. 4. Minimum transmit power vs. SNR

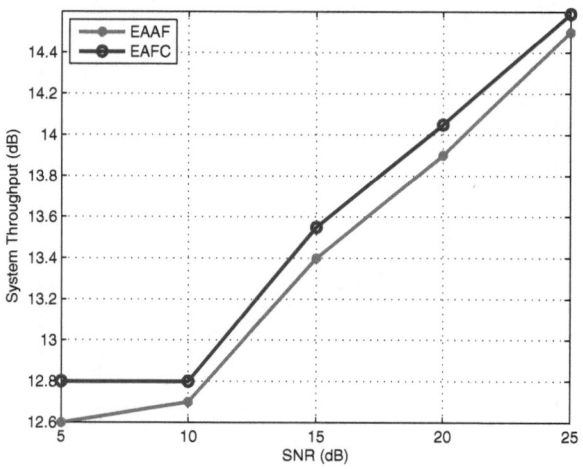

Fig. 5. System throughput vs. SNR

Then, we compare the capacity of the proposed algorithm and EAFC under different SNR with 4 users. The comparison result is shown in Figure 5. It is clear that the proposed algorithm's performance out-performs the other algorithms.

From above comparison, the performance of the proposed algorithm is better than EFAC. Then, we compare the performance of system fairness of EAAF and EAFC through comparing the variance of all user capacity. From Fig.6, the variance obtained from EFFA is significantly less than EAFC.

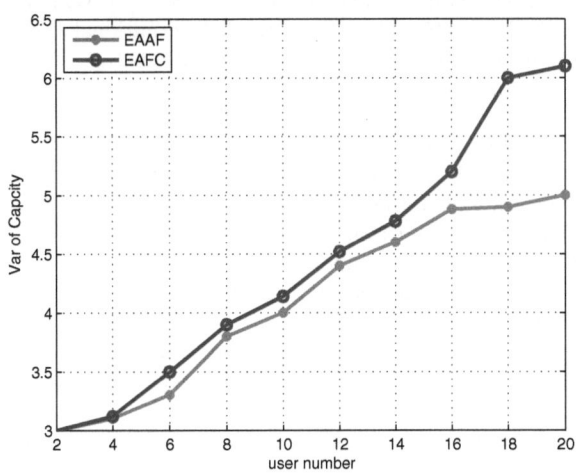

Fig. 6. Var of Capacity vs. number of users

5   Conclusion

In this paper, we have presented the OFDM resource allocation model for TD-LTE system by taking into account both of the cross time slot interference and the system fairness. Accordingly, An evolutionary algorithm based on adaptive fairness threshold has been developed for TD-LTE resource allocation, featuring a better trade-off between the system throughput and system fairness. Experiments have shown the promising results.

Acknowledgments. This work was supported by the Natural Science Foundation of Guangdong Province (S2011030002886, S2012010008813), the National Natural Science Foundation of China (NSFC) under grant: 61272366, and the projects of Science and Technology of Guangdong Province (2012B091100033).

References

1. Wang, W., Zhou, K., Zhang: Effects of synchronization Errors on the Performance of TDD/OFDMA Crossed Timeslot. In: Communications and Networking in China, pp. 870–875. IEEE press, Beijing (2008)
2. Jeong, D.G., Jeon, W.S.: Time Slot Allocation Algorithm with QoS Provision in Downlink OFDMA Systems. IEEE Communications Letters 4, 59–61 (2000)
3. Wie, S.-H., Cho, D.-H.: Time Slot Allocation Scheme based on a Region Division in CDMA/TDD Systems. In: Vehicular Technology Conference (VTC Spring), pp. 2445–2449. IEEE press, Budapest (2001)
4. Choi, Y.S., Sohn, I., Lee, K.B.: A Novel Decentralized Time Slot Allocation Algorithm in Dynamic TDD System. Cell 4, 1268–1272 (2006)

5. Yan, H.-C., Liu, H.-L., Wang, Q.: Dynamic Resource Allocation of TD-LTE System Based on Modified Quantum. In: 2013 Eighth International Conference on Computational Intelligence and Security (CIS), pp. 2726–2737. IEEE press, Leshan (2014)

6. Sadeque, S., Vaughan, I.A.R.: Impact of Individual and Joint Optimizations in Multi Cuser OFDM Resource Allocation by Modified PSO. In: 2011 24th Canadian Conference on Electrical and Computer Engineering (CCECE), pp. 001233–001237. IEEE press, Niagara Falls (2011)

7. Maw, M.S., Sasase, I.: Effcient.: Efficient Resource Allocation for Multiuser MIMO-OFDM Uplink System to Guarantee the Proportional Data Rate Fairness among Users in a System. In: 2011 1st International Symposium on Access Spaces (ISAS), pp. 132–137. IEEE press, Yokohama (2011)

8. Seo, H., Lee, B.G.: A Proportional-Fair Power Allocation Scheme for Fair and Efficient Multiuser OFDM System. In: Global Telecommunications Conference, pp. 3737–3741. IEEE press, Dallas (2004)

9. Mohanram, C., Bhashyam, S.: A Sub-optimal Joint Subcarrier and Power Allocation Algorithm for Multiuser OFDM. IEEE Communications Letters 9, 685–687 (2011)

10. Mahmoud, A.S., Al-Rayyah, A.Y.: Adaptive Power Allocation Algorithm to Support Absolute Proportional Rates Constraint for Scalable OFDM Systems. In: 2010 IEEE 71st Vehicular Technology Conference (VTC 2010-Spring), Taipei, pp. 1–4 (2010)

11. Seo, H., Lee, B.G.: A Proportional-Fair Power Allocation Scheme for Fair and Efficient Multiuser OFDM System. In: IEEE Global Telecommunications Conference, GLOBECOM 2004, Dallas, TX, pp. 3737–3741 (2004)

12. Mohanram, C., Bhashyam, S.: A Sub-optimal Joint Subcarrier and Power Allocation Algorithm for Multiuser OFDM. IEEE Communications Letters 9, 685–687 (2011)

13. Shen, Z., Andrews, J.G., Evans, B.L.: Adaptive resource allocation in multiuser OFDM systems with proportional rate constraints. IEEE Transactions on Wireless Communications 4, 2726–2737 (2005)

14. Wang, C., Chen, C.: A low-complexity iterative power allocation scheme for multiuser OFDM systems. In: Vehicular Technology Conference, VTC Spring 2008, pp. 1935–1939. IEEE press, Singapore (2008)

15. Odhah, N.A., Dessouky, M.I., Al-Hanafy, W.E., Abd El-Samie, F.E.: Greedy Power Allocation Algorithm for Proportional Resource Allocation in Multi-User OFDM Systems. In: 2012 29th National Radio Science Conference (NRSC), pp. 421–428. IEEE press, Cairo (2012)

16. Ho, W.W.L., Liang, Y.-C.: Optimal Resource Allocation for Multiuser MO-OFDM Systems With User Rate Constraints. IEEE Transactions on Vehicular Technology 58, 1190–1203 (2009)

17. Perea-Vega, D., Frigon, J.-F., Girard, A.: Near-Optimal and Efficient Heuristic Algorithms for Resource Allocation in MISO-OFDM Systems. In: 2010 IEEE International Conference on Communications, pp. 1–6. IEEE press, Cape Town (2010)

18. Lee, K.-D., Yum, T.-S.P.: On Pareto-Efficiency Between Profit and Utility in OFDM Resource Allocation. IEEE Transactions on Communications 58, 3277–3285 (2010)

19. Wang, X., Giannakis, G.B.: Resource Allocation for Wireless Multiuser OFDM Networks. IEEE Transactions on Information Theory 57, 4359–4372 (2011)

Enhancing Heuristics for Order Acceptance and Scheduling Using Genetic Programming

John Park[1], Su Nguyen[1,2], Mengjie Zhang[1], and Mark Johnston[1]

[1] Evolutionary Computation Research Group, Victoria University of Wellington,
PO Box 600, Wellington, New Zealand
[2] International University - VNU HCMC, Vietnam
{John.Park,Su.Nguyen,Mengjie.Zhang}@ecs.vuw.ac.nz,
Mark.Johnston@msor.vuw.ac.nz

Abstract. Order acceptance and scheduling (OAS) in make-to-order manufacturing systems is a NP-hard problem for which finding optimal solutions for problem instances can be challenging. Because of this, several heuristic approaches have been proposed in the literature to find near-optimal solutions to OAS. Many previous heuristic approaches are very effective, but require careful design and developing new heuristics can be difficult. Genetic Programming (GP) has been used to generate reusable and efficient heuristics in OAS and shows promising results. However, in terms of solution quality, the evolved heuristics are still less competitive as compared to highly customised heuristics designed by human experts. To overcome these limitations, this paper proposes two new Particle Swarm Optimisation (PSO) approaches to OAS. Afterwards, GP evolved rules are combined with an existing Tabu Search (TS) heuristic and with the proposed PSO algorithms as hybrid approaches to OAS. The experimental results show that these PSO approaches are competitive with effective heuristics such as TS. In addition, TS heuristic greatly benefits from evolved rules, whereas PSO approaches do not benefit.

1 Introduction

Order acceptance and scheduling (OAS) in make-to-order manufacturing systems is an optimisation problem that deals with the decisions of both accepting/rejecting customer *orders* and then scheduling the accepted orders. OAS models environments where the manufacturer has to deal with limited production capacity and high demand. This means that it is likely that the manufacturer cannot meet all of the demands. As a result, the manufacturer needs to determine the orders to accept or reject (acceptance decisions), and how it can process all the accepted orders (scheduling decisions). This paper focuses on OAS in single machine environments with sequence dependent setup times [1]. The problems consist of a set of N orders. Order $j \in \{1, \ldots, N\}$ in the set has release time r_j, processing time p_j, due date d_j, weight/penalty w_j, maximum revenue e_j, and deadline $\overline{d_j}$. In addition, if an order j is processed immediately after order i it has setup time s_{ij}. Order j has the setup time s_{0j} if it is the first

G. Dick et al. (Eds.): SEAL 2014, LNCS 8886, pp. 723–734, 2014.

order to be processed. From completing an order, a profit P_j is gained by the manufacturing system. The profit P_j is equal to the maximum revenue e_j if the order is completed before the due date d_j. Otherwise, the profit P_j is reduced by the time difference between the completion time and due date d_j, denoted as tardiness T_j, multiplied by the weight w_j. If an order is completed after its deadline $\overline{d_j}$, then the profit P_j is zero, and the order is rejected. Overall, this gives us $P_j = \max\{e_j - w_j T_j, 0\}$. The objective function of this problem is to maximise the total profit $TPR = \sum_{j \in A} P_j$, where A is the set of accepted orders.

Ghosh [2] proved that OAS is NP-hard, and finding optimal solutions for OAS problems is very challenging [3]. Therefore, many heuristic approaches have been proposed to find near-optimal solutions. Rom and Slotnick [4] proposed a hybrid approach of finding solutions using Genetic Algorithm (GA) and improving it with a local search heuristic. Oguz et al. [1] developed a Simulated Annealing method (ISFAN) for OAS with sequence dependent setup times, and found that ISFAN can find good solutions for large problem instances. Cesaret et al. [3] then developed a tabu search (TS) method for solving OAS problems that outperforms ISFAN in most problem instances. Lin and Ying [5] proposed an Artificial Bee Colony (ABC) approach with iterated greedy (IG) local search heuristic. Genetic Programming (GP) [6] has been applied to OAS [7][8], where GP is used to evolve reusable heuristics. A literature survey of OAS can be found in Slotnick [9].

One limitation of GP is that the evolved rules usually are not as effective as highly customised heuristics for OAS [7]. Nguyen et al. [10] showed that the quality of the solution generated by GP evolved rules can be improved further by using GA to improve the quality of the solutions. Other hybrid approaches have been proposed for OAS [3][5], but they do not use reusable heuristics. Park et al. [8] have showed that evolved rules can outperform standard rules. In addition, Particle Swarm Optimisation (PSO) [11], which has been applied to other scheduling problems [12], has not been applied to OAS.

1.1 Goal

The goal of this paper is to develop new hybrid approaches to OAS that combine GP evolved rules in conjunction with various heuristic approaches. By doing this, we hope to improve the quality of the solutions that would otherwise be generated either by the evolved rules or the heuristic approaches individually. In addition, we propose novel PSO approaches to OAS, where PSO is applied to OAS for the first time. In summary, the three objectives of this paper are as follows.

(a) Developing two new PSO algorithms for OAS.
(b) Combining an existing TS approach with GP evolved heuristics.
(c) Combining the two PSO algorithms with GP evolved heuristics.

1.2 Organisation

The organisation of this paper is as follows. Section 2 covers the PSO algorithms, the GP representations that are used to generate the evolved rules, and how the evolved rules are used with TS and PSO. Section 3 covers the parameter values used to train the GP methods and for TS and PSO. Section 4 covers the results and evaluations of the approaches. Finally, Section 5 gives conclusions and future research directions.

2   Proposed Methods

In this section, we describe the proposed PSO algorithms that are used for OAS. Afterwards, we describe the GP representations used to generate the evolved rules for OAS, and how they are combined with the TS and PSO algorithms.

2.1   PSO for OAS

PSO is a swarm intelligence algorithm initially proposed by Kennedy and Eberhart [11]. Each particle i in a swarm of size S initially has a random position $\mathbf{x}_i$ and a random velocity $\mathbf{v}_i$. For subsequent iterations, the particle's velocity is influenced by the local best solution $\mathbf{p}_i$ that individual particle i has discovered, and the global best solution $\mathbf{g}$ found by the swarm. Coefficients ω, φ_g, and φ_p are used to decide how the current velocity, the global best position and the local best position affect the velocity update. The final solution generated by PSO is $\mathbf{g}$ after a certain number of iterations T. In addition, for our proposed PSO approaches, a local search heuristic proposed by Lin and Ying [5] is also applied to each particle i at every iteration. This is discussed further below. Overall, this gives us the pseudocode in Algorithm 1 (r_g and r_p are random numbers within the interval $[0, 1]$). The two proposed PSO methods are based on this algorithm. The first PSO method is denoted as Continuous Priority based PSO (CPSO). The second PSO method is denoted as Discrete Permutation based PSO (DPSO), and is a derivation of the discrete permutation based algorithm proposed by Rameshkumar et al. [12].

Calculating Fitness of Particle in OAS Problems. The position $\mathbf{x}_i$ of the particle i undergoes two steps to transform it to a solution. The first step, denoted as raw, converts the particle's position to a permutation of the list of all orders. The second step, denoted as acc, converts this permutation to a list of accepted orders such that the solution is feasible. The fitness of the particle is the total profit TPR.

The first step, raw, converts the position $\mathbf{x}_i$ of particle i to a raw solution π_i. A raw solution is defined as the permutation of all the orders in the problem, i.e., a permutation of $\{1, \ldots, N\}$. It is the sequence in which orders are next considered for processing. The definition of raw for CPSO and DPSO is defined further below. The second step, acc, converts the raw solution π_i to a solution of accepted orders A_i which does not contain any orders that are rejected. If the projected completion time C_j of order j is greater than deadline $\bar{d}_j$, then the order is rejected and the next order at position $k + 1$ is checked. This continues until all orders have either been processed or rejected. For example, suppose that for a particle i the position $\mathbf{x}_i$ is converted to a raw solution $\pi_i = (6, 1, 2, 7, 3, 5, 4, 8)$.

```
Data: S, T, N, v_min, v_max, x_min, x_max
Result: The final global best solution g
/* Initialise the particles */
t ← 0
for each particle i = 1 to S do
    randomly initialise x_i and v_i
    p_i ← x_i
end
set g to the position with maximum f value
/* Move the particles around the search space */
while t < T do
    for i = 0 to S do
        x_i ← apply the local search procedure
        v_i ← ωv_i + φ_g r_g(g − x_i) + φ_p r_p(p_i − x_i)
        x_i ← x_i + v_i
        update p_i if f(x_i) > f(p_i)
        update g if f(x_i) > f(g)
    end
    t ← t + 1
end
```

Algorithm 1. The pseudocode for the PSO algorithms for OAS problems

This means that order 6 may be processed first, followed by order 1, 2, etc. If order 3 is projected to complete past its deadline after processing order 7, then order 3 is rejected and order 5 is checked next. If orders $3, 4, 8$ are rejected, then the final solution is given by $A_i = (6, 1, 2, 7, 5)$. After we get the list of accepted orders, we can calculate the total profit TPR.

Local Search Procedure. Cesaret et al. [3] incorporates local search heuristics in their TS algorithm to improve the search in the solution space. Therefore, the two proposed PSO algorithms use a local search heuristic proposed by Lin and Ying [5] at each iteration to potentially find better solutions located around position $\mathbf{x}_i$ for particle i. The local search procedure is carried out in two steps. The first step finds neighbourhood solution using the iterated greedy (IG) algorithm [13]. The IG algorithm first removes α number of orders from the raw solution π_i in a destructive phase. The removed orders are then reinserted back into the schedule to maximise the objective function value for the partial solutions. If the neighbourhood solution is within some threshold θ_{IG} of the global best solution, a more in-depth search is carried out [5]. In the in-depth search, the position of every order is interchanged with each other, and the solution of the new position is compared against the current particle's solution [5].

Continuous Priority Based PSO. For a problem instance with N orders, the position of each particle i in CPSO algorithm is a vector of size N which is defined as $\mathbf{x}_i = [x_{i1}, \ldots, x_{iN}]$. For $j = 1, \ldots, N$, x_{ij} is a real value bounded within the interval $[x_{min}, x_{max}]$. Likewise, the velocity is also a vector of size N of real values which is defined as $\mathbf{v}_i = [v_{i1}, \ldots, v_{iN}]$, and $v_{ij} \in [v_{min}, v_{max}]$ for

$j = 1, \ldots, N$. Positions and velocities are simply updated using regular vector operation between each index of the vectors, e.g., the position is updated as $\mathbf{x}'_i = \mathbf{x}_i + \mathbf{v}_i = [x_{i1} + v_{i1}, \ldots, x_{iN} + v_{iN}]$.

The transformation function raw that converts $\mathbf{x}_i$ for particle i to the raw solution π_i is defined as follows. The value x_{ij} at each index j of the position vector correspond to the "priority" of processing order j. Therefore, the values in each of the indices are rearranged in a descending order, and the values in each index of the raw solution π_i corresponds to the index of $\mathbf{x}_i$ before it is rearranged. An example of this is shown in Figure 1.

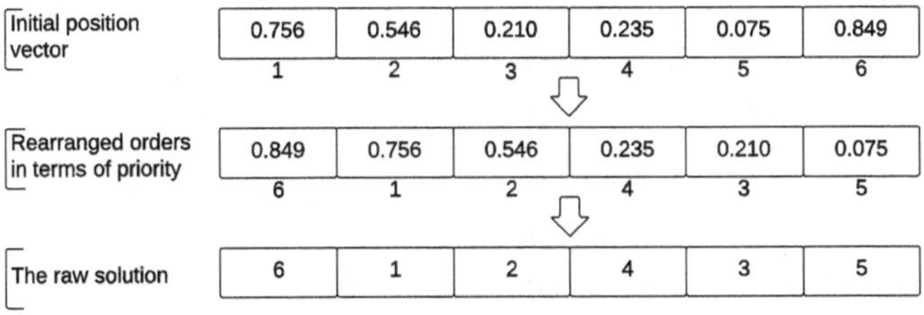

Initial position vector	0.756	0.546	0.210	0.235	0.075	0.849
	1	2	3	4	5	6

Rearranged orders in terms of priority	0.849	0.756	0.546	0.235	0.210	0.075
	6	1	2	4	3	5

The raw solution	6	1	2	4	3	5

Fig. 1. Procedure for computing π_i from $\mathbf{x}_i$ for particle i

Discrete Permutation Based PSO. DPSO is based on the algorithm proposed by Rameshkumar et al. [12]. For a problem instance with N orders, the position $\mathbf{x}_i$ of each particle i in DPSO algorithm is a permutation of numbers $\{1, \ldots, N\}$. This means that each number j, representing order j, only occurs once in $\mathbf{x}_i$ to a unique index k which represents the position the order is inserted. Therefore, transformation function raw to convert position to raw solution is simply defined as $\pi_i = raw(\mathbf{x}_i) = \mathbf{x}_i$.

Velocity $\mathbf{v}_i$ of particle i is represented as a list of "swaps" between values in positions k_1 and k_2 in $\mathbf{x}_i$. This results in the new position being a slightly different permutation of the numbers $\{1, \ldots, N\}$. Upon initialisation, $\mathbf{v}_i$ is a series of random swaps up to length $L = 2$. When the velocity update is applied, the operators between velocity, position and constants are defined as follows:

Subtraction (position - position) Operator: Let $\mathbf{x}_1$ and $\mathbf{x}_2$ be the two positions. Then $\mathbf{x}_2 - \mathbf{x}_1$ is a list of all the "swaps" that can be applied sequentially to get from $\mathbf{x}_1$ to $\mathbf{x}_2$. This results in a new velocity $\mathbf{v}$.

Addition (position + velocity) Operator: Let $\mathbf{x}$ and $\mathbf{v}$ be the position and velocity respectively. Then $\mathbf{x} + \mathbf{v}$ applies the list of swaps in $\mathbf{v}$ to values in $\mathbf{x}$ in the order that the swaps appear. Any equivalent or opposing swaps adjacent to each other will be cancelled out. This results in a new position $\mathbf{x}_{new}$.

Addition (velocity + velocity) Operator: Let $\mathbf{v}_1$ and $\mathbf{v}_2$ be the two velocities. Then $\mathbf{v}_1 + \mathbf{v}_2$ appends the list of swap between indices from $\mathbf{v}_2$ to the end of $\mathbf{v}_1$, such that the new velocity $\mathbf{v}_{new}$ is equivalent to applying $\mathbf{v}_1$ first before $\mathbf{v}_2$.

Multiplication (coefficient × velocity) Operator: Let c and $\mathbf{v}$ be the co-efficient and the velocity. c is a learning coefficient on $\mathbf{v}$, where the new velocity $\mathbf{v}_{new}$ contains $\max\{\lceil c|\mathbf{v}|\rceil, |\mathbf{v}|\}$ number of swaps. The swaps from $\mathbf{v}$ are chosen randomly without replacement and inserted into $\mathbf{v}_{new}$, until $\mathbf{v}_{new}$ has either $c|\mathbf{v}|$ number of swaps or all the swaps from $\mathbf{v}$ (if $c \geq 1$).

In DPSO, r_g and r_p is removed from the velocity update. This is because the **coefficient** × **velocity** operator chooses random swaps for the new velocity. Instead, the velocity update for DPSO is $\mathbf{v}_i \leftarrow \omega\mathbf{v}_i + \varphi_g(\mathbf{g} - \mathbf{x}_i) + \varphi_p(\mathbf{p}_i - \mathbf{x}_i)$.

For example, suppose that DPSO with coefficients $\omega = 0.1, \varphi_g = 0.4, \varphi_p = 0.4$ is applied to a problem instance with $N = 5$ orders. Suppose that $\mathbf{v}_i^k = [(1,3),(4,1)]$, $\mathbf{x}_i^k = [3,2,4,5,1]$, $\mathbf{g}^k = [5,2,1,4,3]$ and $\mathbf{p}_i^k = [3,2,4,5,1]$ for particle i at iteration k. To find the difference between $\mathbf{g}^k$ and $\mathbf{x}_i^k$, we find the list of swaps that can be applied sequentially that convert $\mathbf{x}_i^k$ to $\mathbf{g}^k$. Starting from order 5 on index 1 of $\mathbf{g}^k$, since order 5 is at index 4 of the position vector $\mathbf{x}_i^k$, we get the swap $(1,4)$ to bring order 5 from index 4 to index 1. Afterwards, we swap order 4 at index 3 with order 1 at index 5 to get swap $(3,5)$. Finally, we swap order 3 with order 4, now at indices 4 and 5 respectively, to get the swap $(4,5)$. When applied sequentially to $\mathbf{x}_i^k$, we get $\mathbf{x}_i^k = [3,2,4,5,1] + (1,4) = [5,2,4,3,1]$, then $[5,2,4,3,1]+(3,5) = [5,2,1,3,4]$, and finally $[5,2,1,3,4]+(4,5) = [5,2,1,4,3] = \mathbf{g}^k$. On the other hand, since there is no difference between $\mathbf{p}^k$ and $\mathbf{x}_i^k$, an empty list of swaps is returned. In addition, since $(\mathbf{g}^k - \mathbf{x}_i^k)$ is multiplied by 0.4, we keep $\lceil 0.4 * 3\rceil = 2$ swaps from the list which are chosen randomly. Therefore, at iteration $k + 1$, we have the velocity and the position updated as follows:

$$\mathbf{v}_i^{k+1} = 0.1 \times [(1,3),(4,1)] + 0.4 \times ([5,2,1,4,3] - [3,2,4,5,1]) +$$
$$0.4 \times ([3,2,4,5,1] - [3,2,4,5,1])$$
$$= [(4,1)] + 0.4 \times [(1,4),(3,5),(4,5)] + 0.4 \times []$$
$$= [(4,1),(1,4),(4,5)] = [(4,5)]$$
$$\mathbf{x}_i^{k+1} = [3,2,4,5,1] + [(4,5)] = [3,2,4,1,5]$$

2.2 GP Method

For this paper, we use two existing tree-based GP methods for OAS. They are denoted as GPOAS [7] and GPSR [8]. Both methods use a single priority rule to assign priority values to the remaining orders that are yet to be processed. Out of those remaining orders, only the orders that are released before the earliest projected completion time are considered. This subset of orders are called active orders. From the active orders, GP programs in GPOAS picks the order with the highest priority, whereas GP programs in GPSR pick an order based on probabilities dependent on the priorities. The output from GPOAS is then used in a standard dispatching rule (DR) [7], whereas the output from GPSR is used in a stochastic dispatching rule (SDR) [8].

Combining Evolved DRs with TS. One of the modification made to the TS algorithms is how the initial solutions are generated. Cesaret et al. [3] proposed a DR that uses *Revenue-load ratio* $RLR1_j$ as priority rules for each order j to generate an initial solution. In this paper, we use the DRs evolved using GP to generate the initial solution before carrying out the neighbourhood search done by TS algorithm. This allows the evolved heuristics to be used in conjunction with TS. GP evolved rules generally perform better than the $RLR1_j$ priority rule [7]. This can potentially allow the TS algorithm to perform better with the higher quality initial solution. For example, using the $RLR1$ based priority on a problem instance of size $N = 8$ may produce an initial solution $(6, 1, 8, 2, 7)$ with revenue 101, and using evolved DR may produce an initial solution $(6, 1, 2, 7, 5)$ with revenue 112. The s_{best} generated from the TS algorithm staring from each initial solution can be $s_{best} = (6, 1, 8, 3, 7)$ with revenue 110 for $RLR1$ and $s_{best} = (6, 1, 2, 7, 4, 5)$ with revenue 121 for GP evolved rule respectively.

Combining Evolved DRs with PSO. The following procedure is applied to PSO to incorporate the rules evolved by GP. After initialising the particles in the swarm in PSO, evolved DR is used to generate the initial global best position that the particles could converge to. To do this, a basic kernel procedure is carried out in two steps as described below.

Firstly, the list of accepted orders A that is generated by the evolved DR is converted to a raw solution π. This is done by appending the rejected orders randomly onto the end of A, i.e., if $\overline{A}$ is the set of rejected orders in random ordering, then $\pi = (A, \overline{A})$. This ensures that $A = acc(\pi)$.

Secondly, a fresh position $\mathbf{x}_{dr}$ is generated as position $\mathbf{x}_i$ would be generated initially for particle i in the swarm. From $\mathbf{x}_{dr}$, π_{dr} is computed using *raw*. By comparing π_{dr} with π from A, we get the list of swaps $\mathbf{W}$ that can be applied sequentially to convert π_{dr} to π, and apply the swaps in $\mathbf{W}$ to indices in $\mathbf{x}_{dr}$. For example, suppose that $N = 8$, and we have the raw solution from the DR as $\pi = (2, 1, 6, 8, 4, 5, 3, 7)$ and the fresh position and the corresponding raw solution as $\mathbf{x}_{dr} = (2.4, 1.9, 0.5, -2.1, -1.7, 2.9, -1.1, -2.5)$ and $\pi_{dr} = (6, 1, 2, 7, 3, 5, 4, 8)$. Then the list of swaps is $[(1, 3), (4, 8), (5, 7)]$ and $\mathbf{x}_{dr}$ is updated by adding the swaps to $\mathbf{x}_{dr}$, giving us $\mathbf{x}_{dr} = (0.5, 1.9, 2.4, -2.5, 1.1, 2.9, -1.7, -2.1)$.

3   Parameter Settings

This section describes the parameters used for training the GPOAS and the GPSR for the evolved rules. In addition, the parameters settings for the TS and the PSO approaches are covered.

3.1   GP Parameter Settings

The GP parameter settings are based on the original parameter settings used to evolve rules using GPOAS [7] and GPSR [8]. The population size is set to 1024. Crossover, mutation and reproduction rate are set to 80%, 10%, and 10% respectively. The number of generations is set at 51 and the maximum depth of a GP tree at 8. Tournament selection of size 7 is used to select individuals that will reproduce to the next generation.

3.2   TS Parameter Settings

The key parameters for the TS proposed by Cesaret et al. [3] are the tabu tenure, the threshold value, and the termination criterion, where the tabu tenure is the maximum size of the tabu list. These are kept at the same value as a benchmark, i.e., the maximum size of the tabu list is $\lceil N/6 \rceil$, TS moves to a new solution if its revenue value is greater or equal to $\theta_{TS} = 0.998$ times the current solution, and the TS algorithm terminates after $\epsilon = 50$ iterations without improvement.

3.3   PSO Parameter Settings

For the two proposed PSO algorithms, parameter tuning on the size of the swarm and the number of iterations was carried out to ensure that the swarm of particles had sufficient number of iterations to converge on a solution. After some parameter tuning, the population sizes of both PSO approaches are set at $S = 50$, and the number of iterations at $T = 100$.

Since the values in the position and velocity vectors for CPSO are relative to each other, we kept the minimum and maximum values for both values fairly small. Initially, the velocity constraints were set as $v_{min} = -1$ and $v_{max} = 1$, and the position constraints as $x_{min} = -3$ and $x_{max} = 3$. This meant that less than 5 components of the position vector was at the upper and lower bound for initial tests. Finally, we have the coefficients for the velocity updates as $\omega = 0.7$, $\varphi_g = 1.5$ and $\varphi_p = 1.5$, which is based of Clerc's constriction factor [14]. On the other hand, minimum/maximum values for positions and velocities are not set for DPSO, as positions are permutations of orders and velocities are list of swaps. Also ω, φ_g and φ_p are set at 0.1, 0.4 and 0.4 for DPSO after some parameter tuning. Finally, for the local search heuristic we used the same α and threshold value θ_{IG} specified by Lin and Ying [5] in their Artificial Bee Colony (ABC) approach to OAS, which were $\alpha = 3$ and $\theta_{IG} = 0.01$.

4   Results

To train the rules generated by the GP methods and to evaluate the algorithms, we use the dataset introduced by Oguz et al. [1]. This dataset consists of subsets of data that are generated using three parameter values: the number of orders n, the tardiness factor τ and the due date range R. Each subset consists of 10 problem instances. For each problem instance, an upper bound UB is provided from the linear programming (LP) relaxations of the mixed integer linear programming (MILP) models [1]. The rules are evolved from GPOAS [7] and GPSR [8] by taking the first 5 problem instances from the subset generated from $n = 100$, $\tau = 0.9$ and $R = 0.9$ as training. This is done 30 times, and the best performing rule out of the entire dataset is used to be combined with the TS and PSO algorithms. The TS and PSO algorithms that use the evolved DRs and SDRs are denoted with suffixes '-DR' and '-SDR' respectively. Finally, these are compared against the existing benchmark TS [3] algorithm for the evaluation.

Table 1. Evaluation of the modified TS and PSO algorithms to the benchmark meta-heuristics using the *best* deviation

τ	R	TS			TS-DR			TS-SDR			CPSO			CPSO-DR			CPSO-SDR			DPSO			DPSO-DR			DPSO-SDR		
		a	b	c	a	b	c	a	b	c	a	b	c	a	b	c	a	b	c	a	b	c	a	b	c	a	b	c
0.1	0.1	1	2	3	1	2	2	1	1	2	2	2	3	2	3	3	2	2	3	2	2	3	2	3	4	2	2	3
	0.3	1	2	3	1	1	2	1	1	2	1	2	3	2	2	3	1	2	2	1	2	2	2	2	3	1	2	2
	0.5	0	1	2	0	1	1	0	1	1	1	1	1	1	1	2	1	1	1	1	1	1	1	1	2	1	1	1
	0.7	0	0	0	0	0	0	0	0	0	0	0	1	0	0	1	0	0	1	0	0	1	0	0	1	0	0	1
	0.9	0	0	0	0	0	0	0	0	0	0	0	0	0	0	1	0	0	1	0	0	0	0	0	1	0	0	1
0.3	0.1	1	2	3	2	2	3	1	2	3	2	3	4	2	3	4	2	3	4	2	3	3	2	3	4	2	3	4
	0.3	1	2	5	2	3	5	1	2	4	2	3	5	2	3	6	1	3	5	2	3	5	2	4	6	1	3	5
	0.5	1	2	3	1	2	3	1	2	2	2	3	4	2	3	4	1	2	3	2	3	3	2	3	4	1	2	3
	0.7	0	1	2	1	1	2	0	1	1	0	1	2	1	2	3	0	1	2	0	1	2	1	2	3	0	1	2
	0.9	0	0	2	0	1	2	0	0	1	0	1	2	0	1	3	0	1	2	0	1	2	0	1	3	0	1	2
0.5	0.1	2	4	6	2	4	4	2	3	4	4	4	6	4	6	8	3	5	7	3	4	6	5	6	7	3	5	6
	0.3	2	3	5	3	4	5	2	3	4	4	4	5	4	6	7	3	4	6	4	4	5	5	6	7	3	4	6
	0.5	2	3	4	3	4	5	2	3	4	3	5	6	4	6	7	3	4	5	3	5	6	4	5	7	3	4	5
	0.7	1	2	4	1	3	4	1	2	3	2	3	5	2	4	5	1	2	4	2	4	5	2	4	5	1	2	4
	0.9	0	2	4	1	2	4	1	1	3	2	3	5	2	4	6	1	2	5	2	3	5	2	4	6	1	2	5
0.7	0.1	2	4	6	3	5	6	3	4	6	5	6	7	6	7	10	4	6	8	5	6	7	5	7	10	4	6	8
	0.3	3	6	10	3	6	9	2	5	8	3	7	9	4	8	11	3	6	10	3	7	9	4	8	11	3	6	10
	0.5	4	6	12	4	6	13	3	6	14	5	7	14	6	8	15	4	8	16	5	7	14	5	8	15	4	8	16
	0.7	3	7	13	4	8	11	3	6	9	4	8	11	5	9	11	5	8	12	4	8	10	5	9	11	5	8	12
	0.9	5	8	12	5	8	11	4	7	9	5	8	10	6	9	12	4	8	11	5	8	10	5	9	12	4	8	11
0.9	0.1	7	9	11	6	8	12	7	8	12	7	9	11	8	10	12	7	10	13	7	9	11	8	10	12	7	9	13
	0.3	7	13	17	8	13	15	8	11	14	8	11	13	7	12	14	10	14	18	7	11	14	8	12	15	7	12	14
	0.5	10	15	18	11	15	19	11	13	18	10	13	18	11	16	22	10	13	17	10	13	18	11	14	18	7	14	18
	0.7	10	15	19	11	15	17	10	13	16	10	13	15	10	13	15	12	16	19	10	13	15	10	13	16	11	14	16
	0.9	11	15	22	6	14	21	5	13	19	6	13	19	7	13	19	8	16	23	6	13	18	8	13	18	7	14	19

* a b c denotes the best, average and worst % deviation from the upper bound

** x-DR/SDR represents a x algorithm which uses evolved DR/SDR rule

*** Highlighted cells show the algorithm for which it has the lowest average deviation from the upper bound for the ⟨τ, R⟩

4.1 Best Deviation Evaluation

TS-DR, TS-SDR, CPSO and DPSO are run 30 times for each problem instance and the best results are recorded. For each run, we calculate the percentage deviation from the upper bound UB as $UB = 100 \times (UB - TPR)/UB$. The best deviation is then used as the representative solution for the particular problem instance, and is used to compare the algorithms against the benchmark TS algorithm. The results are shown in Table 1, where a, b and c represent the best, average and worst deviation of the 10 instances.

The results show lower deviations from the upper bound for the TS algorithm that uses the evolved DR and the evolved SDR, especially for orders with high τ and R. This is most likely due to the evolved rules for DRs and SDRs being much more effective than the $RLR1$ priority based DR. Starting from a better initial solution with the evolved rules than $RLR1$ may mean that the TS algorithm has potential to find a better final solution. Therefore, we can conclude that combining TS with rules evolved from GP is an effective method of generating good solutions for OAS.

In addition, we can see that CPSO and DPSO are very competitive as compared to the default TS algorithm. From the table, we can see that it has a lower deviation from the upper bound for high τ and R values and having a comparable deviation for lower τ and R values. In addition, CPSO algorithms are marginally faster than TS algorithm with the local search, with the CPSO algorithm being around 1.3 times faster in terms of average computation time compared to the TS algorithm. This shows that PSO is a very viable method of optimisation for OAS problems.

However, contrary to our expectations, CPSO and DPSO that use evolved DR and SDR to generate the initial global best position performs *worse* than CPSO and DPSO without the use of DRs or SDRs for high τ and R values. One possible reason is that the corresponding solution for the global best position potentially has too high fitness value, which means that particles will prematurely converge to the global best position and get stuck in a local optima. On the other hand, in the default PSO the global best position would shift around quite a bit as new solutions are found, allowing the particles to wander around more. For low τ and R values this is not an issue, since the solutions found by the evolved DRs are already very close to the upper bound on the problem instance. However, since OAS problem instances with high τ and R values are more difficult to find good solutions for, the default PSO that changes global best position often can be better than having DR augmented PSO that has a good initial global best position.

4.2   Average Deviation Evaluation

The algorithms are also tested on their stability by analysing the *average* result for each problem instance. TS, CPSO and DPSO are evaluated 30 times for each problem instance. The results of this is shown in Table 2, where $\bar{x} \pm s$ represents the mean and standard deviation of the deviations over the 10 instances. The version of meta-heuristics TS, CPSO and DPSO with the best average deviation under 5% significance level is highlighted in the table, and the rule with the best deviation under 5% significance level of all other rules for a particular problem subset is marked with a † in the table.

For the results of the *average* deviation evaluation, we can see that TS-SDR outperforms the other algorithms for subsets of low to mid τ and R values. This shows that TS is a fairly robust algorithm and can consistently output good results for various situations. However, for subsets of high τ and R values, CPSO outperforms every other algorithm. This means it is likely that swarm intelligence techniques are effective on difficult and volatile OAS problems. This is further exemplified by the fact that the ABC algorithm proposed by Lin and Ying [5] performs very well for problem subsets of high τ and R values.

Similar to the results of the best deviation evaluation (Section 4.1), CPSO and DPSO augmented with evolved DRs and SDRs do not perform as well as the basic CPSO and DPSO algorithms for high τ and R values. However, there are some promising results for problem subsets of low τ and R values, as CPSO-SDR and DPSO-SDR perform slightly better for low to mid τ and R values. With adjustments to how the evolved DRs and SDRs are incorporated into the PSO algorithm, it is likely to construct new PSO-DR hybrid model that is generally improved for all types of OAS problem instances.

Table 2. Evaluation of modified TS and PSO algorithm with various initial solutions using *average* deviations

τ R	TS	TS-DR	TS-SDR	CPSO	CPSO-DR	CPSO-SDR	DPSO	DPSO-DR	DPSO-SDR
0.10.1	4.1±0.2	2.1±0.1	2.1±0.1	3.4±0.2	2.6±0.0	2.8±0.1	3.4±0.2	2.6±0.0	2.8±0.1
0.3	3.5±0.3	1.8±0.1	1.8±0.1	2.6±0.1	2.3±0.1	2.2±0.1	2.6±0.1	2.3±0.1	2.2±0.1
0.5	2.2±0.3	1.3±0.1	1.1±0.1†	1.5±0.1	1.6±0.1	1.5±0.1	1.5±0.1	1.6±0.1	1.4±0.1
0.7	1.3±0.2	0.5±0.1	0.5±0.1	0.7±0.1	0.7±0.1	0.6±0.0	0.7±0.1	0.7±0.1	0.5±0.1
0.9	0.5±0.1	0.3±0.0	0.2±0.1†	0.2±0.1	0.3±0.0	0.3±0.1	0.2±0.0	0.3±0.0	0.3±0.0
0.30.1	4.9±0.3	2.9±0.1	2.5±0.1†	4.0±0.2	3.3±0.0	3.3±0.1	4.0±0.2	3.3±0.0	3.3±0.1
0.3	4.8±0.3	3.5±0.1	2.6±0.1†	3.9±0.2	4.1±0.1	3.2±0.1	3.9±0.2	4.1±0.1	3.2±0.1
0.5	4.2±0.3	2.6±0.1	2.1±0.1†	3.6±0.2	3.2±0.1	2.6±0.1	3.6±0.1	3.3±0.1	2.6±0.1
0.7	3.1±0.2	1.8±0.1	1.1±0.1†	2.3±0.2	2.3±0.1	1.6±0.1	2.3±0.2	2.3±0.1	1.6±0.1
0.9	2.2±0.2	1.1±0.1	0.7±0.1†	2.0±0.1	1.6±0.1	1.2±0.0	1.9±0.1	1.7±0.1	1.1±0.1
0.50.1	6.9±0.3	4.7±0.2	4.4±0.2†	6.1±0.2	6.4±0.1	6.0±0.2	6.1±0.2	6.4±0.1	6.0±0.2
0.3	7.0±0.4	4.8±0.1	4.0±0.1†	5.9±0.3	6.2±0.0	5.0±0.1	5.9±0.2	6.2±0.0	5.0±0.1
0.5	6.6±0.4	4.9±0.1	3.8±0.2†	6.0±0.3	6.1±0.1	4.7±0.1	6.0±0.2	6.1±0.1	4.7±0.1
0.7	5.3±0.3	3.4±0.1	2.5±0.1†	4.9±0.2	4.4±0.0	3.1±0.1	4.9±0.3	4.4±0.0	3.1±0.1
0.9	4.7±0.3	2.9±0.1	2.0±0.1†	4.0±0.2	4.4±0.1	2.8±0.1	4.1±0.3	4.4±0.1	2.8±0.1
0.70.1	8.2±0.3	5.8±0.2	5.6±0.2†	7.6±0.3	7.2±0.1	7.1±0.2	7.6±0.3	7.2±0.1	7.1±0.2
0.3	9.0±0.4	6.8±0.2	6.3±0.2†	8.4±0.2	8.4±0.2	7.6±0.1	8.3±0.3	8.4±0.1	7.6±0.2
0.5	10.1±0.5	7.5±0.2	7.2±0.2†	8.9±0.3	9.1±0.2	8.9±0.2	8.8±0.3	9.1±0.2	8.8±0.2
0.7	11.9±0.6	9.5±0.2	7.8±0.2†	9.8±0.4	10.9±0.3	9.5±0.2	9.9±0.2	10.9±0.2	9.5±0.2
0.9	12.7±0.6	9.6±0.3	7.9±0.2†	9.6±0.3	10.6±0.3	9.5±0.2	9.6±0.3	10.6±0.3	9.5±0.2
0.90.1	11.8±0.5	10.3±0.3	9.7±0.2†	10.4±0.3	11.6±0.4	11.3±0.2	10.6±0.3	11.5±0.2	11.1±0.2
0.3	16.7±0.6	14.4±0.3	13.3±0.2	13.1±0.3†	13.6±0.3	15.3±0.2	13.2±0.2	13.6±0.4	13.8±0.3
0.5	18.6±0.6	16.7±0.3	15.3±0.3	14.8±0.3†	15.3±0.2	17.7±0.2	14.8±0.3	15.3±0.4	16.1±0.3
0.7	18.5±0.7	16.8±0.3	15.0±0.3	14.9±0.3	15.4±0.4	17.2±0.2	14.8±0.3†	15.4±0.3	15.8±0.3
0.9	18.1±0.5	15.6±0.3	14.5±0.2	14.6±0.3	15.1±0.3	17.0±0.2	14.6±0.3	15.1±0.4	15.2±0.2

* $\bar{x} \pm s$ represents the mean and standard deviation of the rule over the 10 instances.
** Highlighted cell means the variant of TS/CPSO/DPSO algorithm is better than the other variants for the $\langle \tau, R \rangle$ under Z-test of 5% significance level.
*** Cells marked with a cross means the corresponding algorithm is better than the other algorithms for the $\langle \tau, R \rangle$ under Z-test of 5% significance level.

5 Conclusions

Overall, the results show improvements to an existing TS algorithm by the use of GP evolved rules generated from GP. The hybrid system outperforms the standard TS proposed by Cesaret et al. [3] that uses the $RLR1$ heuristic to generate the initial solution. Furthermore, it is possible to evolve heuristics for problems other than OAS to generate initial solutions for TS and other meta-heuristic algorithms.

In addition, it is shown that two different PSO approaches can be applied to OAS in a single machine environment and that these are competitive with state of art meta-heuristics such as TS [3]. However, the current method of incorporating evolved heuristics into PSO is still not as good as using the algorithms by themselves. For future work, it will be interesting to try different approaches to incorporating an evolved heuristic into PSO.

References

1. Oğuz, C., Salman, F.S., Yalçın, Z.B.: Order acceptance and scheduling decisions in make-to-order systems. International Journal of Production Economics 125(1), 200–211 (2010)
2. Ghosh, J.B.: Job selection in a heavily loaded shop. Computers & Operations Research 24(2), 141–145 (1997)
3. Cesaret, B., Oğuz, C., Salman, F.S.: A tabu search algorithm for order acceptance and scheduling. Computers & Operations Research 39(6), 1197–1205 (2012)
4. Rom, W.O., Slotnick, S.A.: Order acceptance using genetic algorithms. Computers & Operations Research 36(6), 1758–1767 (2009)
5. Lin, S., Ying, K.: Increasing the total net revenue for single machine order acceptance and scheduling problems using an artificial bee colony algorithm. Journal of the Operational Research Society (2), 293–311 (2012)
6. Koza, J.R.: Genetic Programming: On the Programming of Computers by Means of Natural Selection. MIT Press (1992)
7. Park, J., Nguyen, S., Zhang, M., Johnston, M.: Genetic programming for order acceptance and scheduling. In: Proceedings of IEEE Congress on Evolutionary Computation (CEC), vol. 7831, pp. 1005–1012 (2013)
8. Park, J., Nguyen, S., Johnston, M., Zhang, M.: Evolving stochastic dispatching rules for order acceptance and scheduling via genetic programming. In: Cranefield, S., Nayak, A. (eds.) AI 2013. LNCS, vol. 8272, pp. 478–489. Springer, Heidelberg (2013)
9. Slotnick, S.A.: Order acceptance and scheduling: A taxonomy and review. European Journal of Operational Research 212(1), 1–11 (2011)
10. Nguyen, S., Zhang, M., Johnston, M., Tan, K.C.: Learning reusable initial solutions for multi-objective order acceptance and scheduling problems with genetic programming. In: Krawiec, K., Moraglio, A., Hu, T., Etaner-Uyar, A.Ş., Hu, B. (eds.) EuroGP 2013. LNCS, vol. 7831, pp. 157–168. Springer, Heidelberg (2013)
11. Kennedy, J., Eberhart, R.: Particle swarm optimization. In: IEEE International Conference on Neural Networks, vol. 4, pp. 1942–1948. IEEE (1995)
12. Rameshkumar, K., Suresh, R., Mohanasundaram, K.: Discrete particle swarm optimization (dpso) algorithm for permutation flowshop scheduling to minimize makespan. In: Wang, L., Chen, K., S. Ong, Y. (eds.) ICNC 2005. LNCS, vol. 3612, pp. 572–581. Springer, Heidelberg (2005)
13. Ruiz, R., Sttzle, T.: A simple and effective iterated greedy algorithm for the permutation flowshop scheduling problem. European Journal of Operational Research 177(3), 2033–2049 (2007)
14. Eberhart, R.C., Shi, Y.: Particle swarm optimization: developments, applications and resources. In: Proceedings of 2001 Congress on Evolutionary Computation, vol. 1, pp. 81–86. IEEE (2001)

Application of Adaptive Streaming Technology in Remotely Driven Electric Vehicles

Kyaw Ko-Ko-Htet[1], Arun-Shankar Narayanan[1] Tan Kok-Kiong[1],
and Chandran Nair[2]

[1] Department of Electrical and Computer Engineering, Faulty of Engineering,
National University of Singapore,
21 Lower Kent Ridge Road, 119077
{A0035642,A0035708,KKTan}@nus.edu.sg
[2] National Instruments Singapore (Pte) Ltd, No. 2 International Business Park
The Strategy Tower 1, #06-03, 609930
Chandran.Nair@ni.com

Abstract. The need for live video transfer across networks is increasingly crucial for modern day applications, ranging from video-based surveillance systems to disarming explosives in a minefield using robots. Some of these applications, for e.g., a tele-operated electric vehicle (EV) as focused in this paper, involve real-time video streaming across wireless networks, which is bandwidth-intensive and delay-sensitive at the same time. This necessitates the need for a mechanism to adapt the data rate in order to suit the network condition and to ensure hassle-free functioning of the intended application. This paper proposes an adaptive streaming mechanism which involves bandwidth estimation based on transmission feedback and systematic adaptation of frame rate and frame quality for a safe drive in a remotely driven EV. The algorithm is implemented in a physical tele-operated EV and experiments are carried out to verify its adaptation performance.

Keywords: Adaptive, Video Streaming, Remote Driving.

1 Introduction

Streaming live video over the internet has always been challenging due to the limited bandwidth, transmission delays and packet losses. Real-time video transmission typically has a minimum bandwidth requirement and is more sensitive to end-to-end delay as compared to other data transmissions due to its continuity requirement. Any potential loss of packets can make the presentation displeasing to the human eyes at the receiving end. Thus there are specific requirements to constraint the transmission delay as well as packet loss to a minimum level in video transmission applications [1]. Furthermore, the internet traffic load condition varies drastically over time, which is detrimental to video transmission. Thus, it is a major challenge to design an efficient video delivery system that can both maximize users' perceived quality of service while achieving high resource utilization of the Internet [2][3].

G. Dick et al. (Eds.): SEAL 2014, LNCS 8886, pp. 735–746, 2014.
© Springer International Publishing Switzerland 2014

Although challenges are prevalent, the need for live video transfer across networks is increasingly crucial for modern day applications[4]. Examples include those used in disaster response systems or real time telemetry of video imagery using unmanned aerial vehicles [5]. Many of these applications require wireless transfer of video that makes the development even more complex due to the dynamic nature of the wireless network environments which can provide only limited, time-varying quality of service for delay-sensitive, bandwidth-intensive, and loss-tolerant multimedia applications [6]. In such regards, achieving a good quality video transmission with the limited network conditions is a key.

Some of the applications mentioned above are delay-sensitive while some others are more tolerant towards minor delays. Due to the fluctuating nature of wireless networks, it is important to have some mechanism to adapt the data rate in order to suit the network condition or else the network might be underutilized or may even potentially cause the streaming to cease, which is detrimental to the application. Many researchers have been focusing on improving the quality of video delivery across the internet by implementing novel algorithms for optimal utilisation of effective bandwidth. Wu et al. summarised the earlier approaches to transfer real-time video over the internet in [1]. More recently, Song and Golubchik devised an adaptive video streaming solution to deliver better video quality among other similar technologies [7]. However, Song gave emphasis to video quality rather than the efficient usage of bandwidth. There are also mainstream products such as Microsoft IIS Smooth Streaming, Adobe Flash Dynamic Streaming and Apple HTTP Adaptive Bitrate Streaming. However, all of these products are intended for use in broadcast for mass entertainment purposes [8].

This paper discusses the application of adaptive streaming technology, introduced in the earlier works [8]-[9] by the same authors, in remotely driven electric vehicles (EV). Earlier works were focusing on a telemedicine application using a low-bandwidth videoconferencing solution and here, the authors introduce the technology in EVs. It is a solution proposed for last mile transportation problem to provide a cost-effective and sustainable door-to-door transportation using a shared fleet of vehicles which will be driven back to a common base by a remote driver located in a central station using audio-visual feedback transmitted from the vehicle [10]. The algorithm used for data rate adaptation in the dynamic bandwidth condition is explained in the next section followed by more details on the use of this algorithm in the last mile transport system using the EVs.

2   Adaptive Video Streaming Algorithm for Remote Driving

Remote driving involves long distance wireless communication in which the available bandwidth is limited and highly dynamic. In this system, a driver located in a remote location should be able to manoeuvre the EV from random localities to a central hub from where the EVs will be made available to the public. The remote driver should also be able to bring the EV from the central hub to the passengers' locations as well upon request [10]. The entire operation will be based on audio-visual feedback from the cameras placed on the vehicle, transmitted through wireless medium in real-time. Due to the dynamics in wireless networks, available bandwidth can be highly volatile and some transmitted data packets might be lost. Although Transmission Control Protocol (TCP) can provide re-transmission of lost data, it significantly increases transmission latency up to a few seconds. Since remote driving is a time-sensitive application, dropping data is preferred compared to waiting for delayed data. Therefore, User Datagram Protocol (UDP), which can provide a higher data throughput for transmission of large amount data with minimal latency and timely transmission of the most recent data [11] [12], is more suitable for this application. The main drawback of UDP is lack of network congestion control and this usually yields packet losses. Additional mechanism is required to be implemented at the application layer in order to avoid sending more data than the allowed bandwidth, and thus lessening network congestion and massive packet losses.

In this application, there is a bi-directional parallel communication in which video and sensor information are sent from EV to RS while control commands are being sent from RS to EV as shown in Fig. 1. This two-way communication can be utilized as a continuous feedback in monitoring network congestion, packet drop rate and bandwidth condition.

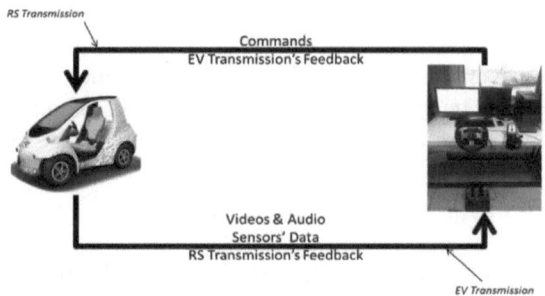

Fig. 1. Bi-directional parallel communication between EV and RS

Motion Joint Photographic Experts Group (MJPEG) compression algorithm is used for this remote driving application. MJPEG has advantages of clear individual images, fast image stream recovery in the event of packet loss, low latency, less processing overhead and better live viewing [13].

Typical frame rate provided by a standard digital video camera is 25-30 frames per second (FPS). For MJPEG compression technique, doubling the FPS leads to doubling the bandwidth usage and thus, it is essential to stream the video at the right FPS count which is sufficient enough for the remote driver to safely manoeuvre the vehicle. The right value of FPS is dependent on the distance travelled by the vehicle between each frame. At a low vehicle speed, low fps is good enough, while high fps is essential at a high speed, since the distance covered is directly proportional to the vehicle speed. As per previous studies conducted on the effect of lower frame rates on human perception, it is pointed out that a frame rate of 10-15 is sufficient for almost all the different tasks to be performed unless the requirement is for sophisticated tasks; such as speech reading based on video frames. The study also pointed out that the frame rate reduction from 25 to 15 is often perceptually similar in appearance to the viewers [14].

The proposed video streaming algorithm consists of three steps. The first step is adapting the frame rate to the value in between 10 to 15 FPS, which are sufficient for the human sight to create the sensation of visual continuity [15]. The second step is estimating available bandwidth by using the continuous feedback from RS. The third step is computing the compression parameter, Q value, such that the transmission rate is within the available bandwidth to avoid massive packet losses.

2.1   Frame Rate Adaptation

As mentioned earlier, the minimal required frame rate should be systematically identified such that the safety of the remote driving operation is not compromised. Consider the situation in which the remote vehicle is travelling at a speed (v) while a camera on the vehicle is capturing at a constant frame rate (f_r). The distance (d_1) which EV travels between two consecutive frames is given by:

$$d_1 = \frac{v}{f_r} \tag{1}$$

If the remote driver has to clearly see up to a distance (d) from the camera, the value of d should be greater than d_1, as the driver should be able to see a point before the vehicle reaches there. The above condition can be written as:

$$d > d_1 = \frac{v}{f_r} \tag{2}$$

Eqn. 2 shows that, for maintaining a constant value for d, higher vehicle speed requires higher frame rate. Thus, the minimal required frame rate (f_r), to satisfy this condition, can be written as:

$$f_r > \frac{v}{d} \tag{3}$$

Additionally, the remote driver requires certain number (n) of frames in order to make vision perception in terms of the vehicle speed. This number has to be based on the vehicle speed as well as the viewable distance from the camera image, typically with a minimum of two to derive a speed estimate. Nonetheless, the frame rate should not be less than 10 to maintain visual continuity. Thus, the required frame rate ($\acute{f}_r$) for perception and judgement of speed can be written as:

$$\acute{f}_r = max(10, \frac{v}{d}n) \tag{4}$$

Eqn. 4 ensures that the remote driver receives n numbers of frames which consist of a particular position/object before the EV reaches that point and this provides enough frames for the remote driver to react appropriately in case of any potential road hazards. In general, low speed operations, such as parking and passing through narrow passages, require high quality image to have better view of obstacles and surrounding. Therefore, above equation can provide a systematic approach in defining a minimal safe frame rate.

The viewable distance (d) from camera is dependent on its view angle as well as the vehicle's turning angle. Fig. 2 shows a four-wheel vehicle turning at a radius of r.

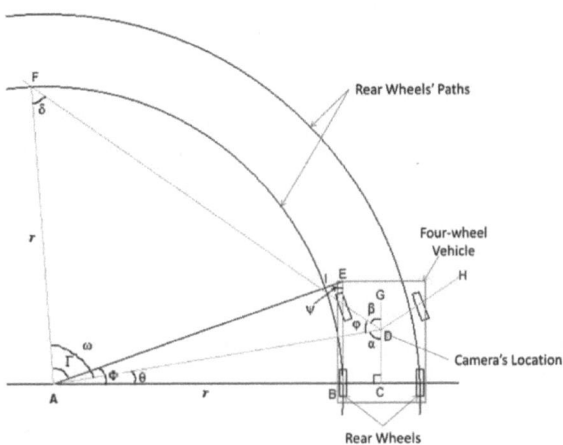

Fig. 2. 2-D view of a four-wheel vehicle turning at a radius of r

Here, the camera is located at the point 'D' and its viewing angle is given by the angle 'FDH'. This angle can cover the vehicle's displacement (s_{FI}) between the point 'I' and the point 'F' which is given by:

$$s_{FI} = r\Gamma \tag{5}$$

where, $\Gamma = (\frac{\pi}{2} + \beta - tan^{-1}\frac{l_{BE}}{r} - sin^{-1}\frac{l_{AD}\sin\varphi}{r})$, the angle formed amongst 'F', 'A', and 'I', $l_{AD} = \sqrt{l_{CD}^2 + (r + l_{BC})^2}$, $\varphi = \frac{\pi}{2} - \beta + tan^{-1}\frac{l_{CD}}{r+l_{BE}}$, $l_{BE} = $ the distance between the centre of the left rear wheel 'B' and the front of the vehicle 'E', $r = $ the vehicle circling radius, $l_{CD} = $ the distance between the centre of the two rear wheels 'C' and the camera location 'D', $\beta = $ half of the viewing angle of the camera, and $l_{BC} = $ the distance between the centre of the left rear wheel 'B' and the centre of the two rear wheels 'C'.

As there is a maximum distance, d_{max}, that the remote camera can see clearly, Eqn. 4 can be rewritten as:

$$\hat{f}_r = max(10, \frac{v}{min(d_{max}, S_{FI})}n) \tag{6}$$

2.2 Estimating Available Bandwidth

In the remote driving, the available data bandwidth is limited and dynamic at the same time since wireless medium is utilised for the data transfer. In order to monitor the available data bandwidth, bi-directional parallel communication (Fig. 1) between EV and RS can be utilized as a continuous feedback. The RS first monitor frame drop rate (FDR) and received data rate (RDR). FDR is used to distinguish over-utilization and under-utilization of the available data bandwidth, and RDR is used to measure the existing BW utilization. Upon detecting frame drops, the BW will be assumed to be over-utilized and the received data rate will be estimated as the maximum available bandwidth. Likewise, when all frames are being received (i.e. $FDR = 0$), BW will be assumed to be under-utilized and the available bandwidth is estimated as received data rate plus Additional Increment Value (AIV). The Estimated Bandwidth (EBW) is computed as shown in Eqn. 7.

$$EBW = \begin{cases} RDR & \text{if } FDR > 0 \\ RDR + AIV & \text{otherwise} \end{cases} \tag{7}$$

2.3 Computing Compression Parameter

The estimated BW (EBW) is periodically fed back to the EV to calculate suitable frame size (f_s) based on the suitable frame rate (Eqn. 6) as shown in Eqn. 8.

$$f_s = \frac{EBW}{f_r} \tag{8}$$

The compression parameter, Q value, is selected such that desired f_s is achieved for each frame. Based on the relationship between Q value and f_s for 10 different images, captured from the same camera mounted on the EV while operated under real-road environment, f_s can be estimated as shown in Eqn. 9. The overall flowchart for this adaptation algorithm is shown in Fig. 3.

$$f_s = 453.51Q + 5449.47 \tag{9}$$

3   Experiment and Verification

The above algorithm is implemented in the existing remote driving system and the details on the implementation are provided in this section. To verify the performance of the algorithm, three different experiments have been carried out with; 1) under-utilisation of BW, 2) over-utilisation of BW, and 3) adaptive algorithm.

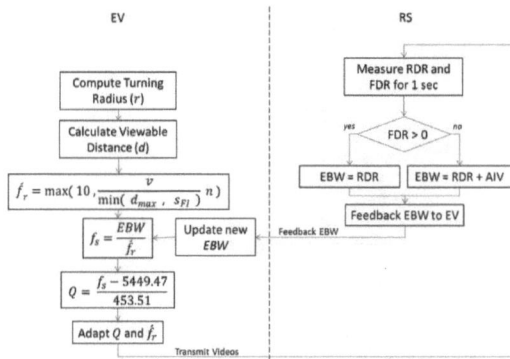

Fig. 3. Overall Flowchart for Adaptation Algorithm

3.1 Hardware Setup

The remote driving system consists of three major sub-systems: EV, Wireless Network and RS. The overall experiment setup is shown in Fig. 4. Cameras are mounted on EV and are connected to the PXI controller (PCI extensions for Instrumentation) [16]. Videos from these cameras will be transmitted from the controller to the RS via a wireless network. The remote driver will be able to control the EV based on the video streams received at the RS.

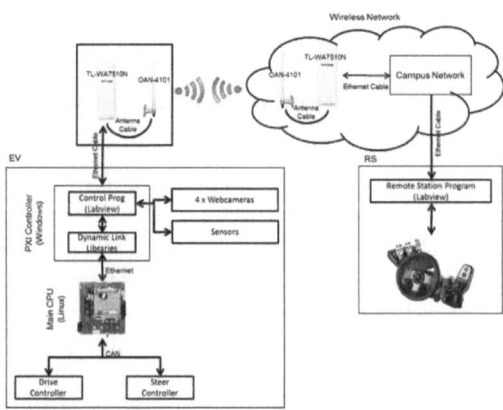

Fig. 4. Block Diagram of Overall System Setup

Electric Vehicle. The EV used is a single-seater vehicle manufactured by Toyota Tsusho and the major components of the EV and its dimensions are shown in Fig. 5.

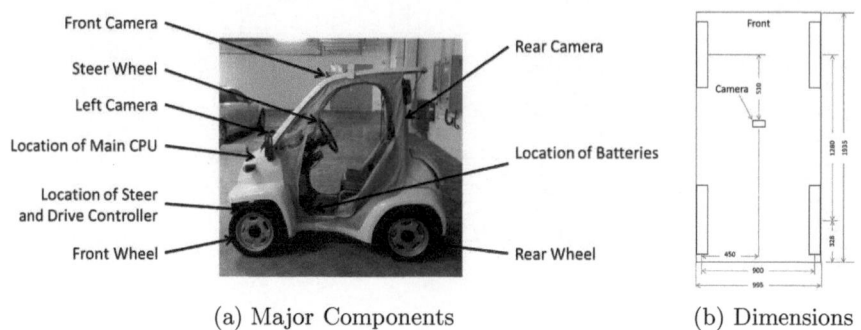

(a) Major Components   (b) Dimensions

Fig. 5. Electric Vehicle

Wireless Communication. As shown in Fig. 4, a pair of TP-link TL-WA7510N with external antenna OAN-4101 is used; one as the wireless access point (AP) and another as the wireless transducer. This network typically can cover up to a radius of 500m. Moreover, the remote station can be located at any place within the campus network.

Remote Station. Fig. 1 shows the remote driving station arrangement in which, Logitech G27 racing wheel assembly is connected to the station and consisting of steering wheel, steel paddle shifter module, and accelerator, brake and clutch pedals. It has 900-degrees steering angle. The remote driver uses it to provide drive commands to manoeuvre the EV.

3.2 Algorithm Implementation Based on Electric Vehicle's Dimension

The bicycle model, which is a common simplification of an Ackermann steered vehicle [17] used for geometric path tracking, is utilized to estimate rotation radius of the EV from the steering angle. The vehicle's trajectory is modelled as a bicycle model in which its steering angle is parallel to that of the vehicle, its rear wheel exists exactly in the middle point between two rear wheels of the vehicle, and its front wheel exists exactly in the middle point between two front wheels as shown in Fig. 6a. Its geometric model [17] is shown in Fig. 6b.

Based on the model, the geometric relationship between rotation radius R and steering angle δ can be written as:

$$R = \frac{L}{tan\delta} \qquad (10)$$

where L is the distance between the front and the rear wheels, 1.28m in this case. With the information from fig. 6, r in Eqn. 5 can be calculated by:

$$r = R - l_{BC} \qquad (11)$$

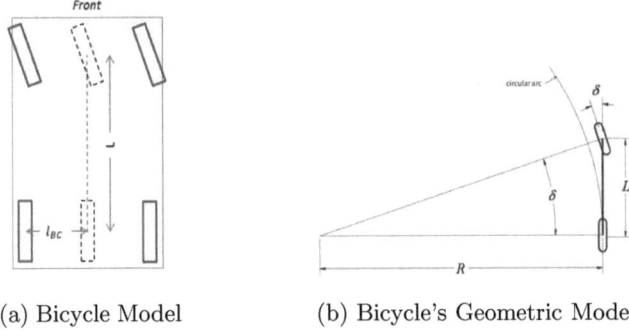

(a) Bicycle Model   (b) Bicycle's Geometric Model

Fig. 6. Geometric Modeling

where $l_{BC} = 0.45$m in this case. The maximum viewable distance, d_{max}, from the vehicle based on the camera image is empirically measured to be 20m. The resultant relationship between steering angle and viewable distance of vehicle path, which is the denominator part of Eqn. 6, $min(d_{max}, s_{FI})$, is shown in Fig. 7. When the steering angle is small, i.e. when the EV moves on a straight road, the whole 20m is visible in the frame. However, as the angle gets bigger, i.e. when the EV is passing through a bent road, the viewing distance gets shorter as a result as shown in Fig. 7.

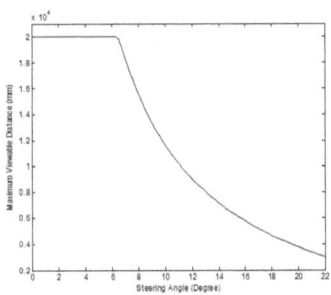

Fig. 7. Steering Angle Vs Viewable Distance of Vehicle's Path

3.3   Experiment Results

Bandwidth Under-Utilization Scenario. In this experiment, frame rate and Q value are set at 10 fps and 15 respectively in order to keep low bandwidth utilization during the remote driving at the test circuit. The average bandwidth utilization is at 125 kilo bytes per second (kBps) and the resulting measurements are shown in Fig. 8a. In the figure, received data rate can be seen varying, possibly due to the difference in frame size after going through the compression

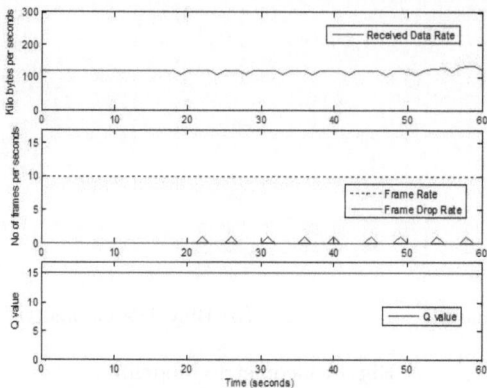

(a) Under-Utilized BW Condition

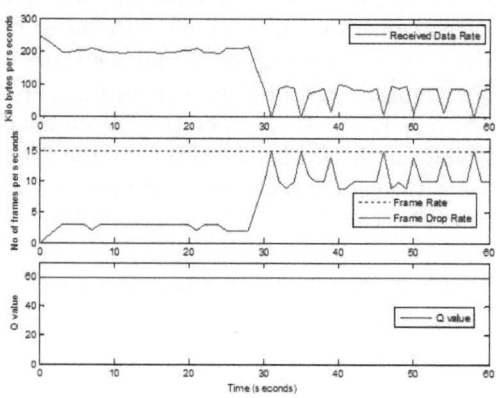

(b) Over-Utilized BW Condition

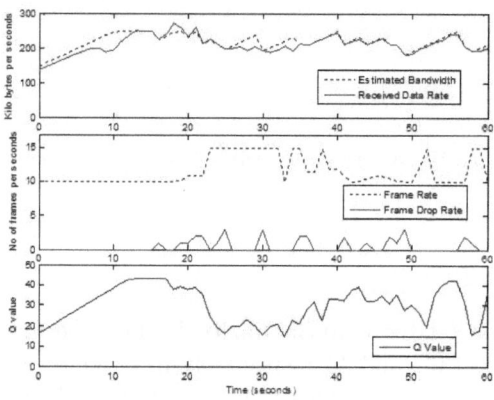

(c) Adaptive Algorithm

Fig. 8. Experiment Results

algorithm as well as intermittent frame dropping. Due to low frame dropping and less congestion, the bandwidth can be perceived as under-utilized. However, the Q-value in this case is too low to deliver good quality images for safely manoeuvring the EV through congested road conditions and thus not suitable in all situations.

Bandwidth Over-Utilization Scenario. In this experiment, both frame rate and Q-value are set at 15 fps and 60 respectively in order to simulate over bandwidth utilization scenario. The resulting measurements are shown in Fig. 8b. During second half of the experiment, frame drop rate becomes close to frame transmission rate and thus, only a small percentage of frames are received by the receiver. This arises due to the transmission bandwidth exceeding the available wireless bandwidth and such a situation is highly undesirable, especially for time critical applications such as remote driving. However, it cannot be avoided in wireless transmission, where available bandwidth is highly volatile, and this demands the use of dynamic bandwidth adaptation with changing conditions.

Dynamic Bandwidth Utilization with Adaptive Algorithm. In this experiment, both frame rate and Q value are adapted based on the described algorithm. The resulting measurements are shown in below figure. The frame rate is adjusted based on EV's speed and viewable distance from the camera. Available bandwidth is estimated based on the frame drop rate and Q value is adapted such that bandwidth utilization is approximately equal to the available bandwidth. In this way, frame drop rate is controlled to avoid the over bandwidth utilized situation while optimally utilising the available network bandwidth, as shown in Fig. 8c. The received data rate is able to adapt to the estimated network bandwidth as well as Q value is seen to adapt accordingly to accommodate frame rate variation due to variation in speed and steering angle while maintaining a low level of frame drop rate.

4   Conclusion

In this paper, adaptive video streaming algorithm is discussed in which the frame rate is adapted based on vehicle travel speed and steering angle. Bandwidth estimation is carried out by using frame drop rate via continuous feedback loop from RS. Based on the estimated bandwidth, compression parameter 'Q' is adjusted to maintain bandwidth utilization just below the available bandwidth, thus optimally utilising the existing network conditions. Implementation details are provided and experiments are carried out to verify its performance. Experiment results prove that, this algorithm is able to adapt the video data according to the dynamics of wireless network efficiently, and thus provide a safe solution for achieving acceptable performance in remote driving applications. This algorithm performance can be further improved by integrating genetic algorithms in 'Q' estimation.

References

1. Wu, D., Hou, Y.T., Zhang, Y.-Q.: Transporting real-time video over the internet: Challenges and approaches. Proceedings of the IEEE 88(12), 1855–1877 (2000)
2. Bolot, J.-C., Turletti, T.: Experience with control mechanisms for packet video in the internet. ACM SIGCOMM Computer Communication Review 28(1), 4–15 (1998)
3. Wu, D., Hou, Y.T., Zhu, W., Lee, H.-J., Chiang, T., Zhang, Y.-Q., Chao, H.J.: On end-to-end architecture for transporting mpeg-4 video over the internet. IEEE Transactions on Circuits and Systems for Video Technology 10(6), 923–941 (2000)
4. Regazzoni, C.S., Fabri, G., Vernazza, G.: Advanced video-based surveillance systems. Springer (1999)
5. Rodriguez, P.A., Geckle, W.J., Barton, J.D., Samsundar, J., Gao, T., Brown, M.Z., Martin, S.R.: An emergency response uav surveillance system. In: AMIA Annual Symposium Proceedings, vol. 2006, p. 1078. American Medical Informatics Association (2006)
6. Van Der Schaar, M., et al.: Cross-layer wireless multimedia transmission: challenges, principles, and new paradigms. IEEE Wireless Communications 12(4), 50–58 (2005)
7. Song, K., Golubchik, L.: Towards user-oriented live video streaming. In: 2010 Proceedings of 19th International Conference on Computer Communications and Networks (ICCCN), pp. 1–6. IEEE (2010)
8. Arun, S.N., Lam, W.C., Tan, K.K.: Innovative solution for a telemedicine application. In: 2012 International Conference for Internet Technology And Secured Transactions, pp. 778–783. IEEE (2012)
9. Narayanan, A.S., Kiong, T.K.: Video conferencing solution for medical applications in low-bandwidth networks. In: 2013 Australasian Telecommunication Networks and Applications Conference (ATNAC), pp. 195–200. IEEE (2013)
10. Kyaw, K.K.H., Narayanan, A.S., Kiong, T.K.: Mitigation of vehicle distribution in an ev sharing scheme for last mile transportation (submitted for publication)
11. Shue, C., Haggerty, W., Dobbins, K.: Osi connectionless transport services on top of udp: Version 1. Technical report, RFC 1240 (1991), http://www.ietf.org/rfc/rfc1240.txt
12. Postel, J.: User datagram protocol. Isi (1980)
13. So-In, C.: Understanding jpeg and applications (2004)
14. Chen, J.Y., Thropp, J.E.: Review of low frame rate effects on human performance. IEEE Transactions on Systems, Man and Cybernetics, Part A: Systems and Humans 37(6), 1063–1076 (2007)
15. Read, P., Meyer, M.-P.: Restoration of motion picture film. Butterworth-Heinemann (2000)
16. National Instrument. PXI Platform - National Instrument, http://www.ni.com/pxi/ (accessed June 6, 2014)
17. Snider, J.M.: Automatic steering methods for autonomous automobile path tracking. Robotics Institute, Pittsburgh, PA, Tech. Rep. CMU-RITR-09-08 (2009)

Optimising Wi-Fi Installations
Using a Multi-Objective Evolutionary Algorithm

Lyndon While* and Chris McDonald

Computer Science & Software Engineering, The University of Western Australia
{lyndon.while,chris.mcdonald}@uwa.edu.au

Abstract. We describe a multi-objective evolutionary algorithm that designs multi-device Wi-Fi installations optimised for three criteria: minimised cost, maximised coverage, and minimised service refusal. At the heart of the system is a detailed simulator for Wi-Fi installations, and a simple parallel evaluation scheme to allow these simulations to be performed in reasonable time. We show that the algorithm can derive good installations for two real-world maps requiring respectively around ten and fifty access points. The fine-grained connectivity and mobility models used in the simulator allow us to derive results that are more realistic than do previous methods.

Keywords: evolutionary algorithms, multi-objective optimisation, Wi-Fi.

1 Introduction

A Wi-Fi network[1] is any wireless local area network that corresponds to the IEEE's 802.11 standards[2]. Wi-Fi installations have become increasingly ubiquitous since the technology was originally patented by the Australian CSIRO in the 1990s[3]: MarketsandMarkets state that the global Wi-Fi market is worth US$40 billion in 2013, and they forecast that it will reach US$93 billion by 2018[4]. It has become common for cities around the world to offer city-wide (or CBD-wide) municipal Wi-Fi service: [5] lists over a hundred cities that provide such service, and many more that are planning to.

The limited range and capacity of individual Wi-Fi devices means that all significant Wi-Fi installations are provided using multiple access points (APs) working together. However, designing such an installation for a given area is a difficult task for several reasons:

- suitable device-mounting points may be hard to find or sparsely distributed;
- the effective range of individual devices is affected by many factors, such as the walls of buildings, trees, and other nearby objects;
- patterns of access vary during any given time period, due to the varying uses of different parts of the area;
- users' mobility makes it difficult to decide where devices should be installed.

---

* Corresponding author.

G. Dick et al. (Eds.): SEAL 2014, LNCS 8886, pp. 747–759, 2014.

[6] discusses the major issues involved, and the design principles commonly used to address them.

The principal contribution of this paper is a new approach to designing the layout of Wi-Fi installations that utilise multiple APs to provide service. We use a multi-objective evolutionary algorithm (MOEA) to select a set of AP locations that optimises the coverage and service provided by a Wi-Fi installation, whilst simultaneously minimising its cost. Experiments show that the algorithm is able to derive a range of excellent layouts for two challenging real-world maps. The algorithm uses detailed simulations as part of its fitness calculations, which allows us to generate more-useful layouts than previous methods; and it incorporates a simple farming scheme for performing fitness calculations in parallel, so it is scalable to much larger problem instances. Also the algorithm is easily adaptable to optimise for other objectives, e.g. the robustness of an installation wrt equipment failure.

The rest of the paper is structured as follows. Section 2 describes previous approaches to designing Wi-Fi installations, and necessary background in multi-objective optimisation. Section 3 describes the design of our multi-objective EA, and of the network simulator used in its fitness calculations. Section 4 describes the results achieved by our algorithm, and Section 5 concludes the paper.

2   Background

This section describes previous approaches to designing Wi-Fi installations, with particular emphasis on attempts to optimise for physical coverage or temporal connectivity. It also describes basic concepts and terminology in multi-objective optimisation.

2.1   Previous Work

The problem of optimally locating finite resources is well studied. Most work on Wi-Fi placement reports on *indoor* set-ups where most clients are stationary, and where mobile clients have little impact on the network design. Many packages enable a building's interior to be defined, and predicted Wi-Fi signal heatmaps to be rendered. The primary exception is Zirari *et al.*[7], who focus on positioning indoor APs in shopping malls and public buildings to provide not just strong network access, but also accurate localisation and tracking of mobile clients.

Wang *et al.*[8] employ the statistical mobility patterns of network users at a university college, in an attempt to provide continuous Wi-Fi connectivity. However, their mobility models were derived by studying an *already deployed* network, and it is unclear if the mobility patterns apply to other environments. In contrast, Zheng *et al.*[9] employ an unspecified approximation algorithm to balance the bandwidth and connectivity demands of users traveling in vehicles and using a roadside Wi-Fi network, to determine worst-case guarantees.

Wang and Kao[10] apply a genetic algorithm (GA) to the placement of APs with heterogeneous costs and capacities, but they assume that client demand

on the network is uniform across an obstacle-free region, and do not model
mobile client movement. Wang and Chen[11] apply a GA to the determination
of locations and heights of antennæ in a cellular network, but their work also
assumes a known, fixed network demand. Agustin *et al.*[12] also apply a GA to
the design of city-wide Wi-Fi, but they restrict their work to static users and
they exploit pre-installed infrastructure to reduce costs.

Bulut and Szymanski[13] employ integer linear programming to position APs
for bandwidth offloading in a 3G cellular network. They employ real user mobility
traces to model the distribution and location of network traffic requests, and
report the percentage of offloading before the 3G network must be used.

The principal advantage that our work offers compared to these previous
works derives from the detailed simulations that we employ to generate realistic
solutions for mobile users.

2.2 Multi-objective Optimisation

In a multi-objective optimisation problem, potential solutions are assessed according to two or more independent quantities. The characteristic of good solutions is that improving in one objective can be achieved only by worsening in at
least one other objective. An algorithm for solving such problems returns a set
of solutions offering different trade-offs between the various objectives.

Consider a problem where a solution x is mapped to a fitness vector $\overline{f_x}$. x
dominates y iff $\overline{f_x}$ is at least as good as $\overline{f_y}$ in every objective, and is better in
at least one objective. x is *non-dominated* wrt a set X iff there is no solution in
X that dominates x. X is a *non-dominated set* iff every solution in X is non-
dominated wrt X. The set of fitness vectors corresponding to a non-dominated
set is a *non-dominated front*. x is *Pareto optimal* iff x is non-dominated wrt the
set of all possible solutions, and the *Pareto optimal set* is the set of all Pareto
optimal solutions. Multi-objective optimisation aims to find (or approximate)
this Pareto optimal set.

With multiple objectives there is only a partial order on solutions, which
causes problems for selection in an EA. The usual solution is to define a ranking
on solutions: one popular scheme[14] defines the *rank* of a solution x wrt a set
X to be the number of solutions in X that dominate x. Selection is then based
on ranks: a lower rank implies a better solution.

Precise definitions of all these terms can be found in [15].

3 Methodology

The key decisions required to apply evolutionary optimisation to a given problem
are the way that solutions are represented, and the design of the associated
variation operators; the choice of objectives to optimise for, and how they are
quantified; and the selection of various parameters and other components of
the algorithm. This section describes these parts of our algorithm, and also the
design, operation, and use of the network simulator in its fitness calculations.

3.1   Representation

We assume a homogeneous deployment where the same device is deployed at all APs (extra service can be provided by placing devices close together). This approach is commonly used by large organisations (such as our own university) that wish to minimise purchasing costs and to streamline operations and device replacement. Thus an installation is specified simply as a set of locations in the area in question. In most situations, most potential device locations are on the outside of buildings. However, the locations available for APs on a given building may be limited for many different reasons, e.g.:

- structural: a suitable device-mounting point must be available;
- architectural: some buildings are difficult to modify;
- regulatory: some buildings cannot be modified for legal reasons;
- aesthetic: the look of a building may be important;
- weather-related: the devices must be protected from the effects of weather;
- security-related: the devices must be safe from e.g. theft and vandalism.

Thus our algorithm assumes that the area in question has been pre-analysed and that a set of *locs* suitable locations has been identified for the installation of devices. These locations are denoted by the indices $0 \ldots locs - 1$, and a solution is represented simply as a non-repeating subset of these indices.

This linear representation has an obvious shortcoming. If Locations $k - 1$ and k are on one building but $k + 1$ is on the next building, then mutating a gene with value k up or down by 1 could have significantly different effects, but the algorithm would be unaware of this difference. However the substantial advantage of the linear representation is its generality: it can be applied without modification to any situation.

3.2   Variation Operators

Variation operators determine how the solutions at one generation are used to generate new solutions to potentially join the population. The principal tasks are pairing up solutions as "parents", combining parents in "crossover" to produce children, and "mutating" the children to create new genetic material.

Parental Selection. is usually biased towards better solutions in the population: together with non-random survival, this is one of the mechanisms by which good genetic material is preserved and propagated. We use a fairly elitist approach: the population is split into thirds xs, ys, zs (from best to worst), then

- solutions in xs are paired point-wise with a random permutation of xs; and
- solutions in xs are paired point-wise with a random permutation of ys; and
- solutions in ys are paired point-wise with a random permutation of zs.

Thus the respective thirds of the population get three, two, and one chance(s) to generate children, and n parents collectively generate $2n$ children. We select randomly n of these to evaluate and to consider for subsequent generations.

Crossover is performed with a simple 1-point scheme. Given two solutions x and y, we choose randomly a location L and we build two children: the first contains locations in x that are less than L plus locations from y that are at least as big as L, and the second contains locations in y that are less than L plus locations from x that are at least as big as L. This scheme has the important property that if both x and y are feasible, then their children are guaranteed to be feasible. Note it is irrelevant whether L is in either x, or y, or both.

Mutation takes one of three forms, chosen randomly for each solution.

- The deletion of an AP: one of the existing APs is removed.
- The addition of an AP: a new AP is inserted that bisects one of the "gaps" between consecutive APs in the solution, within a certain distance.
- The modification of an AP: one of the existing APs is moved to a new location not currently in the solution, within a certain distance.

Both crossover and mutation disallow solutions that have fewer than a certain number of APs, determined by the area of the map in question. Such solutions may rank well in the population, due to their limited cost, but they are unlikely to be useful, due to their lack of coverage. They are discarded and replaced immediately.

3.3   Objectives

The system optimises three objectives, all being minimised.

Percentage of the Map with No Service. For each solution we calculate the percentage of the area where wireless service is *not* available.

Percentage of Requests Unfulfilled. Even in areas where service is available, users sometimes experience *contention*, where a particular AP is fully-occupied. The rate of contention is tricky to estimate because it depends on both the quality of the installation, and the pattern of users' accesses. We simulate a typical number of users (for a given map) over some period of time, and we calculate the percentage of their requests which are unfulfilled. The simulator is described in Section 3.5.

Number of Devices Required. We count the number of devices that a solution deploys, and we use this value as a proxy for its cost. It might be desirable to incorporate other factors too, such as the different cost of mounting devices at various locations, or the implied cost of using less-secure locations.

The first of these objectives acts as a so-called "constraint objective". Solutions offering insufficient coverage are allowed in the population, but they are regarded as infeasible: only solutions with an "uncoverage" value close to zero are regarded as adequate solutions to the problem. It has been demonstrated elsewhere[16–18] that this is a good way to deal with such "soft" constraints.

3.4   Other Algorithm Details

This section describes the other significant details of the operation of the MOEA, specifically to do with selection, initialisation, and termination.

Ranking Procedure and Selection. The algorithm uses a fairly standard ranking procedure for performing selection. Each solution x is given a rank equal to the number of other solutions that dominate x, then ranks are promoted until the required population-size is reached. Normally the final rank promoted must be broken to match the population target n, but we employ a scheme whereby the population can grow and shrink in the range $[n, 2n]$. This mostly eliminates the need to break ranks, limiting it to cases where we are choosing at least $n + 1$ solutions from the final rank, which tends to promote diversity in the population.

This variable-population-size scheme has been used previously with some success[19]. The parallel fitness calculations (Section 3.6) mitigate against any significant performance degradation which can arise when using this scheme.

Initialisation. We create each solution in the initial population randomly: we determine upper and lower bounds on the number of APs in each solution based on the area of the map, then for each solution we select some number k in this range, and we select randomly k of the possible AP positions.

Clearly, more-systematic methods of initialisation are possible; e.g. a uniform distribution of the k AP positions, or a deliberate over-provisioning of APs, which may aid the search. We plan to investigate the benefits of such possibilities.

Termination. We run the algorithm for a fixed number of generations. We plan to investigate other possibilities, such as running until improvement ceases.

3.5   The Network Simulator

The network simulator sits at the heart of our system: it is responsible initially for analysing the given map to determine candidate AP positions, and subsequently for calculating the three objectives for each solution generated by the MOEA.

The map provides the locations of static objects (e.g. buildings and trees) that influence the effectiveness of an AP placement, and walkways linking the buildings. We employ a polygon inflation algorithm to determine and enumerate candidate AP positions every 2m around the outside walls of the building.

Given a solution, the system first determines the area covered by the APs. The map is divided into 50cm-square cells (chosen based on typical walking speeds), and a simple ray-tracing approach calculates the signal-strength arriving at each cell. This is calculated using a link budget equation which includes the Wi-Fi transmission power, antenna gains, device sensitivity, and a free-space path-loss model of signal propagation accounting for transmission frequency and distance[20]. For large maps the ray-tracing can take several seconds, so the simulator pre-calculates the "footprint" of signals from each AP at each cell. The signal-strength at each cell is the maximum of all signals from each AP.

The second metric is calculated by the *cnet* network simulator[21, 22]. *cnet* has been developed over twenty years to support generic wide-area networks, IEEE802.3 LANs, and IEEE802.11b/g WLANs. *cnet* is invoked with a topology file that defines the parameters of the simulation, including the locations of the APs, the number of mobile devices, Wi-Fi transmission characteristics, and the duration of the simulation. *cnet* executes the Floyd-Warshall algorithm to determine the shortest paths between all venues and entry/exit points along the walkways [23, 24]. For maps with several hundred locations this calculation takes only milliseconds, but for larger maps pre-calculation is employed. The mobility model has each device select a random entry point to the map and 2–6 venues to visit: then it generates a route (along walkways) that visits each venue and leaves via the same entry point. Devices move at a constant rate in the range 0.5–3.5m/s, they pause at each venue as if sitting in a lecture or tutorial, and they leave to be at their next venue at the beginning of each hour. Stationary devices are assumed to use APs installed inside venues.

Between venues, each mobile device tries to maintain an association with an AP. Each AP transmits beacon frames each $1024\mu s$, advertising itself to prospective clients. Clients record the signal-strength of all APs within range, and try to associate with each AP, in order of signal-strength, until successful.

Each simulated AP has a carrying capacity of twelve clients, modelled on the recommended capacity for a Cisco Aironet 1140 device supporting 802.11b for casual web-browsing. An AP breaks its association with a client if it has not received any traffic for 10s, when it assumes that the client has moved out of range. As we model only connectivity and actual wireless network data, our clients transmit "heartbeat" packets to maintain their associations. A client disassociates from an AP if it has not received a beacon frame for 30s, or when it arrives at its next venue or leaves the map.

The overall run-time for a particular simulation depends primarily on the area of the map, the number of installed APs, and the number of mobile devices. It is roughly linear in each of the last two.

3.6 Parallelising Calls to the Simulator

The MOEA invokes multiple copies of the simulator on different machines via a simple "farming" scheme, reducing the run-time of the system. In each generation, the MOEA creates a file describing the children which were produced at that generation and which need to be evaluated. The *farmer* program distributes these jobs across a collection of Linux workstations, and it tracks when jobs start and finish, when machines become unavailable, when jobs need to be restarted, and all other bookkeeping information required by the system. When the remote evaluation of all jobs is complete, *farmer* writes the results to an output file for the MOEA.

Because the *farmer* is invoked separately in each generation, most of the lost opportunities for parallelism in the system occur near the end of an invocation, when it is waiting for a few jobs to finish. The *farmer* tries to reduce this factor by invoking big jobs (typically those with more APs) first.

4   Results

We report two experiments, both based on parts of UWA's Crawley campus in Perth, Western Australia[25]. Fig. 1 shows the map used. *North.map* contains roughly the northern half of the campus, and *CSSE.map* contains a subset of this, chosen to illustrate the algorithm working on a small-scale problem. The numbers of mobile devices used in the simulations are estimates of the numbers of internet-active users on the wireless network at any given time. The other principal variable in the simulation is the patterns of movement and internet access of these devices. In these experiments we use a fixed seed for the simulation which ignores these issues: we postpone to future work the analysis of the noise inherent in such simulations.

Table 1 gives the settings used for the experiments, and the timings that resulted.

Table 1. The settings for and timings from the two experiments

CSSE.map	Setting	*North.map*
6,020	area of map (m^2)	71,446
213	number of possible AP positions	1,237
5	minimum number of APs allowed	25
12–18	number of APs at initialisation	45–66
2	simulated time (hours)	2
10	number of mobile devices	100
50 (max 100)	population size	100 (max 200)
100	number of generations	100
30 seconds	average run-time per simulation	23 minutes
66 seconds	maximum run-time per simulation	49 minutes
36 seconds	average run-time per generation	59 minutes
21.2	average degree of parallelism	23.0

Fig. 2 shows the front from one run with *CSSE.map*, and Fig. 3 shows the combined front from two runs with *North.map*. Both graphs plot the percentage of unfulfilled requests vs. the number of APs for different minimum levels of coverage of the map in question: basically coverage is being used as a soft constraint with cut-offs set at various values. We make several observations.

- Both graphs suggest that 99% coverage of the relevant map is possible at reasonable cost and while fulfilling a high percentage of users' requests. The rightmost 99%+ solution from Fig. 3 is illustrated in Fig. 1: generally it shows an even spread of APs, but also it shows a "doubling-up" of APs at key locations such as the Octagon Lecture Theatre, the University Club, and the Arts Building, all on the right-hand side of the map. 99.9% coverage is also achievable, at higher cost and with more service refusals.
- For a given cut-off value for coverage, there is a tendency for the percentage of unfulfilled requests to increase as the number of APs decreases, as expected.

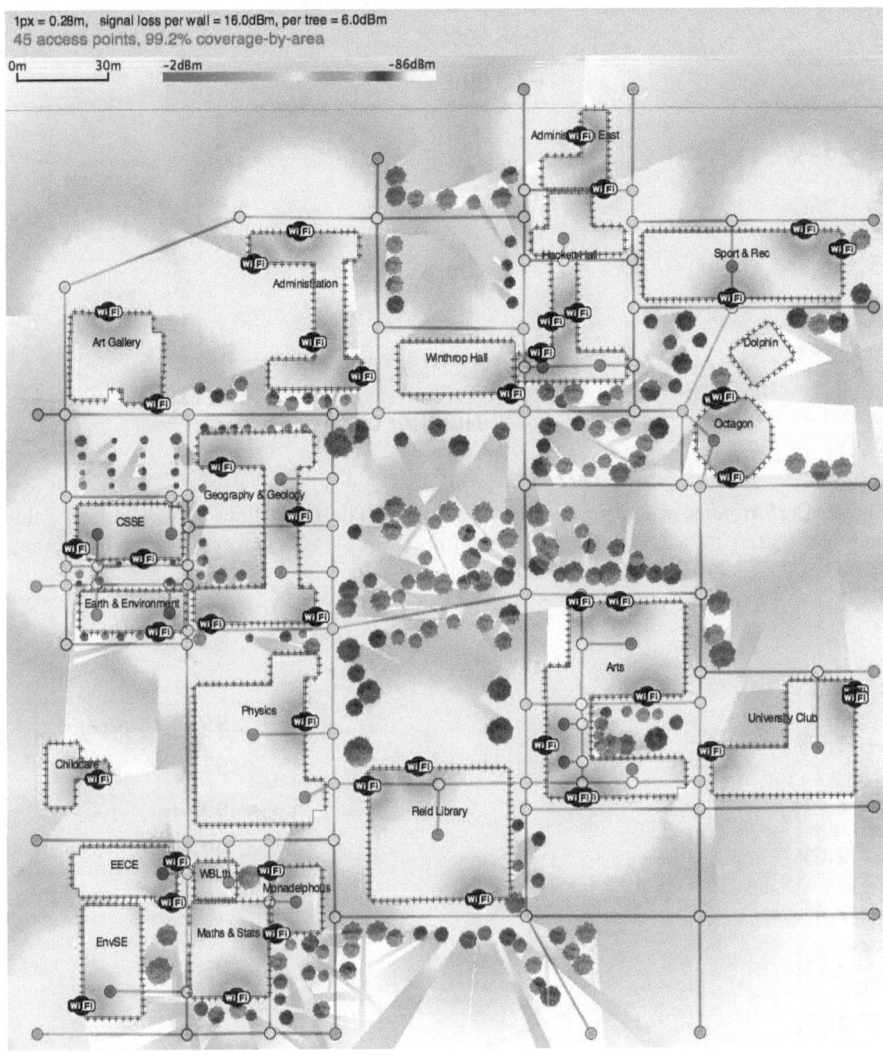

Fig. 1. *North.map* contains the northern part of UWA Crawley campus. Several buildings and many trees are shown. Grey lines denote walkways, yellow dots denote junctions, and green dots denote entrances/exits to this part of the campus. The blue marks on the edges of buildings denote the 1,237 possible AP locations available to the algorithm: the 45 AP locations marked constitute the rightmost 99%+ solution from Fig. 3. *CSSE.map* is a subset of *North.map* that contains the CSSE Building and the two adjacent science buildings, and associated paths: it has 213 possible AP locations.

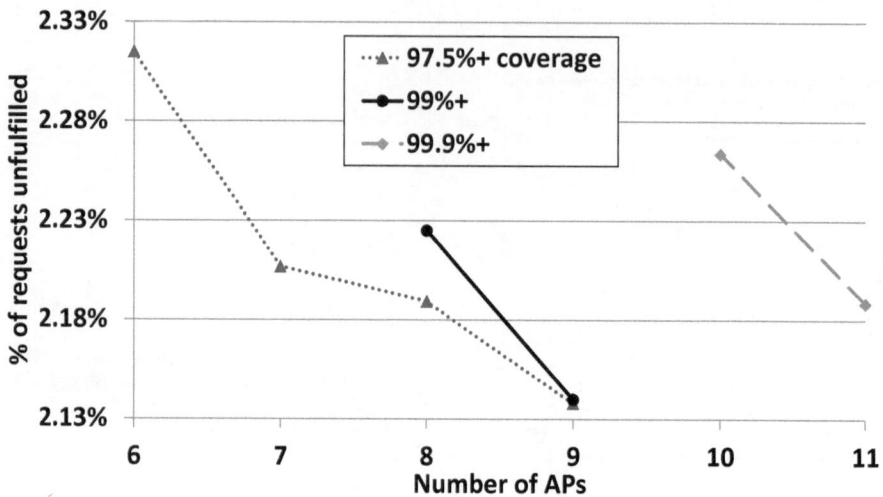

Fig. 2. Performance when processing *CSSE.map*. The three lines show the best solutions generated in one run for various numbers of APs, where coverage of the map exceeds 97.5%, 99%, and 99.9% respectively.

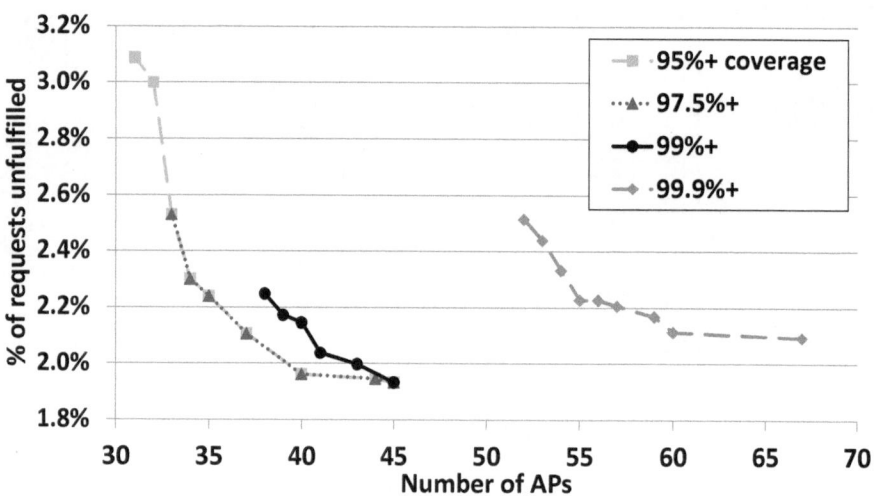

Fig. 3. Performance when processing *North.map*. The four lines combine the best solutions generated in two runs for various numbers of APs, where coverage of the map exceeds 95%, 97.5%, 99%, and 99.9% respectively. The rightmost 99%+ solution is illustrated in Fig. 1.

- Comparing lines for different cut-off values, higher coverage either requires more devices, as expected, or (if the number of devices is unchanged) it results in more requests going unfulfilled. This last point needs some explanation. For a fixed number of APs, achieving higher coverage means that they must be distributed more evenly: close devices have overlapping coverage areas. But an even distribution of APs is unlikely to match the distribution of users' requests, leaving more requests unfulfilled.
- It seems to be impossible to get the percentage of unfulfilled requests below a certain level, about 2% in these experiments. It will always be the case that sometimes the number of users' requests in an area exceeds the offered capacity of the installed APs: this problem is exacerbated by the fact that sometimes after a device stops needing an AP with which it is associated, the AP may not realise for some time that it could be servicing another device.

It is crucial to note that these patterns and trade-offs are exposed as a *result* of running the algorithm: the user does not need to specify anything in advance. This is a major strength of the multi-objective approach.

We note that whilst it would be desirable to compare the performance of our solutions with that of UWA's existing Wi-Fi installation, this is difficult to do in a fair way because the existing system uses indoor APs.

5   Conclusions

We have described a multi-objective evolutionary algorithm that optimises a multi-device Wi-Fi installation for a given map according to three criteria: minimising the number of APs used, minimising the area of the map that lacks coverage, and minimising the proportion of users' requests that are unfulfilled. We have shown that the MOEA delivers good results for two challenging real-world maps that respectively need around ten and fifty APs.

Future work will include improving the simulations that sit at the heart of the system, and optimising for new objectives, such as the robustness of an installation wrt device failure. One issue to be investigated is the noise in the simulations: standard approaches to noise are all expensive[26, 27], so we plan to try a new solution using our parallel evaluation scheme.

References

1. Wikipedia. Wi-Fi (2013), `en.wikipedia.org/wiki/Wi-Fi`
2. IEEE. IEEE^{TM} 802.11 Wireless Local Area Networks (2013),
   `grouper.ieee.org/groups/802/11`
3. Google Patents. Wireless LAN: US 5487069 A (1996),
   `http://www.google.com/patents/US5487069`
4. MarketsandMarkets. Global Wi-Fi Market worth $93.23 Billion By 2018 (2013),
   `www.marketsandmarkets.com/PressReleases/global-wi-fi.asp`
5. Wikipedia. Municipal wireless network (2013),
   `http://en.wikipedia.org/wiki/Municipal_wireless_network`

6. Aerohive Networks. High-Density Wi-Fi Design Principles (2012),
   http://www.aerohive.com/pdfs/Areohive-Whitepaper-Hi-
   DensityPrinciples.pdf
7. Zirari, S., Canalda, P., Mabed, H., Spies, F.: Wi-Fi Access Point Placement within
   Stand-alone. In: 4th ICCE 2012 Hybrid and Combined Wireless Positioning System, pp. 279–284 (2012)
8. Wang, T., Jia, W., Xing, G., Li, M.: Exploiting Statistical Mobility Models for
   Efficient Wi-Fi Deployment. IEEE Transactions on Vehicular Technology 62(1),
   360–373 (2013)
9. Zheng, Z., Sinha, P., Kumar, S.: Sparse WiFi Deployment for Vehicular Internet
   Access with Bounded Interconnection Gap. IEEE/ACM Transactions on Networking 20(3), 956–969 (2012)
10. Wang, C.-S., Kao, L.-F.: The Optimal Deployment of Wi-Fi Wireless Access Points
    Using the Genetic Algorithm. In: ICGEC, pp. 542–545 (2012)
11. Wang, C.-S., Chen, Y.-D.: Base Station Deployment with Capacity and Coverage
    in WCDMA Systems Using Genetic Algorithm at Different Height. In: ICGEC,
    pp. 546–549 (2012)
12. Agustin-Blas, E., Salcedo-Sanz, S., Vidales, P., Urueta, G., Portilla-Figueras, A.,
    Solarski, M.: A Hybrid Grouping Genetic Algorithm for Citywide Ubiquitous WiFi
    Access Deployment. In: IEEE CEC, pp. 2172–2179 (2009)
13. Bulut, E., Szymanski, B.K.: WiFi Access Point Deployment for Efficient Mobile
    Data Offloading. In: 1st ACM International Workshop on Practical Issues and
    Applications in Next Generation Wireless Networks, PINGEN 2012, pp. 45–50
    (2012)
14. Fonseca, C.M., Fleming, P.J.: Genetic Algorithms for Multiobjective Optimization:
    Formulation, Discussion and Generalization. In: 5th ICGA, pp. 416–423 (1993)
15. Coello Coello, C.A., Lamont, G., Van Veldhuizen, D.: Evolutionary Algorithms for
    Solving Multi-objective Problems. Springer (2007)
16. Hingston, P., Barone, L., Huband, S., While, L.: Multi-level Ranking for Con-
    strained Multi-objective Evolutionary Optimisation. In: Runarsson, T.P., Beyer,
    H.-G., Burke, E.K., Merelo-Guervós, J.J., Whitley, L.D., Yao, X. (eds.) PPSN IX.
    LNCS, vol. 4193, pp. 563–572. Springer, Heidelberg (2006)
17. Huband, S., Tuppurainen, D., While, L., Barone, L., Hingston, P., Bearman, R.:
    Maximising Overall Value in Plant Design. Minerals Engineering 19(15), 1470–1478
    (2006)
18. While, L., Hingston, P.: Usefulness of Infeasible Solutions in Evolutionary Search:
    an Empirical and Mathematical Study. In: IEEE CEC, pp. 1363–1370 (2013)
19. While, L., Barone, L., Hingston, P., Huband, S., Tuppurainen, D., Bearman, R.: A
    Multi-objective Evolutionary Algorithm Approach for Crusher Optimisation and
    Flowsheet Design. Minerals Engineering 17(11), 1063–1074 (2004)
20. Tse, D., Viswanath, P.: Fundamentals of Wireless Communication. Cambridge
    University Press (2005)
21. McDonald, C.: Network Simulation using User-level Context Switching. In: Aus-
    tralian UNIX Users' Group Conference, pp. 1–10 (1993)
22. McDonald, C.: cnet network simulator, v3.2.4 (2013),
    http://www.csse.uwa.edu.au/cnet
23. Floyd, R.W.: Algorithm 97: Shortest Path. Communications of the ACM 5(6), 345
    (1962)

24. Warshall, S.: A Theorem on Boolean Matrices. Journal of the ACM 9(1), 11–12 (1962)
25. The University of WA. Campus map (2013),
    http://www.uwa.edu.au/contact/map
26. Di Pietro, A., While, L., Barone, L.: Applying Evolutionary Algorithms to Problems with Noisy, Time-consuming Fitness Functions. In: IEEE CEC, pp. 1254–1261 (2004)
27. Di Pietro, A., While, L., Barone, L.: On the Behaviours of Evolutionary Strategies for Problems with Varying Noise Strength. In: IEEE CEC, pp. 2772–2779 (2008)

Automated Design of Architectural Layouts Using a Multi-Objective Evolutionary Algorithm

Darcy Chia and Lyndon While*

Computer Science & Software Engineering, The University of Western Australia
lyndon.while@uwa.edu.au

Abstract. The internal layouts of buildings in video games are usually designed by hand, but the increasingly expansive and realistic nature of virtual worlds introduces scalability issues which make manual design methods impractical. We present a new methodology that uses a multi-objective evolutionary algorithm to automatically generate building layouts. The method accepts highly versatile input constraints, encoding layouts using a flexible binary tree representation and evaluating them on a range of criteria to ensure authentic results. Tests demonstrate that the method works well for a variety of problem instances representing an apartment, a family house, and an office floor; the ability to generate solutions for different types of buildings and to incorporate non-rectangular spaces shows greater versatility than many previous methods.

Keywords: Architectural design, multi-objective optimisation, evolutionary algorithms.

1 Introduction

People spend much of their time in built environments, both residential and commercial buildings. Traditionally, building design has been performed by architects, engineers, and other experts, in consultation with a client — a resource-intensive process justified by the importance of obtaining good solutions and the lack of adequate alternatives. The same approach has normally been used in online worlds: however with the recent rise in popularity of open-world video games such as the Grand Theft Auto[1] and Left 4 Dead[2] series, in which players can freely roam expansive virtual environments that include entire cities, manually creating unique designs in such contexts is an expensive, if not infeasible, option. It has been estimated that in games like the World of Warcraft[3] series, the manual creation of game content (such as weapons and buildings) absorbs up to 40% of the budget[4, 5]. Thus the automatic generation of realistic and diverse building designs is clearly a desirable goal. Yet game developers continue to model buildings by hand[6], at least partly because current alternatives introduce undesirable compromise. Tutenel *et al.*[7] note that buildings incidental to a game's plot frequently have only their exterior facade rendered, with no interior whatsoever. Such buildings cannot be entered, restricting what a player can and

---

* Corresponding author.

G. Dick et al. (Eds.): SEAL 2014, LNCS 8886, pp. 760–772, 2014.

cannot explore. Other games use a small set of interiors replicated city-wide[7], lessening the variety and authenticity of the virtual world.

This paper explores the use of a multi-objective evolutionary algorithm (MOEA) to generate building floor plans autonomously. The MOEA uses a tree representation of building interiors that describes a recursive subdivision of a given space, and that allocates each derived subspace to a room in the building. This representation has been described previously[8], but it has never been used to generate floor plans in a truly autonomous manner. We propose new genetic operators over this representation, a flexible input system that allows a user to specify (via hard and soft constraints) what rooms they want and the relationships between them, and nine fitness criteria that measure how closely a design matches the specification. The MOEA returns a set of designs that satisfy the specification, with varying trade-offs across the soft constraints. In contrast to most previous work, the MOEA is able to generate floor plans with non-rectangular rooms, and we demonstrate that it can generate plans for various types of buildings, whereas previous work has tended to focus on a single type of building only. We report tests representing a small apartment, a family house, and an office floor: the MOEA delivers a range of promising designs for each. The results show that the method has the potential to become a valuable part of a game developer's toolkit for creating authentic virtual worlds. It may also help architects as a design tool in a traditional setting, although the perceived need for this is less.

The rest of the paper is structured as follows. Section 2 describes relevant background material and previous work. Section 3 specifies the problem that we address and the options that the system currently offers to users. Section 4 describes the details of our multi-objective approach, and Section 5 describes the experiments performed and the results achieved. Section 6 concludes the paper.

2 Background

2.1 Previous Approaches to Automating Architectural Design

Schwarz et al.[9, 10] were the first to automatically generate and evaluate building floor plans. They use two weighted, directed constraint graphs, one whose nodes represent walls in the X direction, the other in the Y direction: thus all rooms are rectangular. The search space is sets of edges in the graphs, and a branch-and-bound technique is used to find good solutions based on criteria such as total area and construction cost.

Much previous work distinguishes between the topology and the geometry of a floor plan. A topology refers to room adjacencies, ignoring spatial coordinates or dimensions; whereas a geometry specifies the position and shape of each room. Medjdoub and Yannou[11] use integer constraint programming to determine feasible topologies (thus reducing the search space of geometries), then they input these to a branch-and-bound algorithm to derive optimal geometrical solutions. Their approach is guaranteed to find optimal solutions, but it is suited only to solving small- or medium-sized problems. Michalek et al.[12] similarly use a genetic algorithm to explore the space of topologies, then gradient-based quadratic programming combined with simulated annealing to find geometrical solutions.

The fitness of a topology is taken to be the fitness of the best geometry that can be generated from that topology. Other work which finds only topological solutions and ignores geometry[13, 14] may serve as a useful starting point for an architect to refine manually, or as useful input to other methods.

A *shape grammar*[15] is based around rules that define transformations between shapes. Shape grammars are a successful and popular tool for modelling the 3D textual features of building facades in video games[16]. They have also been used to generate floor plans[17–19], but their success in this area is limited to particular styles that do not reflect typical modern-day architecture.

Merrell *et al.*[20] train a *Bayesian network* to learn designs using a database of pre-existing floor plans. They are able to generate designs that resemble real-world floor plans, which would presumably seem authentic in a virtual setting, but they assume the availability of existing plans that conform to the design goals. These designs are fed into a stochastic optimisation algorithm which can slide segments of walls horizontally or vertically and can swap the positions of rooms, evaluating fitness using the dimensions and shape of each room.

Quiroz *et al.*[8] use an *interactive genetic algorithm* to generate floor plans. They encode a floor plan as a binary tree, where each internal node divides a rectangular space either horizontally or vertically, and each leaf defines the type of room that a space represents; and they use volunteers to subjectively and collaboratively evaluate plans throughout the evolutionary process. The lack of autonomy and the reliance on human input make this approach ill-suited to the task of generating a large number of building layouts for virtual worlds. Note that a similar encoding scheme is used in our new method (see Section 4.1).

Procedural techniques generate floor plans via a set of steps, each guided by heuristics and random numbers for design variation. These methods are usually fast, but they are restricted to a particular type of building, they tend not to optimise global fitness, and they consider only a limited number of constraints. Marson and Musse[21] generate layouts using squarified treemaps[22], which iteratively divide a space, keeping the subspaces as square as possible. This use of subdivided rectangles resembles the encoding used by our MOEA, but the method is otherwise wholly different. Once rooms are assigned fixed coordinates, a final step generates a corridor to ensure that all rooms are connected, but this corridor necessarily reduces the area of some rooms. Their examples are limited to small six-room apartments, and it is unclear how well the method scales.

Tutenel *et al.*[7] propose a framework that can theoretically *integrate interior layout generation and exterior 3D rendering techniques* to create a complete, cohesive building. Such a process could operate either inside-out or outside-in; either way, consistency is clearly important, e.g. to avoid windows or doors that overlap multiple rooms. A semantic moderator coordinates the 2D interior layout algorithm and the 3D exterior rendering algorithm, without either algorithm knowing how the other operates. We envisage that our new method could be used as a component of such a framework, generating the interior layout for a building while allowing other, specialised algorithms to render the 3D model.

2.2   Multi-objective Optimisation

In a multi-objective optimisation problem, potential solutions are assessed according to two or more independent quantities. The characteristic of good solutions is that improving in one objective can be achieved only by worsening in at least one other objective. An algorithm for solving such problems returns a set of solutions offering different trade-offs between the various objectives.

Consider a fitness function that maps a solution x into a vector $\overline{f_x}$. x *dominates* y iff $\overline{f_x}$ is at least as good as $\overline{f_y}$ in every objective, and is better in at least one. x is *non-dominated* wrt a set X iff there is no solution in X that dominates x. X is a *non-dominated set* iff every $x' \in X$ is non-dominated wrt X. The set of fitness vectors corresponding to a non-dominated set is a *non-dominated front*. x is *Pareto optimal* iff x is non-dominated wrt the set of all possible solutions, and the *Pareto optimal set* is the set of all Pareto optimal solutions. Multi-objective optimisation aims to find (or approximate) this Pareto optimal set.

With multiple objectives there is only a partial order on solutions, which causes problems for selection in an evolutionary algorithm. The usual solution is to define a ranking on solutions: one popular scheme[23] defines the *rank* of a solution x wrt a set X to be the number of solutions in X that dominate x. Selection is then based on ranks: a lower rank implies a better solution.

Precise definitions of all these terms can be found in [24].

3   Problem Specification

Building designs vary a lot, depending on many factors, e.g. the purposes for which the building will be used, the amount of space available, the number and types of rooms required, and the desires and tastes of the client(s). As such, an architectural design tool needs to allow a user to provide a wide range of details in the specification of a problem instance, otherwise the utility of the tool will be limited. Our system allows a user to specify the footprint available, a list of *room types*, e.g. Lounge, Bedroom, Hall, Office, *etc*, and then for each type:

- the number of those rooms required;
- whether the rooms can be used as thoroughfares between other rooms (e.g. a lounge may be a thoroughfare, whereas a bedroom is less likely to be so);
- whether the rooms need to be publicly accessible, i.e. reachable using only thoroughfares (e.g. a bedroom is usually accessible, but not its ensuite);
- a minimum and maximum width, length, and area;
- a maximum aspect ratio;
- a minimum length of external wall, usually to allow for windows or access.

Note that if a building is required to have (say) lounges with different specs, that can be accommodated by having separate types LoungeA and LoungeB.

Secondly, for each pair of room types, a user can specify minimum and maximum ratios of their respective areas; e.g. they may say that a lounge must be at least 20% bigger than a home office, but no more than twice as big.

Finally, the user can specify that certain room types must be adjacent.

Fig. 1 shows an example specification. Other examples are given in [25].

	Number	Th'fare	Access	Width	Length	Area	Max AR	Min wall
Living	1	yes	req	2.5–	3–	9–45	2	1m south
Home office	1	no	req	2.0–	2–	6–30	2	1m any
Kitchen	1	no	req	2.0–	2–	7–35	3	1m any
Bed	2	no	req	2.0–	3–	7–35	2	1m any
Bath	1	no	req	1.5–	2–	4–20	4	0m
Hall	1	yes	req	1.0–1.5	1–	1–50	7	0m

	Living	Office	Kitchen	Bed	Bath	Hall
Living	n/a	1.2–2.0	1.0–1.7	1.0–1.8	2.0–	2.0–
Home office		n/a	0.5–1.4	0.4–1.0	1.0–	1.5–
Kitchen			n/a	0.5–2.0	1.25–	1.0–
Bed				0.7–1.43	1.5–	1.4–
Bath					n/a	0.0–
Hall						n/a

Footprint available: $7 \times 7m$.
Required adjacencies: Kitchen and Living; both Beds and Hall; Bath and Hall.

Fig. 1. The complete specification for an apartment design problem. The first table gives the constraints for the individual room types, the second table gives the area ratio constraint for each pair of room types (empty cells can be inferred from those given), and the final statements give the footprint and adjacencies. Units used are metres.

4   Methodology

This section describes the details of our algorithm: the genetic representation and variation operators used, the objectives and their quantification, and details of initialisation, termination, and archiving.

4.1   Representation

The genetic representation that we use is a binary tree, where

- the root node defines the dimensions of the entire space available;
- each internal node defines a split direction (either H (splitting the X dimension) or V (splitting the Y dimension)), and a proportion $0 < p < 1$: e.g. the node $V, 0.4, l, r$ indicates a split parallel to the X-axis 40% of the way "down" the current space, with children l and r;
- each leaf indicates which room that node's space belongs to.

This representation is very similar to that used in [8] and superficially similar to that used in [21]. It offers several features that are useful in evolving floor plans:

- it is flexible enough to represent any sub-division of the space whose walls are parallel to the axes;
- it allows the representation of non-rectangular rooms in a natural way, simply by having multiple leaves with the same label;
- it is easy to mutate in various natural ways.

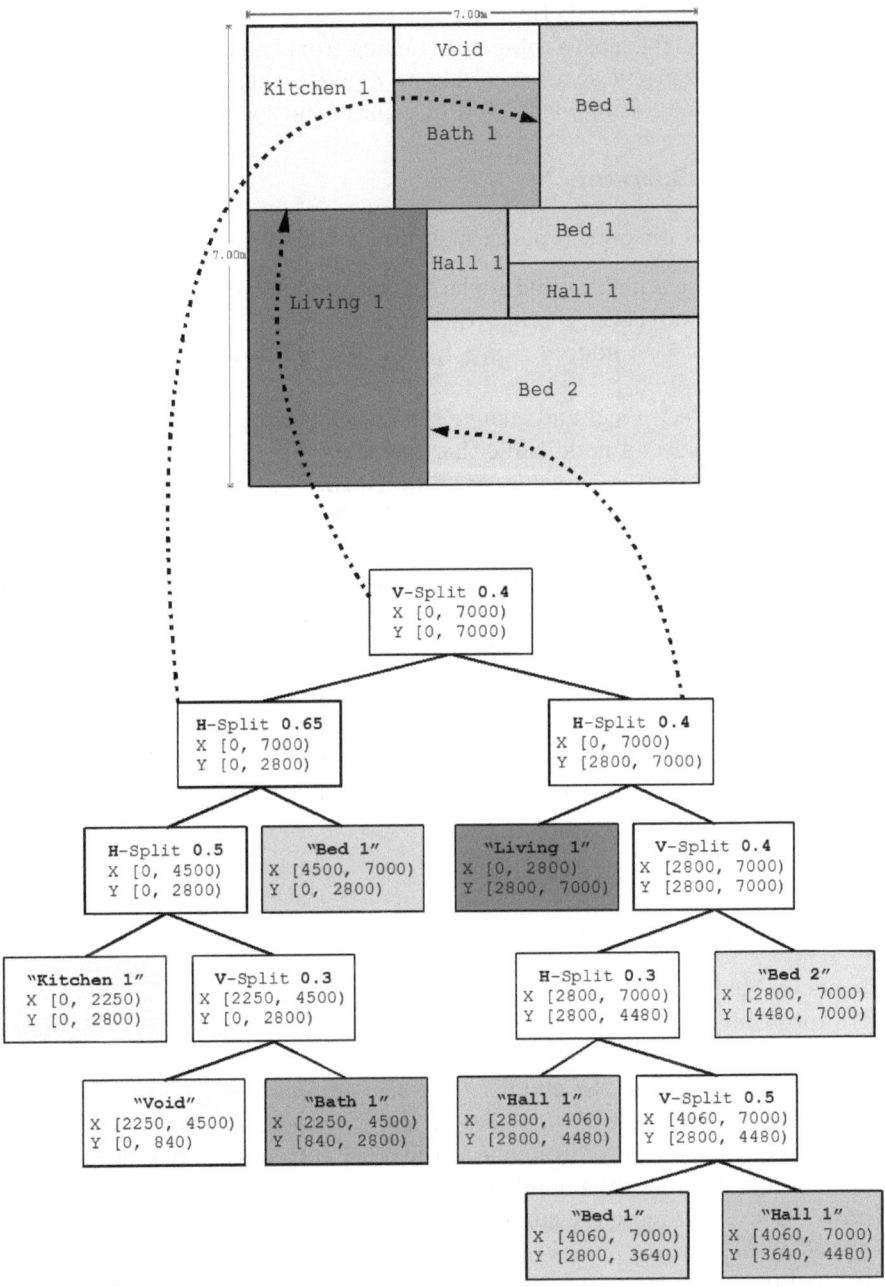

Fig. 2. An example floor plan, and a binary tree representation of that floor plan. Each node illustrated gives the X and Y ranges of the space available to that node, with the origin at the top-left corner of the space. The arrows indicate the correspondence between the tree and the three top-level divisions. Note that Hall 1 and Bed 1 are each associated with two leaves in the tree, hence they are non-rectangular.

Fig. 2 illustrates how a tree is turned into a floor plan. The hierarchy of splits is applied to the available space top-down starting from the root node. Each node d, p, l, r splits the current space according to d and p, then l is applied above or left of the split, and r is applied below or right of the split.

4.2 Variation Operators

We employ several different types of mutation, selected probabilistically.

Insert leaf: select a node n and replace n with a new node d, p, n, l, where d, p, and l are a direction, a proportion, and a new leaf.

Remove leaf: select a node $n = d, p, l, x$ where l is a leaf, and replace n with x.

Change leaf: select a leaf and change the associated room type.

Change node: select a node and either toggle its direction between H and V, or mutate its proportion value by up to ± 0.25.

Rotate subtree: select a node and replace it with one that generates the same floor plan, but rotated either 90°, 180°, or 270°.

Reflect subtree: select a node and replace it with one that generates the same floor plan, but reflected either horizontally, vertically, or both.

Swap subtrees: select two internal nodes and swap their positions.

Swap rooms: select two plan components a and b, and change all leaves of type a to type b and *vice versa*.

[25] gives more details of these mutations and examples illustrating their use. No crossover is currently used: using crossover did not improve the results.

4.3 Objectives

We employ nine objectives to test whether a floor plan meets a specification.

Absence: Minimise the number of missing rooms.

Disjointedness: Minimise the number of disjoint rooms: a room is disjoint if it has two or more non-adjacent areas.

Adjacencies: Minimise the number of broken adjacencies.

Connectivity: Minimise the number of non-connected rooms: two rooms are connected if you can get from one to the other traversing only thoroughfares.

Externality: Minimise the number of rooms with insufficient external wall.

Redundancy: Minimise the number of redundant halls: a hall is redundant if it is adjacent to two or fewer rooms, or if it does not improve room connectivity.

Dimensions: Minimise $\sqrt{(\sum_{r \in rooms} \sum_{d \in dimensions} f(r, d)^2)}$, where $f(r, d)$ is the proportion by which r is outside its set range in width, length, or area.

Aspect ratio: Minimise $\sum_{r \in rooms} g(r)$, where $g(r)$ is the amount by which r exceeds its maximum aspect ratio. $g(r)$ is adjusted for non-rectangular r[25].

Proportionality: Minimise $\sqrt{(\sum_{r, r' \in rooms} h(r, r')^2)}$, where $h(r, r')$ is the amount by which $area(r)/area(r')$ is outside the set range.

The first six of these act as hard constraint objectives: only solutions with a value of 0 for all of these are regarded as feasible/satisfactory. It has been shown elsewhere that this is a good way to derive solutions that satisfy such constraints[26–28]. The other three are effectively soft constraints: we would like to minimise their values, but it may not always be possible to derive an optimal solution.

4.4 Other Algorithm Details

Selection: We use standard Pareto ranking[23] in selection: where we need to break ranks, we favour the hard-constraint objectives over the soft constraints, in the order listed in Section 4.3. Thus feasible solutions are favoured, which tends to increase the proportion of such solutions over time.

Initialisation: All initial solutions are generated using a simple heuristic procedure. Starting from a single (root) node, insert leaves repeatedly with a room not already added, until all of the components are present. Spaces are normally split along their longer axis.

Termination: Currently, the algorithm is run for a pre-defined number of generations. Other termination criteria could be used. A post-processing step is applied to each solution, where adjacent halls are merged into one. Other useful post-processing steps could also be imagined, for example lining up walls that miss each other by only a small amount (instances of this are discussed in Section 5).

5 Results

We report tests of our algorithm on three problem instances of different types:

- a small apartment with seven rooms;
- a family house with twelve rooms;
- an office floor with about twenty rooms, and more-complex access.

Fig. 1 gives the input specification for the apartment design problem; due to space limitations here, the others can be found in [25].

Table 1 shows the settings and performance figures for each experiment. Normally with a MOEA we are interested in identifying the solutions which perform best against the defined objectives. However with this problem we are interested in the MOEA returning a range of solutions that satisfy the specifications of a given problem instance. Thus apart from defining a feasible solution in the usual way to be one which returns 0 for all of the hard constraints, we also specify tolerances on the soft constraints, and we define a *usable* solution to be one which is feasible and which returns values for the other constraints which are all within these tolerances. The maximum values allowed here were 1 for Dimensions and Aspect Ratio, and 0.5 for Proportionality.

Table 1 shows that for all three problems, the MOEA was able to find usable solutions in a reasonable amount of time, although naturally this time increases with the complexity of the problem specification. The system could be used in at least three different modes:

Table 1. Settings and results for the experiments. All figures are averages for ten runs.

	Apartment	House	Office
Population	500	750	1,000
Generations	500	1,000	1,500
Execution time (mins)	2.2	14.0	52.0
Usable solns. generated	122	62	23
Generation of first usable soln.	97	444	1,102
Time to first usable soln. (mins)	0.46	5.7	37

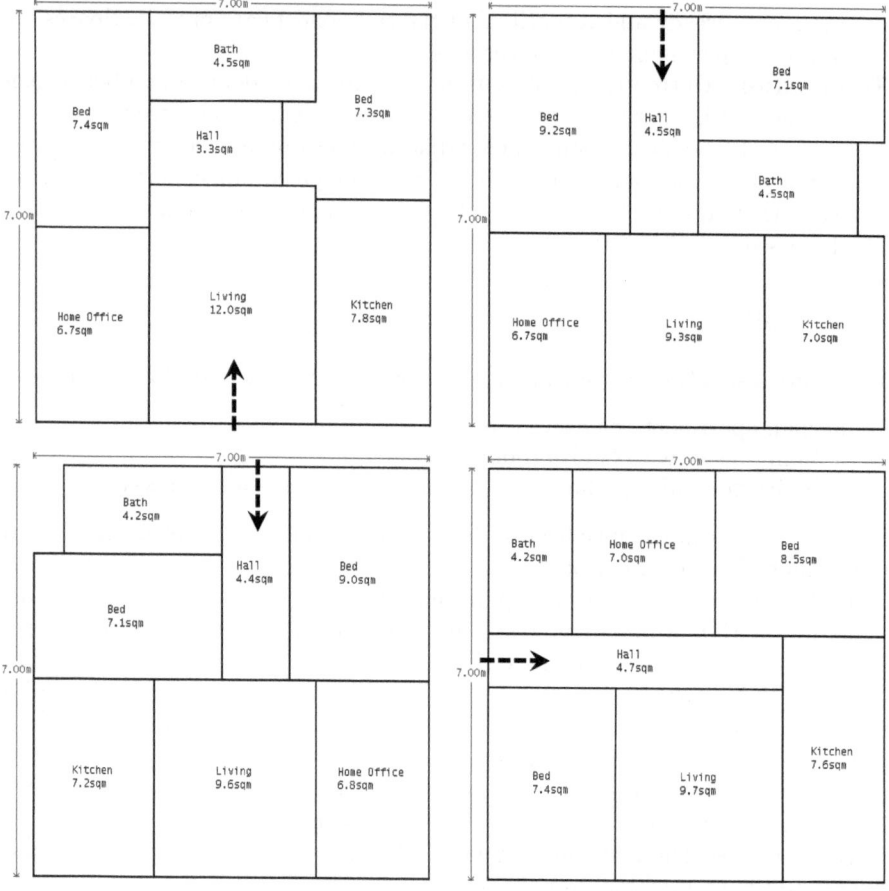

Fig. 3. Four usable solutions returned in one run for the apartment problem. Principal access is through the hall, or directly into the living room.

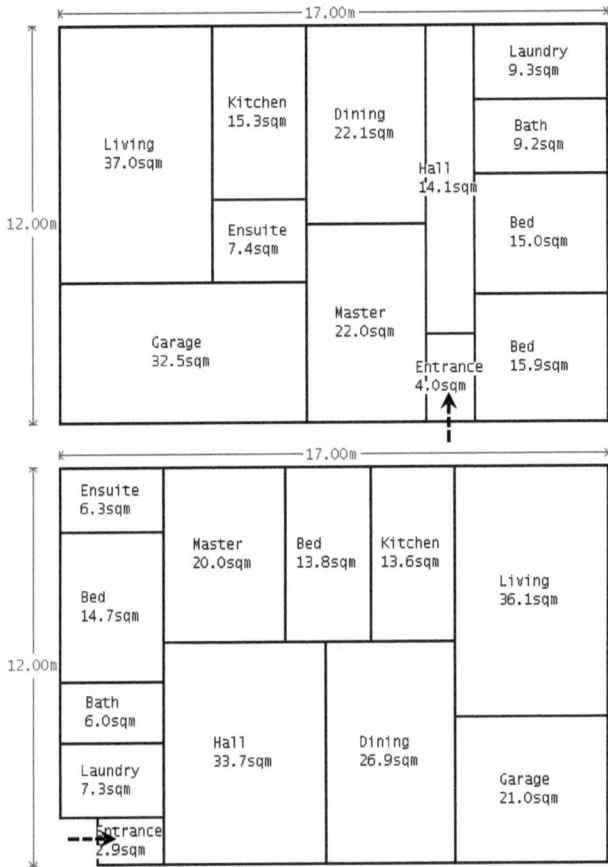

Fig. 4. Two usable solutions returned in one run for the family house problem. Principal access is through the entrance.

- one run of the MOEA, which attempts to derive a range of usable solutions that satisfy a given specification;
- several (much shorter) runs with the same specification, each of which stops after deriving one or a few usable solutions;
- several (again short) runs with slightly varying specifications, again each stopping after deriving a small number of usable solutions.

Note that even if the MOEA is stopped after returning only one usable solution, this layout could trivially be used to generate alternative solutions which may be interesting, using combinations of rotation and reflection, or possibly by swapping rooms with similar dimensions. Also note that in the one run scenario, some usable solutions are likely to be fairly similar, if they share ancestry.

Figs. 3–5 show examples of the layouts derived by the MOEA.

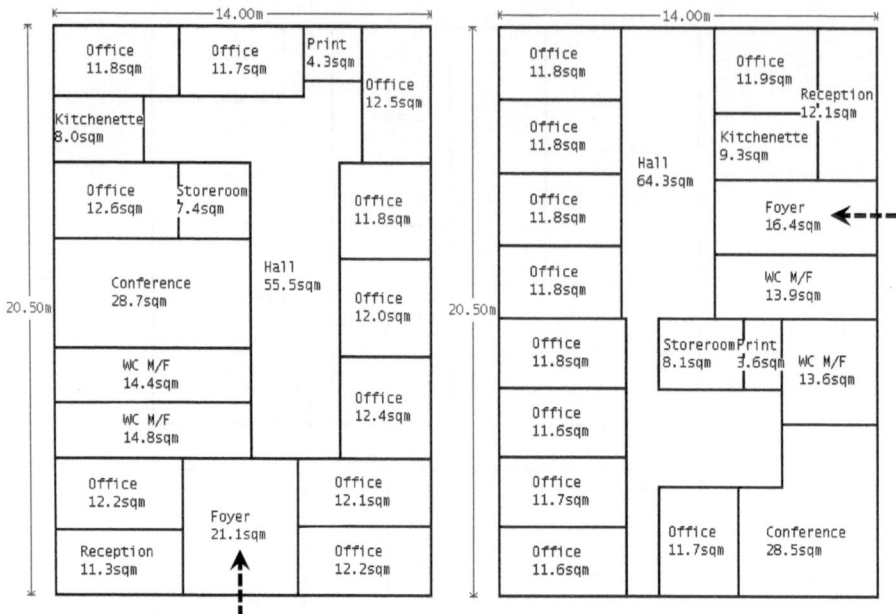

Fig. 5. Two usable solutions returned in one run for the office floor problem. Principal access is through the foyer.

Each figure shows a selection of the usable solutions returned in one run for the relevant problem. The solutions show the variation that can be achieved even within just one run, and some of them illustrate the ability of the MOEA to design non-rectangular rooms, especially the halls in Fig. 5. Some of the layouts also illustrate the improvements that could be achieved using more post-processing, e.g. aligning walls that miss each other marginally, or evening up rooms in cases like the toilets and some of the slightly-different offices in Fig. 5.

6   Conclusions

We have described a multi-objective evolutionary algorithm that autonomously designs floor layouts for buildings according to a user-defined specification. The MOEA uses a tree representation that defines a recursive subdivision of the provided space, assessing layouts against six hard constraints and three soft constraints to determine usability. We have shown that the system works well on three different problems, for a range of buildings. This type of system will help developers to efficiently design large numbers of buildings for virtual worlds.

Future work will focus on enhancing the specifications available to user; on applying post-processing to fine-tune the solutions returned by the MOEA; and on speeding up the MOEA to investigate the possibility of generating virtual building layouts "on-the-fly" as players proceed through a game.

References

1. Rockstar Games. Grand Theft Auto: The Official Site (2013), http://www.rockstargames.com/grandtheftauto (viewed June 30, 2014)
2. Valve. Left 4 Dead Blog (2013), http://www.14d.com (viewed June 30, 2014)
3. Blizzard Entertainment. World of Warcraft (2014), http://www.worldofwarcraft.com (viewed June 30, 2014)
4. Irish, D.: The Game Producer's Handbook. Thomson Course Technology (2005)
5. Hendrikx, M., Meijer, S., Van der Velden, J., Iosup, A.: Procedural Content Generation for Games: a Survey. ACM Transactions on Multimedia Computing, Communications and Applications 9(1), 1 (2013)
6. Lopes, R., Tutenel, T., Smelik, R., de Kraker, K., Bidarra, R.: A Constrained Growth Method for Procedural Floor Plan Generation. In: 11th International Conference on Intelligent Games and Simulation, pp. 13–20 (2010)
7. Tutenel, T., Smelik, R., Lopes, R., de Kraker, K., Bidarra, R.: Generating Consistent Buildings: a Semantic Approach for Integrating Procedural Techniques. IEEE Transactions on Computational Intelligence and AI in Games 3(3), 274–288 (2011)
8. Quiroz, J., Louis, S., Banerjee, A., Dascalu, S.: Towards Creative Design using Collaborative Interactive Genetic Algorithms. In: IEEE CEC, pp. 1849–1856 (2009)
9. Schwarz, A., Berry, D., Shaviv, E.: Representing and Solving the Automated Building Design Problem. Computer-Aided Design 26(9), 689–698 (1994)
10. Schwarz, A., Berry, D., Shaviv, E.: On the Use of the Automated Building Design System. Computer-Aided Design 26(10), 747–762 (1994)
11. Medjdoub, B., Yannou, B.: Separating Topology and Geometry in Space Planning. Computer-Aided Design 32(1), 39–61 (2000)
12. Michalek, J., Choudhary, R., Papalambros, P.: Architectural Layout Design Optimization. Engineering Optimization 34(5), 461–484 (2002)
13. Wong, S., Chan, K.: EvoArch: an Evolutionary Algorithm for Architectural Layout Design. Computer-Aided Design 41(9), 649–667 (2009)
14. Damski, J., Gero, J.: An Evolutionary Approach to Generating Constraint-based Space Layout Topologies. In: CAAD Futures, pp. 855–864 (1997)
15. Stiny, G., Gips, J.: Shape Grammars and the Generative Specification of Painting and Sculpture. Information Processing 71, 1460–1465 (1972)
16. Leblanc, L., Houle, J., Poulin, P.: Component-based Modeling of Complete Buildings. In: Graphics Interface, pp. 87–94 (2011)
17. Stiny, G., Mitchell, W.: The Palladian Grammar. Environment and Planning B 5(1), 5–18 (1978)
18. Koning, H., Eizenberg, J.: The Language of the Prairie: Frank Lloyd Wright's Prairie Houses. Environment and Planning B 8(3), 295–323 (1981)
19. Granadeiro, V., Pisa, L., Duarte, J., Correia, J., Leal, V.: A General Indirect Representation for Optimization of Generative Design Systems by Genetic Algorithms: Application to a Shape Grammar-based Design System. Automation in Construction 35, 374–382 (2013)
20. Merrell, P., Schkufza, E., Koltun, V.: Computer-generated Residential Building Layouts. ACM Transactions on Graphics 29(6), 181 (2010)
21. Marson, F., Musse, S.: Automatic Real-time Generation of Floor Plans based on Squarified Treemaps Algorithm. International Journal of Computer Games Technology 2010, 624817 (2010)
22. Bruls, M., Huizing, K., van Wijk, J.: Squarified Treemaps. In: Joint Eurographics and IEEE TCVG Symposium on Visualization, pp. 33–42 (1999)

23. Fonseca, C.M., Fleming, P.J.: Genetic Algorithms for Multiobjective Optimization: Formulation, Discussion and Generalization. In: 5th ICGA, pp. 416–423 (1993)
24. Coello Coello, C.A., Lamont, G., Van Veldhuizen, D.: Evolutionary Algorithms for Solving Multi-objective Problems. Springer (2007)
25. Chia, D.: Generation of Interior Building Layouts using a Genetic Algorithm. Honours dissertation, Computer Science & Software Engineering. In: UWA (2014)
26. Hingston, P., Barone, L., Huband, S., While, L.: Multi-level Ranking for Constrained Multi-objective Evolutionary Optimisation. In: Runarsson, T.P., Beyer, H.-G., Burke, E.K., Merelo-Guervós, J.J., Whitley, L.D., Yao, X. (eds.) PPSN 2006. LNCS, vol. 4193, pp. 563–572. Springer, Heidelberg (2006)
27. Huband, S., Tuppurainen, D., While, L., Barone, L., Hingston, P., Bearman, R.: Maximising Overall Value in Plant Design. Minerals Eng. 19(15), 1470–1478 (2006)
28. While, L., Hingston, P.: Usefulness of Infeasible Solutions in Evolutionary Search: an Empirical and Mathematical Study. In: IEEE CEC, pp. 1363–1370 (2013)

An Approach for Real-Time Frame Size Adaptation in M-JPEG Streams

Kyaw Ko-Ko-Htet and Tan Kok-Kiong

Department of Electrical and Computer Engineering, Faulty of Engineering,
National University of Singapore,
21 Lower Kent Ridge Road, 119077
{A0035642,KKTan}@nus.edu.sg
http://www.ece.nus.edu.sg/

Abstract. The growth of the video traffic proliferates quickly over the internet as well as wireless networks. With that growth, different video compression standards have been introduced over the years. Among them, *Motion Joint Photographic Expects Group* (M-JPEG) has advantages of avoiding frame-to-frame error propagation and achieving low coding/decoding latency. Due to these advantages, M-JPEG is widely adopted in video-capture devices, wireless IP cameras and industrial real-time applications. On the other hand, due to its nature of dynamic frame sizes, its stream's bit-rate is generally varying and different solutions have been proposed to regulate the bit-rate. As these solutions still persist drawbacks such as having high regulating error, this paper aims to propose an improved approach which can regulate individual frame size of M-JPEG stream in real-time. Experiments indicate that the proposed approach has a straight forward implementation and yet, outperforming in regulating frame size of M-JPEG compared to existing solutions.

Keywords: Adaptation, M-JPEG, Video Streaming.

1 Introduction

Multimedia streaming proliferates quickly reaching 66 percent of all consumer-internet traffic in 2013 and this amount is predicted to reach 79 percent by 2018 [1]. The growth of the video traffic is directly resulting from the increasing variety of applications, such as distributed multimedia applications, medical applications, and central monitoring and control system, all performing video streaming over the internet [2]. Also, an increasing amount of multimedia contents have been streamed over wireless communication channels such as mobile networks and Wi-Fi.

These multimedia contents are generally compressed at servers before transmission to reduce network bandwidth utilization. To fulfill diverse application requirements for multimedia compression, different video compression standards have been introduced over the years. Among them, M-JPEG is a video compression format in which each video frame or interlaced field of a digital video

G. Dick et al. (Eds.): SEAL 2014, LNCS 8886, pp. 773–784, 2014.

sequence is compressed separately as a JPEG image. Originally developed for multimedia PC applications, M-JPEG is now widely used in video-capture devices such as digital cameras, IP cameras, and webcams; and also in non-linear video editing systems [3].

M-JPEG is perhaps the simplest existing video coder; it simply codes each frame in JPEG format and transmits the coded frames sequentially while many other more recent codecs such as MPEG and H.264 use the concept of Group of Pictures (GOP), providing better compression rate as well as smoother transitions between frames [3]. However, due to GOP, these codecs require at least 3 frames to implement the required motion coding, resulting in long latency, which may be undesirable in certain applications. Also, these codecs do not offer much performance margin in videos with lots of objects or scene movements such as traffic videos since they do not provide much data saving due to greater frame-to-frame differences. Also, as the frame-to-frame error is propagated, a single dropped frame can disturb the decoding of subsequent frames. These factors imply the more recent video codecs are not suitable for videos that are taken in emergency or adhoc scenes such as battle fields and post-disaster areas, and also for videos that are transmitted though lossy medium. On the other hand, M-JPEG has the ability to start decoding at any particular frame and localize the transmission errors by using frame-by-frame processing [4] [5]. Due to this advantage, M-JPEG is widely adopted in video-capture devices and wireless IP cameras.

Apart from the video-capture devices, due to its low coding/decoding latency, M-JPEG is also incorporated into a growing number of industrial multimedia information processing applications such as object tracking [6], automated inspection [7], machine vision [8], and vehicle guidance system [9–11]. These industrial media applications can be classified into two broad categories [12]; supervised multimedia control subsystems [13] and multimedia embedded systems (MES). In the first category, the emphasis is fundamentally on the media processing quality, while the real-time constraints are generally soft. However, the applications in the latter category are more demanding in terms of real-time requirements in addition to the media processing quality. Typically, these applications are complex and heterogeneous, encompassing several real-time activities in addition to the media processing ones. Thus, the interference infiltrating into by the multimedia handling components must be limited and predictable [14].

Many MES applications are distributed and they rely on real-time network protocols to provide real-time communication services. However, multimedia traffic such as video streaming generally use a variable bit rate (VBR) traffic source which leads to conflicts with the operational framework of conventional real-time protocols, which usually offer constant-bit-rate (CBR) channels to the applications (e.g., ATM, PROFINET-IRT, Interbus, ControlNet or flexible time-triggered networks (FTT) [15–18]). Fitting a VBR source to a CBR channel may lead to either waste of bandwidth or rejection of frames [14]. Thus, regulating the size of individual frame is vital in fitting the M-JPEG stream into CBR channel of real-time protocols.

Furthermore, achieving CBR stream is crucial not only for these real-time applications but also many other applications such as video servers and medical collaboration systems. This is because the speed bursts dynamically demands different quality-of-service (QoS) grades from servers and networks during a session and it may create undesired interferences to the resource availability on heterogeneous terminals and networks. As there is a great need for reducing speed burst of M-JPEG streams or regulating bit rate of M-JPEG stream, a number of solutions have been proposed recently. However, there is no common standard for M-JPEG streaming for controlling the bit rate in particular. This paper proposes a novel and intelligent approach which can regulate individual frame size of M-JPEG video stream in real-time.

2   Literature Review

As there is a great need to moderate the burstiness of M-JPEG streams and regulate individual frame size of M-JPEG streams, different solutions to control bit-rate of M-JPEG stream have been proposed. These solutions can be classified into three different groups. The first group of solutions typically propose smoothening the bit-rate transmission by setting optimal levels in the buffers; the second one mostly adjusts frame rate in order to achieve desired data rates and the third one generally adapt compression quality (*Q-value*) of M-JPEG compression to attain constant frame size for individual frames. The method used by the first group has a distinct advantage in providing constant image quality and frame rate and thus, its solutions usually aim to smoothen the burstiness in video servers [19–21]. However, buffering usually induce long latency and thus, they are not suitable for many other applications requiring real-time streaming. The method used in the second group can achieve less latency but, due to its frame-rate variation, it is not suitable for applications which are time-sensitive to meet stringent real-time requirements. In such cases, the third group of approaches, which are capable of sustaining constant frame rates as well as regulating frame sizes by adapting the Q-value, are more suitable. In the literature, three different solutions which can regulate frame sizes of individual M-JPEG frames, by adapting the Q-value, are proposed by Nishantha et al.[22], Derin et al. [23], and Silvetre-Blanes et al. [14].

Nishantha et al. [22] proposed a strategy which can regulate the frame size according to the network bandwidth by manipulating the Q value of each individual frame while keeping the frame-rate constant. Increasing and decreasing the value of Q is carried out according to the observation that the frame-size is virtually continuous with respect to Q when it is within the operating range (i.e., the value between 20 and 80). In this approach, the magnitude of the increment or decrement (i.e., ΔQ) is adopted as proposed in Eqn. 1.

$$\Delta Q = \frac{(100 - Q)}{10} \tag{1}$$

Derin et al. [23] recommended an adaptive algorithm which maintains three parameters, namely *QuantScaleCoeff, AggrQScaleFactor*, and *MildQScaleFactor* to perform the adaptations. *QuantScaleCoeff* (Quantization Scaling Coefficient) provides average compression factors between Q value and frame size. *AggrQScaleFactor* (Aggressive Quantization Scaling Factor) is the constant by which previous value of *QuantScaleCoeff* will be multiplied/ divided to obtain its current value in case of aggressive scaling. *MildQScaleFactor* (Mild Quantization Scaling Factor) is the constant to which the previous value of *QuantScaleCoeff* will be multiplied/ divided to obtain its current value in case of mild scaling. Scaling is based on the difference between actual frame size and estimated frame size. When the difference is relatively large, *AggrQScaleFactor* will be used, otherwise *MildQScaleFactor* will be used. When the difference is positive, *QuantScaleCoeff* will be multiplied by either *AggrQScaleFactor* or *MildQScaleFactor* to estimate new *QuantScaleCoeff*. When the difference is negative, *QuantScaleCoeff* will be divided by either *AggrQScaleFactor* or *MildQScaleFactor*.

Silvestre-Blanes et al. [14] suggested the adaptation of Q based on the $R(Q)$ frame bandwidth model shown in Eqn. 2, where α and β are parameters of a curve in which λ regulates the curvature and $\bar{Q} = 100 - Q$ is the compression level which varies symmetrically with respect to the quantification factor

$$R(Q) = \alpha + \frac{\beta}{\bar{Q}^\lambda} \tag{2}$$

Each frame has its own model (α, β, λ). However, α and λ are kept at constant values, based on the assumption that the images acquired in a monitoring application with fixed cameras have a strong similarity between them. The β value is adapted as in Eqn. 3. [24]

$$\beta(i + 1) = \Delta R(i)\bar{Q}^\lambda + \beta(i) \tag{3}$$

where $\Delta R(i)$ is the difference between targeted R value and actual R value. This approach is constrained by the assumption that images should be acquired from fixed cameras and must have a strong similarity. Either shifting of the camera or changing of the background requires re-estimation of the α and λ values. The approach proposed by Nishantha et al. [22] allows shifting of the camera or changing of background. However, it utilizes the increment/ decrement adaptation and thus, unnecessary delays are induced during the dynamic adaptation process in which there are sudden jumps in the desired frame-size level. The approach proposed by Derin et al. [23] makes use of average compression factors between Q value and frame size and thus, it can provide faster adaptation in the case of dynamic adaptations. However, this approach requires two parameters (i.e. *AggrQScaleFactor* and *MildQScaleFactor*) to be tuned properly. Improper tuning of these parameters can result in either a significantly large overshoot or a significantly long settling time; both of which are undesirable for dynamic adaptation. Therefore, this paper aims to provide a novel approach in which the dynamic adaptation can be achieved without tuning any parameter and yet it is able to perform better than the approaches mentioned above.

3   Algorithms of Proposed Approach

In regulating the frame size of M-JPEG compression, the relationship between the frame size and the Q value of JPEG compression is key as the right Q values can provide the target frame sizes for individual frames. However, this relationship is not stationary and it varies based on the contents of each frame. Fig.1 shows the frame size variation of 10 random frames of a video, which was captured from a camera mounted on a moving electric vehicle (EV), with different Q values.

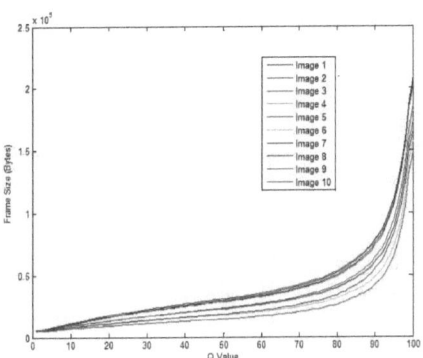

Fig. 1. Frame size vs Q value of 10 different images

According to Fig.1, it can be observed that, although the shapes of these plots are similar, their magnitudes are different. The ratio $(K_n(Q))$ between these plots with respect to that of a reference image, which is the 5th image in this case, can be calculated by using Eqn. 4 and the resulting ratios of above 10 plots are shown in Fig. 2a.

$$K_n(Q) = \frac{S_n(Q)}{S_r(Q)} = \frac{S_n(Q)}{S_5(Q)} \tag{4}$$

where $S_n(Q)$ and $S_r(Q)$ are the frame size of the nth image and the reference image respectively, with the compression quality of Q. According to Fig.2a, the ratios can be modeled as Eqn. 5.

$$\tilde{K}_n(Q) = \begin{cases} (C_n - 1)\frac{Q+10}{30} + 1 & (1 \leq Q < 21) \\ C_n & (21 \leq Q < 81) \\ (C_n - 1)\frac{110-Q}{30} + 1 & (81 \leq Q \leq 100) \end{cases} \tag{5}$$

where $C_n = \frac{1}{60}\sum_{Q=21}^{80} K_n(Q)$ is $\tilde{K}_n$ the estimated value of K_n. The model of the Fig. 2a using Eqn. 5 can be plotted as shown in Fig. 2b.

To study the variation of C_n, a video is captured from a moving EV at 30 fps and the corresponding C_n value of its individual frames is plotted as shown in

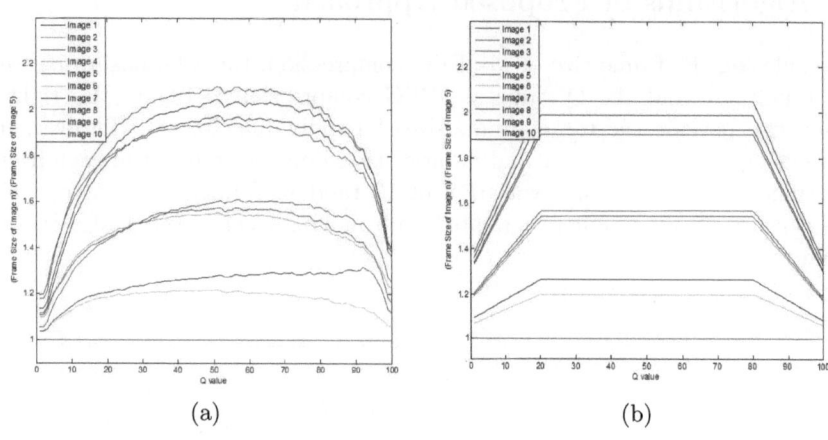

Fig. 2. (a)Frame size ratios vs Q values of 10 different images and (b) its model

Fig.3a. The difference (D_n) between C_n and C_{n-1} can be calculated by Eqn. 6 and the trend of D_n is shown in Fig. 3b.

$$D_n = C_n - C_{n-1} \qquad (6)$$

According to the experiment, $Max(|D_n|) = 0.15664$, $Avg(|D_n|) = 0.006164$ and $Var(|D_n|) = 7.6985 * 10^{-5}$ where $1 \le n \le 7600$. As $Avg(|D_n|) << C_n$ and even $Max(|D_n|) << C_n$, C_{n+1} can be estimated by using Eqn. 7

$$\tilde{C}_{n+1} = C_n \qquad (7)$$

where $\tilde{C}_{n+1}$ is the estimated value of C_{n+1}. In this way, $\tilde{S}_{n+1}(Q)$ can be estimated by using Eqn.8 which is derived from Eqn. 4 and 5.

$$\tilde{S}_{n+1}(Q) = \tilde{K}_n(Q)S_r(Q) = \begin{cases} \{(C_n - 1)\frac{Q+10}{30} + 1\}S_r(Q) & (1 \le Q < 21) \\ C_n S_r(Q) & (21 \le Q < 81) \\ \{(C_n - 1)\frac{110-Q}{30} + 1\}S_r(Q) & (81 \le Q \le 100) \end{cases} \qquad (8)$$

where $\tilde{S}_{n+1}(Q)$ is the estimated value of $S_{n+1}(Q)$.

4   Applying Algorithms of Proposed Approach

Let the first frame be the reference frame and assume that the relationship between the frame size and the Q value (i.e., $S_1(q)|_{1 \le q \le 100}$) is known. Assume that the nth frame size $(S_n(q_n))$, which is compressed with the value of q_n, is known as well. $K_n(q_n)$ can be calculated by using Eqn. 4 with the first frame as a reference frame. That $K_n(q_n)$ is approximately equal to $\tilde{K}_n(q_n)$ and the value

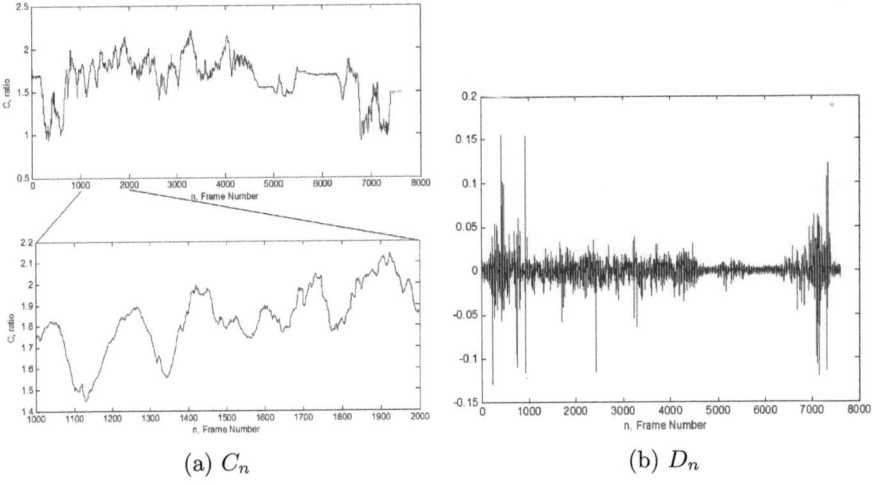

(a) C_n (b) D_n

Fig. 3. Trends of individual frames of video captured from a moving vehicle

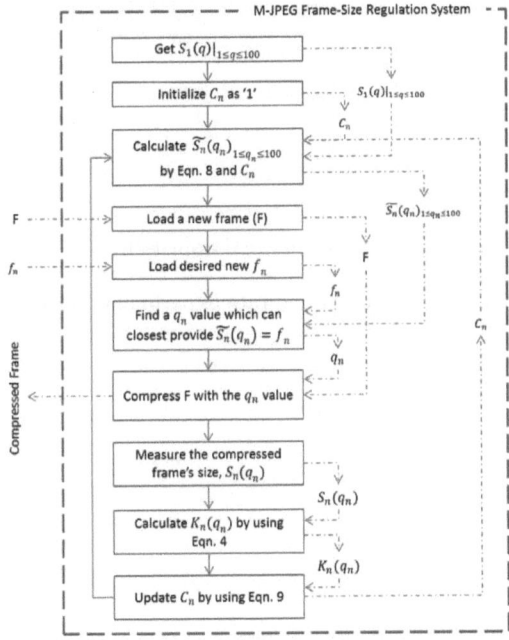

Fig. 4. Flowchart of proposed approach (*dotted arrows show data-flow*)

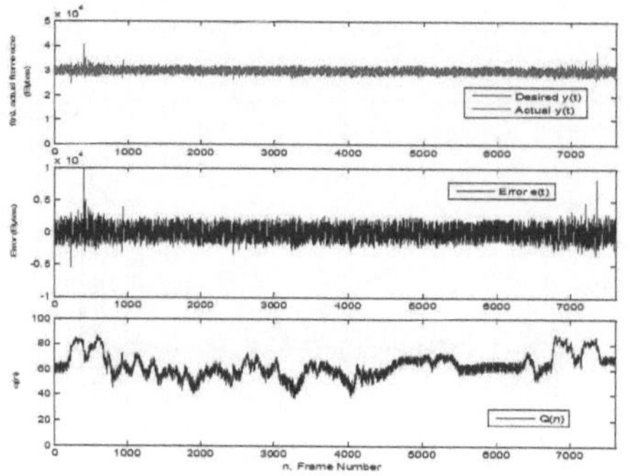

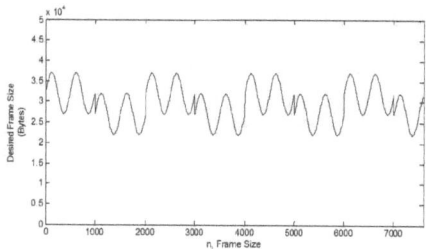

(a) with approach [22] 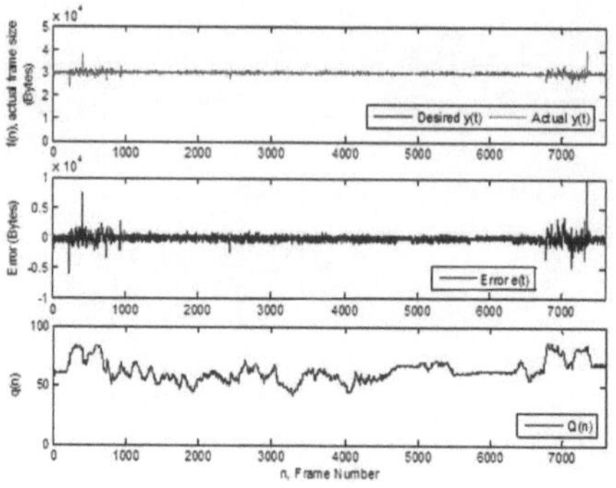   (b) with approach [23]

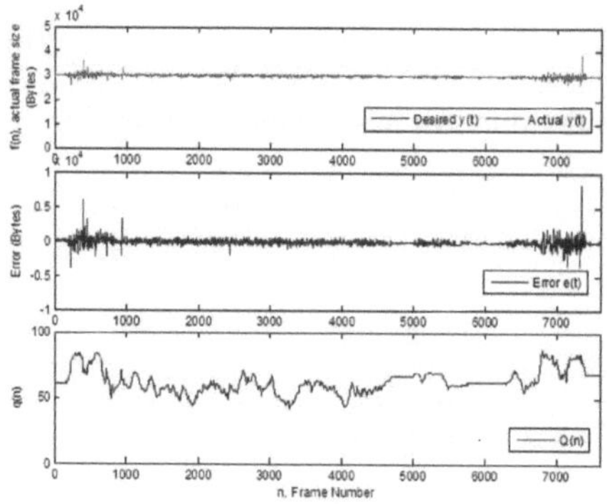

(c) with proposed approach

Fig. 5. Constant frame rate experiments' results

Fig. 6. Desired frame size for dynamic frame rate experiment

of C_n can be calculated by using Eqn. 9 which is derived from Eqn. 5.

$$
C_n = \begin{cases}
\frac{30(\tilde{K}_n(q_n)-1)}{q_n+10} + 1 & (1 \le q_n < 21) \\
\tilde{K}_n(q_n) & (21 \le q_n < 81) \\
\frac{30(\tilde{K}_n(q_n)-1)}{110-q_n} + 1 & (81 \le q_n \le 100)
\end{cases} \tag{9}
$$

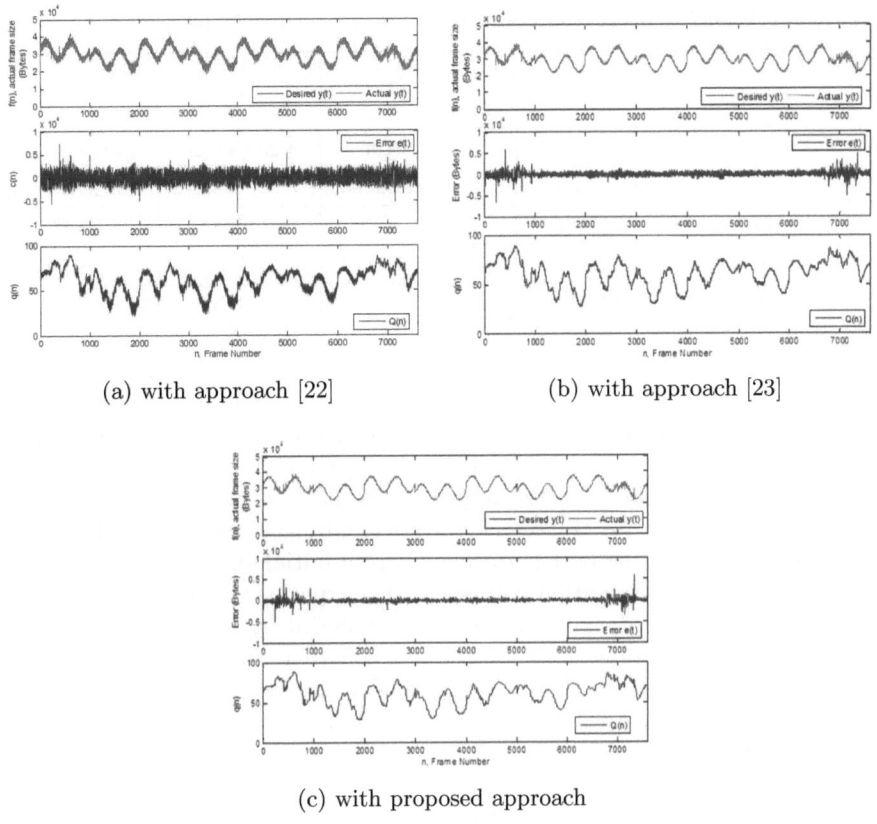

(a) with approach [22]   (b) with approach [23]

(c) with proposed approach

Fig. 7. Dynamic frame rate experiments' results

After knowing C_n, the value for $\tilde{S}_{n+1}(q_{n+1})|_{1 \leq q_{n+1} \leq 100}$ can be calculated by using Eqn. 8. To get the next desired frame size (f_{n+1}), the suitable compression value (q_{n+1}) can be predicted by using the estimated information of $\tilde{S}_{n+1}(q_{n+1})|_{1 \leq q_{n+1} \leq 100}$. This q_{n+1} value can be used to compress the $(n+1)^{th}$ frame and then, the actual size of the compressed frame ($S_{n+1}(q_{n+1})$) can be measured. Again, $K_{n+1}(q_{n+1})$ and C_{n+1} can be calculated by using Eqn. 4 and 9 respectively. In this way, the subsequent values for q (i.e., q_{n+1}, q_{n+2}, etc.) which can provide suitable value of frame size, f (i.e., f_{n+1}, f_{n+2}, etc.) can be estimated.

The above mentioned M-JPEG Frame-Size Regulation algorithm can be implemented according to the following flowchart shown in Fig. 4.

5   Evaluation of the Proposed Approach

In this section, the proposed approach is experimentally compared with the other two similar approaches by Nishantha et al. [22] and Derin et al. [23].

Table 1. RMS error comparison among different approaches

	Constant frame size experiment		Dynamic frame size experiment	
	RMS error (bytes/frames)	RMS error in % of frame size	RMS error (bytes/frames)	RMS error % of frame size
[22] Approach	1211.39	4.04%	1298.83	4.40%
[23] Approach	528.86	1.76%	537.64	1.82%
Proposed Approach	362.70	1.21%	374.26	1.27%

The same set of 7600 video frames used for above C_n trend experiment is used as input frames. As these video frames are captured from a moving camera, the approach in [14], being constraint to stationary camera scenarios, is not used in the comparison. For the approach in [23], optimal parameter values *AggrQScaleFactor* and *MildQScaleFactor* are empirically selected as 0.1 and 0.02 respectively based on the best performance from initial experiment runs.

This performance evaluation consists of two experiment set: constant-frame-size experiment set and dynamic-frame-size experiment set. In the first set of experiments, desired frame sizes are set at a constant value of 30000 bytes and the experiment results are shown in Fig. 5. For the second set of experiments, desired frame sizes are dynamically changing as shown in Fig. 6. Its experiment results are shown in Fig. 7.

From the graph plots, it is clear that there are significant differences in performance while using the three methods for streaming adaptation. To quantify these differences, the RMS errors (i.e., the difference between desired frame size and actual frame size) from both set of experiments are compared in Table 1. Analyzing the data proves that the proposed approach clearly outperforms the other two approaches and it offers a better prediction of Q value in achieving desired individual frame sizes in both the constant-frame-size experiment and the dynamic-frame-size experiment.

6   Conclusion

A novel approach for closely predicting the Q value with minimal latency in time-sensitive M-JPEG streams is proposed in this paper. The proposed approach offers the distinct advantage of no parameter setting requirement enabling a straight forward deployment. Experiment runs were conducted to evaluate the performance of the proposed algorithm in predicting the Q-value for a video sequence from a moving electric vehicle and they verified that the approach outperforms the other known approaches in term of minimizing error in both fixed and dynamic frame-size adaptation for M-JPEG streams.

References

1. Cisco. Cisco Visual Networking Index: Forecast and Methodology (2013-2018),
   `http://www.cisco.com/c/en/us/solutions/collateral/service-provider/`
   `ip-ngn-ip-next-generation-network/white_paper_c11-481360.html`
   (accessed July 16, 2014)
2. Arun, S.N., Lam, W.C., Tan, K.K.: Innovative solution for a telemedicine application. In: 2012 International Conference for Internet Technology And Secured
   Transactions, pp. 778–783. IEEE (2012)
3. Chen, L., Shashidhar, N., Liu, Q.: Scalable secure mjpeg video streaming. In: 2012
   26th International Conference on Advanced Information Networking and Applications Workshops (WAINA), pp. 111–115. IEEE (2012)
4. Qian, L., Jones, D.L., Ramchandran, K., Appadwedula, S.: A general joint source-
   channel matching method for wireless video transmission. In: Proceedings of Data
   Compression Conference, DCC 1999, pp. 414–423. IEEE (1999)
5. Matsuo, M., Ito, R., Kurosaki, M., Sai, B., Kuroki, Y., Miyazaki, A., Ochi, H.:
   Wireless transmission of jpeg 2000 compressed video. In: 2011 13th International
   Conference on Advanced Communication Technology (ICACT), pp. 1020–1024.
   IEEE (2011)
6. Vadakkepat, P., Lim, P., De Silva, L.C., Jing, L., Ling, L.L.: Multimodal approach
   to human-face detection and tracking. IEEE Transactions on Industrial Electronics 55(3), 1385–1393 (2008)
7. Kumar, A.: Computer-vision-based fabric defect detection: a survey. IEEE Transactions on Industrial Electronics 55(1), 348–363 (2008)
8. Cho, C.-S., Chung, B.-M., Park, M.-J.: Development of real-time vision-based fabric inspection system. IEEE Transactions on Industrial Electronics 52(4), 1073–
   1079 (2005)
9. Xie, W.-F., Li, Z., Tu, X.-W., Perron, C.: Switching control of image-based visual
   servoing with laser pointer in robotic manufacturing systems. IEEE Transactions
   on Industrial Electronics 56(2), 520–529 (2009)
10. Hwang, C.-L., Shih, C.-Y.: A distributed active-vision network-space approach
    for the navigation of a car-like wheeled robot. IEEE Transactions on Industrial
    Electronics 56(3), 846–855 (2009)
11. Motai, Y., Kosaka, A.: Hand–eye calibration applied to viewpoint selection for
    robotic vision. IEEE Transactions on Industrial Electronics 55(10), 3731–3741
    (2008)
12. Gomez-Molinero, F.: Real-time requirement of media control applications. In: 19th
    Euromicro Conference on Real-Time Systems, ECRTS 2007, pp. 4–4. IEEE (2007)
13. Rinner, B., Wolf, W.: An introduction to distributed smart cameras. Proceedings
    of the IEEE 96(10), 1565–1575 (2008)
14. Silvestre-Blanes, J., Almeida, L., Marau, R., Pedreiras, P.: Online qos management
    for multimedia real-time transmission in industrial networks. IEEE Transactions
    on Industrial Electronics 58(3), 1061–1071 (2011)
15. Almeida, L., Pedreiras, P., Fonseca, J.A.G.: The ftt-can protocol: Why and how.
    IEEE Transactions on Industrial Electronics 49(6), 1189–1201 (2002)
16. Pedreiras, P., Gai, P., Almeida, L., Buttazzo, G.C.: Ftt-ethernet: a flexible real-
    time communication protocol that supports dynamic qos management on ethernet-
    based systems. IEEE Transactions on Industrial Informatics 1(3), 162–172 (2005)
17. Decotignie, J.-D.: The many faces of industrial ethernet [past and present]. IEEE
    Industrial Electronics Magazine 3(1), 8–19 (2009)

18. Jasperneite, J., Imtiaz, J., Schumacher, M., Weber, K.: A proposal for a generic real-time ethernet system. IEEE Transactions on Industrial Informatics 5(2), 75–85 (2009)
19. Feng, W.-C., Rexford, J.: Performance evaluation of smoothing algorithms for transmitting prerecorded variable-bit-rate video. IEEE Transactions on Multimedia 1(3), 302–312 (1999)
20. Fukuda, K., Wakamiya, N., Murata, M., Miyahara, H.: Qos mapping between users preference and bandwidth control for video transport. In: Building QoS into Distributed Systems, pp. 291–302. Springer (1997)
21. Rexford, J., Sen, S., Basso, A.: A smoothing proxy service for variable-bit-rate streaming video. In: Global Telecommunications Conference, GLOBECOM 1999, vol. 3, pp. 1823–1829. IEEE (1999)
22. Nishantha, D., Hayashida, Y., Hayashi, T.: Application level rate adaptive motion-jpeg transmission for medical collaboration systems. In: Proceedings of 24th International Conference on Distributed Computing Systems Workshops, pp. 64–69. IEEE (2004)
23. Derin, O., Ramankutty, P.K., Meloni, P., Cannella, E.: Towards self-adaptive kpn applications on noc-based mpsocs. Advances in Software Engineering 2012, 11 (2012)
24. Silvestre, J., Almeida, L., Marau, R., Pedreiras, P.: Dynamic qos management for multimedia real-time transmission in industrial environments. In: IEEE Conference on Emerging Technologies and Factory Automation, ETFA, pp. 1473–1480. IEEE (2007)

Automatic Melody Generation Considering User's Evaluation Using Interactive Genetic Algorithm

Mio Takano and Yuko Osana

Tokyo University of Technology,
1401-1 Katakura, Hachioji, Tokyo, Japan
`osana@stf.teu.ac.jp`

Abstract. In this research, an automatic melody generation system considering user's evaluation by interactive genetic algorithm is proposed. In the proposed automatic melody generation system, initial population are generated using the automatic melody generation system by genetic algorithm considering melody blocks of plural melodies, and melodies are generated considering user's evaluation using interactive genetic algorithm. In this system, the trained sample melodies are divided into some melody blocks. Here, melody blocks mean verse, bridge, chorus and so on. And some new melodies are generated considering melody features in each block. The features on rhythm and pitch in each melody block of the sample melodies are trained in some N-gram models, and they are used in order to calculate fitness in the melody generation by genetic algorithm.

1 Introduction

Since the first approach to the automatic composition in 1957, a lot of methods for the automatic composition have been proposed[1]–[3]. As one of these methods, we have proposed the automatic melody generation method using N-gram model and genetic algorithm[3]. In this method, the features on sample melodies are trained using N-gram models[5] per melody blocks. Here, melody block means verse, bridge, chorus and so on. And melodies which have similar features to trained sample melodies can be generated using genetic algorithm[4]. However, in this system, the user's evaluation is not considered in melody generation using genetic algorithm.

In this research, an automatic melody generation system considering user's evaluation by interactive genetic algorithm is proposed. In the proposed automatic melody generation system, initial population are generated using the automatic melody generation system by genetic algorithm considering melody blocks of plural melodies, and melodies are generated considering user's evaluation using interactive genetic algorithm. In this system, the trained sample melodies are divided into some melody blocks, some new melodies are generated considering melody features in each melody block. The features on rhythm and pitch in each melody block of the sample melodies are trained in some N-gram

G. Dick et al. (Eds.): SEAL 2014, LNCS 8886, pp. 785–797, 2014.

models, and they are used in order to calculate fitness in the melody generation by genetic algorithm.

2 Automatic Melody Generation System Using Genetic Algorithm Considering Melody Blocks of Plural Melodies

In this section, the automatic melody generation system using genetic algorithm considering melody blocks of plural melodies which is used in the proposed melody generation system is explained.

In this system, the trained sample melodies are divided into some melody blocks manually, and some new melodies are generated considering melody features in each block. The features on rhythm and pitch in each melody block of sample melodies are trained using some N-gram models, and they are used in order to calculate fitness in the melody generation by the genetic algorithm.

2.1 Flow of Automatic Melody Generation System by Genetic Algorithm Considering Melody Blocks of Plural Melodies

Step 1: Initial Population Generation

In this automatic melody generation system, initial individuals are generated randomly using features on rhythm and pitch in each melody block of training sample melodies.

Step 2: Fitness Calculation

The fitness of each individual is calculated. In this system, the fitness of the individuals which have similar feature to the trained sample melodies becomes high.

Step 3: Selection

Based on fitness calculated in **Step 2**, individuals used in **Step 4** (crossover) are selected by the roulette selection and the elite preserve strategy.

Step 4: Crossover

New individuals are generated from the parents which are selected in **Step 3** by the multi-point crossover. In this system, in order not to generate the melodies which whose notes and rests whose length and position are unnatural, a crossover point is chosen from places other than the middle of a triplet or a rest.

Step 5: Mutation

In order to maintain genetic diversity, the mutation is carried out.

Step 6: Repeat

Steps 2 $\sim$ 5 are repeated T_{max} times.

2.2   Melody Expression in Individuals

In this melody generation system, generated melodies are expressed as individuals. The individual is composed of two parts; (1) rhythm and (2) tone. Rhythm and tone are expressed per unit length. In this system, unit length (that is, minimum length of note/rest) is set to 1/3 length of the demisemiquaver (sixteenth note). The rhythm part is expressed by 0 (rest), 1 (the beginning of the sound) and 2 (the state that the sound continue) per unit length. The tone part is expressed by pitch names and octave such as C_4, A_5.

2.3   Initial Population Generation

In this melody generation system, based on features in each melody block of all training melodies, initial individuals are generated. The initial population generation process is composed of two steps; (1) generation of rhythm (phonetic value) and (2) assignment of pitch to each sound.

(1) Generation of Rhythm (Phonetic Value). First, the rhythm of initial individual is generated based on the Markov model of state transition sequences in each melody block.

In the Markov model of the melody block k (M_R^k), the probability where the state q_i continues after the state q_{i-1}, $P^k(q_i|q_{i-1})$ is calculated as

$$P^k(q_i|q_{i-1}) = \frac{N^k(q_{i-1}^i)}{N^k(q_{i-1})}(q_i \in C_{R_2}) \tag{1}$$

where $N^k(q_{i-1})$ is the number of state sequences in the melody block k in all trained sample melodies and $N^k(q_{i-1}^i)$ is the number of state sequences in the melody block k in all trained sample melodies. q_i is the rhythem state per two bars. C_{R_2} shows the set of rhythm states per two block and it is described as

$$C_{R_2} = \{b_1, b_2, e_1, e_2, s_1, \cdots, s_{96}, r_1, \cdots, r_{96}\} \tag{2}$$

where b_1 and b_2 are the beginning of the block composed of two bar. b_1 show the case when the first sound begins at the beginning of the block. In contrast, b_2 shows the case when the first sound begins at the end of the previous block. e_1 and e_2 are the end of the block. e_1 shows the case when the last sound ends in the block, and e_2 shows the case when the last sound continue to the next block. s_x shows the sound which begins at the position x $(x = 0 \sim 96)$, r_x shows the rest which begins at the position x in each two bar.

The first state in the melody block which is next to the state b_1 or b_2 is generated using the probability which is given by Eq.(1), and the remain rhythm state sequences are generated randomly using the Markov model M_R^k.

(2) Assignment of Pitch to Each Sound. Second, pitch is assigned to each sound based on the occurrence probability of pitch in each position.

The probability of the sound whose pitch is o at the position i in the melody block k, $P^k(s_i^o)$ is given by

$$P^k(s_i^o) = \frac{N^k(s_i^o)}{N^k(s_i)} \tag{3}$$

where $N^k(s_i)$ is the number of sounds whose pitch is o and begin at the position i in the melody block k in all trained sample melodies, and $N^k(s_i^o)$ is the number of the sounds which begin at the position i in the melody block k in all trained sample melodies. The pitch of each sound is determined based on the probability $P^k(s_i^o)$.

2.4   Fitness

In this automatic melody generation system, the fitness of the individual is calculated as summation of (1)rhythm transition, (2) pitch transition, (3) pitch and length transition, (4) transition of the number of sounds per bar, (5) rhythm similarity between phrases, (6) rate of scale unique sounds, (7) pitch difference between two consecutive sounds and (8) distribution of rest length.

Fitness on Transition of Rhythm, Pitch and the Number of Sounds per Bar. In this system, N-gram models on rhythm, tone and the number of sounds per bar are used in order to calculate fitness.

The N-gram model on the feature f is given by

$$V_f(g) = \sum_{k=1}^{K} \frac{1}{N_f(k)} \sum_{j=1}^{N_f(k)} \left(\frac{1}{N_f(g,k,j)} \right.$$

$$\left. \times \sum_{i=1}^{N_f(g,k,j)} P^k_{m(g,k,f,j,i)}(S_f(g,k,j)_i | S_f(g,k,j)_{i-N+1}^{i-1}) \right) \tag{4}$$

where f is given by

$$f \in \{R_2, R_4, R_{last}, T, T_{last}, S, S_{last}, B\}. \tag{5}$$

Here, R_2 is rhythm per two bars, R_4 is rhythm per four bars, R_{last} is rhythm in the last two bars of each melody block, T is pitch in the whole melody block, T_{last} is pitch in the last two bars, S is pitch and length of sound in the whole melody block, S_{last} is pitch and length of sound in last two bars, and B is the number of sounds in a bar.

$N_f(k)$ is the number of state sequence blocks on the feature f in the melody block k of the generated melody and it is given by

$$N_f(k) = \begin{cases} \lfloor N^B(k)/2 \rfloor, & (f = R_2) \\ \lfloor N^B(k)/4 \rfloor, & (f = R_4) \\ 1, & (\text{otherwise}). \end{cases} \tag{6}$$

Here, $N^B(k)$ is the number of bars in the melody block k of the generated melodies and this is determined based on the trained sample melodies. If the numbers of bars in the melody block k of the trained sample melodies are not same, the average number of bars is used. $\lfloor N^B(k)/2 \rfloor$ shows the number of blocks per two bars in the melody block k, and $\lfloor N^B(k)/4 \rfloor$ shows the number of blocks per four bars in the melody block k.

$N_f(g, k, j)$ is the number of states on the feature f of the state sequence block j in the melody block k of the generated melody expressed by the individual g and it is given by

$$
N_f(g, k, j) = \begin{cases}
N_2^S(g, k, j) + 2, & (f = R_2) \\
N_4^S(g, k, j) + 2, & (f = R_4) \\
N_{last}^S(g, k) + 2, & (f = R_{last}) \\
N^S(g, k), & (f = T, S) \\
N_{last}^S(g, k), & (f = T_{last}, S_{last}) \\
N^B(k), & (f = B)
\end{cases}
\tag{7}
$$

where $N_2^S(g, k, j)$ is the number of sounds in the block (composed of two bars) j of the melody block k expressed by the individual g, $N_4^S(g, k, j)$ is the number of sounds in the block (composed of four bars) j of the melody block k expressed by the individual g, $N_{last}^S(g, k)$ is the number of sounds in the last two bars of the melody block k expressed by the individual g, $N^S(g, k)$ is the number of sounds in the melody block k expressed by the individual g, and $N_{last}^S(g, k)$ is the number of sounds in last two bars of the melody block k expressed by the individual g.

$S_f(g, k, j)_i$ is the ith state on the feature f of the state sequence block j in the melody block k of the individual g, and it is given by

$$
S_f(g, k, j)_i \in \begin{cases}
C_{R2} \to C_{R2}, & (f = R_2, R_{last}) \\
C_{R4} \to C_{R4}, & (f = R_4) \\
C_T, & (f = T, T_{last}) \\
C_S, & (f = S, S_{last}) \\
C_B, & (f = B)
\end{cases}
\tag{8}
$$

$$C_{R2} = \{b_1, b_2, e_1, e_2, s_1, \cdots, s_{96}, r_1, \cdots, r_{96}\} \tag{9}$$

$$C_{R4} = \{b_1, b_2, e_1, e_2, s_1, \cdots, s_{192}, r_1, \cdots, r_{192}\} \tag{10}$$

$$C_T = \{X_d \mid X = \{C, C\sharp, D, D\sharp, E, F, F\sharp, G, G\sharp, A, A\sharp, B, R\}, \ d = \{3, \cdots, 6\}\} \tag{11}$$

$$C_S = \{X_d^l \mid X = \{C, C\sharp, D, D\sharp, E, F, F\sharp,$$
$$G, G\sharp, A, A\sharp, B, R\}, \ d = \{3, \cdots, 6\}, \ l = \{1, \cdots, 48\}\} \tag{12}$$

$$C_B = \{1, \cdots, 48\} \tag{13}$$

$S_f(g, k, j)_{i-N+1}^i$ is the state sequence from $i - N + 1$-th state to the $i - 1$-th state on the feature f in the block j of the melody block k expressed by the individual g.

And, in Eq.(4), N^{MB} is the number of melody blocks of the generated melodies and it is determined based on the trained sample melodies. And,

$P_m^k(S_f(g, k, j)_i \mid S_f(g, k, j)_{i-N+1}^i)$ is the probability where the state on the feature f, $S_f(g, k, j)_i$ appears after the state sequence $S_f(g, k, j)_{i-N+1}^i$ in the melody block k in the trained sample melody m and it is estimated in the N-gram model on the feature f of the melody block k in the trained sample melody m ($M_f^k(m)$). In this system, the N-gram model for the trained melody m whose probability is maximum is selected as the N-gram model which is employed to calculate fitness in principle.

$P_{m(g,k,f,j,i)}^k(S_f(g, k, j)_i \mid S_f(g, k, j)_{i-N+1}^{i-1})$ is the probability where the state on the feature f, $S_f(g, k, j)^i$ appears after the state sequence on the feature f, $S_f(g, k, j)_{i-N+1}^{i-1}$ in block j in the melody block k of the gene g and it is estimated in the N-gram model on the feature f of the melody block k in the trained sample melody $m(g, k, f, j, i)$, $M_f^k(m(g, k, f, j, i))$.

In this system, when plural melodies are used as the training sample melodies, the N-gram model whose occurrence probability is highest is selected in order to calculate the fitness. However, in this selection method, different N-gram models are often selected for the consecutive state sequences, as a result, the fitness of the individuals which express unnatural melodies sometimes become high.

In this system, $m(g, k, f, j, i)$ is determined by

$$m(g, k, f, j, i) = \operatorname*{argmax}_m P_m^k(S_f(g, k, j)_i \mid S_f(g, k, i)_{i-N+1}^{i-1}) \tag{14}$$

After N-gram model which used in the calculation of fitness selected by this equation, $m(g, k, f, j, i)$ is determined as follows finally.

$$m(g, k, f, j, i) \begin{cases} m(g, k, f, j, i - 1) \\ (m(g, k, f, j, i - 1) = m(g, k, f, j, i + 1) \\ \text{and } (m(g, k, f, j, i - 2) = m(g, k, f, j, i - 1) \\ \text{or } (m(g, k, f, j, i + 2) = m(g, k, f, j, i + 1))) \\ m(g, k, f, j, i)(\text{otherwise}) \end{cases} \tag{15}$$

That is, in this system, if the following conditions are satisfied, the N-gram model which is selected for the previous and next state sequence is chosen.

1. N-gram model which learns the feature from different melodies are selected for the previous and next state sequences.
2. N-gram model which learns the feature from different melodies are selected for the previous previous and/or next next state sequences.

Fitness on Rhythm Similarity between Phrases. In actual melodies, rhythm of the phrase for every four bars is often mutually similar. So, in this system, the rhythm similarity between phrases are used as the fitness.

The fitness on rhythm similarity between phrases of the individual g, $V_{RS}(g)$ is given by

$$V_{RS}(g) = \sum_{k=1}^{N^{MB}} f_{RS}\left(\frac{1}{\lfloor N^B(k)/4 \rfloor C_2} \sum_{i=1}^{\lfloor N^B(k)/4 \rfloor - 1} \sum_{j=i+1}^{\lfloor N^B(k)/4 \rfloor}\right.$$

$$\left(L_{ij}(g, k) + a \cdot 1 \left(n_{ij}(g, k)/N_i(g, k) + n_{ji}(g, k)/N_j(g, k) \right) \right) \quad (16)$$

where $\lfloor N^B(k)/4 \rfloor$ is the number of phrases in the melody block k, and $\lfloor N^B(k)/4 \rfloor C_2$ means the combination number of two phrases. In Eq.(16), average similarity in all considerable combination of phrases are calculated. Here, $L_{ij}(g, k)$ is the rate of sounds (including rests) whose length and its start position are same in the phrases i and j.

$N_i(g, k)$ shows the number of sounds (including rests) in the phrase i of the melody block k of the individual g, $N_j(g, k)$ is the number of sounds (including rests) in the phrase j of the melody block k of the individual g, $n_{ij}(g, k)$ is the number of sounds (including rests) in the phrase i which begin at the same position (time) to the phrase j, $n_{ji}(g, k)$ is the number of sounds (including rests) in the phrase j which begin at the same position (time) to the phrase i, and a is the weighting coefficient.

Fitness on Rate of Scale Unique Sounds. The most of the melodies in each key consist of their own scale unique sounds. For example, scale unique sounds of C major are C, D, E, F, G, A and B, and these of G major are G, A, B, C, D, E and $F\sharp$.

The fitness on rate of scale unique sounds in the individual g, $V_{US}(g)$ is calculated as

$$V_{US}(g) = \sum_{k=1}^{N^{MB}} f_{US}^k \left(\frac{N^{US}(g, k)}{N^S(g, k)} \right) \quad (17)$$

where $N^S(g, k)$ shows the number of sounds in the melody block k which is expressed by the individual g. $N^{US}(g, k)$ shows the number of the scale unique sounds in the melody block k which is expressed by the individual g. $f_{US}^k(\cdot)$ is the function which is given by

$$f_{US}^k(u) = \begin{cases} 1 & (\theta_{US}^k < u) \\ \dfrac{u}{\theta_{US}^k} & \text{(otherwise)} \end{cases} \quad (18)$$

where θ_{US}^k shows the threshold for the rate of scale unique sounds in the melody block k. And it is determined as follows:

$$\theta_{US}^k = \min_m \left\{ N^{US}(m, k)/N^S(m, k) \right\} \quad (19)$$

where $N^S(m, k)$ shows the number of sounds in the melody block k of the trained sample melody m. $N^{US}(m, k)$ shows the number of scale unique sounds in the melody block k of the trained sample melody m.

Fitness on Pitch Difference between Two Consecutive Sounds. The fitness on pitch difference between two consecutive sounds of the individual g, $V_D(g)$ is given as

$$V_D(g) = \sum_{k=1}^{N^{MB}} \frac{1}{2D_{max}+1} \sum_{i=-D_{max}}^{D_{max}} f_{D_i}^k \left(\frac{N_i^D(g,k)}{N^D(g,k)} \right) \tag{20}$$

where D_{max} is the maximum pitch difference between two sounds, $N^D(g,k)$ is the number of pitch differences between two sounds in the melody block k which is expressed by the individual g, and $N_i^D(g,k)$ is the number of pitch differences between two sounds whose value is i in the melody block k of the individual g. $f_{D_i}^k(\cdot)$ is the function which is given as

$$f_{D_i}^k(u) = \begin{cases} \dfrac{1-u}{1-\theta_{D2}^i(k)} & (\theta_{D2}^i(k) < u) \\[2ex] 1 & (\theta_{D1}^i(k) \le u \le \theta_{D2}^i(k)) \\[2ex] \dfrac{u}{\theta_{D1}^i(k)} & (\theta_{D1}^i(k)) \end{cases} \tag{21}$$

where $\theta_{D1}^i(k)$ and $\theta_{D2}^i(k)$ are the thresholds on the pitch difference i between two consecutive sounds in the melody block k and $\theta_{D1}^i(k) \le \theta_{D2}^i(k)$. These are determined by

$$\theta_{D1}^i(k) = \min_m \left\{ \frac{N_i^D(m,k)}{N^D(m,k)} \right\} \tag{22}$$

$$\theta_{D2}^i(k) = \max_m \left\{ \frac{N_i^D(m,k)}{N^D(m,k)} \right\} \tag{23}$$

where $N^D(m,k)$ shows the number of pitch differences between two consecutive sounds in the melody block k of the trained sample melody m. $N_i^D(m,k)$ shows the number of pitch differences whose value is i in the melody block k of the trained sample melody m.

Fitness on Length of Rests. The fitness on length of rests in the individual g, $V_{RE}(g)$ is given by

$$V_{RE}(g) = \sum_{k=1}^{N^{MB}} \frac{1}{48} \sum_{i=1}^{48} f_{RE} \left(\left(P^k(N_l^R) - P^k(N_l^R(g,k)) \right)^2 \right) \tag{24}$$

where $P^k(N_l^R)$ is the rate of rests whose length is l in the melody block k of the trained sample melodies, and $P(N_l^R(g,k))$ is the rate of rests whose length is l in the melody block k which is expressed by the individual g. These are calculated by

$$P^k(N_l^R) = \frac{\sum\limits_{m} N_l^R(m,k)}{\sum\limits_{m} N^R(m,k)} \tag{25}$$

$$P^k(N_l^R(g)) = \frac{N_l^R(g)}{N^R(g)} \tag{26}$$

where $N_l^R(m,k)$ is the number of rests whose length is l in the melody block k of the trained sample melody m, $N^R(m,k)$ is the number of rests in the melody block k of the trained sample melody m, $N_l^R(g,k)$ is the number of rests whose length is l in the melody block k which is expressed by the individual g and $N^R(g,k)$ is the number of rests in the melody block k which is expressed by the individual g. In Eq.(24), $f_{RE}(\cdot)$ is the function which is given by

$$f_{RE}(u) = \begin{cases} 1 & (u < \theta_{RE1}) \\ 0.5 & (\theta_{RE1} \le u < \theta_{RE2}) \\ 0 & (\text{otherwise}) \end{cases} \tag{27}$$

where θ_{RE1}, θ_{RE2} are the thresholds and $\theta_{RE1} < \theta_{RE2}$.

3 Automatic Melody Generation Reflecting User's Evaluation Using Interactive Genetic Algorithm

In the proposed automatic melody generation system reflecting user's evaluation using interactive genetic algorithm, initial population are generated by the automatic melody generation system by genetic algorithm considering melody blocks of plural melodies described in **2**, and melodies are generated considering user's evaluation using interactive genetic algorithm.

3.1 Melody Expression by Individuals

In the proposed automatic melody generation system, generated melodies are expressed as individuals which is composed of two parts; (1) rhythm and (2) pitch in the same manner used in the system which is used in the generation of the initial population described in **2**.

3.2 Initial Population Generation

In the proposed automatic melody generation system, initial individuals are generated by the automatic melody generation system using genetic algorithm considering melody blocks of plural melodies described in **2**.

3.3 User's Evaluation

Trends of user's evaluation are analyzed and fitness are calculated based on analyzed results in the proposed automatic melody generation system. Here, the melodies are divided into some phrases and a user evaluates these phrases in 5 $(-2 \sim 2)$ degrees.

3.4 Fitness

In the proposed automatic melody generation system, the fitness which are used in the automatic composition system using genetic algorithm considering melody blocks of plural melodies described in **2** are used. In addition, the fitness considering user's evaluation are also used.

In the proposed system, three fitness considering user's evaluation (A) rhythm transition (whole melodies), (B) pitch transition (whole melodies) and (C) pitch and length transition (whole melodies) are used.

These are calculated using N-gram models which learns features of the melodies which was evaluated by a user. The N-gram model on the feature f is given as

$$
V_f(g) = \sum_{k=1}^{N^{MB}} \frac{1}{N_f(k)} \sum_{j=1}^{N_f(k)} \left(\frac{1}{N_f(g,k,j)} \right.
$$

$$
\left. \sum_{i=1}^{N_f(g,k,j)} \left(e^+ (S_f(g,k,j)^i_{i-N+1}) + e^- (S_f(g,k,j)^i_{i-N+1}) \right) \right) \tag{28}
$$

where f is given by

$$
f \in \{IR_2, IT, IS\}. \tag{29}
$$

where IR_2 is the rhythm transition per two bars, IT is the pitch transition, and IS is the the pitch and length transition.

$N_f(k)$ is the number of state sequence blocks on the feature f of the melody block k and is given by

$$
N_f(k) = \begin{cases} \lfloor N^B(k)/2 \rfloor & (f = IR_2) \\ 1 & (\text{otherwise}) \end{cases} \tag{30}
$$

where $N^B(k)$ is the number of bars in the melody block k.

$N_f(g,k,j)$ is the number of states on the feature f of the state sequence block j of the melody block k in the generated melody expressed by the individual g and it is given by

$$
N_f(g,k,j) = \begin{cases} N_2^S(g,k,j) + 2 & (f = IR_2) \\ N^S(g,k) & (f = IT, IS) \end{cases} \tag{31}
$$

where $N_2^S(g,k,j)$ is the number of sounds in the block composed of two bars j of the melody block k which is expressed by the individual g, $N^S(g,k)$ is the number of sounds in the melody block k expressed by the individual g.

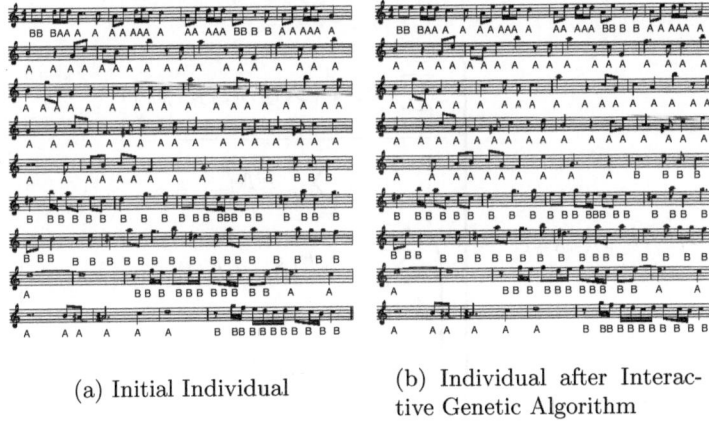

(a) Initial Individual	(b) Individual after Interactive Genetic Algorithm

Fig. 1. An Example of Generated Melody

$S_f(g, k, j)_i$ is the i-th state on the feature f of the state sequence block j of the melody block k in the individual g, and is given by

$$S_f(g, k, j)_i \in \begin{cases} C_{R_2} \to C_{R_2} & (f = IR_2) \\ C_T & (f = IT) \\ C_S & (f = IS) \end{cases} \tag{32}$$

$S_f(g, k, j)^i_{i-N+1}$ shows the state sequece from the $i - N + 1$-th state to the i-th state on the feature f of the state sequence block j of the melody block k in the individual g.

$e^+(S_f(g, k, j)^i_{i-N+1})$ and $e^-(S_f(g, k, j)^i_{i-N+1})$ are the trend of user's evaluation which are given by

$$e^+(S_f(g, k, j)^i_{i-N+1}) = \begin{cases} \dfrac{\sum\limits_{E=1}^{2} EN_k^E(S_f(g, k, j)^i_{i-N+1})}{\sum\limits_{E=1}^{2} EN_k^E} \\ \left(\dfrac{\sum\limits_{E=1}^{2} EN_k^E(S_f(g, k, j)^i_{i-N+1})}{\sum\limits_{E=1}^{2} EN_k^E} > \theta_f \right) \\ 0 \quad \text{(otherwise)} \end{cases} \tag{33}$$

$$e^-(S_f(g,k,j)^i_{i-N+1}) = \begin{cases} \dfrac{\displaystyle\sum_{E=-2}^{-1} EN_k^E(S_f(g,k,j)^i_{i-N+1})}{\displaystyle\sum_{E=-2}^{-1} EN_k^E} \\[2em] \left(\dfrac{\displaystyle\sum_{E=-2}^{-1} EN_k^E(S_f(g,k,j)^i_{i-N+1})}{\displaystyle\sum_{E=-2}^{-1} EN_k^E} > \theta_f\right) \\[2em] 0 \quad \text{(otherwise)} \end{cases} \tag{34}$$

where E is the evaluation by a user. In this system, an user evaluates each phrases as $-2, -1, 0, 1$ or 2. N_k^E is the number of phrases in the melody block k which are evaluated as E, $N_k^E(S_f(g,k,j)^i_{i-N+1})$ shows the number of melodies which include state sequece $N_k^E(S_f(g,k,j)^i_{i-N+1})$ and the its evaluation is E.

4   Computer Experiment Results

Figure 1 shows an example of generated melody of the proposed method which learns the features of two sample melodies (Japanese Animation Songs). Fig. 1(a) shows an example of the generated melody as an initial individual, and A and B show the N-gram model which trained sample melody A or B was used for the calculation of fitness.

Figure 2 shows the transition of the average and maximum fitness. In this experiment, the fitness in first 30 generatations is calculatesd without interactive genetic algorithm.

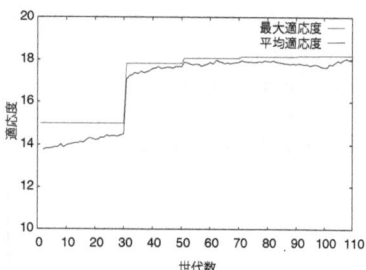

Fig. 2. Fitness

5   Conclusions

In this research, the automatic melody generation system reflecting user's evaluation using interactive genetic algorithm has been proposed. We generated some melodies using the proposed system, and confirmed that the some melodies reflecting the features in each melody block of the trained sample melodies and user's evaluation can be generated.

References

1. Burns, K.H.: The history and development of algorithms in music composition, 1957–1993. Ball State University (1994)
2. Tomari, M., Sato, M., Osana, Y.: Automatic composition based on genetic algorithm and N-gram model. In: Proceedings of IEEE Conference on System, Man and Cybernetics, Singapore (2008)
3. Takano, M., Osana, Y.: Automatic composition system using genetic algorithm and N-gram model considering melody blocks. In: Proceedings of IEEE Congress on Evolutionary Computation, Brisbane (2012)
4. Goldberg, D.E.: Genetic Algorithms in Search, Optimization, and Machine Learning. Addison-Wesley Longman Publishing (1989)
5. Manning, C.D., Schütze, H.: Foundations of Statistical Natural Language Processing. MIT Press (1999)

Object Recognition by Stochastic Metric Learning

Oliver Batchelor and Richard Green

Department of Computer Science
University of Canterbury
Christchurch, New Zealand
oliver.batchelor@pg.canterbury.ac.nz, richard.green@canterbury.ac.nz

Abstract. Descriptors extracted from deep neural networks have been shown to be very discriminative, for example networks such as those trained on the large, very general ImageNet dataset have been used to extract descriptors robust for a variety of image classification tasks. Such retrieval systems utilize feature locality, for example Approximate Nearest Neighbour. Our goal is to use such descriptors as part of a large scale object instance recognition and retrieval system. We propose using deep nonlinear metric learning on Convolutional Neural Networks to learn features with good locality. In particular we worked with two related methods, Neighborhood Components Analysis (NCA) and the related Mean square Error's Gradient Minimization (MEGM).

We utilize a nonlinear form of MEGM as an alternative to NCA and propose some stochastic sampling methods to apply these (normally batch) methods to larger datasets with minibatch Stochastic Gradient Descent (SGD). On a larger scale we found the methods difficult to train, failing to converge or generalizing very badly depending on training method or parameters. This led us to go back to a smaller dataset and examine the factors which lead to good generalization with this form of training.

We found on a small subset of the RGB-D dataset, surprisingly stochastic sampling methods generalized much better with small batch sizes, which acted as a form of regularization. When trained with larger batches, or as a full batch, the dataset was overfit. Given the correct parameters, descriptors extracted performed well at the Nearest Neighbour task and exceeded the performance of those extracted by applying standard supervised training.

1 Introduction

Deep convolutional neural networks in combination with modern GPUs and large image datasets have shown strong performance on image classification tasks [1], and has been applied to related problems such as object detection [2], image segmentation [3] and image retrieval [4].

1.1 Descriptors from Deep Neural Networks

Using descriptors derived from the hidden layers of a neural networks trained using supervised learning, for the purpose of other learning tasks is a relatively

G. Dick et al. (Eds.): SEAL 2014, LNCS 8886, pp. 798–809, 2014.

new idea. These descriptors have been shown to be robust even for quite unrelated tasks [5,4]. The ImageNet dataset [6] is a popular source for pretraining, and pre-trained models exist such as the OverFeat network [2] or the DeCAF feature extractor [5]).

A standard technique in training a Convolutional Neural Netowork (CNN) is to augment the dataset by applying transformations, more data typically gives better generalization. In [7], a CNN was trained on single images which were warped in many different ways. Features obtained from the network were then used in popular classification benchmarks achieving good results. For many years local image descriptors such as Scale Invariant Feature Transform (SIFT) [8] have been used for matching and indexing images, a recent comparison [9] (though perhaps not a fair one) showed that using a CNN for matching tasks performed better than SIFT by a margin similar to the improvement given by SIFT to raw pixel data.

The final layer of a standard deep neural network as used in supervised classification consists of a set of linear classifiers, as such those descriptors are suitable for classification using other linear classifiers such as SVMs. Nearest Neighbor suffers from high dimensionality and noisy or irrelevant dimensions, as such the descriptors produced by a CNN may not be suitable for comparison by distance. For that reason we have looked towards metric learning to directly optimise the descriptors for the purpose of Nearest Neighbor classification.

1.2   Deep Metric Learning

Metric learning has often been used for object recognition and image classification [10,11] (and many others), and especially face recognition, for example [12]. Although most efforts often have focued on mahalanobis distance metric learning (a form of distance metric learning linear transformation), deep metric learning has had some attention [13,11,14,15]. At the expense of much larger computation cost, deep metric learning has been shown to perform much better than its linear counterparts. We use gradients from metric learning to drive Stochastic Gradient Descent (SGD) on a deep CNN.

1.3   Training

Metric learning comes with its own set of challenges, it has often been formulated as batch training method because each example potentially interacts with every other example. In practice descriptors from examples far apart don't interact with each other at all, so approximations can be made as we discuss later. This runs into issues relating to high dimensional spaces, namely the "curse of dimensionality". In high dimensional spaces, such as those we deal with in this paper, if the points (descriptors) were uniformly spaced then on average the number of neighbours increases with dimension.

The interaction between points decays with distance (for example exponentially with NCA). We can use approximations around the local neighbourhood of examples which can be used to create an SGD training proceedure. Using an approximation to the nearest k neighbours is a popular approach, seen in [16,17] (of many). Clustering (amongst other sampling methods) is discussed in [18] such as Farthest Point or Random Projection clustering, the downside of such clustering is that it is hard to control the size of a batch.

Many training methods focus on interaction between pairs of (similar/dissimilar) examples or triples (example, more similar, less similar), for example DrLIM [10] where a spring analogy is used to create an attraction between similar pairs and a repulsion between dissimilar pairs, the advantage with this kind method is that it can be used without explicit class labels, but just a similar/dissimilar annotation.

2 Deep Metric Learning

A deep neural network (in our case a CNN) is used to to create an embedding into a lower dimensional space, creating descriptors which can be compared with their euclidean distance (and classified with nearest neighbour) where the euclidean distance of the raw pixels is both expensive and not a good measure of the distance of the semantic similarity of image content.

We examine non linear versions of two methods NCA [19], the closely related, but less well known MEGM [17]. NCA optimizes a continuous version of the Leave One Out (LOO) performance, it uses a softmax over weights which decay exponentially with distance. The NCA score can be interpreted as the probability that a descriptor will pick another descriptor of the correct class as its nearest neighbour.

The probability, p_{ij}, of one descriptor selecting another descriptor, as its neighbour is defined as a softmax function over weights W_{ij}. The indexes i and j refer to input examples x_i and x_j, and corresponding vector valued output of a CNN $f(x_i)$ and $f(x_j)$ which are the descriptor vectors.

$$p_{ij} = \frac{W_{ij}}{\sum_{k \neq i} W_{ik}} g \tag{1}$$

Then the total probability, p_i, of a point selecting any neighbour with another with its *correct* class is defined as the sum of those neighbour probabilities p_{ij} which have the same class:

$$p_i = \sum_{j:c_j = c_i} p_{ij} \tag{2}$$

Where C_i is the class label of example i. We use a gaussian kernel for the weighting as [17] do.

$$W_{ij} = exp(\frac{-\|f(x_i) - f(x_j)\|_2^2}{2\sigma^2}), W_{ii} = 0 \tag{3}$$

Then the function to be maximized, is the total sum of the probabilities of all descriptors being correctly classified.

$$\mathcal{E}_{nca} = \sum_i p_i \tag{4}$$

Where NCA optimises directly on the probability p_i above, MEGM instead computes for each class $\hat{y}_{ti}$ as a prediction that a descriptor will take class t, where the only difference is that $c_j = t$ as opposed to $c_j = c_i$:

$$\hat{y}_{ti} = \frac{\sum_{j:c_j=t} W_{ij}}{\sum_{k\neq i} W_{ik}} \tag{5}$$

The prediction $\hat{y}_{ti}$ can then be compared with y_{ti} (1 where $t = c_i$, 0 otherwise), it then minimizes the Mean Squared Error (MSE) between prediction and true class label:

$$\mathcal{E}_{megm} = \sum_i \sum_t (y_{ti} - \hat{y}_{ti})^2 \tag{6}$$

Intuitively MEGM can be seen to penalize the case where two classes compete for the same region more so than when one class competes against examples of many different classes, where as NCA would treat the two cases approximately equally. These loss functions can be used to drive gradient descent on a CNN by standard backpropogation. The derivative for MEGM is shown in the appendix, section 7.

We compute the gradient over the outputs of each minibatch and apply backpropogation as usual to find the derivative with respect to the weights of the network. It can be noted that the output, and derivative for MEGM is more expensive to compute because of the additional per class summation, so it would not be suitable with an extremely large number of classes. In practice a large number of the terms can be factored out and precomputed, as well as computing the difference summations in terms of matrix multiplication.

Note the parameter σ was not in the original NCA, and is initialized to the average distance to the nearest neighbours of the initial descriptor output before training. We use it to prevent the weights initializing to zero when the distance between descriptors is large.

Where α controls the tradeoff. When $\alpha\mathcal{E}_{mse} > \mathcal{E}_{nca}$ the descriptors all collapse into the same point.

3   SGD for Metric Learning

Our main proposal is in using minibatch SGD, and applying it to metric learning methods which have been designed as batch learning methods. Metric learning as shown above as a batch method, scales at $O(n^2)$ for n examples. Given that we wish to apply these approaches to large datasets containing hundreds of thousands or millions of images we are forced to consider approximations. The typical method for training a CNN on large numbers of images is using SGD, because it is fast, simple and scales to handle large datasets easily.

The most obvious approximation is to just truncate the influence to the nearest k neighbours as the weight exponentially decays with the square distance. We hypothesised this would lead to the best approximation, however there are many ways of truncating the neighbourhoods. This lends itself to clustering methods and is more complicated than the alternative, which is to sample batches randomly, but use a large enough batch to include several examples of each class.

We propose the following approaches for sampling batches for SGD:

1. **Random Shuffled Batches.** Randomly shuffle the dataset and divide up into batches of a fixed size, exactly how batches would normally be selected for supervised learning. Each batch contains (almost certainly) different numbers of examples from each class.
2. **Stratified Random Batches.** Pick batches by selecting a number of examples from each class, to ensure the same number of examples of each class are represented in each batch.
3. **K-neighbourhoods around Random Points.** Before each training epoch, run the model forward through the training set to obtain descriptors for each example. Select N examples at random and pick the batch as the batch sized neighbourhood (in descriptor space) of each selected example.

3.1 Issues of Scale

The parameter σ is largely optional in theory, as the scale can be factored into the final fully connected weight matrix (or previous layers). We make use of σ for the purposes of numerical stability upon initialization, during training the density of points adjusts itself to fit this parameter. It can also be observed that for different values of σ that the distance between descriptors adjust themselves to fit the new parameter over a few iterations of training.

3.2 Adding Mean Square Error

We experimented with adding the square distance between members of the same class as a means of adding some bias to the loss functions after suspecting that metric learning methods decribed above were overfitting, to force the output distribution to be more simple.

$$\mathcal{E}_{mse} = \sum_i \sum_{j:j_c=i_c} \frac{\|f(x_i) - f(x_j)\|_2^2}{\sigma^2} \tag{7}$$

$$\mathcal{E}_{total} = \mathcal{E}_{nca} + \alpha \mathcal{E}_{mse} \tag{8}$$

3.3 kNN Implementation

We use a brute force k-Nearest Neighbor (kNN) algorithm running on CUDA [20], computing the distance matrix using matrix multiplication followed by using an insertion sort to select the k neighbours of lowest distance. This approach is not scaleable to large datasets, and smarter clustering algorithms will eventually need to be used, however the time for evaluating kNN on the datasets we experimented with are still dominated by the cost of computing descriptors from examples.

4   CNN Architecture

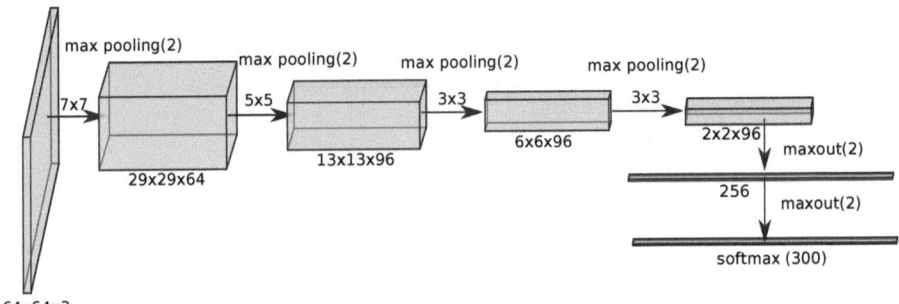

Fig. 1. Convolutional network configuration used for 64×64 rgb images with supervised learning

We used a fairly standard convolutional neural network of six layers, with four layers of convolution and max pooling using rectified linear activation functions, two fully connected layers using maxout [21] units as an activation method, shown in Figure 1. Dropout [22] with a rate of 0.5 is used when training on inputs to the two fully connected layers. For metric learning, the last linear layer and softmax are removed, leaving four convolutional layers and a single fully connected layer giving descriptors of size 256.

Dropout and Maxout have been shown to be beneficial in a supervised learning scenario for the purposes regularization. In the standard supervised training scenario Dropout is of great practical use because it (to some degree) prevents overfitting, and mostly does away with the need for early stopping. However we found that it prevented good generalization when used with metric learning approaches.

4.1   Data Augmentation

In all cases we used randomized data augmentation of the test set by applying random distortions to ensure the network never saw exactly the same image twice, and to increase its tolerance to small changes in lighting, translation and rotation. The parameters of the data augmenmtation can be seen in Figure 1. Without the data augmenmtation supervised training produces substantially

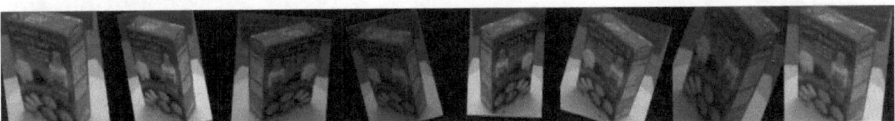

Fig. 2. Example of image distortions resulting from transformations of a single source image

worse generalization than without in both supervised and metric learning approaches. In all experiments testing was performed on non-distorted images and trained with distorted images.

Table 1. Ranges of parameters used for image distortion

scale (uniform)	1 ± 0.2
squash	1 ± 0.2
rotation (rads)	$\pm \frac{\pi}{16}$
translation(x, y) (% image size)	$\pm 5\%$
brightness (additive)	$\pm 20\%$
contrast (multiplicative)	$1 \pm 0.2\%$
gaussian pixel noise	$\sigma = 2 \pm 2$
flip horizontal (probability)	0.5

4.2 Dataset

We experimented with the University of Washington RGB-D dataset primarily because it has a standard test set for instance recongition and a large number of published results, it contains 300 objects from 50 classes. Each object has 3 sequences of rotations at 30, 45 and 60 degrees elevation, each rotation sequence contains approximately 150 images. For instance recognition the sequence at 30 and 60 degrees are used for training and the sequence at 45 degrees is used for testing.

We used a cut down version of the RGB-D dataset for a number of experiments, with 50 objects and 100 images of each object to train on, and 50 images per object to test. The images were randomly selected.

We used a resolution of 64×64 on the RGB-D images, we used the cropped version of the data, our proceedure for resizing was to load all of the images in a sequence and if any image was at a higher resolution than 72×72 we resized the image to 72×72 and all other images in the sequence by the same ratio. The images were then distorted (see Figure 1) and finally centered (modulo translation) on a 64×64 with a black background.

Fig. 3. Example of image distortions resulting from transformations of a single source image

5   Experiments

In all cases (unless otherwise specified) we use a minibatch size of 256, with standard SGD learning rate set to 10^{-2} for supervised learning, and 10^{-5} for metric learning methods. We experimented with other learning rates for metric learning, in some cases a lower learning rate of 10^{-6} was used when the higher rate caused divergence.

We manually divided the training rate by a factor of 10 when the training set accuracy plateaus for supervised learning. Supervised learning methods greatly benefit from reducing the learning rate after time, however we did not notice any benefit to the metric learning methods. Metric learning methods we stopped at 70 epochs, or earlier if they were not converging. Overall we found MEGM to give similar, but slightly better test set accuracy than NCA, we performed most experiments with MEGM for consistency.

5.1   Overall Comparison

We compare the testing classification error between methods, the final accuracy is reported as the test set classification accuracy average over the last 5 iterations. Test set accuracy is percentage accuracy for supervised learning, and for metric learning by $k = 5$ nearest neighbours and selecting the most common class.

Table 2. Summary of training methods

Method	Sampling	Batch	Test accuracy	Train epochs
Initialization			64.0	0
Supervised		256	90.6	40
	5NN		89.0	40
NCA		batch	71.2	50
	random	256	94.0	70
MEGM		batch	74.4	50
	random	128	95.0	70
		256	90.5	70
		512	81.4	70
MEGM	stratified	128	**95.4**	70
		256	94.6	70
		512	87.1	70
MEGM	neighbourhoods	256	80.4	70
MEGM + MSE	stratified	128	**95.3**	70

We sought to compare the sampling approximation to the batch method. As a batch method, clearly SGD is not the ideal training method. We were surprised to see that despite the loss function smoothly decreasing (as can be seen in Figure 5), training failed to generalize well to the training set Figure 4. We anticipated the batch metric learning methods would work best as batch

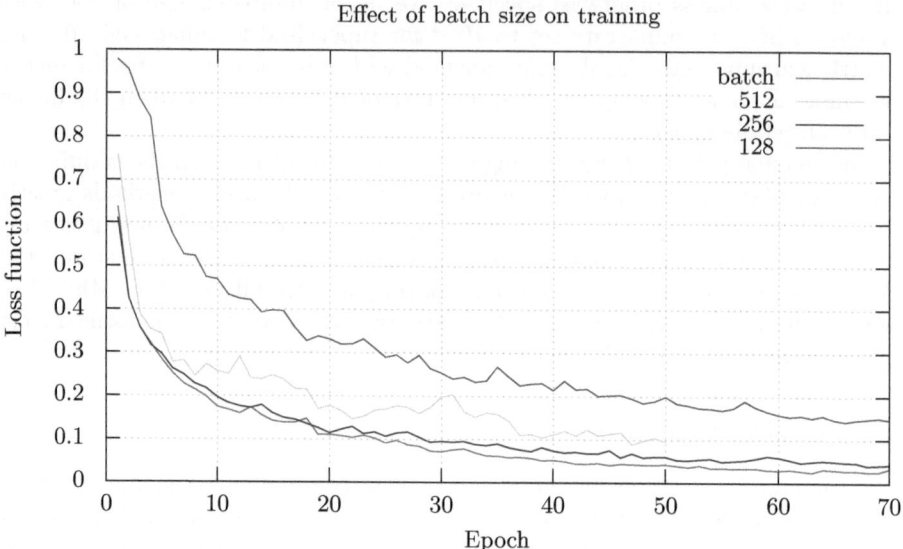

Fig. 4. Loss function for different batch sizes, MEGM loss

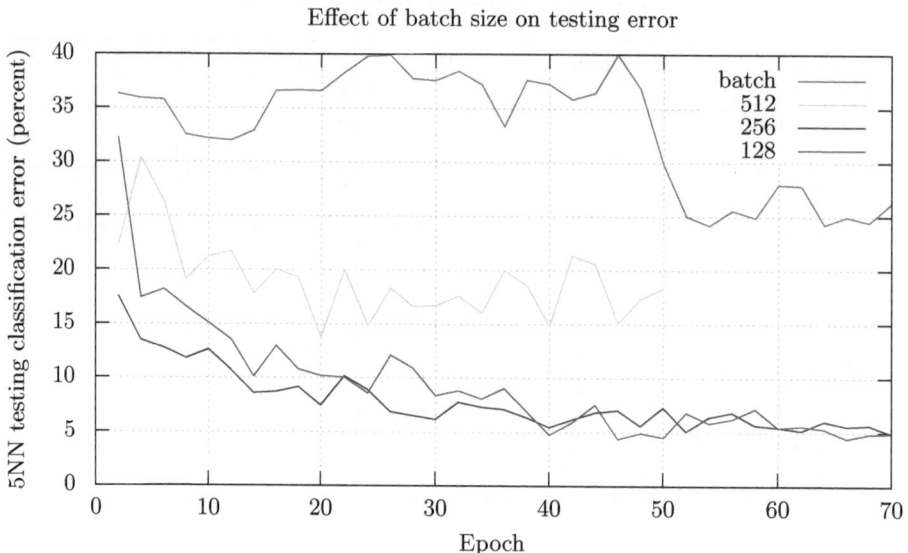

Fig. 5. Testing error for different batch sizes, MEGM loss

methods or with larger batches (as closer approximations to the batch method). We can see that is not the case, and both the batch method and SGD with the larger batch size (512) both failed to generalize well. The same pattern occurred for NCA, as well as using the stratified sampling method. The reason for such overfitting is that we believe the two metric learning methods to have not enough bias to force the network to learn something. NCA allows highly complex and multi-modal distributions with many local minima which, provided the local neighbourhood structure fits, are not penalized by its loss function. Smaller batch sizes however act as a regularization, forcing the descriptors outputs to a simpler form.

5.2 Sampling Method

We compared the three different sampling methods, most noticeably the k-neighbourhood sampling method did not converge well. The loss function can be seen to oscillate wildly and the does not reach a local minimum (reducing the learning rate did not help), and as can be seen in Figure 2 did not produce good generalization to the test set. Reducing the learning rate did not seem to help in this case. In the same figure the results of adding in a MSE term to the loss function can be shown to provide a slightly faster convergence rate while reaching the same testing classification error.

6 Conclusion

We discovered that metric learning with NCA and MEGM can produce good results under the right conditions. Used as a minibatch method, they're sensitive to parameters such as the batch size. Large batch sizes caused significant overfitting, while small batch sizes produced the best generalization, and adding MSE increased convergence rate considerably. Of the proposed sampling methods, random batches and stratified sampling worked much better than neighbourhood sampling which did not converge well.

We validated the proposed idea (at least in the small scale dataset) that the metric learning approach can be used to produce better descriptors than standard supervised learning, despite the toy size dataset. Nonlinear MEGM generalized a little better than NCA on this particular dataset, with similar properties.

We are of the opinion that when either of these metric learning methods do not provide enough bias when combined with deep neural networks. They allow complex (and potentially multi-modal) distributions in the output descriptors, as long as the local neighbourhood structure matches the labelling. We believe this prevents good generalization in our experiments when the batch sizes were larger.

Pairwise interactions complicate the implementation and we believe contribute largely to sensitivity of the training process, so make choosing the correct sampling method much more difficult in practice. A simpler alternative we will investigate in future is to chose a fixed descriptor to represent each class like Nearest

Class Mean (NCM) [23], avoiding the pairwise interaction as well as forcing the neural network to produce a more general metric.

7   Appendix

We adjusted the NCA derivative found in [13] to give the derivative for MEGM output for the ith training case and $t^t h$ class:

$$\frac{\partial \mathcal{E}_{megm}}{\partial f(x_{ti})} = -2\bigg(\sum_{j:c_i=c_j} m_{ti}p_{ij}\Big(d_{ij} - \sum_z p_{iz}d_{iz}\Big)\bigg)$$

$$+ 2\bigg(\sum_{j:c_i=c_j} m_{tj}p_{ji}d_{ji} - \sum_z \Big(\sum_{q:c_z=c_q} p_{zq}\Big)m_{tz}p_{zi}d_{zi}\bigg) \qquad (9)$$

Where err_{ti} is short hand for the partial derivative $\hat{y}_{ti}$ with respect to MSE, and $d_{ij} = f(x_i) - f(x_j)$ is shorthand for the difference between the descriptor vectors. The formula differs from the NCA derivative only by the err_{ti} term.

$$m_{ti} = \frac{\partial \mathcal{E}_{megm}}{\partial \hat{y}_{ti}} = -2(y_{ti} - \hat{y}_{ti}) \qquad (10)$$

References

1. Krizhevsky, A., Sutskever, I., Hinton, G.: ImageNet classification with deep convolutional neural networks. In: Advances in Neural Information Processing Systems (NIPS), pp. 1–9 (2012)
2. Sermanet, P., Eigen, D., Zhang, X., Mathieu, M., Fergus, R., LeCun, Y.: OverFeat: Integrated Recognition, Localization and Detection using Convolutional Networks. arXiv preprint, 1–16 (2013)
3. Masci, J., Giusti, A., Cirean, D., Fricout, G., Schmidhuber, J.: A Fast Learning Algorithm for Image Segmentation with Max-Pooling Convolutional Networks. arXiv preprint (2013)
4. Razavian, A.S., Azizpour, H., Sullivan, J., Carlsson, S.: CNN Features off-the-shelf: an Astounding Baseline for Recognition. arXiv preprint arXiv:1403.6382 (2014)
5. Donahue, J., Jia, Y., Vinyals, O.: Decaf: A deep convolutional activation feature for generic visual recognition. In: International Conference in Machine Learning, vol. 32 (2014)
6. Socher, R.: ImageNet: A large-scale hierarchical image database. In: 2009 IEEE Conference on Computer Vision and Pattern Recognition, pp. 248–255 (2009)
7. Dosovitskiy, A., Springenberg, J., Brox, T.: Unsupervised feature learning by augmenting single images. arXiv preprint arXiv:1312.5242, 1–7 (2013)
8. Lowe, D.G.: Distinctive Image Features from Scale-Invariant Keypoints. International Journal of Computer Vision 60, 91–110 (2004)
9. Fischer, P., Dosovitskiy, A., Brox, T.: Descriptor Matching with Convolutional Neural Networks: a Comparison to SIFT, 1–10 (2014)

10. Hadsell, R., Chopra, S., LeCun, Y.: Dimensionality Reduction by Learning an Invariant Mapping. In: 2006 IEEE Computer Society Conference on Computer Vision and Pattern Recognition (CVPR 2006), vol. 2, pp. 1735–1742 (2006)
11. Min, M.R., Stanley, D., Yuan, Z.: Large-Margin kNN Classification Using a Deep Encoder Network. In: Ninth IEEE International Conference on Data Mining, ICDM 2009 (2009)
12. Kostinger, M., Hirzer, M.: Large scale metric learning from equivalence constraints. In: 2012 IEEE Conference on Computer Vision and Pattern Recognition (CVPR) (2012)
13. Salakhutdinov, R., Hinton, G.: Learning a nonlinear embedding by preserving class neighbourhood structure. In: AI and Statistics (2007)
14. Weston, J.: Deep Learning via Semi-Supervised Embedding (2009)
15. Min, M., Maaten, L.: Deep supervised t-distributed embedding. In: Proceedings of the 27th International Conference on Machine Learning, ICML 2010 (2010)
16. Mensink, T., Verbeek, J., Perronnin, F., Csurka, G.: Metric Learning for Large Scale Image Classification: Generalizing to New Classes at Near-Zero Cost. In: Fitzgibbon, A., Lazebnik, S., Perona, P., Sato, Y., Schmid, C. (eds.) ECCV 2012, Part II. LNCS, vol. 7573, pp. 488–501. Springer, Heidelberg (2012)
17. Zaidi, N.A., Squire, D.M., Suter, D.: A gradient-based metric learning algorithm for k-NN classifiers. In: Li, J. (ed.) AI 2010. LNCS, vol. 6464, pp. 194–203. Springer, Heidelberg (2010)
18. Oneat, D.T.: Fast low-rank metric learning. Masters Thesis, University of Edinburgh (2011)
19. Goldberger, J., Roweis, S., Hinton, G., Salakhutdinov, R.: Neighbourhood components analysis. Advances in Neural Information Processing Systems 17 (2004)
20. Garcia, V., Debreuve, E., Barlaud, M.: Fast k nearest neighbor search using GPU. In: 2008 IEEE Computer Society Conference on Computer Vision and Pattern Recognition Workshops, vol. 2, pp. 1–6. IEEE (2008)
21. Springenberg, J., Riedmiller, M.: Improving Deep Neural Networks with Probabilistic Maxout Units. arXiv preprint arXiv:1312.6116, 1–9 (2013)
22. Hinton, G.E., Srivastava, N., Krizhevsky, A., Sutskever, I., Salakhutdinov, R.R.: Improving neural networks by preventing co-adaptation of feature detectors, 1–18 (2012)
23. Mensink, T., Verbeek, J., Perronnin, F., Csurka, G.: Large Scale Metric Learning for Distance-Based Image Classification (2012)

Automatic Resolution Selection for Edge Detection Using Genetic Programming

Wenlong Fu[1], Mark Johnston[1], and Mengjie Zhang[2]

[1] School of Mathematics, Statistics and Operations Research
Victoria University of Wellington, PO Box 600, Wellington, New Zealand
[2] School of Engineering and Computer Science
Victoria University of Wellington, PO Box 600, Wellington, New Zealand
wenlong.fu@msor.vuw.ac.nz, {mark.johnston,mengjie.zhang}@vuw.ac.nz

Abstract. When Genetic Programming is applied to edge detection, the computational cost is generally expensive. When a set of natural images are used to train edge detectors, using their high resolutions is more expensive than using their low resolutions. However, from existing reports, it is hard to find the influence on performance from using different sampling techniques on low resolutions. In this paper, we propose a GP system to automatically select the resolutions of a single training image to train edge detectors. The results of the experiments show that the GP system can effectively evolve edge detectors based on automatic resolution selection.

Keywords: Genetic Programming, Edge Detection, Image Analysis, Resolution Selection.

1 Introduction

Edge detection has been widely applied to image processing and computer vision [1]. Many techniques for edge detection have been developed based on moving windows [1, 2]. The image gradient has been popularly employed for edge feature extraction, such as the Sobel edge detector [3] and the Canny edge detector [4]. In general, edges extracted from one image with high resolution have higher detection accuracy than the extracted results from the image with low resolution. However, using an image with low resolution to extract edges is generally faster than using the image with high resolution. While considering the importance of remaining detection accuracy, it is desirable to investigate how to detect edges from images with low resolution.

Genetic programming (GP) has been used for object detection and image analysis since the 1990s [5]. In [6], GP was successfully employed to evolve detectors using a shifting function to select pixels instead of using a window. In [7], GP was used to design specific edge detectors based on an energy function, rather than using ground truth. The evolved edge detectors [6, 7] have fast detection speed on raw images. However, it is not clear whether GP can be used to automatically select different low degrees of resolutions on raw images while maintaining the

G. Dick et al. (Eds.): SEAL 2014, LNCS 8886, pp. 810–821, 2014.

detection accuracy. It is therefore worthwhile to investigate automatic resolution selection using GP.

1.1   Goals

The overall goal of this paper is to investigate automatically evolving edge detectors based on automatic resolution selection using GP. In this paper, we will conduct an initial investigation on how to train edge detectors based on automatic resolution selection using a single training image. A GP system for resolution selection is proposed. In order to select different low level resolutions of raw images, different sampling methods are proposed and are considered as functions. Specifically, we would like to investigate the following research objectives.

- Whether GP evolved edge detectors can be used to effectively extract edges when automatic resolution selection is used.
- Whether GP evolved edge detectors have similar detection performance to the edge detectors from human design.

1.2   Organisation

In the remainder of the paper, Section 2 briefly describes the background. Section 3 introduces the proposed GP system for resolution selection. After presenting the experimental design in Section 4, Section 5 describes the results with discussions. Section 6 gives conclusions and future work directions.

2   Background

This section describes edge detection and the related work using GP for edge detection.

2.1   Edge Detection

In general, there are three stages in edge detection. The first stage is to filter noise; the second stage is to extract edge features; and the last stage is to obtain binary edge maps. Since edge points are usually detected based on the extracted edge features, the second stage, namely feature extraction, is very important in edge detection [1, 21].

In order to extract edge features, local pixel information is commonly utilised, such as Gaussian-based edge detection [2]. In Gaussian-based edge detection, multiple scales have been employed for extracting edge features [1, 2, 11]. Similar to human eyes, Gaussian filters with large scales mainly focus on boundary information, but Gaussian filters with small scales mainly focus on the details of edges. From using wavelet sampling on raw images, multiple resolutions can be used to represent these raw images [9]. In different resolution levels in the wavelet domain, edge information (usually in the high frequency domain) is included [1, 10]. It is desired to extract edge information from low resolution raw images after a specific transformation. Meanwhile, a sampling technique for the transformation would be required. Wavelet-based techniques have been successfully applied to edge detection [1].

2.2   Related Work to GP for Edge Detection

The existing works using GP for edge detection are based on images at their raw resolution. This subsection only focuses on the existing work on low-level edge detection. In low-level edge detection, GP has been used to explicitly search for raw pixels based on local windows [15] or implicitly search for raw pixels and a set of raw pixels based on full images [6–8].

There is some work mainly focusing on low-level edge detection [6, 7, 12, 13]. Four macros were suggested to be used for searching for pixels when GP was employed for image processing [20]. Existing edge detectors have been approximated by GP. The Canny edge detector was approximated by applying shifting functions and other common functions in GP [13]. Also, the Sobel edge detector was approximated by GP [14]. The edge detectors evolved by GP based on a 13×13 moving window were compared with the Canny edge detector [15]. In order to obtain edge detectors based on ground truth of natural images, search operators using single pixels and sets of pixels were considered as functions in [6–8], and the evolved edge detectors outperform some existing edge detectors, such as the Sobel edge detector. Considering edges as one-dimensional signals, one-dimensional step filters were designed by GP based on a set of pre-defined edge signals and noise [12].

Additionally, in digital circuit design, edge detectors using a 4×4 window were evolved by GP [18]. Morphological operators were used to evolve binary edge detectors by GP [16, 17]. An ant was designed to find food, namely "edge point", inspired by the problem of artificial ants searching for food in GP [19].

In summary, the existing work on low-level edge detection mainly utilises pixels on images at their raw resolution. However, there is no reports on the detection performance influence of GP edge detectors when the detected edges come from different levels of resolutions.

3   The Method

This section introduces the proposed GP system for resolution selection. The sets of functions and terminals are given first, then the fitness function is described.

3.1   Functions and Terminals

From wavelet domain transformation, low-level resolutions with high frequency parts include edge information [1, 10]. When a sampling technique is used to obtain an image with low resolution, the edge information would be used for extracting edge features in the low resolution levels (not the raw image). However, it is not clear how to design sampling techniques to obtain images (smaller than their raw images) including edge information. We consider sampling techniques as functions so that a GP system can automatically select the size of the images with low resolution.

In wavelet domain transformation, a raw image is separated into different small sub-sections. In our proposed sampling functions, we only select one sub-section as a low resolution image. Considering images from the horizontal or vertical direction to sample pixels, the width of the one-dimensional window is $2 * w_n + 1$ on a signal with length L. Based on the restruction condition from the Nyquist-Shannon sampling theorem [22], the size of the sampled signal is fixed to L/w_n in this paper. For a sampling window, s is used to represent the weights. Here, sampling is based on pixels in a sampling window. Parameter w_n is used to control the sampling scale. Only $w_n = 2$ and $w_n = 3$ are used for an initial investigation in this paper. When $w_n = 2$, $s = [1, 1, 0, -1, -1]$ is used. When $w_n = 3$, $s = [1, 1, 1, 0, -1, -1, -1]$ is used. Equations (4) and (5) are used to represent the sampled value in the new sampled image based on the horizontal and vertical directions. Here, I' is the new image after sampling, x' and y' are the new positions in the new image, I is the raw image, and x and y are the raw positions in the raw image I.

$$Thresh(x) = \max\{0, x\} \tag{1}$$

$$x' = \lfloor \frac{x}{w_n} \rfloor \tag{2}$$

$$y' = \lfloor \frac{y}{w_n} \rfloor \tag{3}$$

$$I'_x(x', y') = Thresh(\sum_{i=-w_n}^{w_n} \sum_{j=-w_n}^{w_n} (s(j) * I(x+i, y+j))) \tag{4}$$

$$I'_y(x', y') = Thresh(\sum_{i=-w_n}^{w_n} \sum_{j=-w_n}^{w_n} (s(i) * I(x+i, y+j))) \tag{5}$$

When $w_n = 2$, a sampled image I' will have a quarter of pixels from image I. The weight vector s is used to only obtain high frequency information from image I.

Considering the low frequency domain, we also design a sampling technique to obtain smaller images. Similar to $I'_x(x', y')$ and $I'_y(x', y')$, the edge information, rather than the raw pixel values, will be sampled.

$$I'_{avg}(x', y') = Thresh(I(x, y) - \frac{\sum_{i=-w_n}^{w_n} \sum_{j=-w_n}^{w_n} (I(x+i, y+j))}{4 * w_n^2 + 4 * w_n + 1}) \tag{6}$$

Note that $I(x, y)$ only represents the image grayscale level in this paper, and only positive values are considered in the image I'. In order to combine sampling results, four functions (see Equations (7) to (10)) are also added into the function set. Here, a result of 1 is produced for a 0 divisor, a and b are pixel grayscale levels from two input images. When two input images are not the same size, the larger image will be shrunk to the size of the smaller one.

$$Add(a, b) = \frac{(a + b)}{2} \tag{7}$$

$$Sub(a, b) = \frac{|a - b|}{2} \tag{8}$$

$$Mul(a, b) = \frac{2ab}{\sqrt{a^2 + b^2}} \tag{9}$$

$$Div(a, b) = \min\{\frac{a}{b}, \frac{b}{a}\} \tag{10}$$

To summarise, the function set includes $\{I'_x, I'_y, I'_{avg}, Add, Sub, Mul, Div\}$. For the terminal set, only the raw image I is considered.

3.2 Fitness Function

The Figure of Merit (FOM) has been previously used as the fitness function in GP for edge detection [23]. FOM considers offset distances from predicted edge points to true edge points. It has lower computational cost than a detection accuracy measure method when predicted edge points are needed to match to true edge points [25]. We directly employ FOM as the fitness function in this paper. Here, we only allow predicted edge points with one pixel offset from a true edge point. Note that the outputs of GP edge detectors are not binary and the size of an output image might be different from the original size of a test image. There are two processes before calculating FOM: obtaining binary edges and scaling an output to the size of the test image.

FOM is defined in Equation (11), where, N_T is the number of true edge points in a single image, N_P is the number of predicted edge points in the image, Set_P is the set of all predicted edge points, α is a weighting factor for detection localisation, and $d(j)$ is the distance from a predicted edge point j to the nearest true edge point in a ground truth edge map. Considering the overlap of a 3×3 window, α is usually set $\frac{1}{9}$.

$$FOM = \frac{1}{\max\{N_T, N_P\}} \sum_{j \in Set_P} \frac{1}{1 + \alpha d^2(j)} \tag{11}$$

The binary result for output $O(x, y)$ is obtained based on Equation (12), where "*mean*" is the average of the output O on all pixels and "*std*" is the standard deviation. Position (x, y) is considered as an edge point when $B(x, y)$ is larger than 0.

$$B(x, y) = Thresh(O(x, y) - (mean + 0.005 * std)) \tag{12}$$

4 The Design of the Experiment

In general, a single image is not sufficient to train GP edge detectors. However, when some degrees of prior domain knowledge are considered in a GP system,

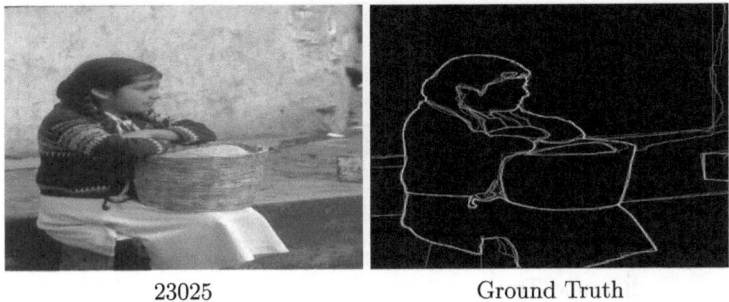

23025   Ground Truth

Fig. 1. Training image 23025 and its ground truth.

Table 1. Training Results (FOM) from Evolved Edge Detectors and Two Simple Designed Edge Detectors

	FOM (mean $\pm$ standard deviation)	p-value
GP	0.3785 ± 0.0070	—
$ED_{w_n=2}$	0.3505	$0.0000 \downarrow$
$ED_{w_n=3}$	0.3509	$0.0000 \downarrow$

a single image has the potential to get good edge detectors [24]. In this paper, we select image 23025 from the Berkeley Segmentation Dataset (BSD) [25] as the single training image. All BSD images with ground truth provided are taken from throughout the world and have size 481×321 pixels. There are 100 test images in the BSD 300 [25]. For fairness of judgement of edges, the ground truth are combined from five to ten persons as graylevel images. Figure 1 shows the training images 23025 and its ground truth. Note that the images in Figure 1 are stretched.

Based on common settings [6, 23, 24] and initial experiments, parameter values used in this paper are: population size 200; maximum generations 60; maximum depth (of a program) 7; and probabilities for mutation 0.15, crossover 0.80 and elitism (reproduction) 0.05. There are 30 independent runs for the experiment.

5   Results and Discussions

In order to compare with the edge detectors designed by humans, we design two edge detectors via using I'_x and I'_y based on $w_n = 2$ and $w_n = 3$. Similar to the simplified Sobel edge detector [3], the designed edge detector is shown in Equation (13).

$$ED_{w_n} = Add(I'_x + I'_y) \tag{13}$$

5.1   Training Results

Table 1 gives the mean and standard deviation of the FOM values of the training results from GP. Here, a t-test with overall significant level 0.05 is employed, and

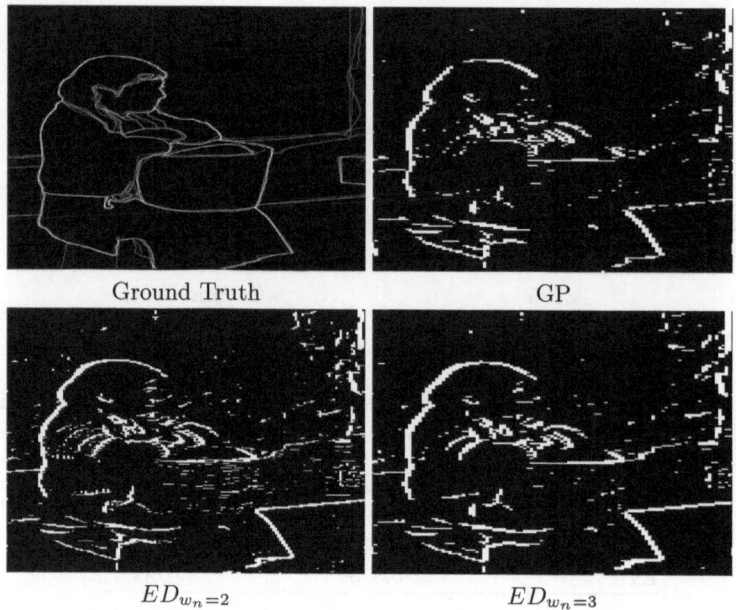

Ground Truth   GP

$ED_{w_n=2}$ $ED_{w_n=3}$

Fig. 2. Detected results on training image 23025

↓ indicates that the relevant result is significantly worse than the results from GP. Comparing to the designed edge detectors $ED_{w_n=2}$ and $ED_{w_n=3}$, the results from GP are significantly better, in terms of FOM. Also, the standard deviation of the results from GP is small, and therefore, the performance of evolving edge detectors is stable, in terms of FOM in the training stage. From the table, it seems that GP has some ability to effectively train edge detectors when the ground truth is provided. Additionally, there is almost no influence between using $w_n = 2$ and $w_n = 3$, in terms of FOM.

Figure 2 shows the detected results from a GP edge detector, $ED_{w_n=2}$ and $ED_{w_n=3}$. Note that the training results from GP are very close, and the standard deviation of the 30 independent runs is only 0.0079. From the visual results, it is found that the detected results from GP have fewer false positives than the results from $ED_{w_n=2}$ and $ED_{w_n=3}$. Since the detected results are based on low resolutions, all detected results are affected by noise.

5.2 Test Performance

Table 2 gives the overall test results from the GP evolved edge detectors on the 100 BSD test images. Here, ↑ represents that the relevant result is significantly better than the results from GP. As we can see, the test results from GP are decreased. From the 100 BSD test images, $ED_{w_n=2}$ and $ED_{w_n=3}$ also have lower FOM values than the single training image 23025. Based on the comparisons among GP, $ED_{w_n=2}$ and $ED_{w_n=3}$, it seems that GP can evolve edge detectors

Table 2. Test Results (FOM) from Evolved Edge Detectors and Two Simple Designed Edge Detectors on the 100 BSD Test Images

	FOM (mean $\pm$ standard deviation)	FOM (best)	p-value
GP	0.2452 ± 0.0079	0.2579	$-$
$ED_{w_n=2}$	0.2393	$-$	$0.0003 \downarrow$
$ED_{w_n=3}$	0.2557	$-$	$0.0000 \uparrow$

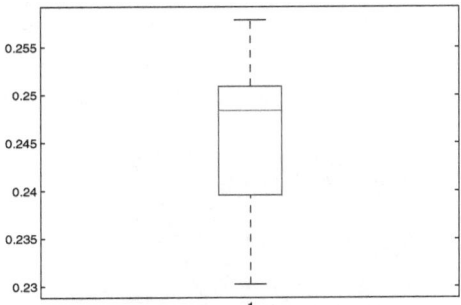

Fig. 3. Box-plot for the test results of the GP evolved edge detectors on the 100 BSD test images

which outperform than $ED_{w_n=2}$ when only a single image is used as the training data. Overall, $ED_{w_n=3}$ has the best test performance, in terms of FOM. It seems that different sampling techniques affect the detected results. Since only a single training image is used, GP automatically selected sampling techniques which can be helpful to find good results on the training image.

Figure 3 shows the box-plot of the 30 independent test results from GP. The best performance from GP is higher than the test performance of $ED_{w_n=3}$. Table 2 gives the best test result of the GP evolved edge detectors is 0.2579. However, most of the edge detectors have lower performance than $ED_{w_n=3}$.

Figure 4 shows the detected results from the best GP edge detector, $ED_{w_n=2}$ and $ED_{w_n=3}$ on an example test image 102016. Similar to the training result, the GP edge detector has fewer false positives than the results from $ED_{w_n=2}$ and $ED_{w_n=3}$. Note that there is no post-processing, so the detected results are not the final binary edge maps. Following the suggestion from [21], this paper only conducts the initial investigation in the feature extraction stage.

From the training and test results, especially comparing to $ED_{w_n=2}$, GP has potential to effectively select sampling techniques to evolve good edge detectors. However, it remains to investigate how to improve the overall performance of GP evolved edge detectors.

Note that the test time is not discussed in this paper. The detection speeds of all edge detectors on each test image are shorter than 0.01 second on a platform with 2.1 GHz CPU and using C++.

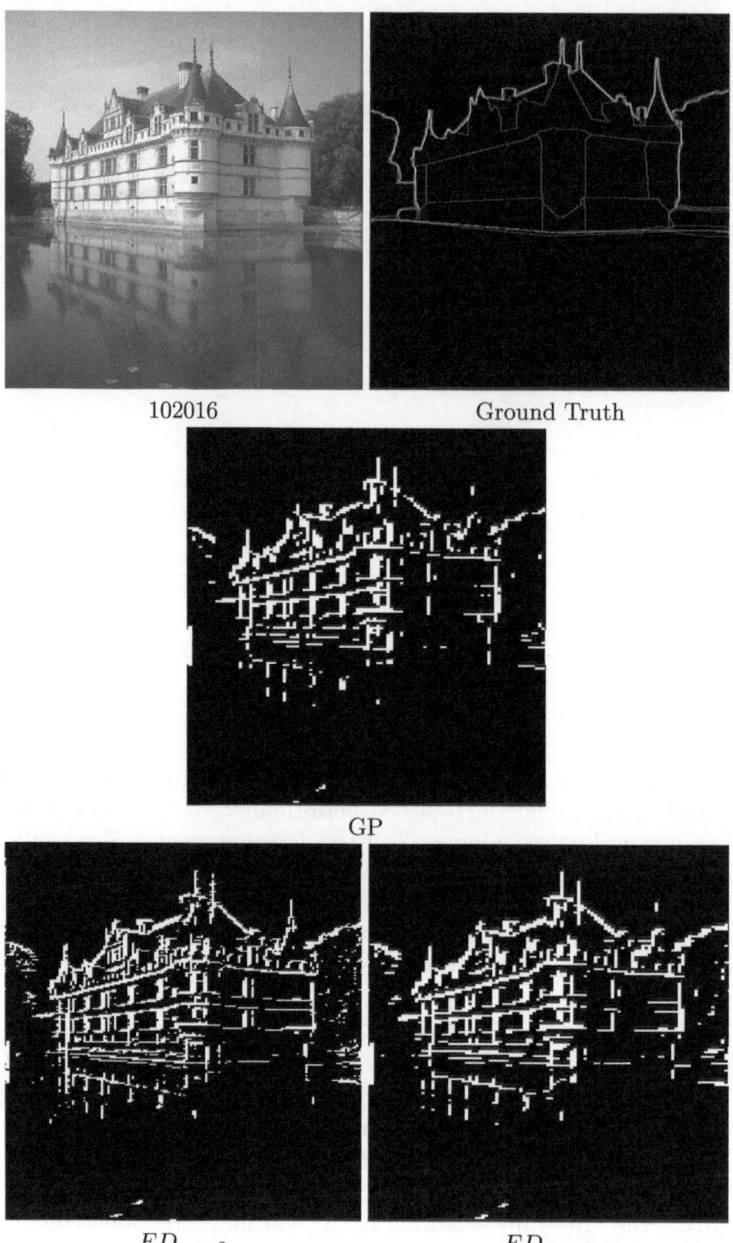

Fig. 4. The detected results on an example image 102061 from the best GP edge detector, $ED_{w_n=2}$ and $ED_{w_n=3}$

5.3   Further Discussion

Similar to one-shot learning [26], only a single training image is used to train edge detectors and the evolved edge detectors are used for a large set of different images. However, there is no prior knowledge from the BSD training dataset, such as the distribution of the single image from the 200 BSD training images. Although there is no prior knowledge from the training data, specific edge knowledge can be considered in the proposed GP system. The proposed sampling techniques mainly focus on high frequency domain (I'_x and I'_y) and difference among pixels (I'_{avg}). Note that edge information mainly exists in high frequency domain. Therefore, it is a potential reason that a single image can be effectively used to evolve edge detectors in the proposed GP system.

6   Conclusions

This paper aimed to investigate automatic selection of sampling techniques using GP for edge detection. Different sampling functions were introduced in the proposed GP system for automatically evolving edge detectors using sampling techniques. The 100 BSD test images were used to test performance on the evolved edge detectors. From the results, GP has potential to evolve good edge detectors via selecting from the pre-defined sampling techniques.

From the detected results, GP evolved edge detectors can be used to effectively extract edges and outperform a simple edge detector, similar to the Sobel edge detector. However, this study is an initial investigation. How to effectively improve the performance of evolved edge detectors and analyse evolved edge detectors would be done in the future work.

References

1. Papari, G., Petkov, N.: Edge and line oriented contour detection: state of the art. Image Vision Computation 29(2–3), 79–103 (2011)
2. Basu, M.: Gaussian-based edge-detection methods: a survey. IEEE Transactions on Systems, Man, and Cybernetics, Part C: Applications and Reviews 32(3), 252–260 (2002)
3. Ganesan, L., Bhattacharyya, P.: Edge detection in untextured and textured images: a common computational framework. IEEE Transactions on Systems, Man, and Cybernetics, Part B: Cybernetics 27(5), 823–834 (1997)
4. Canny, J.: A computational approach to edge detection. IEEE Transactions on Pattern Analysis and Machine Intelligence 8(6), 679–698 (1986)
5. Krawiec, K., Howard, D., Zhang, M.: Overview of object detection and image analysis by means of genetic programming techniques. In: Proceedings of Frontiers in the Convergence of Bioscience and Information Technologies, pp. 779–784 (2007)
6. Fu, W., Johnston, M., Zhang, M.: Genetic programming for edge detection: a global approach. In: Proceedings of the 2014 IEEE Congress on Evolutionary Computation, pp. 254–261 (2011)

7. Fu, W., Johnston, M., Zhang, M.: Unsupervised learning for edge detection using genetic programming. In: Proceedings of 2014 IEEE Congress on Evolutionary Computation, pp. 117–124 (2014)
8. Fu, W., Johnston, M., Zhang, M.: Genetic programming for edge detection using blocks to extract features. In: Proceedings of the 14th Annual Conference on Genetic and Evolutionary Computation, pp. 855–862 (2012)
9. Chan, R.H., Chan, T.F., Shen, L., Shen, Z.: Wavelet algorithms for high-resolution image reconstruction. SIAM Journal on Scientific Computing 24(4), 1408–1432 (2003)
10. Zhang, L., Bao, P.: Edge detection by scale multiplication in wavelet domain. Pattern Recognition Letters 23, 1771–1784 (2002)
11. Grigorescu, C., Petkov, N., Westenberg, M.A.: Contour and boundary detection improved by surround suppression of texture edges. Journal of Image and Vision Computing 22(8), 583–679 (2004)
12. Harris, C., Buxton, B.: Evolving edge detection with genetic programming. In: Proceedings of the 1st Annual Conference on Genetic Programming, pp. 309–314 (1996)
13. Ebner, M.: On the edge detectors for robot vision using genetic programming. In: Processings of the Horst-Michael Groβ, Workshop SOAVE 97-Selbstorganisation von Adaptivem Verhalten, pp. 127–134 (1997)
14. Hollingworth, G.S., Smith, S.L., Tyrrell, A.M.: Design of highly parallel edge detection nodes using evolutionary techniques. In: Proceedings of the Seventh Euromicro Workshop on Parallel and Distributed Processing, pp. 35–42 (1999)
15. Zhang, Y., Rockett, P.: Evolving optimal feature extraction using multi-objective genetic programming: a methodology and preliminary study on edge detection. In: Proceedings of the 8th European Conference on Genetic Programming, pp. 795–802 (2005)
16. Wang, J., Tan, Y.: A novel genetic programming based morphological image analysis algorithm. In: Proceedings of the 13th European Conference on Genetic Programming, pp. 979–980 (2010)
17. Quintata, M., Poli, R., Claridge, E.: Morphological algorithm design for binary images using genetic programming. Genetic Programming and Evolvable Machines 7(1), 81–102 (2006)
18. Golonek, T., Grzechca, D., Rutkowski, J.: Application of genetic programming to edge detector design. In: Proceedings of the International Symposium on Circuits and Systems, pp. 4683–4686 (2006)
19. Bolis, E., Zerbi, C., Collet, P., Louchet, J., Lutton, E.: A GP artificial ant for image processing: Preliminary experiments with EASEA. In: Miller, J., Tomassini, M., Lanzi, P.L., Ryan, C., Tetamanzi, A.G.B., Langdon, W.B. (eds.) EuroGP 2001. LNCS, vol. 2038, pp. 246–255. Springer, Heidelberg (2001)
20. Poli, R.: Genetic programming for image analysis. In: Proceedings of the First Annual Conference on Genetic Programming, pp. 363–368 (1996)
21. Moreno, R., Puig, D., Julia, C., Garcia, M.A.: A new methodology for evaluation of edge detectors. In: Proceedings if the 16th IEEE International Conference on Image Processing (ICIP), pp. 2157–2160 (2009)
22. Marks II, R.J.: Introduction to Shannon Sampling and Interpolation Theory. Springer (1991)
23. Fu, W., Johnston, M., Zhang, M.: Figure of merit based fitness functions in genetic programming for edge detection. In: Bui, L.T., Ong, Y.S., Hoai, N.X., Ishibuchi, H., Suganthan, P.N. (eds.) SEAL 2012. LNCS, vol. 7673, pp. 22–31. Springer, Heidelberg (2012)

24. Fu, W., Johnston, M., Zhang, M.: Is a single image sufficient for evolving edge features by genetic programming? In: Proceedings of the 16th European Conference on the Applications of Evolutionary Computation (2014)
25. Martin, D.R., Fowlkes, C.C., Malik, J.: Learning to detect natural image boundaries using local brightness, color, and texture cues. IEEE Transactions on Pattern Analysis and Machine Intelligence 26(5), 530–549 (2004)
26. Li, F., Rob, F., Pietro, P.: One-Shot Learning of Object Categories. IEEE Transactions on Pattern Analysis and Machine Intelligence 28(4), 594–611 (2006)

Evolutionary Feature Combination Based Seed Learning for Diffusion-Based Saliency

Syed S. Naqvi, Will N. Browne, and Christopher Hollitt

Evolutionary Computation Research Group
Victoria University of Wellington New Zealand
{syed.saud.naqvi,will.browne,christopher.hollitt}@ecs.vuw.ac.nz

Abstract. Diffusion-based saliency detection is a graph-based technique in which the optimal saliency map is computed by saliency propagation over the graph using diffusion of saliency values from one node to another. This is achieved by computing the product of a propagation matrix and a saliency seed vector. The saliency seeds stored in the saliency seed vector contain important prior saliency information usually obtained from a bottom-up saliency model or certain heuristics. Finding the optimal saliency seeds is vital for efficient saliency propagation during the diffusion process. In this work, we propose to investigate the performance of an evolutionary feature combination technique for learning the optimal seeds for diffusion-based saliency detection. We achieve this by adapting an evolutionary feature combination system (having good object detection performance) for the task of seed generation, for diffusion-based saliency, termed as IGASeed. We present quantitative and qualitative comparison of our proposed IGASeed system with the state-of-the-art heuristic and learning approaches for seed prediction. Our results show that our IGASeed technique performs better than most state-of-the-art models and comparable to the best seed learning model with lower computational cost.

1 Introduction

Saliency detection is the task of identifying the most important or useful part of a scene. It has been successfully used in several vision problems such as image compression [8], object recognition [13] and content based image retrieval [4]. The prime motivation for almost all saliency models is either to predict human fixation or to detect salient objects. The focus of this work is on salient object detection. Furthermore, we are interested in formulating the problem of object saliency as identifying the segments of an image that belong to the salient object and discriminating them from the background segments.

Recently, a class of models based on saliency propagation has been introduced that predict the likelihood of a segment belonging to a generic object. These models include energy based models [7], conditional random fields [10], random walk models [7] and manifold ranking models [14]. In all these models a graph-based solution is sought, where the optimal solution comprises of

G. Dick et al. (Eds.): SEAL 2014, LNCS 8886, pp. 822–834, 2014.

a product of a propagation matrix (capturing node similarity information) and a seed vector, which contains saliency information. While the diffusion process for these approaches vary according to their different propagation matrices, the problem of determining the saliency seeds remain common. The seed prediction problem for diffusion based object saliency has not received much attention in prior works, though a few heuristic approaches exist in the literature [14], [9]. Recently, a seed learning method based on optimal feature combination has been reported by Lu et al. [11].

Based on the results of [11], we have two major observations:

- A seed learning system having fewer highly discriminative features may achieve similar performance as compared to a system with large number of features. According to [11], a bottom-up approach (SalSeedProp) performed comparably to a large-scale feature combination seed learning method (OptSeedProp) that incorporates 178 features.
- Learning the best feature combination process may have a more important role in seed learning as compared with the number of features included. Again the results of [11] demonstrated that a feature combination based implementation with uniform weights (MeanSeedProp) having 178 features, performed worse than SalSeedProp [11].

These observations enable us to formulate the following research questions:

- Should a seed learning system incorporate a large number of features with low discrimination power or fewer highly discriminative features?
- How does the optimization of the feature combination process for seed learning compare to the number of features included in the learning system?

The overall goal of this paper is to learn optimal seeds for diffusion-based object saliency. In order to achieve this goal, our previously introduced IGA system of [12] will be adapted to investigate its effectiveness in predicting seeds for a diffusion-based saliency framework, we compare its performance with the state-of-the-art heuristic and learning approaches for seed prediction. We anticipate that the process of efficiently combining features in the best possible manner for producing object saliency is parallel to the problem of seed prediction in diffusion based saliency. The specific objectives of this work are:

- To adapt the IGA system (having considerably lower number of features as compared with OptSeedProp) in the diffusion-based saliency framework to devise a new system, IGASeed. As the IGASeed system utilizes multiple learned schemes, we will address both the questions simultaneously.
- To compare the quantitative performance and the computational complexity of the newly developed IGASeed system with the seed learning system (OptSeedProp) and other state-of-the-art seed prediction approaches.

2   Background

2.1   Diffusion-Based Object Saliency

The process of diffusion based saliency computation involves the product of an optimal seed vector $\mathbf{s}$ and a propagation matrix $\mathbf{A}$. The propagation matrix $\mathbf{A}$ is responsible for capturing the internal relation between nodes and therefore guides the diffusion process. The propagation matrix $\mathbf{A}$ captures similarities between nodes on a graph by the help of an affinity matrix $\mathbf{W}$. Higher affinities between nodes improve the chance of saliency propagation between the nodes during diffusion. A variety of graph based approaches are reported in literature each with a different diffusion mechanism. A few common examples of propagation matrices for diffusion in different frameworks are reported below:

- Quadratic energy models [7]: $\mathbf{A} = (\mathbf{K} + \alpha(\mathbf{D} - \mathbf{W}))^{-1}\mathbf{K}$
- Random walks [7]: $\mathbf{A} = (\mathbf{I} - \alpha\mathbf{W}\mathbf{D}^{-1})^{-1}$
- Manifold Ranking [14]: $\mathbf{A} = (\mathbf{I} - \alpha\mathbf{D}^{-1/2}\mathbf{W}\mathbf{D}^{-1/2})^{-1}$,

where $\mathbf{D}$ is a diagonal matrix computed by summing the rows of the affinity matrix $\mathbf{W}$. The parameter α controls the fitting constraint of the regularization framework. k_i are the weights for the fitting constraint, where $K = \mathrm{diag}\{k_1, k_2, \ldots, k_N\}$. Although having slightly different formulation of the propagation matrix, the above methods work on the same principle. In all frameworks the optimal solution $\mathbf{y}^*$ for diffusion has the form $\mathbf{y}^* = \mathbf{A}\mathbf{s}$.

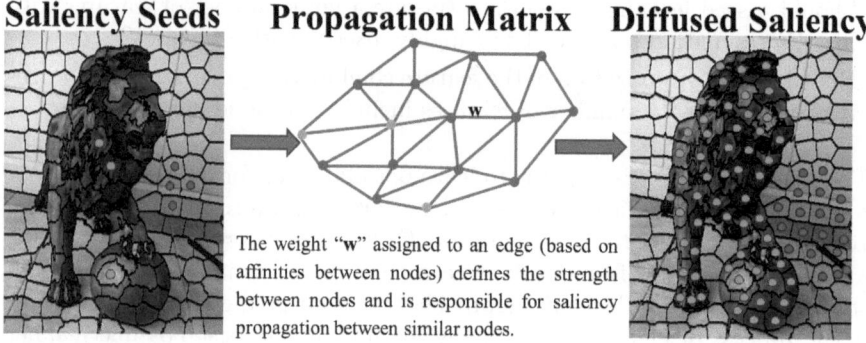

Saliency Seeds   Propagation Matrix   Diffused Saliency

The weight "**w**" assigned to an edge (based on affinities between nodes) defines the strength between nodes and is responsible for saliency propagation between similar nodes.

Fig. 1. The saliency diffusion process. Given an initial set of saliency seeds and a propagation matrix; saliency is propagated between nodes through a diffusion process. The green dots represent the effective object saliency seeds. The red dots depict the background seeds having lower saliency values.

2.2   The IGA System

The IGA system is an evolutionary feature combination based system introduced in [12]. The IGA system learns multiple combination schemes depending upon image type. The feature combination using IGA involves three important variables, i.e., feature weight vector ($\mathbf{w_n}$), normalization operation ($\mathcal{N}$) and an integration operation ($\circ$). The objective is to find the best possible solution (combination scheme) in terms of $\mathbf{w_n}$, $\mathcal{N}$ and $\circ$, such that the difference between the ideal classification accuracy and the computed accuracy is minimized. It uses multiple Genetic Algorithms (GA) to find the optimal combination schemes for different sets of images. The work uses nine saliency features including both pixel and segment based features (for details please refer to [12]).

Before giving an overview of the algorithm behind the IGA system, we would like to describe a few notations for better understanding of the algorithm. The complete feature set and the ground truth sets for training are denoted as G and $\mathcal{F}$, where $\mathcal{F}_i$ and $\mathcal{G}_i$ are feature and ground truth groups, respectively, formed after autonomous feature grouping. $\mathcal{F}_i$ encapsulate feature vectors $\mathbf{f}_i \in \mathbb{R}^{D \times n}$.

The learning process of the IGA system is comprised of the following steps:

1. Find the k nearest neighbors of the first element in $\mathcal{F}$.
2. Form groups $\mathcal{F}_i$ and $\mathcal{G}_i$ using the neighbors from step 1.
3. Delete $\mathcal{F}_i$ and $\mathcal{G}_i$ from $\mathcal{F}$ and $\mathcal{G}$.
4. Repeat steps 1 to 3 untill $\mathcal{F} = \{\}$.
5. For each $\mathcal{F}_i$ find the best possible combination scheme $\lambda_\mathbf{i} = \{\mathbf{w_n}, \mathcal{N}, \circ\}$.
6. Construct a memory set $\mathcal{M}_i$ to store $\{\mathcal{F}_i, \lambda_\mathbf{i}\}$.

In order to test a particular image using the learned model, IGA takes the following procedure:

1. Compute a test feature vector $\mathbf{f_t}$ for a test image.
2. Compute a distance d_i between $\mathbf{f_t}$ and a group $\mathcal{F}_i$.
3. Repeat step 2 to obtain a distance vector $\mathbf{d} = \left[d_1, d_2, \ldots, d_{\frac{N}{k}}\right]$.
4. Choose a memory set $\mathcal{M}_i$ from $\mathcal{M}$ using the smallest distance d_i from $\mathbf{d}$.
5. Use the corresponding optimal combination scheme $\lambda^* = \lambda_\mathbf{i}$ to compute the saliency map.

For specific details of the GA and further details of the IGA system, the reader is directed to [12].

2.3   Optimal Seed Learning System (OptSeedProp)

Recently, a diffusion-based object saliency prediction system was introduced in [11] (OptSeedProp), which learns the set of optimal seeds for object saliency. OptSeedProp attempts to learn the optimal seeds by combining a large number of mid-level features and a bottom-up saliency feature in an optimization framework. The goal is to be able to predict the best seeds by learning the optimal weighted feature combination. The graph is constructed in a similar fashion as in [14]. In order to capture the similarities of two superpixel nodes, a boosted decision tree is used to classify whether the nodes belong to the same object or not. The propagation matrix used by OptSeedProp is given as:

$$\mathbf{A} = (\mathbf{I} - \mathbf{D}^{-1}\mathbf{W})^{-1}. \qquad (1)$$

OptSeedProp uses gradient descent to learn the optimal weight vector $\mathbf{w}$ by maximizing the discriminant saliency criteria between object and background superpixels. OptSeedProp uses a single bottom-up feature and 177 superpixel based mid-level features to learn the optimal seeds. For details, please see [11].

3 From Saliency to Saliency Seeds (IGASeed System)

In the novel IGASeed system, the process of seed based saliency starts by computing features for an input image as shown in Figure 2. The computed features are then combined in the feature fusion block. The combination scheme is selected from the multiple combination schemes learned by the IGA system depending upon the feature distance of the input image from feature groups. We use the same settings for feature grouping and genetic algorithm design as reported in [12]. The evolutionary feature combination based saliency obtained after the fusion process is then used to produce seeds for the diffusion process. In order to map the pixel-level saliency (obtained in the previous step) to segment-level saliency, superpixel information obtained from the image (using SLIC algorithm [2]) is overlayed on the pixel-level saliency (such that pixels inside a superpixel can have different values), and centroids $c_1, ..., c_N$ for each superpixel are computed. These centroids are then used to construct a segment-based saliency or saliency seeds (where each superpixel has similar pixel intensities). In order to form the propagation matrix $\mathbf{A}$, the input image $\mathbf{I}$ is first segmented into superpixels. These superpixels form nodes as the image is represented as a graph $G = (V, E)$. We build a k-regular graph, where each node is connected to its immediate neighbors and to nodes sharing common boundaries with its

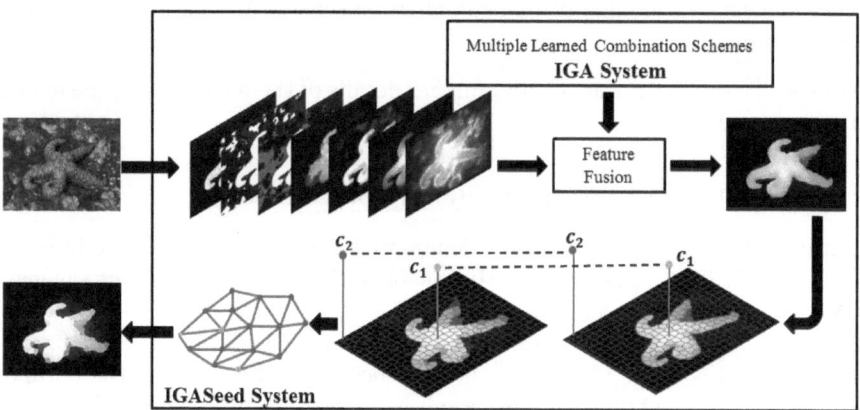

Fig. 2. The process of obtaining saliency seeds from feature combination based saliency and saliency propagation

neighboring nodes. Afterwards, a sparse affinity matrix $\mathbf{W}$ capturing the relation between nodes is constructed as follows [15]:

$$w_{ij} = \exp\left(\frac{-\|x_i - x_j\|^2}{2\sigma^2}\right), \tag{2}$$

where x_i and x_j are the centroids of the nodes and σ is a parameter that controls the strength of the relation between two nodes. The propagation matrix $\mathbf{A}$ is then computed by the following diffusion scheme which produced the best results in our experiments.

$$\mathbf{A} = \frac{\mathbf{I}}{\mathbf{D} - \alpha\mathbf{W}}. \tag{3}$$

The number of superpixels N in our implementation is set to 200. Following [15], α and σ^2 are chosen to be 0.99 and 0.1.

4   Experimental Design

4.1   Data Sets

We select the well-known benchmark database called segmentation evaluation database (SED) [3] from the salient object detection literature. SED is divided into two datasets SED1 and SED2, based on image type and the level of difficulty. SED1 contains a total of 100 single object foreground images, while SED2 contains 100 images each having two different foreground objects. The ground truth is provided in the database, having either two or three classes and annotations from three different human subjects for each image. For saliency evaluation, the ground truth is first processed by pixel level voting to obtain a binary ground truth map with 1's representing an object and 0's for background.

4.2   Selected Models for Comparison

Baseline Seed Prediction Models. In this work, two baseline seed predicting heuristic methods are used to produce saliency seeds.

- **Background Seeds (BS):** Following the heuristic approach of [14], i.e, using boundary nodes as queries for ranking of nodes. We use all boundary nodes from the four sides of an image as seeds to compute a saliency map using the diffusion process.
- **Reconstruction Error Based Seeds (SRE):** We use the sparse reconstruction errors obtained by using the background nodes as basis functions for sparse representation. The sparse reconstruction errors are then used as saliency seeds for computing diffusion based saliency. Sparse reconstruction errors using background templates were first used by [9] as part of the Bayesian system to predict saliency.

OptSeedProp. For OptSeedProp [11], we use the numerical results from the paper as the implementation is not publically available. Consequently, we are not able to report receiver operating characteristic (ROC) and average precision (AP) curves for OptSeedProp.

Deterministic Models. RC [5], Gof [6] and FTS [1] are used for comparison. We use the author provided implementations for these models to compute the reported results.

4.3   Performance Evaluation

For a fair comparison, we evaluate the performance of models using the quantitative measures reported in [11], i.e., the average area under the receiver operating characteristic curve (AUC) and the average precision (AP). In order to compute AUC and AP, ROC and precision curves must be generated for a set of thresholds for each saliency map. We use the standard benchmark method reported in [1] to generate ROC and precision curves for a defined set of thresholds. According to this benchmark, a saliency map is segmented at discrete thresholds within a range of [0,255], and compared with the ground truth to compute either the true positive rate, false positive rate or the precision at each value of the threshold. The AUC is then computed by finding the area under the ROC curve and AP is obtained by averaging the precision values for each image.

5   Quantitative Results

This section presents the performance comparison of selected seed based and state-of-the-art deterministic models with our proposed IGASeed system.

5.1   Performance Comparisons with Seed Based Techniques

Table 1 presents the comparison of seed based techniques with our proposed IGASeed system.

Table 1. Object detection performance with seed based techniques: AUC/AP

AUC/AP	BS	SRE	IGASeed	OptSeedProp
SED1	0.758/0.767	0.801/0.795	0.930/0.850	**0.953/0.891**
SED2	0.727/0.716	0.760/0.757	0.840/0.777	**0.906/0.806**

For SED1 the OptSeedProp approach performs better than all other models with our proposed IGASeed technique producing the second best results. The high performance of OptSeedProp is due to the high number of features that it incorporates (i.e., 178 features in total) at the cost of high computational load. On the other hand our IGASeed technique achieves comparable results using only nine features presenting a low computational cost solution. Our IGASeed system learns multiple combination schemes to enhance its performance at the cost of high computation time at the training stage. However the time required to compute the saliency for a test image is considerably faster than the OptSeed-Prop model (for details please see Section 5.3). OptSeedProp performs better than our IGASeed system on the SED2 dataset. However it is to be noted that our IGASeed system produces comparable results. The analogous performance of our IGASeed system indicates that the process of optimally combining features carries more importance than including a large number of features, for quality seed prediction with the added advantage of low computational cost. Our IGASeed technique outperforms the primitive heuristic seed prediction models without adding considerable computational time.

5.2   Performance Comparisons with Deterministic Approaches

This section presents the comparison of our IGASeed technique with the state-of-the-art deterministic models as shown in Table 2. It can be observed that our IGASeed system achieves considerably better results than all other models on the SED1 dataset. RC [5] produces the second best results.

Table 2. Object detection performance comparison with deterministic techniques: AUC/AP

AUC/AP	FTS	Gof	RC	IGASeed
SED1	0.686/0.5656	0.822/0.641	0.889/0.820	**0.930/0.850**
SED2	0.791/0.722	0.832/0.682	**0.925/0.806**	0.840/0.777

We anticipate that the higher performance of our IGASeed system as compared with the state-of-the-art models is due to its richness in terms of features and its effective combination scheme at the cost of additional computational time. It is to be noted that despite using more time as compared to FTS and RC, the time used by our IGASeed system to produce a saliency map is comparable to most state-of-the-art saliency models. RC produces the best performance amongst both diffusion and non-diffusion based systems including OptSeedProp on SED2 dataset with our IGASeed system performing the second best. This might indicate that the current seed prediction based methods are not exceptionally good at scenes with multiple objects and more robust diffusion processes are required to deal with multiple object scenarios.

To complement the numerical results in Tables 1 and 2, we present the ROC and thresholded precision curves in Figure 3. In compliance with the numerical results, the ROC and precision curves show that our IGASeed system performs better than all the models for the SED1 dataset. Our IGASeed maintains its higher true positive rate for all thresholds and highest precision values for majority of threshold levels. In terms of the ROC curve, RC and Gof produce comparable results, while RC maintains higher true positive rate for lower thresholds.

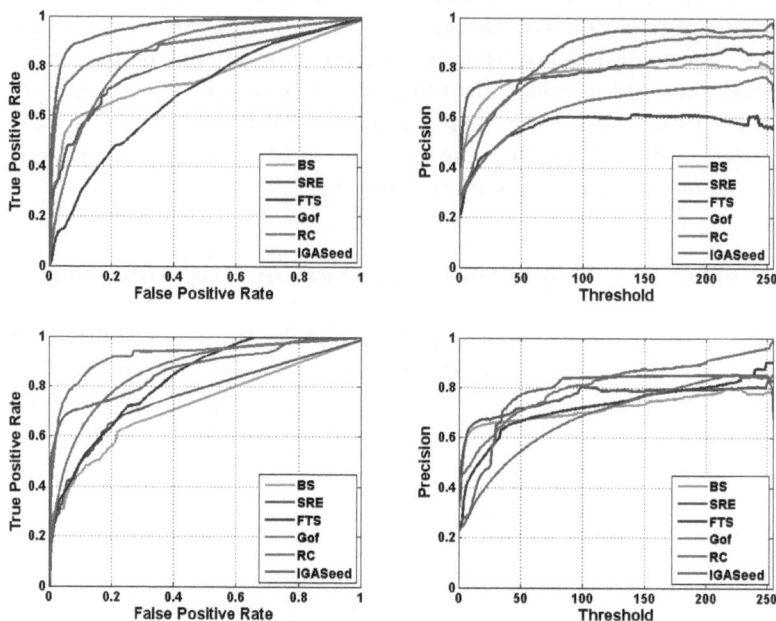

Fig. 3. ROC and thresholded precision curves for all models. Top row shows the results for SED1 and bottom row presents the results for SED2 dataset respectively. Our IGASeed system is shown in red. This figure is best viewed in color.

For SED2 the close performance of our IGASeed and the Gof model in terms of AUC can be observed from the ROC curve. RC performs the best in capturing more area under the curve. In terms of the precision curve, our IGASeed system performs similar to RC, however RC scores higher average precision at higher thresholds boosting its overall score.

It can be observed that the performance of BS and SRE models in terms of precision at low thresholds is remarkably good. A probable explanation of this performance could be the sparser maps that these models usually produce. The sparse maps when thresholded at low levels score high precision but fail to maintain it on higher thresholds, assigning the salient regions low values in most cases.

We performed an unpaired two-tailed t-test to determine the statistical significance of our IGASeed system. In terms of AUC, our IGASeed system was found to be statistically significant over all models with a p-value < 0.0001 for the SED1 dataset. On the AUC comparison for SED2, it significantly outperformed

BS, SRE and FTS with p-values of <0.0001, <0.0001 and 0.014, respectively and performed similar to Gof with a p-value of 0.554. RC performed better than IGASeed with a p-value < 0.0001 on SED2 in terms of AUC.

In terms of AP on the SED1 dataset, our IGASeed system was found to be statistically significant over all models with a p-value 0.0026 for RC and <0.0001 for all other models. On SED2 for AP comparison, it performed better than BS and FTS with a p-value of 0.01, while performed similar to SRE with a p-value 0.231 and significantly lower than RC with a p-value of 0.0168. We are not able to report the statistical significance results for the OptSeedProp system as they do not report the variance results in their paper.

5.3   Comparison of Computational Time

Table 3 present the computational time requirements for our IGASeed and Opt-SeedProp systems. Comparing the two models using complexity results is not trivial as there are several different modalities involved, which make the comparison infeasible. Therefore we limit our discussion to the reported timings in seconds for the scope of this work. Here we neglect the time for weighted addition of features for both approaches, as it is negligible. The reported timings are average of 10 selected images of size 300×400 from the SED1 dataset and they were computed on a i7 vpro 3.2 GHz processor with 8 GB of RAM.

The complexities for individual features are reported for presenting the overall context. In terms of computational complexity of individual features, the upper bound on a feature in our IGASeed system is greater than the upper bound on any individual feature in the OptSeedProp model. However evaluating bounds on computational complexity of features in isolation will make the comparison unfair as it is the sum of timings that governs the completion time.

It can be noted that our IGASeed techniques is an order of times faster than its counterpart in terms of the total average time spent in seconds (as shown in brackets in Table 3). Although the individual timings for OptSeedProp features are smaller (due to superpixel level feature computation for all features) as compared with our IGASeed features. However, the number of superpixel level features of a type along with the number of scales is very high for the OptSeedProp technique, boosting the overall computation time.

6   Qualitative Results

The better performance of our IGASeed system in terms of visual quality is demonstrated by the saliency maps for the images in rows 2 and 6 (Figure 4), for which almost all the models struggle to produce high quality saliency maps. For the fifth row image, our IGASeed system uniformly highlights both salient objects as compared with other models.

BS and SRE occasionally produce highly sparse saliency maps failing to highlight the salient object in a uniform manner and perform poorly on parts of the salient regions. This may be due to the low number of representative seeds selected by these models. FTS appears to filter important information on a few

Table 3. Computational time requirements for our IGASeed system and OptSeedProp system. F1-F9 are the nine features used by our IGASeed system. F1-F177 are features used by the OptSeedProp system. We only report timing and complexity information for representative features for OptSeedProp system. All timings reported here are in seconds.

IGA				OptSeedProp			
Feature	Time (14.78)	Complexity	Remarks	Feature	Time (52.36)	Complexity	Remarks
F1	5.36	$O(N)$	N: number of pixels in image	F1-25	22.7	$O(n_p^2)$	n_p: number of super-pixels in image
F2	0.67	$O(n_c^2)$	n_c: number of clusters (k-means)	F26-125	21	$O(n_b^2)$	n_b: number of boundary superpixels in image at each side
F3	0.67	$O(N_c)$	N_c: number of pixels in a cluster	F161-163	3.15	$O(n_p^2)$	n_p: number of super-pixels in image
F4	0.25	$O(N)$	N: number of pixels in image	F164-166	3.15	$O(n_p^2)$	n_p: number of super-pixels in image
F5+F6	7.08	$O(N_p)$	N_p: number of pixels in a region	F167-168	0.4	$O(n_s^2)$	n_s: number of selected superpixel regions
F7	0.18	$O(n_p^2)$	n_p: number of super-pixels in image	F169	0.91	$O(n_p^2)$	n_p: number of super-pixels in image
F8	0.18	$O(n_p^2)$	n_p: number of super-pixels in image	F173-177	1	$O(n_b^2)$	n_b: number of boundary superpixels in image at each side
F9	0.39	$O(n_r)$	n_r: number of reduced superpixels in image				

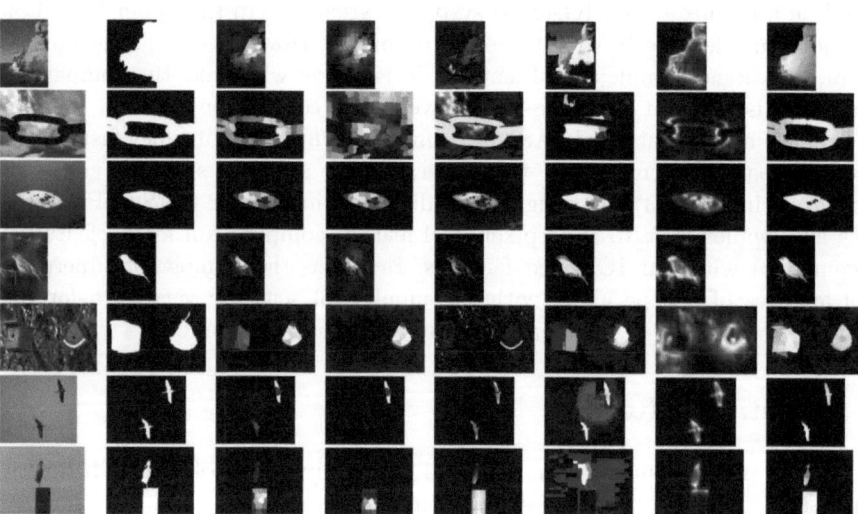

Fig. 4. Visual comparison of all models on selected images from SED1 and SED2. Left to right (in rows): input, ground truth, SRE, BS, FTS, RC, Gof and our proposed IGASeed.

images, resulting in assigning similar scores to the objects and the background. RC includes unwanted noise in images, which seldomly affects its quantitative performance due to its low intensity in the saliency maps as compared with the intensity of salient object regions.

7 Conclusions and Future Work

In this work, we have investigated the following scenarios for seed learning: 1) high number of less informative features versus low number of highly discriminative features and 2) optimal feature combination vs the number of features included. The novel IGASeed system, despite using considerably fewer features with a better feature combination scheme, achieved comparable results to the OptSeedProp on both the datasets. The good performance of our IGA system for seed prediction demonstrates the potential of a reduced feature implementation with rich feature combination schemes, as compared with the large scale feature combination implementation of OptSeedProp. The IGASeed system has the added advantage of low computational cost. Our results encourage us to investigate richer learning approaches with the best performing selected features for improving seed learning for diffusion-based object saliency.

References

1. Achanta, R., Hemami, S., Estrada, F., Susstrunk, S.: Frequency-tuned salient region detection. In: IEEE CVPR, pp. 1597–1604 (2009)
2. Achanta, R., Shaji, A., Smith, K., Lucchi, A., Fua, P., Susstrunk, S.: Slic superpixels compared to state-of-the-art superpixel methods. IEEE Trans. Pattern Anal. Mach. Intell. 34(11), 2274–2282 (2012)
3. Alpert, S., Galun, M., Basri, R., Brandt, A.: Image segmentation by probabilistic bottom-up aggregation and cue integration. In: IEEE CVPR, pp. 1–8 (2007)
4. Chen, T., Cheng, M.M., Tan, P., Shamir, A., Hu, S.M.: Sketch2photo: Internet image montage. In: ACM SIGGRAPH Asia, pp. 124:1–124:10 (2009)
5. Cheng, M.M., Zhang, G.X., Mitra, N., Huang, X., Hu, S.M.: Global contrast based salient region detection. In: IEEE CVPR, pp. 409–416 (2011)
6. Goferman, S., Zelnik-Manor, L., Tal, A.: Context-aware saliency detection. IEEE Trans. on Pattern Anal. Mach. Intell. 34(10), 1915–1926 (2012)
7. Gopalakrishnan, V., Hu, Y., Rajan, D.: Random walks on graphs for salient object detection in images. IEEE Trans. Image Process. 19(12), 3232–3242 (2010)
8. Itti, L.: Automatic foveation for video compression using a neurobiological model of visual attention. IEEE Trans. Image Process. 13(10), 1304–1318 (2004)
9. Li, X., Lu, H., Zhang, L., Ruan, X., Yang, M.H.: Saliency detection via dense and sparse reconstruction. In: IEEE ICCV, pp. 2976–2983 (2013)
10. Liu, T., Yuan, Z., Sun, J., Wang, J., Zheng, N., Tang, X., Shum, H.Y.: Learning to detect a salient object. IEEE Trans. Pattern Anal. Mach. Intell. 33(2), 353–367 (2011)
11. Lu, S., Mahadevan, V., Vasconcelos, N.: Learning optimal seeds for diffusion-based salient object detection. In: IEEE CVPR (2014)

12. Naqvi, S., Browne, W., Hollitt, C.: Genetic algorithms based feature combination for salient object detection, for autonomously identified image domain types. In: IEEE CEC (2014)
13. Rutishauser, U., Walther, D., Koch, C., Perona, P.: Is bottom-up attention useful for object recognition? In: IEEE CVPR, vol. 2, pp. II-37–II-44 (2004)
14. Yang, C., Zhang, L., Lu, H., Ruan, X., Yang, M.H.: Saliency detection via graph-based manifold ranking. In: IEEE CVPR, pp. 3166–3173 (2013)
15. Zhou, D., Bousquet, O., Lal, T.N., Weston, J., Schlkopf, B.: Learning with local and global consistency. In: NIPS, pp. 321–328 (2004)

Analysis of Hybrid Classification Approach
to Differentiate Dense and Non-dense Grass Regions

Sujan Chowdhury[1], Brijesh Verma[1], and David Stockwell[2]

[1] Central Queensland University, Australia
{s.chowdhury2,b.verma}@cqu.edu.au
[2] Queensland Transport and Main Roads, Australia
David.R.Stockwell@tmr.qld.gov.au

Abstract. Vegetation classification from satellite and aerial images is a common research area for fire risk assessment and environmental surveys for decades. Recently classification from video data obtained by vehicle mounted video in outdoor environments is receiving considerable attention due to the large number of real-world applications. However this is a very challenging task and requires novel research techniques. This paper presents an analysis of hybrid classification approach to distinguish vegetation in particularly the type of roadside grasses from videos recorded by the Queensland transport and main roads. The proposed framework can distinguish dense and non-dense grass regions from roadside video data. While most of the recent works focuses on infrared images, proposed approach uses image texture feature for vegetation region classification. Analysis of hybrid approach using texture feature and multiple classifiers is the main contribution of this research work. The classifiers include: Support Vector Machine (SVM), Neural Network (NN), k-Nearest Neighbor (k-NN), AdaBoost and Naïve Bayes. The different images were created from video data containing roadside vegetation in various conditions for training and testing purposes. The hybrid classification approach has been analysed on roadside data obtained and results are discussed.

Keywords: Hybrid Classification, Neural Networks, Vegetation Analysis, Feature Extraction, Support Vector Machine, k-Nearest Neighbor, AdaBoost, Naïve Bayes.

1   Introduction

Research on distinguishing vegetation area especially different types of vegetation regions has been recognized very important topic in the remote sensing field. However, the recognition of vegetation from video data recorded by ground vehicle is quite new even up to date. Surprisingly, there is no known method that can effectively identify the vegetation region from roadside data due to its similar spectral signature with respect the color, shape as well as the regions are not properly distributed in the field [1]. Hence, this research tries to implement a new idea for roadside vegetation classification. Instead of classifying different types of grasses, initially this paper

G. Dick et al. (Eds.): SEAL 2014, LNCS 8886, pp. 835–846, 2014.
© Springer International Publishing Switzerland 2014

focuses on dense and non-dense region identification. Although a variety of vegetation can be found on the roadside with respect to color, structure, this paper consider only how they distributed on the field.

Designing and implementing automatic image classification algorithms within autonomous system is very young and unexplored area of research. Meanwhile several approaches on weed classification have been developed using various classifiers such as k-Nearest Neighbour [2], Adaptive boost [3], Artificial Neural Network [4], Support Vector Machine [5] and Wavelet [6]. The increased use of automation on roadside data is the main reason for motivating into the research on detection and classification of vegetation region from roadside [7] [8] [9].

In 2004, tree species classification technique from high resolution forest imagery was developed by Kanda et al. [10]. In this study, they used GLCM as texture feature. Although many features can be generated from GLCM, according to the literature they selected homogeneity as their feature vector. Qian Yu et al. [11] studied detailed vegetation classification in high spatial resolution airborne Digital Airborne Imaging System (DAIS) imagery using 52 features where nine GLCMs features have been used. In 2007, Ghazali et al. [12] introduced a 2 Dimensional Discrete Wavelet Transform (2D-DWT) based feature extraction technique to investigate the characteristic of narrow and broad weed. According to [13], GLCM and FFT have been used to recognize types of weeds as either narrow or broad. Recognition results showed that for offline images, FFT achieve 89.2% and 91% correct classification rate for narrow and broad weed recognition where as 81% and 81.5% classification rates were recorded for the GLCM. Whereas for recorded video FFT scores 80.60% and 81.10% for narrow and broad weed recognition, respectively. On the other hand, for GLCM obtained accuracy was 70.4% for the narrow category and 72.5% for the broad. Hence for real time application FFT has been chosen. In addition, Ghazali et al. [14] obtained above 80% accuracy by using a combination of statistical grey-level co-occurrence matrix (GLCM), structural approach Fast Fourier Transform (FFT), and scale-invariant feature transform (SIFT) features in a real-time weeds control system for an oil palm plantation. Apart from this, Wu et al. [15] utilized GLCM and histogram statistics-based texture features extracted from four spatial orientations, horizontal, left diagonal, vertical and right diagonal corresponding to 0°, 45°, 90° and 135° respectively for weed and corn seedling recognition with SVM classifier. Zhengrong et al. [16] investigated state-of-art texture descriptors such as Gray-Level Co-Occurrence Matrix (GLCM), Local Binary Patterns (LBP) for object-based vegetation species classification and evaluate the performance. In this work, 10 spectral and texture feature descriptors were evaluated using SVM by means of classification accuracy. In contrast, some researchers [17] used 3 texture features (GLCM, Gabor Wavelet (GW), Uniform LBP (ULBP)) to classify vegetation species. The evaluation results suggest selecting appropriate feature and classification algorithm for different categories. Overall, the classification accuracies of all classifiers and texture features are not as good as expected. In 2003, Tang et al. [18] performed texture based weed classification using low-level Gabor wavelets-based feature extraction algorithm to classify images into broadleaf and grass categories. In this research, three species of broadleaf weeds (common cocklebur, velvetleaf, and ivyleaf morning glory) and two grasses (giant foxtail and crabgrass) that are common in Illinois were studied. Although it can classify effectively, a serious drawback of this method was that sample images were limited to 40 images with 20 samples for each class. Another drawback with the method was the processing time as each weed image needs to perform four frequency levels. Mustapha and Mustafa [19] developed an algorithm to extract texture based features based on GW to categorize broad and narrow leaf weeds, which achieved an accuracy of 88.17%.

Another class of weed classification technique was proposed by Ishak et al.[20] which utilized the combination of a Gabor wavelet (GW) and Gradient Field of Distribution (GFD) to extract a new set of feature vector. In their work, an Artificial Neural Network (ANN) has been applied for classification purpose. A total of 400 images of 200 grasses and 200 broadleaf weeds with different lighting conditions were used to test the effectiveness and listed accuracy was 93.75%.

In recent years, some researchers use LIDAR (Light Detection and Ranging) sensors in order to calculate texture feature from 3D structure. In [21] Nguyen et al. combines 2D-3D information for vegetation detection. In terms of 2D feature they consider the mean and standard deviation of brightness, color and histogram. The limitation of their approach is time consuming because of mapping 2D and 3D information from two different sensors by using coarse calibration method [21]. The approach can be efficiently used when considering time is not a criteria [22]. Most of the researchers use texture feature and color features in order to detect grass region [23] [24] . They extract the color component from YUV color space and enhance the image by changing the color or brightness of the pixels. Their proposed approach works well when the grass color is green [25], but in real scenario in Queensland road there is no specific color of the grass. So it's difficult to use color as a feature vector for those vegetation area identification. The aim of this paper is to combine the NN, SVM, k-NN, AdaBoost and Naïve Bayes together with a texture feature for dense and non-dense area classification.

Among the reported research on vegetation detection mainly focused on specific texture: grass-field [23] and weed [26] as well as with specific color: green leaf, green grass [25]. Some research works focused on specific structure such as foliage and needle tree [27]. From the literature cited above, it is clear that only few research works were done that works on every types of vegetation commonly found on roadside. In this paper pixel-based method for roadside vegetation detection is presented which is applied on roadside video data collected by moving vehicle.

The rest of this paper is organized as follows: in the third section proposed approach is introduced with texture feature selection and classifier training. Experimental results and analysis of proposed approach is presented in the fourth section. The conclusions and future research directions are presented in the last section of this paper.

2   Proposed Approach

The main focus of this research is based on the video data collected by Queensland transport and main roads (TMR) from different parts of Queensland. Using four video cameras fitted on different positions, data are collected automatically from the roadside. Directions include left, right, front and rear. This research especially focuses on left side video data. For this study ten meter from roadside are considered. The video camera was set to automatic exposure and focus. Presently, all frames shown in Fig.1 were taken from the video. All the videos are saved in .avi format for further processing. The data size around 250 GB with resolution 1632 X 624. Frame rate for the video was 5 and codec used for this purpose was Motion JPEG Video (MJPG). For this research purpose database contains 110 images extracted from video sequences, which are manually segmented to differentiate between dense and non-dense region. No additional equipment needed as this study only focus on visible spectrum. The foundation of the proposed hybrid technique is based on that the vegetation features can be learnt and distinguished by the fusion of different base classifiers. The proposed technique consists of 5 stages such as data acquisition, extraction of selected features, training of base classifiers and classification with trained base classifiers, majority voting and calculation of accuracy. A scenario of the proposed technique is shown in Fig. 2.

Fig. 1. Different Types of Dense and Non-dense Region

At the very beginning of the proposed technique, the research focuses on deciding which color spaces would be suitable for this kind of region differentiation. The study starts with several color spaces like RGB, gray, HSV, YUV YCbCr that area commonly used in vegetation detection. RGB color space is initially rejected because from RGB image we cannot differentiate between grass color and soil color. As we lose information on gray image, this option is also not good choice. Though HSV color space show good performance on green vegetation, it is not always true under most different environment conditions. Better accuracy is achieved using YCbCr color space as its work with brightness of the pixel as well as blue and red difference between the pixels.   On the basis of the conclusion, candidate image is converted into YCbCr color model and then convert it into binary level. The frame size of each image after scaling was 384 X 384.

2.1   Feature Extraction

For human being it seems to be very simple to understand the difference between dense and sparse region based on its typical colors, textures and geometric distributions. But to train machine and get decision from it is quite challenging. The most important challenge is choosing appropriate features.

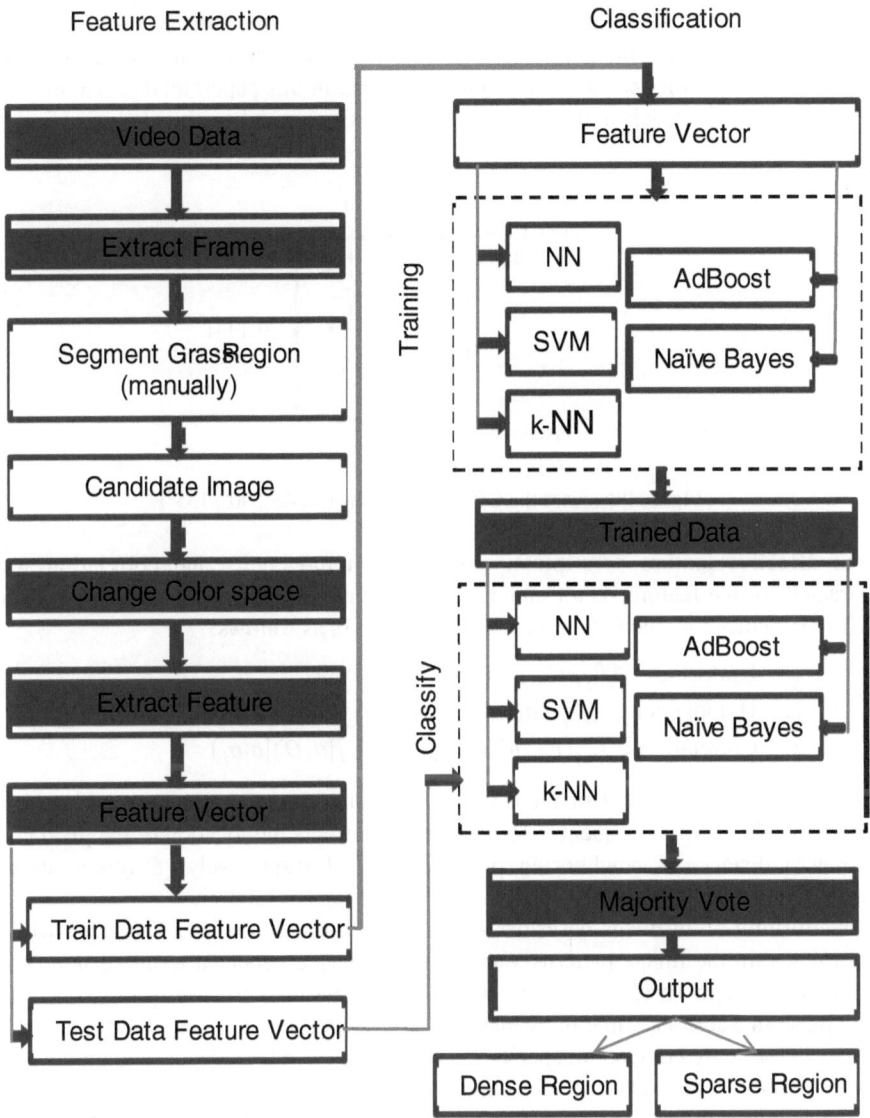

Fig. 2. Flow Chart of the Proposed System Implementation

Texture feature consider as a powerful source of information and have been intensively used in image classification [15]. A good number of methods exist in order to extract texture feature from image. Among them gray-level co-occurrence matrix (GLCM), Local Binary Patterns (LBP) are the widely used texture descriptors for weed and leaf classification [15]. As dense and sparse region classification is similar kind of classification, we adopt the idea of using GLCM as a texture feature for our purpose. But the key difference of using GLCM is the way that we using the pixel for our purpose. Usually texture feature tries to extract the pattern within the adjacent areas. It also integrates the information on direction, adjacent interval and change amplitude for analyzing structure or pattern. In this paper, texture feature were generated from 32 levels in 0^0, 45^0, 90^0, 135^0 directions respectively (shown in Fig. 3).

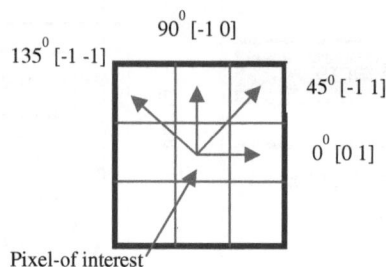

Fig. 3. Illustrates the Offset Directions (0^0, 45^0, 90^0, 135^0)

Four effective texture descriptors: uniformity, homogeneity and correlation were extracted for the feature vector and a small portion of the scenario is shown in Table 1. The formulas for calculating the above parameters as follows:

1. Uniformity: $\sum_{i,j}^{N} p(i,j|d,\theta)^2$
2. Homogeneity: $\sum_{i,j}^{N}((p(i,j|d,\theta))|(1 + |i - j|))$
3. Correlation: $\sum_{i,j}^{n}(i - \mu i)(j - \mu j)(p(i,j|d,\theta)|\sigma_i\sigma_j)$

Where N is the number of intensity levels in the image. An element p (i, j ld, θ) represents the relative frequency. Here i and j represents the location of the pixel (x,y) and at a distance d neighboring pixel gray level respectively. θ represents the orientation angle.

Uniformity of an image describes the random of image texture. The value will be high if within the image patterns vary greatly and vice versa. In terms of non-dense region the pattern varies greatly, so the value will be high (shown in the 3rd column for sparse in Table 1), while the value is small (shown in the 2nd column for sparse in Table 1) for dense compare to sparse. Therefore, uniformity added as a feature vector. Homogeneity represents the closeness within the image pixels. Low value of homogeneity presents the image pixels are close enough within their pattern, whereas high value of homogeneity presents the big variation within the pixel values. From the table 1 in 4th column the value is small for dense and value is a little bit high for sparse. Hence, homogeneity considered as a feature vector. Another important feature also added as a feature vector for differentiation of dense and non-dense region is known as correlation. This feature represents the linear dependency of an image. If pixel values vary greatly, correlation value will be small, while the value will be large if all the elements in the matrix are equal. The value of correlation will be big if the texture areas of an image are similar in certain direction.

Table 1. Texture Feature Vector

Image	Feature 1		Feature 2	
	Dense	Sparse	Dense	Sparse
1	0.513361377	0.666191085	0.412776971	0.515697507
2	0.579373343	0.657258439	0.507681093	0.517547735
3	0.518351603	0.626493744	0.497914291	0.516757606
4	0.588950211	0.623618024	0.638743607	0.495329285
5	0.534794282	0.620239126	0.484430235	0.459876482
6	0.446476271	0.709995826	0.34273133	0.514854551
7	0.439425231	0.617367189	0.334520367	0.431400026
8	0.511729273	0.657870402	0.405405136	0.478781784
9	0.529651143	0.62778723	0.455379704	0.462826774
10	0.526651609	0.639158566	0.397081212	0.480070446

2.2   Hybrid Classification

After extracting features, the next phase is to train the machine using the extracted features. A wide variety of machine learning algorithms have been addressed in many published works for classification tasks. In this paper, we try to address those algorithms used for our purpose and describe the reason for choosing them. At the very beginning of our work, we choose three classifiers and now extended to five: Support Vector Machine (SVM), Neural Network (NN), k-Nearest Neighbor (k-NN), AdaBoost, and Naïve Bayes Classifier.

Classification begins with SVM as it widely used for weed, crops and leaf classification [1]. It separate the clusters by drawing a hyper plane, in such a way that feature vectors with dense region are on one side of the plane and non-dense region feature vectors will be on the other side of the plane. In terms of computational complexity and memory requirements SVM achieved optimum performance. Hence SVM added as a classifier. The reason for choosing k-NN as a base classifier because of its overall recognition accuracy and time complexity [28]. A more impressive work has been presented in [29] to classify the weed according to their types based on the feature obtained using combination of Gradient Field Distribution (GFD) and Grey Level Co-occurrence Matrix (GLCM). BP-ANN was used as a base classifier and ANN has produced good generalization accuracy in many other applications, hence the idea of adding NN as a base classifier. Adaboost has been used to classify 3D aerial lidar data into four categories: road, grass, buildings, and trees using five features [30] and achieved over 92% accuracy. So the performance of the algorithm attracts us to choose Adaboost as a base classifier. A study on Bayes feature fusion has been presented for three types of image classification: "soil" (class 1), "tall vegetation" (class 2), and "grass" (dry or green – class 3) [31]. Hence we added "naive Bayes" as our base classifier. By adding more classifiers, we want to analyze how performance varies by increasing the number of base classifiers.

3   Experimental Results and Discussions

This section provides experimental results for the classification accuracy in order to analyze the hybrid technique by varying the number of classifiers. The feature extraction technique together with classification technique was applied on the dense and non-dense regions. 110 images have been used for training and testing to check the classification performance. Firstly, images were segmented manually and prepared for training and testing. From the test sets, features were extracted and then corresponding classifiers were applied for classification. Figs. 4 and 5 show the original image, image after changing the color space, and the dense region after coloring respectively.

Fig. 4. Dense a) Original Image b) Histogram Image c) Coloring Dense Region

Fig. 5. Non-dense a) Original Image b) Histogram Image c) Coloring Dense Region

Two types of image classification have been presented: "dense region" (class 1), "non-dense region vegetation" (class 2). The class "1" represents dense region and "2" represents non-dense region. Five machine learning algorithms have been used (SVM, NN, k-NN, AdaBoost, Naïve Bayes) as base classifiers. The overall performance of the proposed approach for classification of dense and non-dense region is depicted in Table 2 and results are discussed.

Initially, with different kernel functions classification accuracies of SVM were compared and the most suitable kernel function for dense and non-dense region classification has been chosen. The best accuracy is listed in Table 2 and it shows that using RBF-based kernel function recognition accuracy obtained was 90% for training and 85% on test data.

Table 2. Classification Results using SVM, NN, k-NN, Adaboost, Naive Bayes and Hybrid

Technique/Approach	Function	Train Accuracy (%)	Test Accuracy (%)
SVM	'rbf'	90	85
NN	H.U=12, Iterations= 3500, RMS error=0.0001	90	85
k -NN	$k = 7$	85	80
AdaBoost	Learning cycle=100, Learners='Discriminants'	87	82
Naïve Bayes	fitNaiveBayes	90	85
Hybrid Approach (SVM, NN, k-NN)	SVM: rbf NN : H.U= 12, Iterations= 3500 k -NN: k =7	95	90
Hybrid Approach (SVM,NN,k-NN, AdaBoost,Naïve Bayes)	SVM: rbf NN : H.U= 12, Iterations= 3500 k-NN: k=7 AdaBoost: Learning cycle=100, Learners='Discriminants' Naïve Bayes: fitNaiveBayes	95	90

The second classifier used in the proposed hybrid approach is a NN classifier with different number of hidden units, iterations, and RMS errors. At every stage we manually changed the parameter and checked the classification accuracy. Finally, the best parameters are chosen which give accuracy around 90% for training and 85% for testing dataset same as SVM. The obtained parameter accuracy is listed in Table 2.

The third classifier used in the proposed hybrid approach is k-NN with the closest feature vector and results obtained are listed in Table 2. From the above table it shows the highest accuracies on training and test datasets were obtained using k =7. We ran experiments with different value of k, but observed degradation of accuracy using *k*-NN.   Results obtained using the *k*-NN is lower than the other two classifiers which were 85% for training and 80% for testing.

The fourth classifier used is AdaBoost, which shows similar performance like SVM and NN in terms of accuracy. Table 2 shows that, accuracy achieved by AdaBoost is 87% for training and 82% for testing using 100 learning cycle and 'Discriminant' in learners.

The fifth and final classifier used for classification was Naïve Bayes with fitNaiveBayes and predict function. Accuracy obtained using this classifier is similar to SVM and NN, which is 90% for training and 85% for testing.

Finally based on the decision obtained using different classifiers, the results are fused using majority voting with 3 classifiers and 5 classifiers. The results with different classifiers and hybrid results (after majority voting) are summarized in Table 2. The principle of majority voting is that decision is based on majority wins. Thus if majority of classifiers predict the region as dense then the region is classified as dense and vice versa.   The results shown in Table 2 indicate that hybrid approach achieved the highest classification accuracy when RBF kernel function was chosen for SVM; number of hidden unit and iterations were chosen 12 and 3500 respectively for NN and the value of *k* was chosen as 7 for *k*-NN, 100 learning cycle and 'Discriminant' learners was chosen for AdaBoost. Table 2 shows that best results obtained using the combination of feature vector and hybrid approach were 95% (training) and 90% (testing) respectively. From the above analysis it is obvious that by just increasing the number of classifiers doesn't increase the accuracy of hybrid approach. So choosing appropriate classifiers and using them efficiently should be the main focus of the future research.

4   Conclusion

This paper presents and analyses a hybrid approach with texture features for dense and non-dense regions classification. The proposed hybrid approach increments the number of diverse classifiers and analyses the results to see whether there is any change in accuracy or not.

After conducting experiments, it was found that the results using hybrid classification are better than the results using individual classifiers. In comparison of the results obtained with individual classifiers, the hybrid classification approach achieved 95% and 90% accuracy respectively on training and test data. The analysis showed that the accuracy didn't increase when more than three classifiers are added.

In this research the candidate images were manually segmented. More research will be conducted on segmentation to crop the region automatically. The experiments were conducted using a dense and non-dense grass region which is most important factor in identifying roadside fire. In real scenario on roadside, in addition to grasses there are also presence of other vegetation like trees and shrubs. Future research will be directed towards using a large dataset with different types of vegetation extracted from the video data and using evolutionary algorithms for optimization of classifiers.

References

1. Tellaeche, A., Pajares, G., Burgos-Artizzu, X.P., Ribeiro, A.: A computer vision approach for weeds identification through Support Vector Machines. Applied Soft Computing 11, 908–915 (2011)
2. Cho, T.-H., Conners, R.W., Araman, P.A.: A comparison of rule-based, k nearest neighbor, and neural net classifiers for automated industrial inspection. In: Proceedings of the IEEE/ACM International Conference on Developing and Managing Expert System Programs, pp. 202–209 (1991)
3. Liu, M.: Fingerprint classification based on Adaboost learning from singularity features. Pattern Recognition 43, 1062–1070 (2010)
4. Petrová, J., Moravec, H., Slavıková, P., Mudrová, M., Procházka, A.: Neural network in object classification using matlab. Network 12, 10 (2012)
5. Yang, H.-Y., Wang, X.-Y., Wang, Q.-Y., Zhang, X.-J.: LS-SVM based image segmentation using color and texture information. Journal of Visual Communication and Image Representation 23, 1095–1112 (2012)
6. Rehman, A., Gao, Y., Wang, J., Wang, Z.: Image classification based on complex wavelet structural similarity. Signal Processing: Image Communication 28, 984–992 (2012)
7. Harbas, I., Subasic, M.: Detection of roadside vegetation using features from the visible spectrum. In: 37th International Convention on Information and Communication Technology, Electronics and Microelectronics (MIPRO), pp. 1204–1209 (2014)
8. Nguyen, D.V., Kuhnert, L., Thamke, S., Schlemper, J., Kuhnert, K.D.: A novel approach for a double-check of passable vegetation detection in autonomous ground vehicles. In: 2012 15th International IEEE Conference on Intelligent Transportation Systems (ITSC), pp. 230–236 (2012)
9. Nguyen, D.V., Kuhnert, L., Kuhnert, K.D.: Spreading algorithm for efficient vegetation detection in cluttered outdoor environments. Robotics and Autonomous Systems 60, 1498–1507 (2012)
10. Kanda, F., Kubo, M., Muramoto, K.-I.: Watershed segmentation and classification of tree species using high resolution forest imagery. In: Proceedings of 2004 IEEE International Geoscience and Remote Sensing Symposium, IGARSS 2004, pp. 3822–3825 (2004)
11. Yu, Q., Gong, P., Clinton, N., Biging, G., Maggi Kelly, A., Schirokauer, D.: Object-based detailed vegetation classification with airborne high spatial resolution remote sensing imagery. Photogrammetric Engineering & Remote Sensing 72(7), 799–811 (2006)
12. Ghazali, K.H., Mansor, M.F., Mustafa, M.M., Hussain, A.: Feature extraction technique using discrete wavelet transform for image classification. In: 5th Student Conference on Research and Development, SCOReD 2007, pp. 1–4 (2007)
13. Mustafa, M.M., Hussain, A., Ghazali, K.H., Riyadi, S.: Implementation of image processing technique in real time vision system for automatic weeding strategy. In: 2007 IEEE International Symposium on Signal Processing and Information Technology, pp. 632–635 (2007)
14. Ghazali, K.H., Razali, S., Mustafa, M.M., Hussain, A.: Machine vision system for automatic weeding strategy in oil palm plantation using image filtering technique. In: 3rd International Conference on Information and Communication Technologies: From Theory to Applications, ICTTA 2008, pp. 1–5 (2008)
15. Wu, L., Wen, Y.: Weed/corn seedling recognition by support vector machine using texture features. African Journal of Agricultural Research 4, 840–846 (2009)

16. Li, Z., Hayward, R., Zhang, J., Jin, H., Walker, R.: Evaluation of spectral and texture features for object-based vegetation species classification using support vector machines. In: International Archives of the Photogrammetry, Remote Sensing and Spatial Information Sciences (Part A), ISPRS, Vienna, Austria, pp. 122–127 (2010)

17. Li, Z., Liu, Y., Hayward, R., Walker, R.: Empirical comparison of machine learning algorithms for image texture classification with application to vegetation management in power line corridors. In: International Archives of the Photogrammetry, Remote Sensing and Spatial Information Sciences (Part A), ISPRS, Vienna, Austria, pp. 128–133 (2010)

18. Tang, L., Tian, L.F., Steward, B.L.: Classification of broadleaf and grass weeds using Gabor wavelets and an artificial neural network. Transactions of the ASAE 46, 1247 (2003)

19. Mustapha, A., Mustafa, M.M.: Development of a real-time site sprayer system for specific weeds using gabor wavelets and neural networks model. In: Proceedings of the Malaysia Science and Technology Congress, Malaysia, pp. 406–413 (2005)

20. Ishak, A.J., Hussain, A., Mustafa, M.M.: Weed image classification using gabor wavelet and gradient field distribution. Computers and Electronics in Agriculture 66, 53–61 (2009)

21. Nguyen, D.V., Kuhnert, L., Jiang, T., Thamke, S., Kuhnert, K.D.: Vegetation detection for outdoor automobile guidance. In: 2011 IEEE International Conference on Industrial Technology (ICIT), pp. 358–364 (2011)

22. Nguyen, D.V., Kuhnert, L., Kuhnert, K.D.: Structure overview of vegetation detection. A novel approach for efficient vegetation detection using an active lighting system. Robotics and Autonomous Systems 60, 498–508 (2012)

23. Zafarifar, B., de With, P.H.N.: Grass field detection for TV picture quality enhancement. In: International Conference on Consumer Electronics, ICCE 2008 Digest of Technical Papers, pp. 1–2 (2008)

24. Herman, S., Janssen, J., Bellers, E., Wendorf, J.: Automatic segmentation-based grass detection for real-time video. Google Patents (2004)

25. Gu, Y.J., Zhong, J.: Grass Detection Based on Color Features. In: 2010 Chinese Conference on Pattern Recognition (CCPR), pp. 1–5 (2010)

26. Sabeenian, R.S., Palanisamy, V.: Texture based weed detection using multi resolution combined statistical & spatial frequency (MRCSF). International Journal of Information Technology and Computer Science (IJITCS) 5, 253 (2009)

27. Jean-Francois Lalonde, N.V., Huber, D.F., Hebert, M.: Natural terrain classification using three-dimensional ladar data for ground robot mobility. Journal of Field Robotics 23, 839–861 (2006)

28. Chen, Y., Lin, P., He, Y., Xu, Z.: Classification of broadleaf weed images using Gabor wavelets and Lie group structure of region covariance on Riemannian manifolds. Biosystems Engineering 109, 220–227 (2011)

29. Juraiza Ishak, A., Mustafa, M.M., Hussain, A.: Gradient field distribution and grey level co-occurrence matrix techniques for automatic weed classification. In: 5th International Symposium on Mechatronics and Its Applications, ISMA 2008, pp. 1–5 (2008)

30. Lodha, S.K., Fitzpatrick, D.M., Helmbold, D.P.: Aerial lidar data classification using AdaBoost. In: Sixth International Conference on 3-D Digital Imaging and Modeling, 3DIM 2007, pp. 435–442 (2007)

31. Shi, X., Manduchi, R.: A study on bayes feature fusion for image classification. In: Conference on Computer Vision and Pattern Recognition Workshop, CVPRW 2003, p. 95 (2003)

Image Segmentation: A Survey of Methods Based on Evolutionary Computation

Yuyu Liang, Mengjie Zhang, and Will N. Browne

School of Engineering and Computer Science
Victoria University of Wellington, P.O. Box 600, Wellington 6140, New Zealand
{yuyu.liang,mengjie.zhang,will.browne}@ecs.vuw.ac.nz

Abstract. Image segmentation is mainly used as a preprocessing step in problems of image processing and computer vision. Its performance has a great influence on subsequent tasks. Evolutionary Computation (EC) techniques have been introduced to the area of image segmentation due to their high search capacity. However, there are rarely comprehensive surveys on EC based image segmentation methods, which can enable researchers to get a quick understanding of this area and compare the existing methods. Therefore, this paper provides an overview of EC based image segmentation methods, and discusses the remaining issues in this area. It is observed that among all EC techniques, four of them (genetic algorithms, genetic programming, differential equation and partial swarm optimization) are more frequently used and GAs are the most popular technique. It is noted that low generalization capacity and computational complexity are two common problems in EC techniques applied to image segmentation.

Keywords: Evolutionary Computer Vision, Image Segmentation, Evolutionary Computation, Genetic Algorithms, Genetic Programming.

1 Introduction

Image segmentation is a process of partitioning pixels of an image to different regions based on specific information, which are normally intensity, texture or color. For a segmented image, pixels in one region are similar to each other according to a homogeneity criterion, yet pixels in different regions are heterogeneous [1–3]. It is a major step in both image processing and computer vision systems. The segmented images are often the input to high-level image tasks, such as feature extraction, object detection, image recognition and classification. Since this process divides an image into several homogeneous regions and helps find regions of interest, images become easier to manipulate and more meaningful for the high-level tasks [1].

However, partitioning an image typically has a high computation cost. The enumeration of all possible solutions is an exhaustive process, which forms a huge search space [3]. To address this problem, the search methods selected are crucial for algorithms' efficiency. Evolutionary Computation (EC) algorithms

G. Dick et al. (Eds.): SEAL 2014, LNCS 8886, pp. 847–859, 2014.

solve problems using natural selection, and have been shown to be powerful search methods. Therefore, EC-based image segmentation is an active research area now.

When using EC techniques to conduct image segmentation, most related works combine them with other classic segmentation algorithms such as threshold, region growing, clustering and partial differential equations [4–9]. In these hybrid methods, EC takes the role of optimizing parameters or minimizing/ maximizing objective functions. In contrast, other works [10–14] apply EC techniques to generate segmentation algorithms from a subset of basic image operators such as filters, histogram equalization and threshold. Therefore, we group EC-based image segmentation techniques into two branches: hybrid segmentation techniques combining EC and classic mehtods (Section 3); image segmentation operators evolved by EC (Section 4).

The overall goal of this paper is to discuss EC-based image segmentation methods, analyze the improvements over existing methods and identify problems and challenges. The specific objectives are:

- Overview the existing hybrid methods that combine EC techniques with the classic methods for image segmentation;
- Discuss the current EC methods, evolving image segmentation operators from primitive image processing operators;
- Identify the main challenges, issues and future work directions.

The rest of the paper is organized as follows. Section 2 introduces EC and image segmentation. Section 3 surveys existing works on methods combining EC techniques and classic segmentation methods. Section 4 focuses on the EC methods that evolve image segmentation algorithms. In Section 5, conclusions are drawn.

2   A Brief Review of Evolutionary Computation and Image Segmentation

As a subfield of artificial intelligence, EC normally addresses continuous optimization or combinatorial optimization problems. It can be mainly categorized into three groups [15, 16]: evolutionary algorithms, swarm intelligence and others (as shown in Fig. 1). EC algorithms have a similar framework [15], which is presented in Fig. 2. They have been successfully applied to many problems in image processing and computer vision, including edge detection, image segmentation, object detection, classification and recognition.

Image segmentation is an important step for many problems in image processing and computer vision. It is an active area with many alogrithms, methods and techniques proposed. According to papers [1, 2], most segmentation techniques can fall into one of the following categories.

- Threshold based methods [4, 17–19];
- Region based methods (region growing, region splitting, and region merging) [5, 20];

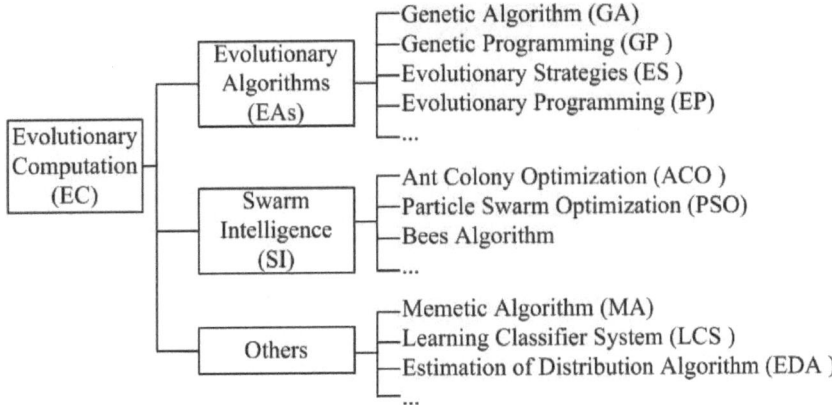

Fig. 1. EC Algorithms

- Edge based methods [6];
- Clustering based methods [7, 21–23];
- Neural network based methods [8];
- Partial Differential Equation (PDE) based methods (snakes or active contour model (ACM), level sets, mumford shah and so on) [24, 25].

The six categories of image segmentation techniques provide a clue to investigate hybrid methods of EC with classic algorithms. The research on investigating hybrid methods can thus be divided into these six branches: EC-Threshold based, EC-Region based, EC-Edge based, EC-Clustering based, EC-Neural Network based and EC-PDE based image segmentation.

3 Hybrid Image Segmentation Techniques of EC and Classic Mehtods

Based on the research conducted by Khan [1], combining two or more segmentation techniques together can lead to better performance. EC algorithms have been introduced to the area of image segmentation, forming hybrid methods with many classic segmentation algorithms. Since image segmentation is a costly combination task, it is related to a huge search space. As powerful search techniques [15], EC is suitable for complex search tasks in image segmentation.

3.1 EC-Threshold Based Techniques

Thresholding is a simple and popular image segmentation method, which has been widely used in practical applications. For threshold-based algorithms, the selection of the optimal threshold is the biggest challenge, and still remains a problem [4, 17]. Thresholds are found by analyzing the image histogram in most

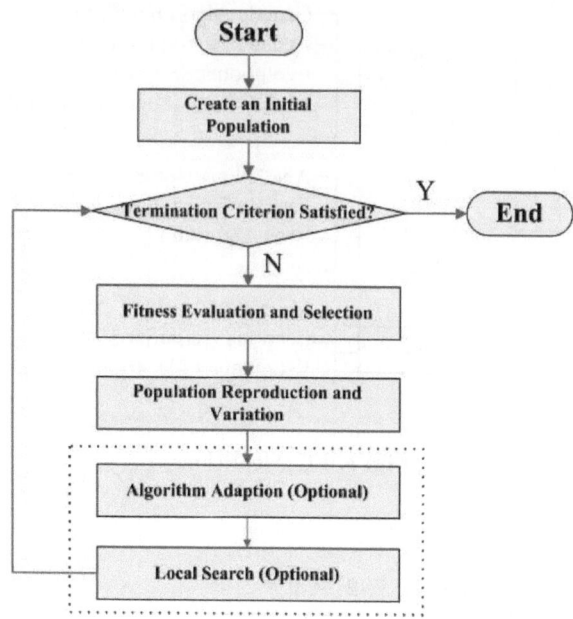

Fig. 2. The Framework of EC Algorithms

related algorithms. In this process, an objective function related to threshold values is often minimized or maximized. EC techniques have been used to find the optimal threshold [4, 17–19].

Banimelhem and Yahya [4] utilizes a thresholding technique to conduct image segmentation. To find optimal thresholds, a genetic algorithm is employed due to its search capability in practical applications. The segmentation problem has been transformed into an optimization problem in this paper by searching for optimal thresholds. The proposed method is tested on four images, but there is no evaluation on the segmentation performance.

Kanungo et al. [17] proposes a threshold-based approach using GA-Crowding to select the optimal threshold from histogram images. A crowding algorithm, which is firstly proposed by De Jong in 1975, is a multimodal function optimizer [26]. Under ideal conditions, the histograms of images with two classes have a deep and sharp valley between two peaks. The bottom of the valley is the threshold [27]. In this paper, the crowding algorithm is used to find the peaks on histogram images, and then GA is applied to locate the valley between the peaks. This technique can not only operate with bimodal features, but also multimodal features. Although it can deal with bimodal features well, it performs poorly for images with histograms of tri-modal features.

PSO is used to select multilevel thresholds in image segmentation [18]. The goal is to maximize the Otsu and Kapur objective functions. Otsu and Kapur are two optimal thresholding methods based on between-class variance and entropy criterion respectively. Both of them are efficient for bi-level thresholding. Even though these methods can be extended to multi-level thresholding, the computation time increases exponentially with the number of thresholds. A GA based multilevel thresholding algorithm is also implemented as a comparison in this paper. Compared with GAs based method, the proposed PSO based method achieves better results in terms of solution quality, convergence and robustness.

Pei et al. [19] combines Otsu with Differential Evolution for image segmentation. It employs the Differential Evolution technique to search for the optimal threshold. This system avoids the weakness of Otsu, which is based on maximum between-cluster variance and cannot manipulate low signal-to-noise well. Moreover, it is faster than the 2D maximum between-cluster variance method, which has a high computation complexity.

3.2   EC-Region Based Techniques

For region-based methods, segmentation is usually conducted through grouping neighboring pixels with similar intensities, such as region-growing and split-and-merge methods [28].

Al-Faris et al. [5] combine PSO and Seeded Region Growing (SRG) to segment images. SRG confronts two problms that are the selection of seeds and the similarity criteria. This paper chooses K randomly seeds when initializing the particles of the swarm. Thus SRG is combined with PSO, and then is evolved by PSO. The new algorithm is tested on the dataset – RIDER Breast MRI [29]. When compared with Support Vector Machine (SVM), K Nearest Neighbors (KNN), Fuzzy C-means (FCM), Bayesian, and Improved Self-Training (IMPST) techniques, there are two highlights. One is that the classification performance of the proposed algorithm improves greatly. The other is that parameters, such as threshold value, the suspected region window and seed pixel, are automatically selected.

3.3   EC-Edge Based Techniques

In an image, edges are the local changes in intensity, so edge detection is a process of finding discontinuity in intensity or the first derivative of intensity [30]. Edge detection techniques have been used as one kind of segmentation technique.

Diazi et al. [6] proposes a segmentation method using edge detection. Their system consists of three steps: calculate depth gradients and orientation gradients; find an edge map from the gradients using a GA; label pixels. Even though the proposed method can locate thin and closed edges and reduce false edges by employing a GA, it is sensitive to noise, and the edge map cannot describe image surfaces well.

3.4   EC-Clustering Based Techniques

Clustering is a process of organizing data into clusters that have high intra-cluster and low inter-cluster similarity. It is clear that intra-cluster similarity should be maximized and inter-cluster similarity should be minimized. Based on this idea, objective functions are defined. Clustering-based works introduce EC to optimize objective functions [7, 21–23].

Shirakawa and Tomoharu [7] presents a multi-objective evolutionary algorithm to optimize two clustering objectives (overall deviation and edge value). The overall deviation is the summed distance between data items and their cluster center, and the connectivity evaluates the degree to which neighbor pixels have been put in the same cluster. The proposed method can determine the number of clusters automatically. The results of four test images (pepper, sailboat, terra, paprika) are considered relatively good by the authors.

Maulik et al. [21–23] research EC-based fuzzy clustering methods for image segmentation, and the basic discipline is similar. Take paper [21] as an example, an improved differential evolution is applied to optimize multiple parameters for fuzzy clustering (XB and J_m). J_m calculates the global cluster variance,while the XB index is a combination of global and local situations. When tested on six data sets from the UCI dataset repository, this method outperforms a multiobjective version of classical differential evolution based fuzzy clustering technique (MODEFC), NSGA-II [23] and the FCM algorithm. The difference between papers [21–23] lies in the selection of EC techniques. In the papers [22, 23], GA is utilized to conduct parameter optimization, while DE is applied in the paper [21]. The authors have not compared them directly, so it is not clear which one is better for image segmentation.

3.5   EC-Neural Network Based Techniques

Neural networks adapted to image segmentation are such methods, which processes small areas of an image using a network or a set of networks [31]. After the processing, a decision mechanism is defined to mark the areas of an image based on the category recognized by the neural network.

Bilotta [8] proposes a method of using GAs to explore a CNN (Cellular Neural Network) to detect and segment lesions contained within magnetic resonance imaging (MRI) images automatically. Since it does not need any manual segmentation, this system is fully automatical. Tested on 11 patients with 20 slices each, CNN can segment most of the lesions and thus it is a useful tool to evaluate MR lesions.

3.6   EC-PDE Based Techniques

The PDE methods used for segmentation are mainly Snakes (or Active Contour Model), Level Set and Mumford-Shah model [24]. Several EC techniques have been combined with PDEs to conduct image segmentation.

Payel et al. [9] utilizes GAs to evolve level-set functions using texture and shape information. In the evolutionary process, the contours (level-set functions) are formed through minimizing an energy function by the gradient descent method. The proposed system is used to segment the prostate region on pelvic and MRI images. The GAs-based level-set algorithm proposed in the paper does not need to compute derivatives in curve evolution, reducing computational complexity. In addition, it is domain-independent and can deal with multiple features for segmentation.

In paper [32], GA is introduced to optimize parameters of level sets. Specifically, segmentation parameters are encoded in each individual of the GA. Individual fitness is calculated by comparing the segmentation result with the reference segmentation. This system is tested on computed tomography (CT) images and achieves good performance, but reference segmentations and contours are needed.

Roulu et al. [33] presents an evolutionary snakes to segment nuclei in histopathological images. This evolutionary algorithm uses mutation and crossover mechanisms to search solution space, yet does not use the natural selection process for the selection of the best individual. In this paper, the snake algorithm is transformed to a combinatory optimization problem. An evolutionary algorithm is proposed to search the solution space. The defined fitness function of evolutionary algorithm reduces the search space, generating a fast convergence.

Cruz-Aceves et al. [34] propose a new method using differential evolution to guide the evolution of multiple active contours (MACDE). This system is tested on synthetic images containing complex objects, Gaussian noise, and deep concavities, and datasets of sequential computed tomography and real MRI images as well. Evaluation is based on the comparison of segmentation results and reference images. MACDE is better than original ACM in terms of accuracy, robustness, and efficiency. However, reference segmentations from experts need to be provided, which increases the workload in applying this method.

Wang et al. [35] intoduces PSO to overcome drawbacks of the classic snake model, such as high level of sensitivity to noise and local optimization. The combination of PSO and snakes performs better in image segmentation than snakes alone. However, the related parameters of the system cannot be adjusted automatically, and convergence speed is low.

4   Image Segmentation Operators Evolved by EC

In areas of artificial intelligence, sysmbolic processing, and machine learning, many problems can be considered as discovering a computer program [36], including image segmentation. Among EC techniques, GP and GA are used by existing works to generate segmentation algorithms. GAs are heuristic search methods that mimic the process of natural selection and uses fixed-length strings to represent possible solutions. GP is a specialization of GA, which commonly uses parse trees to replace strings for the representation[10]. Driven by a fitness function defined to evaluate a solution's ability to deal with a given task, GA and GP seek to optimize a population of computer programs.

Research [10–13] is aimed to apply GP to evolve image operators, such as filters and classifiers, to conduct image segmentation. For example, Poli [10] uses the evolved filters to build pixel-classification-based segmentation methods. In those papers, segmentation is regarded as a classification task on the pixel level. Similarly, Brumby et al. [14] utilizes a GA to evolve image segmentation algorithm from certain primitive image operators.

Poli [10] proposes a method using GP to evolve filters to detect features of interest and conduct pixel-classification-based segmentation. In order to make the filters feasible and efficient, this paper focuses on choosing a suitable terminal set, a function set and a fitness function. The dataset contains real medical images (head MR images, X-ray coronarograms blood vessels). The method is measured with sensitivity and specificity defined in (1) and (2). It achieves 61.5% in sensitivity and 99.2% specificity, while neural network only achieves 31.7% and 92.2% respectively.

$$Sensitivity = |TP|/(|TP| + |FN|) . \tag{1}$$

$$Specificity = |TN|/(|TN| + |FP|) . \tag{2}$$

where $|TP|$(True Positive) and $|FN|$(False Negative) mean points belonging to the objects of interest are correctly and incorrectly detected by an evolved program respectively; $|FP|$(False Positive) and $|TN|$(True Negative) represent points belonging to the non-objects of interest are correctly and incorrectly detected by the evolved program respectively.

In paper [11], GP is employed to evolve texture classifiers to conduct texture image segmentation containing both binary textures and multiple textures. Table 1 shows how to use the generated image operators to do segmentation. The system is fast with high accuracy, and can deal with complex shapes. There are also some weaknesses: a) it needs prior information—the number of texture classes the images have; b) the training time is quite long (from several hours to several days); c) it is problem-specific.

One common problem in segmentation is that algorithms are usually not generalized, since they use priori information of subjects which limits them to specific problems [12]. To solve this problem, Singh et al. [12] utilizes GP to generate image segmentation programs from primitive image-operators (e.g. filters, edge detectors). There are two advantages of the proposed algorithm: it does not need priori information as other automatic image segmentation algorithms require; it produces simple MATLAB-based programs, which are easy to be understood and used by other researchers. This technique is compared with GENIE Pro. (GENtic Image Exploitation) [14], and consistently produces better results.

Roberts [13] also employs GP to evolve image operators with image inputs. The highlight in this paper is introducing a caching mechanism to overcome the high time consumption. Due to the evolutionary process where the evaluation of individuals costs much time. By storing the results of subtree evaluations, the caching system does not need to re-evaluate the trees, which are copied from the previous generations. Tested on the mole images with skin lesions, the method achieves 92.3% of average sensitivity and 97.2% of average specificity.

Brumby et al.[14] propose a software, Genetic Imagery Exploitation (GENIE), which is an image segmentation and classification tool. This software uses GA to

Table 1. A Flow Chart of Pixel-classification-based Segmentation. (Adapted from [11])

Input:	test image (I); window size (n); step size for moving the window (v in vertical direction; h in horizontal direction)
Output: segmented image	

1. Use the generated operator to sweep I:
a) Start from the top-left of I.
b) Get a subimage with the size of n*n; use the operator to classify it and get a label.
c) Label the pixels of the subimage with the result of b).
d) Move the window to the right horizontally with h pixels and repeat b) and c), till the window reach the right boundary.
e) Move the window down vertically with v pixels and repeat b), c) and d), till the window reach the down-right corner of I.

2. Generate the output with the labels of each pixel.
a) Use some mechanisms(e.g. voting) to decide the final label of each pixel.
b) Assign pixels with the same label the same color or intensity, and output the result image.

evolve image-processing algorithms from low-level image operators. In this paper, primitive image operators are described and the chromosomal representation of an algorithm is presented. This software is tested on an airborne simulator picture of the Gulf of Mexico, and it achieves an accuracy of over 98%.

This section summarizes the research using GP and GA to evolve image operators. In these four papers [10–14], GP and GA are used in a different way with the EC techniques presented in Section 3. The hybrid techniques combining EC and classic segmentation algorithms mainly utilize EC to optimize parameters or objective functions required by classic algorithms. In comparison, using GP and GA to generate image segmentation operators is creating segmentation programs, which can be directly applied to image segmentation.

Table 2. Summary of EC-based Methods for Image Segmentation

EC-based Methods	Specific Techniques
Hybrid Methods	GAs-Threshold [4, 17], PSO-Threshold [18], DE-Threshold [19] PSO-Region [5] GAs-Edge [6] GAs-Clustering [22, 23], DE-Clustering [21] GAs-Neural Nerwork [8] GAs-Level Sets [9, 32], DE-ACM [34], PSO-Snake [35]
Evolution of Image Operator	GP [10–13], GA [14]

5 Conclusions

In this paper, various techniques of EC based image segmentation have been discussed. They fall into two groups, including hybrid segmentation techniques involving EC and classic mehtods, and EC-evolved image segmentation operators. Existing approaches of EC based image segmentation methods presented in this paper have been summarized in Table 2.

Based on the survey conducted in this paper, it is observed that:

- Most EC-based image segmentation techniques are based on the GAs, GP, DE or PSO techniques. These EC techniques are used to deal with parameter optimization or pixel-level problems in image segmentation.
- GAs have been the most popular methods in image segmentation. This is because many classic segmentation algorithms require parameter optimization that is suited to GA. Traditional search methods can often be stuck in local optima, while GAs are often able to handle large and complex search space and to obtain the global optimum. Moreover, GP and GA can be used in a different way with other EC techniques for image segmentation. Due to the characteristic of generating programs, they are employed to generate image segmentation operators.
- According to the identified weaknesses of the systems presented by the papers we studied, problems in EC-based image segmentation are the same as certain open issues in the whole area. Most of the image segmentation methods are still domain-specific, and need prior knowledge to fulfill the segmentation task. Thus, the generalization capacity of systems are low. Another big challenge lies in computational efficiency, which is a barrier for real applications. Actually, the computation complexity is often because of the evaluation process utilizing a costly fitness function.

One interesting future work would be to focus on GP-based image segmentation. GP is a promising method in image processing and computer vision, and it has been widely used for image classification and feature construction [37–41]. With the expectation that normally high-level features have low dimensions and can improve performance for the complex image segmentation tasks, We will consider applying GP to conduct feature construction in image segmentation tasks.

References

1. Khan, W.: Image Segmentation Techniques: A Survey. Journal of Image and Graphics 1(4), 166–170 (2013)
2. Senthilkumaran, N., Rajesh, R.: Edge detection techniques for image segmentation a survey of soft computing approaches. International Journal of Recent Trends in Engineering 1(2), 250–254 (2009)
3. Bhandarkar, S.M., Zhang, H.: Image Segmentation Using Evolutionary Computation. IEEE Transactions on Evolutionary Computation, 1–21 (1999)

4. Banimelhem, O., Yahya, Y.A.: Multi-Thresholding Image Segmentation Algorithm Using Genetic Algorithm. In: World Congress in Computer Science, Computer Engineering, and Applied Computing (2011)

5. Al-Faris, Q.A., Ngah, U.K., Isa, N.A.M., Shuaib, I.L.: Breast MRI Tumour Segmentation using Modified Automatic Seeded Region Growing Based on Particle Swarm Optimization Image Clustering. In: Online Conference on Soft Computing in Industrial Applications Anywhere on Earth, Online Version, pp. 1–11 (2012)

6. Diazi, I., Branch, J., Boulancer, P.: A Genetic Algorithm to Segment Range Image by Edge Detection. In: International Conference on Industrial Electronics and Control Applications, pp. 7–14 (2005)

7. Shirakawa, S., Tomoharu, N.: Evolutionary image segmentation based on multiobjective clustering. In: IEEE Congress on Evolutionary Computation. IEEE (2009)

8. Bilotta, E., Cerasa, A., Pantano, P., Quattrone, A., Staino, A., Stramandinoli, F.: A CNN Based Algorithm for the Automated Segmentation of Multiple Sclerosis Lesions. In: Di Chio, C., Cagnoni, S., Cotta, C., Ebner, M., Ekárt, A., Esparcia-Alcazar, A.I., Goh, C.-K., Merelo, J.J., Neri, F., Preuß, M., Togelius, J., Yannakakis, G.N. (eds.) EvoApplicatons 2010, Part I. LNCS, vol. 6024, pp. 211–220. Springer, Heidelberg (2010)

9. Payel, G., Melanie, M., James, T., Arthur, H.: A Genetic Algorithm-based Algorithm-Level Set Set Curve Evolution for Prostate Segmentation on Pelvic CT and MRI Images. In: Biomedical Image Analysis and Machine Learning Technologies: Applications and Techniques, pp. 127–149. IGI Global (2010)

10. Riccardo, P.: Genetic programming for feature detection and image segmentation. In: Fogarty, T.C. (ed.) AISB-WS 1996. LNCS, vol. 1143, Springer, Heidelberg (1996)

11. Song, A., Ciesielski, V.: Fast Texture Segmentation using Genetic Programming. In: The 2003 Congress on Evolutionary Computation, pp. 2126–2133. IEEE (2003)

12. Singh, T., Nawwaf, K., Mohmmad, D., Rabab, W.: Genetic Programming based Image Segmentation with Applications to Biomedical Object Detection. In: Proceedings of the 11th Annual Conference on Genetic and Evolutionary Computation, pp. 1123–1130. ACM (2009)

13. Roberts, M.E.: The Effectiveness of Cost Based Subtree Caching Mechanisms in Typed Genetic Programming for Image Segmentation. In: Raidl, G.R., Cagnoni, S., Cardalda, J.J.R., Corne, D.W., Gottlieb, J., Guillot, A., Hart, E., Johnson, C.G., Marchiori, E., Meyer, J.-A., Middendorf, M. (eds.) EvoIASP 2003, EvoWorkshops 2003, EvoSTIM 2003, EvoROB/EvoRobot 2003, EvoCOP 2003, EvoBIO 2003, and EvoMUSART 2003. LNCS, vol. 2611, pp. 444–454. Springer, Heidelberg (2003)

14. Brumby, P., Theiler, P., Perkins, J., Harvey, R., Szymanski, J., Bloch, J.: Investigation of Image Feature Extraction by a Genetic Algorithm. In: Proceedings of SPIE, pp. 24–31 (1999)

15. Zhang, J., Zhan, Z.H., Lin, Y., Chen, N., Gong, Y.J., Zhong, J.H., Shi, Y.H.: Evolutionary Computation Meets Machine Learning: A Survey. IEEE Computational Intelligence Magazine 6(4), 68–75 (2011)

16. Evolutionary Algorithm, http://en.wikipedia.org/wiki/Evolutionary_algorithm

17. Kanungo, P., Nanda, P.K., Samal, U.C.: Image segmentation using thresholding and genetic algorithm (2006)

18. Duraisamy, S.P., Kayalvizhi, R.: A New Multilevel Thresholding Method Using Swarm Intelligence Algorithm for Image Segmentation. Journal of Intelligent Learning Systems and Applications 2(03), 126 (2010)

19. Pei, Z., Zhao, Y., Liu, Z.: Image segmentation based on Differential Evolution algorithm. In: International Conference on Image Analysis and Signal Processing, IASP 2009. IEEE Press (2009)
20. Kaganami, H.G., Beij, Z.: Region Based Detection versus Edge Detection. IEEE Transactions on Intelligent Information Hiding and Multimedia Signal Processing, 1217–1221 (2009)
21. Saha, I., Maulik, U., Bandyopadhyay, S.: An Improved Multi-objective Technique for Fuzzy Clustering with Application to IRS Image Segmentation. In: Giacobini, M., Brabazon, A., Cagnoni, S., Di Caro, G.A., Ekárt, A., Esparcia-Alcázar, A.I., Farooq, M., Fink, A., Machado, P. (eds.) EvoWorkshops 2009. LNCS, vol. 5484, pp. 426–431. Springer, Heidelberg (2009)
22. Maulik, U., Bandyopadhyay, S.: Fuzzy Partitioning using a Real-coded Variable-length Genetic Algorithm for Pixel Classification. IEEE Transaction on Geoscience and Remote Sending 41(5), 1075–1081 (2003)
23. Bandyopadhyay, S., Maulik, U., Anirban, M.: Multiobjective genetic clustering for pixel classification in remote sensing imagery. IEEE Transactions on Geoscience and Remote Sensing 45(5), 1506–1511 (2007)
24. Jiang, X., Zhang, R., Nie, S.: Image Segmentation Based on PDEs Model: a Survey. In: 2009 3rd International Conference on Bioinformatics and Biomedical Engineering, pp. 1–4 (2009)
25. Chan, T.F., Shen, J., Vese, L.: Variational PDE models in image processing. Notices of AMS 50(1), 14–26 (2003)
26. De Jong, K.A.: An analysis of the behavior of a class of genetic adaptive systems. Dissertation Abstracts International. 36(10) (1975)
27. Prewitt, J., Mendelsohn, M.L.: The analysis of cell images. Annals of the New York Academy of Sciences 128(3), 1035–1053 (1966)
28. Jiao, L.: Evolutionary-based image segmentation methods. Image Segmentation (10), 180–224 (2011)
29. RIDER Breast MRI. National Biomedical Imaging Archive (NBIA), U.o. Michigan, Editor 2007, U.S. National Cancer Institute (2011)
30. Senthilkumaran, N., Rajesh, R.: Edge Detection Techniques for Image Segmentation-A Survey of Soft Computing Approaches. International Journal of Recent Trends in Engineering 1(2), 250–254 (2009)
31. Pathegama, M., Göl, Ö.: Edge-end pixel extraction for edge-based image segmentation. Transactions on Engineering, Computing and Technology 2, 213–216 (2004)
32. Drio, O., Raul, F.: Liver Segmentation using Level Sets and Genetic Algorithms. In: Fourth International Conference on Computer Vision Theory and Applications, pp. 154–159 (2009)
33. Roula, A., Bouridane, A., Kurugollu, F.: An evolutionary snake algorithm for the segmentation of nuclei in histopathological images. In: 2004 International Conference on Image Processing. IEEE (2004)
34. Cruz-Aceves, I., Avina-Cervantes, G., Lopez-Hernandez, M., Rostro-Gonzalez, H., Garcia-Capulin, H., Torres-Cisneros, M., Guzman-Cabrera, R.: Multiple Active Contours guided by Differential evolution for Medical image segmentation. In: Computational and Mathematical Methods in Medicine (2013)
35. Wang, K., Guo, Q., Zhuang, D., Chu, H., Fu, B.: Application of Snake Model based on PSO in The Image Segmentation. In: The Sixth World Congress on Intelligent Control and Automation, WCICA 2006, pp. 9637–9640 (2006)
36. Koza, J.: Genetic Programming as a Means for Programming Computers by Natural Selection. Statistics and Computing Journal (1993)

37. Tackett, W.A.: Genetic Programming for Feature Discovery and Image Discrimination. In: ICGA, pp. 303–311 (1993)
38. Belpaeme, T.: Evolving visual feature detectors. In: Floreano, D., Mondada, F. (eds.) ECAL 1999. LNCS, vol. 1674, pp. 266–270. Springer, Heidelberg (1999)
39. Eshatian, K., Hang, M., Ndreae, P.: A Filter Approach to Multiple Feature Construction for Symbolic Learning Classifiers using Genetic Programming. IEEE Transactions on Evolutionary Computation 16(5), 645–661 (2012)
40. Smart, W., Zhang, M.: Classification Strategies for Image Classification in Genetic Programming. In: Proceeding of Image and Vision Computing Conference, pp. 402–407. Palmerston North, New Zealand (2003)
41. Jabeen, H., Baig, R.: Review of classification using genetic programming. International Journal of Engineering Science and Technology 2(2), 94–103 (2010)

Author Index

Abbass, Hussein A.   395
AbdAllah, AbdelMonaem F.M.   1
Abdullah, Salwani   554
Aguirre, Hernán   143, 167, 239
Ahmed, Soha   593
Al-Sahaf, Harith   335
Alvarez, Isidro M.   383
Arajy, Yahya Z.   554
Armas, Rolando   167

Batchelor, Oliver   798
Bonyadi, Mohammad Reza   431
Browne, Will N.   383, 822, 847

Chai, Jinze   529
Chen, Wei-Neng   37
Cheung, Yiu-ming   713
Chia, Darcy   760
Chowdhury, Sujan   835
Ciesielski, Vic   311, 371, 542
Consoli, Pietro A.   359
Cresswell-Miley, Cain   347

da Silva, Alexandre Sawczuk   180
Dau, Hoang Anh   311, 542
Derbel, Bilel   143
Dick, Grant   491
Drozdik, Martin   143
Drugan, Madalina M.   299

Ernst, Andreas T.   691
Essam, Daryl L.   1

Fu, Wenlong   516, 810

Gallagher, Marcus   119, 131, 455
Gong, Yue-Jiao   644
Green, Richard   798

Hamano, Kouta   58
Haqqani, Mohammad   228
Hollitt, Christopher   822
Huang, Ye   192

Isasi, Pedro   299

Jia, Ya-Hui   37
Jiang, He   216
Johnston, Mark   335, 618, 656, 668, 723, 810

Katzgraber, Helmut G.   13
Kazimipour, Borhan   479
Kifah, Saif   554
Kirley, Michael   443
Kok-Kiong, Tan   735, 773
Ko-Ko-Htet, Kyaw   735, 773
Köppen, Mario   58

Lai, Xiaochen   216
Li, Jing-Jing   37
Li, Ming   529
Li, Ping   582
Li, Xiaodong   228, 479, 631, 691
Li, Zhen-hua   713
Liang, Yuyu   847
Liefooghe, Arnaud   143
Liu, Hai-Lin   713
Liu, Ivy   569
Liu, Jing   680
Liu, Xing   582
Lozano, Jose A.   94
Lu, Xiaofen   82

Ma, Hui   106, 180
Manderick, Bernard   299
Marshall, Richard J.   618, 668
McDonald, Chris   747
McDonald, Ross B.   13
Mei, Yi   631, 691
Mendiburu, Alexander   94
Menzel, Stefan   82
Michalewicz, Zbigniew   431
Minku, Leandro L.   359
Mishra, Krishna Manjari   119
Morgan, Rachael   455
Morino, Yuya   323

Nagao, Tomoharu   467
Nair, Chandran   735
Naqvi, Syed S.   822

Narayanan, Arun-Shankar   735
Neshatian, Kourosh   347
Neumann, Frank   251, 419
Nguyen, Anh Quang   251, 419
Nguyen, Hoai Bach   569
Nguyen, Su   656, 723

Ohnishi, Kei   58
Olhofer, Markus   25
Osana, Yuko   323, 785
Owen, Caitlin   263
Oyama, Akira   239

Park, John   723
Peng, Lifeng   593
Piao, Yong   216

Qin, A.K.   479
Qiu, Xin   192

Ren, Zhilei   216

Sabar, Nasser R.   703
Salim, Flora D.   542
Santana, Roberto   13, 94
Sarker, Ruhul A.   1
Sato, Hiroyuki   274
Sato, Yuji   155
Shafi, Kamran   395
Shang, Lin   582
Shirakawa, Shinichi   467
Smalikho, Olga   25
Song, Andy   311, 371, 542, 691, 703
Stockwell, David   835
Sun, Wencheng   216
Sundar, Shyam   48

Takano, Mio   785
Tan, Kay Chen   192, 656
Tan, Kok Kiong   287
Tanaka, Kiyoshi   143, 167, 239

Tang, Ke   82
Tran, Binh   503, 605
Tsutsui, Shigeyoshi   70
Tumer, Kagan   204, 407

Verel, Sebastien   143
Verma, Brijesh   835
von der Osten, Friedrich Burkhard   443

Wagner, Markus   251, 431
Wang, Lixia   680
Wang, Liya   529
Wang, Qiang   713
Whigham, Peter A.   263
While, Lyndon   747, 760

Xie, Feng   371, 542
Xie, Jing   691
Xu, Jian-Xin   192
Xue, Bing   503, 516, 569, 593, 605

Yang, Shiqin   155
Yao, Xin   82, 359, 631
Yazawa, Yuki   239
Ying-Lin,   644
Yliniemi, Logan   204, 407
Yu, Fan   529
Yu, Shuang   287
Yu, Xinghuo   228
Yu, Yang   106

Zhan, Zhi-Hui   644
Zhang, Ge-Yi   644
Zhang, Jun   37, 644, 713
Zhang, Mengjie   106, 180, 335, 383, 503,
    516, 569, 593, 605, 618, 656, 668, 723,
    810, 847
Zheng, Yu   529
Zhou, Mingxing   680
Zhou, Zhe   582